Theory of Nuclear Structure

To my
wife Dipti
and
son Ritankar

Theory of NUCLEAR STRUCTURE

M. K. Pal

Saha Institute of Nuclear Physics
Calcutta

S&Æ SCIENTIFIC AND ACADEMIC EDITIONS

Library of Congress Catalog Card Number: 83–4603
ISBM: 0–442-27524–2

First published by Affiliated East–West Press, 1981
104 Nirmal Tower 26, Barakhamba Road, New Delhi 110 001

Published by Van Nostrand Reinhold Company Inc.
135 West 50th Street New York, N.Y. 10020

15 14 13 12 11 10 9 8 7 6 5 4 3 2 1

Library of Congress Cataloging in Publication Data

Pal, M K. (Manoj Kumar), 1932–
Theory of Nuclear Structure

Includes bibliography and index.
1. Nuclear structure. I. Title
Qc793.3.S8P34 1983 539.7'4 83–4603
ISBN: 0-442–27524–2

Contents

Foreword

In this text, the theory of nuclear structure, including a study of the nuclear models, has been presented with the help of the basic ingredients of the two-nucleon interaction as revealed by the study of two-nucleon data and the widely accepted reaction matrix formalism of nuclear many-body theory. This approach represents one type of expert outlook on nuclear theory. There also exists an equally powerful lobby that advocates discarding any reference to the actual two-nucleon interaction in structural work in view of the uncertainties of the present-day many-body models. Although the first type of outlook has been favoured in presenting the subject, the actual mathematical formalism in the relevant chapters has been kept sufficiently general, offering the reader a choice between the calculated two-body reaction matrix and the empirically determined effective potential. In so doing, a compromise has been effected in presenting an integrated approach to the subject without impairing the usefulness of the book to the student who should know most of the important developments in the subject without prejudice to any particular school of physicists.

The book, which is intended to serve as a text at the masters level, is so written that the student will easily grasp, on his own, about three-quarters of each chapter; the remaining content may, in general, be best assimilated under the guidance of a teacher. It also contains enough up-to-date material to be useful to the research physicist in nuclear structure in the first year of his studies. For the benefit of such a reader, who needs a quick review of the subject with references of the latest work, every chapter has sufficient coverage; care has been taken to make this coverage fairly representative rather than fully comprehensive.

Each chapter begins with a qualitative introduction and proceeds from fundamental concepts to develop the subject, paying close attention to pedagogic details. Some specialized topics of quantum mechanics, frequently used in nuclear theory, are covered in the appendices. A few problems, chosen with the purpose of giving working practice, are set at the end of every chapter and every appendix. The only prerequisite to an understanding of the content is a grounding in elementary nuclear physics and nonrelativistic quantum mechanics.

The volume has resulted from lectures delivered over the past several years at Saha Institute of Nuclear Physics, Calcutta; University of Delhi; The Australian National University, Canberra; and University of Oregon, Eugene. Chapter VI and Appendix G very closely follow the lecture on the theory of vibration in nuclei delivered at the International Centre for Theoretical Physics, Trieste, during January-March 1969. I gratefully acknowledge the permission granted by the Centre and by the International Atomic Energy Agency to use this material.

I would like to thank Mrs. Melinda Piluso and Mrs. Kay Knighten who typed a part of the manuscript in Eugene, and Mr. Jayanta Datta and Mr. Shibapada Bhattacharyya who completed it in Calcutta.

January 1981
Saha Institute of Nuclear Physics
Calcutta

M. K. Pal

Atomic and Nuclear Constants

1 Angstrom unit $= 1$ AU $= 10^{-8}$ cm

1 fermi $= 1$ fm $= 10^{-13}$ cm

Velocity of light $= c = 2.99793 \times 10^{10}$ cm/sec

Charge of electron $= e = 4.80286 \times 10^{-10}$ esu (electrostatic unit)

Energy acquired by electron in passing through potential difference of 1 volt $= 1$ eV $=$ 1.6021×10^{-12} erg

1 million electronvolt $= 1$ MeV $= 1.6021 \times 10^{-6}$ erg

1 kilo electronvolt $= 1$ keV $= 1.6021 \times 10^{-9}$ erg

Planck's constant h divided by $2\pi = \hbar = 1.054 \times 10^{-27}$ erg sec $= 6.5817 \times 10^{-22}$ MeV sec

$\hbar c = 1.9732 \times 10^{-11}$ MeV cm

Fine-structure constant $= \dfrac{e^2}{\hbar c} = 1/137.037$

Mass (m) of electron times $c^2 = mc^2 = 0.510976$ MeV

Mass (M_p) of proton times $c^2 = M_pc^2 = 938.211$ MeV $= 1836.12mc^2$

1 atomic mass unit $=$ one-sixteenth of mass of O^{16} atom $= 931.141$ MeV$/c^2$

Bohr radius $= a_H = \dfrac{\hbar^2}{me^2} = 0.52917$ AU

Classical electron radius $= r_e = \dfrac{e^2}{mc^2} = 2.8175$ fm

Electron Compton wavelength $= \lambda_c = \dfrac{\hbar}{mc} = 3.8612 \times 10^{-11}$ cm

Compton wavelength of π-meson $= \dfrac{\hbar}{m_\pi c} = 1.4132$ fm

1 nuclear magneton $= 1$ nM $= \dfrac{e\hbar}{2M_pc} = 3.1524 \times 10^{-18}$ MeV/gauss

I Two-Nucleon Problem

1. INTRODUCTION

A. GENERAL SURVEY

An atom is made up of a positively charged nucleus surrounded by negatively charged electrons rotating under the influence of the attractive Coulomb field. The radius of a typical electron orbit is of the order of the Angstrom unit (A.U.), whereas that of the nucleus is several fermis (fm).

All the chemical properties of atoms and the formation of molecules through the bonding of atoms are determined by the behaviour of atomic electrons. The energy emitted in atomic spectroscopy as electromagnetic radiation, and also in various chemical reactions, is the result of the transition of electrons between different energy levels, and the typical magnitude of such energy release in a single process is of the order of a few electron volts (eV). The study of the atomic energy levels arising from the motion of electrons in the nuclear Coulomb field, and under the influence of their mutual Coulomb interaction, forms the subject matter of atomic spectroscopy.

The atomic nucleus consists of two kinds of particles, namely, proton and neutron, denoted by the common name nucleon. Both have approximately the same mass, each being about 1840 times the electron mass. Therefore, almost all the mass of an atom is located in its nucleus. The neutron is chargeless, whereas the proton bears a positive charge, equal in magnitude to the negative charge (e) of an electron. In a nucleus, the total number of protons is called the atomic number (denoted by Z), and the total number of nucleons (i.e., protons and neutrons) is known as the mass number (denoted by A); in an electrically neutral atom, the atomic number Z gives the number of electrons orbiting the nucleus. The Z- and A-value are usually shown respectively as subscript and superscript to the chemical symbol of an element; for example, ${}_6C^{12}$ signifies the carbon atom containing 12 nucleons of which 6 are protons.

The fact that there is a massive nucleus inside an atom, carrying positive charge and occupying a very small volume, was first established by Rutherford in 1911 in his celebrated scattering experiments. The complete picture of the nucleus, comprising neutrons and protons, emerged in 1932 with the discovery of the neutron by Chadwick.

The radioactivity of uranium atoms, which was historically the earliest nuclear phenomenon, was first observed by Becquerel in 1896. This phenomenon entails the spontaneous emission of radiation by the nuclei of radioactive atoms. As a matter of fact, many heavy atoms such as polonium, radium, and radon gas, which are all radioactive, were first chemically isolated and identified by Pierre and Marie Curie in 1898 while they were studying the radiations emitted by these substances.

Radiations from radioactive substances are of three types: (i) α-radiation, consisting of ${}_2He^4$ atoms charged with two units of electricity due to the absence of the two orbital

1 (45-122/1977)

electrons; (ii) β-radiation, consisting of electrons or positrons; and (iii) γ-radiation, or ordinary electromagnetic radiation, with a much larger frequency, i.e., a much shorter wavelength than visible light or X-rays. The typical energy carried by a single α- or β-particle, or a single γ-ray quantum (photon), is of the order of a million electronvolt (MeV), i.e., a million times larger than the typical energy release in chemical reactions or the processes analyzed in atomic spectroscopy.

Although radioactivity was first discovered in naturally occurring substances, whose atoms are heavier than $_{82}Pb^{208}$, many artificially prepared elements also exhibit this phenomenon. The chemical isolation and identification of all such substances come within the field of nuclear chemistry. The study of elements heavier than uranium ($Z = 92$), called transuranic elements, is an essential part of this subject. Such research has extended the periodic table to include elements of atomic number higher than 92, the highest atomic number element discovered so far being 107.

Rutherford and his collaborators used the α-particles emitted by radioactive substances as projectiles, and examined their scattering on various atoms. The positive charge of an atom, causing scattering through the Coulomb field, was assumed to be located at the centre of the atom. The angular distribution of the scattered α-particles was worked out by the application of the concepts of classical mechanics (see Goldstein[1]). A quantum mechanical derivation yields the same classical formula, and can be found in any standard text (see, for example, Schiff[2]). For a given angle of scattering of an α-particle, the distance of its closest approach to the point charge Ze can be calculated in terms of its initial kinetic energy. As long as this distance is greater than the actual extent of the positive charge Ze in space, the point-charge formula should agree with experimental observation. The distance for which significant departure from the Rutherford formula has been observed gives a rough idea of the radius of the positive charge distribution in an atom, i.e., the radius of the nucleus. Rutherford established in his experiments that the nuclear radius is much smaller than 10^{-12} cm. More precise experiments, performed later, have shown that the radius is usually a few fm. The fact that nucleons in a nucleus are confined to such a small region of space is convincing proof of the presence of very strong attractive and strong nuclear forces holding them together.

Several kinds of interaction are known in present-day physics: (i) gravitational interaction between different celestial bodies; (ii) weak interaction, which gives rise to the emission of β-particles; (iii) electromagnetic interaction, which exists between charged particles by virtue of their charge, current, and magnetic moments; and (iv) strong interaction, which binds the nucleons in a nucleus. The weak and strong interactions govern a host of subatomic phenomena; a comprehensive study of these processes, involving not only electron, neutron, and proton but also many other elementary particles, belongs to the realm of elementary particle physics or, in a broader sense, of high-energy physics.

The attractive nuclear interaction between neutrons or between a neutron and a proton gives rise to an attractive (negative) potential energy, whereas the Coulomb repulsion between protons results in a repulsive (positive) potential energy. Since the Coulomb force is very much weaker than the nucleon-nucleon force, there is always a net attractive nuclear potential energy (V). The motion of nucleons inside the nucleus also gives a positive kinetic energy (T). If the nucleus is a bound system, then the sum $T + V\ (=E)$ is necessarily negative. The magnitude $|E|$ of the attractive total energy E is called the *binding energy*, and is related to the actual mass of the nucleus and the masses of nucleons as now described.

According to the special theory of relativity, mass and energy are equivalent; the amount of energy equivalent to the mass M is given by Mc^2, where c is the velocity of light. Thus, a nucleus of Z protons and N neutrons ($Z + N = A$) has a total mass-energy equal to $(ZM_p + NM_n)c^2$, where the subscripts p and n refer to proton and neutron, respectively. The total energy of the system is obtained by adding $E(=T + V)$ to this mass-energy. If the actual mass of the nucleus is $M(Z, N)$, then the total energy is $M(Z, N)c^2$. Equating the two energy values, we obtain

$$(ZM_p + NM_n)c^2 + T + V = M(Z, N)c^2$$

or, equivalently,

$$E(=T + V) = [M(Z, N) - ZM_p - NM_n]c^2.$$

Our earlier observation that the energy E is negative therefore implies that the actual mass of a nucleus $M(Z, N)$ is less than the sum of the masses of Z protons and N neutrons composing the nucleus. For determining E, $M(Z, N)$ may be replaced by the mass of the electrically neutral atom (i.e., the nucleus plus the Z electrons around the nucleus), and M_p by M_H, the mass of a hydrogen atom (i.e., a proton plus an orbital electron). The expression for E computed with these masses is, however, equal to $(T + V)$ for the nucleons plus $(T + V)$ for the electrons. The magnitude of the second $(T + V)$ is the binding energy of the electrons, and represents only about a millionth of the nuclear binding energy. Thus, atomic masses can be used, with negligible error, in the foregoing expression for E.

The importance of mass measurements was realized very early in the history of nuclear physics. Since the measurements by Aston in 1927, mass spectrometers and spectrographs have been gradually refined to attain perfection, and provide accurate values of the binding energy of nuclei. These instruments also separate the various *isotopes* of the same nuclear species whose existence was first established by J. J. Thomson in 1913. Thomson showed that, for a given proton number, it is possible to have several nuclei with a different number of neutrons. All such nuclei have the same positive charge (Z) but different mass (A). The neutral atoms of these nuclei carry the same number (Z) of electrons, and hence possess identical *chemical* properties; they are called different isotopes of the same chemical element. Similarly, all nuclei having different Z but the same mass number A have very similar *nuclear* properties and are known as *isobars*.

In a nucleus, the binding energy *per nucleon* begins at about 1 MeV for a deuteron (${}_1H^2$), very quickly reaches approximately 7 MeV for ${}_2He^4$, and then stays at a more or less stable value of 8 MeV for heavier elements. For the nuclei towards the end of the periodic table, the binding energy again drops to smaller values. The constancy of the binding energy *per nucleon* for most nuclei in the periodic table is called the *saturation* property of nuclear binding energy. According to the saturation phenomenon, the *total* binding energy is proportional to A, the mass number. There is a corresponding saturation phenomenon for the nuclear volume. It has been observed in various methods of measuring that the nuclear radius is roughly given by $r_0A^{1/3}$, where r_0 is an approximate constant lying between 1 fm and 1.2 fm. If the nucleus is assumed to be a sphere, the nuclear volume is then seen to be proportional to A. The same result can be alternatively stated in terms of density: since both the total mass and volume are proportional to A, the density is predicted to be constant. The constancy of nuclear density is also frequently referred to as a saturation phenomenon.

The initial quick rise of the binding energy per nucleon from H^2 to He^4 is due to the *very short range* (of the order of fm) of the nucleon-nucleon force. In H^2, the two nucleons tend to pull each other into the range of this force so that they can gain maximum attractive potential energy. But then, according to the uncertainty principle, confinement within a small distance implies a large kinetic energy. So, if the particles succeed in pulling each other into the range of the force, the resulting large positive kinetic energy may even unbind the system. Thus, H^2 is inclined to be a rather loosely bound structure, with the relative wavefunction of the two nucleons projecting a tail appreciably outside the range of the force; the kinetic energy corresponding to this somewhat spread-out structure is sufficiently small to make the deuteron a bound system with a small binding energy of about 2 MeV. As the number of nucleons is four in He^4, the kinetic energy tends to increase by a factor of 2 (the ratio of the number of particles), whereas the same factor for the potential energy is 6 (the ratio of the number of pairs). Therefore, the increased number of bonds between the particles can pull all the particles into the range of the nucleon-nucleon force; the consequent gain in kinetic energy cannot swamp the gain in potential energy because of the unequal multiplicative factors. Since all the four nucleons now stay within the nuclear force range, the binding energy per nucleon is able to reach about 7 MeV.

The short-ranged attractive potential that causes the quick rise of binding energy from H^2 to He^4 cannot, without any further modification, cause the saturation phenomenon. A potential that is always attractive is inclined to draw all the particles in a nucleus within the force range, and hence the binding energy of each pair becomes equally effective. Thus, the total binding energy tends to be proportional to the total number of pairs, $\frac{1}{2}A(A-1)$, or roughly to A^2 for large A, and the binding energy per nucleon, instead of being a constant, increases linearly with A.

The observed saturation may be produced either (i) through the exchange-dependence of the two-nucleon potential or (ii) by the possible existence of a repulsive core region of the two-nucleon potential. Historically, the earliest ideas on exchange were propounded by Heisenberg (1932) and Majorana (1933), who proposed a space-spin-exchange mechanism and a space-exchange mechanism, respectively. A pure spin-exchange mechanism was at the same time put forward by Bartlett. According to the exchange mechanism, two-nucleon states having opposite behaviour under the exchange have potentials with opposite signs. For instance, in the case of a space-exchange Majorana potential, the spatially symmetric states of the two nucleons (i.e., even orbital angular momentum states) remain unchanged under space exchange and hence they have an attractive potential, whereas the antisymmetric spatial states of the two nucleons change sign under space exchange and hence have a repulsive Majorana potential. The spin-exchange Bartlett potential is similarly attractive and repulsive in respectively the spin-triplet (symmetric under spin exchange) and spin-singlet (antisymmetric under spin exchange) states. Similar observations hold for the space-spin Heisenberg exchange potential. In addition to these three exchange-dependent potentials, there can be some non-exchange potential which is the same for all states; this is known as the Wigner potential. Detailed explanations of the foregoing behaviours under exchange are given in Sections 2 and 3. The fact that an exchange-dependent potential is attractive in some two-nucleon states and repulsive in others now implies that all the nucleons in a nucleus cannot condense into a small volume; the repulsion between many of the pairs tends to balance the dragging-in effect of the attractive pairs.

In case the two-nucleon potential has a repulsive core region, whenever the two nucleons approach each other closer than the core radius, the strong repulsion prevents the formation of the collapsed state. There is very strong evidence that both the exchange mechanism and the repulsive core simultaneously exist in the nuclear potential and together help to bring about saturation. Detailed calculations on these effects are presented in Chapter II. The concept of the repulsive core was initially adopted by analogy with the van der Waals force between molecules. This force is repulsive at short distances and attractive at large distances; as a result, liquids and solids in bulk are always characterized by the saturation property, i.e., they have a constant density, and their volume is therefore proportional to the total number of molecules present in the block of the sample. The idea of the exchange mechanism was also borrowed by Heisenberg from molecular physics in analogy with the property of the force between two hydrogen atoms in a hydrogen molecule; it can be proved, with the help of the wavefunctions of the two electrons in the two hydrogen atoms, that their interaction is attractive and repulsive, respectively, in states that are symmetric and antisymmetric under exchange. In the nuclear case, the exchange mechanism was simply *assumed* by Heisenberg and no attempt was made to justify it from more fundamental concepts.

Basic concepts, however, later emerged with the formulation of the meson theory of nuclear forces. The initial proposal by Yukawa in 1935 was dependent on the presence of a meson of mass equal to approximately 200 times the electron mass. The first meson, called the μ-meson, was identified in 1937 in cosmic rays. It was, however, realized that the μ-meson does not at all interact strongly with nucleons. The discovery in 1947 of the π-meson, having a mass somewhat higher than Yukawa's prediction, established the correctness of his proposition.

Nucleons interact by exchanging π-mesons. A derivation of the resultant potential can be obtained by applying a quantum field theory of the π-mesons. According to this theory, π-mesons are the quanta of the meson field, and can be created or destroyed in interaction with the nucleon field whose quanta are the nucleons. The process is analogous to the emission or absorption of light by a charged particle in motion, where the emitted or absorbed 'particle' is a photon, the quantum of the electromagnetic field. The predicted potential depends on the type of meson field used. For example, the meson could be a scalar particle (spin zero, parity even) or a pseudoscalar particle (spin zero, parity odd). In recent years, other types of mesons, e.g., the vector mesons (spin one, parity odd), have also been discovered. Various exchange potentials and the tensor potential have been derived by applying different types of meson field. More recently, the repulsive core potential has been shown to be associated with the exchange of two new mesons (ω, ϕ).

The standard approach in conventional nuclear theory, which is initially followed in our study, is to treat the two-nucleon interaction in a phenomenological manner, that is, *assume* it to be derivable from a potential, and then set up a reasonable form for it in agreement with general theoretical considerations and experimental data. Most experimental data providing direct information on the potential concern the two-nucleon systems: neutron-proton (n-p), proton-proton (p-p), and neutron-neutron (n-n). The n-p system produces a bound state, the ${}_1H^2$ nucleus, whereas no bound state exists in the p-p and n-n systems.

A considerable amount of data is available on the scattering of a proton by a proton or by a neutron. Such data have been compiled for incident proton energy, ranging from a few MeV to a few hundred MeV. For each energy value, measurements are made of the angular distribution of the scattered particle (represented by the differential cross-section—see Section 7A for precise definition), the total cross-section (integral of the differential cross-section over all angles), and several quantities dependent on the polarization of the spin of the scattered nucleon. An analysis of the properties of the deuteron, and the available n-p and p-p scattering data in terms of a two-body potential, is presented in detail in the rest of Chapter I. What needs emphasis here is that even the simplest data on the two-nucleon systems reveal that the force is really dependent on the type of two-nucleon state under consideration. In the n-p system, the state that corresponds to the ground state of ${}_1H^2$ is a predominantly spin-triplet (symmetric under spin exchange), S-state, i.e., state of orbital angular momentum zero (symmetric under space exchange). Hence, this state is symmetric under space-spin exchange. On the other hand, in n-n and p-p systems, because the two nucleons are identical, the Pauli exclusion principle demands that an allowed state be antisymmetric under space-spin exchange. Were there no state-dependence of the two-nucleon force, the interaction in an allowed state of n-n and p-p systems would have been the same as that in ${}_1H^2$, and n-n and p-p systems could then have a bound state. Similar evidence of state-dependence is found in low-energy n-p scattering data. For instance, it is shown in Section 11B that these data clearly indicate different scattering potentials for the spin-singlet and spin-triplet S-state.

Two very important observable properties of the nucleons are their spin and magnetic moment. Each nucleon has spin $\frac{1}{2}$ (in units of $\hbar$), like the electron, and it is therefore expected to satisfy the Dirac equation. According to the Dirac theory, a particle with charge e and of mass M has a magnetic moment $e\hbar/(2Mc)$. If we use the charge and mass of the proton in this quantity, then the resultant magnetic moment is called 1 nuclear magneton (nM), which serves as the unit in specifying the nucleon and nuclear magnetic moments. The proton is therefore expected to have a magnetic moment of 1 nM, and the neutron a magnetic moment equal to zero (since $e = 0$ for a neutron). Both these conclusions do not agree with experimental data. The observed moment of proton, equal to 2.78 nM, and that of neutron, equal to -1.93 nM, are therefore anomalous from the point of view of the Dirac theory. These anomalous values are explained in terms of the additional contribution made by the charged 'meson cloud' surrounding the nucleons.

Taking the nuclear dimension to be about 10^{-12} cm, we can use the uncertainty relation to estimate the nucleon momentum. The average kinetic energy of nucleons inside the nucleus obtained in this way is approximately 25 MeV, which is very small as compared with the rest mass-energy of a nucleon ($\approx$930 MeV). Therefore, in treating the nucleons theoretically, the complete sophistication of the relativistic Dirac equation is not necessary. The nonrelativistic Schrödinger equation, supplemented by the Pauli theory of spin, is sufficient for our purposes.

Each of the nucleons in a nucleus possesses, according to the foregoing description, an intrinsic spin $\frac{1}{2}$; every one of them also has angular momentum due to its rotational motion. All such angular momenta and intrinsic spins couple with one another and produce many possible values, in general, of the total angular momentum (sometimes called "spin" at the risk of confusion) of the entire nucleus. Depending on the nature of the interaction between nucleons, the state corresponding to one of these values has the least energy and therefore represents the ground state of the nucleus; the others signify excited states of the same nucleus. Since the orbital angular momentum has an integral quantum number, whereas the intrinsic spin is $\frac{1}{2}$, each nucleon has a total angular momentum that is half-integral (as per the quantum mechanical angular momentum coupling rule). Therefore, the states of a nucleus having an even number of nucleons always have integral spin, whereas those of an odd-mass nucleus have half-integral spin.

Several experimental techniques are currently used to determine the spins of the ground and excited states of nuclei. Many of these states also have a measurable magnetic dipole moment and an electric quadrupole moment. The latter arises whenever there is a lack of spherical symmetry of the nuclear charge distribution. As a consequence of simple angular momentum conservation (explained in detail in Section 6), states of angular momentum zero can have neither a dipole nor a quadrupole moment, whereas states of angular momentum $\frac{1}{2}$ cannot have any quadrupole moment. The ground-state moments of nuclei have been, whenever they are nonvanishing, very systematically measured through almost the entire periodic table. Comparatively recently, techniques have been developed and applied to the measurement of moments of excited states as well.

The compilation of data on the ground and excited states of nuclei falls within the scope of nuclear spectroscopy. The excited states of nuclei, studied in such work, are formed by bombarding target nuclei with energetic projectiles. Usually, electron, proton, and nuclei of such elements as H^2, H^3, He^3, and He^4, are energized in high-energy machines, known as nuclear accelerators, and used as projectiles to probe the interior of a nucleus. Heavier nuclei, for example, C^{12} and O^{16}, have also been employed as probes in recent times. Energetic electrons have a special advantage over other projectiles: the de Broglie wavelength for energy of a few hundred MeV is very much smaller in an electron than in a massive projectile because of the small rest mass of the electron. Therefore, high-energy electrons can probe more minute details of the interior of a nucleus. For studying the charge distribution in a nucleus, electrons are ideally suited. In many nuclear studies, the neutron, which is chargeless, is also considered a better probe than the aforementioned nuclei, all of which have positive charge. This is because the neutron, unlike such charged particles, is not repelled by the target nucleus, and so can penetrate quite deep inside a nucleus. However, neutrons cannot be accelerated in a machine because the acceleration is achieved by an electric field which affects only charged particles.

Each projectile can, after striking the target nucleus, give rise to nuclear reactions; that is, a different nucleon or a light nucleus may be ejected from the target, and, since the total charge and the mass of the target plus the incident projectile have to be conserved, the product nucleus remaining in the target is also different from the initial one. Due to momentum conservation, the product nucleus suffers a recoil, which, if sufficiently large in magnitude, enables it to move out of the target material. Experimental studies are usually made on the differential cross-section, total cross-section, and angular correlation between the outgoing particle and recoiling product nucleus. The polarization of the spin of the outgoing particle and of the recoiling nucleus, and sometimes the correlation in their polarizations, are also examined. Changing a nuclear species by bombarding with an energetic projectile is referred to as *nuclear transmutation*. The expression *nuclear reaction*, as used in nuclear physics, has a wider connotation than nuclear transmutation. The elastic and inelastic scattering of the incident projectile by the target nucleus fall within the definition of nuclear reaction. In scattering, the target nucleus and incident projectile remain unchanged; they can, however, exchange energy. If the projectile is scattered without delivering any energy to the target nucleus, then the scattering is called elastic, but if it delivers some energy when it gets scattered, then the scattering is said to be inelastic. In inelastic scattering, therefore, the target nucleus is left in a state of excitation by virtue of the extra energy it acquires from the projectile. During transmutation also, the product nucleus may be in its ground state or any of its excited states. The excited states usually decay to the ground state by emitting γ-rays. Sometimes, even the ground state of the product nucleus, originally resulting from the reaction, may not be stable in the sense that it may emit a β-particle (an electron or a positron) and get converted into a different nucleus. The examination of the β- and γ-rays associated with the product nuclei provides valuable data on their energy levels. This type of β- and γ-radioactivity, caused by bombarding with particles accelerated in the laboratory, is called *induced* radioactivity to distinguish it from the radioactivity of naturally occurring elements beyond Pb^{208}.

All the early nuclear transmutation experiments, following the discovery of the nucleus by Rutherford, employed α-particles from polonium (a naturally occurring radioactive element) as projectiles. With the use of Al^{27} as the target nucleus, it was established that the outgoing protons form two separate energy groups, corresponding to two different energy levels of the final nucleus Si^{30}; the γ-ray connecting these two levels was also observed. The same experiment, carried out with a thick target, proved that α-particles of several different energies were very effective in producing the reaction. Though the incident α-particles lost energy in the thick target and provided a continuous spectrum of energy to the target nucleus, the reaction took place for only a few select values of the energy. This phenomenon was interpreted as *resonance effect* in nuclear reactions. In other experiments, where Be^9 and B^{10} nuclei were used as targets, the outgoing particle was found to be a very penetrating radiation, like γ-rays. Chadwick correctly identified this particle as the neutron, whose high penetrating power is due to the lack of charge and consequent absence of electromagnetic interaction with matter.

A few years before these experiments with Al^{27}, several groups of workers started building nuclear accelerators. The success of their efforts was more or less simultaneous. In 1932, Cockroft and Walton developed their high-voltage machine and, with protons accelerated to a few hundred keV, caused the first artificial nuclear transmutation of Li^7, yielding two α-particles. The group headed by Lawrence in Berkeley succeeded in running their first cyclotron, a high-energy machine in which the charged particle is made to move in a circle of gradually increasing radius by confining it with the help of a perpendicular magnetic field and imparting energy to it, at each half-turn of its cycle, with an alternating electric field. About the same time, Van de Graaff launched his experimental electrostatic high-voltage generator. Machines using the cyclic principle of cyclotrons, and the high-voltage principle of the Van de Graaff generator, were simultaneously developed during the last three decades. Moreover, many ingenious innovations have been introduced to deliver ever higher energies accompanied by high-beam currents (caused by the motion of accelerated charged particles). These machines are now an almost indispensable part of experimental work in nuclear physics.

Special machines have also been built to accelerate electrons for nuclear physics research. Energetic projectiles from such accelerators have provided a wealth of experimental data on the energy levels of nuclei, and on the exact mechanism of nuclear reactions. A projectile may cause a nuclear reaction in various ways: for example, it may enter a nucleus, get absorbed, and the resultant nucleus (known as the *compound nucleus*) may disintegrate in several different ways; or the projectile may simply interact with only one or two particles near the surface of the nucleus and eject them while colliding with the nucleus; or the projectile itself may break up while interacting with the target nucleus, and one of its fragments may leave as the emergent particle, the others getting absorbed by the nucleus. A study of all such reaction mechanisms forms an interesting and active area of modern nuclear physics.

To complete this general survey, we mention two other nuclear processes, which are also regarded as special types of nuclear reaction. These processes are related to the elements at the beginning and end of the periodic table whose binding energies are smaller than the average value of about 8 MeV per nucleon for the rest of the nuclei. Let (Z_1, N_1) and (Z_2, N_2) be the number of protons and neutrons of two nuclei. Let us also consider another nucleus with the number of protons and neutrons as (Z, N), where $Z_1 + Z_2 = Z$ and $N_1 + N_2 = N$. We denote the masses of these three nuclei by M and their binding energies by B. Then

$$M(Z, N)c^2 = (ZM_p + NM_n)c^2 - B(Z, N),$$
$$M(Z_1, N_1)c^2 = (Z_1M_p + N_1M_n)c^2 - B(Z_1, N_1),$$
$$M(Z_2, N_2)c^2 = (Z_2M_p + N_2M_n)c^2 - B(Z_2, N_2).$$

Adding the last two expressions and subtracting from the first, we obtain

$$[M(Z, N) - M(Z_1, N_1) - M(Z_2, N_2)]c^2 = B(Z_1, N_1) + B(Z_2, N_2) - B(Z, N).$$

If nuclei 1 and 2 belong to the beginning of the periodic table and nucleus (Z, N) resulting from their fusion lies in the middle region, where the elements have normal binding energy, then

$$B(Z_1, N_1) + B(Z_2, N_2) < B(Z, N),$$

and hence, from the foregoing equation,

$$M(Z, N) < M(Z_1, N_1) + M(Z_2, N_2).$$

Since the sum of the rest mass-energies of nuclei 1 and 2 is greater than the rest mass-energy of the final nucleus, the excess rest mass-energy of the initial state must have been *evolved* during the fusion process. On the other hand, if nucleus (Z, N) belongs to the end of the periodic table and undergoes fission into two nearly equal fragments, which lie in the middle region, then $B(Z_1, N_1) + B(Z_2, N_2) > B(Z, N)$, and hence, from the same equation, $M(Z, N) > M(Z_1, N_1) + M(Z_2, N_2)$. Therefore, in the *fission* of the heavy nucleus (Z, N) into two smaller fragments, the rest mass-energy of the initial state is greater than the rest mass-energy of the final state; hence, the excess rest mass-energy will have been *evolved* in the fission process.

The fusion process (often referred to as the thermonuclear process) acts as the source of energy of the stars and, on a 'miniature' scale, of the H-bomb. The release of energy at a controlled rate by means of the fusion process is a coveted goal, and is currently the focal point of considerable research work in thermonuclear physics. In the fission process, a heavy nucleus such as U^{235} undergoes fission when struck by a slow neutron. The two medium-heavy nuclei, formed as a result of the fission, contain many more neutrons than do stable nuclei with the same Z-value. Therefore, the fission fragments emit some of these extra neutrons while trying to reach a stable state. Neutrons so emitted, when suitably slowed down, can once again cause fission of more U^{235} nuclei. Thus, the fission becomes a chain-reacting process and can be used as a self-sustaining process of energy output. In the so-called 'atom' bomb, this process is utilized in an uncontrolled manner so that a tremendous amount of energy is released within a very short time, thereby causing an 'explosion'. On the other hand, in a nuclear reactor, the same principle is applied with a controlled rate of energy release. In a power reactor, the controlled power output is converted into electrical energy. A reactor is an extremely useful machine for nuclear research purposes. It acts as a source of slow neutrons of tremendous flux. These neutrons can be used to bombard many nuclei and produce new isotopes, which can be used for studies in nuclear spectroscopy; many such isotopes serve clinical purposes also.

By far the biggest source of slow neutrons for experimental work is the nuclear reactor. Higher-energy neutrons are obtained by bombarding suitable targets with charged particles from accelerators.

The foregoing survey is intended to acquaint the reader with the scope of nuclear physics. The data on cross-section, angular distribution, polarization, etc., of nuclear reactions provide, as already mentioned, an insight into the reaction mechanism; in addition, they help us to

2 (45-122/1977)

test the wavefunctions of the various nuclear states involved in the reaction. The latter type of information is rather difficult to extract unambiguously from the reaction analysis because of the uncertainty about the exact reaction mechanism. In recent years, considerable progress has been made in this direction. Nuclear spectroscopic data, obtained from an analysis of β-, γ-decay, and the measurement of spins and moments offer a more direct test of nuclear wavefunctions. Theoretically, the wavefunctions of a nucleus containing many nucleons are quite difficult to calculate. In the first place, the exact nature of nuclear forces is uncertain. Secondly, the quantum mechanical calculation of a system of more than a few nucleons is almost impossible to carry out exactly. The approximate methods employed in nuclear physics are based on models of the nucleus, the choice of the model being guided by experimental data. For instance, the shell and collective models of nuclei have been very successfully applied to explain various kinds of experimental data over different regions of the periodic table. A deeper understanding of the origin, success, and interrelationship of these models has also been recently achieved. All these studies are grouped under the broad head of *nuclear structure*, which is the theme of this text. The field of nuclear structure is intimately connected with the two-nucleon forces because all the detailed structural characteristics of a heavy nucleus have to be understood in terms of the basic interactions between pairs of nucleons.

The study of the nuclear two-body problem provides the requisite information on two-nucleon forces. The three- and four-body systems, namely, H^3, He^3, and He^4, can be treated by fairly straightforward quantum mechanical methods, and hence they offer a valuable testing ground for nuclear forces derived from two-body data. As we go beyond the mass number four, the increasing complexity of the many-nucleon system makes it prohibitive to directly apply the many-body Schrödinger equation and compels recourse to the structural models mentioned earlier. A detailed presentation of nuclear structure theory, commencing with two-nucleon interaction and a description of nuclear models, forms the subject matter of our study. The calculation of an effective interaction for heavier nuclei from the actual two-nucleon interaction and a deeper understanding of structural models in terms of this effective interaction are also included. The book ends with the more sophisticated and up-to-date theoretical work of recent years, which attempts to arrive at a unified conceptual framework of the different structural models.

B. GENERAL CONCEPTS OF SYMMETRY

Some of the basic properties of the two-nucleon potential already discussed in Section 1A are: (i) the range, which is very short and assumed to be smaller than the spatial extent of the deuteron nucleus, and hence lies between 1 fm and 2 fm; (ii) the repulsive core whose radius is stipulated at much less than 1 fm; and (iii) the exchange-dependence. In Section 3, we shall construct very general forms of the two-nucleon potential based on these concepts and make certain broad physical *assumptions* on its symmetry characteristics. These characteristics also impose restrictions on the two-nucleon wavefunction, which can be inferred in a general manner without even solving the two-nucleon Schrödinger equation. These general characteristics of the wavefunctions are discussed in Section 2. To facilitate the reader's understanding of these symmetry arguments, we now present some general quantum mechanical results. Those who are already familiar with these concepts may proceed directly to Section 2.

The symmetry requirements we impose are of several kinds. One type corresponds to a *continuous* operation, such as the requirement of symmetry under the rotation of the coordinate system through *any* angle. Another corresponds to a *discrete* operation, such as the reflection of the coordinate system. Obviously, reflection is not a continuous process because the final orientation of the coordinate axes cannot be reached from the initial orientation by going through a continuous succession of stages. A *scalar* quantity remains invariant under these transformations of coordinate axes, whereas a pseudoscalar quantity remains unchanged under rotation but changes sign under reflection. Now, since the Hamiltonian H of a system of particles is the operator that stands for the total energy of the system, we expect, from physical considerations, that its expression does not depend on the choice of the coordinate axes; in other words, H should be a scalar under the combined symmetry operations of rotation and reflection. Exception to the invariance requirement of the Hamiltonian under reflection is known in other areas of physics (see Section 2).

Yet another type of symmetry requirement arises from the identity of particles. If P_{12} is the operator that exchanges the space and spin coordinates of the two indistinguishable particles, for instance, two neutrons or two protons, then the expression of their Hamiltonian H must remain unchanged under the operation of P_{12}. If we are considering a neutron and a proton, then, of course, the particles are distinguishable through their charge, rendering this particular symmetry requirement unnecessary. But there is a mathematical way of considering these two particles also as indistinguishable, in which case the definition of P_{12} has to be somewhat more generalized. This mathematical artifice is described in Section 2C. P_{12}, like reflection, is not a continuous symmetry operation.

A fourth kind of symmetry requirement, used in nuclear physics and also other branches of physics, concerns the reversal of the sign of time, and is usually called the time-reversal operation. This too is a discrete symmetry operation. Those who are not familiar with it should refer to the detailed mathematical description in Appendix D before going any further.

Let us denote the operator for any of the four symmetry operations by the symbol $\mathcal{S}$. The kinetic energy of each nucleon is given by $\frac{1}{2}Mv^2$, where $\mathbf{v} = d\mathbf{r}/dt$ is the velocity of the nucleon, and M is its mass. Under rotation, the square of any vector remains unchanged, and hence the kinetic energy is invariant. Under reflection, the vector $\mathbf{r}$ changes sign, i.e., $\mathbf{v}$ changes sign, but v^2, and hence the kinetic energy, remains unchanged. Under the time-reversal operation, t changes to $-t$, by definition, and so $\mathbf{v}$ changes to $-\mathbf{v}$, but once again $\frac{1}{2}Mv^2$ remains unchanged. The total kinetic energy of the two nucleons is given by $(\frac{1}{2}M_1v_1^2 + \frac{1}{2}M_2v_2^2)$, where the subscripts 1 and 2 label the two nucleons. Under P_{12}, the label 1 changes to 2 and 2 changes to 1, maintaining the same expression of the total kinetic energy. Thus, the kinetic energy is indeed invariant under all the foregoing symmetry operations. Therefore, the requirement of the invariance of H under any of the symmetry operations $\mathcal{S}$ is equivalent to imposing the same requirement on the two-body potential V.

Let us now consider the general requirement of the invariance of H under $\mathcal{S}$. This is described by the operator equation

$$\begin{aligned} \mathcal{S}H\mathcal{S}^{-1} &= H, \\ \mathcal{S}H &= H\mathcal{S}, \\ [\mathcal{S}, H] &= 0. \end{aligned} \tag{I.1}$$

The pair of square brackets denotes the *commutator* of the two operators and stands for $\mathcal{S}H - H\mathcal{S}$. It is known from elementary quantum mechanics that if two operators commute, they have simultaneous eigenstates. The construction of such states is now described.

If Ψ is an eigenfunction of H, the result (I.1) implies that $\mathcal{S}\Psi$ is also an eigenfunction of

H belonging to the same eigenvalue. The proof follows. Since $H\mathcal{S} = \mathcal{S}H$,

$$H\mathcal{S}\Psi = \mathcal{S}H\Psi = E\mathcal{S}\Psi,$$

where

$$H\Psi = E\Psi.$$

These two equations establish our assertion. Now there are two ways in which $\mathcal{S}\Psi$ and Ψ can have the same eigenvalue of H. First, $\mathcal{S}\Psi$ can be the state Ψ itself multiplied by a constant, in which case the constant is clearly the eigenvalue of $\mathcal{S}$ in the state Ψ, and the latter is indeed a simultaneous eigenfunction of H and $\mathcal{S}$. The second possibility arises if $\mathcal{S}\Psi$ is not simply proportional to Ψ, but represents a different wavefunction, say, Φ. In this case, we conclude that the energy eigenvalue E of H has a degeneracy. The states Ψ and Φ are degenerate with respect to H; neither of them is an eigenfunction of $\mathcal{S}$ on its own. If we continue this process, the second state Φ itself may produce another state when operated upon by $\mathcal{S}$, which must also be degenerate with Ψ and Φ. In this way, we may encounter a situation where a set of states, generated by the operation of $\mathcal{S}$, degenerate with respect to the eigenvalue of H. In general, none of these states is, on its own, an eigenfunction of $\mathcal{S}$, but suitable linear combinations of these states, which also have definite eigenvalues of $\mathcal{S}$, may be constructed.

Since the eigenstates of the symmetry operators, described earlier, are quite simple, we get a lot of information on the eigenfunctions of H without even using the Schrödinger equation.

Let us first consider the case where $\mathcal{S}$ stands for an *infinitesimal* rotation through an angle ϵ about *any* axis $\mathbf{n}$. Since rotation is a *continuous* operation, the consideration of an infinitesimal rotation is necessary and sufficient. From (BI.4) in Appendix B, the operator generating this rotation is given by

$$\mathbb{1} - i\epsilon\mathbf{n}\cdot\mathbf{J},$$

where $\mathbf{J}$ is the total angular momentum of the system. The commutator of this operator with H is proportional to $[\mathbf{J}, H]$, and hence, from (I.1), we obtain

$$[\mathbf{J}, H] = 0.$$

Since this equation is true for each component of $\mathbf{J}$,

$$[J^2, H] = 0.$$

The properties of the angular momentum operator $\mathbf{J}$ and its components are given in Appendix A (Section II). Most readers may already know these results from elementary quantum mechanics. Those who are not familiar with them should study Appendix A (Sections I–IV) at this point. Since J_x, J_y, J_z do not all commute with one another, but J^2 and any one component of $\mathbf{J}$ (say, J_z) do, we conclude that J^2, J_z, and H have simultaneous eigenfunctions. The eigenfunctions of J^2 and J_z are called angular momentum eigenfunctions. If $\mathbf{J}$ denotes the total angular momentum, then it is obtained by vectorially summing [see Appendix A (Section IV)] the orbital angular momentum and spin angular momentum, $\mathbf{L}$ and $\mathbf{S}$, respectively. The construction of the eigenfunctions of J^2 and J_z, in terms of orbital angular momentum eigenfunctions [the spherical harmonics of Appendix A (Section I)], and the spin eigenfunctions, with the help of the Clebsch-Gordon coefficients, are discussed in Appendix A (Section IV). Thus, the eigenfunctions of J^2 and J_z, which are completely known, are simultaneous eigenfunctions of H as well. The quantum numbers labelling these eigenfunctions [see Appendix A (Section II)] are j and m, where the eigenvalue of J^2 is $j(j+1)$ and that of J_z is m. Therefore, these quantum numbers label the eigenfunctions of H too. Any operator that commutes with H, for example, J^2 and J_z, thus provides its characteristic quantum number as a labelling index of the eigenfunction of H. Such a quantum number is subsequently referred to as a *good* quantum number.

For a Hamiltonian that is rotationally invariant, it is also possible to prove that all Ψ^j_m with $m = -j$ to $m = +j$ are degenerate with respect to the eigenvalue of H. Since H commutes with all the three components of **J**, (I.1) is true for $J_\pm = J_x \pm iJ_y$, i.e., the step-up and step-down operators. Hence, if Ψ^j_m is an eigenfunction of H, $(J_x \pm iJ_y)\Psi^j_m$ also is an eigenfunction belonging to the same eigenvalue of H. However, these two operators produce the states $\Psi^j_{m\pm1}$. Thus, the eigenvalue of H for the state Ψ^j_m does not depend on m if H is rotationally invariant.

We next consider the discrete symmetry operations. It is implied in the very nature of these operations that, if any of them is performed twice in succession, the Hamiltonian and the states should remain unaltered. For example, reflecting the coordinates twice means returning to the initial orientation of the axes. Therefore, under $\mathcal{S}^2$, where $\mathcal{S}$ is one of the discrete symmetry operators, the description of the state must remain unchanged. Since the wavefunctions Ψ and $e^{i\delta}\Psi$, where δ is a real phase parameter, represent the same physical state, we can, in general, have

$$\mathcal{S}^2\Psi = e^{i\delta}\Psi. \tag{I.2a}$$

If this has to hold for every arbitrary state Ψ, then δ should not depend on the Ψ chosen, and hence all the states Ψ are eigenstates of $\mathcal{S}^2$ belonging to the same eigenvalue. Without any loss of generality we may choose $e^{i\delta} = 1$. The case of time-reversal operation is somewhat special and is considered later in this section.

With the present choice $e^{i\delta} = 1$, we have

$$\mathcal{S}^2\Psi = \Psi,$$

$$\mathcal{S}^2 = \mathbb{1} \tag{I.2b}$$

in view of the arbitrariness of Ψ. Since (I.2b) can be factorized as

$$(\mathcal{S} - \mathbb{1})(\mathcal{S} + \mathbb{1}) = 0,$$

it follows that, corresponding to the two solutions $\mathcal{S} = \pm\mathbb{1}$ of this equation, there exist two *eigenvalues* of the operator $\mathcal{S}$, equal to ±1. Thus, if Ψ itself is a simultaneous eigenstate of H and $\mathcal{S}$, then there can be two kinds of state Ψ corresponding to the two eigenvalues of $\mathcal{S}$. They satisfy

$$\mathcal{S}\Psi = \pm\Psi.$$

When $\mathcal{S}$ is the reflection operator, these two kinds of eigenfunction Ψ are said to have *even* (+ sign) and *odd* (− sign) *parity*.

If $\mathcal{S}$ is now identified with the exchange operator P_{12}, then, in general, $P_{12}\Psi$ and Ψ will be different wavefunctions. According to our earlier discussion, these wavefunctions are degenerate with respect to the eigenvalue of H. The linear combinations $(\mathbb{1} \pm P_{12})\Psi$ have the same eigenvalue of H but different eigenvalues of P_{12}, namely, $+1$ and -1. The combination $(\mathbb{1} - P_{12})\Psi$, which has the eigenvalue -1, is said to be *antisymmetric* under exchange, whereas the other combination is called *symmetric*. According to the Pauli exclusion principle, two nucleons cannot occupy the same single-particle state. If ϕ and ϕ' denote two single-particle wavefunctions, then $\phi(1)\phi'(2)$, where 1 and 2 refer to nucleon coordinates, is a special form for the two-body wavefunction Ψ. The antisymmetric combination $(\mathbb{1} - P_{12})\Psi$ for this special kind of Ψ is obviously given by

$$\phi(1)\phi'(2) - \phi(2)\phi'(1).$$

If $\phi = \phi'$, this antisymmetric combination identically vanishes. Thus, the Pauli exclusion principle is incorporated into the two-nucleon wavefunction by working with the *antisymmetric* two-nucleon states.

Finally, we consider the case of the time-reversal operator T. It is shown in Appendix D that, according to a standard phase convention,

$$T\Psi^j_m = (-1)^{j-m}\Psi^j_{-m}.$$

Therefore,

$$T^2\Psi^j_m = (-1)^{j-m}T\Psi^j_{-m} = (-1)^{j-m}(-1)^{j+m}\Psi^j_m = (-1)^{2j}\Psi^j_m.$$

Thus, according to the phase convention adopted here, $e^{i\delta}$ of (I.2a) is, in the case of time-reversal, chosen equal to $+1$ when j is an integer, and equal to -1 when j is half-integral. These two cases correspond to wavefunctions of an even and odd number of nucleons, respectively. In the first case, T has the eigenvalues ± 1, as proved earlier, whereas in the second case, T^2 has the eigenvalue -1 and it is impossible to find an eigenstate of the operator T itself.

The states Ψ and $T\Psi$ have the same eigenvalue of H and, by a suitable linear combination of the two, we can construct two states that correspond to the two eigenvalues of T in the case of even nuclei. These are given by

$$(\mathbb{1} \pm T)\Psi.$$

This assertion can be verified by applying T directly on these two linear combinations. The fact that Ψ and $T\Psi$ have the same eigenvalue of H when H is invariant under T represents a nontrivial result only if we are considering a special Hamiltonian H that is not invariant under a completely general rotation. In that case, Ψ^j_m and $(-1)^{j-m}\Psi^j_{-m}$ are proved to be degenerate by the present method; but we have already proved a more general result, namely, the wavefunctions Ψ^j_m for all m between $-j$ and j are degenerate for a rotationally invariant H. In the case of the two-nucleon Hamiltonian, we assume a complete rotational invariance; hence, its time-reversal invariance does not give any extra information on the two-nucleon wavefunction. In Chapters III, IV, and V, we shall consider a one-body type Hamiltonian of the whole nucleus given by

$$\sum_{i=1}^{A} (T_i + \mathcal{V}_i),$$

where the summation goes over all the nucleons, T denotes the kinetic energy, and $\mathcal{V}$ stands for an average potential in which the nucleons move. In Chapter III, $\mathcal{V}$ is regarded as a spherically symmetric potential, and hence invariant under a general rotation. In this case, therefore, the single-nucleon wavefunctions corresponding to the single-nucleon Hamiltonian $T + \mathcal{V}$ are of the type ψ^j_m. In Chapters IV and V, we shall deal with more general average potentials, which may have either a symmetry with respect to any rotation about one special axis, or no symmetry at all under continuous rotation about any axis. In both these cases, the single-particle eigenfunctions ψ_m (or ψ) and $T\psi_m$ (or $T\psi$) of H are degenerate in energy.

2. GENERAL SYMMETRY PROPERTIES OF TWO-NUCLEON HAMILTONIAN AND TWO-NUCLEON STATES

As pointed out in Section 1, our main aim in studying the two-nucleon problem (i.e., the deuteron, and two-nucleon scattering) is to obtain information on the two-nucleon interaction. Because of the complexity of this interaction, arising out of its state-dependence, it is useful to begin with very general symmetry requirements and proceed to derive the maximum number of unambiguous conclusions from experimental data. We shall start by assuming the validity of five properties of the interaction.

(i) The interaction does not depend on the choice of coordinate axes; rather, it remains invariant under the rotation of these axes. This requires the interaction to be a scalar [or a pseudoscalar, which is ruled out by property (iii)] in ordinary space, and hence the eigenfunctions of the two-nucleon system will correspond to definite values of the total angular momentum quantum number j.

(ii) The interaction is charge-independent. That is, apart from the Coulomb interaction between a pair of protons, the interaction in p-p, n-n, and n-p systems is the same. This statement requires further qualification. Because of the requirement of the Pauli exclusion principle with respect to space-spin coordinates in p-p and n-n systems (i.e., systems of two *identical* Fermions), the states possible for them are possible also for the n-p system, but not vice versa. Charge-independence holds only for such states; the extra states characteristic of only the n-p system may exhibit different interactions. Charge-independence is a very simple hypothesis, which has been found to work in the analysis of two-body data. Some high-energy data may be explained by assuming different interactions for n-p and p-p systems; they can also be explained by retaining the charge-independence hypothesis. Thus, the hypothesis, though not very unambiguously established, is favoured for its simplicity.

(iii) The interaction remains invariant under a reflection of the coordinate system. This rules out a pseudoscalar nature for the Hamiltonian, and implies that the eigenfunctions of the two-nucleon system can be classified according to even or odd parity.

That parity is not conserved in weak interactions is now well-known. No compelling evidence on the violation of parity conservation has yet been obtained in strong interaction physics. Experiments have set a limit to the extent to which parity may still be violated in nucleon-nucleon interaction. This limit is extremely small.

(iv) The interaction remains invariant under Galilean transformation, according to which a physical theory remains unchanged if a transformation is made from one coordinate system to another having a uniform relative velocity. Considered in the new coordinate frame, each of the momenta $\mathbf{k}_1, \mathbf{k}_2, \ldots$ of the two nucleons in the old frame acquires an extra additive momentum, say, $\boldsymbol{\varkappa}$, arising from the motion of the new coordinate frame. Thus, Galilean invariance demands, for the matrix elements of the two-nucleon potential V in momentum space, the equality

$$\langle \mathbf{k}_1', \mathbf{k}_2' | V | \mathbf{k}_1, \mathbf{k}_2 \rangle = \langle \mathbf{k}_1' + \boldsymbol{\varkappa}, \mathbf{k}_2' + \boldsymbol{\varkappa} | V | \mathbf{k}_1 + \boldsymbol{\varkappa}, \mathbf{k}_2 + \boldsymbol{\varkappa} \rangle.$$

Using the definitions of total and relative momenta, namely,

$$\mathbf{K} = \mathbf{k}_1 + \mathbf{k}_2, \qquad \mathbf{k} = \tfrac{1}{2}(\mathbf{k}_1 - \mathbf{k}_2),$$

we can write the equality also as

$$\langle \mathbf{K}', \mathbf{k}' | V | \mathbf{K}, \mathbf{k} \rangle = \langle \mathbf{K}' + 2\boldsymbol{\varkappa}, \mathbf{k}' | V | \mathbf{K} + 2\boldsymbol{\varkappa}, \mathbf{k} \rangle. \tag{I.3a}$$

Expressing the matrix element on the left-hand side explicitly in terms of coordinate space integrals, we obtain

$$\langle \mathbf{K}', \mathbf{k}'| V |\mathbf{K}, \mathbf{k}\rangle = \int d^3R' \int d^3r' \int d^3R \int d^3r \langle \mathbf{K}', \mathbf{k}' \mid \mathbf{R}', \mathbf{r}'\rangle\langle \mathbf{R}', \mathbf{r}'| V |\mathbf{R}, \mathbf{r}\rangle\langle \mathbf{R}, \mathbf{r} \mid \mathbf{K}, \mathbf{k}\rangle$$

$$= \frac{1}{(2\pi)^6} \int d^3R' \int d^3r' \int d^3R \int d^3r \langle \mathbf{R}', \mathbf{r}'| V |\mathbf{R}, \mathbf{r}\rangle \exp(-i\mathbf{K}'\cdot\mathbf{R}') \exp(i\mathbf{K}\cdot\mathbf{R})$$

$$\times \exp(-i\mathbf{k}'\cdot\mathbf{r}') \exp(i\mathbf{k}\cdot\mathbf{r}).$$

Here, $\mathbf{R}$ and $\mathbf{r}$ denote the centre-of-mass and the relative coordinate vectors

$$\mathbf{R} = \tfrac{1}{2}(\mathbf{r}_1 + \mathbf{r}_2), \qquad \mathbf{r} = \mathbf{r}_1 - \mathbf{r}_2,$$

which are canonically conjugate to $\mathbf{K}$ and $\mathbf{k}$, respectively. The factor $(2\pi)^{-6}$ comes from the $(2\pi)^{-3/2}$ normalization constant of each momentum eigenfunction. In the same manner, the right-hand side of (I.3a) also can be expressed in terms of coordinate integrals; the only difference we will have in that case is the occurrence of the additional factor $\exp[i2\boldsymbol{\varkappa}\cdot(\mathbf{R} - \mathbf{R}')]$ in the integrand. In order that the two expressions be equal for *any* $\boldsymbol{\varkappa}$, as demanded by (I.3a), we should have this additional factor in the integrand equal to unity, i.e., $\mathbf{R} = \mathbf{R}'$. This equality is guaranteed if the coordinate space matrix element of V vanishes whenever $\mathbf{R} \neq \mathbf{R}'$. Therefore, Galilean invariance requires

$$\langle \mathbf{R}', \mathbf{r}'| V |\mathbf{R}, \mathbf{r}\rangle = \delta(\mathbf{R}, \mathbf{R}')\langle \mathbf{R}, \mathbf{r}'| V |\mathbf{R}, \mathbf{r}\rangle. \tag{I.3b}$$

(v) The interaction is invariant under translation of the origin of the coordinate system through *any* distance $\mathbf{a}$. According to this requirement, we have

$$\langle \mathbf{r}_1', \mathbf{r}_2'| V |\mathbf{r}_1, \mathbf{r}_2\rangle = \langle \mathbf{r}_1' + \mathbf{a}, \mathbf{r}_2' + \mathbf{a}| V |\mathbf{r}_1 + \mathbf{a}, \mathbf{r}_2 + \mathbf{a}\rangle,$$

$$\langle \mathbf{R}', \mathbf{r}'| V |\mathbf{R}, \mathbf{r}\rangle = \langle \mathbf{R}' + \mathbf{a}, \mathbf{r}'| V |\mathbf{R} + \mathbf{a}, \mathbf{r}\rangle. \tag{I.3c}$$

In close analogy with what followed (I.3a), we now express the left- and right-hand side of (I.3c) in terms of momentum integrals. The reader will then be convinced that (I.3c) can be satisfied if $\mathbf{K}' = \mathbf{K}$, i.e.,

$$\langle \mathbf{K}', \mathbf{k}'| V |\mathbf{K}, \mathbf{k}\rangle = \delta(\mathbf{K}, \mathbf{K}')\langle \mathbf{K}, \mathbf{k}'| V |\mathbf{K}, \mathbf{k}\rangle. \tag{I.3d}$$

Next, by combining the requirements (I.3b) and (I.3d), we show that the matrix element of V on the right-hand side of (I.3b) is, in fact, *independent* of $\mathbf{R}$ and the matrix element of V on the right-hand side of (I.3d) is independent of $\mathbf{K}$. To clarify this, let us introduce momentum integrals and write

$$\langle \mathbf{R}, \mathbf{r}'| V |\mathbf{R}, \mathbf{r}\rangle = \frac{1}{(2\pi)^6} \int d^3K' \int d^3k' \int d^3K \int d^3k \langle \mathbf{R}, \mathbf{r}' \mid \mathbf{K}', \mathbf{k}'\rangle\langle \mathbf{K}', \mathbf{k}'| V |\mathbf{K}, \mathbf{k}\rangle\langle \mathbf{K}, \mathbf{k} \mid \mathbf{R}, \mathbf{r}\rangle.$$

On the right-hand side, using (I.3d) in the matrix element of V and carrying out the $\mathbf{K}'$-integration, we obtain

$$\langle \mathbf{R}, \mathbf{r}'| V |\mathbf{R}, \mathbf{r}\rangle = \frac{1}{(2\pi)^6} \int d^3K \int d^3k' \int d^3k \exp(i\mathbf{k}'\cdot\mathbf{r}') \exp(-i\mathbf{k}\cdot\mathbf{r})\langle \mathbf{K}, \mathbf{k}'| V |\mathbf{K}, \mathbf{k}\rangle.$$

As the integral on the right-hand side is not in any way dependent on $\mathbf{R}$, we have verified our earlier assertion. Therefore, we finally write

$$\langle \mathbf{R}', \mathbf{r}'| V |\mathbf{R}, \mathbf{r}\rangle = \delta(\mathbf{R}, \mathbf{R}')\langle \mathbf{r}'| V |\mathbf{r}\rangle. \tag{I.3e}$$

Proceeding in an analogous manner (the proof is left to the reader), we easily establish

$$\langle \mathbf{K}', \mathbf{k}'| V |\mathbf{K}, \mathbf{k}\rangle = \delta(\mathbf{K}, \mathbf{K}')\langle \mathbf{k}'| V |\mathbf{k}\rangle. \tag{I.3f}$$

A potential with the general coordinate space representation, as shown in (I.3e), is called *nonlocal.* In the special case of a *local* potential, we have, by the definition of locality, $\mathbf{r} = \mathbf{r}'$ in the matrix elements of V, and then (I.3e) reduces to

$$\langle \mathbf{R}', \mathbf{r}'| V |\mathbf{R}, \mathbf{r}\rangle = \delta(\mathbf{R}, \mathbf{R}')\delta(\mathbf{r}, \mathbf{r}')V(\mathbf{r}), \tag{I.3g}$$

where the diagonal matrix element $\langle \mathbf{r}| V |\mathbf{r}\rangle$ has been appropriately written as a function of $\mathbf{r}$. In our study on the two-nucleon problem, we shall *assume* a local potential, though a particular type of nonlocal potential is mentioned in Section 12.

The foregoing general requirements lead us to certain general characteristics of the two-nucleon eigenstates which are now described in detail.

A. SPATIAL STATES

If $\mathcal{E}$ denotes the total energy, and ψ the two-body wavefunction, then the Schrödinger equation is given by

$$[-\frac{\hbar^2}{2M}(\nabla_1^2 + \nabla_2^2) + V]\psi = \mathcal{E}\psi.$$

Transforming $\mathbf{r}_1$ and $\mathbf{r}_2$ to $\mathbf{r}$ and $\mathbf{R}$, we get

$$(-\frac{\hbar^2}{2\mu}\nabla^2 - \frac{\hbar^2}{2\mathcal{M}}\nabla_{\mathbf{R}}^2 + V)\psi = \mathcal{E}\psi,$$

where the reduced mass $\mu = \frac{1}{2}M$, the total mass $\mathcal{M} = 2M$, $\mathbf{\nabla}$ is the gradient operator for the relative coordinate $\mathbf{r}$, and $\mathbf{\nabla}_{\mathbf{R}}$ is the same for the centre-of-mass coordinate $\mathbf{R}$. It is clear from this equation that, since the spatial dependence of the potential V is through the relative coordinate $\mathbf{r}$ alone, the equation is easily separable into relative and centre-of-mass motion. If e denotes the kinetic energy of the centre-of-mass in a state Φ, then

$$-\frac{\hbar^2}{2\mathcal{M}}\nabla_{\mathbf{R}}^2\Phi = e\Phi$$

and the Schrödinger equation simplifies to

$$(-\frac{\hbar^2}{M}\nabla^2 + V)\Psi = E\Psi, \tag{I.3h}$$

where $\psi = \Phi\Psi$, Ψ is the wavefunction for the relative motion, and $E(=\mathcal{E} - e$, i.e., the total energy minus the centre-of-mass energy) is clearly the energy of relative motion (kinetic plus potential). In the centre-of-mass coordinate system, the centre-of-mass is stationary, i.e., $e = 0$, and hence the centre-of-mass wavefunction Φ is a constant.

A spatial wavefunction Ψ of (I.3h) for a given orbital angular momentum quantum number l of relative motion can be written as (see Schiff[2])

$$\frac{u_l(r)}{r}Y_m^l(\theta, \phi),$$

where $u_l(r)$ is the pure radial function and $Y_m^l(\theta, \phi)$ is the orbital angular momentum function. Under the parity operation $\mathbf{r} \rightarrow -\mathbf{r}$, this wavefunction is multiplied by $(-1)^l$. Thus, even-parity eigenfunctions will have even values of l, and odd-parity eigenfunctions will have odd

3 (45-122/1977)

values of l. Since the total angular momentum quantum number j ($\mathbf{j} = \mathbf{l} + \mathbf{S}$, where $\mathbf{S}$ is the spin of the two-nucleon system) is required to be a good quantum number, the spatial eigenfunctions can be mixtures of either even or odd values of l consistent with a given j.

The effect of interchanging the spatial coordinates $\mathbf{r}_1$ and $\mathbf{r}_2$ is to change $\mathbf{r} \to -\mathbf{r}$. Thus, the space-exchange operation P_r is equivalent to the parity operation. It follows from what has been said that the two-nucleon eigenstates will have definite symmetry characteristic under spatial exchange: the even-parity function will be symmetric, and the odd-parity function, antisymmetric.

B. SPIN STATES

The construction of the spin functions $\chi_M^S(\sigma_1, \sigma_2)$ having specified values of the total spin quantum number S and its projection M is straightforward. If the individual spin functions [see Appendix A (Section III)] are denoted by $\chi_\mu^{1/2}(\sigma)$, where $\frac{1}{2}$ is the spin quantum number and $\mu = \pm\frac{1}{2}$ are its projection, then the coupling of the two spin $\frac{1}{2}$ particles can be done with the help of the Clebsch-Gordon coefficients [see Appendix A (Section IV)]:

$$\chi_M^S(\sigma_1, \sigma_2) = \sum_{\mu, \mu'} \begin{bmatrix} \frac{1}{2} & \frac{1}{2} & S \\ \mu & \mu' & M \end{bmatrix} \chi_\mu^{1/2}(\sigma_1)\chi_{\mu'}^{1/2}(\sigma_2). \tag{I.4a}$$

The two possible values of S are 0 and 1, and the corresponding wavefunctions given by (I.4a) can be explicitly written down, using the values of the Clebsch-Gordon coefficients listed in Appendix A (Table AVII.1). We thus have

$$\chi_0^0(\sigma_1, \sigma_2) = \frac{1}{\sqrt{2}}[\chi_{1/2}^{1/2}(\sigma_1)\chi_{-1/2}^{1/2}(\sigma_2) - \chi_{-1/2}^{1/2}(\sigma_1)\chi_{1/2}^{1/2}(\sigma_2)] \qquad \text{(singlet)}, \tag{I.4b}$$

$$\chi_{-1}^1(\sigma_1, \sigma_2) = \chi_{-1/2}^{1/2}(\sigma_1)\chi_{-1/2}^{1/2}(\sigma_2),$$

$$\chi_0^1(\sigma_1, \sigma_2) = \frac{1}{\sqrt{2}}[\chi_{1/2}^{1/2}(\sigma_1)\chi_{-1/2}^{1/2}(\sigma_2) + \chi_{-1/2}^{1/2}(\sigma_1)\chi_{1/2}^{1/2}(\sigma_2)] \qquad \text{(triplet)}, \tag{I.4c}$$

$$\chi_1^1(\sigma_1, \sigma_2) = \chi_{1/2}^{1/2}(\sigma_1)\chi_{1/2}^{1/2}(\sigma_2).$$

It is clear from these expressions that, under the exchange $\sigma_1 \leftrightarrow \sigma_2$ of the spin coordinates of the two nucleons, the singlet function changes sign and the three triplet functions remain unchanged. If we denote the spin-exchange operator by P_σ, then this result can be stated as

$$P_\sigma \chi_M^S(\sigma_1, \sigma_2) = (-1)^{1+S}\chi_M^S(\sigma_1, \sigma_2). \tag{I.5a}$$

It should be noted that σ, used as the spin coordinate in the arguments of these spin functions, is a two-valued variable, and is therefore not treated as a vector. On the other hand, when the spin *operator* is required, we shall denote it by the vector $\boldsymbol{\sigma}$ with the components σ_x, σ_y, σ_z.

The spin functions $\chi_\mu^{1/2}(\sigma)$ with $\mu = +\frac{1}{2}$ and $\mu = -\frac{1}{2}$ are often referred to as the "up" and "down" spin states, respectively. Some authors denote the "up" state by α and the "down" state by β. The notation in this text, though more complicated, has the advantage of being consistent with the general angular momentum notation used in Appendix A (Section II).

C. ISOTOPIC SPIN STATES

Two-nucleon wavefunctions depend also on the charge of the two nucleons. A convenient way to take this dependence into account is to define an artificial spin operator, called the *isotopic spin* $\boldsymbol{\tau}$. In analogy with the ordinary spin operator $\boldsymbol{\sigma}$, the z-component of $\boldsymbol{\tau}$ is required to have

two possible values $+1$ and -1, which may be related to the neutron and the proton, respectively. The opposite convention could also have been consistently followed; but the fact that stable heavy nuclei have a neutron excess implies that the convention we are following has the advantage of giving positive values of the projection of the total isotopic spin in such nuclei.

The individual particle isospin states may be denoted by $\rho_{\mu}^{1/2}(\tau)$, where $\mu = +\frac{1}{2}$ (neutron) and $\mu = -\frac{1}{2}$ (proton). We should recall here that $\boldsymbol{\sigma} = 2\mathbf{s}$, where $\mathbf{s}$ is the actual spin, and $s_z = \pm\frac{1}{2}$. Similarly, $\boldsymbol{\tau} = 2\mathbf{t}$, where $\mathbf{t}$ is the actual isospin, and $t_z = \pm\frac{1}{2}$.

The construction of two-nucleon states of total isospin T and projection M_T is similar to that in (I.4a). The symmetry under exchange of the two nucleons also follows in an analogous manner, i.e., the states $T = 1$ are symmetric, and the state $T = 0$ is antisymmetric. If P_τ is the exchange operator, then the relation analogous to (I.5a) is given by

$$P_\tau \rho_{M_T}^T(\tau_1, \tau_2) = (-1)^{1+T} \rho_{M_T}^T(\tau_1, \tau_2). \tag{I.5b}$$

The main difference between a nucleon's intrinsic spin $\boldsymbol{\sigma}$ and isospin $\boldsymbol{\tau}$ is that the components of $\boldsymbol{\sigma}$ are associated with the *actual* x-, y-, z-direction of the coordinate system in the *physical* space, whereas the components of $\boldsymbol{\tau}$ in the three directions have no relation whatsoever to the physical space. Therefore, we shall refer to τ_x, τ_y, τ_z as the components of $\boldsymbol{\tau}$ along three mutually perpendicular directions in the *fictitious isospin space*. The only physically significant quantity about the isospin of the nucleon is the value of τ_z, which specifies whether a given nucleon is a neutron or a proton. The z-component of a two-body $\mathbf{T}$, namely, T_z, is simply the sum of $\frac{1}{2}\tau_z$ of the two nucleons; hence, the value M_T of T_z is also a physically significant quantity, elucidating which of the systems—the p-p, the n-n, and the n-p—is being considered. The values of M_T for these three systems are respectively -1, $+1$, and 0. The projections -1 and $+1$ can result only from $T = 1$, whereas the projection 0 can result from both $T = 1$ and $T = 0$. Thus, n-n and p-p systems have the state $T = 1$, whereas the n-p system has both the states $T = 1$ and $T = 0$.

The requirement of *charge-independence* establishes that any space-spin state allowed in n-n, n-p, and p-p systems has the same interaction in all the three systems. The implication, in terms of isotopic spin, is that the interaction in the states $T = 1$ does not depend on the value of the projection quantum number M_T of the state. However, as the state $T = 0$ is present only in the n-p system, its interaction may differ from that in the state $T = 1$, without violating the charge-independence hypothesis.

Another hypothesis, which is less restrictive, is known as *charge symmetry*. According to this hypothesis, n-n and p-p systems must have the same nuclear interaction, but the n-p system has a different interaction even in the same space-spin state. This therefore requires that, of the three substates of $T = 1$, $M_T = \pm 1$ have the same interaction, whereas $M_T = 0$ may have a different interaction.

Let us consider the implication of charge-independence a little further. Since the interaction does not depend on the value of M_T, i.e., the value of the projection of T along the z-axis in the isospin space, it follows that the z-axis can be chosen in *any* direction without altering the interaction. In other words, the interaction must be invariant with respect to *any* arbitrary rotation of the coordinate axes in the isospin space. Therefore, as explained in Appendix B, the isospin part of the interaction must be a *scalar* (rank zero) with respect to rotations in the isospin space. In analogy with the result derived in Section 1B for the ordinary angular momentum operator $\mathbf{J}$ and rotation in the physical space, we can establish that the isospin part of the two-nucleon eigenfunction will have T and M_T as good quantum numbers.

After isospin has been introduced, any two nucleons can be treated as identical Fermions, and a generalized Pauli exclusion principle demanded for them under the combined exchange of space, spin, and isospin coordinates. The usual requirement of antisymmetry under the exchange of space and spin coordinates for p-p and n-n systems is now replaced by the antisymmetry under exchange of space, spin, and isospin coordinates for p-p, n-n, as well as n-p systems. Therefore, this principle requires

$$P_r P_\sigma P_\tau \Psi(\mathbf{r}, \sigma_1, \sigma_2, \tau_1, \tau_2) = -\Psi(\mathbf{r}, \sigma_1, \sigma_2, \tau_1, \tau_2), \tag{I.6}$$

where Ψ is an eigenfunction of the two-body system. Since the eigenfunction has already been shown to have definite symmetry characteristics with respect to P_r and P_τ, it follows from (I.6) that it should have definite symmetry characteristic also under P_σ. The product $P_r P_\sigma P_\tau$ is required to have the value -1, and hence the combinations allowed are

(i) $P_\tau = +1$

$P_r = 1, \qquad P_\sigma = -1,$

$P_r = -1, \qquad P_\sigma = 1;$

(ii) $P_\tau = -1$

$P_r = 1, \qquad P_\sigma = 1,$

$P_r = -1, \qquad P_\sigma = -1.$

The requirement of definite spin symmetry means that the eigenfunctions are characterized by definite values of S. We should note that this follows from the Pauli principle, and not from any *direct* invariance requirement of the two-nucleon Hamiltonian. On the other hand, the fact that T and parity need to be good quantum numbers follows from *direct* invariance requirements.

The allowed states, in accordance with (i) and (ii), are

(i) $T = 1$

$l =$ even, $\qquad S = 0,$

$l =$ odd, $\qquad S = 1;$

(ii) $T = 0$

$l =$ even, $\qquad S = 1,$

$l =$ odd, $\qquad S = 0.$

The different values of orbital angular momentum are usually denoted by spectroscopic nomenclature:

$$S\ (l = 0), \quad P\ (l = 1), \quad D\ (l = 2), \quad F\ (l = 3), \quad G\ (l = 4), \quad H\ (l = 5), \ldots .$$

The states $S = 0$ and $S = 1$ are usually denoted by their multiplicity: spin-singlet and spin-triplet. Similarly, the states $T = 0$ and $T = 1$ are called isospin-singlet and isospin-triplet.

Although l is not a conserved quantum number, j is. For a given j, l can have values ranging from $l = j + S$ to $l = |j - S|$ [see Appendix A (Section IV)]. Thus, in a spin-singlet state, only one value, namely, $l = j$, is allowed. On the other hand, in a spin-triplet state, there are three possible values of l for a given j, namely, $l = j$ and $l = j \pm 1$. The parity of the state $l = j$ is obviously opposite to that of the states $l = j \pm 1$. The two states $l = j \pm 1$ will, in general, mix with each other in a spin-triplet eigenfunction of parity $(-1)^{j+1}$.

We now list states of low orbital angular momentum [the triplet states that mix with each other are connected by a + sign; the spin multiplicity $(2S + 1)$ is given as a superscript and the total angular momentum j as a subscript to the letters $S, P, D, \ldots$, which stand for orbital angular momenta]:

$$\begin{aligned} &{}^1S_0,\ {}^1D_2,\ {}^1G_4,\ \ldots, \\ &{}^3P_0,\ {}^3P_1,\ ({}^3P_2 + {}^3F_2),\ {}^3F_3,\ ({}^3F_4 + {}^3H_4),\ \ldots, \end{aligned} \tag{I.7}$$

$$\begin{aligned} &{}^1P_1,\ {}^1F_3,\ {}^1H_5,\ \ldots, \\ &({}^3S_1 + {}^3D_1),\ {}^3D_2,\ ({}^3D_3 + {}^3G_3),\ {}^3G_4,\ \ldots . \end{aligned} \tag{I.8}$$

The states in (I.7) have $T = 1$, and those in (I.8) have $T = 0$. However, the T-value is automatically determined from the parity, and spin S by the generalized Pauli exclusion principle, and therefore need not be explicitly mentioned.

It has already been noted that n-n and p-p systems have only the states $T = 1$, whereas the n-p system has both the states $T = 1$ and $T = 0$. The n-p system therefore has twice as many states as the n-n system or the p-p system. In the foregoing list, only the states (I.7) occur in n-n and p-p systems, whereas both (I.7) and (I.8) occur in the n-p system. Since $T = 1$ is symmetric under exchange, the wavefunctions for the states (I.7) are antisymmetric under space-spin exchange; that is, they satisfy the ordinary Pauli exclusion principle. Since n-n and p-p systems correspond to $T = 1$ alone, the corresponding wavefunctions satisfy the ordinary Pauli exclusion principle, as expected. On the other hand, since both the states $T = 1$ and $T = 0$ are possible for the n-p system, this system has states that are both symmetric and antisymmetric under ordinary space-spin exchange. In other words, the n-p wavefunctions span the entire space of the space-spin functions, as they should. It is clear then that the isospin formalism has not altered any physics of the problem. It has simply separated the states common to the n-n, p-p, and n-p systems from the states that are present in the n-p system alone. It has further put the states common to the three systems on an identical footing, thereby delineating their charge-independence.

3. GENERAL FORMS OF TWO-NUCLEON INTERACTION

The charge-independence hypothesis limits the dependence of the two-nucleon interaction on isospin operators to two possible forms, namely,

$$\begin{aligned} &\mathbb{1}, \\ &\boldsymbol{\tau}_1 \cdot \boldsymbol{\tau}_2, \end{aligned} \tag{I.9}$$

because these are the only two independent scalar quantities in the isospin space that can be formed with $\boldsymbol{\tau}_1$ and $\boldsymbol{\tau}_2$.

Rotation-reflection invariance also restricts space-spin dependence to scalar expressions formulated from the variables $\mathbf{r}$, $\mathbf{p}$, $\boldsymbol{\sigma}_1$, and $\boldsymbol{\sigma}_2$, where $\mathbf{p}$ is the relative momentum. To limit the total number of possible expressions, we introduce two more requirements, which are generally valid.

(i) The two-nucleon interaction is invariant under time-reversal. Even in weak interactions, where parity violation has been established, the time-reversal invariance is still regarded as valid. Strong interactions too have so far been found to satisfy this particular invariance requirement.

(ii) The two-nucleon interaction is invariant under the exchange of the nucleons. Since the permissible isospin expressions (I.9) satisfy this requirement, the space-spin part of the interaction should also do so. Just because the Hamiltonian has this invariance property, the eigenfunctions turn out to be of definite symmetry under particle exchange.

In order to further limit the possible forms of interaction, an assumption is usually made about its momentum dependence. The restrictive nature of this assumption has been increasingly relaxed as data on high-energy differential cross-section, polarization, and correlation of spins have accumulated with the passage of time. The various stages of this assumption are now described.

(i) The interaction does not depend on $\mathbf{p}$. This assumption implies a strictly static two-nucleon interaction.

(ii) The interaction is only linearly dependent on $\mathbf{p}$. This may be a good approximation as long as we restrict ourselves to low-energy data. We shall presently see that, under this assumption, the only permitted velocity-dependent interaction is the spin-orbit coupling $(\mathbf{r} \times \mathbf{p})\cdot\mathbf{S}$.

(iii) The interaction may have a general dependence on p^2. Obviously, this sort of velocity dependence may be necessary for the explanation of high-energy data. In Sections 11 and 12, we shall examine the current experimental situation, and study to what extent such a general velocity dependence may be required by the data.

We now proceed to construct general expressions with space-spin variables. [Some mathematical preliminaries needed for this purpose are given in Appendix B (Section II).] We shall first use assumption (ii) about momentum dependence, as was originally done by Wigner and Eisenbud[3]. These general expressions held ground until Okubo and Marshak[4] extended them by introducing assumption (iii) about momentum dependence, which is presented towards the end of this section.

Since $\sigma_\alpha^2 = \mathbb{1}$ and $\sigma_\alpha \sigma_\beta = i\sigma_\gamma$ (α, β, γ being a cyclic permutation of the x-, y-, z-component), it is not worthwhile to use bilinear terms in the components of $\boldsymbol{\sigma}_1$ alone, or $\boldsymbol{\sigma}_2$ alone. The only independent operators that can be formed with one $\boldsymbol{\sigma}_1$ and one $\boldsymbol{\sigma}_2$ are given by

$$\mathbb{1}, \quad \boldsymbol{\sigma}_1\cdot\boldsymbol{\sigma}_2 \qquad \text{(scalar)}, \tag{I.10a}$$

$$\boldsymbol{\sigma}_1 \pm \boldsymbol{\sigma}_2, \quad \boldsymbol{\sigma}_1 \times \boldsymbol{\sigma}_2 \qquad \text{(axial vector)}, \tag{I.10c}$$

$$\mathcal{S}_{\alpha\beta} = \tfrac{1}{2}(\sigma_{1\alpha}\sigma_{2\beta} + \sigma_{1\beta}\sigma_{2\alpha}) - \tfrac{1}{3}\delta_{\alpha\beta}\boldsymbol{\sigma}_1\cdot\boldsymbol{\sigma}_2 \qquad \text{(tensor)}. \tag{I.10d}$$

The operator $\boldsymbol{\sigma}$, being an angular momentum, is itself an axial vector (analogous to the orbital angular momentum $\mathbf{r} \times \mathbf{p}$). The components of $\boldsymbol{\sigma}_1$ and $\boldsymbol{\sigma}_2$, and hence those of $\boldsymbol{\sigma}_1 \times \boldsymbol{\sigma}_2$, do not change sign under reflection. Thus, $\boldsymbol{\sigma}_1 \times \boldsymbol{\sigma}_2$ is also an axial vector.

A. NO HIGHER THAN LINEAR DEPENDENCE ON p

With $\mathbf{r}$ and $\mathbf{p}$ we can construct the Hermitean quantities (remembering that we shall not go beyond linear dependence on $\mathbf{p}$)

$$\mathbf{r}^2, \quad (\mathbf{r}\cdot\mathbf{p} + \mathbf{p}\cdot\mathbf{r}) \qquad \text{(scalar)}, \tag{I.11a}$$

$$\mathbf{r}, \quad \mathbf{p} \qquad \text{(vector)}, \tag{I.11b}$$

$$\mathbf{r} \times \mathbf{p} \qquad \text{(axial vector)}, \tag{I.11c}$$

$$R_{\alpha\beta} = r_\alpha r_\beta - \tfrac{1}{3}\delta_{\alpha\beta} r^2, \quad (\mathbf{r}, \mathbf{p})^{(2)} + (\mathbf{p}, \mathbf{r})^{(2)} \qquad \text{(tensor)}. \tag{I.11d}$$

We have constructed only Hermitean expressions because these will be employed in constructing the potential that is required to be a Hermitean operator. In constructing Hermitean

combinations, we kept in mind that the components of **r** and **p** are individually Hermitean, but the product of any component of **r** and any component of **p** is not Hermitean since **r** and **p** do not commute with each other. The *symmetrized* combinations, appearing in the second operator in (I.11a) and (I.11d), are easily verified as being Hermitean. In expressing the last-named operator, we have used the notation given in (BII.6) of Appendix B, namely, $(\mathbf{r}, \mathbf{p})^{(2)}$, which denotes a second-rank tensor constructed with the vectors **r** and **p** in that order.

In order to form scalar expressions, we are permitted to take the scalar product of an expression from (I.10) and the corresponding expression from (I.11). The number of permitted expressions will be severely limited if the requirements include invariance under both time-reversal and particle exchange. Under time-reversal,

$$\boldsymbol{\sigma} \rightarrow -\boldsymbol{\sigma}, \qquad \mathbf{r} \rightarrow \mathbf{r}, \qquad \mathbf{p} \rightarrow -\mathbf{p},$$

and, under particle exchange, both **r** and **p** change sign. We further wish to multiply, at the end, the resultant expressions by a general spatial function $V(r/a)$, where a is called the range parameter. Therefore, it is not necessary to retain terms containing r^2 alone while forming general expressions.

If we keep all these facts in mind, the different forms that will result are

$$\begin{array}{ccc} \text{isospin} & & \text{space-spin} \\ \left.\begin{matrix} \mathbb{1} \\ \boldsymbol{\tau}_1 \cdot \boldsymbol{\tau}_2 \end{matrix}\right\} & \times V(r/a) \times & \left\{\begin{matrix} \mathbb{1} \\ \boldsymbol{\sigma}_1 \cdot \boldsymbol{\sigma}_2 \\ (\mathbf{r} \times \mathbf{p}) \cdot (\boldsymbol{\sigma}_1 + \boldsymbol{\sigma}_2) \\ \mathcal{T} = 3\boldsymbol{\sigma}_1 \cdot \hat{r}\, \boldsymbol{\sigma}_2 \cdot \hat{r} - \boldsymbol{\sigma}_1 \cdot \boldsymbol{\sigma}_2 \end{matrix}\right. \end{array} \tag{I.12}$$

The third space-spin expression is obviously the spin-orbit coupling term $\mathbf{l} \cdot \mathbf{S}$, whereas the last one, $\mathcal{T}$, is the celebrated tensor force operator. The latter results from the scalar multiplication of $\mathcal{S}_{\alpha\beta}$ in (I.10d) and $R_{\alpha\beta}$ in (I.11d) as follows:

$$\begin{aligned} \sum_{\alpha\beta} R_{\alpha\beta} \mathcal{S}_{\alpha\beta} &= \sum_{\alpha\beta} (r_\alpha r_\beta - \tfrac{1}{3}\delta_{\alpha\beta} r^2)[\tfrac{1}{2}(\sigma_{1\alpha}\sigma_{2\beta} + \sigma_{1\beta}\sigma_{2\alpha}) - \tfrac{1}{3}\delta_{\alpha\beta}\boldsymbol{\sigma}_1 \cdot \boldsymbol{\sigma}_2] \\ &= \tfrac{1}{2} \sum_\alpha \sigma_{1\alpha} r_\alpha \sum_\beta \sigma_{2\beta} r_\beta + \tfrac{1}{2} \sum_\beta \sigma_{1\beta} r_\beta \sum_\alpha \sigma_{2\alpha} r_\alpha - \frac{r^2}{3} \sum_\alpha \sigma_{1\alpha}\sigma_{2\alpha} \\ &\quad - \frac{\boldsymbol{\sigma}_1 \cdot \boldsymbol{\sigma}_2}{3} \sum_\alpha r_\alpha r_\alpha + \tfrac{1}{9} r^2 \boldsymbol{\sigma}_1 \cdot \boldsymbol{\sigma}_2 \sum_\alpha \delta_{\alpha\alpha} \\ &= (\boldsymbol{\sigma}_1 \cdot \mathbf{r})(\boldsymbol{\sigma}_2 \cdot \mathbf{r}) - \tfrac{1}{3} r^2 \boldsymbol{\sigma}_1 \cdot \boldsymbol{\sigma}_2. \end{aligned} \tag{I.13}$$

Since the arbitrary radial function $V(r/a)$ is already present in (I.12), we divide (I.13) by r^2. A further normalization by a factor of 3 (to eliminate the occurrence of the fraction $\frac{1}{3}$) leads to the expression of $\mathcal{T}$ given in (I.12), where $\hat{r}$ represents the unit vector in the direction of **r**.

The resultant forms in (I.12) can be divided into two broad categories: (i) those that depend on r, but not on (θ, ϕ), i.e., the direction of **r**, and (ii) those that have a general dependence on r, θ, ϕ, i.e., the magnitude as well as the direction of **r**. Types (i) and (ii) are called *central* and *noncentral* potentials, respectively.

The general form of a central potential is found from (I.12) to be a linear combination of $\mathbb{1}$, $\boldsymbol{\sigma}_1 \cdot \boldsymbol{\sigma}_2$, $\boldsymbol{\tau}_1 \cdot \boldsymbol{\tau}_2$, and $\boldsymbol{\sigma}_1 \cdot \boldsymbol{\sigma}_2 \boldsymbol{\tau}_1 \cdot \boldsymbol{\tau}_2$, each operator being multiplied by a suitable radial shape function $V(r/a)$. The range parameter a may, in general, be different for the different operators. These spin-isospin operators clearly make the potential state-dependent.

We have

$$\boldsymbol{\sigma}_1 \cdot \boldsymbol{\sigma}_2 = \begin{cases} 1 & \text{(spin-triplet)} \\ -3 & \text{(spin-singlet)} \end{cases} \tag{I.14a}$$

$$\boldsymbol{\tau}_1 \cdot \boldsymbol{\tau}_2 = \begin{cases} 1 & \text{(isospin-triplet)} \\ -3 & \text{(isospin-singlet)} \end{cases} \tag{I.14b}$$

and hence

$$\boldsymbol{\sigma}_1 \cdot \boldsymbol{\sigma}_2 \boldsymbol{\tau}_1 \cdot \boldsymbol{\tau}_2 = \begin{cases} -3 & \text{(space-symmetric, i.e., even } l \text{ states)} \\ 9 & \text{(space-antisymmetric, i.e., odd } l \text{ states, with spin-singlet)} \\ 1 & \text{(space-antisymmetric, i.e., odd } l \text{ states, with spin-triplet)} \end{cases} \tag{I.15}$$

The specification of space symmetry in (I.15) is possible in view of the total antisymmetry of the states in space, spin, and isospin.

The operators in (I.14) and (I.15) are very conveniently expressed in terms of three other operators, which have more physical content. Let us consider the operator

$$P_\sigma = \tfrac{1}{2}(\mathbb{1} + \boldsymbol{\sigma}_1 \cdot \boldsymbol{\sigma}_2). \tag{I.16a}$$

It is clear from (I.14a) that

$$P_\sigma = \begin{cases} +1 & \text{(spin-triplet)} \\ -1 & \text{(spin-singlet)} \end{cases}$$

Thus, P_σ achieves the same task as exchanging the spin coordinates in the spin-triplet (symmetric) and spin-singlet (antisymmetric) states. In other words, it is the spin-exchange operator or the Bartlett operator introduced in (I.5a). Since carrying out the exchange twice means getting back to the initial situation, we have $P_\sigma^2 = \mathbb{1}$.

In the same way as in the case of spin, the operator

$$P_\tau = \tfrac{1}{2}(\mathbb{1} + \boldsymbol{\tau}_1 \cdot \boldsymbol{\tau}_2) \tag{I.16b}$$

serves the purpose of exchanging the isospin coordinates in the two-body isospin functions, as described by (I.5b). As with P_σ^2, so too here we have $P_\tau^2 = \mathbb{1}$. According to the requirement of total antisymmetry of the space, spin, and isospin functions, expressed in (I.6), and in view of the fact that the square of the exchange operators is $\mathbb{1}$, we have $-P_\tau = P_r P_\sigma$. Hence, instead of calling P_τ the isospin exchange operator, we could identify $-P_\tau$ as the space-spin exchange operator or the Heisenberg exchange operator P_{H}.

With the help of $P_r = -P_\sigma P_\tau$ and (I.16a) and (I.16b), we express the space-exchange operator or the Majorana operator P_{M} as

$$P_{\text{M}} = P_r = -\tfrac{1}{4}(\mathbb{1} + \boldsymbol{\sigma}_1 \cdot \boldsymbol{\sigma}_2)(\mathbb{1} + \boldsymbol{\tau}_1 \cdot \boldsymbol{\tau}_2). \tag{I.16c}$$

The three exchange operators P_r, P_σ, and P_τ can conveniently replace the operators in (I.14) and (I.15) in the general expression of the central potential. The general expression also contains a term with the operator $\mathbb{1}$; this term does not cause any exchange of two-nucleon coordinates, and is often called the Wigner potential.

If we now consider the noncentral potentials in (I.12), namely, the spin-orbit and the tensor terms, we notice that, apart from the radial shape function $V(r/a)$, they can be multiplied only by a linear combination of $\mathbb{1}$ and $\boldsymbol{\tau}_1 \cdot \boldsymbol{\tau}_2$, which can obviously be expressed in terms of $\mathbb{1}$ and P_τ. This simpler exchange-dependence arises from the fact that the noncentral forces are effective only in spin-triplet states, as now explained.

The spin-orbit force $\boldsymbol{l} \cdot \mathbf{S}$ is diagonal in the spin quantum number S, has a value zero in the singlet state, and a nonvanishing value in the triplet state (see Section 4). The tensor force behaves as a second-rank tensor in the spin-space, as is apparent from its construction. Hence, it can give a matrix element connecting a triplet state with a triplet state, but none that connects a singlet state with a singlet state. Operating on a singlet ($S = 0$) state, the second-rank tensor tends to lead to a state of spin 2, but such a state does not exist for the two-nucleon system; hence, the result for a singlet-singlet or a singlet-triplet matrix element of the tensor force is zero. On the other hand, operating on a spin-triplet state ($S = 1$), the second-rank tensor could give rise to spin states with $S = 1, 2, 3$ (follows from coupling rules); since the states $S = 2, 3$ do not exist, the only nonvanishing matrix element is triplet-triplet (see Section 4 for explicit evaluation of the matrix element).

Since the noncentral force is effective only in the spin-triplet state, it is clear that a spin-exchange operator is unnecessary for specifying its exchange character. In other words, P_σ always has the value unity for noncentral forces, and hence the operators $\mathbb{1}$, P_σ, P_τ, $P_\sigma P_\tau$ $(= -P_r)$ reduce to only $\mathbb{1}$ and P_τ. For the spin-triplet state, as soon as the value of P_τ is specified, the value of P_r is determined from the total antisymmetry, $P_r = -P_\tau$. Thus, the exchange-dependence of noncentral forces could have been alternatively specified in terms of $\mathbb{1}$ and P_r, i.e., by specifying the noncentral force separately for the odd and even l states.

B. GENERAL DEPENDENCE ON p

If the two-nucleon potential has a general dependence on $\mathbf{p}$, the list (I.11) has to be extended as

$$r^2, \quad p^2, \quad (\mathbf{r}\cdot\mathbf{p} + \mathbf{p}\cdot\mathbf{r}), \quad l^2 \qquad \text{(scalar)}, \tag{I.17a}$$

$$\mathbf{r}, \quad \mathbf{p} \qquad \text{(vector)}, \tag{I.17b}$$

$$\boldsymbol{l} = \mathbf{r} \times \mathbf{p} \qquad \text{(axial vector)}, \tag{I.17c}$$

$$\left.\begin{array}{ll} \mathcal{R}_{\alpha\beta} = r_\alpha r_\beta - \frac{1}{3}\delta_{\alpha\beta} r^2, & \mathcal{P}_{\alpha\beta} = p_\alpha p_\beta - \frac{1}{3}\delta_{\alpha\beta} p^2 \\ \mathcal{L}_{\alpha\beta} = l_\alpha l_\beta - \frac{1}{3}\delta_{\alpha\beta} l^2, & (\mathbf{r}, \mathbf{p})^{(2)} + (\mathbf{p}, \mathbf{r})^{(2)}. \end{array}\right\} \qquad \text{(tensor)} \tag{I.17d}$$

The rules for combining (I.17) with (I.10) are the same as stated after the expressions (I.11). The shape function $V(r/a)$, however, now gets extended to a more general function of the scalar quantities appearing in (I.17a), namely, $V(r^2, p^2, l^2)$. The dependence on the other scalars $\mathbf{r}\cdot\mathbf{p}$ and $\mathbf{p}\cdot\mathbf{r}$ has to be through $(\mathbf{r}\cdot\mathbf{p})(\mathbf{p}\cdot\mathbf{r})$ and $(\mathbf{p}\cdot\mathbf{r})(\mathbf{r}\cdot\mathbf{p})$ because of the requirement of hermiticity and time-reversal invariance. Since, however,

$$l^2 = (\mathbf{r} \times \mathbf{p})\cdot(\mathbf{r} \times \mathbf{p}) = \tfrac{1}{2}(r^2p^2 + p^2r^2) - (\mathbf{r}\cdot\mathbf{p}\mathbf{p}\cdot\mathbf{r} + \mathbf{p}\cdot\mathbf{r}\mathbf{r}\cdot\mathbf{p}) + 3,$$

it is clear that the dependence of V on r^2, p^2, and l^2 also accounts for the completely general dependence on $\mathbf{r}\cdot\mathbf{p}$ and $\mathbf{p}\cdot\mathbf{r}$. It may seem, at first sight, that two more tensors—those obtained by coupling ($\mathbf{r}$ and $\boldsymbol{l}$) and ($\mathbf{p}$ and $\boldsymbol{l}$)—are missing from (I.17d). But it should be remembered that, since $\boldsymbol{l}$ is an axial vector and $\mathbf{r}$ and $\mathbf{p}$ are polar vectors, these two tensors will have a different transformation property under reflection. They can only produce pseudoscalars when multiplied by (I.10d).

4 (45-1122/1977)

The complete list of potentials that results in the case of a general dependence on **p** is

$$
\begin{array}{ll}
\text{isospin} & \text{space-spin}
\end{array}
$$

$$
\left.\begin{array}{l}\mathbb{1}\\ \boldsymbol{\tau}_1\cdot\boldsymbol{\tau}_2\end{array}\right\}\times V(r^2, p^2, l^2)\times\left\{\begin{array}{ll}
\mathbb{1} & \text{(I.18a)}\\
\boldsymbol{\sigma}_1\cdot\boldsymbol{\sigma}_2 & \text{(I.18b)}\\
(\mathbf{r}\times\mathbf{p})\cdot(\boldsymbol{\sigma}_1+\boldsymbol{\sigma}_2) & \text{(I.18c)}\\
\mathcal{T}=3\boldsymbol{\sigma}_1\cdot\mathbf{r}\boldsymbol{\sigma}_2\cdot\mathbf{r}-\boldsymbol{\sigma}_1\cdot\boldsymbol{\sigma}_2 & \text{(I.18d)}\\
3\boldsymbol{\sigma}_1\cdot\boldsymbol{l}\boldsymbol{\sigma}_2\cdot\boldsymbol{l}-l^2\boldsymbol{\sigma}_1\cdot\boldsymbol{\sigma}_2 & \text{(I.18e)}\\
3\boldsymbol{\sigma}_1\cdot\hat{\mathbf{p}}\boldsymbol{\sigma}_2\cdot\hat{\mathbf{p}}-\boldsymbol{\sigma}_1\cdot\boldsymbol{\sigma}_2 & \text{(I.18f)}
\end{array}\right.
$$

(I.18e) and (I.18f) result from $\sum_{\alpha\beta} \mathcal{L}_{\alpha\beta}S_{\alpha\beta}$ and $\sum_{\alpha\beta} \mathcal{P}_{\alpha\beta}S_{\alpha\beta}$, respectively. $\hat{\mathbf{p}}$ is the unit vector in the direction of **p**. It is shown in Section 10B that only one of these two expressions is necessary for the general description of two-nucleon processes (which necessarily conserve energy, and are therefore described as on-energy-shell processes); the other one, though redundant for the description of on-energy-shell processes, becomes a requisite for producing the off-energy-shell behaviour of the potential. Such processes are important in the three- and many-nucleon system, where a pair of colliding nucleons does not necessarily have to conserve energy by themselves, since they are able to transfer energy and momentum to a neighbouring nucleon.

For our purposes, we arbitrarily choose the term (I.18e) and omit (I.18f) for on-energy-shell processes. This new term in the potential is sometimes referred to as the quadratic spin-orbit coupling term because it can be related to $(\boldsymbol{l}\cdot\mathbf{S})^2$ as

$$
\begin{aligned}
2(\boldsymbol{l}\cdot\mathbf{S})^2 &= \tfrac{1}{2}\boldsymbol{l}\cdot(\boldsymbol{\sigma}_1+\boldsymbol{\sigma}_2)\boldsymbol{l}\cdot(\boldsymbol{\sigma}_1+\boldsymbol{\sigma}_2)\\
&= \tfrac{1}{2}(\boldsymbol{\sigma}_1\cdot\boldsymbol{l}\boldsymbol{\sigma}_2\cdot\boldsymbol{l}+\boldsymbol{\sigma}_2\cdot\boldsymbol{l}\boldsymbol{\sigma}_1\cdot\boldsymbol{l})+\tfrac{1}{2}(\boldsymbol{\sigma}_1\cdot\boldsymbol{l}\boldsymbol{\sigma}_1\cdot\boldsymbol{l}+\boldsymbol{\sigma}_2\cdot\boldsymbol{l}\boldsymbol{\sigma}_2\cdot\boldsymbol{l}).
\end{aligned}
\tag{I.19}
$$

The second term on the right-hand side of (I.19) can be further simplified as

$$
\begin{aligned}
\tfrac{1}{2}(\boldsymbol{\sigma}_1\cdot\boldsymbol{l}\boldsymbol{\sigma}_1\cdot\boldsymbol{l}+\boldsymbol{\sigma}_2\cdot\boldsymbol{l}\boldsymbol{\sigma}_2\cdot\boldsymbol{l}) &= \tfrac{1}{2}\sum_{\alpha,\beta}(\sigma_{1\alpha}l_\alpha\sigma_{1\beta}l_\beta+\sigma_{2\alpha}l_\alpha\sigma_{2\beta}l_\beta)\\
&= \tfrac{1}{2}\sum_{\alpha}(l_\alpha l_\alpha+l_\alpha l_\alpha)+\tfrac{1}{2}\sum_{\alpha\neq\beta}(\sigma_{1\alpha}\sigma_{1\beta}l_\alpha l_\beta+\sigma_{2\alpha}\sigma_{2\beta}l_\alpha l_\beta)\\
&= l^2+\tfrac{1}{2}\sum_{(\alpha\beta)}(\sigma_{1\alpha}\sigma_{1\beta}l_\alpha l_\beta+\sigma_{1\beta}\sigma_{1\alpha}l_\beta l_\alpha+\sigma_{2\alpha}\sigma_{2\beta}l_\alpha l_\beta+\sigma_{2\beta}\sigma_{2\alpha}l_\beta l_\alpha)\\
&= l^2-\tfrac{1}{2}\sum_{\gamma}(\sigma_{1\gamma}l_\gamma+\sigma_{2\gamma}l_\gamma)=l^2-\tfrac{1}{2}(\boldsymbol{\sigma}_1+\boldsymbol{\sigma}_2)\cdot\boldsymbol{l}\\
&= l^2-\boldsymbol{l}\cdot\mathbf{S}.
\end{aligned}
\tag{I.20}
$$

The summation $(\alpha\beta)$ is only over the pairs (xy), (yz), (zx), and α, β, γ is a cyclic permutation of the x-, y-, z-component. We have used $\sigma_\alpha\sigma_\beta = i\sigma_\gamma$ and $[l_\alpha, l_\beta] = il_\gamma$ to arrive at the simplified expression (I.20). Using (I.20) in (I.19), we have

$$
\begin{aligned}
\boldsymbol{\sigma}_1\cdot\boldsymbol{l}\boldsymbol{\sigma}_2\cdot\boldsymbol{l} &= \tfrac{1}{2}(\boldsymbol{\sigma}_1\cdot\boldsymbol{l}\boldsymbol{\sigma}_2\cdot\boldsymbol{l}+\boldsymbol{\sigma}_2\cdot\boldsymbol{l}\boldsymbol{\sigma}_1\cdot\boldsymbol{l})\\
&= (\boldsymbol{l}\cdot\mathbf{S})^2-l^2+\boldsymbol{l}\cdot\mathbf{S}(\boldsymbol{l}\cdot\mathbf{S}+1).
\end{aligned}
\tag{I.21a}
$$

The value of the operator $\boldsymbol{l}\cdot\mathbf{S}$ is given in (I.27), where it is clear that if $l = j$, then $\boldsymbol{l}\cdot\mathbf{S} = 0$ for $S = 0$, and $\boldsymbol{l}\cdot\mathbf{S} + 1 = 0$ for $S = 1$. On the other hand, if $l = j \pm 1$ and $S = 1$, then

$$
(\boldsymbol{l}\cdot\mathbf{S})(\boldsymbol{l}\cdot\mathbf{S}+1) = l^2.
$$

Using these results in (I.21a), we get

$$\boldsymbol{\sigma}_1\cdot \boldsymbol{l}\boldsymbol{\sigma}_2\cdot \boldsymbol{l} = (\boldsymbol{l}\cdot \mathbf{S})^2 - l^2\delta_{lj}. \tag{I.21b}$$

This completes the demonstration of the correspondence of the extra term in the potential to a quadratic spin-orbit coupling.

Several authors have used the quadratic spin-orbit coupling term $\boldsymbol{\sigma}_1\cdot \boldsymbol{l}\boldsymbol{\sigma}_2\cdot \boldsymbol{l}$ in various linear combinations with $(l^2\boldsymbol{\sigma}_1\cdot\boldsymbol{\sigma}_2)$. The particular combination used in (I.18e) makes it a second-rank tensor in the separate orbital and spin spaces. Since other linear combinations do not preserve this definite tensorial property, they may give rise to matrix elements for the spin-singlet state as well.

Hamada and Johnston[5] used the linear combination

$$\boldsymbol{\sigma}_1\cdot\boldsymbol{\sigma}_2 l^2 - \tfrac{1}{2}(\boldsymbol{\sigma}_1\cdot \boldsymbol{l}\boldsymbol{\sigma}_2\cdot \boldsymbol{l} + \boldsymbol{\sigma}_2\cdot \boldsymbol{l}\boldsymbol{\sigma}_1\cdot \boldsymbol{l}) = (\delta_{lj} + \boldsymbol{\sigma}_1\cdot\boldsymbol{\sigma}_2)l^2 - (\boldsymbol{l}\cdot\mathbf{S})^2, \tag{I.22}$$

whereas Breit et al[6] worked with the operator

$$l^2 - (\boldsymbol{l}\cdot\mathbf{S})^2 - \boldsymbol{l}\cdot\mathbf{S}, \tag{I.23}$$

which is equivalent to $l^2\delta_{lj}$.

The Okubo-Marshak potentials, given by (I.18), are far too general because they contain the shape function $V(r^2, p^2, l^2)$. Such a shape function can be attributed different values for different energies and orbital angular momenta. For simpler potentials, given by (I.12), we had shape functions of the type $V(r/a)$, and it was the exchange operators that varied the potential in different spin, isospin states. In the present case, we have a much wider freedom to choose a potential for *every* l-value, and then to make it different for different energies as well. A phenomenological potential in which all this freedom is exploited can obviously fit any amount of experimental data, and hence becomes meaningless as phenomenology. In practice, therefore, the simple shape function $V(r/a)$ is used also with the Okubo-Marshak forms of the potential. Green[7] tried to apply an energy-dependent shape function of a simple type, namely,

$$V(r^2, p^2) = p^2 V(r) + V(r)p^2; \tag{I.24}$$

this is usually used to replace the hard core of the shape function. Section 5A gives more information on the choice of the shape function in the phenomenological two-nucleon potential; it is also shown in Section 12 that the potential is required to have a strongly repulsive core region followed by a weaker attractive tail, in order that the two-nucleon scattering data can be well-reproduced. When the core is idealized by an *infinitely* repulsive region, it is called the hard core of the two-nucleon potential. The velocity-dependent potential (I.24) was designed in the course of attempts to fit the scattering data without introducing a hard core in the potential.

4. MATRIX ELEMENTS OF TWO-NUCLEON POTENTIAL

Our ultimate aim is to write the Schrödinger equation with the two-nucleon potential arrived at in Section 3. Using general invariance arguments, we have already shown that each of the eigenfunctions will have a spin-angle part and an isospin part, which can be described by a linear sum of functions of the type $|lSjm, TM_T\rangle$, where the sum is over l, consistent with the value of j and the parity of the eigenfunction. These functions are completely known: the isospin part labelled by TM_T is the function $\rho^T_{M_T}(\tau_1, \tau_2)$ defined in Section 2C, and the spin-angle functions labelled $lSjm$ are obtained by coupling [see Appendix A (Section IV)] the spherical harmonic $Y^l(\theta, \phi)$ with the spin functions $\chi^S(\sigma_1, \sigma_2)$ described in Section 2B. The only unknown part of the wavefunction is the pure radial part.

For $S = 0$ (spin-singlet), $l = j$ is the only possible value of l, whereas for $S = 1$ (spin-triplet), l can have three possible values: j and $j \pm 1$. The last two have parity opposite to that of $l = j$. Thus, the eigenfunction for $S = 1$ and parity $\pi = -(-1)^j$ is the only complicated coupled wavefunction; it has the structure

$$\frac{u}{r}|l = j - 1,\ S = 1, jm,\ TM_T\rangle + \frac{w}{r}|l = j + 1,\ S = 1, jm,\ TM_T\rangle; \tag{I.25}$$

but the eigenfunction for $l = j$ ($S = 0$ or $S = 1$) and parity $\pi = (-1)^j$ is uncoupled and has the simple expression

$$\frac{u}{r}|l = j,\ Sjm,\ TM_T\rangle. \tag{I.26}$$

The radial functions u and w are the only unknown parts of the two-nucleon wavefunction. Hence, our ultimate aim is to convert the Schrödinger equation into pure radial differential equations for u and w. Obviously, (I.26) will yield an uncoupled equation for u, whereas (I.25) will give rise to two coupled equations for u and w. (These equations are obtained in Section 5.) In order to eliminate the standard spin-angle and isospin parts of the wavefunctions from the Schrödinger equation, we need to multiply the equation from the left by the complex conjugate of these wavefunctions, integrate over the angles, and sum over the spin and isospin. This procedure gives rise to the matrix elements $\langle l'Sjm,\ TM_T|\ V\ |lSjm,\ TM_T\rangle$ in the potential term of the Schrödinger equation. Thus, we must know the values of these matrix elements of the potential V before we can explicitly express the radial Schrödinger equations. As groundwork for Section 5, we shall now derive these matrix elements of V.

A. CENTRAL POTENTIAL

The central potential contains the exchange operators P_σ, P_τ, P_r, which obviously cannot give a matrix element connecting different values of l. [It is important to note the implication: that, with a pure central potential, the coupling of states in (I.25) would have broken up, and we would have obtained states labelled by a single l-value.]

The matrix element for $P_\tau = (-1)^{T+1}$ and that for $P_\sigma = (-1)^{S+1}$. The space-exchange operator P_r, which is equivalent with the parity operation, has the matrix element π, the parity of the states. We should recall that reflection invariance was required in constructing V in Section 3, and the matrix elements are therefore nonvanishing only when both the states have the same parity. Thus, there is no ambiguity when writing the value of P_r.

B. NONCENTRAL POTENTIAL

The noncentral potential consists of the linear spin-orbit coupling, the quadratic spin-orbit coupling, and the tensor operator. The exchange operator for these components is a linear combination of $\mathbb{1}$ and P_τ. The value of P_τ is known as soon as the isospin T of the state is specified. Therefore, it suffices to evaluate the matrix elements of the operators $\boldsymbol{l}\cdot\mathbf{S}$, $(\boldsymbol{l}\cdot\mathbf{S})^2$, and $\mathcal{T}$ between the states $|lSjm\rangle$ and $\langle l'Sjm|$.

It is simple to evaluate the matrix elements of $\boldsymbol{l}\cdot\mathbf{S}$, once again diagonal in l and nonvanishing only for $S = 1$. Using

$$j^2 = l^2 + S^2 + 2\boldsymbol{l}\cdot\mathbf{S}$$

or

$$\boldsymbol{l}\cdot\mathbf{S} = \tfrac{1}{2}(j^2 - l^2 - S^2),$$

we obtain

$$\langle l'Sjm, TM_T| V|lSjm, TM_T\rangle = \tfrac{1}{2}\delta_{ll'}[j(j+1) - l(l+1) - S(S+1)]. \tag{I.27}$$

Since $\boldsymbol{l}\cdot\mathbf{S}$ has only diagonal matrix elements, the matrix elements of $(\boldsymbol{l}\cdot\mathbf{S})^2$ will simply be the square of the matrix element (I.27). This, with the help of the identity (I.21b), enables us to arrive at the matrix element of the quadratic spin-orbit coupling term.

What now remains is the evaluation of the matrix elements of the tensor operator $\mathcal{T}$. Before we proceed to this evaluation, we shall express the operator $\mathcal{T}$ in terms of the components of the spherical harmonic Y^2 and the components of the second-rank tensor $(\boldsymbol{\sigma}_1, \boldsymbol{\sigma}_2)^{(2)}$ defined by [see (BII.6) in Appendix B]

$$(\boldsymbol{\sigma}_1, \boldsymbol{\sigma}_2)^{(2)}_M = \sum_m \begin{bmatrix} 1 & 1 & 2 \\ m & M-m & M \end{bmatrix} (\sigma_1)_m(\sigma_2)_{M-m}. \tag{I.28}$$

The expression for $\mathcal{T}$, given in (I.12) and (I.13), is in terms of cartesian components of the second-rank tensors $R_{\alpha\beta}$ and $\mathcal{S}_{\alpha\beta}$. The expression we are aiming at now is in terms of spherical components of second-rank tensors. Such an expression will allow us to handle the matrix elements by the general results given in Appendix B (Section III).

We define the tensor y^l in terms of the spherical harmonic Y^l (here, and subsequently, $[a]$ stands for $2a+1$) as

$$y^l_m = \sqrt{\frac{4\pi}{[l]}}\, Y^l_m. \tag{I.29}$$

Then, from the spherical harmonics in Table AI.1 (see Appendix A) and the definition (I.11d) of $R_{\alpha\beta}$, we get

$$\begin{aligned} r^2y^2_0 &= \frac{r^2}{2}(3\cos^2\theta - 1) = \tfrac{3}{2}(z^2 - r^2) = \tfrac{3}{2}R_{zz}, \\ r^2y^2_{\pm1} &= \mp\sqrt{\tfrac{3}{2}}\,r^2\cos\theta\sin\theta\, e^{\pm i\phi} = \mp\sqrt{\tfrac{3}{2}}\,z(x \pm iy) = \mp\sqrt{\tfrac{3}{2}}(R_{zx} \pm iR_{yz}), \\ r^2y^2_{\pm2} &= \tfrac{1}{2}\sqrt{\tfrac{3}{2}}\,r^2\sin^2\theta\, e^{\pm 2i\phi} = \tfrac{1}{2}\sqrt{\tfrac{3}{2}}(x \pm iy)(x \pm iy) = \tfrac{1}{2}\sqrt{\tfrac{3}{2}}(R_{xx} - R_{yy} \pm 2iR_{xy}). \end{aligned} \tag{I.30}$$

Similarly, using the definitions (I.28) and (I.10d), and the values of the Clebsch-Gordon coefficients given in Table AVII.2 (see Appendix A), we obtain

$$\begin{aligned} (\boldsymbol{\sigma}_1, \boldsymbol{\sigma}_2)^{(2)}_0 &= \sqrt{\tfrac{3}{2}}\,\mathcal{S}_{zz}, \\ (\boldsymbol{\sigma}_1, \boldsymbol{\sigma}_2)^{(2)}_{\pm1} &= \mp(\mathcal{S}_{zx} \pm i\mathcal{S}_{yz}), \\ (\boldsymbol{\sigma}_1, \boldsymbol{\sigma}_2)^{(2)}_{\pm2} &= \tfrac{1}{2}(\mathcal{S}_{xx} - \mathcal{S}_{yy} \pm 2i\mathcal{S}_{xy}). \end{aligned} \tag{I.31}$$

We next take the scalar products of the components (I.30) and (I.31) and get

$$\begin{aligned} r^2y^{(2)}\cdot(\boldsymbol{\sigma}_1, \boldsymbol{\sigma}_2)^{(2)} &= \sum_{M=-2}^{2} (-1)^M r^2y^2_M(\boldsymbol{\sigma}_1, \boldsymbol{\sigma}_2)^{(2)}_{-M} \\ &= \tfrac{3}{2}\sqrt{\tfrac{3}{2}}R_{zz}\mathcal{S}_{zz} + \sqrt{\tfrac{3}{2}}[(R_{zx} + iR_{yz})(\mathcal{S}_{zx} - i\mathcal{S}_{yz}) + \text{h.c.}] \\ &\quad + \tfrac{1}{4}\sqrt{\tfrac{3}{2}}[(R_{xx} - R_{yy} + 2iR_{xy})(\mathcal{S}_{xx} - \mathcal{S}_{yy} - 2i\mathcal{S}_{xy}) + \text{h.c.}] \\ &= \sqrt{\tfrac{3}{2}}[\tfrac{1}{2}(R_{xx}\mathcal{S}_{xx} + R_{yy}\mathcal{S}_{yy}) + \tfrac{3}{2}R_{zz}\mathcal{S}_{zz} - \tfrac{1}{2}(R_{xx}\mathcal{S}_{yy} + R_{yy}\mathcal{S}_{xx}) \\ &\quad + 2(R_{zx}\mathcal{S}_{zx} + R_{yz}\mathcal{S}_{yz} + R_{xy}\mathcal{S}_{xy})], \end{aligned} \tag{I.32}$$

where h.c. stands for Hermitean conjugate. Since $R_{xx} + R_{yy} + R_{zz} = 0$ and $\mathcal{S}_{xx} + \mathcal{S}_{yy} + \mathcal{S}_{zz} = 0$, we may write

$$\begin{aligned} \tfrac{1}{2}R_{zz}\mathcal{S}_{zz} &= \tfrac{1}{2}(R_{xx} + R_{yy})(\mathcal{S}_{xx} + \mathcal{S}_{yy}) \\ &= \tfrac{1}{2}(R_{xx}\mathcal{S}_{xx} + R_{yy}\mathcal{S}_{yy}) + \tfrac{1}{2}(R_{xx}\mathcal{S}_{yy} + R_{yy}\mathcal{S}_{xx}). \end{aligned}$$

Using this result in (I.32), we get

$$r^2 y^{(2)} \cdot (\boldsymbol{\sigma}_1, \boldsymbol{\sigma}_2)^{(2)} = \sqrt{\tfrac{3}{2}}[(R_{xx}S_{xx} + R_{yy}S_{yy} + R_{zz}S_{zz}) + 2(R_{zx}S_{zx} + R_{yz}S_{yz} + R_{xy}S_{xy})]$$
$$= \sqrt{\tfrac{3}{2}} \sum_{\alpha,\beta} R_{\alpha\beta}S_{\alpha\beta} = \sqrt{\tfrac{3}{2}} \frac{r^2}{3} \mathcal{T}$$

or

$$\mathcal{T} \equiv \frac{3}{r^2} \sum_{\alpha,\beta} R_{\alpha\beta}S_{\alpha\beta} = \sqrt{6} y^{(2)} \cdot (\boldsymbol{\sigma}_1, \boldsymbol{\sigma}_2)^{(2)}. \tag{I.33}$$

The desired expression of the tensor operator is given by (I.33).

To evaluate the matrix elements of (I.33), we first use the result (BIII.15) from Appendix B and get

$$\langle l, S = 1\ jm | \mathcal{T} | l', S = 1\ jm \rangle = \sqrt{6}(-1)^{l+1-j} \sqrt{[l][1]}\, W(l'1l1; j2) \langle l \| y^2 \| l' \rangle$$
$$\times \langle (\tfrac{1}{2}\ \tfrac{1}{2})S = 1 \| (\boldsymbol{\sigma}_1, \boldsymbol{\sigma}_2)^{(2)} \| (\tfrac{1}{2}\ \tfrac{1}{2})S = 1 \rangle. \tag{I.34}$$

We have incorporated the detailed structure of the coupled spin state [which was understood on the left-hand side of (I.34)] in the reduced matrix element of the spin operator. This reduced matrix element can be evaluated by using, from Appendix B, first (BIII.16), and then (BIII.6) for $\langle \tfrac{1}{2} \| \boldsymbol{\sigma} \| \tfrac{1}{2} \rangle$ and, from Appendix A, (AVI.5) for the evaluation of the $9j$-symbol [the symmetry property of the $9j$-symbol described in Appendix A (Section VI) is needed for its conversion into the right form]. We thus obtain

$$\langle (\tfrac{1}{2}\ \tfrac{1}{2})1 \| (\boldsymbol{\sigma}_1, \boldsymbol{\sigma}_2)^{(2)} \| (\tfrac{1}{2}\ \tfrac{1}{2})1 \rangle = 2\sqrt{\tfrac{5}{3}}. \tag{I.35}$$

The reduced matrix element of $y^{(2)}$ is evaluated by first using (BIII.3) from Appendix B, and then applying Table AVII.4 (see Appendix A) for the evaluation of the Clebsch-Gordon coefficient:

$$\langle l \| y^{(2)} \| l' \rangle = \sqrt{\frac{[l']}{[l]}} \begin{bmatrix} l' & 2 & l \\ 0 & 0 & 0 \end{bmatrix} = \begin{cases} -\sqrt{\dfrac{l(l+1)}{[l-1][l+1]}}, & l = l' \\[2ex] \sqrt{\tfrac{3}{2}}\sqrt{\dfrac{l(l-1)}{[l][l-1]}}, & l = l' + 2 \end{cases} \tag{I.36}$$

We then convert the W-function of (I.34) into $W(l'l11; 2j)$ by applying its symmetry properties [see (AV.6) in Appendix A], and ultimately take recourse to Table AVII (see Appendix A) for its evaluation. The final results are given by

$$\sqrt{6}(-1)^{l+1-j} W(l'1l1; j2)\sqrt{[l][1]} = \begin{cases} -\sqrt{\tfrac{3}{5}}\sqrt{\dfrac{[l-1][l+1]}{l(l+1)}}, & l = l' = j \\[2ex] \sqrt{\tfrac{3}{5}}\sqrt{\dfrac{(l+1)[l+1]}{l(l-1)}}, & l = l' = j + 1 \\[2ex] \sqrt{\tfrac{3}{5}}\sqrt{\dfrac{l[l-1]}{(l+1)[l+1]}}, & l = l' = j - 1 \\[2ex] 3\sqrt{\tfrac{2}{5}}\sqrt{\dfrac{[l]}{[l-1]}}, & l = j + 1,\ l' = j - 1 \end{cases} \tag{I.37}$$

Collecting the results given in (I.35) to (I.37) and putting in (I.34), we get the matrix elements of the tensor operator as

$l \backslash l'$	j
j	2

(I.38a)

$l \backslash l'$	$j-1$	$j+1$
$j-1$	$-\frac{2(j-1)}{2j+1}$	$\frac{6\sqrt{j(j+1)}}{2j+1}$
$j+1$	$\frac{6\sqrt{j(j+1)}}{2j+1}$	$-\frac{2(j+2)}{2j+1}$

(I.38b)

As an application of these results, let us consider the matrix elements of $\mathcal{T}$ between the states $^3S_1$-$^3D_1$ and $^3D_1$-$^3D_1$. Using $j = 1$, we get [from (I.38b)]

$$\langle {}^3D_1|\mathcal{T}|{}^3D_1\rangle = -2, \tag{I.39a}$$

$$\langle {}^3D_1|\mathcal{T}|{}^3S_1\rangle = 2\sqrt{2}. \tag{I.39b}$$

These matrix elements enter the Schrödinger equation of the deuteron (see Section 5).

5. TWO-NUCLEON SCHRÖDINGER EQUATION

The reduced mass of the two-nucleon system is $\frac{1}{2}M$, where M is the mass of a single nucleon. The Schrödinger equation in the centre-of-mass coordinate system is therefore given by

$$\left(-\frac{\hbar^2}{M}\nabla^2 + V\right)\Psi = E\Psi, \tag{I.40}$$

where the potential V and the wavefunction Ψ are functions of r, θ, ϕ, σ_1, σ_2, τ_1, and τ_2, and E is the energy in the centre-of-mass system. E is negative and equal to $-\epsilon$ (ϵ is the binding energy of the deuteron and equal to 2.226 MeV) when we consider the bound-state problem, but positive and equal to $\frac{1}{2}E_{\text{lab}}$ in the scattering case, E_{lab} being the energy, measured in the laboratory, of the incident nucleon (usually, the proton from a high-energy accelerator) hitting a stationary target nucleon.

We define the quantities

$$\begin{aligned} v &= -\frac{M}{h^2}V, \\ k^2 &= \frac{M}{h^2}E, \\ \gamma^2 &= \frac{M}{h^2}\epsilon \end{aligned} \tag{I.41}$$

such that k^2 is positive for the scattering state and negative for the bound state. In the latter state, γ^2 is positive and $k = i\gamma$.

It has been pointed out in Section 4 that the eigenfunctions Ψ are labelled by the quantum number Sjm, TM_T and belong to one of two categories: (i) uncoupled wavefunction with a unique value of l, namely, $l = j$, corresponding to both $S = 0$ and $S = 1$; (ii) coupled wavefunctions corresponding to $l = j \pm 1$ for $S = 1$. The structures of the coupled and uncoupled wavefunctions have been given in (I.25) and (I.26), respectively.

A. UNCOUPLED RADIAL EQUATION

We consider first the Schrödinger equation (I.40) for an uncoupled state given by (I.26) by using the standard result

$$\nabla^2[\frac{u}{r}Y^l_m(\theta, \phi)] = \frac{1}{r}[\frac{d^2}{dr^2} - \frac{l(l+1)}{r^2}]uY^l_m(\theta, \phi), \tag{I.42}$$

which also implies

$$\nabla^2 \frac{u}{r}|lSjm, TM_T\rangle = \frac{1}{r}[\frac{d^2}{dr^2} - \frac{l(l+1)}{r^2}]u|lSjm, TM_T\rangle. \tag{I.43}$$

(I.43) follows from (I.42) if the angular momentum coupled state $|lSjm\rangle$ is broken up in terms of the spherical harmonic Y^l and the spin functions χ^S.

We substitute $(u/r)|lSjm, TM_T\rangle$ for Ψ in (I.40), use the result (I.43), and then multiply the resultant equation from the left by the spin-angle-isospin function $\langle lSjm, TM_T|$. This procedure yields, in the uncoupled case ($l = j$, $S = 1$, or $S = 0$), the radial equation

$$\frac{d^2u}{dr^2} - \frac{j(j+1)}{r^2}u - \langle jSjm, TM_T|\, v\, |jSjm, TM_T\rangle u + k^2u = 0. \tag{I.44}$$

The potential V consists of the shape function $V(r/a)$, a linear combination of the various exchange operators, and a combination of the noncentral operator $\boldsymbol{l}\cdot\mathbf{S}$, $(\boldsymbol{l}\cdot\mathbf{S})^2$, $\mathcal{T}$. The matrix elements of the exchange operators and the noncentral operators have been evaluated in Section 4. When these explicit matrix elements are used in the potential term in (I.44), this term becomes a number multiplied by the shape function $V(r/a)$ and the radial wavefunction u.

The shape function is usually written as

$$V(r/a) = -V_0 f(r/a), \tag{I.45a}$$

where V_0 is a strength parameter and $f(r/a)$ determines the radial shape of the potential. Since the nucleon-nucleon potential binds the nucleons within a very small radius, namely, that of the order of a few fm, the functional form of $f(r/a)$ and the range parameter a are so chosen that the potential quickly tends to zero as r increases beyond a few fm. Some of the commonly used shape functions are

$$\begin{aligned}
f(r/a) &= \begin{cases} 1 & \text{for } r \leqslant a \\ 0 & \text{for } r > a \end{cases} && \text{(square-well);} \\
f(r/a) &= e^{-r/a} && \text{(exponential well);} \\
f(r/a) &= e^{-r/a}/(r/a) && \text{(Yukawa well);} \\
f(r/a) &= e^{-r^2/a^2} && \text{(Gaussian well);} \\
f(r/a) &= e^{-r/a}/(1 - e^{-r/a}) && \text{(Hülthen well).}
\end{aligned} \tag{I.45b}$$

It is clear from these expressions that the Yukawa well tends to infinity at $r = 0$, and that the Gaussian well has the value unity and zero derivative at $r = 0$. These shape functions are represented in Fig. I.1 with the overall attractive negative sign of (I.45a). A pure central potential of the square-well, and exponential and Hülthen types yields analytic solutions of the S-state

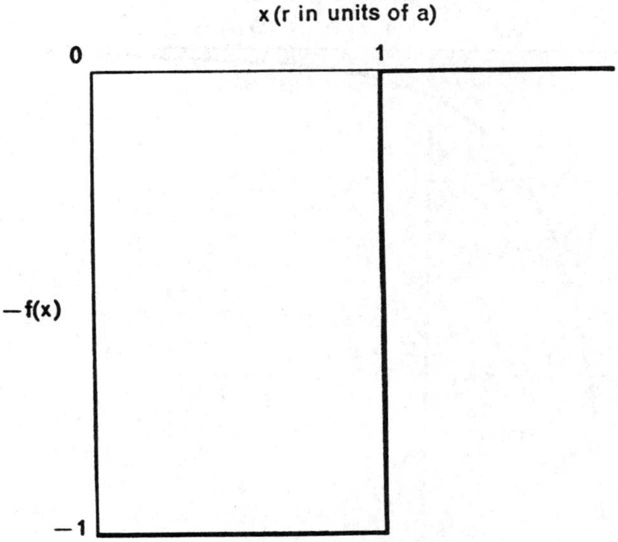

(a) Square-well

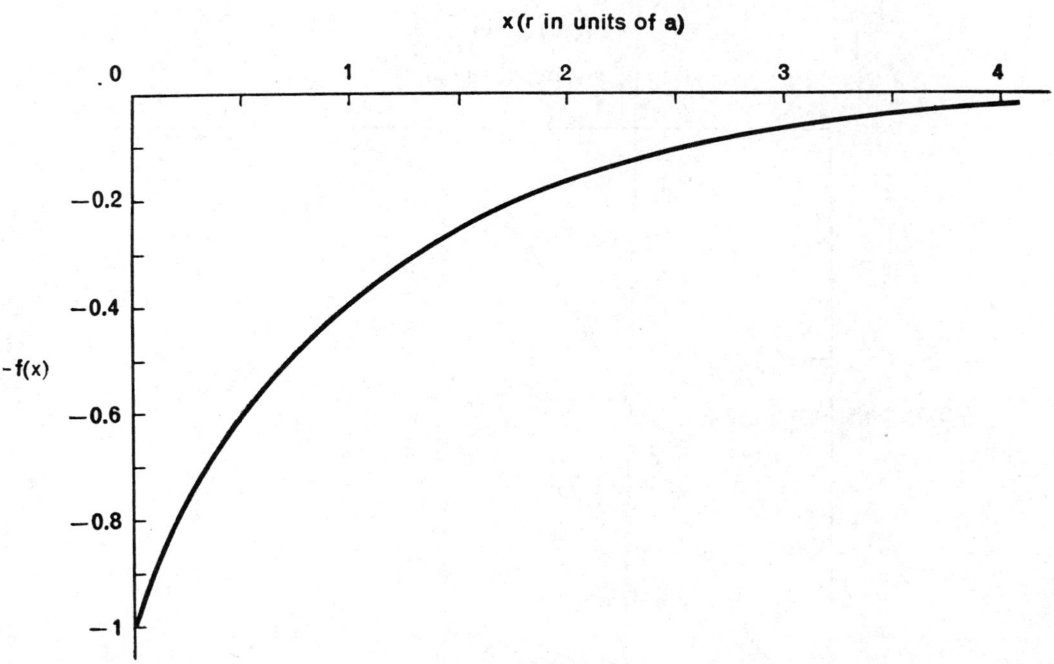

(b) Exponential

Fig. I.1 Several standard shapes of two-nucleon potential (cont.).

5 (45-122/1977)

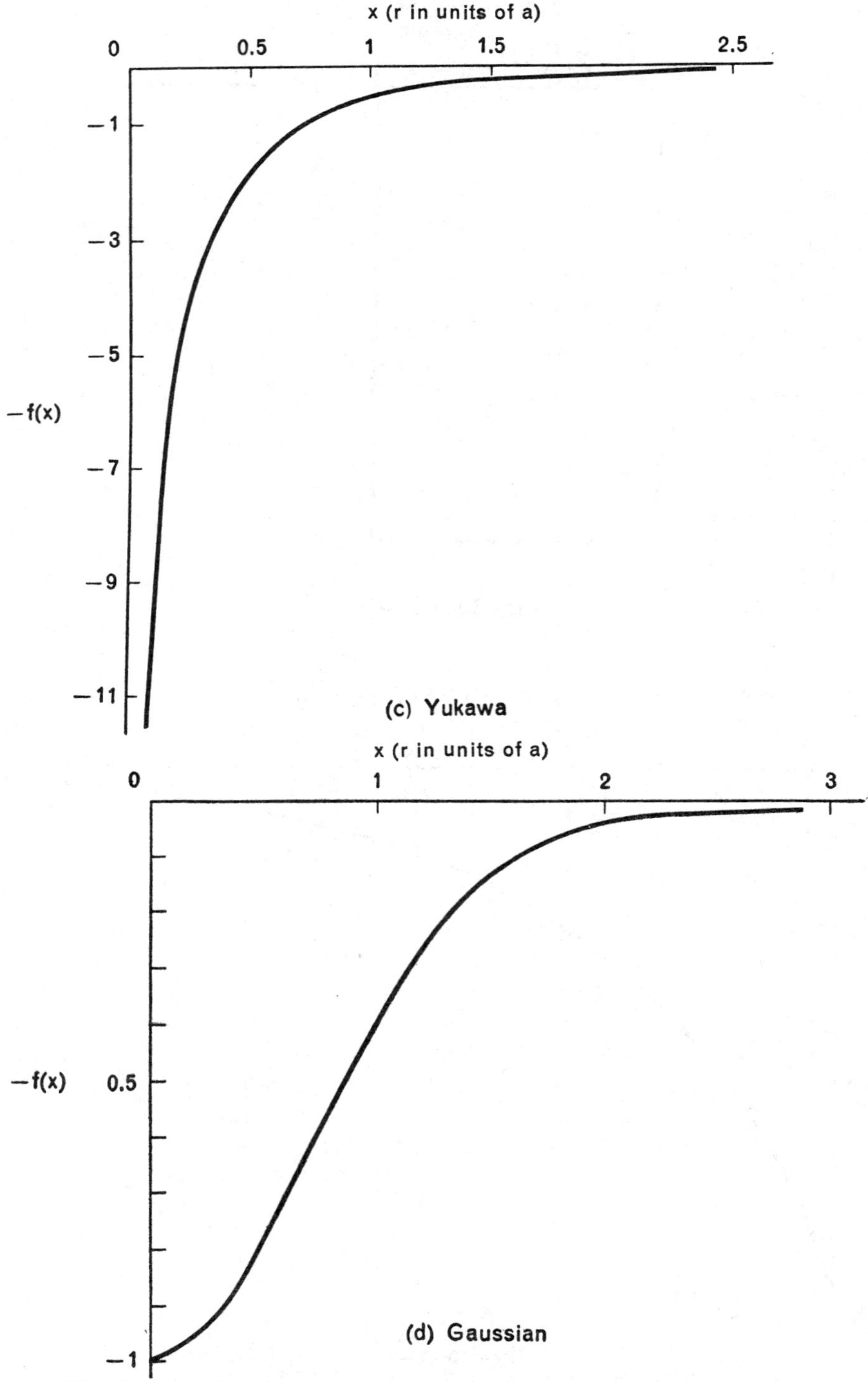

Fig. I.1 Several standard shapes of two-nucleon potential (cont.).

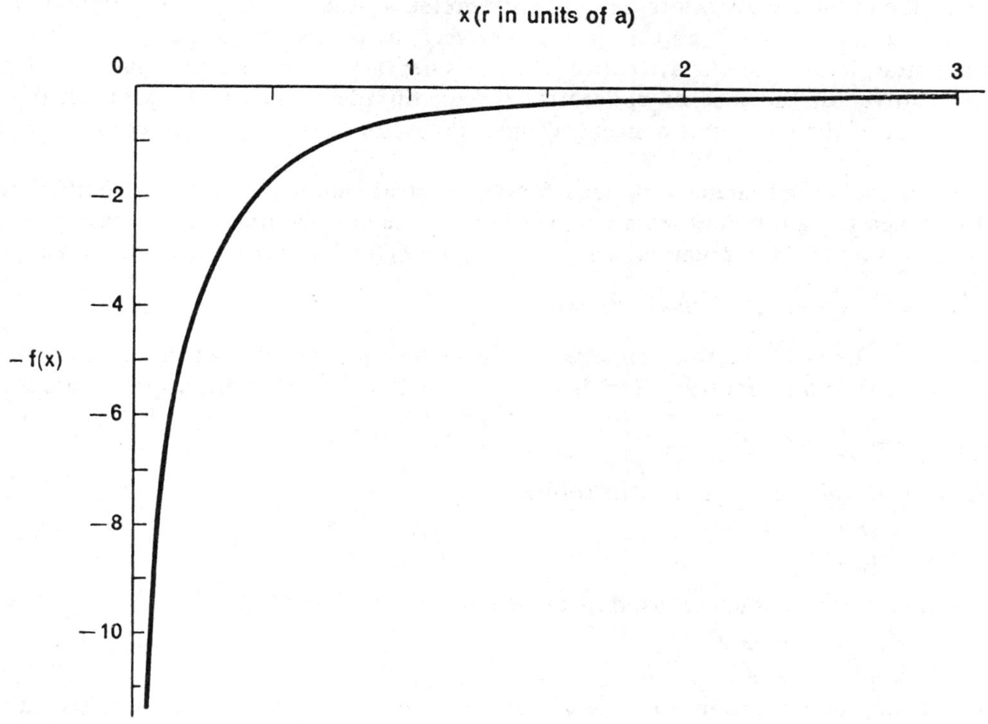

(e) Hülthen

Fig. I.1 Several standard shapes of two-nucleon potential.

Schrödinger equation. It is left to the reader to work out these solutions. The Hülthen well resembles the Yukawa well at $r \to 0$ and an exponential well at $r \to \infty$, and hence the exact S-state wavefunction of the deuteron obtained with the Hülthen well is frequently used in nuclear physics. This wavefunction is given by

$$u(r) = e^{-\gamma r}(1 - e^{-r/a}) \tag{I.46a}$$

in the internal region, and by

$$u(r) = e^{-\gamma r} \tag{I.46b}$$

in the asymptotic region. The extent of the deuteron wavefunction is clearly related to γ^{-1}, which is sometimes called the deuteron radius. γ^{-1} is much larger than the potential range a, and hence, for larger r, $e^{-r/a}$ decays much quicker than $e^{-\gamma r}$. Thus, (I.46a) automatically passes on to (I.46b) in the asymptotic region. An interesting conclusion that can be derived from the analytic solutions of the S-state deuteron equation (easiest to verify in the square-well case) for the three potentials—square-well, exponential, and Hülthen—is that V_0a^2 is roughly a constant. In other words, the same binding energy of the deuteron can be produced with an entire set of potentials of varying strength V_0, provided the range parameter a is suitably adjusted.

The depth parameter $-V_0$, multiplied by the number that results from the matrix elements of the exchange operators, may be interpreted as the depth of the potential for the type of state (specified by T and S or, alternatively, by parity and S) being considered. The central potential in any state is attractive when this depth is negative and repulsive when this depth is positive. In the case of noncentral forces, this depth has to be multiplied by the matrix element of the particular noncentral operator before deciding whether it is attractive or repulsive.

Finally, the radial shape functions for the central, tensor, and $l\cdot S$ potentials may in general be taken to be different from one another. We denote the linear combination of shape functions that occur in the potential term of (I.44) by $\mathcal{F}(r)$ and write the equation as

$$\frac{d^2u}{dr^2} - \frac{j(j+1)}{r^2}u + \mathcal{F}(r)u + k^2u = 0. \tag{I.47}$$

The asymptotic forms of the solution u can be immediately studied. Let us consider the case when $r \to 0$. If the potential term $\mathcal{F}(r)$ is less singular than r^{-2}, then the approximate equation

$$\frac{d^2u}{dr^2} - \frac{l(l+1)}{r^2}u = 0, \qquad l = j,$$

holds for $r \to 0$. This equation has the solution

$$u(r) = \begin{cases} r^{-l} \\ r^{l+1} \end{cases}$$

If the solution has to be well-behaved at the origin, then

$$u(r) = r^{l+1}, \qquad r \to 0, \tag{I.48}$$

is the only acceptable solution.

The asymptotic solution for $r \to \infty$ can be similarly worked out. In this case, the potential term $\mathcal{F}(r)$, which contains decreasing exponential functions, and hence diminishes much faster than r^{-n} (n is positive) as $r \to \infty$, can be dropped from the equation. The equation we then have to solve is

$$\frac{d^2u}{dr^2} - \frac{l(l+1)}{r^2}u + k^2u = 0, \qquad l = j. \tag{I.49}$$

This second-order differential equation has two linearly independent solutions, namely,

$$\frac{u}{r} = \begin{cases} j_l(kr) \\ n_l(kr) \end{cases}$$

which are called the spherical Bessel function, spherical Neumann function, respectively. These functions are defined by Schiff[2] who also lists many of their properties, including explicit forms for a few low-order functions ($l = 0, 1, 2$) [see Appendix C (Section I)]. For large r, in particular, they behave as

$$\begin{aligned} j_l(kr) &= \sin(kr - \tfrac{1}{2}l\pi)/(kr), \qquad r \to \infty, \\ n_l(kr) &= -\cos(kr - \tfrac{1}{2}l\pi)/(kr), \qquad r \to \infty. \end{aligned} \tag{I.50}$$

Here these solutions are valid for (I.47) only as $r \to \infty$, and cannot be extrapolated to the origin; it is nevertheless useful to remember certain properties for $r \to 0$:

$$\begin{aligned} j_l(kr) &= \frac{(kr)^l}{(2l+1)!!}, \qquad r \to 0, \\ n_l(kr) &= -\frac{(2l-1)!!}{(kr)^{l+1}}, \qquad r \to 0. \end{aligned} \tag{I.51}$$

It is therefore clear that only $j_l(kr)$ is well-behaved at the origin, whereas $n_l(kr)$ has a very strong singularity at $r = 0$.

The general solution of the radial Schrödinger equation (I.47), for large r, may be obtained by taking a general linear combination of the two functions (I.50). In the bound-state case, $k = i\gamma$, and hence the linear combinations behave as follows:

$$h_l^{(1)}(i\gamma r) = j_l(i\gamma r) + in_l(i\gamma r) \rightarrow \frac{e^{-\gamma r}}{r}, \qquad r \rightarrow \infty, \tag{I.52a}$$

$$h_l^{(2)}(i\gamma r) = j_l(i\gamma r) - in_l(i\gamma r) \rightarrow \frac{e^{\gamma r}}{r}, \qquad r \rightarrow \infty. \tag{I.52b}$$

These are called the spherical Hankel functions of the first and second kind, respectively. Since, as $r \rightarrow \infty$, $h_l^{(1)}$ is well-behaved but $h_l^{(2)}$ blows up, the bound-state solution of (I.47) behaves asymptotically as $h_l^{(1)}(i\gamma r)$, defined by (I.52a).

On the other hand, if the scattering state is considered, then any general linear combination

$$A_l[\cos \delta_l j_l(kr) - \sin \delta_l n_l(kr)] \tag{I.53a}$$

may be taken as the asymptotic form for $r \rightarrow \infty$ of the solution u/r. A_l and δ_l are the two constants that determine the arbitrary superposition of the two solutions (I.50). For $r \rightarrow \infty$, the asymptotic form becomes

$$u = \frac{1}{k} A_l \sin (kr - \tfrac{1}{2} l\pi + \delta_l), \qquad r \rightarrow \infty. \tag{I.53b}$$

The constant δ_l is called the phase shift of the partial wave l.

Once the asymptotic forms of the function u are known, it is possible to solve (I.47) numerically for all r. For the purposes of numerical integration, the essential trait of (I.47) can be simply described as

$$\frac{d^2u}{dr^2} = \bar{f}u, \tag{I.54}$$

where $\bar{f}$ is a known function of r. Using a Taylor expansion for the points $(a + \Delta)$ and $(a - \Delta)$, we obtain

$$u(a + \Delta) + u(a - \Delta) = 2u(a) + \Delta^2 u''(a) + \frac{\Delta^4}{12} u^{\mathrm{IV}}(a)$$

or

$$u(a + \Delta) = -u(a - \Delta) + 2u(a) + \Delta^2 \bar{f}(a)u(a) + \frac{\Delta^4}{12} u^{\mathrm{IV}}(a). \tag{I.55a}$$

We have retained terms up to the fourth-order derivative in the expansion, and, in the last step, have substituted the second-order derivative from the differential equation (I.54) itself. Next, we make similar Taylor expansions for u'', retain terms up to u^{IV} as before, and obtain

$$u''(a + \Delta) + u''(a - \Delta) = 2u''(a) + \Delta^2 u^{\mathrm{IV}}(a)$$

or

$$\Delta^2 u^{\mathrm{IV}}(a) = -2\bar{f}(a)u(a) + \bar{f}(a + \Delta)u(a + \Delta) + \bar{f}(a - \Delta)u(a - \Delta).$$

Using this expression in (I.55a), we get the final integration formula

$$u(a+\Delta) = [1 - \frac{\Delta^2}{12}\bar{f}(a+\Delta)]^{-1}[\{\frac{\Delta^2}{12}\bar{f}(a-\Delta) - 1\}u(a-\Delta) + \{\frac{5\Delta^2}{6}\bar{f}(a) + 2\}u(a)]. \quad \text{(I.55b)}$$

If u is known at the two points $(a - \Delta)$ and a, then this formula enables the extension of u to the point $(a + \Delta)$. Other integration formulas, available in texts on numerical analysis (see, for example, Buckingham[8]), achieve different degrees of accuracy; however, if the step-length Δ chosen is sufficiently small, (I.55b) is usually adequate.

If we use the asymptotic form (I.48) to evaluate u at $r = 0$, and Δ, then (I.55b) would enable us to calculate it step by step at all subsequent points. In the case of a scattering state, the log-derivative of the solution thus obtained can be matched with the log-derivative of the asymptotic form (I.53b) at a suitably chosen large value of r (the criterion being that at the matching radius the potential term of the radial equation is very much smaller than the energy term). This procedure determines the phase shift δ_l.

Conversely, we could start with the asymptotic forms $rj_l(kr)$ and $rn_l(kr)$ at a suitably chosen large value of r, the criterion being the same as the one just described. We can then use (I.55b) with a negative value of Δ and integrate inwards step by step towards the origin. The two solutions u_1 and u_2, obtained by starting with $rj_l(kr)$ and $rn_l(kr)$, respectively, will be linearly independent since they have different asymptotic forms at $r \to \infty$. We can then determine a linear combination of these two solutions so that the boundary condition at $r = 0$ is satisfied; that is, we require

$$(u_1 + cu_2)\Big|_{r=0} = 0 \quad \text{(I.56a)}$$

and determine c. Once c is determined, the required solution u is obtained, except for an overall normalization. The log-derivative of u can then be used, as before, with the log-derivative of (I.53b) for calculating the phase shift δ_l.

The latter method is specially suited when the shape function of the potential possesses an infinitely repulsive core (hard core) of radius r_c. Here the wavefunction u is required to be zero at $r = r_c$. Therefore, we can integrate u_1 and u_2 from asymptotically large r to $r = r_c$, and then require at $r = r_c$ the boundary condition

$$(u_1 + cu_2)\Big|_{r=r_c} = 0. \quad \text{(I.56b)}$$

The phase shift is determined by the same procedure.

If we wish to integrate the Schrödinger equation in the bound-state case (i.e., $k^2 = -\gamma^2$), we have only one asymptotic expression, namely, (I.52a), to start with. The solution u, now obtained by using (I.55b) backwards, will not automatically satisfy the boundary condition (I.56a) at $r = 0$ (no hard core) or (I.56b) at $r = r_c$ (hard core). In either case, one depth parameter V_0 belonging to the potential has to be varied until the required boundary condition is reproduced to the desired accuracy. This feature is essentially connected with the eigenvalue-eigenfunction nature of the bound-state problem. Physically, this means that the potential has a unique strength in order that it can produce a bound state $|lSjm\rangle$ with the required binding energy γ^2 (i.e., ϵ). Of course, in the two-nucleon case, there is only one bound state, namely, the ground state of the deuteron, which is actually the coupled $(^3S_1 + {}^3D_1)$-state. Therefore, it will be required that we solve the bound-state equation only in the case of the coupled differential equations (to be derived in Section 5B). Our coverage of the solution of the uncoupled equation for the bound-state case is merely of general mathematical interest.

Finally, it is worthwhile to note that we have suggested starting the solutions with the asymptotic forms, without caring to specify the constant such a solution could be multiplied with. This is because we are solving a homogeneous differential equation, and any solution of it that is multiplied by an arbitrary constant still remains a solution of the equation. In other words, an arbitrary multiplying constant continues to be undetermined in the solution of a homogeneous equation. This, however, is no hindrance, not even in the process of evaluating the phase shift from the calculated solution. For this purpose, we need only the log-derivatives of this solution and the asymptotic expression, and the log-derivative of any function f being $(1/f)(df/dr)$ is clearly independent of any constant preceding f. The overall undetermined constant can be specified only by requiring a certain normalization for u. In the bound-state case, the normalization is

$$\int_0^\infty u^2\,dr = 1,$$

but, in the scattering case, the normalization is such that the ingoing wave part e^{-ikr}/r of u/r in the asymptotic region $(r \to \infty)$ is equal to the ingoing part of the incident wave e^{ikz}. This case is considered in further detail in Section 7B.

B. COUPLED RADIAL EQUATIONS

To obtain the coupled radial equations, we put the coupled solution (I.25) in the Schrödinger equation (I.40) and then use the general result (I.43). We thus obtain

$$\frac{1}{r}[(\frac{d^2}{dr^2} - \frac{\hat{l}^2}{r^2}) - v + k^2](u\,|l = j - 1,\, S = 1, jm,\, TM_T\rangle + w\,|l = j + 1,\, S = 1, jm,\, TM_T\rangle) = 0.$$

We observe that the operator $\hat{l}^2$, operating on the angular momentum eigenfunction, acquires the appropriate value $l(l + 1)$. We multiply this equation from the left first by $\langle l = j - 1,\, S = 1,\, jm,\, TM_T|$ and then by $\langle l = j + 1,\, S = 1, jm,\, TM_T|$ and use the orthonormality of these functions to simplify the results. This procedure yields the two coupled radial equations

$$[\frac{d^2u}{dr^2} - \frac{j(j-1)}{r^2} + k^2]u - \langle j - 1,\, S = 1, jm,\, TM_T|\, v\,|j - 1,\, S = 1, jm,\, TM_T\rangle u$$
$$- \langle j - 1,\, S = 1, jm,\, TM_T|\, v\,|j + 1,\, S = 1, jm,\, TM_T\rangle w = 0, \qquad \text{(I.57a)}$$

$$[\frac{d^2}{dr^2} - \frac{(j+1)(j+2)}{r^2} + k^2]w - \langle j + 1,\, S = 1, jm,\, TM_T|\, v\,|j + 1,\, S = 1, jm,\, TM_T\rangle w$$
$$- \langle j + 1,\, S = 1, jm,\, TM_T|\, v\,|j - 1,\, S = 1, jm,\, TM_T\rangle u = 0. \qquad \text{(I.57b)}$$

The matrix elements of v, which are nondiagonal in l, are equal because of the hermiticity of V. The radial function resulting from these nondiagonal matrix elements of v is denoted by $\mathcal{H}(r)$. Similar functions resulting from the diagonal matrix elements of v in the states $l = j - 1$ and $l = j + 1$ are represented as $\mathcal{F}(r)$ and $\mathcal{G}(r)$, respectively. Then the coupled radial equations can be written as

$$\frac{d^2u}{dr^2} - \frac{j(j-1)}{r^2}u + k^2u + \mathcal{F}(r)u + \mathcal{H}(r)w = 0, \qquad \text{(I.58a)}$$

$$\frac{d^2w}{dr^2} - \frac{(j+1)(j+2)}{r^2}w + k^2w + \mathcal{G}(r)w + \mathcal{H}(r)u = 0. \tag{I.58b}$$

Once again, any solution $\begin{pmatrix} u \\ w \end{pmatrix}$ of (I.58) can be multiplied by a common arbitrary constant, and the resultant $\begin{pmatrix} u \\ w \end{pmatrix}$ will still be a solution of these homogeneous equations. This overall constant can be fixed by a normalization condition similar to that in the uncoupled case.

When there is no hard core in the potential functions, u and w are required to be well-behaved at $r = 0$, and their functional dependence on r for $r \to 0$ will still be given by (I.48), with $l = j \mp 1$. On the other hand, in the presence of a hard core of radius r_c, both u and w should vanish at $r = r_c$.

The asymptotic behaviour for $r \to \infty$ is similar to that of the uncoupled case because, when $r \to \infty$, the potential functions $\mathcal{F}(r)$, $\mathcal{G}(r)$, $\mathcal{H}(r)$ all tend to zero as a decreasing exponential and (I.58a) and (I.58b) become uncoupled and identical to (I.49) with $l = j \mp 1$.

The ground state of the deuteron is a special case of (I.58a) and (I.58b), with $k^2 = -\gamma^2$ (bound state) and $j = 1$. u and w are now the radial functions for the $^3S_1$- and $^3D_1$-state, respectively. The centrifugal term $j(j-1)$ becomes zero in (I.58a). For $r \to \infty$, the asymptotic forms of u/r and w/r are given by (I.52a) with $l = 0$ and $l = 2$, respectively. With the help of (CI.8) from Appendix C, we obtain

$$u(r) = \gamma r h_0^{(1)}(i\gamma r) = \gamma r[j_0(i\gamma r) + in_0(i\gamma r)] = e^{-\gamma r}, \qquad r \to \infty, \tag{I.59a}$$

$$w(r) = \gamma r h_2^{(1)}(i\gamma r) = e^{-\gamma r}[1 + \frac{3}{\gamma r} + \frac{3}{(\gamma r)^2}], \qquad r \to \infty. \tag{I.59b}$$

Since an overall multiplying constant is always undetermined with these functions, we have multiplied the asymptotic functions $rh_l^{(1)}(i\gamma r)$ by an extra γ to yield somewhat simpler final forms.

The coupled equations (I.58a) and (I.58b) may be expressed in simpler notation as

$$\frac{d^2u}{dr^2} = \bar{f}u + \bar{h}w, \tag{I.60a}$$

$$\frac{d^2w}{dr^2} = \bar{h}u + \bar{g}w, \tag{I.60b}$$

where $\bar{f}$, $\bar{g}$, and $\bar{h}$ are known functions of r. The integration formulas for u and w can now be obtained by following steps similar to those involved in the derivation of (I.55b). The final result for u is given by

$$\begin{aligned} u_+ = {} & [(1-g_+)(1-f_+) - h_+^2]^{-1}[\{-(1-g_+)(1-f_-) + h_+h_-\}u_- \\ & + \{(1-g_+)(10f+2) + 10hh_+\}u \\ & + \{(1-g_+)h_- - (1-g_-)h_+\}w_- + \{10h(1-g_+) + h_+(10g+2)\}w]. \end{aligned} \tag{I.61}$$

The result for w_+ is obtained from (I.61) by the interchanges $f \leftrightarrow g$ and $u \leftrightarrow w$. Here the functions f, g, h are $\Delta^2/12$ times $\bar{f}$, $\bar{g}$, $\bar{h}$, respectively, where Δ is the integration step-length. The functions with the subscripts + and − are to be evaluated at the points $a + \Delta$ and $a - \Delta$, respectively, whereas the unsubscripted functions are to be evaluated at the point a. The backward integration formula is obtained from (I.61) by interchanging the subscripts + and −.

The integration of the deuteron radial functions should begin with the asymptotic forms of

$u(r)$ and $w(r)$, given by (I.59a) and (I.59b). An overall normalization constant for these functions will remain undetermined, but the relative weight of the two can be adjusted. Therefore, the solutions should commence with $u(r)$ of (I.59a) and $\eta w(r)$, where η is an arbitrary mixing parameter and $w(r)$ is given by (I.59b). The backward integration formula is used to integrate up to $r = 0$ (no hard core) or $r = r_c$ (hard core). The procedure is repeated for two different values of η so that there are two linearly independent solutions, namely, $\begin{pmatrix} u_1 \\ w_1 \end{pmatrix}$ and $\begin{pmatrix} u_2 \\ w_2 \end{pmatrix}$, corresponding to the values η_1 and η_2. We now use arbitrary superposition of these two solutions so as to satisfy the boundary condition at $r = 0$ (no hard core) or $r = r_c$ (hard core). The required equations

$$(u_1 + cu_2)\Big|_{r=0 \text{ or } r=r_c} = 0, \tag{I.62a}$$

$$(w_1 + cw_2)\Big|_{r=0 \text{ or } r=r_c} = 0 \tag{I.62b}$$

obviously cannot be satisfied only by the choice of a single constant c. This fact is connected with the eigenvalue-eigenfunction nature of the problem. We have to change a depth parameter V_0 in the potential until both the boundary conditions (I.62a) and (I.62b) are satisfied by the choice of the single constant c. The depth V_0 so determined is the right depth of the potential so that the deuteron can be bound with the right binding energy. The final $u_1 + cu_2$ and $w_1 + cw_2$ are the required radial functions u and w, respectively. The overall normalization constant for them is determined by requiring

$$\int_0^\infty (u^2 + w^2)\, dr = 1. \tag{I.63}$$

In the scattering case, it is true that (I.58a) and (I.58b) get uncoupled for $r \to \infty$ because $\mathcal{H}(r) \to 0$ as a decreasing exponential function. Each of the partial waves $l = j - 1$ and $l = j + 1$ could therefore be required to satisfy boundary conditions of the type (I.53b), obtained in the uncoupled case. However, if the aim were to determine the phase shifts δ_{j-1} and δ_{j+1} by solving u and w from the coupled equations and then matching their log-derivatives with those of the appropriate asymptotic expressions, we would have ended up with complex phase shifts. This is because the partial waves $j \pm 1$ are coupled to each other through the scattering potential, and hence the flux of the beam for each partial wave is not individually conserved. The flux from each partial wave leaks into the other. If the phase shift is complex, then the imaginary part of the phase shift $i\epsilon_l$ gives rise to a factor exp $(-2\epsilon_l)$ in the amplitude of the outgoing wave (e^{ikr}). [This fact becomes explicit from (I.86)–(I.88) in Section 7B.] Thus, the imaginary part of the complex phase shift takes care of the damping of the intensity of the outgoing wave compared to that of the ingoing wave, the damping itself being a consequence of the loss of flux.

It is possible, however, to work with a special mixture of the partial waves $l = j - 1$ and $l = j + 1$ in the coupled case so that the special eigenwaves (mixtures of $l = j \pm 1$) are scattered without any loss of flux. The phase shifts for these special mixtures are therefore real and are referred to as the eigenphase shifts of the coupled states. We shall describe the eigenphase shift formalism in Section 7B, and then further discuss the solution of the coupled equations in the scattering case.

Before concluding this section, it would be worthwhile for the reader to be acquainted with the nature of the bound-state (Fig. I.2a) and scattering-state (Fig. I.2b) radial functions u

6 (45–122/1977)

plotted as a function of r in the case of a central potential. In both figures, the potential function has been indicated by lines labelled V on the negative side of the ordinate. The radial wavefunction for a potential without any hard core is also shown. When there is no hard core, the potential starts at $r = 0$; but, in the presence of a hard core, the attractive potential ends at $r = r_c$, and then, from $r = r_c$ to $r = 0$, there is an infinitely large potential of positive (repulsive) value. The radial functions are represented by solid lines on the positive side of the ordinate. The thicker line corresponds to the case of the hard core, and the thinner line to

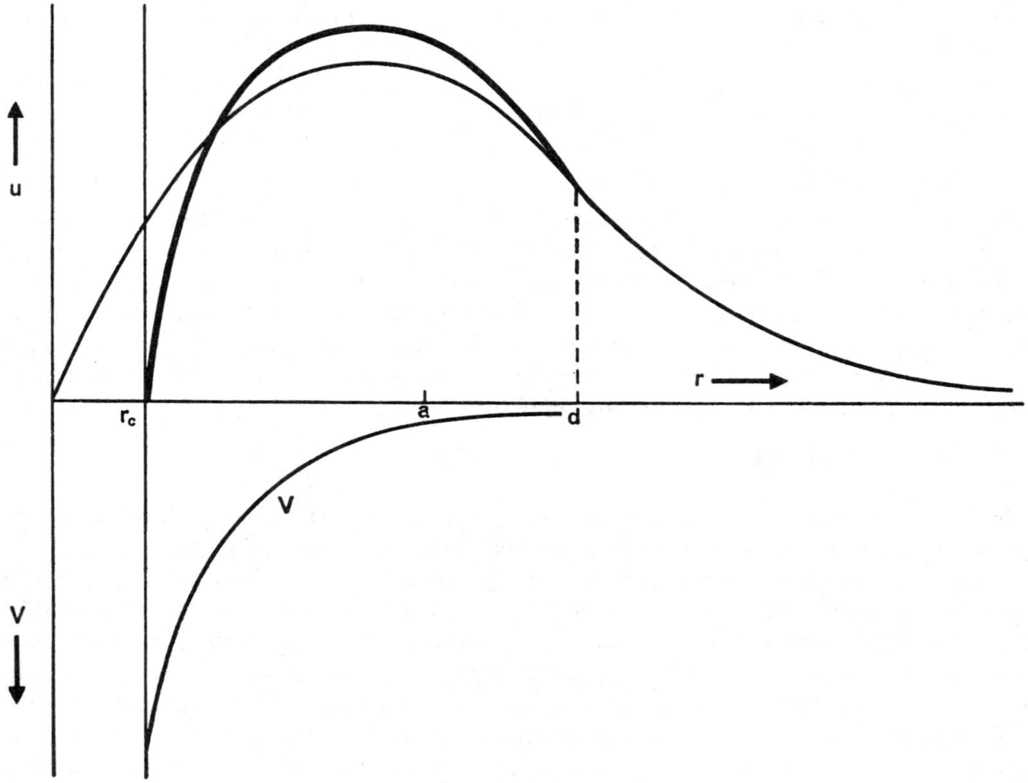

(a) Bound-state solution

Fig. I.2 Hard-core potential (followed by attractive well-behaved potential) and corresponding radial wavefunction (cont.).

the case of no hard core. When there is no hard core, the internal solutions start at a zero value for $r = 0$, whereas for the hard-core potential they start at a zero value for $r = r_c$, as shown in Fig. I.2. These internal functions then grow up and finally decrease and smoothly join the asymptotic forms at a point $r = d$ (which is somewhat larger than the range parameter a), where the potential $V(r/a)$ has decayed to a very much smaller value than the energy E that goes into the Schrödinger equation. It is clear from Fig. I.2a that, in the bound-state case, the internal solution must pass through a maximum and start decreasing before the asymptotic distance d is reached. This is so because, if it is still growing at $r = d$, it cannot smoothly match on to the exponentially decreasing tail of the wavefunction.

In the case of a scattering state of low energy, the internal solution is very similar to the bound-state solution because, inside the range of the potential, the potential term in the Schrödinger equation dominates the energy term. For high-energy scattering, however, the internal scattering solution distinctly differs from the bound-state solution. In the asymptotic region, the scattering solution is a sine wave with a phase shift δ. It is clear that an internal solution, which is still increasing at $r = d$, or has already started decreasing at $r = d$, has no difficulty in matching on to a sine wave. Only the latter situation is shown in Fig. I.2b. The

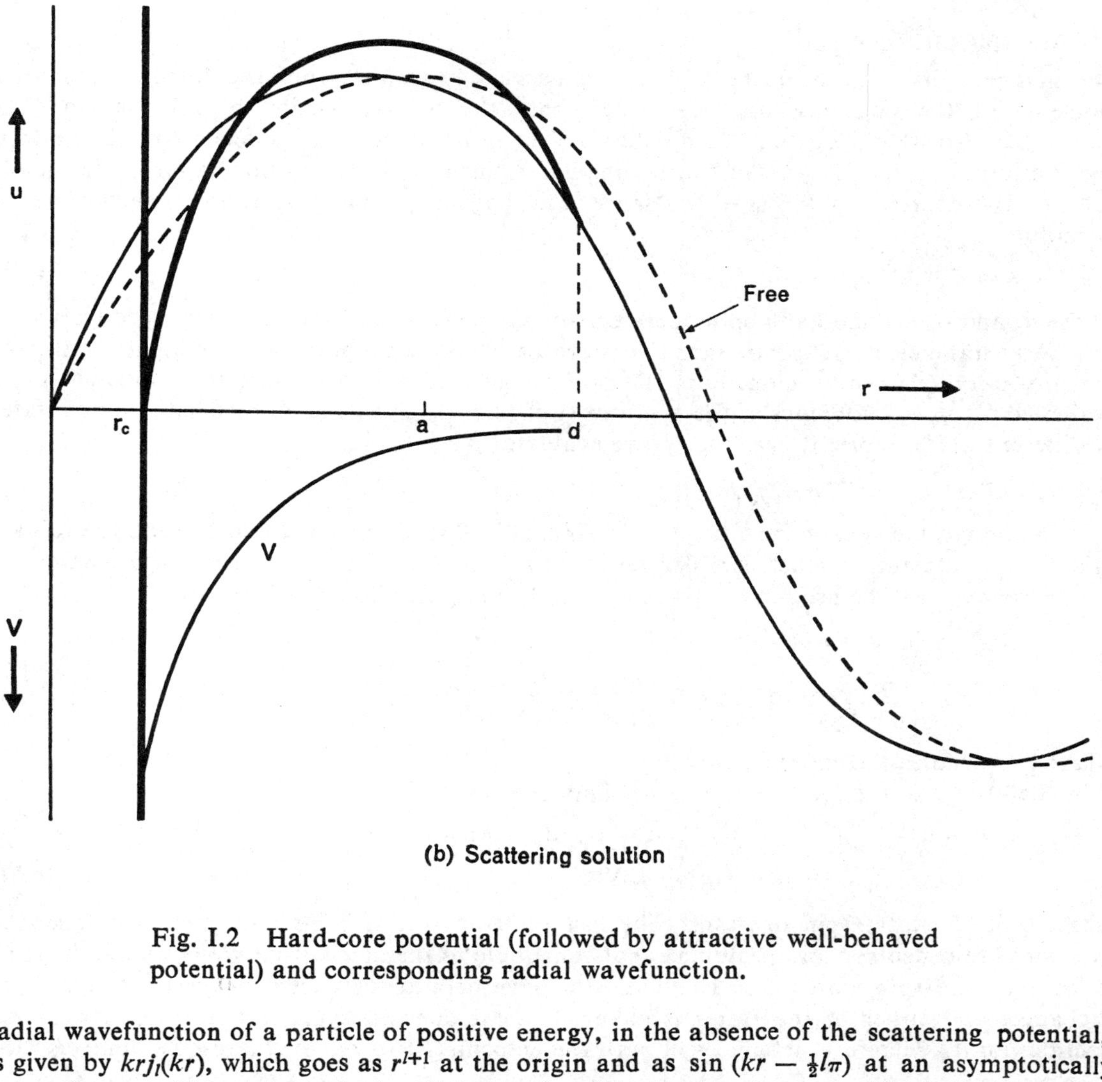

(b) Scattering solution

Fig. I.2 Hard-core potential (followed by attractive well-behaved potential) and corresponding radial wavefunction.

radial wavefunction of a particle of positive energy, in the absence of the scattering potential, is given by $krj_l(kr)$, which goes as r^{l+1} at the origin and as $\sin(kr - \frac{1}{2}l\pi)$ at an asymptotically large distance. This unperturbed radial wavefunction is also shown in Fig. I.2b (the curve labelled "Free"). Obviously, the displacement of this curve with respect to the actual asymptotic curve at large distances determines the phase shift.

6. STATIC ELECTROMAGNETIC MOMENTS OF DEUTERON

Having obtained the complete wavefunction

$$\frac{u}{r}\,|l=0,\,S=1,\,j=1,\,m\rangle+\frac{w}{r}\,|l=2,\,S=1,\,j=1,\,m\rangle \tag{I.64}$$

for the deuteron in Section 5B, we are now in a position to compute (i) the magnetic dipole moment and (ii) the electric quadrupole moment of the deuteron, which are the two static electromagnetic moments that the deuteron ground state possesses. The experimental values for (i) and (ii) are 0.8573 nM and 2.77 millibarns (mb; 1 barn $= 10^{-24}$ cm^2), respectively.

A. MAGNETIC MOMENT

The magnetic moment operator $\boldsymbol{\mu}$ has two parts: one results from the intrinsic magnetic moments of the nucleons, and the other from the orbital motion of the proton. Let $\mu_n(=-1.9128$ nM) and $\mu_p(=2.7925$ nM) denote the intrinsic magnetic moments of the neutron and the proton, respectively. The orbital angular momentum of the proton about the centre-of-mass of the deuteron is $\frac{1}{2}\mathbf{r}\times\mathbf{p}=\frac{1}{2}\boldsymbol{l}$. Hence, the magnetic moment operator in units of nM is given by

$$\boldsymbol{\mu}=\mu_n\boldsymbol{\sigma}_n+\mu_p\boldsymbol{\sigma}_p+\tfrac{1}{2}\boldsymbol{l}, \tag{I.65}$$

where $\boldsymbol{\sigma}_n$ and $\boldsymbol{\sigma}_p$ are the Pauli spin operators for the neutron and the proton, respectively.

Any static electromagnetic moment is evaluated, by convention, for the highest value of the projection of angular momentum. Thus, for a state $|lSjm\rangle$, the expectation value has to be evaluated for $m=j$. Obviously, the operator (I.65) cannot give a matrix element between states of different l. Therefore, it is enough if we evaluate

$$\langle lSjm=j|\,\mu_z\,|lSjm=j\rangle.$$

Although the general results given in Appendix B (Section III) could be used to evaluate this matrix element, a much simpler result, which holds for only the *expectation value of a vector operator*, can be used in the present case. The required result is given by

$$\begin{aligned}\langle A_z\rangle&=\langle \mathbf{A}\cdot\mathbf{J}\rangle\langle J_z\rangle/\langle J^2\rangle\\&=\frac{m}{j(j+1)}\langle \mathbf{A}\cdot\mathbf{J}\rangle,\end{aligned} \tag{I.66}$$

where $\langle\ \rangle$ denotes the expectation value.

Before we use the results (I.66), we simplify $\boldsymbol{\mu}$ as

$$\begin{aligned}\boldsymbol{\mu}&=\tfrac{1}{2}(\mu_n+\mu_p)(\boldsymbol{\sigma}_n+\boldsymbol{\sigma}_p)+\tfrac{1}{2}(\mu_n-\mu_p)(\boldsymbol{\sigma}_n-\boldsymbol{\sigma}_p)+\tfrac{1}{2}\boldsymbol{l}\\&=(\mu_n+\mu_p)\mathbf{S}+\tfrac{1}{2}\boldsymbol{l}+\tfrac{1}{2}(\mu_n-\mu_p)(\boldsymbol{\sigma}_n-\boldsymbol{\sigma}_p),\end{aligned} \tag{I.67}$$

where $\mathbf{S}$ is the total spin operator. The last term $(\boldsymbol{\sigma}_n-\boldsymbol{\sigma}_p)$, being antisymmetric under the exchange of the neutron and proton, cannot contribute to the expectation value we require; this is because the spin states $S=1$ and $S=0$ have definite symmetry properties under spin exchange, and either of the states is changed to the opposite symmetry state after $(\boldsymbol{\sigma}_n-\boldsymbol{\sigma}_p)$ operates on it. Therefore, it suffices if, with the help of (I.66), we evaluate $\langle S_z\rangle$ and $\langle l_z\rangle$ for $m=j$.

Since $\boldsymbol{l}=\mathbf{J}-\mathbf{S}$, we have

$$l^2=J^2+S^2-2\mathbf{S}\cdot\mathbf{J}$$

or

$$\langle \mathbf{S}\cdot\mathbf{J}\rangle = \tfrac{1}{2}\langle J^2 + S^2 - l^2\rangle$$
$$= \tfrac{1}{2}[j(j+1) + S(S+1) - l(l+1)].$$

Similarly, $\mathbf{S} = \mathbf{J} - \boldsymbol{l}$ leads to

$$\langle \boldsymbol{l}\cdot\mathbf{J}\rangle = \tfrac{1}{2}[j(j+1) + l(l+1) - S(S+1)].$$

Using these results, together with (I.66) and (I.67) for $m = j$, we obtain

$$\langle lSjm = j|\,\mu_z\,|lSjm = j\rangle = \frac{1}{2(j+1)}[(\mu_n + \mu_p + \tfrac{1}{2})j(j+1) + (\mu_n + \mu_p - \tfrac{1}{2})\{S(S+1) - l(l+1)\}]. \tag{I.68}$$

There being no radial operator in μ_z, the functions u/r and w/r give rise to the simple radial integrals $\int_0^\infty u^2\,dr$ and $\int_0^\infty w^2\,dr$, which multiply respectively the $^3S_1$- and $^3D_1$-matrix element of μ_z calculated from (I.68). These radial integrals are obviously the probabilities of the $^3S_1$- and $^3D_1$-state in the deuteron, and the normalization of the deuteron wavefunction expressed by (I.63) means that the sum of these probabilities is equal to unity. We denote these probabilities by p_S and p_D, noting that $p_S + p_D = 1$.

Evaluating (I.68) for $S = 1, j = 1$ with respectively $l = 0$, $l = 2$, we get

$$\mu(^3S_1) = \mu_n + \mu_p = 0.8797\ \text{nM}, \tag{I.69a}$$

$$\mu(^3D_1) = -\tfrac{1}{2}(\mu_n + \mu_p) + \tfrac{3}{4} = 0.3101\ \text{nM}. \tag{I.69b}$$

Because the observed magnetic moment of the deuteron is very close to the value given in (I.69a), we conclude that the deuteron ground state is predominantly $^3S_1$. Since the S-state does not have any orbital angular momentum, the magnetic moment in this state results entirely from the intrinsic magnetic moments of the nucleons; further, in the spin-triplet state, the spins of the two nucleons are aligned parallel, and hence their individual magnetic moments should add up. This gives a physical understanding of the result (I.69a). Since the $^3D_1$-state magnetic moment (I.69b) is less than the $^3S_1$-state moment, it is possible to reduce the magnetic moment from the $^3S_1$-state value by mixing it with a small part of the $^3D_1$-state. The probability p_D of the $^3D_1$-state in the deuteron ground state can be calculated if the magnetic moment is made to agree with the observed value. Thus,

$$0.8573 = 0.8797p_S + 0.3101p_D$$
$$= 0.8797 - 0.5696p_D$$

or

$$p_D = 0.0224/0.5696 = 0.039 = 3.9\ \%.$$

One test of an acceptable deuteron wavefunction, from the point of view of its magnetic moment, is therefore given by

$$\int_0^\infty w^2\,dr = p_D = 0.039. \tag{I.70}$$

There remain, however, several uncertainties relating to the correct expression for the magnetic moment operator itself. The fact that there are meson exchange currents in a nucleus implies the existence of additional magnetic moment caused by such currents. The velocity-dependent forces also give rise to extra contribution to the magnetic moment. The origin of the latter is now demonstrated for the spin-orbit coupling $\mathbf{l}\cdot\mathbf{S}$.

It is well-known (see, for example, Schiff[2]) that, in the presence of an electromagnetic field, the changes in the Hamiltonian of a system are obtained by altering the momentum of the system $\mathbf{p} \rightarrow \mathbf{p} - (e\mathbf{A}/c)$, where e is the charge of the system and $\mathbf{A}$ the vector potential of the electromagnetic field. For a constant steady magnetic field $\mathbf{H}$, the vector potential $\mathbf{A}$ is given by

$$\mathbf{A} = \tfrac{1}{2}(\mathbf{H} \times \mathbf{r})$$

as can be verified by forming $\nabla \times \mathbf{A}$ and checking that the result is indeed $\mathbf{H}$. The spin-orbit coupling $V(r/a)(\mathbf{r} \times \mathbf{p})\cdot\mathbf{S}$ therefore gives rise to an extra term in the Hamiltonian:

$$-\frac{e}{c}V(r/a)(\mathbf{r} \times \mathbf{A})\cdot\mathbf{S} = -\frac{e}{2c}V(r/a)\mathbf{r} \times (\mathbf{H} \times \mathbf{r})\cdot\mathbf{S}$$

$$= -\frac{e}{2c}V(r/a)[r^2\mathbf{S} - (\mathbf{r}\cdot\mathbf{S})\mathbf{r}]\cdot\mathbf{H}.$$

Since the energy of a magnetic dipole of moment $\boldsymbol{\mu}$ in the presence of $\mathbf{H}$ is given by $-\boldsymbol{\mu}\cdot\mathbf{H}$, we identify the extra magnetic moment arising from the spin-orbit term of the Hamiltonian as

$$\boldsymbol{\mu} = \frac{e}{2c}V(r/a)[r^2\mathbf{S} - (\mathbf{r}\cdot\mathbf{S})\mathbf{r}]. \tag{I.71}$$

Because of the uncertainties associated with the magnetic moment operator, the estimate (I.70) of the D-state probability is not very reliable. A certain latitude of variation has to be allowed in this quantity while checking the acceptability of the deuteron wavefunctions calculated for various parameters of the two-nucleon potential.

B. ELECTRIC QUADRUPOLE MOMENT

General Concepts of Electric Multipole Moments

The quadrupole moment of a nucleus arises from a special kind of deformation of its charge distribution. Let us first consider the case of a set of point charges bounded in a region of space, enclosed by a surface represented, for simplicity, by the closed curve in Fig. I.3, and

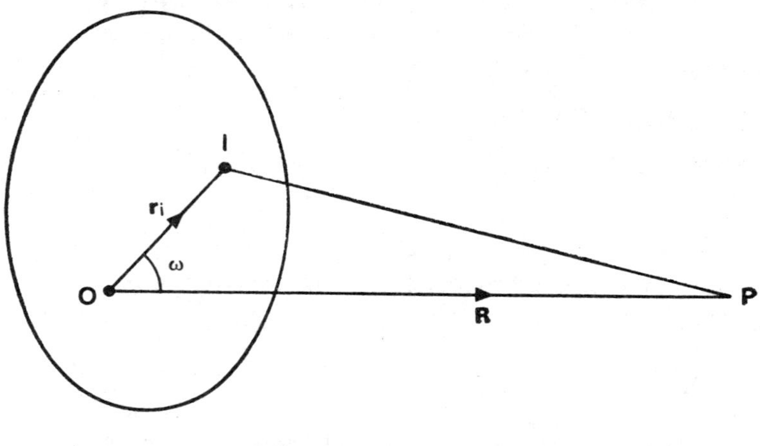

Fig. I.3 Electrostatic potential at P due to distribution of charge enclosed in a surface represented by closed curve.

calculate the potential at a point P outside the charge distribution. Taking O as the origin of a coordinate system and I as a typical point charge inside the charge distribution, we can denote the distances OI and OP by the vectors $\mathbf{r}_i$ and $\mathbf{R}$, respectively. The distance IP is then given by $(R^2 + r_i^2 - 2Rr_i \cos\omega)^{1/2}$, where ω is the angle between $\mathbf{R}$ and $\mathbf{r}_i$. Therefore, the potential v_i at the point P due to the charge e, located at the point I, is given by

$$v_i = \frac{e}{(R^2 + r_i^2 - 2Rr_i \cos\omega)^{1/2}} = e \sum_{K=0}^{\infty} \frac{r_<^K}{r_>^{K+1}} P_K(\cos\omega). \tag{I.72a}$$

Here $r_<$ and $r_>$ are the smaller and the larger of the two quantities R and r_i. This result follows from the definition of the Legendre polynomial P_K (see Whittaker and Watson[9]). If (θ_i, ϕ_i) and (Θ, Φ) represent the polar coordinates of $\mathbf{r}_i$ and $\mathbf{R}$ in the chosen coordinate system, then $P_K(\cos\omega)$ can be written by the spherical harmonic addition theorem [see (BI.21) in Appendix B] as

$$P_K(\cos\omega) = \frac{4\pi}{[K]} \sum_{q=-K}^{K} Y_q^K(\theta_i, \phi_i) Y_q^{K*}(\Theta, \Phi) = \sum_{q=-K}^{K} y_q^K(\theta_i, \phi_i) y_q^{K*}(\Theta, \Phi).$$

If the point P is extremely remote from the charge distribution, then, in (I.72a), $r_<$ and $r_>$ can be replaced by r_i and R, respectively. Then we have

$$v_i = \sum_{K=0}^{\infty} \sum_{q=-K}^{K} \frac{e}{R^{K+1}} y_q^{K*}(\Theta, \Phi) r_i^K y_q^K(\theta_i, \phi_i).$$

To obtain the potential v at the point P due to the entire charge distribution, this expression has to be summed over all the point charges contained in the distribution. That gives

$$v = \sum_i v_i = \sum_{K=0}^{\infty} \sum_{q=-K}^{K} \frac{e}{R^{K+1}} y_q^{K*}(\Theta, \Phi) \Omega_q^K, \tag{I.72b}$$

where

$$\Omega_q^K = \sum_i r_i^K y_q^K(\theta_i, \phi_i). \tag{I.72c}$$

The quantity (I.72c) is called the q-component of the K-th moment of the charge distribution. It is clear from (I.72b) that the potential due to a charge distribution at an external point can be determined if all these moments (i.e., Ω_q^K for all values of K and q) of the charge distribution are known.

If we regard the distribution of charge as continuous, specified by the charge density $e\rho$, then the summation in (I.72c) obviously requires to be replaced by an integration over the region of space V in which the charge is contained, i.e.,

$$\Omega_q^K = \int_V \rho(\mathbf{r}) r^K y_q^K(\theta, \phi)\, d^3r. \tag{I.72d}$$

The density $\rho(\mathbf{r})$ at the point $\mathbf{r}$ can, in general, be a function of all the three polar coordinates r, θ, ϕ, and it is therefore expressible as a complete summation over the spherical harmonics as

$$\rho(\mathbf{r}) = \sum_{l=0}^{\infty} \sum_{m=-l}^{l} C_{lm}(r) Y_m^{l*}(\theta, \phi). \tag{I.73}$$

If this expression of $\rho(\mathbf{r})$ is inserted in (I.72d), then, from the orthogonality of the spherical harmonics, only the term $C_{Kq}(r) Y_q^{K*}(\theta, \phi)$ of $\rho(\mathbf{r})$ contributes to the moment Ω_q^K and yields

$$\Omega_q^K = \sqrt{\frac{4\pi}{[K]}} \int_0^{\infty} C_{Kq}(r) r^{K+2}\, dr.$$

The first term of $\rho(\mathbf{r})$ corresponds to $l = 0$, $m = 0$ for which the spherical harmonic is the constant $1/\sqrt{4\pi}$. Therefore, the monopole part ($l = 0$) of the charge density does not depend on the angles θ, ϕ, and this part represents a spherically symmetric quantity. All higher terms of $\rho(\mathbf{r})$ depend on the angles through the higher-order spherical harmonics, and hence they represent the deformation of the charge-density distribution. In particular, the quadrupole moment arises from the ($l = 2$)-terms ($m = -2$ to $m = +2$) of $\rho(\mathbf{r})$, which are said to represent the quadrupole-type deformation of the charge distribution.

According to the principles of quantum mechanics, the moment Ω_q^K of a nucleus is the expectation value of the operator (I.72c) with respect to the nuclear wavefunction Ψ; the summation i goes over all the protons because they are the particles that carry the nuclear charge. The resultant expression

$$\langle \Omega_q^K \rangle = \langle \Psi | \sum_{i=1}^{Z} r_i^K y_q^K(\theta_i, \phi_i) | \Psi \rangle \tag{I.74}$$

can be very easily converted into the classical form (I.72d) with a suitable definition of $\rho(\mathbf{r})$, as is now shown. By using the property of the δ-function, any function $f(\mathbf{r}_i)$ can be written as

$$f(\mathbf{r}_i) = \int d^3r\, f(\mathbf{r})\delta(\mathbf{r} - \mathbf{r}_i).$$

Identifying the function $f(r_i)$ with $r_i^K Y_q^K(\theta_i, \phi_i)$, we obtain

$$\langle \Omega_q^K \rangle = \int d^3r\, r^K Y_q^K(\theta, \phi) \langle \Psi | \sum_{i=1}^{Z} \delta(\mathbf{r} - \mathbf{r}_i) | \Psi \rangle.$$

Comparing this with (I.72d), we obtain the quantum mechanical expression for $\rho(\mathbf{r})$ as

$$\rho(\mathbf{r}) = \langle \Psi | \sum_{i=1}^{Z} \delta(\mathbf{r} - \mathbf{r}_i) | \Psi \rangle. \tag{I.75}$$

A nuclear state Ψ occurring in (I.74) has a definite angular momentum J and projection M. From the general result obtained in Appendix B (Section III), we conclude that the matrix element of Y_q^K between two states of angular momentum J and projection M is nonvanishing if $q = 0$ and if K lies between 0 and $2J$. Further, since the parity of the state Ψ is the same on the two sides of the matrix element in (I.75), K can only be even integers, i.e., $0, 2, 4, \ldots, 2J$ or $(2J - 1)$. The last value is $2J$ or $2J - 1$ depending on whether J is an integer or a half-integer.

Specialization to Electric Quadrupole Moment

We now specialize our discussion to the *static* quadrupole moment. It is enough, according to the preceding discussion, to consider the ($q = 0$)-component of the operator. From (I.72c), Ω_0^2 is given by

$$\Omega_0^2 = \sum_{i=1}^{Z} r_i^2 y_0^2(\theta_i) = \sum_{i=1}^{Z} r_i^2 \tfrac{1}{2}(3\cos^2\theta_i - 1).$$

The ($q = 0$)-component of y_q^2 does not depend on the ϕ-angle, and the explicit expression of y_0^2 used here is from Table AI.1 (see Appendix A). Instead of using the operator Ω_0^2, as just defined, it is customary in nuclear physics to use the operator Q_0, which is equal to $2\Omega_0^2$. We shall abide by this convention, and the *static* quadrupole moment operator of a nuclear state will imply Q_0, rather than Ω_0^2. Some alternative expressions of Q_0 are

$$
\begin{aligned}
Q_0 &= \sum_{i=1}^{Z} r_i^2(3\cos^2\theta_i - 1) \\
&= \sum_{i=1}^{Z} (3z_i^2 - r_i^2) \\
&= 2\sum_{i=1}^{Z} r_i^2 y_0^2(\theta_i) \\
&= \sqrt{\frac{16\pi}{5}} \sum_{i=1}^{Z} r_i^2 Y_0^2(\theta_i).
\end{aligned} \tag{I.76a}
$$

According to the discussion following (I.75), a state of $J = 0$ or $J = \frac{1}{2}$ cannot have any quadrupole moment. In order that a state may have a nonvanishing static quadrupole moment, its angular momentum has to be 1 or higher. The nonvanishing value of the quadrupole moment itself may be positive or negative. An examination of the expressions (I.76a) reveals that the positive and negative values correspond to the expectation value of $3z^2$ being respectively larger and smaller than the expectation value of r^2 $(=x^2 + y^2 + z^2)$. This means that, in the former case, the extent of the deformed charge distribution in the z-direction is greater than that in the x- and y-direction. The situation is obviously the opposite in the case of a negative quadrupole moment. In both the cases, the deformed charge distribution has a spheroidal shape. The z-axis is the symmetry axis of the spheroid, i.e., the cross-section of the deformed charge distribution perpendicular to the z-axis has a circular shape. The spheroidal shapes corresponding to positive and negative quadrupole moments are shown in Fig. I.4 and

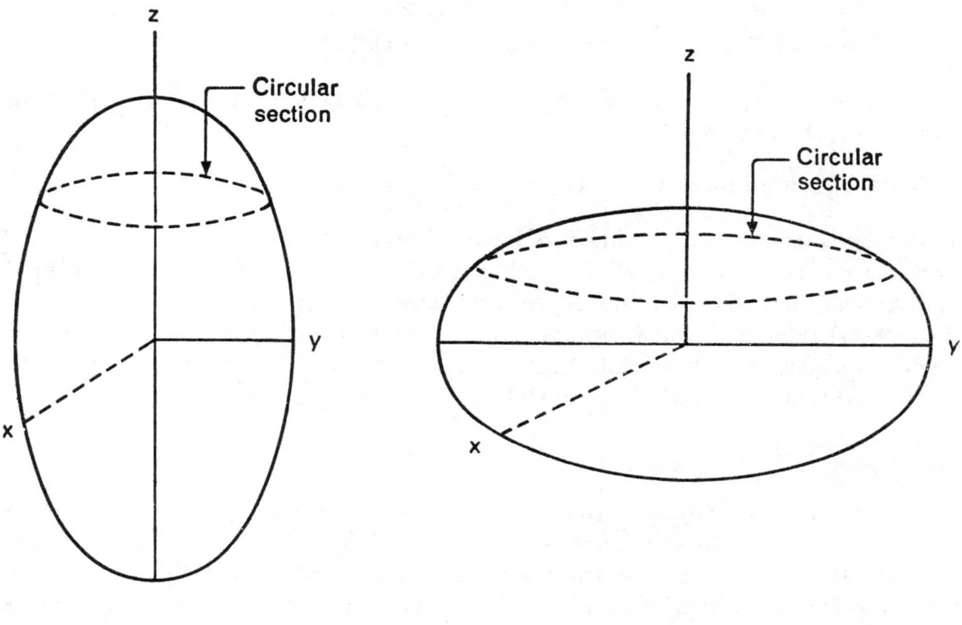

(a) Prolate (b) Oblate

Fig. I.4 Spheroidal charge distribution of prolate and oblate shapes.

7 (45-122/1977)

are known as prolate (Fig. I.4a) and oblate (Fig. I.4b). In both these cases, the section in the xy-plane is circular.

In the case of the deuteron, the positive value of the quadrupole moment therefore signifies a prolate spheroidal shape of the charge distribution. Its quadrupole moment operator is given, according to (I.76a), by

$$Q_0 = 3z_p^2 - r_p^2 = r_p^2(3\cos^2\theta - 1) = \tfrac{1}{4}r^2(3\cos^2\theta - 1) = \tfrac{1}{2}r^2 y_0^2(\theta). \qquad \text{(I.76b)}$$

Here we have used the fact that the distance $\mathbf{r}_p$ of the proton from the centre-of-mass is $\frac{1}{2}\mathbf{r}$.

We shall first evaluate the general matrix element of (I.76b) between the states $\langle l, S, j, m = j|$ and $|l', S, j, m = j\rangle$, and then give specific values to l, l' in order to obtain the two types of contribution of the wavefunction (I.64). The highest projection $m = j$ is chosen because static electromagnetic moments are always quoted for this projection by way of convention. Using, from Appendix B, (BIII.1) and (BIII.14) in succession, we obtain

$$\langle lSjj|\, Q_0\, |l'Sjj\rangle = \begin{bmatrix} j & 2 & j \\ j & 0 & j \end{bmatrix} \langle lSj|\, |Q|\, |l'Sj\rangle, \qquad \text{(I.77a)}$$

where

$$\langle lSj|\, |Q|\, |l'Sj\rangle = \tfrac{1}{2}\left(\int_0^\infty R_l R_{l'} r^4\, dr\right)(-1)^{l'-l} U(Sl'j2; jl) \sqrt{\frac{[l']}{[l]}} \begin{bmatrix} l' & 2 & l \\ 0 & 0 & 0 \end{bmatrix}. \qquad \text{(I.77b)}$$

R_l and $R_{l'}$ are the radial wavefunctions corresponding to the states l and l'.

We first take $l' = 0$ and $l = 2$ in (I.77b) and use the fact that both the Clebsch-Gordon coefficient and the U-function become unity in this case. Further, since $R_l = w/r$ and $R_{l'} = u/r$, we get

$$({}^3S_1\text{-}{}^3D_1)\text{-reduced matrix element} = \frac{1}{2\sqrt{5}} \int_0^\infty (uw) r^2\, dr. \qquad \text{(I.78a)}$$

Similarly, putting $S = 1$, $l' = l = 2$, and $j = 1$ in (I.77b) and using (I.36) and Table AVII.6 (see Appendix A), we obtain

$$({}^3D_1\text{-}{}^3D_1)\text{-reduced matrix element} = -\frac{1}{2\sqrt{10}} \int_0^\infty w^2 r^2\, dr. \qquad \text{(I.78b)}$$

The actual matrix elements are obtained by multiplying the results (I.78) by the Clebsch-Gordon coefficient occurring in (I.77a), which has the value $1/\sqrt{10}$ for $j = 1$. We finally note that, in the expectation value of the quadrupole moment operator with respect to the state (I.64), the $({}^3D_1\text{-}{}^3D_1)$-matrix element occurs once, whereas the $({}^3S_1\text{-}{}^3D_1)$-matrix element occurs twice. Hence, we get the quadrupole moment q of the deuteron by adding twice the result (I.78a) to the result (I.78b), and then multiplying the sum by $1/\sqrt{10}$. Thus,

$$q = \frac{\sqrt{2}}{10} \int_0^\infty uwr^2\, dr - \frac{1}{20} \int_0^\infty w^2 r^2\, dr. \qquad \text{(I.78c)}$$

The tensor potential is so adjusted that the resultant w accurately reproduces the observed quadrupole moment. The binding energy ϵ and the quadrupole moment q can be well-reproduced for several sets of values of the parameters in the potential. These sets can then be somewhat narrowed down by also checking the value of the magnetic moment. However, as already mentioned, the magnetic moment leaves some latitude in view of the uncertainty in the operator itself. As a result, the choice of the parameters in the potential cannot be reduced to a unique set.

will imply Q_0, rather than Ω_0^2. Some alternative expressions of Q_0 are

$$
\begin{aligned}
Q_0 &= \sum_{i=1}^{Z} r_i^2(3\cos^2\theta_i - 1) \\
&= \sum_{i=1}^{Z} (3z_i^2 - r_i^2) \\
&= 2\sum_{i=1}^{Z} r_i^2 y_0^2(\theta_i) \\
&= \sqrt{\frac{16\pi}{5}} \sum_{i=1}^{Z} r_i^2 Y_0^2(\theta_i). \qquad \text{(I.76a)}
\end{aligned}
$$

According to the discussion following (I.75), a state of $J = 0$ or $J = \frac{1}{2}$ cannot have any quadrupole moment. In order that a state may have a nonvanishing static quadrupole moment, its angular momentum has to be 1 or higher. The nonvanishing value of the quadrupole moment itself may be positive or negative. An examination of the expressions (I.76a) reveals that the positive and negative values correspond to the expectation value of $3z^2$ being respectively larger and smaller than the expectation value of r^2 $(=x^2 + y^2 + z^2)$. This means that, in the former case, the extent of the deformed charge distribution in the z-direction is greater than that in the x- and y-direction. The situation is obviously the opposite in the case of a negative quadrupole moment. In both the cases, the deformed charge distribution has a spheroidal shape. The z-axis is the symmetry axis of the spheroid, i.e., the cross-section of the deformed charge distribution perpendicular to the z-axis has a circular shape. The spheroidal shapes corresponding to positive and negative quadrupole moments are shown in Fig. I.4 and

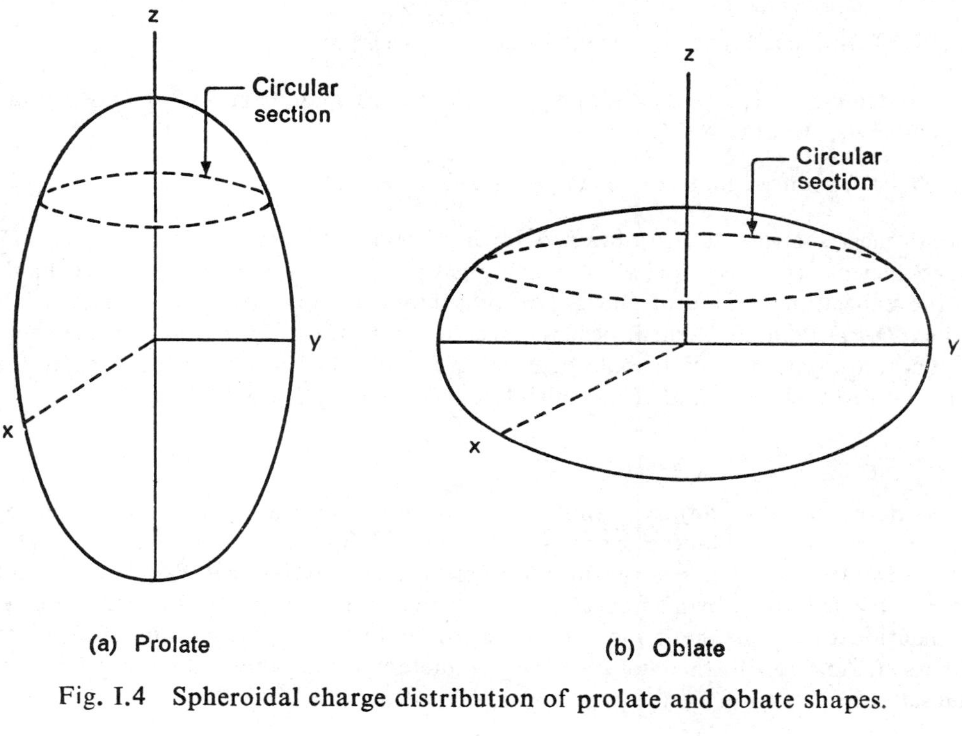

(a) Prolate (b) Oblate

Fig. I.4 Spheroidal charge distribution of prolate and oblate shapes.

7 (45-122/1977)

are known as prolate (Fig. I.4a) and oblate (Fig. I.4b). In both these cases, the section in the xy-plane is circular.

In the case of the deuteron, the positive value of the quadrupole moment therefore signifies a prolate spheroidal shape of the charge distribution. Its quadrupole moment operator is given, according to (I.76a), by

$$Q_0 = 3z_p^2 - r_p^2 = r_p^2(3\cos^2\theta - 1) = \tfrac{1}{4}r^2(3\cos^2\theta - 1) = \tfrac{1}{2}r^2 y_0^2(\theta). \tag{I.76b}$$

Here we have used the fact that the distance $\mathbf{r}_p$ of the proton from the centre-of-mass is $\frac{1}{2}\mathbf{r}$.

We shall first evaluate the general matrix element of (I.76b) between the states $\langle l, S, j, m = j|$ and $|l', S, j, m = j\rangle$, and then give specific values to l, l' in order to obtain the two types of contribution of the wavefunction (I.64). The highest projection $m = j$ is chosen because static electromagnetic moments are always quoted for this projection by way of convention. Using, from Appendix B, (BIII.1) and (BIII.14) in succession, we obtain

$$\langle lSjj|\, Q_0\, |l'Sjj\rangle = \begin{bmatrix} j & 2 & j \\ j & 0 & j \end{bmatrix} \langle lSj|\, |Q|\, |l'Sj\rangle, \tag{I.77a}$$

where

$$\langle lSj|\, |Q|\, |l'Sj\rangle = \tfrac{1}{2}\left(\int_0^\infty R_l R_{l'} r^4\, dr\right)(-1)^{l'-l}U(Sl'j2;jl)\sqrt{\frac{[l']}{[l]}}\begin{bmatrix} l' & 2 & l \\ 0 & 0 & 0 \end{bmatrix}. \tag{I.77b}$$

R_l and $R_{l'}$ are the radial wavefunctions corresponding to the states l and l'.

We first take $l' = 0$ and $l = 2$ in (I.77b) and use the fact that both the Clebsch-Gordon coefficient and the U-function become unity in this case. Further, since $R_l = w/r$ and $R_{l'} = u/r$, we get

$$({}^3S_1\text{-}{}^3D_1)\text{-reduced matrix element} = \frac{1}{2\sqrt{5}}\int_0^\infty (uw)r^2\, dr. \tag{I.78a}$$

Similarly, putting $S = 1$, $l' = l = 2$, and $j = 1$ in (I.77b) and using (I.36) and Table AVII.6 (see Appendix A), we obtain

$$({}^3D_1\text{-}{}^3D_1)\text{-reduced matrix element} = -\frac{1}{2\sqrt{10}}\int_0^\infty w^2 r^2\, dr. \tag{I.78b}$$

The actual matrix elements are obtained by multiplying the results (I.78) by the Clebsch-Gordon coefficient occurring in (I.77a), which has the value $1/\sqrt{10}$ for $j = 1$. We finally note that, in the expectation value of the quadrupole moment operator with respect to the state (I.64), the $({}^3D_1\text{-}{}^3D_1)$-matrix element occurs once, whereas the $({}^3S_1\text{-}{}^3D_1)$-matrix element occurs twice. Hence, we get the quadrupole moment q of the deuteron by adding twice the result (I.78a) to the result (I.78b), and then multiplying the sum by $1/\sqrt{10}$. Thus,

$$q = \frac{\sqrt{2}}{10}\int_0^\infty uwr^2\, dr - \tfrac{1}{20}\int_0^\infty w^2 r^2\, dr. \tag{I.78c}$$

The tensor potential is so adjusted that the resultant w accurately reproduces the observed quadrupole moment. The binding energy ϵ and the quadrupole moment q can be well-reproduced for several sets of values of the parameters in the potential. These sets can then be somewhat narrowed down by also checking the value of the magnetic moment. However, as already mentioned, the magnetic moment leaves some latitude in view of the uncertainty in the operator itself. As a result, the choice of the parameters in the potential cannot be reduced to a unique set.

7. SCATTERING FORMALISM FOR TWO-NUCLEON SYSTEM

A. DEFINITION AND KINEMATICS

Now we examine the scattering of a proton by a neutron caused by the nucleon-nucleon potential between the two particles. The scattering of a proton by a proton (discussed in Section 8) is further complicated by the presence of the Coulomb potential. However, the definitions and kinematic considerations that follow apply to both n-p and p-p scattering.

In a typical experimental set-up, the neutron is contained in a target material inside a scattering chamber, and the proton beam from an accelerator is scattered by the target. In Fig. I.5, N denotes the target neutron, and the protons are incident on it along the direction

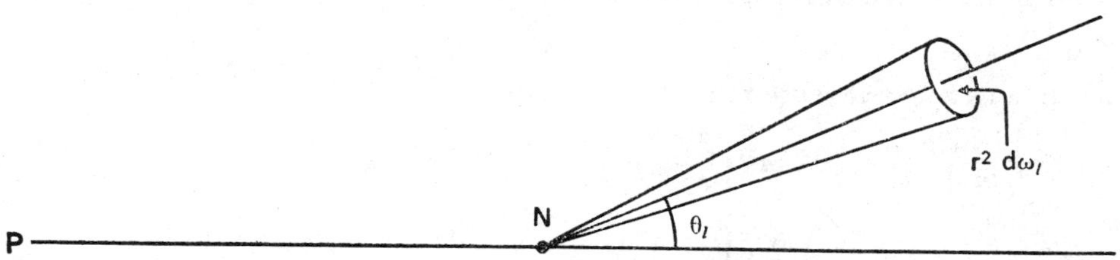

Fig. I.5 Set-up for measuring differential scattering cross-section.

PN, which we choose as the z-axis of the coordinate system *fixed in the laboratory*. These protons are scattered in all possible directions. We consider the scattering in the direction NP', which makes an angle θ_l with the direction of the incident beam. A detector, set up at a distance r from the scattering centre, subtends a finite solid angle $d\omega_l$ at N. It counts the scattered protons proceeding per unit time along the direction θ_l, through the solid angle $d\omega_l$ (i.e., across the element of area $r^2\, d\omega_l$) with a certain efficiency. An independent knowledge of this efficiency enables the calculation of the total number of protons N_S scattered per unit time within the solid angle $d\omega_l$ in the direction θ_l. What is obtained in this way is obviously not the scattering due to one target neutron, but rather that due to all the neutrons contained within a volume of the target defined by its thickness t along the z-direction, and area A marked by the cross-section of the incident beam. The total number of target neutrons N_T responsible for the scattering is therefore given by $At\rho$, where ρ is the density of neutrons in the target material. Similarly, the total number of incident protons can be specified in terms of the number of particles N_P incident per unit time per unit area of the beam. This latter quantity is called the flux of particles. If we divide N_S first by N_T and then by N_P, and also by the solid angle $d\omega_l$ then we get the probability per unit time of scattering of the proton at an angle θ_l within unit solid angle, for a single scattering centre, and for the unit incident flux of protons. This experimental quantity is known as the differential cross-section of scattering for the angle θ_l and is denoted by either $\sigma(\theta_l)$ or $d\sigma/d\omega$. The integral of this quantity over all the angles is called the total scattering cross-section.

In the foregoing definition, we have used the angles measured in the laboratory coordinate system. A similar definition of the differential cross-section can be given also for the centre-of-mass coordinate system. If $\sigma(\theta)$ denotes the differential cross-section in the centre-of-mass system, then, in view of the fact that θ_l and θ are related in a definite way, $\sigma(\theta_l)$ and $\sigma(\theta)$ can also be related to each other once and for all. The theory of the scattering process is always worked out in the centre-of-mass coordinate system, and hence $\sigma(\theta)$, and not $\sigma(\theta_l)$, is the quantity that will be calculated directly by means of the quantum mechanical theory. On other hand, $\sigma(\theta_l)$ is the quantity measured in experiments in the laboratory. Therefore, relation just mentioned is essential for a comparison of theory and experiment.

The derivation relating the various laboratory quantities with the corresponding centre-of-mass (CM) quantities is given in Schiff[2]. Only the relevant results are quoted here.

Considering, in the laboratory, a projectile of mass M_1 moving with initial velocity $\mathbf{v}_1$ and incident on a fixed target particle of mass M_2, we have

$$E_{\text{lab}} = \tfrac{1}{2}M_1 v_1^2, \tag{I.79a}$$

$$E_{\text{CM}} = \frac{M_2}{M_1 + M_2} E_{\text{lab}} \tag{I.79b}$$

such that, for nucleon-nucleon scattering ($M_1 = M_2$),

$$E_{\text{CM}} = \tfrac{1}{2}E_{\text{lab}}. \tag{I.79c}$$

The scattering angles θ_l and θ are related by the equation

$$\tan\theta_l = \frac{\sin\theta}{m + \cos\theta}, \qquad m = \frac{M_1}{M_2}, \tag{I.79d}$$

and hence

$$\sigma(\theta_l) = \frac{(1 + m^2 + 2m\cos\theta)^{3/2}}{1 + m\cos\theta}\sigma(\theta). \tag{I.79e}$$

For the case of nucleon-nucleon scattering, $m = 1$, and we have

$$\theta = 2\theta_l, \qquad \sigma(\theta) = \tfrac{1}{4}\sec\theta_l\,\sigma(\theta_l). \tag{I.79f}$$

In Section 7B, we shall describe the quantum mechanical calculation of $\sigma(\theta)$ in terms of an incident plane wave e^{ikz} along the z-direction, and a scattered outgoing wave of the type $r^{-1}e^{ikr}M(\theta)$, where M is called the scattering amplitude.

The probability current of any quantum mechanical wavefunction Ψ is given by (see Schiff[2])

$$\text{Re}\,\Psi^*(-i\hbar\mu^{-1}\nabla)\Psi,$$

where Re denotes "the real part of", and μ signifies the reduced mass of the system (=half of the nucleon mass for nucleon-nucleon scattering). For the incident wave, we use

$$\Psi_{\text{inc}} = e^{ikz}$$

to obtain the probability current along the z-direction given by

$$\hbar k\mu^{-1} = v,$$

where v is the velocity. Thus, the incident wave represents a flux of v particles per unit area per unit time.

In the same way, the scattered wave

$$\Psi_{\text{scatt}} = r^{-1}e^{ikr}M(\theta)$$

represents a probability current $r \to \infty$ along the radial direction given by

$$r^{-2}v|M(\theta)|^2.$$

Therefore, the total number of scattered particles passing through the area $r^2\,d\omega$ is

$$v\,d\omega\,|M(\theta)|^2.$$

Since we are now considering a case where there is only one scattering centre, our earlier definition of the differential cross-section asserts that

$$\frac{d\sigma}{d\omega} = \frac{v|M(\theta)|^2}{v} = |M(\theta)|^2. \tag{I.79g}$$

This expression for the differential scattering cross-section in terms of the scattering amplitude is used in our subsequent discussions.

B. SCATTERING THEORY

Assuming the direction of the incident particle to be the z-axis of the centre-of-mass coordinate system, we get the incident wavefunction at a large distance from the scattering centre given by the plane wave e^{ikz}. If the incident spin state corresponds to S and projection m, then

$$\Psi_{\text{inc}} = e^{ikz}\chi_m^S = \sum_{l=0} i^l\sqrt{4\pi[l]}\,j_l(kr)Y_0^l\chi_m^S. \tag{I.80a}$$

The partial-wave expansion of e^{ikz} used here is a standard result and is given in Appendix C (Section I). We now couple the spherical harmonic Y_0^l with the spin function χ_m^S and introduce the angular momentum coupled wavefunction $|lSjm\rangle$, which has already been used for the description of the two-nucleon spin-angle wavefunction. We also use the asymptotic form of $j_l(kr)$ for $r \to \infty$, given by (I.50). We thus obtain

$$\Psi_{\text{inc}} = \sum_{j=0} \sum_{l=|j-S|}^{j+S} i^l\sqrt{4\pi[l]}\begin{bmatrix} l & S & j \\ 0 & m & m \end{bmatrix}\frac{1}{kr}\sin\left(kr - \tfrac{1}{2}l\pi\right)|lSjm\rangle. \tag{I.80b}$$

The amplitude of $-[e^{-i\{kr-(1/2)l\pi\}}/2ikr]|lSjm\rangle$ in the incident wave is obviously given by i^lC_{lSjm}, where

$$C_{lSjm} = \sqrt{4\pi[l]}\begin{bmatrix} l & S & j \\ 0 & m & m \end{bmatrix}. \tag{I.81a}$$

Similarly, the amplitude of $(e^{ikr}/r)|lSjm\rangle$ in (I.80b) is given by

$$C_{lSjm}/(2ik). \tag{I.81b}$$

The incident wavefunction Ψ_{inc} is a solution of the free Schrödinger equation, that is, (I.40) without the potential term. As discussed in Section 5, the actual solution Ψ of the Schrödinger equation (I.40) can also be decomposed into the states $|lSjm\rangle$, of which the states $l = j$ ($S = 1$ or $S = 0$) are uncoupled and the states $l = j \pm 1$ ($S = 1$) are coupled. By definition, the scattered wave Ψ_{scatt} is given by

$$\Psi_{\text{scatt}} = \Psi - \Psi_{\text{inc}}, \tag{I.82}$$

and it should be a purely outgoing wave (i.e., the radial part should be of the type e^{ikr}/r). Therefore, the ingoing wave part (e^{-ikr}/r) of Ψ should be exactly equal to that of Ψ_{inc}, so that the ingoing part vanishes on the right-hand side of (I.82).

Uncoupled States

We first determine the part of Ψ_{scatt} that contains the uncoupled states $|lSjm\rangle$ with $l = j$. The radial part of these states, for $r \to \infty$, has already been given in (I.53b) as

$$\frac{u}{r} = \frac{A_l}{kr}\sin\left(kr - \tfrac{1}{2}l\pi + \delta_l^S\right), \qquad r \to \infty. \tag{I.83a}$$

The coefficient of $[-e^{-i\{kr-(1/2)l\pi\}}/(2ikr)]$ in this wavefunction is clearly given by

$$A_l \exp(-i\delta_l^S).$$

Equating this with the ingoing amplitude of the state $|lSjm\rangle$ of Ψ_{inc}, namely, $i^l C_{lSjm}$, given by (I.81a), we get

$$A_l = i^l \exp(i\delta_l^S) C_{lSjm}, \qquad l = j.$$

The coefficient of $(e^{ikr}/r)|lSjm\rangle$, in (I.83a), is similarly seen to be

$$\frac{A_l}{2ik} \exp(-i\tfrac{1}{2}l\pi) \exp(i\delta_l^S) = \frac{1}{2ik} C_{lSjm} \exp(2i\delta_l^S). \tag{I.83b}$$

If we subtract (I.81b)—which gives the same amplitude in Ψ_{inc}—from (I.83b), we obtain, according to (I.82), the amplitude of the $|l = j, Sjm\rangle$ part of the scattered wave. The full expression for this scattered partial wave is therefore given by

$$\frac{1}{2ik} C_{lSjm}[\exp(2i\delta_l^S) - 1]\frac{e^{ikr}}{r}|lSjm\rangle, \qquad l = j. \tag{I.84a}$$

If we now break up the angular momentum coupling in $|lSjm\rangle$ and write it as

$$|lSjm\rangle = \sum_{m'} \begin{bmatrix} l & S & j \\ m-m' & m' & m \end{bmatrix} Y^l_{m-m'}(\theta, \phi)\chi^S_{m'}(\sigma_1, \sigma_2), \tag{I.84b}$$

then it is clear that the incident beam of the spin projection m can be scattered, in general, to all possible values of the spin projection m'. The coefficient of $(e^{ikr}/r)\chi^S_{m'}$ in (I.84a) is called the scattering amplitude in the uncoupled state $l = j$, corresponding to the scattering from an initial state of given spin projection m to a final state of spin projection m'. We shall denote this quantity by $M^{jSj}_{m',m}$, and fully express it with the help of (I.84a) and (I.84b) as

$$M^{jSj}_{m',m}(\theta, \phi) = \frac{1}{2ik}\sqrt{4\pi[j]} \begin{bmatrix} j & S & j \\ o & m & m \end{bmatrix} \begin{bmatrix} j & S & j \\ m-m' & m' & m \end{bmatrix} [\exp(2i\delta_j^S) - 1] Y^j_{m-m'}(\theta, \phi). \tag{I.85}$$

The sum of (I.85) over $j = 0 \rightarrow j = \infty$ gives the total scattering amplitude corresponding to all the uncoupled states.

Before evaluating the scattering amplitude in the coupled case, we shall so alter the preceding method of derivation that it will allow immediate generalization to the coupled case. Since our aim is to expand the total wave in the ingoing and outgoing waves, we could have straightaway started with the general asymptotic form

$$u = Ae^{-i[kr-(1/2)l\pi]} - Be^{i[kr-(1/2)l\pi]}, \qquad r \rightarrow \infty. \tag{I.86}$$

The amplitude of the outgoing wave should obviously be related to that of the ingoing wave through the scattering process itself. This physical fact is incorporated in (I.86) by requiring

$$B = \mathcal{S}A, \tag{I.87}$$

where $\mathcal{S}$ is called the scattering operator or the scattering matrix. Here $\mathcal{S}$ relates the single quantity A to the single quantity B, and hence it is a very simple one-dimensional matrix (i.e., a single scalar quantity). Comparing (I.83a) with (I.86), we obtain

$$A = -A_l \exp(-i\delta_l^S)/(2ik),$$

$$B = -A_l \exp(i\delta_l^S)/(2ik),$$

and hence

$$\mathcal{S} = B/A = \exp(2i\delta_l^S), \tag{I.88}$$

where δ_l^S is the phase shift for the uncoupled state $l = j$. Thus, the asymptotic form (I.83a), used earlier, is exactly equivalent to the form

$$u = A[e^{-i\{kr-(1/2)l\pi\}} - \exp(2i\delta_l^S)e^{i\{kr-(1/2)l\pi\}}], \qquad r \to \infty.$$

Coupled Case

In the coupled case, we start with asymptotic forms of the coupled radial functions u and w, written in analogy with (I.86) as

$$u = A_1 e^{-i[kr-(1/2)(j-1)\pi]} - B_1 e^{i[kr-(1/2)(j-1)\pi]}, \qquad r \to \infty, \tag{I.89a}$$

$$w = A_2 e^{-i[kr-(1/2)(j+1)\pi]} - B_2 e^{i[kr-(1/2)(j+1)\pi]}, \qquad r \to \infty. \tag{I.89b}$$

Since the waves $l = j \pm 1$ are coupled to each other, we cannot require B_1 to be determined entirely from A_1 by means of the scattering process. The same observation is true for A_2 and B_2. Because of the coupling of the two partial wave channels, an ingoing wave in the channel $l = j - 1$ or $l = j + 1$ gets scattered to both the channels $l = j - 1$ and $l = j + 1$. Thus, in the present case, we can impose the more generalized requirement, namely, that the amplitude vector $\begin{pmatrix} B_1 \\ B_2 \end{pmatrix}$ of the outgoing wave be related to $\begin{pmatrix} A_1 \\ A_2 \end{pmatrix}$ of the ingoing wave through a 2×2 scattering matrix $\mathcal{S}$:

$$\begin{pmatrix} B_1 \\ B_2 \end{pmatrix} = \mathcal{S}\begin{pmatrix} A_1 \\ A_2 \end{pmatrix}. \tag{I.90}$$

Now since

$$B_i = \sum_{j=1}^{2} \mathcal{S}_{ij} A_j,$$

we find that either B_1 or B_2 is given as a linear sum of both A_1 and A_2, as desired.

There being no flux loss from the combined channels, we require

$$\sum_{i=1}^{2} A_i^* A_i = \sum_{i=1}^{2} B_i^* B_i$$

or

$$(A_1^* \quad A_2^*)\begin{pmatrix} A_1 \\ A_2 \end{pmatrix} = (B_1^* \quad B_2^*)\begin{pmatrix} B_1 \\ B_2 \end{pmatrix}.$$

Using (I.90) and its Hermitean conjugate on the right-hand side of this equation, we get

$$\mathcal{S}^\dagger \mathcal{S} = \mathbb{1},$$

i.e., $\mathcal{S}$ is required to be a unitary matrix. [It should be noted that, in the simple uncoupled case, $\mathcal{S} = \exp(2i\delta_j^S)$ was unitary. This is because, during the scattering process, there is no loss of flux from the wave $l = j$ since it is uncoupled.]

It is also assumed that time-reversal invariance is another general requirement on the $\mathcal{S}$-matrix. In the coupled case, the $\mathcal{S}$-matrix will ultimately be computed by solving the coupled Schrödinger equation, in which the scattering agency, namely, the two-nucleon potential, has already been chosen to satisfy this general invariance requirement (see Section 3). The $\mathcal{S}$-matrix computed from such potentials will also satisfy time-reversal invariance. As a result of this invariance property, the $\mathcal{S}$-matrix becomes symmetric. The matrix element $\langle l=j-1|\,\mathcal{S}\,|l=j+1\rangle$ denotes scattering from an initial state $l=j+1$ to a final state $l=j-1$. Under time-reversal, the initial and final states get reversed in their roles, and the matrix element is transformed to $\langle l=j+1|\,\mathcal{S}\,|l=j-1\rangle$. The invariance under time-reversal therefore requires these two matrix elements to be equal, i.e., $\mathcal{S}$ to be symmetric.

We next prove that the 2×2 symmetric unitary matrix $\mathcal{S}$ can be diagonalized by a *real orthogonal* matrix $\mathcal{O}$. If

$$\mathcal{O}^{-1}\mathcal{S}\mathcal{O}=D, \tag{I.91a}$$

where D is the diagonal form of $\mathcal{S}$, then

$$\mathcal{S}\mathcal{O}=\mathcal{O}D \tag{I.91b}$$

or

$$\tilde{\mathcal{O}}\tilde{\mathcal{S}}=\tilde{D}\tilde{\mathcal{O}} \qquad (\sim \text{ denotes a transposed matrix})$$

or

$$\tilde{\mathcal{O}}\mathcal{S}=D\tilde{\mathcal{O}} \qquad (\tilde{\mathcal{S}}=\mathcal{S} \text{ due to symmetry, and } \tilde{D}=D \text{ due to the diagonal nature of } D).$$

But, from (I.91a),

$$\mathcal{O}^{-1}\mathcal{S}=D\mathcal{O}^{-1},$$

and hence a comparison of the last two equations establishes

$$\mathcal{O}^{-1}=\tilde{\mathcal{O}}, \tag{I.92a}$$

i.e., $\mathcal{O}$ is an orthogonal matrix. The fact that $\mathcal{O}$ is real, at the same time, follows from the unitarity of $\mathcal{S}$ as now detailed. We first use the fact that any eigenvector of a matrix is also an eigenvector of the inverse matrix, and the eigenvalues are reciprocals of each other. Thus, if $|\alpha\rangle$ is an eigenvector of $\mathcal{S}$ belonging to the eigenvalue d_α, i.e.,

$$\mathcal{S}|\alpha\rangle=d_\alpha|\alpha\rangle, \tag{I.93}$$

then

$$\mathcal{S}^{-1}|\alpha\rangle=(d_\alpha)^{-1}|\alpha\rangle$$

automatically holds. Therefore,

$$\langle\beta|\,\mathcal{S}^{-1}\,|\alpha\rangle=(d_\alpha)^{-1}\langle\beta\,|\,\alpha\rangle. \tag{I.94a}$$

But

$$\langle\beta|\mathcal{S}^{\dagger}\equiv\langle\beta|\mathcal{S}^{-1}=d_\beta^*\langle\beta| \qquad [\text{follows from the Hermitean conjugate of (I.93) with } \alpha\to\beta],$$

and hence

$$\langle\beta|\,\mathcal{S}^{-1}\,|\alpha\rangle=d_\beta^*\langle\beta\,|\,\alpha\rangle. \tag{I.94b}$$

Subtracting (I.94a) from (I.94b), we get

$$(d_\beta^*-d_\alpha^{-1})\langle\beta\,|\,\alpha\rangle=0.$$

When $\beta=\alpha$, this equation yields

$$d_\alpha^*=(d_\alpha)^{-1}. \tag{I.95}$$

Thus, the diagonal elements d_α of the diagonal matrix D should be of the form $\exp(2i\delta^\alpha)$, where δ^α is real. Further, (I.95) means

$$D^\dagger = D^{-1},$$

that is, D is also unitary. Now, taking the inverse of (I.91b), we get

$$\mathcal{O}^{-1}\mathcal{S}^{-1} = D^{-1}\mathcal{O}^{-1}$$

or

$$\mathcal{O}^{-1}\mathcal{S}^\dagger = D^\dagger\mathcal{O}^{-1}.$$

Similarly, the Hermitean conjugate of (I.91b) yields

$$\mathcal{O}^\dagger\mathcal{S}^\dagger = D^\dagger\mathcal{O}^\dagger.$$

A comparison of these two results establishes

$$\mathcal{O}^{-1} = \mathcal{O}^\dagger, \tag{I.92b}$$

i.e., $\mathcal{O}$ is unitary. If we combine (I.92a) with (I.92b), we find

$$\tilde{\mathcal{O}} = \mathcal{O}^{-1} = \mathcal{O}^\dagger \equiv \tilde{\mathcal{O}}^*,$$

that is, the matrix $\mathcal{O}$ is also real.

The most general form of a real orthogonal 2×2 matrix may be described by a single real parameter ϵ as

$$\mathcal{O} = \begin{pmatrix} \cos\epsilon & -\sin\epsilon \\ \sin\epsilon & \cos\epsilon \end{pmatrix}.$$

The orthonormality of the rows and columns of this matrix can be immediately checked. The form of the diagonal matrix D has already been established as

$$D = \begin{pmatrix} \exp(2i\delta_j^\alpha) & 0 \\ 0 & \exp(2i\delta_j^\beta) \end{pmatrix},$$

where δ_j^α and δ_j^β are real parameters. From (I.91a) and these general forms of $\mathcal{O}$ and D, it becomes obvious that the symmetric unitary scattering matrix $\mathcal{S}$ can be parametrized in terms of the three real quantities ϵ, δ_j^α, and δ_j^β as

$$\mathcal{S} = \mathcal{O}D\mathcal{O}^{-1} = \begin{pmatrix} \cos\epsilon & -\sin\epsilon \\ \sin\epsilon & \cos\epsilon \end{pmatrix}\begin{pmatrix} \exp(2i\delta_j^\alpha) & 0 \\ 0 & \exp(2i\delta_j^\beta) \end{pmatrix}\begin{pmatrix} \cos\epsilon & \sin\epsilon \\ -\sin\epsilon & \cos\epsilon \end{pmatrix}$$

$$= \begin{pmatrix} \exp(2i\delta_j^\alpha)\cos^2\epsilon + \exp(2i\delta_j^\beta)\sin^2\epsilon & [\exp(2i\delta_j^\alpha) - \exp(2i\delta_j^\beta)]\sin\epsilon\cos\epsilon \\ [\exp(2i\delta_j^\alpha) - \exp(2i\delta_j^\beta)]\sin\epsilon\cos\epsilon & \exp(2i\delta_j^\alpha)\sin^2\epsilon + \exp(2i\delta_j^\beta)\cos^2\epsilon \end{pmatrix}. \tag{I.96}$$

The parameter ϵ is called the mixing parameter, and the two δ's the eigenphase shifts. (The reason for these nomenclatures will soon be clear.)

(I.91b) implies that the two columns of $\mathcal{O}$ are the two eigenvectors of $\mathcal{S}$ belonging to the two eigenvalues given by the two diagonal elements of D. Thus,

$$\mathcal{S}\begin{pmatrix} \cos\epsilon \\ \sin\epsilon \end{pmatrix} = \exp(2i\delta_j^\alpha)\begin{pmatrix} \cos\epsilon \\ \sin\epsilon \end{pmatrix}, \tag{I.97a}$$

$$\mathcal{S}\begin{pmatrix} -\sin\epsilon \\ \cos\epsilon \end{pmatrix} = \exp(2i\delta_j^\beta)\begin{pmatrix} -\sin\epsilon \\ \cos\epsilon \end{pmatrix}. \tag{I.97b}$$

8 (45-122/1977)

Comparing these expressions with (I.90), we get two special choices of A_1 and A_2 for which B_1 and B_2 are simply proportional to A_1 and A_2, respectively. The eigenwaves, obtained from (I.89a) and (I.89b) corresponding to these special choices, are called α- and β-wave, respectively. They can be represented in the space of $|j-1, 1, jm\rangle$ and $|j+1, 1, jm\rangle$ by the two-dimensional vectors

$$\begin{pmatrix} u_\alpha \\ w_\alpha \end{pmatrix} = \begin{pmatrix} \cos \epsilon[e^{-i\{kr-(1/2)(j-1)\pi\}} - \exp(2i\delta_j^\alpha)e^{i\{kr-(1/2)(j-1)\pi\}}] \\ \sin \epsilon[e^{-i\{kr-(1/2)(j+1)\pi\}} - \exp(2i\delta_j^\alpha)e^{i\{kr-(1/2)(j+1)\pi\}}] \end{pmatrix} \quad (\alpha\text{-wave}), \tag{I.98a}$$

$$\begin{pmatrix} u_\beta \\ w_\beta \end{pmatrix} = \begin{pmatrix} -\sin \epsilon[e^{-i\{kr-(1/2)(j-1)\pi} - \exp(2i\delta_j^\beta)e^{i\{kr-(1/2)(j-1)\pi\}}] \\ \cos \epsilon[e^{-i\{kr-(1/2)(j+1)\pi\}} - \exp(2i\delta_j^\beta)e^{i(kr-(1/2)(j+1)\pi\}}] \end{pmatrix} \quad (\beta\text{-wave}). \tag{I.98b}$$

Scattering amplitude for coupled states The coefficient of $(1/r) \exp[-i(kr - \frac{1}{2}l\pi)]|lSjm\rangle$ in Ψ_{inc} is given by [see (I.80b)]

$$A_l = -\frac{i^l}{2ik}C_{lSjm}, \tag{I.99}$$

where C_{lSjm} is defined by (I.81a). Considering the two coupled l-values, $l_1 = j-1$ and $l_2 = j+1$ for a given j, we can write the corresponding part of the total wavefunction Ψ as

$$\sum_{i=1}^{2} \frac{1}{r}[A_i \exp\{-i(kr - \tfrac{1}{2}l_i\pi)\} - B_i \exp\{i(kr - \tfrac{1}{2}l_i\pi)\}]|l_iSjm\rangle. \tag{I.100}$$

The coefficients A_1 and A_2 in (I.100) can be obtained from (I.99) by assigning the l-values of l_1 and l_2, respectively. This is because the ingoing wave part of Ψ must be equal to that of Ψ_{inc}, as stated after (I.82).

The outgoing spherical wave $(1/r)e^{ikr}$ in the total wavefunction has an amplitude due to these two coupled l-values, given by the B-term of (I.100), namely,

$$-\sum_{i=1}^{2} B_i \exp(-\tfrac{1}{2}il_i\pi)|l_iSjm\rangle. \tag{I.101}$$

Substituting

$$B_i = \sum_{k=1}^{2} S_{ik}A_k$$

according to the definition (I.90), and then writing A_k explicitly from (I.99), we obtain

$$-\sum_{i=1}^{2} B_i \exp(-\tfrac{1}{2}il_i\pi)|l_iSjm\rangle = \frac{1}{2ik}\sum_{i=1}^{2}\sum_{k=1}^{2} S_{ik}C_{l_kSjm}i^{l_k-l_i}|l_iSjm\rangle. \tag{I.102a}$$

The coefficient of $(1/r)e^{ikr}$ due to these two l-values in Ψ_{inc} is given, according to (I.81b), as

$$\frac{1}{2ik}\sum_{i=1}^{2} C_{l_iSjm}|l_iSjm\rangle. \tag{I.102b}$$

According to (I.82), the coefficient of $(1/r)e^{ikr}$ in Ψ_{scatt} due to the two coupled states $l = j \pm 1$ is obtained by subtracting (I.102b) from (I.102a). Let $M_{m'm}^{Sj}$ denote the scattering amplitude due to these two coupled states, corresponding to the transition from the initial spin state χ_m^S to the final spin state $\chi_{m'}^S$. The component $\chi_{m'}^S$ contained in the state $|l_iSjm\rangle$ is clearly given by [see (I.84b)]

$$\langle Sm' \mid l_iSjm\rangle = \begin{bmatrix} l_i & S & j \\ m-m' & m' & m \end{bmatrix} Y_{m-m'}^{l_i}(\theta, \phi). \tag{I.103}$$

Using this expression and subtracting (I.102b) from (I.102a) as described, we obtain

$$M^{Sj}_{m'm} = \frac{1}{2ik}\sum_{i=1}^{2}\left\{\sum_{k=1}^{2}(\mathcal{S}_{ik} - \delta_{ik})i^{l_k - l_i}C_{l_kSjm}\right\}\langle Sm' \mid l_iSjm\rangle$$
$$= \frac{\sqrt{4\pi}}{2ik}\sum_{i=1}^{2}\sum_{k=1}^{2}(\mathcal{S}_{ik} - \delta_{ik})i^{l_k - l}\sqrt{[l_k]}\begin{bmatrix} l_k & S & j \\ 0 & m & m \end{bmatrix}\begin{bmatrix} l_i & S & j \\ m-m' & m' & m \end{bmatrix}Y^{l_i}_{m-m'}(\theta, \phi). \tag{I.104}$$

The matrix elements $\mathcal{S}_{ik}$ are to be substituted from (I.96).

Differential and Total Cross-Section

To put together all the results so far derived in this section, we refer back to (I.85), and delineate the uncoupled amplitude first for $S = 1$, and then for $S = 0$, as

$$M^{j,S=1,j}_{m',m}(\theta, \phi) = \frac{\sqrt{4\pi}}{2ik}\sqrt{[j]}[\exp(2i\delta^{(1)}_j) - 1]\begin{bmatrix} j & 1 & j \\ 0 & m & m \end{bmatrix}\begin{bmatrix} j & 1 & j \\ m-m' & m' & m \end{bmatrix}Y^{j}_{m-m'}(\theta, \phi), \tag{I.105}$$

$$M^{j,S=0,j}_{0,0}(\theta, \phi) = \frac{\sqrt{4\pi}}{2ik}\sqrt{[j]}[\exp(2i\delta^{(0)}_j) - 1]Y^j_0(\theta, \phi). \tag{I.106}$$

In the singlet case, the only possible value of m and m' is zero, and both the Clebsch-Gordon coefficients are equal to unity; this accounts for the simpler appearance of (I.106).

The total singlet scattering amplitude is obtained by summing (I.106) over $j = 0 \to j = \infty$, whereas the total triplet scattering amplitude is given by first adding (I.104) and (I.105) and then summing over $j = 0 \to j = \infty$:

$$M^{S=0}_{0,0}(\theta, \phi) = \sum_{j=0}^{\infty} M^{j,S=0,j}_{0,0}(\theta, \phi), \tag{I.107a}$$

$$M^{S=1}_{m',m}(\theta, \phi) = \sum_{j=0}^{\infty}[M^{j,S=1,j}_{m',m}(\theta, \phi) + M^{S=1,j}_{m',m}(\theta, \phi)]. \tag{I.107b}$$

Therefore, the singlet and triplet differential cross-sections corresponding to these scattering amplitudes are given, according to (I.79g), by

$$\left(\frac{d\sigma}{d\omega}\right)_{S=0} = \left|\sum_{j=0}^{\infty} M^{j,S=0,j}_{0,0}(\theta, \phi)\right|^2, \tag{I.108a}$$

$$\left(\frac{d\sigma}{d\omega}\right)_{S=1,m} = \sum_{m'=-1}^{1}\left|\sum_{j=0}^{\infty} M^{j,S=1,j}_{m',m}(\theta, \phi) + M^{S=1,j}_{m',m}(\theta, \phi)\right|^2. \tag{I.108b}$$

In the triplet case, since the result can be in all possible values of m', the differential cross-section for $|Sm\rangle \to |Sm'\rangle$ scattering, which is the square of the modulus of (I.107b), has been summed over m', that is, between $m' = -1$ and $m' = +1$. (I.108b) gives the triplet differential cross-section corresponding to an initial spin projection m, which has therefore been used as a label on the left-hand side.

To obtain $d\sigma/d\omega$ in the case where no measurement is made on the spin of the initial state, we must multiply (I.108a) by the probability $p(0, 0)$ of the initial singlet state, and (I.108b) by the probability $p(1, m)$ of the initial triplet state $|S = 1, m\rangle$. The second expression has obviously to be summed over all m-values, from $m = -1$ to $m = +1$, and then added to the singlet expression. For a completely unpolarized incident beam, the four initial spin states

$|0, 0\rangle$ and $|1, m\rangle$ ($m = -1$ to $m = +1$) occur with equal probability, and hence $p(0, 0) = p(1, m) = \frac{1}{4}$. Therefore,

$$\frac{d\sigma}{d\omega} = \tfrac{1}{4}\left(\frac{d\sigma}{d\omega}\right)_{S=0} + \tfrac{1}{4}\sum_m \left(\frac{d\sigma}{d\omega}\right)_{S=1,m}. \tag{I.109}$$

The total cross-section is obtained by integrating the differential cross-section over all angles. The singlet total cross-section is obviously given by

$$\begin{aligned}\sigma_{S=0} &= \frac{4\pi}{k^2}\sum_{j=0}^{\infty}\sum_{j'=0}^{\infty}\sqrt{[jj']}\exp[i(\delta_j^{(0)}-\delta_{j'}^{(0)})\sin\delta_j^{(0)}\sin\delta_{j'}^{(0)} \\ &\quad\times\int_0^{\pi}\sin\theta\,d\theta\int_0^{2\pi}d\phi\;Y_0^j(\theta,\phi)Y_0^{j'}(\theta,\phi) \\ &= \frac{4\pi}{k^2}\sum_{j=0}^{\infty}[j]\sin^2\delta_j^{(0)}.\end{aligned} \tag{I.110}$$

This result follows from the orthonormality of spherical harmonics, according to which the integral over angles is simply $\delta_{j,j'}$. Thus, the singlet total cross-section does not contain any interference between states of different angular momenta. In the first step of (I.110), $[jj']$ denotes $[j][j'] = (2j+1)(2j'+1)$.

The computation of the triplet total cross-section is a little more complicated: it involves examining the structure of each term in the triplet amplitude, and noting that the dependence on m and m' is through factors of the type

$$\begin{bmatrix} l & 1 & j \\ 0 & m & m\end{bmatrix}\begin{bmatrix} l_1 & 1 & j \\ m-m' & m' & m\end{bmatrix} Y_{m-m'}^{l_1}(\theta,\phi).$$

Thus, the m, m' summation of any general term in the square modulus, and the subsequent angle-integration, yields

$$\begin{aligned}&\sum_{m,m'}\begin{bmatrix} l & 1 & j \\ 0 & m & m\end{bmatrix}\begin{bmatrix} l' & 1 & j' \\ 0 & m & m\end{bmatrix}\begin{bmatrix} l_1 & 1 & j \\ m-m' & m' & m\end{bmatrix}\begin{bmatrix} l_1' & 1 & j' \\ m-m' & m' & m\end{bmatrix} \\ &\quad\times\int_0^{\pi}\sin\theta\,d\theta\int_0^{2\pi}d\phi\;Y_{m-m'}^{l_1}(\theta,\phi)Y_{m-m'}^{l_1'*}(\theta,\phi) \\ &= \delta_{l_1,l_1'}\sum_m\begin{bmatrix} l & 1 & j \\ 0 & m & m\end{bmatrix}\begin{bmatrix} l' & 1 & j' \\ 0 & m & m\end{bmatrix}\sum_{m'}\begin{bmatrix} l_1 & 1 & j \\ m-m' & m' & m\end{bmatrix}\begin{bmatrix} l_1' & 1 & j' \\ m-m' & m' & m\end{bmatrix} \\ &= \delta_{l_1,l_1'}\delta_{j,j'}\sum_m\begin{bmatrix} j & 1 & l \\ -m & m & 0\end{bmatrix}\begin{bmatrix} j & 1 & l' \\ -m & m & 0\end{bmatrix}\frac{[j]}{\sqrt{[l][l']}} \\ &= \delta_{l_1,l_1'}\delta_{j,j'}\delta_{l,l'}\frac{[j]}{[l]}.\end{aligned} \tag{I.111}$$

We have used the orthonormality of spherical harmonics, and that of the Clebsch-Gordon coefficients [see (AIV.3) in Appendix A], and also the last symmetry property of the Clebsch-Gordon coefficient [see (AIV.4) in Appendix A].

The various Krönecker deltas in (I.111) require that, in the triplet total cross-section, there cannot be any interference (i) between terms of different j (due to the presence of $\delta_{j,j'}$), (ii) between terms with different spherical harmonics for the same j (because of $\delta_{l_1,l_1'}$), and (iii) between the two terms of (I.104) for each spherical harmonic (due to $\delta_{l,l'}$). Using all these facts, we get

$$\sigma_{S=1} = \frac{4\pi}{k^2}\sum_{j=0}^{\infty}[j]\sin^2\delta_j^{(1)} + \frac{\pi}{k^2}\sum_{j=0}^{\infty}[j]\{|(\mathcal{S}_{11}-1)|^2 + |(\mathcal{S}_{22}-1)|^2 + |\mathcal{S}_{12}|^2 + |\mathcal{S}_{21}|^2\}.$$

This can be further simplified by noting that the expression within the braces is the trace of the matrix $(\mathcal{S} - \mathbb{1})(\mathcal{S}^\dagger - \mathbb{1})$ and that the trace of $\mathcal{S}$ is the sum of its eigenvalues. Thus,

$$\sigma_{S=1} = \frac{4\pi}{k^2}\sum_{j=0}^{\infty}[j]\{\sin^2\delta_j^{(1)} + \sin^2\delta_j^{\alpha} + \sin^2\delta_j^{\beta}\}. \tag{I.112}$$

Recalling the probability factor of $\frac{1}{4}$ preceding the two terms in (1.109), we finally obtain

$$\sigma = \frac{\pi}{k^2}\sum_{j=0}^{\infty}[j]\{\sin^2\delta_j^{(0)} + \sin^2\delta_j^{(1)} + \sin^2\delta_j^{\alpha} + \sin^2\delta_j^{\beta}\}. \tag{I.113}$$

C. CALCULATION OF PHASE SHIFTS AND MIXING PARAMETER

Computing the differential and total scattering cross-section with the help of (I.109) and (I.113) becomes possible only when we know the phase shifts and the mixing parameter ϵ. The calculation of the phase shift $\delta_j^{(S)}$ in the uncoupled case has been described in Section 5. We need only explain the method of computing δ_j^{α}, δ_j^{β}, and ϵ from the solution of the coupled radial Schrödinger equations (I.57).

We first notice that, for a given j, the parity of the coupled states is $-(-1)^j$ and, since $S = 1$, the value of T automatically becomes fixed from the total antisymmetry of the states. Therefore, although T has not so far been mentioned in our description of scattering formalism, there is no difficulty in fixing this quantum number in the coupled equations. These equations can be solved by the numerical integration method described in Section 5 [see (I.61)], starting with the asymptotic forms for the solutions at large r, and then integrating inwards step by step.

To find out the asymptotic forms needed to commence the integration, we rewrite u_α, w_α, u_β, w_β, given by (I.98a), as

$$\begin{aligned} u_\alpha &= -2i\cos\epsilon\exp(i\delta_j^{\alpha})[\sin\{kr - \tfrac{1}{2}(j-1)\pi\}\cos\delta_j^{\alpha} \\ &\quad + \cos\{kr - \tfrac{1}{2}(j-1)\pi\}\sin\delta_j^{\alpha}], \qquad r\to\infty, \\ w_\alpha &= -2i\sin\epsilon\exp(i\delta_j^{\alpha})[\sin\{kr - \tfrac{1}{2}(j+1)\pi\}\cos\delta_j^{\alpha} \\ &\quad + \cos\{kr - \tfrac{1}{2}(j+1)\pi\}\sin\delta_j^{\alpha}], \qquad r\to\infty. \end{aligned} \tag{I.114}$$

Replacing δ_j^{α} by δ_j^{β} and ϵ by $(\epsilon + \pi/2)$ in these expressions, we get the results for u_β and w_β. It is clear from these asymptotic expressions that we can obtain two solutions by starting the numerical integration first with

$$\begin{pmatrix} u_1 \\ w_1 \end{pmatrix} = \begin{pmatrix} \sin\{kr - \frac{1}{2}(j-1)\pi\} \\ \eta\sin\{kr - \frac{1}{2}(j+1)\pi\} \end{pmatrix}, \qquad r\to\infty, \tag{I.115a}$$

and then with

$$\begin{pmatrix} u_2 \\ w_2 \end{pmatrix} = \begin{pmatrix} \cos\{kr - \frac{1}{2}(j-1)\pi\} \\ \eta\cos\{kr - \frac{1}{2}(j+1)\pi\} \end{pmatrix}, \qquad r\to\infty, \tag{I.115b}$$

where η is an adjustable parameter. It is clear from the asymptotic form (I.114) that the solutions u_1, u_2 and w_1, w_2 obtained in this way should be superposed in the same ratio.

That is,

$$u = u_1 + Cu_2,$$
$$w = w_1 + Cw_2 \qquad \text{(I.116)}$$

should be assumed to be a general solution. The reason we had to take the parameter η to be the same in (I.115a) and (I.115b) is also clear from the nature of (I.114). In the expressions (I.114), the ratio of the coefficients of the sine terms of u_α and w_α is equal to the ratio of the coefficients of the cosine terms. The same remark is true for the β-solution. For the α-solution, η is obviously tan ϵ, whereas for the β-solution, η is $-\cot \epsilon$, so that $\eta_\alpha \eta_\beta = -1$.

After obtaining the complete solutions u_1, w_1 and u_2, w_2 by integrating the coupled equations up to $r = 0$ (no hard core) or $r = r_c$ (hard core), we require a zero boundary value for the linear combinations (I.116) at $r = 0$ or $r = r_c$. The two boundary conditions for u and w have to be assured by using the parameter η and the arbitrary constant C. The value of η, say, η_α, thus found will be equal to tan ϵ, and the mixing parameter ϵ can therefore be determined. Similarly, the constant C yields the value of δ_j^α through $C = \tan \delta_j^\alpha$.

But the scattering formalism we have developed guarantees a second solution of η. If this solution is labelled η_β, then we know that $\eta_\alpha \eta_\beta = -1$. Therefore, after one η is ascertained, the entire procedure can be repeated, replacing η by $-\eta^{-1}$. If the numerical accuracy is adequate, we expect to find a single constant C' (different from the previous C) with which $(u_1 + C'u_2)$ and $(u_2 + C'u_2)$ automatically satisfy their boundary conditions at $r = 0$ or $r = r_c$. The second solution C' obtained in this way gives δ_j^β.

The reader should note that the nomenclatures α and β are very arbitrary. Whatever η and C were first found in our numerical search were taken to give the α-solution. This is a consequence of the arbitrariness in the formalism itself, which clearly states that the α-solution goes over to the β-solution, and vice versa, when ϵ is replaced by $(\epsilon + \pi/2)$. We can, however, introduce a convention to secure agreement between the values of ϵ, δ_j^α, δ_j^β obtained by different workers. At low energies (i.e., small k^2), the difference in the centrifugal barrier term $l(l+1)/r^2$ for $l = j - 1$ and $l = j + 1$ tends to produce two pure states. Both $\epsilon = 0$ and $\epsilon = \pi/2$ can produce such pure solutions. However, $\epsilon = 0$ gives pure solutions for which the α- and β-wave correspond to $l = j - 1$ and $l = j + 1$, respectively. The convention that is followed can now be stated: the solution of ϵ that goes to zero (instead of $\pi/2$) at low energy is the usual choice. This clearly has the effect of calling the predominantly $(l = j - 1)$-state the α-state, and the predominantly $(l = j + 1)$-state the β-state.

8. SPECIAL CONSIDERATION OF p-p SCATTERING

The scattering formulas derived in Section 7 are applicable only to the case of n-p scattering, and not to p-p scattering, for two reasons. First, because of the Coulomb interaction between the two protons, the Schrödinger equation is modified with an extra term corresponding to the Coulomb potential e^2/r. Since the Coulomb potential goes to zero slower than the centrifugal term $l(l+1)/r^2$ as $r \to \infty$, the asymptotic equation should contain also the Coulomb term. The asymptotic solutions are therefore no longer given by $j_l(kr)$ and $n_l(kr)$, but by the radial solutions of the Schrödinger equation for the Coulomb potential. The scattering in the p-p system thus contains a pure Coulomb part, a pure nuclear part, and an interference between the two types of scattering.

The second reason is the Pauli exclusion principle, and the consequent requirement of antisymmetry with respect to the exchange of space and spin coordinates. In Section 7, we have used an incident wave $e^{ikz}\chi_m^S$, which contains all partial waves, $l = 0 \to l = \infty$, corresponding to both the spin values S (=1, 0). Because the incident wave has all partial waves, the total wave Ψ and the scattered wave Ψ_{scatt} also contain all possible l-waves corresponding to a given spin S. In the case of p-p scattering, this is not true. Because of the Pauli principle, we would have to use a symmetric (l = even) spatial part in Ψ_{scatt} corresponding to $S = 0$ and an antisymmetric (l = odd) spatial part for $S = 1$.

A. PURE COULOMB SCATTERING

We consider first the case of *pure* Coulomb scattering. This gives the Coulomb wavefunctions that have to be used at asymptotically large distances in the case of our p-p scattering problem arising from the Coulomb plus the pure nucleon-nucleon potential.

The Schrödinger equation for the repulsive Coulomb potential e^2/r can be written as

$$(\nabla^2 + k^2 - \frac{1}{\beta r})\Psi(\mathbf{r}) = 0, \tag{I.117}$$

where $\beta^{-1} = Me^2/\hbar^2$. We solve this equation in two ways: first, obtain the solution of (I.117) straightaway; and second, obtain the solution in spherical polar coordinates, thereby implying a partial wave separation.

Method 1 We write the solution as

$$\Psi(\mathbf{r}) = e^{ikz}\mathcal{F}(\zeta_-) \quad \text{with} \quad \zeta_- = r - z, \tag{I.118}$$

which is dictated by a specific consideration. Since the incident nucleon is along the z-axis, and the Coulomb potential is spherically symmetric, there is a cylindrical symmetry about the z-axis. Hence, if we try to regard $\mathcal{F}$ as a function of the three parabolic coordinates, namely, $\zeta_\pm = r \pm z$ and the azimuthal angle ϕ, then the axial symmetry around z implies that there cannot be any dependence on ϕ. $\mathcal{F}$ can, in general, be a function of both $(r - z)$ and $(r + z)$, but only in the former case do we get $\Psi(\mathbf{r})$ of (I.118) as a sum of an incident wave and an *outgoing* spherical wave (see the subsequent derivation for a proof of this statement), which is the desired form in scattering formalism.

If we substitute (I.118) in (I.117), then the resultant equation for $\mathcal{F}$ is found to be a confluent hypergeometric equation [see Appendix C (Section II)] for the variable $ik\zeta_-$ and the parameters $a = -i\alpha$, $c = 1$. The solution of this equation, well-behaved at $r \to 0$, is given by

$$\mathcal{F} = F(-i\alpha, 1; ik\zeta_-),$$

where F is the confluent hypergeometric function represented by the series (CII.3a) of Appendix C and also by the integral representation (CII.5a) and (CII.5b) of Appendix C. The constant $\alpha = 1/(2\beta k) = e^2/(\hbar v)$, where v (the velocity in the centre-of-mass coordinate system) $= 2\hbar k/M$ (since $\hbar k$ is the momentum and $\frac{1}{2}M$ the reduced mass). In this case, since the variable $ik\zeta_-$ is positive-imaginary, we choose the upper sign in the exponential occurring in the asymptotic form (CII.5c) of Appendix C. Thus, we obtain

$$\Psi(\mathbf{r}) = e^{ikz}F(-i\alpha, 1; ik\zeta_-)$$

$$= \frac{\Gamma(1)}{\Gamma(1 + i\alpha)}e^{\pi\alpha}(ik\zeta_-)^{+i\alpha}e^{ikz} + \frac{\Gamma(1)}{\Gamma(-i\alpha)}(ik\zeta_-)^{-1-i\alpha}e^{ikr}, \qquad r \to \infty$$

$$= \frac{e^{(\pi/2)\alpha}}{\Gamma(1+i\alpha)}[e^{i(kz+\alpha \ln k\zeta_-)} - i\alpha\frac{\Gamma(1+i\alpha)}{\Gamma(1-i\alpha)}i^{-1}\frac{e^{i(kr-\alpha \ln k\zeta_-)}}{k\zeta_-}]$$

$$= \frac{e^{(\pi/2)\alpha}}{\Gamma(1+i\alpha)}[e^{i(kz+\alpha \ln k\zeta_-)} - \frac{e^2}{Mv^2}\operatorname{cosec}^2\frac{\theta}{2}\, e^{-i\alpha \ln \sin^2(\theta/2)+2i\eta_0}\frac{e^{i(kr-\alpha \ln 2kr)}}{r}], \tag{I.119}$$

where

$$\frac{\Gamma(1+i\alpha)}{\Gamma(1-i\alpha)} = e^{2i\eta_0}. \tag{I.120}$$

The first term $e^{i(kz+\alpha \ln k\zeta_-)}$ corresponds to the incident wave with Coulomb distortion, and the second term to the outgoing spherical wave with Coulomb distortion represented by $\alpha \ln 2kr$. If we normalize the incident wave to unit amplitude by multiplying (I.119) by $e^{(-\pi/2)\alpha}\Gamma(1+i\alpha)$, then the coefficient of the second term is, by definition, the scattering amplitude

$$f_{\text{coul}}(\theta) = -\frac{e^2}{Mv^2}\operatorname{cosec}^2\frac{\theta}{2}\, e^{-i\alpha \ln \sin^2(\theta/2)+2i\eta_0}. \tag{I.121}$$

It is clear from this result that the differential cross-section $(d\sigma/d\omega)_{\text{coul}} = |f_{\text{coul}}(\theta)|^2$ is the familiar Rutherford expression

$$(\frac{d\sigma}{d\omega})_{\text{coul}} = (\frac{e^2}{Mv^2})^2 \operatorname{cosec}^4\frac{\theta}{2}. \tag{I.122}$$

In the case of p-p scattering, however, this is not the correct expression for Coulomb scattering because of the antisymmetry requirement under space-spin exchange. We therefore need to form the symmetric and antisymmetric combinations to go with $S = 0$ and $S = 1$, respectively:

$$F_{\text{coul}}^{\text{singlet}}(\theta) = f_{\text{coul}}(\theta) + f_{\text{coul}}(\pi-\theta)$$
$$= -\frac{e^2}{Mv^2}e^{2i\eta_0}(\operatorname{cosec}^2\frac{\theta}{2}\, e^{-i\alpha \ln \sin^2(\theta/2)} + \sec^2\frac{\theta}{2}\, e^{-i\alpha \ln \cos^2(\theta/2)}) \quad \text{(singlet)}, \tag{I.123}$$

$$F_{\text{coul}}^{\text{triplet}}(\theta) = f_{\text{coul}}(\theta) - f_{\text{coul}}(\pi-\theta)$$
$$= -\frac{e^2}{Mv^2}e^{2i\eta_0}(\operatorname{cosec}^2\frac{\theta}{2}\, e^{-i\alpha \ln \sin^2(\theta/2)} - \sec^2\frac{\theta}{2}\, e^{-i\alpha \ln \cos^2(\theta/2)}) \quad \text{(triplet)}. \tag{I.124}$$

Since $F_{\text{coul}}^{\text{singlet}}(\theta)$ accompanies the spin state $|S=0, m_s=0\rangle$, and $F_{\text{coul}}^{\text{triplet}}(\theta)$ the spin state $|S=1, m_s\rangle$, $(m_s = \pm 1, 0)$, the orthogonality of the spin functions establishes that there is no interference between singlet and triplet scattering in the expression of $(d\sigma/d\omega)_{\text{coul}}$. Further, since the potential is central and spin-independent, the initial spin projection m_s remains unchanged in the scattering process. Each term in the square modulus of the scattering amplitude has to be multiplied by $P(S, m_s)$, the probability of the spin state $|S, m_s\rangle$ in the initial beam, and then summed over m_s. For an initial unpolarized beam, this procedure gives a factor of $\frac{1}{4}$ in the singlet state, and $\frac{3}{4}$ in the triplet state. Thus,

$$(\frac{d\sigma}{d\omega})_{\text{coul}} = \tfrac{1}{4}|F_{\text{coul}}^{\text{singlet}}(\theta)|^2 + \tfrac{3}{4}|F_{\text{coul}}^{\text{triplet}}(\theta)|^2. \tag{I.125}$$

For future reference, we finally write the Coulomb solution $\Psi(\mathbf{r})$ with the correct normalization [mentioned after (I.120)] as

$$\Psi(\mathbf{r}) = e^{(-\pi/2)\alpha}\Gamma(1+i\alpha)e^{ikz}F(-i\alpha, 1; ik\zeta_-). \tag{I.126}$$

Method 2 Since the potential of (I.117) is central, the angular part of the solution of (I.117) is the spherical harmonic $Y^l_m(\theta, \phi)$, where l is the orbital angular momentum and m its projection. The radial Schrödinger equation for the radial function, $u_l(r)/r$, is given by

$$\frac{d^2u_l}{dr^2} - \frac{l(l+1)}{r^2}u_l + k^2u_l - \frac{1}{\beta r}u_l = 0. \tag{I.127}$$

The solution of this equation, for $r \to 0$, is r^{l+1}. For $r \to \infty$, if we exclude both the potential and centrifugal terms, the solution behaves approximately as e^{ikr}. Therefore, we try a solution in the form

$$\frac{u_l(r)}{r} = r^l e^{ikr}\mathcal{F}_l(r). \tag{I.128}$$

Substituting (I.128) in (I.127), we obtain, for $\mathcal{F}_l(r)$,

$$\frac{d^2\mathcal{F}_l}{d\rho^2} + \frac{1}{\rho}(c-\rho)\frac{d\mathcal{F}_l}{d\rho}\frac{a}{\rho}\mathcal{F}_l = 0,$$

where $\rho = -2ikr$, $c = 2(l+1)$, and $a = l + 1 + i\alpha$. This is the confluent hypergeometric equation given in Appendix C (Section II), and hence the solution well-behaved at $r = 0$ is

$$\mathcal{F}_l(r) = F(l+1+i\alpha, 2l+2; -2ikr). \tag{I.129}$$

To explore the asymptotic form of this solution at $r \to \infty$, we make use of the lower sign of the exponential in (CII.5c) of Appendix C since the variable ρ is negative-imaginary. Thus,

$$\begin{aligned}\frac{u_l(r)}{r} &= r^l e^{ikr}\Big[\frac{\Gamma(2l+2)}{\Gamma(l+1-i\alpha)}e^{-i\pi(l+1+i\alpha)}(-2ikr)^{-l-1-i\alpha} \\ &\quad + \frac{\Gamma(2l+2)}{\Gamma(l+1+i\alpha)}(-2ikr)^{-l-1+i\alpha}e^{-2ikr}\Big], \qquad r\to\infty \\ &= \frac{\Gamma(2l+2)}{|\Gamma(l+1+i\alpha)|}\frac{e^{(\pi/2)\alpha}}{(2k)^l}\frac{1}{kr}\sin\left(kr - \frac{l\pi}{2} - \alpha\ln 2kr + \eta_l\right),\end{aligned} \tag{I.130}$$

where

$$\Gamma(l+1+i\alpha) = \Gamma^*(l+1-i\alpha) = |\Gamma(l+1+i\alpha)|\exp(i\eta_l).$$

The Coulomb phase shift η_l is thus defined by

$$\exp(2i\eta_l) = \frac{\Gamma(l+1+i\alpha)}{\Gamma(l+1-i\alpha)}. \tag{I.131}$$

In order to make this solution closely analogous to $j_l(kr)$, we choose the normalization appropriately and obtain the resultant normalized solution, $F_l(kr)$, as

$$\begin{aligned}F_l(kr) &= \frac{\Gamma(l+1+i\alpha)\exp(-i\eta_l)}{\Gamma(2l+2)}(2k)^l e^{(-\pi/2)\alpha}[r^l e^{ikr}F(l+1+i\alpha, 2l+2; -2ikr)] \\ &= \frac{1}{kr}\sin\left(kr - \tfrac{1}{2}l\pi - \alpha\ln 2kr + \eta_l\right), \qquad r\to\infty.\end{aligned} \tag{I.132}$$

In view of the symmetry about the z-axis, only Y^l_0 can enter the expression for the complete wavefunction, which can now be written as

$$\Psi(\mathbf{r}) = \sum_{l=0}^{\infty} C_l F_l(kr) Y^l_0(\theta). \tag{I.133}$$

9 (45-122/1977)

The normalization constant C_l for the partial waves can be so chosen that (I.133) becomes identical to (I.126). From (I.133), we get

$$C_lF_l(kr) = \int_0^{2\pi} d\phi \int_0^{\pi} \sin\theta \, d\theta \, \Psi(\mathbf{r}) Y_0^l(\theta)$$

$$= \sqrt{\pi[l]} \int_{-1}^{+1} d(\cos\theta) \, \Psi(\mathbf{r}) P_l(\cos\theta), \tag{I.134}$$

where the Legendre polynomial $P_l(\cos\theta)$ is related to the spherical harmonic Y_0^l by $P_l(\cos\theta) = (4\pi/[l])^{1/2} Y_0^l$. Using $\Psi(\mathbf{r})$ from (I.126) in (I.134), we can directly evaluate C_l. The resultant value of C_l should be independent of the value of r used in (I.134). For our purpose, the value of $r \to 0$ is very convenient. We thus obtain

$$C_l r^l = \sqrt{\pi[l]} \exp(i\eta_l) \frac{\Gamma(1+i\alpha)}{\Gamma(l+1+i\alpha)} \frac{\Gamma(2l+2)}{(2k)^l} \underset{r\to 0}{\mathrm{Lt}} \int_{-1}^{+1} dx \, e^{-ikr(1-x)}$$

$$\times F(-i\alpha, 1; ikr(1-x)) P_l(x) \tag{I.135}$$

or

$$C_l = i^l \sqrt{4\pi[l]} \exp(i\eta_l). \tag{I.136}$$

[The intervening steps of this derivation are given in Appendix C (Section III).] Therefore, we now get two equivalent expressions for $\Psi(\mathbf{r})$, namely,

$$\Psi(\mathbf{r}) = \sum_{l=0}^{\infty} i^l \sqrt{4\pi[l]} \exp(i\eta_l) F_l(kr) Y_0^l(\theta) \tag{I.137}$$

$$= e^{(-\pi/2)\alpha} \Gamma(1+i\alpha) e^{ikz} F(-i\alpha, 1; ik(r-z)). \tag{I.126}$$

The ingoing spherical wave part of (I.137) is now automatically equal to that of (I.126), and hence that of the incident wave contained in (I.126). The scattering amplitude is given by the coefficient of the outgoing spherical wave part $r^{-1}e^{+i(kr-\alpha \ln 2kr)}$ of $(\Psi - \Psi_{\text{inc}})$. Thus, another expression of the Coulomb amplitude, which is identically equal to (I.121), is given by

$$f_{\text{coul}}(\theta) = \frac{1}{2ik} \sum_{l=0}^{\infty} \sqrt{4\pi[l]} \exp(2i\eta_l) - 1] Y_0^l(\theta). \tag{I.138}$$

The antisymmetrization ($S = 1$) and symmetrization ($S = 0$) of this amplitude and the computation of $(d\sigma/d\omega)$ have already been described.

The Coulomb solution $F_l(kr)$, given by (I.132), is regular at the origin, and hence is the only solution of interest for the treatment of *pure* Coulomb scattering. There is, however, a second solution of the confluent hypergeometric equation [see Appendix C (Section II)], which could have been used, instead of (I.129), to derive a second linearly independent Coulomb solution. In the present case, the parameter c of the confluent hypergeometric equation is $(2l + 2)$, which has positive integral values 2, 4, 6, ... Therefore, the solution $G(a, c; z)$ given by (CII.6a) in Appendix C should be used to write the second solution. With the same kind of normalization as $F_l(kr)$, the new solution for $u_l(r)/r$, to be denoted by $G_l(kr)$, is given as

$$G_l(kr) = i\frac{\Gamma(l+1+i\alpha)\exp(i\eta_l)}{\Gamma(2l+2)}(2k)^l e^{(-\pi/2)\alpha} r^l e^{ikr} G(l+1+i\alpha, 2l+2; -2ikr)$$

$$= -\frac{1}{kr}\cos(kr - \tfrac{1}{2}l\pi - \alpha \ln 2kr + \eta_l), \qquad r \to \infty. \tag{I.139}$$

The asymptotic form of $G_l(kr)$ is very similar to that of $n_l(kr)$. The extra i in the normalization should be noted.

The solution $G_l(kr)$ is useful in the study of scattering by the Coulomb plus a short-range force. The solution near the origin, in such a case, is the solution of the Schrödinger equation with both kinds of potential, but at a large distance, where the short-range interaction goes to zero, the solution should be written as the most general solution of the Coulomb equation, i.e., a linear combination of $F_l(kr)$ and $G_l(kr)$.

B. COULOMB PLUS NUCLEAR SCATTERING

Uncoupled State and Singlet Scattering

For the uncoupled state $l = j$, the Schrödinger equation (I.47) gets modified to

$$\frac{d^2u_l}{dr^2} - \frac{l(l+1)}{r^2}u_l + \mathcal{F}(r)u_l + k^2u_l - \frac{1}{\beta r}u_l = 0, \qquad l = j. \tag{I.140}$$

The equation for $r \to \infty$ is obtained from (I.140) by omitting the term $\mathcal{F}(r)u_l$ since $\mathcal{F}(r)$, containing the nucleon-nucleon potential, goes to zero as $r \to \infty$. The two linearly independent solutions of the asymptotic equation are given by $rF_l(kr)$ and $rG_l(kr)$ (see Section 8A). Hence, the most general asymptotic solution of (I.140) is given by

$$\begin{aligned}\frac{u_l}{r} &= A_l[\cos\delta_l\, F_l(kr) - \sin\delta_l\, G_l(kr)], \qquad r \to \infty\\ &= \frac{A_l}{kr}\sin(kr - \tfrac{1}{2}l\pi - \alpha\ln 2kr + \eta_l + \delta_l), \qquad r \to \infty.\end{aligned} \tag{I.141}$$

The asymptotic solution contains the Coulomb phase shift η_l plus the nuclear phase shift δ_l. The phase shift δ_l can be computed by the numerical method described in Section 5. The procedure entails solving (I.140) numerically with two different initial forms for u_l at $r \to \infty$, namely, $rF_l(kr)$ and $rG_l(kr)$. The correct linear combination of the two independent solutions is determined by satisfying the boundary condition at $r = 0$ (no hard core) or $r = r_c$ (hard core). δ_l is then obtained by equating the log-derivative of this general solution with that of the asymptotic form (I.141).

To arrive at the new expression for the scattering amplitude, we have to go through the derivation given in Section 7 with the modifications needed by the Coulomb potential and by the Pauli principle. Now the incident wave for $r \to \infty$ is the same as that in the pure Coulomb case [we have derived two alternative forms for the latter in Section 8A; see (I.119) and the normalization of (I.126)]:

$$\begin{aligned}\Psi_{\text{inc}} &= \exp[i\{kz + \alpha\ln k(r - z)\}], \qquad r \to \infty\\ &= \sum_{l=0}^{\infty} i^l\sqrt{4\pi[l]}\,Y_0^l(\theta)\frac{1}{2ikr}[-\exp\{-i(kr - \tfrac{1}{2}l\pi - \alpha\ln 2kr)\}\\ &\quad + \exp\{i(kr - \tfrac{1}{2}l\pi - \alpha\ln 2kr)\}].\end{aligned} \tag{I.142}$$

The partial-wave expansion is useful now. This expansion follows from the fact that the ingoing wave part of Ψ_{inc} must be the same as that of Ψ given by (I.137). The amplitude of the ingoing wave $-(2ikr)^{-1}\exp[-i(kr - \frac{1}{2}l\pi - \alpha\ln 2kr)]$ in Ψ_{inc} corresponding to a spin state χ_m^S is still i^lC_{lSjm}, where C_{lSjm} is obtained from the expression (I.81a). Similarly, the amplitude of the outgoing part $e^{i(kr-\alpha\ln 2kr)}/r$ is identical to (I.81b).

As regards the total u_l/r, the asymptotic form (I.141) now replaces (I.83a). The coefficient of $-\exp[-i(kr - \frac{1}{2}l\pi - \alpha \ln 2kr)]/(2ikr)$ in (I.141) is given by

$$A_l \exp[-i(\eta_l + \delta_l^S)],$$

and hence if we carry out all the subsequent steps, (I.84a) gets replaced by

$$\frac{1}{2ik} C_{lSjm} [\exp(2i\eta_l + 2i\delta_l^S) - 1] \frac{e^{i(kr-\alpha \ln 2kr)}}{r} |lSjm\rangle.$$

The coefficient of $(1/r)e^{i(kr-\alpha \ln 2kr)}\chi^S_{m'}$ in this expression gives $M^{jSj}_{m',m}(\theta, \phi)$, and the only difference between the new expression and (I.85) is the extra Coulomb phase shift added to δ_j^S:

$$M^{jSj}_{m',m}(\theta, \phi) = \frac{1}{2ik}\sqrt{4\pi[j]}\begin{bmatrix} j & S & j \\ 0 & m & m \end{bmatrix}\begin{bmatrix} j & S & j \\ m-m' & m' & m \end{bmatrix}\{\exp(2i\eta_j + 2i\delta_j^S) - 1\}$$
$$\times Y^j_{m-m'}(\theta, \phi). \tag{I.143}$$

For singlet ($S = 0$) scattering, the projections $m = m' = 0$, the Clebsch-Gordon coefficients are unity, and the symmetrization of the amplitude gives only even values of the orbital angular momentum $l(=j)$ and a multiplicative factor of 2. Thus, the singlet amplitude is given by

$$M^{j,0,j}_{0,0}(\theta) = \frac{1}{ik}\sum_{j=0\,(\text{even})}^{\infty} \sqrt{4\pi[j]}[\{\exp(2i\eta_j) - 1\} + \exp(2i\eta_j)\{\exp(2i\delta_j^{(0)}) - 1\}]Y^j_0(\theta)$$

$$= F^{\text{singlet}}_{\text{coul}}(\theta) + \frac{1}{ik}\sum_{j=0\,(\text{even})}^{\infty} \sqrt{4\pi[j]}\exp(2i\eta_j)[\exp(2i\delta_j^{(0)}) - 1]Y^j_0(\theta). \tag{I.144}$$

$F^{\text{singlet}}_{\text{coul}}(\theta)$ is the symmetrized Coulomb amplitude given by (I.123). It has been proved in Section 8A that the closed expression (I.123) is identically equal to the partial wave sum for the term $[\exp(2i\eta_j) - 1]$. The second term in (I.144) contains the nuclear phase shift δ_j, which decreases with increasing j. Hence, while computing the scattering amplitudes, the second term need be evaluated for only a few low partial waves, depending on the energy of the nucleon being scattered. Since such a procedure is erroneous for the Coulomb scattering, it has been necessary to separate the pure Coulomb scattering part in (I.144), carry out the sum over $j = 0 \to j = \infty$ (even), and replace it by the closed expression (I.123).

The triplet uncoupled amplitude, given by (I.143) with $S = 1$, is now considered along with the triplet coupled amplitudes.

Coupled States and Triplet Scattering

The coupled differential equations for $l = j - 1$ and $l = j + 1$ obtained from (I.58) by inserting the Coulomb term are

$$\frac{d^2u}{dr^2} - \frac{j(j-1)}{r^2}u + k^2u + F(r)u + H(r)w - \frac{1}{\beta r}u = 0, \tag{I.145a}$$

$$\frac{d^2w}{dr^2} - \frac{(j+1)(j+2)}{r^2}w + k^2w + G(r)w + H(r)u - \frac{1}{\beta r}w = 0. \tag{I.145b}$$

For $r \to \infty$, the nuclear potential terms $F(r)$, $G(r)$, $H(r)$ go to zero, and the equations get uncoupled. The asymptotic wavefunctions for $r \to \infty$ are therefore pure Coulomb solutions of the form $\sin(kr - \frac{1}{2}l\pi - \alpha \ln 2kr + \eta_l)$ and $\cos(kr - \frac{1}{2}l\pi - \alpha \ln 2kr + \eta_l)$ with $l = j \pm 1$. Alternatively, the asymptotic solutions can be written as a sum of ingoing and outgoing distorted Coulomb waves as

$$\begin{aligned} u = A_1 \exp [-i\{kr - \tfrac{1}{2}(j-1)\pi - \alpha \ln 2kr + \eta_{j-1}\}] \\ - B_1 \exp [i\{kr - \tfrac{1}{2}(j-1)\pi - \alpha \ln 2kr + \eta_{j-1}\}], \qquad r \to \quad , \end{aligned} \tag{I.146a}$$

$$\begin{aligned} w = A_2 \exp [-i\{kr - \tfrac{1}{2}(j+1)\pi - \alpha \ln 2kr + \eta_{j+1}\}] \\ - B_2 \exp [i\{kr - \tfrac{1}{2}(j+1)\pi - \alpha \ln 2kr + \eta_{j+1}\}], \qquad r \to \infty. \end{aligned} \tag{I.146b}$$

Obviously, these forms replace (I.89a) and (I.89b). The definition of the $\mathcal{S}$-matrix, its unitarity, symmetry, and diagonalization (which defines the mixing parameter ϵ and the eigenphase shifts δ_j^α and δ_j^β) given in Section 7 will obviously remain unmodified. The final expressions (I.98a) and (I.98b) for the α- and β-wave, however, contain the more complicated ingoing and outgoing exponential terms occurring in (I.146a) and (I.146b).

The coefficient of $(1/r)e^{-i[kr-(1/2)l\pi-\alpha \ln 2kr]}|lSjm\rangle$ in $\Psi_{\rm inc}$ [see (I.142)] is still expressed by (I.99). However, this has now to be equated with the coefficients $A_i \exp(-i\eta_i)$ [see (I.146)], where $i = 1, 2$ correspond as before to $l_1 = j - 1$ and $l_2 = j + 1$. Thus,

$$A_i = -\frac{i^{l_i}}{2ik} C_{l_iSjm} \exp(i\eta_i). \tag{I.147}$$

The coefficients B_i are still expressed by (I.101).

The coefficient of $(1/r)e^{i(kr-\alpha \ln 2kr)}$ in Ψ, due to these two coupled waves, is given, according to (I.146), by

$$-\sum_{i=1}^{2} B_i \exp(-\tfrac{1}{2}il_i\pi) \exp(i\eta_i)|l_iSjm\rangle, \tag{I.148}$$

which replaces (I.101). The same coefficient contained in $\Psi_{\rm inc}$ is given, according to (I.142), by (I.102b). Now, repeating the steps following (I.101) with the help of (I.142) and (I.102b), we can easily verify that the expression (I.104) for $M_{m'm}^{Sj}$ changes only through the replacement of $\mathcal{S}_{ik}$ by $\bar{\mathcal{S}}_{ik}$, where

$$\bar{\mathcal{S}}_{ik} = \exp(i\eta_i)\mathcal{S}_{ik} \exp(i\eta_k) \tag{I.149a}$$

or

$$\bar{\mathcal{S}} = D_c \mathcal{S} D_c, \tag{I.149b}$$

$$D_c = \begin{pmatrix} \exp(i\eta_{j-1}) & 0 \\ 0 & \exp(i\eta_{j+1}) \end{pmatrix}. \tag{I.149c}$$

The matrix elements $\mathcal{S}_{ik}$ are still expressed by (I.96). The parameters ϵ, δ_j^α, δ_j^β occurring in (I.96) are now to be determined by solving (I.145). The procedure is similar to that described in Section 7C. The initial forms for large r, as given by (I.115), should now be modified by adding the appropriate Coulomb phase shift η_l in each of the sine and cosine functions.

Now we consider the spatial antisymmetrization of the p-p ($S = 1$) scattering. As a result of this antisymmetrization, only odd spherical harmonics can contribute with a multiplicative factor 2. Hence, the contribution (I.143) from the uncoupled triplet state $l = j$ should be accepted with a factor of 2 only when j is an odd integer, whereas the contribution from the coupled states $l = j \pm 1$, given by two times (I.104) with $\bar{\mathcal{S}}$ replacing $\mathcal{S}$, is to be reckoned when j is an even integer.

The various l-values that result from the summation over j are given by $l = 1;\ 1;\ (1, 3);\ 3;\ (3, 5);\ 5;\ (5, 7);\ \ldots$, where the l's produced from different $j(= 0, 1, 2, \ldots)$ are separated by semicolons, and the coupled states are the ones enclosed within the parentheses. The first $l = 1$

arises from $j=0$, and it is uncoupled because its partner $l=j-1$ corresponds to a negative l, which is not allowed. It should be noted that only odd l's appear (due to spatial antisymmetry) and that each l occurs from three neighbouring j-values $j=l-1$, $j=l$, and $j=l+1$. This knowledge can be utilized to rewrite the part of the triplet amplitude that comes from the diagonal matrix elements of the scattering matrix in terms of the triplet Coulomb amplitude given by (I.124), and the part that contains the nuclear scattering. This separation is analogous to (I.144) for singlet scattering, and the purpose is the same as that described after that equation.

The diagonal part is given by

$$\frac{\sqrt{4\pi}}{ik}\sum_{j=0}^{\infty}\sum_{l}\sqrt{[l]}(\bar{\mathcal{S}}_{l,l}^{(j)}-1)\begin{bmatrix} l & 1 & j \\ 0 & m & m \end{bmatrix}\begin{bmatrix} l & 1 & j \\ m-m' & m' & m \end{bmatrix} Y_{m-m'}^{l}(\theta,\phi),$$

where $l=j$ when j is an odd integer, and $l=j\pm1$ when j is an even integer; the scattering matrix, in general, depends on j, and hence j has been superscribed on $\bar{\mathcal{S}}$. The diagonal matrix elements of $\bar{\mathcal{S}}$ are given by

$$\bar{\mathcal{S}}_{j,j}^{(j)}=\exp\,[2i(\eta_j+\delta_j^{(1)})]=\exp\,(2i\eta_l)\mathcal{S}_{l,l}^{(j)} \qquad (l=j),$$

$$\bar{\mathcal{S}}_{l,l}^{(j)}=(D_c)_{l,l}\mathcal{S}_{l,l}^{(j)}(D_c)_{l,l}=\exp\,(2i\eta_l)\mathcal{S}_{l,l}^{(j)} \qquad (l=j\pm1),$$

where $\mathcal{S}_{l,l}$ $(l=j\pm1)$ are given in Section 7 [see (I.96)]. We now make use of the counting of the states in terms of odd l, as just outlined. The series is then seen to be equivalent to

$$\frac{\sqrt{4\pi}}{ik}\sum_{l=1\,(\text{odd})}^{\infty}\sum_{j}^{l,\,l\pm1}[l]^{1/2}\{\exp\,(2i\eta_l)\mathcal{S}_{l,l}^{(j)}-1\}Y_{m-m'}^{l}\begin{bmatrix} l & 1 & j \\ 0 & m & m \end{bmatrix}\begin{bmatrix} l & 1 & j \\ m-m' & m' & m \end{bmatrix}. \qquad \text{(I.150)}$$

It should be noted that the Coulomb phase shifts do not have any explicit j-dependence because the Coulomb potential is a pure spatial function. η is determined as soon as l is specified; the three different j-states, namely, l and $l\pm1$, corresponding to the same l, have the same phase shift η_l. Thus, if we consider $[\exp\,(2i\eta_l)-1]$ instead of $[\exp\,(2i\eta_l)\mathcal{S}_{l,l}^{(j)}-1]$, the sum over j concerns the product of the Clebsch-Gordon coefficients alone. The use of (AIV.3b) in Appendix A clarifies that this sum is $\delta_{m,m'}$. Hence, we obtain, for $[\exp\,(2i\eta_l)-1]$,

$$\frac{\sqrt{4\pi}}{ik}\sum_{l=1\,(\text{odd})}^{\infty}[l]^{1/2}[\exp\,(2i\eta_l)-1]Y_0^{l}(\theta),$$

irrespective of the value of the initial spin projection m. We have already noticed in Section 8A that this partial wave sum is equal to the closed expression (I.124). Thus, after separating the pure Coulomb scattering, the triplet scattering amplitude is given by

$$\begin{aligned} M_{m',m}^{1,j}(\theta,\phi) &= F_{\text{coul}}^{\text{triplet}}(\theta)+\frac{1}{ik}\sum_{l=1\,(\text{odd})}\sqrt{4\pi[l]}\,Y_{m-m'}^{l}(\theta,\phi) \\ &\times\sum_{j}^{l,\,l\pm1}\exp\,(2i\eta_l)(\mathcal{S}_{l,l}^{(j)}-1)\begin{bmatrix} l & 1 & j \\ 0 & m & m \end{bmatrix}\begin{bmatrix} l & 1 & j \\ m-m' & m' & m \end{bmatrix} \\ &-\frac{\sqrt{4\pi}}{ik}\sum_{j=0}^{\infty}\sum_{l\neq l'}^{j\pm1}\sqrt{[l]}\begin{bmatrix} l & 1 & j \\ 0 & m & m \end{bmatrix}\begin{bmatrix} l' & 1 & j \\ m-m' & m' & m \end{bmatrix}Y_{m-m'}^{l'}(\theta,\phi) \\ &\times\exp\,[i(\eta_{j-1}+\eta_{j+1})]\mathcal{S}_{l,l'}^{(j)}. \end{aligned} \qquad \text{(I.151)}$$

The first and second terms on the right-hand side of (I.151) result from the diagonal parts of

$\bar{\mathcal{S}}$ given by (I.150), and the last term from the nondiagonal elements of the $\bar{\mathcal{S}}$-matrix. From (I.96) and the definition $\bar{\mathcal{S}} = D_c \mathcal{S} D_c$, we have

$$\bar{\mathcal{S}}_{j-1,\,j+1} = \bar{\mathcal{S}}_{j+1,\,j-1} = \exp\,[i(\eta_{j-1} + \eta_{j+1})]\mathcal{S}^{(j)}_{j-1,j+1}$$
$$= \exp\,[i(\eta_{j-1} + \eta_{j+1})] \sin \epsilon \cos \epsilon[\exp\,(2i\delta_j^\alpha) - \exp\,(2i\delta_j^\beta)].$$

As j and l increase, the phase shifts δ_j^α and δ_j^β approach zero, and hence the contribution of the last term in (I.151) also gets ever smaller. Therefore, this term, as well as the second term of (I.151) (discussed in connection with singlet scattering), does not present any problem for numerical evaluation.

The procedure for squaring the scattering amplitudes, the subsequent summation and averaging over m', m, and the computation of the differential and total cross-sections from those results have been covered in Section 7.

C. BAR PHASE SHIFTS

In Section 7, we parametrized the unitary symmetric scattering matrix in terms of ϵ, δ_j^α, and δ_j^β. An alternative parametrization in terms of three new parameters $\bar{\epsilon}$, $\bar{\delta}_j^\alpha$, and $\bar{\delta}_j^\beta$ is also possible, if we use the unitary diagonal matrix

$$D = \begin{pmatrix} \exp\,(i\bar{\delta}_j^\alpha) & 0 \\ 0 & \exp\,(i\bar{\delta}_j^\beta) \end{pmatrix}$$

and another unitary symmetric matrix

$$\mathcal{E} = \begin{pmatrix} \cos 2\bar{\epsilon} & i \sin 2\bar{\epsilon} \\ i \sin 2\bar{\epsilon} & \cos 2\bar{\epsilon} \end{pmatrix}$$

to form the matrix $D\mathcal{E}D$. This matrix is obviously unitary and symmetric, and is given in terms of three independent real parameters. Hence,

$$\mathcal{S} = D\mathcal{E}D \tag{I.152}$$

is a perfectly acceptable alternative parametrization of the scattering matrix. The new barred parameters can be related to the old parameters by equating (I.152) to (I.96).

Introducing $\Delta_\pm = \delta_j^\alpha \pm \delta_j^\beta$ and a similar definition for the barred quantities, we get

$$\mathcal{S} = \begin{pmatrix} \exp\,(i\Delta_+)\{\cos \Delta_- + i \cos 2\epsilon \sin \Delta_-\} & i \exp\,(i\Delta_+) \sin \Delta_- \sin 2\epsilon \\ \cdots & \exp\,(i\Delta_+)\{\cos \Delta_- - i \cos 2\epsilon \sin \Delta_-\} \end{pmatrix}$$

$$= \begin{pmatrix} \exp\,(i\bar{\Delta}_+) \exp\,(i\bar{\Delta}_-) \cos 2\bar{\epsilon} & i \exp\,(i\bar{\Delta}_+) \sin 2\bar{\epsilon} \\ \cdots & \exp\,(i\bar{\Delta}_+) \exp\,(-i\bar{\Delta}_-) \cos 2\bar{\epsilon} \end{pmatrix},$$

where the dotted element is equal to the other nondiagonal element from the symmetry of the $\mathcal{S}$-matrix.

The relationships we are seeking are therefore given by

$$\exp\,(i\bar{\Delta}_+) \exp\,(i\bar{\Delta}_-) \cos 2\bar{\epsilon} = \exp\,(i\Delta_+)\{\cos \Delta_- + i \cos 2\epsilon \sin \Delta_-\}, \tag{I.153a}$$

$$\exp\,(i\bar{\Delta}_+) \exp\,(-i\bar{\Delta}_-) \cos 2\bar{\epsilon} = \exp\,(i\Delta_+)\{\cos \Delta_- - i \cos 2\epsilon \sin \Delta_-\}, \tag{I.153b}$$

$$\exp\,(i\bar{\Delta}_+) \sin 2\bar{\epsilon} = \exp\,(i\Delta_+) \sin \Delta_- \sin 2\epsilon. \tag{I.153c}$$

Taking the ratio of (I.153a) and (I.153b), we get

$$\exp(2i\bar{\Delta}_-) = \frac{\cos\Delta_- + i\cos 2\epsilon \sin\Delta_-}{\cos\Delta_- - i\cos 2\epsilon\sin\Delta_-}. \tag{I.154}$$

Similarly, from (I.153b) and (I.153c),

$$\tan 2\bar{\epsilon} = \frac{\sin\Delta_- \sin 2\epsilon}{\cos\Delta_- - i\cos 2\epsilon\sin\Delta_-}\exp(-i\bar{\Delta}_-)$$

$$= \frac{\sin\Delta_- \sin 2\epsilon}{(\cos^2\Delta_- + \cos^2 2\epsilon \sin^2\Delta_-)^{1/2}}.$$

The inverse of the square-root of (I.154) has been substituted for $\exp(-i\bar{\Delta}_-)$ in order to arrive at the last step. A very simple result for $\sin 2\bar{\epsilon}$ that follows from the foregoing equation is

$$\sin 2\bar{\epsilon} = \frac{\sin\Delta_- \sin 2\epsilon}{(\sin^2\Delta_- \sin^2 2\epsilon + \cos^2\Delta_- + \cos^2 2\epsilon \sin^2\Delta_-)^{1/2}} = \sin\Delta_- \sin 2\epsilon. \tag{I.155a}$$

Hence, from (I.153c) we get

$$\exp(i\bar{\Delta}_+) = \exp(i\Delta_+)$$

or

$$\bar{\Delta}_+ = \Delta_+. \tag{I.155b}$$

Using this result in (I.153a) and (I.153b) and subtracting the two equations, we derive

$$\cos 2\bar{\epsilon}\sin\bar{\Delta}_- = \cos 2\epsilon \sin\Delta_-$$

or

$$\sin\bar{\Delta}_- = \tan 2\bar{\epsilon}/\tan 2\epsilon. \tag{I.155c}$$

The value of $\sin\Delta_-$ from (I.155a) has been used to arrive at the last step. The expressions (I.155) enable us to relate the barred quantities to the unbarred ones.

The introduction of the bar phase shifts in the analysis of nucleon-nucleon scattering was initially motivated by the convenience of relating the parameters of the nuclear $\mathcal{S}$-matrix to those of the total $\bar{\mathcal{S}}$-matrix. It is useful to recall the result for the p-p scattering $\bar{\mathcal{S}}$-matrix, which is related to the nuclear $\mathcal{S}$-matrix by $\bar{\mathcal{S}} = D_c\mathcal{S}D_c$. Now, if $\mathcal{S}$ is parametrized in terms of barred quantities, then

$$\mathcal{S} = D_c D\mathcal{E}DD_c$$

$$= D_t\mathcal{E}D_t,$$

where the new diagonal matrix D_t is given by

$$D_t = D_c D = \begin{pmatrix}\exp(i\eta_{j-1}) & 0 \\ 0 & \exp(i\eta_{j+1})\end{pmatrix}\begin{pmatrix}\exp(i\bar{\delta}_j^\alpha) & 0 \\ 0 & \exp(i\bar{\delta}_j^\beta)\end{pmatrix}$$

$$= \begin{pmatrix}\exp[i(\eta_{j-1} + \bar{\delta}_j^\alpha)] & 0 \\ 0 & \exp[i(\eta_{j+1} + \bar{\delta}_j^\beta)]\end{pmatrix}.$$

Thus, in terms of the barred parameters, the passage from $\mathcal{S}$ to $\bar{\mathcal{S}}$ is very easy. Both have the same type of formal expression, and the set of parameters $(\bar{\epsilon}, \bar{\delta}_j^\alpha, \bar{\delta}_j^\beta)$ simply changes to $(\bar{\epsilon}, \bar{\delta}_j^\alpha + \eta_{j-1}, \bar{\delta}_j^\beta + \eta_{j+1})$. Since the Coulomb phase shifts are known, the transformation from $\mathcal{S}$ to $\bar{\mathcal{S}}$ is trivial.

9. EFFECTIVE-RANGE FORMALISM FOR LOW-ENERGY SCATTERING

Semiclassical arguments may be used to show that, at very low energy, it is the S-state alone that contributes to scattering, and then, as energy increases, ever higher angular momentum states start participating in the process. This is a consequence of the short range of the nucleon-nucleon force. If this range is denoted by a and the momentum by $\mathbf{p}$, then the maximum angular momentum state that can be affected by the scattering potential is obviously given by pa. Equating the square of this quantity with $l(l+1)\hbar^2$, we can very easily evaluate the energy at which a given l-state begins to acquire importance. Very simple computation places this energy for the ($l=1$)-state at approximately 10 MeV. In this section, we consider scattering below this energy, and try to explore the energy-dependence of the S-wave scattering. For p-p scattering, it is enough if we deal with scattering in the $^1S_0$-state, but for n-p scattering, both $^1S_0$- and $^3S_1$-scattering have to be considered. In n-p scattering, since the tensor force mixes the $^3S_1$- with the $^3D_1$-state, strictly speaking we should consider the α-state [mixed ($^3S_1 + {}^3D_1$)-eigenstate of the scattering matrix with the $^3S_1$-state *predominant*; see Section 7] and *not* the pure $^3S_1$-state.

A. n-p SCATTERING

We first consider low-energy n-p scattering. From (I.110) and using only the $j=0$ term, we get, for $^1S_0$-scattering,

$$\sigma_{S=0} = \frac{4\pi}{k^2 \operatorname{cosec}^2 \delta_0^{(0)}} = \frac{4\pi}{k^2 + k^2 \cot^2 \delta_0^{(0)}}. \tag{I.156}$$

An identical expression holds for $^3S_1$-scattering with $\delta_0^{(0)}$ replaced by the appropriate phase shift. The total low-energy n-p cross-section is then obtained by adding $\frac{1}{4}\sigma_{S=0}$ and $\frac{3}{4}\sigma_{S=1}$. Thus, the examination of the energy-dependence of low-energy n-p scattering is reduced to the study of the quantity $k \cot \delta$ for the $^1S_0$- and $^3S_1$-state. (We omit the labels from δ to simplify the notation.)

To show that $k \cot \delta$ is an even function of k, i.e., a function of k^2, we use $\phi_l^{(1)}$ and $\phi_l^{(2)}$ to denote two linearly independent solutions of the asymptotic equation (I.49). Instead of identifying $\phi_l^{(1)}$ and $\phi_l^{(2)}$ as $krj_l(kr)$ and $krn_l(kr)$, respectively, we can alternatively take

$$\begin{aligned}\phi_l^{(1)} &= -ikr[j_l(kr) - in_l(kr)]e^{-i(1/2)l\pi} \\ &= -ikrh_l^{(2)}(kr)e^{-i(1/2)l\pi} \\ &= e^{-ikr}, \qquad r \to \infty, \\ \phi_l^{(2)} &= ikr[j_l(kr) + in_l(kr)]e^{i(1/2)l\pi} \\ &= ikrh_l^{(1)}(kr)e^{i(1/2)l\pi} \\ &= e^{ikr}, \qquad r \to \infty.\end{aligned}$$

The solutions $u_l^{(1)}$ and $u_l^{(2)}$ of the Schrödinger equation (I.47) which behave, for $r \to \infty$, as $\phi_l^{(1)}$ and $\phi_l^{(2)}$, respectively, are henceforth labelled $f_l(k, r)$ and $f_l(-k, r)$. If $f_l(k, r)$ is a solution of the Schrödinger equation, $f_l(-k, r)$ too automatically is a solution because the Schrödinger equation contains k^2. Further, since $f_l(k, r)$ behaves as e^{-ikr} at $r \to \infty$, $f_l(-k, r)$ automatically behaves as e^{ikr} at $r \to \infty$. Therefore, $f_l(-k, r)$ is indeed the same as the independent solution $u_l^{(2)}$.

We now wish to construct the exact solution $u_l(r)$ by taking a linear combination of

10 (45-122/1977)

$f_l(k, r)$ and $f_l(-k, r)$ so that it automatically satisfies the boundary condition at $r \to \infty$; the linear combination will be chosen such that

$$u_l(r) = (\text{constant}) \times r^{l+1}, \qquad r \to 0,$$

and hence $u_l'(r) = r^l$, $r \to 0$, with a suitable normalization. The Wronskian (denoted by W) of $f_l(k, r)$ and $f_l(-k, r)$ can be shown, from (I.49), to be a constant, independent of r [see Problem 3 (Chapter I)]. Thus, it can be evaluated at $r \to \infty$ by using their asymptotic forms, which yield

$$f_l(k, r)\frac{d}{dr}f_l(-k, r) - f_l(-k, r)\frac{d}{dr}f_l(k, r) \equiv W[f_l(k, r), f_l(-k, r)] = 2ik.$$

This fact, together with the choice

$$f_l(k) = \operatorname*{Lt}_{r\to 0} r^l f_l(k, r),$$

establishes that

$$u_l(r) = -\frac{1}{2ik}[f_l(-k)f_l(k, r) - f_l(k)f_l(-k, r)] \tag{I.157}$$

indeed satisfies $u_l'(r) = r^l$, $r \to 0$.

If we compare (I.157) for $r \to \infty$ with (I.86), and use (I.87) and (I.88) for the $\mathcal{S}$-matrix, we get

$$\exp(2i\delta_l) = \mathcal{S}_l(k) = \frac{f_l(k)}{f_l(-k)}e^{il\pi} = \operatorname*{Lt}_{r\to 0}\frac{f_l(k, r)}{f_l(-k, r)}e^{il\pi}. \tag{I.158}$$

Now, specializing this equation to ($l = 0$)-scattering and (in order to simplify the notation) no longer using 0 as a subscript, we obtain

$$\frac{f(k)}{f(-k)} = e^{2i\delta} = \frac{e^{i\delta}}{e^{-i\delta}} = \frac{\cos\delta + i\sin\delta}{\cos\delta - i\sin\delta}$$

or

$$i\frac{f(k) + f(-k)}{f(k) - f(-k)} = \frac{\cos\delta}{\sin\delta} = \cot\delta.$$

Hence,

$$k\cot\delta = N(k^2)/[D(k^2)], \tag{I.159}$$

where

$$N(k^2) = f(k) + f(-k), \tag{I.160a}$$

$$D(k^2) = \frac{1}{ik}[f(k) - f(-k)]. \tag{I.160b}$$

It should be noted that both the right-hand sides of (I.160) are even functions of k and therefore N, D, and hence $k\cot\delta$, are functions of k^2.

By using the asymptotic forms for $f(k, r)$ and $f(-k, r)$, we derive from (I.157) the asymptotic form $\bar{u}(r)$ of $u(r)$ as

$$\bar{u}(r) = -\frac{1}{2ik}[f(-k)e^{-ikr} - f(k)e^{ikr}]. \tag{I.161}$$

This function is a solution of the Schrödinger equation without the potential term. If we extrapolate it to the origin, and normalize it by requiring $\bar{u}(r = 0) = 1$, then we get

$$\bar{u}(r = 0) = 1 = \frac{1}{2ik}[f(k) - f(-k)]. \tag{I.162}$$

Hence, with this particular normalization of the asymptotic function $\bar{u}$, we get

$$k \cot \delta = \tfrac{1}{2}[f(k) + f(-k)] = \tfrac{1}{2}N(k^2). \tag{I.163}$$

If we now make a Taylor expansion of $k \cot \delta$ about $k^2 = 0$, we obtain

$$k \cot \delta = -a^{-1} + \tfrac{1}{2}r_0 k^2 - P r_0^3 k^4 + \ldots, \tag{I.164}$$

where

$$-a^{-1} = \operatorname*{Lt}_{k^2 \to 0} k \cot \delta = \operatorname*{Lt}_{k^2 \to 0} \tfrac{1}{2}N(k^2), \tag{I.165a}$$

$$r_0 = \operatorname*{Lt}_{k^2 \to 0} \frac{d}{d[k^2]} N(k^2), \tag{I.165b}$$

$$-P r_0^3 = \tfrac{1}{4} \operatorname*{Lt}_{k^2 \to 0} \frac{d^2}{d[k^2]^2} N(k^2). \tag{I.165c}$$

The constants a, r_0, and P are called the scattering length, effective range, and shape parameter, respectively. In what follows, they are related to physically more transparent quantities.

First, we consider the Schrödinger equation for $\bar{u}$, namely,

$$\frac{d^2\bar{u}}{dr^2} + k^2\bar{u} = 0, \tag{I.166}$$

and the limit of $k^2 \to 0$. The resultant equation

$$\frac{d^2\bar{u}_0}{dr^2} = 0$$

(the subscript now refers to the value of k^2 or, equivalently, energy) has the solution

$$\bar{u}_0 = A + Br,$$

where A and B are integration constants. If this has to be normalized to unity at $r = 0$, then $A = 1$; and let us further replace B by another constant $-a^{-1}$. Then

$$\bar{u}_0 = 1 - a^{-1}r,$$

which obviously represents a straight line crossing the abscissa and ordinate at a and 1, respectively. Hence, a has a very simple interpretation: it is that value of r for which the zero-energy asymptotic solution has zero value, provided it is normalized to unity at the origin.

We now wish to show that the same $-a^{-1}$ occurs in (I.164). To do this, we require $\bar{u}_0(r = a) = 0$ in (I.161) and obtain

$$0 = \operatorname*{Lt}_{k^2 \to 0} -\frac{1}{2ik}[f(-k)e^{-ika} - f(k)e^{ika}]$$

or

$$\operatorname*{Lt}_{k \to 0} \frac{f(k)}{f(-k)} = \operatorname*{Lt}_{k \to 0} \frac{e^{-ika}}{e^{ika}}$$

or

$$\operatorname*{Lt}_{k \to 0} \frac{f(k) + f(-k)}{\dfrac{1}{2ik}[f(k) - f(-k)]} = \operatorname*{Lt}_{k \to 0} -\frac{2k \cos ka}{\sin ka}.$$

Using the normalization (I.162), we get

$$\operatorname*{Lt}_{k \to 0} \tfrac{1}{2}[f(k) + f(-k)] = \operatorname*{Lt}_{k \to 0} -\frac{k \cos ka}{\sin ka}$$

$$= \underset{k\to 0}{\mathrm{Lt}} - \frac{k[1 - \frac{(ka)^2}{2} + \ldots]}{ka - \frac{(ka)^3}{6} + \ldots}$$

$$= -a^{-1}.$$

This proves the assertion just made, and relates the scattering length to a simple property of the zero-energy asymptotic solution.

Figure I.6 shows the difference in the sign of the scattering lengths in two typical situa-

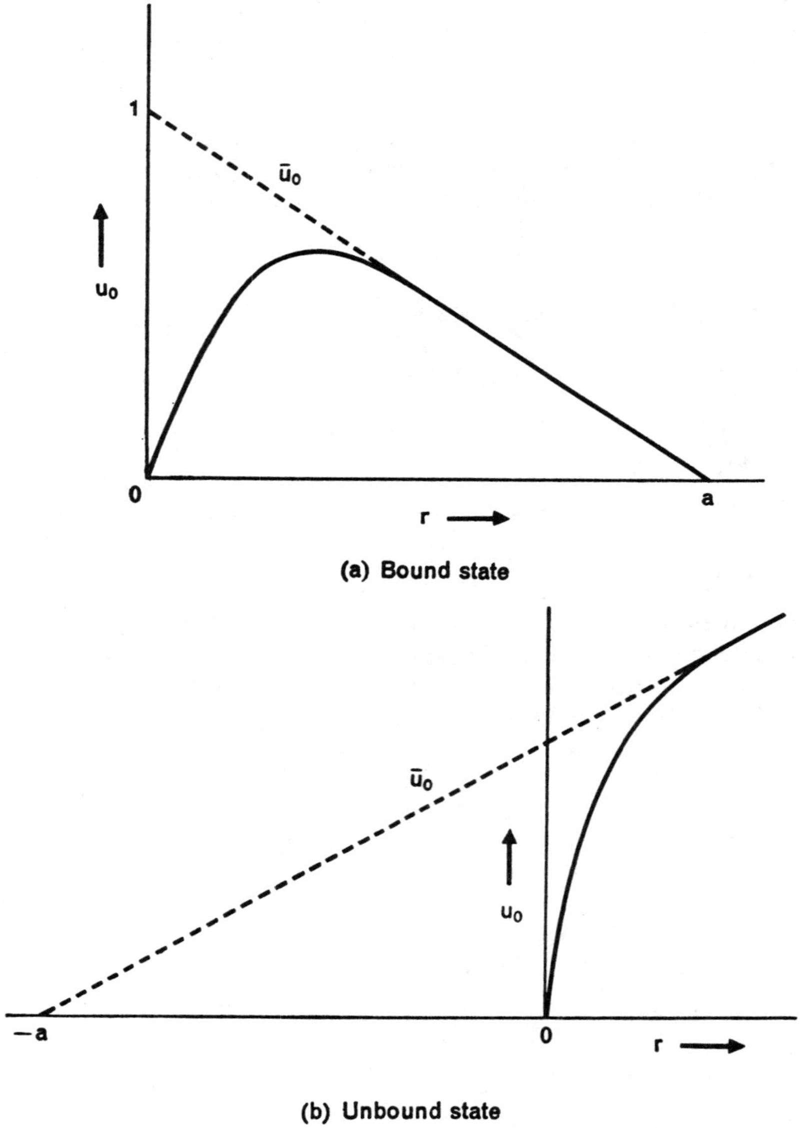

Fig. I.6 Radial wavefunctions.

tions for the zero-energy wavefunction u_0. If the potential is strong enough and can produce a bound state, then the solution has to bend down before the asymptotic region is reached such that a decaying asymptotic solution can be smoothly matched on to it. This situation is shown in Fig. I.6a. The asymptotic straight line solution now crosses the abscissa on the positive side of r, and hence the scattering length a is positive.

The second situation for a weak potential which cannot produce a bound state is shown in Fig. I.6b. Here the internal exact solution u_0 does not bend down in the asymptotic region, and hence the smooth matching of a decaying asymptotic solution becomes impossible. The zero-energy straight line asymptotic solution now crosses the abscissa on the negative side of r, and hence the scattering length is negative. Since the $^1S_0$-state in the deuteron is unbound, we expect to obtain a negative scattering length for low-energy n-p $^1S_0$-scattering. On the other hand, since $^3S_1$ is a bound state, the corresponding n-p scattering length should be positive. (See Section 11 for the experimental values.)

The next step entails relating the effective range r_0 and the shape parameter P given by (I.165b) and (I.165c) to the wavefunctions u and $\bar{u}$. We start with the Schrödinger equation for u:

$$\frac{d^2u}{dr^2} + k^2u + f(r)u = 0. \tag{I.167}$$

The asymptotic solution for this equation at $r \to \infty$ is taken in the form $C \sin (kr + \delta)$, where δ is the phase shift, and C an arbitrary normalization constant. For the solution $\bar{u}(r)$ of (I.166), we use the same form, and further normalize it to unity at $r = 0$, which gives $C = (\sin \delta)^{-1}$. Thus,

$$\bar{u}(r) = \frac{\sin (kr + \delta)}{\sin \delta}, \tag{I.168}$$

and its log-derivative $K(r)$, evaluated at the origin, is given by

$$K(0) = \frac{1}{\bar{u}}\frac{d\bar{u}}{dr}\bigg|_{r=0} = k \cot \delta. \tag{I.169}$$

The exact solution u of (I.167) has to satisfy the two boundary conditions

$$u(0) = 0, \tag{I.170a}$$

$$\underset{r\to\infty}{\text{Lt}}\, u(r) = \underset{r\to\infty}{\text{Lt}}\, \bar{u}(r). \tag{I.170b}$$

We now multiply (I.167) by u from the left and integrate between zero and infinity; similarly, (I.166) is multiplied by $\bar{u}$ and integrated between an adjustable parameter $\bar{r}$ (to be chosen later) and infinity. Subtracting the two results, we obtain

$$0 = L = \int_0^\infty dr\, u[\frac{d^2u}{dr^2} + k^2u + f(r)u] - \int_{\bar{r}}^\infty dr\, \bar{u}[\frac{d^2\bar{u}}{dr^2} + k^2\bar{u}]. \tag{I.171}$$

By a partial integration, we get the result

$$\int_0^\infty dr\, u\frac{d^2u}{dr^2} - \int_{\bar{r}}^\infty dr\, \bar{u}\frac{d^2\bar{u}}{dr^2} = u\frac{du}{dr}\bigg|_0^\infty - \bar{u}\frac{d\bar{u}}{dr}\bigg|_{\bar{r}}^\infty - \int_0^\infty (\frac{du}{dr})^2\, dr + \int_{\bar{r}}^\infty (\frac{d\bar{u}}{dr})^2\, dr.$$

In view of (I.170), the first and second terms of the right-hand side of the foregoing equation contribute only

$$(\bar{u}\frac{d\bar{u}}{dr})_{\bar{r}} \equiv \bar{u}^2(\bar{r})K(\bar{r}).$$

Collecting these results and putting in (I.171), we have

$$0 = L = \int_0^\infty dr\,[-(\frac{du}{dr})^2 + (k^2 + f)u^2] - \int_{\bar{r}}^\infty dr\,[-(\frac{d\bar{u}}{dr})^2 + k^2\bar{u}^2] + \bar{u}^2(\bar{r})K(\bar{r}). \quad \text{(I.172)}$$

We next prove, with the help of (I.172), that K is stationary under variations of u and $\bar{u}$, subject to the boundary conditions (I.170). Since L is always required to be zero, $\delta L = 0$. Hence,

$$0 = \delta L = \bar{u}^2(\bar{r})\delta K(\bar{r}) + 2\bar{u}(\bar{r})K(\bar{r})\delta u(\bar{r})$$
$$+ 2\int_0^\infty dr\,[-\frac{du}{dr}\frac{d}{dr}(\delta u) + (k^2 + f)u\delta u] - 2\int_{\bar{r}}^\infty [-\frac{d\bar{u}}{dr}\frac{d}{dr}(\delta\bar{u}) + k^2\bar{u}\delta\bar{u}]. \quad \text{(I.173)}$$

Integrating by parts the first term of the first integral, we obtain

$$-\frac{du}{dr}\delta u\Big|_0^\infty + \int_0^\infty dr\,\frac{d^2u}{dr^2}\delta u.$$

The second term, together with the rest of the integrand, vanishes by virtue of (I.167). Similar partial integration of the first term of the second integral and the subsequent use of (I.166) leaves only

$$-\frac{d\bar{u}}{dr}\delta\bar{u}\Big|_{\bar{r}}^\infty.$$

The boundary conditions (I.170) demand that

$$\delta u(0) = 0,$$

$$\operatorname*{Lt}_{r\to\infty} \delta u(r) = \operatorname*{Lt}_{r\to\infty} \delta\bar{u}(r).$$

Hence,

$$-2\frac{du}{dr}\delta u\Big|_0^\infty + 2\frac{d\bar{u}}{dr}\delta\bar{u}\Big|_{\bar{r}}^\infty = -2\bar{u}(\bar{r})K(\bar{r})\delta\bar{u}(\bar{r}).$$

Using all these results in (I.173), we obtain

$$0 = \delta L = \bar{u}^2(\bar{r})\delta K(\bar{r}),$$

which proves the stationary nature of $K(\bar{r})$ with respect to variations in the wavefunctions.

We now choose $\bar{r} = 0$. (A second possible choice of $\bar{r}$ is discussed later in this section.) Then, by virtue of (I.169) and the normalization $\bar{u}(0) = 1$, we have

$$\delta(k \cot \delta) = 0;$$

obviously, the first δ denotes variation in $k \cot \delta$ corresponding to variations in the wavefunctions, whereas the second δ represents the phase shift.

We are now in a position to return to (I.172) with $\bar{r} = 0$, and rewrite it as

$$k \cot \delta = \int_0^\infty dr\,[(\frac{du}{dr})^2 - (\frac{d\bar{u}}{dr})^2 + k^2(\bar{u}^2 - u^2) - fu^2], \quad \text{(I.174)}$$

and to consider variations in u and $\bar{u}$ due to a small variation of k^2, i.e.,

$$\delta u = \delta k^2 \frac{du}{dk^2}$$

$$\delta \bar{u} = \delta k^2 \frac{d\bar{u}}{dk^2}.$$

Because $k \cot \delta$ is stationary with respect to δu and $\delta \bar{u}$, as just proved, we know that there cannot be any change in $k \cot \delta$ arising from variations in the wavefunctions. But since there is an *explicit* k^2-dependence in the integral on the right-hand side of (I.174), there will be a variation in $k \cot \delta$, when energy is varied, which is given by the differentiation of the k^2-term. Thus,

$$\frac{d}{dk^2}(k \cot \delta) = \int_0^\infty dr\,(\bar{u}^2 - u^2). \tag{I.175}$$

This general and *exact* relation, specialized to the case $k^2 \to 0$, now enables us to write down $\frac{1}{2}r_0$ in terms of the wavefunctions. We have

$$\tfrac{1}{2}r_0 = \operatorname*{Lt}_{k^2\to 0} \frac{d}{dk^2}(k \cot \delta) = \int_0^\infty dr\,(\bar{u}_0^2 - u_0^2), \tag{I.176}$$

where the label zero on the wavefunctions once again refers to the value of k^2.

The next higher term (the k^4-term) in the expansion (I.164) of $k \cot \delta$ as a function of k^2 can also be evaluated from (I.175). The coefficient of this term, $-Pr_0^3$, is obviously given by

$$-Pr_0^3 = \operatorname*{Lt}_{k^2\to 0} \tfrac{1}{2}\frac{d}{dk^2}\left[\frac{d}{dk^2}(k \cot \delta)\right] = \int_0^\infty dr\,\left(\bar{u}\frac{d\bar{u}}{dk^2} - u\frac{du}{dk^2}\right)_{k^2\to 0}. \tag{I.177}$$

A much simpler method for deriving the same results is the one due to Bethe[10]. The more involved derivation of the effective-range expression (I.176), which we have just covered, is due to Raphael[11]; this method becomes interesting when, following Raphael, we choose a different value of the adjustable parameter $\bar{r}$ later in this section.

In Bethe's manner of derivation of the effective-range formalism, we start with (I.166) and (I.167) for two different energies k_1^2 and k_2^2. The corresponding wavefunctions are also labelled by the subscripts 1 and 2. Multiplying the equation for u_1 by u_2 from the left, and vice versa, and then subtracting the two equations and integrating between 0 and ∞, we get

$$\int_0^\infty \left(u_2\frac{d^2u_1}{dr^2} - u_1\frac{d^2u_2}{dr^2}\right) dr = (k_2^2 - k_1^2)\int_0^\infty u_1u_2\,dr.$$

Since the integrand on the left-hand side is

$$\frac{d}{dr}\left(u_2\frac{du_1}{dr} - u_1\frac{du_2}{dr}\right),$$

we obtain

$$\left(u_2\frac{du_1}{dr} - u_1\frac{du_2}{dr}\right)\Bigg|_0^\infty = (k_2^2 - k_1^2)\int_0^\infty u_1u_2\,dr.$$

Handling in the same manner the equations for $\bar{u}_1$ and $\bar{u}_2$, we get

$$\left(\bar{u}_2\frac{d\bar{u}_1}{dr} - \bar{u}_1\frac{d\bar{u}_2}{dr}\right)\Bigg|_0^\infty = (k_2^2 - k_1^2)\int_0^\infty \bar{u}_1\bar{u}_2\,dr.$$

Since u_1, u_2 become equal to $\bar{u}_1$, $\bar{u}_2$, respectively, when $r \to \infty$, the left-hand sides of these two equations are equal at the upper limit ∞. At the lower limit, $u_1 = u_2 = 0$, and hence only the $(\bar{u}_1, \bar{u}_2)$-term contributes. Thus, when we subtract the two equations, we obtain

$$\left(-\bar{u}_2\frac{d\bar{u}_1}{dr} + \bar{u}_1\frac{d\bar{u}_2}{dr}\right)_{r=0} = (k_2^2 - k_1^2)\int_0^\infty (\bar{u}_1\bar{u}_2 - u_1u_2)\, dr. \tag{I.178}$$

We now put $k_1^2 = 0$ and $k_2^2 = k^2$, and make use of the normalization $\bar{u}(0) = 1$. (I.178) thus yields

$$k \cot \delta = k^2\int_0^\infty (\bar{u}\bar{u}_0 - uu_0)\, dr + \left(\frac{d\bar{u}_0}{dr}\right)_{r=0}. \tag{I.179}$$

From the expression of $\bar{u}_0$ given after (I.166), we get

$$\left(\frac{d\bar{u}_0}{dr}\right)_{r=0} = -a^{-1}.$$

Further, in the integrand of (I.179), we make use of the Taylor expansions

$$\bar{u} = \bar{u}_0 + k^2\left(\frac{d\bar{u}}{dk^2}\right)_{k^2\to 0} + \ldots,$$

$$u = u_0 + k^2\left(\frac{du}{dk^2}\right)_{k^2\to 0} + \ldots .$$

Then (I.179) is easily seen to yield the expansion (I.164) of $k \cot \delta$ in powers of k^2; and the parameters $\frac{1}{2}r_0$ and $-Pr_0^3$ are given by the expressions (I.176) and (I.177), respectively.

In the expansion of (I.164), only the first two terms are expected to be important at low energy (i.e., small k^2). Unless the data are very accurate, it is difficult to obtain any information on the third term, containing P, from an analysis of the low-energy scattering cross-section. The two parameters a and r_0, which are derived from a direct analysis of the data, can be reproduced by a short-range potential of any reasonable shape because any such potential contains two adjustable parameters, namely, the depth and the range. Hence, unless the data are very reliable, the low-energy analysis will give information only on the range and depth of the potential, and no information at all on its radial shape. Only if we can extract information on the parameter P of (I.164) is there any hope of distinguishing between different potential shapes. This explains the nomenclature used for P. At present, experimental data have sufficient accuracy to make an evaluation of P possible for the ^1S-state. The numerical values of a, r_0, and P are given in Section 11.

We next describe an alternative choice of the parameter $\bar{r}$, appearing in (I.172). Instead of the normalization $\bar{u}(0) = 1$, we now choose $\bar{u}(\bar{r}) = 1$. Further, since $\bar{u}(r)$ is proportional to $\sin(kr + \delta)$, the log-derivative $K(\bar{r})$ is given by $k \cot(k\bar{r} + \delta)$. Therefore, since $K(\bar{r})$ is stationary under changes in the wavefunctions, we get

$$\frac{d}{dk^2}[k \cot(k\bar{r} + \delta)] = \int_{\bar{r}}^\infty \bar{u}^2\, dr - \int_0^\infty u^2\, dr.$$

This equation replaces (I.175), which was obtained for $\bar{r} = 0$. We may now choose $\bar{r}$ in such a way that this expression becomes equal to zero for $k^2 \to 0$. The advantage of such a choice is that the k^2-term is absent in the expansion of $k \cot(k\bar{r} + \delta)$, and we then have

$$k \cot(k\bar{r} + \delta) = A + \tfrac{1}{2}Ck^4 + \ldots . \tag{I.180}$$

There are three parameters A, C, and $\bar{r}$ in this expression, just as there were a, r_0, and P in the earlier effective-range expansion (I.164). But since the k^2-term is absent in (I.180), $k \cot (k\bar{r} + \delta)$ is less energy-dependent than $k \cot \delta$, and hence this new expansion is expected to hold over a larger energy range.

The equation determining $\bar{r}$ is given by

$$\int_{\bar{r}}^{\infty} \bar{u}_0^2 \, dr - \int_0^{\infty} u_0^2 \, dr = 0. \tag{I.181a}$$

Making use of the definition (I.176) of r_0 in this expression, we get

$$\tfrac{1}{2} r_0 - \int_0^{\bar{r}} \bar{u}_0^2 \, dr = 0, \tag{I.181b}$$

which yields, with the help of $\bar{u}_0$ given after (I.166),

$$\tfrac{1}{2} r_0 = \bar{r} - a^{-1}\bar{r}^2 + \tfrac{1}{3} a^{-2}\bar{r}^3. \tag{I.182a}$$

We can similarly relate the scattering length to the new parameters. (I.180) gives the expression

$$k \cot \delta = \frac{k + (A + \frac{1}{2}Ck^4) \cot k\bar{r}}{\cot k\bar{r} - (Ak^{-1} + \frac{1}{2}Ck^3)}. \tag{I.182b}$$

Taking the limit $k^2 \to 0$ on the right-hand side, and remembering that $\operatorname{Lt}_{k^2 \to 0} k \cot k\bar{r} = \bar{r}^{-1}$, we obtain

$$-a^{-1} = \operatorname*{Lt}_{k^2 \to 0} (k \cot \delta) = \frac{A}{1 - A\bar{r}}. \tag{I.182c}$$

If we expand the right-hand side of (I.182b) in powers of k^2 and pick up the coefficient of k^4, the result should be the same as $-Pr_0^3$ in (I.164). The relationship between P and the new Raphael parameters obtained in this way is given by

$$-P = \tfrac{1}{24}(1 + x + \tfrac{1}{3}x^2)^{-3}[\tfrac{1}{3}x^3 - \tfrac{4}{5}x^2 + (1 + x)^2(1 + \beta)], \tag{I.182d}$$

where

$$x = \frac{A\bar{r}}{1 - A\bar{r}} = -a^{-1}\bar{r}, \qquad \beta = \frac{3}{2}\frac{C}{(\bar{r})^3}.$$

The Raphael derivation is interesting for a second reason: it provides the basis of a semi-empirical approach due to Feshbach and Lomon[12] for the analysis of two-nucleon scattering data. These authors argue that, instead of analyzing two-nucleon data in terms of the phase shifts of various states, it may be possible to carry out the analysis in terms of a boundary condition at a suitably chosen radius. The Raphael derivation, presented here, elucidates that the log-derivative $K(\bar{r})$ at a radius $\bar{r}$ is stationary with respect to variations in the wavefunction and, moreover, that a certain $\bar{r}$ can be chosen such that, at this radius, K has a fairly weak energy-dependence.

Finally, we indicate what happens to the effective-range formalism if the nuclear potential has a hard core. In such a case, the actual wavefunction u vanishes at $r = r_c$, the core radius, and hence (in Bethe's derivation) the lower limit on the integrals involving u should be changed from 0 to r_c. Thus, the basic equation (I.178) is still valid, provided we change the

11 (45-122/1977)

right-hand side to

$$(k_2^2 - k_1^2)\left(\int_0^\infty \bar{u}_1 \bar{u}_2 \, dr - \int_{r_c}^\infty u_1 u_2 \, dr\right). \tag{I.183a}$$

A similar alteration has to be introduced in the integral on the right-hand side of (I.179). *The final result is clearly an expansion of k cot δ in the old form (I.164)*, where a^{-1} has the same interpretation, but r_0 and P have slightly altered definitions. Instead of (I.176), we obviously obtain

$$\tfrac{1}{2}r_0 = \int_0^\infty dr \, \bar{u}_0^2 - \int_{rc}^\infty dr \, u_0^2; \tag{I.183b}$$

and instead of (I.177), we have,

$$-Pr_0^3 = \int_0^\infty dr \, (\bar{u}\frac{d\bar{u}}{dk^2})_{k^2\to0} - \int_{r_c}^\infty dr \, (u\frac{du}{dk^2})_{k^2\to0}. \tag{I.183c}$$

As far as the Raphael formalism is concerned, the expression (I.180) is still valid, but (I.181a) determining $\bar{r}$ obviously changes to

$$\int_{\bar{r}}^\infty \bar{u}_0^2 \, dr - \int_{r_c}^\infty u_0^2 \, dr = 0. \tag{I.183d}$$

But, in view of the new equation (I.183b) of $\frac{1}{2}r_0$, (I.183d) and (I.181b) become identical. Therefore, the relationship between the Raphael parameters A, C, $\bar{r}$ and the parameters a, r_0, P remains unchanged.

Relationship of Triplet Scattering Length and Effective Range with Deuteron Binding Energy

If the deuteron is assumed to be a pure $^3S_1$-state, then its wavefunction u_g and the corresponding asymptotic function $\bar{u}_g$ satisfy (I.167) and (I.166), respectively, with k^2 replaced by $-\gamma^2$ [see (I.41) for definition]. Therefore, (I.178) holds with the substitution $k_1^2 = -\gamma^2$, $u_1 = u_g$, and $\bar{u}_1 = \bar{u}_g$. As before, we put $k_2^2 = k^2$, and hence $u_2 = u$, $\bar{u}_2 = \bar{u}$. Thus,

$$(-\bar{u}\frac{d\bar{u}_g}{dr} + \bar{u}_g\frac{d\bar{u}}{dr})_{r=0} = (k^2 + \gamma^2)\int_0^\infty (\bar{u}\bar{u}_g - uu_g)\, dr. \tag{I.184}$$

The asymptotic wavefunction of the deuteron is given by

$$\bar{u}_g = e^{-\gamma r} \tag{I.185}$$

and it clearly has the desired normalization $\bar{u}_g(r = 0) = 1$. Further,

$$(\frac{d\bar{u}_g}{dr})_{r=0} = -\gamma, \tag{I.186}$$

and hence (I.184) yields

$$k \cot \delta + \gamma = (k^2 + \gamma^2)\int_0^\infty (\bar{u}\bar{u}_g - uu_g)\, dr. \tag{I.187}$$

Letting $k^2 \to 0$ and using the definition (1.165a), we then have

$$-a^{-1} + \gamma = \gamma^2\int_0^\infty (\bar{u}\bar{u}_g - uu_g)\, dr. \tag{I.188}$$

This equation is exact and can now be approximated by noting that, outside the range of the potential, u, u_g are equal to $\bar{u}$, $\bar{u}_g$, respectively. Hence, the integrand on the right-hand side is nonvanishing only within the potential range where the potential term of the Schrödinger equation dominates the energy term. Therefore, dropping the energy term as an approximation, we replace the integral on the right-hand side of (I.188) by

$$\int_0^\infty (\bar{u}\bar{u}_g - uu_g)\,dr \approx \int_0^\infty (\bar{u}_0^2 - u_0^2)\,dr \equiv \tfrac{1}{2}r_0, \tag{I.189}$$

where the subscripts on u, $\bar{u}$ refer to zero energy, and the definition (I.176) has been used. Inserting (I.189) back in (I.188), we obtain the desired relationship

$$a^{-1} = \gamma - \tfrac{1}{2}r_0\gamma^2. \tag{I.190}$$

B. p-p SCATTERING

Low-energy p-p scattering takes place entirely in the $^1S_0$-state. The corresponding wavefunctions $\bar{u}$, u satisfying (I.127) and (I.140) with $l = 0$ are

$$\frac{d^2\bar{u}}{dr^2} + k^2\bar{u} - \frac{1}{\beta r}\bar{u} = 0, \tag{I.191a}$$

$$\frac{d^2u}{dr^2} + k^2u - \frac{1}{\beta r}u + \mathscr{F}(r)u = 0. \tag{I.191b}$$

We consider the two energies k_1^2 and k_2^2 and treat these two equations in the same manner as in Bethe's derivation (see Section 9A). The final equation is once again identical to (I.178).

To proceed, we need explicit values of u and du/dr for the limit $r \to 0$. The evaluation of these quantities, although quite straightforward with the help of Section 8 and Appendix C (Section II), involves a lot of complicated algebraic manipulations, which are now undertaken. We finally choose the normalization of $\bar{u}$ so that $\bar{u}(r = 0)$ becomes unity. But, for the present, we disregard the normalization constant in the asymptotic solution (I.141) and rewrite it, for $l = 0$, as

$$\bar{u} = kr[\cot\delta\, F_0(kr) - G_0(kr)]. \tag{I.192}$$

The normalization omitted here is clearly $A_0/\sin\delta$, and an extra multiplying factor k has been introduced for algebraic convenience. The expressions for $krF_0(kr)$ and $-krG_0(kr)$ are now written with the help of (I.132) and (I.139). Thus,

$$krF_0(kr) = \frac{\Gamma(1 + i\alpha)}{\Gamma(2)} \exp(-i\eta_0)e^{-(\pi/2)\alpha}e^{ikr}krF(1 + i\alpha, 2; -2ikr),$$

$$-krG_0(kr) = -i\frac{\Gamma(1 + i\alpha)}{\Gamma(2)} \exp(-i\eta_0)e^{-(\pi/2)\alpha}e^{ikr}krG(1 + i\alpha, 2; -2ikr).$$

Here $\Gamma(2) = 1$ and

$$\begin{aligned}\Gamma(1 + i\alpha)\exp(-i\eta_0)e^{-(\pi/2)\alpha} &= |\Gamma(1 + i\alpha)| \exp(i\eta_0)\exp(-i\eta_0)e^{-(\pi/2)\alpha}\\ &= |\Gamma(1 + i\alpha)|e^{-(\pi/2)\alpha}\\ &= [e^{-\pi\alpha}\Gamma(1 + i\alpha)\Gamma^*(1 + i\alpha)]^{1/2}.\end{aligned}$$

The result for $\Gamma(1 + i\alpha)$ stated after (I.130) has been used here. Now

$$\begin{aligned}e^{-\pi\alpha}\Gamma(1 + i\alpha)\Gamma^*(1 + i\alpha) &= \Gamma(1 + i\alpha)\Gamma(1 - i\alpha)e^{-\pi\alpha}\\ &= e^{-\pi\alpha}i\alpha\Gamma(i\alpha)\Gamma(1 - i\alpha) = \frac{i\pi\alpha e^{-\pi\alpha}}{\sin(i\pi\alpha)} = \frac{2\pi\alpha}{e^{2\pi\alpha} - 1}.\end{aligned}$$

Here we have applied the standard result

$$\Gamma(x)\Gamma(1 - x) = \pi/\sin(\pi x).$$

We introduce the symbol C^2 such that

$$C^2 = \frac{2\pi\alpha}{e^{2\pi\alpha} - 1}. \tag{I.193a}$$

Therefore,

$$\Gamma(1 + i\alpha)\exp(-i\eta_0)e^{-(\pi/2)\alpha} = C. \tag{I.193b}$$

Using this result in the expressions for $krF_0(kr)$ and $-krG_0(kr)$, we obtain

$$krF_0(kr) = Ce^{ikr}krF(1 + i\alpha, 2; -2ikr), \tag{I.194a}$$

$$-krG_0(kr) = -iCe^{ikr}krG(1 + i\alpha, 2; -2ikr). \tag{I.194b}$$

We now further simplify (I.194) by using the expansions (CII.3a) and (CII.7) from Appendix C for the series F and G. Since our ultimate aim is to evaluate these functions and their derivatives at $r = 0$, it is enough, in the power series expansions, to retain terms in r up to the first power; all higher powers of r may be omitted because even in the derivative they produce terms proportional to r and higher powers of r, which vanish at $r = 0$. We have

$$F(1 + i\alpha, 2; -2ikr) = 1 + \frac{1 + i\alpha}{2}(-2ikr) + \dots$$
$$= 1 - (1 + i\alpha)ikr + \dots .$$

Therefore, retaining terms up to the first power in r, we obtain

$$krF_0(kr) = Ckr(1 + ikr)[1 - (1 + i\alpha)ikr] + \dots$$
$$= Ckr. \tag{I.195a}$$

The next term in the expansion is clearly given by

$$(Ckr)(ikr)(1 - 1 - i\alpha) = (Ckr)r/(2\beta),$$

where $1/\beta = 2\alpha k$. $G(1 + i\alpha, 2; -2ikr)$ can be simplified in a similar way. Using (CII.7) from Appendix C and writing the last term first, we get

$$G(1 + i\alpha, 2; -2ikr) = -\frac{e^{-2\pi i(1+i\alpha)} - 1}{2\pi i}[\{2\ln(-2ikr) + \pi\cot\pi(1 + i\alpha) + i\pi\}$$
$$- 2\{\psi(1) + \psi(2) - \psi(1 + i\alpha)\}]$$
$$- 2e^{-i\pi(1+i\alpha)}\frac{1}{\Gamma(1 + i\alpha)}(-2ikr)^{-1}\frac{1}{\Gamma(1 - i\alpha)}. \tag{I.195b}$$

We have used the lower signs in (CII.7) because $z = -2ikr$ is in the lower-half plane. While applying (CII.7), we have retained only the first term of $F(1 + i\alpha, 2; -2ikr)$, which is unity, and the first term of the summation over r. This is because (I.194b) has a multiplicative factor kr, and our aim is to retain terms in r up to the first power. We shall further simplify (I.195b) before putting it in (I.194b). The last term on the right-hand side is, in view of (I.193b), equal to

$$\frac{i}{C^2}(kr)^{-1}.$$

In the expression within the square brackets,

$$i\pi + 2\ln(-2ikr) = i\pi + 2\ln(e^{-i(\pi/2)}2kr) = i\pi - i\pi + 2\ln(2kr)$$

$$= 2\ln\frac{r}{\beta} - 2\ln\alpha,$$

where β is defined after (I.195a).

Next, we use (CII.8) from Appendix C and the standard results

$$\psi(1+x) = \frac{1}{x} + \psi(x),$$

$$\psi(1-x) = \psi(x) + \pi\cot(\pi x)$$

to obtain the simplification

$$\psi(1) + \psi(2) = 1 + 2\psi(1) = 1 - 2\gamma,$$

$$\pi\cot\pi(1+i\alpha) + 2\psi(1+i\alpha) = \psi(1+i\alpha) + \psi(-i\alpha)$$

$$= \frac{1}{i\alpha} + \psi(i\alpha) + \psi(-i\alpha)$$

$$= \frac{1}{i\alpha} - 2\gamma + \sum_{S=0}^{\infty}\frac{2}{S+1} - \sum_{S=0}^{\infty}\left(\frac{1}{S+i\alpha} + \frac{1}{S-i\alpha}\right)$$

$$= \frac{1}{i\alpha} - 2\gamma + 2\alpha^2\sum_{S=1}^{\infty}\frac{1}{S(S^2+\alpha^2)}.$$

Here γ (the Eulerian constant) $= 0.5772\ldots$. Using all these results in (I.195b), we obtain

$$G(1+i\alpha, 2; -2ikr) = \frac{i}{C^2}(kr)^{-1} - \frac{e^{2\pi\alpha}-1}{\pi i}$$
$$\times\left[\ln\frac{r}{\beta} - \ln\alpha + \frac{1}{2i\alpha} + \gamma - 1 + \alpha^2\sum_{S=1}^{\infty}\frac{1}{S(S^2+\alpha^2)}\right] + \ldots .$$

Substituting in (I.194b) and retaining terms in r up to the first power, we obtain the final expression

$$-krG_0(kr) = C^{-1}(1+ikr) + \frac{C(e^{2\pi\alpha}-1)}{\pi}kr$$
$$\times\left[\ln\frac{r}{\beta} - \ln\alpha + \gamma + \alpha^2\sum_{S=1}^{\infty}\frac{1}{S(S^2+\alpha^2)} - 1 - ik\beta\right] + \ldots . \qquad \text{(I.195c)}$$

Using (I.195a) and (I.195c) in (I.192), and now multiplying by an additional normalization constant C, we get

$$\bar{u}(r) = C^2(\cot\delta)kr + (1+ikr) + \frac{C^2(e^{2\pi\alpha}-1)}{\pi}kr$$
$$\times\left[\ln\frac{r}{\beta} - \ln\alpha + \gamma + \alpha^2\sum_{S=1}^{\infty}\frac{1}{S(S^2+\alpha^2)} - 1 - ik\beta\right] + \ldots$$
$$= \frac{\pi}{e^{2\pi\alpha}-1}\frac{r}{\beta}\cot\delta + 1 + \frac{r}{\beta}\left[\ln\frac{r}{\beta} - \ln\alpha + \gamma - 1 + \alpha^2\sum_{S=1}^{\infty}\frac{1}{S(S^2+\alpha^2)}\right]. \qquad \text{(I.196)}$$

We have used the explicit value of C^2, given by (I.193a), to arrive at the final form. Since $r \ln r \to 0$ when $r \to 0$, the value of (I.196) at $r = 0$ is obviously equal to unity. This explains why we chose the normalization constant equal to C from (I.193). Thus,

$$\bar{u}(r = 0) = 1. \tag{I.197a}$$

The evaluation of $d\bar{u}/dr$ at $r = 0$ from (I.196) is also quite straightforward. We have

$$\left(\frac{d\bar{u}}{dr}\right)_{r=0} = \beta^{-1}\frac{\pi \cot \delta}{e^{2\pi\alpha} - 1} + \beta^{-1}\left[-\ln \alpha + \gamma + \alpha^2 \sum_{S=1}^{\infty} \frac{1}{S(S^2 + \alpha^2)}\right] + \beta^{-1} \operatorname*{Lt}_{r\to 0} \ln \frac{r}{\beta}. \tag{I.197b}$$

It should be recalled (see Section 8) that $\beta^{-1}(= 2\alpha k = Me^2/\hbar^2)$ is independent of scattering energy, and $\alpha[= e^2/(\hbar v)$, where v is the velocity in the centre-of-mass coordinate system] depends on the value of k^2. Therefore, while applying (I.197) to the basic expression (I.178), we must append subscripts 1 and 2 to α and the phase shift δ, but not to β and γ. The last term of (I.197b), because of its energy independence, cancels out in the difference indicated on the left-hand side of (I.178), despite its blowing up at $r \to 0$. Thus,

$$\beta^{-1}(K_2 - K_1) = (k_2^2 - k_1^2)\int_0^\infty (\bar{u}_1\bar{u}_2 - u_1u_2)\, dr, \tag{I.198a}$$

where

$$K = \frac{\pi \cot \delta}{e^{2\pi\alpha} - 1} - \ln \alpha + \alpha^2 \sum_{S=1}^{\infty} \frac{1}{S(S^2 + \alpha^2)} - \gamma. \tag{I.198b}$$

It should be noted that the last term of K is a constant and cancels out in the difference $(K_2 - K_1)$; hence, whether $-\gamma$ in the definition of K is selected instead of $+\gamma$ appearing in (I.197b) is immaterial. Since the present definition of K, containing the constant $-\gamma$, is conventional, we abide by it.

As in Section 9A, so too here we identify k_1^2 with zero energy and put $k_2^2 = k^2$ and, from (I.198a), obtain

$$\beta^{-1}K = \beta^{-1} \operatorname*{Lt}_{k^2\to 0} K + k^2 \int_0^\infty (\bar{u}\bar{u}_0 - uu_0)\, dr. \tag{I.199a}$$

This equation is still exact. The effective-range expansion is obtained by expanding u and $\bar{u}$ in a power series of k^2, which yields, as in Section 9A,

$$\beta^{-1}K = -a^{-1} + \tfrac{1}{2}r_0k^2 + \ldots, \tag{I.199b}$$

where the scattering length a is given by

$$-a^{-1} = \beta^{-1} \operatorname*{Lt}_{k^2\to 0} K \tag{I.199c}$$

and the effective range r_0 by the familiar expression

$$r_0 = 2\int_0^\infty (\bar{u}_0^2 - u_0)\, dr. \tag{I.199d}$$

Thus, the effective-range expansion of p-p $^1S_0$-scattering is an expansion of $\beta^{-1}K$ in powers of k^2, K being defined by (I.198b). This is in contrast to the n-p $^1S_0$-scattering, where the expanded quantity was much simpler and equal to $k \cot \delta$.

10. POLARIZATION IN TWO-NUCLEON SCATTERING

A. GENERAL CONCEPTS AND DEFINITIONS

Polarization of Spin $\frac{1}{2}$ Particle

A spin $\frac{1}{2}$ particle is described by a two-component spinor wavefunction. We learn in Appendix A (Section III) that the spin of a particle described by the wavefunction $\begin{pmatrix} a \\ b \end{pmatrix}$ points in the direction of the unit vector **n**, specified by the angles (θ, ϕ), such that

$$a = \cos\frac{\theta}{2}e^{-i(\phi/2)},$$
$$b = \sin\frac{\theta}{2}e^{i(\phi/2)}. \tag{I.200}$$

This statement actually means that the operator $(\boldsymbol{\sigma}\cdot\mathbf{n})$ has the value unity for the state $\begin{pmatrix} a \\ b \end{pmatrix}$.

A *single* spin $\frac{1}{2}$ particle, described by any spinor wavefunction, is therefore *completely* polarized in the direction **n** whose (θ, ϕ)-angles are given by (I.200). Since $(\boldsymbol{\sigma}\cdot\mathbf{n})$ has the value unity, we conclude that $\boldsymbol{\sigma}$ has the value **n** for the spin state $\begin{pmatrix} a \\ b \end{pmatrix}$. Therefore, the expectation value $\langle\boldsymbol{\sigma}\rangle$ for the spin state $\begin{pmatrix} a \\ b \end{pmatrix}$ can be taken to be the definition of polarization for this state because this expectation value has the *magnitude* unity and the direction **n**.

The concept of partial polarization or zero polarization applies to a *beam* of particles. Suppose we have a beam of spin $\frac{1}{2}$ particle consisting of N particles, specified by the spin states $\begin{pmatrix} a^{(r)} \\ b^{(r)} \end{pmatrix}$ with $r = 1, \ldots, N$. For the r-th particle, the expectation value of $\boldsymbol{\sigma}$ defines the direction, say, $\mathbf{n}^{(r)}$. In general, this direction may vary for different particles. The polarization of the entire beam is defined to be the average of $\langle\boldsymbol{\sigma}\rangle$ over the whole beam, i.e., by

$$N^{-1}\sum_{r=1}^{N}\mathbf{n}^{(r)}.$$

We shall denote the average of the expectation value of any quantity for the whole beam of particles by an overhead bar. Thus, the definition of the polarization of a beam of spin $\frac{1}{2}$ particle is given by

$$\begin{aligned}\mathscr{P} = \overline{\langle\boldsymbol{\sigma}\rangle} &= N^{-1}\sum_{r=1}^{N}(a^{(r)}b^{(r)})\boldsymbol{\sigma}\begin{pmatrix} a^{(r)} \\ b^{(r)} \end{pmatrix} \\ &= N^{-1}\sum_{r=1}^{N}\mathbf{n}^{(r)},\end{aligned} \tag{I.201}$$

where the polar angles $\theta^{(r)}$, $\phi^{(r)}$ specifying the unit vector $\mathbf{n}^{(r)}$ are given in terms of the state amplitudes $a^{(r)}$, $b^{(r)}$ by (I.200).

If the beam consists of a very large number of particles having the directions $\mathbf{n}^{(r)}$ *oriented completely at random*, then the summation in (I.201) can obviously be replaced by an integration over the entire range of the (θ, ϕ)-angles. The components of **n**,

$$n_x = \sin\theta\cos\phi,$$

$$n_y = \sin\theta \sin\phi,$$

$$n_z = \cos\theta,$$

all give zero when integrated over

$$\int_0^{2\pi} d\phi \int_{-1}^{+1} d(\cos\theta).$$

Thus, the definition (I.201) yields *zero* polarization for a beam of spin $\frac{1}{2}$ particle having individual spin alignments **n** completely at random in space.

If the directions are not completely random, the vector addition of the N unit vectors in (I.201) will obviously yield a single vector, whose magnitude is given by $N\mathcal{P}$, where $\mathcal{P} \leqslant 1$. Clearly, $\mathcal{P}$ can be equal to 1 only if all the vectors point in the same direction. Therefore, the polarization $\boldsymbol{\mathcal{P}}$, as defined by (I.201), now has a magnitude $\mathcal{P}$ less than or equal to 1 and a resultant direction determined by the vector addition.

Density Matrix

A very convenient quantity for carrying out the averaging over all the particles just mentioned is the statistical density matrix. We denote it by ρ, and shall precisely define it in the next paragraph. In the discussion on polarization, the states drawing our attention are the spin states of the particles, and hence the density matrix for our purpose is a matrix in the spin-space. The reader should, however, remember that the concept of a density matrix is broad-based and not restricted to spin states. In the definition that follows, the notation is therefore kept general.

Let a set of basic states for the specification of the states of the individual particles be denoted by $|i\rangle$, $i = 1, 2, \ldots, n$. In terms of these states, we can specify the state $|\psi^{(r)}\rangle$ of the r-th particle as

$$|\psi^{(r)}\rangle = \sum_{i=1}^{N} C_i^{(r)}|i\rangle. \tag{I.202}$$

The operator ρ corresponding to the density matrix is now defined by

$$\rho = N^{-1} \sum_{r=1}^{N} |\psi^{(r)}\rangle\langle\psi^{(r)}| \tag{I.203a}$$

such that its matrix element between the basic states $\langle i|$ and $|j\rangle$ becomes

$$\begin{aligned}\langle i|\,\rho\,|j\rangle &= N^{-1} \sum_{r=1}^{N} \langle i\,|\,\psi^{(r)}\rangle\langle\psi^{(r)}\,|\,j\rangle \\ &= N^{-1} \sum_{r=1}^{N} C_i^{(r)} C_j^{(r)*}.\end{aligned} \tag{I.203b}$$

The final step follows from (I.202) by using the orthonormality of the basic states $|i\rangle$.

The usefulness of this density matrix is apparent from what follows on the average of the expectation value $\langle\Omega\rangle$ of any operator Ω denoting a physical observable. By definition, $\langle\Omega\rangle$ for the r-th particle is given by

$$\frac{\langle\psi^{(r)}|\,\Omega\,|\psi^{(r)}\rangle}{\langle\psi^{(r)}\,|\,\psi^{(r)}\rangle} = \frac{\sum_{i,j} C_i^{(r)} C_j^{(r)*} \langle j|\,\Omega\,|i\rangle}{\sum_i |C_i^{(r)}|^2}. \tag{I.204}$$

The denominator accounts for the fact that the state $\psi^{(r)}$ may not be normalized to unity.

However, we assume the normalization constant necessary for this purpose to be independent of r. Our task now is to carry out the averaging of the expectation value over all the particles in the beam. We thus obtain

$$\overline{\langle\Omega\rangle} = N^{-1} \sum_{r=1}^{N} \frac{\langle\psi^{(r)}|\,\Omega\,|\psi^{(r)}\rangle}{\langle\psi^{(r)}\mid\psi^{(r)}\rangle} = \frac{N^{-1} \sum\limits_{i,j} \sum\limits_{r=1}^{N} C_i^{(r)} C_j^{(r)*} \langle j|\,\Omega\,|i\rangle}{N^{-1} \sum\limits_{i} \sum\limits_{r=1}^{N} |C_i^{(r)}|^2}$$

$$= \frac{\sum\limits_{i,j} \langle i|\,\rho\,|j\rangle\langle j|\,\Omega\,|i\rangle}{\sum\limits_{i} \langle i|\,\rho\,|i\rangle} = \frac{\mathrm{Tr}\,\rho\Omega}{\mathrm{Tr}\,\rho}. \tag{I.205}$$

The first step here merely employs the definition of the average value. The second step makes use of (I.204) with the fact that the normalization of $\langle\psi^{(r)}\mid\psi^{(r)}\rangle$ in the denominator is independent of r. In the third step, the definition (I.203b) has been applied. In the final step, "Tr" denotes the trace, i.e., the sum of the diagonal elements of the corresponding matrices. The result (I.205) is very important, and it shows that a knowledge of the n^2 matrix elements of the density matrix ρ is enough to carry out a *statistical* averaging over all the particles of an ensemble.

In fact, the total number of independent matrix elements of ρ is even less than n^2. From (I.203b), it is obvious that

$$\langle j|\,\rho\,|i\rangle = \langle i|\,\rho\,|j\rangle^*,$$

i.e., the matrix ρ is Hermitean. Thus, the n diagonal matrix elements and half of the $n(n-1)$ nondiagonal matrix elements are really independent, so that, for carrying out a statistical average, the number of matrix elements we need to specify is

$$n + \tfrac{1}{2}n(n-1) = \tfrac{1}{2}n(n+1).$$

Further, we have already assumed $\langle\psi^{(r)}\mid\psi^{(r)}\rangle$ to be independent of r. Let

$$\langle\psi^{(r)}\mid\psi^{(r)}\rangle \equiv \sum_{i=1}^{n} |C_i^{(r)}|^2 = C,$$

and hence

$$\mathrm{Tr}\,\rho = \sum_{i=1}^{n} N^{-1} \sum_{r=1}^{N} |C_i^{(r)}|^2$$

$$= N^{-1} \sum_{r=1}^{N} \sum_{i=1}^{n} |C_i^{(r)}|^2 = C. \tag{I.206a}$$

In particular, if all the states $\psi^{(r)}$ are normalized to unity, then

$$\mathrm{Tr}\,\rho = 1, \tag{I.206b}$$

$$\overline{\langle\Omega\rangle} = \mathrm{Tr}\,\rho\Omega. \tag{I.206c}$$

The subsidiary condition (I.206a) or (I.206b) imposes a further restriction on the n diagonal elements of ρ.

Having introduced the general concepts of a density matrix, we now start specializing it to our problem of spin polarization in nucleon-nucleon scattering. We deal first with the simpler case of a beam of spin $\frac{1}{2}$ particle without bringing in the target nucleon for the present. The density matrix now is a 2×2 matrix in the spin-space, and hence we can, in general, express

12 (45-122/1977)

it in terms of the 2×2 unit matrix $\mathbb{1}$, and the three Pauli spin matrices σ_x, σ_y, σ_z of Appendix A (Section III). We introduce the general symbol S_μ ($\mu = 1, \ldots, 4$) to denote these four matrices. From the properties of the Pauli matrices given in Appendix A (Section III), we obtain

$$\mathrm{Tr}\; S_\mu S_\nu = 2\delta_{\mu\nu}. \tag{I.207}$$

The general expression of the density matrix is given by

$$\rho = \sum_{\mu=1}^{4} C_\mu S_\mu,$$

where the coefficients C_μ can be formally expressed in terms of ρ itself. We multiply both sides of this equation by S_ν and take the trace using the result (I.207). Then

$$C_\nu = \tfrac{1}{2}\, \mathrm{Tr}\; \rho S_\nu.$$

Therefore,

$$\begin{aligned} \rho &= \tfrac{1}{2} \sum_{\mu=1}^{4} (\mathrm{Tr}\; \rho S_\mu) S_\mu \\ &= \tfrac{1}{2}(\mathrm{Tr}\; \rho)\mathbb{1} + \tfrac{1}{2}(\mathrm{Tr}\; \rho \boldsymbol{\sigma})\cdot\boldsymbol{\sigma} \\ &= \tfrac{1}{2}(\mathrm{Tr}\; \rho)(\mathbb{1} + \overline{\langle\boldsymbol{\sigma}\rangle}\cdot\boldsymbol{\sigma}) \\ &= \tfrac{1}{2}(\mathrm{Tr}\; \rho)(\mathbb{1} + \boldsymbol{\mathscr{P}}\cdot\boldsymbol{\sigma}). \end{aligned} \tag{I.208}$$

In the second step, the four operators $\mathbb{1}$ and σ_x, σ_y, σ_z for S_μ have been explicitly used and the summation carried out over μ. In the third and final steps, the definitions (I.205) and (I.201), respectively, have been applied.

In the case of nucleon-nucleon scattering, both the incident and target particles have spin $\frac{1}{2}$, and hence the combined spin-space is four-dimensional. The density matrix ρ is therefore a 4×4 matrix and can, in general, be expanded in terms of the 16 matrices

$$\mathbb{1}, \quad \boldsymbol{\sigma}_1, \quad \boldsymbol{\sigma}_2, \quad \boldsymbol{\sigma}_1\boldsymbol{\sigma}_2, \tag{I.209}$$

where 1 and 2 label respectively the incident and target nucleons. These matrices have been obtained by a generalization of the aforementioned 2×2 case. Each of the vectors $\boldsymbol{\sigma}_1$, $\boldsymbol{\sigma}_2$ stands for three components, and the quantity $\boldsymbol{\sigma}_1\boldsymbol{\sigma}_2$ actually represents the nine components $\sigma_{1\alpha}\sigma_{2\beta}$, where each of α and β can be any of the components x, y, z; $\mathbb{1}$ is the four-dimensional unit matrix. The four-dimensional matrix corresponding to $\sigma_{1\alpha}$ is obtained by taking the direct product of this 2×2 Pauli matrix with the 2×2 unit matrix for the target nucleon. The same observation holds for any component of $\boldsymbol{\sigma}_2$. Finally, any component $\sigma_{1\alpha}\sigma_{2\beta}$ is represented by the direct product of the corresponding Pauli matrices. We do not require *explicit* matrix forms of these operators in our treatment. It will suffice to remember the result analogous to (I.207) for the new operators (I.209). We shall continue using the general notation S_μ to denote any one of them, but we must remember now that μ goes over the set of 16 (instead of four) operators. The new result for the trace, which follows from the trace of the components of $\boldsymbol{\sigma}_1$ and $\boldsymbol{\sigma}_2$, is given by

$$\mathrm{Tr}\; S_\mu S_\nu = 4\delta_{\mu\nu}, \tag{I.210}$$

and hence the general expression

$$\rho = \sum_{\mu=1}^{16} C_\mu S_\mu$$

now reduces in the usual way to

$$\rho = \tfrac{1}{4} \sum_{\mu=1}^{16} \mathrm{Tr}\,(\rho S_\mu) S_\mu \tag{I.211a}$$

$$= \tfrac{1}{4}(\mathrm{Tr}\,\rho)(\mathbb{1} + \mathscr{P}_1 \cdot \boldsymbol{\sigma}_1 + \mathscr{P}_2 \cdot \boldsymbol{\sigma}_2 + \sum_{\alpha,\beta} \overline{\langle \sigma_{1\alpha}\sigma_{2\beta} \rangle} \sigma_{1\alpha}\sigma_{2\beta}). \tag{I.211b}$$

In the last term, each of the summations over α, β goes over the x-, y-, z-component, $\mathscr{P}_1$ and $\mathscr{P}_2$ are respectively the polarizations of the incident and target nucleons, and the average value of the product $\sigma_{1\alpha}\sigma_{2\beta}$ appearing in the last term is called the polarization correlation function.

A relation such as (I.211b) is valid for the spin density matrix corresponding to the incident as well as the scattered two-nucleon states. It is also obvious that some terms of this expansion may be absent because of the lack of polarization of the target nucleons or incident nucleons or both.

According to the scattering theory (covered in Section 7), the coefficients $C_i^{(r)}$ of the expression (I.202) for the scattered outgoing wave are related to the corresponding coefficients in the expansion of the incident wave through the scattering amplitude M (which was found to be a matrix in the two-nucleon spin-space). It is therefore clear from the definition (I.203b) that the density matrix for the scattered state is related to that for the incident state through the same matrix M. We now proceed to obtain this relationship.

Density matrix for incident and scattered states The scattering theory was worked out in Section 7 by starting with the incident wave

$$\Psi_{\mathrm{inc}} = e^{ikz}\chi_m^S.$$

Corresponding to this incident wave, the scattered wave of spin projection m' was described as

$$r^{-1}e^{ikr}M_{m',m}^S(\theta, \phi)\chi_{m'}^S,$$

where M is the scattering amplitude. In general, the scattered wave was found to consist of waves of all spin projection, and hence given by the sum

$$\Psi_{\mathrm{scatt}} = r^{-1}e^{ikr} \sum_{m'} M_{m',m}^S \chi_{m'}^S. \tag{I.212}$$

Since our aim in this section is to consider polarization, we shall start with an incident wave in which the various spin states occur with *different* amplitudes a_m^S. Thus,

$$\Psi_{\mathrm{inc}} = e^{ikz} \sum_{S,m} a_m^S \chi_m^S. \tag{I.213}$$

The scattered wave corresponding to this incident wave is obviously obtained by multiplying (I.212) by a_m^S and carrying out the summation over the spin quantum numbers. Thus,

$$\Psi_{\mathrm{scatt}} = r^{-1}e^{ikr} \sum_{S,m} \sum_{m'} M_{m',m}^S a_m^S \chi_{m'}^S$$

$$= r^{-1}e^{ikr} \sum_{S,m'} b_{m'}^S \chi_{m'}^S, \tag{I.214a}$$

where

$$b^S_{m'} = \sum_m M^S_{m',m} a^S_m. \tag{I.214b}$$

Comparing (I.213) and (I.214a) with (I.202), we conclude that the basic states in the expansion of the incident wave are $e^{ikz}\chi^S_m$ with the coefficients a^S_m, whereas the basic states in the expansion of the outgoing scattered wave are $r^{-1}e^{ikr}\chi^S_m$ with the coefficients b^S_m. The functions e^{ikz} and $r^{-1}e^{ikr}$ merely take account of the spatial dependence of the incident and scattered waves. The information on the incident and scattered density matrices is contained, according to (I.203b), in the coefficients a^S_m and b^S_m. It should be noted that the summation label i of (I.202) now corresponds to the two quantum numbers S and m; the values (0, 0), (1, 1), (1, 0), and (1, −1) of these quantum numbers span the four-dimensional spin-space of the two nucleons. For the sake of simplicity, we use the compact notation i to denote the two quantum numbers (S, m) when we write, according to the definition (I.203b), the density matrices for the incident and scattered states as

$$\langle i|\, \rho_{\text{inc}}\, |j\rangle = N^{-1} \sum_{r=1}^{N} a_i^{(r)} a_j^{(r)*}, \tag{I.215a}$$

$$\langle i|\, \rho_{\text{scatt}}\, |j\rangle = N^{-1} \sum_{r=1}^{N} b_i^{(r)} b_j^{(r)*}, \tag{I.215b}$$

where N is the total number of particles and (r) labels the r-th particle. According to our simpler notation, the definition (I.214b) can be rewritten as

$$b_i = \sum_k M_{ik} a_k. \tag{I.216}$$

It should be noted that, although the summation over k here implies a sum over all the four values of (S, m), the scattering amplitude M itself does not have any component connecting states of different S-quantum number. Hence, this new equation is identical to (I.214b). Using (I.216) in (I.215b), we get

$$\begin{aligned}
\langle i|\, \rho_{\text{scatt}}\, |j\rangle &= N^{-1} \sum_{r=1}^{N} \sum_k M_{ik} a_k^{(r)} \sum_l M_{jl}^* a_l^{(r)*} \\
&= \sum_{k,l} M_{ik} \Big[N^{-1} \sum_{r=1}^{N} a_k^{(r)} a_l^{(r)*} \Big] (M^\dagger)_{lj} \\
&= \sum_{k,l} M_{ik} \langle k|\, \rho_{\text{inc}}\, |l\rangle (M^\dagger)_{lj} \\
&= \langle i|\, M \rho_{\text{inc}} M^\dagger\, |j\rangle.
\end{aligned}$$

In simplifying this expression, we have used the definition (I.215a). Thus, we have proved a very convenient result for later implementation, namely,

$$\rho_{\text{scatt}} = M \rho_{\text{inc}} M^\dagger. \tag{I.217}$$

The differential cross-section, which has been calculated in Section 7 in terms of the square modulus of the matrix elements of M, is now expressed in an alternative manner, using ρ_{scatt} and ρ_{inc}. From the expression (I.213), the flux of incident particles per unit area per unit time is given by (see the flux calculation in Section 7)

$$v \sum_{S,m} |a^S_m|^2 = v \sum_i |a_i|^2,$$

where v is the incident velocity, and i the abbreviated notation for (S, m). The statistical average of this flux over all the particles in the incident beam can be written as

$$N^{-1} \sum_{r=1}^{N} \sum_{i} |a_i^{(r)}|^2 = v \operatorname{Tr} \rho_{\text{inc}}.$$

In a similar manner, the statistical average of the outgoing flux through the area $r^2\, d\omega$ for the scattered state (I.214a) is

$$N^{-1} v\, d\omega \sum_{r=1}^{N} \sum_{i} |b_i^{(r)}|^2 = v\, d\omega \operatorname{Tr} \rho_{\text{scatt}}.$$

Therefore, the differential cross-section, according to the definition of Section 7, is given by

$$\frac{d\sigma}{d\omega} \equiv \sigma(\theta) = \frac{\operatorname{Tr} \rho_{\text{scatt}}}{\operatorname{Tr} \rho_{\text{inc}}}. \tag{I.218}$$

This is the second important relationship that will be useful for our study in this section.

Let us examine how the general result (I.218) agrees with the procedure followed in Section 7 for an unpolarized incident beam. Putting together (I.107), (I.108), and (I.109), we notice that the differential cross-section of an unpolarized beam was defined by

$$\frac{d\sigma}{d\omega} = \tfrac{1}{4} \sum_{S,m} \sum_{m'} |M^S_{m',m}|^2,$$

where $\frac{1}{4}$ is the probability of each spin state in the incident wave, i.e., the statistical average of the quantity $|a^S_m|^2$ was taken to be $\frac{1}{4}$, irrespective of the values of S and m. Let us make use of this fact in our general definition (I.218). Denoting the statistical average by an overhead bar, we get

$$\frac{d\sigma}{d\omega} = \frac{\sum_{S,m} \overline{|b^S_m|^2}}{\sum_{S,m} \overline{|a^S_m|^2}} = \sum_{S,m} \overline{\left|\sum_{m'} M^S_{m,m'} a^S_{m'}\right|^2}$$

$$= \tfrac{1}{4} \sum_{S,m} \sum_{m'} |M^S_{m,m'}|^2.$$

In simplifying this expression, we have replaced the statistical average of each $|a^S_m|^2$ in the denominator by $\frac{1}{4}$, and obtained unity by summing over (S, m). In the numerator, the square of each $|a^S_{m'}|^2$ has, when averaged over all the particles, been replaced by $\frac{1}{4}$, and the statistical average of the cross terms $a^{S*}_m a^S_{m'}$ has been assumed to be zero.

B. GENERAL EXPRESSION OF SCATTERING AMPLITUDE

The scattering amplitude derived in Section 7 is in terms of spherical harmonics and Clebsch-Gordon coefficients. We now express the same result in an alternative *formal* manner. Before we consider the general expression of M for nucleon-nucleon scattering, we examine a spin $\frac{1}{2}$ particle scattered on a zero spin target. This case is much simpler and demonstrates the principles involved in the method.

In the case of a spin $\frac{1}{2}$ particle scattered on a zero spin target, the spin-space is two-dimensional, and hence the scattering amplitude matrix M is expressible in terms of the four 2×2 matrices $\mathbb{1}$, σ_x, σ_y, σ_z, used earlier in this section. However, it has to be a scalar quantity (because its *magnitude* should not depend on the orientation of the coordinate frame), and we must therefore try to construct scalar expressions with these four operators and the kinematic variables for the scattering process. The latter are given by the incident and the outgoing momentum vectors $\mathbf{k}_i$ and $\mathbf{k}_f$. Each of these is a polar vector and, for elastic scattering, both have the same *magnitude* k. The angle between them is the scattering angle θ. We can use them to construct the different types of quantities

$$\begin{aligned}
&k_i^2 = k^2, \qquad k_f^2 = k^2, \qquad \mathbf{k}_i \cdot \mathbf{k}_f = k^2 \cos\theta && \text{(scalar);}\\
&\mathbf{P} = \frac{\mathbf{k}_i + \mathbf{k}_f}{|\mathbf{k}_i + \mathbf{k}_f|}, \qquad \mathbf{K} = \frac{\mathbf{k}_f - \mathbf{k}_i}{|\mathbf{k}_f - \mathbf{k}_i|} && \text{(vector);}\\
&\mathbf{n} = (\mathbf{k}_i \times \mathbf{k}_f)/|\mathbf{k}_i \times \mathbf{k}_f| && \text{(axial vector);}\\
&\text{second-rank tensor.}
\end{aligned} \tag{I.219a}$$

The second-rank tensor constructed with $\mathbf{k}_i$ and $\mathbf{k}_f$ has not been explicitly written down because it will not be able to combine with any term from $\mathbb{1}$ to $\boldsymbol{\sigma}$. As there are no independent bilinear combinations of the components of a *single* spin operator $\boldsymbol{\sigma}$ (see Section 3), we have only

$$\begin{aligned}
&\mathbb{1} \quad \text{(scalar),}\\
&\boldsymbol{\sigma} \quad \text{(axial vector).}
\end{aligned} \tag{I.219b}$$

We are permitted to take the scalar product of these two matrices with the corresponding quantities in (I.219a). Thus, the scattering amplitude is given by

$$M = g(k^2, \cos\theta)\mathbb{1} + h(k^2, \cos\theta)\boldsymbol{\sigma}\cdot\mathbf{n}, \tag{I.220}$$

where g and h are scalar functions of the kinematic variables k^2 and $\cos\theta$ appearing in the scalar expression of (I.219a). The vector $\mathbf{n}$, by definition, is the unit vector perpendicular to the plane of $\mathbf{k}_i$, $\mathbf{k}_f$, i.e., the scattering plane.

It should be noted that the requirement of a scalar expression of M guarantees its invariance under the rotation and reflection of the coordinate frame. Since the scattering potential V has the additional property of time-reversal invariance, we should ascertain whether or not (I.220) satisfies this requirement. We know that $\boldsymbol{\sigma}$, being an angular momentum, changes sign when $t \to -t$. Under the time-reversal operation, the particle starts moving from the final state towards the initial state, i.e., $\mathbf{k}_f \to -\mathbf{k}_i$ and $\mathbf{k}_i \to -\mathbf{k}_f$. Thus,

$$\mathbf{n} \to (\mathbf{k}_f \times \mathbf{k}_i)/|\mathbf{k}_f \times \mathbf{k}_i| = -\mathbf{n}.$$

Therefore, (I.220) satisfies the time-reversal invariance.

We now extend the foregoing considerations to the case of nucleon-nucleon scattering. Since we have two spin operators $\boldsymbol{\sigma}_1$, $\boldsymbol{\sigma}_2$, we can construct all the expressions given by (I.10). The list (I.219a) remains unchanged, but by $\mathbf{k}_i$ and $\mathbf{k}_f$ we now understand the initial and final momenta of two-nucleon scattering in the centre-of-mass coordinate system. Further, the second-rank tensors need to be explicitly written. Since the tensor $\mathcal{S}_{\alpha\beta}$ of (I.10) is bilinear in $\boldsymbol{\sigma}$, it does not change sign under reflection. Hence, we must construct second-rank tensors of $\mathbf{n}$ with $\mathbf{n}$, of $\mathbf{P}$ with $\mathbf{P}$, of $\mathbf{K}$ with $\mathbf{K}$, or of $\mathbf{P}$ with $\mathbf{K}$. Combining $\mathbf{n}$ with either $\mathbf{K}$ or $\mathbf{P}$ gives a tensor that obviously changes sign under reflection. We also examine the behaviour of M under time-reversal. The tensor $\mathcal{S}_{\alpha\beta}$ is invariant under this operation. Under the same operation,

$$\mathbf{P} = -\mathbf{P}, \qquad \mathbf{K} = \mathbf{K}, \qquad \mathbf{n} = -\mathbf{n}.$$

Therefore, the second-rank tensor constructed with $\mathbf{K}$ and $\mathbf{P}$ changes sign under time-reversal and is unable to combine with $\mathcal{S}_{\alpha\beta}$. The only admissible second-rank tensors are thus given by

$$\mathcal{K}_{\alpha\beta} = K_\alpha K_\beta - \tfrac{1}{3}\delta_{\alpha\beta},$$
$$\mathcal{P}_{\alpha\beta} = P_\alpha P_\beta - \tfrac{1}{3}\delta_{\alpha\beta}, \tag{I.221}$$
$$\mathcal{N}_{\alpha\beta} = n_\alpha n_\beta - \tfrac{1}{3}\delta_{\alpha\beta}.$$

The square of each of the unit vectors **K**, **P**, **n** is unity, and hence the corresponding factor is absent from the $\delta_{\alpha\beta}$-terms. The scalar products of these tensors with the spin tensor of (I.10) yield (in the manner of Section 3) terms in M of the type

$$\boldsymbol{\sigma}_1\cdot\mathbf{A}\boldsymbol{\sigma}_2\cdot\mathbf{A} - \tfrac{1}{3}\boldsymbol{\sigma}_1\cdot\boldsymbol{\sigma}_2, \tag{I.222a}$$

where **A** = **K**, **P**, **n**.

The axial vectors of (I.10) can combine with **n** of (I.219a). But since $(\boldsymbol{\sigma}_1 \times \boldsymbol{\sigma}_2)$ does not change sign under time-reversal, whereas **n** does, the only admissible terms of M obtained in this way are of the type

$$(\boldsymbol{\sigma}_1 \pm \boldsymbol{\sigma}_2)\cdot\mathbf{n}. \tag{I.222b}$$

Finally, scalar terms of (I.219a) and (I.10) yield terms of the type

$$(\mathbb{1} \quad \text{or} \quad \boldsymbol{\sigma}_1\cdot\boldsymbol{\sigma}_2)(\text{function of } k^2 \text{ and } \cos\theta). \tag{I.222c}$$

Before putting together (I.222a), (I.222b), and (I.222c) for the general expression of M, we point out another very important fact. It is clear from the definition of the unit vectors **K**, **P**, and **n**, given in (I.219a), that both **K** and **P** belong to the plane of $\mathbf{k}_i$ and $\mathbf{k}_f$ (the scattering plane), and **n** is perpendicular to this plane. Further, in the case of elastic scattering, $k_i = k_f = k$, and hence

$$\mathbf{K}\cdot\mathbf{P} = 0,$$
$$\mathbf{P}\times\mathbf{K} = \mathbf{n}.$$

Thus, **K** and **P** are perpendicular to each other, and a right-handed screw turned from **P** to **K** goes towards the normal direction **n**. Therefore, the directions **P**, **K**, and **n** define a set of right-handed orthogonal coordinate axes, and we can label them the x'-, y'-, z'-direction of a new coordinate frame. From this it is clear that all the three terms of the type

$$\boldsymbol{\sigma}_1\cdot\mathbf{A}\boldsymbol{\sigma}_2\cdot\mathbf{A}$$

with **A** = **K**, **P**, **n** are not independent. Their sum is equal to

$$\sigma_{1x'}\sigma_{2x'} + \sigma_{1y'}\sigma_{2y'} + \sigma_{1z'}\sigma_{2z'},$$

i.e., $\boldsymbol{\sigma}_1\cdot\boldsymbol{\sigma}_2$. We can either accept all three of these terms and omit the $(\boldsymbol{\sigma}_1\cdot\boldsymbol{\sigma}_2)$-term of (I.222c) or take any two of these terms and include a $(\boldsymbol{\sigma}_1\cdot\boldsymbol{\sigma}_2)$-term in M. Let us adopt the first alternative and write the general expression of M as

$$\begin{aligned} M = {} & g_0(k^2, \cos\theta)\mathbb{1} + f(k^2, \cos\theta)(\boldsymbol{\sigma}_1 - \boldsymbol{\sigma}_2)\cdot\mathbf{n} \\ & + h_0(k^2, \cos\theta)(\boldsymbol{\sigma}_1 + \boldsymbol{\sigma}_2)\cdot\mathbf{n} + h_K(k^2, \cos\theta)(\boldsymbol{\sigma}_1\cdot\mathbf{K})(\boldsymbol{\sigma}_2\cdot\mathbf{K}) \\ & + h_P(k^2, \cos\theta)(\boldsymbol{\sigma}_1\cdot\mathbf{P})(\boldsymbol{\sigma}_2\cdot\mathbf{P}) + h_n(k^2, \cos\theta)(\boldsymbol{\sigma}_1\cdot\mathbf{n})(\boldsymbol{\sigma}_2\cdot\mathbf{n}). \end{aligned}$$

At this point, we would wish to incorporate another symmetry requirement in M, which follows from the corresponding symmetry of the two-body potential. The symmetry we contemplate is the invariance under the exchange of the two nucleons. This, of course, presupposes that we are using the language of isospin to describe the two nucleons, and the nucleons are then completely equivalent in our description.

Each of the momenta $\mathbf{k}_i$ and $\mathbf{k}_f$, being the *relative* momentum of the two nucleons, must change sign under an exchange of the nucleons. Therefore, $\mathbf{K} \to -\mathbf{K}$, $\mathbf{P} \to -\mathbf{P}$ under this symmetry operation. However, $\mathbf{n}$ being bilinear in $\mathbf{k}_i$ and $\mathbf{k}_f$ does not change, i.e., $\mathbf{n} \to \mathbf{n}$. An examination of the different terms of M now reveals that the f-term changes sign under the exchange of the two nucleons, whereas all the other terms remain unchanged. We therefore discard the f-term and rewrite the general expression of M as

$$M = g_0 \mathbb{1} + h_0(\boldsymbol{\sigma}_1 + \boldsymbol{\sigma}_2)\cdot\mathbf{n} + h_K \boldsymbol{\sigma}_1\cdot\mathbf{K}\boldsymbol{\sigma}_2\cdot\mathbf{K} + h_P \boldsymbol{\sigma}_1\cdot\mathbf{P}\boldsymbol{\sigma}_2\cdot\mathbf{P} + h_n \boldsymbol{\sigma}_1\cdot\mathbf{n}\boldsymbol{\sigma}_2\cdot\mathbf{n}. \tag{I.223}$$

Each of the coefficients g, h here is a function of k^2 and $\cos\theta$, and we must remember that a general expression of this kind is valid separately for the isospin ($T = 0$)- and ($T = 1$)-state scattering amplitude. Therefore, the n-p and p-p systems need respectively ten and five complex quantities (the g- and h-coefficients) to describe the scattering amplitude at any energy and angle (k^2 and $\cos\theta$). However, all the ten real numbers required to describe the five complex quantities for each T cannot be determined from scattering measurements. Since the square modulus of the scattering amplitude occurs in any measurable quantity, an overall phase parameter is always left undetermined. Thus, nine independent scattering measurements at every energy and angle are required to enable us to write down M for p-p scattering to the extent of an overall phase. However, all the nine functions of angle obtained by performing experiments at all angles are not independent because the scattering amplitude matrix M has to satisfy certain restrictions, depending on the unitary property of the $\mathcal{S}$-matrix. Taking these constraints into account, we can show that *five* independent types of measurement are needed to specify the scattering amplitude for a given value of energy. Such measurements will then, in principle, allow a unique determination of the M-matrix or, equivalently, the phase shifts. In practice, however, there may be additional constraints, depending on the value of the experimental energy. For example, at very low energy there is a constraint on the angular momentum, and only the S-state ($l = 0$) scattering is of consequence. Therefore, only one value of the phase shift suffices for each isospin state. Further, the variation of this single phase shift with energy (as long as the energy is low) is fairly well described in terms of two parameters, namely, effective range and scattering length.

Finally, we try to express (I.223) in a more compact manner. Collecting the terms that are independent of $\boldsymbol{\sigma}_1$, we define their sum $g(k^2, \cos\theta)$ as

$$g(k^2, \cos\theta) = g_0(k^2, \cos\theta)\mathbb{1}_2 + h_0\boldsymbol{\sigma}_2\cdot\mathbf{n}. \tag{I.224a}$$

Similarly, the coefficients of all terms dependent on $\boldsymbol{\sigma}_1$ are put together, and we define

$$\mathbf{h}(k^2, \cos\theta) = h_0(k^2, \cos\theta)\mathbf{n}\mathbb{1}_2 + h_K(k^2, \cos\theta)(\boldsymbol{\sigma}_2\cdot\mathbf{K})\mathbf{K} + h_P(k^2, \cos\theta)(\boldsymbol{\sigma}_2\cdot\mathbf{P})\mathbf{P} + h_n(k^2, \cos\theta)(\boldsymbol{\sigma}_2\cdot\mathbf{n})\mathbf{n}. \tag{I.224b}$$

With these definitions, M is clearly expressible as

$$M = g(k^2, \cos\theta)\mathbb{1}_1 + \mathbf{h}(k^2, \cos\theta)\cdot\boldsymbol{\sigma}_1. \tag{I.225}$$

Each of the expressions (I.224a) and (I.224b) is a matrix in the space of the target nucleon (2). The subscripts 2 and 1 on the unit matrices in (I.224) and (I.225) indicate that these are 2×2 unit matrices in the spin-space of the corresponding nucleons. The expression for $\mathbf{h}$ is not only a 2×2 matrix in the spin-space of the target nucleon, but each term in it also behaves as an axial vector quantity under the rotation and reflection of the coordinate axes. This explains the fact that in (I.225) the second term becomes a scalar upon multiplication by the axial vector $\boldsymbol{\sigma}_1$. The *formal* similarity of the expression (I.225) and the expression (I.220) for a spin $\frac{1}{2}$ particle scattered on a zero spin target should be noted.

C. PRINCIPLES OF POLARIZATION EXPERIMENTS

We are now in a position to work out the expressions that are necessary for a complete understanding of the different types of polarization measurement.

Differential Cross-Section of Polarized and Unpolarized Beams

First, using (I.218) for $\sigma(\theta)$, then (I.217) for the ρ_{scatt} of the numerator, and finally (I.211a) to replace the ρ_{inc} of the numerator, we obtain

$$\begin{aligned}\sigma(\theta) &= \frac{\text{Tr}\ \rho_{\text{scatt}}}{\text{Tr}\ \rho_{\text{inc}}} = \frac{\text{Tr}\ M\rho_{\text{inc}}M^\dagger}{\text{Tr}\ \rho_{\text{inc}}} \\ &= \frac{1}{4\ \text{Tr}\ \rho_{\text{inc}}} \sum_\mu (\text{Tr}\ \rho_{\text{inc}}\, S_\mu)(\text{Tr}\ MS_\mu M^\dagger) \\ &= \tfrac{1}{4} \sum_\mu \overline{\langle S_\mu \rangle}_{\text{inc}} (\text{Tr}\ MS_\mu M^\dagger). \end{aligned} \tag{I.226a}$$

In the final step, we have used (I.205) to obtain the statistical average of an expectation value.

Using the sixteen operators S_μ explicitly in this expression, we derive the result

$$\begin{aligned}\sigma(\theta) &= \tfrac{1}{4}\ \text{Tr}\ MM^\dagger + \tfrac{1}{4}\mathcal{P}_1^{\text{inc}}\boldsymbol{\cdot}(\text{Tr}\ M\boldsymbol{\sigma}_1 M^\dagger) + \tfrac{1}{4}\mathcal{P}_2^{\text{inc}}\boldsymbol{\cdot}(\text{Tr}\ M\boldsymbol{\sigma}_2 M^\dagger) \\ &\quad + \tfrac{1}{4} \sum_{\alpha,\beta} \overline{\langle \sigma_{1\alpha}\sigma_{2\beta}\rangle}\ (\text{Tr}\ M\sigma_{1\alpha}\sigma_{2\beta}M^\dagger), \end{aligned} \tag{I.226b}$$

where $\mathcal{P}_1^{\text{inc}}$ and $\mathcal{P}_2^{\text{inc}}$ denote the polarization of the two nucleons in the initial state (i.e., the state Ψ_{inc}). (I.226b) is the general expression for the differential cross-section when both the incoming and target nucleons may have a polarization. This expression can be easily specialized to the cases described hereafter.

Case 1 When both the incoming and target nucleons are unpolarized, we have

$$\mathcal{P}_1^{\text{inc}} = \mathcal{P}_2^{\text{inc}} = 0, \qquad \overline{\langle \sigma_{1\alpha}\sigma_{2\beta}\rangle} = 0.$$

Representing the differential cross-section for this special case by $\sigma_0(\theta)$, we obtain, from (I.226b),

$$\begin{aligned}\sigma_0(\theta) &= \tfrac{1}{4}\ \text{Tr}\ MM^\dagger \\ &= \tfrac{1}{4}\ \text{Tr}\ [(g\mathbb{1}_1 + \mathbf{h}\boldsymbol{\cdot}\boldsymbol{\sigma}_1)(\mathbb{1}_1 g^\dagger + \boldsymbol{\sigma}_1\boldsymbol{\cdot}\mathbf{h}^\dagger)] \\ &= \tfrac{1}{2}\ \text{Tr}_2\ (gg^\dagger + \mathbf{h}\boldsymbol{\cdot}\mathbf{h}^\dagger). \end{aligned} \tag{I.227}$$

Here we have first used the expression (I.225) for M, and its Hermitean conjugate. We have then evaluated the traces of the spin operators belonging to nucleon 1, using the standard results given in Appendix A (Section III). In the resultant expression, we have yet to evaluate the trace in the space of nucleon 2, as indicated by Tr_2, because the original trace was in the four-dimensional combined spin-space of the two nucleons.

To further simplify (I.227), we make use of (I.224) and recall that $\boldsymbol{\sigma}_2\boldsymbol{\cdot}\mathbf{P} = \sigma_{2x'}$, $\boldsymbol{\sigma}_2\boldsymbol{\cdot}\mathbf{K} = \sigma_{2y'}$, and $\boldsymbol{\sigma}_2\boldsymbol{\cdot}\mathbf{n} = \sigma_{2z'}$. Therefore, in evaluating Tr_2, we can again use the standard results

13 (45–122/1977)

given in Appendix A (Section III). In this way, we obtain

$$\sigma(\theta) = |g_0|^2 + 2|h_0|^2 + |h_K|^2 + |h_P|^2 + |h_n|^2. \tag{I.228}$$

The right-hand side of this equation is a function of k^2 and cos θ. There is no dependence of this expression on the azimuthal angle ϕ of the direction $\mathbf{k}_f$ of scattering. This is a consequence of the lack of polarization of the incident beam.

Case 2 When the target nucleon is unpolarized, but the incident nucleon has a polarization, we have

$$\mathscr{P}_2^{\text{inc}} = 0, \qquad \langle\overline{\sigma_{1\alpha}\sigma_{2\beta}}\rangle = 0.$$

Therefore (I.226b) yields

$$\begin{aligned}\sigma(\theta) &= \tfrac{1}{4}\,\text{Tr}\, MM^\dagger + \tfrac{1}{4}\mathscr{P}_1^{\text{inc}}\boldsymbol{\cdot}(\text{Tr}\, M\boldsymbol{\sigma}_1 M^\dagger)\\ &= \sigma_0(\theta)(1 + \mathscr{P}_1^{\text{inc}}\boldsymbol{\cdot}\mathbf{a}),\end{aligned} \tag{I.229a}$$

where $\mathbf{a}$ is given by

$$\mathbf{a} = \frac{\text{Tr}\, M\boldsymbol{\sigma}_1 M^\dagger}{\text{Tr}\, MM^\dagger}. \tag{I.229b}$$

To simplify (I.229b), we need the value of Tr $M\boldsymbol{\sigma}_1 M^\dagger$, which is evaluated, as before, in two steps: first, Tr_1 is taken in the spin-space of nucleon 1, using M and $M^\dagger$ from (I.225); and then Tr_2 of the resultant expression is evaluated. After completing the first step, we obtain

$$\text{Tr}\, M\boldsymbol{\sigma}_1 M^\dagger = 2\,\text{Tr}_2\,(g\mathbf{h}^\dagger + \mathbf{h}g^\dagger - i\mathbf{h}\times\mathbf{h}^\dagger). \tag{I.230}$$

The term $(\mathbf{h}\times\mathbf{h}^\dagger)$ arises from the trace of $(\mathbf{h}\boldsymbol{\cdot}\boldsymbol{\sigma}_1)\boldsymbol{\sigma}_1(\mathbf{h}^\dagger\boldsymbol{\cdot}\boldsymbol{\sigma}_2)$, the detailed evaluation of which is left to the reader [see Problem 4 (Chapter I)]. It can be further verified that $\text{Tr}_2\,(\mathbf{h}\times\mathbf{h}^\dagger) = 0$. Using (I.224) and evaluating $\text{Tr}_2\,(g\mathbf{h}^\dagger + \mathbf{h}g^\dagger)$ in detail, we have

$$\sigma(\theta) = \sigma_0(\theta)(1 + a\mathscr{P}_1^{\text{inc}}\boldsymbol{\cdot}\mathbf{n}), \tag{I.231a}$$

where

$$a = \frac{|\text{Tr}\, M\boldsymbol{\sigma}_1 M^\dagger|}{\text{Tr}\, MM^\dagger} \tag{I.231b}$$

$$= \frac{2\,\text{Re}\,(g_0 + h_n)h_0^*}{|g_0|^2 + 2|h_0|^2 + |h_K|^2 + |h_P|^2 + |h_n|^2}. \tag{I.231c}$$

The quantity a is called the asymmetry parameter for the present. The reason for this nomenclature becomes obvious from the discussion that follows.

We recall that $\mathbf{n}$ is the unit vector perpendicular to the plane of $\mathbf{k}_i$ and $\mathbf{k}_f$. The direction of $\mathbf{k}_i$ is the z-axis, and $\mathbf{k}_f$ makes an angle θ with this direction (Fig. I.7). It is possible to rotate $\mathbf{k}_f$ around $\mathbf{k}_i$ on the surface of a cone, keeping the angle θ fixed. This amounts to taking the ϕ-angle of the direction $\mathbf{k}_f$ through its entire range, i.e., 0 to 2π. While rotating $\mathbf{k}_f$, the direction of $\mathbf{n}$ keeps changing continuously. Therefore, for a fixed θ, the direction $\mathbf{n}$ is related to the ϕ-angle of $\mathbf{k}_f$. Let us consider the very simple cases of $\mathbf{k}_f$ pointing "Left" and "Right", with respect to the incident direction $\mathbf{k}_i$, as shown in Fig. I.7. Both the directions $\mathbf{k}_f$ and $\mathbf{k}_i$ are assumed to be contained in the plane of the paper. In the case of $\mathbf{k}_f$ in the left direction, a *right-handed* screw driven from the direction $\mathbf{k}_i$ towards $\mathbf{k}_f$ moves upwards; therefore, the corresponding $\mathbf{n}$ is perpendicular to the plane of the paper, and it points upwards. On the other hand, in the case of $\mathbf{k}_f$ in the right direction, a *right-handed* screw driven from $\mathbf{k}_i$ towards $\mathbf{k}_f$ moves downwards; hence, $\mathbf{n}$ is still perpendicular to the plane of the paper, but it points

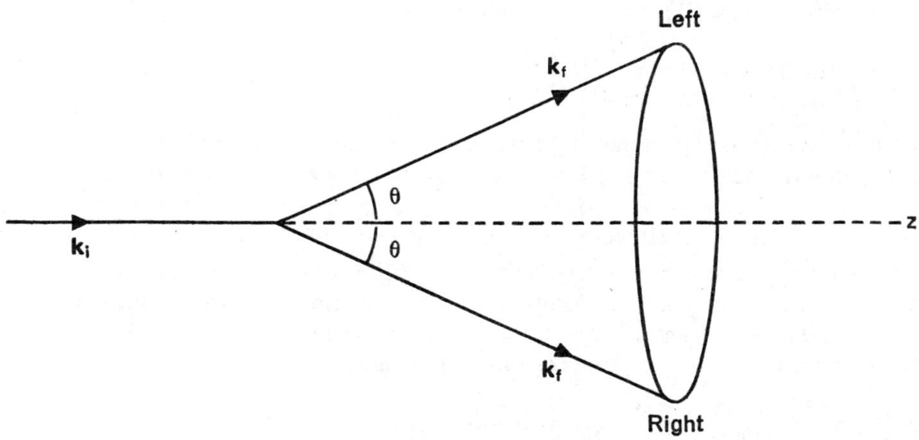

Fig. I.7 Set-up for measuring left-right asymmetry in scattering of polarized nucleons.

downwards. Therefore, the directions $\mathbf{n}$ in the two cases are exactly opposite to each other, and hence, according to (I.231a), the second term in the expressions of $\sigma_R(\theta)$ and $\sigma_L(\theta)$ (R denoting right, and L left) are exactly equal in magnitude, but opposite in sign. If $\mathbf{n}$ describes the *upward* normal, then

$$\sigma_L(\theta) = \sigma_0(\theta)(1 + a\mathcal{P}_1^{\text{inc}}\cdot\mathbf{n}),$$

$$\sigma_R(\theta) = \sigma_0(\theta)(1 - a\mathcal{P}_1^{\text{inc}}\cdot\mathbf{n}).$$

Therefore,

$$\frac{\sigma_L(\theta) - \sigma_R(\theta)}{\sigma_L(\theta) + \sigma_R(\theta)} = a\mathcal{P}_1^{\text{inc}}\cdot\mathbf{n}. \tag{I.232}$$

For a known polarization of the incident beam, this expression enables us to determine the asymmetry parameter a. It is clear from the foregoing details that a nonvanishing value of a gives rise to the azimuthal asymmetry in the scattering of a polarized beam, and hence the nomenclature. The measured asymmetry is proportional to the asymmetry parameter a of the scatterer, and also to the component of the incident polarization along the normal to the scattering plane. Therefore, if the scattering plane contains the direction of incident polarization, no asymmetry is observed. It is further obvious that the notation $\sigma(\theta)$ on the left-hand side of (I.231a) is a little misleading because it conceals the ϕ-dependence of the quantity. We, however, choose to keep the notation simple, at the risk of some confusion.

The foregoing derivation can also be specialized to the case of nucleon scattering on a spinless target. The expression (I.226a) is obviously valid, but the summation μ now goes over the four operators $\mathbb{1}$ and $\boldsymbol{\sigma}$ of the nucleon. Therefore, (I.226b) changes to the simpler expression

$$\sigma(\theta) = \tfrac{1}{2}\,\text{Tr}\,MM^\dagger + \tfrac{1}{2}\mathcal{P}^{\text{inc}}\cdot(\text{Tr}\,M\boldsymbol{\sigma}M^\dagger), \tag{I.233}$$

where the trace is in the two-dimensional spin-space of the nucleon. For the matrix M, we now have to use the simple expression (I.220). Corresponding to the two cases treated earlier for nucleon-nucleon scattering, we now have

$$\sigma_0(\theta) = \tfrac{1}{2}\,\mathrm{Tr}\, MM^\dagger = |g|^2 + |h|^2 \qquad \text{(case 1—unpolarized incident nucleon)}, \qquad \text{(I.234)}$$

$$\sigma(\theta) = \sigma_0(\theta)(1 + a\mathcal{P}^{\mathrm{inc}}\cdot\mathbf{n}) \qquad \text{(case 2—polarized incident nucleon)} \qquad \text{(I.235a)}$$

with

$$a = \frac{2\,\mathrm{Re}\, gh^*}{|g|^2 + |h|^2}. \qquad \text{(I.235b)}$$

As before, the asymmetry parameter at an angle θ can be determined from the left-right azimuthal asymmetry at that angle; the expression of the asymmetry is still given by (I.232).

Polarization as a Result of Scattering

We now derive an expression for the polarization of the scattered nucleon beam. Since we are interested in the polarization of the beam *after* the scattering has taken place, we must use, in the definition (I.205), the density matrix for the final state, i.e., $\rho_{\mathrm{scatt}} = M\rho_{\mathrm{inc}}M^\dagger$. Therefore, the expression we obtain for the polarization of the scattered beam is

$$\mathcal{P}_1^{\mathrm{scatt}} = \frac{\mathrm{Tr}\,(M\rho_{\mathrm{inc}}M^\dagger\boldsymbol{\sigma}_1)}{\mathrm{Tr}\,\rho_{\mathrm{inc}}} \times \frac{\mathrm{Tr}\,\rho_{\mathrm{inc}}}{\mathrm{Tr}\,\rho_{\mathrm{scatt}}} = [\sigma(\theta)]^{-1}\frac{\mathrm{Tr}\,(M\rho_{\mathrm{inc}}M^\dagger\boldsymbol{\sigma}_1)}{\mathrm{Tr}\,\rho_{\mathrm{inc}}}$$

or

$$\sigma(\theta)\mathcal{P}_1^{\mathrm{scatt}} = \frac{1}{4\,\mathrm{Tr}\,\rho_{\mathrm{inc}}}\sum_\mu (\mathrm{Tr}\,\rho_{\mathrm{inc}}S_\mu)(\mathrm{Tr}\, MS_\mu M^\dagger\boldsymbol{\sigma}_1).$$

In the final step, we have used the expression (I.211a) for ρ_{inc}. Using the definition of $\overline{\langle S_\mu\rangle}$ in the foregoing equation, we get

$$\sigma(\theta)\mathcal{P}_1^{\mathrm{scatt}} = \tfrac{1}{4}\sum_\mu \overline{\langle S_\mu\rangle}_{\mathrm{inc}}(\mathrm{Tr}\, MS_\mu M^\dagger\boldsymbol{\sigma}_1)$$

$$= \tfrac{1}{4}\,\mathrm{Tr}\, MM^\dagger\boldsymbol{\sigma}_1 + \tfrac{1}{4}\mathcal{P}_1^{\mathrm{inc}}\cdot\mathrm{Tr}\, M\boldsymbol{\sigma}_1M^\dagger\boldsymbol{\sigma}_1. \qquad \text{(I.236)}$$

In the final step, we have used the explicit forms of S_μ, and dropped the terms containing $\mathcal{P}_2^{\mathrm{inc}}$ and $\overline{\langle\sigma_{1\alpha}\sigma_{2\beta}\rangle}$ [see (I.226b)] on the understanding that we shall always consider initially *unpolarized targets*.

We first specialize (I.236) to the case of an unpolarized incident beam, i.e., $\mathcal{P}_1^{\mathrm{inc}} = 0$. In this case, $\sigma(\theta)$ reduces to $\sigma_0(\theta)$ and we obtain

$$\sigma_0(\theta)\mathcal{P}_1^{\mathrm{scatt}} = \tfrac{1}{4}\,\mathrm{Tr}\, MM^\dagger\boldsymbol{\sigma}_1$$

or

$$\mathcal{P}_1^{\mathrm{scatt}} = \frac{\mathrm{Tr}\, MM^\dagger\boldsymbol{\sigma}_1}{\mathrm{Tr}\, MM^\dagger}. \qquad \text{(I.237a)}$$

$\mathrm{Tr}\, MM^\dagger\boldsymbol{\sigma}_1$ can be evaluated in the same way as $\mathrm{Tr}\, M\boldsymbol{\sigma}_1M^\dagger$. It is easy to verify by such a direct evaluation that

$$\mathrm{Tr}\, MM^\dagger\boldsymbol{\sigma}_1 = 2\,\mathrm{Tr}_2\,(g\mathbf{h}^\dagger + \mathbf{h}g^\dagger + i\mathbf{h}\times\mathbf{h}^\dagger). \qquad \text{(I.237b)}$$

But since the last term of (I.237b) is zero, we obtain the very important result

$$\mathrm{Tr}\, MM^\dagger\boldsymbol{\sigma}_1 = \mathrm{Tr}\, M\boldsymbol{\sigma}_1M^\dagger.$$

A comparison of (I.229b) and (I.237a) therefore reveals that $\mathcal{P}_1^{\mathrm{scatt}}$ and $\mathbf{a}$ are identical. If we write

$$\mathcal{P}_1^{\mathrm{scatt}} = \mathcal{P}\mathbf{n}, \qquad \text{(I.238)}$$

then the magnitude $\mathscr{P}$ of the polarization is given by the asymmetry parameter a of (I.231b) and (I.231c).

The result (I.238) also holds for the scattering of a nucleon on a zero spin target. In this case, $\mathscr{P}$ is equal to (I.235b) for the asymmetry parameter.

We next consider the more general case described by (I.236), where the initial beam has a polarization $\mathscr{P}_1^{\text{inc}}$. According to (I.237a) and (I.238),

$$\tfrac{1}{4}\,\text{Tr}\; MM^\dagger\boldsymbol{\sigma}_1 = \sigma_0(\theta)\mathscr{P}_1^{\text{scatt}} = \sigma_0(\theta)\mathscr{P}\mathbf{n}. \tag{I.239}$$

We next simplify the second term of (I.236) by explicitly evaluating the trace. A straightforward evaluation of Tr_1 yields

$$\tfrac{1}{4}\mathscr{P}_1^{\text{inc}}\boldsymbol{\cdot}\text{Tr}\;(M\boldsymbol{\sigma}_1M^\dagger\boldsymbol{\sigma}_1) = \tfrac{1}{2}\,\text{Tr}_2\,[\mathscr{P}_1^{\text{inc}}(gg^\dagger - \mathbf{h}\boldsymbol{\cdot}\mathbf{h}^\dagger) + 2\,\text{Im}\,\{(\mathscr{P}_1^{\text{inc}}\times\mathbf{h})g^\dagger\} + 2\,\text{Re}\,\{(\mathscr{P}_1^{\text{inc}}\boldsymbol{\cdot}\mathbf{h})\mathbf{h}^\dagger\}]. \tag{I.240}$$

Here Re and Im denote real and imaginary parts of the respective quantities. $\frac{1}{2}\,\text{Tr}_2$ of $gg^\dagger$ and $\mathbf{h}\boldsymbol{\cdot}\mathbf{h}^\dagger$ were evaluated when simplifying (I.227), and added to obtain (I.228). Now, subtracting the two, we get

$$\tfrac{1}{2}\mathscr{P}_1^{\text{inc}}\,\text{Tr}_2\,(gg^\dagger - \mathbf{h}\boldsymbol{\cdot}\mathbf{h}^\dagger) = [|g_0|^2 - (|h_K|^2 + |h_P|^2 + |h_n|^2)]\mathscr{P}_1^{\text{inc}}. \tag{I.241a}$$

Evaluating Tr_2 in a similar manner, we obtain

$$\tfrac{1}{2}\,\text{Tr}_2\,\mathbf{h}g^\dagger = (h_0g_0^* + h_0^*h_n)\mathbf{n},$$

and hence

$$\tfrac{1}{2}\,\text{Tr}_2\,[2\,\text{Im}\,(\mathscr{P}_1^{\text{inc}}\times\mathbf{h})g^\dagger] = 2\,\text{Im}\,[h_0^*(h_n - g_0)](\mathscr{P}_1^{\text{inc}}\times\mathbf{n}). \tag{I.241b}$$

Similarly,

$$\tfrac{1}{2}\,\text{Tr}_2\,\mathbf{h}\mathbf{h}^\dagger = (|h_0|^2 + |h_n|^2)\mathbf{n}\mathbf{n} + |h_K|^2\mathbf{K}\mathbf{K} + |h_P|^2\mathbf{P}\mathbf{P}.$$

Therefore,

$$\tfrac{1}{2}\,\text{Tr}_2\,[2\,\text{Re}\,(\mathscr{P}_1^{\text{inc}}\boldsymbol{\cdot}\mathbf{h})\mathbf{h}^\dagger] = 2(\mathbf{n}\boldsymbol{\cdot}\mathscr{P}_1^{\text{inc}})(|h_0|^2 + |h_n|^2)\mathbf{n} + 2|h_K|^2(\mathbf{K}\boldsymbol{\cdot}\mathscr{P}_1^{\text{inc}})\mathbf{K} + 2|h_P|^2(\mathbf{P}\boldsymbol{\cdot}\mathscr{P}_1^{\text{inc}})\mathbf{P}. \tag{I.241c}$$

The third term on the right-hand side of (I.240) has been expressed in (I.241c) in terms of the unit vectors **K**, **P**, and **n**. It is quite trivial to express (I.241a) and (I.241b) also in terms of these unit vectors. In (I.241a), we write

$$\mathscr{P}_1^{\text{inc}} = (\mathscr{P}_1^{\text{inc}}\boldsymbol{\cdot}\mathbf{K})\mathbf{K} + (\mathscr{P}_1^{\text{inc}}\boldsymbol{\cdot}\mathbf{P})\mathbf{P} + (\mathscr{P}_1^{\text{inc}}\boldsymbol{\cdot}\mathbf{n})\mathbf{n}.$$

In (I.241b), the cross product $(\mathscr{P}_1^{\text{inc}}\times\mathbf{n})$ is contained in the plane perpendicular to **n**, i.e., the plane of **K** and **P**, and hence

$$\mathscr{P}_1^{\text{inc}}\times\mathbf{n} = -(\mathscr{P}_1^{\text{inc}}\boldsymbol{\cdot}\mathbf{P})\mathbf{K} + (\mathscr{P}_1^{\text{inc}}\boldsymbol{\cdot}\mathbf{K})\mathbf{P}.$$

Using all these results in (I.240), we obtain

$$\begin{aligned}\tfrac{1}{4}\mathscr{P}_1^{\text{inc}}\boldsymbol{\cdot}&\text{Tr}\;(M\boldsymbol{\sigma}_1M^\dagger\boldsymbol{\sigma}_1)\\ &= (|g_0|^2 + 2|h_0|^2 + |h_n|^2 - |h_K|^2 - |h_P|^2)(\mathscr{P}_1^{\text{inc}}\boldsymbol{\cdot}\mathbf{n})\mathbf{n}\\ &\quad + [(|g_0|^2 + |h_K|^2 - |h_P|^2 - |h_n|^2)(\mathscr{P}_1^{\text{inc}}\boldsymbol{\cdot}\mathbf{K}) + 2\,\text{Im}\,\{h_0^*(g_0 - h_n)\}(\mathscr{P}_1^{\text{inc}}\boldsymbol{\cdot}\mathbf{P})]\mathbf{K}\\ &\quad + [(|g_0|^2 + |h_P|^2 - |h_K|^2 - |h_n|^2)(\mathscr{P}_1^{\text{inc}}\boldsymbol{\cdot}\mathbf{P}) - 2\,\text{Im}\,\{h_0^*(g_0 - h_n)\}(\mathscr{P}_1^{\text{inc}}\boldsymbol{\cdot}\mathbf{K})]\mathbf{P}.\end{aligned} \tag{I.242}$$

We next consider the geometry shown in Fig. I.8. $\mathbf{k}_i$ and $\mathbf{k}_f$ are the incident and outgoing momenta (each of magnitude k) in the centre-of-mass frame. The angle θ between them is the scattering angle in the centre-of-mass system. The momenta $(\mathbf{k}_i + \mathbf{k}_f)$ and $(\mathbf{k}_f - \mathbf{k}_i)$, constructed from geometrical considerations, are also shown in the diagram, and the unit vectors in these directions are labelled **P** and **K**, respectively. From geometry, the angle between **P** and $\mathbf{k}_i$ is given by $\theta/2$. In Section 7, we derived the relationship between the scattering angle θ_l in

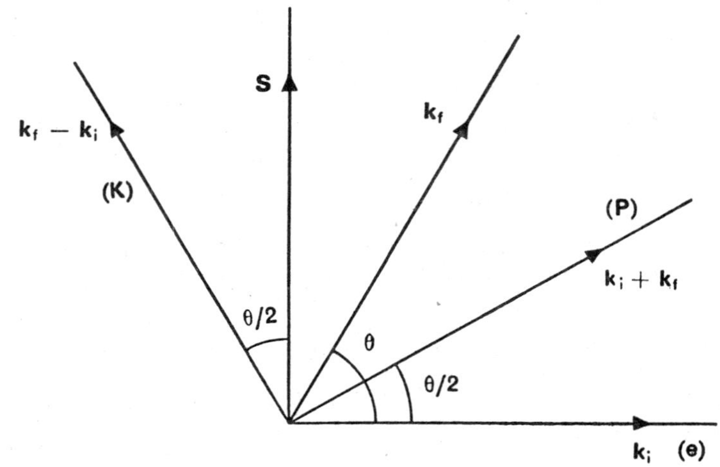

Fig. I.8 Two possible sets of coordinate axes in description of polarization phenomena.

the laboratory system and the same (θ) in the centre-of-mass system, namely, $\theta_l = \theta/2$. Since the incident momentum $\mathbf{k}_i$ has the same direction in the laboratory and the centre-of-mass system, we conclude that the final momentum $(\mathbf{k}_f)_{lab}$ in the laboratory system points in the same direction as **P**. Therefore, the last term in (I.242) gives the component of $\mathcal{P}_1^{scatt}$ in the direction of the outgoing momentum in the laboratory frame. The second term similarly gives the component of $\mathcal{P}_1^{scatt}$ in the scattering plane, but perpendicular to the direction of the outgoing momentum in the laboratory frame. The first term, together with (I.239), gives the component $\mathcal{P}_1^{scatt}$ in a direction normal to the scattering plane. The coefficients of the three terms in (I.242) contain the components of the incident polarization along the directions **n**, **P**, and **K**. However, it is more convenient to consider the components of $\mathcal{P}_1^{inc}$ along the incident momentum in the laboratory frame (the same as the direction of $\mathbf{k}_i$), along the perpendicular to this direction in the scattering plane (the direction S in Fig. I.8), and along the direction **n** normal to the scattering plane. It is apparent that the unit vector **n**, the unit vector **e** along the direction of $\mathbf{k}_i$, and the unit vector S define a right-handed coordinate system. Our aim is achieved if we express $(\mathcal{P}_1^{inc}\cdot\mathbf{K})$ and $(\mathcal{P}_1^{inc}\cdot\mathbf{P})$, appearing in (I.242), in terms of $(\mathcal{P}_1^{inc}\cdot\mathbf{e})$ and $(\mathcal{P}_1^{inc}\cdot\mathbf{S})$. From the geometry defined in Fig. I.8, it easily follows that

$$\begin{aligned} \mathcal{P}_1^{inc}\cdot\mathbf{K} &= -(\mathcal{P}_1^{inc}\cdot\mathbf{e}) \sin\frac{\theta}{2} + (\mathcal{P}_1^{inc}\cdot\mathbf{S}) \cos\frac{\theta}{2}, \\ \mathcal{P}_1^{inc}\cdot\mathbf{P} &= (\mathcal{P}_1^{inc}\cdot\mathbf{e}) \cos\frac{\theta}{2} + (\mathcal{P}_1^{inc}\cdot\mathbf{S}) \sin\frac{\theta}{2}. \end{aligned} \tag{I.243}$$

We first substitute (I.243) in (I.242), and then substitute the resultant expression and (I.239) in (I.236). Thus,

$$\frac{\sigma(\theta)}{\sigma_0(\theta)}\mathscr{P}_1^{\text{scatt}} = (\mathscr{P} + D\mathscr{P}_1^{\text{inc}}\cdot\mathbf{n})\mathbf{n} + (A\mathscr{P}_1^{\text{inc}}\cdot\mathbf{e} + R\mathscr{P}_1^{\text{inc}}\cdot\mathbf{S})\mathbf{K} + (A'\mathscr{P}_1^{\text{inc}}\cdot\mathbf{e} + R'\mathscr{P}_1^{\text{inc}}\cdot\mathbf{S})\mathbf{P}. \tag{I.244}$$

The quantities D, A, R, A', and R' are respectively given by

$$\sigma_0(\theta)D(\theta) = |g_0|^2 + 2|h_0|^2 + |h_n|^2 - |h_K|^2 - |h_P|^2 = \sigma_0(\theta) - 2(|h_K|^2 + |h_P|^2)$$

or

$$\sigma_0(\theta)[1 - D(\theta)] = 2(|h_K|^2 + |h_P|^2); \tag{I.245a}$$

$$\sigma_0(\theta)A(\theta) = -(|g_0|^2 + |h_K|^2 - |h_P|^2 - |h_n|^2)\sin\frac{\theta}{2} + 2\,\text{Im}\,[h_0^*(g_0 - h_n)]\cos\frac{\theta}{2}; \tag{I.245b}$$

$$\sigma_0(\theta)R(\theta) = (|g_0|^2 + |h_K|^2 - |h_P|^2 - |h_n|^2)\cos\frac{\theta}{2} + 2\,\text{Im}\,[h_0^*(g_0 - h_n)]\sin\frac{\theta}{2}; \tag{I.245c}$$

$$\sigma_0(\theta)A'(\theta) = (|g_0|^2 + |h_P|^2 - |h_K|^2 - |h_n|^2)\cos\frac{\theta}{2} + 2\,\text{Im}\,[h_0^*(g_0 - h_n)]\sin\frac{\theta}{2}; \tag{I.245d}$$

$$\sigma_0(\theta)R'(\theta) = (|g_0|^2 + |h_P|^2 - |h_K|^2 - |h_n|^2)\sin\frac{\theta}{2} - 2\,\text{Im}\,[h_0^*(g_0 - h_n)]\cos\frac{\theta}{2}. \tag{I.245e}$$

The parameter $\mathscr{P}$ (equal to the asymmetry parameter a) has already occurred in the expression (I.237a) of the polarization of a scattered beam, when the initial beam was unpolarized; it has also occurred as the asymmetry parameter in the expression (I.231a) of the differential cross-section of an initially polarized beam. This parameter is usually measured in a double scattering experiment, where the first scattering is the nucleon-nucleon scattering, and the second scattering on a spinless C^{12} target serves the purpose of an analyzer. The detailed geometry and layout of the experiment are described in Section 10D. The parameters D, A, and R occurring in (I.244) are called respectively the depolarization, asymmetry, and rotation parameters. They are measured in different layouts of triple scattering experiments, in which the second scattering is the nucleon-nucleon scattering, and the first and last scatterings are on C^{12} targets, serving the purpose of polarizer and analyzer, respectively, of the nucleon beam. The layout and geometry of these experiments are also described in Section 10D.

Spin-Correlation Parameters

Before describing the set-up of double and triple scattering experiments, we consider the principles involved in the measurement of the correlation of spin polarization of the two outgoing nucleons (i.e., the scattered nucleon and the recoiling target nucleon). For the sake of simplicity, we consider an incident unpolarized beam (i.e., $\mathscr{P}_1^{\text{inc}} = 0$) and an initially unpolarized target (i.e., $\mathscr{P}_2^{\text{inc}} = 0$, and $\overline{\langle\sigma_{1\alpha}\sigma_{2\beta}\rangle}_{\text{initial}} = 0$). The correlation in spin is described by the various components of $\overline{\langle\boldsymbol{\sigma}_1\boldsymbol{\sigma}_2\rangle}_{\text{final}}$, where 2 is the target nucleon. This quantity is obtained by a procedure analogous to that for the derivation of (I.236). Since we have assumed an unpolarized target, only the analogue of the first term of (I.236) is present, and is given by

$$\sigma_0(\theta)\overline{\langle\boldsymbol{\sigma}_1\boldsymbol{\sigma}_2\rangle}_{\text{final}} = \tfrac{1}{4}\,\text{Tr}\,MM^\dagger\boldsymbol{\sigma}_1\boldsymbol{\sigma}_2. \tag{I.246}$$

The extra operator here being $\boldsymbol{\sigma}_2$, the trace in the spin-space of the first particle is that of $MM^\dagger\boldsymbol{\sigma}_1$, which is expressed by (I.237b). Therefore,

$$\tfrac{1}{4}\,\text{Tr}\,MM^\dagger\boldsymbol{\sigma}_1\boldsymbol{\sigma}_2 = \tfrac{1}{2}\,\text{Tr}_2\,[(g\mathbf{h}^\dagger + \mathbf{h}g^\dagger + i\mathbf{h}\times\mathbf{h}^\dagger)\boldsymbol{\sigma}_2]$$

$$= \tfrac{1}{2}\,\mathrm{Tr}_2\,[2\,\mathrm{Re}\,(g\mathbf{h}^\dagger)\boldsymbol{\sigma}_2] + \frac{i}{2}\,\mathrm{Tr}_2\,[(\mathbf{h}\times\mathbf{h}^\dagger)\boldsymbol{\sigma}_2]. \tag{I.247a}$$

From (I.224), we obtain

$$\tfrac{1}{2}\,\mathrm{Tr}_2\,g\mathbf{h}^\dagger\boldsymbol{\sigma}_2 = (|h_0|^2 + g_0h_n^*)\mathbf{nn} + g_0(h_K^*\mathbf{KK} + h_P^*\mathbf{PP}) + ih_0(h_P^*\mathbf{PK} + h_K^*\mathbf{KP}),$$

and hence 2 Re of this quantity is given by

$$[2|h_0|^2 + 2\,\mathrm{Re}\,(g_0h_n^*)]\mathbf{nn} + 2\,\mathrm{Re}\,(g_0h_K^*)\mathbf{KK} + 2\,\mathrm{Re}\,(g_0h_P^*)\mathbf{PP} + 2\,\mathrm{Im}\,(h_0h_K^*)\mathbf{KP} - 2\,\mathrm{Im}\,(h_0h_P^*)\mathbf{PK}. \tag{I.247b}$$

The second term of (I.247a) can be evaluated by rewriting $\mathbf{h}$ in the form

$$\mathbf{h} = h_0\mathbf{n} + \sum_i h_i\sigma_{2i}\mathbf{e}_i,$$

where the unit vectors $\mathbf{e}_i$ ($i = x', y', z'$) are in the directions $\mathbf{P}$, $\mathbf{K}$, $\mathbf{n}$. Since $\mathbf{h}$ consists of two types of terms, the h_0's and h_i's, it is clear that $(\mathbf{h}\times\mathbf{h}^\dagger)\boldsymbol{\sigma}_2$ will give rise to four different types of terms. The traces of these terms can be evaluated by the techniques already described, and thus we can obtain

$$\frac{i}{2}\,\mathrm{Tr}_2\,(\mathbf{h}\times\mathbf{h}^\dagger)\boldsymbol{\sigma}_2 = 2\,\mathrm{Im}\,(h_0h_K^*)\mathbf{PK} - 2\,\mathrm{Im}\,(h_0h_P^*)\mathbf{KP} - 2\,\mathrm{Re}\,(h_Kh_n^*\mathbf{PP} + h_Ph_n^*\mathbf{KK} + h_Ph_K^*\mathbf{nn}). \tag{I.247c}$$

Using (I.247) in (I.246), we get

$$\langle\boldsymbol{\sigma}_1\boldsymbol{\sigma}_2\rangle_{\text{final}} = C_{nn}\mathbf{nn} + C_{KP}\mathbf{KP} + C_{PK}\mathbf{PK} + C_{PP}\mathbf{PP} + C_{KK}\mathbf{KK}, \tag{I.248}$$

where

$$\begin{aligned}
\sigma_0(\theta)C_{nn} &= 2|h_0|^2 + 2\,\mathrm{Re}\,(g_0h_n^* - h_Ph_K^*),\\
\sigma_0(\theta)C_{KP} &= 2\,\mathrm{Im}\,[h_0(h_K^* - h_P^*)],\\
C_{PK} &= C_{KP},\\
\sigma_0(\theta)C_{PP} &= 2\,\mathrm{Re}\,(g_0h_P^* - h_Kh_n^*),\\
\sigma_0(\theta)C_{KK} &= 2\,\mathrm{Re}\,(g_0h_K^* - h_Ph_n^*).
\end{aligned} \tag{I.249}$$

In (I.248), the two vectors in each term correspond to the directions of polarization of the first and second particle. Thus, C_{nn} describes the correlation in their polarizations perpendicular to the scattering plane. Before we discuss the directions of polarization specified by the other terms, we again refer to Fig. I.8, and the description already given on the direction of motion of the outgoing particle 1 in the *laboratory coordinate system*. It should be recalled that this motion is along $\mathbf{P}$. Therefore, as far as the polarization of the scattered particle (here particle 1) is concerned (i.e., the first vector in each term), $\mathbf{P}$ specifies a polarization parallel to its motion, and $\mathbf{K}$ a polarization in the scattering plane perpendicular to its direction of motion. To investigate similar facts on the recoiling target particle (here particle 2), we recall the results arrived at in Section 7: (i) the recoiling particle in the centre-of-mass system moves in a direction opposite to that of the scattered particle; and (ii) the angle giving the direction of a particle in the laboratory system is half of the same angle measured in the centre-of-mass system. It is therefore clear from Fig. 1.8 that the recoiling particle moves in the direction of $-\mathbf{k}_f$ in the centre-of-mass system, and $-\mathbf{K}$ in the laboratory system. Therefore, whenever the second vector in (I.248) is $\mathbf{K}$, it describes the polarization of the recoiling particle opposite to the direction of its motion in the *laboratory frame*, and whenever it is **P**, it describes the polarization in the scattering plane perpendicular to the direction of motion. We shall see in Section 10D that the polarization of a particle in the direction of its motion (*along* or *opposite* the direction) is more difficult to measure than the polarization in directions perpendicular to the motion of the particle. Therefore, the coefficients C_{nn} and C_{KP} are easier to measure than the other coefficients in (I.248).

D. DESCRIPTION OF DOUBLE AND TRIPLE SCATTERING EXPERIMENTS

In Section 7, the layout of a single scattering experiment giving the differential cross-section $\sigma_0(\theta)$ of an unpolarized beam was described. The new quantities $\mathcal{P}$, D, R, A, . . . , introduced in Section 10C, are measured in double and triple scattering experiments. Before describing the set-up for their measurements in the case of nucleon-nucleon scattering, let us explain the principles of a polarizer and analyzer (usually a C^{12} target), frequently used in these experiments. The key equations for our purpose are (I.232) and (I.238), and the reader would do well to get familiar with them at this stage.

Polarizer and Analyzer

Let us consider an unpolarized beam scattered on a C^{12} target (spinless). After the scattering, the beam is polarized perpendicular to the scattering plane [see (I.238) and the paragraph following it], the magnitude $\mathcal{P}_C$ of the polarization (C refers to C^{12}) being equal to the asymmetry parameter a_C of (I.235b). This scattering is shown in Fig. I.9, where C_1 is the

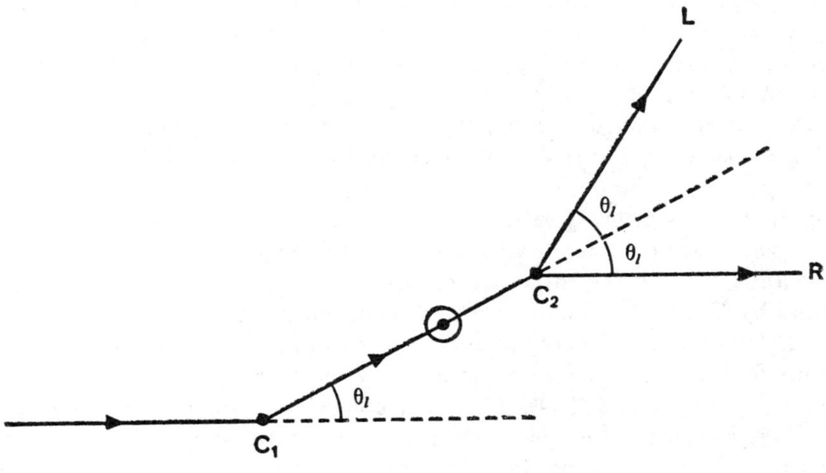

Fig. I.9 Set-up for measuring polarizing power of a target.

carbon scatterer and the arrows indicate the directions of the incident and outgoing nucleon (both in the plane of the paper), making an angle θ_l in the laboratory. The polarization of the outgoing beam is perpendicular to the plane of the figure and points upwards. This is indicated by the dot enclosed in a circle on the outgoing beam. This beam is next scattered on a second C^{12} target (C_2 in Fig. I.9). We measure the left-right asymmetry of the final beam in the two directions (also in the plane of the paper) shown after C_2. According to the statement following (I.235b), this asymmetry is given by (I.232), i.e., $a_2\mathcal{P}_1^{\text{inc}}\cdot\mathbf{n}_2$, where $\mathcal{P}_1^{\text{inc}}$ is the polarization

14 (45-122/1977)

of the beam incident on C_2 and is equal to $\mathcal{P}_C\mathbf{n}_1$. Here subscript 1 refers to the first scattering and subscript 2 to the second. The direction of the polarization ($\mathbf{n}_1$) and the normal $\mathbf{n}_2$ to the second scattering plane both point upwards with respect to the plane of the paper, and hence $\mathbf{n}_1 \cdot \mathbf{n}_2 = 1$. Therefore, the measured left-right asymmetry is equal to $a_2\mathcal{P}_C$. Since the asymmetry parameter is equal in magnitude to $\mathcal{P}_C$, the measured asymmetry gives $\mathcal{P}_C^2$ or the magnitude of $\mathcal{P}_C$. This is usually referred to as the polarizing power of the carbon target. In the experiment just described, the first scattering caused the polarization with the polarizing power equal to $\mathcal{P}_C$, whereas the second scattering served as the analyzer of this polarization through its asymmetry parameter a_2. Therefore, a is also referred to as the analyzing power of the target, which is equal in magnitude to the polarizing power of the same target.

Measurement of $\mathcal{P}$

In the set-up described in Fig. I.9, let us substitute the first scatterer by a nucleon. Then the first scattering causes a polarization of the outgoing nucleon beam equal to $\mathcal{P}\mathbf{n}_1$, where $\mathcal{P}$ now refers to nucleon-nucleon scattering and is the quantity to be measured. The second scattering on C^{12} serves the purpose of the analyzer and, according to (I.232) and the concluding statement of our discussion on polarizer and analyzer, the measured left-right asymmetry of the final beam determines $a_C\mathcal{P} = \mathcal{P}_C\mathcal{P}$. Since $\mathcal{P}_C$ is already known, we obtain $\mathcal{P}$ from this double scattering experiment.

Measurement of D

The quantities $D, R, A, \ldots$ occur in the expression of the polarization of the outgoing beam corresponding to an *incident polarized* beam. Therefore, in this case, we need a polarized nucleon beam before the nucleon-nucleon scattering can be caused. The first scattering is usually done on a C^{12} target, which serves the purpose of polarizing the beam; the second scattering is the nucleon-nucleon scattering under investigation; the various components of the polarization of the outgoing beam are then analyzed by the third scatterer, which once again is a C^{12} target.

The simplest triple scattering set-up for measuring D is shown in Fig. I.10. Here all the three scattering planes are coincident with the plane of the paper, and the single normal direction $\mathbf{n}$ therefore applies to all the three scatterings. The nucleon scatterer is indicated by N, and the C^{12} scatterers by C_1 and C_3. The incident nucleon beam on C_1 is unpolarized, whereas the outgoing beam C_1N is polarized perpendicular to the paper given by $\mathcal{P}_1^{\text{inc}} = \mathcal{P}_C\mathbf{n}$. Since the normal direction for the second scattering is the same as that of the first scattering, we now have only the ($\mathcal{P}_1^{\text{inc}}\cdot\mathbf{n}$)-term in (I.244). The other components in the scattering plane, namely, $\mathcal{P}_1^{\text{inc}}\cdot\mathbf{e}$ and $\mathcal{P}_1^{\text{inc}}\cdot\mathbf{S}$, are zero in the present set-up of the experiment. Therefore, the polarization of the outgoing beam NC_3 is predicted to be

$$\frac{\sigma(\theta)}{\sigma_0(\theta)}\mathcal{P}_1^{\text{scatt}} = (\mathcal{P} + D\mathcal{P}_C)\mathbf{n}$$

in a direction normal to the scattering plane. This is indicated by the dot enclosed in a circle on the line NC_3. According to (I.235a) and the fact that $a = \mathcal{P}$, we obtain

$$\sigma(\theta) = \sigma_0(\theta)(1 + \mathcal{P}\mathcal{P}_C).$$

Therefore,

$$\mathcal{P}_1^{\text{scatt}} = \frac{\mathcal{P} + D\mathcal{P}_C}{1 + \mathcal{P}\mathcal{P}_C}\mathbf{n}. \tag{I.250a}$$

The left-right asymmetry in the third scattering is therefore given by

$$\mathcal{P}_C \mathcal{P}_1^{scatt} \cdot \mathbf{n} = \frac{\mathcal{P} + D\mathcal{P}_C}{1 + \mathcal{P}\mathcal{P}_C}.$$

Since $\mathcal{P}$ and $\mathcal{P}_C$ are already known, this experiment enables us to determine the parameter D.

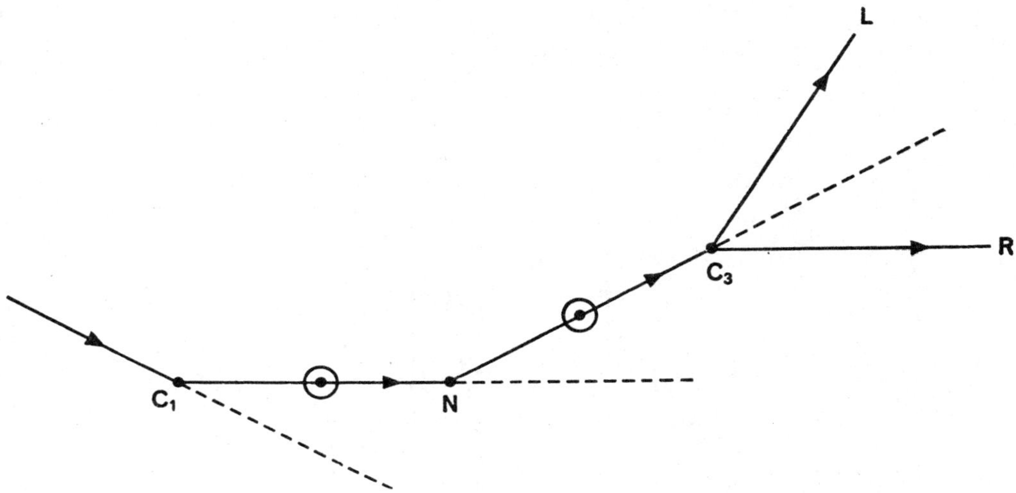

Fig. I.10 Triple scattering set-up for measuring depolarization parameter.

If the incident beam in nucleon-nucleon scattering had a complete polarization (i.e., $\mathcal{P}_C\mathbf{n} = \mathbf{n}$ or $\mathcal{P}_C = 1$), then, according to (I.250a),

$$\mathcal{P}_1^{scatt} = \frac{\mathcal{P} + D}{1 + \mathcal{P}}\mathbf{n}.$$

Therefore, the outgoing beam can remain completely polarized only if $D = 1$, which is the maximum value of D. For any value of $D < 1$, the outgoing beam is found to be only partially polarized. This explains the nomenclature *depolarization parameter.*

Measurement of *R*

Figure I.11 shows the set-up for R measurement. The subsequent scattering planes are crossed with respect to each other such that $\mathbf{n}_1 \cdot \mathbf{n}_2 = \mathbf{n}_2 \cdot \mathbf{n}_3 = 0$. Let the first scattering plane be horizontal. Through the outgoing direction C_1N we draw the vertical plane in solid lines containing the third scatterer C_3. As before, the second scatterer N is a nucleon target. The third scattering plane has been drawn through the line NC_3. The left-right asymmetry of the final scattering in this plane is the quantity measured in this experiment. The polarization of the beam C_1N is given by $\mathcal{P}_1^{inc} = \mathcal{P}_C\mathbf{n}_1$; but $\mathbf{n}_1$, which is in the vertical direction, is now contained in the plane of the second scattering. The direction $\mathbf{n}_2$, normal to the second scattering plane, is horizontal, and hence $\mathcal{P}_1^{inc}$ does not have any normal component with respect to the *second scattering* plane. In (I.244), the direction $\mathbf{e}$ is along the incident laboratory momentum, and $\mathbf{S}$ is perpendicular to this direction in the scattering plane. Therefore, in the present case, $\mathcal{P}_1^{inc}$ points in the direction of $\mathbf{S}$ for the second scattering. Further, since $\mathbf{n}_2$ and $\mathbf{n}_1$ are perpendicular,

$$\sigma(\theta) = \sigma_0(\theta)(1 + \mathcal{P}\mathcal{P}_C \mathbf{n}_1 \cdot \mathbf{n}_2) = \sigma_0(\theta). \tag{I.250b}$$

Therefore, (I.244) yields, for the polarization of the beam NC_3, the expression

$$\mathcal{P}_1^{\text{scatt}} = R\mathcal{P}_C\mathbf{K} + R'\mathcal{P}_C\mathbf{P}, \tag{I.250c}$$

where **P** is in the direction of NC_3 (the outgoing direction in the laboratory) and **K** is in the vertical plane but perpendicular to the outgoing direction. Since the third scattering plane

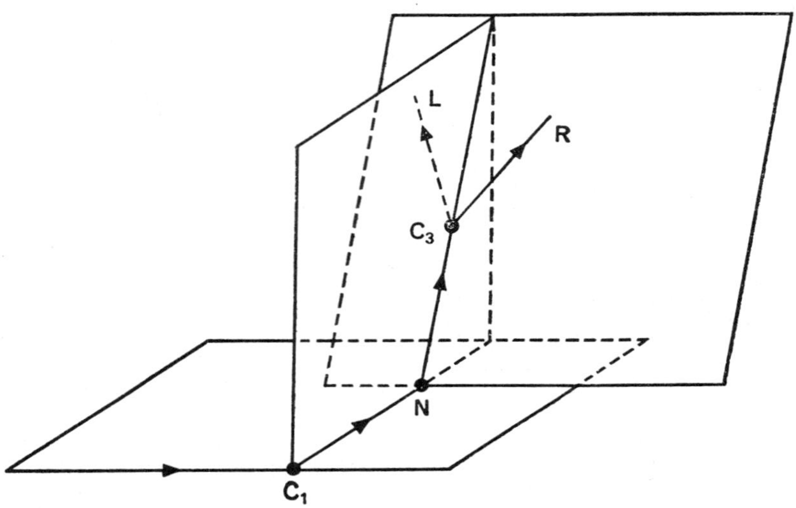

Fig. I.11 Triple scattering set-up for measuring parameter R.

must pass through NC_3, it is clear that the normal to this plane, $\mathbf{n}_3$, is always perpendicular to **P**. Therefore, in the left-right asymmetry measurement in the third scattering plane, the term **P** does not occur. According to (I.232), this asymmetry is given by

$$\mathcal{P}_C\mathcal{P}_1^{\text{scatt}} \cdot \mathbf{n}_3 = \mathcal{P}_C^2(R\mathbf{K} + R'\mathbf{P}) \cdot \mathbf{n}_3 = \mathcal{P}_C^2 R\mathbf{K} \cdot \mathbf{n}_3 = \mathcal{P}_C^2 R. \tag{I.250d}$$

Since the third scattering plane is crossed with respect to the second, the normal $\mathbf{n}_3$ must be contained in the vertical plane. By definition, it should also be perpendicular to any line in the scattering plane, and hence to NC_3. Therefore, it follows that **K** is coincident with $\mathbf{n}_3$, i.e., $\mathbf{K} \cdot \mathbf{n}_3 = 1$. This explains the final result, which enables us to determine R from the measured asymmetry and known $\mathcal{P}_C$.

Measurement of A

The measurement of the parameter A is very similar in principle to that of R. We now need an external device to rotate the polarization $\mathcal{P}_C\mathbf{n}_1$ of the beam C_1N with respect to its direction of motion. Such a rotation can be achieved by applying a magnetic field perpendicular to the direction of $\mathbf{n}_1$ and C_1N. The field rotates the direction of motion C_1N through an amount δ, and the direction of spin by an amount Δ. The two quantities δ and Δ are not equal because of the anomalous magnetic moment of the nucleon. Therefore, by adjusting the applied field, it is possible to rotate the polarization to the direction of motion. In the final condition, the rotated direction of C_1N is obviously in a different plane. Therefore, in Fig. I.12, we do not show the first scattering plane. C_1N is the rotated beam, and the small arrow above it indicates

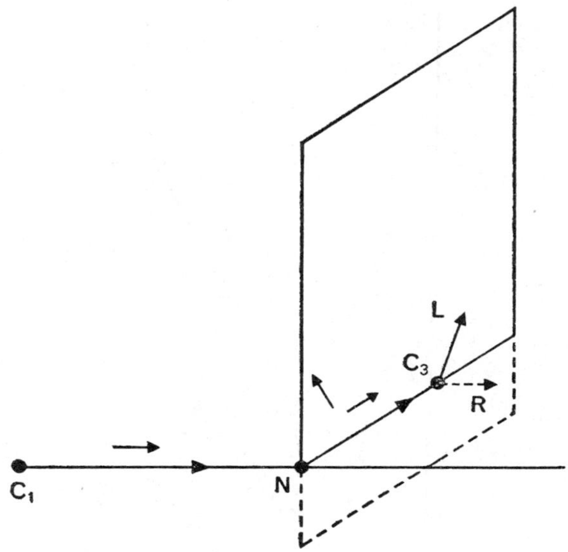

Fig. I.12 Triple scattering set-up for measuring parameter A.

the direction of polarization. The scattered beam NC_3 and incident C_1N are assumed to be in the plane of the paper. The third scattering on C_3 is in the vertical plane, and the left-right asymmetry is measured as shown. Since the polarization of the incident beam is along the direction **e**, only two terms of (I.244) are nonvanishing, and so we get

$$\mathcal{P}_1^{\text{scatt}} = \mathcal{P}_C(A\mathbf{K} + A'\mathbf{P}). \tag{I.250e}$$

$\sigma(\theta)$ is still equal to $\sigma_0(\theta)$, as shown in (I.250b), because the direction of incident polarization continues to be perpendicular to the normal $\mathbf{n}_2$. This fact has been used in writing (I.250e). As before, the left-right asymmetry measurement in the crossed plane rejects the component **P** of (I.250e) and yields a result equal to $\mathcal{P}_C^2 A$, which allows the determination of A.

Measurements of R' and A' require the use of magnetic fields between the second and third scatterers. Such experiments are not usually done, and hence we shall refrain from giving any further description.

Measurement of Correlation Parameters C_{nn} and C_{KP}

The disposition of apparatus for the C_{nn} experiment is shown in Fig. I.13. The incident and all the scattered beams are contained in the plane of the paper. N is the nucleon scatterer. The other two C^{12} scatterers, C_1 and C_2, are in the path of the scattered and recoiling nucleons, which are detected *in coincidence* after the first scattering. The incident beam on N is unpolarized. According to (I.249), the polarization correlation of the beams NC_1 and NC_2 is expressed in the general form

$$\overline{\langle \boldsymbol{\sigma}_1\boldsymbol{\sigma}_2 \rangle}_{\text{final}} = \sum_{i,j} C_{ij}\mathbf{e}_i\mathbf{e}_j,$$

where $\mathbf{e}_i$ ($i = 1, 2, 3$) are the unit vectors along **P**, **K**, and **n** corresponding to the first scattering. The nonvanishing coefficients C_{ij} are given by C_{nn}, C_{KP}, C_{PK}, C_{PP}, and C_{KK}. Since the left-right asymmetry is measured after scattering on C_1 and C_2 in the same plane as the first

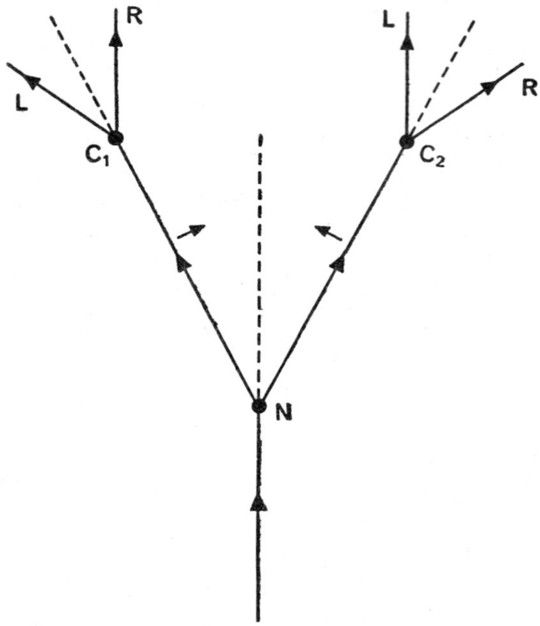

Fig. I.13 Set-up for measuring correlation parameters C_{nn} and C_{KP}.

scattering, all the normals point in the same direction, i.e., perpendicular to the plane of the paper. Left-right asymmetry measurements therefore detect only the *nn*-component of the polarization correlation. The coefficient C_{nn} is given by

$$\mathscr{P}_C^2 C_{nn} = \frac{N_{1L} + N_{2R} - N_{1R} - N_{2L}}{N_{1L} + N_{2R} + N_{1R} + N_{2L}}, \tag{I.251}$$

where the N's denote the counts in the counters specified by the subscripts.

C_{KP} denotes the correlation coefficient for the polarization of NC_1 and NC_2 in directions perpendicular to them in the plane of the paper (indicated by the small arrows in Fig. I.13). These components can be measured from the asymmetry in the crossed planes through NC_1 and NC_2. The coefficient C_{KP} is given by an expression similar to (I.251).

E. EXPRESSIONS IN TERMS OF MATRIX ELEMENTS OF *M*

The purpose of measuring the single, double, and triple scattering parameters is to analyze them in terms of phase shifts at each energy, and extract a set of phase shifts that is as unique as possible. The expressions of the parameters given so far are in terms of the subscripted *g*- and **h**-coefficients. For completeness, we now indicate how these coefficients are expressed in terms of the elements $M^S_{m',m}$ of the scattering amplitude-matrix. Once this is done, the expressions for $M^S_{m',m}$ (see Sections 7 and 8) allow the results to be cast in terms of the phase shifts and the spherical harmonics.

We start with the expression (I.223) for M. Taking the trace of both sides, we are left with only the g_0-term. Then

$$g_0 = \tfrac{1}{4}\,\mathrm{Tr}\,M = \tfrac{1}{4}\sum_{S,m} \langle S, m|\,M\,|S, m\rangle$$

$$= \tfrac{1}{4}\sum_{S,m} M^S_{m,m}. \tag{I.252}$$

Similarly, if we multiply both sides of (I.223) by $(\boldsymbol{\sigma}_1 + \boldsymbol{\sigma}_2)\cdot\mathbf{n}$ and then take the trace, only the corresponding term on the right-hand side of (I.223) yields a factor of 8. Therefore,

$$h_0 = \tfrac{1}{8}\,\mathrm{Tr}\,M(\boldsymbol{\sigma}_1 + \boldsymbol{\sigma}_2)\cdot\mathbf{n}$$

$$= \tfrac{1}{8}\sum_{S,m,m'} \langle S, m|\,M\,|S, m'\rangle\,\langle S, m'|\,(\boldsymbol{\sigma}_1 + \boldsymbol{\sigma}_2)\cdot\mathbf{n}\,|S, m\rangle. \tag{I.253a}$$

In further simplifying the right-hand side of (I.253a), we must recall (see Section 7) that the states $|S, m\rangle$, used in the matrix elements of M, have their projection quantum numbers referred to the incident direction $\mathbf{k}_i$ as the z-axis. We choose the direction of $\mathbf{n}$ as the y-axis, and then the direction $\mathbf{S}$ in Fig. 1.8 clearly becomes the x-axis. Therefore,

$$(\boldsymbol{\sigma}_1 + \boldsymbol{\sigma}_2)\cdot\mathbf{n} = 2S_y = \frac{1}{i}(S_+ - S_-),$$

where $\mathbf{S}$ is the two-nucleon spin operator. Using this form in (I.253a), we have

$$h_0 = \frac{i}{8}\sum_m \left[M^{(1)}_{m,m-1}\sqrt{(1+m)(2-m)} - M^{(1)}_{m,m+1}\sqrt{(1-m)(2+m)}\right]. \tag{I.253b}$$

Since $S_\pm$ give a vanishing result for the singlet $(S = 0)$-state, only the triplet matrix elements appear in (I.253b).

In a manner similar to h_0, any of the coefficients $h_A(\mathbf{A} = \mathbf{P}, \mathbf{K}, \mathbf{n})$ is given by

$$h_A = \tfrac{1}{4}\,\mathrm{Tr}\,M(\boldsymbol{\sigma}_1\cdot\mathbf{A})(\boldsymbol{\sigma}_2\cdot\mathbf{A})$$

$$= \tfrac{1}{4}\sum_{S,m,m'} \langle S, m|\,M\,|S, m'\rangle\,\langle S, m'|\,(\boldsymbol{\sigma}_1\cdot\mathbf{A})(\boldsymbol{\sigma}_2\cdot\mathbf{A})\,|S, m\rangle. \tag{I.254}$$

From Fig. I.8,

$$\boldsymbol{\sigma}\cdot\mathbf{P} = \sigma_z \cos\frac{\theta}{2} + \sigma_x \sin\frac{\theta}{2},$$

$$\boldsymbol{\sigma}\cdot\mathbf{K} = -\sigma_z \sin\frac{\theta}{2} + \sigma_x \cos\frac{\theta}{2},$$

and, according to our choice of y-axis,

$$\boldsymbol{\sigma}\cdot\mathbf{n} = \sigma_y.$$

These expressions enable us to evaluate the second factor in (I.254) in terms of the matrix elements of σ_x, σ_y, σ_z. Further simplification of (I.254) is left to the reader.

The expressions h_P, h_K, h_n are, however, not independent. Their sum is obviously given by

$$h_P + h_K + h_n = \tfrac{1}{4}\,\mathrm{Tr}\,M\boldsymbol{\sigma}_1\cdot\boldsymbol{\sigma}_2$$

$$= \tfrac{1}{4}\sum_{S,m} \langle S, m|\,\boldsymbol{\sigma}_1\cdot\boldsymbol{\sigma}_2\,|S, m\rangle M^S_{m,m}$$

$$= -\tfrac{3}{4}M^{(0)}_{0,0} + \tfrac{1}{4}\sum_m M^{(1)}_{m,m}. \tag{I.255}$$

In the case of p-p scattering, the M-matrix refers to the state $T = 1$ uniquely, and the $M^{(0)}$- and $M^{(1)}$-component, occurring in all the equations from (I.252) to (I.255), are respectively the symmetrized and antisymmetrized amplitudes mentioned in Section 8.

In the case of n-p scattering, if we use the $M^{(0)}$- and $M^{(1)}$-component (given in Section 7) in all the equations from (I.252) to (I.255), we undoubtedly get the subscripted g- and h-coefficients (in terms of phase shifts), but now each such coefficient can be split up into the $(T = 1)$- and $(T = 0)$-component. Corresponding to the split

$$M = \tfrac{1}{2}[M(T = 1) + M(T = 0)]$$

of the M-matrix, each g- and h-coefficient is now given by

$$\begin{aligned} g_i &= \tfrac{1}{2}[g_i(T = 1) + g_i(T = 0)], \\ h_i &= \tfrac{1}{2}[h_i(T = 1) + h_i(T = 0)]. \end{aligned} \tag{I.256}$$

If we ignore Coulomb effects for very high-energy scattering, then g_i and h_i for $T = 1$, appearing in (I.256), are the same as those derived for p-p scattering. The factor of $\frac{1}{2}$ is necessary to secure this equality because the symmetrization and antisymmetrization of the scattering amplitudes for p-p scattering in the states $S = 0$ and $S = 1$ give rise to an extra factor of 2 whereas this factor is absent for the $(T = 1)$-component of the n-p scattering amplitude. The $(T = 0)$-component in (I.256) is related to the $(l = \text{odd})$-terms of the scattering amplitude for $S = 0$ and the $(l = \text{even})$-terms for $S = 1$.

The expressions derived in Section 10C for the various scattering parameters obviously hold for n-p scattering too. But we could use the explicit split of the g- and h-coefficients in their $(T = 1)$- and $(T = 0)$-component, and then apply the fact that the $(T = 1)$-components are the same as for p-p scattering. An interesting relation that can be derived in this way for the polarization $\mathcal{P}(\theta)$ for the pure $(T = 0)$-state is

$$2[\sigma_0(\theta)\mathcal{P}(\theta)]_{\text{n-p}} - 2[\sigma_0(\pi - \theta)\mathcal{P}(\pi - \theta)]_{\text{n-p}} = [\sigma_0(\theta)\mathcal{P}(\theta)]_{T=1} + [\sigma_0(\theta)\mathcal{P}(\theta)]_{T=0}. \tag{I.257}$$

The derivation is set as an exercise (see Problem 7 at the end of this chapter).

11. ANALYSIS OF TWO-NUCLEON DATA

A. LOW-ENERGY p-p SCATTERING DATA

The nuclear part of p-p scattering is in the $^1S_0$-state. The Coulomb scattering comes from all angular momentum states, and is given by the expressions (I.123) and (I.124) for spin-singlet and spin-triplet states, respectively. For triplet spin, the lowest angular momentum state is P, and hence we ignore the nuclear scattering in the triplet state at low energy, and consider only the Coulomb scattering of (I.124). For the singlet state, we use the expression (I.144), retaining only the $(j = 0)$-term. The differential cross-section is therefore given by

$$\begin{aligned} \frac{d\sigma}{d\omega} &= \tfrac{1}{4}\left|F^{\text{singlet}}_{\text{coul}}(\theta) + \frac{2}{k}\exp\,[i(2\eta_0 + \delta_0)]\sin\delta_0\right|^2 + \tfrac{3}{4}|F^{\text{triplet}}_{\text{coul}}(\theta)|^2 \\ &= \tfrac{1}{4}|F^{\text{singlet}}_{\text{coul}}(\theta)|^2 + \tfrac{3}{4}|F^{\text{triplet}}_{\text{coul}}(\theta)|^2 \\ &\quad + \frac{\sin\delta_0}{k}\,\text{Re}\,[F^{\text{singlet}}_{\text{coul}}(\theta)\exp\{-i(2\eta_0 + \delta_0)\}] + \frac{1}{k^2}\sin^2\delta_0 \end{aligned}$$

$$= \left(\frac{e^2}{Mv^2}\right)^2\left[\text{cosec}^4\frac{\theta}{2} + \sec^4\frac{\theta}{2} - \text{cosec}^2\frac{\theta}{2}\sec^2\frac{\theta}{2}\cos\left(2\alpha\ln\tan\frac{\theta}{2}\right)\right]$$

$$- \frac{e^2}{Mv^2}\frac{\sin\delta_0}{k}\left[\text{cosec}^2\frac{\theta}{2}\cos\left(2\alpha\ln\sin\frac{\theta}{2} + \delta_0\right) + \sec^2\frac{\theta}{2}\cos\left(2\alpha\ln\cos\frac{\theta}{2} + \delta_0\right)\right]$$

$$+ \frac{1}{k^2}\sin^2\delta_0. \tag{I.258}$$

Because of the symmetrization of the singlet amplitude and the antisymmetrization of the triplet amplitude, the square modulus of each of these remains unchanged when $\theta \to (\pi - \theta)$. Therefore, p-p scattering (at all energies) is symmetric about the 90° centre-of-mass angle, and it is enough to plot the experimental data between 0° and 90°. The first term in the expression (I.258) gives the pure Coulomb scattering and blows up as $\theta \to 0$. It diminishes in magnitude very quickly as θ increases and reaches an extremely small value for $\theta = 90°$. The last term, which denotes pure nuclear scattering at low energy, is independent of angle (because it represents *S*-state scattering), and it predominates over the pure Coulomb scattering at large angles. In the intermediate region of angles, the second term, representing the interference between the Coulomb and nuclear scattering, shows up in a very striking way; see the dip in the typical angular distribution curve (solid line) shown in Fig. I.14. (The dashed extrapola-

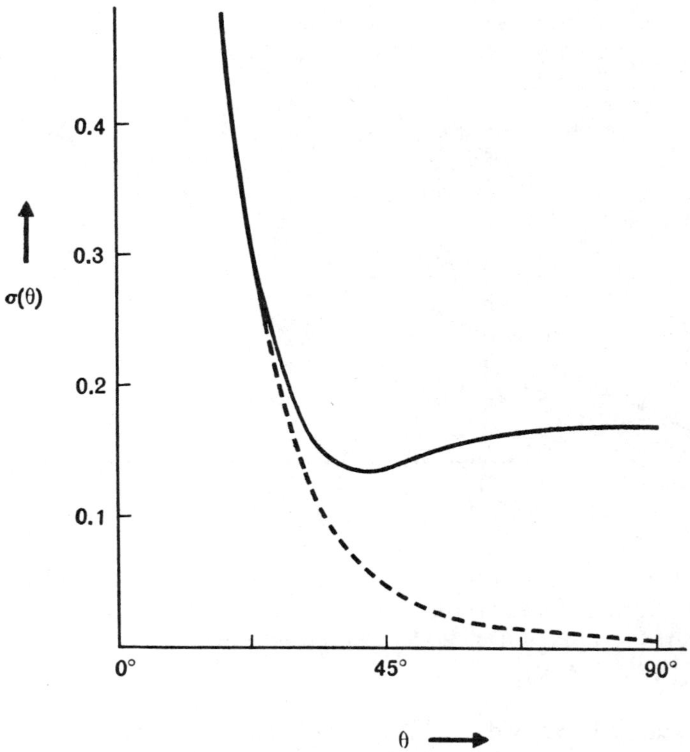

Fig. 1.14 Typical p-p differential cross-section curve. (Following Blatt, J. M., and Weisskopf, V. F., Theoretical Nuclear Physics, Wiley, New York, 1952, p 90.)

tion represents the pure Coulomb scattering.) It is clear from this diagram that at intermediate angles the cross-section is lower than pure nuclear scattering of larger angles. Thus, the interference term is established to be negative in sign, which in turn indicates that the phase shift δ_0 is positive. The cosine terms inside the square brackets remain positive even if δ_0 is negative, but sin δ_0 changes sign, and hence the interference term would have become positive had δ_0 been negative.

An analysis of the experimental data in terms of (I.258) determines the phase shift δ_0 at each energy. Therefore, it is quite straightforward to plot the quantity K of (I.198b) as a function of energy (k^2). This is done in Fig. I.15, where the lower curve corresponds to the

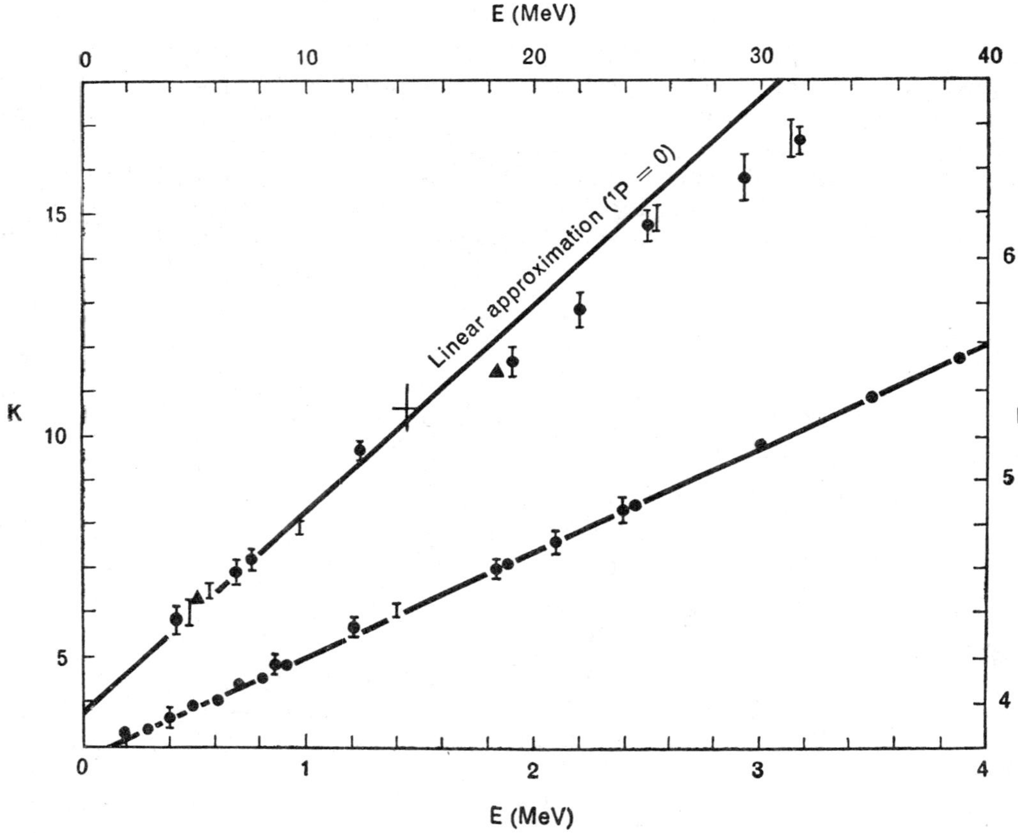

Fig. I.15 Plot of K for p-p scattering as function of incident laboratory energy of proton. [From Breit, G., and Gluckstern, R., *Ann. Rev. Nucl. Sc.*, **2,** 365 (1953); also Gammel, J. L., and Thaler, R. M., *Progress in Elementary Particle and Cosmic Ray Physics*, **5,** 120 (1960).]

values of the abscissa and the ordinate shown at respectively the bottom and the right-hand edge, the upper curve corresponds to a higher energy range (drawn at the top), and the corresponding values of K are given on the left-hand side. If the terms up to k^2 in the expression (I.199b) are important, we expect that K as a function of energy will appear as a straight line. Any departure from a straight line at higher energies indicates the presence of the shape-dependent term $-Pr_0^3k^4$ in the effective-range expansion of K. The experimental curve is found to be a straight line up to about 18 MeV. (It should be noted that, in conformity with customary practice, all energies quoted in this section are for the laboratory system, whereas all angles are for the centre-of-mass system.) Above 18 MeV, experimental points lie below the theoretical straight line, showing a positive value for the shape-dependent parameter P. This conclusion, however, is not very trustworthy because, at this energy, there can be some scattering in the ^3P-states, which have not been considered in the effective range theory and in (I.258). For the singlet state, the straight line part of the upper curve yields

$$a = -7.68 \text{ fm}, \qquad r_0 = 2.65 \text{ fm}. \tag{I.259}$$

B. LOW-ENERGY n-p SCATTERING DATA

Low-energy n-p scattering takes place in the $^1S_0$-state and in the α-state of coupled $^3S_1 + {}^3D_1$, i.e., the coupled state in which $^3S_1$ is predominant. Because of the very small weight of the $^3D_1$-state, the measured angular distribution is spherically symmetric. For this reason, the fit to the data is usually discussed in terms of the total scattering cross-section. Using the effective range expansion for $k \cot \delta$ in (I.156) and its counterpart for the $^3S_1$-state, we obtain

$$\sigma = \tfrac{1}{4}\frac{4\pi}{k^2 + (-a_s^{-1} + \tfrac{1}{2}r_{0s}k^2)^2} + \tfrac{3}{4}\frac{4\pi}{k^2 + (-a_t^{-1} + \tfrac{1}{2}r_{0t}k^2)^2}, \tag{I.260a}$$

the subscripts s and t denoting the singlet and triplet state quantities. In the limit of zero energy ($k^2 \to 0$), this expression reduces to

$$\sigma(k^2 \to 0) = \tfrac{1}{4}(4\pi a_s^2) + \tfrac{3}{4}(4\pi a_t^2). \tag{I.260b}$$

The experimental quantity to be compared with this expression is obtained by extrapolating the low-energy part of the curve of σ versus k^2 to zero energy.

The experimental data are shown in Fig. I.16. The *extrapolated value* at zero energy is found to be (20.36 ± 0.10) barns. The *actual* experimental data for energies comparable to 1 eV are shown in Fig. I.17a. It is found that the observed cross-section at energies less than 1 eV quickly increases to about four times the aforementioned extrapolated value. This quick increase is an effect of chemical binding, which is discussed later in this section. For the present, we confine ourselves to the fit to the extrapolated zero-energy value.

The zero-energy triplet cross-section $4\pi a_t^2$ in (I.260b) can be very easily estimated by making use of the expression (I.190) for the triplet scattering length. The maximum value of the right-hand side is clearly γ, and hence the minimum estimate for a_t is γ^{-1}, i.e., the deuteron radius ($=4.31$ fm). The minimum value of the right-hand side of (I.190) corresponds to the largest estimate of the effective range r_{0t}. Since the deuteron is a very loosely bound system, its radius γ^{-1} can be taken to be the upper limit of r_{0t}, reducing the right-hand side to $\frac{1}{2}\gamma$. Thus, the maximum estimate of a_t is $2\gamma^{-1}$. Using these maximum and minimum estimated values of a_t, we predict that the zero-energy triplet cross-section lies within the range

$$2.33 \text{ barns} \leqslant \sigma_t\ (k^2 \to 0) \leqslant 9.32 \text{ barns},$$

which is very much smaller than the extrapolated experimental value quoted earlier. This disagreement has to be understood in terms of the singlet scattering $4\pi a_s^2$ in (I.260b). This was first pointed out by Wigner, who suggested that the neutron-proton interaction is different in spin-singlet and spin-triplet even-parity states. Using the limiting values of the triplet cross-section in (I.260b), together with the experimental value of 20.36 barns on the left-hand side, we can easily estimate the limiting values of the singlet scattering cross-section $4\pi a_s^2$. These limits are given by

$$74.6 \text{ barns} \geqslant \sigma_s(k^2 \to 0) \geqslant 53.6 \text{ barns}.$$

The large *magnitude* of the singlet scattering length, obtained in this way, can be interpreted

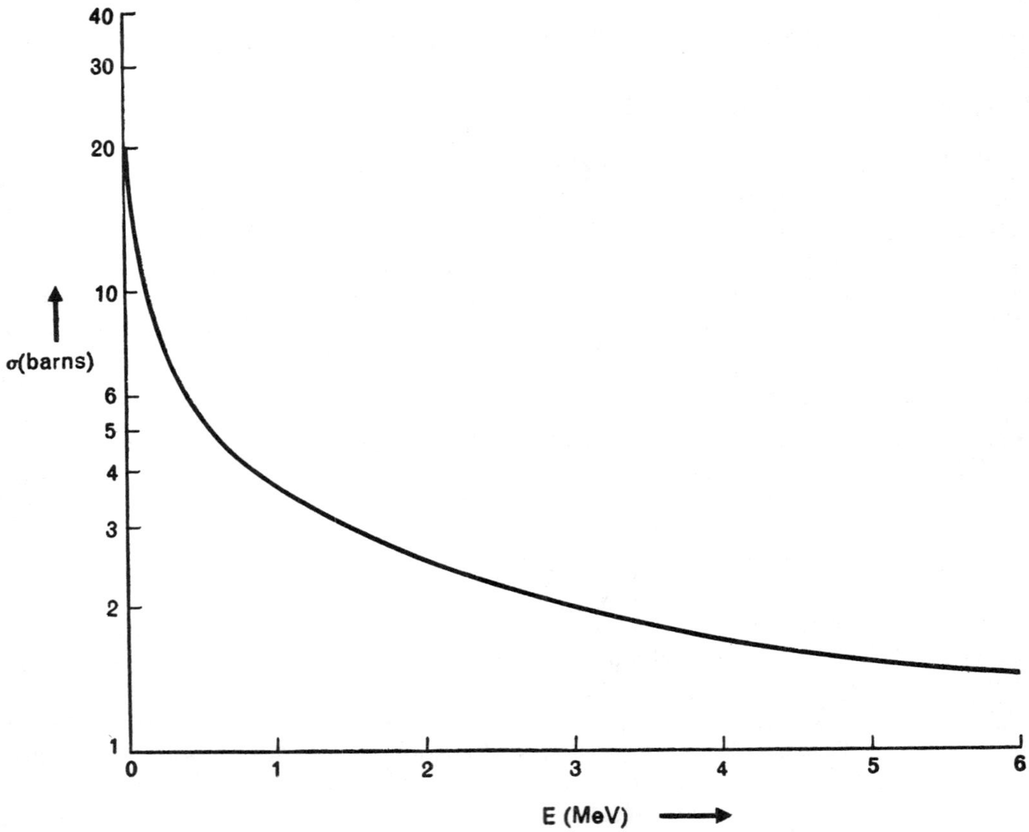

Fig. I.16 Neutron-proton scattering cross-section at low energy. (Following Blatt, J. M., and Weisskopf, V. F., Theoretical Nuclear Physics, Wiley, New York, 1952, p 70.)

in terms of the scattering length diagram (Fig. I.6). It is clear from Fig. I.6a that the more the internal wavefunction bends down, the closer to the origin the point a moves. The stronger binding implies a smaller positive value of the scattering length. A very large positive value of a therefore means a very weakly bound system. The unbound case, depicted in Fig. I.6b, shows that, if the internal solution just becomes horizontal when the asymptotic region is reached, i.e., if the system is just unbound, the dashed curve 'meets' the abscissa on the negative side at $a = -\infty$. The more and more unbound the system is, the closer does the intercept of the dashed curve move to the origin from the negative side. Thus, a very large negative value of the scattering length implies an unbound system having a 'virtual state' just above zero energy. The value of the zero-energy cross-section implies only that a_s^2 is large, and fails to disclose the very vital information on the sign of a_s. The latter determines, as just explained, whether the $^1S_0$-state of the n-p system is very weakly bound or corresponds to a virtual state just above zero energy. To glean this information, we have obviously to design an experiment in which terms linear in a_s can affect the result. These experiments entail the *coherent* scattering of neutrons by (i) the two protons belonging to ortho- and para-hydrogen molecules or (ii) the protons at the lattice points of a crystal or (iii) the protons belonging to a liquid 'mirror', which causes a total reflection of the incident neutron beam. All these experiments measure the coherent scattering length f defined by

$$f = \tfrac{1}{2}(3a_t + a_s). \tag{I.261}$$

The best value for this quantity is obtained from the experiment described in (iii) and is given by $-(3.78 \pm 0.02)$ fm. Combining this with the expression for zero energy, namely, (I.260b), and the experimental value of $\sigma(k^2 \to 0)$, we easily derive

$$a_t = (5.38 \pm 0.02)\ \text{fm}, \qquad a_s = -(23.69 \pm 0.06)\ \text{fm}. \tag{I.262a}$$

Of course, the fact that a has a positive sign (because the deuteron is bound) must be applied in order that the solutions a_t and a_s of (I.260b) and (I.261) become unique. Having obtained a_t, we can straightaway calculate r_{0t} from (I.190) by using the known value of the deuteron radius, namely,

$$r_{0t} = (1.70 \pm 0.03)\ \text{fm}. \tag{I.262b}$$

Finally, r_{0s} can be determined, *in principle*, from (I.260a) by applying the already known values of a_s, a_t, r_{0t} and the experimental curve of σ versus energy. In practice, however, the observed data are insensitive to the choice of r_{0s}; with the accuracy of the available data it appears that any value of r_{0s} between 1.5 fm and 3.5 fm is quite satisfactory. It is usual to assume charge-independence of nucleon-nucleon interaction and take r_{0s} to be the same as the ^1S-effective range for p-p scattering.

Before describing the principle of the coherent scattering experiments, let us discuss how low-energy n-p scattering is affected by the chemical binding of the target proton belonging to a heavy molecule. Such chemical binding is responsible for the quick rise of the cross-section in Fig. I.17a at energies less than 1 eV.

Effect of Chemical Binding

The binding energy of the hydrogen atom belonging to a target molecule is usually of the order of 1 eV. As long as the energy of the incident neutron is much larger than this binding energy, the recoil of the target proton is sufficiently large to tear it off the molecule. Thus, as far as n-p scattering is concerned, the proton can still be regarded as free, and we expect to get a cross-section of the order of 23 barns. This is clear from the range near 1 eV in Fig. 1.17a. As the energy of the incident neutron gets smaller, the recoil of the proton is not sufficient to dislocate it from the molecule, and then the binding to the molecule begins to affect the results through a rise in the cross-section. The bound proton is normally in its lowest energy state, and the small recoil energy tends to excite it to one of its excited (still *bound*) states of vibration. When the recoil energy is smaller than the energy $\hbar\omega$ of the first excited vibrational state, the molecule, together with the proton, remains in the same energy state, and the incident neutron undergoes elastic scattering. As the recoil energy gets higher and higher, successive states of vibration of the proton can be excited. This phenomenon is marked by the consecutive bumps in the cross-section curve of Fig. I.17b, where σ_{free} is the cross-section due to an unbound proton. The bumps coincide with the excitation of successive excited states of vibration of the prot

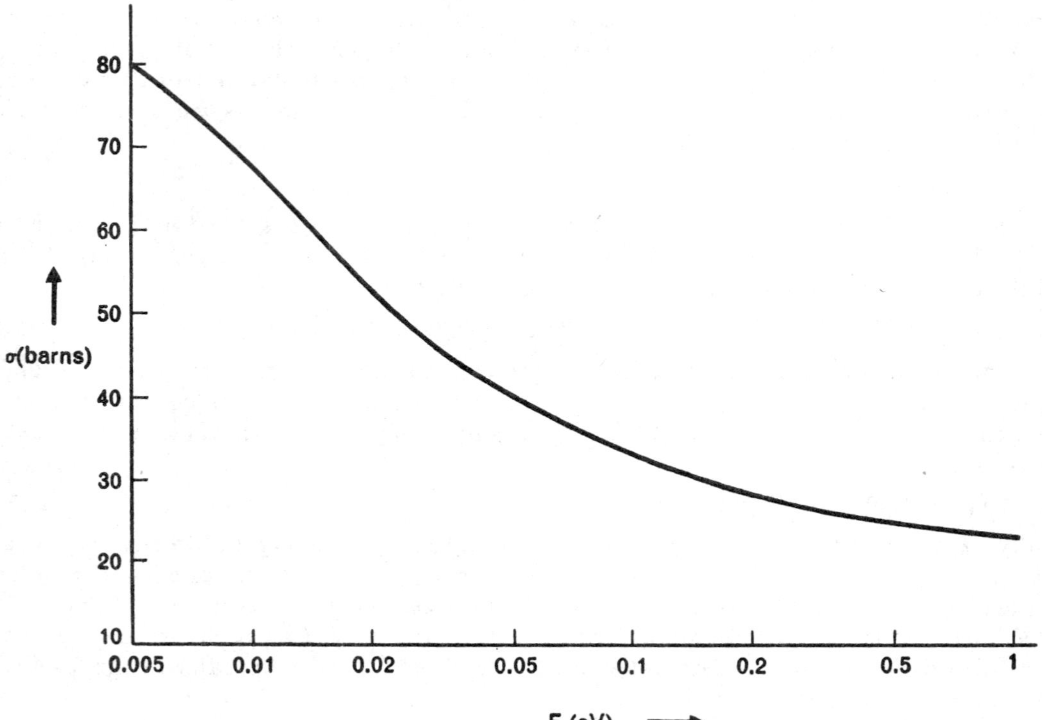

(a) Increase in cross-section with decreasing energy (following Blatt, J. M., and Weisskopf, V. F., Theoretical Nuclear Physics, Wiley, New York, 1952, p 72)

Fig. I.17 Cross-section for neutron scattering by protons bound in a molecule (cont.).

Let us consider the case of extremely small energy when the proton remains in its lowest state, and show that the cross-section for such energy is indeed four times the free proton cross-section. This effect arises from the fact that the reduced mass of a free neutron-proton system is different from that of the system consisting of a neutron and a proton bound in a molecule. Denoting the mass of the molecule by M_{m} and that of the nucleon by M, we get the reduced mass μ' of the latter system as

$$\mu' = \frac{MM_{\text{m}}}{M + M_{\text{m}}}, \qquad \text{(I.263a)}$$

which is $\approx M$ when $M_{\text{m}} \gg M$ (i.e., the proton is bound in a very heavy molecule). On the other hand, the reduced mass of the free neutron-proton system is

$$\mu = \tfrac{1}{2}M.$$

Thus, *for a very heavy molecule*, the ratio of the two reduced masses is expressed as

$$\mu'/\mu = 2.$$

It can be shown that the cross-section of scattering by the bound proton is given, to a very

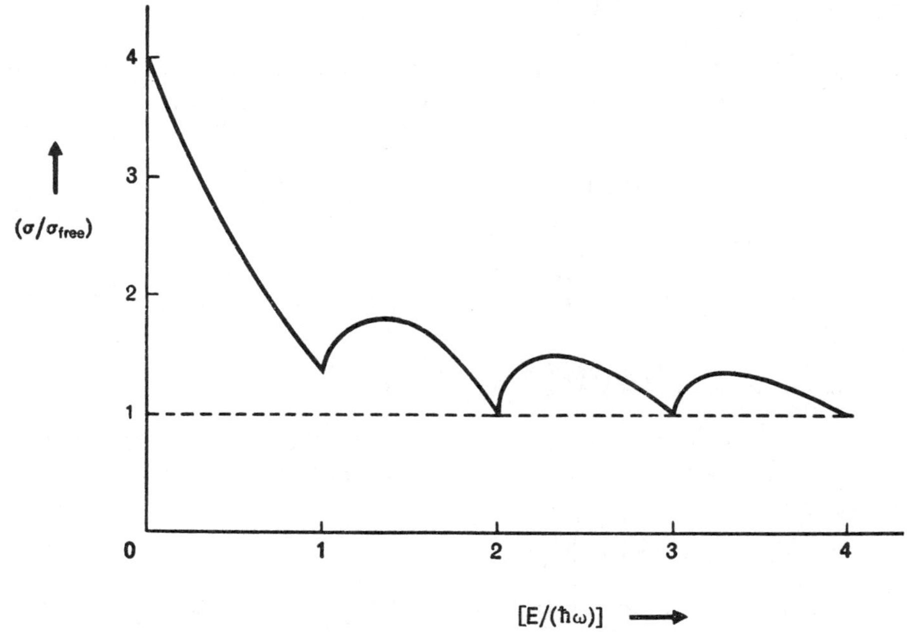

(b) Cross-section of scattering (σ) by a bound proton, plotted against energy (E) (following Bethe, H. A., and Morrison, P., Elementary Nuclear Theory, 2nd edn., Wiley, New York, 1956, p 63)

Fig. I.17 Cross-section for neutron scattering by protons bound in a molecule.

good approximation, by the Born approximation formula using a pseudo-potential, and that the final result is

$$\sigma = \left(\frac{\mu'}{\mu}\right)^2 4\pi a^2, \tag{I.263b}$$

where a is the scattering length. This expression establishes the desired result for a heavy molecule. An outline of the pseudo-potential method, originally due to Fermi, can be obtained from Blatt and Weisskopf[13] whose work is the basis for most of Section 11B.

Coherent Scattering of Neutrons

Let us consider the scattering of a neutron beam by a set of N scattering centres (protons or any other nucleus) located at the points $\mathbf{r}_1, \mathbf{r}_2, \ldots, \mathbf{r}_i, \ldots, \mathbf{r}_N$. The origin of the coordinates is taken at a fixed point O with respect to which an incident or an outgoing neutron has the coordinate vector $\mathbf{r}_n$. The *relative* coordinate of the incident neutron with respect to the scattering centre at $\mathbf{r}_i$ is thus $(\mathbf{r}_n - \mathbf{r}_i)$. According to the discussion in Section 7, the scattering amplitude $f(\theta_i)$ due to the i-th scattering centre, where θ_i is the angle made by the relative coordinate vector $(\mathbf{r}_n - \mathbf{r}_i)$ with the incident momentum $\mathbf{k}$, is defined by the asymptotic wave-function

$$\exp\,[i\mathbf{k}\cdot(\mathbf{r}_n - \mathbf{r}_i)] + f(\theta_i)\,\frac{\exp\,(ik|\mathbf{r}_n - \mathbf{r}_i|)}{|\mathbf{r}_n - \mathbf{r}_i|}.$$

The second term represents the scattering due to the i-th scattering centre, when the incident wave is given by the first term. Therefore, if the incident wave is written as

$$\Psi_{\text{inc}} = \exp\,(i\mathbf{k}\cdot\mathbf{r}_n), \tag{I.264a}$$

the corresponding scattered wave due to the i-th centre will be

$$\Psi^{(i)}_{\text{scatt}} = f(\theta_i)\exp\,(i\mathbf{k}\cdot\mathbf{r}_i)\,\frac{\exp\,(ik|\mathbf{r}_n - \mathbf{r}_i|)}{|\mathbf{r}_n - \mathbf{r}_i|}. \tag{I.264b}$$

Considering the scattering by all the scattering centres, we obtain Ψ_{scatt} corresponding to the incident wave (I.264a):

$$\Psi_{\text{scatt}} = \sum_{i=1}^{N} f(\theta_i)\exp\,(i\mathbf{k}\cdot\mathbf{r}_i)\,\frac{\exp(ik|\mathbf{r}_n - \mathbf{r}_i|)}{|\mathbf{r}_n - \mathbf{r}_i|}. \tag{I.264c}$$

If the origin O is fixed near the scattering target, then all the $\mathbf{r}_i$'s will have small magnitudes. To compute the scattering cross-section, we need (see Section 7) the expression for Ψ_{scatt} for $r_n \to \infty$. Writing

$$|\mathbf{r}_n - \mathbf{r}_i| = (r_n^2 + r_i^2 - 2\mathbf{r}_i\cdot\mathbf{r}_n)^{1/2} \approx r_n - \mathbf{r}_i\cdot\mathbf{e}_n,$$

where $\mathbf{e}_n$ is the unit vector in the direction of $\mathbf{r}_n$, we obtain from (I.264c) the asymptotic form

$$\Psi_{\text{scatt}} = \sum_{i=1}^{N} f(\theta_i)\exp\,[ik\mathbf{r}_i\cdot(\mathbf{e}_k - \mathbf{e}_n)]\,\frac{\exp\,(ikr_n)}{r_n}, \qquad r_n \to \infty. \tag{I.265}$$

Here $\mathbf{e}_k$ is the unit vector in the incident direction, i.e., in the direction of $\mathbf{k}$. As in Section 7, we compute the scattering flux flowing across the area $r_n^2\,d\omega_n$, which is

$$\frac{\hbar k}{M}\Big|\sum_{i=1}^{N} f(\theta_i)\exp\,[ik\mathbf{r}_i\cdot(\mathbf{e}_k - \mathbf{e}_n)]\Big|^2\,d\omega_n.$$

The incident flux per unit area corresponding to the expression (I.264a) is equal to

$$\hbar k/M.$$

The ratio of these two expressions yields

$$\frac{\text{outgoing flux per unit solid angle}}{\text{incident flux per unit area}}$$

$$= \Big|\sum_{i=1}^{N} f(\theta_i)\exp\,[ik\mathbf{r}_i\cdot(\mathbf{e}_k - \mathbf{e}_n)]\Big|^2 \tag{I.266a}$$

$$= \sum_{i=1}^{N} |f(\theta_i)|^2 + \sum_{i\neq j}^{N} f(\theta_i)f^*(\theta_j)\exp\,[ik(\mathbf{e}_k - \mathbf{e}_n)\cdot(\mathbf{r}_i - \mathbf{r}_j)]. \tag{I.266b}$$

If the N scattering centres are randomly distributed in the target material, then the phase angle $k\mathbf{r}_i\cdot(\mathbf{e}_k - \mathbf{e}_n)$ corresponding to the different scattering centres will have completely random values. Thus, the $(i \neq j)$-terms in the expression (I.266b) will, under such circumstances, sum up to a negligible value. The expression (I.266b) then reduces to

$$\sum_{i=1}^{N} |f(\theta_i)|^2 = N|f(\theta)|^2. \tag{I.267}$$

Under such conditions, the scattering by the different scattering centres is said to be incoherent with respect to each other. In Section 7, when we computed the differential cross-section per unit solid angle and for one scattering centre, the proportionality of the scattering to the total number of scattering centres was tacitly assumed. We have now provided an understanding of that tacit assumption; according to (I.267), the proportionality to N follows if the scattering by the different scattering centres in the target material is incoherent, which is so when they are randomly distributed.

If the scattering centres belong to a crystalline target material, then their regular arrangement causes a definite relationship in the phase angle $k\mathbf{r}_i\cdot(\mathbf{e_k} - \mathbf{e_n})$ of (I.266a) for different scattering centres. This phase relationship is very much akin to what obtains in the case of X-ray scattering by crystals. The Bragg or Laue conditions for coherent X-ray scattering by a crystal in certain preferred directions are well-known. We shall show that the phase relationship demanded by (I.266a) for coherent scattering by different centres is identical to that derived by Kittel[14] for X-ray scattering. It is clear from (I.266b) that if

$$k(\mathbf{e_k} - \mathbf{e_n})\cdot(\mathbf{r}_i - \mathbf{r}_j)$$

is an integral multiple of 2π for all pairs (i, j) of scattering centres, then the second term for different i and j add up coherently and the expression becomes

$$\sum_{i=1}^{N} |f(\theta_i)|^2 + \sum_{i\neq j}^{N} f(\theta_i)f^*(\theta_j) = |\sum_{i=1}^{N} f(\theta_i)|^2. \tag{I.268}$$

Here the scattering amplitudes due to different scattering centres have added up coherently, and the square of the modulus is taken *after* the coherent addition. In a crystal, $(\mathbf{r}_i - \mathbf{r}_j)$ for any i and j is of the form $(n_1\mathbf{a} + n_2\mathbf{b} + n_3\mathbf{c})$, where (n_1, n_2, n_3) are integers and $\mathbf{a}$, $\mathbf{b}$, $\mathbf{c}$ are the basic translation vectors under which the crystal remains invariant. Comparing our results with Kittel's, we note that our $(\mathbf{r}_i - \mathbf{r}_j)$ is his quantity $\mathbf{r}$, our $\mathbf{e_k}$ is his incident unit vector $\mathbf{s}_0$, and our $\mathbf{e_n}$ is his outgoing unit vector $\mathbf{s}$. The wave number k in our case is indeed $2\pi/\lambda$, where λ is the De Broglie wavelength of the incident neutron. Thus, the phase angle $k(\mathbf{e_k} - \mathbf{e_n})\cdot(\mathbf{r}_i - \mathbf{r}_j)$ is identical to Kittel's expression (2.3) for X-ray scattering and the subsequent equations (2.4) to (2.8), derived by Kittel[14], therefore hold good in our case.

Let the N scattering centres consist of nuclei of different types $\alpha, \beta, \ldots$. If the number of nuclei of type α is N_α and the scattering amplitude of neutron on these nuclei is f_α, then, along a given direction θ, the expression (I.268) contributes

$$|\sum_{\alpha} N_\alpha f_\alpha(\theta)|^2. \tag{I.269}$$

Since the energy of the incident neutron in these experiments must be very low—too low to dislodge a nucleus from the crystal—, we have also to correct each f_α for the binding effect discussed in this section. The reduced mass μ_α of the incident neutron of mass M, in the case of scattering on an unbound nucleus of type α having mass MA_α, is

$$\mu_\alpha = \frac{M^2 A_\alpha}{MA_\alpha + M} = \frac{A_\alpha}{A_\alpha + 1} M.$$

Taking the crystal to be infinitely heavy compared to the incident neutron, we obtain the

16 (45-122/1977)

reduced mass μ' for scattering by nuclei bound in the crystal as

$$\mu' = M.$$

Therefore, the binding effect is incorporated in the expression (I.269) by multiplying each f_α by the corresponding (μ'/μ_α), i.e., $(A_\alpha + 1)/A_\alpha$. A further modification of (I.269) is necessary in the terms that correspond to scattering by nuclei having a nonvanishing spin I_α. If I_α is nonvanishing, then the combined system of the incident neutron and the target nucleus α has two possible values of total angular momentum: $(J_\alpha = I_\alpha + \frac{1}{2})$ and $(J_\alpha = I_\alpha - \frac{1}{2})$. In the case $I_\alpha = 0$, only one value of J_α, namely, $\frac{1}{2}$, is possible, and hence this case is excluded from the present discussion When $I_\alpha \neq 0$, the two values of J_α usually give rise to different scattering amplitudes; let us denote these by $f_\alpha^{(+)}$ and $f_\alpha^{(-)}$ for $(J_\alpha = I_\alpha + \frac{1}{2})$ and $(J_\alpha = I_\alpha - \frac{1}{2})$, respectively. The number of scattering nuclei N_α also subdivides into two sets $N_\alpha^{(+)}$ and $N_\alpha^{(-)}$ corresponding to the two values of J_α. If the incident neutron is unpolarized, and the target nucleus spin I_α is oriented in all possible directions with equal statistical probability, then the fraction of the total number of nuclei N_α that will produce $(J_\alpha = I_\alpha + \frac{1}{2})$ or $(J_\alpha = I_\alpha - \frac{1}{2})$ with the incident neutron spin is clearly given by the corresponding statistical probability of these states. The number of substates with a given value of J_α is $(2J_\alpha + 1)$ and the total number of substates taking into account both values of J_α is

$$[2(I_\alpha + \tfrac{1}{2}) + 1] + [2(I_\alpha - \tfrac{1}{2}) + 1] = 2(2I_\alpha + 1).$$

The ratio of the two, i.e., $(1/2)(2J_\alpha + 1)/(2I_\alpha + 1)$, thus determines the fraction of nuclei that produces the prescribed J_α with the incident neutron. We therefore write

$$N_\alpha^{(+)} = \frac{I_\alpha + 1}{2I_\alpha + 1} N_\alpha, \qquad N_\alpha^{(-)} = \frac{I_\alpha}{2I_\alpha + 1} N_\alpha.$$

The considerations of spin, in the case of $I_\alpha \neq 0$, thus replace each term $N_\alpha f_\alpha(\theta)$ of (I.269) by the expression

$$N_\alpha^{(+)} f_\alpha^{(+)}(\theta) + N_\alpha^{(-)} f_\alpha^{(-)}(\theta) = \frac{N_\alpha}{2I_\alpha + 1}[(I_\alpha + 1) f_\alpha^{(+)}(\theta) + I_\alpha f_\alpha^{(-)}(\theta)]. \tag{I.270}$$

In the special case when the scattering nucleus in the crystal is a proton, we have $I_\alpha = \frac{1}{2}$; $f_\alpha^{(+)}$ and $f_\alpha^{(-)}$ then denote respectively the triplet and singlet n-p scattering amplitudes. The numerical factors with these scattering amplitudes in (I.270), leaving aside N_α, then become the familiar factors $\frac{3}{4}$ and $\frac{1}{4}$. The binding effect gives the additional factor $(A_\alpha + 1)/A_\alpha$, which is equal to 2 for the proton $(A_\alpha = 1)$.

The expression (I.269), with the modifications for the binding effect and nuclear spin, as outlined, gives the intensity of coherent scattering of the neutron beam along the favoured directions θ (the so-called Bragg angles). Those interested in the details of the coherent neutron scattering experiment by crystals containing hydrogen should refer to Bethe and Morrison[10]. The final number that is deduced from such an experiment in the case of coherent scattering by a proton is the coherent scattering amplitude given by

$$f_H = 2(\tfrac{3}{4} f_t + \tfrac{1}{4} f_s), \tag{I.271}$$

where the numerical factor 2 takes account of the binding effect and the subscripts t and s stand for triplet and singlet scattering, respectively. As explained after the general expression (I.270), the scattering amplitudes f_t and f_s correspond to the superscripts (+) and (−), respectively, in that expression. In view of the low energy of the neutrons in these experiments, f_t

and f_s can be replaced by the corresponding zero-energy quantities. Comparing the wavefunction in the limit $k \to 0$, namely,

$$\Psi(\mathbf{r}) = 1 - a/r, \qquad k \to 0,$$

with the general expression defining the scattering amplitude

$$\Psi(\mathbf{r}) = e^{ikz} + f(\theta)\frac{e^{ikr}}{r},$$

we find, in the limit of $k \to 0$, that the second expression reduces to the first, provided we make the identification

$$f = -a. \tag{I.272}$$

Thus, the zero-energy scattering amplitude is merely the negative of the scattering length. We therefore rewrite (I.271) as

$$f_{\mathrm{H}} = -2(\tfrac{3}{4}a_t + \tfrac{1}{4}a_s). \tag{I.273}$$

The value, determined for this quantity in the coherent scattering experiment, is

$$f_{\mathrm{H}} = 3.9 \text{ fm}. \tag{I.274a}$$

The measurement of the low-energy total cross-section, described after (I.260b), yields

$$\tfrac{1}{4}a_s^2 + \tfrac{3}{4}a_t^2 = 162 \text{ fm}^2. \tag{I.274b}$$

(I.274a) and (I.274b) can be compatible with each other only if a_s differs greatly from a_t. Furthermore, since a_t is positive, a_s has to be negative so that the positive sign of f_{H} is reproduced. The final values of a_s and a_t, derived from (I.274a) and (I.274b), have already been quoted in (I.262a).

It is possible to obtain the coherent scattering amplitude by reflecting a neutron beam on a liquid hydrocarbon-reflecting surface. Phenomena such as reflection and refraction are most aptly described in terms of a bulk property—the refractive index μ—of the medium through which wave propagation takes place. In order to understand why the refractive index concept is associated with a coherence of the amplitude of the waves scattered by various points in the medium, we first point out a property of the total scattering cross-section σ, which is called the *optical theorem*. According to this theorem (see Problem 5 at the end of this chapter),

$$\sigma = \frac{4\pi}{k} \operatorname{Im} f(0), \tag{I.275}$$

where $f(0)$ denotes the scattering amplitude in the forward direction. Thus, we notice that even though the scattering by the various points in the medium is incoherent, in the sum of the cross-section of scattering by the various points in the medium, the scattering amplitudes in the forward direction add up in a coherent manner. Unfortunately, the optical theorem relates only the imaginary part of this coherent sum to the measured total cross-section. Our aim is therefore to obtain a relation similar to the optical theorem that will express a measurable quantity in terms of the real part of the scattering amplitude in the forward direction. We shall soon see that such a relationship exists for the real part of the refractive index μ.

To calculate the refractive index μ, we consider, in a semiclassical manner, the case of a plane wave incident along the z-axis from the left on a sheet of refractive material of thickness Δt and index of refraction μ. Let the origin of the coordinates be at O (Fig. I.18). Our aim is to find an expression for the transmitted wave at the point P located to the right of the sheet. In the absence of refractive material, the wave at P is e^{ikz}.

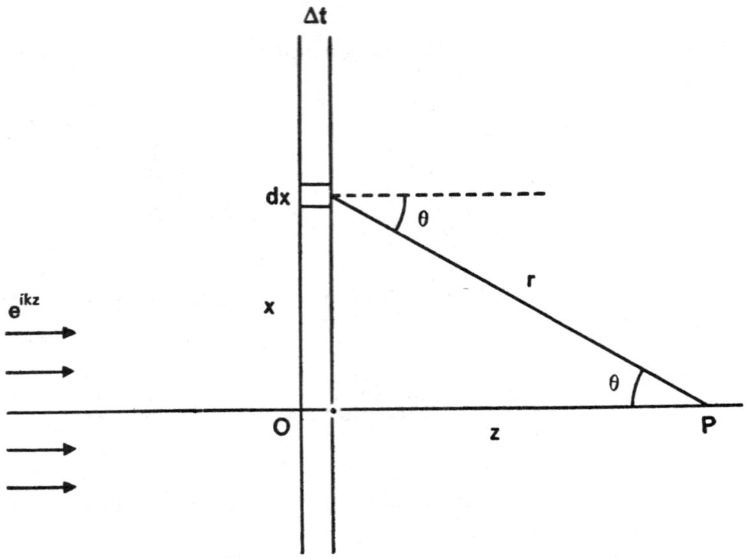

Fig. I.18 Scattering by a target of thickness Δt. (Beam incident from left and transmitted beam observed at P.)

According to the definition of the refractive index μ, the presence of refractive material simply changes the path length Δt to $\mu\Delta t$; hence, the wave at P in the presence of the material is obtained by replacing z in the expression of the plane wave by the path length $\mu\Delta t + (z - \Delta t)$. Therefore,

$$\text{wave at } P = e^{ik[(z-\Delta t)+\mu\Delta t]} = e^{-k(\text{Im}\,\mu)\Delta t} e^{ikz} e^{ik[(\text{Re}\,\mu)-1]\Delta t}. \tag{I.276}$$

Here the refractive index μ has been taken to be complex in general. The expression (I.276) states that the imaginary part of μ gives rise to an absorption of the wave in the refractive material, and its real part is responsible for causing a change in the phase angle of the plane wave. The absorption can very simply be related to the total scattering cross-section σ; in fact, the relationship to be derived is at the root of determining σ from a transmission experiment. Let I_0 be the number of neutrons incident on unit area of the target material and I the number transmitted through it in the forward direction. The loss of particles in the transmitted beam must be equal to the total number of neutrons scattered away from the forward beam. If σ is the total cross-section and N the total number of scattering centres per unit volume of the target, then the number of neutrons scattered away is given by $\sigma I_0 N\Delta t$. Therefore,

$$I = I_0 - \sigma I_0 N\Delta t = I_0(1 - \sigma N\Delta t) \approx I_0 e^{-\sigma N\Delta t}. \tag{I.277}$$

The measurement of the transmitted intensity I and the direct intensity I_0 (measured by removing the target from the beam path), together with a knowledge of N and Δt, thus determines σ. This is the principle of the transmission experiment just referred to. In our case, what is important is the comparison of (I.277) with (I.276). According to the expression (I.276), the transmitted intensity at P, which is proportional to the square of the modulus of this expression, contains the absorptive factor $\exp[-2k(\text{Im}\,\mu)\Delta t]$. A comparison of this factor with the absorptive factor of (I.277) allows the identification

$$2k(\text{Im}\,\mu) = \sigma N \quad \text{or} \quad \text{Im}\,\mu = \sigma N/(2k). \tag{I.278}$$

This expression has already given us a relationship between the imaginary part of the refractive index and the cross-section σ. It will very soon be related to the optical theorem (I.275).

Since our main aim is to obtain an expression of (Re μ) with the help of the corresponding factor in (I.276), let us directly compute the wave at P by adding the incident wave e^{ikz} with all the scattered spherical waves reaching P from the various points in the target. Let us consider the scattering centres contained within the radii x and $(x + dx)$ of the target material with O as the centre. Each of these centres produces a spherical wave $f(\theta)e^{ikr}/r$ at P, where r is the radial distance of P from these scattering centres and $f(\theta)$ is the scattering amplitude. The total number of scattering centres contained within these radii is given by $N\Delta t 2\pi x\, dx$. Thus, integrating over the entire target, we obtain

$$\begin{aligned}
\text{wave at } P &= e^{ikz} + 2\pi N\Delta t \int_0^\infty \frac{e^{ikr}}{r} x\, dx\, f(\theta) \\
&= e^{ikz} + 2\pi N\Delta t \int_z^\infty f(\theta)e^{ikr}\, dr \\
&= e^{ikz} - 2\pi N\Delta t f(0) \frac{e^{ikz}}{ik} \\
&\approx e^{ikz} e^{i[(2\pi N/k)f(0)\Delta t]}.
\end{aligned} \tag{I.279}$$

In arriving at the second step, we have noted that z has a fixed value, and hence the relation $r^2 = z^2 + x^2$ leads to $r\, dr = x\, dx$. According to this relation, the limits $x = 0$ and $x = \infty$ clearly change to $r = z$ and $r = \infty$. In the third step, we have carried out the integration with the usual convergence factor $e^{-\epsilon r}$, where ϵ is a *positive* infinitesimal quantity; this device makes the contribution to the integral from its upper limit equal to zero. In carrying out the integration, we have also used the fact that, for the very low-energy neutron scattering under consideration, the scattering is entirely in the S-state, and hence the scattering amplitude is independent of θ and can be extracted from the integral. However, at the end of the integration, we have noted that the nonvanishing contribution to the integral is from the lower limit $r = z$, for which the angle θ is equal to zero. Hence, in the final expression, the constant value of the S-state scattering amplitude has been specifically labelled with the angle $\theta = 0$.

In a more general derivation, without assuming pure S-state scattering, we have first to express $f(\theta)$ as a function of r, say, $F(r)$, by using the relation $\cos\theta = z/r$ or $\theta = \cos^{-1}(z/r)$ (see Fig. I.18). An integration by parts then yields the third step of (I.279) as the first term; in the other term, we find the integral of $e^{ikr}(dF/dr)$. Usually, the second term is ignored by assuming a slow variation of F with r.

We next compare the expression (I.279) with (I.276) and obtain

$$k(\text{Im}\,\mu) = \frac{2\pi N}{k}\,\text{Im}\,f(0), \qquad k[(\text{Re}\,\mu) - 1] = \frac{2\pi N}{k}\,\text{Re}\,f(0). \tag{I.280}$$

The first of these relations, together with (I.278), leads to

$$\sigma = \frac{2k}{N}\frac{2\pi N}{k^2}\,\text{Im}\,f(0) = \frac{4\pi}{k}\,\text{Im}\,f(0),$$

i.e., the optical theorem (I.275). The second equation in (I.280) gives the desired relationship between the real part of the refractive index and the real part of the scattering amplitude. Replacing the latter by $-a$, where a is the scattering length, we finally obtain

$$(\text{Re } \mu) - 1 = -\frac{2\pi N}{k^2} a. \tag{I.281a}$$

We have so far assumed only one type of scattering nucleus in the target having scattering length a for neutron scattering and number density N per unit volume of the target material. If we have several types of nuclei for which these quantities are given by a_α and N_α, then the summation over the different types yields

$$(\text{Re } \mu) - 1 = -\frac{2\pi}{k^2} \sum_\alpha N_\alpha a_\alpha. \tag{I.281b}$$

Thus, the real part of the refractive index is indeed related to the coherent sum of the scattering amplitudes in the forward direction.

For a measurement of (Re μ), however, a total reflection type experiment is carried out. It is clear from (I.281b) that (Re μ) is less than unity if the coherent sum $\sum_\alpha N_\alpha a_\alpha$ is positive. In the way earlier indicated for Bragg reflection in crystals, each $N_\alpha f_\alpha(0)$ in (I.281b) has to be replaced by $N_\alpha^{(+)} f_\alpha^{(+)}(0) + N_\alpha^{(-)} f_\alpha^{(-)}(0)$ with the help of (I.270). Thus, when the scattering nucleus α is a proton, we have two terms—one with a positive scattering *amplitude* $f_\alpha^{(-)}(0)$ (because the singlet scattering *length* a_s is negative) and the other with a negative scattering *amplitude* $f_\alpha^{(+)}(0)$ (because the triplet scattering *length* a_t is positive). Because $|a_s|$ is very much larger than a_t, the expression $N_\alpha^{(+)} f_\alpha^{(+)}(0) + N_\alpha^{(-)} f_\alpha^{(-)}(0)$ for scattering by protons turns out to be positive. According to (I.281b), therefore, (Re μ) is larger than 1 if the target material is pure liquid hydrogen. In that case, we know that there cannot be any total reflection of the neutron beam when it is incident from air on to the liquid surface. Fortunately, for most heavier nuclei, the scattering length is positive, and hence by taking liquids that contain such nuclei, in addition to hydrogen, we can easily make (Re μ) < 1. Under such circumstances, the total reflection of the neutron beam will take place at an angle of incidence i_c for which the corresponding angle of refraction θ_c into the liquid is 90°. Thus, (Re μ) will be given by

$$\text{Re } \mu \equiv \frac{\sin i_c}{\sin \theta_c} = \sin i_c.$$

Usually, a convenient liquid hydrocarbon is used as the target material, for which (Re μ) turns out to be very close to unity. As a result, i_c (which is measured from the normal to the reflecting surface) is nearly equal to $\pi/2$. The slight departure from $\pi/2$ is denoted by δ_c. Thus, $i_c = (\pi/2 - \delta_c)$, and δ_c is the so-called glancing angle, i.e., the angle made by the incident beam with the reflecting surface. In terms of the *small* angle δ_c, we have

$$(\text{Re } \mu) = \sin i_c = \sin\left(\frac{\pi}{2} - \delta_c\right) = \cos \delta_c \approx 1 - \tfrac{1}{2}\delta_c^2.$$

When we combine this expression with (I.281b), the expression we derive for the coherent sum of the scattering amplitudes in terms of the *measured* glancing angle δ_c is

$$\delta_c^2 = -\frac{4\pi}{k^2} \sum_\alpha N_\alpha f_\alpha(0) = \frac{4\pi}{k^2} \sum_\alpha N_\alpha a_\alpha. \tag{I.281c}$$

The scattering amplitude of carbon is taken to be known from an independent measurement of the neutron scattering cross-section on a carbon target. The concentration of C- and H-nuclei in the liquid hydrocarbon then enables the use of (I.281c) and the derivation of the amplitude $f_{\rm H}$ for hydrogen, as defined in (I.273).

The third type of experiment for the determination of $f_{\rm H}$ consists of scattering neutrons on hydrogen molecules, which are of two types: (i) ortho-hydrogen, in which the protons of the two hydrogen atoms in the molecule have a total spin $S = 1$, and (ii) para-hydrogen, in which the two protons have a total spin $S = 0$. Since the hydrogen molecules are randomly oriented in a hydrogen target, the scattering of neutrons by different molecules is incoherent. On the other hand, the scattering amplitudes of the two protons in each molecule coherently add up with each other as follows. We denote the incident neutron by the subscript n and the two protons by the subscripts 1 and 2. The factor (μ'/μ) for the binding effect in the scattering by each proton is clearly equal to $\frac{4}{3}$ (it should be checked that $\mu' = \frac{2}{3}M$ and $\mu = \frac{1}{2}M$, M being the nucleon mass). The spin state of the incident neutron with any of the target protons can be either singlet or triplet. The projection operators for these states for the proton i $(i = 1, 2)$ are given by

$$\Lambda_{\rm s}^{(i)} = \tfrac{1}{4}(1 - \boldsymbol{\sigma}_{\rm n}\cdot\boldsymbol{\sigma}_i), \qquad \Lambda_{\rm t}^{(i)} = \tfrac{1}{4}(3 + \boldsymbol{\sigma}_{\rm n}\cdot\boldsymbol{\sigma}_i).$$

It can be easily checked that $\Lambda_{\rm s}$ produces zero acting on a triplet state, and unity acting on a singlet state; the operator $\Lambda_{\rm t}$ does just the opposite. In this sense, they are the projection operators, as claimed. The scattering amplitude of the neutron scattering due to this proton is then expressed as

$$\tfrac{4}{3}(f_{\rm s}\Lambda_{\rm s}^{(i)} + f_{\rm t}\Lambda_{\rm t}^{(i)}),$$

$\frac{4}{3}$ being the binding factor. Summing over the two protons in the molecule, we then obtain the scattering amplitude f due to a molecule:

$$\begin{aligned} f_{\rm mol} &= \tfrac{4}{3}\sum_i (f_{\rm s}\Lambda_{\rm s}^{(i)} + f_{\rm t}\Lambda_{\rm t}^{(i)}) \\ &= -\tfrac{4}{3}\sum_i (a_{\rm s}\Lambda_{\rm s}^{(i)} + a_{\rm t}\Lambda_{\rm t}^{(i)}) \\ &= -\tfrac{2}{3}[(a_{\rm s} + 3a_{\rm t}) + (a_{\rm t} - a_{\rm s})\boldsymbol{\sigma}_{\rm n}\cdot\tfrac{1}{2}(\boldsymbol{\sigma}_1 + \boldsymbol{\sigma}_2)] \\ &= -\tfrac{2}{3}[(a_{\rm s} + 3a_{\rm t}) + (a_{\rm t} - a_{\rm s})\boldsymbol{\sigma}_{\rm n}\cdot\mathbf{S}]. \end{aligned} \tag{I.282}$$

Here $\mathbf{S}$ is the total spin operator for the two protons in the hydrogen molecule, and the values of the quantum number S for ortho- and para-hydrogen have already been quoted. The expression (I.282) is the operator that stands for the scattering amplitude. To find its value for the two kinds of hydrogen molecules, we introduce the spin operator

$$\mathcal{S} = \mathbf{s}_{\rm n} + \mathbf{S} = \tfrac{1}{2}\boldsymbol{\sigma}_{\rm n} + \mathbf{S},$$

which represents the total spin of the incident neutron and the two protons in the target molecule. Squaring this vector equation, we easily obtain

$$\mathcal{S}^2 = s_{\rm n}^2 + S^2 + \boldsymbol{\sigma}_{\rm n}\cdot\mathbf{S}.$$

Thus, the expectation value of $\boldsymbol{\sigma}_{\rm n}\cdot\mathbf{S}$, occurring in (I.282), will be

$$\langle\boldsymbol{\sigma}_{\rm n}\cdot\mathbf{S}\rangle = \langle(\mathcal{S}^2 - s_{\rm n}^2 - S^2)\rangle = \mathcal{S}(\mathcal{S}+1) - \tfrac{3}{4} - S(S+1). \tag{I.283}$$

Here $\mathcal{S}$ and S are the quantum numbers corresponding to the spins $\mathcal{S}$ and $\mathbf{S}$, respectively. According to the angular momentum coupling rules, $\mathcal{S}$ has only one value, namely, $\mathcal{S} = \frac{1}{2}$, in the case of para-hydrogen for which $S = 0$. Therefore, for the scattering by the para-hydrogen molecule, (I.283) is zero, and (I.282) yields the value

$$f_{\text{mol}}^{\text{para}} = -\tfrac{2}{3}(a_s + 3a_t). \tag{I.284a}$$

However, in the case of ortho-hydrogen ($S = 1$), the coupling rule gives two possible values of the quantum number $\mathcal{S}$, namely, $\mathcal{S} = \frac{1}{2}$ and $\mathcal{S} = \frac{3}{2}$. In the two cases, we have

$$\langle \boldsymbol{\sigma}\cdot\mathbf{S}\rangle_{\text{ortho}} = \begin{cases} -2 & \text{for } \mathcal{S} = \frac{1}{2} \\ 1 & \text{for } \mathcal{S} = \frac{3}{2} \end{cases}$$

Using these values in (I.282), we obtain

$$f_{\text{mol}}^{\text{ortho}}(\mathcal{S} = \tfrac{1}{2}) = -\tfrac{2}{3}(a_t + 3a_s), \tag{I.284b}$$

$$f_{\text{mol}}^{\text{ortho}}(\mathcal{S} = \tfrac{3}{2}) = -\tfrac{8}{3}a_t. \tag{I.284c}$$

In the scattering experiment, the quantum number $\mathcal{S}$, however, is not measured. Hence, we have to first form the cross-section with these scattering amplitudes and then average over this quantum number. The averaging factor $(2\mathcal{S} + 1)/[2(2S + 1)]$ is equal to $\frac{1}{3}$ and $\frac{2}{3}$ for respectively $\mathcal{S} = \frac{1}{2}$ and $\mathcal{S} = \frac{3}{2}$. In this manner,

$$\begin{aligned} \sigma_{\text{ortho}} &= 4\pi\tfrac{4}{9}[\tfrac{1}{3}(a_t + 3a_s)^2 + \tfrac{2}{3}16a_t^2] \\ &= 4\pi\tfrac{4}{9}[(a_s + 3a_t)^2 + 2(a_t - a_s)^2]. \end{aligned} \tag{I.285a}$$

In the same way, (I.284a) yields

$$\sigma_{\text{para}} = 4\pi\tfrac{4}{9}(a_s + 3a_t)^2 = \sigma_{\text{ortho}} - \frac{32\pi}{9}(a_t - a_s)^2. \tag{I.285b}$$

Thus, σ_{ortho} is larger than σ_{para}, provided a_t and a_s are different. This fact was proved in early experiments which established Wigner's hypothesis, namely, that the nucleon-nucleon potential is spin-dependent, thereby making the singlet and triplet scattering lengths different. The experimental values of σ_{para} and σ_{ortho} are nearly 4 barns and 125 barns, respectively. It is clear from (I.285) that these values enable us to determine a_t and a_s with the additional information that a_t is positive.

C. TWO-NUCLEON SCATTERING IN MEDIUM-ENERGY RANGE

Scattering data in the range 10 MeV to about 500 MeV is treated in the medium-range category. Above this upper limit, meson production starts taking place and the scattering ceases to be purely elastic. High-energy scattering ($\approx$BeV) is not dealt with here.

The reader should first realize that data do not exist as a continuous function of energy in our specified energy range. The various energy regions where extensive data are available are determined by the working capacity of the accelerators around the world. Thus, one of the earliest and most popular energies for p-p scattering has been in the range 310–350 MeV. Experiments have also been carried out in regions around 10–25 MeV, 30 MeV, 40 MeV, 50 MeV, 70 MeV, 95 MeV, 130–190 MeV, 170–260 MeV, and 400–460 MeV.

The second point to note is that the p-p data are more abundant and more accurate than the n-p data; this is because of purely experimental reasons connected with the difficulty in obtaining mono-energetic neutron beams. The n-p experiments can also be performed with proton beams, but then the target neutron must belong to a nucleus such as the deuteron; in this case, the extraction of the scattering due to the neutron alone from the composite scattering of the neutron and proton in the deuteron target would have to rely on theoretical models. Such experiments and analyses of n-p scattering have also been conducted.

The earliest measurements were on the total and differential cross-sections only. The first polarization measurements were those for 310 MeV p-p scattering. A large number of data has been summarized by Wilson[15] and Hess[16].

Several representative curves with data at various energies are shown in Fig. I.19 (n-p) and Fig. I.20 (p-p). Some characteristic features deserve special mention.

n-p Scattering

(i) In the range 10–200 MeV, the total cross-section falls rapidly, according to the inverse power of energy, from the zero-energy value of 20.36 barns. In the range 200–400 MeV, it remains nearly constant and then rises due to the extra inelastic (meson production) cross-section.

(ii) The differential cross-section curve is almost symmetric about the 90° centre-of-mass angle. The slight departure from symmetry is evident in two ways: (a) the backward (180°) scattering is somewhat larger than the forward (0°) scattering and gets enhanced with the increase in energy; and (b) the position of the minimum of the differential cross-section shifts to angles larger than 90° with increasing energy.

p-p Scattering

(i) The total scattering cross-section initially falls with energy (72 mb at 40 MeV and 23 mb at 140 MeV), then remains nearly constant till about 350 MeV, beyond which there is an increase due to inelastic processes.

(ii) In the differential cross-section curve, Coulomb scattering is predominant at small angles followed by the Coulomb nuclear interference region, beyond which the nearly pure nuclear scattering is more or less isotropic up to 90° (the curve is symmetric about this angle due to the identity of the two particles). With increasing energy, the extent of the Coulomb and Coulomb nuclear interference region shrinks towards smaller angles, and the isotropy of nearly pure nuclear scattering correspondingly gets extended up to 20°; the interference region extends from 20° to 50°, whereas the isotropic nuclear scattering holds in the range 50°–90°. At 345 MeV, on the other hand, the isotropy extends down to 15°.

(iii) In the range 300–450 MeV, the p-p differential cross-section at 90° is nearly four times that for n-p scattering at the same energy.

12. DIFFERENT FORMS OF TWO-NUCLEON POTENTIAL

The near symmetry about 90° of the n-p differential cross-section curve shows the existence of an exchange-dependence which is approximately of the Serber type, i.e., a potential that exists for the even states only. The magnitude of the differential cross-section at 180° being somewhat higher than that at 0° indicates that the strength of the Majorana exchange is slightly larger than that of the non-exchange component. In other words, there is a small odd-state potential whose strength has a sign opposite to that of the even-state potential.

The shift of the minimum of the n-p differential cross-section curve from 90° to higher angles is indirect evidence of the presence of a noncentral potential. To appreciate this, let us consider the scattering at an energy where only the *S*-, *P*-, and *D*-state scattering are important. The scattering in the even states *S* and *D* is predominant, and it is a sum of terms

17 (45-122/1977)

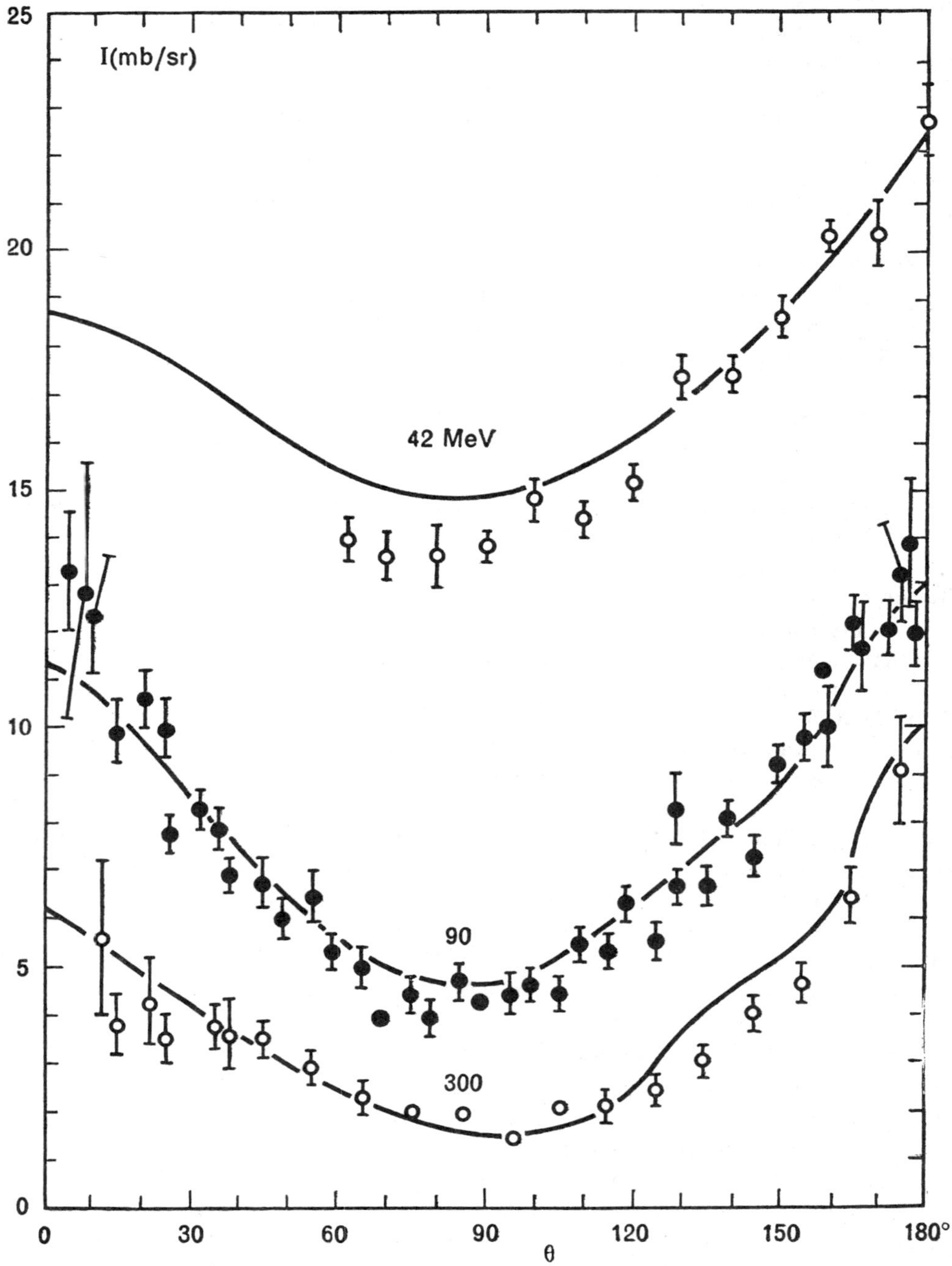

Fig. I.19 Scattering data for n-p scattering. [Following Hamada, T., and Johnston, I. D., *Nucl. Phys.*, **34**, 382 (1962).]

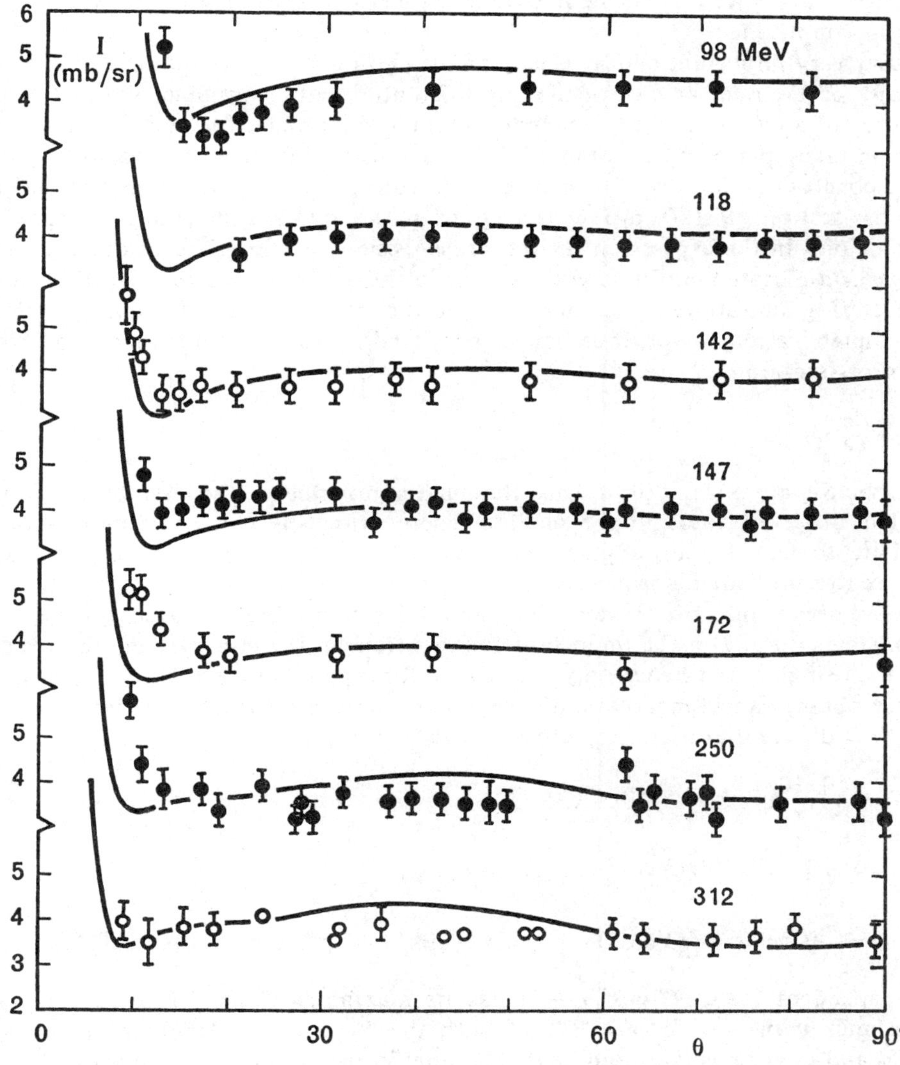

Fig. I 20 Scattering data for p-p scattering. [Following Hamada, T., and Johnston, I. D., *Nucl. Phys.*, **34**, 382 (1962).]

containing Y_0^0 and $Y_0^2(\theta)$, where θ is the scattering angle. Such a function of θ is obviously symmetric about 90° with its minimum located at that angle.

Now let us consider the small *P*-state scattering which has to be added to the scattering in even states. This brings in a term proportional to $-Y_0^1(\theta)$, i.e., $-\cos\theta$, the minus sign being a consequence of the opposite sign of the odd-state potential. The small *P*-state scattering therefore reduces the amplitude slightly at $\theta = 0$, increases it by an equal amount at $\theta = \pi$, and leaves the amplitude at $\theta = \pi/2$ unaltered. Thus, the minimum of the differential cross-section curve remains at $\theta = \pi/2$. This conclusion would not have been true had there been noncentral forces leading to the presence of spherical harmonics $Y^l_m(\theta, \phi)$ with $m \neq 0$ in the scattering amplitude.

Another very important characteristic of the two-nucleon potential is established by the near isotropy of the nuclear part of the p-p differential scattering. Since the S-state scattering is independent of angle, it may be concluded at first sight that, for some mysterious reason, p-p scattering takes place only in the S-state even when the energy rises to a few hundred MeV. This conclusion, however, is immediately ruled out by examining the magnitude of the p-p cross-section. S-state p-p scattering is in the spin-singlet state; the spin averaging gives a factor of $\frac{1}{4}$ in the expression for the cross-section, whereas the symmetrization of the *angle-independent* S-state amplitude with respect to the angles θ and $(\pi - \theta)$ gives a factor of 2 in the scattering amplitude and a factor of 4 in the cross-section. The spin-averaging factor and the symmetrization factor thus cancel out, leading to the expression of the S-state *total* scattering cross-section

$$\sigma = \frac{4\pi}{k^2} \sin^2 \delta_0, \tag{I.286}$$

where δ_0 is the S-state phase shift. Thus, the maximum value of the S-state p-p cross-section is $4\pi/k^2$. The observed value in the medium energy range is twice this maximum estimate, proving that, though higher angular momentum states contribute to p-p scattering, the differential scattering remains isotropic.

To delve deeper into the mystery, let us consider scattering at an energy where angular momentum states up to D make important contribution. In the p-p system, the even states S and D are spin-singlet, and hence only the central force contributes to the scattering in these states. Once again, cancelling the spin averaging and symmetrization factors, we have the combined S, D differential scattering cross-section given by

$$\begin{aligned}\sigma(\theta) &= \frac{\pi}{k^2}\left|\sum_l^{0,\,2} (2l + 1) \sin \delta_l P_l(\cos \theta)\right|^2 \\ &= \frac{\pi}{k^2}\left|[\sin \delta_0 + \tfrac{5}{2}(3 \cos^2 \theta - 1) \sin \delta_2]\right|^2 \\ &= \frac{\pi}{k^2}[\sin^2 \delta_0 + 5(3 \cos^2 \theta - 1) \sin \delta_0 \sin \delta_2 + \tfrac{25}{4}(3 \cos^2 \theta - 1)^2 \sin^2 \delta_2].\end{aligned} \tag{I.287}$$

The angle-dependent factor $(3 \cos^2 \theta - 1)$ has its maximum value $+2$ for $\theta = 0$ and $\theta = \pi$, and its minimum value -1 for $\theta = \pi/2$. If both the S- and D-state potential are attractive, then both δ_0 and δ_2 are positive, and hence the interference term makes a positive contribution at $\theta = 0$ and $\theta = \pi$, and a negative contribution at $\theta = \pi/2$. The other angle-dependent term in (I.287), which represents the contribution of the D-state, is positive at all angles and for any sign of δ_2. Then, at $\theta = \pi/2$, $\sigma(\theta)$ is predicted to have its minimum. On the other hand, if δ_0 and δ_2 have opposite signs, then, in the interference term, $\sin \delta_0 \sin \delta_2$ is negative, and hence the contribution of this term now becomes negative at $\theta = 0$ and $\theta = \pi$, and positive at $\theta = \pi/2$. The increase of the cross-section, due to the interference term at $\pi/2$, and the decrease at $\theta = 0$ and at $\theta = \pi$ then have the effect of nullifying the angle-dependence of the last term and establishing a near isotropic distribution. The clue to the isotropy therefore lies in the opposite sign of δ_0 and δ_2. Jastrow[17] suggested an ingenious way of achieving this end, namely, by considering a potential that has a repulsive core region of very small extent followed by an attractive tail. If the repulsive core has a sufficiently small radius, then, at the energies of a few hundred MeV, only the S-state gets affected by it; as a result, the S-state acquires a negative phase shift because the repulsive core dominates the attractive tail. On the other hand, each higher angular momentum state corresponds, semiclassically speaking, to a certain impact parameter between the two nucleons; as long as this parameter is larger than the core radius, the phase shift gets determined by the attractive outer potential, and has a positive sign.

The idea of the hard core was first suggested by Jastrow. More or less simultaneously, Case and Pais[18] proposed the spin-orbit potential as a remedy to the problem of near isotropy of the p-p scattering. A simple Born approximation estimate shows that the scattering due to the spin-orbit potential vanishes at $\theta = 0$ and $\theta = \pi$. The equality of the scattering at $\theta = 0$ and $\theta = \pi$ follows, of course, from the symmetry of the p-p scattering about $\theta = \pi/2$. The Born approximation expression further shows that the cross-section is nonvanishing at $\theta = \pi/2$ and builds up as we proceed from $\theta = 0$ towards $\theta = \pi/2$. The scattering due to the central potential of an approximately Serber type is peaked at $\theta = 0$ and $\theta = \pi$ with a minimum near $\theta = \pi/2$. Thus, the scattering due to the spin-orbit potential, which helps to build up the cross-section at $\pi/2$, has the effect of leading to near isotropic scattering. However, after the initial suggestion by Case and Pais, several authors worked out exact scattering calculations and came to the conclusion that the spin-orbit potential cannot produce the observed isotropy in the entire energy range up to several hundred MeV. On the other hand, the hard core of the potential, as suggested by Jastrow, is more effective in this respect.

As soon as the earliest polarization data started coming in, it became established that the spin-orbit potential plays a crucial role in explaining the data and is an indispensable part of the two-nucleon interaction. As a result of all these qualitative and semiquantitative facts, different brands of the two-nucleon potential developed over the years. All these are characterized by central, tensor, and spin-orbit components and rely on the hard core in the radial shape for the near isotropy of the p-p scattering. As for the detailed radial form of the well-behaved potential outside the hard core, most workers have used shapes predicted with the help of the meson theory. Some details of a few of these phenomenological potentials are now given.

Phenomenological Potentials with Hard Core

Several brands of two-nucleon potential were suggested by Gammel and Thaler[19], and by Brueckner, Gammel, and Thaler. About the same time, Signell and Marshak[20] produced a phenomenological potential by adopting, as a starting point, the Gartenhaus potential derived on the basis of the meson theory, and then adding to it the hard core and the spin-orbit potentials.

The two most popular and successful brands of the two-nucleon potential having a hard core are due to Hamada and Johnston[5] and to Breit et al[6]. These potentials reproduce all the two-body scattering data (including the polarization parameters) as a function of energy over the energy range of several hundred MeV. The Yale potential of Breit and his collaborators was specially designed to reproduce the phase shifts, in various two-nucleon states, as smooth functions of energy. As a first step, the phase parameters (the phase shifts, and the mixing parameter in the case of coupled states) were determined as a function of the energy by direct fit to all the scattering and polarization data (in the manner described in Section 10). The setting-up of the potential with its parameters adjusted to reproduce the phase parameters may be regarded as the second step in this type of work. The first step, namely, the determination of the phase parameter as a function of energy, is rather sophisticated and has been practised very efficiently by several groups of workers, including the Livermore and Yale teams. The actual procedure, now almost standardized, entails expressing the scattering amplitude as the sum over partial waves up to a certain maximum orbital angular momentum l_{max}. The usual value chosen for l_{max} is 5. The contribution of all the higher partial waves is then taken to be represented by the one-pion exchange contribution (OPEC) to the scattering amplitude. This idea, initially suggested by Moravcsik[21], helped to reduce the ambiguity in the determination of the phase shifts. As already mentioned, the 310 MeV p-p data received the earliest attention. Before Moravcsik's suggestion, the entire scattering was represented by a partial wave sum, with l going to a fairly large value. As a result, the number of phase parameters was too many to be unambiguously determined. In the earliest analyses of this type (see Stapp[22]), five alternative sets of phase parameters produced approximately similar statistical fit to the 310 MeV p-p data. When the data were reanalyzed by using the OPEC plus a fewer number of phase parameters, the latter became determined uniquely. In view of this success, the Yale group, while setting up the phenomenological potential, took the one-pion exchange potential (OPEP) as a given component of the potential and then determined the rest of the potential by fitting the energy-dependent phase parameters up to l_{max}. More details on this potential are given later in this section.

The Hamada-Johnston (HJ) two-nucleon potential is written as

$$V = V_c + V_t \mathcal{T} + V_{lS} \boldsymbol{l} \cdot \mathbf{S} + V_{ll} l_{12}, \tag{I.288a}$$

where $\mathcal{T}$ is the tensor operator occurring in (I.12), and the quadratic spin-orbit coupling l_{12} is defined by (I.22). The radial shape functions V_c, V_t, V_{lS}, and V_{ll} are given by

$$\begin{aligned} V_c &= 0.08(\tfrac{1}{3}\mu)(\boldsymbol{\tau}_1 \cdot \boldsymbol{\tau}_2)(\boldsymbol{\sigma}_1 \cdot \boldsymbol{\sigma}_2) Y(x)[1 + a_c Y(x) + b_c Y^2(x)], \\ V_t &= 0.08(\tfrac{1}{3}\mu)(\boldsymbol{\tau}_1 \cdot \boldsymbol{\tau}_2) Z(x)[1 + a_t Y(x) + b_t Y^2(x)], \\ V_{lS} &= \mu G_{lS} Y^2(x)[1 + b_{lS} Y(x)], \\ V_{ll} &= \mu G_{ll} x^{-2} Z(x)[1 + a_{ll} Y(x) + b_{ll} Y^2(x)]. \end{aligned} \tag{I.288b}$$

Here μ is the pion mass ($=139.4$) in MeV and x is the internucleon distance measured in units of the pion Compton wavelength ($=1.415$ fm); the nucleon mass M is taken to be 6.73μ; and the functions $Y(x)$ and $Z(x)$ are defined by

$$Y(x) = \frac{e^{-x}}{x}, \qquad Z(x) = \left(1 + \frac{3}{x} + \frac{3}{x^2}\right) Y(x). \tag{I.288c}$$

The values of the parameters a_c, b_c, a_t, b_t, . . . , as determined from the detailed fit to the scattering data, are given in Table I.1. These radial shapes of the potential are used outside the hard core of radius $x_c = 0.343$.

The HJ potential, as originally proposed (see Table I.1), would lead to *bound* triplet ($l = J$)-odd states which are known to be non-existent. Therefore, the triplet ($l = J$)-odd state potential has subsequently been modified as follows. It has been defined as -0.26744μ in the region $x_c < x \leqslant 0.487$ and by (I.288) for $x > 0.487$.

The Yale potential of Breit et al[6] is very similar to the HJ potential. Here the OPEP (to be labelled V_{OPEP}) is explicitly used and the quadratic spin-orbit coupling is written in a somewhat different form. The entire two-nucleon potential is expressed as

$$V = V_{\text{OPEP}} + V_c + V_t \mathcal{T} + V_{lS} \boldsymbol{l} \cdot \mathbf{S} + V_q [Q_{12} - (\boldsymbol{l} \cdot \mathbf{S})^2], \tag{I.289a}$$

Table I.1 Parameters of Hamada-Johnston[5] potential defined by (I.288)

Spin-Parity State	Parameters								
	a_c	b_c	a_t	b_t	G_{LS}	b_{LS}	G_{LL}	a_{LL}	b_{LL}
Singlet-even	8.7	10.6					−0.000891	0.2	−0.2
Triplet-odd	−9.07	10.6	−1.29	0.55	0.1961	−7.12	−0.000891	−7.26	6.92
Triplet-even	6.0	−1.0	−0.5	0.2	0.0743	−0.1	0.00267	1.8	−0.4
Singlet-odd	−8.0	12.0					−0.00267	2.0	6.0

where the operator multiplying V_q is the negative of (I.23). The OPEP is given by

$$V_{\text{OPEP}} = \left(\frac{g^2}{12}\right)\mu\left(\frac{\mu}{M}\right)^2 \boldsymbol{\tau}_1 \cdot \boldsymbol{\tau}_2 \frac{e^{-x}}{x}\left[\boldsymbol{\sigma}_1 \cdot \boldsymbol{\sigma}_2 + \left(1 + \frac{3}{x} + \frac{3}{x^2}\right)\mathcal{T}\right], \tag{I.289b}$$

where x, μ, and M are the quantities defined in the context of the HJ potential, and the coupling constant g^2 is given by $g^2/14 = 0.94$ in the singlet-even states and unity elsewhere. In computing the meson Compton wavelength, the neutral pion mass was used for singlet-even and triplet-odd states [i.e., $(T = 1)$-states], whereas for singlet-odd and triplet-even states [i.e., $(T = 0)$-states] a weighted mean of charged and neutral pion masses was used in the proportion 2-to-1. The hard-core radius was taken to be $x_c = 0.35$. All the depths V_c, V_t, . . . were in the form

$$V = \sum_{n=1}^{7} a_n e^{-2x}/x^n. \tag{I.289c}$$

The values of the parameters a_n in the various spin-parity states and for the different types of V (i.e., V_c, V_t, . . .) are listed in Table I.2.

Phenomenological Potentials without Hard Core

Soft-core potential Both the HJ potential and the Yale potential have been very extensively used in nuclear many-body calculations and structural work (see Chapter II). Though these potentials have been highly successful in achieving a good fit to the two-nucleon data, they are deficient in producing good results in the many-body calculations. It is felt that, in the latter calculations, a two-nucleon potential with a *softer* repulsive core (instead of the *infinitely* repulsive core) would be more efficient. With this aim in view, a soft-core potential was later introduced by Reid[23] by fitting the two-nucleon phase parameters up to an energy of about 350 MeV. Bressel et al[24] earlier derived a soft-core potential from similar analyses. Reid's soft-core potential was a sequel to a hard-core potential, determined by him earlier from the two-nucleon data. The main characteristic that distinguishes the Reid potential from the HJ and the Yale potential is that Reid determined the potential in each two-nucleon state *independent* of the other states. It may appear that this approach would produce one potential for each of the infinite number of two-nucleon states, and hence the data-fitting in this manner would become patently meaningless. In practice, however, this is not so. Since the highest energy ($\approx$350 MeV) considered in the analysis is rather low, Reid confined himself to a few

Table I.2 Parameters a_n (in MeV) for Yale potential (see Breit et al[6]) in (I.289c)

Spin-Parity State	V-Type	n							
		0	1	2	3	4	5	6	7
Singlet-even	c	21.925	−19.6303	−194.782	66.4334	−15.2873	−14.5395	1.115	0
	q	0.3333	0.5	0.1	−2.0	−5.1083	−0.2333	0.2	0
Triplet-odd	c	0	−14.28	18.72	−17.3	0	−5.3	−1.3	0
	t	1.5	50.8984	−83.3812	8.7693	−3.1988	1.6172	0.52	0
	IS	0	0	0	0	−2.9794	−76.4565	43.8285	−7.4186
Singlet-odd	c	0	−96.0	−71.001	18.0	8.0	125.8	5.01	0
Triplet-even	c	0	−47.667	−18.47	−1.0	3.55	0	0	0
	t	0	17.3933	7.775	13.535	3.0	−1.4971	0	0
	IS	0	0	0	0	14.35	7.4875	0	0
	q	0	0	5.3333	0	−13.5917	−7.4167	−1.6667	0

two-nucleon states, namely, those with $J \leqslant 2$. The HJ and the Yale potentials with their extra quadratic spin-orbit term have, in fact, as much freedom as Reid's potential had when confined to the specific states. The potentials in the low angular momentum states, considered by Reid, are more or less adequate in determining the low-energy behaviour of nuclei, i.e., the binding energy of nuclei considered as a many-body system (see Chapter II) and other structural aspects dealt with in Chapters III-VI.

In the singlet and the uncoupled triplet states (which have $l = J$), Reid used a central potential, and for the coupled triplet states he used a potential having central, tensor, and spin-orbit ($l \cdot S$, and *not* quadratic) components. In the p-p system, confining ourselves to ($J \leqslant 2$)-states, we have only two singlet-even states, ${}^1S_0$ and ${}^1D_2$. Reid chose different central potentials in these two states. In fact, the quadratic spin-orbit coupling term in the HJ and the Yale potentials does precisely this job. In these cases, the tensor and $l \cdot S$ do not contribute in the singlet states, but the quadratic spin-orbit coupling terms (in addition to the central force) do. From (I.22) and (I.23), it is easy to verify that the potential in the singlet states in the HJ and Yale cases is respectively given by

$$V(\mathrm{HJ}) = V_c - 2l(l+1)V_{ll} \tag{I.290a}$$

and

$$V(\mathrm{Yale}) = V_c - l(l+1)V_q. \tag{I.290b}$$

The second term in each expression vanishes in the S-state, and makes a repulsive contribution (taking V_c, V_{ll}, V_q to be attractive) in the D-state. Thus, the quadratic spin-orbit coupling has the effect of making the ${}^1S_0$- and ${}^1D_2$-potential different, which is Reid's assumption. Securing the difference in the potential of the two states was, in fact, the main motivation of Hamada and Johnston and Breit and his colleagues when they introduced the quadratic spin-orbit coupling. The ${}^1S_0$-phase shift at the various energies determines the central potential V_c, leading to ${}^1D_2$-phase shifts which are found to be overattractive in comparison with what is given by the direct phase shift analysis. The second terms in (I.290) provide the repulsive contribution necessary to bring down the ${}^1D_2$-phase shifts to their observed values.

It can be verified from (I.22) and (I.23) that, in the triplet coupled states, the quadratic spin-orbit term of the Breit potential vanishes, whereas that in the HJ potential becomes proportional to $l \cdot S$. Therefore, both these potentials agree with the form used by Reid for the triplet coupled states.

Having enumerated the main features of the Reid potential, we now list those of his soft-core potential. Any reader who would like to have information on the hard-core potential should refer to Reid[23].

In Reid's soft-core potential, $h = 10.463$ MeV; x has the definition stated after (I.288b) with the inverse of the meson Compton wavelength taken as 0.7 fm^{-1}; and all the numbers representing the depth of the various components are in MeV.

($T = 1$)-states

$$V({}^1S_0) = -h(e^{-x} - 1650.6e^{-4x} + 6484.2e^{-7x})x^{-1},$$

$$V({}^1D_2) = -h(e^{-x} - 12.322e^{-2x} - 1112.6e^{-4x})x^{-1},$$

$$V({}^3P_0) = -h\left[\left(1 + \frac{4}{x} + \frac{4}{x^2}\right)e^{-x} - \left(\frac{16}{x} + \frac{4}{x^2}\right)e^{-4x}\right]x^{-1} + (27.133e^{-2x} - 790.74e^{-4x} + 20662e^{-7x})x^{-1},$$

$$V(^3P_1) = h[(1 + \frac{2}{x} + \frac{2}{x^2})e^{-x} - (\frac{8}{x} + \frac{2}{x^2})e^{-4x}]x^{-1} + (-135.25e^{-2x} + 472.81e^{-3x})x^{-1},$$

$$V(^3P_2 - {}^3F_2) = V_c + V_t\mathcal{T} + V_{lS}\mathbf{l}\cdot\mathbf{S}$$

with

$$V_c = (\tfrac{1}{3}he^{-x} - 933.48e^{-4x} + 4152.1e^{-6x})x^{-1},$$

$$V_t = h[(\frac{1}{3} + \frac{1}{x} + \frac{1}{x^2})e^{-x} - (\frac{4}{x} + \frac{1}{x^2})e^{-4x}]x^{-1} - 34.925e^{-3x}/x, \tag{I.291a}$$

$$V_{lS} = -2074.1e^{-6x}/x.$$

(T = 0)-states

$$V(^1P_1) = (3he^{-x} - 634.39e^{-2x} + 2163.4e^{-3x})x^{-1},$$

$$V(^3D_2) = -3h[(1 + \frac{2}{x} + \frac{2}{x^2})e^{-x} - (\frac{8}{x} + \frac{2}{x^2})e^{-4x}]x^{-1} - 220.12e^{-2x}/x + 871e^{-3x}/x,$$

$$V(^3S_1 - {}^3D_1) = V_c + V_tS_{12} + V_{lS}\mathbf{l}\cdot\mathbf{S}$$

with

$$V_c = (-he^{-x} + 105.468e^{-2x} - 3187.8e^{-4x} + 9924.3e^{-6x})x^{-1},$$

$$V_t = -h[(1 + \frac{3}{x} + \frac{3}{x^2})e^{-x} - (\frac{12}{x} + \frac{3}{x^2})e^{-4x}]x^{-1} + 351.77e^{-4x}/x - 1673.5e^{-6x}/x, \tag{I.291b}$$

$$V_{lS} = 708.91e^{-4x}/x - 2713.1e^{-6x}/x.$$

Velocity-dependent potential and nonlocal potential Two types of potential other than the soft-core potential have been used to obviate the difficulties of dealing with the hard core. Of these, one is characterized by an explicit quadratic momentum dependence

$$\frac{p^2}{M}W(r) + W(r)\frac{p^2}{M}, \tag{I.292}$$

where $\mathbf{p}$ is the momentum operator, and $W(r)$ denotes the radial dependence. This form was originally proposed by Peierls, and investigated in detail by Razavy et al[25] and Green[26].

The second type of potential is characterized by its nonlocality. A nonlocal potential is represented in the coordinate space by a function such as $V(\mathbf{r}, \mathbf{r}')$, which is nonvanishing even when $\mathbf{r} \neq \mathbf{r}'$. The local potential, so far dealt with, is a special case of the nonlocal potential and is represented by $V(\mathbf{r}, \mathbf{r}')\delta(\mathbf{r}, \mathbf{r}')$. A separable nonlocal potential, which has the special form

$$V(\mathbf{r}, \mathbf{r}') = f(\mathbf{r})f(\mathbf{r}'), \tag{I.293}$$

has been widely used. Tabakin[27] studied such a potential in detail by taking a sum of two separable terms [as in (I.293)] in each two-nucleon state. Further extension of the separable form has been considered, especially in dealing with the three-body problem. For good reviews of the separable potential and its applications, see Mitra[28] and Levinger[29].

13. ONE-BOSON EXCHANGE POTENTIAL (OBEP)

We have stated in Section 12 that the one-pion exchange contribution (OPEC) to the scattering amplitude has been explicitly included in the two-body data analysis, and the data then give us information on the phase parameters of states up to a certain maximum orbital angular momentum l_{max}. Correspondingly, in the analysis in terms of a two-body potential, many authors (e.g.,

Breit et al[6]) included the one-pion exchange potential (OPEP) explicitly and then set up the rest of the potential phenomenologically. Qualitatively speaking, the motivation behind such approaches is as follows. The potential acting between a pair of particles due to the exchange of a meson has a range of the order of the meson Compton wavelength, which is inversely proportional to the meson mass. Since the π-meson is the lightest boson that can be exchanged between a pair of nucleons, the OPEP determines the long-range part (beyond the pion-Compton wavelength) of the two-nucleon potential. If we want information on the two-nucleon potential at medium and short ranges, we are then faced with the computation of the potential arising from the exchange of heavier bosons and two, three, . . . pions. Since this computation is comparatively more difficult, Breit and his coworkers were satisfied with determining it phenomenologically.

A breakthrough has been achieved in recent years with the realization that multi-meson systems most often have strongly correlated resonance states behaving as a single boson. It is therefore speculated that some of these multi-meson resonances, when exchanged between two nucleons, may dominate the medium- and short-range behaviour of the two-nucleon potential. The potential computed in this way is called the one-boson exchange potential (OBEP).

Besides the exchange of one π-meson, other exchanges have been explicitly considered in the OBEP: (i) the ρ-meson (two-pion system with quantum numbers $J = 1$, parity $P = -1$, isospin $T = 1$) having a mass ≈ 750 MeV (pion mass ≈ 135 MeV); (ii) the ω-meson ($J = 1$, $P = -1$, $T = 0$, mass ≈ 780 MeV) which is a three-pion resonance; (iii) the ϕ-meson which has the same quantum numbers as that of ω, a mass ≈ 1020 MeV, and is a resonance of the $K\bar{K}$ system; (iv) the so-called σ-meson ($J = 0$, $P = +1$, $T = 0$) which is postulated to occur in the two-pion system; and (v) the η-meson which has all the quantum numbers of the π-meson, except for isospin $T = 0$ (π-meson has $T = 1$).

The exchange of the ω- and ϕ-meson gives rise to a repulsive potential in all states, which becomes very strong at about 0.2 fm, the Compton wavelength of ϕ. This effect is therefore comparable with the strongly repulsive core of the phenomenological two-nucleon potential. The exchange of η-meson creates a very weak potential, and may be neglected.

The σ-meson is currently in an experimentally unconfirmed state. In the OBEP theory, its mass and coupling strength go as parameters, which are then determined by detailed fit to the two-nucleon data. The introduction of the σ-meson is regarded as a convenient parametrization of the effect of the two-pion system in the S-state ($J = 0$). The two-pion wavefunction has to be symmetric under the exchange of spatial and isospin coordinates (they have no spin); since the S-state is symmetric, the isospin state too has to be symmetric, and the only symmetric state that can be obtained by coupling two unit isospin states is the state with $T = 0$. Thus, the quantum numbers are quite unique. Some authors have included a heavier meson (called the s-meson) with the same quantum numbers as those of the σ-meson.

The main difference amongst workers on the meson-theoretic two-nucleon potential lies in their manner of treatment of the two-pion system. The approach, just described, in which the effect of the two-pion system is parametrized through one or two isoscalar ($T = 0$) scalar ($J = 0$) mesons is one such treatment. In another development of the theory, the effect of the S-state of the two-pion system is parametrized through the scattering length and effective range. In yet another attempt, the effect of the two-pion continuum is considered in more detail, and the resultant potential taken into account explicitly. The potential generated in any of these ways is attractive and counteracts the repulsive effect of the ω- and ϕ-exchange at medium range.

Various authors also differ in the details of their method of computing the potential. Broadly speaking, conventional field-theoretical techniques and the dispersion theoretic method are the two principal methods of solving the problem. They are not detailed here.

The properties of each boson that enter the expression of the corresponding potential are the coupling constant of the boson field to the nucleon field (this parameter determines the strength of the potential), and the mass of the boson (the inverse of which, as mentioned earlier, determines the range of the potential). With the exception of the scalar-isoscalar meson, the masses of the other bosons, considered in the OBEP, are known from experimental data. Most of the resonance states have a finite width, and in that sense the mass is not very unique. A slight latitude of variation can therefore be tolerated in the mass parameters. In the detailed analysis by Scotti and Wong[30], the ρ-meson mass was thus taken to be an additional adjustable parameter. With the exception of the π-meson, the value of the coupling constant of these bosons is not known too accurately from experimental data. Moreover, in the case of the vector bosons (i.e., $J = 1$), each boson is characterized by two coupling constants corresponding to the vector- and tensor-type coupling. Thus, we need, in principle, two coupling constants for each ρ, ω, and ϕ. In practice, there is evidence that the tensor coupling in the cases ω and ϕ is comparatively weak, and hence these three bosons give rise, in a practical data analysis, to four (instead of six) adjustable strength parameters. The η-meson gives rise to one more, and the σ-meson to two, namely, the coupling strength and the mass. Considering the ρ-meson mass, we have so far counted eight adjustable parameters. Three more cut-off parameters enter through an exponential convergence factor needed in each vector meson potential. One more parameter was introduced to give the correct k^{2l}-type behaviour for small k to the partial wave amplitude of angular momentum l. Thus, the scattering analysis based on the OBEP has twelve parameters, much smaller in number than those contained in the phenomenological potentials. Yet the fit to the two-body data is statistically as good (perhaps slightly better) as that achieved with the phenomenological potentials.

The number of parameters in the OBEP can be considerably reduced by taking advantage of the interrelationship amongst the various coupling strengths postulated by the SU_3- and SU_6-symmetry requirements. Ball et al[31] thus reduced the total number of parameters to four: the mass and coupling strength of σ; a single coupling strength that relates the strengths for ρ, ω, and ϕ; a single parameter that describes the cut off needed for convergence; and the correct k^{2l}-type threshold behaviour. The fit achieved in this way is, once again, highly satisfactory.

The expressions for the potential due to the exchange of the different mesons are now quoted from the work of Bryan and Scott[32]. We first deal with the isoscalar ($T = 0$) mesons, of which we have four at our disposal: a scalar ($J = 0, P = +1$) meson σ; a pseudoscalar ($J = 0, P = -1$) meson η; and two vector ($J = 1, P = -1$) mesons, ω and ϕ. The expression of the interaction Lagrangian $\mathcal{L}_{\text{int}}$, which describes the coupling of the boson field (ϕ) to the nucleon field (ψ), becomes relevant in defining the coupling constant (g). Therefore, in each case, the expression of the Lagrangian is followed by that of the OBEP. The potential in every case is in units of the rest mass (m) energy of the corresponding boson, and the radial distance x is in units of the Compton wavelength of the same particle. M is the nucleon mass. The functions $f(x)$, $J(x)$, and $g(x)$ are defined by

$$f(x) = e^{-x}x^{-1}, \qquad J(x) = x^{-1}\frac{d}{dx}f(x), \qquad g(x) = 1 + 3x^{-1} + 3x^{-2}.$$

(i) $T = 0$, $J = 0$, $P = +1$ (isoscalar, scalar meson),

$$\mathcal{L}_{\text{int}} = (4\pi)^{1/2} g \bar{\psi}\psi\phi,$$

$$V = g^2\left(1 - \tfrac{1}{8}\frac{m^2}{M^2}\right)\left[-f(x) + \tfrac{1}{2}\frac{m^2}{M^2}J(x)\boldsymbol{l}\cdot\mathbf{S}\right]. \tag{I.294a}$$

(ii) $T = 0$, $J = 0$, $P = -1$ (isoscalar, pseudoscalar meson),

$$\mathcal{L}_{\text{int}} = (4\pi)^{1/2} g \bar{\psi}\gamma_5\psi\phi \qquad (\gamma_5 \text{ is the usual Dirac matrix}),$$

$$V = \frac{g^2}{12}\left(\frac{m}{M}\right)^2 f(x)[\boldsymbol{\sigma}_1\cdot\boldsymbol{\sigma}_2 + g(x)\mathcal{T}] \qquad (\mathcal{T} \text{ is the usual tensor force operator}). \tag{I.294b}$$

(iii) $T = 0$, $J = 1$, $P = -1$ (isoscalar, vector meson),

$$\mathcal{L}_{\text{int}} = (4\pi)^{1/2}[g\bar{\psi}\gamma^{\mu}\psi\phi_{\mu} + \{f/(2m)\}$$

$$\times \bar{\psi}\sigma^{\mu\nu}\psi\{\frac{\partial}{\partial x_{\nu}}\phi^{\mu} - \frac{\partial}{\partial x_{\mu}}\phi^{\nu}\}]$$

(γ^{μ}, $\sigma^{\mu\nu}$ are the usual Dirac matrices—see, for example, Schweber et al[33]—, g is the vector coupling constant, and f the tensor coupling constant),

$$V = g^2\Big[\Big\{1 + \frac{\tfrac{1}{2}(m/M)f}{1 + \tfrac{1}{4}(m/M)^2 g}\Big\}^2\Big(1 + \tfrac{1}{4}\frac{m^2}{M^2}\Big)f(x)$$

$$+ \Big\{1 + \frac{1 + \tfrac{1}{8}(m/M)^2 f}{\tfrac{1}{2}(m/M)g}\Big\}^2 \tfrac{1}{12}\frac{m^2}{M^2}\{2\boldsymbol{\sigma}_1\cdot\boldsymbol{\sigma}_2 - g(x)\mathcal{T}\}f(x)$$

$$+ \Big\{1 + \frac{4(m/M)f}{\tfrac{3}{2}(m/M)^2 g} + \frac{f^2}{g^2}\Big\}\tfrac{3}{2}\frac{m^2}{M^2}J(x)\boldsymbol{l}\cdot\mathbf{S}\Big]. \tag{I.294c}$$

The expressions for the case of the ($T = 1$)-mesons are obtained from the foregoing ones by consistently replacing ϕ in the Lagrangian by $\boldsymbol{\tau}\cdot\boldsymbol{\varphi}$, where $\boldsymbol{\tau}$ is the isospin operator for the nucleon, and $\boldsymbol{\varphi}$ a vector in the isospin space. The expressions for the OBEP are derived from (I.294) by multiplying each expression by $\boldsymbol{\tau}_1\cdot\boldsymbol{\tau}_2$, where the subscripts refer to the two nucleons. The three new expressions then describe respectively the OBEP due to the exchange of an isovector scalar meson, isovector pseudoscalar meson (π), and isovector vector meson (ρ). It should be noted that the pion-exchange potential obtained in this way agrees with the expression (I.289b) for the OPEP. Bryan and Scott used an isovector scalar meson in addition to the usual σ-meson (isoscalar scalar); they treated the coupling constant and mass of this meson as parameters and determined for them the values $g^2 = 6.5$, $m = 770$ MeV, whereas the corresponding values for the σ-meson were $g^2 = 9.4$, $m = 560$ MeV.

The OBEP just quoted has an (r^{-3})-behaviour at the origin, and hence it is cut off to zero inside a small sphere about the origin. The cut-off radius chosen by Bryan and Scott was 0.6 fm.

A review article on the OBEP by the late Erkelenz[34] and another by Brown[35] are recommended to the reader. More recently, Green and Haapakoski[36] more competently computed the $^1S_0$-potential by considering the effect of the N^*-resonance, known as $\Delta_{3,3}$ (1230 MeV) [subscripts 3, 3 stand for $2J$ and $2T$, respectively], which plays a dominant role in π-N scattering. The OBEP due to the σ-exchange mainly contributes to the $^1S_0$-potential and, as already mentioned, this part of the OBEP is unreliable due to the unsatisfactory experimental situation of the σ-resonance. The recent work considering the $\Delta_{3,3}$ is yet another approach to the simulation of the two-pion effects.

REFERENCES

1. Goldstein, H., Classical Mechanics, Addison-Wesley, Reading, Mass., 1950, p 84.
2. Schiff, L. I., Quantum Mechanics, 2nd edn., McGraw-Hill, New York, 1955, pp 97–99, 117.
3. Wigner, E., and Eisenbud, L., *Phys. Rev.*, **56**, 214 (1939).
4. Okubo, S., and Marshak, R. E., *Ann. Phys.* (N.Y.), **4**, 166 (1958).
5. Hamada, T., and Johnston, I. D., *Nucl. Phys.*, **34**, 382 (1962).
6. Lassila, K. E., Hull, M. H., Ruppel, H. M., McDonald, F. A., and Breit, G., *Phys. Rev.*, **126**, 881 (1962).
7. Green, A. M., *Nucl. Phys.*, **33**, 218 (1962); *ibid*, **47**, 671 (1963).
8. Buckingham, R. A., Numerical Methods, Pitman, London, 1962.
9. Whittaker, E. T., and Watson, G. N., A Course of Modern Analysis, Cambridge University Press, London, 1958.
10. Bethe, H. A., and Morrison, P., Elementary Nuclear Theory, 2nd edn., Wiley, New York, 1956, p 56.
11. Raphael, R., *Phys. Rev.*, **102**, 905 (1956).
12. Feshbach, H., and Lomon, E. L., *Phys. Rev.*, **102**, 891 (1956).
13. Blatt, J. M., and Weisskopf, V. F., Theoretical Nuclear Physics, Wiley, New York, 1952, p 74.
14. Kittel, C., Introduction to Solid State Physics, 2nd edn., Wiley, New York, 1956, pp 44–48.
15. Wilson, R., The Nucleon-Nucleon Interaction, Interscience, New York, 1962.
16. Hess, W. N., *Revs. Mod. Phys.*, **30**, 368 (1958).
17. Jastrow, R., *Phys. Rev.*, **79**, 389 (1950); *ibid*, **81**, 165 (1951).
18. Case, K. M., and Pais, A., *Phys. Rev.*, **80**, 203 (1950).
19. Gammel, J. L., and Thaler, R. M., *Progress in Elementary Particle and Cosmic Ray Physics*, **5**, 99 (1960).
20. Signell, P., and Marshak, R. E., *Phys. Rev.*, **109**, 1229 (1958).
21. Moravcsik, M. J., *Ann. Rev. Nucl. Sc.*, **10**, 291 (1960); *Rep. Prog. Phys.*, **35**, 587 (1972); Cziffra, P., MacGregor, M. H., Moravcsik, M. J., and Stapp, H. P., *Phys. Rev.*, **114**, 880 (1959).
22. Stapp, H. P., Ypsilantis, T. J., and Metropolis, N., *Phys. Rev.*, **105**, 302 (1957).
23. Reid, R. V., Jr., *Ann. Phys.* (N.Y.), **50**, 411 (1968).
24. Bressel, C., Kerman, A. K., and Lomon, E. L., *Bull. Am. Phys. Soc.*, **10**, 584 (1966).
25. Razavy, M., Field, G., and Levinger, J. S., *Phys. Rev.*, **125**, 269 (1962).
26. Green, A. M., *Nucl. Phys.*, **47**, 671 (1963).
27. Tabakin, F., *Ann. Phys.* (N.Y.), **30**, 51 (1964).
28. Mitra, A. N., *Adv. Nucl. Phys.*, **3**, 1 (1969).

29. Levinger, J. S., Nuclear Physics, Springer Tracts in Modern Physics, Vol 71, ed. by G. Höhler, Springer-Verlag, Berlin, 1974, p 88.

30. Scotti, A., and Wong, D. Y., *Phys. Rev.*, **138**, B145 (1965).

31. Ball, J. S., Scotti, A., and Wong, D. Y., *Phys. Rev.*, **142**, 1000 (1966).

32. Bryan, R. A., and Scott, B. L., *Phys. Rev.*, **135**, B434 (1964).

33. Schweber, S. S., Bethe, H. A., and deHoffman, F., Mesons and Fields, Vol 1, Row, Peterson and Co., Evanston, Illinois, 1956.

34. Erkelenz, K., *Phys. Rep.*, **13C**, 191 (1974).

35. Brown, G. E., in Proc. Fifth Int. Conf. on High-Energy Physics and Nuclear Structure, Uppsala (1973), ed. by G. Tibell, North-Holland, Amsterdam, 1974.

36. Green, A. M., and Haapakoski, P., *Nucl. Phys.*, **A221**, 429 (1974).

PROBLEMS

1. (a) Consider the tensor operator $\mathcal{T}$ defined by (I.12) and evaluate the expectation value of $\mathcal{T}^2$ in a two-body triplet state of projection unity [i.e., χ_1^1 of (I.4c)].

(b) Give arguments to show that the $^3D_1$-state wavefunction can be obtained by applying $\mathcal{T}$ on the $^3S_1$-state wavefunction

$$\Psi(^3D_1) = N\mathcal{T}\Psi(^3S_1).$$

Assume $\Psi(^3S_1)$ to be normalized to unity and evaluate the normalization constant N such that $\Psi(^3D_1)$ is also similarly normalized.

(c) Evaluate the matrix elements of $\mathcal{T}$ connecting $\Psi(^3S_1)$ with $\Psi(^3D_1)$, and $\Psi(^3D_1)$ with $\Psi(^3D_1)$, using the results of Problems 1a and 1b. Check the results with the help of (I.38b).

2. (a) Analytically solve the S-state radial Schrödinger equation for a bound state of energy $-\epsilon$ by using the exponential and Hülthen potentials of (I.45b). [Hint: Use the variable transformations $Z = e^{-r/(2a)}$ (exponential) and $Z = 1 - e^{-r/a}$ (Hülthen).]

(b) Solve the S-state radial Schrödinger equation with a square-well potential of depth V_0 and range a. Find the value of the product V_0a^2 when the state is just bound.

3. The Wronskian of any two functions f and g of a variable r is defined as

$$W(f, g) = f\frac{dg}{dr} - g\frac{df}{dr}.$$

Show that the Wronskian of the two independent solutions $f_l(k, r)$ and $f_l(-k, r)$ of (I.47), as defined in Section 9A, is independent of r. [Hint: Multiply the differential equation for $f_l(k, r)$ by $f_l(-k, r)$ and, vice versa, subtract and integrate.]

4. (a) Obtain (I.230) after evaluating the trace of the spin operators of the first particle.

(b) Show that $\mathrm{Tr}_2\,(\mathbf{h} \times \mathbf{h}^\dagger) = 0$ by detailed evaluation.

(c) Establish (I.231c) by evaluating the trace of (I.231b) in detail.

5. Consider, for the scattering amplitude of a central potential, the partial-wave expression

$$f(\theta) = \frac{1}{2ik}\sum_{l=0}^{\infty}(2l + 1)[\exp(2i\delta_l) - 1]P_l(\cos\theta),$$

where P_l is the Legendre polynomial. Compute the corresponding total cross-section σ and directly show

$$\sigma = \frac{4\pi}{k}\,\mathrm{Im}\, f(0),$$

where Im denotes the "imaginary part of" and $f(0)$ denotes the scattering amplitude for $\theta = 0$.

6. Use the Born approximation expression (EI.1c) from Appendix E to compute the scattering amplitude separately for a central, tensor, and spin-orbit potential.

7. Derive (I.257).

II Nuclear Binding Energy and Many-Body Theory

14. NUCLEAR BINDING ENERGY

The definition and preliminary facts of nuclear binding energy have been presented in Section 1. In this chapter, we shall calculate the nuclear binding energy, starting with the nuclear many-body Hamiltonian which is a sum of the kinetic energy and the interaction between all pairs of nucleons [see (II.3)]. The mathematical apparatus of the many-body perturbation theory is applied for this purpose. The actual development of the many-body formalism begins only in Section 15. Here we prepare the ground for a systematic discussion on nuclear binding energy in terms of physically distinguishable contributions. Such a discussion is invariably based on an empirical mass formula. The most celebrated of this category was, for a long time, the Bethe-Weizsäcker (BW) formula. Many modifications to this formula have been suggested by various authors at different times; however, only recently a more versatile formula has been designed to encompass several features of nuclear structure not contained in the BW version. The new mass formula was proposed by Myers and Swiatecki (MS) and contains mainly the effect of nuclear deformation and shell structure in addition to the features of the BW formula.

The actual nuclear binding energy in the BW mass formula is split up into five different components, each with a definite physical origin. The mass formula is written as

$$M(Z, N)c^2 = (M_p Z + M_n N)c^2 + E(Z, N), \tag{II.1}$$

where $M(Z, N)$ is the mass of the nucleus made up of Z protons and N neutrons; M_p and M_n are the masses of a proton and a neutron, respectively; and the quantity $E(Z, N)$ is the negative of the binding energy of the nucleus. For a stable nucleus, E is negative and hence the binding energy is positive. As just mentioned, the quantity E can be broken up into five different components, each of which we shall now discuss. The main purpose of such a mass formula is to delineate the dependence of the various terms on the relevant variables N, Z, and A. Whenever necessary, we shall also make use of an expression similar to (II.1) for the *atomic* mass, in which case the proton mass M_p is replaced by the mass of the hydrogen atom M_H.

Volume Energy

If we consider a *hypothetical* nucleus made up of an equal number of protons and neutrons, and having no surface and no Coulomb interaction amongst its protons, what sort of binding energy should we expect? The answer to this question is at the basis of the concept of the volume energy term. Theoretically, it is the expectation value of a nuclear many-body Hamiltonian, containing only the kinetic energy and the typical nuclear interaction amongst the nucleons; the Coulomb term of the actual nuclear Hamiltonian must be dropped, according to the definition just given, and the nuclear wavefunctions to be used in the computation of the

19 (45-122/1977)

expectation value must correspond to a constant density everywhere (a fall-off of the density implies the existence of a nuclear surface). Such a hypothetical nucleus is called the infinite nuclear matter, and our explicit calculation later shows that the expectation value just defined is proportional to $A(=N+Z)$. This, as explained in Section 1, is the saturation property of the typical nuclear forces in a nucleus. The first term of the binding energy formula is therefore a term proportional to A and is defined by

$$\text{volume energy} = -a_V A, \tag{II.2a}$$

which provides the *predominantly* attractive contribution to the total energy $E(Z, N)$; this is because the entire effect of the attractive nucleon-nucleon forces is contained in this particular term. In the empirical formula, of which (II.2a) is merely one term, the constant a_V is determined by fitting known nuclear masses, and is found to have a value approximately 15 MeV. A more precise value is given in Table II.1. From what has just been said, it is clear that in a *calculation* of nuclear binding energy based on the hypothetical nuclear matter concept the final calculated value has to be compared with only the volume energy term.

The binding energy of an actual nucleus differs from (II.2a) because of the limitations involved in the concept of the 'hypothetical' nucleus. The limitations must therefore be removed by adding several correction terms to (II.2a). One such term must take into account the effect of the nuclear surface for an actual nucleus. Other terms must allow for the Coulomb repulsion amongst the protons, and the fact that most medium-heavy and heavy nuclei have $N > Z$, and not $N = Z$, as implied in the concept of the volume energy term. These terms are now introduced.

Surface Energy

In the empirical formula, all that needs insertion is the dependence of the surface energy term on the quantities N, Z, and A. The subsequent argument shows that it depends on only A explicitly and is proportional to $A^{2/3}$. The way the nuclear surface affects binding energy is by preventing each nucleon near the surface from having neighbours on *all* sides to interact with. A nucleon deep inside the nucleus interacts with all the nucleons contained in a *sphere* of radius equal to the range of the nuclear force and thus contributes to the total binding energy its fullest share, which, according to the concept of the volume energy term, is roughly 15 MeV per nucleon. A nucleon near the surface, on the contrary, interacts with only the nucleons contained in a *part* of the *sphere* just described because the remainder of the sphere, with the nucleon at its centre, is outside the nuclear surface and contains no nucleons to interact with. As a result, it contributes less than its share of the volume energy per nucleon. Thus, the correction to the *negative* volume energy due to the surface effect must be a term with a *positive* sign and this must be proportional to the total surface area of the nucleus. In this sense, this particular correction term is akin to the surface tension effect of a liquid drop. Since the radius of a nucleus is given by $r_0 A^{1/3}$, the surface area is proportional to $A^{2/3}$, and hence we write the surface energy term as

$$\text{surface energy} = +a_S A^{2/3}. \tag{II.2b}$$

If the nuclear shape is spherical, then the surface area is $4\pi r_0^2 A^{2/3}$ and, in this case, the empirical parameter a_S can be taken to define the nuclear surface tension S as

$$a_S = 4\pi r_0^2 S. \tag{II.2b$'$}$$

The surface area of a deformed nucleus, however, is different, and the expression (II.2b) therefore changes. In the MS mass formula, this effect of deformation has been included.

Coulomb Energy

Once again, if we want the dependence of the Coulomb energy on the quantities Z, N, and A, this can be very easily inferred from the classical ideas of a charged liquid drop. For a spherical drop of radius R having a total charge Ze, the classical expression for the Coulomb energy is given by

$$\frac{3}{5}\frac{(Ze)^2}{R}.$$

The simplest way to derive this classical expression is to consider the Coulomb potential on a spherical shell of thickness dr at radius r. The charge contained in this shell is $4\pi r^2\rho\, dr$, where ρ is the charge per unit volume; the Coulomb potential on this shell due to the sphere of radius r is given by $(4\pi/3)\rho r^3/r = (4\pi/3)\rho r^2$. Therefore, the Coulomb energy gained in putting the charge on the shell around the sphere of radius r, which is a product of the charge and the potential, is expressed as

$$\left(\frac{4\pi}{3}\right)^2 3\rho^2 r^4\, dr.$$

The Coulomb energy gained in building up the charge on the entire sphere, starting with $r = 0$ and ending with $r = R$, is then denoted by

$$\left(\frac{4\pi}{3}\right)^2 3\rho^2 \int_0^R r^4\, dr = \frac{3}{5}\frac{(Ze)^2}{R}$$

because

$$\frac{4\pi}{3}\rho R^3 = Ze.$$

The same expression holds for the major part of the Coulomb energy, worked out quantum mechanically by employing correct many-body wavefunctions. Minor corrections arise due to the effect of the Pauli exclusion principle. We shall ignore these finer details, and use in our mass formula

$$\text{Coulomb energy} = \frac{3}{5}\frac{e^2}{r_0}\frac{Z^2}{A^{1/3}} = a_C\frac{Z^2}{A^{1/3}}. \qquad \text{(II.2c)}$$

The value of the parameter a_C is therefore essentially determined by the quantity r_0, which is known from experimental data on nuclear radii. Taking $r_0 = 1.2$ fm, we obtain $a_C = 0.71$ MeV. The Coulomb parameter in the mass formula is thus the easiest to determine.

Symmetry Energy

The symmetry energy part of the binding energy formula represents the addition that is necessary to account for the unequal number of neutrons and protons in many stable nuclei. The dependence of this term on the quantities N, Z, and A is not obvious. In Section 21B, we note that, with model nuclear wavefunctions, the correction behaves as $Af(T_z/A)$, where $T_z = \frac{1}{2}(N - Z)$. For all stable nuclei, the quantity (T_z/A) is very small, and hence while writing the correction term it is quite adequate to take account of only the leading term in the function f

of T_z/A. It will also be shown that the leading term is quadratic. Therefore, we write

$$\text{symmetry energy} = a_V x_V A^{-1} T_z^2 = \tfrac{1}{4} a_V x_V \frac{(N-Z)^2}{A}, \tag{II.2d}$$

where a_V is given by (II.2a); the symmetry energy is therefore determined by the new independent parameter x_V.

Combining (II.2a) and (II.2d), we can write the volume and this symmetry energy term (the mass formula as

$$-a_V A(1 - x_V I^2), \tag{II.2d'}$$

where $I = A^{-1}T_z$. In this sense, the symmetry energy hitherto referred to can be described as the *volume* symmetry energy. This terminology is in keeping with the MS mass formula.

In the MS mass formula, the authors introduce an additional *surface* symmetry energy term along with the surface energy term (II.2b) and write the combined terms as

$$+a_S A^{2/3}(1 - x_S I^2). \tag{II.2d''}$$

However, while choosing numerical values of the parameters, they put $x_V = x_S$.

The quantity T_z used in the expressions (II.2d) is the third component of the isospin **T** for the nucleus. The concept of isospin for the individual nucleons and for the two-nucleon system (after coupling the two individual isospins) has been introduced in Chapter I. In keeping with the ideas there, the concept can be extended to a nucleus made up of many nucleons. The total isospin **T** is a *vector* sum of the individual isospin $\frac{1}{2}$ of the nucleons. The quantum number T in the eigenvalue $T(T+1)$ of T^2 can therefore have many possible values consistent with the coupling rules. (For instance, it had the two values 1 and 0 in the two-nucleon system.) The situation for the third component of **T**, however, is very simple. Since the component is additive, T_z is obtained by adding the third component of isospin of the individual nucleons. According to our convention, each neutron and each proton contributes respectively $+\frac{1}{2}$ and $-\frac{1}{2}$ to this component, and therefore

$$T_z = \tfrac{1}{2}(N - Z),$$

as defined earlier. Since T_z is the third component of **T**, its value, according to angular momentum rules, must range from $+T$ to $-T$, integrally spaced. Therefore, we conclude that the *minimum* value of T *for a given nucleus* is equal to the value of its T_z, i.e., $\frac{1}{2}(N - Z)$. The maximum value of T for the same nucleus is $\frac{1}{2}A$ because A individual vectors of magnitude $\frac{1}{2}$ have been coupled to obtain T.

Pairing Energy

In order to understand the origin of the pairing energy term, let us first study the question of stability of nuclei against β-decay. In β^--decay, the charge of the nucleus (Z) increases by one, whereas it decreases by one in β^+-decay and electron capture. In all these processes, the mass number A of the nucleus does not change. If we consider atomic masses, and if $M(A, Z) > M(A, Z+1)$, then the nucleus (A, Z) can obviously undergo β^--decay spontaneously and change itself to the nucleus $(A, Z+1)$. On the other hand, if the atomic mass $M(A, Z) > M(A, Z-1)$, then the nucleus undergoes spontaneous β^+-emission or K-capture and changes into the nucleus $(A, Z-1)$.

So, for stability against β-decay, the atomic mass $M(A, Z)$ has to be smaller than both $M(A, Z+1)$ and $M(A, Z-1)$. Since our purpose now is to study the stability of nuclei with a given value of A and different values of Z, we replace N by $(A - Z)$ in (II.1) and (II.2d) and note that, for a given A, the dependence of M on Z is quadratic, and hence $M(A, Z)$ plotted as a function of Z is a parabola (the so-called *mass parabola*), as shown in Fig. II.1a.

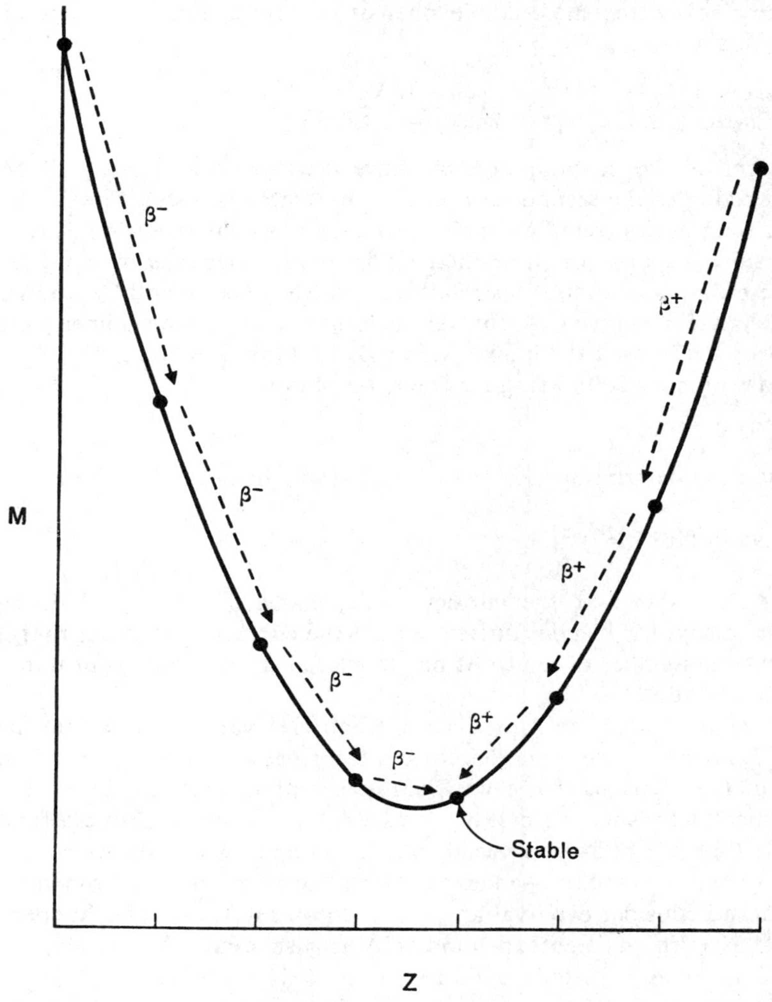

(a) Mass parabola without pairing energy term

Fig. II.1 Mass parabola and single-particle energy levels (cont.).

All the nuclei in this figure decay (as indicated by the arrows) to their neighbours, except the one at the minimum point, which is predicted to be stable. Recalling what was said after (II.1) about atomic masses, we derive

$$c^2 \frac{\partial M(A, Z)}{\partial Z}\bigg|_{\text{given } A} = (M_H - M_n)c^2 + \frac{2a_C}{A^{1/3}}Z - a_V x_V \frac{A - 2Z}{A}.$$

In obtaining this expression, we have used (II.2a)–(II.2d) for the binding energy E. It is clear that the volume and surface energies have made no contribution to it, whereas the Coulomb and symmetry energies have given rise to the last two terms. If we equate the expression to zero, we clearly get the value of Z that corresponds to the minimum point of the parabola. This Z therefore gives us the most stable nucleus—stable against all types of β-decay—with the given A. In this way,

$$\frac{Z}{A} = \frac{a_V x_V + (M_n - M_H)c^2}{2a_V x_V + 1.42A^{2/3}} = \frac{a_V x_V + 0.79}{2a_V x_V + 1.42A^{2/3}}.$$

In the second term of the denominator, we have used the value of a_C (in MeV) mentioned after (II.2c); the value of the second term in the numerator is also in MeV. The point to note here is that this relationship remains unchanged even after the inclusion of the pairing energy term (to be introduced soon), provided the pairing energy depends on only A and not explicitly on Z. Therefore, we can use this relation, with the experimentally known Z-value of the most stable nucleus for a given A, for the determination of the symmetry energy parameter $a_V x_V$. In his work, Bethe used the known value $Z = 80$ for $A = 200$ (${}_{80}Hg^{200}$ is the most stable nucleus with mass number 200); using this fact, we obtain

$$a_V x_V = 93.2 \text{ MeV},$$

and then the equation determining Z for the most stable nucleus of a given A reads

$$\frac{Z}{A} = (1.98 + 0.015A^{2/3})^{-1}.$$

It is clear that Z/A starts at the correct value, namely, $\approx 1/2$, and then decreases as A increases; that is to say, for heavier nuclei, we get the correct prediction: that Z has to be less than $\frac{1}{2}A$ and that the number of neutrons has to exceed the number of protons in order that a stable nucleus is formed.

Now, *in general*, this formula predicts a fractional value of Z at the minimum of the mass parabola; the nearest integer is therefore the correct prediction for Z. There is nothing in the derivation that tells us that even Z-values are predicted more often than odd Z-values or vice versa. Since there is no preference for odd or even Z in the formula for the most stable one, we conclude that, for odd-mass nuclei, stable isotopes with odd or even Z are, statistically speaking, equally abundant—an expectation borne out by experimental facts. On the other hand, when we consider even values of A, if the most stable one happens to have even Z, then, since A is even, its neutron number N is also even. Conversely, if this particular nucleus has odd Z, then N also is odd. Experimentally, however, we know that even-even nuclei are much more stable than odd-odd nuclei. The stability of the latter is so poor that, with the exception of H^2, Li^6, B^{10}, and N^{14}, there is no other odd-odd stable nucleus in the periodic table. It is this shortcoming of a mass formula with only the four terms (II.2a) to (II.2d) that necessitates the addition of the pairing energy term. How such a term provides the remedy for even A nuclei is qualitatively demonstrated in Fig. II.1b. We deliberately insert an extra attractive and a repulsive contribution in the even-even and odd-odd nuclei, respectively, whereas no such extra contribution is made to odd-mass nuclei. Thus,

$$\text{pairing energy} = \begin{cases} +\delta f(A) & \text{(odd-odd)} \\ 0 & \text{(odd-mass)} \\ -\delta f(A) & \text{(even-even)} \end{cases} \tag{II.2e}$$

where δ is a constant parameter, and the function f depends on only A. It is clear that the pairing energy term keeps the mass parabola for odd-mass nuclei unaltered (see Fig. II.1a). But it produces two parabolas for even A nuclei, as depicted in Fig. II.1b, where the upper

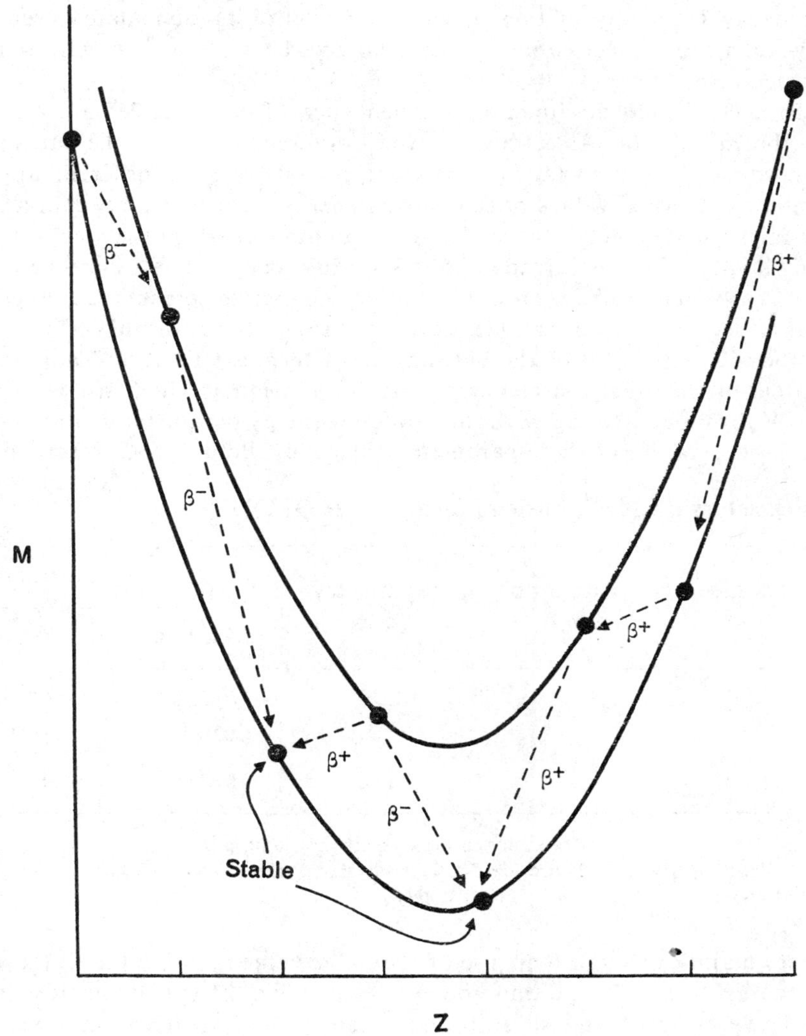

(b) Mass parabolas (solid lines) with inclusion of pairing energy term in even-mass nuclei

Fig. II.1 Mass parabola and single-particle energy levels (cont.).

one corresponds to odd-odd nuclei and the lower one to even-even nuclei. The *upward* and *downward* shifts in the two cases are caused respectively by the *repulsive* and *attractive* contributions from (II.2e). As is shown by the arrows, now the odd-odd nucleus (Z) at the minimum point of the upper curve can undergo β-decay to its neighbours $Z \pm 1$ on the even-even curve. In the case of the even-even curve, not only is the nucleus near the minimum stable, but some of its immediate neighbours on either side are also expected to be stable, provided each such member (Z) on the even-even curve is below its neighbours $Z \pm 1$ on the odd-odd curve. This is because the decay from any of these even-even nuclei to its immediate lower neighbour on the same curve can take place only through double-beta decay, which is an extremely improbable process.

Both the constant δ and the functional dependence of A in (II.2e) are determined by a detailed fit to the mass data. As a result, several different forms for the pairing energy term are currently in vogue. It is important to note that, in spite of their different appearances, all these terms predict numerical values of the pairing energy term that lie within a fraction of an MeV, even for fairly heavy nuclei. Since the total binding energy of a nucleus with a mass number in the vicinity of 100 is already about 800 MeV, such a small discrepancy in the pairing energy term really makes, from author to author, negligible percentage change in the nature of the actual fit to the binding energy data secured through their formulas.

Some of the different forms of the pairing energy term are listed in Table II.1, which also specifies the values of the other parameters in the mass formula. The first row gives the parameters of the BW formula, starting with the Coulomb energy parameter a_C in which $r_0 = 1.2$ fm has been used. Earlier values of these parameters listed by Bethe‡ and Fermi differ from the

Table II.1 Parameters (in MeV) of mass formula [see (II.2)]

Volume (a_V)	Surface (a_S)	Coulomb (a_C)	Symmetry ($a_V x_V$)	Pairing		
				Author	$f(A)$	δ
−15.75	17.8	0.71	93.2	Fermi*	$A^{-3/4}$	33.5
				WB**	A^{-1}	135
−15.677	18.56	0.717	112.2	MS†	$A^{-1/2}$	11

*Fermi, E., Nuclear Physics, University of Chicago Press, Chicago, 1950, p 6.
**Blatt, J. M., and Weisskopf, V. F., Theoretical Nuclear Physics, Wiley, New York, 1952, p 232.
†Myers, W. D., and Swiatecki, W. J., *Nucl. Phys.*, **81**, 1 (1966).

values listed here mainly through their use of the older value $r_0 = 1.5$ fm in the evaluation of the Coulomb energy term. The second row gives the values of the parameters in the MS formula (which, however, contains several additional terms); the volume symmetry energy parameter differs from the value of that in the BW formula because of the presence of the additional surface symmetry energy term in the MS formula. For the symmetry energy term, we have listed the complete parameter $a_V x_V$; if a value of x_V is required, it can be very easily derived. The Myers-Swiatecki paper lists the parameter $\frac{1}{4}x_V$ (and calls it κ) and quotes its value as 1.79, which has been used in obtaining 112.2 (in the second row). We have already explicitly shown how the values of a_C and $a_V x_V$ are obtained in practice. Since, by the definition (II.2e), the pairing energy term for odd-mass nuclei is zero, in principle, the value of a_V and a_S in the

‡Bethe, H. A., *Revs. Mod. Phys.*, **8**, 165 (1936).

first row could be obtained by fitting the experimental masses of any two odd-mass nuclei. The values -15.75 and 17.8 listed in the table have, however, been computed from a least-squares fit to the masses of a large number of nuclei, rather than only two.

Stability of Nuclei

The mass formula is very useful in studying the stability of nuclei against the emission of various particles such as β, α, and a nucleon. The case of β-decay has already been used while introducing the concept of the pairing energy. The criteria for the stability against α- or nucleon emission can be worked out as follows. The initial nucleus has the mass $M(Z, N)$, and, if it has to emit an α-particle having two neutrons and two protons, the residual nucleus is $(Z-2, N-2)$. If the final mass $M(Z-2, N-2)+M(\alpha)$ is smaller than $M(Z, N)$, then clearly the decay is energetically permitted. Thus, the criterion for α-stability is given by

$$M(Z, N) - M(Z-2, N-2) < M(\alpha),$$

where $M(\alpha)$ is the mass of the α-particle. The nucleon masses cancel out from the two sides and the inequality can be rewritten as

$$E(Z, N) - E(Z-2, N-2) < E(\alpha)$$

or

$$E(Z-2, N-2) + E(\alpha) - E(Z, N) > 0 \qquad [(Z, N) \text{ is stable}].$$

Since the energies are negative, we can write the condition for *instability*, in terms of the binding energies $|E|$, as

$$|E(Z-2, N-2)| + |E(\alpha)| - |E(Z, N)| > 0 \qquad [(Z, N) \text{ is unstable}].$$

If this condition is satisfied, then the positive value of the expression gives the kinetic energy of the emitted α-particle. In a practical computation, the binding energy of the α-particle is taken from experimental data, and that of the two heavy nuclei with the mass numbers $A(=N+Z)$ and $A-4$ is computed from the mass formula, using the relationship between Z and A which corresponds to the minimum of the mass parabola for a given value of A. The kinetic energies of the α-particles emitted by naturally occurring radioactive elements (such as Ra with $A = 226$) computed in this manner are of the right order of magnitude. As a matter of fact, the kinetic energy of the emitted α as a function of increasing A starts becoming positive for A around 150, that is, much before the natural α-emitters are encountered. However, the computed kinetic energy for such nuclei is very small, and therefore the probability of penetration through the Coulomb barrier turns out to be almost prohibitive.

In the same manner as for α-emission, the condition of *instability* against neutron emission is obtained as

$$|E(Z, N-1)| - |E(Z, N)| > 0.$$

Here the term analogous to $E(\alpha)$ in the preceding expression is absent because the nucleon does not have any intrinsic binding energy. For all *stable* elements, this condition of *instability* is, as expected, not satisfied.

Discussion on Mass Formula in Terms of Single-Particle Model

A few remarks are now made to provide some insight into the terms of the mass formula. We already know that the volume binding energy term is a sum of the kinetic energy of the nucleons and their potential energy of interaction *through the nuclear forces*. The former is

20 (45-122/1977)

positive whereas the latter is attractive. Now, in both the kinetic energy term and the potential energy term, it becomes energetically advantageous (i.e., helps in reducing the energy) for a nucleus with a given A to make $N = Z = \frac{1}{2}A$. When $N \neq Z$, the kinetic energy can be shown to increase quadratically with $(N - Z)/A$ (this is explicitly shown in Section 21B). For the part of the potential energy arising from *nuclear forces*, a similar observation holds. In the case of potential energy, it is not very easy to demonstrate this fact explicitly; it can, however, be made plausible by qualitative arguments in terms of the occupancy of single-particle orbitals by the nucleons in a nucleus. There is evidence (see Section 24) in nuclei that nucleons, in the lowest order of approximation, can be ascribed single-particle quantum states corresponding to the energy levels in an average potential field. Now, if this is accepted as an established fact, then it is easy to understand why two neutrons (with opposite spin) and two protons like to occupy each single-particle spatial state of the average potential well. This produces the best possible overlap in the spatial part of their wavefunction, and hence allows them to make their potential energy of interaction as strong as possible. As a matter of fact, we have noted in Section 1 that nuclear forces do show a saturation effect in groups of two neutrons and two protons. Therefore, as long as we consider only the nuclear forces, the minimum of the attractive potential energy is secured if $N = Z = \frac{1}{2}A$ = even integer. This condition can be satisfied only if A is a multiple of 4; we shall call such nuclei $4n$-nuclei, where n is an integer. If A is even, then there is a possibility that it may also be of the $(4n + 2)$-type. In the latter case, the preceding criterion requiring $N = Z =$ *even integer* cannot be satisfied; however, it is clear that, for a minimum of energy, the nucleus would like to choose its neutron and proton numbers in such a way that the saturated single-particle spatial states, each with two protons and two neutrons, are the maximum possible. In other words, only two nucleons may be left over for a partially filled single-particle state. For these last two nucleons, nuclear interaction energy becomes more attractive if one of them is a neutron and the other a proton (because n-p interaction is stronger than n-n or p-p interaction). Thus, for all even-A nuclei, there is a tendency to make the neutron and proton numbers equal. [The observant reader will note that the most stable $(4n + 2)$ type nucleus has been predicted to be odd-odd (i.e., odd-neutron and odd-proton number) according to these simple considerations. As we have already learnt, the pairing energy effects cause departure from this tendency.] As soon as there is a departure from this condition, the number of completely filled single-particle spatial states decreases, and hence we expect the attractive potential energy to be weaker; the energy then increases compared with that in the $N = Z$ case. Since the situation $N = Z$ corresponds to a minimum, any departure from this condition must give rise to a *positive* term in energy quadratic in $(N - Z)/A$; a linear term would necessarily imply that $N = Z$ was not the condition for a minimum. All this qualitative description throws some light on the mechanism of the volume binding energy, and the origin of the symmetry energy in the case $N \neq Z$.

Let us now examine the consequence of the Coulomb force. To start with, the Coulomb energy is small for light nuclei. But as we go on to heavier and heavier nuclei, for a given A, the value $Z = \frac{1}{2}A$, determined by the considerations on volume and symmetry energy terms, no longer remains energetically advantageous. Because of the large Coulomb energy, the nucleus then tends to make Z smaller than $\frac{1}{2}A$ and thereby reduces a repulsive contribution to its energy. However, we have seen that as soon as N becomes unequal to Z the repulsive symmetry energy term comes into play. Therefore, the tendency of the nucleus to reduce Z does not go unchecked. The competition between the Coulomb and symmetry energy terms determines the value of $Z(<\frac{1}{2}A)$ for which the minimum of energy can be obtained.

Finally, we ask whether the origin of the pairing energy term also can be understood from the single-particle picture invoked for the volume and symmetry energy. We have seen that, as a result of the competition between the Coulomb and symmetry energy, stable heavy and medium-heavy nuclei have $N \neq Z$, and hence there are at least a few single-particle spatial states, each with less than its full quota of two neutrons and two protons. A stable even-even nucleus (N, Z) has levels up to a certain maximum filled with two neutrons and two protons, and the remaining $(N - Z)$ neutrons go in pairs to levels lying immediately above the completely filled ones, as shown in Fig. II.1c where circles with oppositely directed arrows denote nucleons

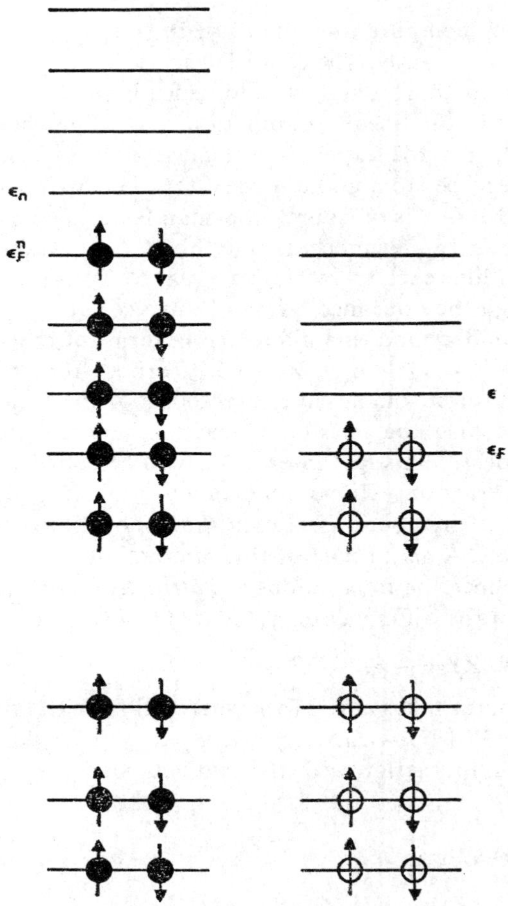

(c) Single-particle energy levels of neutrons and protons

Fig. II.1 Mass parabola and single-particle energy levels.

with opposite spin orientation. The odd-mass nucleus $(N, Z-1)$ is therefore obtained by removing a proton from the *topmost filled* proton level (ϵ_F, say, which is called the fermi level). The odd-mass nucleus $(N, Z+1)$ is similarly obtained by putting a proton in the first available level above ϵ_F, for instance, the level ϵ which already contains two neutrons.

Let us now consider the mass difference

$$c^2[M(N, Z+1) - 2M(N, Z) + M(N, Z-1)].$$

Since the masses of the individual nucleons cancel out, this expression is also equal to

$$E(N, Z+1) - 2E(N, Z) + E(N, Z-1).$$

The mass data can be used to compute this quantity. In this special combination, the volume energy terms clearly cancel out exactly. The cancellation of the surface, symmetry, and Coulomb terms is not exact; however, for fairly large A (and hence large N, Z and large neutron excess), the difference of these terms in the linear combination of binding energies (as just specified) is expected to be very small. Yet the experimental mass data on the even-even and odd-mass nuclei require this difference to be from about $\frac{1}{2}$ MeV to a few MeV. According to the complete mass formula, this observed result is of course the manifestation of the pairing energy term. Since the attractive pairing energy term occurs only in $E(N, Z)$, and not in the odd-mass values $E(N, Z+1)$, we expect the difference to be almost equal to $2\delta f(A)$; a slight departure from this interpretation occurs through the noncancellation of surface, symmetry, and Coulomb energy terms. Now let us try to understand this difference in terms of the single-particle picture just given. The difference $E(N, Z-1) - E(N, Z)$ clearly arises from the missing single-particle energy (ϵ_F) of the removed proton, the missing interaction of the latter with the *two* neutrons belonging to its orbit, and also the missing interaction between the removed proton with its partner proton in the same orbit. This last interaction will be called the pairing interaction and denoted by $-G$; the n-p interaction will be denoted by $-\bar{G}$. It is clear that, for simplicity, we are assuming the *residual* interaction between the removed proton and nucleons belonging to other orbits to be negligible. A major part of this interaction, however, determines the single-particle energy ϵ, ϵ_F, ... Since the interactions are *attractive*, the *missing* interactions give rise to *positive* contributions and the difference $E(N, Z-1) - E(N, Z)$ is then

$$E(N, Z-1) - E(N, Z) = -\epsilon_F + 2\bar{G} + G.$$

The factor 2 in the second term takes care of the interaction with *two* neutrons. In the same manner, the difference $E(N, Z+1) - E(N, Z)$ arises partly from the single-particle energy (ϵ) of the added proton, and its interaction with the two neutrons in its orbit. Thus,

$$E(N, Z+1) - E(N, Z) = \epsilon - 2\bar{G}.$$

Adding these two results, we obtain

$$E(N, Z+1) - 2E(N, Z) + E(N, Z-1) = (\epsilon - \epsilon_F) + G.$$

In general, the single-particle levels ϵ and ϵ_F near the proton fermi level ϵ_F are very close to each other, unless of course the particular nucleus has protons in completely closed shells. (See Section 23 for the meaning of "closed shells".) In the latter case, $(\epsilon - \epsilon_F)$ is the difference between the highest level of the closed shells and the lowest level of the next higher shell; this difference is usually of the order of 10 MeV or more. Inside a shell, however, the various single-particle levels are very close together, and the pairing interaction G predominates over $(\epsilon - \epsilon_F)$ in the difference of the binding energies. This is why the terminology "pairing energy" is used to denote the term $\pm\delta f(A)$ of the mass formula.

In the manner just described, we can consider also the nuclei $(N \pm 1, Z)$, and compute from the mass data

$$c^2[M(N+1, Z) - 2M(N, Z) + M(N-1, Z)]$$

or, equivalently,

$$E(N+1, Z) - 2E(N, Z) + E(N-1, Z).$$

According to Fig. II.1c, the nucleus $(N-1, Z)$ is obtained by removing a neutron from the last occupied neutron level (energy $= \epsilon_F^n$), whereas the nucleus $(N+1, Z)$ is obtained by putting an extra neutron in the level (ϵ_n) just above ϵ_F^n. The difference $E(N-1, Z) - E(N, Z)$ comes from the missing pairing interaction $-G_n$ between the removed neutron and its partner in the same orbit, and its missing single-particle energy ϵ_F^n. On the other hand, the difference $E(N+1, Z) - E(N, Z)$ results entirely from the extra single-particle energy ϵ_n. Thus,

$$E(N-1, Z) - E(N, Z) = -\epsilon_F^n + G_n,$$

$$E(N+1, Z) - E(N, Z) = \epsilon_n,$$

and hence

$$E(N+1, Z) - 2E(N, Z) + E(N-1, Z) = (\epsilon_n - \epsilon_F^n) + G_n.$$

Once again, this expression is also identical to the one for the protons. Unless the neutrons in the nucleus denoted by (N, Z) belong to completely closed neutron shells, $(\epsilon_n - \epsilon_F^n)$ is small, and hence the difference in the binding energies considered here is roughly equal to the pairing energy between a neutron pair.

In the same manner as just described, we can now obtain the expressions

$$E(N+1, Z-1) - E(N, Z) = \epsilon_n - \epsilon_F + G + 2\bar{G},$$

$$E(N-1, Z+1) - E(N, Z) = \epsilon - \epsilon_F^n - 2\bar{G} + G_n$$

or

$$E(N+1, Z-1) - 2E(N, Z) + E(N-1, Z+1) = [(\epsilon - \epsilon_F) + G] + [(\epsilon_n - \epsilon_F^n) + G_n]$$
$$\approx G + G_n.$$

Here the special linear combination of the binding energies can be related to the mass data on the even-even nucleus (N, Z) and the odd-odd nuclei $(N+1, Z-1)$ and $(N-1, Z+1)$ as

$$c^2[M(N+1, Z-1) - 2M(N, Z) + M(N-1, Z+1)].$$

In most cases, the odd-odd nuclei are unstable, and hence the quantity cannot be directly computed from mass data. However, the β-decay energies indirectly give the quantities $c^2[M(N+1, Z-1) - M(N, Z)]$ and $c^2[M(N-1, Z+1) - M(N, Z)]$. Once again, according to the foregoing expressions, the special linear combination depends mainly on the pairing interaction between two protons and between two neutrons. This interpretation is in agreement with the semi-empirical mass formula, according to which the aforementioned linear combination of masses depends predominantly on the pairing energy term and is equal to $4\delta f(A)$.

The foregoing description, which is in terms of a set of single-particle levels and the pairing interaction between identical nucleons in the same spatial state, but having opposite spin directions, is a somewhat oversimplified version of the actual situation. The pairing interaction as defined here is made more general in Chapters III and VI. The present simple interaction between identical nucleons in the same spatial states but opposite spin states was first considered by Bethe (see Table II.1). Later, Mayer generalized the concept in connection with the *jj*-coupling shell model (see Section 24). According to the more recent nuclear structure theory (Section 41), the generalized pairing interaction leads to expressions for all the aforementioned mass differences, which can be obtained by consistently replacing quantities of the type $[(\epsilon - \epsilon_F) + G]$ and $[(\epsilon_n - \epsilon_F^n) + G_n]$ by $[(\epsilon - \epsilon_F)^2 + \Delta^2]^{1/2}$ and $[(\epsilon_n - \epsilon_F^n)^2 + \Delta_n^2]^{1/2}$, respectively. The quantity Δ or Δ_n is called the energy gap parameter and is related to the corresponding strength G or G_n of the pairing interaction through (VI.95). In this type of nuclear structure theory (described in Section 41), the effect of pairing in a set (rather than in just one) of close-lying single-particle levels has been carefully considered. If ϵ is taken nearly equal to ϵ_F, and the pairing theory concerns a pair in a single level, then (VI.95) leads to $\Delta \approx \frac{1}{2}G$, a result with which the mass differences (already worked out according to our simple pairing picture) is in perfect agreement.

15. BASIC MATHEMATICAL APPARATUS FOR MANY-BODY THEORY

In Chapter I, we have obtained fairly reliable information on the two-nucleon interaction, and described several alternative two-nucleon potentials which explain all the two-body data quite accurately. In a nucleus made up of more than two nucleons, the total nucleon interaction is obviously a sum of the interaction between all the pairs of nucleons. In addition, there could be typical many-body interactions that are not describable in terms of two-body interactions. However, it is always worthwhile to start with the simplest ideas, and bring in more complicated concepts only when the experimental data absolutely demand them. Nuclear data have so far been quite well-understood in terms of two-body interactions alone, i.e., without resorting to more complicated three- or more-than-three-body forces.

For a nucleus comprising A nucleons, we shall therefore work with the Hamiltonian

$$H = \sum_{i=1}^{A} T(i) + \sum_{i<j}^{A} V(ij), \tag{II.3}$$

where T denotes the single-particle kinetic energy operator, and V the two-nucleon potential. The labels enclosed within the parentheses refer to particle coordinates. The restriction $i < j$ in the second sum takes care of the fact that the interaction has to be summed, counting each pair (ij) only once. The Schrödinger equation

$$H\Psi(1, 2, \ldots, A) = E\Psi(1, 2, \ldots, A) \tag{II.4}$$

cannot, in general, be solved in a straightforward manner. Apart from the difficulties of the hard-core potential, the wavefunction itself contains $3A$ spatial coordinates, and an equal number of spin and isospin coordinates. When $A > 2$, (II.4), in which many coordinates are coupled to each other through the two-body potential, cannot be solved exactly. We must therefore take recourse to approximate methods.

In the quantum mechanics of all many-body systems, we start with an approximate Hamiltonian

$$H_0 = \sum_{i=1}^{A} [T(i) + \mathcal{V}(i)], \tag{II.5}$$

where $\mathcal{V}$ is a suitably defined one-body potential. Solving the Schrödinger equation corresponding to this Hamiltonian is easy because it consists of one-body terms that *do not* couple the coordinates of different nucleons. First, we find a set of single-particle eigenvalues

$\epsilon_\alpha, \epsilon_\beta, \ldots$ and the corresponding single-particle functions $\phi_\alpha, \phi_\beta, \ldots$ by solving the one-body Schrödinger equation

$$(T + \mathcal{V})\phi_\alpha = \epsilon_\alpha \phi_\alpha; \tag{II.6}$$

then an A-particle product wavefunction

$$\Phi_{\alpha\beta\ldots\zeta}(1, 2, \ldots, A) = \phi_\alpha(1)\phi_\beta(2)\ldots\phi_\zeta(A) \tag{II.7a}$$

obviously satisfies the many-body equation

$$H_0\Phi_{\alpha\beta\ldots\zeta}(1, 2, \ldots, A) = E_0\Phi_{\alpha\beta\ldots\zeta}(1, 2, \ldots, A) \tag{II.7b}$$

with

$$E_0 = \epsilon_\alpha + \epsilon_\beta + \ldots + \epsilon_\zeta. \tag{II.7c}$$

The many-body eigenfunctions of (II.5) are therefore products of single-particle wavefunctions of the type (II.7a), and the eigenvalue E_0 is the sum of the corresponding single-particle energies. The many-body eigenfunctions of the simple Hamiltonian H_0 are labelled by A sets of single-particle quantum numbers $\alpha\beta\ldots\zeta$. (It should be noted that each of the letters $\alpha, \beta\ldots$ is an abbreviation for the full set of quantum numbers needed to specify a single-particle state.) There can evidently be several such many-body wavefunctions corresponding to the various ways of choosing the A single-particle states $\alpha\beta\ldots\zeta$.

Our ultimate aim, however, is to solve (II.3), and hence we rewrite it as

$$H = H_0 + H_I, \tag{II.8}$$

where

$$H_I = \sum_{i<j}^{A} V(ij) - \sum_{i=1}^{A} \mathcal{V}(i). \tag{II.9}$$

Since the solutions of H_0 are already known, the many-body calculation now reduces to an approximate treatment of the residual interaction H_I by perturbation techniques. The rapidity of convergence of the perturbation treatment depends on the strength of the *residual* interaction, which can be controlled by a judicious choice of the one-body potential $\mathcal{V}$. A choice of $\mathcal{V}$ from such a criterion obviously relates the one-body potential to the given two-body potential, averaged over one of the particles.

For the present, we do not consider the niceties in the choice of $\mathcal{V}$, but only assume that a one-body potential exists such that a perturbation treatment of (II.9) becomes meaningful. When we do need to use explicit forms for the single-particle wavefunctions ϕ_α, we shall use semi-empirical forms for $\mathcal{V}$, dictated partly by certain types of experimental data and partly by the convenience of solving the one-body equation (II.6). These model one-body potentials and wavefunctions are described in Section 17. In many structural calculations, we need not be overconcerned about refining $\mathcal{V}$; rather we can, in good faith, extensively use the model single-particle wavefunctions. Should a refined calculation of $\mathcal{V}$ be desired, these model single-particle wavefunctions may be used as basis wavefunctions. Such calculations are dealt with in Section 35.

Before we proceed with the development of the many-body perturbation theory, we need to incorporate in the many-body unperturbed wavefunction (II.7a) a very important requirement, namely, antisymmetry under the exchange of any two nucleons demanded by the Pauli exclusion principle. We therefore now present the details of this antisymmetrization, and the method of calculating the matrix elements of one-body and two-body symmetric operators using the antisymmetric many-body wavefunctions. In view of the form (II.9) of the residual interaction, such matrix elements appear in the perturbation calculation.

A. ANTISYMMETRIZATION

Let us consider the many-body Hamiltonian (II.3) or (II.5). If we make any permutation of the particle indices, these Hamiltonians remain unchanged. If P is the operator denoting the permutation, then the invariance property is expressed by

$$PH_0P^{-1} = H_0$$

or

$$PH_0 = H_0P. \tag{II.10}$$

The same expressions hold for H as well. Although we restrict our discussion to the eigenfunctions of H_0, our remarks are true for the eigenfunctions of H too.

Operating with P on (II.7b), we obtain

$$PH_0\Phi_{\alpha\beta\ldots\zeta}(1, 2, \ldots, A) = E_0P\Phi_{\alpha\beta\ldots\zeta}(1, 2, \ldots, A),$$

and hence, using (II.10) on the left-hand side,

$$H_0P\Phi_{\alpha\beta\ldots\zeta}(1, 2, \ldots, A) = E_0P\Phi_{\alpha\beta\ldots\zeta}(1, 2, \ldots, A).$$

Thus, if $\Phi_{\alpha\beta\ldots\zeta}(1, 2, \ldots, A)$ is an eigenfunction of H_0 with the eigenvalue E_0, then $P\Phi_{\alpha\beta\ldots\zeta}(1, 2, \ldots, A)$ is also an eigenfunction of H_0 belonging to the same eigenvalue; therefore, any linear combination of all the $P\Phi_{\alpha\beta\ldots\zeta}(1, 2, \ldots, A)$, where P is either $\mathbb{1}$ or any other general permutation of the particle indices, is also an eigenfunction of H_0 belonging to the eigenvalue E_0. This result therefore provides us with the method of antisymmetrizing (II.7a) without interfering with the property requiring the resultant function to be still an eigenfunction of H_0 belonging to the known eigenvalue E_0. The procedure entails superposing all $P\Phi_{\alpha\beta\ldots\zeta}(1, 2, \ldots, A)$ in such a way that the sum becomes antisymmetric under the exchange of any two nucleon indices.

To arrive at the required linear sum, we consider first the simpler cases of two- and three-nucleon systems. The results indicate the requirements of the general A-nucleon case.

Two Nucleons

The normalized antisymmetric wavefunction is given by

$$\Phi_{\alpha\beta} = \frac{1}{\sqrt{2}}[\phi_\alpha(1)\phi_\beta(2) - \phi_\alpha(2)\phi_\beta(1)], \tag{II.11a}$$

which can be written in terms of permutation operators as

$$\Phi_{\alpha\beta} = \frac{1}{\sqrt{2}}[\mathbb{1} - P(1, 2)][\phi_\alpha(1)\phi_\beta(2)]. \tag{II.11b}$$

Here $P(1, 2)$ exchanges the nucleon indices. In the two-nucleon case, there are only two possible permutations of the two nucleons, namely, keeping them unchanged (which corresponds to $\mathbb{1}$) or interchanging them [which corresponds to $P(1, 2)$]. Thus, (II.11b) is indeed a linear sum of $P\Phi_{\alpha\beta}(1, 2)$, with the two possible permutations $\mathbb{1}$ and $P(1, 2)$, and the coefficients $1/\sqrt{2}$ and $-1/\sqrt{2}$, respectively. [In our notation, P denotes any general permutation, and $P(i, j)$ stands for a *single* exchange of the labels i and j.]

We use $\mathcal{A}$ to denote the antisymmetrizing operator

$$\mathcal{A} = \frac{1}{\sqrt{2}}[\mathbb{1} - P(1, 2)] = \frac{1}{\sqrt{A!}} \sum_P \pi(P)P, \tag{II.12}$$

where $A(=2)$ is the number of nucleons, and $\pi(P)$ is the parity of the permutation P, i.e., $\pi(P) = +1$ when P involves an even number of particle *exchanges* and $\pi(P) = -1$ when P involves an odd number of particle *exchanges*. In the present case, it is equal to $+1$ for $\mathbb{1}$, and -1 for $P(1, 2)$. Since the square of the exchange operator is $\mathbb{1}$, we get

$$\mathcal{A}^2 = \tfrac{1}{2}[\mathbb{1} - P(1, 2)][\mathbb{1} - P(1, 2)] = \mathbb{1} - P(1, 2) = \sqrt{2}\mathcal{A} = \sqrt{A!}\mathcal{A}. \tag{II.13}$$

It should be observed that the wavefunction (II.11a) could also be written as a determinant

$$\Phi_{\alpha\beta} = \frac{1}{\sqrt{2}} \begin{vmatrix} \phi_\alpha(1) & \phi_\beta(1) \\ \phi_\alpha(2) & \phi_\beta(2) \end{vmatrix}. \tag{II.14}$$

In this form, the antisymmetry of the wavefunction under an exchange of particle indices becomes very transparent. The exchange corresponds to interchanging two rows of the determinant, and it is a well-known property of determinants that such an exchange multiplies the determinant by -1. It should be noted that the particle states correspond to various columns of the determinant, and the coordinate labels 1, 2, . . . correspond to various rows. This could be reversed, if desired, because a determinant and its transpose are identical.

The normalization factor also is quite easy to understand: the total number of terms in the antisymmetric wavefunction is 2, and the terms themselves are obviously orthogonal to each other. The square of each term, when integrated, gives unity. Hence, without the normalization factor, the square of the wavefunction integrates to 2; this explains why the normalization factor in (II.11a) or (II.14) is $1/\sqrt{2}$.

Three Nucleons

We next consider an antisymmetric wavefunction of three nucleons corresponding to the set of single-particle states $(\alpha\beta\gamma)$. In analogy with (II.14), we write the wavefunction as

$$\Phi_{\alpha\beta\gamma} = \frac{1}{\sqrt{3!}} \begin{vmatrix} \phi_\alpha(1) & \phi_\beta(1) & \phi_\gamma(1) \\ \phi_\alpha(2) & \phi_\beta(2) & \phi_\gamma(2) \\ \phi_\alpha(3) & \phi_\beta(3) & \phi_\gamma(3) \end{vmatrix}. \tag{II.15}$$

Since there are 3! terms in the expansion of the determinant, which are orthogonal to each other, the normalization factor here is $1/\sqrt{3!}$. A determinant can be very conveniently expanded in terms of permutation operators operating on the product of the diagonal elements. In the present case, we obtain

$$\Phi_{\alpha\beta\gamma} = \frac{1}{\sqrt{A!}} \sum_P \pi(P)P[\phi_\alpha(1)\phi_\beta(2)\phi_\gamma(3)]. \tag{II.16}$$

The antisymmetrizer is therefore defined, in general, by the last expression in (II.12). The square of $\mathcal{A}$ can also be shown, in general, to have the last form given in (II.13).

We digress here to demonstrate that $\mathcal{A}^2$ in the three-nucleon case is indeed $\sqrt{A!}\mathcal{A}$. This will familiarize the reader with the elementary manipulations of permutation operators. If we begin with the set of numbers (1, 2, 3) and rearrange them as (e_1, e_2, e_3), then the permutation

21 (45-122/1977)

corresponding to this operation can be symbolically written as $\begin{pmatrix} 1 & 2 & 3 \\ e_1 & e_2 & e_3 \end{pmatrix}$, implying that 1 is to be replaced by e_1, 2 by e_2, and 3 by e_3. Obviously, e_1, e_2, e_3 differ from one another and lie within the set of numbers (1, 2, 3). According to this notation, the six (=3!) different ways of rearranging 1, 2, 3 are given by

$$\mathbb{1} = \begin{pmatrix} 1 & 2 & 3 \\ 1 & 2 & 3 \end{pmatrix}, \quad P(1,3)P(1,2) = \begin{pmatrix} 1 & 2 & 3 \\ 2 & 3 & 1 \end{pmatrix}, \quad P(2,3)P(1,2) = \begin{pmatrix} 1 & 2 & 3 \\ 3 & 1 & 2 \end{pmatrix};$$
$$P(1,2) = \begin{pmatrix} 1 & 2 & 3 \\ 2 & 1 & 3 \end{pmatrix}, \quad P(2,3) = \begin{pmatrix} 1 & 2 & 3 \\ 1 & 3 & 2 \end{pmatrix}, \quad P(3,1) = \begin{pmatrix} 1 & 2 & 3 \\ 3 & 2 & 1 \end{pmatrix}. \tag{II.17}$$

The operators have been deliberately subdivided into two sets. Those in the second row correspond to *single* exchanges of two numbers, the third remaining unchanged. Hence, for all these operators, $\pi(P) = -1$. On the other hand, the operators in the first row correspond to *zero* exchange (the $\mathbb{1}$-operator), and the product of *two* consecutive exchanges. Hence, all these have $\pi(P) = +1$. The $\pi(P) = +1$ and $\pi(P) = -1$ groups are respectively called even and odd permutations. It should be observed that when we write a product such as $P(1,3)P(1,2)$ it is understood that the operations indicated by the exchange operators have to be carried out in sequence *starting from the right*.

Since doing the same exchange twice in succession means getting back to the initial arrangement, we have the square of any exchange operator equal to $\mathbb{1}$. If we take account of this fact, it is quite easy to see that the product of any two of these six permutation operators will be a third one belonging to the set. If we multiply any two operators from the first row, we get one from the first row because the product of two even permutations is necessarily even. Similarly, if we multiply any two from the second row, we once again get one belonging to the first row because the product now corresponds to *two* successive exchanges. On the other hand, if we multiply one from the first row by one from the second row, we get one belonging to the latter because now the product corresponds to an odd number of exchanges. Thus, it is clear that if

$$PP' = P'', \tag{II.18a}$$

where each of these operators belongs to the set of six operators, then we have

$$\pi(P)\pi(P') = \pi(P''). \tag{II.18b}$$

Let us now write the results of multiplying all the pairs of operators (II.17) as in Table II.2. Here the rows and columns are defined by Roman numerals, which stand for the six operators in the order shown in (II.17). We use the convention that the operator corresponding to the column is placed in the product on the right. The result of any such product has been trivially constructed by faithfully following the definitions; for example, to construct II × III we first write III as $\begin{pmatrix} 1 & 2 & 3 \\ 3 & 1 & 2 \end{pmatrix}$ and then recall that II changes $1 \to 2$, $2 \to 3$, and $3 \to 1$, and hence the second row of III changes, after operation by II, to the order (1 2 3); therefore,

$$\text{II} \times \text{III} = \begin{pmatrix} 1 & 2 & 3 \\ 1 & 2 & 3 \end{pmatrix} = \mathbb{1}.$$

Table II.2 Multiplication of permutation operators for three particles

		Even			Odd		
		I	II	III	IV	V	VI
Even	I	I	II	III	IV	V	VI
	II	II	III	I	VI	IV	V
	III	III	I	II	V	VI	IV
Odd	IV	IV	V	VI	I	II	III
	V	V	VI	IV	III	I	II
	VI	VI	IV	V	II	III	I

A very important result, explicitly demonstrated in Table II.2, is now stated: scanning any row or column, we find all the six operators (II.17) occurring once; in other words, if $P'' = PP'$, and we choose a given P' and allow P to run over all the six operators (or equivalently keep P fixed and allow P' to change), then P'' once describes the whole set in (II.17) in a certain changed order.

This result, together with (II.18b), allows us to simplify $\mathcal{A}^2$. We have

$$\mathcal{A}^2 = \frac{1}{A!} \sum_{P,\,P'} \pi(P)\pi(P')PP'. \tag{II.19}$$

For a given P', consider the entire sum over P. From the result just stated and (II.18b),

$$\sum_{P} \pi(P)\pi(P')PP' = \sum_{P''} \pi(P'')P''.$$

Since this result is independent of P', the summation over the latter in (II.19) gives a numerical factor, which is equal to the total number of P' operators ($=A!$). Thus,

$$\mathcal{A}^2 = \sum_{P''} \pi(P'')P'' = \sqrt{A!}\,\mathcal{A}. \tag{II.13}$$

The property of the product PP' used here is generally true for the permutations in an A-nucleon system even when $A > 3$. Thus, the property (II.13), derived for three nucleons and earlier for two nucleons, is valid in general.

A Nucleons

The generalization of these results to A nucleons ($A > 3$) is quite obvious. We have to choose a set of A single-nucleon states ($\alpha, \beta, \ldots, \zeta$) and construct a determinantal wavefunction

$$\Phi_{\alpha\beta\gamma\ldots\zeta} = \frac{1}{\sqrt{A!}} \begin{vmatrix} \phi_\alpha(1) & \phi_\beta(1) & \ldots & \phi_\zeta(1) \\ \phi_\alpha(2) & \phi_\beta(2) & \ldots & \phi_\zeta(2) \\ \vdots & \vdots & & \vdots \\ \phi_\alpha(A) & \phi_\beta(A) & \ldots & \phi_\zeta(A) \end{vmatrix}, \tag{II.20a}$$

which is identically equal to

$$\Phi_{\alpha\beta\gamma\ldots\zeta} = \mathcal{A}[\phi_\alpha(1)\phi_\beta(2)\ldots\phi_\zeta(A)] \tag{II.20b}$$

with

$$\mathcal{A} = \frac{1}{\sqrt{A!}}\sum_P \pi(P)P. \tag{II.21}$$

The result (II.13) for $\mathcal{A}^2$ is still valid.

B. MATRIX ELEMENTS

Because of the indistinguishability of the nucleons, any dynamical variable that we wish to evaluate with these many-body wavefunctions is necessarily symmetric in all the nucleon variables. Important examples are the operators

$$\sum_{i=1}^{A} T(i), \qquad \sum_{i<j}^{A} V(ij), \qquad \sum_{i=1}^{A} \mathcal{V}(i).$$

Moreover, we shall restrict ourselves to one-body (for example, T and $\mathcal{V}$) and two-body (for example, V) type operators.

One-Body Type Symmetric Operator

Let us denote a general operator of the one-body type by

$$F = \sum_{i=1}^{A} f(i). \tag{II.22}$$

Since each term of this operator contains only one nucleon index, this operator can, at the most, change the state of one nucleon when it operates on a many-body wavefunction $\Phi_{\alpha\beta\ldots\zeta}$. Therefore, only two kinds of nonvanishing matrix elements are obtained: (i) the diagonal matrix element connecting $\Phi_{\alpha\beta\ldots\zeta}$ with itself (here no single-nucleon state is changed by F); and (ii) the nondiagonal matrix element connecting $\Phi_{\alpha\beta\ldots\mu\ldots\zeta}$ with $\Phi_{\alpha\beta\ldots\mu'\ldots\zeta}$, where one single-particle state μ changes to μ' through the operator F, and all the other single-particle states remain unaltered. The results for these two cases are now worked out.

Since the operator F is symmetric in the nucleon indices, it remains unchanged if any permutation operator operates on it, e.g., for the three-nucleon case,

$$P(1, 2)[f(1) + f(2) + f(3)] = f(2) + f(1) + f(3) = F.$$

Therefore, for any permutation operator P, we have

$$PF = FP.$$

We use this result, together with the form (II.20b) for the wavefunctions, and the result (II.13) for $\mathcal{A}^2$, and get

$$\begin{aligned}
&\langle\Phi_{\alpha\beta\ldots\zeta}|\, F\, |\Phi_{\alpha\beta\ldots\zeta}\rangle \\
&= \frac{1}{A!}\langle\phi_\alpha(1)\phi_\beta(2)\ldots\phi_\zeta(A)|\sum_P \pi(P)PF\sum_{P'} \pi(P')P'\, |\phi_\alpha(1)\phi_\beta(2)\ldots\phi_\zeta(A)\rangle \\
&= \langle\phi_\alpha(1)\phi_\beta(2)\ldots\phi_\zeta(A)|\, F\sum_{P''} \pi(P'')P''\, |\phi_\alpha(1)\phi_\beta(2)\ldots\phi_\zeta(A)\rangle.
\end{aligned}$$

Since all the single-particle wavefunctions $\alpha, \beta, \ldots, \zeta$ are different (clearly the determinantal wavefunction would vanish otherwise), only the $(P'' = \mathbb{1})$-term can give a nonvanishing result in the foregoing expression. For any other P'', one or more particle indices correspond to different state labels on the two sides, and hence give zero through orthogonality. Thus, the *diagonal* matrix element is given by

$$\begin{aligned}\langle\Phi_{\alpha\beta\ldots\zeta}|\, F\, |\Phi_{\alpha\beta\ldots\zeta}\rangle &= \langle\phi_\alpha(1)\phi_\beta(2)\ldots\phi_\zeta(A)|\sum_{i=1}^{A} f(i)\,|\phi_\alpha(1)\phi_\beta(2)\ldots\phi_\zeta(A)\rangle\\ &= \langle\phi_\alpha|\, f\, |\phi_\alpha\rangle + \langle\phi_\beta|\, f\, |\phi_\beta\rangle + \ldots + \langle\phi_\zeta|\, f\, |\phi_\zeta\rangle\\ &= \sum_{\xi}\langle\xi|\, f\, |\xi\rangle. \end{aligned} \tag{II.23}$$

Here ξ stands for any of the single-particle states belonging to the set $(\alpha\beta\ldots\zeta)$. The particle labels have been omitted in the last step because they are dummy integration variables, as far as each single-particle matrix element occurring here is concerned.

Using the foregoing procedure and arguments, we can express the nondiagonal matrix element (in which one single-particle state differs on the two sides) as

$$\begin{aligned}&\langle\Phi_{\alpha\beta\ldots\mu\ldots\zeta}|\, F\, |\Phi_{\alpha\beta\ldots\mu'\ldots\zeta}\rangle\\ &= \langle\phi_\alpha(1)\phi_\beta(2)\ldots\phi_\mu(m)\ldots\phi_\zeta(A)|\sum_{i=1}^{A} f(i)\,|\phi_\alpha(1)\phi_\beta(2)\ldots\phi_{\mu'}(m)\ldots\phi_\zeta(A)\rangle\\ &= \langle\mu|\, f\, |\mu'\rangle. \end{aligned} \tag{II.24}$$

The last step follows from the fact that only the $f(m)$-term can contribute a nonvanishing result, whereas all the other $f(i)$, $i \neq m$, give zero because $\phi_\mu^*(m)\phi_{\mu'}(m)$ integrates to zero.

Two-Body Type Symmetric Operator

We denote a general operator of the two-body type by

$$G = \sum_{i<j}^{A} g(ij).$$

Since each term here is a function of two-nucleon variables, at the most, the states of two nucleons can change when G operates on a many-nucleon wavefunction. Therefore, there are now three different types of nonvanishing matrix elements: (i) the diagonal matrix elements connecting $\Phi_{\alpha\beta\ldots\zeta}$ with itself; (ii) the nondiagonal matrix element where one single-particle state differs on the two sides, i.e., connecting $\Phi_{\alpha\beta\ldots\mu\ldots\zeta}$ with $\Phi_{\alpha\beta\ldots\mu'\ldots\zeta}$; and (iii) the nondiagonal matrix element where two single-particle states differ on the two sides, i.e., connecting $\Phi_{\alpha\beta\ldots\mu\ldots\nu\ldots\zeta}$ with $\Phi_{\alpha\beta\ldots\mu'\ldots\nu'\ldots\zeta}$. We assume that μ, μ' and ν, ν' occupy the same columns in the two determinantal wavefunctions. This does not destroy any generality because, if they do not occupy the same column, they can be made to do so by exchanging a suitable number of columns of the determinants, giving either a multiplicative $+1$ or -1, depending on the number of necessary exchanges.

We proceed exactly as in the case of F, and use the fact that $PG = GP$ by virtue of the complete symmetry of G. Thus, we arrive at the result

$$\langle\Phi_{\alpha\beta\ldots\zeta}|\, G\, |\Phi_{\alpha\beta\ldots\zeta}\rangle = \langle\phi_\alpha(1)\phi_\beta(2)\ldots\phi_\zeta(A)|\, G \sum_{P''}\pi(P'')P''\, |\phi_\alpha(1)\phi_\beta(2)\ldots\phi_\zeta(A)\rangle.$$

If we pick up the term $g(1, 2)$ from G, then it is clear that, for a nonvanishing result, we cannot permute the particle indices 3, 4, . . . , A; they have to be retained where they appear in the left-hand state. If this restriction is imposed, then there are clearly only two operators, namely, $P'' = \mathbb{1}$ and $P(1, 2)$, that can contribute to the aforementioned result. Therefore, the $g(1, 2)$-term gives

$$\begin{aligned}&\langle\phi_\alpha(1)\phi_\beta(2)|\, g(1, 2)[\mathbb{1} - P(1, 2)]\, |\phi_\alpha(1)\phi_\beta(2)\rangle\\&\quad = \langle\phi_\alpha(1)\phi_\beta(2)|\, g(1, 2)\, |\phi_\alpha(1)\phi_\beta(2)\rangle - \langle\phi_\alpha(1)\phi_\beta(2)|\, g(1, 2)\, |\phi_\alpha(2)\phi_\beta(1)\rangle\\&\quad = (\alpha\beta|\, g\, |\alpha\beta).\end{aligned}$$

The two-body matrix element within the parentheses appearing in the last step is an abbreviation denoting the *direct* minus the *exchange* matrix elements of the preceding step. The nomenclatures *direct* and *exchange* are self-explanatory in view of the fact that they arise from $\mathbb{1}$ and $P(1, 2)$, respectively. The contribution of all the pair operators $g(i, j)$ occurring in G can be similarly evaluated. The final result (obviously a sum over all the pairs of states occurring in Φ) is

$$\langle\Phi_{\alpha\beta\ldots\zeta}|\, G\, |\Phi_{\alpha\beta\ldots\zeta}\rangle = \sum_{\xi<\eta} (\xi\eta|\, g\, |\xi\eta), \tag{II.25}$$

where $(\xi\eta)$ is any pair belonging to the set $(\alpha\beta \ldots \zeta)$, and the sum is over all such pair states.

Proceeding in an analogous manner, we get

$$\begin{aligned}&\langle\Phi_{\alpha\beta\ldots\mu\ldots\zeta}|\, G\, |\Phi_{\alpha\beta\ldots\mu'\ldots\zeta}\rangle\\&\quad = \langle\phi_\alpha(1)\phi_\beta(2)\ldots\phi_\mu(m)\ldots\phi_\zeta(A)|\, G \sum_{P''} \pi(P'')P''\, |\phi_\alpha(1)\phi_\beta(2)\ldots\phi_{\mu'}(m)\ldots\phi_\zeta(A)\rangle.\end{aligned}$$

In this case also, for any $g(i, j)$ occurring in G, only $P'' = \mathbb{1}$ and $P(i, j)$ can contribute. However, there is a further restriction because of $\mu \neq \mu'$. If either of (i, j) is not equal to m, then $\phi_\mu^*(m)\phi_{\mu'}(m)$ integrates to zero. Hence, only terms of the type $g(i, m)$ with $P'' = \mathbb{1}$ and $P(i, m)$ contribute. Thus, for the first kind of nondiagonal matrix element, we get

$$\langle\Phi_{\alpha\beta\ldots\mu\ldots\zeta}|\, G\, |\Phi_{\alpha\beta\ldots\mu'\ldots\zeta}\rangle = \sum_{\xi} (\xi\mu|\, g\, |\xi\mu'), \tag{II.26}$$

where the summation over ξ goes over the entire set of one-nucleon states $\alpha\beta \ldots \zeta$ (leaving out the particular state μ or μ').

The second kind of nondiagonal matrix element where the two states μ, ν differ on the two sides can also be obtained from similar arguments. In this case, however, if $\phi_\mu(m)$, $\phi_{\mu'}(m)$ and $\phi_\nu(n)$, $\phi_{\nu'}(n)$ occur in the two wavefunctions, then only the $g(m, n)$-term of G with $P'' = \mathbb{1}$ and $P(m, n)$ can obviously contribute a nonvanishing result, and we get

$$\langle\Phi_{\alpha\beta\ldots\mu\ldots\nu\ldots\zeta}|\, G\, |\Phi_{\alpha\beta\ldots\mu'\ldots\nu'\ldots\zeta}\rangle = (\mu\nu|\, g\, |\mu'\nu'). \tag{II.27}$$

It should be remembered that each of the two-particle antisymmetric matrix elements (enclosed within the parentheses) occurring in (II.25)–(II.27) is given by the direct minus exchange matrix elements

$$(\xi\eta|\, g\, |\xi\eta) = \langle\phi_\xi(1)\phi_\eta(2)|\, g(1, 2)\, |\phi_\xi(1)\phi_\eta(2)\rangle - \langle\phi_\xi(1)\phi_\eta(2)|\, g(1, 2)\, |\phi_\eta(1)\phi_\xi(2)\rangle. \tag{II.28}$$

We finally stress that the results in (II.23)–(II.27), together with the definition (II.28), are very important and are repeatedly used in subsequent chapters. These results should be memorized, even if the detailed bookkeeping needed for this derivation is ignored by a pragmatic reader who may not consider such details important.

16. SECOND-QUANTIZATION TECHNIQUE

Another powerful technique, which is extensively used in the many-body theory for proper antisymmetrization and the evaluation of matrix elements, is called the method of second quantization. The reader should refer to the text on quantum mechanics by Landau and Lifshitz* for the basic concepts of this method. We deal with the results here in a pragmatic manner without going into any detailed proof.

Let us introduce an operator $C_\alpha^\dagger$ which, operating on any state $|\Phi\rangle$, produces an extra particle in the state α, in addition to what is already contained in $|\Phi\rangle$. Therefore, if $|\Phi\rangle$ stands for a state with no particle (called the "vacuum state"), the $C_\alpha^\dagger|\Phi\rangle$ represents the state $|\alpha\rangle$ of a single particle. On the other hand, if $|\Phi\rangle$ denotes a determinantal state of A particles, then $C_\alpha^\dagger|\Phi\rangle$ is a state of $(A + 1)$ particles, the last state added to what is already contained in $|\Phi\rangle$ being the state α. $C_\alpha^\dagger$ is called the creation operator for the single-particle state α. Similarly, the Hermitean conjugate operator C_α removes a single particle in the state α operating on any state $|\Phi\rangle$. Thus, if $|\Phi\rangle$ is an A-particle state, $C_\alpha|\Phi\rangle$ is the state of $(A - 1)$ particles, which is obtained by eliminating the state α contained in $|\Phi\rangle$. It therefore follows that, if the state α is not initially present in $|\Phi\rangle$, then $C_\alpha|\Phi\rangle$ must be zero. The operator C_α is called the destruction or annihilation operator for the single-particle state α.

If we keep the Pauli exclusion principle in mind, then $C_\alpha^\dagger|\Phi\rangle$ must be zero if the state α is already present in $|\Phi\rangle$, because we are trying to produce another particle in the state α in contravention of the exclusion principle. The two results

$$C_\alpha^\dagger|\Phi\rangle = 0 \qquad \text{if } \alpha \text{ is an occupied state in } \Phi, \tag{II.29a}$$

$$C_{\alpha'}|\Phi\rangle = 0 \qquad \text{if } \alpha' \text{ is not an occupied state in } \Phi \tag{II.29b}$$

are very useful in the algebraic manipulations of the creation and destruction operators.

We next state the anticommutation properties of these operators, which also aid in the subsequent algebraic manipulations. $(AB + BA)$, which is called the anticommutator of any two operators A and B, is denoted by the symbol $\{A, B\}$. With this notation, we have

$$\{C_\alpha, C_\beta\} = \{C_\alpha^\dagger, C_\beta^\dagger\} = 0, \tag{II.30a}$$

$$\{C_\alpha, C_\beta^\dagger\} = \delta_{\alpha\beta}. \tag{II.30b}$$

(II.30a) states that whenever we interchange two destruction or two creation operators we get a minus sign. It easily follows that the result $\{C_\alpha^\dagger, C_\beta^\dagger\} = 0$ guarantees the Pauli exclusion principle, as now proved. We have

$$C_\alpha^\dagger C_\beta^\dagger + C_\beta^\dagger C_\alpha^\dagger = 0$$

or

$$C_\alpha^\dagger C_\beta^\dagger = -C_\beta^\dagger C_\alpha^\dagger.$$

Therefore,

$$C_\alpha^\dagger C_\beta^\dagger|\Phi\rangle = -C_\beta^\dagger C_\alpha^\dagger|\Phi\rangle.$$

Making $\alpha = \beta$, we get

$$C_\alpha^\dagger C_\alpha^\dagger|\Phi\rangle = -C_\alpha^\dagger C_\alpha^\dagger|\Phi\rangle.$$

Since a quantity can be equal to the negative of itself only if it is zero, we have established the desired result that two particles cannot occupy the same state.

We proceed to state without proof the expression of the operators in the new language.

*Landau, L. D., and Lifshitz, E. M., Quantum Mechanics, 2nd edn., Pergamon, Oxford, 1958, p 227.

Any one-body type operator is given by

$$F = \sum_{i=1}^{A} f_i = \sum_{\alpha,\beta} \langle\alpha| f |\beta\rangle C_\alpha^\dagger C_\beta, \tag{II.31a}$$

and a two-body type operator by

$$G = \sum_{i<j=1}^{A} g_{ij} = \tfrac{1}{2} \sum_{\alpha,\beta,\gamma,\delta} \langle\alpha\beta| g |\gamma\delta\rangle C_\alpha^\dagger C_\beta^\dagger C_\delta C_\gamma. \tag{II.31b}$$

It should be noted that the matrix element of g occurring in (II.31b) is *not* the antisymmetrized matrix element (the latter could have been used by changing the factor $\frac{1}{2}$ to $\frac{1}{4}$). The second point to note is that $|\gamma\delta\rangle$ in the matrix element occurs in the reverse order in the corresponding operator $C_\delta C_\gamma$.

The Hamiltonian is therefore given by

$$H = \sum_{\alpha,\beta} \langle\alpha| T |\beta\rangle C_\alpha^\dagger C_\beta + \tfrac{1}{2} \sum_{\alpha,\beta,\gamma,\delta} \langle\alpha\beta| V |\gamma\delta\rangle C_\alpha^\dagger C_\beta^\dagger C_\delta C_\gamma. \tag{II.32}$$

We now wish to establish the results (II.23)–(II.28) by using the second-quantization technique. For the one-body operator of (II.31a), we get

$$\langle\Phi| \sum_{i=1}^{A} f_i |\Phi\rangle = \sum_{\alpha,\beta} \langle\alpha| f |\beta\rangle\langle\Phi| C_\alpha^\dagger C_\beta |\Phi\rangle. \tag{II.33}$$

According to (II.29b),

$$C_\beta|\Phi\rangle = 0$$

if β is not occupied in Φ. The Hermitean conjugate of (II.29b) further demands

$$\langle\Phi|C_\alpha^\dagger = 0$$

if α is not occupied in Φ. Therefore, the unrestricted summation over α and β in (II.33) is actually a restricted summation over the states occupied in Φ. Further,

$$C_\alpha^\dagger C_\beta = -C_\beta C_\alpha^\dagger + \delta_{\alpha\beta}$$

according to (II.30b). Therefore, (II.33) simplifies to

$$\sum_{\alpha,\beta}^{\text{occ}} \delta_{\alpha\beta}\langle\alpha| f |\beta\rangle\langle\Phi \mid \Phi\rangle - \sum_{\alpha,\beta}^{\text{occ}} \langle\alpha| f |\beta\rangle\langle\Phi| C_\beta C_\alpha^\dagger |\Phi\rangle.$$

Since α is required to be an occupied state in Φ, $C_\alpha^\dagger|\Phi\rangle = 0$ according to (II.29a). Therefore, the second term in this expression is zero. Using the $\delta_{\alpha\beta}$-factor in the first term, we finally get the very simple result

$$\langle\Phi| \sum_{i=1}^{A} f_i |\Phi\rangle = \sum_{\alpha}^{\text{occ}} \langle\alpha| f |\alpha\rangle.$$

The expression $\langle\Phi \mid \Phi\rangle$ has been put equal to unity because $|\Phi\rangle$ is a normalized state. This result agrees with (II.23).

Before proceeding to derive (II.24) by the present technique, let us adopt a simple way of expressing the many-body state $\Phi_{\alpha\beta\ldots\mu'\ldots\xi}$ by taking the state $\Phi_{\alpha\beta\ldots\mu\ldots\xi}$ as the reference state. We first note that one of the single-particle states, namely, μ, has been removed from the reference state and substituted by the single-particle state μ' to obtain the state $\Phi_{\alpha\beta\ldots\mu'\ldots\xi}$. According to the definition of the creation and destruction operators, the removal and substitution, just described, is produced by $C^\dagger_{\mu'}C_\mu$, and hence

$$|\Phi_{\alpha\beta\ldots\mu'\ldots\zeta}\rangle = C^\dagger_{\mu'}C_\mu|\Phi_{\alpha\beta\ldots\mu\ldots\zeta}\rangle. \quad \text{(II.34)}$$

Therefore,

$$\langle\Phi_{\alpha\beta\ldots\mu\ldots\zeta}|\,F\,|\Phi_{\alpha\beta\ldots\mu'\ldots\zeta}\rangle = \sum_{\xi,\eta}\langle\xi|\,f\,|\eta\rangle\langle\Phi_{\alpha\beta\ldots\mu\ldots\zeta}|\,C^\dagger_\xi C_\eta C^\dagger_{\mu'}C_\mu\,|\Phi_{\alpha\beta\ldots\mu\ldots\zeta}\rangle.$$

This expression is simplified by permuting $C^\dagger_{\mu'}$ to the extreme left, where it produces zero, acting on the left-hand state by virtue of the Hermitean conjugate of (II.29b). In this way, we are left with the matrix element of $\delta_{\eta\mu'}C^\dagger_\xi C_\mu$, where $\delta_{\eta\mu'}$ was acquired while transposing C_η and $C^\dagger_{\mu'}$. We once again take C_μ to the left and use the Hermitean conjugate of (II.29a). Thus, the final simplified result contains $\delta_{\eta\mu'}\delta_{\xi\mu}$. Using all this, we reproduce the expression (II.24).

The derivation of the expressions (II.25)–(II.27), using the second-quantization technique and similar arguments, is left as an exercise.

17. MODELS FOR SINGLE-NUCLEON WAVEFUNCTION AND POTENTIAL

It has been mentioned in Section 15 that although the one-body potential $\mathcal{V}$ should, in principle, be determined by averaging the known two-body potential, in practice, a model $\mathcal{V}$ and model single-particle wavefunctions are very frequently used in nuclear structural calculations. We shall describe several such models in this section.

A. PLANE-WAVE STATES

The simplest single-particle wavefunction is the plane wave

$$\phi_{\mathbf{k}}(\mathbf{r}) = \frac{1}{\sqrt{\Omega}}\exp(i\mathbf{k}\cdot\mathbf{r}), \quad \text{(II.35a)}$$

where $\mathbf{k} = \hbar^{-1}\mathbf{p}$ is the wave number, $\mathbf{p}$ the momentum, and Ω the volume of a large box in which the nucleons are assumed to be enclosed. The presence of the box makes the eigenvalue $\hbar\mathbf{k}$ of the momentum variable $\mathbf{p}$ discrete and the wavefunctions (II.35a) satisfy the orthogonality relation

$$\langle\phi_{\mathbf{k}}\,|\,\phi_{\mathbf{k}'}\rangle = \frac{1}{\Omega}\int_\Omega d^3r\,\exp[i(\mathbf{k}'-\mathbf{k})\cdot\mathbf{r}] = \delta_{\mathbf{k},\mathbf{k}'}, \quad \text{(II.35b)}$$

where $\delta_{\mathbf{k},\mathbf{k}'}$ is the Krönecker delta. When $\Omega\to\infty$, $\mathbf{k}$ becomes continuous and the orthogonality relation should change to

$$\langle\phi_{\mathbf{k}}|\,\phi_{\mathbf{k}'}\rangle = \delta(\mathbf{k},\mathbf{k}'), \quad \text{(II.35c)}$$

where $\delta(\mathbf{k},\mathbf{k}')$ is the Dirac delta function. This new equation demands that

$$\phi_{\mathbf{k}}(\mathbf{r}) = \frac{1}{(2\pi)^{3/2}}\exp(i\mathbf{k}\cdot\mathbf{r}) \quad \text{(II.36)}$$

when $\Omega\to\infty$ and $\mathbf{k}$ is continuous. The normalization constant $(2\pi)^{-3/2}$ follows from representing the one-dimensional Dirac delta function as

$$\delta(a, a') = (2\pi)^{-1}\int_{-\infty}^{\infty}\exp[i(a-a')x]\,dx.$$

The model developed with the single-particle plane-wave states is often called the *fermi-gas model*. The plane wavefunctions have also to be used extensively in the study of an idealized many-nucleon system, called infinite nuclear matter. This hypothetical system is assumed to be of infinite extent without a surface. Since one point in the infinite medium is physically equivalent to any other, there is no natural choice of the origin of coordinates. If the origin is shifted through a distance $\mathbf{a}$, then the wavefunction $\phi(\mathbf{r})$ changes to $\phi(\mathbf{r}-\mathbf{a})$. These two states should be equivalent to each other, and hence $\phi(\mathbf{r}-\mathbf{a})$ can at the most differ from $\phi(\mathbf{r})$ through a phase. We first consider an infinitesimal displacement $\boldsymbol{\epsilon}$ and, by Taylor expansion to the first order in $\boldsymbol{\epsilon}$, obtain

$$\phi(\mathbf{r}) \rightarrow \phi(\mathbf{r}-\boldsymbol{\epsilon}) = \phi(\mathbf{r}) - \boldsymbol{\epsilon}\cdot\nabla\phi(\mathbf{r}) = (1 - \frac{i}{\hbar}\boldsymbol{\epsilon}\cdot\mathbf{p})\phi(\mathbf{r}),$$

where $\mathbf{p} = -i\hbar\nabla$ is the momentum operator. Thus, the infinitesimal displacement operator is given by

$$1 - \frac{i}{\hbar}\boldsymbol{\epsilon}\cdot\mathbf{p}.$$

It is worthwhile to compare this result with the infinitesimal rotation operator (BI.4) in Appendix B, and to note that the role of the angular momentum operator has here been taken by the linear momentum. If we now consider a finite displacement $\mathbf{a}$, then, following the same procedure as in the case of rotation, we can prove

$$\phi(\mathbf{r}-\mathbf{a}) = \exp\,[(-i/\hbar)\mathbf{a}\cdot\mathbf{p}]\phi(\mathbf{r}).$$

By asserting that $\phi(\mathbf{r}-\mathbf{a})$ and $\phi(\mathbf{r})$ differ only through a phase, we get

$$\phi(\mathbf{r}-\mathbf{a}) = \exp\,[-i\delta(\mathbf{a})]\phi(\mathbf{r}),$$

where $\delta(\mathbf{a})$ is a real phase, dependent on the displacement $\mathbf{a}$. Comparing these two expressions, we conclude that $\phi(\mathbf{r})$ must be an eigenfunction of the momentum operator $\mathbf{p}$, i.e., it must be a plane wave of the type (II.36). In this case, the phase $\delta(\mathbf{a})$ is obviously given by $\mathbf{k}\cdot\mathbf{a}$.

Since there are two kinds of nucleons, each with two spin states, a given momentum state $\mathbf{k}$ gets fully occupied, according to the Pauli principle, with two neutrons (spin up and spin down) and two protons. The model ground state of the many-nucleon system constructed with the plane-wave states is thus a determinant in which the momentum states from $k = 0$ to a certain maximum k_F (called the fermi momentum) are occupied; each momentum state has four spin-isospin states and accepts two neutrons and two protons. The total number of momentum states up to $k_F\,(=\hbar^{-1}p_F)$ in a box of volume Ω is given by

$$\frac{\Omega}{(2\pi\hbar)^3}\int_0^{p_F} 4\pi p^2\,dp = \frac{\Omega k_F^3}{6\pi^2}.$$

Multiplying this by 4 (the spin-isospin multiplicity), we get the total number of occupied states (equal to the total number of nucleons)

$$A = \frac{2}{3\pi^2}\Omega k_F^3. \tag{II.37}$$

It is known from experimental data that the radius R of a nucleus comprising A nucleons is proportional to $A^{1/3}$ and is expressed as

$$R = r_0 A^{1/3},$$

where $r_0 \approx 1.07$ fm for heavy nuclei ($A > 40$). Thus, $\Omega = (4\pi/3)r_0^3 A$, and hence

$$k_F = \frac{(9\pi)^{1/3}}{2} r_0^{-1} \approx 1.52 r_0^{-1}. \tag{II.38}$$

The density of nucleons is obviously given by

$$\rho = \frac{A}{\Omega} = \frac{3}{4\pi} r_0^{-3}. \tag{II.39}$$

The smaller the value of r_0, the smaller is the volume Ω accessible to the A nucleons, and hence the larger is the density ρ.

We shall work out two more mathematical results for the plane-wave states in order to facilitate our discussion on infinite nuclear matter. The first one concerns the summation over all momentum states that occur as intermediate states in the development of the many-body perturbation theory. The completeness relation gives

$$\mathbb{1} = \sum_{\mathbf{k}} |\phi_{\mathbf{k}}\rangle \langle\phi_{\mathbf{k}}|.$$

Multiplying with $\langle \mathbf{r}|$ from the left and $|\mathbf{r}'\rangle$ from the right, and using (II.35a), we get

$$\delta(\mathbf{r}, \mathbf{r}') = \frac{1}{\Omega} \sum_{\mathbf{k}} \exp [i\mathbf{k}\cdot(\mathbf{r} - \mathbf{r}')].$$

On the other hand, if we consider $\Omega \to \infty$ and continuous $\mathbf{k}$, we get, from (II.36),

$$\delta(\mathbf{r}, \mathbf{r}') = \frac{1}{(2\pi)^3} \int d^3k \exp [i\mathbf{k}\cdot(\mathbf{r} - \mathbf{r}')].$$

Therefore, the prescription for replacing the $\mathbf{k}$ summation of $\exp [i\mathbf{k}\cdot(\mathbf{r} - \mathbf{r}')]$ by an integration is defined by

$$\sum_{\mathbf{k}} \to \frac{\Omega}{(2\pi)^3} \int d^3k. \tag{II.40}$$

The second result we wish to arrive at is the transformation of the two-particle product wavefunction $\exp (i\mathbf{k}_1\cdot\mathbf{r}_1) \exp (i\mathbf{k}_2\cdot\mathbf{r}_2)$ to plane-wave states of the centre-of-mass coordinate $\mathbf{R}$, and relative coordinate $\mathbf{r}$, defined as

$$\begin{aligned} \mathbf{r} &= \mathbf{r}_1 - \mathbf{r}_2, \\ \mathbf{R} &= \tfrac{1}{2}(\mathbf{r}_1 + \mathbf{r}_2). \end{aligned} \tag{II.41}$$

The canonically conjugate momenta are given by

$$\begin{aligned} \mathbf{k} &= \tfrac{1}{2}(\mathbf{k}_1 - \mathbf{k}_2), \\ \mathbf{K} &= \mathbf{k}_1 + \mathbf{k}_2 \end{aligned} \tag{II.42}$$

such that

$$\mathbf{k}_1\cdot\mathbf{r}_1 + \mathbf{k}_2\cdot\mathbf{r}_2 = \mathbf{K}\cdot\mathbf{R} + \mathbf{k}\cdot\mathbf{r}.$$

Hence, the required transformation is given by

$$\exp (i\mathbf{k}_1\cdot\mathbf{r}_1) \exp (i\mathbf{k}_2\cdot\mathbf{r}_2) = \exp (i\mathbf{K}\cdot\mathbf{R}) \exp (i\mathbf{k}\cdot\mathbf{r}). \tag{II.43}$$

We could also use a more symmetric coordinate transformation

$$\bar{\mathbf{r}} = \frac{1}{\sqrt{2}}(\mathbf{r}_1 - \mathbf{r}_2),$$
$$\bar{\mathbf{R}} = \frac{1}{\sqrt{2}}(\mathbf{r}_1 + \mathbf{r}_2) \tag{II.44}$$

and the canonically conjugate momenta

$$\bar{\mathbf{k}} = \frac{1}{\sqrt{2}}(\mathbf{k}_1 - \mathbf{k}_2),$$
$$\bar{\mathbf{K}} = \frac{1}{\sqrt{2}}(\mathbf{k}_1 + \mathbf{k}_2). \tag{II.45}$$

These new coordinates and momenta once again yield a simple relation

$$\exp(i\mathbf{k}_1\cdot\mathbf{r}_1)\exp(i\mathbf{k}_2\cdot\mathbf{r}_2) = \exp(i\bar{\mathbf{K}}\cdot\bar{\mathbf{R}})\exp(i\bar{\mathbf{k}}\cdot\bar{\mathbf{r}}). \tag{II.46}$$

B. WOODS-SAXON POTENTIAL

In a real finite nucleus, the density varies in the surface region, and the use of single-particle plane-wave states is not a very suitable approximation. We may naively argue that the shape of the overall potential $\mathcal{V}$, offered by a nucleus to a neighbouring nucleon, should somewhat resemble the shape of its density distribution. The validity of this statement is more firmly established in Section 35 when we deal with the self-consistent determination of $\mathcal{V}$ by the averaging of the given two-body potential. Therefore, any experiment that gives information on the density distribution in nuclei provides the guideline for an empirical choice of $\mathcal{V}$. We shall shortly discuss the results of high-energy electron scattering experiments for this purpose.

Direct information on $\mathcal{V}$ may also be obtained by analyzing the data on the scattering of a single nucleon from various nuclei. Such an analysis is done on the casual assumption that all the nucleus does to the nucleon is to provide an overall potential, arising from the two-body interaction between the oncoming nucleon and the nucleons inside the nucleus. In practice, the nucleon usually produces a few different kinds of nuclear reactions after it hits the nucleus: (i) it may be directly scattered (shape elastic scattering) or (ii) it may be absorbed by the nucleus, leading to the formation of the compound nucleus, and then (a) re-emitted (compound nuclear scattering) or (b) the compound nucleus may break up, emitting some other particle (deuteron, alpha, . . .).

If the potential that is required to describe the collision of a nucleon and a nucleus be taken as real, then it can produce only elastic scattering and no absorption of the nucleon by the nucleus. This is clear from the theory of scattering by a real potential (considered in Section 7) in connection with nucleon-nucleon scattering. We learnt there that the outgoing amplitude is related to the ingoing amplitude through a unitary $\mathcal{S}$-matrix, and, as a result, the total outgoing flux is the same as the total ingoing flux. This situation can be altered by considering a complex potential for which the imaginary part produces an absorption of the ingoing flux. (See Problem 6 at the end of this chapter.)

There is a great deal of information on the complex potential (also called the optical potential, which gives the name optical model to this simple model of nucleon-nucleus scattering) emerging from the analysis of experimental data. We write the potential as $\mathcal{V} + i\mathcal{W}$, where $\mathcal{V}$ is the real part, and $\mathcal{W}$ the imaginary part of the optical potential. The imaginary part produces absorption of the nucleon, and hence contains information on the compound elastic scattering as well as the processes emitting other particles. On the other hand, the real part produces shape elastic scattering, and represents the potential we are directly interested in for our structure theory. This potential produces a discrete set of bound states, and also continuum states which can be directly calculated by solving the one-body Schrödinger equation (II.6).

We here summarize the information on $\mathcal{V}$ by stating that it consists of a pure spatial part and a single-particle spin-orbit coupling term $\boldsymbol{l}\cdot\mathbf{s}$. For the sake of simplicity, we do not bring in the $(\boldsymbol{l}\cdot\mathbf{s})$-term for the present; it is dealt with in Section 24 when we consider the single-particle equation in more detail in connection with the nuclear shell model. For now we are satisfied with the pure spatial part of $\mathcal{V}$, which is found to have the shape

$$\mathcal{V}(r) = \frac{\mathcal{V}_0}{1 + \exp[(r - R)/a]}. \tag{II.47}$$

This potential is usually called the Woods-Saxon potential (see Woods and Saxon[1]), and looks like the solid line (for a fairly large R) in Fig. II.2. The parameter $\mathcal{V}_0$ gives the strength of

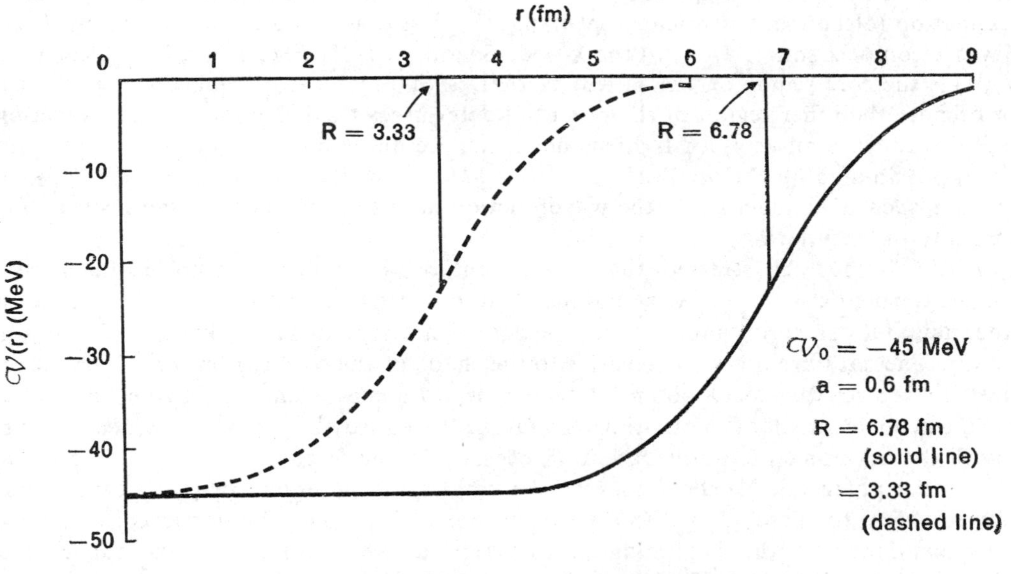

Fig. II.2 Woods-Saxon type average nuclear potential.

the potential, and is attractive (i.e., negative in sign); the radius parameter R determines the extent of the core region of nearly uniform density, and the fall-off distance in the surface region is determined by the diffuseness parameter a. The usual range of the values of a is $0.5 \rightarrow 0.7$ fm, and that of r_0 (where $R = r_0 A^{1/3}$) is $1.15 \rightarrow 1.35$ fm. This interpretation of the parameters R and a is clear from the considerations now described. At $r = 0$, the potential is given by

$$\frac{\mathcal{V}_0}{1 + \exp(-R/a)}.$$

As long as R is fairly large, this value is roughly equal to $\mathcal{V}_0$. As r increases, the term $\exp[(r - R)/a]$ in the denominator also increases, but as long as $r \ll R$, the term 1 in the denominator predominates, and there is very little change in the value of $\mathcal{V}$. Thus, the extent of the core region of more or less constant magnitude depends on how big R is. When $r = R$, the potential drops to $\mathcal{V}_0/2$; from then on, since r is larger than R, the quantity $(r - R)/a$ becomes positive, and the exponential starts growing fairly fast as r increases. Very soon, 1 gets small compared with the exponential term, and the potential becomes approximately $\mathcal{V}_0 \exp[-(r - R)/a]$ $(r \gg R)$ which decays at a rate determined by the magnitude of a.

Although the diffuseness parameter varies very little from nucleus to nucleus, the radius parameter R, being proportional to $A^{1/3}$, changes considerably from light to heavy nuclei. For a nucleus with a rather small R, the Woods-Saxon potential looks like the dashed curve in Fig. II.2, where the surface region is very pronounced as compared with the very small core. It is clear from the shape of these curves that the optical potential for a heavy nucleus can be very nearly approximated by a suitably defined square-well, whereas for lighter nuclei a finite depth potential varying as r^2 (the harmonic oscillator potential) is a fairly good representation. Since the square-well has a very sharp edge, it produces more pronounced diffraction of the incident nucleon wave than the diffused surface Woods-Saxon potential. Therefore, as far as the fit to nucleon scattering data is concerned, the square-well may not be a very good approximation (although, historically speaking, the first optical potential set up by Feshbach et al[2] was a complex square-well) to the Woods-Saxon potential. However, in nuclear structure theory, it is the core region of the potential that is of greater consequence than the surface region because the inner region of the potential determines the behaviour of the wavefunction at small distances. Similarly, for lighter nuclei, the oscillator potential of finite depth may be further approximated by the oscillator potential $\frac{1}{2}M\omega^2 r^2$, which grows to ∞ when $r \to \infty$. We expect that such a change alters the wavefunction near the periphery of the nucleus, but not considerably in the interior.

Finally, it must be stressed that the potential $\mathcal{V}$, determined from the analysis of nucleon scattering data, is not very unique. Without appreciably spoiling the fit, the depth and the radius R can simultaneously be changed within certain limits, keeping $\mathcal{V}_0 R^2$ roughly a constant. The data are not very sensitive to the shape of the core region too. There have been attempts to replace the local potential (which is, to some extent, energy-dependent) by an energy-independent nonlocal potential. As far as structure theory is concerned, we usually work with the harmonic oscillator potential, occasionally with the Woods-Saxon, and seldom with any other potential. Needless to say, the single-particle equation (II.6) cannot be analytically solved in the case of the Woods-Saxon potential; it has to be numerically handled.

As mentioned at the beginning of this section, we should also take the clue to an empirical $\mathcal{V}$ by making it conform to the shape of the density distribution inside a nucleus. This is an empirical way of making the one-body potential consistent with the density (in lieu of a fully self-consistent determination of $\mathcal{V}$ from the given V). At present, there is a vast amount of experimental data on nuclear charge distribution. Because of the Coulomb repulsion between the protons, and the fact that in heavier nuclei there is a neutron excess, the density distribution of neutrons may be somewhat different from the charge distribution in a nucleus (which gives only the proton density distribution). However, since the neutron excess tends to spread the neutrons over a larger volume, and the Coulomb repulsion also has a similar effect on the protons, we assume that the charge distribution, revealed in electron scattering experiments, is a fairly reliable guide to the density distribution in a nucleus. At any rate, the spirit of our present discussion is empirical, and all we want to check is that the density distribution roughly agrees with the shape of $\mathcal{V}$ given in this section.

The differential cross-section $\sigma(\theta)$ for the scattering of an electron of energy E by a nucleus of charge Ze, assumed to be located at a point, is given by

$$\sigma(\theta) = \frac{Z^2e^4}{4E^2}\frac{\cos^2(\theta/2)}{\sin^4(\theta/2)} = \sigma_M(\theta). \tag{II.48}$$

This is the well-known Mott formula for electron scattering by a point charge. As long as the electron energy is small, and it does not penetrate *into* the nuclear charge distribution, this formula describes the electron scattering excellently. However, when the electron energy is large ($\sim 10^2$ MeV), the electron penetrates into the nuclear charge distribution, and (II.48) becomes invalid. Because of the high energy of the electron, the Born approximation equation

$$\sigma(\theta) = \sigma_M(\theta) \times |F_{if}(\theta)|^2 \tag{II.49}$$

has been extensively applied; in (II.49), the form factor $F_{if}(\theta)$, corresponding to an initial nuclear state Ψ_i and a final nuclear state Ψ_f, is given by

$$F_{if}(\theta) = \frac{1}{Z}\langle\Psi_f|\sum_p \exp(i\mathbf{q}\cdot\mathbf{r}_p)|\Psi_i\rangle. \tag{II.50}$$

This formula takes into account the possibility of inelastic scattering corresponding to an excitation of the nucleus from its ground state. The summation p goes over the protons only, and $\mathbf{q}$ is the momentum transfer of the electron, given by $\mathbf{q} = \mathbf{k}_i - \mathbf{k}_f$, where $\mathbf{k}_i$ and $\mathbf{k}_f$ are the initial and final momenta of the electron in units of $\hbar$.

In the case of elastic scattering, both Ψ_i and Ψ_f are equal to Ψ_0, the ground state of the nucleus, and $k_i = k_f = k$. Then $q = 2k \sin \frac{1}{2}\theta$ with $k = E/(\hbar c)$, if relativistic electron energy is assumed. Since

$$\exp(i\mathbf{q}\cdot\mathbf{r}_p) = \int d^3r \exp(i\mathbf{q}\cdot\mathbf{r})\delta(\mathbf{r}, \mathbf{r}_p),$$

we may write, for elastic scattering,

$$F(\theta) = \int d^3r \exp(i\mathbf{q}\cdot\mathbf{r})\rho(\mathbf{r}), \tag{II.51}$$

where

$$\rho(\mathbf{r}) = Z^{-1}\langle\Psi_0|\sum_p \delta(\mathbf{r}, \mathbf{r}_p)|\Psi_0\rangle. \tag{II.52}$$

The quantity $\rho(\mathbf{r})$ is the proton probability density inside the nucleus at the point $\mathbf{r}$.

The observed differential cross-section $\sigma(\theta)$, divided by the Mott cross-section $\sigma_M(\theta)$, gives the experimental $|F(\theta)|^2$ as a function of θ. Various shapes of the charge density $\rho(\mathbf{r})$ are used to obtain the best fit to the experimental curve. For a nucleus having a spherically symmetric density $\rho(r)$, the angle-integration in (II.51) can easily be carried out and $F(\theta)$ reduces to

$$F(\theta) = \frac{4\pi}{q}\int_0^\infty r\rho(r)\sin(qr)\,dr. \tag{II.53}$$

The normalization (EI.6) in Appendix E now becomes

$$1 = 4\pi\int_0^\infty \rho(r)r^2\,dr.$$

If we expand $\sin(qr)$ in (II.53) for small q in powers of q, and use the foregoing normalization,

we obtain

$$F(\theta) = 1 - \tfrac{1}{6}q^2\langle r^2\rangle + \cdots, \tag{II.54}$$

where

$$\langle r^2\rangle = 4\pi\int_0^\infty \rho(r)r^4\,dr \tag{II.55}$$

is the mean-square radius of the charge distribution. Thus, an extrapolation of the form factor to $q \to 0$ (i.e., small θ, in view of $q = 2k \sin \tfrac{1}{2}\theta$) gives us information on the mean-square radius.

In principle, the relation (II.51) can be inverted by using the standard result

$$\int \exp\,[i\mathbf{q}\cdot(\mathbf{r} - \mathbf{r}')]\,d^3q = (2\pi)^3\delta(\mathbf{r} - \mathbf{r}').$$

We multiply (II.51) by $\exp\,(-i\mathbf{q}\cdot\mathbf{r}')$, integrate over $\mathbf{q}$, and get

$$\rho(\mathbf{r}') = \frac{1}{(2\pi)^3}\int d^3q\,\exp\,(-i\mathbf{q}\cdot\mathbf{r}')F(\mathbf{q}). \tag{II.56}$$

Since the dependence of $F(\theta)$ on θ is through the momentum transfer $\mathbf{q}$, we have used $F(\mathbf{q})$ in (II.56). For a spherical density $\rho(r)$, we have just seen that $F(\mathbf{q})$ is actually a function of the magnitude q, and hence the angle-integration in (II.56) can be carried out, yielding

$$\rho(r) = \frac{1}{2\pi^2 r}\int_0^\infty qF(q)\sin\,(qr)\,dq. \tag{II.57}$$

It is clear from (II.57) that the charge density $\rho(r)$ could be uniquely obtained from the measured form factor had it been possible for the experiment to give $F(q)$ for all q between 0 and ∞. Since this is almost impossible, the data on $F(q)$ is analyzed usually with various assumed shapes of $\rho(r)$ containing one or two parameters. It is clear from (II.54) that the low-energy data can be analyzed only in terms of one parameter, namely, the mean-square radius; as we go higher and higher in q, the higher and higher moments $\langle r^{2n}\rangle$ of the charge distribution $\rho(r)$ appear in $F(q)$. Thus, when the experiments are done with fairly high energies, quite a few moments of the charge distribution, appearing in an expansion of $F(q)$ in powers of q, grow significant, and hence an analysis of the data in terms of an entire density distribution $\rho(r)$ becomes meaningful.

Some details on the derivation of (II.49) and the shapes of the charge distributions of various nuclei and their root-mean-square radii are given in Appendix E. The data on the root-mean-square radii are very useful in fixing a parameter in the assumed model $\mathcal{V}(r)$ in nuclear structure calculations. We summarize the results here by saying that a fermi-type charge distribution

$$\rho(r) = \frac{\rho_0}{1 + \exp\,[(r - R)/a]} \tag{II.58}$$

explains the electron scattering data fairly well. It should be observed that this shape is the same as that of the Woods-Saxon potential $\mathcal{V}(r)$. The observation on the shrinkage of the core region (the radius R) as we proceed to lighter nuclei is true for the charge distribution too; the skin thickness however (determined by the diffuseness parameter a) stays more or less constant. Once again, in many nuclei, changes in the shape of $\rho(r)$ in the core region within reasonable limits do not appreciably alter the fit to $F(\theta)$. For very light nuclei, the charge distributions calculated with wavefunctions of an infinite oscillator potential $\frac{1}{2}M\omega^2r^2$ (see Section 17C for these wavefunctions) are found to reproduce the data fairly well.

C. ISOTROPIC HARMONIC OSCILLATOR POTENTIAL

The model single-particle potential most extensively used in nuclear structure study is the isotropic harmonic oscillator potential

$$\mathcal{V}(r) = \tfrac{1}{2}M\omega^2(x^2 + y^2 + z^2) = \tfrac{1}{2}M\omega^2r^2, \tag{II.59}$$

where M is the nucleon mass, and ω the angular frequency of the oscillator; if α is the spring constant, then $\omega^2 = \alpha/M$. Since the three harmonic oscillators occurring in (II.59) in the directions x, y, z, have the same spring constant, $\mathcal{V}(r)$ is called the three-dimensional *isotropic* oscillator potential.

The single-particle Schrödinger equation, with this potential, in terms of cartesian coordinates can be very easily separated into three harmonic oscillator equations for the three directions x, y, z. The equation for the x-direction, for example, is

$$\left(-\frac{h^2}{2M}\frac{d^2}{dx^2} + \tfrac{1}{2}M\omega^2x^2\right)\phi(x) = E_x\phi(x) \tag{II.60}$$

whose eigenvalues and eigenfunctions (see Schiff[3]) are

$$E_x = \hbar\omega(n_x + \tfrac{1}{2}), \qquad n_x = 0, 1, 2, \ldots, \tag{II.61a}$$

$$\phi_{n_x}(x) = [b\sqrt{\pi}2^{n_x}n_x!]^{-1/2}H_{n_x}(\xi)\exp(-\tfrac{1}{2}\xi^2). \tag{II.61b}$$

Here $\xi = x/b$, and $b^2 = \hbar/(M\omega)$. The Hermite polynomial $H_{n_x}(\xi)$ can be computed from the results

$$H_0(\xi) = 1, \qquad H_1(\xi) = 2\xi, \qquad H_{n+1}(\xi) = 2\xi H_n(\xi) - 2nH_{n-1}(\xi). \tag{II.62}$$

Similar equations hold for E_y, $\phi(y)$ and E_z, $\phi(z)$. The energy E and the full three-dimensional wavefunction are obviously given by

$$E = E_x + E_y + E_z = \hbar\omega(n_x + n_y + n_z + \tfrac{3}{2}) \tag{II.63}$$

and

$$\phi(\mathbf{r}) = \phi(x)\phi(y)\phi(z). \tag{II.64}$$

The energy E is determined by the total oscillator quantum number λ, defined as

$$\lambda = n_x + n_y + n_z. \tag{II.65}$$

In general, for a given λ (which can have positive integral values), (n_x, n_y, n_z) can have various possible sets of values consistent with (II.65). Thus, there is a great deal of degeneracy in each eigenvalue.

In nuclear structure work, it is more convenient to use the solutions of the three-dimensional harmonic oscillator equation in spherical polar coordinates (r, θ, ϕ). Since the potential (II.59) is spherically symmetric, the angle-dependence of the wavefunctions is given by spherical harmonics; hence, we write

$$\phi(r) = \frac{R(r)}{r}Y^l_m(\theta, \phi). \tag{II.66}$$

The radial function $R(r)$ satisfies the usual radial Schrödinger equation

$$\left[\frac{d^2}{dr^2} - \frac{l(l+1)}{r^2} + k^2 - v(r)\right]R = 0, \tag{II.67}$$

where

$$k^2 = \frac{2ME}{\hbar^2},$$

$$v(r) = \frac{2M}{\hbar^2}\mathcal{V}(r) = (\frac{M\omega}{\hbar})^2 r^2.$$

We now examine the dimension of $\hbar/(M\omega)$. This can be written as $[\hbar/(Mc)](c/\omega)$. The nucleon Compton wavelength $[\hbar/(Mc)]$ has the dimension of length. The velocity of light c has the dimension of (length) $\times$ (time)$^{-1}$, and ω has the dimension of (time)$^{-1}$. Hence, $\hbar/(M\omega)$, which has the dimension of (length)^2, is from now on denoted by b^2, where b is called the oscillator parameter and has the dimension of length

$$b = (\frac{\hbar}{M\omega})^{1/2}. \tag{II.68}$$

We also introduce the dimensionless radial coordinate

$$\rho = \frac{r}{b}$$

and convert the radial equation (II.67) into an equation in ρ such that

$$[\frac{d^2}{d\rho^2} - \frac{l(l+1)}{\rho^2} + 2\mathcal{E} - \rho^2]R = 0, \tag{II.69}$$

where

$$\mathcal{E} = b^2k^2 = \frac{\hbar}{M\omega}\times\frac{ME}{\hbar^2} = \frac{E}{\hbar\omega} \tag{II.70}$$

is the energy measured in units of $\hbar\omega$.

For $\rho \to 0$, the nature of the solution is determined by the first two terms of (II.69), and is therefore given by

$$R(\rho) = \begin{cases}\rho^{l+1} \\ \rho^{-l}\end{cases} \qquad \rho \to 0 \tag{II.71}$$

We accept the first one because the other one, for $l \neq 0$, blows up at the origin. Similarly, the solution for $\rho \to \infty$ is determined by the first and last terms in (II.69), and is described by

$$R(\rho) = \begin{cases}e^{-(1/2)\rho^2} \\ e^{(1/2)\rho^2}\end{cases} \qquad \rho \to \infty$$

Once again, the first one is accepted from the condition of well behaviour. The R we therefore substitute in (II.69) is

$$R(\rho) = \rho^{l+1}e^{-(1/2)\rho^2}F(\rho); \tag{II.72}$$

and we obtain, for $F(\rho)$,

$$\frac{d^2F}{d\rho^2} + 2(\frac{l+1}{\rho} - \rho)\frac{dF}{d\rho} + [2\mathcal{E} - (2l+3)]F = 0. \tag{II.73}$$

The solution to this equation can be obtained in the form of a series

$$F(\rho) = \sum_{k=0}^{\infty} c_k\rho^{k+\beta}, \tag{II.74}$$

where the indicial parameter β and the recurrence relation for the coefficients c_k can be obtained in the usual way. It can finally be verified that the series defined by (II.74) is the confluent hypergeometric series [see Appendix C (Section II)] with $\rho^2 = z$, $\frac{1}{2}(l + \frac{3}{2} - \mathcal{E}) = a$, and $(l + \frac{3}{2}) = c$. Thus,

$$F(\rho) = F(\tfrac{1}{2}(l + \tfrac{3}{2} - \mathcal{E}), l + \tfrac{3}{2}; \rho^2). \tag{II.75}$$

This result could have been obtained also by putting $\rho^2 = z$ in (II.73) and checking that the resultant differential equation is identical to (CII.1) of Appendix C.

We now use the solution (II.75) in (II.72) and investigate what happens when $\rho \to \infty$. In order that $R(\rho)$ can go as $e^{-(1/2)\rho^2}$ when $\rho \to \infty$, the confluent hypergeometric series must terminate after a finite number of terms. This happens when the quantity a of $F(a, c; z)$ is a negative integer or zero. In the present case, we thus obtain

$$\tfrac{1}{2}(l + \tfrac{3}{2} - \mathcal{E}) = -n, \qquad n = 0, 1, 2, \ldots, \tag{II.76}$$

or

$$E_{nl} = \hbar\omega\mathcal{E} = \hbar\omega(2n + l + \tfrac{3}{2}). \tag{II.77}$$

Thus, in terms of the new quantum numbers n, l, the total oscillator quantum number λ is given by

$$\lambda = 2n + l. \tag{II.78}$$

Once again, for each integral λ, the quantum numbers n, l may acquire various possible values, giving rise to considerable degeneracy.

We now collect all these results and write out the radial function, multiplied by a normalization constant $\bar{N}_{nl}$, as

$$R_{nl}(\rho) = \bar{N}_{nl}\rho^{l+1} \exp(-\tfrac{1}{2}\rho^2)F(-n, l + \tfrac{3}{2}; \rho^2).$$

The present F has a finite number of terms and is related to the Laguerre polynomials by

$$L_n^a(z) = \frac{[\Gamma(n + a + 1)]^2}{n!\Gamma(a + 1)}F(-n, a + 1; z). \tag{II.79}$$

Therefore,

$$R_{nl}(\rho) = N_{nl}\rho^{l+1} \exp(-\tfrac{1}{2}\rho^2)L_n^{l+1/2}(\rho^2).$$

The normalization constant N_{nl} is determined by requiring

$$1 = \int_0^\infty R_{nl}^2(r)\,dr = b\int_0^\infty R_{nl}^2(\rho)\,d\rho$$

and using

$$\int_0^\infty z^a e^{-z} L_m^a(z) L_n^a(z)\,dz = \delta_{m,n}\frac{[\Gamma(n + a + 1)]^3}{n!}. \tag{II.80}$$

In this way, we obtain

$$R_{nl}(r) = b^{-1/2}\left\{\frac{2n!}{[\Gamma(n + l + \frac{3}{2})]^3}\right\}^{1/2}\rho^{l+1} \exp(-\tfrac{1}{2}\rho^2)L_n^{l+1/2}(\rho^2) \tag{II.81a}$$

$$= b^{-1/2}\left\{\frac{2\Gamma(n + l + \frac{3}{2})}{n!}\right\}^{1/2}\rho^{l+1} \exp(-\tfrac{1}{2}\rho^2) \sum_{k=0}^{n} (-1)^k\Gamma^{-1}(k + l + \tfrac{3}{2})\binom{n}{k}\rho^{2k}. \tag{II.81b}$$

The form (II.81b) follows from (II.81a) by using (II.79), and (CII.3b) from Appendix C. It should be observed that the complete radial function

$$\frac{R_{nl}(r)}{r} = \frac{R_{nl}(\rho)}{b\rho}$$

is obtained from (II.81) by replacing $b^{-1/2}$ by $b^{-3/2}$, and the factor ρ^{l+1} by ρ^{l}. Many authors use the full radial function (let us call it $\mathcal{R}_{nl}$), instead of the R_{nl}, while quoting expressions for the oscillator radial functions. The reader is warned against confusing R_{nl} and $\mathcal{R}_{nl}$ through oversight.

It should be noted that, for $n = 0$, the sum over k in (II.81b) reduces to $\Gamma^{-1}(l + \frac{3}{2})$, and hence

$$R_{0l}(r) = b^{-1/2}\left[\frac{2}{\Gamma(l + \frac{3}{2})}\right]^{1/2}\rho^{l+1} \exp\left(-\tfrac{1}{2}\rho^2\right). \tag{II.82a}$$

Similarly, for $n = 1$, the two terms in the sum can be easily evaluated to get

$$R_{1l}(r) = b^{-1/2}\left[\frac{2l + 3}{\Gamma(l + \frac{3}{2})}\right]^{1/2}\rho^{l+1} \exp\left(-\tfrac{1}{2}\rho^2\right)\left[1 - \frac{2}{2l + 3}\rho^2\right]. \tag{II.82b}$$

The expressions (II.82) could also have been obtained from (II.81a) by using

$$L_0^a(z) = \Gamma(a + 1), \qquad L_1^a(z) = \Gamma(a + 2)[(a + 1) - z]. \tag{II.83a}$$

To compute the radial functions R_{nl} with $n \geqslant 2$, the easiest method is to use the recurrence relation

$$L_{n+1}^a(z) = \frac{a + n + 1}{n + 1}[(a + 2n + 1 - z)L_n^a(z) - (a + n)^2 L_{n-1}^a(z)] \tag{II.83b}$$

and evaluate $L_n^{l+1/2}(\rho^2)$ for $n \geqslant 2$, step by step, starting with (II.83a). The use of (II.81a) then allows the computation of a whole set of $R_{nl}(r)$, for a given l.

We now get back to (II.78) for the total oscillator quantum number λ, and observe that all the degenerate oscillator states belonging to a given λ have the same parity. This is because $2n$ is always an even integer, and l is therefore odd or even depending on whether λ is odd or even. All the oscillator states up to $\lambda = 7$ are listed in Table II.3.

Table II.3 Isotropic harmonic oscillator states

Value of λ	Degenerate States (nl)
0	(00)
1	(01)
2	(02), (10)
3	(03), (11)
4	(04), (12), (20)
5	(05), (13), (21)
6	(06), (14), (22), (30)
7	(07), (15), (23), (31)

According to our definition of n, its lowest value is zero. An equally popular convention of defining this radial quantum number gives the n-value as one higher than ours, and hence starts with unity instead of zero. This definition is preferred by many authors because here n directly gives the total number of radial nodes in the wavefunction R_{nl} (the node at zero is not counted). Once again, the reader is warned against confusing these two different definitions of n.

Product of Two Oscillator Functions and Moshinsky Transformation

We shall express the product of the oscillator functions of two particles in terms of oscillator functions corresponding to the relative and centre-of-mass coordinates defined by (II.41) and (II.44). The expression we are aiming at will be the analogue of (II.46), derived for the plane-wave functions. In the present case, however, the relationship is a little more complicated in the sense that there is a finite number of terms in the transformation, and the calculation of the transformation coefficients is a nontrivial algebraic problem.

We first start with the oscillator Hamiltonian for two particles:

$$H(1, 2) = -\frac{\hbar^2}{2M}(\nabla_1^2 + \nabla_2^2) + \tfrac{1}{2}M\omega^2(r_1^2 + r_2^2). \tag{II.84a}$$

If we use the transformation (II.41), we get

$$H(1, 2) = -\frac{\hbar^2}{2\mathcal{M}}\nabla_{\mathbf{R}}^2 + \tfrac{1}{2}\mathcal{M}\omega^2R^2 - \frac{h^2}{2\mu}\nabla_{\mathbf{r}}^2 + \tfrac{1}{2}\mu\omega^2r^2,$$

where $\mathcal{M} = 2M$ and $\mu = \frac{1}{2}M$. Thus, the Hamiltonian has clearly separated into oscillator Hamiltonians for the centre-of-mass coordinate $\mathbf{R}$, and the relative coordinate $\mathbf{r}$. The same remark is true if we use the transformation (II.44), in which case we get

$$H(1, 2) = -\frac{\hbar^2}{2M}\nabla_{\bar{\mathbf{R}}}^2 + \tfrac{1}{2}M\omega^2\bar{R}^2 - \frac{\hbar^2}{2M}\nabla_{\bar{\mathbf{r}}}^2 + \tfrac{1}{2}M\omega^2\bar{r}^2. \tag{II.84b}$$

An advantage of (II.84b) is that the same mass M appears in the centre-of-mass ($\bar{\mathbf{R}}$) and relative ($\bar{\mathbf{r}}$) Hamiltonians. The definition (II.68) of the oscillator parameter b that appears in the solutions of the harmonic oscillator equation tells us that this parameter remains unaltered if we use $\bar{\mathbf{R}}$ and $\bar{\mathbf{r}}$; on the other hand, if we use $\mathbf{R}$ and $\mathbf{r}$, then the centre-of-mass oscillator has the parameter $b_{\text{CM}} = (1/\sqrt{2})b$, and $b_{\text{rel}} = \sqrt{2}b$. From now on, we shall use the set $\bar{\mathbf{R}}$ and $\bar{\mathbf{r}}$, although the entire work can be done with $\mathbf{R}$ and $\mathbf{r}$ also. If a change to ($\mathbf{R}$, $\mathbf{r}$) is required, the reader is reminded to change the oscillator parameters appropriately.

Since the expressions (II.84) describe the same Hamiltonian $H(1, 2)$, we conclude that both $\phi_{n_1l_1m_1}(\mathbf{r}_1)\phi_{n_2l_2m_2}(\mathbf{r}_2)$ and $\phi_{\mathcal{N}\mathcal{L}\mathcal{M}}(\bar{\mathbf{R}})\phi_{nlm}(\bar{\mathbf{r}})$ form a complete set of wavefunctions of two particles moving in the harmonic oscillator potential. Each ϕ here is a harmonic oscillator wavefunction for the coordinate shown within the parentheses. It is therefore guaranteed that any one product wavefunction $\phi_{n_1l_1m_1}(\mathbf{r}_1)\phi_{n_2l_2m_2}(\mathbf{r}_2)$ should be expressible in terms of a complete set of harmonic oscillator functions $\phi_{\mathcal{N}\mathcal{L}\mathcal{M}}(\bar{\mathbf{R}})\phi_{nlm}(\bar{\mathbf{r}})$. There are certain conservation requirements that will, however, limit the sum to a finite number of terms.

Instead of describing this transformation in terms of the product wavefunctions, we define it in terms of the angular momentum coupled wavefunctions

$$\left[\phi_{n_1l_1}(\mathbf{r}_1), \phi_{n_2l_2}(\mathbf{r}_2)\right]_M^L = \sum_{m_1m_2}\begin{bmatrix} l_1 & l_2 & L \\ m_1 & m_2 & M \end{bmatrix}\phi_{n_1l_1m_1}(\mathbf{r}_1)\phi_{n_2l_2m_2}(\mathbf{r}_2),$$

$$\left[\phi_{\mathcal{N}\mathcal{L}}(\bar{\mathbf{R}}), \phi_{nl}(\bar{\mathbf{r}})\right]^{L'}_{M'} = \sum_{\mathcal{M}m} \begin{bmatrix} \mathcal{L} & l & L' \\ \mathcal{M} & m & M' \end{bmatrix} \phi_{\mathcal{N}\mathcal{L}\mathcal{M}}(\bar{\mathbf{R}})\phi_{nlm}(\bar{\mathbf{r}}),$$

where L is the total orbital angular momentum quantum number and M its projection. The same definitions apply to L' and M'. The advantage of using these coupled wavefunctions in the transformation is that now we have two more conserved quantities: L and M should be the same as L' and M' from angular momentum conservation. We therefore write the transformation as

$$\left[\phi_{n_1l_1}(\mathbf{r}_1), \phi_{n_2l_2}(\mathbf{r}_2)\right]^{L}_{M} = \sum_{\mathcal{N}\mathcal{L},\, nl} \langle \mathcal{N}\mathcal{L}, nl : L| \,|n_1l_1, n_2l_2 : L\rangle \left[\phi_{\mathcal{N}\mathcal{L}}(\bar{\mathbf{R}}), \phi_{nl}(\bar{\mathbf{r}})\right]^{L}_{M}, \tag{II.85}$$

where the transformation coefficients (to be determined by detailed algebra) are denoted by the symbol $\langle \,||\, \rangle$. The various quantities occurring inside the transformation bracket are self-explanatory. As already mentioned, the sum on the right-hand side is found to contain a finite number of terms from physical conservation requirements, which we shall now discuss.

The total energy of the two oscillator functions on the left-hand side of (II.85) is given by

$$\hbar\omega(2n_1 + l_1 + 2n_2 + l_2 + 3) = \hbar\omega(\lambda_1 + \lambda_2 + 3).$$

From the requirement of conservation of the total energy, this number should be equal to the energy contained in each term on the right-hand side, namely,

$$\hbar\omega(2\mathcal{N} + \mathcal{L} + 2n + l + 3) = \hbar\omega(\Lambda + \lambda + 3).$$

Thus,

$$\lambda_1 + \lambda_2 = \Lambda + \lambda. \tag{II.86a}$$

There is a parity conservation requirement too. But since $(\lambda_1 + \lambda_2)$ determines the parity of the left-hand side and $(\Lambda + \lambda)$ the parity of the right-hand side, (II.86a) already guarantees parity conservation. The second nontrivial requirement is provided by the angular momentum coupling on the right-hand side, which demands that $\mathcal{L}$ and l be able to couple and produce the *given* L of the left-hand side. This therefore gives

$$(\mathcal{L} + l) \geqslant L \geqslant |\mathcal{L} - l|. \tag{II.86b}$$

To illustrate how the expressions (II.86) limit the number of terms in (II.85), let us consider the case of $(n_1l_1) = (0p)$, $(n_2l_2) = (1d)$, and $L = 3$. We now have $\lambda_1 + \lambda_2 = 5$. Therefore, any possible break-up of 5 into two positive integers would give, by (II.86a), the values of Λ and λ. Thus, the allowed combinations are

$$\Lambda = 0, \lambda = 5; \quad \lambda = 0, \Lambda = 5; \quad \Lambda = 1, \lambda = 4;$$
$$\lambda = 1, \Lambda = 4; \quad \Lambda = 2, \lambda = 3; \quad \lambda = 2, \Lambda = 3.$$

Now, if we consider Table II.3 which gives the oscillator states for various λ (applicable to Λ also), then, corresponding to the set $\Lambda = 0, \lambda = 5$, we get the three states

$$[\phi_{00}(\bar{\mathbf{R}}), \phi_{05}(\bar{\mathbf{r}})], \quad [\phi_{00}(\bar{\mathbf{R}}), \phi_{13}(\bar{\mathbf{r}})], \quad [\phi_{00}(\bar{\mathbf{R}}), \phi_{21}(\bar{\mathbf{r}})].$$

We now apply (II.86b) to these three states. Since L has been specified to be 3, only the second state is acceptable; the first and third states can produce the total angular momentum 5 and 1, respectively, and hence are not acceptable. In the same way, we can deal with the remaining five sets.

Having demonstrated how to count the $\mathcal{N}\mathcal{L}, nl$ that appear in (II.85), we are left with the task of obtaining the transformation coefficients themselves. These transformation coefficients were first defined by Talmi[4], and later by Moshinsky[5], who prepared extensive numerical tables for them. Appendix F gives the necessary algebraic formulas for the computation of these coefficients. For very extensive computations involving many oscillator functions, it is advisable to program these formulas for the computing machine, and calculate and store them for subsequent use. If only a few coefficients are needed in a small calculation, reference to the tables prepared by Moshinsky and Brody[6] will yield quicker results. We shall use the term *harmonic oscillator brackets* to denote these transformation coefficients.

18. PERTURBATION THEORY IN THE LOWEST TWO ORDERS

We have already discussed the many-body eigenvalues and eigenfunctions of the unperturbed Hamiltonian H_0. Denoting the unperturbed ground-state wavefunction by Φ_0, and the corresponding energy eigenvalue by $E_0^{(0)}$, we obtain

$$H_0\Phi_0 = E_0^{(0)}\Phi_0. \tag{II.87}$$

The ground state Φ_0 is a determinantal wavefunction of the type (II.20a) and is specified by a set of single-particle states $\alpha, \beta, \gamma, \ldots, \zeta$, which are called *occupied* states from now on. An excited state Φ_n of this Hamiltonian belonging to the energy eigenvalue $E_n^{(0)}$ satisfies the equation

$$H_0\Phi_n = E_n^{(0)}\Phi_n. \tag{II.88}$$

The excited states $\Phi_m, \Phi_n, \ldots$ are also determinantal wavefunctions, but they obviously differ from the ground state in having one or more single-particle states different from those *occupied* in Φ_0.

The situation can be discussed with the help of the single-particle energy levels given in Fig. II.3. The states $\alpha, \beta, \gamma, \ldots, \zeta$ starting from the bottom of the single-particle spectrum are *occupied* in Φ_0; there is a large number of states above the highest occupied level (called the *fermi level, fermi surface,* or *fermi energy*), which are left unoccupied in the ground state. The excited states Φ_m are obtained by lifting any number of particles from the occupied to the unoccupied states. If only one particle is lifted from the *occupied* state μ to the *unoccupied* state μ', then obviously the difference between the energy of this Φ_m and the energy of Φ_0 is given by

$$E_m^{(0)} - E_0^{(0)} = \epsilon_{\mu'} - \epsilon_\mu, \tag{II.89a}$$

where $\epsilon_{\mu'}$, ϵ_μ are the energies of the corresponding single-particle states. If we compare the structure of such a Φ_m with Φ_0, then we notice that Φ_m has a vacancy in the state μ and an additional occupancy of the state μ'; otherwise they are identical. For this reason, such a state is often called a one-hole one-particle (1h-1p) state, where the vacant state μ is called the *hole* state and the extra occupied state μ' is called the *particle* state. In the same way, the higher excited states Φ_m may have 2p-2h, 3p-3h, ... The excitation energy of an np-nh state is given, in general, by the sum of the energies (ϵ_p) of the n-particle states minus the sum of the energies (ϵ_h) of the n-hole states, i.e.,

$$E_m^{(0)} - E_0^{(0)} = \sum_p \epsilon_p - \sum_h \epsilon_h. \tag{II.89b}$$

Having explicitly introduced the unperturbed wavefunctions and the energies, we now proceed to work out the perturbation formalism for the ground state in the lowest two orders. Denoting the ground-state energy in the order k by $E_0^{(k)}$, we write the standard expressions (see, for example, Schiff[3]) as

$$E_0^{(0)} = \langle\Phi_0|\, H_0\, |\Phi_0\rangle, \tag{II.90a}$$

$$E_0^{(1)} = \langle\Phi_0|\, H_{\rm I}\, |\Phi_0\rangle, \tag{II.90b}$$

$$E_0^{(2)} = \sum_n \frac{\langle\Phi_0|\, H_{\rm I}\, |\Phi_n\rangle \langle\Phi_n|\, H_{\rm I}\, |\Phi_0\rangle}{E_0^{(0)} - E_n^{(0)}}, \tag{II.90c}$$

where the summation n runs over all the excited unperturbed states, and H_0 and $H_{\rm I}$ are given by (II.8) and (II.9), respectively.

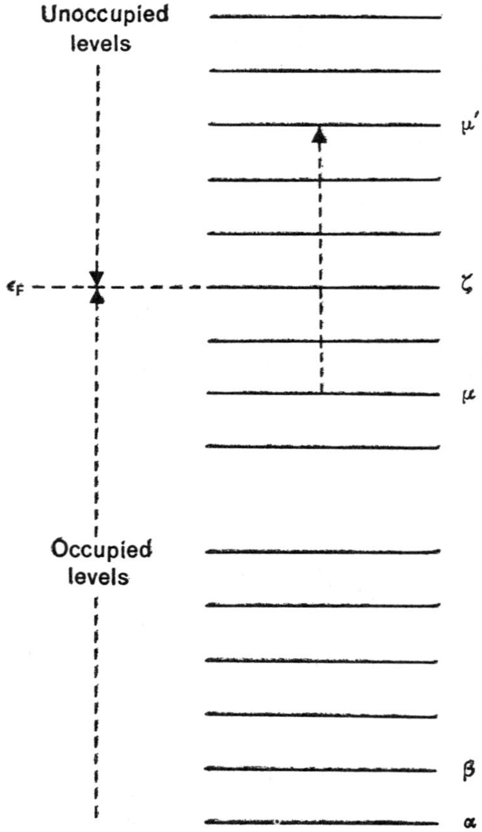

Fig. II.3 Fermi energy and occupied and unoccupied single-particle levels.

Further simplification is obtained with the help of (II.23)–(II.27). We have

$$E_0^{(1)} = \langle\Phi_0|\, \sum_{i<j} V_{ij}\, |\Phi_0\rangle - \langle\Phi_0|\, \sum_i \mathcal{V}_i\, |\Phi_0\rangle$$

$$= \sum_{\xi<\eta}^{\rm occ} (\xi\eta|\, V\, |\xi\eta) - \sum_{\xi}^{\rm occ} \langle\xi|\, \mathcal{V}\, |\xi\rangle, \tag{II.91a}$$

where occ denotes the set of single-particle states occupied in Φ_0. The unperturbed energy

(II.90a) is given by

$$E_0^{(0)} = \langle \Phi_0 | \sum_i (T_i + \mathcal{V}_i) | \Phi_0 \rangle = \sum_{\xi}^{\text{occ}} \langle \xi | T + \mathcal{V} | \xi \rangle. \tag{II.91b}$$

It is convenient to combine the equations (II.91) and then work out the right-hand side of the resultant expression as

$$E_0^{(0)} + E_0^{(1)} = \sum_{\xi}^{\text{occ}} \langle \xi | T | \xi \rangle + \sum_{\xi<\eta}^{\text{occ}} (\xi\eta | V | \xi\eta). \tag{II.92}$$

To simplify the matrix element in (II.90c), we first write

$$\langle \Phi_0 | H_I | \Phi_n \rangle = \langle \Phi_0 | \sum_{i<j} V_{ij} | \Phi_n \rangle - \langle \Phi_0 | \sum_i \mathcal{V}_i | \Phi_n \rangle. \tag{II.93}$$

In the first term, a nonvanishing result is obtained (i) when the excited state Φ_n differs from Φ_0 through the state of one particle, i.e., Φ_n is a 1h-1p type state with respect to Φ_0, and also (ii) when it differs from Φ_0 through two single-particle states, i.e., when it is a state of the 2h-2p type. On the other hand, the second term of (II.93) is nonvanishing only when Φ_n is of the type (i). We shall work out these two cases separately.

Case (i)

$$\langle \Phi_0 | H_I | \Phi_n \rangle = \sum_{\xi}^{\text{occ}} (\xi\mu | V | \xi\mu') - \langle \mu | \mathcal{V} | \mu' \rangle, \tag{II.94a}$$

$$E_n^{(0)} - E_0^{(0)} = \epsilon_{\mu'} - \epsilon_{\mu}, \tag{II.94b}$$

where μ and μ' are the hole and particle states, respectively. The average potential $\mathcal{V}$ can be so chosen that (II.94a) vanishes, i.e.,

$$\langle \mu | \mathcal{V} | \mu' \rangle = \sum_{\xi}^{\text{occ}} (\xi\mu | V | \xi\mu') \tag{II.95}$$

should be satisfied. The average potential $\mathcal{V}$, defined in this way, is in agreement with the Hartree-Fock definition of the average potential in atomic physics. In the case of infinite nuclear matter, where the single-particle states are the momentum eigenstates, it is easy to show that each term of (II.94a) vanishes individually. We shall therefore put the contribution of the 1h-1p type Φ_n in (II.90c) equal to zero.

Case (ii)

$$\langle \Phi_0 | H_I | \Phi_n \rangle = \langle \Phi_0 | \sum_{i<j} V_{ij} | \Phi_n \rangle = (\mu\nu | V | \mu'\nu'), \tag{II.96a}$$

$$E_n^{(0)} - E_0^{(0)} = \epsilon_{\mu'} + \epsilon_{\nu'} - \epsilon_{\mu} - \epsilon_{\nu}, \tag{II.96b}$$

where μ, ν are the two hole states and μ', ν' are the corresponding particle states.

Substituting (II.96) in (II.90c), we obtain

$$E_0^{(2)} = - \sum_{\mu<\nu} \sum_{\mu'<\nu'} \frac{|(\mu\nu | V | \mu'\nu')|^2}{\epsilon_{\mu'} + \epsilon_{\nu'} - \epsilon_{\mu} - \epsilon_{\nu}}, \tag{II.97}$$

where the summation label n has been appropriately replaced by the hole pair and the particle pair of Φ_n.

The expressions (II.92) and (II.97) are now applied to the case of the infinite nuclear matter. It is obvious that the matrix elements of V appearing in these expressions would be infinitely large if we wish to use a realistic two-nucleon potential with a hard core; even a soft-core potential with a very large repulsive core would cause difficulties because then each matrix element of V is very large and the convergence of the perturbation terms in successive orders is very poor. For this reason, we deliberately use for V a weak well-behaved potential

$$V = V_0 f(r/a)(g_W P_W + g_M P_M + g_\sigma P_\sigma + g_H P_H) \tag{II.98a}$$

with the strength parameters g satisfying

$$g_W + g_M + g_\sigma + g_H = 1. \tag{II.98b}$$

The operator P with the subscripts W, M, σ, and H denotes respectively Wigner, Majorana, Bartlett, and Heisenberg exchange defined in Section 3A; in particular, P_W is the same as the unity operator, and $P_M = P_r$, $P_H = -P_\tau$. The depth parameter V_0 is typically taken to be 40 MeV to 50 MeV with a range parameter $a \approx 1.4$–1.5 fm, and the shape function $f(r/a)$ is any of the well-behaved forms in (I.45b).

The subsequent perturbation calculations are only of academic interest; nevertheless, they acquaint the reader with several important concepts and mathematical manipulations of the infinite nuclear matter theory. The formulation of a more detailed and sophisticated many-body perturbation theory applicable to the realistic two-nucleon potential is dealt with in Section 19.

Kinetic Energy

The single-particle state $|\xi\rangle$, in the case of infinite nuclear matter, actually stands for $|\mathbf{k}, m_s, m_\tau\rangle$, where m_s, m_τ are the projection quantum numbers of spin and isospin. Thus, (II.35a) leads to

$$\begin{aligned}\sum_\xi^{\text{occ}} \langle\xi| T |\xi\rangle &= \sum_{\mathbf{k}, m_s, m_\tau}^{\text{occ}} \langle\mathbf{k}, m_s, m_\tau| \left(-\frac{\hbar^2}{2M}\nabla^2\right) |\mathbf{k}, m_s, m_\tau\rangle \\ &= 4 \sum_{\mathbf{k}}^{\text{occ}} \frac{\hbar^2 k^2}{2M} \frac{1}{\Omega} \int_\Omega d^3r \exp(-i\mathbf{k}\cdot\mathbf{r}) \exp(i\mathbf{k}\cdot\mathbf{r}) \\ &= 4 \sum_{\mathbf{k}}^{\text{occ}} \frac{\hbar^2 k^2}{2M}.\end{aligned}$$

The summation is replaced by integration with the help of (II.40), and finally the result (II.37) is used to obtain

$$\begin{aligned}\langle T\rangle &= \frac{4\Omega}{(2\pi)^3} \int^{\text{occ}} d^3k \frac{\hbar^2 k^2}{2M} \\ &= \frac{4\Omega}{(2\pi)^3} 4\pi \frac{\hbar^2}{2M} \int_0^{k_F} k^4\, dk \\ &= \frac{\hbar^2}{2M} \frac{2\Omega k_F^5}{5\pi^2} = A \tfrac{3}{5} \frac{\hbar^2 k_F^2}{2M}\end{aligned} \tag{II.99a}$$

or

$$\frac{\langle T\rangle}{A} = \tfrac{3}{5} \frac{\hbar^2 k_F^2}{2M}. \tag{II.99b}$$

Thus, the average kinetic energy per nucleon is equal to three-fifths of the maximum kinetic energy of a fermi-sea nucleon.

Potential Energy

A two-body state such as $|\xi\eta\rangle$ is now more explicitly written as $|\mathbf{k}_1, m_s, m_\tau; \mathbf{k}_2, m_s', m_\tau'\rangle$, where the semicolon separates the state labels of the two nucleons. The antisymmetrized matrix element of V between any two states is then given by

$$(\bar{\mathbf{k}}_1, \bar{m}_s, \bar{m}_\tau; \bar{\mathbf{k}}_2, \bar{m}_s', \bar{m}_\tau'|\, V\,|\mathbf{k}_1, m_s, m_\tau; \mathbf{k}_2, m_s', m_\tau')$$
$$= \langle\bar{\mathbf{k}}_1, \bar{m}_s, \bar{m}_\tau; \bar{\mathbf{k}}_2, \bar{m}_s', \bar{m}_\tau'|V(|\mathbf{k}_1, m_s, m_\tau; \mathbf{k}_2, m_s', m_\tau'\rangle - |\mathbf{k}_2, m_s', m_\tau'; \mathbf{k}_1, m_s, m_\tau\rangle). \tag{II.100}$$

The Wigner term of (II.98a) does not exchange anything, and hence, for this term, (II.100) yields

$$V_0 g_W[\delta_{\bar{m}_s, m_s}\delta_{\bar{m}_\tau, m_\tau}\delta_{\bar{m}_s', m_s'}\delta_{\bar{m}_\tau', m_\tau'}\langle\bar{\mathbf{k}}_1, \bar{\mathbf{k}}_2|\, f(r/a)\,|\mathbf{k}_1, \mathbf{k}_2\rangle$$
$$- \delta_{\bar{m}_s, m_s'}\delta_{\bar{m}_\tau, m_\tau'}\delta_{\bar{m}_s', m_s}\delta_{\bar{m}_\tau', m_\tau}\langle\bar{\mathbf{k}}_1, \bar{\mathbf{k}}_2|\, f(r/a)\,|\mathbf{k}_2, \mathbf{k}_1\rangle]. \tag{II.101}$$

On the other hand, with any of the exchange potentials of (II.98a) we have first to carry out the appropriate exchange in the two states enclosed within the parentheses in (II.100) and then write down the result analogous to (II.101). For example, because the Majorana operator exchanges the spatial part of the wavefunction, we first interchange $\mathbf{k}_1 \leftrightarrow \mathbf{k}_2$ and then obtain $V_0 g_M$ $\times$ [expression (II.101) with the momentum matrix elements interchanging places]. The Bartlett and Heisenberg terms can be handled in the same manner. The first one interchanges $m_s \leftrightarrow m_s'$ in (II.100), and the second interchanges both $m_s \leftrightarrow m_s'$ and $\mathbf{k}_1 \leftrightarrow \mathbf{k}_2$.

The matrix elements occurring in (II.101) can, by using (II.43), be simplified to

$$\langle\bar{\mathbf{k}}_1, \bar{\mathbf{k}}_2|\, f(r/a)\,|\mathbf{k}_1, \mathbf{k}_2\rangle = \langle\bar{\mathbf{K}}, \bar{\mathbf{k}}|\, f(r/a)\,|\mathbf{K}, \mathbf{k}\rangle$$
$$= \delta_{\mathbf{K},\bar{\mathbf{K}}}\langle\bar{\mathbf{k}}|\, f(r/a)\,|\mathbf{k}\rangle$$
$$= \frac{1}{\Omega}\delta_{\mathbf{K},\bar{\mathbf{K}}}\int d^3r \exp[i(\mathbf{k} - \bar{\mathbf{k}})\cdot\mathbf{r}]f(r/a)$$
$$= \frac{1}{\Omega}\delta_{\mathbf{K},\bar{\mathbf{K}}}F(|\mathbf{k} - \bar{\mathbf{k}}|), \tag{II.102a}$$

where

$$F(q) = \int d^3r \exp(i\mathbf{q}\cdot\mathbf{r})f(r/a) \tag{II.102b}$$

is the Fourier transform of the shape function. Using the Yukawa shape for F, we can explicitly work out the integral in (II.102b) and obtain

$$F(q) = \frac{4\pi a^3}{1 + q^2a^2}. \tag{II.102c}$$

The δ-function in (II.102a) has been obtained from $\langle\bar{\mathbf{K}}\,|\,\mathbf{K}\rangle$ by the normalization condition (II.35b) of the plane waves, and it expresses the requirement of total momentum conservation of the interacting pair of nucleons. All these results are now applied to the potential energy term of (II.92) to obtain

$$\sum_{\xi<\eta}^{\text{occ}} (\xi\eta|V|\xi\eta) = \tfrac{1}{2} \sum_{\mathbf{k}_1, m_s, m_\tau}^{\text{occ}} \sum_{\mathbf{k}_2, m'_s, m'_\tau}^{\text{occ}} (\mathbf{k}_1, m_s, m_\tau; \mathbf{k}_2, m'_s, m'_\tau| V |\mathbf{k}_1, m_s, m_\tau; \mathbf{k}_2, m'_s, m'_\tau)$$

$$= \tfrac{1}{2} V_0 \sum_{\mathbf{k}_1, \mathbf{k}_2}^{\text{occ}} \sum_{m_s, m'_s} \sum_{m_\tau, m'_\tau} [(g_W - \delta_{m_s, m'_s}\delta_{m_\tau, m'_\tau} g_M + \delta_{m_s, m'_s} g_\sigma - \delta_{m_\tau, m'_\tau} g_H)$$

$$\times \langle \mathbf{k}_1, \mathbf{k}_2| f(r/a) |\mathbf{k}_1, \mathbf{k}_2\rangle$$

$$- (\delta_{m_s, m'_s}\delta_{m_\tau, m'_\tau} g_W - g_M + \delta_{m_\tau, m'_\tau} g_\sigma - \delta_{m_s, m'_s} g_H)$$

$$\times \langle \mathbf{k}_1, \mathbf{k}_2| f(r/a) |\mathbf{k}_2, \mathbf{k}_1\rangle]. \qquad \text{(II.103)}$$

Here we have replaced the pair sum ($\xi < \eta$) by half times the independent summations over ξ and η; this is legitimate because each of the extra terms corresponding to $\xi = \eta$ entering through this process is identically zero by virtue of the cancellation of the direct and the exchange terms of the antisymmetrized matrix element. The sum over the projection quantum numbers in (II.103) can be immediately performed, taking due account of the Krönecker deltas. Further, from (II.102), and the definition (II.42) of the relative momentum, we obtain

$$\langle \mathbf{k}_1, \mathbf{k}_2| f(r/a) |\mathbf{k}_1, \mathbf{k}_2\rangle = \frac{1}{\Omega} F(0) = \frac{4\pi a^3}{\Omega}, \qquad \text{(II.104a)}$$

$$\langle \mathbf{k}_1, \mathbf{k}_2| f(r/a) |\mathbf{k}_2, \mathbf{k}_1\rangle = \frac{1}{\Omega} F(2|\mathbf{k}|) = \frac{1}{\Omega} \frac{4\pi a^3}{1 + 4k^2 a^2}. \qquad \text{(II.104b)}$$

In this way, the final simplified form for (II.103) is found to be

$$\langle V\rangle = \frac{2V_0}{\Omega} \sum_{\mathbf{k}_1, \mathbf{k}_2}^{\text{occ}} [g_1 F(0) + g_2 F(2k)], \qquad \text{(II.105a)}$$

where

$$g_1 = 4g_W - g_M + 2g_\sigma - 2g_H,$$
$$g_2 = -g_W + 4g_M - 2g_\sigma + 2g_H. \qquad \text{(II.105b)}$$

The momentum summation is easy to carry out in the first term. It should be observed that four times $\sum_{\mathbf{k}_1}$ or $\sum_{\mathbf{k}_2}$ is each equal to A, the total number of nucleons, and that (A/Ω) is the density ρ given by (II.39). Therefore,

$$\text{first term of (II.105a)} = \tfrac{1}{8} V_0 g_1 A\rho F(0) = \frac{V_0}{3\pi} g_1 A \kappa_F^3, \qquad \text{(II.106)}$$

where $\kappa_F = k_F a$. In the second term, the summations are much more involved to carry out because of the presence of the momentum-dependent quantity $F(2k)$. We first replace the sums by the integrations following (II.40). Then

$$\frac{1}{\Omega} \sum_{\mathbf{k}_1, \mathbf{k}_2}^{\text{occ}} F(2k) = \frac{\Omega}{(2\pi)^6} \int^{\text{occ}} d^3k_1 \int^{\text{occ}} d^3k_2\, F(2k)$$

$$= \frac{\Omega}{(2\pi)^6} \underset{(k_1 \leqslant k_F,\ k_2 \leqslant k_F)}{\int d^3K \int d^3k}\, F(2k). \qquad \text{(II.107)}$$

Here we have converted the integrations over $\mathbf{k}_1$, $\mathbf{k}_2$ into those over the total and relative momenta **K** and **k** and have specified the restrictions beneath the integral so that only the occupied states are taken into account.

Now, from the definitions of **K**, **k** in terms of $\mathbf{k}_1$, $\mathbf{k}_2$, we get

$$\mathbf{k}_1 = \tfrac{1}{2}\mathbf{K} + \mathbf{k}, \quad \text{i.e.,} \quad k_1^2 = \tfrac{1}{4}K^2 + k^2 + Kk\cos\theta,$$

$$\mathbf{k}_2 = \tfrac{1}{2}\mathbf{K} - \mathbf{k}, \quad \text{i.e.,} \quad k_2^2 = \tfrac{1}{4}K^2 + k^2 - Kk\cos\theta,$$

where θ is the angle between the vectors **K** and **k**. Since each of k_1 and k_2 should be less than or equal to k_F, we have, from these two relations,

$$\cos\theta \leqslant \frac{k_F^2 - (\frac{1}{4}K^2 + k^2)}{Kk}, \tag{II.108a}$$

$$\cos\theta \geqslant -\frac{k_F^2 - (\frac{1}{4}K^2 + k^2)}{Kk}. \tag{II.108b}$$

These relations imply that, for the given K, k, and k_F, the integration over the angles of **K** in (II.107) is not unrestricted. As a matter of fact, if $k_F^2 - (\frac{1}{4}K^2 + k^2)$ is not positive, the requirements (II.108) become mutually contradictory, and no permissible value of $\cos\theta$ exists. Hence, the first requirement is

$$k_F^2 - (\tfrac{1}{4}K^2 + k^2) \geqslant 0$$

or

$$K \leqslant 2(k_F^2 - k^2)^{1/2}. \tag{II.109a}$$

For the specified k_F and k, (II.109a) gives the maximum value of K, which yields a non-vanishing result in (II.107). When this condition is satisfied, we can distinguish between the two possibilities

$$\frac{k_F^2 - (\frac{1}{4}K^2 + k^2)}{Kk} \geqslant 1$$

or

$$K \leqslant 2(k_F - k) \tag{II.109b}$$

and

$$\frac{k_F^2 - (\frac{1}{4}K^2 + k^2)}{Kk} \leqslant 1$$

or

$$K \geqslant 2(k_F - k). \tag{II.109c}$$

In the first case, all the values of $\cos\theta$ between -1 and $+1$ are obviously permitted by (II.108). But, in the second case, $\cos\theta$ cannot go over the full range; it goes from its minimum value given by the right-hand side of (II.108b) to its maximum value given by the right-hand side of (II.108a).

Using in (II.107) all the considerations just specified, we have the expression (II.107) given by

$$\frac{1}{\Omega}\sum_{\mathbf{k}_1,\mathbf{k}_2}^{\text{occ}} F(2k) = \frac{\Omega}{(2\pi)^6}4\pi\int_0^{k_F} k^2\,dk\,F(2k)$$
$$\times 4\pi\left[\int_0^{2(k_F-k)} K^2\,dK + \int_{2(k_F-k)}^{2(k_F^2-k^2)^{1/2}} K^2\,dK\,\frac{k_F^2 - (\frac{1}{4}K^2 + k^2)}{Kk}\right].$$

In the first term, included within the square brackets, the integration limits of $\cos\theta$ are such that $\cos\theta$ can go over its full range of values, and hence the integration over the angles of **K** has yielded a factor 4π placed just outside the square brackets. On the other hand, in the second term within the square brackets, the integration limits of $\cos\theta$ impose a restriction on the maximum and minimum values of $\cos\theta$, just mentioned, and hence the angle-integration yields

$$2\pi[(\cos\theta)_{\max} - (\cos\theta)_{\min}] = 4\pi\frac{k_F^2 - (\frac{1}{4}K^2 + k^2)}{Kk}. \qquad \text{(II.110a)}$$

This explains the structure of the second term. It should finally be noted that each of $\mathbf{k}_1$ and $\mathbf{k}_2$ varies in *magnitude* from $0 \to k_F$, and hence $(\mathbf{k}_1 - \mathbf{k}_2)$ varies in *magnitude* from $0 \to 2k_F$; this explains the limits of the k-integration.

Carrying out the K-integrations explicitly, we obtain

$$\frac{1}{\Omega}\sum_{\mathbf{k}_1,\mathbf{k}_2}^{\text{occ}} F(2k) = \frac{\Omega}{(2\pi)^6}4\pi\int_0^{k_F} k^2\,dk\,F(2k)\frac{32\pi}{3}k_F^3\left(1 - \frac{3}{2}\frac{k}{k_F} + \frac{1}{2}\frac{k^3}{k_F^3}\right). \qquad \text{(II.110b)}$$

Using the explicit expression of $F(2k)$ from (II.104b), we can complete the momentum integration in (II.110b). The integral encountered is given by

$$\int_0^1 dx\,\frac{x^2(1 - \frac{3}{2}x + \frac{1}{2}x^3)}{1 + 4\kappa_F^2x^2},$$

where $x = k/k_F$ and $\kappa_F = k_Fa$. First, we write the integrand I in the form

$$\begin{aligned}4\kappa_F^2 I &= \frac{(1 + 4\kappa_F^2x^2)(1 - \frac{3}{2}x + \frac{1}{2}x^3) - (1 - \frac{3}{2}x + \frac{1}{2}x^3)}{1 + 4\kappa_F^2x^2}\\
&= (1 - \tfrac{3}{2}x + \tfrac{1}{2}x^3) - \frac{1 - \frac{3}{2}x + \frac{1}{2}x^3}{1 + 4\kappa_F^2x^2}\\
&= (1 - \tfrac{3}{2}x + \tfrac{1}{2}x^3) - \frac{1}{1 + 4\kappa_F^2x^2} - \frac{(8\kappa_F^2)^{-1}(1 + 4\kappa_F^2x^2)x}{1 + 4\kappa_F^2x^2} + \left(\frac{1}{8\kappa_F^2} + \frac{3}{2}\right)\frac{x}{1 + 4\kappa_F^2x^2}\\
&= 1 - \left(\frac{1}{8\kappa_F^2} + \frac{3}{2}\right)x + \tfrac{1}{2}x^3 - \frac{1}{1 + 4\kappa_F^2x^2} + \left(\frac{1}{8\kappa_F^2} + \frac{3}{2}\right)\frac{1}{8\kappa_F^2}\frac{8\kappa_F^2x}{1 + 4\kappa_F^2x^2}.\end{aligned}$$

A term-by-term integration is now straightforward. Collecting all these results with (II.106), we obtain the final simplified form for (II.105a) as

$$\frac{\langle V\rangle}{A} = \frac{V_0}{3\pi}g_1\kappa_F + \frac{V_0}{4\pi}g_2\left[3\kappa_F - \frac{1}{2\kappa_F} - 4\tan^{-1}(2\kappa_F) + \left(\frac{3}{2\kappa_F} + \frac{1}{8\kappa_F^3}\right)\ln(4\kappa_F^2 + 1)\right]. \qquad \text{(II.111)}$$

The binding energy per nucleon up to the first order in the perturbation theory is given by the sum of (II.99b) and (II.111), which is a function of the fermi momentum k_F or, equivalently, a function of the nuclear radius parameter r_0 or the density of nucleons ρ. The basic idea in infinite nuclear matter theory is that, although we go to the limit $\Omega \to \infty$, A too simultaneously becomes very large in such a way that the density ρ, which is very simply given by the radius parameter r_0, still remains a finite well-determined number. The calculated value of the binding energy per nucleon can be plotted as a function of the parameter r_0 or, equivalently, ρ or k_F. If the theory is satisfactory, then this curve should display a minimum, and the equilibrium density and binding energy of the nucleus should correspond to this minimum. The purpose of the theoretical calculation is to check whether the binding energy versus the r_0-curve does indeed show a minimum of right magnitude at the correct r_0. In many cases, the theoretical curve may not even show a minimum. If, for example, the curve monotonically decreases as $r_0 \to 0$, then the state of lowest binding energy corresponds to $r_0 = 0$; such a theory therefore predicts that the nucleus energetically prefers to be in a collapsed state of zero radius, and hence it is obviously wrong.

With this general discussion on the type of result to expect, let us now closely examine our binding energy expression (II.99b) plus (II.111). The nature of the binding energy versus the r_0-curve depends, to a large extent, on the type of exchange mixture being used in the two-nucleon potential. If we consider a mixture of Wigner and Majorana forces alone, then $g_1 = 4g_W - g_M$ and $g_2 = -g_W + 4g_M$. If we normalize $g_W + g_M$ to unity, then we can parametrize g_1 and g_2 by one parameter α chosen from

$$g_W = \tfrac{1}{2}(1 + \alpha), \qquad g_M = \tfrac{1}{2}(1 - \alpha),$$

and hence

$$g_1 = \tfrac{1}{2}(3 + 5\alpha), \qquad g_2 = \tfrac{1}{2}(3 - 5\alpha).$$

The binding energy (E/A) curves (see Brueckner[7]) for several values of α are shown in Fig. II.4. The nature of the curves can be understood in this way: when r_0 becomes small, k_F gets very large, and hence the κ_F^3-term in (II.111) dominates the other terms; this term will give an attractive contribution (since the attractive V_0 is present as an overall factor) if $g_1 \equiv \tfrac{1}{2}(3 + 5\alpha)$ is positive, i.e., as long as $\alpha > -0.6$. For values of $\alpha \leqslant -0.6$, the positive kinetic energy and the repulsive potential energy terms add up, giving a curve above the r_0-axis. For values of α sufficiently larger than -0.6, the attractive potential energy term overtakes the kinetic energy term and produces a negative energy. But, unfortunately, the curves do not display a genuine minimum. From what has already been said, these monotonically decreasing curves, as $r_0 \to 0$, predict a collapsed state.

The situation improves if we use an exchange-dependence for which $g_1 = 0$. Such calculations have been done by Swiatecki[8] for the Yukawa and also the Gaussian potential $f(r/a) = e^{-r^2/a^2}$, using a $(\boldsymbol{\tau}_1 \cdot \boldsymbol{\tau}_2)$-type exchange-dependence. Huby[9] did a similar calculation with a Yukawa potential of the type $V_0 \boldsymbol{\tau}_1 \cdot \boldsymbol{\tau}_2 (g^2 + f^2 \boldsymbol{\sigma}_1 \cdot \boldsymbol{\sigma}_2) e^{-r/a}/(r/a)$, where g^2 and f^2 are two strength parameters. It should be observed that the exchange-dependence of Huby's potential is described by

$$g_W = f^2 - g^2, \qquad g_M = -4f^2, \qquad g_\sigma = -2f^2, \qquad g_H = 2(f^2 - g^2)$$

such that

$$g_1 = 0, \qquad g_2 = -3(g^2 + 3f^2).$$

Huby chose $V_0 = -13.4$ MeV, $f^2 = 0.595$, $g^2 = 0.258$, and $a = 1.75$ fm. The results for the binding energy are shown in Fig. II.5. The curve a corresponds to our first-order expression (II.111), and indeed shows a minimum. The values of r_0 and E/A corresponding to the minimum, however, greatly vary from the observed results. Huby calculated the correction to E/A due to the second-order perturbation term. (We end this section with the details of the second-order calculation.) His final results (that is, first order plus second order) are shown by the curve b. Although the saturation binding energy improves to some extent, its value and that of r_0 still do not agree with the observed numbers.

Various other radial shapes of the two-body potential have also been tried in this kind of

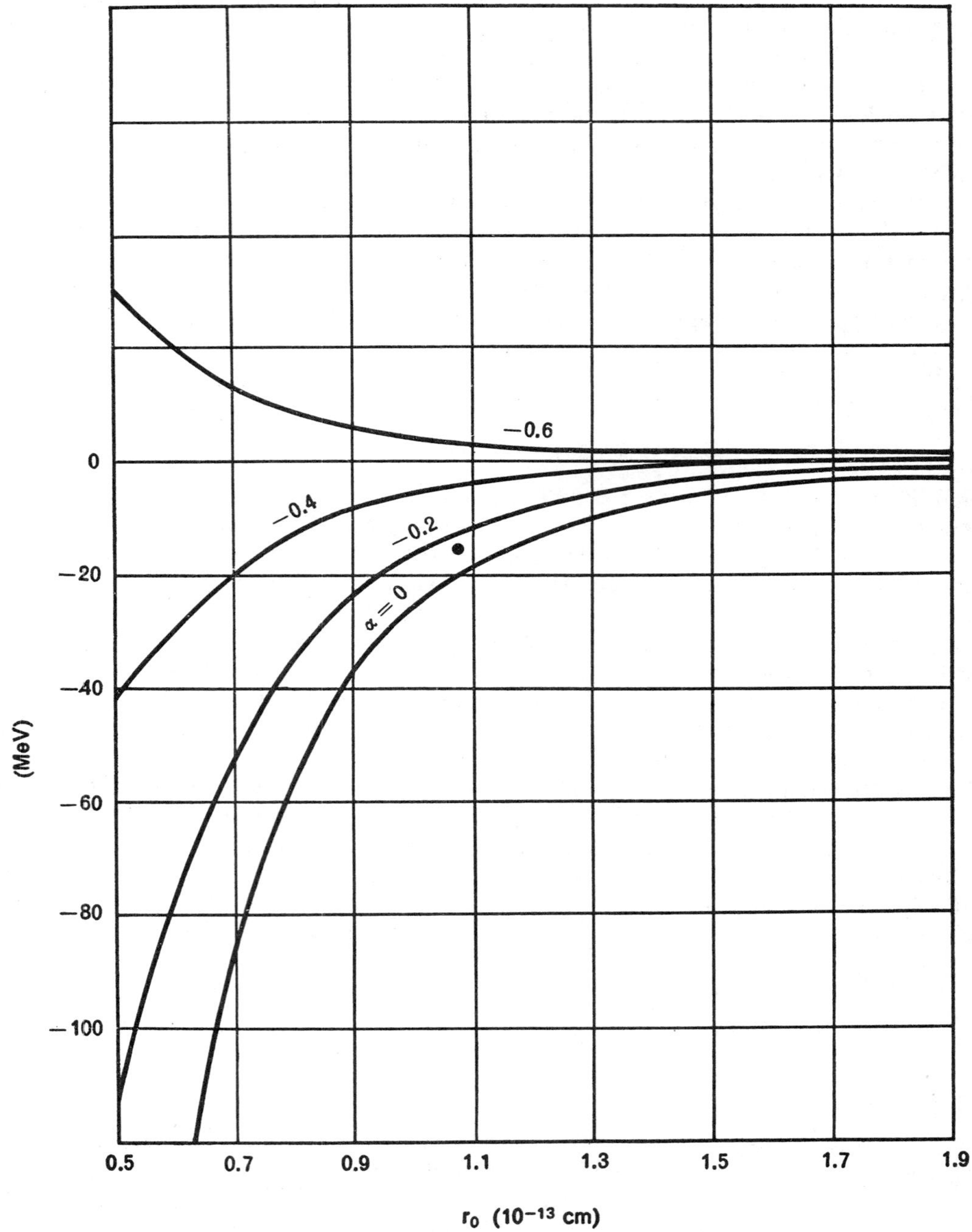

Fig. II.4 Binding energy per nucleon as function of nuclear radius parameter r_0 for various mixtures of Wigner and Majorana exchanges of purely central two-nucleon potential having Yukawa shape. (Following Brueckner, K. A., in The Many-Body Problem, ed. by C. Dewitt, Dunod, Paris, 1959, p 113.)

calculation. The Gaussian shape used by Swiatecki is an example. This simply entails recalculating $F(0)$ and $F(2|\mathbf{k}|)$ of (II.104), using the appropriate shape function $f(r/a)$, and then repeating all the steps leading to (II.111). A very interesting potential has been used by Bethe[10]. Rather than choosing a potential with a complicated kind of exchange-dependence, he uses a

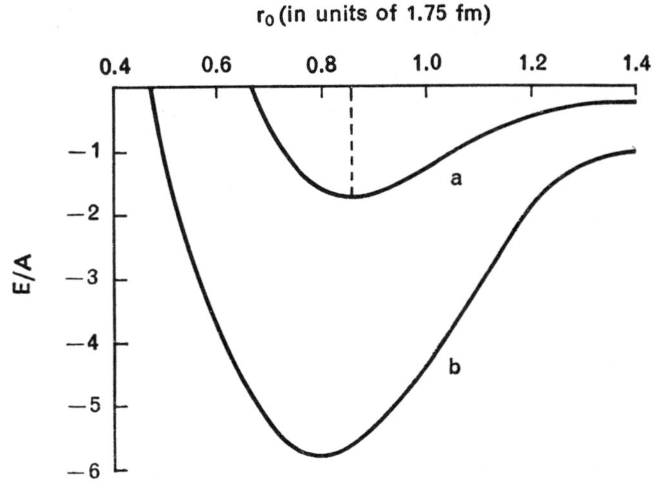

Fig. II.5 Results of binding energy calculation with a well-behaved central potential. [From Huby, R., *Proc. Phys. Soc.* (London), **A62**, 62 (1949).]

simple model potential, which is assumed to exist only in the $S(l=0)$-state of the relative motion of two interacting nucleons. Further, the radial shape is assumed to be an exponential $f(r/a) = e^{-r/a}$. For this potential, we have to work with the Wigner term of (II.103) with $g_W = 1$, and then, in computing the Fourier transform F of (II.102a), use only the S-wave part of the relative wavefunction exp ($i\mathbf{k}\cdot\mathbf{r}$). In this way, we finally obtain the result

$$\frac{\langle V \rangle}{A} = \frac{3V_0}{2\pi}[(\text{II.111})], \tag{II.112}$$

where [(II.111)] stands for the quantity enclosed within the square brackets in (II.111). Using a range parameter equal to 0.706 fm, and also keeping V_0 just sufficient to produce a resonance at zero energy in the two nucleon S-state, Bethe obtains a minimum in the binding energy curve at $k_F a = 2.92$. This corresponds to a value of r_0, which is more than three times smaller than the observed value, and a saturation nuclear density about thirty times too large.

The most realistic calculation of the potential energy using the perturbation theory has been that by Tabakin[11], who employed the separable nonlocal potential mentioned in Section 12. The two curves in Fig. II.6 show the binding energy per nucleon in the first and second order. In the first order, saturation is obtained at $k_F = 1.6\ \text{fm}^{-1}$ with $E/A = -8$ MeV, whereas with the inclusion of the second-order energy these values are given by $1.8\ \text{fm}^{-1}$ and -14.1 MeV, respectively. It should be observed that the value of k_F is much higher than the proper saturation value.

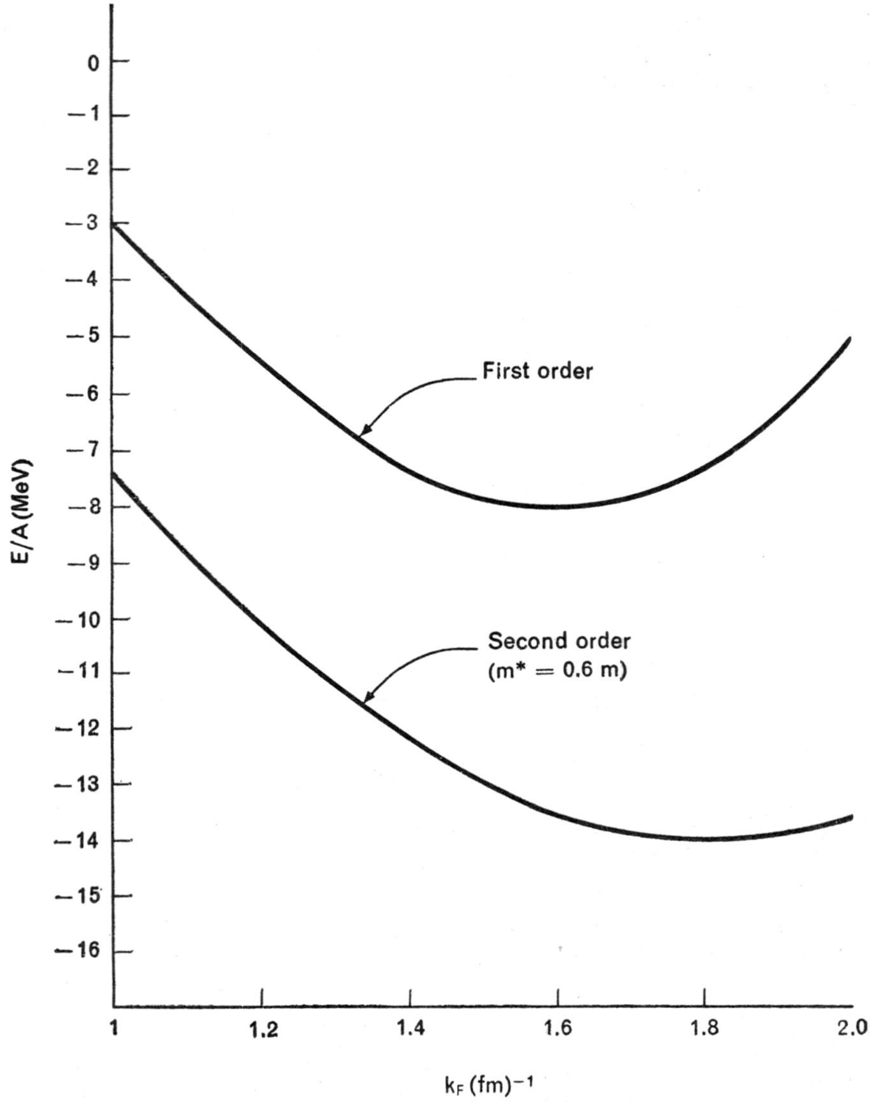

Fig. II.6 Results of binding energy calculation with a separable nonlocal potential. [Following Tabakin, F., *Ann. Phys.* (N.Y.), **30**, 51 (1964).]

Second-Order Calculation

The expression (II.97) has to be evaluated in detail. We have already seen from (II.102a) that the total two-body momentum must be the same for the initial and final states in a matrix element of V. Denoting the initial momenta of the two fermi-sea particles by $\mathbf{k}_1$, $\mathbf{k}_2$ and the momenta of the same particles when they go out of the fermi sea by $\bar{\mathbf{k}}_1$, $\bar{\mathbf{k}}_2$, we then obtain

$$\mathbf{k}_1 + \mathbf{k}_2 = \bar{\mathbf{k}}_1 + \bar{\mathbf{k}}_2 = \mathbf{K}, \qquad \tfrac{1}{2}(\mathbf{k}_1 - \mathbf{k}_2) = \mathbf{k}, \qquad \tfrac{1}{2}(\bar{\mathbf{k}}_1 - \bar{\mathbf{k}}_2) = \bar{\mathbf{k}}.$$

Further,

$$\epsilon(\mathbf{k}_1) + \epsilon(\mathbf{k}_2) = \frac{\hbar^2}{2M}(k_1^2 + k_2^2) = \frac{\hbar^2 K^2}{4M} + \frac{\hbar^2 k^2}{M},$$

$$\epsilon(\bar{\mathbf{k}}_1) + \epsilon(\bar{\mathbf{k}}_2) = \frac{\hbar^2}{2M}(\bar{k}_1^2 + \bar{k}_2^2) = \frac{\hbar^2 K^2}{4M} + \frac{\hbar^2 \bar{k}^2}{M}.$$

Therefore, the energy denominator in (II.97) is given by

$$\epsilon(\bar{\mathbf{k}}_1) + \epsilon(\bar{\mathbf{k}}_2) - \epsilon(\mathbf{k}_1) - \epsilon(\mathbf{k}_2) = \frac{\hbar^2}{M}(\bar{k}^2 - k^2). \tag{II.113}$$

For the evaluation of the numerator, we begin by noting that the direct minus exchange matrix element of V can be written in a compact form as

$$\begin{aligned}&(\mathbf{k}_1, m_s, m_\tau; \mathbf{k}_2, m_s', m_\tau'|\, V\, |\bar{\mathbf{k}}_1, \bar{m}_s, \bar{m}_\tau; \bar{\mathbf{k}}_2, \bar{m}_s', \bar{m}_\tau') \\ &\quad = \langle \mathbf{k}_1, m_s, m_\tau; \mathbf{k}_2, m_s', m_\tau'|\, V(\mathbb{1} - P_M P_\sigma P_\tau)\, |\bar{\mathbf{k}}_1, \bar{m}_s, \bar{m}_\tau; \bar{\mathbf{k}}_2, \bar{m}_s', \bar{m}_\tau'\rangle. \end{aligned} \tag{II.114}$$

Writing out V in the form (II.98a), where P_W is actually the unity operator and $P_H = P_M P_\sigma$, we obtain

$$\begin{aligned} V(\mathbb{1} - P_M P_\sigma P_\tau) &= V_0 f(r/a)(g_W \mathbb{1} + g_M P_M + g_\sigma P_\sigma + g_H P_M P_\sigma - g_W P_M P_\sigma P_\tau - g_M P_\sigma P_\tau \\ &\qquad - g_\sigma P_M P_\tau - g_H P_\tau) \\ &= V_0 f(r/a)[\mathcal{P}(P_\sigma, P_\tau) + \bar{\mathcal{P}}(P_\sigma, P_\tau) P_M], \end{aligned} \tag{II.115a}$$

where

$$\mathcal{P} = g_W \mathbb{1} + g_\sigma P_\sigma - g_M P_\sigma P_\tau - g_H P_\tau, \tag{II.115b}$$

$$\bar{\mathcal{P}} = g_M \mathbb{1} + g_H P_\sigma - g_W P_\sigma P_\tau - g_\sigma P_\tau. \tag{II.115c}$$

Using (II.115a) in (II.114), we obtain

$$\begin{aligned}&(\mathbf{k}_1, m_s, m_\tau; \mathbf{k}_2, m_s', m_\tau'|\, V\, |\bar{\mathbf{k}}_1, \bar{m}_s, \bar{m}_\tau; \bar{\mathbf{k}}_2, \bar{m}_s', \bar{m}_\tau') \\ &\quad = V_0 \langle \mathbf{k}_1, m_s, m_\tau; \mathbf{k}_2, m_s', m_\tau'| f(r/a) \\ &\qquad \times (\mathcal{P}|\bar{\mathbf{k}}_1, \bar{m}_s, \bar{m}_\tau; \bar{\mathbf{k}}_2, \bar{m}_s', \bar{m}_\tau'\rangle + \bar{\mathcal{P}}|\bar{\mathbf{k}}_2, \bar{m}_s, \bar{m}_\tau; \bar{\mathbf{k}}_1, \bar{m}_s', \bar{m}_\tau'\rangle). \end{aligned} \tag{II.116}$$

The summations in (II.97) now entail a summing over the occupied momentum states ($\mathbf{k}_1, \mathbf{k}_2$) with a factor of $\frac{1}{2}$, a similar summing over the unoccupied momentum states ($\bar{\mathbf{k}}_1, \bar{\mathbf{k}}_2$) with a factor of $\frac{1}{2}$, and a summing over $m_s, m_\tau, \bar{m}_s, \bar{m}_\tau, m_s', m_\tau', \bar{m}_s', \bar{m}_\tau'$. Because the energy denominator (II.113) is independent of spin-isospin states, the sums over the eight m-quantum numbers can be carried out on the square of the modulus of (II.116) alone. While doing so, we note that

$$\sum_{\bar{m}} |\bar{m}_s, \bar{m}_\tau; \bar{m}_s', \bar{m}_\tau'\rangle\langle \bar{m}_s, \bar{m}_\tau; \bar{m}_s', \bar{m}_\tau'| = \mathbb{1} \tag{II.117}$$

from the completeness of the spin isospin states, where the sum is over all the four quantum numbers with an overhead bar. Finally, each of the sums over $m_s, m_\tau, m_s', m_\tau'$ will be a trace of the operator concerned in the corresponding spin-isospin space. When, for example, $P_\sigma = \frac{1}{2}(\mathbb{1} + \boldsymbol{\sigma}_1 \cdot \boldsymbol{\sigma}_2)$ is summed over m_s (the spin states of the first particle), we have

$$\sum_{m_s} \langle m_s|\, P_\sigma\, |m_s\rangle = \mathrm{Tr}_1\, P_\sigma = \mathbb{1}_2 \tag{II.118}$$

because $\text{Tr}_1\, \boldsymbol{\sigma}_1 = 0$ and $\text{Tr}_1\, \mathbb{1} = 2\mathbb{1}_2$. Here $\mathbb{1}_2$ is the unity operator in the spin-space of the second particle. Upon subsequent summation over m_s', (II.118) yields 2. Thus,

$$\sum_{m_s, m_s'} \langle m_s, m_s'|\, P_\sigma\, |m_s, m_s'\rangle = 2.$$

Further summation over m_τ, m_τ' of this quantity clearly yields an additional factor of 4. Denoting the complete sum over all the four quantum numbers by the symbol $\text{Tr}_{1,2}$, we obtain

$$\text{Tr}_{1,2}\, P_\sigma \equiv \sum_{m_s, m_s'} \sum_{m_\tau, m_\tau'} \langle m_s, m_\tau;\, m_s', m_\tau'|\, P_\sigma\, |m_s, m_\tau;\, m_s', m_\tau'\rangle = 8. \tag{II.119a}$$

In the same manner,

$$\text{Tr}_{1,2}\, P_\tau = 8, \tag{II.119b}$$

$$\text{Tr}_{1,2}\, P_\sigma P_\tau = 4 \tag{II.119c}$$

and, of course,

$$\text{Tr}_{1,2}\, \mathbb{1} = 16. \tag{II.119d}$$

Using all these results in (II.116), we get

$$\sum_m \sum_{\bar m} |(\mathbf{k}_1, m_s, m_\tau;\, \mathbf{k}_2, m_s', m_\tau'|\, V\, |\bar{\mathbf{k}}_1, \bar m_s, \bar m_\tau;\, \bar{\mathbf{k}}_2, \bar m_s', \bar m_\tau')|^2$$

$$= |V_0|^2\, \text{Tr}_{1,2}\, (\mathcal{P}^2 |\langle \mathbf{k}_1, \mathbf{k}_2|\, f\, |\bar{\mathbf{k}}_1, \bar{\mathbf{k}}_2\rangle|^2 + \bar{\mathcal{P}}^2 |\langle \mathbf{k}_1, \mathbf{k}_2|\, f\, |\bar{\mathbf{k}}_2, \bar{\mathbf{k}}_1\rangle|^2$$
$$+ 2\mathcal{P}\bar{\mathcal{P}} \langle \mathbf{k}_1, \mathbf{k}_2|\, f\, |\bar{\mathbf{k}}_1, \bar{\mathbf{k}}_2\rangle \langle \bar{\mathbf{k}}_2, \bar{\mathbf{k}}_1|\, f\, |\mathbf{k}_1, \mathbf{k}_2\rangle). \tag{II.120}$$

The $\text{Tr}_{1,2}$ operation for the operators $\mathcal{P}^2$, $\bar{\mathcal{P}}^2$ and $\mathcal{P}\bar{\mathcal{P}}$ can be carried out with the help of (II.119) and the explicit forms (II.115b) and (II.115c) of $\mathcal{P}$ and $\bar{\mathcal{P}}$. The two matrix elements of f can also be substituted from (II.102), which contain the two Fourier transforms $F(\mathbf{k} - \bar{\mathbf{k}})$ and $F(\mathbf{k} + \bar{\mathbf{k}})$. The details of this algebra will not be discussed.

The final expression of the second-order energy can be obtained, provided we know how to carry out a sum of the type

$$\sum_{\mathbf{k}_1, \mathbf{k}_2}^{\text{occ}} \sum_{\bar{\mathbf{k}}_1, \bar{\mathbf{k}}_2}^{\text{unocc}} (\bar k^2 - k^2)^{-1} \mathcal{F}(\mathbf{k}, \bar{\mathbf{k}}) \delta(\mathbf{K}, \bar{\mathbf{K}}),$$

where $\mathcal{F}(\mathbf{k}, \bar{\mathbf{k}})$ can be one of the three functions $|F(\mathbf{k} + \bar{\mathbf{k}})|^2$, $|F(\mathbf{k} - \bar{\mathbf{k}})|^2$, and $F(\mathbf{k} + \bar{\mathbf{k}})F(\mathbf{k} - \bar{\mathbf{k}})$. The sums can be converted into integrals over $\mathbf{k}_1$, $\mathbf{k}_2$, $\bar{\mathbf{k}}_1$, $\bar{\mathbf{k}}_2$ in the usual way; the integration variables can then be changed to $\mathbf{K}$, $\bar{\mathbf{K}}$, $\mathbf{k}$, and $\bar{\mathbf{k}}$ with appropriate restrictions on the magnitudes k_1, k_2 ($\leqslant k_F$) and $\bar k_1$, $\bar k_2$ ($\geqslant k_F$). In view of the $\delta(\mathbf{K}, \bar{\mathbf{K}})$-factor, we are left with the integrations over $\mathbf{K}$, $\mathbf{k}$, and $\bar{\mathbf{k}}$. To tackle these integrals, subject to the restrictions just mentioned, we have to follow a procedure similar to that described for the momentum summation in the calculation of the first-order energy. Here it is convenient to keep $\mathbf{K}$ fixed and first carry out the integration over the angles of $\mathbf{k}$ and $\bar{\mathbf{k}}$.

It is easy to see that the restrictions (II.108) apply to the angle θ between the vectors $\mathbf{k}$ and $\mathbf{K}$; however, the restrictions (II.109) now have to be reinterpreted as those on the magnitude of k for a given value of K. Thus, (II.109a) is replaced by

$$k \leqslant (k_F^2 - \tfrac{1}{4}K^2)^{1/2}, \tag{II.121a}$$

which fixes the maximum value of the k-integration. Similarly, (II.109b) and (II.109c) are replaced by

$$k \leqslant k_{\mathrm{F}} - \tfrac{1}{2}K, \tag{II.121b}$$

$$k \geqslant k_{\mathrm{F}} - \tfrac{1}{2}K. \tag{II.121c}$$

(II.121b) permits all the values of cos θ between -1 and $+1$, whereas (II.121c) allows cos θ to vary between $-P_k(K)$ and $+P_k(K)$, where $P_k(K)$ is given by the right-hand side of (II.108a).

Considerations similar to those used in the case of **k** lead to some conclusions for the angles of $\bar{\mathbf{k}}$. First, the minimum value of the $\bar{k}$-integration is determined by the condition

$$\bar{k} \geqslant (k_{\mathrm{F}}^2 - \tfrac{1}{4}K^2)^{1/2}. \tag{II.122a}$$

Further, when

$$\bar{k} \geqslant k_{\mathrm{F}} + \tfrac{1}{2}K, \tag{II.122b}$$

the angle-integration can be carried over the full range of -1 to $+1$. On the other hand, if

$$\bar{k} \leqslant k_{\mathrm{F}} + \tfrac{1}{2}K, \tag{II.122c}$$

the quantity cos $\bar{\theta}$ can vary between $-Q_{\bar{k}}(K)$ and $+Q_{\bar{k}}(K)$, where

$$Q_{\bar{k}}(K) = \frac{(\tfrac{1}{4}K^2 + \bar{k}^2) - k_{\mathrm{F}}^2}{K\bar{k}}. \tag{II.123}$$

These restrictions on the angles of **k** and $\bar{\mathbf{k}}$ can be very lucidly specified in terms of Fig. II.7. The two circles represent fermi spheres, each of radius k_{F}. The separation between the centres C and C' of the two spheres is equal to **K**. The intersection of the two spheres is a lens-shaped

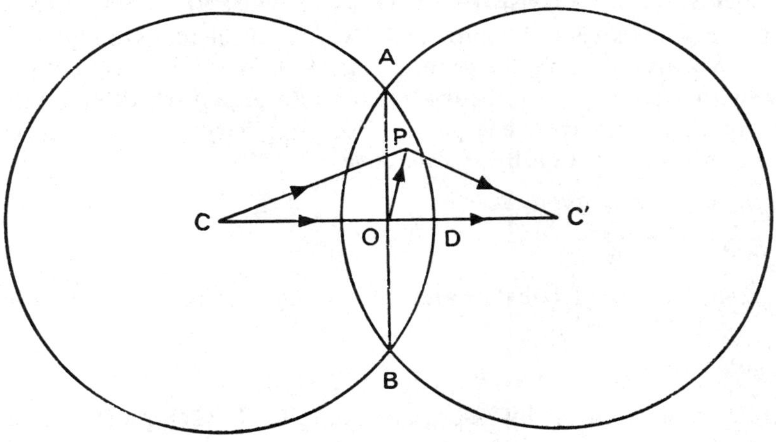

Fig. II.7 Two intersecting fermi spheres showing region of integration in momentum space required in evaluating second-order contribution to binding energy.

region in momentum space. We assert that the momentum vectors, corresponding to all points in this lens-shaped region, measured from the centre O, give the permissible values of the initial relative momentum **k**. To elucidate this, let us consider any point P and join it to the points O, C, and C'. It is clear that the vector $\overrightarrow{CP}$ is equal to $\overrightarrow{CO} + \overrightarrow{OP}$, that is, $\tfrac{1}{2}\mathbf{K} + \mathbf{k}$.

Similarly, the vector $\overrightarrow{PC'}$ is given by $\overrightarrow{OC'} - \overrightarrow{OP}$, that is, $\frac{1}{2}\mathbf{K} - \mathbf{k}$. Thus, $\overrightarrow{CP}$ and $\overrightarrow{PC'}$ are respectively $\mathbf{k}_1$ and $\mathbf{k}_2$. According to our construction, as long as P is a point in the lens-shaped region, the length of the vectors $\overrightarrow{CP}$ and $\overrightarrow{PC'}$ will always be less than the radius k_F of the two spheres. Thus, the restrictions $k_1 \leqslant k_F$ and $k_2 \leqslant k_F$ are correctly incorporated if all the momentum vectors measured from O, and contained in the lens-shaped region, define the permissible domain of integration of $\mathbf{k}$. It is clear that when we complete the sphere of radius $OD = (k_F - \frac{1}{2}K)$ with O as the centre, this sphere is contained entirely within the lens. Hence, from $k = 0$ to $(k_F - \frac{1}{2}k)$, the angle-integration of $\mathbf{k}$ is over the full range of spherical geometry. The two points A and B are evidently at a distance $k = (k_F^2 - \frac{1}{4}K^2)^{1/2}$ from O, specifying the maximum value of k to be used in the integration. It is clear that, in the regions of the lens between the points A and B and the sphere of radius $(k_F - \frac{1}{2}K)$, the angle θ (measured with respect to $\mathbf{K}$, i.e., $\overrightarrow{CC'}$ as the z-axis) is bound between a minimum and a maximum value. Obviously, the minimum and maximum values of θ for a given k are obtained respectively on the right and left surfaces of the lens. The minimum and maximum values of $\cos\theta$, resulting from the geometrical construction, can easily be found to agree with their values [stated after (II.121c)].

In Fig. II.7, the region outside the boundary of the two fermi spheres gives the permitted domain of integration of $\bar{\mathbf{k}}$. The proof of this statement and the division of the $\bar{k}$-integration into the two ranges (II.122b) and (II.122c) also follow from geometrical considerations based on Fig. II.7. The procedure is the same as that described for the k-integration.

19. GOLDSTONE'S LINKED-CLUSTER PERTURBATION EXPANSION

We now present a sophisticated and complete derivation of the many-body perturbation theory (see Goldstone[12]). We wish to solve the ground state Ψ and the corresponding energy E of a Hamiltonian H with a perturbation theory based on an unperturbed Hamiltonian H_0, and its ground state Φ having unperturbed energy E_0. For simplicity, Φ is shown without the subscript 0. Denoting the perturbation by H_I, we have

$$H\Psi = (H_0 + H_I)\Psi = E\Psi, \tag{II.124a}$$

$$H_0\Phi = E_0\Phi. \tag{II.124b}$$

Multiplying (II.124a) from the left by Φ^* and integrating over all the particle coordinates, we get

$$\langle\Phi|\, H_0\, |\Psi\rangle + \langle\Phi|\, H_I\, |\Psi\rangle = E\langle\Phi \mid \Psi\rangle.$$

In the first term, we replace $\langle\Phi|H_0$ by $\langle\Phi|E_0$, according to (II.124b), and obtain

$$(E - E_0)\langle\Phi \mid \Psi\rangle = \langle\Phi|\, H_I\, |\Psi\rangle$$

or

$$E - E_0 = \frac{\langle\Phi|\, H_I\, |\Psi\rangle}{\langle\Phi \mid \Psi\rangle}. \tag{II.125}$$

The perturbation approach entails expressing Ψ in a perturbative series based on the unperturbed Φ; then (II.125) yields the correction to the energy in various orders.

To develop the perturbation expansion of Ψ in a formal way, we follow a time-dependent approach and resort to artificial 'switching on' of the perturbation H_I very slowly, commencing with the unperturbed H_0 at time $t = -\infty$. We write

$$H_I^{\epsilon}(t) = e^{\epsilon t} H_I, \qquad \epsilon > 0. \tag{II.126}$$

We ultimately let ϵ *go to zero through positive values.* We notice from (II.126) that, at $t \to -\infty$, because of the positive value of ϵ, the perturbation $H_I^{\epsilon}(t) \to 0$, and hence we are able to satisfy our earlier assumption that only H_0 is present at $t = -\infty$. Further, as the time increases from $(t = -\infty)$ to $(t = 0)$, the interaction $H_I^{\epsilon}(t)$ gradually increases through its dependence on the exponential factor, and ultimately, at $t = 0$, it acquires the full value H_I of the interaction. So, it is only at $t = 0$ that the wavefunction $\Psi(t)$ evolves to the correct solution of H given in (II.124a). The gradual development of the interaction from zero value at $t = -\infty$ to its full value at $t = 0$ is usually called the *adiabatic* switching process.

We shall work with the Hamiltonian

$$H_{\epsilon}(t) = H_0 + H_I^{\epsilon}(t), \tag{II.127a}$$

solve the time-dependent Schrödinger equation

$$H_{\epsilon}(t)\Psi_{\epsilon}(t) = i\hbar \frac{\partial \Psi_{\epsilon}(t)}{\partial t}, \tag{II.127b}$$

and then evaluate $\Psi_{\epsilon}(t)$ at $t = 0$ in the limit $\epsilon \to +0$. The boundary condition to be imposed on $\Psi_{\epsilon}(t)$ is obtained from the fact that, at $t \to -\infty$, the wavefunction has to correspond to the Hamiltonian H_0.

The *stationary state* wavefunction Φ, used in (II.124b), is the *time-independent* wavefunction for the Hamiltonian H_0, and the corresponding time-dependent wavefunction $\Phi(t)$ in the Schrödinger picture is given by

$$\Phi(t) = \exp\left(-\frac{i}{\hbar} E_0 t\right)\Phi. \tag{II.127c}$$

The boundary condition on $\Psi_{\epsilon}(t)$ can therefore be stated as

$$\underset{\epsilon \to +0}{\mathrm{Lt}} \; \underset{t \to -\infty}{\mathrm{Lt}} \; \Psi_{\epsilon}(t) = \underset{t \to -\infty}{\mathrm{Lt}} \; \Phi(t). \tag{II.127d}$$

Further, the stationary state solution Ψ (time-independent) of the complete Hamiltonian has the time-dependent expression

$$\Psi(t) = \exp\left(-\frac{i}{\hbar} E t\right)\Psi \tag{II.127e}$$

in the Schrödinger picture. At time $t = 0$, (II.127e) becomes identical to the time-independent function Ψ. Hence, our tacit assumption in the adiabatic switching procedure is that we expect the validity of the equation

$$\underset{\epsilon \to +0}{\mathrm{Lt}} \; \Psi_{\epsilon}(t = 0) = \Psi(t = 0) = \Psi. \tag{II.127f}$$

There are examples of Hamiltonians where the adiabatic switching procedure leads to a state Ψ, which is *not* the *lowest* energy state of the complete Hamiltonian. In these cases, the *lowest* energy of the exact Hamiltonian cannot ever be obtained in a perturbative manner from the unperturbed ground state We omit such cases from our consideration.

Instead of trying to obtain a formal solution of (II.127b) subject to the condition (II.127d), we first convert the equation into a different representation, known as the *interaction representation*. The wavefunction $\hat{\Psi}_{\epsilon}(t)$ of this representation is related to the wavefunction

$\Psi_\epsilon(t)$ of the Schrödinger representation through the unitary transformation

$$\hat{\Psi}_\epsilon(t) = \exp\left(\frac{i}{\hbar}H_0 t\right)\Psi_\epsilon(t). \tag{II.128a}$$

The same transformation $\exp[(i/\hbar)H_0 t]$ has to be applied to any other Schrödinger wavefunction. Any operator Ω of the Schrödinger representation then *automatically* transforms in the interaction representation into $\hat{\Omega}(t)$ given by

$$\hat{\Omega}(t) = \exp\left(\frac{i}{\hbar}H_0 t\right)\Omega \exp\left(-\frac{i}{\hbar}H_0 t\right).$$

In particular, the Hamiltonian operator $H_\epsilon(t)$ of (II.127a) transforms into (since H_0 is invariant under the foregoing transformation)

$$\hat{H}_\epsilon(t) = H_0 + \hat{H}_\mathrm{I}^\epsilon(t) \tag{II.128b}$$

with

$$\begin{aligned}\hat{H}_\mathrm{I}^\epsilon(t) &= \exp\left(\frac{i}{\hbar}H_0 t\right)H_\mathrm{I}^\epsilon(t)\exp\left(-\frac{i}{\hbar}H_0 t\right)\\ &= e^{\epsilon t}\exp\left(\frac{i}{\hbar}H_0 t\right)H_\mathrm{I}\exp\left(-\frac{i}{\hbar}H_0 t\right).\end{aligned} \tag{II.128c}$$

Making the substitutions (II.128a) and (II.128b) in (II.127b), we obtain

$$i\hbar\frac{\partial\hat{\Psi}_\epsilon(t)}{\partial t} = \hat{H}_\mathrm{I}^\epsilon(t)\hat{\Psi}_\epsilon(t). \tag{II.129}$$

Applying the unitary transformation $\exp[(i/\hbar)H_0 t]$ on the Schrödinger wavefunction $\Phi(t)$ of (II.127c), we easily obtain

$$\hat{\Phi}(t) = \exp\left(-\frac{i}{\hbar}E_0 t\right)\exp\left(\frac{i}{\hbar}E_0 t\right)\Phi = \Phi. \tag{II.130a}$$

Thus, the unperturbed wavefunction becomes time-independent in the interaction representation, and hence the boundary condition (II.127d) reduces to

$$\underset{\epsilon\to+0}{\mathrm{Lt}}\ \underset{t\to-\infty}{\mathrm{Lt}}\ \hat{\Psi}_\epsilon(t) = \underset{t\to-\infty}{\mathrm{Lt}}\ \hat{\Phi}(t) \equiv \Phi. \tag{II.130b}$$

Further, according to (II.128a), the exact wavefunction in both representations becomes identical at $t = 0$. Therefore, the adiabatic assumption (II.127f) can be rewritten as

$$\underset{\epsilon\to+0}{\mathrm{Lt}}\ \hat{\Psi}_\epsilon(t=0) = \underset{\epsilon\to+0}{\mathrm{Lt}}\ \Psi_\epsilon(t=0) = \Psi. \tag{II.130c}$$

We can immediately carry out an integration with respect to time and express the solution of (II.129) as

$$\hat{\Psi}_\epsilon(t) = C - \frac{i}{\hbar}\int^t dt'\ \hat{H}_\mathrm{I}^\epsilon(t')\hat{\Psi}_\epsilon(t'). \tag{II.131a}$$

The occurrence of t on the integral sign means that the integral has to be evaluated as an indefinite one, and then we have to put $t' = t$ in the result. In (II.131a), C is the integration constant, which can be determined by imposing the condition (II.130b). Thus,

$$\hat{\Psi}_\epsilon(t) = \Phi - \frac{i}{\hbar}\int_{-\infty}^{t} dt'\, \hat{H}_\mathrm{I}^\epsilon(t')\hat{\Psi}_\epsilon(t'), \tag{II.131b}$$

and, by imposing (II.130c), we obtain from this equation

$$\Psi = \underset{\epsilon\to+0}{\mathrm{Lt}}\ \hat{\Psi}_\epsilon(t=0) = \Phi + \underset{\epsilon\to+0}{\mathrm{Lt}}\ (-\frac{i}{\hbar})\int_{-\infty}^{0} dt'\, \hat{H}_\mathrm{I}^\epsilon(t')\hat{\Psi}_\epsilon(t'). \tag{II.132a}$$

This formal expression of Ψ allows us to develop it in a perturbation series by carrying out an iterative procedure on the integral: we substitute for $\hat{\Psi}_\epsilon(t')$ from (II.131b), that is,

$$\hat{\Psi}_\epsilon(t') = \Phi - \frac{i}{\hbar}\int_{-\infty}^{t'} dt''\, \hat{H}_\mathrm{I}^\epsilon(t'')\hat{\Psi}_\epsilon(t'')$$

and then again substitute a similar expression for $\hat{\Psi}_\epsilon(t'')$ in the integral, and so on. In this way,

$$\begin{aligned}\Psi &= \Phi + \underset{\epsilon\to+0}{\mathrm{Lt}}\left[(-\frac{i}{\hbar})\int_{-\infty}^{0} dt'\, \hat{H}_\mathrm{I}^\epsilon(t')\Phi + (-\frac{i}{\hbar})^2\int_{-\infty}^{0} dt'\int_{-\infty}^{t'} dt''\, \hat{H}_\mathrm{I}^\epsilon(t')\hat{H}_\mathrm{I}^\epsilon(t'')\Phi + \ldots\right] \\ &= \underset{\epsilon\to+0}{\mathrm{Lt}}\ U_\epsilon\Phi,\end{aligned} \tag{II.132b}$$

where

$$U_\epsilon = \mathbb{1} + \sum_{n=1}^{\infty}\left[(-\frac{i}{\hbar})^n\int_{-\infty}^{0} dt_1\int_{-\infty}^{t_1} dt_2\ldots\int_{-\infty}^{t_{n-1}} dt_n\, \hat{H}_\mathrm{I}^\epsilon(t_1)\hat{H}_\mathrm{I}^\epsilon(t_2)\ldots\hat{H}_\mathrm{I}^\epsilon(t_n)\right]. \tag{II.132c}$$

Using (II.132b) in the master equation (II.125), we obtain

$$E - E_0 = \underset{\epsilon\to+0}{\mathrm{Lt}}\ \frac{\langle\Phi|\ \hat{H}_\mathrm{I}^\epsilon\ (t=0)U_\epsilon\ |\Phi\rangle}{\langle\Phi|\ U_\epsilon\ |\Phi\rangle}, \tag{II.133}$$

where H_I has been substituted from (II.128c) in terms of its interaction representation at $t = 0$.

At this stage, we introduce the second quantized expression (II.32) for the many-body Hamiltonian. We add and subtract the average one-body potential $\mathcal{V}$ in the manner of (II.5) and (II.9), and then obtain

$$H_0 = \sum_{\alpha,\beta}\langle\alpha|\ (T + \mathcal{V})\ |\beta\rangle C_\alpha^\dagger C_\beta = \sum_\alpha \epsilon_\alpha C_\alpha^\dagger C_\alpha, \tag{II.134a}$$

$$H_\mathrm{I} = \tfrac{1}{2}\sum_{\alpha,\beta,\gamma,\delta}\langle\alpha\beta|\ V\ |\gamma\delta\rangle C_\alpha^\dagger C_\beta^\dagger C_\delta C_\gamma - \sum_{\alpha,\beta}\langle\alpha|\ \mathcal{V}\ |\beta\rangle C_\alpha^\dagger C_\beta. \tag{II.134b}$$

Here ϵ_α is the eigenvalue of $(T + \mathcal{V})$ in the single-particle state $|\alpha\rangle$. In order to obtain the expression of $\hat{H}_\mathrm{I}$, which is in the interaction representation, we need $\hat{C}_\alpha^\dagger(t)$, $\hat{C}_\beta(t)$, ... also in the interaction representation. By definition,

$$\hat{C}_\alpha^\dagger(t) = \exp(\frac{i}{\hbar}H_0 t)C_\alpha^\dagger \exp(-\frac{i}{\hbar}H_0 t),$$

and hence, by straightforward differentiation, we obtain

$$\begin{aligned} i\hbar\frac{\partial\hat{C}_\alpha^\dagger(t)}{\partial t} &= -\exp(\frac{i}{\hbar}H_0 t)[H_0, C_\alpha^\dagger]\exp(-\frac{i}{\hbar}H_0 t) \\ &= -\epsilon_\alpha\hat{C}_\alpha^\dagger(t).\end{aligned}$$

The last step follows from a direct evaluation of the commutator with the help of (II.134a) for H_0, and the anticommutation relation of the creation and destruction operators. A direct time-integration of this equation yields

$$\hat{C}_\alpha^\dagger(t) = \exp\left(\frac{i}{\hbar}\epsilon_\alpha t\right)C_\alpha^\dagger. \tag{II.135a}$$

In obtaining this form, we have also imposed the condition $\hat{C}_\alpha^\dagger\,(t=0) = C_\alpha^\dagger$ which follows from the definition of $\hat{C}_\alpha^\dagger(t)$. The Hermitean conjugate of (II.135a) guarantees

$$\hat{C}_\beta(t) = \exp\left(-\frac{i}{\hbar}\epsilon_\beta t\right)C_\beta. \tag{II.135b}$$

With the help of (II.135), we therefore obtain

$$H_{\rm I}^\epsilon(t) = e^{\epsilon t}\Big[\frac{1}{2}\sum_{\alpha,\beta,\gamma,\delta}\langle\alpha\beta|\,V\,|\gamma\delta\rangle C_\alpha^\dagger C_\beta^\dagger C_\delta C_\gamma \exp\left\{\frac{i}{\hbar}(\epsilon_\alpha+\epsilon_\beta-\epsilon_\gamma-\epsilon_\delta)t\right\}$$
$$-\sum_{\alpha,\gamma}\langle\alpha|\,\mathcal{V}\,|\gamma\rangle C_\alpha^\dagger C_\gamma \exp\left\{\frac{i}{\hbar}(\epsilon_\alpha-\epsilon_\gamma)t\right\}. \tag{II.136}$$

With these expressions, we shall first examine the structure of

$$\underset{\epsilon\to+0}{\rm Lt}\frac{U_\epsilon|\Phi\rangle}{\langle\Phi|\,U_\epsilon\,|\Phi\rangle} \tag{II.137}$$

occurring in (II.133), and then the linked-cluster expansion of Goldstone will follow.

When we insert (II.136) and (II.132c) in (II.137), we observe a set of exponential time factors corresponding to each time $t_1, t_2, \ldots, t_n$, and then the groups of creation and destruction operators, such as $(C_{\alpha_i}^\dagger C_{\beta_i}^\dagger C_{\delta_i} C_{\gamma_i})$ and $(C_{\alpha_i}^\dagger C_{\gamma_i})$, appearing in a continued product; the subscript i here refers to the subscript to the t to which the particular group of creation and destruction operators belongs. When writing the continued product of these operators, we must take into account the order in which the various t_i's occurred in (II.132c).

If Φ is a many-body state in which single-particle states up to a certain maximum energy are occupied, then we have

$C_\gamma|\Phi\rangle = 0$ (if γ is a particle state above the occupied set of states), (II.138a)

$C_\alpha^\dagger|\Phi\rangle = 0$ (if α is a hole state, i.e., one of the states already occupied in Φ). (II.138b)

On the other hand, $C_\gamma|\Phi\rangle$ and $C_\alpha^\dagger|\Phi\rangle$ stand for a 1h and a 1p state if γ is an occupied state and α an unoccupied state. The *contraction* of two such operators is, by definition, the expectation value of their product in the state Φ. We denote the contraction by the symbol $\langle\ \ \rangle$. Thus,

$$\langle C_\alpha^\dagger C_\gamma\rangle \equiv \langle\Phi|\,C_\alpha^\dagger C_\gamma\,|\Phi\rangle = n_\gamma\delta_{\alpha\gamma}, \tag{II.138c}$$

according to the property (II.138a) of C_γ. Here n_γ is the occupation probability of a state, i.e., $n_\gamma = 1$ if γ is an occupied state in Φ, otherwise it is zero. In a similar manner, the property (II.138b) of $C_\alpha^\dagger$ yields

$$\langle C_\gamma C_\alpha^\dagger\rangle = (1-n_\alpha)\delta_{\alpha\gamma}. \tag{II.138d}$$

We now state Wick's theorem in a simplified manner, which best suits our purpose. According to this simplified version of the theorem, the product of any number of creation and destruction operators is equivalent to the result

$$\begin{aligned} ABCD\ldots = {} & N(ABCD\ldots) \\ & + \langle AB\rangle N(CD\ldots) + \ldots \\ & + \langle AB\rangle\langle CD\rangle N(EF\ldots) + \ldots . \end{aligned} \tag{II.139}$$

Here we have used the general symbols $A, B, \ldots$ to denote the creation and destruction operators. The letter N stands for the 'normal product' and indicates an arrangement of the operators, enclosed within the parentheses following N, in which all 'particle' and 'hole' *creation* operators (i.e., $C_\alpha^\dagger$ if α is an unoccupied state, and C_γ if γ is an occupied state) have been permuted to the *left* of the 'particle' and 'hole' *destruction* operators (C_α, $C_\gamma^\dagger$, ...); in *securing this arrangement, the sign of the permutation must be included.* The second line of (II.139) says that we must continue contracting *every* pair of operators, and multiply it by the normal product of the remaining operators; here also, we must include a minus sign if the final arrangement of operators is reached through an odd number of permutations. For example, the term $\langle AC\rangle N(BD\ldots)$ should appear with a minus sign. The third line of (II.139) shows that we must next contract two pairs of operators (in all possible ways), keeping an account of the sign due to the permutations, and multiply the contraction by the normal product of the remaining operators. The process has to be continued until all the operators are contracted in pairs in all possible ways.

In view of Wick's theorem and the properties (II.138) of Φ, we conclude that, in the expression of $U_\epsilon|\Phi\rangle$, terms that contribute are those that have *only* particle and hole *creation* operators in the normal products, and terms that have no normal products at all (i.e., the fully contracted terms). The last type of terms are obviously the only ones that contribute to the denominator of (II.137). The most convenient way to keep track of all nonvanishing terms is to draw diagrams.

A contraction $\langle C_{\alpha_i}^\dagger C_{\gamma_j}\rangle$ of the creation operator taken from time t_i and the destruction operator taken from time $t_j\,(t_i > t_j)$ is nonvanishing, according to (II.138c), only if $\alpha_i = \gamma_j$, and the state α_i or the state γ_j is occupied in Φ. This is represented in Fig. II.8 by the line

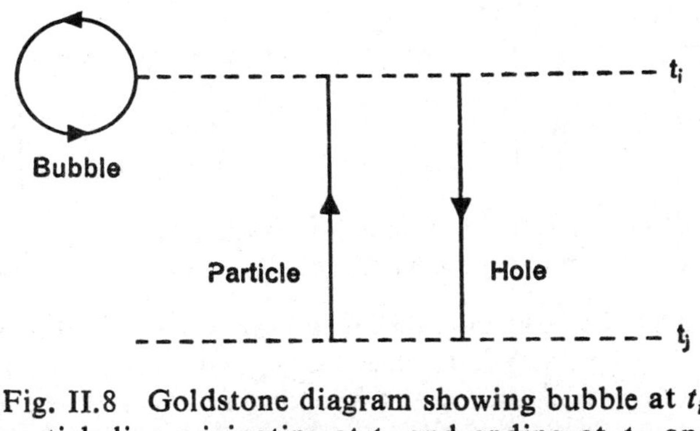

Fig. II.8 Goldstone diagram showing bubble at t_i, particle line originating at t_j and ending at t_i, and hole line going from t_i to t_j.

with the downward-pointing arrow showing a propagation from time t_i to time t_j (*from* the creation *to* the destruction operator). Such a line moving downwards therefore represents the propagation of a 'hole' state. According to our convention, a line going *out* from some point denotes *creation* at that point and a line going *towards* a point signifies *destruction* at that point. This explains why, in the representation of $\langle C^{\dagger}_{\alpha_i} C_{\gamma_j} \rangle$, the arrow is from t_i to t_j. In Fig. II.8, we have taken the direction of time flow from the bottom towards the top. Therefore, time t_i, which is greater than t_j, is shown by a horizontal line above that for t_j. In the same manner, we can represent the contraction $\langle C_{\gamma_i} C^{\dagger}_{\alpha_j} \rangle$ of the destruction operator taken from time t_i and the creation operator taken from time $t_j (t_i > t_j)$ by the line with the upward-pointing arrow. According to (II.138b), the state γ_i or the state α_j must now stand for a 'particle' state; the direction of the arrow has been fixed, as before, by noting where the particle is created and where it is destroyed. A third kind of nonvanishing contraction is that of two operators belonging to the same time t_i, i.e., $\langle C^{\dagger}_{\alpha_i} C_{\gamma_i} \rangle$. From the expression of $H_I(t_i)$, it is clear that the creation operator in such a case always occurs to the left of the destruction operator, and hence, according to (II.138a), $\alpha_i (= \gamma_i)$ must be an occupied state. This process is represented by the "Bubble" in Fig. II.8 where both the creation and the destruction are found at t_i, as required.

The normal products of the operators correspond to the destruction or the creation of particles and holes from the state Φ. An operator $C^{\dagger}_{\alpha_i}$, contained in a normal product, stands for the creation of a particle (if α is an unoccupied state), and for the destruction of a hole (if α is an occupied state) taking place at time t_i. Similarly, the operator C_{γ_i} present in a normal product stands for the creation of a hole state at time t_i or the destruction of a particle state at the same time, depending on whether γ is an occupied or an unoccupied state. On a diagram, the creation and destruction operators for particles and holes appear as free lines running from the edges of the diagram to the horizontal line standing for the relevant time. An arrow that points downwards will be a 'hole' and one that points upwards will be a 'particle'; the movement of the arrow towards or away from the time line indicates whether a creation or a destruction has taken place. In this way, we arrive at the possibilities shown in Fig. II.9.

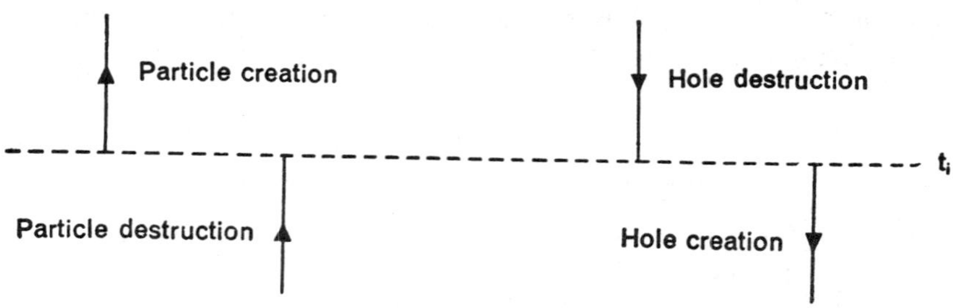

Fig. II.9 Goldstone diagram showing particle creation, particle destruction, hole creation, and hole destruction at t_i.

According to (II.132c), we have interactions taking place at n different times $t_1, t_2, \ldots, t_n$ in a process of order n. At each instant of time, the V- or $\mathcal{V}$-interaction is represented along the corresponding time line. Each V-line has two ends, each end (to be called a *vertex*) with one particle line or one hole line going in and another going out; this is because of $C^{\dagger}_{\alpha} C^{\dagger}_{\beta} C_{\delta} C_{\gamma}$ with each V-interaction. On the other hand, each $\mathcal{V}$-interaction (because of its $C^{\dagger}_{\alpha} C_{\gamma}$) has at one end one particle line or one hole line going in and another going out; the other end of the $\mathcal{V}$-interaction is however left free and *deliberately* terminated by a cross to emphasize the fact that it stands for a $\mathcal{V}$-interaction.

The rule for associating the matrix elements of V and $\mathcal{V}$ with the interaction lines is as follows. The states (or state) that correspond(s) to ingoing lines must be written as *initial* states (or state) (*right*-hand side of the matrix element) and those going out as final states (left-hand side of the matrix element). In the matrix element of V, the correct correspondence between the first state on the right and the first state on the left as the ingoing line and the *corresponding* outgoing line must be maintained; the same correspondence must also be kept between the second single-particle state on the right and the one on the left. Thus, the direct and exchange terms of a two-body matrix element correspond to different diagrams. For example, the two diagrams of Fig. II.10 correspond to the direct and exchange matrix elements (in the first order) $\langle\alpha\beta| V |\alpha\beta\rangle$ and $\langle\alpha\beta| V |\beta\alpha\rangle$, respectively. As required by the correspondence just mentioned, the arrows in the diagrams representing the exchange matrix element $\langle\alpha\beta| V |\beta\alpha\rangle$ indicate $\alpha \rightarrow \beta$ and $\beta \rightarrow \alpha$.

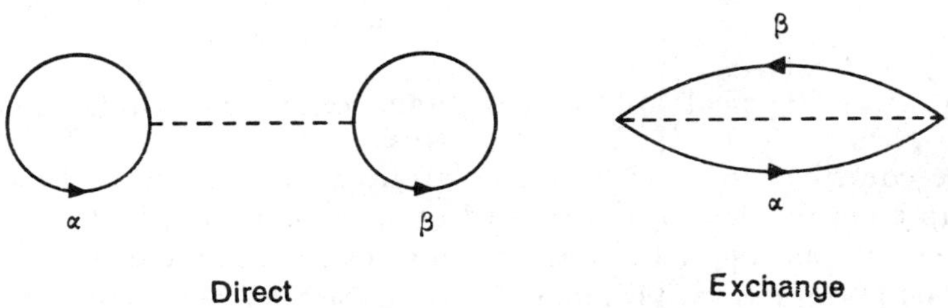

Fig. II.10 Direct and exchange diagrams of first order in two-body potential V.

Now that we have stated all the rules for diagrammatic representation, we note that, in the expression $U_\epsilon\Phi$, the bottom of the diagram cannot have any free particle line or free hole line because we are starting with Φ as the initial state. In the n-th order, there will be n-interaction lines at times $t_1, t_2, \ldots, t_n$ with $t_1 > t_2 > \ldots > t_n$. Finally, at the top of the diagram, there will be several free particle and free hole lines (in the terms containing normal products). There will also be some diagrams with no external lines (the fully contracted terms); such diagrams obviously correspond to processes contained in $\langle\Phi| U_\epsilon |\Phi\rangle$ in which we start with Φ and end with Φ. In addition, there will be diagrams in which there are physically isolated parts, some of which may not have any external lines. Such *disconnected* parts having *no external* lines are called *unlinked* parts. Once again, the unlinked parts obviously represent processes contained in $\langle\Phi| U_\epsilon |\Phi\rangle$.

We must now consider the time-integrations in a diagram having one or more unlinked parts. For an understanding of the proof, just one such unlinked part in a diagram (see Fig. II.11) is adequate. Suppose the times at which the interactions take place in this unlinked part are given by $t_1', t_2', \ldots, t_j'$, where $t_1' > t_2' > \ldots > t_j'$; and, on the other hand, the times at which the interactions take place in the other parts of the diagram (which have some external lines at the top by definition) are $t_1, t_2, \ldots, t_i$ with $t_1 > t_2 > \ldots > t_i$. We have to admit all possible ordering of the times in the set $(t_1', t_2', \ldots, t_j')$ with respect to those in the set $(t_1, t_2, \ldots, t_i)$. In diagrammatic terms, this process simply means a physical displacement, up or down, of the unlinked part *as a whole* with respect to the rest of the diagram. If we carry on independent time-integrations within the two sets and *multiply* the results, it is obvious that the effect of all such physical displacements of the unlinked part will be correctly taken care of. Therefore, we

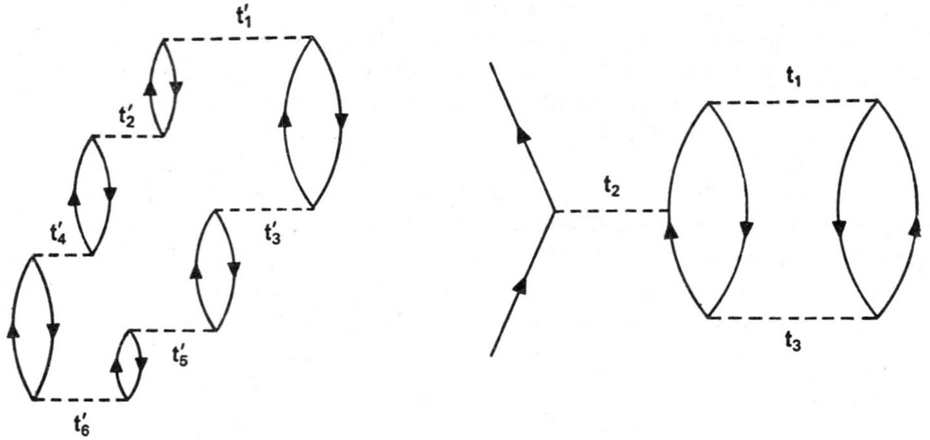

Fig. II.11 Typical Goldstone diagram containing unlinked part.

conclude that the contribution to $U_\epsilon \Phi$ of any diagram having unlinked parts is the ground-state-to-ground-state matrix element represented by such an unlinked part multiplied by the expression for the rest of the diagram. It is therefore clear that the quantity $\langle \Phi | U_\epsilon | \Phi \rangle$ in the denominator of (II.137) results as a common factor in the numerator and, after the cancellation of these common quantities from the numerator and the denominator, only the *linked* diagrams in the expression for $U_\epsilon \Phi$ remains. By definition, a linked diagram is any diagram that is not unlinked. Therefore, it will be characteristic of all these diagrams in $U_\epsilon \Phi$ to have external lines. It is important to note that physically *disconnected* parts, each with *external lines*, will still be considered a *linked* diagram in $U_\epsilon \Phi$.

Now let us examine the final expression (II.133) for the energy. The *linked-cluster* diagram of $U_\epsilon | \Phi \rangle$ now has to be multiplied by $\langle \Phi | H_I(t=0)$ to obtain the energy. The interaction $H_I(t=0)$ can take care of only two outgoing and two ingoing lines (the V-part), or only one outgoing and one ingoing line (the $\mathcal{V}$-part), at time $t=0$. Further, as a result of this last interaction, we want all external lines to disappear from the top of the diagram leading to the final state $\langle \Phi |$. Therefore, not all linked diagrams contained in $U_\epsilon | \Phi \rangle$, but rather only those having two or four external lines, ultimately contribute to the energy. It is essential to note that two *disconnected* parts having external lines were counted as *linked* diagrams in $U_\epsilon \Phi$; however, after $\langle \Phi | H_I(t=0)$ multiplies from the left, the last H_I interaction establishes *connection* between the disconnected parts of $U_\epsilon | \Phi \rangle$. Thus, in the contribution to the energy, we pick up all *ground-state-to-ground-state diagrams* that are physically *connected* to one another. It is intuitively obvious that any such connected ground-state-to-ground-state diagram will be comprised of a set of closed polygons of particle and hole lines. In each closed polygon, the corners represent interaction vertices, where interaction lines connect it to other polygons. The directions of the arrows on the lines making a polygon must be continuous through the successive vertices. Each such polygon is called a *loop*. The sides of the polygons, which are hole lines or particle lines connecting two different times, correspond to various contractions of pairs, as already explained. The number of permutations needed to go from the initial arrangement of the creation and destruction operators to the one that corresponds to the pairs contracted in a loop is always one more than the number of hole lines belonging to the loop. Therefore, the sign associated with any ground-state-to-ground-state connected diagram can be taken to be $(-1)^{l+h}$, where l is the number of loops and h the number of hole lines. If there are also $\mathcal{V}$-interaction lines, each such line will carry an additional minus sign due to the fact that $-\mathcal{V}$ occurs in H_{I}. Hence, the final rule for the sign of a diagram is given by $(-1)^{l+h+n(\mathcal{V})}$, where $n(\mathcal{V})$ is the number of $\mathcal{V}$-interaction lines.

The diagrams drawn before the time-integrations have been called Goldstone diagrams by many authors. Let us now explicitly consider the time-integrations in the connected ground-state-to-ground-state Goldstone diagram. According to (II.136) and (II.132c), the integration over t_n in the n-th order term gives (we shall include one $-i/\hbar$ factor with each time-integration)

$$-\frac{i}{\hbar}\int_{-\infty}^{t_{n-1}} dt_n \exp(\epsilon t_n) \exp[-\frac{i}{\hbar}(-\epsilon_{\alpha_n}-\epsilon_{\beta_n}+\epsilon_{\gamma_n}+\epsilon_{\delta_n})t_n]$$
$$= (\epsilon_{\gamma_n}+\epsilon_{\delta_n}-\epsilon_{\alpha_n}-\epsilon_{\beta_n})^{-1}\exp(\epsilon t_{n-1})\exp[-\frac{i}{\hbar}(-\epsilon_{\alpha_n}+\epsilon_{\beta_n}+\epsilon_{\gamma_n}+\epsilon_{\delta_n})t_{n-1}], \tag{II.140a}$$

$$-\frac{i}{\hbar}\int_{-\infty}^{t_{n-1}} dt_n \exp(\epsilon t_n)\exp[\frac{i}{\hbar}(\epsilon_{\alpha_n}-\epsilon_{\gamma_n})t_n]$$
$$= (\epsilon_{\gamma_n}-\epsilon_{\alpha_n})^{-1}\exp(\epsilon t_{n-1})\exp[\frac{i}{\hbar}(\epsilon_{\alpha_n}-\epsilon_{\gamma_n})t_{n-1}]. \tag{II.140b}$$

(II.140a) results for the V-interaction term and (II.140b) for the $\mathcal{V}$-interaction term.

In a ground-state-to-ground-state diagram, there are no *free* particle or *free* hole lines; all creation and destruction operators have been contracted pairwise, and hence the particle and hole lines on such a diagram necessarily run from one interaction vertex to another. The creation operator $C_{\alpha_n}^{\dagger}$ or $C_{\beta_n}^{\dagger}$ in $H_{\text{I}}(t_n)$ must have contracted with a destruction operator which is to its left and belongs to another time t_i $(i \neq n)$. According to (II.138d), such a contraction requires α_n to be a particle state running from time t_n to time t_i. In the same way, the operator C_{γ_n} or C_{δ_n} must have contracted with a creation operator to its left, i.e., either belonging to the same time t_n or a different time t_i $(i \neq n)$. In either case, according to (II.138c), γ_n and δ_n have to be hole states; in the first case, it represents a "bubble" running from t_n to t_n, whereas in the second case, the hole line runs from t_i to t_n. In view of these identifications of the labels α_n, β_n, γ_n, δ_n, we conclude that the energy denominators appearing in either (II.140a) or (II.140b) are equal to the energy of the holes minus the energy of the particles present in the time interval between t_{n-1} and t_n.

Let us next examine the integration over t_{n-1}. At this stage, we have to multiply (II.140a) or (II.140b) by the time factors from $H_{\text{I}}(t_{n-1})$ and then integrate. For an understanding of the details, it is adequate to consider only the V-interaction part. Collecting (II.140a) with the V-part of $H_{\text{I}}(t_{n-1})$, we obtain the time integral

$$-\frac{i}{\hbar}\int_{-\infty}^{t_{n-2}} dt_{n-1}\exp(2\epsilon t_{n-1})\exp[-\frac{i}{\hbar}(-\epsilon_{\alpha_n}-\epsilon_{\beta_n}+\epsilon_{\gamma_n}+\epsilon_{\delta_n}-\epsilon_{\alpha_{n-1}}-\epsilon_{\beta_{n-1}}$$
$$+\epsilon_{\gamma_{n-1}}+\epsilon_{\delta_{n-1}})t_{n-1}]. \tag{II.141}$$

We have to consider two possibilities: any of the particle lines α_n, β_n or the hole lines γ_n, δ_n may either meet an interaction vertex at t_{n-1} or continue unhindered in the time interval t_{n-1} to t_{n-2}. In the first case, any line that meets an interaction vertex corresponds to a contraction of a creation and a destruction operator, one taken from time t_{n-1} and the other from time t_n. The single-particle labels on these two operators are therefore required to be the same; the corresponding energies in (II.141), one appearing with a plus sign and the other with a minus sign, then cancel each other. What remains in the exponent is the energy of the hole states minus the energy of the particle states, these holes and particles now being those present in the diagram between the time interval t_{n-1} and t_{n-2}. The time-integration in (II.141) thus once again produces an energy denominator, $(\sum_h \epsilon_h - \sum_p \epsilon_p)$. Continuing in this manner with subsequent time-integrations over $t_{n-2}, t_{n-3}, \ldots$, we shall always arrive at the foregoing rule for the energy denominators.

We can now summarize all the results of the diagram technique. To find the contribution to the ground-state energy in any order n of the perturbation theory, we need to draw all connected diagrams without any external lines containing n-interactions. To calculate the contribution of any of these diagrams, we have to write down the matrix elements of V or $\mathcal{V}$ at each interaction vertex, placing the ingoing single-particle states on the right and the corresponding outgoing ones on the left; for every intermediate state, i.e., the portion of the diagram between two interaction lines, we have to write an energy denominator equal to the sum of hole energies minus the sum of particle energies, taking into account the hole and particle lines present in that particular intermediate state; finally, we have to write the sign of the diagram in front of the resultant expression; it is plus or minus depending on whether the sum of the number of hole lines, the number of closed loops, and the number of $\mathcal{V}$-interactions in the diagram is even or odd. The rule for the final summation over the hole and particle lines is given at the end of this section.

As an illustration of the diagrammatic method, we consider the three first-order diagrams. The first of these diagrams contains the $\mathcal{V}$-interaction, shown by the cross in Fig. II.12a. The

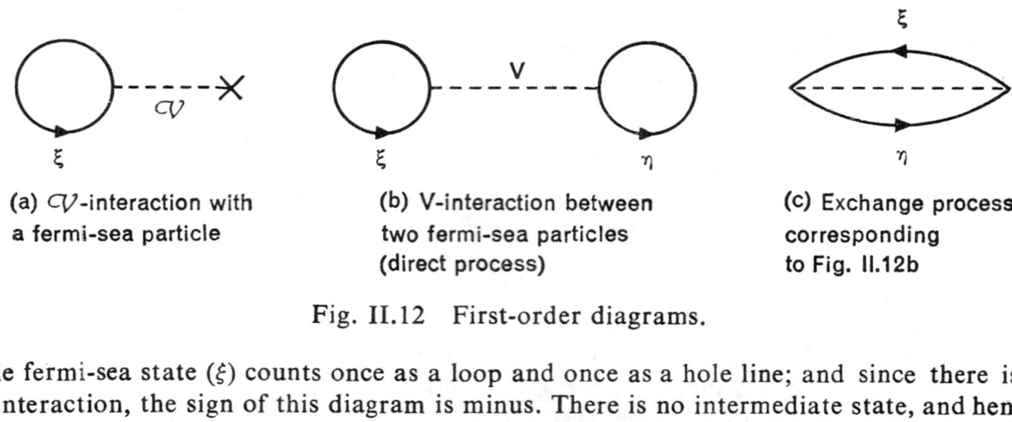

(a) $\mathcal{V}$-interaction with a fermi-sea particle

(b) V-interaction between two fermi-sea particles (direct process)

(c) Exchange process corresponding to Fig. II.12b

Fig. II.12 First-order diagrams.

single fermi-sea state (ξ) counts once as a loop and once as a hole line; and since there is one $\mathcal{V}$-interaction, the sign of this diagram is minus. There is no intermediate state, and hence no energy denominator. The matrix element at the interaction vertex is $\langle\xi|\,\mathcal{V}\,|\xi\rangle$. Thus, the result is given by

$$\text{Fig. II.12a} = -\langle\xi|\,\mathcal{V}\,|\xi\rangle. \qquad \text{(II.142a)}$$

Figure II.12b has two loops, two hole lines, and no $\mathcal{V}$-interaction, and hence the sign is plus. No energy denominator is required, and the matrix element of the V-interaction is $\langle\xi\eta|\,V\,|\xi\eta\rangle$. Thus,

$$\text{Fig. II.12b} = +\langle\xi\eta|\,V\,|\xi\eta\rangle. \qquad \text{(II.142b)}$$

Figure II.12c has only one loop, two hole lines, and no $\mathcal{V}$-interaction, and hence its sign is minus. Once again, no energy denominator is needed, and the final result with the matrix element of V is given by

$$\text{Fig. II.12c} = -\langle \eta\xi | V | \xi\eta \rangle. \tag{II.142c}$$

Clearly then, Fig. II.12c is the exchange counterpart of Fig. II.12b, and together they produce

$$\text{Fig. II.12b} + \text{Fig. II.12c} = (\xi\eta | V | \xi\eta). \tag{II.142d}$$

To illustrate a more complicated case, let us consider the second-order diagrams. Figure II.13a contains two loops, two hole lines, and no $\mathcal{V}$-interaction. Hence, its sign is plus.

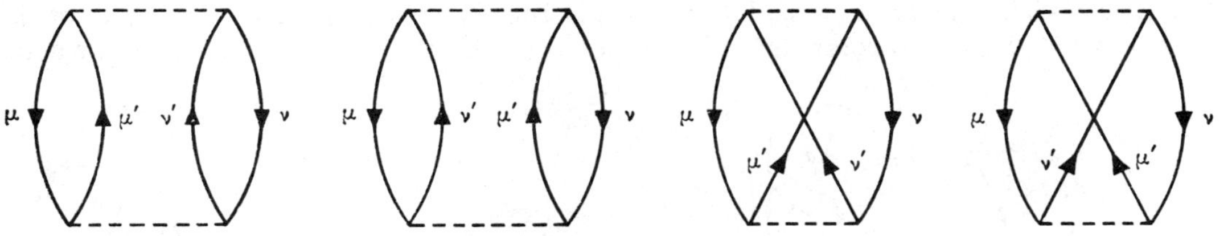

(a) Direct (c) Direct (b) Exchange (d) Exchange

Fig. II.13 Second-order diagrams with V-interaction.

The energy denominator is given by $\epsilon_\mu + \epsilon_\nu - \epsilon_{\mu'} - \epsilon_{\nu'}$, and hence

$$\text{Fig. II.13a} = \frac{\langle \mu\nu | V | \mu'\nu' \rangle \langle \mu'\nu' | V | \mu\nu \rangle}{\epsilon_\mu + \epsilon_\nu - \epsilon_{\mu'} - \epsilon_{\nu'}}. \tag{II.143a}$$

The only difference in Fig. II.13b is that it contains only one loop and, at the upper interaction line, $\mu' \to \nu$ and $\nu' \to \mu$; the corresponding matrix element is therefore $\langle \nu\mu | V | \mu'\nu' \rangle$, and *not* $\langle \mu\nu | V | \mu'\nu' \rangle$ as in Fig. II.13a. Thus,

$$\text{Fig. II.13b} = -\frac{\langle \nu\mu | V | \mu'\nu' \rangle \langle \mu'\nu' | V | \mu\nu \rangle}{\epsilon_\mu + \epsilon_\nu - \epsilon_{\mu'} - \epsilon_{\nu'}}. \tag{II.143b}$$

For a given pair of occupied states (μ, ν) and a given pair of particle states (μ', ν'), it is also possible to label the diagrams alternatively, as shown in Figs. II.13c and II.13d. Writing their contribution as just outlined and then adding all the four diagrams, we easily obtain

$$\text{Fig. II.13a}+\text{Fig. II.13b}+\text{Fig. II.13c}+\text{Fig. II.13d} = \frac{(\mu\nu | V | \mu'\nu')(\mu'\nu' | V | \mu\nu)}{\epsilon_\mu + \epsilon_\nu - \epsilon_{\mu'} - \epsilon_{\nu'}}. \tag{II.143c}$$

The rule for summing over the hole and particle lines is now stated. After having written down the contribution of a diagram with an arbitrary labelling of the lines, we have obviously to sum each hole line over all the occupied fermi-sea states and each particle line over all the single-particle states above the fermi sea. However, when a set of q hole lines is completely equivalent, for example, the pair (ξ, η) in Figs. II.12b and II 12c and Fig. II.13, independent sums over all of them are redundant and must be divided by $q!$, which is the number of ways in which one set of labelling of the q equivalent hole lines can be permuted amongst themselves. In the same manner, the summations over p equivalent particle lines in any part of a diagram must be divided by $p!$.

27 (45-122/1977)

If we apply this summation rule to Fig. II.12, we obtain the complete first-order contribution as

$$-\sum_{\xi}\langle\xi|\,\mathcal{V}\,|\xi\rangle+\frac{1}{2!}\sum_{\xi}\sum_{\eta}(\xi\eta|\,V\,|\xi\eta),$$

which is the same as (II.91a). The expression (II.92) then follows in the same manner. Similarly, the summation rule produces from (II.143c) a complete second-order contribution, which exactly agrees with (II.97).

20. TWO-BODY REACTION MATRIX

Examples of all possible first-order diagrams have been shown in Fig. II.12. In the second order, it is possible to draw three diagrams other than those in Fig. II.13. Examples of these are given in Fig. II.14. However, it is easy to see that these diagrams correspond to a lh-lp

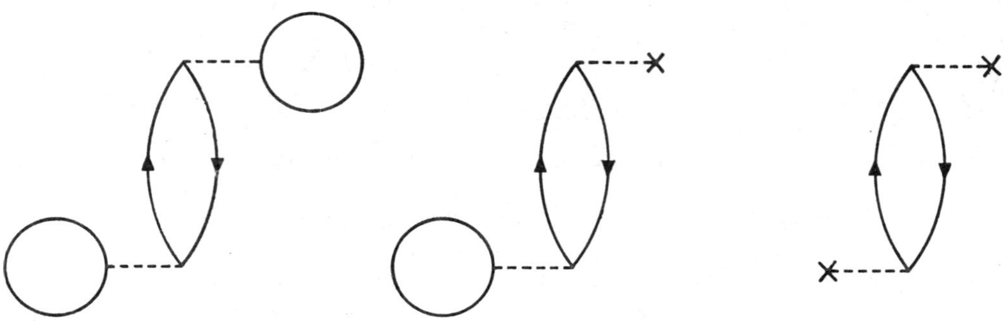

Fig. II.14 Second-order diagrams with lh-lp type intermediate state.

type intermediate state, and hence their contributions vanish by the arguments presented after (II.95).

Some representative third-order diagrams are shown in Fig. II.15. Similarly, other diagrams in higher orders can be detailed.

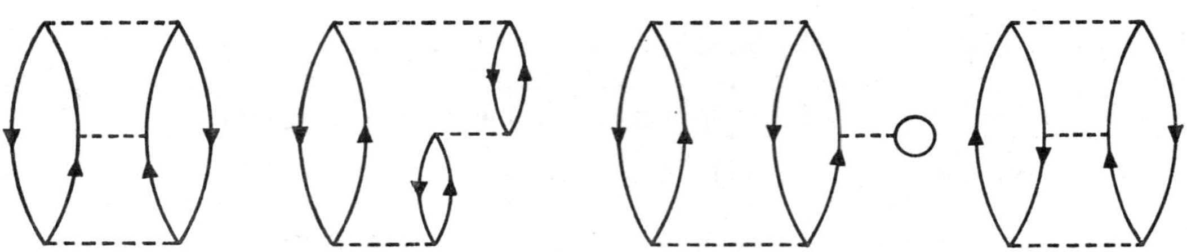

(a) Entering into sum for t-matrix (see Fig. II.16)

(b) Ring

(c) Particle-bubble

(d) Hole-particle interaction

Fig. II.15 Representative third-order diagrams.

Now, in the nuclear case, where there is a strongly repulsive core region of the two-nucleon potential, any of the single interaction vertex for the V-interaction will be very big and, in fact, infinitely large, if the core is idealized by the hard sphere. How then do these diagrams help us? To discover the answer, we have to use the known physical properties of the given many-body system to reckon the comparative importance of these diagrams. In the nuclear case, the strong short-range part of the interaction very strongly correlates the motion of a pair of nucleons, and the probability of a third nucleon or more nucleons coming *simultaneously* close enough to a given correlated pair is very small. This, therefore, gives us the clue that all the diagrams describing the interactions of a given pair of nucleons to all orders of the perturbation theory must be given more importance than diagrams that need the presence of more than two fermi-sea nucleons.

In the discussions that follow, we imply the existence of the exchange diagrams, and argue in terms of the direct diagrams alone. The first-order diagram Fig. II.12b, the second-order diagram Fig. II.13a, and the third-order diagram Fig. II.15a, all describe the interaction between a given pair of fermi-sea nucleons. In the first order, they interact directly; in the second order, they have interacted, risen above the fermi sea, and then finally interacted to return to their initial states inside the fermi sea; in the third order, the intermediate state particles have interacted once more, still staying outside the fermi sea, and, after the final interaction, have returned to the holes inside the fermi sea. In this way, more and more interaction lines can be inserted within the pair of intermediate state particles, but care should be taken to always keep them outside the fermi sea. The infinite number of this particular class of perturbation theory diagrams is shown in Fig. II.16. Brueckner and his collaborators first

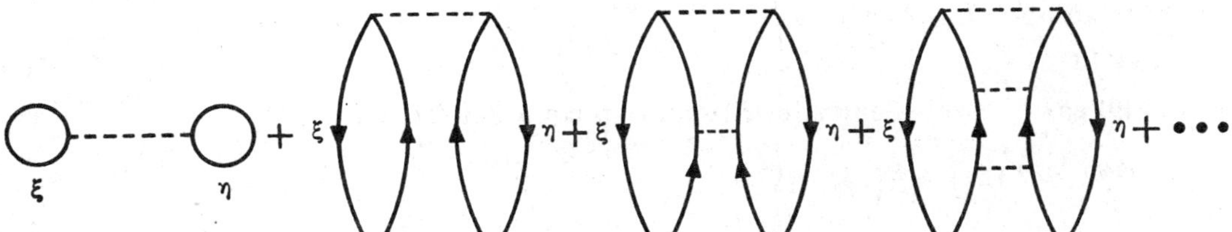

Fig. II.16 Sum of ladder diagrams defining effective potential t between two fermi-sea particles ξ and η.

proposed that this set of diagrams be summed, in the nuclear case, as exactly as possible. The result then defines the reaction matrix t between nucleon-pair states. The derivation follows.

Applying the rules given in Section 19, we can write the contribution of the diagrams in Fig. II.16, for a given pair of occupied states $(\xi\eta)$, as

$$\cdots \sum_{\xi''<\eta''} \sum_{\xi'<\eta'} \frac{(\xi\eta|\, V \ldots |\xi''\eta'')(\xi''\eta''|\, V\, |\xi'\eta')(\xi'\eta'|\, V\, |\xi\eta)}{\ldots (\epsilon_\xi + \epsilon_\eta - \epsilon_{\xi''} - \epsilon_{\eta''})(\epsilon_\xi + \epsilon_\eta - \epsilon_{\xi'} - \epsilon_{\eta'})}. \tag{II.144}$$

It should be noted that, in conformity with our earlier observation, we have implied the presence of the relevant exchange diagrams, and hence the direct minus exchange matrix elements, represented within the parentheses, occur in (II.144).

Let us define a projection operator Q, which projects outside the fermi sea, and an energy denominator $e(w)$ by

$$\frac{Q}{e(w)} = \sum_{\xi'<\eta'} \frac{|\xi'\eta')(\xi'\eta'|}{w - \epsilon_{\xi'} - \epsilon_{\eta'}}, \tag{II.145}$$

where w is a given number and ξ', η' are single-particle states outside the fermi sea. Then the expression (II.144) can be written as

$$(\xi\eta|\, V \frac{Q}{e(w_0)} V \frac{Q}{e(w_0)} V \ldots V \frac{Q}{e(w_0)} V\,|\xi\eta) \tag{II.146a}$$

with

$$w_0 = \epsilon_\xi + \epsilon_\eta. \tag{II.146b}$$

The complete sum of all these diagrams may be used to define a matrix element of the reaction matrix t as

$$(\xi\eta|\, t\,|\xi\eta) \equiv (\xi\eta|\, V\,|\xi\eta) + (\xi\eta|\, V\frac{Q}{e(w_0)} V\,|\xi\eta) + (\xi\eta|\, V\frac{Q}{e(w_0)} V\frac{Q}{e(w_0)} V\,|\xi\eta) + \ldots . \tag{II.147}$$

In general, we then define the operator t by

$$t(w) = V + V\frac{Q}{e(w)}V + V\frac{Q}{e(w)}V\frac{Q}{e(w)}V + \ldots . \tag{II.148}$$

(II.148) can be formally rewritten as

$$\begin{aligned} t(w) &= V + V\frac{Q}{e(w)}[V + V\frac{Q}{e(w)}V + V\frac{Q}{e(w)}V\frac{Q}{e(w)}V + \ldots] \\ &= V + V\frac{Q}{e(w)}t(w). \end{aligned} \tag{II.149}$$

If we define an operator Ω such that

$$t = V\Omega, \tag{II.150}$$

then (II.149) can be equivalently rewritten as an equation for Ω as

$$V\Omega(w) = V[\mathbb{1} + \frac{Q}{e(w)}V\Omega(w)]$$

or

$$\Omega(w) = \mathbb{1} + \frac{Q}{e(w)}V\Omega(w) = \mathbb{1} + \frac{Q}{e(w)}V + \frac{Q}{e(w)}V\frac{Q}{e(w)}V + \ldots . \tag{II.151}$$

Letting (II.151), with $w = w_0$ of (II.146b), operate on the unperturbed two-nucleon state $|\xi\eta)$, we obtain

$$\begin{aligned} \Omega(w_0)|\xi\eta) &= |\xi\eta) + \frac{Q}{e(w_0)}V\Omega(w_0)|\xi\eta) \\ &= |\xi\eta) + \frac{Q}{e(w_0)}V|\xi\eta) + \frac{Q}{e(w_0)}V\frac{Q}{e(w_0)}V|\xi\eta) + \ldots . \end{aligned} \tag{II.152}$$

The resultant state, as given by the right-hand side of (II.152), comprises the unperturbed pair state and all possible two-particle excitations above the fermi sea resulting from the interactions occurring between the pair as many times as possible. It should be noted that the projection operator Q restricts the two-particle states, after any interaction between the pair, to states outside the fermi sea. This is an effect of the Pauli exclusion principle because all the states inside the fermi sea are already occupied by other nucleons, and hence these are not available to the interacting pair when they scatter. The state $\Omega(w_0)|\xi\eta)$ is therefore the exact state of the pair corresponding to the unperturbed state $|\xi\eta)$. This exact state of the pair, embedded in the many-body medium, differs in two respects from the state of a pair left all by itself: (i) the restriction due to the Pauli principle as just described, and (ii) the initial energy w_0 [contained in $e(w_0)$], which is not equal to the sum of the free-particle energies of the pair. (The important consequences of these differences are described in detail in Section 22.)

Using the symbol $|\xi\eta)_{\text{exact}}$, we then rewrite (II.152) as

$$|\xi\eta)_{\text{exact}} \equiv \Omega(w_0)|\xi\eta) = |\xi\eta) + \frac{Q}{e(w_0)}V|\xi\eta)_{\text{exact}}. \tag{II.153}$$

This equation, due originally to Bethe and Goldstone[13], can be converted into the coordinate space and then solved for the exact wavefunction. The details are given in Section 22. The point to note here is that, in the core region, V becomes infinitely large, the exact wavefunction goes to zero, and hence their product still remains a meaningful small finite quantity. If the core is only strongly repulsive, and not exactly infinite in magnitude, the exact wavefunction also remains extremely small in that region, and once again the contribution to their product from the core region is small. It should be observed that, by virtue of the definition (II.150) and the definition of $|\xi\eta)_{\text{exact}}$, any matrix element of t is given by

$$(\xi'\eta'|\ t\ |\xi\eta) \equiv (\xi'\eta'|\ V\Omega\ |\xi\eta) = (\xi'\eta'|\ V\ |\xi\eta)_{\text{exact}}. \tag{II.154}$$

Since the product $V|\xi\eta)_{\text{exact}}$ has become a meaningful small quantity in the core region of V, the matrix elements of t, as computed from (II.154), are of reasonable magnitude.

Now, in terms of the t-matrix elements, the entire contribution of the infinite set of V-diagrams shown in Fig. II.16 has reduced simply to a single matrix element $(\xi\eta|\ t\ |\xi\eta)$ of t. This is illustrated in Fig. II.17, where the bubbles represent the fermi-sea states ξ, η as before,

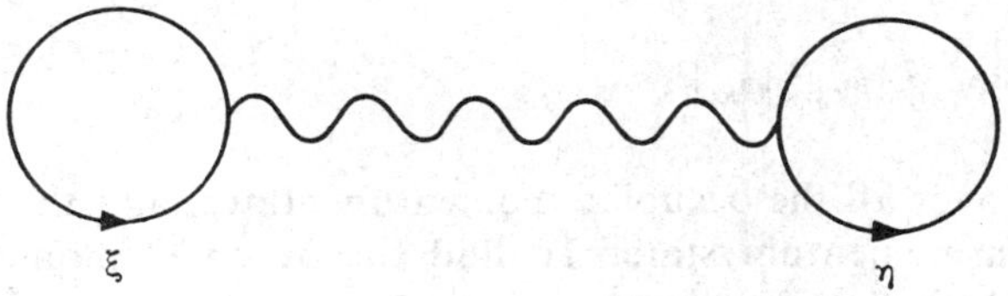

Fig. II.17 First-order diagram in effective potential t (wavy line).

and the wavy line denotes the reaction operator or the effective two-body potential t. If all the diagrams apart from those used in the definition of this matrix element of t are ignored, then the many-body ground-state energy is clearly given by

$$E_0 = \sum_{\xi}^{\text{occ}} \langle\xi|\ T\ |\xi\rangle + \sum_{\xi<\eta} (\xi\eta|\ t\ |\xi\eta). \tag{II.155}$$

21. CALCULATION OF SURFACE AND SYMMETRY ENERGY

A. SURFACE ENERGY (Semi-infinite Nuclear Matter)

Idealized nuclear matter of *infinite* extent does not have any surface (i.e., no boundary region marked by a fall-off in the density), and hence it has no surface energy. For the purpose of calculating the surface energy, therefore, we usually modify the idealized model as follows. We keep the extent of the nucleus in two directions, say, x and y, infinite, but introduce a wall in the z-direction at $z = 0$, placing all the nucleons to the right of this wall, i.e., in the region

$z = 0$ to $z = +\infty$. Such an idealized system is often referred to as the *semi-infinite* nuclear matter. The solid wall at $z = 0$ may actually be an infinitely repulsive potential, extending from $z = -\infty$ to $z = 0$ and preventing nucleons from entering this region of space. We may also introduce in the region $z = -\infty$ to $z = 0$ a repulsive potential of a different shape; one such example, considered later in this section, is a linearly rising repulsive potential which becomes infinitely large at $z = -\infty$.

First, let us consider the case of the infinitely repulsive potential in the region $z = -\infty$ to $z = 0$. Since the motion in the z-direction extends to $+\infty$, the quantum states are still of the continuum type. Corresponding to the momentum k_z, we have the wavefunction for such a state, satisfying the correct boundary condition (namely, it has to be zero at $z = 0$), given by

$$\phi_{k_z} = \underset{L\to\infty}{\mathrm{Lt}} \left(\frac{2}{L}\right)^{1/2} \sin k_z z, \tag{II.156a}$$

where L is the length of the region in which the nucleons move along any coordinate axis, and goes to the limit infinity for the model just described. The quantum states for the x- and y-direction are still described by

$$\phi_{k_x} = \underset{L\to\infty}{\mathrm{Lt}} \left(\frac{1}{L}\right)^{1/2} \exp\,(ik_x x), \tag{II.156b}$$

$$\phi_{k_y} = \underset{L\to\infty}{\mathrm{Lt}} \left(\frac{1}{L}\right)^{1/2} \exp\,(ik_y y). \tag{II.156c}$$

The nucleon density computed with these wavefunctions is clearly independent of x and y, but depends on the coordinate z. At any point (x, y, z), it is given by

$$\rho(z) = \sum_{k_x,k_y,k_z}^{\mathrm{occ}} 4|\phi_{k_x}|^2\,|\phi_{k_y}|^2\,|\phi_{k_z}|^2, \tag{II.157}$$

where the summation is over all the occupied momentum states, and the factor 4 takes care of the four nucleons in each momentum state. To find the occupied momentum states, we first note that the allowed values of k_x, k_y, k_z are

$$k_x L = 2\pi n_x, \qquad k_y L = 2\pi n_y, \qquad k_z L = 2\pi n_z,$$

where n_x, n_y are integers, both positive and negative, but n_z has only *positive* integral values. The permitted values of k_x, k_y, as already noted, follow from the imposition of periodic boundary conditions and the fact that the exponential functions for the same positive and negative integer values of n_x or n_y represent *independent* eigenstates. To obtain the allowed values of k_z, we have imposed the periodic boundary condition (which becomes the same as the standing wave boundary condition because $\phi_{k_z} = 0$ at $z = 0$); we must however note that $\sin\,[(2\pi n_z/L)z]$ and $\sin\,[(-2\pi n_z/L)z]$ are no longer independent wavefunctions (they differ only through a trivial phase factor -1), and hence we have restricted n_z to positive integer values. In the momentum space, therefore, the (k_z = negative)-half of the space is not allowed. All the occupied states are now contained inside a *hemisphere* of radius k_F in the 'upper' half, i.e., (k_z = positive)-half of the momentum space. Keeping this fact in mind, we have to convert the momentum summation in (II.157) into an integration in the limit $L \to \infty$. We denote the volume L^3 by Ω and obtain

$$\rho(z) = \frac{8}{\Omega} \sum_{k_x,k_y,k_z}^{\mathrm{occ}} \sin^2 k_z z = \frac{8}{\Omega}\,\frac{\Omega}{(2\pi)^3} \int_{\mathrm{hemisphere}} d^3k\, \sin^2\,(kz\cos\theta),$$

where θ is the polar angle of $\mathbf{k}$ in the momentum space, and the momentum integration is over the occupied hemisphere of radius k_F. Carrying out the integration explicitly, we get

$$\begin{aligned}\rho(z) &= \frac{2}{\pi^2}\int_0^{k_F} k^2\,dk \int_0^{\pi/2} \sin\theta\,d\theta \sin^2(kz\cos\theta)\\ &= \frac{1}{\pi^2}\int_0^{k_F} k^2\,dk \int_0^1 d\eta\,[1-\cos(2kz\eta)], \qquad \eta=\cos\theta\\ &= \frac{1}{\pi^2}\int_0^{k_F} k^2\,dk\,[1-\frac{\sin(2kz)}{2kz}]\\ &= \rho(\infty)[1-\frac{3}{2k_F z}j_1(2k_F z)],\end{aligned} \tag{II.158a}$$

where

$$\rho(\infty) = \frac{1}{3\pi^2}k_F^3, \tag{II.158b}$$

and the spherical Bessel function j_1 [see Appendix C (Section I)] is given by

$$j_1(\xi) = \frac{\sin\xi}{\xi^2} - \frac{\cos\xi}{\xi}. \tag{II.158c}$$

For large z, the function $j_1(2k_F z)$ behaves as $\sin(2k_F z - \pi/2)$, which always remains finite in magnitude, and hence the second term of (II.158a) goes to zero for $z \to \infty$ by virtue of the factor $2k_F z$ in the denominator. Thus, $\rho(\infty)$ is the value of the nucleon density at a fairly large distance from the surface. As we move towards $z = 0$, the second term of (II.158a) makes itself felt, gives rise to small oscillations in the density, and is followed by a smooth fall-off to $\rho(z) = 0$ at $z = 0$. The value at $z = 0$ follows from the result

$$j_1(\xi) = \tfrac{1}{3}\xi, \qquad \xi \to 0.$$

It should be noted that, for a given k_F, the expression (II.158b) for $\rho(\infty)$ is exactly one-half of the density of infinite nuclear matter [see (II.37)]. This is understandable in the present case where the occupied momentum states belong to a hemisphere and not a sphere.

Because of the fall-off in the density $\rho(z)$ in the vicinity of $z = 0$, the amount of nuclear matter $\int_0^\infty \rho(z)\,dz$, contained in a cylinder with the axis along z and having unit cross-section, is less than $\int_0^\infty \rho(\infty)\,dz$. However, the same amount at density $\rho(\infty)$ will be contained in a cylinder extending from $z = a$ to $z = \infty$, where a is to the right of $z = 0$. Thus, the location $z = a$ is defined by

$$\int_0^\infty \rho(z)\,dz = \int_a^\infty \rho(\infty)\,dz. \tag{II.159}$$

As shown in Fig. II.18, this definition ensures that the cross-hatched areas with different shades are equal. In this figure, $\rho(z)$, in units of $\rho(\infty)$, has been plotted as a function of $\zeta(=k_F z)$. Since the density $\rho(z)$ starts falling off from a point to the right of $z = a$ and eventually reaches zero value at $z = 0$, the point $z = a$ may be taken to be the 'location' of the surface of semi-infinite nuclear matter. The nuclear density is seen to fall off gradually across this surface. A sharp fall-off in the density at a well-defined z-value is ruled out from energetic considerations; such a sharp fall-off in the manner of a step function implies that the momentum of the particles at the particular point z is completely uncertain.

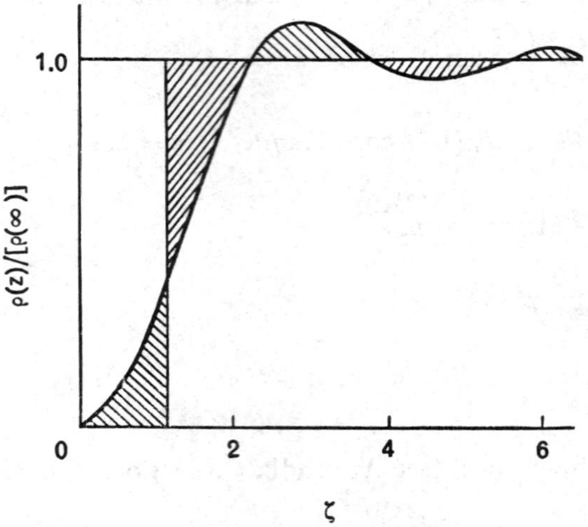

Fig. II.18 Density of semi-infinite nuclear matter plotted as function of distance along z-axis.

In order to compute the surface energy, we proceed as follows. Let $E(z)$ be the energy per unit volume at the point (x, y, z) of the semi-infinite nuclear matter. Then the amount of energy contained in a cylinder of unit cross-section and axis lying along z is given by

$$\int_0^\infty dz\, E(z). \tag{II.160a}$$

The same quantity of nuclear matter at density $\rho(\infty)$ is contained inside the cylinder of length a to ∞, where a is defined by (II.159). If the energy density of infinite nuclear matter is denoted by $E(\infty)$ (which is independent of any coordinate), then the total energy of infinite nuclear matter inside the cylinder is

$$\int_a^\infty dz\, E(\infty) = \frac{E(\infty)}{\rho(\infty)} \int_0^\infty \rho(z)\, dz, \tag{II.160b}$$

where the definition (II.159) has been used to substitute for $\int_a^\infty dz$. By definition, the total energy (II.160a) of the semi-infinite nuclear matter must be equal to the volume energy of an equal quantity of infinite nuclear matter, i.e., (II.160b), plus the surface energy. Therefore,

$$\text{surface energy per unit area} = \int_0^\infty dz\, [E(z) - \frac{E(\infty)}{\rho(\infty)}\rho(z)]$$

$$= \int_0^\infty dz\, \mathcal{E}(z)\rho(z), \tag{II.161a}$$

where

$$\mathcal{E}(z) = \frac{E(z)}{\rho(z)} - \frac{E(\infty)}{\rho(\infty)}. \tag{II.161b}$$

The actual computation therefore reduces to that of $E(z)$.

The kinetic energy part, $T(z)$, of $E(z)$ can be very easily obtained. We have

$$T(z) = -\frac{\hbar^2}{2M} \sum_{k_x,k_y,k_z}^{\text{ccc}} 4(\phi^*_{k_x}\phi^*_{k_y}\phi^*_{k_z})\nabla^2(\phi_{k_x}\phi_{k_y}\phi_{k_z}),$$

where the factor 4 takes care of the four nucleons in each occupied momentum state. In a straightforward manner, similar to the calculation of $\rho(z)$, we obtain

$$\begin{aligned} T(z) &= \frac{\hbar^2}{2M}\frac{8}{\Omega}\frac{\Omega}{(2\pi)^3}\int_{\text{hemisphere}} d^3k\, k^2 \sin^2(kz\cos\theta) \\ &= \frac{\hbar^2}{2M}\frac{1}{\pi^2}\int_0^{k_F} k^4\, dk\, [1 - \frac{\sin(2kz)}{2kz}] \\ &= \tfrac{3}{5}\frac{\hbar^2 k_F^2}{2M}\rho(\infty)(1 + \frac{5\cos\xi}{\xi^2} - \frac{15\sin\xi}{\xi^3} - \frac{30\cos\xi}{\xi^4} + \frac{30\sin\xi}{\xi^5}), \end{aligned} \tag{II.162}$$

where $\xi = 2k_F z$. The factor preceding the square brackets is clearly the kinetic energy per unit volume of the infinite nuclear matter at density $\rho(\infty)$.

The potential energy part, $V(z)$, of $E(z)$ is, however, a little more involved to derive. If $\langle V\rangle$ is the total potential energy, then $V(z)$, by definition, is given by

$$\langle V\rangle = \int V(z)\, dx\, dy\, dz. \tag{II.163}$$

Therefore, the calculation of $V(z)$ relies on expressing the potential energy $\langle V\rangle$ as an integral of the type in (II.163). The expression (II.105a) for $\langle V\rangle$ is still valid. However, the single-particle momentum eigenfunction, to be used in (II.104), is no longer given by $(1/\sqrt{\Omega})\exp(i\mathbf{k}\cdot\mathbf{r})$; it is to be obtained from (II.156) and is given by

$$\phi_{\mathbf{k}}(\mathbf{r}) = \phi_{k_x}(x)\phi_{k_y}(y)\phi_{k_z}(z). \tag{II.164}$$

We denote the two-particle coordinates by $\mathbf{r}$ and $\mathbf{r}'$ (instead of $\mathbf{r}_1$ and $\mathbf{r}_2$) and express the matrix elements in (II.104) as integrals over these coordinates. Substituting in (II.105a) and then comparing with (II.163), we obtain

$$V(z) = \frac{g_1}{2}\rho(\mathbf{r})\int d^3r'\, \rho(\mathbf{r}')f(|\mathbf{r}-\mathbf{r}'|) + \frac{g_2}{2}\int d^3r'\, |\rho(\mathbf{r},\mathbf{r}')|^2 f(|\mathbf{r}-\mathbf{r}'|), \tag{II.165}$$

where

$$\rho(\mathbf{r}) = \sum_{\mathbf{k}}^{\text{occ}} 4|\phi_{\mathbf{k}}(\mathbf{r})|^2, \tag{II.166a}$$

$$\rho(\mathbf{r},\mathbf{r}') = \sum_{\mathbf{k}}^{\text{occ}} 4\phi^*_{\mathbf{k}}(\mathbf{r})\phi_{\mathbf{k}}(\mathbf{r}'). \tag{II.166b}$$

In general, (II.165) defines a function V of all the three coordinates contained in $\mathbf{r}$. In the present case of semi-infinite nuclear matter, $\rho(\mathbf{r})$, as defined by (II.166a), which is the same as (II.157) and (II.158a), depends only on z; the exchange density $\rho(\mathbf{r},\mathbf{r}')$ can also be explicitly worked out, and the second term of (II.165) can be shown to depend only on z. The explicit evaluation will now be done, taking a Gaussian form for $f(|\mathbf{r}-\mathbf{r}'|)$ for the sake of simplicity, i.e.,

$$\begin{aligned} f(|\mathbf{r}-\mathbf{r}'|) &= \exp[-(|\mathbf{r}-\mathbf{r}'|)^2/a^2] \\ &= \exp[-(x-x')^2/a^2]\exp[-(y-y')^2/a^2]\exp[-(z-z')^2/a^2]. \end{aligned}$$

Since $\rho(\mathbf{r}')$ depends only on z', we have, for the integral in the first term of (II.165),

$$
\begin{aligned}
[V(z)]_{\text{direct term}} &= \tfrac{1}{2}g_1\rho(\mathbf{r}) \int d^3r'\, \rho(\mathbf{r}')f(|\mathbf{r}-\mathbf{r}'|) \\
&= \int_{-\infty}^{\infty} \exp[-(x-x')^2/a^2]\,dx' \int_{-\infty}^{\infty} \exp[-(y-y')^2/a^2]\,dy' \\
&\quad \times \int_{-\infty}^{\infty} dz' \exp[-(z-z')^2/a^2]\rho(z') \qquad\qquad (\text{II}.167) \\
&= \pi a^2 \int_{-\infty}^{\infty} dz' \exp[-(z-z')^2/a^2]\rho(z').
\end{aligned}
$$

This integral can be evaluated after inserting $\rho(z')$ from (II.158a); the result is clearly a function of z alone. The factor $\rho(\mathbf{r})$ also depends only on z, and, as a result, the first term of (II.165) has been established to be a function of z. To evaluate the second term of this expression, we first work out the exchange density with the help of (II.166b), (II.164), and (II.156) as

$$
\begin{aligned}
\rho(\mathbf{r},\mathbf{r}') &= \operatorname*{Lt}_{L\to\infty} \frac{8}{L^3} \sum_{\mathbf{k}}^{\text{occ}} \sin k_z z \sin k_z z' \exp[-ik_x(x-x')] \exp[-ik_y(y-y')] \\
&= \frac{8}{\Omega}\frac{\Omega}{(2\pi)^3} \int_{\text{hemisphere}} d^3k \sin k_z z \sin k_z z' \exp[-ik_x(x-x')]\exp[-ik_y(y-y')] \\
&= \frac{1}{\pi^3}\int_0^{k_F} k_\perp\, dk_\perp \int_0^{2\pi} d\theta \exp[-i\mathbf{k}_\perp\cdot(\mathbf{r}_\perp-\mathbf{r}'_\perp)] \int_0^{(k_F^2-k_\perp^2)^{1/2}} dk_z \sin k_z z \sin k_z z'.
\end{aligned}
$$

Here $\mathbf{k}_\perp$ is the projection of $\mathbf{k}$ on the (x, y)-plane having the components k_x, k_y and making an angle θ with the x-direction. The last integral over k_z gives rise to a function of z, z', and $k_\perp$ for which we introduce the symbol

$$
F(k_\perp, z, z') = \int_0^{(k_F^2-k_\perp^2)^{1/2}} dk_z \sin k_z z \sin k_z z'. \qquad (\text{II}.168)
$$

With this notation, the square of the modulus of the exchange density which enters (II.165) can be written as

$$
\begin{aligned}
|\rho(\mathbf{r},\mathbf{r}')|^2 &= \frac{1}{\pi^3}\int_0^{k_F} k_\perp\, dk_\perp \int_0^{k_\perp} k'_\perp\, dk'_\perp \int_0^{2\pi} d\theta \int_0^{2\pi} d\theta' \exp[-i(\mathbf{k}_\perp-\mathbf{k}'_\perp)\cdot(\mathbf{r}_\perp-\mathbf{r}'_\perp)] \\
&\quad \times F(k_\perp, z, z')F(k'_\perp, z, z').
\end{aligned}
$$

We have to multiply this expression by the Gaussian shape function and integrate over x', y', z'. The integrals over x', y' are clearly of the form now specified, and can be worked out as

$$
\begin{aligned}
&\int_{-\infty}^{\infty} dx' \int_{-\infty}^{\infty} dy'\, e^{-(x-x')^2/a^2} e^{-(y-y')^2/a^2} \exp[-i(\mathbf{k}_\perp-\mathbf{k}'_\perp)\cdot(\mathbf{r}_\perp-\mathbf{r}'_\perp)] \\
&= \int_{-\infty}^{\infty} dx'\, e^{-(x-x')^2/a^2} \exp[-i(k_x-k'_x)(x-x')] \int_{-\infty}^{\infty} dy'\, e^{-(y-y')^2/a^2} \\
&\quad \times \exp[-i(k_y-k'_y)(y-y')] \\
&= \pi a^2 \exp[-(k_x-k'_x)^2/4] \exp[-(k_y-k'_y)^2/4] \\
&= \pi a^2 \exp[-\tfrac{1}{4}(\mathbf{k}_\perp^2+\mathbf{k}_\perp'^2-2\mathbf{k}_\perp\cdot\mathbf{k}'_\perp)].
\end{aligned}
$$

It is clear from this expression that the second term of (II.165), i.e., the space-exchange term, is not in any way dependent on x and y. We finally obtain

$$[V(z)]_{\text{exchange term}} = \tfrac{1}{2}g_2\left(\frac{a}{\pi}\right)^2 \int_0^{k_F} k_\perp\, dk_\perp \int_0^{k_F} k'_\perp\, dk'_\perp \int_0^{2\pi} d\theta \int_0^{2\pi} d\theta' \int_{-\infty}^{\infty} dz'$$
$$\times e^{-(z-z')^2/a^2} F(k_\perp, z, z') F(k'_\perp, z, z') \exp\left[-\tfrac{1}{4}(k_\perp^2 + k'^2_\perp - 2k_\perp k'_\perp \cos\phi)\right],$$

where ϕ is the angle between $\mathbf{k}_\perp$ and $\mathbf{k}'_\perp$, i.e., $\phi = \theta - \theta'$. Of the two angle-integrations over θ and θ', one can immediately be carried out yielding (2π), whereas the other becomes an integration over ϕ; and, according to the integral representation of the Bessel function, the latter integral yields

$$\int_0^{2\pi} d\phi \exp\left(\tfrac{1}{2}k_\perp k'_\perp \cos\phi\right) = 2\pi J_0\left(-\frac{i}{2}k_\perp k'_\perp\right) = 2\pi \sum_{n=0}^{\infty} \frac{(\tfrac{1}{4}k_\perp k'_\perp)^{2n}}{(n!)^2}.$$

Using all these results, we obtain

$$[V(z)]_{\text{exchange term}} = \tfrac{1}{2}g_E \int_{-\infty}^{\infty} dz'\, e^{-(z-z')^2/a^2} \sum_{n=0}^{\infty} [\mathcal{F}_n(z, z')]^2, \tag{II.169a}$$

where

$$\mathcal{F}_n(z, z') = 2a \int_0^{k_F} dk_\perp\, k_\perp \frac{(\tfrac{1}{4}k_\perp)^{2n}}{n!} F(k_\perp, z, z'). \tag{II.169b}$$

In the special case we are considering here, $F(k_\perp, z, z')$, given by (II.168), can be explicitly worked out; therefore, $\mathcal{F}_n(z, z')$ of (II.169b) can also be evaluated in a fairly straightforward manner. Usually, a few of the lowest terms in the summation over n of (II.169a) suffice.

A better way of calculating the surface energy is to take a potential in the region $z = -\infty$ to $z = 0$, and then do a variational calculation for the minimum of nuclear energy with respect to some parameter in the potential. We consider only a linear potential (though this is not very realistic)

$$\mathcal{V}(z) = -\frac{\hbar^2 k_F^2}{2M}\frac{z}{z_0} \quad \text{for } -\infty < z < 0. \tag{II.170}$$

This potential has the value zero at $z = 0$, always stays repulsive for negative values of z, and becomes infinitely repulsive at $z \to -\infty$. The parameter z_0 determines its slope, and is clearly equal to the *magnitude* of z at which the potential becomes equal to the fermi energy. In this case, the wavefunction (II.156a) has to be replaced by the solution of the one-dimensional Schrödinger equation corresponding to the linear potential (II.170). These wavefunctions are given in texts on quantum mechanics, for example, Landau and Lifshitz[14]. The calculation of $\rho(z)$, $T(z)$, and $V(z)$ has then to be repeated by replacing everywhere in our earlier work in this section the wavefunction ϕ_{k_z} of (II.156a) by the appropriate wavefunction of the linear potential. This procedure leads to fairly detailed computations, and can be carried out by an enterprising reader with the help of a publication by Swiatecki[15] on which our discussion in this section is based. Since the wavefunction for the z-coordinate now extends up to $z = -\infty$ to the left, and vanishes only at that point [because $\mathcal{V}(-\infty) = \infty$], the tailing-off of the density $\rho(z)$ continues right up to $z = -\infty$. Thus, in the definition (II.161a) of surface energy, the integration limit $z = 0$ has to be replaced by $z = -\infty$.

The computation of surface energy in the presence of an average nuclear potential $\mathcal{V}(z)$ can, however, be carried out in an *approximate* manner if *the variation of the potential with distance is rather slow*. Under such circumstances, we *assume* that the earlier concepts of semi-infinite nuclear matter, based on a set of occupied states characterized by the momenta $k = 0 \rightarrow k = k_F$, are still valid in a *local* sense; that is, we now assume that, at each point $\mathbf{r}$ in space, depending on the local density $\rho(\mathbf{r})$, a set of states characterized by the momenta $k = 0 \rightarrow k = k_{max}(\mathbf{r})$ are occupied. The maximum occupied momentum $k_{max}(\mathbf{r})$ at the point $\mathbf{r}$ is taken to be related to $\rho(\mathbf{r})$ in the same way as k_F, in the case of semi-infinite nuclear matter, is related to the density $\rho(\infty)$, i.e., (II.158b). We thus have

$$\rho(\mathbf{r}) = \frac{1}{3\pi^2} k^3_{max}(\mathbf{r}). \tag{II.171}$$

Since, however, the density is not given, but the potential $\mathcal{V}(\mathbf{r})$ is, we have to make a further assumption relating $\rho(\mathbf{r})$ with $\mathcal{V}(\mathbf{r})$. Of course, given $\mathcal{V}(\mathbf{r})$, we can, in principle, find out the single-particle occupied states and from these *calculate* $\rho(\mathbf{r})$. The purpose of the present approximate method is to avoid detailed computations, and hence we use a simplified approach, which is now explained. In the absence of the potential, a set of single-particle states up to the fermi energy $\epsilon_F [= \hbar^2 k_F^2/(2M)]$ are occupied. We assume that the characterization of the occupied states remains the same in the presence of the potential. However, at a point where the potential energy is $\mathcal{V}(\mathbf{r})$, the zero-momentum state starts at a total energy ϵ given by

$$0 = T(\mathbf{r}) = \epsilon - \mathcal{V}(\mathbf{r}), \quad \text{i.e.,} \quad \epsilon = \mathcal{V}(\mathbf{r}). \tag{II.172}$$

Therefore, the lowest state occupied at the point $\mathbf{r}$, according to this simple picture, has an energy given by (II.172). The highest occupied state of energy ϵ_F corresponds in the same way to a local momentum $k_{max}(\mathbf{r})$ given by

$$\frac{\hbar^2 k^2_{max}(\mathbf{r})}{2M} = T_{max}(\mathbf{r}) = \epsilon_F - \mathcal{V}(\mathbf{r}) = \frac{\hbar^2 k_F^2}{2M} - \mathcal{V}(\mathbf{r}). \tag{II.173}$$

This simple relationship, along with (II.171), can thus be applied to calculate $\rho(\mathbf{r})$ from the given $\mathcal{V}(\mathbf{r})$, obtaining

$$\rho(\mathbf{r}) = \frac{1}{3\pi^2}[k_F^2 - \frac{2M}{\hbar^2}\mathcal{V}(\mathbf{r})]^{3/2}. \tag{II.174}$$

Since the spirit of this approximate calculation is to borrow the expressions for nuclear matter and use $k_{max}(\mathbf{r})$, instead of k_F, in those expressions for the calculations of local quantities at the point $\mathbf{r}$, we go on maintaining the same procedure in the evaluation of the kinetic energy density $T(\mathbf{r})$ and the potential energy density $V(\mathbf{r})$. In this way,

$$T(\mathbf{r}) = [\tfrac{3}{5}\frac{\hbar^2 k^2_{max}(\mathbf{r})}{2M}]\rho(\mathbf{r}) = \tfrac{3}{5}[k_F^2 - \frac{2M}{\hbar^2}\mathcal{V}(\mathbf{r})]^{5/2}\frac{1}{3\pi^2}\frac{\hbar^2}{2M}. \tag{II.175}$$

Here the first factor within the square brackets stands for the kinetic energy per nucleon, and hence multiplication by the density $\rho(\mathbf{r})$ of nucleons produces the kinetic energy per unit volume. The derivation of the expression for $V(\mathbf{r})$, obeying the assumptions of the present simplified model, is left as an exercise.

In the case of semi-infinite nuclear matter, the dependence of $\rho(\mathbf{r})$, $T(\mathbf{r})$, $V(\mathbf{r})$ on $\mathbf{r}$ is only through z. In the case of the linear potential (II.170), $\rho(z)$, according to (II.174), starts falling off from $\rho(\infty)$ as we go to the left of $z = 0$, and it reaches the value zero at the point $z = -z_0$, where $\mathcal{V}(z)$ has the value $[\hbar^2 k_F^2/(2M)]$. In the definition of the surface energy (II.161a), the integration should, as remarked after (II.170), extend from $z = -\infty$ to $z = +\infty$ in the presence of a potential $\mathcal{V}(z)$. However, in the simple model calculation being considered now, it suffices if we carry out the integration between $z = -z_0$ and $z = 0$; this is because $\rho(z)$ is zero to the left of $z = -z_0$, and $E(z)/\rho(z)$ is equal to $E(\infty)/\rho(\infty)$ to the right of $z = 0$, making $\mathcal{E}(z)$ equal to zero in this region. For convenience, we shift the origin to the point $-z_0$ and use the symbol ζ to denote the length along the z-axis measured from this origin (i.e., $\zeta = z_0 + z$). In the case of the linear potential (II.170), we then have

$$\rho(\zeta) = \frac{k_F^3}{3\pi^2}\frac{\zeta^{3/2}}{z_0^{3/2}} = \rho(\infty)\left(\frac{\zeta}{z_0}\right)^{3/2}, \tag{II.176a}$$

$$T(\zeta) = \left(\frac{3}{5}\frac{\hbar^2 k_F^2}{2M}\right)\left(\frac{k_F^3}{3\pi^2}\right)\left(\frac{\zeta}{z_0}\right)^{5/2} = T(\infty)\left(\frac{\zeta}{z_0}\right)^{5/2}. \tag{II.176b}$$

As an exercise, the interested reader may write down the expression for $V(\zeta)$ for a Yukawa or a Gaussian potential. The integration of $\mathcal{E}(z)\rho(z)$ [see (II.161a)] in the range $z = -z_0$ to $z = 0$ becomes an integration from 0 to z_0 in terms of the new variables ζ. The final expression for the surface energy depends on the parameter z_0 of the potential. The parameter z_0 may be chosen by minimizing the surface energy. The numerical agreement of such a simple calculation, based on the first-order perturbation expression of V for a well-behaved Gaussian or Yukawa potential, is, however, not very encouraging. A more detailed calculation for the same two-body potential based on the exact wavefunctions of the linear potential, as mentioned after (II.170), also gives unsatisfactory agreement.

For a fairly comprehensive list of references to more satisfactory calculations of surface energy, the reader is referred to a paper on this topic by Day[16] which is based on the two-body reaction matrix.

B. SYMMETRY ENERGY

For the calculation of symmetry energy, we go back to a nucleus of infinite extent, but take a neutron number (N) that is different from the proton number (Z). The total nucleon number ($N + Z$) is denoted by A, and we assume that the departure of N and Z from $\frac{1}{2}A$ is small. Thus,

$$N \equiv \tfrac{1}{2}A + T_z, \qquad Z \equiv \tfrac{1}{2}A - T_z \tag{II.177a}$$

with

$$T_z = N - \tfrac{1}{2}A = N - \tfrac{1}{2}(N + Z) = \tfrac{1}{2}(N - Z),$$

and T_z is assumed to be small as compared with A. Since the nucleus is of infinite extent, the single-particle states are the plane-wave states characterized by the momenta $\mathbf{k}$ of the particles. However, the total number of occupied states for neutrons, equal to N, is different from that for protons, equal to Z. In other words, the fermi momentum for the neutrons, (k_F^N), is different from that for the protons, (k_F^P). Considering two neutrons or two protons with opposite spins in each momentum state, and taking Ω to be the nuclear volume, we easily have

$$\frac{2\Omega}{(2\pi)^3}\frac{4\pi}{3}(k_F^N)^3 = N, \qquad \frac{2\Omega}{(2\pi)^3}\frac{4\pi}{3}(k_F^P)^3 = Z. \tag{II.177b}$$

Substituting for N and Z from (II.177a), we get

$$(k_F^N)^3 = \frac{3\pi^2 A}{2\Omega}\left(1 + \frac{T_z}{2A}\right), \qquad (k_F^P)^3 = \frac{3\pi^2 A}{2\Omega}\left(1 - \frac{T_z}{2A}\right). \tag{II.178a}$$

The first factor in each expression of (II.178a) is the original $(k_F)^3$ for the idealized nuclear matter which has an equal number of neutrons and protons. Thus,

$$k_F^N = k_F(1+\delta)^{1/3}, \qquad k_F^P = k_F(1-\delta)^{1/3}, \tag{II.178b}$$

where

$$\delta = \frac{T_z}{2A} = \tfrac{1}{2}I, \tag{II.178c}$$

I being the quantity in (II.2d′) defining symmetry energy.

The part of the symmetry energy that arises from the kinetic energy of the system of nucleons is easy to calculate. According to our derivation in Section 18, the kinetic energy per particle of an infinite many-body system is three-fifths of the fermi energy. Hence, the total kinetic energy in the present case is given by

$$\begin{aligned}\langle T\rangle &= \tfrac{3}{5}\frac{\hbar^2(k_F^N)^2}{2M}N + \tfrac{3}{5}\frac{\hbar^2(k_F^P)^2}{2M}Z \\ &= \tfrac{3}{5}\frac{\hbar^2}{2M}\frac{2\Omega}{(2\pi)^3}\frac{4\pi}{3}[(k_F^N)^5 + (k_F^P)^5] \end{aligned}\tag{II.179a}$$

$$\begin{aligned} &= \tfrac{3}{5}\frac{\hbar^2}{2M}\frac{2\Omega}{(2\pi)^3}k_F^5\frac{4\pi}{3}[(1+\delta)^{5/3} + (1-\delta)^{5/3}] \\ &= (\tfrac{3}{5}\frac{\hbar^2}{2M}k_F^2)(\frac{4\Omega}{8\pi^3}\frac{4\pi}{3}k_F^3)(1+\tfrac{5}{9}\delta^2). \end{aligned}\tag{II.179b}$$

Therefore,

$$\frac{\langle T\rangle}{A} = \tfrac{3}{5}\epsilon_F(1+\tfrac{5}{9}\delta^2) = \tfrac{3}{5}\epsilon_F + \tfrac{1}{3}\epsilon_F\delta^2. \tag{II.179c}$$

We have here used (II.177) to replace N and Z, and then (II.178b) for k_F^N and k_F^P. Finally, we have retained only up to the quadratic term in δ, and, when writing (II.179c), have replaced the second factor within the parentheses in (II.179b) by its value A. ϵ_F is the fermi energy $[\hbar^2 k_F^2/(2M)]$.

The second term in (II.179c), when compared with (II.2d′), gives the contribution to the symmetry energy. Before working out the contribution of the potential energy $\langle V\rangle$ to the symmetry-energy term, let us do some schematic work. Let the total binding energy E, which is a function of δ through its dependence on k_F^N and k_F^P, be expanded in a power series of δ as

$$E(\delta) = E(0) + \delta(\frac{dE}{d\delta})_0 + \tfrac{1}{2}\delta^2(\frac{d^2E}{d\delta^2})_0 + \ldots . \tag{II.180a}$$

The first term here is clearly the energy of the idealized nuclear matter with $N = Z$ and the fermi momentum k_F. By our assumption, we have a minimum of $E(\delta)$ corresponding to $\delta = 0$. Therefore,

$$(\frac{dE}{d\delta})_0 = 0,$$

$$(\frac{d^2E}{d\delta^2})_0 > 0.$$

The coefficient of the symmetry-energy term in E is therefore required to be positive from very general considerations. (II.180a) thus becomes

$$E(\delta) = E(0) + \tfrac{1}{2}\delta^2\left(\frac{d^2E}{d\delta^2}\right)_0 + \ldots, \tag{II.180b}$$

and we have to compute the second derivative of E with respect to δ for $\delta = 0$ (indicated by the subscript zero outside the close parenthesis) in order to know the coefficient of the symmetry-energy term.

As already remarked, the dependence of E on δ comes through k_F^N and k_F^P, and hence we write

$$\frac{dE}{d\delta} = \frac{dk_F^N}{d\delta}\frac{\partial E}{\partial k_F^N} + \frac{dk_F^P}{d\delta}\frac{\partial E}{\partial k_F^P}.$$

Therefore,

$$\left(\frac{d^2E}{d\delta^2}\right)_{\delta=0} = \left(\frac{dk_F^N}{d\delta}\right)^2_{\delta=0}\left\{\frac{\partial^2 E}{\partial(k_F^N)^2}\right\}_{\delta=0} + \left(\frac{\partial k_F^P}{d\delta}\right)^2_{\delta=0}\left\{\frac{\partial^2 E}{\partial(k_F^P)^2}\right\}_{\delta=0}$$

$$+\left(\frac{d^2k_F^N}{d\delta^2}\right)_{\delta=0}\left(\frac{\partial E}{\partial k_F^N}\right)_{\delta=0} + \left(\frac{d^2k_F^P}{d\delta^2}\right)_{\delta=0}\left(\frac{\partial E}{\partial k_F^P}\right)_{\delta=0}.$$

From (II.178b), we get

$$\left(\frac{dk_F^N}{d\delta}\right)^2_{\delta=0} = \left(\frac{dk_F^P}{d\delta}\right)^2_{\delta=0} = \tfrac{1}{9}k_F^2,$$

$$\left(\frac{d^2k_F^N}{d\delta^2}\right)_{\delta=0} = \left(\frac{d^2k_F^P}{d\delta^2}\right)_{\delta=0} = -\tfrac{2}{9}k_F.$$

Using these results, we finally have

$$\left(\frac{d^2E}{d\delta^2}\right)_0 = \frac{k_F^2}{9}\left[\left\{\frac{\partial^2 E}{\partial(k_F^N)^2}\right\}_0 + \left\{\frac{\partial^2 E}{\partial(k_F^P)^2}\right\}_0\right]$$

$$- \tfrac{2}{9}k_F\left[\left(\frac{\partial E}{\partial k_F^N}\right)_0 + \left(\frac{\partial E}{\partial k_F^P}\right)_0\right]. \tag{II.181}$$

Working out this expression for the kinetic energy part of E with the help of (II.179a), we can very easily verify that the result agrees with (II.179b) and (II.179c). It is advantageous, however, to first note that, because E has a minimum at $\delta = 0$, i.e., $k_F^N \to k_F$ and $k_F^P \to k_F$, the second line of (II.181) is zero. To compute the symmetry energy, we thus need the contribution of kinetic and potential energy to only the first line of (II.181). The contribution from the kinetic energy can easily be verified to be $\tfrac{2}{3}\epsilon_F\delta^2$ (instead of $\tfrac{1}{3}\epsilon_F\delta^2$) appearing in (II.179c). Our task is therefore completed if we can write down $[\partial^2\langle V\rangle/\{\partial(k_F^N)^2\}]_0$ and $[\partial^2\langle V\rangle/\{\partial(k_F^P)^2\}]_0$. In the expression for potential energy, the quantities k_F^N and k_F^P occur as upper limits of momentum integrations, and the derivative can be easily written down with the general result for any x and any function $f(x)$, namely,

$$\frac{d}{dx}\int_0^x f(\xi)\,d\xi = f(x).$$

The details may be worked out by an enterprising reader.

22. CALCULATION OF REACTION MATRIX AND NUCLEAR BINDING ENERGY

A. INFINITE NUCLEAR MATTER

The definition of the two-body reaction matrix has been given in (II.149). The point we would like to emphasize here is that the quantity w is a fixed numerical parameter as far as the t-matrix equation is concerned, and this is why this parametric dependence of the various quantities has been explicitly shown in (II.149). In the calculation of the energy of infinite nuclear matter, we need [see (II.155)] matrix elements of t of the type $(\xi\eta|\, t\, |\xi\eta)$, where ξ, η are fermi-sea states; for such matrix elements of t, the parameter w is equal to w_0 of (II.146b). We may also encounter cases where w is not necessarily the starting energy of the pair of nucleons interacting through the t-operator; this will happen if other nucleons, excited out of the fermi sea, are present as spectators when the two nucleons under consideration are interacting. In such a case, w may greatly differ from the sum of the starting energies of the initial nucleons. The corresponding t-matrix is said to be *off-energy shell*, whereas the t-matrix in which the parameter w is equal to the initial energy of the interacting pair is called *on-energy shell*.

The matrix elements of $t(w)$ are given by (II.154), and can be calculated after the correlated two-body state $|\xi\eta)_{\text{exact}}$ has been calculated from the unperturbed initial state $|\xi\eta)$ of the pair of nucleons by means of the Bethe-Goldstone equation (II.153) (see Bethe and Goldstone[13]), which we rewrite in terms of the two-body wavefunction as

$$\Psi_{\xi\eta} = \Phi_{\xi\eta} + \frac{Q}{e(w)} V \Psi_{\xi\eta}. \tag{II.182}$$

The second term, upon iteration, becomes

$$\left[\frac{Q}{e(w)}V + \frac{Q}{e(w)}V\frac{Q}{e(w)}V + \ldots\right]\Phi_{\xi\eta}, \tag{II.183}$$

a form which, although useless in obtaining $\Psi_{\xi\eta}$ in the case of a strong-core potential, is well suited for reckoning the good quantum numbers labelling $\Phi_{\xi\eta}$, which serve for the labelling of $\Psi_{\xi\eta}$ as well. The potential V to the extreme right in each term of (II.183) occurs in a matrix element $\langle\xi'\eta'|\, V\, |\xi\eta\rangle$ in view of the form (II.145) for Q/e. Thus, the quantum numbers of $\Phi_{\xi\eta}$ conserved in the operation with V are carried forward to the labelling of the state $|\xi'\eta'\rangle$ of the Q-operator. The same observation is true of the matrix elements of subsequent V-operators, as we proceed towards the left. It is therefore convenient to label $\Phi_{\xi\eta}$ with quantum numbers conserved by the two-nucleon potential V. It is important to realize that, since $\Phi_{\xi\eta}$ is a two-nucleon state, the discussions in Section 2 are of help in reckoning the labelling quantum numbers. Thus, the two-nucleon spin quantum number S, and the isospin quantum number T and its projection M_T can be very profitably used in specifying $\Phi_{\xi\eta}$. Similarly, the centre-of-mass wavefunction in the spatial part of $\Phi_{\xi\eta}$ appears unmodified in $\Psi_{\xi\eta}$ in spite of the repeated operation of V because V does not depend on the centre-of-mass coordinate $\mathbf{R}$ of the two nucleons. (II.182) can thus be very easily reduced to an equation connecting the unperturbed wavefunction of the relative coordinate $\mathbf{r}$ with the corresponding exact wavefunction. This is done explicitly in what follows. After the Bethe-Goldstone equation for the wavefunction of relative motion has been obtained in this way, that wavefunction can be broken up into states of various orbital angular momentum l (in the manner of Chapter I), and the good quantum number S can be coupled with l to obtain the good quantum numbers J and M. All these statements become more explicit as we proceed.

In the case of infinite nuclear matter, the occupied plane-wave states (including spin and isospin parts) are denoted by the quantum numbers $(\mathbf{k}_1, \sigma_1, \tau_1)$ and $(\mathbf{k}_2, \sigma_2, \tau_2)$, instead of ξ and η, and the unoccupied states ξ', η' above the fermi sea by $(\mathbf{k}_1', \sigma_1', \tau_1')$ and $(\mathbf{k}_2', \sigma_2', \tau_2')$, respectively. σ and τ together with their subscripts denote the projection quantum number for spin and isospin, respectively. The summations over ξ', η' in (II.145) then become integrations over the momenta $\mathbf{k}_1'$, $\mathbf{k}_2'$ and summations over the spin and isospin projection quantum numbers:

$$\sum_{\xi'<\eta'}^{\text{unocc}} \rightarrow \tfrac{1}{2} \sum_{\substack{\sigma_1', \tau_1' \\ \sigma_2', \tau_2'}} \int_{k_1'>k_F} d^3k_1' \int_{k_2'>k_F} d^3k_2' \left[\frac{\Omega}{(2\pi)^3}\right]^2. \tag{II.184}$$

The single-particle energies $\epsilon(\xi')$, $\epsilon(\eta')$, . . . and the energies of the occupied states $\epsilon(\xi)$, $\epsilon(\eta)$, . . . correspond to the energies of a self-consistent single-particle Hamiltonian $(T + \mathcal{V})$. For infinite nuclear matter, the single-particle wavefunctions are required to be the plane waves $\exp(i\mathbf{k}\cdot\mathbf{r})$ from very general symmetry considerations, and hence the average potential $\mathcal{V}$ must be diagonal in $\mathbf{k}$. In other words, $\exp(i\mathbf{k}\cdot\mathbf{r})$ must be an eigenfunction of $\mathcal{V}$, i.e., $\mathcal{V}$ must be a function of the momentum $\mathbf{k}$ only. Since $\mathcal{V}$ has to be a scalar, and k^2 $(=\mathbf{k}\cdot\mathbf{k})$ is the only scalar we can form from $\mathbf{k}$, we conclude that the potential energy $\mathcal{V}$ of the particles in the average field is a function of k^2 only. Expanding $\mathcal{V}$ in powers of k^2 and writing the kinetic energy T explicitly, we obtain

$$\epsilon(k) \equiv \frac{\hbar^2 k^2}{2M} + \mathcal{V}(k^2) = \frac{\hbar^2 k^2}{2M} + (A + Bk^2 + Ck^4 + \ldots). \tag{II.185a}$$

If we retain quantities up to only the quadratic term, we get the energy $\epsilon(k)$ in the so-called *effective-mass approximation*

$$\epsilon(k) \approx \left(\frac{\hbar^2 k^2}{2M} + Bk^2\right) + A = \frac{\hbar^2 k^2}{2M^*} + A, \tag{II.185b}$$

where

$$(2M^*)^{-1} = (2M)^{-1} + (B/\hbar^2). \tag{II.185c}$$

In practice, the t-matrix equation has to be solved with an initially specified value of $\epsilon(k)$, and then, at the end of the t-matrix calculation, the energy of a particle of momentum $\mathbf{k}$ and spin-isospin projection σ, τ has to be recalculated from the definition

$$\epsilon(k) = \frac{\hbar^2 k^2}{2M} + \mathcal{V}(k^2) = \frac{\hbar^2 k^2}{2M} + \sum_{\mathbf{k}_1, \sigma_1, \tau_1}^{\text{occ}} (\mathbf{k}_1, \sigma_1, \tau_1; \mathbf{k}, \sigma, \tau |\, t\, |\mathbf{k}_1, \sigma_1, \tau_1; \mathbf{k}, \sigma, \tau). \tag{II.185d}$$

The momentum summation can be converted into an integration in the usual manner. The recalculated values of $\epsilon(k)$ must agree with the initially prescribed values. In general, a few iterations of the whole procedure are needed [i.e., start with the recalculated $\epsilon(k)$, recompute t, and then again calculate $\epsilon(k)$ from the new t] until self-consistency is obtained in the $\epsilon(k)$. Considerable experience has been gained by now from calculations done in the past, and a good prescription [due to Bethe (see Rajaraman and Bethe[17])] for the starting energies is given by (II.185b) with the following choice of parameters:

$$\frac{M^*}{M} = 0.7, \qquad A = -112 \text{ MeV} \qquad \text{(occupied states)}, \tag{II.185e}$$

$$\frac{M^*}{M} = 1, \qquad A = 0 \qquad \text{(unoccupied states)}. \tag{II.185f}$$

29 (45–122/1977)

A few remarks about these prescriptions are pertinent here. First, the energy spectrum (II.185f) of the unoccupied ('particle') states consists, according to (II.185b), of the kinetic energy $[\hbar^2k^2/(2M)]$ only; there is no potential energy $\mathcal{V}(k^2)$ for these states. Second, the spectrum of the occupied ('hole') states gives rise to negative energies (i.e., bound states) up to the fermi momentum k_F, the usual value for which is about 1.35 fm^{-1}. As a result, *at* the fermi momentum, the hole state is very much below the particle state, the latter having a positive energy as per (II.185f). Thus, there is an energy gap in the single-particle energy spectrum at the fermi surface.

As a step towards solving the Bethe-Goldstone equation (II.182), we first separate the centre-of-mass motion. In the present case of infinite nuclear matter, we have, for the unperturbed state,

$$\Phi_{\mathbf{k}_1,\mathbf{k}_2}(\mathbf{r}_1,\mathbf{r}_2)=\frac{1}{\sqrt{\Omega}}\exp(i\mathbf{k}_1\cdot\mathbf{r}_1)\frac{1}{\sqrt{\Omega}}\exp(i\mathbf{k}_2\cdot\mathbf{r}_2)$$

or

$$\Phi_{\mathbf{K},\mathbf{k}}(\mathbf{R},\mathbf{r})=\frac{1}{\sqrt{\Omega}}\exp(i\mathbf{K}\cdot\mathbf{R})\frac{1}{\sqrt{\Omega}}\exp(i\mathbf{k}\cdot\mathbf{r}), \tag{II.186a}$$

where

$$\begin{aligned}\mathbf{R}&=\tfrac{1}{2}(\mathbf{r}_1+\mathbf{r}_2), &\quad \mathbf{r}&=\mathbf{r}_1-\mathbf{r}_2,\\ \mathbf{K}&=\mathbf{k}_1+\mathbf{k}_2, &\quad \mathbf{k}&=\tfrac{1}{2}(\mathbf{k}_1-\mathbf{k}_2).\end{aligned} \tag{II.186b}$$

For simplicity, we have omitted the spin-isospin part of the wavefunction, and have avoided an explicit antisymmetrization under the exchange of the two particles. These complications are introduced at a later stage in this section [see (II.196) and (II.197a)]. As far as the separation of the centre-of-mass motion is concerned, these omissions do not alter the results in any way. According to the discussion following the expression (II.183), the centre-of-mass part $(1/\sqrt{\Omega})\exp(i\mathbf{K}\cdot\mathbf{R})$ in (II.186) remains unmodified in the exact wavefunction $\Psi_{\mathbf{k}_1,\mathbf{k}_2}$. Thus, we write

$$\Psi_{\mathbf{k}_1,\mathbf{k}_2}(\mathbf{r}_1,\mathbf{r}_2)\equiv\Psi_{\mathbf{K},\mathbf{k}}(\mathbf{R},\mathbf{r})=\frac{1}{\sqrt{\Omega}}\exp(i\mathbf{K}\cdot\mathbf{R})\psi_{\mathbf{k}}(\mathbf{r}), \tag{II.187}$$

where the correlated wavefunction $\psi_{\mathbf{k}}(\mathbf{r})$ for the relative motion has to be obtained by solving the Bethe-Goldstone equation. Before writing the resultant equation, we note that $w(=w_0)$ for the on-energy shell t-matrix is a *negative* quantity, and, according to Bethe's prescription [(II.185b) and (II.185f)],

$$\begin{aligned}\epsilon(k_1')+\epsilon(k_2')&=\frac{\hbar^2}{2M}(k_1'^2+k_2'^2)\\ &=\frac{\hbar^2}{2M}(\tfrac{1}{2}K^2+2k'^2)=\frac{\hbar^2K^2}{4M}+\frac{\hbar^2k'^2}{M}.\end{aligned} \tag{II.188a}$$

Using these results, we obtain

$$e(w_0)=w_0-\frac{\hbar^2K^2}{4M}-\frac{\hbar^2k'^2}{M}=-\frac{\hbar^2}{M}(\gamma^2+k'^2), \tag{II.188b}$$

where

$$-\frac{\hbar^2}{M}\gamma^2=w_0-\frac{\hbar^2K^2}{4M}$$

is a given *negative* quantity (because w_0 is *negative*). We have here used the fact that the centre-of-mass momentum $(\mathbf{k}_1' + \mathbf{k}_2')$ for the intermediate state is the same as $\mathbf{K}\ (=\mathbf{k}_1 + \mathbf{k}_2)$ for the initial state. Both w_0 and K are given numbers at the time of tackling the Bethe-Goldstone equation.

Using (II.188b), (II.187), (II.186), and (II.184) in (II.182), we have

$$\frac{1}{\sqrt{\Omega}} \exp(i\mathbf{K}\cdot\mathbf{R})\psi_{\mathbf{k}}(\mathbf{r}) = \frac{1}{\sqrt{\Omega}} \exp(i\mathbf{K}\cdot\mathbf{R})\frac{1}{\sqrt{\Omega}} \exp(i\mathbf{k}\cdot\mathbf{r}) - \left[\frac{\Omega}{(2\pi)^3}\right]^2\frac{M}{\hbar^2}$$
$$\times \int d^3K' \int_{(k_1', k_2' > k_F)} d^3k' \langle \mathbf{R}, \mathbf{r} \mid \mathbf{K}', \mathbf{k}'\rangle\langle \mathbf{K}', \mathbf{k}'| V |\mathbf{K}, \psi_{\mathbf{k}}\rangle(\gamma^2 + k'^2)^{-1}. \tag{II.189}$$

The coordinate representation of $|\mathbf{K}, \psi_{\mathbf{k}}\rangle$ with respect to $\langle\mathbf{R}, \mathbf{r}|$ is, by definition, the wavefunction on the left-hand side of (II.189). The integration over $d^3k_1'\, d^3k_2'$ has been rewritten in (II.189) in terms of $d^3K'\, d^3k'$, where $\mathbf{K}'$ and $\mathbf{k}'$ are related to $\mathbf{k}_1'$, $\mathbf{k}_2'$ in the manner of (II.186b). In the matrix element of V, occurring in (II.189), only $\langle\mathbf{k}'|$ and $|\psi_{\mathbf{k}}\rangle$ participate, and the centre-of-mass states in the matrix element give rise to the simple orthogonality integral

$$\langle \mathbf{K}' \mid \mathbf{K}\rangle = \frac{1}{\Omega} \int d^3R \exp(-i\mathbf{K}'\cdot\mathbf{R}) \exp(i\mathbf{K}\cdot\mathbf{R})$$
$$= \frac{(2\pi)^3}{\Omega}\delta(\mathbf{K}, \mathbf{K}'),$$

where δ is the Dirac delta function. Integration over d^3K', occurring in (II.189), can then be trivially carried out, which converts $\langle\mathbf{R}, \mathbf{r} \mid \mathbf{K}', \mathbf{k}'\rangle$ into $\langle\mathbf{R}, \mathbf{r} \mid \mathbf{K}, \mathbf{k}'\rangle$, i.e., $(1/\sqrt{\Omega}) \exp(i\mathbf{K}\cdot\mathbf{R}) \times (1/\sqrt{\Omega}) \exp(i\mathbf{k}'\cdot\mathbf{r})$. Using all these results, and cancelling out the common factor $(1/\sqrt{\Omega}) \exp(i\mathbf{K}\cdot\mathbf{R})$, we finally obtain, from (II.189), the desired equation

$$\psi_{\mathbf{k}}(\mathbf{r}) = \frac{1}{\sqrt{\Omega}} \exp(i\mathbf{k}\cdot\mathbf{r}) - \frac{\Omega}{(2\pi)^3} \int_{(k_1', k_2' > k_F)} d^3k' \frac{1}{\sqrt{\Omega}} \exp(i\mathbf{k}'\cdot\mathbf{r})(\gamma^2 + k'^2)^{-1}\langle\mathbf{k}'| \frac{M}{\hbar^2}V|\psi_{\mathbf{k}}\rangle.$$

Writing the matrix element of V in the obvious integral form, namely,

$$\langle\mathbf{k}'| V |\psi_{\mathbf{k}}\rangle = \frac{1}{\sqrt{\Omega}} \int d^3r' \exp(-i\mathbf{k}'\cdot\mathbf{r}')V(\mathbf{r}')\psi_{\mathbf{k}}(\mathbf{r}'),$$

we rewrite the foregoing equation for $\psi_{\mathbf{k}}(\mathbf{r})$ as

$$\psi_{\mathbf{k}}(\mathbf{r}) = \frac{1}{\sqrt{\Omega}} \exp(i\mathbf{k}\cdot\mathbf{r}) - \int d^3r'\, G(\mathbf{r}, \mathbf{r}')v(\mathbf{r}')\psi_{\mathbf{k}}(\mathbf{r}'), \tag{II.190a}$$

where Green's function $G(\mathbf{r}, \mathbf{r}')$ is given by

$$G(\mathbf{r}, \mathbf{r}') = -\frac{1}{(2\pi)^3} \int_{(k_1', k_2' > k_F)} d^3k' \frac{\exp[i\mathbf{k}'\cdot(\mathbf{r} - \mathbf{r}')]}{\gamma^2 + k'^2} \tag{II.190b}$$

and

$$v(\mathbf{r}') = -\frac{M}{\hbar^2}V(\mathbf{r}').$$

It is worthwhile to remember that both (II.190a) and (II.190b) have an implicit dependence on $\mathbf{K}\ (=\mathbf{k}_1' + \mathbf{k}_2')$, although this is not explicitly shown in the notation here. The reader should also observe that we have written (II.190a) for the correlated wavefunction corresponding to the *normalized* unperturbed wavefunction; this equation could have been written also with the omission of the normalization factor $1/\sqrt{\Omega}$, in which case $\psi_{\mathbf{k}}(\mathbf{r})$ would be interpreted as the correlated wavefunction corresponding to the *unnormalized* unperturbed wavefunction $\exp(i\mathbf{k}\cdot\mathbf{r})$.

It is now interesting to compare (II.190) with the equations that give the exact wavefunction and Green's function (see Schiff[3]) for the case of two-body scattering *when the two particles do not have any other particles to interact with.* The Bethe-Goldstone wavefunction (II.190a) and the corresponding Green's function (II.190b) describe a pair of particles *embedded in the nuclear many-body medium.* Although the Bethe-Goldstone wavefunction and Green's function are formally very similar to the scattering wavefunction and the corresponding Green's function, there are very important fundamental differences between the two which will now be described. First, the momentum integration in the scattering Green's function is unrestricted, whereas that in (II.190b) has the restrictions indicated under the integral sign. Second, the integrand of the scattering Green's function has a singularity at $k' = \pm k$, which are *on* the path of the momentum integration; it is the prescription for choosing the integration contour near the singular point $k' = +k$ that leads to the outgoing spherical wave form for the scattering Green's function, and eventually the same type of scattered spherical wave in the scattering wavefunction for $r \to \infty$. In the Bethe-Goldstone case, on the other hand, this important singularity at the point $k' = k$ is absent; the integrand of (II.190b) has poles at the purely imaginary points $k' = \pm i\gamma$, which are *not* on the integration contour along the real axis. Thus, the asymptotic behaviour of the Bethe-Goldstone wavefunction for $r \to \infty$ is quite different from the spherical outgoing wave-scattering solution. As a matter of fact, by *ignoring* the restriction on the $\mathbf{k}'$-integration in (II.190b), we can immediately obtain an *approximate* expression for the Bethe-Goldstone Green's function by the contour integration procedure. The result can be easily verified to be

$$\bar{G}(\mathbf{r}, \mathbf{r}') = -\frac{1}{4\pi}\frac{\exp(-\gamma|\mathbf{r} - \mathbf{r}'|)}{|\mathbf{r} - \mathbf{r}'|}. \tag{II.191}$$

To remind the reader that this Green's function is an approximation to (II.190b), we have used the overhead bar. When this approximate Green's function is used in (II.190a), we immediately notice that, for $r \to \infty$, the second term of (II.190a) decays as $e^{-\gamma r}$, and we obtain the somewhat surprising result

$$\psi_{\mathbf{k}}(\mathbf{r}) = \frac{1}{\sqrt{\Omega}}\exp(i\mathbf{k}\cdot\mathbf{r}), \qquad r \to \infty. \tag{II.192}$$

In other words, the exact Bethe-Goldstone wavefunction, instead of behaving as scattered waves with phase shifts, acts exactly like the unperturbed plane wave at large values of r. This property is called the 'healing' of the exact wavefunction (see Gomes et al[18]). The exact wavefunction departs from the unperturbed wavefunction only at small values of r; it acquires, as it were, a 'wound' at small r which 'heals up' beyond $r \approx 1$ fm, where the exact wavefunction and the unperturbed wavefunction become identical. The healing property which has been established here with the approximate Green's function (II.191) has been found to be true even in more accurate calculations of the Bethe-Goldstone wavefunction.

Let us now return to the exact equation (II.190b) and study the difficulties involved in evaluating it. The restriction on the integral, as we know, has arisen from the Pauli exclusion principle, which required us to take the intermediate particle states $\mathbf{k}_1'$ and $\mathbf{k}_2'$ (satisfying the total momentum conservation, $\mathbf{k}_1' + \mathbf{k}_2' = \mathbf{K} = \mathbf{k}_1 + \mathbf{k}_2$) outside the fermi sphere of radius k_F. It has been shown in Section 18 that, to satisfy this requirement, we have to break up the integration over dk' into the ranges shown in (II.122) and, in each range, carry out the integration over the angles of $\mathbf{k}'$ within the appropriate limits of $\cos\theta$. Not only do these integrations make the final results rather complicated, but, what is more, they point to the fact that Green's function depends on the magnitude as well as the *direction* of the vector $\mathbf{K}$. Thus, the Bethe-Goldstone wavefunction would have to be computed for many given *angles* and magnitudes of $\mathbf{K}$. To avoid this complicated calculation, it is usual to introduce an approximation that removes the dependence of Green's function on the *direction* of $\mathbf{K}$. This approximation is called the *angle-averaged approximation* for the Pauli operator Q. It entails doing the angle-integration in (II.190b) over the *full* range by introducing into the integrand the factor $Q(k', K)$ given by

$$Q(k', K) = \begin{cases} 0 & \text{if } 0 \leqslant k' < (k_F^2 - \frac{1}{4}K^2)^{1/2} \\ \dfrac{(\frac{1}{4}K^2 + k'^2) - k_F^2}{Kk'} & \text{if } (k_F^2 - \frac{1}{4}K^2)^{1/2} \leqslant k' < (k_F + \frac{1}{2}K) \\ 1 & \text{if } k' \geqslant k_F + \frac{1}{2}K \end{cases} \qquad \text{(II.193)}$$

It is clear that the quantity $Q(k', K)$ is obtained by carrying out the integrations over $d(\cos\theta)\, d\phi$ for the various domains of k', shown in (II.122), within the appropriate limits of $\cos\theta$ and ϕ $(0 \to 2\pi)$ and then dividing the result by 4π. In this sense, (II.193) indeed expresses the restriction due to the Pauli principle in an 'angle-averaged' manner. The approximate Green's function in this case is given by

$$\begin{aligned} G_{av}(\mathbf{r}, \mathbf{r}') &= -\frac{1}{(2\pi)^3} \int d^3k' \frac{\exp[i\mathbf{k}'\cdot(\mathbf{r} - \mathbf{r}')]}{\gamma^2 + k'^2} Q(k', K) \\ &= -\frac{1}{2\pi^2} \frac{1}{|\mathbf{r} - \mathbf{r}'|} \int_0^\infty dk'\, Q(k', K) \frac{k' \sin k'|\mathbf{r} - \mathbf{r}'|}{\gamma^2 + k'^2}. \end{aligned} \qquad \text{(II.194)}$$

The corresponding Bethe-Goldstone wavefunction can be obtained from (II.190a). Several calculations in this approximation have been done by Brueckner and his collaborators[19]. These authors, however, treated the particle energies somewhat differently. Instead of using only the kinetic energies in accordance with Bethe's prescription, Brueckner and his colleagues used self-consistent particle energies, such as (II.185d), in which $(\mathbf{k}, \sigma, \tau)$ is now taken to denote a particle state. However, in evaluating the second term of (II.185d), we have to be careful. Since t depends on w, we have to prescribe a value of w that is suitable when a particle above the fermi surface is interacting with an occupied fermi-sea particle. Since particles have to be excited above the fermi sea in *pairs*, due to the requirement of total momentum conservation, at the stage when a particle interacts with an occupied fermi-sea nucleon, there will always be another spectator particle and its hole in the fermi sea. Thus, as already remarked, there is a case where the parameter w for the t cannot be merely the sum of the starting energies of the two interacting nucleons; the excitation energy due to the spectator particle will also have to be reckoned. This is the so-called off-energy shell effect, and Brueckner and his collaborators considered it in giving a self-consistent definition of the potential energy of particle states. Bethe has discussed this point in detail (see Rajaraman and Bethe[17]) and has stated why the potential energy for the particle states should be dropped.

In the preceding discussions, we have used an integral equation form for the Bethe-

Goldstone equation (II.190). It is possible to very easily convert this equation into an integro-differential equation form. With this aim in view, we first note that $-(\gamma^2 + k'^2)^{-1}$ in Green's function (II.190b) could be produced by the operation of $(\nabla^2 - \gamma^2)^{-1}$ on the integral $\int d^3k' \exp(i\mathbf{k}'\cdot\mathbf{r}) \exp(-i\mathbf{k}'\cdot\mathbf{r}')$ with specified restrictions on the momentum integration. This momentum integral is actually the coordinate space representation of the Pauli operator Q, and is shown as

$$\frac{1}{(2\pi)^3}\int_{(k_1', k_2' > k_F)} d^3k' \exp(i\mathbf{k}'\cdot\mathbf{r}) \exp(-i\mathbf{k}'\cdot\mathbf{r}') = \frac{\Omega}{(2\pi)^3}\int_{(k_1', k_2' > k_F)} d^3k' \langle \mathbf{r} \mid \mathbf{k}'\rangle\langle \mathbf{k}' \mid \mathbf{r}'\rangle$$
$$= \langle \mathbf{r}| \sum_{\substack{\mathbf{k}' \\ (k_1', k_2' > k_F)}} |\mathbf{k}'\rangle\langle \mathbf{k}' \mid \mathbf{r}'\rangle = \langle \mathbf{r}| Q |\mathbf{r}'\rangle.$$

Therefore, we can rewrite (II.190b) as

$$G(\mathbf{r}, \mathbf{r}') = (\nabla^2 - \gamma^2)^{-1}\langle \mathbf{r}| Q |\mathbf{r}'\rangle,$$

and hence (II.190a) yields [with the notation $\phi_{\mathbf{k}}(\mathbf{r}) = (1/\sqrt{\Omega}) \exp(i\mathbf{k}\cdot\mathbf{r})$]

$$(\nabla^2 - \gamma^2)[\phi_{\mathbf{k}}(\mathbf{r}) - \psi_{\mathbf{k}}(\mathbf{r})] = \int d^3r' \langle \mathbf{r}| Q |\mathbf{r}'\rangle\langle \mathbf{r}'| v |\psi_{\mathbf{k}}\rangle$$
$$= \langle \mathbf{r} \mid Qv\psi_{\mathbf{k}}\rangle. \tag{II.195a}$$

This is the promised integro-differential equation, with the differential operator ∇^2 on the left-hand side and the integral over $\mathbf{r}'$ on the right-hand side (Q being the integral operator). It is advantageous for our subsequent study to introduce the 'defect wavefunction' $\chi_{\mathbf{k}}(\mathbf{r})$ defined by

$$\chi_{\mathbf{k}}(\mathbf{r}) = \phi_{\mathbf{k}}(\mathbf{r}) - \psi_{\mathbf{k}}(\mathbf{r}) \tag{II.195b}$$

such that (II.195a) becomes

$$(\nabla^2 - \gamma^2)\chi_{\mathbf{k}}(\mathbf{r}) = \langle \mathbf{r} \mid Qv\psi_{\mathbf{k}}\rangle. \tag{II.195c}$$

This integro-differential equation reduces to a pure differential equation if we now make the assumption of putting $Q = \mathbb{1}$. This approximation, as is clear, amounts to ignoring the restriction on the integration over $\mathbf{k}'$ (due to the Pauli principle) in Green's function. The form of Green's function in this approximation is given by (II.191). The differential equation form will, however, be more convenient to work with. Putting $Q = \mathbb{1}$ in (II.195c), we obtain

$$(\nabla^2 - \gamma^2)\chi_{\mathbf{k}}(\mathbf{r}) = v(\mathbf{r})\psi_{\mathbf{k}}(\mathbf{r}) \tag{II.195d}$$

or

$$[\nabla^2 - \gamma^2 + v(\mathbf{r})]\psi_{\mathbf{k}}(\mathbf{r}) = -(k^2 + \gamma^2)\phi_{\mathbf{k}}(\mathbf{r}). \tag{II.195e}$$

Although $\psi_{\mathbf{k}}(\mathbf{r})$, calculated from this equation, is approximate, we have not introduced a new notation for it for the sake of simplicity. The goodness of the approximation $Q = \mathbb{1}$ involved in this approximate Bethe-Goldstone equation is discussed later in this section.

Before solving the differential equation (II.195d) or (II.195e), we choose to introduce the complications due to the spin-isospin part of the pair wavefunction and its antisymmetrization. We have already sketched this procedure after the expression (II.183). The unperturbed two-particle *spatial* wavefunction can be either symmetrized or antisymmetrized. Omitting the centre-of-mass part, $(1/\sqrt{\Omega}) \exp(i\mathbf{K}\cdot\mathbf{R})$, of the two-nucleon wavefunction, we obtain respectively the symmetric and antisymmetric relative wavefunctions

$$\phi_{\mathbf{k}}^{(s)}(\mathbf{r}) = \frac{1}{\sqrt{2\Omega}}[\exp(i\mathbf{k}\cdot\mathbf{r}) + \exp(-i\mathbf{k}\cdot\mathbf{r})], \tag{II.196a}$$

$$\phi_{\mathbf{k}}^{(a)}(\mathbf{r}) = \frac{1}{\sqrt{2\Omega}}[\exp(i\mathbf{k}\cdot\mathbf{r}) - \exp(-i\mathbf{k}\cdot\mathbf{r})]. \tag{II.196b}$$

Using the familiar partial wave expansions of $\exp(i\mathbf{k}\cdot\mathbf{r})$ and $\exp(-i\mathbf{k}\cdot\mathbf{r})$ from Appendix C (Section I), we are easily convinced that (II.196a) and (II.196b) consist of only even l and odd l, respectively. The combination (S, T) of the spin-isospin functions with which these two kinds of spatial wavefunctions are to be multiplied is already well-known to the reader (see Chapter I). For the sake of brevity, we omit the superscripts (s) and (a) from $\phi_{\mathbf{k}}(\mathbf{r})$, and write the summation over l in the partial wave summations with a prime, reminding the reader thereby that this summation goes over l of the appropriate parity in keeping with the S, T quantum numbers of the spin-isospin functions. Thus, the complete unperturbed wavefunction in its antisymmetrized form is written as

$$\phi_{\mathbf{k}}(\mathbf{r})\chi_{M_S}^S\rho_{M_T}^T = \frac{1}{\sqrt{\Omega}}\sum_l{}' i^l\sqrt{8\pi(2l+1)}(kr)^{-1}f_l(r)Y_0^l(\theta)\chi_{M_S}^S\rho_{M_T}^T. \tag{II.197a}$$

Here χ^S and ρ^T are the spin and isospin functions, and

$$f_l(r) = krj_l(kr). \tag{II.197b}$$

If we had a central potential alone, then this form would have been convenient to work with. The noncentral potential does not conserve the quantum numbers M_S and l, and hence we shall couple Y_0^l and $\chi_{M_S}^S$ in (II.197a) and work with the angular momentum coupled expression of the unperturbed wavefunction

$$\frac{1}{\sqrt{\Omega}}\sum_l{}' i^l\sqrt{8\pi(2l+1)}(kr)^{-1}f_l(r)\sum_J\begin{bmatrix} l & S & J \\ 0 & M_S & M_S\end{bmatrix}|lSJM_S\rangle\rho_{M_T}^T, \tag{II.198}$$

where $|lSJM_S\rangle$ is the spin-angle function defined by (I.80b). We here have $\sqrt{8\pi}$, instead of the usual $\sqrt{4\pi}$, because of the *two* terms in (II.196) and the normalization factor $1/\sqrt{2}$. In the exact two-body wavefunction, all the quantum numbers (with the exception of l) of each of these partial waves are conserved; the radial wavefunction $f_l(r)$ will also change to the exact radial function, say, $u_{l'}^{(l)}(r)$. In this notation, the superscript denotes the partial wave of the unperturbed wavefunction, and the subscript represents the orbital angular momentum of the corresponding exact wavefunction. From the discussions in Chapter I, we know that the possibility of $l' \neq l$ arises only in the case of the coupled spin-triplet states having $l = J \pm 1$. In such a case, commencing with the unperturbed $(l = J + 1)$- *or* $(l = J - 1)$-wave, we reach, in the exact wavefunction, *both* the $(l' = J \pm 1)$-waves. The cases of the spin-singlet state $(l = J)$ and the uncoupled triplet state having $l = J$ are much simpler. In these cases, labelling the exact radial function with the single subscript l is quite adequate. For simplicity of presentation, we shall first discuss the uncoupled states in detail, and then point out the modifications in the case of the coupled states.

For the uncoupled case, the exact radial wavefunction with the quantum number $l\ (=J)$ is defined with the same extraneous numerical factors as in (II.198) as

$$\begin{bmatrix} l & S & l \\ 0 & M_S & M_S\end{bmatrix} i^l\sqrt{8\pi(2l+1)}(kr)^{-1}u_l(r)|lSlM_S\rangle\rho_{M_T}^T. \tag{II.199}$$

The difference between $f_l(r)$ and $u_l(r)$ is called $\zeta_l(r)$, i.e.,

$$\zeta_l(r) = f_l(r) - u_l(r).$$

The l-th partial wave of the uncoupled type in the expansion of the defect wavefunction $\chi_{\mathbf{k}}(\mathbf{r})$ of (II.195b) is therefore written as

$$\begin{bmatrix} l & S & l \\ 0 & M_S & M_S \end{bmatrix} i^l\sqrt{8\pi(2l+1)}(kr)^{-1}\zeta_l(r)|lSlM_S\rangle\rho^T_{M_T}. \tag{II.200}$$

It is now quite straightforward to use the partial wave forms (II.198)–(II.200) in the differential equations (II.195d) and (II.195e) and obtain the corresponding radial equations in the manner of Section 5. We then have

$$\left[\frac{d^2}{dr^2} - \frac{l(l+1)}{r^2} - \gamma^2 + v_{lST}(r)\right]\zeta_l(r) = v_{lST}(r)f_l(r) \tag{II.201a}$$

or

$$\left[\frac{d^2}{dr^2} - \frac{l(l+1)}{r^2} - \gamma^2 + v_{lST}(r)\right]u_l(r) = -(k^2+\gamma^2)f_l(r). \tag{II.201b}$$

Here $v_{lST}(r)$ is the diagonal matrix element of v with respect to the spin angle and isospin function $|lSlM_S\rangle\rho^T_{M_T}$. (For these matrix elements, see Section 4.)

(II.201b) is a *nonhomogeneous* second-order differential equation for the unknown function $u_l(r)$. This is in contrast to the homogeneous equation obtained in Section 5; as a matter of fact, the reader will notice that, except for the nonhomogeneous term $f_l(r)$ on the right-hand side, the left-hand side of this equation is formally identical to the bound-state ($k^2 = -\gamma^2$) two-body equation (I.44). We need two boundary conditions to solve this second-order equation; and we must remember now that the nonhomogeneous term here does not allow us to treat the overall normalization of $u_l(r)$ as arbitrary. One appropriate boundary condition is, of course, given by

$$u_l(r) = 0 \quad \begin{cases} \text{at } r = 0 \text{ (if there is no hard core)} \\ \text{at } r = c \text{ (if there is a hard core of radius } c) \end{cases}$$

The other boundary condition is defined by the healing property, already mentioned, and also apparent from (II.201a). At $r \to \infty$, $v_{lST}(r) \to 0$, and hence this equation demands $\zeta_l(r) \to 0$ as a damped exponential $e^{-\gamma r}$. Therefore, the second boundary condition is given by $u_l(r) = f_l(r)$ at large r, where $v_{lST}(r)$ is negligible. The forward and backward numerical integration formulas for the solution of (II.201b) can be worked out in the manner of Section 5; these formulas will have some extra terms resulting from the nonhomogeneous part of the differential equation. A practical procedure for computing the solution step by step is as follows. (i) Apply the backward integration formula at a fairly large r, taking $u_l(r) = f_l(r)$ and giving an *arbitrary* value to $u_l(r-\Delta)$; integrate towards the origin and stop at a suitable point R to the right of the core; call this solution $u_l^{(e)}$. (ii) Repeat the procedure in (i) with the homogeneous equation, taking $u_l(r)$ and $u_l(r-\Delta)$ at the starting points from the asymptotic form $h_l^{(1)}(i\gamma r)$ of the homogeneous solution; call this solution $\phi_l^{(e)}$. The solution to the right of R (which will eventually be the matching radius) can then be written as $u_l^{(e)} + x\phi_l^{(e)}$, where x is an unknown constant for the present; notice that this form still guarantees the boundary condition $u_l(r) = f_l(r)$ at the chosen value of the starting r. (iii) Use the forward integration formula at $r = c$ or $r = 0$ (depending on whether or not there is a hard core), taking the boundary value zero of u_l at that point and an *arbitrary* value at the next point to the right; integrate up to the point R step by step; call this solution $u_l^{(i)}$. Obtain, similarly, the solution of the homogeneous equation starting at the internal boundary with the correct boundary condition and an arbitrary value at the next point; integrate up to R and denote this solution by $\phi_l^{(i)}$. The general form of the internal solution is then $u_l^{(i)} + y\phi_l^{(i)}$, where y is an unknown constant to be presently determined. Matching the values and derivatives of the inner and outer solutions at R, we obtain two equations for the determination of x and y.

To calculate the contribution of any of the uncoupled states to the matrix elements of t, we recall that we have to write the unperturbed state on the left-hand side, the potential V in the middle, and the exact state on the right-hand side. Moreover, if we want a *diagonal* matrix element, then the exact state on the right must correspond to the unperturbed state on the left. This means that, along with the complex conjugate of the l-wave part of (II.198), we take the l-wave part of (II.199), put V in the middle, and evaluate the matrix element. Finally, adding the contributions of the various partial waves, we obtain

$$8\pi \sum_l{}' (2l+1)k^{-2}\begin{bmatrix} l & S & l \\ 0 & M_S & M_S \end{bmatrix}^2 \int_0^\infty dr\, f_l(r) V_{lST}(r) u_l(r). \tag{II.202}$$

Here $V_{lST}(r)$ is the diagonal matrix element of the two-body potential V [and *not* v which is equal to $-(M/\hbar^2)V$], mentioned after (II.201b). The radial integral here has to be broken up into a core part and an outer part as

$$\int_0^\infty dr\, f_l(r) V_{lST}(r) u_l(r) = \int_c^\infty dr\, f_l(r) V_{lST}(r) u_l(r) + \int_0^c dr\, f_l(r) V_{lST}(r) u_l(r). \tag{II.203a}$$

The first integral can be easily evaluated as soon as $u_l(r)$ has been obtained by solving (II.201b). Inside the hard core, $u_l(r) = 0$, but $V_{lST}(r) \to \infty$, and usually the second integral works out to a finite value which is, however, quite tricky to obtain. We now proceed to demonstrate the procedure.

Multiplying (II.201a) from the left by $f_l(r)$ and integrating from $0 \to c$, we get

$$-\frac{M}{\hbar^2}\int_0^c dr\, f_l(r) V_{lST}(r) u_l(r) = \int_0^c dr\, f_l(r)\left[\frac{d^2}{dr^2} - \frac{l(l+1)}{r^2} - \gamma^2\right]\zeta_l(r).$$

The differential equation for the l-th partial wave $f_l(r)$ of the plane wave is given by

$$\left[\frac{d^2}{dr^2} - \frac{l(l+1)}{r^2} + k^2\right] f_l(r) = 0.$$

Multiplying this equation by $\zeta_l(r)$ from the left, integrating between $0 \to c$, and then subtracting from the preceding equation, we easily obtain

$$\frac{M}{\hbar^2}\int_0^c dr\, f_l(r) V_{lST}(r) u_l(r) = (k^2+\gamma^2)\int_0^c f_l(r)\zeta_l(r)\, dr + (\zeta_l f_l' - f_l \zeta_l')_0^c$$

$$= (k^2+\gamma^2)\int_0^c f_l^2(r)\, dr + f_l(c) u_l'(c). \tag{II.203b}$$

In the final step, we have made use of the fact that $\zeta_l(r)\,[\equiv f_l(r) - u_l(r)]$ is always equal to $f_l(r)$ inside and on the boundary of the core because the exact wavefunction $u_l(r) = 0$ at all such points. However, in the term involving the derivatives f_l', ζ_l', we have exercised caution. There is no contribution to this term from $r = 0$ because $f_l = krj_l(kr)$ is zero at this point. On the other hand, at $r = c$, $\zeta_l(c) = f_l(c)$, but $\zeta_l'(c) \neq f_l'(c)$. This is because the function u_l has a disconti-

30 (45–122/1977)

nuity in its derivative at $r = c$; everywhere on the inner side of the core, $u_l = 0$, and hence $u_l'(c - \epsilon) = 0$, ϵ being an infinitesimal quantity, whereas, on the outer edge of the core, the exact wavefunction usually starts with a finite value of the derivative, making $u_l'(c + \epsilon) \neq 0$. In view of this fact, $\zeta_l'(c + \epsilon) = f_l'(c + \epsilon) - u_l'(c + \epsilon)$, and this explains the second term of (II.203b), which is called the core-boundary term for obvious reasons. The first term of the same expression is then the core-volume term. The expressions (II.203) complete the evaluation of (II.202).

We next discuss the coupled ($l = J \pm 1$)-states of triplet spin. In this case, starting with the unperturbed l-wave ($J + 1$ or $J - 1$) of (II.198), we obtain the corresponding exact wavefunction as a sum over the two coupled orbital states. Therefore, the expression replacing (II.199) is given by

$$\sum_{l'=|J-1|}^{J+1} i^{l'}\sqrt{8\pi(2l+1)}\begin{bmatrix} l & S & J \\ 0 & M_S & M_S \end{bmatrix}(kr)^{-1}u_{l'}^{(l)}(r)|l'SJM_S\rangle \rho_{M_T}^{T}. \tag{II.204}$$

From (II.195e), we then obtain (in the manner of Section 5B) the coupled equations

$$[\frac{d^2}{dr^2} - \frac{l(l+1)}{r^2} - \gamma^2 + v_{llSJT}(r)]u_l^{(l)}(r) + v_{l'lSJT}(r)u_{l'}^{(l)}(r) = -(k^2 + \gamma^2)f_l(r), \tag{II.205a}$$

$$[\frac{d^2}{dr^2} - \frac{l'(l'+1)}{r^2} - \gamma^2 + v_{l'l'SJT}(r)]u_{l'}^{(l)}(r) + v_{ll'SJT}(r)u_l^{(l)}(r) = 0, \tag{II.205b}$$

where l' is the wave coupled to the l-wave of the given J. $v_{l_1l_2SJT}$ denotes the matrix element of v connecting the state $|l_1SJM\rangle\rho_{M_T}^T$ with $|l_2SJM\rangle\rho_{M_T}^T$. The unperturbed radial wavefunction in these coupled equations can be considered a two-component vector $\begin{pmatrix} f_l \\ 0 \end{pmatrix}$ and the corresponding exact wavefunction is $\begin{pmatrix} u_l^{(l)} \\ u_{l'}^{(l)} \end{pmatrix}$. As $r \to \infty$ and $v(r) \to 0$, (II.205a) reduces to

$$[\frac{d^2}{dr^2} - \frac{l(l+1)}{r^2} - \gamma^2]\zeta_l^{(l)}(r) = 0$$

or

$$\zeta_l^{(l)}(r) = h_l^{(1)}(i\gamma r).$$

Here $\zeta_l^{(l)}(r)$ is, by definition, $f_l(r) - u_l^{(l)}(r)$, and since $h_l^{(1)}(i\gamma r)$, according to Appendix C (Section I), goes to zero at $r \to \infty$, we conclude that $u_l^{(l)}(r)$ heals up to $f_l(r)$. In the same manner, (II.205b) tells us that $u_{l'}^{(l)}(r) = h_{l'}^{(1)}(i\gamma r) \to 0$ for $r \to \infty$. Therefore, at large r, the exact wavefunction $\begin{pmatrix} u_l^{(l)}(r) \\ u_{l'}^{(l)}(r) \end{pmatrix}$ indeed heals up to the unperturbed $\begin{pmatrix} f_l \\ 0 \end{pmatrix}$. At the inner boundary (i.e., $r = 0$ or $r = c$), the boundary value of the two radial functions is zero. The numerical integration procedure starting with the inner and outer boundary conditions is left as an exercise.

The contribution to the diagonal matrix element of t for the coupled states of a *given J* can be easily computed with the help of (II.204) and (II.198). We obtain

$$8\pi \sum_{l=|J-1|}^{J+1} \sum_{l'=|J-1|}^{J+1} \sum_{l''} \sqrt{(2l+1)(2l''+1)}k^{-2}\begin{bmatrix} l & S & J \\ 0 & M_S & M_S \end{bmatrix}\begin{bmatrix} l'' & S & J \\ 0 & M_S & M_S \end{bmatrix}$$
$$\times \int_0^\infty dr\, f_{l''}(r)V_{l''l'SJT}(r)u_{l'}^{(l)}(r). \tag{II.206}$$

We have here summed (II.204) over the two possible values of l to obtain the exact wavefunction corresponding to the two unperturbed waves. The unperturbed wave on the left-hand side of the t-matrix element has been written with the summation over l''. The two-body potential V in the middle is thus sandwiched between $\langle l''SJM_S|\rho^T_{M_T}$ and $|l'SJM_S\rangle\rho^T_{M_T}$. The resultant matrix element of V (which is independent of M_S and M_T) has been denoted by the subscripted $V(r)$ in (II.206). One of the Clebsch-Gordon coefficients has come from the exact wavefunction (II.204), whereas the other (having l'') has come from the unperturbed wave.

We now recall that, in computing the binding energy, we need to sum $\langle \mathbf{k}_1, \mathbf{k}_2; SM_STM_T| t |\mathbf{k}_1, \mathbf{k}_2; SM_STM_T\rangle$ over all the possible spin-isospin states (i.e., on the quantum numbers S, M_S, T, and M_T) and then over all the possible occupied pair states $\mathbf{k}_1, \mathbf{k}_2$. First, we consider the summations over S, M_S, T, and M_T. Since the various contributions to the t-matrix element, given by (II.202) and (II.206), are independent of M_T, the summation over this quantum number produces $(2T + 1)$. The summation over M_S can also be readily carried out in (II.202) and (II.206) from the general properties (AIV.3a) and (AIV.4) in Appendix A of the Clebsch-Gordon coefficients

$$\sum_{M_S} \begin{bmatrix} l & S & J \\ 0 & M_S & M_S \end{bmatrix} \begin{bmatrix} l'' & S & J \\ 0 & M_S & M_S \end{bmatrix} = \frac{[J]}{\sqrt{[l][l'']}} \sum_{M_S} \begin{bmatrix} J & S & l \\ -M_S & M_S & 0 \end{bmatrix} \begin{bmatrix} J & S & l'' \\ -M_S & M_S & 0 \end{bmatrix}$$
$$= \frac{[J]}{[l]} \delta_{ll''}.$$

In (II.202), this factor is unity because $l = l' = J$, whereas in (II.206) the summation has a very important simplifying effect: it removes the summation over l'', cancels the factor $\sqrt{(2l + 1)(2l' + 1)}$, and finally replaces (II.206) by

$$8\pi(2T + 1)(2J + 1)k^{-2} \sum_{l, l'} \int_0^\infty dr\, f_l(r) V_{ll'SJT}(r) u_{l'}^{(l)}(r). \tag{II.207a}$$

The factor $(2T + 1)$, resulting from the M_T-summation, has also been included here. For ready reference, we write down also the similar expression resulting from (II.202), which is

$$8\pi(2T + 1)(2J + 1)k^{-2} \int_0^\infty dr\, f_J(r) V_{JSJT}(r) u_J(r). \tag{II.207b}$$

The expressions (II.207) have to be summed over all the values of J, and, for each J, over all the possible values of S and T. While considering the corresponding values of l, l', the requirement of antisymmetry has to be borne in mind. Usually, the summation is carried over all the two-nucleon states beginning with $l = 0$ and going up to a certain maximum l $(= 3, 4)$; states beyond this l are rather unimportant in low-energy phenomena. The t-matrix element obtained after this summation has then to be summed over all the occupied pair states $\mathbf{k}_1$, $\mathbf{k}_2$ so as to obtain the binding energy of infinite nuclear matter. Since the whole calculation described so far has been done with a given value of K, and the diagonal t-matrix element, obtained through (II.207), depends on k and not on the direction of $\mathbf{k}$, the required pair summation can be done in the manner of Section 18.

($Q = \mathbb{1}$)-Approximation

The foregoing treatment of the t-matrix with pure differential equations is based on the neglect of the Pauli exclusion principle, i.e., the ($Q = \mathbb{1}$)-approximation. We now present a semi-quantitative argument, originally due to Bethe (see Rajaraman and Bethe[17]), in support of this approximation.

According to (II.182), the ($Q = \mathbb{1}$)-approximation will be good, provided V is such that $(1/e)V\psi$ *automatically* has small Fourier components inside the fermi sphere. From the definition of the defect wavefunction χ, we have

$$\chi = \phi - \psi = -\frac{1}{e}V\psi.$$

Therefore, we have to compute the Fourier transform of χ, and verify if it has small components for $k \leqslant k_F$.

The customary argument is based on ζ_0, i.e., the ($l = 0$)-component of χ in the case of a pure hard-core potential V of radius c, and no potential for $r > c$. Inside the core radius,

$$\zeta_0 = f_0(r) = \sin kr, \qquad r < c,$$

and outside the core, the equation determining $\zeta_0(r)$, namely, (II.201a), becomes

$$\left(\frac{d^2}{dr^2} - \gamma^2\right)\zeta_0(r) = 0,$$

which has the solution

$$\zeta_0(r) = Ae^{-\gamma r}, \qquad r \geqslant c.$$

From the continuity of $\zeta_0(r)$ at $r = c$, we have

$$A = \sin kc\, e^{\gamma c},$$

and hence

$$\zeta_0(r) = \sin kc\, e^{-\gamma(r-c)}, \qquad r \geqslant c.$$

This defect wavefunction, the unperturbed function $f_0(r)$, and the exact function $u_0(r)[=f_0(r) - \zeta_0(r)]$ are shown in Fig. II.19. The nearly triangular shape of $\zeta_0(r)$, with its vertex at $r = c$, and the known shape of $\sin kr$ immediately indicate that the Fourier transform of $\zeta_0(r)$, which

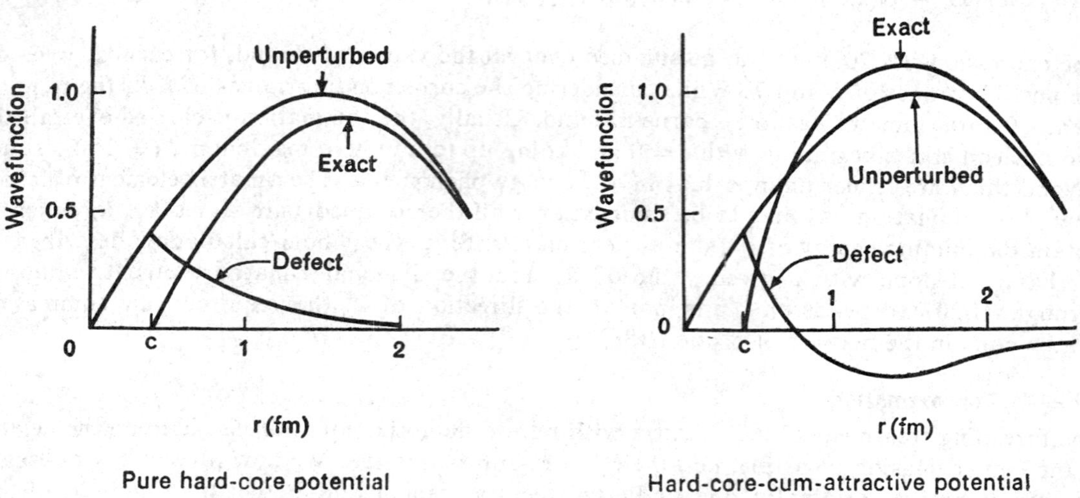

Fig. II.19 Unperturbed, exact, and defect wavefunctions for S-state.

for the S-state becomes the overlap integral

$$\int_0^\infty \sin kr\, \zeta_0(r)\, dr$$

between $\sin kr$ and $\zeta_0(r)$, will clearly have a large value only if the first maximum of $\sin kr$ coincides with the vertex of the 'triangle'. Thus, the Fourier transform of $\zeta_0(r)$ is large when

$$kc = \frac{\pi}{2} \quad \text{or} \quad k = \frac{\pi}{2c} \approx 3.9 \text{ fm}^{-1} \qquad \text{for } c = 0.4 \text{ fm}.$$

When $K = 0$, the relative momentum k is also the momentum of the individual particle. The value 3.9 fm^{-1} for k is just about three times the fermi momentum k_F ($\approx 1.36 \text{ fm}^{-1}$). Thus, the intermediate states in Q, which are important for $(1/e)V\psi$, lie very much beyond the fermi sphere in the case of a hard-core potential V.

This conclusion is valid also in the case of an attractive potential beyond the hard core. The attractive potential has the effect of pulling in the exact wavefunction $u_0(r)$ towards the core, as shown in Fig. II.19. As a result, it overshoots $f_0(r)$ beyond a certain value of r, and therefore $\zeta_0(r)$ becomes slightly negative in that region. However, the nearly triangular shape of $\zeta_0(r)$ is still preserved, and hence the foregoing argument on the overlap with $\sin kr$ still holds good.

Reference Spectrum Method versus Separation Method

The method for calculating the t-matrix with the help of the solutions of pure differential equations depends on two very important assumptions: (i) the quadratic expression for the energies of the particle states, which helped to get the ∇^2-operator; and (ii) the neglect of the Pauli exclusion principle. In his original work, Bethe suggested, in respect of (i), a spectrum of the type $A + Bk^2$ with nonvanishing values of both A and B for the particle states lying between 3 fm^{-1} and 5 fm^{-1} (the range of momenta within which the hard core scatters the particles according to the foregoing discussion). However, in later years, it was realized (see Rajaraman and Bethe[17]) that the free-particle spectrum $[\hbar^2k^2/(2M)]$ is the best choice. This is what we have done here. Bethe's work also contains, as already mentioned, a quadratic assumption on the hole spectrum with prescribed values of A and B. This assumption is, however, not so crucial in determining the equations because the energy w_0 of the occupied pair state appears in the equation as a parameter. However, we have already noted that the hole energies entering w_0 must be self-consistently reproduced from the calculated t-matrix. The quadratic prescription that Bethe has given ensures that the assumed hole energies are close to their correct self-consistent values. Bethe has called the assumed spectrum the "reference spectrum", and the method, as we have described it for the calculation of the t-matrix, is therefore known as the *reference spectrum method.* It is possible to obtain a correction to the t-matrix, calculated by the reference spectrum method, arising from the departure of the reference spectrum from the actual spectrum. This correction is called the *dispersion correction*, and we shall obtain the expression for it later in this section [see the second term of (II.217a)].

Assumption (ii), although not reflected in the name of the method, also plays a very crucial role in its working. As already emphasized, without the neglect of the Pauli principle, the equation of the method would never have reduced to a pure differential equation. We have reproduced Bethe's argument which shows that the neglect of the Pauli principle is a good approximation. Along with the calculation of the dispersion correction, we shall presently show [see the first term of (II.217a)] how to calculate the correction to the approximate t-matrix arising from the Pauli exclusion principle.

Before giving these corrections, we introduce another method, called the *separation method*, which is motivated by the desire to make the Pauli correction smaller than that following from the reference spectrum method. Originally, this new method, as proposed by Moszkowski and Scott[20], had a somewhat different motivation which will also be mentioned as we go along.

In the reference spectrum method, we have used the complete nuclear two-body potential in writing the Bethe-Goldstone equation. While arguing in favour of the ($Q = \mathbb{1}$)-approximation, however, we have discussed the situation for only the hard core of the potential, ignoring the outer attractive potential which has an exponential or a Yukawa-type tail. We know, from the general properties of Fourier transforms, that the longer the distance within which a spatial function is nonvanishing, the shorter are the important momentum values in its Fourier transform. Thus, the decaying tail of the outer attractive potential will have Fourier components for much smaller momenta than the Fourier components of the hard core of radius 0.4 fm. Even if we include a part of the outer potential, say, up to approximately 1 fm, the important Fourier components will still lie fairly outside the fermi sphere. Therefore, we first separate the potential V into

$$V = V_s + V_l,$$

where the short-range part V_s includes the hard core plus a part of the outer potential (up to a distance d), and the long-range part V_l is the rather weak decaying tail of the outer potential. At the stage of writing the approximate Bethe-Goldstone equation with the ($Q = \mathbb{1}$)-approximation, we use only V_s. According to our foregoing discussion, the corresponding t-matrix (let us call it t_s) will then suffer less because of the neglect of the Pauli principle. This approach, however, involves the additional task of computing the effect of V_l, which has been left out of the t_s-matrix. Here Moszkowski and Scott[20] argued that since V_l is a rather weak potential it can be treated by the straightforward perturbation theory (in the manner of Section 18) in the first one or two orders. In addition, they proposed a choice of the *separation distance d* in the division of the potential in a manner that makes the diagonal matrix element of t_s for a 'free' (i.e., not in the nuclear medium) nucleon pair identically equal to zero. They hoped that, with such a choice, most nuclear calculations can then be done only with the weak well-behaved potential V_l.

Having sketched the motivation of Moszkowski and Scott, we now proceed to a general derivation (see Bethe et al[21]) which relates the exact t-matrix with an approximate one (the approximation may be in its treatment of some or all of the operators Q, e, and V) through a formally exact equation. Such an equation then allows us to calculate the correction, i.e., the difference between t and the approximate one. At a stroke, we then achieve the goal of calculating the dispersion correction (due to the approximation on e), the Pauli correction (due to the approximation on Q), and the correction due to the long-range potential in the separation method.

The original equation for the correct t-matrix is given by

$$t = V\Omega = V + V\frac{Q}{e}t \quad \text{or} \quad \Omega = \mathbb{1} + \frac{Q}{e}t; \tag{II.208a}$$

and the equation for the approximate t-matrix, say, t_A, in terms of the approximate operators Q_A, e_A, and V_A is given by

$$t_A = V_A\Omega_A = V_A + V_A\frac{Q_A}{e_A}t_A \quad \text{or} \quad \Omega_A = \mathbb{1} + \frac{Q_A}{e_A}t_A. \tag{II.208b}$$

If we choose the operators Q_A, e_A, and V_A to be Hermitean, then t_A obviously is Hermitean. Thus, the Hermitean conjugate of (II.208b) can be written as

$$\Omega_A^\dagger = \mathbb{1} + t_A \frac{Q_A}{e_A}. \tag{II.208c}$$

Multiplying (II.208a) by $t_A^\dagger$ $(=\Omega_A^\dagger V_A)$ from the left, and (II.208c) by $t(=V\Omega)$ from the right, and then subtracting the two, we obtain

$$\Omega_A^\dagger(V - V_A)\Omega = t - t_A + t_A\left(\frac{Q_A}{e_A} - \frac{Q}{e}\right)t$$

or

$$t = t_A + \Omega_A^\dagger(V - V_A)\Omega + t_A\left(\frac{Q_A}{e_A} - \frac{Q}{e}\right)t. \tag{II.209}$$

This is the required equation connecting t with t_A. The second and third terms on the right-hand side represent the correction terms. The second term is nonvanishing only if we have not used the full potential V in calculating the approximate matrix t_A. For example, it is nonvanishing in the separation method, where $V_A = V_s$, and hence $(V - V_A) = V_l$; this term then gives the correction due to the long-range part of the potential in the separation method. The last term obviously gives the correction due to the Pauli principle and the dispersion correction. In the separation method, the exact Q and exact e are not used in writing the Bethe-Goldstone equation for the short-range potential. Hence, in this method, both the Pauli and dispersion corrections have to be calculated from the last term of (II.209). Now we consider some details about the Bethe-Goldstone equation of the separation method. It is customary in this method, as in the reference spectrum method, to use $Q_A = \mathbb{1}$. In their original work, Moszkowski and Scott[20] used the free-particle energies for the particle as well as the hole states while calculating e_A. Thus,

$$\begin{aligned} e_A &= \frac{\hbar^2}{2M}(k_1^2 + k_2^2 - k_1'^2 - k_2'^2) \\ &= \frac{\hbar^2}{M}(k^2 - k'^2) = \frac{\hbar^2}{M}(k^2 + \nabla^2). \end{aligned} \tag{II.210a}$$

Here $\mathbf{k}_1$, $\mathbf{k}_2$ are the momenta of the occupied states, and $\mathbf{k}_1'$, $\mathbf{k}_2'$ those of the intermediate particle states. The centre-of-mass energy $[\hbar^2/(4M)]K^2$ and $[\hbar^2/(4M)]K'^2$ clearly cancel out because $\mathbf{K} = \mathbf{K}'$. It should, however, be noted that it is not necessary to use exactly this approximate form for e_A in the separation method. For example, the approximate e_A calculated with Bethe's prescription (we call it e_R) in the reference spectrum method, which is

$$e_R = \frac{\hbar^2}{M}(-\gamma^2 + \nabla^2), \tag{II.210b}$$

could have been used equally well. In fact, we expect the corresponding approximate t_s to be better than the Scott-Moszkowski t_s because with e_R the dispersion correction is definitely lower than that for the e_A of (II.210a). The expression (II.210a) was originally chosen in order to make the diagonal matrix element of t_s zero with a special choice of the separation distance d. This will now be shown.

A comparison of (II.210a) with (II.210b), together with the fact that $Q_A = \mathbb{1}$, and $V_A = V_s$, easily establishes that the Scott-Moszkowski equations can be obtained from Bethe's reference spectrum equations (II.201) and (II.205) by replacing γ^2 everywhere by $-k^2$, and V everywhere by V_s. For specific discussions, we consider only the uncoupled case (II.201b) and write the equation incorporating the suggested substitution as

$$[\frac{d^2}{dr^2} - \frac{l(l+1)}{r^2} + k^2 + v_s(r)]u_l(r) = 0. \tag{II.211a}$$

As is to be expected, this equation does not differ from the uncoupled two-nucleon scattering equation (I.44). For convenience of notation, we have omitted the state labels l, S, T from v and used only the subscript s as a reminder that it is related to V_s, and not the full V. Along with this equation, we consider the equation for the l-th partial wave of exp ($i\mathbf{k}\cdot\mathbf{r}$), namely,

$$[\frac{d^2}{dr^2} - \frac{l(l+1)}{r^2} + k^2]f_l(r) = 0, \tag{II.211b}$$

where $f_l(r)$ is given by (II.197b). Multiplying (II.211a) by $f_l(r)$ from the left, (II.211b) by $u_l(r)$ from the left, and subtracting and integrating from $r = 0$ to the separation distance d, we obtain

$$-\int_0^d f_l(r)v_s(r)u_l(r)\,dr = (f_l u_l' - u_l f_l')_0^d$$

$$= f_l(d)u_l'(d) - u_l(d)f_l'(d). \tag{II.212}$$

Since there is no potential beyond $r = d$, the integral on the left-hand side is equal to the corresponding integral between the limits 0 and ∞. According to (II.202), this integral therefore gives the contribution of the partial-wave state l to the diagonal t-matrix. On the right-hand side of (II.212), we have only the contribution from the upper limit d because, at the lower limit, both $u_l(0)$ and $f_l(0)$ vanish. Thus, the contribution to the diagonal element of the t-matrix can be made equal to zero, provided the separation distance d of the potential is so chosen as to satisfy

$$(\frac{u_l'}{u_l})_{r=d} = (\frac{f_l'}{f_l})_{r=d}. \tag{II.213}$$

It is now clear why the Scott-Moszkowski programme was overambitious. In order to make $\langle\mathbf{k}|\, t\, |\mathbf{k}\rangle$ zero, we have to choose the separation distance d for each partial wave, according to the prescription (II.213). Though this is generally impossible for a single value of d, the hurdle can be overcome by introducing different separation distances for different partial-wave states. Once this is done, we should be prepared to keep the d for a particular state fixed while k of the incident wave changes. If d is chosen from (II.213) for a particular k and l, there is no guarantee that for the same d and l (II.213) will hold for all k. In other words, d will turn out to be not only state-dependent but energy-dependent as well.

The Scott-Moszkowski equation (II.211a) tells us that beyond d this equation becomes identical to (II.211b) and has solutions which are, *in general*,

$$f_l(r) = krj_l(kr),$$

$$g_l(r) = krn_l(kr).$$

Therefore, without any extra boundary condition at $r = d$ or beyond this point, the solution $u_l(r)$ will be a linear combination of $f_l(r)$ and $g_l(r)$, and hence at $r \to \infty$ it displays a phase shift with respect to the incident wave $f_l(r)$. This is contrary to the property of healing required of a good Bethe-Goldstone wavefunction. The imposition of the boundary condition (II.213) at the separation distance d precisely serves the purpose of removing the phase shift and guaranteeing the healing property at $r \to \infty$ (in fact, it ensures healing beyond d). Once the log-derivative of the internal wavefunction is matched to the log-derivative of f_l, and the continuity of the wavefunction is ensured at d, i.e., $[u_l(d) = f_l(d)]$, the wavefunction beyond d obviously continues to be $f_l(r)$ at *every* point because of the absence of any potential in this region.

The computation of the solution of (II.211a) and the determination of d from (II.213) are now sketched. Since (II.211a) is homogeneous and of second order, its solution will have an overall normalization constant and a second arbitrary constant. The latter is fixed as soon as the numerical solution is commenced at the core radius with the given boundary value $u_l(c) = 0$. The value $u_l(c + \Delta)$ can be arbitrarily fixed at the beginning of the numerical integration because of the overall arbitrary constant multiplying the solution. As we integrate outwards, we compute the log-derivative (u_l'/u_l) at each point, and check if it agrees with the log-derivative (f_l'/f_l) at the point. The point where this equality holds is our desired separation distance d. Finally, we fix the overall arbitrary constant in front of u_l by ensuring the continuity of the solution at d through the equation $u_l(d) = f_l(d)$.

The question we have to answer now is: can a d that satisfies (II.213) always be found? If not, then the separation method will not work. It is easy to examine this point if we draw the wavefunctions as in Fig. II.20. For simplicity, let us adhere to the S-state. The unperturb-

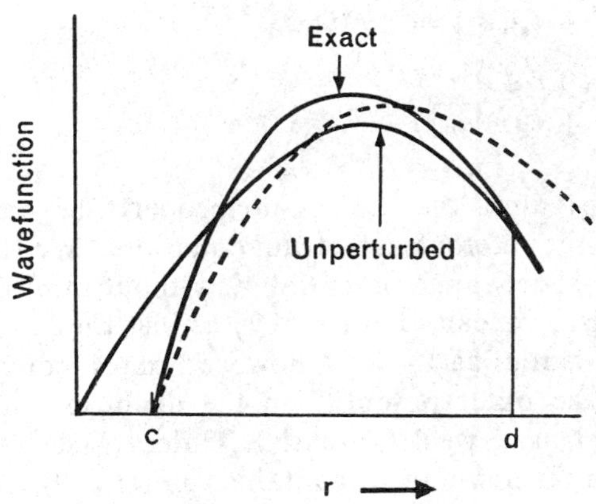

Fig. II.20 Unperturbed and exact wavefunctions in separation method.

ed wavefunction $\sin kr$ starts with a zero value at $r = 0$ and, by the time the core radius is reached, it has already acquired a phase angle kc. On the other hand, if we had a pure hard-core potential, the solution beyond the core radius would be a linear combination of $\sin kr$ and $\cos kr$, i.e., $\sin (kr + \delta)$. Since this wavefunction has to vanish at $r = c$, the phase shift δ is $-kc$. Thus, the exact wavefunction, when it starts at c, is behind the unperturbed wavefunction in phase by an amount kc. In order that its derivative and value can smoothly match on to those of the unperturbed wavefunction at d, in accordance with (II.213) and as shown in Fig. II.20, it has to make up for the handicap in phase at the starting point and acquire the same phase angle as $f_0(r)$ by the time the point d is reached. This is clearly possible only if the exact wavefunction beyond c can 'bend down' faster than the unperturbed wavefunction. We know that an attractive potential beyond c helps in 'pulling in' the exact wavefunction, whereas a repulsive wavefunction 'pushes it out'. In the latter case, the rate of 'bending down' of the wavefunction is obviously slower than that of the unperturbed function, and hence the initial handicap in phase will never be made up. On the other hand, in the case of an attractive potential, the pulling-in effect really causes a faster bending down of the exact wavefunction, as compared with the unperturbed one and, *provided the bending down is fast enough,* the exact wavefunction will catch up with the unperturbed one at some finite d. It is possible to visualize a weak attractive potential for which the pulling-in effect is so small that the complete 'catching-up' in phase may not take place even at $r \to \infty$, or it may take place only at a fairly large value of d. Even in the latter case, the separation method, *as proposed by Scott and Moszkowski*, will not work. This is because, if d turns out to be large, the healing distance for the Scott-Moszkowski wavefunction (equal to d) misses the most important property of the *small* healing distance of a realistic Bethe-Goldstone wavefunction. The case of a weak potential is demonstrated by the dashed line in Fig. II.20.

We now arrive at a more general way of applying the separation method in conjunction with the e_A of the reference spectrum method. We have now to use e_R of (II.210b) as the expression for e_A, and use V_s instead of V in the equations of the reference spectrum method. To be specific, the uncoupled equations (II.201) become

$$\left[\frac{d^2}{dr^2} - \frac{l(l+1)}{r^2} - \gamma^2 + v_s\right]\zeta_l(r) = v_s(r) f_l(r), \tag{II.214a}$$

$$\left[\frac{d^2}{dr^2} - \frac{l(l+1)}{r^2} - \gamma^2 + v_s(r)\right]u_l(r) = -(k^2 + \gamma^2) f_l(r). \tag{II.214b}$$

Because γ^2 has a fairly large value, the healing-up property beyond the potential V_s is now automatic, just as it was in the reference spectrum equations. We thus have the advantage of choosing the extent d of the short-range potential V_s without imposing any constraint such as (II.213). In fact, if we so wish, we can identify $V_s(r)$ as just the hard-core part of the potential and take the entire outer potential as V_l. It is, however, more convenient to include, say, up to approximately 1 fm of the outer potential in the definition of V_s such that the V_l that remains can truly be treated as a weak potential. Unless V_l is fairly weak, the subsequent calculation of the correction term due to it, contained in (II.209), becomes rather involved.

According to (II.214a), we have, for $r \geqslant d$,

$$\zeta_l(r) = h_l^{(1)}(i\gamma r) \to 0, \qquad r \to \infty.$$

The point to note here is that $\zeta_l(r) \to 0$, i.e., $u_l(r)$ heals up to $f_l(r)$ only at a substantially large r; at the separation distance, $\zeta_l(d)$ is given by $h_l^{(1)}(i\gamma d)$. Thus, this separation distance, in contrast with that in the Scott-Moszkowski treatment, is no longer *exactly* equal to the healing distance. The method of obtaining the numerical solution of the radial equation (II.214a) or (II.214b), starting with the form $h_l^{(1)}(i\gamma r)$ for $\zeta_l(r)$ at $r = d$, will be similar to that discussed in relation to the reference spectrum equations. The matching distance will have to be chosen at some point between c and d. The subsequent calculation of the t-matrix (t_s, say) due to the V_s will also be achieved in the manner described there. The t_s calculated in this way will, in general, have nonvanishing diagonal matrix elements, in contrast with the vanishing result achieved in the Scott-Moszkowski treatment with the boundary condition (II.213).

Calculation of Correction Terms

In both the types of separation method (the Scott-Moszkowski version and that obtained by a combination with the reference spectrum), we have to calculate the second term of (II.209), giving the correction due to the long-range part of the potential V_l beyond the separation distance d. In the reference spectrum method employing the *complete* potential V, this term is identically zero. In the diagonal matrix element of t, for example, this correction term contributes

$$\langle\phi_{\mathbf{k}}|\,\Omega_s^{\dagger}V_l\Omega\,|\phi_{\mathbf{k}}\rangle = \langle\psi_{\mathbf{k}}^{s}|\,V_l\,|\psi_{\mathbf{k}}\rangle, \tag{II.215}$$

where $\psi_{\mathbf{k}}^{s}$ is the exact wavefunction for the short-range potential, and $\psi_{\mathbf{k}}$ is the exact wavefunction for the complete problem with the correct Q, V, and e. The latter is, of course, unknown because we have not solved the exact Bethe-Goldstone equation with the correct Q, V, and e. Thus, the correction term has to be evaluated to various orders in a perturbation theory based on the solution $\psi_{\mathbf{k}}^{s}$ as the unperturbed wavefunction. In other words, we need an expansion of Ω in which the first term is Ω_s, and the subsequent terms are obtained in a perturbative manner. With this aim in view, we write

$$\Omega = \mathbb{1} + \frac{Q}{e}V\Omega,$$

$$\Omega_s = \mathbb{1} + \frac{\mathbb{1}}{e_A}V_s\Omega_s,$$

where e_A is either the reference spectrum expression (II.210b) or the Scott-Moszkowski expression (II.210a). Thus,

$$\Omega - \Omega_s = \frac{Q}{e}V\Omega - \frac{\mathbb{1}}{e_A}V_s\Omega_s$$

or

$$\Omega = \Omega_s - \frac{\mathbb{1}}{e_A}V_s\Omega_s + \frac{Q}{e}V\Omega.$$

Operating on $\phi_{\mathbf{k}}$ with this expression and multiplying by V_l from the left, we obtain

$$V_l\Omega\phi_{\mathbf{k}} = V_l\psi_{\mathbf{k}}^{s} + V_l\chi_{\mathbf{k}}^{s} + V_l\frac{Q}{e}(V_s + V_l)\Omega\phi_{\mathbf{k}}.$$

In obtaining the second term, we have replaced $-e_A^{-1}V_s\Omega_s\phi_{\mathbf{k}}$ by $\chi_{\mathbf{k}}^{s}$ from the Bethe-Goldstone equation for the short-range potential. This term, however, is *exactly* zero because V_l is nonvanishing beyond the separation distance, whereas the defect wavefunction $\chi_{\mathbf{k}}^{s}$ is precisely zero in this region. Similarly, in the first term, $\psi_{\mathbf{k}}^{s}$ contributes beyond the separation distance, and, in this region, $\psi_{\mathbf{k}}^{s}$ is *exactly* equal to the unperturbed wavefunction. Thus,

$$V_l\Omega\phi_{\mathbf{k}} = V_l\phi_{\mathbf{k}} + V_l\frac{Q}{e}(V_s + V_l)\Omega\phi_{\mathbf{k}}. \tag{II.216a}$$

This relation is exact.

We now have to evaluate the second term in an iterative manner. For this purpose, we replace $(Q/e)V_s\Omega$ by $(\mathbb{1}/e_A)V_s\Omega_s$. Ignoring the effect of Q in the term containing V_s is justified in the first approximation because of the arguments already given. But then this term vanishes when multiplied by the V_l from the left, as in the argument preceding (II.216a). Thus, the second term of this equation reduces to approximately $V_l(Q/e)V_l\Omega\phi_{\mathbf{k}}$, where the Pauli operator is retained because its effect for the long-range part of the potential cannot be ignored in the first approximation. On the other hand, we can introduce the first approximations $e \approx e_A$ and $\Omega \approx \mathbb{1}$ in its evaluation, the latter approximation being justified by the fact that, in the region where V_l is nonvanishing, $\Omega\phi_{\mathbf{k}}\,(\equiv\psi_{\mathbf{k}})$ is indeed the healed-up function $\phi_{\mathbf{k}}^{s}$. The separation distance may not be the correct healing distance for the exact Bethe-Goldstone wavefunction $\psi_{\mathbf{k}}$; however, if d is chosen not too small, the aforementioned approximation can be made as exact as possible. Thus, (II.216a) in the first approximation can be replaced by

$$V_l\Omega\phi_{\mathbf{k}} = V_l\phi_{\mathbf{k}} + V_l\frac{Q}{e}V_l\phi_{\mathbf{k}}. \tag{II.216b}$$

This expression, as is clear, is already of the second order in the potential V_l. The correctio term (II.215) can therefore be written to this order in V_l as

$$\langle\phi_{\mathbf{k}}|\,\Omega_s^{\dagger}V_l\Omega\,|\phi_{\mathbf{k}}\rangle \approx \langle\phi_{\mathbf{k}}|\,V_l\,|\phi_{\mathbf{k}}\rangle + \langle\phi_{\mathbf{k}}|\,V_l\frac{Q}{e}V_l\,|\phi_{\mathbf{k}}\rangle. \tag{II.216c}$$

Usually, the second term of (II.216c) is evaluated with the angle-averaged approximation for the Pauli operator Q.

We next discuss the Pauli and dispersion corrections contained in the third term of (II.209). These corrections have to be calculated in the reference spectrum method as well as the separation method; further, Q_A in both these methods is $\mathbb{1}$, and hence the correction term becomes

$$t_A\left(\frac{Q}{e} - \frac{\mathbb{1}}{e_A}\right)t.$$

In evaluating it, the first approximation entails substituting t_A for the last t. Then we have to compute

$$t_A\frac{\mathbb{1}}{e_A}e_A\left(\frac{Q}{e} - \frac{\mathbb{1}}{e_A}\right)e_A\frac{\mathbb{1}}{e_A}t_A.$$

We have deliberately inserted the factors $e_A^{-1}e_A$ adjacent to t_A so that, when we use $\langle\phi_{\mathbf{k}}|$ and $|\phi_{\mathbf{k}}\rangle$ at the two ends of this chain of operators for obtaining the matrix element, we easily secure $\langle\chi_{\mathbf{k}}^{A}|$ and $|\chi_{\mathbf{k}}^{A}\rangle$ for $\langle\phi_{\mathbf{k}}|t_A e_A^{-1}$ and $e_A^{-1}t_A|\phi_{\mathbf{k}}\rangle$, respectively. In this manner, the correction to the diagonal matrix element is given by

$$\langle\chi_{\mathbf{k}}^{A}|\,e_A\left(\frac{Q}{e} - \frac{\mathbb{1}}{e_A}\right)e_A\,|\chi_{\mathbf{k}}^{A}\rangle = \langle\chi_{\mathbf{k}}^{A}|\,[-(\mathbb{1} - Q)e_A + (e_A - e)e^{-1}Qe_A]\,|\chi_{\mathbf{k}}^{A}\rangle. \tag{II.217a}$$

The first term in this expression is the Pauli correction. It *removes*, through the presence of the $(\mathbb{1} - Q)$-factor with a *minus* sign, the effect of the fermi-sea states earlier allowed to participate as intermediate states in the calculation of t_A. It should be observed that the computation of this term needs only a knowledge of the approximate e_A for states inside the fermi sea; hence, the momentum integration *inside* the fermi sphere, present through the operator $(\mathbb{1} - Q)$, namely,

$$\int_{(k_1', k_2' \leqslant k_F)} d^3k'\,|\mathbf{k}'\rangle\langle\mathbf{k}'|,$$

can be carried out in the manner of Section 18. Using this expression for $(\mathbb{1} - Q)$ explicitly,

we obtain

$$\text{Pauli correction} = \int_{(k_1',\, k_2' \leqslant k_F)} d^3k' \, \frac{\hbar^2}{M}(k'^2 + \gamma^2) \, |\langle \mathbf{k}' \mid \chi_{\mathbf{k}}^{\mathrm{A}} \rangle|^2. \tag{II.217b}$$

The Fourier transform $\langle \mathbf{k}' \mid \chi_{\mathbf{k}}^{\mathrm{A}} \rangle$ of the defect wavefunction, in the case of a short-range potential, was shown to be peaked in the region 3–5 fm^{-1}, whereas the integration in (II.217b) extends up to k_F ($\approx 1.36\ \mathrm{fm}^{-1}$). Thus, the Pauli correction to t_s will be very small. In the case of t_A calculated with the complete potential V, this correction is somewhat appreciable because the long-range part of the potential then makes the Fourier components $\langle \mathbf{k}' \mid \chi_{\mathbf{k}}^{\mathrm{A}} \rangle$ contained inside the fermi sea comparatively larger.

The second term within the square brackets in (II.217a) represents the dispersion correction due to the departure of e_A from the exact e. The presence of the operator Q tells us that the difference $(e_A - e)$ for the states *outside* the fermi sea participates in building up this correction. If we adopt the attitude that the parameter w_0 in e_A has been made self-consistent and that the correct spectrum for the particle states is indeed the *free*-particle spectrum used in our definition of e_A, then the dispersion correction term is identically zero. This is indeed the situation in the reference spectrum method. In the case of the Scott-Moszkowski type separation method, however, this is not true because here, although the particle energies in e_A are the correct ones, the energy of the occupied states w_0 is not self-consistently determined but replaced by the energy of a free pair. We have noted that, in the Scott-Moszkowski case, $e_A = (-\hbar^2/M)(k'^2 - k^2)$, whereas the 'correct' e has the expression $(-\hbar^2/M)(k'^2 + \gamma^2)$ in which the quantity w_0 entering the definition of γ^2 is the self-consistent energy of the fermi-sea pair. Therefore,

$$e - e_A = -\frac{\hbar^2}{M}(k^2 + \gamma^2).$$

This expression enables us to carry out the momentum integration present in the dispersion term of (II.217a).

Finally, we have to justify the substitution of t_A for t in the foregoing calculation. We have already observed in the discussion following (II.217b) that, in the reference spectrum method using the full potential, the Pauli correction may be somewhat large. Therefore, the substitution of t_A for t in that case may not be a very good approximation. On the other hand, in any version of the separation method, this correction is much smaller. Hence, if we wish to relate the exact reaction matrix t_s^{exact}, *for the short-range potential* V_s, with the corresponding approximate matrix, t_s, calculated by making $Q = \mathbb{1}$ and $e = e_A$, then, in the relation [obtained from (II.209) by making $V = V_A = V_s$]

$$t_s^{\text{exact}} = t_s + t_s\left(\frac{Q}{e} - \frac{\mathbb{1}}{e_A}\right)t_s^{\text{exact}}, \tag{II.218a}$$

the substitution of t_s for t_s^{exact} in the second term is fairly well-justified. The subsequent splitting of this term into the Pauli and dispersion corrections, and their estimates, can then be accomplished by the method just given. Were we to use the reference spectrum expression for e, we would finally conclude

$$t_s^{\text{exact}} \approx t_s. \tag{II.218b}$$

Or, if we are fastidious, we may estimate the Pauli correction according to (II.217b) using $\chi_{\mathbf{k}}^{s}$

in the place of $\chi_{\mathbf{k}}^{A}$, and add this correction to the right-hand side of (II.218b). Finally, the exact t for the full potential and the t_s^{exact} of (II.218a) will be connected by the relation [obtainable from (II.209) by putting $V_A = V_s$ and $Q_A = Q$, $e_A = e$]

$$t = t_s^{\text{exact}} + \Omega_s^{\dagger} V_l \Omega. \tag{II.218c}$$

The evaluation of the correction term in this equation has already been discussed and is given by (II.216c).

B. FINITE NUCLEI

One of the initial attempts at calculating the t-matrix elements for finite nuclei was made by Brueckner et al[22]. In their work, they assumed that the density inside a finite nucleus does not change too rapidly from point to point (which may not be a very good approximation in the surface region of the nucleus), and hence a 'local density approximation' holds good. At each point $\mathbf{R}$ in the nucleus, the density $\rho(\mathbf{R})$ is taken to define a fermi momentum $k_F(\mathbf{R})$ by the relation $k_F^3(\mathbf{R}) = (3\pi^2/2)\rho(\mathbf{R})$. Then all the t-matrix results of infinite nuclear matter are borrowed, and the t-matrix is transformed into the coordinate space by

$$\langle \mathbf{r}|\, t\, |\mathbf{r}'\rangle = \int d^3k \int d^3k' \, \langle \mathbf{r} \mid \mathbf{k}'\rangle \langle \mathbf{k}'|\, t\, |\mathbf{k}\rangle \langle \mathbf{k} \mid \mathbf{r}\rangle.$$

Here $\mathbf{r}$ and $\mathbf{r}'$ are the relative coordinates of the interacting pair of nucleons having the centre-of-mass coordinate $\mathbf{R}$, and $\mathbf{k}$, $\mathbf{k}'$ are the relative momenta. In general, t is nonlocal in the relative coordinate (as just shown). This expression, when evaluated, has also an implicit dependence on $\mathbf{R}$ because the value of $k_F(\mathbf{R})$ required for obtaining $\langle \mathbf{k}'|\, t\, |\mathbf{k}\rangle$ must be the local fermi momentum at the centre-of-mass $\mathbf{R}$ of the interacting pair. If we take a spherically symmetric nucleus, then the density $\rho(\mathbf{R})$ at any point does not depend on the angles of $\mathbf{R}$; it depends only on the magnitude R. Therefore, the corresponding $k_F(\mathbf{R})$ also becomes a function of R alone. Thus, the computation is reduced to taking a set of R-values, starting from the centre of the nucleus to a point sufficiently beyond the nuclear 'radius', and then, corresponding to each value of R, calculating the matrix $\langle \mathbf{r}|\, t\, |\mathbf{r}'\rangle$ for various values of relative coordinates. The latter can then be treated as a nonlocal two-nucleon potential with which a self-consistent calculation (see Chapter V) of the total energy, single-particle energies, wavefunctions, and the density $\rho(R)$ can be done. Such a calculation, although very painstaking, has been done for O^{16} and Ca^{40} with a moderate measure of success.

At about the same time as Brueckner et al, Banerjee and Dutta-Roy[23], Dawson and his collaborators[24], and Eden and Emery[25] also worked independently on calculations of the t-matrix elements for finite nuclei. In all this work, the advantage of using the harmonic oscillator wavefunctions (see Section 17) has been stressed.

After the simplicity of Bethe's reference spectrum method was demonstrated for infinite nuclear matter, there has been a prolific outburst of work on finite nuclei along the same lines, using the harmonic oscillator wavefunctions as the single-particle functions. The studies of Kuo and Brown[26], MacKeller and Becker[27], Shakin et al[28], Grillot and McManus[29], and Barrett et al[30] are typical of the work in this field. The method adopted by different workers varies slightly from one to the other. Our treatment is based primarily on the method for infinite nuclear matter, which has been presented in Section 22A. Formally, this presentation has more in common with the Kuo-Brown work than with the other studies.

As already mentioned, the single-particle wavefunction will be taken as the harmonic oscillator wavefunction. The method of expressing the two-particle harmonic oscillator functions in terms of the centre-of-mass and relative harmonic oscillator functions has been described in Section 17. We now consider a two-body wavefunction of the type $|N\mathcal{L}, nl : L\rangle$, where the quantum numbers $N\mathcal{L}$ define the harmonic oscillator function for the centre-of-mass motion, nl describe the same for the relative motion, and L is the resultant orbital angular momentum. We know that the t-matrix will leave $N\mathcal{L}$ unchanged, and produce the correlated wavefunction $\psi_{nl}(\mathbf{r})$ from the unperturbed harmonic oscillator function $\phi_{nl}(\mathbf{r})$. The Bethe-Goldstone equation has therefore to be set up for a given energy w_0 of the interacting pair, and a given centre-of-mass state $(N\mathcal{L})$ of energy $E_{N\mathcal{L}} = (2N + \mathcal{L} + \frac{3}{2})\hbar\omega$. For the antisymmetry of the state, we have to combine the spin-isospin quantum numbers S, T with the l-values of appropriate parity in the manner of Section 22A. The coupling of l and S to produce the conserved quantum number $\mathcal{J}$ has also to be done, as described there.

To set up the Bethe-Goldstone equation, we once again make the approximation $Q = \mathbb{1}$ and, in the definition of e, take w_0 to be the *given* initial energy of the interacting pair. For the particle energies in the intermediate state, two types of assumption are usually made. According to Bethe's prescription for nuclear matter, these energies must be the unperturbed energies of the particle states. Since we are using the harmonic oscillator wavefunctions as the unperturbed ones, Bethe's prescription may, in this case, be taken to mean the sum of the energies of two particles moving in harmonic oscillator orbits beyond the occupied set of states. If $(n_1'l_1')$ and $(n_2'l_2')$ denote two such harmonic oscillator states, then their energies can be expressed also in terms of the energies of their centre-of-mass state $(N'\mathcal{L}')$ and of their relative state $(n'l')$. According to (II.86a), we have

$$\epsilon_{n_1'l_1'} + \epsilon_{n_2'l_2'} = \epsilon_{N'\mathcal{L}'} + \epsilon_{n'l'}.$$

However, if we have commenced with the given centre-of-mass state $(N\mathcal{L})$, then the state $(N'\mathcal{L}')$ must be taken to be the same as $(N\mathcal{L})$. In this way,

$$e = (w - \epsilon_{N\mathcal{L}}) - \epsilon_{n'l'} = (w - \epsilon_{nl}) - H_0(\mathbf{r}) = -\left[\frac{\hbar^2}{M}\gamma'^2 + H_0(\mathbf{r})\right], \tag{II.219a}$$

where γ'^2 has been defined as

$$-\frac{\hbar^2}{M}\gamma'^2 = w - \epsilon_{nl} \tag{II.219b}$$

and $H_0(\mathbf{r})$ is the harmonic oscillator Hamiltonian for the relative coordinate

$$H_0(\mathbf{r}) = -\frac{\hbar^2}{M}\nabla^2 + \tfrac{1}{4}M\omega^2 r^2. \tag{II.219c}$$

Acting on the intermediate state $|N\mathcal{L}, n'l'\rangle$, $H_0(\mathbf{r})$ reproduces the energy $\epsilon_{n'l'}$, which explains the appearance of $H_0(\mathbf{r})$ in lieu of $\epsilon_{n'l'}$ in the final step of (II.219a).

An alternative form for e follows from taking Bethe's prescription literally and using the *free*-particle energies for particle states as was done in the case of infinite nuclear matter. According to this approximation, therefore,

$$e = w + \frac{\hbar^2}{4M}\nabla_{\mathbf{R}}^2 + \frac{\hbar^2}{M}\nabla^2,$$

where $\nabla_{\mathbf{R}}^2$ refers to the centre-of-mass coordinate and ∇^2 to the relative coordinate. Acting on the intermediate state $|N\mathcal{L}, n'l'\rangle$, the operator $[-\hbar^2/(4M)]\nabla_{\mathbf{R}}^2$ no longer produces the same state multiplied by a number. However, an approximation is made, and the kinetic energy operator for the centre-of-mass motion is replaced by its average value, which is $\frac{1}{2}\epsilon_{N\mathcal{L}}$ in the harmonic oscillator state $(N\mathcal{L})$. Thus, the foregoing expression for e reduces to

$$e = (w - \tfrac{1}{2}\epsilon_{N\mathcal{L}}) + \frac{\hbar^2}{M}\nabla^2 = \frac{\hbar^2}{M}(-\gamma^2 + \nabla^2), \tag{II.220a}$$

where

$$-\frac{\hbar^2}{M}\gamma^2 = w - \tfrac{1}{2}\epsilon_{N\mathcal{L}}. \tag{II.220b}$$

A third alternative form of e is that applied by workers who closely follow the separation method as prescribed by Scott and Moszkowski. In the case of infinite nuclear matter, the Scott-Moszkowski approximation entails replacing the exact starting energy w_0 of the pair by the *free*-particle value. Hence, in this case, the analogous approximation will incur replacing w_0 by $(\epsilon_{N\mathcal{L}} + \epsilon_{nl})$. Making this substitution in (II.219a), we obtain the new approximate form for e as

$$e = (\epsilon_{N\mathcal{L}} + \epsilon_{nl} - \epsilon_{N\mathcal{L}}) - \epsilon_{n'l'} = \epsilon_{nl} - \epsilon_{n'l'} = \epsilon_{nl} - H_0(\mathbf{r}). \tag{II.221}$$

The expression (II.220a) is formally identical to (II.210b), which is the e-operator of Bethe's reference spectrum method for infinite nuclear matter. Thus, if we define, in analogy with (II.195b),

$$\chi_{nl}(\mathbf{r}) = \phi_{nl}(\mathbf{r}) - \psi_{nl}(\mathbf{r}),$$

then the equation for $\chi_{nl}(\mathbf{r})$ will obviously be identical to (II.195d) in which $\chi_{\mathbf{k}}(\mathbf{r})$ and $\psi_{\mathbf{k}}(\mathbf{r})$ are now replaced by $\chi_{nl}(\mathbf{r})$ and $\psi_{nl}(\mathbf{r})$. As in the case of infinite nuclear matter, we here denote the pure radial functions corresponding to $\phi_{nl}(\mathbf{r})$, $\psi_{nl}(\mathbf{r})$, and $\chi_{nl}(\mathbf{r})$ by $f_{nl}(r)$, $u_{nl}(r)$, and $\zeta_{nl}(r)$, respectively. These functions then satisfy (II.201a), according to which $u_{nl}(r)$ heals up to the harmonic oscillator radial function $f_{nl}(r)$. Care should be exercised in using the form (II.201b) because of the occurrence of $-k^2$ in it. This quantity arose, as is obvious from (II.201a), through the replacement of $[(d^2/dr^2) - \{l(l+1)/r^2\}]f_l(r)$ by $-k^2 f_l(r)$. This replacement cannot be made, in the present case, for the harmonic oscillator function $f_{nl}(r)$.

The treatment of the coupled states with (II.220a) for e is also similar to that given for infinite nuclear matter. The coupled equations for $u_{nl}^{(l)}(r)$ and $u_{nl'}^{(l)}(r)$ corresponding to the unperturbed harmonic oscillator function $f_{nl}(r)$ are identical to (II.205), provided we remember to replace the $(-k^2)$-term by $[d^2/dr^2 - \{l(l+1)/r^2\}]f_{nl}(r)$. In this way, the coupled equations in the reference spectrum method are found to be

$$\left[\frac{d^2}{dr^2} - \frac{l(l+1)}{r^2} - \gamma^2 + v_{llS\mathcal{J}T}(r)\right]\chi_{nl}^{(l)}(r) - v_{ll'S\mathcal{J}T}(r)u_{nl'}^{(l)}(r) = v_{llS\mathcal{J}T}(r)f_{nl}(r), \tag{II.222a}$$

$$\left[\frac{d^2}{dr^2} - \frac{l'(l'+1)}{r^2} - \gamma^2 + v_{l'l'S\mathcal{J}T}(r)\right]u_{nl'}^{(l)}(r) - v_{l'lS\mathcal{J}T}(r)\chi_{nl}^{(l)}(r) = v_{l'l'S\mathcal{J}T}(r)f_{nl}(r). \tag{II.222b}$$

A comparison of (II.219a) with (II.220a) will convince the reader that if the former is used, then too the differential equations for the radial functions are very similar. In all cases, we simply have to replace γ^2 by γ'^2 and add a term $(-M/\hbar^2)(\frac{1}{4}M\omega^2 r^2)$, arising from the second term of (II.219c), to the left-hand side of each radial equation.

Finally, the expression (II.221) leads to

$$\psi_{nl}(\mathbf{r}) = \phi_{nl}(\mathbf{r}) - \frac{\mathbb{1}}{H_0(\mathbf{r}) - \epsilon_{nl}} V\psi_{nl}(\mathbf{r}),$$
$$[H_0(\mathbf{r}) - \epsilon_{nl}](\phi_{nl} - \psi_{nl}) = V\psi_{nl}(\mathbf{r}). \tag{II.223}$$

Since $H_0(\mathbf{r})$ is the harmonic oscillator Hamiltonian of which $\phi_{nl}(\mathbf{r})$ is an eigenstate having the eigenvalue ϵ_{nl}, we must have

$$[H_0(\mathbf{r}) - \epsilon_{nl}]\phi_{nl}(\mathbf{r}) = 0.$$

Therefore, (II.223) becomes a homogeneous equation for $\psi_{nl}(\mathbf{r})$ and appears as

$$[H_0(\mathbf{r}) + V(\mathbf{r}) - \epsilon_{nl}]\psi_{nl}(\mathbf{r}) = 0.$$

In terms of the pure radial function $u_{nl}(r)$, we then have

$$\left[\frac{d^2}{dr^2} - \frac{l(l+1)}{r^2} + v_{llS\mathcal{J}T}(r) + U(r) + \mathcal{E}_{nl}\right]u_{nl}(r) = 0, \tag{II.224}$$

where

$$\mathcal{E}_{nl} = \frac{M}{\hbar^2}\epsilon_{nl}, \qquad U(r) = -\frac{M}{\hbar^2}(\tfrac{1}{4}M\omega^2r^2),$$

and $v_{llS\mathcal{J}T}(r)$ is the diagonal matrix element of $(-M/\hbar^2)V(\mathbf{r})$ in the two-nucleon state specified by l, S, $\mathcal{J}$, T. The case of the coupled states can be similarly worked out. The reader can easily verify that the required equations are given by

$$\left[\frac{d^2}{dr^2} - \frac{l(l+1)}{r^2} + v_{llS\mathcal{J}T}(r) + U(r) + \mathcal{E}_{nl}\right]u_{nl}^{(l)}(r) + v_{l'lS\mathcal{J}T}(r)u_{nl'}^{(l)}(r) = 0, \tag{II.225a}$$

$$\left[\frac{d^2}{dr^2} - \frac{l'(l'+1)}{r^2} + v_{l'l'S\mathcal{J}T}(r) + U(r) + \mathcal{E}_{nl}\right]u_{nl'}^{(l)}(r) + v_{ll'S\mathcal{J}T}(r)u_{nl}^{(l)} = 0. \tag{II.225b}$$

The quantity $v_{ll'S\mathcal{J}T}$ is the matrix element of $(-M/\hbar^2)V(\mathbf{r})$ connecting the two-body states $|lS\mathcal{J}T\rangle$ and $|l'S\mathcal{J}T\rangle$.

Since the e in (II.224) and (II.225) closely resembles that of the Scott-Moszkowski work, these equations have been applied only with the short-range part V_s of the two-nucleon potential. One of the most important cases treated by Kuo and Brown concerns the coupled $(^3S_1 + {}^3D_1)$-states. Therefore, we shall discuss (II.225) in further detail, and point out how the solutions determine the separation distance d of the potential.

In the case of the uncoupled equation (II.224), we notice that, for large r, where $v(r) \to 0$, $u_{nl}(r)$ is a solution of the harmonic oscillator equation. However, in general, for large r, the solution can be a linear combination of the well-behaved solution $f_{nl}(r)$ (worked out in Section 17) and the other linearly independent solution of the harmonic oscillator equation. But, if we choose the separation distance d by requiring $u_{nl}(d) = f_{nl}(d)$ and $u'_{nl}(d) = f'_{nl}(d)$, then, beyond d, where only the harmonic oscillator potential is present, the solution $u_{nl}(r)$ indeed continues to be $f_{nl}(r)$ and does not mix with the other linearly independent solution. Thus, the correct healing up of $u_{nl}(r)$ to the harmonic oscillator function $f_{nl}(r)$ is assured. These considerations, however, do not apply so smoothly to (II.225) because of the coupled nature of these equations. In this case, for a given unperturbed wave $f_{nl}(r)$, we have nonvanishing solutions for both $u_{nl}^{(l)}(r)$ and $u_{nl'}^{(l)}(r)$. According to the two equations, however, in the region of r where $v(r) \to 0$, $u_{nl}^{(l)}(r)$ behaves as $f_{nl}(r)$ plus the other linearly independent solution of the harmonic oscillator having

32 (45-122/1977)

the same n and l; the same observation holds for $u_{nl'}^{(I)}(r)$ which, according to (II.225b), must behave as $f_{nl'}(r)$ plus the other linearly independent solution having the same n and l'. By imposing the boundary conditions at the separation distance $r = d$, namely,

$$u_{nl}^{(I)}(d) = f_{nl}(d), \qquad u_{nl}^{(I)\prime} = f_{nl}'(d), \tag{II.226a}$$

we can, of course, make the solution $u_{nl}^{(I)}(r)$ properly heal up to the harmonic oscillator function $f_{nl}(r)$ beyond the point d. The healing-up condition on the function $u_{nl'}^{(I)}(r)$ which is coupled to $u_{nl}^{(I)}(r)$ *should have been* (because the unperturbed wave does not have any component corresponding to l')

$$u_{nl'}^{(I)}(r) = 0 \quad \text{for } r > b, \tag{II.226b}$$

where b is the healing distance, and it *has* to be reasonably small. However, the very nature of this solution at large r, mentioned prior to (II.226a), rules out the automatic validity of (II.226b) happening. This problem has been taken care of in the literature on the subject in an *arbitrary* manner by demanding the boundary condition

$$\frac{1}{u_{nl'}^{(I)}} \frac{d}{dr} u_{nl'}^{(I)} \bigg|_{r=d} = \frac{1}{rh_{l'}(i\gamma r)} \frac{d}{dr}[rh_{l'}(i\gamma r)] \bigg|_{r=d}, \tag{II.226c}$$

where $h_{l'}$ is the spherical Hankel function [see Appendix C (Section I)], and the value of the parameter γ is taken from the reference spectrum expression (II.220b). There is more arbitrariness in this procedure: beyond $r = d$, the solution $u_{nl'}^{(I)}$ is assumed to be $rh_{l'}^{(1)}(i\gamma r)$ which, however, does not follow from (II.225b), after putting $v(r) = 0$.

The coupled equations are integrated out from the core radius with the initial conditions $u_{nl}^{(I)}(c) = 0$, $u_{nl'}^{(I)}(c) = 0$. To start the numerical integration, arbitrary values are given to these functions at $(c + \Delta)$. This is allowed for $u_{nl}^{(I)}$ because there is an arbitrary *common* multiplying constant for $u_{nl}^{(I)}$ and $u_{nl'}^{(I)}$. Giving an arbitrary value to the second function at $(c + \Delta)$, we are, however, doing more than what is permitted. This would have been acceptable were the relative proportion of the two functions arbitrary. But this is not so because the relative mixture of the two functions must be determined by the coupled equations themselves. To remove this arbitrariness, we build up a second numerical solution (to be denoted by an overhead bar), starting with the same boundary values at c and giving the previous value to $u_{nl}^{(I)}(c + \Delta)$ and a different value to $u_{nl'}^{(I)}(c + \Delta)$. The correct solution then is of the form

$$A\left\{\begin{bmatrix} u_{nl}^{(I)} \\ u_{nl'}^{(I)} \end{bmatrix} + C\begin{bmatrix} \bar{u}_{nl}^{(I)} \\ \bar{u}_{nl'}^{(I)} \end{bmatrix}\right\},$$

where A and C are constants to be determined. The linear superposition caused by C makes the final solution completely general, and the constant A plays the role of the overall multiplying constant. We now determine A, C, and d by ensuring the three boundary conditions embodied in (II.226a) and (II.226c).

The logically satisfactory way to resolve the arbitrariness in treating the healing up of $u_{nl'}^{(I)}$ appears to be as follows. Do not use (II.221) in setting up the coupled equations in a separation method; use (II.220a) instead. In that case, the coupled equations, as already discussed, will be given by (II.222); because we would like to separate the potential into a short- and a long-range part, the potential v appearing in these equations will now have to be taken to correspond to the short-range part alone. Beyond the separation distance where this potential goes to zero, we *automatically* have, from (II.222), the healing-up conditions

$$\zeta_{nl}^{(l)}(r) \to rh_l^{(1)}(i\gamma r), \qquad u_{nl'}^{(l)}(r) \to rh_{l'}^{(1)}(i\gamma r),$$

both of which tend to zero for large r as $e^{-\gamma r}$. The large positive value of γ ensures the healing. The coupled equations to be solved, however, are nonhomogeneous. The reader should be able to work out the numerical integration of these equations with the help of the description given after (II.201b). This is left as an interesting exercise.

The coupled equations have been applied in practice to only the $(^3S_1 + {}^3D_1)$-case. The coupling in the triplet states of higher angular momenta has been ignored in almost all computations. In the particular case of $(^3S_1 + {}^3D_1)$, when the unperturbed wave is in the $(l = 0)$-channel, $u_{nl}^{(l)}$ and $u_{nl'}^{(l)}$ correspond respectively to the S- and D-wavefunction. The matrix element $v_{llS\mathcal{J}T}$, as shown in Section 4, then consists only of the central potential, whereas the other diagonal element is made up of the central, tensor, spin-orbit, and quadratic spin-orbit potentials. The nondiagonal elements $v_{ll'S\mathcal{J}T}$ and $v_{l'lS\mathcal{J}T}$ are composed entirely of the tensor potential.

Calculation of Correction Terms

Correction terms can be calculated with equations similar to (II.216c) and (II.217a), provided the plane-wave functions are suitably replaced by the harmonic oscillator functions. In the case of the $^1S_0$- and $^3S_1$-state, for example, the first term of (II.216c) has a contribution from only the long-range part of the *central* potential, for reasons already stated. For the $^1S_0$-state, the noncentral potential never contributes, and hence the second-order correction term becomes much smaller than the first-order correction term. On the other hand, the second-order term has, in the $^3S_1$-state, a very important contribution from the long-range part of the tensor potential. The tensor potential changes the $^3S_1$-state to an intermediate $^3D_1$-state, which then goes back to the $^3S_1$-state, giving a nonvanishing second-order result. However, it is very difficult to satisfactorily treat the Q-operator in the case of finite nuclei. Even if we assume that the intermediate particle states have free-particle energies [as in (II.220a)], the replacement of the intermediate states contained in Q by plane-wave states with momenta larger than k_F would not be correct. Nevertheless, this procedure is followed in the simplest type of calculations. The quantity k_F is calculated from the given A and given Ω $[=(4\pi/3)r_0^3A]$, where r_0 is taken from the experimentally measured 'radius' of the nucleus $R = r_0A^{1/3}$. For Q, the angle-averaged approximation (II.193) is then used. The reader may work out the detailed formulas on his own or with the help of the publication of Kuo and Brown[26]. More careful treatment (for example, Wong[31], and MacKeller and Becker[27]) of the Q-operator also exists, but none is above criticism.

The Pauli and dispersion corrections may also be evaluated by closely following (II.217a). Once again, the $\chi_{\mathbf{k}}^{A}$ has to be replaced by the corresponding harmonic oscillator defect wavefunction, and a workable assumption be introduced for the projection operator Q and $(\mathbb{1} - Q)$. In the simplest calculations, the plane-wave approximation, with angle-averaging, has been used.

C. RESULTS OF CALCULATIONS

Infinite Nuclear Matter

The results of calculations for infinite nuclear matter have been summarized by Bethe[32]. His work is the source for the binding energy per nucleon (in MeV) and the corresponding fermi momentum k_F (in fm^{-1}) quoted in Table II.4. The parameter k_F can be related to the density in the usual way. The code names for the potentials are elucidated in the footnote to the table, and the reader is referred to Section 12 and to the respective original work for details on these potentials.

Table II.4 Calculated binding energy and fermi momentum of infinite nuclear matter

Author(s)	Potential*	Energy/A (MeV)	k_F (fm^{-1})
Brueckner and Gammel[19]	BGT	−15.2	1.02
Brueckner and Masterson[19]	Yale	−8.3	1.19
Sprung and Bhargava[33]	HJ	−10.9	1.38
Ingber[34]	OBEP, Ingber	−14.8	1.31
Haftel and Tabakin[35]	OBEP, BS	−15.0	1.58
Kallio and Day[36]	Reid (HC)	−7.2	1.36
	Reid (SC)	−11.05	1.36
		−11.2 (minimum)	1.44

*BGT = Brueckner-Gammel-Thaler; HJ = Hamada-Johnston; OBEP = one-boson-exchange potential; BS = Bryan-Scott; HC = hard core; SC = soft core.

The authors quoted in the first three lines of Table II.4 used a nonvanishing potential energy in their expression of the particle energies in the intermediate states. The other authors included in this table followed Bethe's prescription for the particle energy.

The first point we notice in Table II.4 is the excellent value of the binding energy and k_F in the case of the BGT potential. In later work on two-body physics, this potential has been found to be unrealistic. The HJ and Yale potentials, both having a hard core, came into vogue in the post-BGT potential days. These potentials, as is clear from Table II.4, underbind the nucleons by a very large amount. In very recent years, the OBEP has become quite successful in explaining two-body data, and the results with the Ingber and Bryan-Scott OBEP shown in the table are quite encouraging.

The general failure of the Yale and HJ potentials in the nuclear binding energy calculation has been attributed to several reasons. (i) The many-body binding energy is very crucially determined by the contribution from the two-nucleon $^3S_1$-state. We have already noted that the central potential gives a direct contribution in this state, whereas the tensor potential contributes in the second order via the $^3D_1$-state. As a result, of the two potentials, the one with the higher central-to-tensor ratio in the triplet state will produce larger binding energy in the nuclear matter calculation. The HJ potential has a large tensor component, and the mixing parameter ϵ for the coupled ($^3S_1 + {}^3D_1$)-state is very large. (ii) The existence of the hard-core repulsion is, according to Bethe, responsible for at least a part of the failure. It has been observed that replacing the hard core by a soft core definitely improves the situation.

This provides much of the motivation of Reid's work on a two-body potential with the soft core. The value −11.2 MeV of the binding energy for this potential is still slightly low. Bethe[32] summarized all the corrections to this value for the binding energy. Rajaraman and

Bethe[17] (see original references in this review article) considered the effect of all higher-order diagrams containing three fermi-sea particles. Dahlbohm[37] made the most detailed estimate of the three-body contribution by using the Reid potential. Day[38] roughly estimated the contribution of diagrams containing four hole lines. The contribution due to three-body forces (if any) has also been discussed by various authors (see Bethe[32]). The final binding energy, obtained after adding all these corrections to -11.05 MeV, has been found (see Bethe[32]) to be -15.4 MeV per nucleon at $k_F = 1.36\ \text{fm}^{-1}$. Thus, the current results for infinite nuclear matter are quite satisfactory.

Finite Nuclei

In the case of finite nuclei, there is, apart from obtaining the binding energy from the t-matrix elements, considerable interest in using these matrix elements in nuclear structure calculations of the type described in subsequent chapters.

Binding energy has been calculated by employing the Hartree-Fock theory (Chapter V), and the t-matrix elements calculated with a harmonic oscillator basis. Once again, the hard-core potentials, Yale and BJ, with which most calculations for finite nuclei have been done, are found to underbind the nuclei by a large amount.

An important effect that needs consideration in structural calculations on finite nuclei is now described. Most finite nuclei contain a set of single-particle states, completely occupied by nucleons, and a few nucleons near the fermi surface in partially filled single-particle states. We shall refer to these nucleons as *valence* nucleons. Now the nuclear structural calculations usually entail taking into account the effect of the residual interactions on these valence nucleons with the tacit assumption that the nucleons sufficiently below the fermi surface, occupying completely filled levels, more or less act as an inert core structure for the nucleus. In the reaction matrix theory, therefore, we have the additional work of computing this effective interaction between any two valence nucleons. The Brueckner reaction matrix t, computed in this chapter, is still strong enough to cause excitation of particles from the core when any of the core particles interacts with any of the valence particles. These effects have to be included in defining an effective interaction G between any two valence nucleons in terms of the reaction matrix t.

The task of replacing the many-body linked-cluster expansion of all the particles by a linked-cluster expansion of the valence particles alone was formally achieved by Bloch and Horowitz[39]. Brandow[40] has made very important contributions in recent years in the formal development of this subject.

On the computational side, Kuo and Brown[26] first emphasized the necessity of constructing the effective interaction G between two valence nucleons by adding to the t-interaction the effect of exciting core particles. They worked out this effect for the two valence nucleons in O^{18} and F^{18} up to the second order in the reaction matrix t. Figure II.21a shows the interaction between the two valence nucleons directly through the t-operator. The two vertical lines represent the two valence nucleons, and the wavy line represents the t-interaction between them. The core excitation process considered by Kuo and Brown is shown in Fig. II.21b. At the first vertex involving t (wavy line), a core nucleon interacts with a valence nucleon and gets excited as a particle (upward arrow) above the core states, leaving a hole (downward arrow) in the core; at the second vertex, this resultant particle interacts with the other valence particle and falls back into the hole that was earlier left behind in the core. This particular process is known as the core-polarization effect. The effective interaction G between the valence nucleons can now be

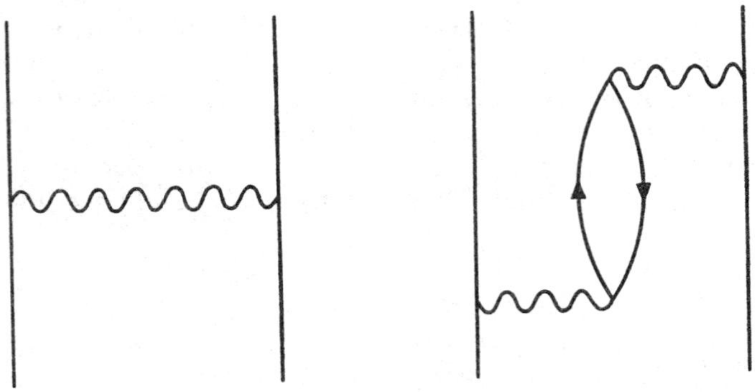

(a) Interaction t only

(b) Interaction via hole-particle pair

Fig. II.21 Two-body effective interaction between two valence nucleons.

defined as shown in Fig. II.22a. Thus, the final diagram of Fig. II.22b, where the saw-toothed line represents G, now stands as a replacement of Figs. II.21a and II.21b containing t.

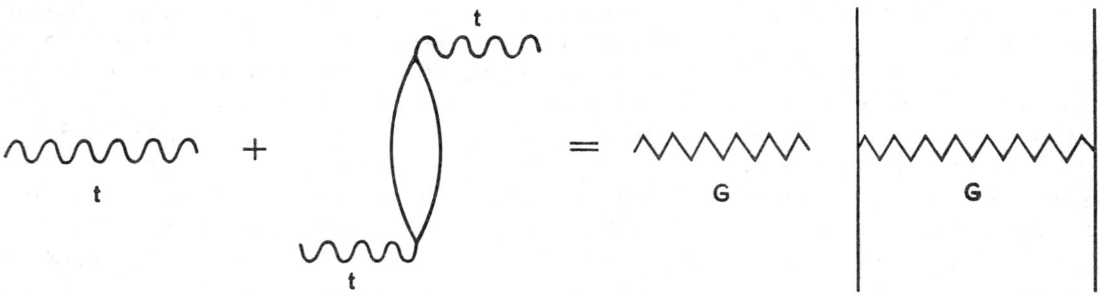

(a) Definition of G

(b) Two terms of Fig. II.21 written as single term with renormalized interaction

Fig. II.22 Renormalized interaction G (saw-toothed line) between pair of valence particles.

Kuo and Brown[26] found the contributions of the first- and second-order diagrams (Fig. II.21a and Fig. II.21b, respectively) in G to be comparable. The computed values reproduced the spectra of O^{18} and F^{18} quite well. However, numerical mistakes were discovered in this work, and subsequent corrections made the agreement poor. Kuo[41] then restored the agreement by including the effect of two more diagrams which are of second order in t. The effect of replacing the hole-particle excitation of the core by the TDA- or RPA-type (see Section 42B) excitation was also investigated by Kuo[42] and found not to change the results appreciably. Once again, numerical mistakes were noticed in his work, this time by Osnes and

Warke[43], who pointed out that the RPA excitation of the core builds up very large corrections to the first order t, and spoils the good fit to the experimental spectra.

On the conceptual level, the fact that the second-order term in the computation of G is comparable to, or even larger in some cases, than the first-order term led to questions on the convergence of the entire procedure. Barrett and Kirson[44] then calculated all the third-order and a few fourth-order terms. They found that nearly all the third-order terms were of the same order of magnitude as the second-order terms.

Selected sums of higher-order diagrams were then attempted by various authors. In fact, the TDA and RPA calculations already belong to this category because the TDA- and RPA-type excitation of the core consists of a chain of successive hole-particle pairs always going upwards (TDA) or both up and down (RPA). Other kinds of summation, which include the effect of 'screening' of the hole-particle pair and the modification of a t-interaction vertex in Fig. II.21b, have also been done by Kirson and Zamick[45]. The numerical results are somewhat embarrassing: (i) there is no convergence of the perturbation series for G in terms of the successive orders in the t-operator; (ii) the values of the G-matrix elements computed by various authors for use in structural calculations on finite nuclei therefore depend very much on which diagrams are included in the computation; (iii) the inclusion of all higher-order diagrams in the best known way by Kirson (including RPA, self-screening, and vertex modification) gives results for G that are very nearly equal to those for the bare t-matrix; and (iv) numerical agreement with the spectra is best when spectroscopic calculations are made with G-matrix elements computed by the simplest Kuo method (i.e., by keeping only the second-order diagrams without even doing the TDA or RPA).

The work on calculating G has been very comprehensively reviewed by Barrett and Kirson[46]. The reader should consult this article for all the original references in this field. A comparatively recent conference proceeding edited by Barrett[47] summarizes the status of the subject up to the middle of 1975.

REFERENCES

1. Woods, R. D., and Saxon, D. S., *Phys. Rev.*, **95**, 577 (1954).
2. Feshbach, H., Porter, C. E., and Weisskopf, V. F., *Phys. Rev.*, **96**, 448 (1954).
3. Schiff, L. I., Quantum Mechanics, 2nd edn., McGraw-Hill, New York, 1955, pp 60, 162–64.
4. Talmi, I., *Helv. Phys. Acta*, **25**, 185 (1952).
5. Moshinsky, M., *Nucl. Phys.*, **13**, 104 (1959); Brody, T. A., Jacob, G., and Moshinsky, M., *Nucl. Phys.*, **17**, 16 (1960).
6. Moshinsky, M., and Brody, T. A., Tables of Transformation Brackets, Monografias del Instituto de Fisica, Mexico City, Mexico, 1960.
7. Brueckner, K. A., in The Many-Body Problem, ed. by C. Dewitt, Dunod, Paris, 1959.
8. Swiatecki, W. J., *Proc. Phys. Soc.* (London), **A63**, 1208 (1950).
9. Huby, R., *Proc. Phys. Soc.* (London), **A62**, 62 (1949).
10. Bethe, H. A., *Phys. Rev.*, **103**, 1353 (1956).
11. Tabakin, F., *Ann. Phys.* (N.Y.), **30**, 51 (1964).
12. Goldstone, J., *Proc. Roy. Soc.* (London), **A239**, 267 (1957).
13. Bethe, H. A., and Goldstone, J., *Proc. Roy. Soc.* (London), **A238**, 551 (1957).

14. Landau, L. D., and Lifshitz, E. M., Quantum Mechanics, Pergamon Press, Oxford, 1958, p 70.
15. Swiatecki, W. J., *Proc. Phys. Soc.* (London), **A64**, 226 (1951); *ibid*, **A68**, 285 (1955).
16. Day, B. D., *Phys. Rev.*, **136**, B1594 (1964).
17. Rajaraman, R., and Bethe, H. A., *Revs. Mod. Phys.*, **39**, 745 (1967).
18. Gomes, L. C., Walecka, J. D., and Weisskopf, V. F., *Ann. Phys.* (N.Y.), **3**, 241 (1958).
19. Brueckner, K. A., and Gammel, J. L., *Phys. Rev.*, **109**, 1023 (1958); Brueckner, K. A., and Masterson, K. A., *Phys. Rev.*, **128**, 2267 (1962).
20. Moszkowski, S. A., and Scott, B. L., *Ann. Phys.* (N.Y.), **11**, 65 (1960).
21. Bethe, H. A., Brandow, B. H., and Petschek, A. G., *Phys. Rev.*, **129**, 225 (1963).
22. Brueckner, K. A., Gammel, J. L., and Weitzner, H., *Phys. Rev.*, **110**, 431 (1958); Brueckner, K. A., Lockett, A. M., and Rotenberg, M., *Phys. Rev.*, **121**, 255 (1961).
23. Banerjee, M. K., and Dutta-Roy, B., *Ann. Phys.* (N.Y.), **7**, 484 (1959).
24. Dawson, J. F., Talmi, I., and Walecka, J. D., *Ann. Phys.* (N.Y.), **18**, 339 (1962); Dawson, J. F., and Walecka, J. D., *Ann. Phys.* (N.Y.), **22**, 133 (1963).
25. Eden, R. J., and Emery, V. J., *Proc. Roy. Soc.* (London), **A248**, 266 (1958).
26. Kuo, T. T. S., and Brown, G. E., *Nucl. Phys.*, **85**, 40 (1966); Kuo, T. T. S., *Nucl. Phys.*, **A90**, 199 (1967); *ibid*, **A103**, 71 (1967).
27. MacKeller, A. D., and Becker, R. L., *Phys. Letters*, **18**, 308 (1965).
28. Shakin, C. M., and Waghmare, Y. R., *Phys. Rev. Letters*, **16**, 403 (1966); Shakin, C. M., Waghmare, Y. R., and Hull, M. H., *Phys. Rev.*, **161**, 1006 (1967); Shakin, C. M., Waghmare, Y. R., Tomaselli, M., and Hull, M. H., *Phys. Rev.*, **161**, 1015 (1967).
29. Grillot, D., and McManus, H., *Nucl. Phys.*, **A113**, 161 (1968).
30. Barrett, B. R., Hewitt, R. G. L., and McCarthy, R. J., *Phys. Rev.*, **C3**, 1137 (1971).
31. Wong, C. W., *Nucl. Phys.*, **A91**, 399 (1967).
32. Bethe, H. A., *Ann. Rev. Nucl. Sc.*, **21**, 93 (1971).
33. Sprung, D. W. L., and Bhargava, P. C., *Ann. Phys.* (N.Y.), **42**, 222 (1967).
34. Ingber, L., *Phys. Rev.*, **174**, 1250 (1966).
35. Haftel, M. I., and Tabakin, F., *Nucl. Phys.*, **A158**, 1 (1970); *Phys. Rev.*, **C3**, 921 (1971).
36. Kallio, A., and Day, B. D., *Nucl. Phys.*, **A124**, 177 (1969).
37. Dahlbohm, T. K., see Bethe, H. A., *Ann. Rev. Nucl. Sc.*, **21**, 137 (1971).
38. Day, B. D., *Phys. Rev.*, **151**, 826 (1966).
39. Bloch, C., and Horowitz, J., *Nucl. Phys.*, **8**, 91 (1958).
40. Brandow, B. H., *Revs. Mod. Phys.*, **39**, 771 (1967).
41. Kuo, T. T. S., *Nucl. Phys.*, **A103**, 71 (1967).
42. Kuo, T. T. S., *Phys. Letters*, **26B**, 63 (1967).
43. Osnes, E., and Warke, C. S., *Phys. Letters*, **30B**, 306 (1969).
44. Barrett, B. R., and Kirson, M. W., *Nucl. Phys.*, **A148**, 145 (1970).
45. Kirson, M. W., and Zamick, L., *Ann. Phys.* (N.Y.), **60**, 188 (1970); Kirson, M. W., *Ann. Phys.* (N.Y.), **66**, 624 (1971).
46. Barrett, B. R., and Kirson, M. W., *Adv. Nucl. Phys.*, **6**, 219 (1973).

47. Barrett, B. R., ed., Effective Interactions and Operators in Nuclei, Springer-Verlag, Berlin, 1975.

PROBLEMS

1. Derive the various types of matrix elements of T and V by second-quantization algebra.

2. Work out the complete series solution for the isotropic harmonic oscillator potential.

3. (a) Calculate the probability $P(k)$ of finding a pair of nucleons in infinite nuclear matter with given spin-isospin and relative momentum of magnitude k. Plot this probability as a function of k.

(b) Calculate the probability $P_K(k)$ of finding a pair of nucleons having relative momentum k in infinite nuclear matter when the total momentum is given as K.

(c) Calculate the probability $P(r)$ of finding a pair of nucleons in infinite nuclear matter separated by a distance r, and plot it as a function of r.

4. In order to derive the dependence of any Goldstone diagram on the density ρ $(=A/\Omega)$, we need to know the dependence on A and Ω of three quantities, namely, (i) the summation over all occupied fermi-sea states of a hole line, (ii) each two-body matrix element of V, and (iii) the summation of each particle line over all the states outside the fermi sea. Show that (i) gives a factor A, (ii) is proportional to Ω^{-1}, and (iii) is proportional to Ω.

5. Apply the results of Problem 4 to show that the contribution of each of the diagrams in Fig. II.16 (defining t) to the binding energy per nucleon (E/A) is proportional to ρ. Show that the same contribution from each of the third-order diagrams in Figs. II.15a and II.15b is proportional to ρ^2. [Hint: While considering the sums mentioned in Problems 4(i) and 4(iii), the reader is advised to bear in mind the momentum conservation and sum over only the independent momenta.]

Note All the diagrams in any order of t, containing three and only three hole lines, can be shown by a generalization of the procedure of solving Problem 5 to give a contribution proportional to ρ^2 in E/A. Rajaraman and Bethe[17] therefore argued that all such diagrams have to be summed (and not considered individually) to find the correction to the contribution of Fig. II.16.

6. (a) Take a central complex potential $V(r) + iW(r)$ and write down the corresponding radial Schrödinger equation for the l-th partial wave and its complex conjugate equation. With the help of these two equations, derive an equation satisfied by the Wronskian of u_l and u_l^*, where (u_l/r) is the radial wavefunction for the l-th partial wave. Show as a consequence of this equation that, if $u_l \to 0$ as a decaying exponential (characteristic of a bound state) at $r \to \infty$, the corresponding eigenvalue of energy has to be complex and the imaginary part of the energy is given by

$$\text{Im } E = \int_0^\infty W(r)|u_l(r)|^2 \, dr.$$

Note If $W(r)$ is attractive, Im E is negative and we can write $E = E_R - \frac{1}{2}i\Gamma$, where Γ is real. Then the corresponding time factor of the wavefunction is given by

$$\exp\left[\frac{-i}{\hbar}(E_R - \tfrac{1}{2}i\Gamma)t\right] = \exp\left(-\frac{\Gamma}{2\hbar}t\right)\exp\left(\frac{-i}{\hbar}E_R t\right).$$

33 (45–122/1977)

Thus, a bound state of the complex potential is in reality only *quasi-bound*, and it decays with time as $\exp[\{-\Gamma/(2\hbar)\}t]$ if the imaginary part of the potential is attractive.

(b) In the scattering case, the energy is a given real quantity. Use this fact in the same Wronskian equation to show

$$[u_l(r), u_l^*(r)]_w = -i\frac{4m}{\hbar^2}\int_0^r W(r')|u_l(r')|^2\, dr',$$

where the bracketed expression with w as subscript on the left-hand side denotes the Wronskian.

Note For a real potential, $W(r') = 0$, and hence $u_l(r) = u_l^*(r)$, i.e., the radial function is real whereas, for a complex potential, this equation demands that $u_l(r)$ be a complex radial function.

(c) Writing

$$u_l(r) = F_l(r) + iG_l(r),$$

we get, if $W(r)$ is attractive everywhere,

$$[G_l(r), F_l(r)]_w > 0 \quad \text{for all } r.$$

Use this condition to prove that the complex phase shift δ_l has a positive imaginary part when $W(r)$ is attractive.

(d) Write, in the usual way,

$$u_l(r) = c_l[e^{-i\{kr-(1/2)l\pi\}} - \mathcal{S}_l(k)e^{i\{kr-(1/2)l\pi\}}], \qquad r \to \infty,$$

where c_l is an overall normalization constant and $\mathcal{S}_l(k) = \exp[2i\delta_l(k)]$ is the scattering function. Prove that, for this $u_l(r)$, the ingoing flux is given by

$$\text{in} = v\sum_{l=0}^{\infty}|c_l|^2,$$

and the outgoing flux by

$$\text{out} = v\sum_{l=0}^{\infty}|c_l|^2|\mathcal{S}_l(k)|^2,$$

where v is the incident velocity in the scattering problem.

Note Thus, if δ_l is real (as in the case of the real potential), $|\mathcal{S}_l(k)|^2 = 1$, and the in and out flux are equal. On the other hand, for a complex potential with an attractive imaginary part, the phase shift δ_l has been proved to have a positive imaginary part, and hence the out flux is less than the in flux, i.e., such a complex potential produces absorption.

III Nuclear Shell Model

23. INTRODUCTION

It is well-known that the electrons in an atom move in single-particle orbits corresponding to the various energy levels of the attractive Coulomb potential set up by the nucleus from the centre of the atom. Comparatively weak Coulomb repulsion between pairs of electrons denotes only a small perturbation to the independent particle behaviour of the electrons. As far as the motion of the electrons in the outer orbits is concerned, the repulsive forces on them, exerted by the electrons moving in orbits closer to the nucleus, get averaged out to an effective overall potential which partially screens the attractive nuclear potential. This feature further reduces the effect of the residual interactions amongst the electrons, and accounts for the success of an independent particle model for the electron motion in atoms. The single-particle energy levels determined by the screened nuclear potential have the characteristic of occurring in groups separated from each other by energy intervals that are large compared with the typical energy difference between the levels of each group. The levels belonging to each group are said to form a *shell*. The states of a many-electron atom are obtained by filling up these levels from the bottom with electrons, in accordance with the Pauli exclusion principle. In an atom where all the levels belonging to a shell are occupied by electrons, we expect to find the energy necessary to excite it to be fairly large, namely, equal to the energy needed to lift an electron from the outermost level to the next level of the shell above it. Such an atom will also 'refuse' to interact chemically with other atoms because it cannot accept an electron in the level of its outermost electrons. This is experimentally the situation with the inert gases of the periodic table. On the other hand, alkali metals which have only one electron in a shell above a set of completely occupied shells would be expected to be chemically very active, and it should be very easy to excite the outermost electron of such an atom to any of the close-lying levels of the same shell. These properties of alkali atoms are also experimentally well-established. In fact, it is well-known that the periodicity of the chemical properties of various atoms occurring in the same columns of the periodic table is precisely explained by the shell structure of the electron configurations.

About the year 1950, evidence of shell structure was obtained for the atomic nucleus also. It was found that the role analogous to that of the inert gases in atomic physics is played in nuclear physics by a set of nuclei having either the neutron or proton number equal to 2, 8, 20, 50, 82, and 126. These nucleon numbers are usually called *magic numbers*. The special properties of such nuclei are also common to two other nucleon numbers in a less marked degree; these nucleon numbers, 28 and 40, are therefore described as *semi-magic*. Some of the anomalous nuclear properties corresponding to the magic and semi-magic numbers are now listed. (For more details, the reader is referred to Mayer and Jensen[1].)

(i) Nuclei having magic N- or Z-values, i.e., neutron or proton values, show anomalous large binding energy as compared with the smoothly varying prediction from the semi-empirical mass formula of Section 14. The high binding energy of Pb^{208}, for example, makes itself perceptible, starting with nuclei that are ten or more mass units below or above it. The higher binding energy also manifests itself in the energy release in α- and β-decay connecting such nuclei.

(ii) The binding energy of the last nucleon in a nucleus, plotted as a function of the nucleon number, displays very large values at the magic numbers.

(iii) The 9th, 51st, and 83rd neutrons in O^{17}, Kr^{87}, and Xe^{137}, respectively, are very loosely bound, making all these nuclei spontaneous neutron emitters.

(iv) The total number of stable nuclei with N or Z equal to the magic numbers is much larger than the average number of nuclei for a nonmagic N- or Z-value in the neighbourhood.

(v) The first excited state of even nuclei having N or Z equal to the magic numbers occurs at an energy that is unusually large as compared with the same excitation energy of neighbouring nuclei.

All these facts, and several others, compiled by Mayer, were accepted as direct evidence on the existence of shell structure in the nucleus. The magic numbers mark the filling up of a group of close-lying levels belonging to a shell; in analogy with atomic physics, we therefore expect these nuclei to be very stable, and harder to excite when compared with their neighbours. The nuclei with one nucleon more than the magic numbers are comparable with the alkali atoms: the last nucleon in them is comparatively loosely bound and easily excitable.

We have seen in Chapter I that the two-nucleon interaction is very strong and has a strongly repulsive core in some two-body states. Moreover, there is no stronger attraction of the nucleons with any object located at the centre of the nucleus. Therefore, the understanding of nuclear shell structure requires different considerations from those relating to electron shell structure. It is clear that the two-nucleon interactions between all pairs average out to a large extent and produce a strong overall potential in which the nucleons move more or less independently. The residual two-nucleon interaction that remains after the extraction of the one-body potential is evidently weaker than the original interaction.

Many important theoretical questions are associated with the somewhat surprising discovery of the shell-model behaviour of the nucleons in a nucleus. For instance, how do the strong two-nucleon interactions average out to a smooth shell-model potential well? And how do the residual pair interactions get diluted? These questions were resolved several years after the shell model was proposed, and after many detailed shell-model calculations were done with an empirical overall potential, and empirical residual two-nucleon interactions, which were assumed to be smooth, well-behaved, and much more diluted than the actual interactions. The answers to the knotty theoretical problems came with the advent of the Brueckner-Goldstone many-body theory described in Chapter II. Initially, this theory was developed for the idealized infinite nuclear matter, but it did provide an understanding of why the nucleons almost always move more or less independent of one another. It was pointed out by Gomes et al[2] that the Pauli exclusion principle makes a fundamental difference between the motion of a pair of nucleons when they are on their own and that of the nucleons when they are embedded in a many-nucleon medium. In the former case, the strong interaction gives rise to a distinctive correlated wavefunction exhibiting large phase shifts at large separation distances. On the other hand, in the latter case, the Pauli principle prevents a given pair of nucleons from scattering to states occupied by other nucleons in the fermi sea; the correlation built up as a result of the scattering to very high momentum states by the repulsive core is confined to a very small separation distance ($\approx$1 fm), after which the wavefunction quickly 'heals up' to the free-nucleon wavefunction without any phase shift. The reader should remember that the effect of three-body or higher than three-body clustering of nucleons is also found to be small in many-body calculations.

Thus, the nucleons in the many-body medium move, almost all the time, as independent particles, despite the strongly repulsive core of the pair interaction. Only when two nucleons approach each other to within a fermi does their motion veer considerably from free-particle motion; the pair wavefunction at such a short distance of approach is so strongly correlated that it vanishes at a distance equal to, but less than, the hard-core radius. The many-body theory, described in Section 22, provides the method of calculating the t-matrix elements between unperturbed wavefunctions, using the actual potential V and the correlated wavefunction ψ. The t-matrix is therefore the effective two-body potential to be used along with the uncorrelated two-nucleon states. With this effective two-nucleon potential, we can do a Hartree-Fock calculation, as is done in atomic physics for an atom with many electrons; such a procedure yields a self-consistent average potential $\mathcal{V}$, which is the overall single-particle potential used in shell-model calculation. The sum of the *effective* two-nucleon potential over all pairs minus the sum of the average potential $\mathcal{V}$ over all individual nucleons should be treated as the residual interaction. The effects that the residual interaction can produce require detailed calculation in nuclear shell theory.

However, it should be remembered that the shell-model wavefunction of a nucleus is at the best a model wavefunction. Along with the effective potential, it does give the correct result in an energy calculation, but it lacks the short-range correlation between any pair of nucleons present in the *actual* wavefunction. This fact further implies that when we use the shell-model wavefunction in calculating the value of any other physical observable we must correspondingly use, in principle, an effective operator for the latter. However if the physical quantity we wish to evaluate does not depend crucially on the short-range correlations of a pair wavefunction, then the transformation to the effective operator may not be needed.

Although we have mentioned that $\mathcal{V}$ should be calculated by the Hartree-Fock method from the effective matrix elements of the actual two-nucleon potential, we shall postpone such a first-principle derivation of $\mathcal{V}$ to Chapter V. In this chapter, we shall *assume*, rather than derive, a form for $\mathcal{V}$ in accordance with the empirical development of the shell model which featured in early years. The potential extensively used in shell-model work is the isotropic harmonic oscillator potential supplemented by a spin-orbit coupling term. The eigenvalues and eigenfunctions of the former have already been calculated in Section 17. In Section 24, we shall consider the effect of adding the spin-orbit coupling term to such a potential. It is then seen that this empirically-determined form of the one-body potential does indeed produce a level system that explains all the observed magic numbers.

As regards the residual two-body potential in detailed shell-model calculations, we shall develop the formalism for such calculations in a way that leaves us the option of using either the t-matrix elements of finite nuclei (described in Section 22) or an empirically-determined well-behaved smooth potential, such as the one applied in the many-body perturbation calculation of Section 18; the addition of a tensor and two-body spin-orbit potential to (II.98a) may also be done, if desired.

24. SINGLE-PARTICLE LEVELS AND MAGIC NUMBERS

We refer to Section 17C for the single-particle eigenvalues and eigenfunctions belonging to the harmonic oscillator potential. Table III.1 gives the various energy levels in order of increasing energy. A given state nl has $(2l + 1)$ degenerate substates corresponding to the values of the projection quantum number $m_l = -l, -l + 1, \ldots, l$. Each such (nlm_l)-substate can accommodate two nucleons of each kind (i.e., neutron or proton) corresponding to the two alignments of spin m_s $(= \pm\frac{1}{2})$. Thus, the Pauli exclusion principle allows $2(2l + 1)$ nucleons of each kind to go to a particular oscillator state nl. This enables us to estimate how many neutrons or protons are needed to fill up the various oscillator energy levels $\lambda = 0, 1, 2, \ldots$.

Table III.1 Shell closure for isotropic harmonic oscillator

Value of λ	Values of nl	Number of Nucleons Needed to Fill Up Shell	Total Number of Nucleons at Shell Closure
0	(00)	2	2
1	(01)	6	8
2	(10)(02)	12	20
3	(11)(03)	20	40
4	(20)(12)(04)	32	72
5	(21)(13)(05)	42	114
6	(30)(22)(14)(06)	56	170

The first energy level $\lambda = 0$ is obviously filled up with 2 nucleons, and the second level $\lambda = 1$ with 6 nucleons. The third level $\lambda = 2$ contains two nl-states, namely, (02) and (10) which are filled up with 10 nucleons and 2 nucleons, respectively. Thus, the third oscillator energy level $\lambda = 2$ is filled up with 12 more nucleons, and so on. The number of nucleons needed to fill up each oscillator energy level λ is shown in the third column of Table III.1. When each oscillator energy level is filled up, we say that a *shell closure* has taken place. The *total* number of nucleons at each shell closure is given in the last column of the table. The entry in each line of this column is a sum of all the numbers in the third column from the first line to the particular line under consideration.

Words such as *energy level*, *state*, and *substate* carry a special connotation. Each energy level is determined by the value of λ and is degenerate in two ways: first, it contains several 'states' of different nl and, second, an nl-state itself is degenerate with respect to the projection quantum numbers m_l and m_s. We have called the states nlm_lm_s the *substates* of the state nl. We shall also very frequently use the word *sublevel* to denote the various nl-states present for a particular energy level. When all the sublevels belonging to a particular oscillator energy level are filled up, only then do we say that we have a *closed-shell* nucleus. In between two successive closed-shell nuclei there will obviously be a few nucleon numbers corresponding to the complete filling-up (i.e., closure) of the various sublevels. These nuclei are said to have a *closed sublevel* or *closed subshell*.

The way the word *shell* is being used here makes it synonymous with the oscillator energy level. This, however, is not strictly true when we consider the spin-orbit coupling. According to the general concept of a shell, introduced in Section 23, the word denotes a collection of states lying very close in energy and separated from a similar group by a comparatively larger energy interval. In the case of the harmonic oscillator potential, therefore, a shell becomes identically specified by a λ-value, and all sublevels of a shell are degenerate. We shall presently see that both these features change with the introduction of the spin-orbit coupling.

Returning to the third column of Table III.1, we expect nuclei having the neutron or proton numbers given here to be particularly stable. Therefore, they should be compared with the magic numbers mentioned in Section 23. We notice that only the first three numbers 2, 8, and 20 (corresponding to He^4, O^{16}, and Ca^{40}) agree with the known magic numbers. The rest of the numbers appearing in this column differ from the observed magic numbers 50, 82, and 126. To explain these higher magic numbers, Mayer suggested that a spin-orbit coupling term

$$V_{ls}(\mathbf{r}) = -U(r)\boldsymbol{l}\cdot\mathbf{s} \tag{III.1}$$

be added to the spherical shell-model potential (e.g., the harmonic oscillator potential). This potential causes a splitting of the $(j = l \pm \frac{1}{2})$-levels, as now shown. Since $j^2 = l^2 + s^2 + 2\boldsymbol{l}\cdot\mathbf{s}$, we have

$$\begin{aligned}\langle n(l\tfrac{1}{2})jm|\ V_{ls}(\mathbf{r})\ |n(l\tfrac{1}{2})jm\rangle &= -\langle U(r)\rangle_{nl}\tfrac{1}{2}[j(j+1) - l(l+1) - \tfrac{3}{4}]\\ &= \begin{cases}\tfrac{1}{2}(l+1)\langle U(r)\rangle_{nl} & \text{for } j = l - \tfrac{1}{2}\\ -\tfrac{1}{2}l\langle U(r)\rangle_{nl} & \text{for } j = l + \tfrac{1}{2}\end{cases}\end{aligned} \tag{III.2}$$

Here $|nl\frac{1}{2}jm\rangle$ is a single-particle state with the radial function R_{nl}/r and a total angular momentum j obtained by coupling the orbital angular momentum l with the spin $\frac{1}{2}$; m is the projection of j along the quantization axis; and $\langle U(r)\rangle_{nl}$ is the radial integral

$$\int_0^\infty R_{nl}^2(r)U(r)\,dr$$

of the shape function $U(r)$ of the spin-orbit potential. (III.2) shows that the state $j = l - \frac{1}{2}$ is pushed up (+ sign), whereas the state $j = l + \frac{1}{2}$ is pushed down (− sign). The splitting of the two j-components of a given l is found to be

$$\Delta E_l = \tfrac{1}{2}(2l+1)\langle U(r)\rangle_{nl}. \tag{III.3}$$

It should be noted that the signs of the energies $j = l \pm \frac{1}{2}$ in (III.2) are determined by the choice of an attractive (−) sign in the definition (III.1). With such a choice, the higher j-level for a given l is found to be below the lower one. We shall shortly see that this is required by the observed magic numbers.

After the introduction of the spin-orbit coupling, various sublevels in an oscillator shell will be specified by the set of quantum numbers nlj. Each such sublevel will be filled up with $(2j + 1)$ nucleons of each kind corresponding to the possible $(2j + 1)$ values of m.

The level $nl = 0s$ for $\lambda = 0$ now changes to $0s_{1/2}$, where the subscript denotes the j-quantum number. It should be observed that, for an s-state ($l = 0$), only one value of j $(= l + \frac{1}{2})$ is possible. This shell is still filled up with $(2j + 1 =)$ 2 nucleons. The next level $0p$ for $\lambda = 1$ is now split up into $0p_{3/2}$ and $0p_{1/2}$, the former occurring below the latter. They are filled up by 4 and 2 nucleons, respectively. Thus, the total number of nucleons necessary to fill up the second shell is still 6, and the total number of nucleons when this shell is closed is still 8. The next shell $\lambda = 2$ contains $1s$ and $0d$, which now split up into $1s_{1/2}$, $0d_{5/2}$, and $0d_{3/2}$. The total number of nucleons to fill these up is still the same as that shown in Table III.1. Important changes start taking place from the next group of sublevels which correspond to $\lambda = 3$, and $nl = 1p$, $nl = 0f$. These levels now split up into $1p_{3/2}$, $1p_{1/2}$ and $0f_{7/2}$, $0f_{5/2}$. If we assume the radial integral in (III.3) to be very weakly dependent on the quantum numbers

n, l, then we expect the splitting, which is proportional to $(2l + 1)$, to be larger for the f-state than for the p-state. Thus, the state $0f_{7/2}$ will be the lowest of the sublevels, and will be filled up with 8 nucleons; the nucleon number $(20 + 8 =)$ 28 therefore marks the closure of the sublevels $0f_{7/2}$ and accounts for the occurrence of this semi-magic number. The remaining sublevels $1p_{3/2}$, $1p_{1/2}$, and $0f_{5/2}$ require 12 nucleons to be filled up, bringing the total number of nucleons to 40, which is also known to be a semi-magic number. In order that the magic number 50 be reproduced, we obviously have to consider the next set of levels belonging to $\lambda = 4$. If we once again invoke the dependence of the spin-orbit splitting on $(2l + 1)$, we expect that the level $0g_{9/2}$ will be the lowest in this set. The spin-orbit splitting for the g-state is so large that this particular level is pushed down from above to the vicinity of the levels contained in $\lambda = 3$. If this hypothesis is accepted, then it is easy to see that the $0g_{9/2}$ requiring 10 nucleons to be filled up precisely explains the existence of the magic number 50. We now notice that the spin-orbit coupling has radically changed the concept of an oscillator shell specified by λ. Instead of the levels $1p$ and $0f$ contained in $\lambda = 3$ defining this particular shell, the level $0g_{9/2}$, which came down from the next higher λ, also teamed up with $1p_{3/2}$, $1p_{1/2}$, $0f_{7/2}$, and $0f_{5/2}$ to redefine the shell. Thus, all the sublevels specified by the quantum numbers nlj and present between two actual magic numbers is the realistic definition of a shell, as is understood by nuclear structure theorists. Very often the expression *major shell* is used to describe the same entity. It should also be observed that all the sublevels belonging to the major shell between the magic numbers 20 and 50 have odd parity, except for the level $0g_{9/2}$ which came down from above. This level has an even parity, and may be called an *intruder* level in this particular major shell.

The idea of spin-orbit splitting can be similarly applied to the sublevels in $\lambda = 5, 6, 7$ for the explanation of the remaining two magic numbers. The sublevel $0g_{7/2}$, together with the rest of the sublevels in $\lambda = 4$, i.e., $2s_{1/2}$, $1d_{5/2}$, and $1d_{3/2}$, requires 20 nucleons, taking the total to 70 nucleons. Then $0h_{11/2}$ comes down from $\lambda = 5$ to the vicinity of these levels, and the extra 12 nucleons filling it up explain the magic number 82. In the same way, the level $0i_{13/2}$ coming down to the vicinity of the sublevels in $\lambda = 6$ will explain the magic number 126. These levels, namely, $0h_{11/2}$ and $0i_{13/2}$, once again are the intruder levels in the respective major shells and have parity opposite to that of the rest of the levels.

Figure III.1 shows the single-particle level spectrum. The harmonic oscillator levels are indicated in the extreme left-hand column; the values of $\lambda = 0, 1, 2, \ldots$ times $\hbar\omega$ and the parity of the sublevels contained in each λ are also specified here. The next column gives an arbitrary splitting of the various degenerate levels nl contained in each λ; this type of splitting could be produced, for example, by replacing the harmonic oscillator potential by a square-well potential or some other more realistic central potential. The third column shows the shell-model levels with the spin-orbit splitting; beyond the nucleon number 40, the migration of the $(j = l + \frac{1}{2})$-level for the highest l to the lower set of levels is very clear here. The next two columns list the count of the nucleons for filling up the various levels. The extreme right-hand column suitably adds up these numbers and demonstrates the reproduction of the magic numbers. The reader would do well to use this diagram in familiarizing himself with the sublevels nlj contained in each major shell, i.e., between two subsequent magic numbers. Too much importance should not be attached to the ordering of the sublevels contained in a major shell, as illustrated in Fig. III.1. The way Mayer did this ordering is explained when we deal, in Section 25, with the ground-state spins of nuclei predicted by the shell model.

To increase the reader's familiarity with some real nuclei, in the context of shell-model

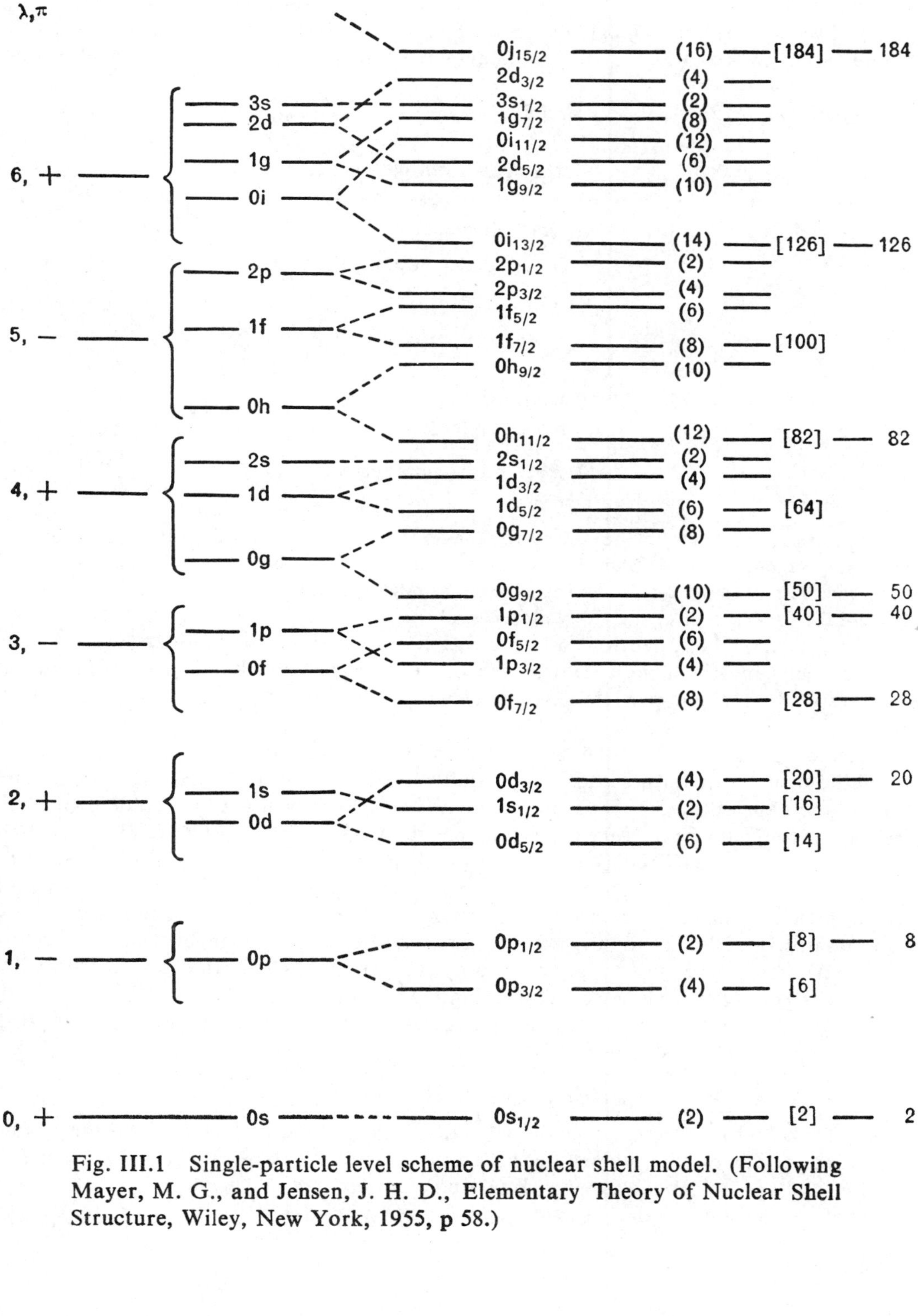

Fig. III.1 Single-particle level scheme of nuclear shell model. (Following Mayer, M. G., and Jensen, J. H. D., Elementary Theory of Nuclear Shell Structure, Wiley, New York, 1955, p 58.)

ideas, we add a few facts:

(i) If we treat the semi-magic numbers 28 and 40 on an equal footing with the other magic numbers, then the nuclei in the first column of Table III.2 occupy a very special position

Table III.2 Closed-shell nuclei

Nucleus	Z, N	Other Nuclei in Neighbourhood
He^4	2, 2	All $0p$-shell nuclei with $4 < A < 16$
O^{16}	8, 8	F, O, and Ne isotopes up to $A = 20$
Ca^{40}	20, 20	K, Sc, and Ca isotopes near $A = 40$
Ca^{48}	20, 28	
Ni^{56}	28, 28	Ni isotopes above $A = 56$
Zr^{90}	40, 50	Y, Zr, and Nb isotopes near $A = 90$
Pb^{208}	82, 126	Tl, Pb, and Bi isotopes near $A = 208$

in nuclear structure physics, in the sense that they are doubly magic, i.e., both have neutrons and protons in closed shells. Strictly speaking, only the first three and the last nuclei are doubly magic. The others involve the semi-magic numbers 28 and 40. Of the latter nuclei, Ca^{48} and Zr^{90} conform very well to a doubly closed-shell picture.

(ii) Up to Ca^{40}, both neutrons and protons fill up the same shell-model levels. Beyond Ca^{40}, *stable* nuclei mostly have a neutron excess, and hence the outermost neutrons in such nuclei correspond to single-particle levels which are usually well above the levels of the protons.

(iii) Nuclei in which either the neutrons or protons correspond to a closed shell, leaving the other kind of nucleons to proceed to a partially filled shell, can be given the special name *single closed-shell* nuclei. Besides the isotopes of the nuclei shown in the first column of Table III.2, the stable Sn isotopes are very important members of this category ($Z = 50$, $N > 50$).

(iv) In detailed shell-model calculations such as those described in this chapter, the closed shells are usually treated as an inert core; only the nucleons in the outermost partially filled sublevels are assumed to determine the various nuclear properties. The complexity of the calculation increases tremendously with a rise in the number of such nucleons. Therefore, detailed calculations have become practicable only for the very light nuclei, and for the nuclei near the closed-shell ones listed in the first column of Table III.2. The third column of the table will familiarize the reader with the names of such nuclei, which are very popular with workers engaged in detailed shell-model work.

25. GROUND-STATE SPIN OF NUCLEI

One of the physical properties of nuclei very lucidly explained by the shell model is the ground-state total angular momentum, sometimes called the ground-state spin at the risk of confusion with the *intrinsic* spin S. When we consider single-nucleon states specified by the quantum numbers *nljm*, the intrinsic spin of each nucleon is already coupled with its orbital angular momentum l to form the individual particle total angular momentum j. Thus, when all the j's of the individual nucleons are coupled, we automatically get the *total* angular momentum of the system. In this manner of coupling the angular momenta, the total intrinsic spin S does not appear anywhere. For obvious reasons, when we couple the individual particle j-values to obtain the total J, the coupling scheme is referred to as *jj-coupling*.

There is an alternative coupling scheme for the shell-model states which uses the single-particle states specified by the quantum numbers nlm_lm_s, rather than $nljm$. In this scheme, all the orbital angular momenta l are coupled to arrive at the total orbital angular momentum L; simultaneously, the intrinsic spins $\frac{1}{2}$ of the particles are coupled to each other to form the total S; finally, L and S are coupled to each other to obtain the total J. This coupling scheme is called *LS-coupling*.

In all our work on shell-model angular momenta, we shall confine our discussion to the jj-coupling picture; only while doing certain calculations do we need to transform our jj-coupled functions to LS-coupled ones.

To be able to predict the ground-state spin, we shall have to make use of the energy levels in Fig. III.1 and consider the possible assignment of the particles to the various levels so that the lowest state for the given number of nucleons can be obtained. In general, such an assignment involves a set of completely occupied levels $n_cl_cj_c$ and a few nucleons distributed in some of the m-sublevels of a partially filled level nlj. Sometimes the last few nucleons may be distributed among more than one partially filled level nlj. A unique prediction of the ground-state spin is possible only in certain simple cases; in more complicated cases, additional physical assumptions are required. We shall start with the simple cases and then proceed to the more complicated ones, outlining at the same time the extra assumption that may be necessary in each case.

A. SPIN OF A COMPLETELY FILLED LEVEL

Let us first consider a level nlj with $(2j + 1)$ particles completely occupying its various m-sublevels. The level is filled up, and we wish to find out the total angular momentum of such a state. Since all the m-sublevels are occupied and m occurs in $\pm$ pairs of equal magnitude ($m = \pm\frac{1}{2}, \pm\frac{3}{2}, \ldots, \pm j$), the total angular momentum projection M, which is the sum of the individual m-values, is equal to zero. The result $M = \Sigma\, m$ follows from the fact that

$$\mathbf{J} = \sum_i \mathbf{j}(i),$$

and hence

$$J_z = \sum_i j_z(i),$$

where $\mathbf{J}$ and $\mathbf{j}$ are the total and individual particle angular momentum *operators*, respectively, and the label i denotes the various nucleons.

If $(2j + 1)$ nucleons are to be distributed amongst the $(2j + 1)$ sublevels of nlj, there is only one way to do it; therefore, there is *only one* determinantal antisymmetric many-body state in such a case. Since the total angular momentum projection of the state is $M = 0$, it can have only the angular momentum $J = 0$. It should be observed that $M = 0$ could have resulted from any higher value of J also, but in that case there should have been $(2J + 1)$ many-body states corresponding to the various projections of the non-zero J. The fact that there is *only one* state in the present case is crucial in uniquely fixing the value of $J = 0$.

To clarify the counting of the many-body states, let us note the difference between a closed-shell case and a partially filled one. Suppose we are considering the case of 3 nucleons in a level *nlj*, where $j = \frac{5}{2}$. This level can accommodate 6 nucleons corresponding to the *m*-values $\pm\frac{1}{2}$, $\pm\frac{3}{2}$, and $\pm\frac{5}{2}$. The 3 nucleons at our disposal can be distributed amongst these 6 sublevels in ($^6C_3$=) 20 ways; therefore, now there are 20 *different* many-body states. Each of the 20 states corresponds to a particular allotment of the projection quantum numbers. For example, $(+\frac{5}{2}, +\frac{3}{2}, +\frac{1}{2})$, $(+\frac{5}{2}, +\frac{3}{2}, -\frac{1}{2})$, $(+\frac{5}{2}, +\frac{3}{2}, -\frac{3}{2})$, and $(+\frac{5}{2}, +\frac{1}{2}, -\frac{1}{2})$ are several such possible assignments. As is clear, each such assignment has a total projection quantum number which is obtained by adding the three numbers. In general, there may be several states with the same total projection; for example, each of the last two assignments just stated has $M = +\frac{5}{2}$. In Section 29, we shall consider this technique in detail to arrive at the various possible values of J in a partially filled level.

B. ONE NUCLEON IN A LEVEL

We next consider the case of a nucleus having several filled levels $n_c l_c j_c$ and one nucleon in an unfilled level *nlj*. According to our previous discussion, the total angular momentum of each of the completely occupied levels $n_c l_c j_c$ is zero. By coupling all these zero angular momenta, we still get a state of total angular momentum zero and projection zero. To this state, we have to finally couple the state of the last nucleon in the level *nlj*. If the projection quantum number to which the last particle belongs is m, then obviously the final total angular momentum J and its projection M are respectively given by j and m, the corresponding quantities for the last nucleon of the nucleus.

C. ONE NUCLEON MISSING FROM A LEVEL

An equally easy case is one where a nucleus has several filled levels $n_c l_c j_c$ and an additional level *nlj* containing $2j$ nucleons. The last level would then have only one nucleon missing from a completely filled status. Let the missing nucleon belong to the sublevel of projection $-m$, and let the projection of the many-body state under consideration be M. If the additional nucleon of projection $-m$ were also present, then we would, once again, have a completely filled level *nlj* with a total projection zero and angular momentum zero. Therefore,

$$M - m = 0 \quad \text{or} \quad M = m.$$

Since the coupling of the many-body state with the angular momentum j of the missing nucleon produces a completely occupied state of total angular momentum zero, it follows from the angular momentum coupling rule that the former has a unique value of $J = j$.

We may therefore conclude that a many-body state with a set of fully occupied levels and one particle of state $(j, -m)$ missing from the level *nlj* corresponds to a total angular momentum j, which is equal to that of the missing particle, but with a projection equal and *opposite* to that of the missing particle.

We now introduce a new nomenclature. A many-body state of the type we are considering will be called a one-*hole* state of angular momentum j and projection m [associated with the absence of a particle in the state $(j, -m)$]. It is shown in Section 30A that there is a similar one-to-one correspondence between the states of A particles in a level *nlj* and those of A holes in the same level.

The instances characterized by a completely filled level, and by one particle or one hole in a level, are simple cases where a unique prediction of the total angular momentum is possible.

Already, these results enable the prediction of the ground-state spins of several nuclei. For example, C^{12} [filled $(0s_{1/2})$- and $(0p_{3/2})$-level] and O^{16} [filled $(0s_{1/2})$-, $(0p_{3/2})$-, and $(0p_{1/2})$-level] have zero spin; C^{13} [one particle in $(0p_{1/2})$- and filled $(0s_{1/2})$-, $(0p_{3/2})$-level] and N^{15} [one hole in $(0p_{1/2})$- and filled $(0s_{1/2})$-, $(0p_{3/2})$-level] have a ground-state spin of $\frac{1}{2}$; O^{17} and F^{17} [one particle in $(0d_{5/2})$-level, filled $(0s_{1/2})$-, $(0p_{3/2})$-, and $(0p_{1/2})$-level] have a ground-state spin of $\frac{5}{2}$. It is easy to find many heavier nuclei also where these simple results are applicable.

Although we have stated the foregoing results in the context of ground-state spin, it is clear, from the way they have been arrived at, that they can be applied to the prediction of spins of excited states also, provided such states belong to the category of the simple cases just mentioned. The example of C^{13} clarifies the situation. In this nucleus, the 6 protons and 6 neutrons fill up the $(0s_{1/2})$- and $(0p_{3/2})$-level, whereas the last neutron occupies the next level $0p_{1/2}$ so that the lowest energy can be obtained. This is therefore the likely composition of the ground state of C^{13}. If we intend to excite such a system, we have several options.

(i) Excite a neutron from the filled $(0p_{3/2})$-level to the $(0p_{1/2})$-level, thereby filling the $(0p_{1/2})$-neutron level and creating a hole in the $(0p_{3/2})$-level.

(ii) Excite a proton in the way suggested in (i); this, however, does not close the $(0p_{1/2})$-level because this level needs two neutrons, or two protons, to fill up, whereas in the present case we have only one neutron and one proton.

(iii) Excite the neutron from the level $0p_{1/2}$- to the higher levels $0d_{5/2}$, $1s_{1/2}, \ldots$.

It is clear that cases (i) and (iii) allow the application of the simple results. In case (i), we predict an excited state $3/2^-$, i.e., spin 3/2 and odd parity (since the hole belongs to a p-state); the parity is the same as that of the ground state [$1/2^-$ because of the $(0p_{1/2})$-particle]. The notation we have just used is standard in nuclear structure; according to it, J^π denotes the angular momentum J and parity π of the state. In case (iii), we have a set of closed levels and a particle in $(0d_{5/2})$- or $(1s_{1/2})$-, . . . level. Thus, the spin and parity of such excited states will be $5/2^+$, $1/2^+, \ldots$. Case (ii), however, is more complicated; it consists of one proton hole in $0p_{3/2}$, and one neutron and one proton in $0p_{1/2}$. The coupling of the three angular momenta present here can generally give rise to states of several possible angular momenta.

D. ARBITRARY NUMBER OF PARTICLES IN A LEVEL—PAIRING ASSUMPTION

Let us now deal with one of the more complicated cases, namely, that of an arbitrary number, k, of particles in a level nlj. The total number of antisymmetric states of such a system is equal to the total number of ways in which k different m-values can be picked out of the total set of $(2j + 1)$ m-values. The number is given by ${}^{2j+1}C_k$. The total projection M for each set of k projection quantum numbers is obtained by summing them. The method of determining the various permissible J-values with the help of these M-values is discussed in Section 29. What we emphasize here is the fact that, for a general system $(nlj)^k$, i.e., one where there are k nucleons in the level nlj, the total angular momentum J can have many possible values. Therefore, strictly speaking, a unique prediction of the ground-state spin cannot be made. The ordering of the various J-states of such a nucleus will depend on exactly what kind of residual interaction is present between the pairs of nucleons. If, however, there is a physical clue to the character of these residual interactions which is of a sufficiently general nature, then it can be utilized to predict which of the various permissible J-states is likely to be the lowest in energy (and hence corresponds to the ground state of the nucleus). This is exactly what is done in making a shell-model prediction of ground-state spins in a complicated case.

We first state the general characteristic of the residual interaction just referred to. A pair of neutrons or protons in a particular j-level can couple their angular momenta to all possible even integers between 0 and $(2j - 1)$; but any reasonable residual *short-range* interaction produces the maximum attractive energy for the 'paired' state of angular momentum zero. From now on, we shall use the expressions *paired state* and *pairing* in a very specialized sense: to denote a pair of nucleons of angular momentum zero. The fact that the residual interaction depresses the paired two-nucleon state to the maximum extent is explicitly demonstrated in Section 28D by working out the matrix elements of the two-body interaction for a δ-function type potential; the result, however, is true for a wide variety of short-range forces. We shall also consider in Section 41D an idealized 'pairing interaction' which is effective only for the paired state, and zero for a two-nucleon state of non-zero angular momentum.

Granted the pairing property of the two-nucleon residual interaction, we find it quite straightforward to predict which of the various J-states of $(nlj)^k$ is lowermost in energy. In the case of an even number of nucleons, k, the minimum energy is obviously obtained when all the nucleons are paired, i.e., they form $k/2$ pairs, each of angular momentum zero. The resultant angular momentum, which is a sum of the angular momenta of all the pairs, is evidently zero. This being true for both an even number of protons and an even number of neutrons, we always expect an even-even nucleus (frequently referred to as *even* nucleus for brevity) to have a ground-state angular momentum zero. No exception to this rule is known in the periodic table.

In the same way as just described, for an odd value of k, the ground state will correspond to $(k - 1)/2$ pairs, each of angular momentum zero, and a last odd nucleon of angular momentum j. By coupling all these angular momenta, we get a total J equal to j, the same as the angular momentum of a single nucleon in the state nlj. If therefore we consider an even-odd or an odd-even nucleus (each called *odd-mass* nucleus), the ground-state spin will be the same as that of the single-particle level to which the last odd particle belongs. Except in rare cases, this rule is also found to be true over the whole periodic table.

The last observation needs clarification. We recall the fact that the observed magic numbers helped to finalize the group of single-particle levels that occur between two magic numbers, i.e., in a major shell. As far as the ordering of the various levels inside a shell is concerned, it could have been well-determined had we precisely known the shell-model potential. Since, however, this potential is empirically determined, it is a part of the empirical approach to fix the ordering of the levels inside a shell from experimental data. If we start distributing the nucleons in a given odd nucleus amongst the shell-model levels starting from the bottom, we shall obviously not be able to decide the precise level to which the last odd nucleon goes because the ordering of the levels as we approach the end is not known in advance. But our predicted ground-state spin according to the pairing hypothesis, together with the experimentally observed value, now enables us to determine the single-particle level to which the last odd nucleon goes. Once this has been ascertained with the help of experimental data on one odd-mass nucleus, then the ground-state spin of several successive odd-mass nuclei can be predicted to be the same; this is because a given nlj-level will be filled with $(2j + 1)$ nucleons, and the spin of all the nuclei having $1, 3, 5, \ldots, 2j$ nucleons in this particular level will be equal to j, according to our pairing hypothesis. As the nucleon number counted from the point where the level nlj started filling up exceeds $(2j + 1)$, we should expect the next few odd nuclei to have a different spin. The observed value of the first one of them helps us identify the succeeding single-particle level, and then the next few nuclei, having the new ground-state spin, corroborate our shell-model ideas. This was the method followed by Mayer in establishing the level sequence inside the various major shells, as illustrated in Fig. III.1.

At several points in the periodic table, there are departures from the scheme of level filling just outlined, and this, once again, can be explained in terms of the pairing ideas. This kind of departure usually occurs when a level of low spin occurs just below a level of much higher spin. If we denote the two spins by j and j', respectively, then, according to the foregoing rule, we expect nuclei having 1, 3, 5, . . . , $2j$ nucleons in the j-level to show a ground-state spin equal to j, and immediately following these nuclei there will be odd-mass nuclei with ground-state spin of j'; the latter nuclei are expected to have a fully-occupied j-level and 1, 3, 5, . . . , $2j'$ particles in the j'-level. Very frequently, however, if j' is much larger than j, these nuclei are also found to have a ground-state spin of j instead of the expected j'. This apparently implies that the Pauli exclusion principle has broken down, and the level j is accommodating more than $(2j + 1)$ nucleons. The pairing hypothesis satisfactorily explains this paradox without violating a basic concept like the Pauli principle.

In Section 41D, we shall use a model pairing interaction to show that the attractive pairing energy is proportional to $(2j + 1)$ [see (VI.94b) with $j_1 = j_1' = j$]. For more realistic two-body potentials, the strict proportionality to $(2j + 1)$ may not be true, but it would be safe to say that the attractive matrix element for the state $J = 0$ gets larger as the single-particle j increases. This result therefore implies that, in the previous example, it may be energetically more advantageous to put nucleons in pairs in the j'-orbit rather than the j-orbit because the attractive pairing energy is larger in the j'-orbit. If the difference in the pairing energy is enough to swamp the difference in the single-particle energies of the pair $[2\epsilon(j') - 2\epsilon(j)]$, then the pair would definitely prefer to go to the j'-level than to the j-level. This situation could arise if the j'-level is just above the j-level in the single-particle level diagram, and j' much larger than j. When this happens, the j'-level tends to fill in pairs, at the cost of a few pairs in the lower j-level; hence, the j-level then remains unfilled for quite a large number of nuclei in the particular region of the periodic table. Since the last odd nucleon does not gain any pairing energy because it does not have a partner, it prefers to go to the j-level, rather than the higher j'-level, when vacancies are available in both levels. This completely explains the observed existence of many odd-mass nuclei with the ground-state spin j, in apparent violation of the Pauli principle.

All the foregoing facts get clarified by the data on ground-state spins of several odd-mass nuclei given in Table III.3. Here we have listed odd-proton nuclei, and hence by our pairing hypothesis the last odd proton determines the ground-state spin. Since the even number of neutrons always couple their angular momenta to zero, no detailed information on the neutrons is tabulated. The last four columns in the table contain the relevant information on the protons. The first entry is a nucleus with 21 protons, which is one proton more than the magic number 20. At $Z = 20$, the major shell with the levels $1s_{1/2}$, $0d_{5/2}$, and $0d_{3/2}$ has been filled up. The next set of levels is $0f_{7/2}$, $0f_{5/2}$, $1p_{3/2}$, $1p_{1/2}$, and $0g_{9/2}$. The nucleus with 21 protons (Sc^{45}) has a ground-state spin of 7/2, showing that the level $0f_{7/2}$ is the lowest of this set of levels. Having found this, we can now predict the ground-state spins of nuclei having 23, 25, and 27 protons, all of which should be 7/2. This agrees with the observed results, except in the case of Mn^{55}. At $Z = 28$, the level $0f_{7/2}$ gets filled up, and hence we expect the nucleus with $Z = 29$ to have a different spin, and indeed we find this to be 3/2, showing that the level immediately above $0f_{7/2}$ is $1p_{3/2}$. The nucleus with $Z = 31$ is then expected to have a spin 3/2, which is true; but then $1p_{3/2}$ fills up with 4 protons, and hence the nucleus with $Z = 33$ would be expected to have a different spin. This, however, is not true.

Table III.3 Ground-state spin of odd-mass nuclei

Z	Nucleus	A	Ground-State Spin	Level of Last Odd Proton (j)	Next Level (j')	Number of Protons	
						j	j'
21	Sc	45	7/2	$0f_{7/2}$	$1p_{3/2}$	1	0
23	V	51	7/2	$0f_{7/2}$	$1p_{3/2}$	3	0
25	Mn	53	7/2	$0f_{7/2}$	$1p_{3/2}$	5	0
25	Mn	55	5/2	$0f_{7/2}$	$1p_{3/2}$	5	0
27	Co	57	7/2	$0f_{7/2}$	$1p_{3/2}$	7	0
27	Co	59	7/2	$0f_{7/2}$	$1p_{3/2}$	7	0
29	Cu	63	3/2	$1p_{3/2}$	$1f_{5/2}$	1	0
29	Cu	65	3/2	$1p_{3/2}$	$1f_{5/2}$	1	0
31	Ga	69	3/2	$1p_{3/2}$	$1f_{5/2}$	3	0
31	Ga	71	3/2	$1p_{3/2}$	$1f_{5/2}$	3	0
33	As	75	3/2	$1p_{3/2}$	$1f_{5/2}$	3	2
35	Br	79	3/2	$1p_{3/2}$	$1f_{5/2}$	3	4
35	Br	81	3/2	$1p_{3/2}$	$1f_{5/2}$	3	4
37	Rb	81	3/2	$1p_{3/2}$	$1f_{5/2}$	3	6
39	Y	89	1/2	$1p_{1/2}$	$0g_{9/2}$	1	0

We notice from Table III.3 that the same ground-state spin of 3/2 is observed for all the odd-proton nuclei up to $Z = 37$. This then is an example of the pairing effect. In order that the semi-magic number 40 can be reproduced, the two levels after $1p_{3/2}$ should be $0f_{5/2}$ and $1p_{1/2}$, and next should come $0g_{9/2}$. Of the first two levels, we choose $0f_{5/2}$ as lower in energy since this has a spin higher than 3/2. Then we interpret the observed ground-state spin 3/2 up to $Z = 37$ by assuming that $0f_{5/2}$ is just above $1p_{3/2}$, and receives protons in pairs from $Z = 33$ onwards, as shown in the last column of the table. The level $1p_{3/2}$ is thus left with 3 protons all the time, and gives rise to the observed spin of 3/2 for all the odd nuclei in this region. At $Z = 37$, the level $0f_{5/2}$ is completely filled with protons, and hence the peculiarity of the pairing effect cannot continue any further. When we add one more pair of protons, one of the pairs necessarily goes to $1p_{3/2}$, filling it up, and the second one goes to $1p_{1/2}$, $0f_{5/2}$ being already fully occupied. This explains the observed spin 1/2 of the last nucleus Y in the table. Similar considerations apply all over the periodic table, except in the regions of permanent deformation (discussed in Chapter IV).

The case of Mn^{55}, as shown in Table III.3, is peculiar, and obviously it cannot be explained even with pairing ideas. There are a few more instances of such delinquent cases: Ne^{21} and Na^{23} [$(0d_{5/2})$-level, $J = 3/2$]; Ti^{47} [$(0f_{7/2})$-level, $J = 5/2$]; and Se^{79} [$(0g_{9/2})$-level, $J = 7/2$].

We now describe a very important consequence of a level of higher angular momentum located just above a level of lower angular momentum filling in pairs. Although the ground state, which is the state of lowest energy, is obtained in this capricious way, we can still have a state with the normal way of filling the levels. The excited state of the odd nucleus obtained thereby will have the last odd nucleon located in the level of higher spin (j'), and hence will have a resultant spin j'. The ground-state spin j is much lower than the spin of this close-lying excited level, and the two states differ in the level-assignment of only one particle. We shall see in Section 27 that the decay of such an excited state to the ground state by the emission of a γ-ray is given essentially by the transition matrix element of a single particle from the state j' to the state j. It is explained in Section 27 that this decay probability is very much cut down due to the small energy of the γ-ray and its large multipolarity (which is related to the difference in the angular momenta of the two states). A long-lived excited state of a nucleus, lying very close to the ground state, is called an *isomeric* state. We therefore expect to observe an isomeric state in all the odd-mass nuclei where pairing effects produce a ground-state spin which differs from that of the topmost single-particle level of very high spin. The abundance of isomerism in certain regions of the periodic table is a well-known fact, and these regions are usually called *islands of isomerism*. The location of these islands in the periodic table agrees very well with the regions where we expect to find a high-spin shell-model level close to a low-spin one.

Finally, we apply our coupling ideas to the prediction of the spins of odd-odd nuclei. The neutrons will have a spin equal to that of the last odd neutron, and a similar observation applies to the protons that are also odd in number. Let these spins be j_n and j_p, respectively. The angular momentum J of the entire nucleus is obtained by coupling these two angular momenta, thereby producing all possible values between $|j_n - j_p|$ and $(j_n + j_p)$. It is therefore not possible to uniquely predict the ground-state spin of an odd nucleus. From a study of the observed spins of odd-odd nuclei, Nordheim tried to establish two empirical rules:

Weak rule (which does not predict an exact value of J) If $j_p = l_p + \frac{1}{2}$ and $j_n = l_n + \frac{1}{2}$ or $j_p = l_p - \frac{1}{2}$ and $j_n = l_n - \frac{1}{2}$, then $J \leqslant (j_n + j_p)$.

Strong rule If $j_p = l_p + \frac{1}{2}$ and $j_n = l_n - \frac{1}{2}$ or $j_p = l_p - \frac{1}{2}$ and $j_n = l_n + \frac{1}{2}$, then $J = |j_n - j_p|$.

Most of the odd-odd nuclei are unstable and the foregoing coupling rules are based on the assignments of spin and parity of such nuclei from β-decay data. However, it must be remembered that there are many observed violations of Nordheim's rules. de-Shalit[3] has tried to understand these coupling rules by introducing a model two-body interaction between the neutron and the proton, and then calculating the energy of all the states of different J.

For odd-odd nuclei having a neutron (proton) hole and proton (neutron) particle, the rule $J = j_n + j_p - 1$ holds fairly well.

26. STATIC ELECTROMAGNETIC MOMENTS OF NUCLEI

The two most important static electromagnetic moments that have been systematically measured for a large number of nuclei are the magnetic dipole and electric quadrupole moments. In recent years, new experimental techniques have been perfected to study these moments (especially the magnetic dipole moment) for comparatively long-lived *excited states* also. A vast amount of data has been made available in this way. Although many detailed calculations remain to be done, the data have been fairly well-understood and systematized in qualitative terms. We shall presently see that very simple shell-model ideas, requiring the moments to be the property of the outermost nucleons, can explain the observed moments in a few cases; in other instances, such ideas have to be supplemented with possible magnetic or electric polarization of the so-called *core* which is composed of the inner nucleons in closed levels. The details of these core-polarization effects are considered in Section 30B. In this section, we shall work out the predictions based on a simple model wavefunction, and point out the inadequacies of such a model. We shall first explain the basic assumptions behind the model wavefunction in the context of the so-called *seniority* wavefunction defined here. The details of the moment calculation follow the definition.

Seniority Wavefunction

The seniority wavefunction has its basis in the pairing ideas introduced in Section 25. We have argued there that the nucleon-nucleon forces energetically favour the formation of 'pairs' of identical nucleons of zero angular momentum. A many-nucleon wavefunction having this paired structure, and *antisymmetric under the exchange of any two nucleons*, can be classified by its seniority. This nomenclature precisely denotes the number of 'unpaired' nucleons in the wavefunction. We recall that, according to our terminology, two identical nucleons coupled to a *non-zero* angular momentum are considered to be unpaired. Thus, the energetically most favoured ground-state wavefunction of an even number (A) of protons or neutrons which consists of $A/2$ pairs is a seniority-*zero* state. If a pair of nucleons is broken and coupled to a non-zero angular momentum, then the resultant wavefunction corresponds to a higher energy (according to pairing concepts), and has two nucleons that are unpaired. Therefore, such a wavefunction belongs to an excited state of the system and has seniority two. By breaking more and more pairs in this way, we get excited states of an even number of neutrons or protons having seniority four, six, ... Obviously, the highest seniority wavefunction is obtained by breaking all the $A/2$ pairs (which gives a wavefunction of seniority A) equal to the number of nucleons. So too, the lowest state of an odd number ($A + 1$) of neutrons or protons, which consists of $A/2$ pairs and one unpaired particle, is a seniority-one state. Higher excited states of the system are obtained by breaking one, two, three, . . . , $A/2$ pairs, which give rise to wavefunctions of seniority three, five, seven, . . . , ($A + 1$), respectively.

The core nucleons belonging to a seniority wavefunction may be considered to be the paired ones, which have a resultant angular momentum zero; therefore, the core is inert as far as angular momentum goes. We should observe that this slightly extends our usual concept of the core of a nucleus having a shell structure. All the sublevels of a nucleus that are completely filled have a total angular momentum zero, and, in this sense, they form the inert core of the nucleus. In the seniority wavefunction, we have, over and above the nucleons in such inert closed levels, *one* or *more* partially filled sublevels that have pairs of nucleons in paired ($J = 0$)-state and a few unpaired nucleons (their number being equal to the seniority). In our extended definition of the core, we have included the paired nucleons of partially filled levels over and above those in the completely filled levels. The point we emphasize here is that the paired nucleons in partially filled levels, although inert as far as angular momentum is concerned, can, in some cases, play a role in the evaluation of the electromagnetic moments. This is because they are required to be antisymmetric under exchange with the unpaired nucleons.

Extreme Single-Particle Model of the Wavefunction

The seniority wavefunction of the ground state of an odd-mass nucleus has one particle outside the core, which is inert with respect to angular momentum. But, because of the antisymmetry of the paired nucleons with the last odd nucleon, the core can affect the value of the moments.

We now introduce a much more simplified model which predicts the same ground-state spin as the seniority wavefunction, but lacks any antisymmetrization between the paired particles and the last odd particle. In this model, we regard the core as an inert object and completely overlook its structure in terms of individual nucleons. The core is, so to say, a mass of inert nuclear matter of zero angular momentum to which a single nucleon is coupled to produce the odd-mass nucleus. The wavefunction of the entire nucleus is therefore a simple product of the single-nucleon wavefunction and the wavefunction of the core, regarded as a collective body of zero spin. The magnetic and the quadrupole moment of the nucleus, calculated with such a wavefunction, are obviously the same as the values obtained for the last odd nucleon. This model is therefore called the extreme *single-particle model.* The collective model of the nucleus, used for the deformed nuclei and described in Section 23, treats the odd-mass nuclei in exactly the same way. It is assumed in that model that the odd nucleon is coupled to a collective nuclear droplet (the core), which is describable in terms of a few collective coordinates. The present model is also the same in spirit, but has a spherical core instead of a deformed one; if we wish to talk in terms of the collective coordinates of the core, then there is only one such, namely, the radius of the spherical object.

Until we give the details on the construction of antisymmetric shell-model wavefunctions (see Sections 29 and 30), we shall restrict our discussion to the extreme single-particle model while evaluating physical quantities such as the static moments in this section and the dynamic transition probabilities in Section 27.

A. MAGNETIC MOMENT

The intrinsic spins of the neutron and the proton, and the orbital angular momentum of the latter, are the most direct sources of the nuclear magnetic moment. The magnetic moment produced by a certain angular momentum (orbital, intrinsic, or total) is usually written as the particular *gyromagnetic ratio* times the angular momentum itself. The gyromagnetic ratio for the orbital momentum is equal to $e/(2Mc)$, where e is the charge of the proton, M the nucleon mass, and c the velocity of light. If Planck's constant $\hbar$, which is present in all angular momenta as the unit, is extracted from the angular momentum, and the gyromagnetic ratio is multiplied with it, then we obtain, in the case of orbital angular momentum $l\hbar$, the corresponding magnetic moment given by

$$\boldsymbol{\mu}_l = \frac{e\hbar}{2Mc}\boldsymbol{l} = g_l\boldsymbol{l}.$$

The quantity preceding $\boldsymbol{l}$ (which is now dimensionless) is called the nuclear magneton (nM). All nuclear magnetic moments are expressed in this unit. In contemporary literature on nuclear structure, the gyromagnetic ratio (as just defined), times $\hbar$, is somewhat inaccurately also called the gyromagnetic ratio; it is denoted by g and is usually quoted in units of nuclear magneton. Thus, the gyromagnetic ratio for orbital motion is given by

$$g_l = \begin{cases} 1 \text{ nM} & \text{(for a proton)} \\ 0 & \text{(for a neutron)} \end{cases} \tag{III.4}$$

In the same manner, the spin magnetic moment is given by

$$\boldsymbol{\mu}_s = g_s\mathbf{s},$$

where g_s is the spin gyromagnetic ratio. In analogy with the spin gyromagnetic ratio of the electron, equal to $e/(mc)$ (m is the electron mass), we would expect the proton g_s to be $e\hbar/(Mc)$, i.e., 2 nM; and since the neutron does not possess any charge, its g_s is expected to be zero. This, however, is not true because we know that these particles possess anomalous intrinsic magnetic moments equal to 2.7925 nM (proton) and −1.9128 nM (neutron). Since the spin is 1/2, the spin gyromagnetic ratio of the nucleons is therefore given by

$$g_s = \begin{cases} 5.5850 \text{ nM} & \text{(for a proton)} \\ -3.8256 \text{ nM} & \text{(for a neutron)} \end{cases} \tag{III.5}$$

With these preliminary definitions, we now write the magnetic moment operator $\boldsymbol{\mu}$ of a nucleon as

$$\boldsymbol{\mu} = g_l \boldsymbol{l} + g_s \mathbf{s}. \tag{III.6}$$

The operator for the whole nucleus is obtained by summing this expression over all the nucleons.

The operators $\frac{1}{2}(1 + \tau_z)$ and $\frac{1}{2}(1 - \tau_z)$ are respectively the projection operators for a neutron ($\tau_z = +1$)- and a proton ($\tau_z = -1$)-state, where τ_z is the z-component of isotopic spin. Hence, we could also combine the definitions (III.4)–(III.6) and write a single expression

$$\boldsymbol{\mu} = \tfrac{1}{2}(1 - \tau_z)(\boldsymbol{l} + 5.585\mathbf{s}) - \tfrac{1}{2}(1 + \tau_z)3.8256\mathbf{s}. \tag{III.7}$$

We could also define a gyromagnetic ratio g_j for the total angular momentum $\mathbf{j}$ of the nucleon such that

$$\boldsymbol{\mu} = g_j \mathbf{j}. \tag{III.8}$$

g_j will, of course, be related to g_l and g_s through this equation and (III.6). In order to obtain this relationship, we make use of the following result for the *expectation value* of any *vector* operator $\mathbf{A}$ in terms of that for the angular momentum $\mathbf{j}$:

$$\langle \mathbf{A} \rangle = \frac{\langle \mathbf{A} \cdot \mathbf{j} \rangle}{\langle j^2 \rangle} \langle \mathbf{j} \rangle.$$

Hence, as far as the expectation value is concerned, (III.6) is equivalent to the expression

$$\boldsymbol{\mu} = \frac{1}{j(j+1)} (g_l \langle \boldsymbol{l} \cdot \mathbf{j} \rangle + g_s \langle \mathbf{s} \cdot \mathbf{j} \rangle) \mathbf{j}.$$

Comparing this result with (III.8), we obtain

$$g_j = \frac{1}{j(j+1)} (g_l \langle \boldsymbol{l} \cdot \mathbf{j} \rangle + g_s \langle \mathbf{s} \cdot \mathbf{j} \rangle). \tag{III.9}$$

$\langle \boldsymbol{l} \cdot \mathbf{j} \rangle$ can be evaluated by making use of the operator identity

$$\mathbf{s} = \mathbf{j} - \boldsymbol{l},$$

i.e.,

$$s^2 = j^2 + l^2 - 2\boldsymbol{l} \cdot \mathbf{j}$$

or

$$\boldsymbol{l} \cdot \mathbf{j} = \tfrac{1}{2}(j^2 + l^2 - s^2).$$

A similar identity can be used to evaluate $\langle \mathbf{s} \cdot \mathbf{j} \rangle$. Finally, we obtain

$$g_j = \tfrac{1}{2}(g_l + g_s) + \frac{1}{2j(j+1)} [l(l+1) - \tfrac{3}{4}](g_l - g_s)$$

$$
= \begin{cases} \dfrac{1}{2(j+1)}[(2j+3)g_l - g_s] & \text{for } l = j + \tfrac{1}{2} \\[2ex] \dfrac{1}{2j}[(2j-1)g_l + g_s] & \text{for } l = j - \tfrac{1}{2} \end{cases} \tag{III.10}
$$

By definition, the magnetic moment is the expectation value of μ_z in a state $|(l\tfrac{1}{2})jm\rangle$ with the *maximum projection* $m = j$. Therefore, according to (III.8), the value of g_j, as given by (III.10), should be multiplied by the expectation value of $\langle j_z \rangle = j$ to obtain the observed single-nucleon magnetic moment μ. Thus, for $l = j - \tfrac{1}{2}$,

$$
\mu = \tfrac{1}{2}[(2j-1)g_l + g_s]
$$

$$
= \begin{cases} \tfrac{1}{2}g_s & \text{(for a neutron)} \\ j - \tfrac{1}{2} + \tfrac{1}{2}g_s & \text{(for a proton)} \end{cases} \tag{III.11a}
$$

and, for $l = j + \tfrac{1}{2}$,

$$
\mu = \begin{cases} -\dfrac{j}{(j+1)}\tfrac{1}{2}g_s & \text{(for a neutron)} \\[2ex] \dfrac{j}{(j+1)}(j + \tfrac{3}{2} - \tfrac{1}{2}g_s) & \text{(for a proton)} \end{cases} \tag{III.11b}
$$

The values of the neutron and proton magnetic moments, as given by (III.11), are called the *Schmidt values*. $\tfrac{1}{2}g_s$ occurring in these expressions is the intrinsic magnetic moment of the respective particles.

Before discussing the agreement of the expressions (III.11) with experimental data, we would like to point out that the Schmidt values are *theoretically exact* for a single particle outside a set of closed levels, and also in the case of a single hole in an otherwise closed level. The requirement of antisymmetry, already mentioned, does not affect these simple results. We shall prove these statements by first considering the case of closed levels and then adding or removing one particle.

Closed Levels

Since there is only one way of assigning sublevels to the nucleons in the case of closed levels, the antisymmetric wavefunction Φ is a unique determinant in which all the projection quantum numbers for the different j's are occupied. According to (II.23), we have

$$
\langle \Phi | \sum_{i=1}^{A} \mu_z(i) | \Phi \rangle = \sum_{nlj} \sum_{m} \langle nljm | \mu_z | nljm \rangle, \tag{III.12}
$$

where the summation nlj goes over all the occupied levels; the summation m proceeds over all the sublevels $-l$ to $+l$; and the summation i in the operator on the left-hand side goes over all the A particles, as mentioned after (III.6).

The matrix element in (III.12) can be written as [see (BIII.1) in Appendix B]

$$
\begin{bmatrix} j & 1 & j \\ m & 0 & m \end{bmatrix} \langle nlj \| \boldsymbol{\mu} \| nlj \rangle.
$$

$\boldsymbol{\mu}$ being a vector operator, its rank is unity, which explains the occurrence of 1 in the Clebsch-Gordon coefficient. The quantum number m now occurs in the Clebsch-Gordon coefficient only.

Since $\begin{bmatrix} j & 0 & j \\ m & 0 & m \end{bmatrix}$ is unity, the summation m yields

$$\sum_m \begin{bmatrix} j & 1 & j \\ m & 0 & m \end{bmatrix} = \sum_m \begin{bmatrix} j & 1 & j \\ m & 0 & m \end{bmatrix}\begin{bmatrix} j & 0 & j \\ m & 0 & m \end{bmatrix}$$

$$= \frac{[j]}{\sqrt{[1]}} \sum_m \begin{bmatrix} j & j & 1 \\ m & -m & 0 \end{bmatrix}\begin{bmatrix} j & j & 0 \\ m & -m & 0 \end{bmatrix} = 0.$$

Here we have used, from Appendix A, the third symmetry property (AIV.4) and the orthogonality property (AIV.3a) of the Clebsch-Gordon coefficients. This work therefore establishes that the *magnetic moment of a state of several closed levels is zero.* From the manner of this derivation, it is clear that the same vanishing result would have followed for any other physical observable which is represented by an operator of rank $K \neq 0$. This general result for a set of closed levels does not, however, really need the detailed proof given here; it simply follows from angular momentum conservation. Since a set of closed levels produces a state $|\Phi\rangle$ of angular momentum $J = 0$, the action on it of an operator of rank K produces a state of angular momentum K only. If $K \neq 0$, obviously its scalar product with $\langle\Phi|$ vanishes because $\langle\Phi|$ has $J = 0$.

One-Particle or One-Hole State

The antisymmetric state of a set of closed levels plus a single particle in the level (j_p, m_p) is also a unique determinant. Hence, according to (II.23), the magnetic moment now is the sum (III.12) over all the occupied levels, which has been proved to be zero, plus the result for the last occupied level (j_p, m_p), which is generally non-zero. Thus, the zero result for the closed levels immediately proves our assertion that the magnetic moment of closed levels plus one particle is given by the Schmidt value for the last particle.

The one-hole state, the hole being in the state $(j, -m)$, is also represented by a unique determinant. Had the state $(j, -m)$ been present in the set of occupied states of the determinant, we would have obtained a zero value for the magnetic moment (according to our discussion for the case of closed levels). Since the contribution of this state is absent, the result should be given by its negative, i.e.,

$$-\langle nlj, -m|\,\mu_z\,|nlj, -m\rangle = -\begin{bmatrix} j & 1 & j \\ -m & 0 & -m \end{bmatrix}\langle nlj\|\,\boldsymbol{\mu}\,\|nlj\rangle$$

$$= \begin{bmatrix} j & 1 & j \\ m & 0 & m \end{bmatrix}\langle nlj\|\,\boldsymbol{\mu}\,\|nlj\rangle$$

$$= \langle nljm|\,\mu_z\,|nljm\rangle. \qquad \text{(III.13)}$$

Thus, the magnetic moment of a hole state of angular momentum j and projection m [recall, from Section 25, that the removal of a particle $(j, -m)$ gives rise to a one-hole state of the same spin but opposite projection] is the same as that of a single particle in the state (j, m). This, incidentally, proves our earlier assertion about the Schmidt value of the magnetic moment for a one-hole state. In arriving at the final step of (III.13), we have made use of the second symmetry relation (AIV.4) from Appendix A to replace the Clebsch-Gordon coefficient with negative projections by the one with positive projections.

More General Case

In a more general case, we may have more than one nucleon in one or more partially filled levels. We have already noted that the antisymmetry requirement between such nucleons usually gives rise to results different from the Schmidt value; such results are considered in Section 30.

Comparison of experimental data and Schmidt values We now go back to the Schmidt expressions (III.11) of the magnetic moment and compare the predictions with experimental data on odd-mass nuclei. The single-particle quantum numbers l and j of the last odd nucleon may be obtained from Section 25.

If the Schmidt values are plotted as a function of j, they obviously give rise to two curves corresponding to $j = l + \frac{1}{2}$ and $j = l - \frac{1}{2}$ for both the neutron and the proton. These are shown in Figs. III.2 (proton) and III.3 (neutron) along with the experimental values for the odd-mass nuclei. We now trace the detailed comparison of experimental data with the Schmidt

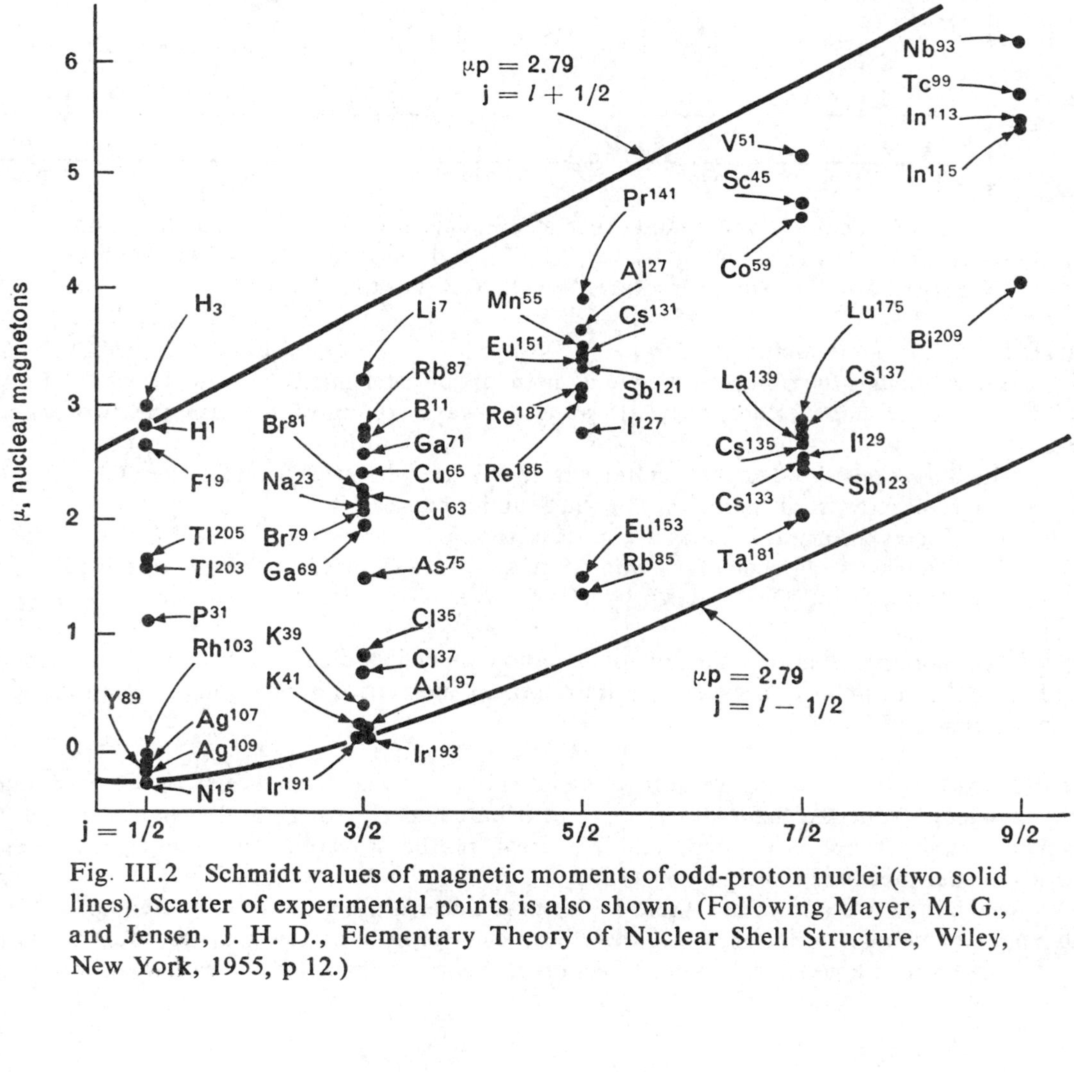

Fig. III.2 Schmidt values of magnetic moments of odd-proton nuclei (two solid lines). Scatter of experimental points is also shown. (Following Mayer, M. G., and Jensen, J. H. D., Elementary Theory of Nuclear Shell Structure, Wiley, New York, 1955, p 12.)

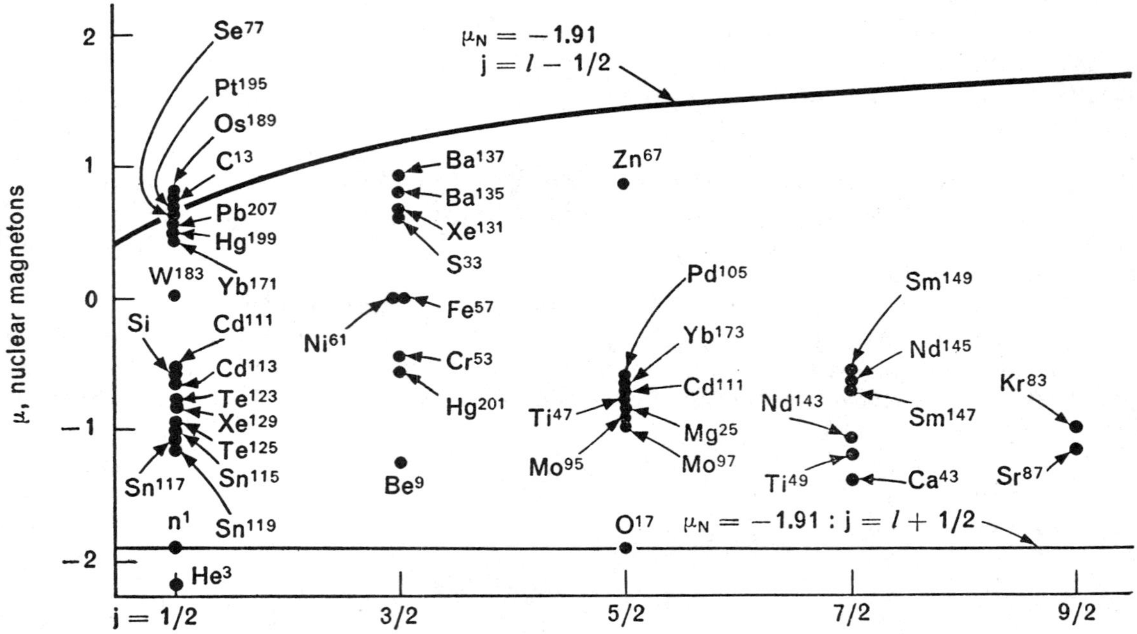

Fig. III.3 Schmidt values and experimental data for magnetic moments of odd-neutron nuclei. (Following Mayer, M. G., and Jensen, J. H. D., Elementary Theory of Nuclear Shell Structure, Wiley, New York, 1955, p 13.)

values by stating some facts:

(i) Although the observed magnetic moments of most nuclei deviate by about 0.5–1.5 nM from the Schmidt values, they are almost always contained between the two Schmidt lines.

(ii) The exceptions to the statement in (i) are observed for H^3, He^3, C^{13}, and N^{15} which depart slightly from the bounds of the Schmidt curves.

(iii) The experimental values are most of the time closer to the correct l-line.

(iv) The closest agreement (within 0.2 nM) to the Schmidt values is obtained for the nuclei named in (ii), and O^{17}, K^{39}, K^{41}, and *all nuclei having the last odd nucleon in a* $(p_{1/2})$-*level.*

(v) Amongst the one-particle or one-hole nuclei, which should obey the Schmidt results, Bi^{209} is an important exception; its observed moment differs from the Schmidt value by more than 1 nM.

Possible explanations of departure from Schmidt values The departure of the observed magnetic moments from the Schmidt values is attributed to (i) error in the expression of the magnetic moment operator used, and (ii) error in the wavefunction, the extreme single-particle model presumably being very crude.

Analyzing error (i), we find there are several uncertain corrections to the expression of the magnetic moment operator, which have been mentioned in connection with our discussion on the deuteron magnetic moment in Section 6. Some of these effects have been only very roughly estimated. They are: (a) the correction due to the meson exchange currents; (b) the departure of the intrinsic moments of the nucleons from their free values when they are embedded in nuclear medium (the anomalous intrinsic moments, being the effects of the meson cloud around the nucleons, can very reasonably reflect the difference between the meson cloud inside a nucleus and that around a free nucleon); and (c) the correction due to the velocity-dependent spin-orbit term (III.1) in the shell-model Hamiltonian, an explicit expression for which has been worked out in Section 6, where the spin-orbit potential is taken as $V(r/a)(\mathbf{r} \times \mathbf{p}) \cdot \mathbf{s}$. A comparison of this form with (III.1) reveals that, in using (I.71) in the present case, $-\hbar^{-1}U(r)$ must be substituted for $V(r/a)$; the factor $\hbar^{-1}$ appears because $\mathbf{r} \times \mathbf{p} = l\hbar$. Thus, the additional magnetic moment due to spin-orbit coupling is represented by the value of the z-component of the operator as

$$\frac{e}{2\hbar c}U(r)[(\mathbf{r} \cdot \mathbf{s})\mathbf{r} - r^2\mathbf{s}],$$

i.e.,

$$\frac{M}{\hbar^2}U(r)r^2\left[\frac{4\pi}{3}(Y^1 \cdot \mathbf{s})Y^1_0 - s_z\right] \text{ nM}, \tag{III.14}$$

where Y^1 is the spherical harmonic of rank unity. (III.14) is in units of nuclear magneton. The reader may satisfy himself that $(M/\hbar^2)U(r)r^2$ is dimensionless; he is further asked to evaluate the expectation value of (III.14) for a state $|nljm = j\rangle$ by using the tensor operator technique of Appendix B (Section III). A numerical estimate of this and the effects (a) and (b) just outlined gives a correction to the magnetic moment of the order of 0.1–0.2 nM. We have already stated that many observed moments differ from the Schmidt values by much more than this order of magnitude, and hence error (ii) is very important.

Analyzing error (ii), we observe that when the wavefunction of the nucleons outside the closed levels is suitably antisymmetrized, a correction to the Schmidt value, commonly referred to as the Pauli correction, is usually obtained. If all these nucleons are in one nlj-level, then the expectation value of μ_z for the antisymmetric state having seniority one happens to be the same as the Schmidt value. If, however, they are distributed among more than one unfilled level, the correct antisymmetrized state gives a result equal to a linear sum of the contribution from the various levels with appropriate coefficients. The most important aspect of the wavefunction, as far as magnetic moment is concerned, is the admixture of small amounts of other states Φ_α ($\alpha \neq 0$) in the predominant state Φ_0 which obeys the lowest seniority shell-model ideas. From the composition of the magnetic moment operator (III.6), it is obvious that it cannot connect two different single-particle states of different l quantum number; it can, however, connect two such states of the same l and different j. If a state Φ_α ($\alpha \neq 0$) be such that only one single-particle state in Φ_α is different from one in Φ_0 and these two single-particle states have the same l but different j, then we can have a fairly large nonvanishing matrix element of the magnetic moment operator between Φ_0 and Φ_α. Let the actual ground-state wavefunction be written as

$$\Psi_0 = c_0\Phi_0 + \sum_\alpha c_\alpha\Phi_\alpha, \tag{III.15}$$

where c_α is much smaller than c_0. Then the magnetic moment for this state is given, to the first order in c_α, by

$$\langle\Psi_0|\sum_{i=1}^{A} \mu_z(i)\,|\Psi_0\rangle = c_0^2\langle\Phi_0|\sum_{i=1}^{A} \mu_z(i)\,|\Phi_0\rangle + 2\sum_{\alpha\neq 0} c_\alpha c_0\langle\Phi_\alpha|\sum_{i=1}^{A} |\,\mu_z(i)\,|\Phi_0\rangle. \tag{III.16}$$

36 (45–122/1977)

The first term represents the magnetic moment of the predominant state Φ_0, and the second term the corrections due to the impurity states Φ_α. We have noted that, for a particular class of Φ_α, the second term can give a fairly significant contribution to the magnetic moment. Arima and Horie[4] and also Blinstoyle and Perks[4] discussed the corrections to the Schmidt values of the magnetic moment due to this kind of small admixture, and demonstrated that most of the observed values can be satisfactorily explained. They proved, in particular, that such a correction is very small for nuclei having the odd particle in a ($p_{1/2}$)-level. This fact is in excellent agreement with the observed situation already mentioned.

Arima and Horie[4] and Blinstoyle and Perks[4] calculated the coefficients c_α by using a δ-function type residual potential between nucleons. In more recent years, Mottelson proposed a 'new look' at the magnetic moments which is similar to the Arima-Horie work in principle, but differs in some details, and finally casts the results in a form that makes some of the physical contents very transparent. In this new work, once again, an admixture of the type (III.15) is considered, but the coefficients c_α are calculated by using a *spin-dependent* two-nucleon interaction. The states Φ_0, Φ_α of an odd-mass nucleus having a total angular momentum $J(=j)$ can be obtained by antisymmetrizing a core wavefunction of even nucleons coupled to a single-particle state of angular momentum i. For states of maximum projection $M = j$, we have

$$|\Phi_0\rangle = |\Phi_c, j : J = j, M = j\rangle_A,$$

$$|\Phi_\alpha\rangle = |\Phi_c', j : J = j, M = j\rangle_A,$$

where A denotes "antisymmetric" states, and the core states Φ_c, Φ_c' are different. Such states are used to manipulate the second term in (III.16) along with the perturbation expression for c_α, and this correction term is finally demonstrated to have the appearance of an expectation value for the single-particle state $|j, m = j\rangle$ of the operator

$$\delta g_s s_z - g_p(Y^2, s)^1_0. \tag{III.17}$$

The quantities δg_s and g_p in this expression depend on the properties of the core wavefunctions Φ_c and Φ_c'. The second operator denotes a coupled tensor of rank 1 and component 0, obtained by coupling the spherical harmonic Y^2 with the spin operator s. Detailed work shows that the core state Φ_c' contained in Φ_α is obtained by causing a spin polarization in the core state Φ_c of Φ_0. Hence, the correction to the magnetic moment operator, represented by (III.17), is called the core-polarization correction. It should be observed that, in this 'new look' method, the correction to the magnetic moment due to errors in the wavefunction is finally transformed to a correction of the operator itself. The final expression of the magnetic moment operator is the result of adding (III.6) and (III.17) and is given by

$$\mu_z = g_l l_z + (g_s + \delta g_s)s_z - g_p(Y^2, s)^1_0. \tag{III.18}$$

The details of the derivation of the core-polarization effect are given in Section 30B.

For the present, we consider (III.18) to be a semiphenomenological expression for the magnetic moment operator and check how far the magnetic moment data can be fitted by manipulation with the parameters δg_s and g_p. We first observe the fact that the operator $(Y^2, s)^1_0$, as far as the single-particle expectation value is concerned, can be replaced by the operator s_z multiplied by an expression dependent on the single-particle quantum numbers l and j. To demonstrate this, we evaluate the expectation values of this operator and s_z. Using (BIII.16) and (BIII.13) from

Appendix B, we obtain

$$\langle l\tfrac{1}{2}jm=j|\,(Y^2,\mathbf{s})^1_0\,|l\tfrac{1}{2}jm=j\rangle=\begin{bmatrix} j & 1 & j \\ j & 0 & j \end{bmatrix}\begin{bmatrix} l & \tfrac{1}{2} & j \\ 2 & 1 & 1 \\ l & \tfrac{1}{2} & j \end{bmatrix}\langle l||Y^2||l\rangle\langle\tfrac{1}{2}||\mathbf{s}||\tfrac{1}{2}\rangle,$$

$$\langle l\tfrac{1}{2}jm=j|\,s_z\,|l\tfrac{1}{2}jm=j\rangle=\begin{bmatrix} j & 1 & j \\ j & 0 & j \end{bmatrix}U(l\tfrac{1}{2}j1;j\tfrac{1}{2})\langle\tfrac{1}{2}||\mathbf{s}||\tfrac{1}{2}\rangle.$$

Comparing these two expressions, we get the desired result

$$(Y^2,\mathbf{s})^1_0\to C(l,j)s_z, \tag{III.19a}$$

where

$$C(l,j)=\begin{bmatrix} l & \tfrac{1}{2} & j \\ 2 & 1 & 1 \\ l & \tfrac{1}{2} & j \end{bmatrix}\langle l||Y^2||l\rangle U(l\tfrac{1}{2}j1;j\tfrac{1}{2})^{-1}. \tag{III.19b}$$

Therefore, as far as the expectation value is concerned, we can rewrite (III.18) as

$$\mu_z=g_l l_z+g_s^{\text{eff}}s_z, \tag{III.20}$$

where

$$g_s^{\text{eff}}=g_s+\delta g_s-C(l,j)g_p. \tag{III.21}$$

Although (III.20) appears very similar to the original magnetic moment operator (III.6), the main difference between the two is the (l,j)-dependence of g_s^{eff} as against the constancy of g_s. It is clear from (III.19b) that the last term of (III.21) is zero for $l=0$. Therefore, the quantity δg_s can be empirically determined by reproducing the magnetic moments of odd nuclei with the last particle in $s_{1/2}$-orbit. The value that produces a good average fit to most of the data is given by $\delta g_s=-0.5$. Thus, the core-polarization effect produces an *intrinsic* gyromagnetic ratio $(g_s+\delta g_s)$, which is quenched with respect to the actual intrinsic gyromagnetic ratio g_s of the nucleons. To arrive at a rough empirical value of the parameter g_p, we may now exploit the fact that most odd nuclei having the last nucleon in the $(p_{1/2})$-level has magnetic moments in fair agreement with the Schmidt value. That is, for $l=1$, $j=\tfrac{1}{2}$, the extra terms δg_s and $C(l,j)g_p$ in (III.21) must very nearly cancel out each other.

Figures III.4 and III.5 show the kind of trend that results from an empirical approach to (III.20) and (III.21). The horizontal solid lines are the predicted values for each j-value shown at the base of the figures. The l-value is also indicated. The ordinates represent g_s^{eff}/g_s, which would have been equal to 1 had the Schmidt values been correct. Most of the experimental points scatter below 1, and it is seen that their average trend for each (l,j) is fairly in agreement with the empirically predicted solid lines. We emphasize, however, that a more detailed calculation of δg_s and g_p is possible by using detailed wavefunctions describing the core polarization. Unless such calculations are done, no comment can be made on the agreement of the core-polarization theory with the observed scatter of the experimental points.

B. ELECTRIC QUADRUPOLE MOMENTS OF NUCLEI

The neutrons, being uncharged, do not contribute anything to the electric quadrupole moment operator of a nucleus. Each of the protons, on the other hand, contributes an amount

$$q_0=3z^2-r^2=2r^2y_0^2(\theta), \tag{III.22}$$

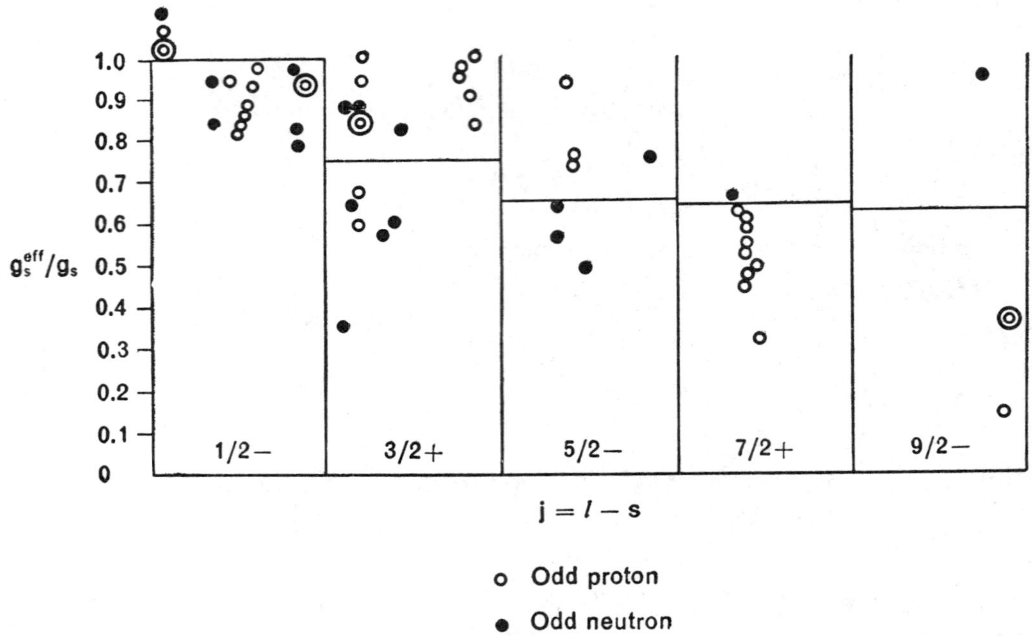

Fig. III.4 Effective gyromagnetic ratio (III.21) in case of $j = l - \frac{1}{2}$ calculated with rough model of core polarization. Scatter of experimental points is also shown. (Following Bodenstedt, E., and Rogers, J. D., in Perturbed Angular Correlations, ed. by E. Karlsson, E. Matthias, and K. Siegbahn, North-Holland, Amsterdam, 1964, p 91.)

where $y_0^2(\theta)$ is the renormalized spherical harmonic of (AI.7a) from Appendix A. Summing this quantity over all the protons, we obtain the quadrupole moment operator of the entire nucleus as

$$Q_0 = \sum_{i=1}^{Z} q_0(i) = 2 \sum_{i=1}^{Z} r_i^2 y_0^2(\theta_i) \tag{III.23a}$$

$$= 2 \sum_{i=1}^{A} \tfrac{1}{2}[1 - \tau_z(i)] r_i^2 y_0^2(\theta_i). \tag{III.23b}$$

Here A is the total nucleon number, and Z the proton number in the nucleus. Since the factor $\frac{1}{2}(1 - \tau_z)$ is equal to one for a proton and zero for a neutron, the summation over the Z protons in (III.23a) is completely equivalent to the summation over all the A nucleons in (III.23b).

We have proved, in connection with our discussion on the magnetic moment of a nucleus having a set of completely filled levels (see Section 26A), that the value of any operator of rank not equal to zero vanishes for such a system. Since the quadrupole moment is given by y_0^2, i.e., a tensor operator of second rank, we conclude that the quadrupole moment of a closed-level nucleus is zero.

In analogy with the proof given for magnetic moment, it therefore follows that the quadrupole moment of a nucleus having one proton outside a set of closed levels is given by the quadrupole moment of this last proton alone. If the state of the last proton be denoted by

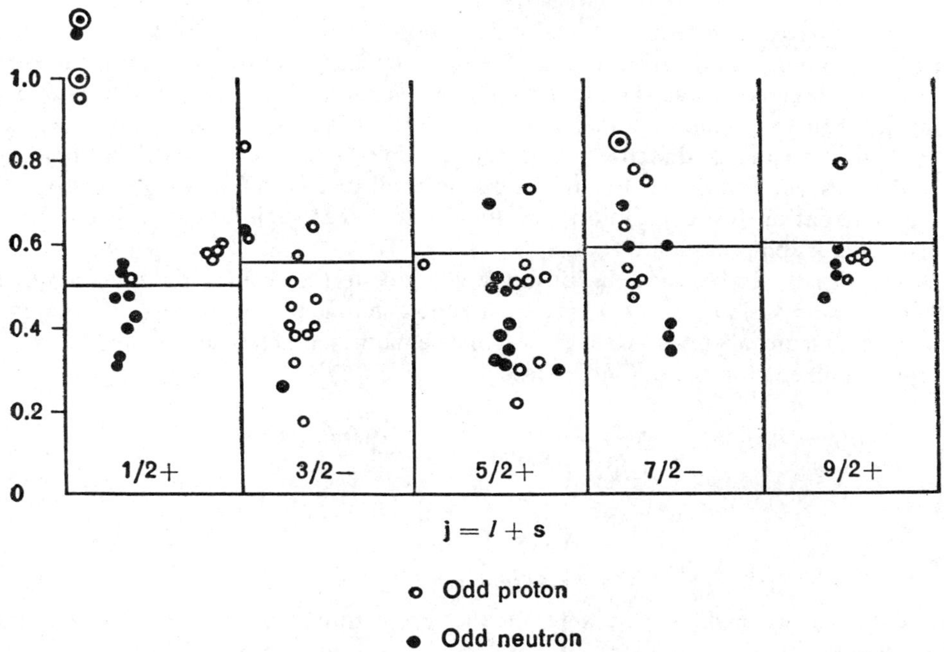

Fig. III.5 Effective gyromagnetic ratio (III.21) in case of $j = l + \frac{1}{2}$ calculated with rough model of core polarization. Scatter of experimental points is also shown. (Following Bodenstedt, E., and Rogers, J. D., in Perturbed Angular Correlations, ed. by E. Karlsson, E. Matthias, and K. Siegbahn, North-Holland, Amsterdam, 1964, p 91.)

$|nl\frac{1}{2}jm\rangle$, then its quadrupole moment, *by definition*, is the expectation value of the operator (III.22) when the state has the maximum projection $m = j$. Thus, we have

$$\langle nl\tfrac{1}{2}jj|\, q_0\, |nl\tfrac{1}{2}jj\rangle = \begin{bmatrix} j & 2 & j \\ j & 0 & j \end{bmatrix} 2\langle l\tfrac{1}{2}j\|\, \mathscr{Y}^2 \|\, l\tfrac{1}{2}j\rangle \left(\int_0^\infty R_{nl}^2 r^2\, dr\right)$$

$$= 2\begin{bmatrix} j & 2 & j \\ j & 0 & j \end{bmatrix}\begin{bmatrix} j & 2 & j \\ \frac{1}{2} & 0 & \frac{1}{2} \end{bmatrix} \int_0^\infty R_{nl}^2 r^2\, dr$$

$$= -\frac{2j-1}{2(j+1)} \int_0^\infty R_{nl}^2 r^2\, dr. \tag{III.24}$$

The value of the reduced matrix element, appearing in this derivation, has been obtained from (BIII.17) of Appendix B, and explicit expressions for the Clebsch-Gordon coefficients have been written down from Table AVII.4 (Appendix A) to arrive at the final step. It is clear that the expression is nonvanishing only if $j > \frac{1}{2}$. In contrast with the expression for the single-particle magnetic moment, we notice that the single-particle quadrupole moment depends on the exact nature of the radial function R_{nl}. The definition of our radial functions, and their expressions for a harmonic oscillator shell-model potential have been given in Section 17C. Since the value of the radial integral in (III.24) is positive definite, and $(2j - 1) > 0$ for $j > \frac{1}{2}$, it follows that the quadrupole moment of a one-particle state is negative. This has a very simple physical interpretation. Angular momentum is, by definition, perpendicular to the plane of motion of the particle. Semiclassically speaking, in the present case, the angular momentum 'points' in the z-direction because it has the maximum possible projection $m = j$ along that axis. Hence, the particle orbit for the state under our consideration is confined to a plane perpendicular to the z-axis, and the charge distribution of the particle will obviously be 'flattened' ('pancake' shape) with respect to the z-direction. In our general discussion on the quadrupole moment of a charge distribution (see Section 6B), we have stated that such a charge distribution possesses a negative quadrupole moment.

We can easily derive the relationship analogous to (III.13) for the quadrupole moment of a proton one-hole state in terms of the quadrupole moment for the proton one-particle state. Following the same arguments as given for magnetic moment, we conclude that the value of quadrupole moment for the one-hole state jm is given by

$$\begin{aligned} -\langle nlj - m|\, q_0\, |nlj - m\rangle &= -\begin{bmatrix} j & 2 & j \\ -m & 0 & -m \end{bmatrix} \langle nlj\|\, q\, \|nlj\rangle \\ &= -\begin{bmatrix} j & 2 & j \\ m & 0 & m \end{bmatrix} \langle nlj\|\, q\, \|nlj\rangle \\ &= -\langle nljm|\, q_0\, |nljm\rangle. \end{aligned} \tag{III.25}$$

In this case, the reversal of the sign of the projection quantum numbers in the Clebsch-Gordon coefficient gives a factor of $(-1)^{j+2-j} = +1$ as against $(-1)^{j+1-j} = -1$ which was obtained for the Clebsch-Gordon coefficient in (III.13). This fact accounts for the difference in sign between the two expressions. We enunciate the result contained in (III.25): the quadrupole moment of a proton hole state is equal and *opposite* to that of the corresponding particle state.

In a more general case, where there is more than one proton in one or more partially filled levels, the evaluation is once again quite involved, due to the antisymmetry requirement in the wavefunction, and can be done with the help of (III.119b) of Section 29A.

As in the case of the magnetic moment, so too here we can predict a value for these complicated cases, provided we make the extreme single-particle assumption for the wavefunction of an odd-proton nucleus. With this assumption, if $|nljm\rangle$ denotes the state of the last odd proton, (III.24) can be used to predict its quadrupole moment.

An interesting fact about the observed quadrupole moments of odd-mass nuclei is that very often a nucleus having an odd *neutron* is found to have a significant quadrupole moment. This fact is clearly in disagreement with the prediction arrived at from the extreme single-particle model. Among the light nuclei, the case of O^{17} is almost a classic example and has always drawn a great deal of attention of theorists. The nuclei O^{17} and F^{17} contain, respectively, one neutron and one proton outside the closed-shell nucleus O^{16}. Therefore, in these cases, the theoretical predictions should be exact. Thus, O^{17} should have a zero quadrupole moment, and F^{17} a quadrupole moment equal to that of a single proton in the $(0d_{5/2})$-level. To make a rough estimate of the latter, we use, for the radial integral in (III.24), an approximate value where the integral denotes the square of the radius of the last proton orbit. Hence, a rough value for it is the square of the nuclear radius, i.e., $(r_0 A^{1/3})^2$. Using a value of r_0 in the range 1.1 fm to 1.4 fm, we get the estimated quadrupole moment of F^{17} within the range -4.6 fm^2 to -7.5 fm^2. No experimental data is available on F^{17}, but the observed quadrupole moment of O^{17} is close to -3 fm^2. That is to say, not only does the odd-neutron nucleus O^{17} have a quadrupole moment, but, in addition, the observed value is comparable with what is expected for the corresponding odd-proton nucleus F^{17}.

The conclusion that an odd-neutron nucleus should have a zero quadrupole moment is valid also under more general circumstances, namely, even when we consider the *antisymmetrized* seniority wavefunctions. Since the neutrons do not contribute any quadrupole moment under any circumstances, the quadrupole moment must arise entirely from the protons. But the lowest seniority antisymmetrized wavefunction for an even number of protons has an angular momentum zero, and hence has a zero quadrupole moment by virtue of the angular momentum selection rule. Therefore, the observed quadrupole moments of odd-neutron nuclei strongly suggest that the 'core' wavefunction of such a nucleus does not strictly have an angular momentum zero; due to the interaction of the odd particle with the core, there must be small admixtures of states of higher angular momentum in the core. Even if such admixtures are small, they can cause a very significant core contribution to the quadrupole moment.

In analogy with the nomenclature already used in the discussion of magnetic moment, we describe the core contribution to the quadrupole moment as the effect of *core polarization*. In the case of magnetic moment, the *spin polarization* of the core was the contributor to the core magnetic moment, whereas here it is the *quadrupole-type polarization* (as is obvious from the derivation that follows) of the proton charge distribution of the core that is responsible for the entire quadrupole moment of odd-neutron nuclei.

Since the core-polarization effect is apparent from the nonvanishing quadrupole moment of many odd-neutron nuclei, there is no reason why such effects should not be present for the odd-proton nuclei as well. In the odd-proton case, the core contribution adds up to the value of the quadrupole moment of the protons in the partially filled levels. We shall now give a general derivation of the core contribution which is applicable to all odd-mass (i.e., both odd-proton and odd-neutron) nuclei.

The structure of the admixed ground-state wavefunction is still given by (III.15), stated in our discussion on similar effects on magnetic moment. The expression (III.16) is now replaced by a new expression in which the quadrupole moment operator q_0 substitutes for μ_z everywhere. The correction to the quadrupole moment arising out of the admixed states Φ_α will be given by matrix elements of the type $\langle\Phi_\alpha|\, Q_0\, |\Phi_0\rangle$, where Φ_α and Φ_0 are the same as those stated after (III.16). Using (BIII.14) from Appendix B, and remembering that Φ_c has zero angular momentum, we obtain

$$\begin{aligned}\langle\Phi_\alpha|\, Q_0\, |\Phi_0\rangle &= \begin{bmatrix} j & 2 & j \\ j & 0 & j \end{bmatrix} \langle\Phi_c'(J')j : J = j|\, |Q|\, |\Phi_c(0), j : J = j\rangle \\ &= \begin{bmatrix} j & 2 & j \\ j & 0 & j \end{bmatrix} U(j0j2; jJ')(-1)^{J'} \langle\Phi_c'(J')|\, |Q|\, |\Phi_c(0)\rangle.\end{aligned}$$

The values within the parentheses following Φ_c and Φ_c' denote the angular momenta of these states. The U-function is equal to 1 when $J' = 2$, and zero otherwise. Since the Clebsch-Gordon coefficient $\begin{bmatrix} 0 & 2 & 2 \\ 0 & 0 & 0 \end{bmatrix}$ is unity, we can replace the reduced matrix element between the core states by the actual matrix element, and obtain

$$\langle\Phi_\alpha|\, Q_0\, |\Phi_0\rangle = \begin{bmatrix} j & 2 & j \\ j & 0 & j \end{bmatrix} \langle\Phi_c'(J' = 2, M' = 0)|\, Q_0\, |\Phi_c(J' = 0, M' = 0)\rangle. \qquad \text{(III.26)}$$

Thus, the correction to the quadrupole moment is given in terms of the matrix elements of Q_0

between the predominant core state Φ_c and the admixed core state Φ_c'. In the case of odd-neutron nuclei having an even number of protons in their lowest seniority state, $\langle \Phi_0 | Q_0 | \Phi_0 \rangle$ is zero, and hence the correction just derived is the entire quadrupole moment. For odd-proton nuclei, $\langle \Phi_0 | Q_0 | \Phi_0 \rangle$ is nonvanishing and may be evaluated by the application of the extreme single-particle model (an approximate method) or the exact seniority wavefunction [see (III.119b) in Section 29A].

The foregoing derivation gives useful information on the nature of the excited core states Φ_c'. First, Φ_c' has to have an angular momentum $J' = 2$. Second, since Q_0 is a single-particle type operator, (III.26) will be nonvanishing only if Φ_c' is obtained from Φ_c by exciting one (*and not more than one*) particle from an occupied state $(nljm)$ to an unoccupied state $(n'l'j'm')$ in which case the matrix element in (III.26) is given by

$$\langle \Phi_c' | Q_0 | \Phi_c \rangle = \langle n'l'j'm' | q_0 | nljm \rangle. \tag{III.27}$$

Therefore, a non-zero core contribution is obtained if the excitation $(nljm) \rightarrow (n'l'j'm')$ be such that (i) $m = m'$; (ii) l, l' have the same parity; (iii) l, l', 2 satisfy the angular momentum coupling requirement; (iv) j, j', 2 satisfy the same requirement as in (iii); and (v) the radial integral $\int_0^\infty R_{nl} R_{n'l'} r^2 \, dr$ is nonvanishing.

If Φ_c stands for a closed-shell core, then Φ_c' is obviously a 1h-1p *proton* state. To satisfy the parity requirement (ii), the lowest energy excitation must be across two major shells. In the case of a nonclosed-shell core, Φ_c has nucleons in the closed levels, and an even number of protons and neutrons in one or more partially filled levels. The neutrons are of no consequence as far as the quadrupole moment is concerned; and hence Φ_c' may be an excited state obtained by promoting one proton from either the fully occupied or the partially filled levels across two major shells. There is a further possibility in this case, namely, the even number of protons in the partially filled levels may now be excited from one such level to another of the same major shell by the Q_0-operator. To consider this aspect of core excitation, we must make a distinction between the odd-neutron and the odd-proton case. In the former, we have an even number of protons in the partially filled levels, and their lowest *antisymmetrized* state is a seniority-zero state of angular momentum zero. The operator Q_0 can therefore cause an excitation to a state of angular momentum two, belonging to higher seniority. In the odd-proton case, the situation is obviously the same, provided we make the extreme single-particle assumption, i.e., if we do not demand antisymmetry between the last odd proton and the even number of protons in the so-called *core*. The effect of this antisymmetrization is to give somewhat modified results.

Since the admixed core state Φ_c' is found to have an angular momentum two, and it results from an excitation of a proton in Φ_c, it is clear that the core contribution to the quadrupole moment is a result of quadrupole-type polarization of the proton charge distribution. This is what we have stated before starting the derivation.

There is a fundamental difference, experimentally and theoretically, between the core-*polarization* effect for odd-mass nuclei, whose normal equilibrium shape is spherical, and the core-*deformation* effect of these nuclei in the rotational region. Experimentally, the latter nuclei are found to have quadrupole moments that are one or two orders of magnitude larger than the single-particle value predicted by (III.24). In the case of odd-mass nuclei in the spherical region, the experimental values differ appreciably from the single-particle estimate but never by an order of magnitude. It is true that the odd-neutron nuclei in the spherical region possess a non-zero quadrupole moment, in contradiction to their single-particle prediction of zero value, but the observed number never drastically exceeds the single-particle estimate made by replacing the odd neutron by a proton; very often the observed value is indeed close to the estimate made by means of this artificial substitution.

We can describe the situation in an alternative manner. Since the quadrupole moment arises from particles each with a charge e, the correction due to the core-polarization effect modifies, so to say, the charge of the last odd particle to an effective charge e_{eff}. When this odd particle is a neutron, e_{eff} is very close to the proton charge e, whereas in the case of an odd proton, e_{eff} is the actual charge e of the particle plus a charge of the same order of magnitude. An empirical way of interpreting the quadrupole moment data of the odd-mass nuclei in the spherical region is to make the estimate from the single-particle expression (III.24) and then multiply it by a parameter x representing the ratio (e_{eff}/e) so that the observed value is reproduced. This type of empirical interpretation therefore simply gives a value of one parameter x connected with the effective charge of the odd neutron or odd proton. The purpose of a more detailed theory would then be to evaluate the core-polarization matrix elements in detail and check whether or not the observed effective charge parameter is obtained.

The trend in the values of the quadrupole moment Q of various nuclei in the periodic table is shown in Fig. III.6. In this figure, Q has been expressed in units of ZR^2, where Z is the nuclear charge and R the nuclear radius. Very useful tables giving the ground-state spins and static electromagnetic moments (magnetic dipole and electric quadrupole) for all odd-mass and odd-odd nuclei, together with the same quantities measured for the excited states of nuclei, may be obtained from the reference cited in the captions to Figs. III.4 and III.5.

27. ELECTROMAGNETIC TRANSITION PROBABILITY

A. SUMMARY OF GENERAL RESULTS FROM GAMMA-DECAY THEORY

The details of the gamma-decay theory are not covered in this book. The summary of the basic concepts and the expressions for the transition probability given in this section should be sufficient for an understanding of the description of electromagnetic transition probability. Shell-model wavefunctions are used in this chapter and other types of wavefunction in the subsequent chapters.

The charge (ρ), current ($\mathbf{j}$), and magnetization ($\mathbf{M}$) densities inside a nucleus act as the source of an electromagnetic field, which emits radiation in the form of photons whose wavelength is very much smaller than that of ordinary light and X-rays. These photons are the nuclear γ-rays. The direction of emission of the photon is perpendicular to the plane of the electric and magnetic fields which are derivable from the vector potential $\mathbf{A}$ through the usual equations

$$\mathbf{H} = \text{curl}\,\mathbf{A}, \qquad \mathbf{E} = -\frac{1}{c}\frac{\partial \mathbf{A}}{\partial t}. \tag{III.28}$$

The vector potential can in general be expanded in terms of the vector spherical harmonics $\mathbf{T}_M^{(l1)L}$, defined by (BII.11) in Appendix B. The component $\mathbf{T}_M^{(L1)L}$ of $\mathbf{A}$ defines a radiation field through (III.28), and the γ-ray photon corresponding to that field is said to be of the *magnetic type* and *multipolarity* 2^L. In the same manner, the components of $\mathbf{A}$ corresponding to $\mathbf{T}_M^{(L\pm1,1)L}$ give rise to γ-ray photons of the *electric type* and the same multipolarity 2^L. It is clear that the electric and magnetic types of radiation of the same multipolarity have opposite parity. It hardly needs to be emphasized that each *type* of radiation comprises both the electric and magnetic *fields* determined by (III.28).

37 (45-122/1977)

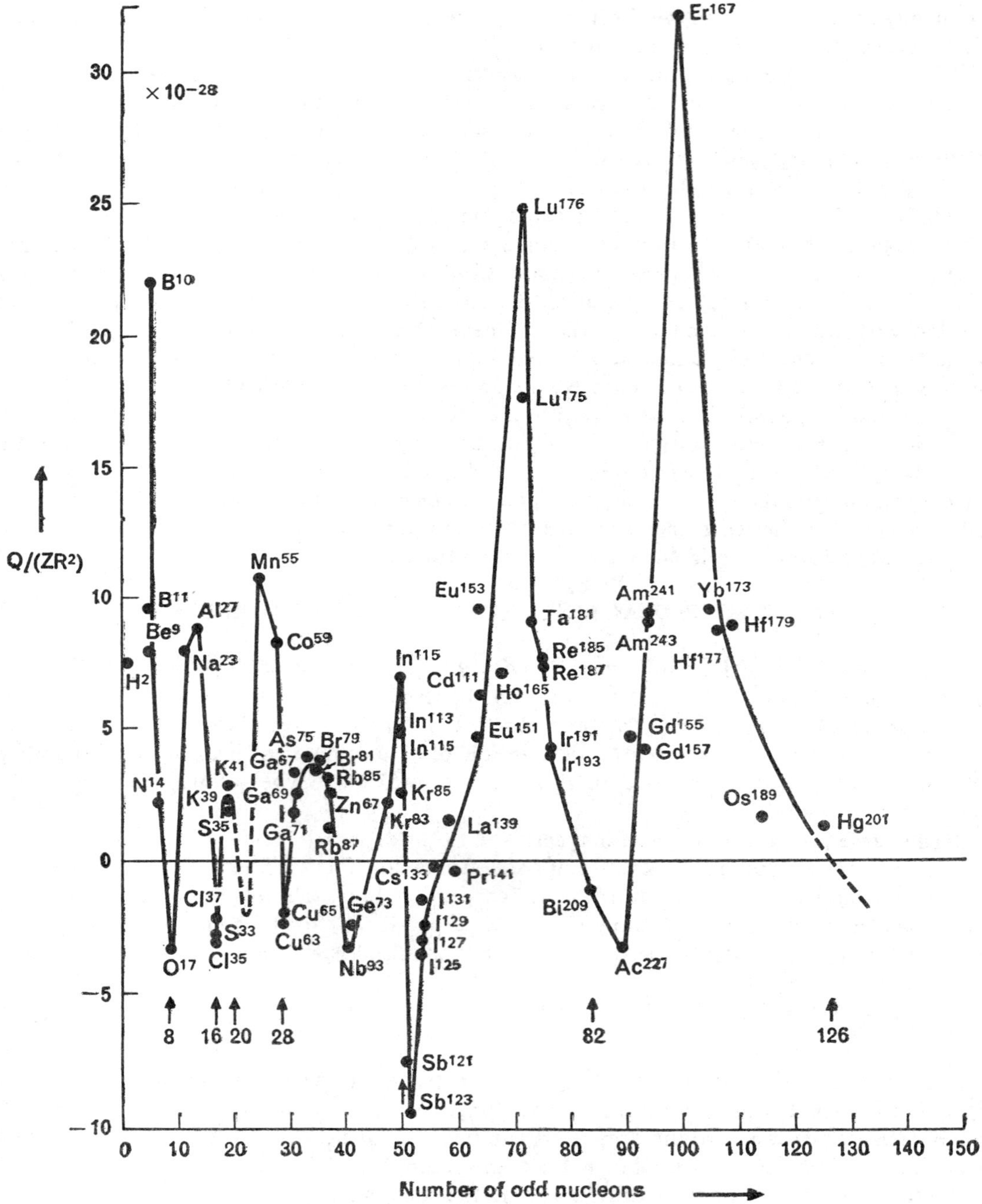

Fig. III.6 Quadrupole moments of nuclei expressed in units of ZR^2. (Following Frauenfelder, H., and Steffen, R. M., in Nuclear Spectroscopy, Part A, ed. by F. Ajzenberg-Selove, Academic Press, New York, 1960, p 548.)

The γ-ray energy $\hbar\omega$ is clearly equal to the energy difference ΔE of the two nuclear levels connected by the gamma-transition. Therefore, the wave number $k(\equiv\omega/c)$ of the photon is given by $k = (\hbar c)^{-1}\Delta E \approx .005\Delta E\ \text{fm}^{-1}$, where ΔE is in MeV. Thus, γ-ray energy in the range 1–10 MeV yields values of k in the range .005–.05 fm^{-1}.

The transition probability per unit time (λ) between the two nuclear states J_i and J_f for a given multipolarity L of the emitted γ-ray is now quoted separately for the electric and magnetic types as

$$\lambda(\mathrm{E}L) = \frac{8\pi(L+1)}{L[(2L+1)!!]^2}\frac{k}{\hbar}\,\Sigma\,|Q_M^L + Q'^L_M|^2, \tag{III.29a}$$

$$\lambda(\mathrm{M}L) = \frac{8\pi(L+1)}{L[(2L+1)!!]^2}\frac{k}{\hbar}\,\Sigma\,|\mathcal{M}_M^L + \mathcal{M}'^L_M|^2. \tag{III.29b}$$

Here EL and ML within the parentheses denote a 2^L-pole transition of the electric and the magnetic type, respectively. The summation symbol stands for an averaging over the projection M_i of the initial state J_i and a summation over the projection M_f of the final state J_f. The matrix elements Q, Q', $\mathcal{M}$, $\mathcal{M}'$ are defined as

$$Q_M^L = e\langle J_f M_f|\sum_{\alpha=1}^{Z}(\bar{r}_\alpha)^L Y_M^L(\theta_\alpha, \phi_\alpha)\,|J_i M_i\rangle, \tag{III.30a}$$

$$Q'^L_M = -\frac{1}{L+1}\frac{e}{2Mc}\langle J_f M_f|\sum_{\alpha=1}^{A}\mu_\alpha(\mathbf{p}\bar{r}^L Y_M^L)_\alpha \cdot (\bar{\mathbf{r}}\times\boldsymbol{\sigma})_\alpha\,|J_i M_i\rangle, \tag{III.30b}$$

$$\mathcal{M}_M^L = \frac{2}{L+1}\frac{e}{2Mc}\langle J_f M_f|\sum_{\alpha=1}^{Z}(\mathbf{p}\bar{r}^L Y_M^L)_\alpha \cdot \boldsymbol{l}_\alpha\,|J_i M_i\rangle, \tag{III.31a}$$

$$\mathcal{M}'^L_M = \frac{e}{2Mc}\langle J_f M_f|\sum_{\alpha=1}^{A}\mu_\alpha(\mathbf{p}\bar{r}^L Y_M^L)_\alpha \cdot \boldsymbol{\sigma}_\alpha\,|J_i M_i\rangle. \tag{III.31b}$$

Here e is the charge of the proton; M is the mass of the nucleon; (r, θ, ϕ) are the nucleon coordinates; $\bar{r} = kr$; α refers to a nucleon inside the nucleus; Z is the total number of protons and A the total number of nucleons in the nucleus; $\mathbf{p} = -i\hbar\boldsymbol{\nabla}$ is the momentum operator of a nucleon; $\boldsymbol{l}$ and $\boldsymbol{\sigma}$ are the orbital angular momentum and spin operators, respectively, of the nucleon and μ is its magnetic moment (in units of nM). The averaging and summation over the projection quantum numbers are necessary because we are here quoting the probability of decay measured in an experiment in which no information on the polarization of the emitted γ-ray is secured. Experiments that provide this additional information are also possible.

We now explore the relative order of magnitudes of (III.30) and (III.31). The first relevant quantity for this purpose is the dimensionless radial coordinate $\bar{r} = kr$. When the nuclear matrix elements are evaluated, we expect $\bar{r}$ to yield a quantity of the order of kR, R being the nuclear radius. Using $R = r_0 A^{1/3}$, with a maximum possible value of $r_0 \approx 1.4$ fm, and the estimate of k already given, we obtain

$$kR \approx .005\Delta E r_0 A^{1/3} \approx 0.21$$

for $A = 216$ and $\Delta E = 5$ MeV. *For lighter nuclei and softer γ-rays, the estimate is much less.* The second relevant quantity in judging the orders of magnitude is $\mathbf{p}/(Mc)$ which occurs in (III.30b) and (III.31). The magnitude of this quantity, $p/(Mc)$, is clearly equal to v/c, where v is the velocity of the nucleon inside the nucleus. A rough estimate of v/c can be obtained from the average kinetic energy of about 25 MeV for the nucleons as

$$\frac{v}{c} = \left(\frac{\frac{1}{2}Mv^2}{\frac{1}{2}Mc^2}\right)^{1/2} \approx \left(\tfrac{25}{500}\right)^{1/2} = 10^{-1}\sqrt{5} \approx 0.22.$$

It should be noticed that, in contrast to kR, this estimate has no relation to the mass of the nucleus or the energy of the γ-ray. For a very massive nucleus and γ-energy of about 5 MeV, $kR \approx v/c$.

An estimate of (III.30a) in units of e is therefore given by $(kR)^L$, whereas that of (III.30b) in the same units is $(v/c)(kR)^{L+1}$. Since kR is most often smaller than 0.2, Q'^L_M is expected to be smaller than Q^L_M by at least a factor of 0.04, and hence it can be dropped from (III.29a). We can further compare Q^L_M for various values of L. Since it is proportional to $(kR)^L$, its magnitude very rapidly decreases with increasing L. The ratio of Q^L_M in two successive orders is *at the most* of the order of 1/5, and hence its square $|Q^L_M|^2$ decreases by a factor of 25 or more as the multipole order is increased by 1. Therefore, it is clear that, in (III.30a), the *lowest value of L* permitted by the two nuclear states is the most important contributor. An estimate of $\lambda(EL)$ made with only the Q^L_M-term of (III.29a) and this lowest value of L should be sufficient for the purpose of theoretical prediction of the transition probability between the two given nuclear states. This therefore explains the necessity and advantage of discussing the nuclear gamma-transitions in terms of the multipole components of the electromagnetic field.

We next examine the orders of magnitude of the expressions (III.31). Both of them are clearly proportional to $(v/c)(kR)^L$. Thus, in (III.29b), both the terms are equally important. They are clearly related to the magnetic moment of the proton orbital motion ($\mathcal{M}^L_M$-term) and the intrinsic magnetic moment of both protons and neutrons ($\mathcal{M}'^L_M$-term). The observation about the decrease in the decay probability by a factor of at least 1/25 as the multipole order increases by one is obviously applicable to the magnetic transitions as well, and hence, for computational purposes, the lowest value of L permitted by the nuclear states in (III.31) suffices.

A comparison of the magnitudes of $\lambda(EL)$ with those of $\lambda(ML)$ is also interesting and useful. It is clear that, in the same order L, the magnetic-type transition which is proportional to $(v/c)(kR)^L$ is less probable than the electric-type transition which is proportional to $(kR)^L$. Hence, (EL) will be still larger than $\lambda(ML')$ with $L' = L + 1$. This fact is applied in the discussion that follows.

We are now in a position to answer the question: which is the lowest order multipole transition that can take place between two *given* nuclear states? The nuclear operators in Q^L_M (Q'^L_M has been dropped for the reasons already explained), $\mathcal{M}^L_M$, and $\mathcal{M}'^L_M$ are each of rank L. This is very obvious in (III.30a), where we have just $\bar{r}^L Y^L_M$ as the operator. It should be observed that, in (III.31), the scalar product in $\mathbf{p} \cdot \boldsymbol{l}$ and $\mathbf{p} \cdot \boldsymbol{\sigma}$ makes them zero-rank operators, and hence the rank of the entire operator is still determined by $\bar{r}^L Y^L_M$. According to the result of Appendix B (Section III), we therefore conclude that $|J_i - J_f| \leqslant L \leqslant (J_i + J_f)$. This therefore explains that the allowed multipolarity will be either $L_{min} = |J_i - J_f|$ or $L_{min} + 1$ (provided $L_{min} + 1 < J_i + J_f$), depending on the parity of the two nuclear states π_i and π_f. The parity of the operator in Q^L_M is obviously given by $(-1)^L$, whereas that of the operators in (III.31) is $(-1)^{L+1}$. This observation follows from the fact that $\mathbf{p} \cdot \boldsymbol{l}$ or $\mathbf{p} \cdot \boldsymbol{\sigma}$ has the additional parity -1 (because $\mathbf{p}$ is a polar vector which changes sign under reflection, whereas the axial vectors $\boldsymbol{l}$ and $\boldsymbol{\sigma}$ remain unchanged under the same operation).

We now distinguish between two alternative cases which depend on the given parity change $\pi_i\pi_f$ of the two nuclear states, namely,

$$\pi_i\pi_f = (-1)^{L_{min}},$$

$$\pi_i\pi_f = (-1)^{L_{min}+1}.$$

In the first case, the electric-type transition of multipolarity L_{min} is obviously allowed, and the lowest magnetic multipole permitted by the parity change is $L_{min} + 1$. But $\lambda(\text{M}, L_{min} + 1)$ is much smaller than $\lambda(\text{E}L_{min})$, according to our earlier findings. Hence, this case, for which the transition is 'pure' ($\text{E}L_{min}$), is very simple. On the other hand, in the second case, the parity requirement is satisfied by $L_{min} + 1$ for the electric-type transition, and L_{min} for the magnetic-type transition. It is very difficult to say which of the two, $\lambda(\text{M}, L_{min})$ and $\lambda(\text{E}, L_{min} + 1)$, is larger. The type of sketchy estimate, made earlier, implies that the two quantities may be of the same order of magnitude. But if a detailed evaluation is made of the $\boldsymbol{l}$- and spin-matrix elements, then $\lambda(\text{M}, L_{min})$ may sometimes be larger than $\lambda(\text{E}, L_{min} + 1)$. There are cases where the electric transition may be much enhanced due to other effects. Thus, in the second case, the observed transition is generally a mixture of $(\text{E}, L_{min} + 1)$ and (M, L_{min}). The transition corresponding to the first case is called *parity-favoured*, and that for the second case is called *parity-unfavoured*. The mixed E2, M1-transition, observed in many nuclear γ-decays, is a good example of the parity-unfavoured situation. This, incidentally, is a case where the E2-matrix element is very much enhanced in many nuclei as a result of collective effects (see Chapter IV).

It is pertinent to make a few observations before proceeding with the evaluation of the γ-decay probabilities. The first point, which the reader may have already noticed, is that our estimate of the relative order of magnitude of the various matrix elements did not include the matrix elements of Y^L_M, $\boldsymbol{l}$, and $\boldsymbol{\sigma}$. While comparing $\lambda(\text{E}L)$ with $\lambda(\text{E}, L + 1)$, the ratio of the matrix elements of Y^L_M and Y^{L+1}_M should also have been a relevant quantity. The same observation holds in a comparison of $\lambda(\text{M}L)$ with $\lambda(\text{M}, L + 1)$, in which case the matrix elements of $\boldsymbol{l}$ and $\boldsymbol{\sigma}$ cancel out in the ratio. For our earlier estimate of these relative magnitudes to be valid, a tacit assumption is implied, namely, the ratio of the matrix elements of Y^L_M and Y^{L+1}_M is a quantity of the order of unity. For our earlier comparison of $\lambda(\text{E}L)$ with $\lambda(\text{M}L)$ to be valid, the same assumption is clearly implied for the matrix elements of $\boldsymbol{l}$ and $\boldsymbol{\sigma}$. It must be remembered that these order-of-magnitude ideas on the matrix elements of $\boldsymbol{l}$, $\boldsymbol{\sigma}$, Y^L_M, $Y^{L+1}_M, \ldots$ are essentially right, although an *exact* evaluation may sometimes add another numerical factor (smaller or greater than unity) to our previous order-of-magnitude estimates. This is precisely the reason (already mentioned) that makes $\lambda(\text{M}, L_{min})$ larger than $\lambda(\text{E}, L_{min} + 1)$ in some cases.

The second point that should attract the reader's attention is that the parities of the nuclear operators of the EL-transition [in (III.30a)] and the ML-transition [in (III.31a)] are given by $(-1)^L$ and $(-1)^{L+1}$, respectively. In our reference to the parity of the 2^L-pole component of the *vector potential* **A**, we have stated that the ML-component of **A**, which is given by the vector spherical harmonic $\mathbf{T}^{(L1)L}_M$, has the parity $(-1)^L$ and that of the EL-component, $\mathbf{T}^{(l1)L}_M$ with $l = L \pm 1$, is given by $(-1)^{L+1}$. It should be observed that these parities of the *vector potential* are just the opposite of the parities of the *nuclear operators* in (III.30a) and (III.31). The reader must not confuse the parities of these two different quantities, and would do well to memorize the results for the *nuclear operators*, which will frequently enter our discussions on structure theory.

The quantities that directly concern us in nuclear theory are the square of the matrix elements Q^L_M and $(\mathcal{M}^L_M + \mathcal{M}'^L_M)$. Accordingly, modified transition probabilities, called reduced transition probabilities and denoted by $B(\mathrm{E}L)$ and $B(\mathrm{M}L)$, are defined by omitting the extraneous factors containing 8π, $L, \ldots$, and the energy of the γ-ray, contained in k^{2L+1} (we recall that an energy-dependent factor k^{2L} is concealed in the *square* of the matrix element of $\bar{r}^L$).

Before we write down the expressions for the reduced transition probabilities, we explicitly carry out the summation and averaging over the projection quantum numbers, introducing the symbols $\Omega^L_M(\mathrm{E})$ and $\Omega^L_M(\mathrm{M})$ to denote the nuclear operators:

$$\Omega^L_M(\mathrm{E}) = e \sum_{\alpha=1}^{A} r^L_\alpha Y^L_M(\theta_\alpha, \phi_\alpha)\tfrac{1}{2}(1 - \tau_z)_\alpha, \tag{III.32a}$$

$$\Omega^L_M(\mathrm{M}) = \frac{e\hbar}{2Mc} \sum_{\alpha=1}^{A} [\nabla r^L Y^L_M(\theta, \phi)]_\alpha \cdot [\frac{2}{L+1} l \tfrac{1}{2}(1 - \tau_z) + \mu\sigma]_\alpha. \tag{III.32b}$$

It should be noticed that, except for the factor k^L contained in $\bar{r}^L$, these operators are the same as those appearing in Q^L_M and $(\mathcal{M}^L_M + \mathcal{M}'^L_M)$. The factor $\frac{1}{2}(1 - \tau_z)$ restricts the corresponding sum over α in (III.32) to the protons. In replacing $\mathbf{p}$ by $-i\hbar\nabla$, we have omitted from (III.32b) the factor $-i$ since it gives rise to unity in the square of the modulus that appears in (III.29b). The omission of k^L from the definitions (III.32) is motivated by the definition of the reduced transition probability which, as mentioned, does not contain the factor k^{2L+1}.

In terms of the operator Ω^L_M, the nuclear matrix element is given by

$$\langle J_f M_f | \, \Omega^L_M \, | J_i M_i \rangle = \begin{bmatrix} J_i & L & J_f \\ M_i & M & M_f \end{bmatrix} \langle J_f | \, |\Omega^L| \, | J_i \rangle. \tag{III.33}$$

Since $M_i + M = M_f$ is required by the Clebsch-Gordon coefficient, we conclude that the component M of the multipole operator Ω^L is determined to be $(M_f - M_i)$ the moment we specify the projections M_i and M_f of the nuclear states. The reduced matrix element being independent of the projection quantum numbers, we can now very easily carry out the summation over M_f and the averaging over M_i implied in the notation Σ. Thus,

$$\Sigma \begin{bmatrix} J_i & L & J_f \\ M_i & M_f - M_i & M_f \end{bmatrix}^2 \equiv \frac{1}{[J_i]} \sum_{M_i} \sum_{M_f} \begin{bmatrix} J_i & L & J_f \\ M_i & M_f - M_i & M_f \end{bmatrix}^2 = \frac{[J_f]}{[J_i]}. \tag{III.34}$$

The factor $[J_i]^{-1} \equiv (2J_i + 1)^{-1}$, together with $\sum_{M_i}$, takes care of the averaging. We have first carried out the summation over M_i, getting unity by the result (AIV.3a) from Appendix A; the subsequent summation of unity over M_f thus yields $[J_f]$. Using the results (III.33) and (III.34) in the definition of the reduced transition probability, we obtain

$$\begin{aligned} B(L) &\equiv \Sigma \, |\langle J_f M_f | \, \Omega^L_M \, | J_i M_i \rangle|^2 \\ &= \frac{[J_f]}{[J_i]} |\langle J_f | \, |\Omega^L| \, | J_i \rangle|^2. \end{aligned} \tag{III.35}$$

The application of the definitions (III.32a) and (III.32b) in this expression would yield respectively $B(\mathrm{E}L)$ and $B(\mathrm{M}L)$.

In nuclear structure theory, we shall very frequently be dealing with the E2- and the M1-transition, and hence we give the explicit expressions of the operators $\Omega^2_M(\mathrm{E})$ and $\Omega^1_M(\mathrm{M})$ for future reference as

$$\Omega^2_M(\mathrm{E}) = e \sum_{\alpha=1}^{A} \tfrac{1}{2}(1 - \tau_z)_\alpha r^2_\alpha Y^2_M(\theta_\alpha, \phi_\alpha), \tag{III.36a}$$

$$\Omega_M^1(\mathrm{M}) = \frac{e\hbar}{2Mc}\sqrt{\frac{3}{4\pi}}\sum_{\alpha=1}^{A}\left[\tfrac{1}{2}(1-\tau_z)l_M + \mu\sigma_M\right]_\alpha. \tag{III.36b}$$

(III.36a) follows directly from (III.32a) by putting $L = 2$. (III.36b), however, needs some simplifying procedure, starting with $L = 1$ in (III.32b). We first notice that rY_M^1 is $\sqrt{3/(4\pi)}(\mathbf{r})_M$, according to (AI.9) from Appendix A. Since ∇ operating on the cartesian components (x, y, z) of $\mathbf{r}$ gives the unit vectors $(\mathbf{e}_x, \mathbf{e}_y, \mathbf{e}_z)$, respectively, it follows from the relationship between the cartesian components and the spherical components $(\mathbf{r})_M$ that

$$\nabla(\mathbf{r})_M = \mathbf{e}_M.$$

Hence,

$$\nabla(\mathbf{r})_M \cdot \mathbf{A} = \mathbf{e}_M \cdot \mathbf{A} = A_M$$

for any vector $\mathbf{A}$. This result has been utilized in expressing the simplified form (III.36b). It should be noticed that, except for the factor $\sqrt{3/(4\pi)}$, the expression for $\Omega^1(\mathrm{M})$ is identical to the static magnetic moment operator (III.6). The factor $e\hbar/(2Mc)$ in (III.36b) is the nuclear magneton, which served as the unit in expressing (III.6). Since $\mathbf{s} = \frac{1}{2}\boldsymbol{\sigma}$, and $\frac{1}{2}g_s$ is, according to (III.5), the intrinsic magnetic moment μ of the nucleon, the spin terms in (III.6) and (III.36b) exactly agree. And g_l being equal to unity for a proton and zero for a neutron, the term $\frac{1}{2}(1-\tau_z)\mathbf{l}$ of (III.36b) is also equal to the term $g_l\mathbf{l}$ of (III.6). A similar comparison of (III.36a) with (III.23b) would convince the reader that the $(M = 0)$-component of the E2-transition operator is related to the static quadrupole moment operator Q_0 through a numerical factor $\sqrt{5/(16\pi)}$.

Finally, we mention a very important fact concerning gamma-decay, namely, that *no monopole* $(L = 0)$*-transition is possible*. According to (III.32b), the M0-operator contains a ∇ acting on the constant $r^0Y_0^0\ (=1/\sqrt{4\pi})$, and hence the operator is identically zero. This result also follows from the fact that the M0-component of the electromagnetic field is derivable from the vector potential $\mathbf{T}_M^{(L1)L}$ with $L = 0$, but this particular vector spherical harmonic does not exist; the orbital angular momentum $l = 0$, after being coupled with spin unity, can give rise to only the resultant $L = 1$, and *not* $L = 0$. To prove the absence of the E0-transition, we examine the expression (III.32a) for $L = 0$. Since $r^LY_M^L$ is then a constant $(=1/\sqrt{4\pi})$, the summation over all the protons yields $\Omega_0^0(\mathrm{E}) = (4\pi)^{-1/2}Ze$; the matrix element of this constant connecting two *different* nuclear states is zero due to their orthogonality. In analogy with the case of the M0-transition, we can also show that the vector potential for the E0-component of the radiation field is identically zero. We have already stated that the vector potential for the EL-component is a suitable mixture of $\mathbf{T}_M^{(l1)L}$ with $l = L \pm 1$. It can be shown that the particular mixture is obtained by taking the curl of $\mathbf{T}_M^{(L1)L}$. Since $\mathbf{T}_M^{(L1)L}$ is zero for $L = 0$, its curl is also zero, and hence the absence of the E0-component of the radiation field follows.

A consequence of the absence of the E0-transition is that an excited state 0^+ of a nucleus cannot decay by gamma-emission to a lower 0^+ state (usually the ground state). According to the angular momentum and parity selection rules, already stated, E0 is the only transition permissible between two such states; but since E0 does not exist, there is no gamma-decay. Usually, an excited state of this type decays by emitting internal conversion electrons or by electron-positron pair emission (provided the energy difference between the two states exceeds the minimum amount necessary for the creation of a pair). In the same way, no gamma-decay can take place between a 0^- and a 0^+ state because M0 is then the only transition permitted by the selection rule.

B. SINGLE-PARTICLE TRANSITION PROBABILITY ACCORDING TO THE SHELL MODEL

Here we shall first evaluate the expression (III.35) for the reduced transition probability, using single-particle shell model wavefunctions as the initial and final states. Since we are considering only one nucleon, the summation over all the particles in the expression for Ω_M^L is omitted. We thus obtain

$$B(L) = \frac{[j']}{[j]} |\langle n'l'\tfrac{1}{2}j'|\,|\Omega^L|\,|nl\tfrac{1}{2}j\rangle|^2, \tag{III.37}$$

where the unprimed and primed quantum numbers refer respectively to the initial and final states of the single nucleon. Using (III.32a) without the summation over α, we get the reduced matrix element for the electric-type transition as

$$\langle n'l'\tfrac{1}{2}j|\,|\Omega^L(\mathrm{E})|\,|nl\tfrac{1}{2}j\rangle = e\langle n'l'\tfrac{1}{2}j'|\,|r^LY^L|\,|nl\tfrac{1}{2}j\rangle$$

$$= e\left(\frac{[L]}{4\pi}\right)^{1/2}\langle r^L\rangle(-1)^L\tfrac{1}{2}[1+(-1)^{l+l'+L}]\begin{bmatrix} j' & L & j \\ \frac{1}{2} & 0 & \frac{1}{2}\end{bmatrix}, \tag{III.38a}$$

where $\langle r^L\rangle$ represents the radial integral

$$\langle r^L\rangle = \int_0^\infty R_{n'l'}r^LR_{nl}\,dr. \tag{III.38b}$$

The reduced matrix element of Y^L has been written from (BIII.17) in Appendix B.

The evaluation of the reduced matrix element of $\Omega^L(\mathrm{M})$ is somewhat more complicated. Each term in (III.32b), without the summation over α, has the form

$$(\nabla r^LY_M^L)\cdot\mathbf{A},$$

where $\mathbf{A}$ is either $\boldsymbol{l}$ or $\boldsymbol{\sigma}$. We explicitly write the scalar product between ∇ and $\mathbf{A}$ in terms of spherical components [see (BII.10) in Appendix B] and obtain

$$(\nabla r^LY_M^L)\cdot\mathbf{A} = \sum_\mu(-1)^\mu(\nabla_\mu r^LY_M^L)A_{-\mu}. \tag{III.39a}$$

The quantity $(\nabla_\mu r^LY_M^L)$ can obviously be expanded in terms of spherical harmonics. Writing the spherical harmonic Y_M^L as a ket $|L, M\rangle$ and then introducing the complete set of states $\sum_{L'M'}|L'M'\rangle\langle L'M'|$ on the left, we get

$$\nabla_\mu r^L|L, M\rangle = \sum_{L', M'}|L'M'\rangle\langle L'M'|\,\nabla_\mu r^L\,|L, M\rangle$$

$$= \sum_{L'}|L', M' = M+\mu\rangle\begin{bmatrix} L & 1 & L' \\ M & \mu & M+\mu\end{bmatrix}\langle L'|\,|\nabla r^L|\,|L\rangle. \tag{III.39b}$$

The Clebsch-Gordon coefficient requires M' to be equal to $(M+\mu)$, and hence the summation over M' has disappeared. The reduced matrix elements of ∇ are nonvanishing for $L' = L\pm 1$, and their values are given by (BIII.9) from Appendix B. We thus obtain

$$\langle L' = L-1|\,|\nabla r^L|\,|L\rangle = -\left(\frac{L}{2L-1}\right)^{1/2}\left(\frac{d}{dr}+\frac{L+1}{r}\right)r^L$$

$$= -\left(\frac{L}{2L-1}\right)^{1/2}(2L+1)r^{L-1},$$

$$\langle L' = L + 1 \| \nabla r^L \| L \rangle = \left(\frac{L+1}{2L+3}\right)^{1/2}\left(\frac{d}{dr} - \frac{L}{r}\right) r^L = 0.$$

Therefore, the summation in (III.39b) reduces to one term, i.e., only $L' = L - 1$. Rewriting the kets as spherical harmonics, we get

$$\nabla_\mu r^L Y_M^L = -\begin{bmatrix} L & 1 & L-1 \\ M & \mu & M+\mu \end{bmatrix} (2L+1)\left[\frac{L}{(2L-1)}\right]^{1/2} r^{L-1} Y_{M+\mu}^{L-1}, \tag{III.39c}$$

which is the desired expression to be substituted in (III.39a).

The summation over the projection quantum number μ can now be carried out. Collecting the μ-dependent quantities, we obtain

$$\begin{aligned} \sum_\mu (-1)^\mu \begin{bmatrix} L & 1 & L-1 \\ M & \mu & M-\mu \end{bmatrix} Y_{M+\mu}^{L-1} A_{-\mu} &= \sum_\mu (-1)^\mu (-1)^{L+1-L+1} (-1)^{1-\mu} \begin{bmatrix} L-1 & 1 & L \\ M+\mu & -\mu & M \end{bmatrix} \\ &\quad \times \left(\frac{2L-1}{2L+1}\right)^{1/2} Y_{M+\mu}^{L-1} A_{-\mu} \\ &= -\left(\frac{2L-1}{2L+1}\right)^{1/2} (Y^{L-1}, \mathbf{A})_M^L, \end{aligned} \tag{III.40a}$$

where

$$(Y^{L-1}, \mathbf{A})_M^L = \sum_\mu \begin{bmatrix} L-1 & 1 & L \\ M+\mu & -\mu & M \end{bmatrix} Y_{M+\mu}^{L-1} A_{-\mu}. \tag{III.40b}$$

We have here used the notation of (BII.8) from Appendix B to denote the resultant tensor of rank L and component M, obtained by compounding the spherical harmonic Y^{L-1} with the vector operator $\mathbf{A}$ ($= \boldsymbol{l}$ or $= \boldsymbol{\sigma}$). In arriving at (III.40a), we have used the symmetry relations (AIV.4), from Appendix A, of the Clebsch-Gordon coefficient. In Section 27A, we have stated that the nuclear operators in $\mathcal{M}_M^L$ and $\mathcal{M}'^L_M$ are tensors of rank L and component M; we have now *explicitly* demonstrated in (III.40a) the forms of these tensors.

Collecting all these results and substituting in (III.39a), we get

$$(\nabla r^L Y_M^L) \cdot \mathbf{A} = [L(2L+1)]^{1/2} r^{L-1} (Y^{L-1}, \mathbf{A})_M^L. \tag{III.40c}$$

(III.32b) therefore yields, for a single particle,

$$\Omega_M^L(\mathrm{M}) = \frac{e\hbar}{2Mc}[L(2L+1)]^{1/2} r^{L-1}\left[\frac{2g_l}{L+1}(Y^{L-1}, \boldsymbol{l})_M^L + \mu (Y^{L-1}, \boldsymbol{\sigma})_M^L\right], \tag{III.41}$$

where $g_l = 1$ for a proton and $g_l = 0$ for a neutron. In the special case of $L = 1$, this expression is easily seen to agree with (III.36b) except for the summation over α. For this verification, we need the identity now given, which follows from the definition (III.40b), together with the values of Y_0^0 and the Clebsch-Gordon coefficient ($=1$):

$$(Y^0, \mathbf{A})_M^1 = \begin{bmatrix} 0 & 1 & 1 \\ 0 & M & M \end{bmatrix} Y_0^0 A_M = (4\pi)^{-1/2} A_M.$$

The reduced matrix element of the operator (III.41) can be evaluated by the formulas in Appendix B (Section III). In terms of the radial integral, defined by (III.38b), we have

$$\begin{aligned} \langle n'l'\tfrac{1}{2}j' \| \Omega^L(\mathrm{M}) \| nl\tfrac{1}{2}j \rangle &= \frac{e\hbar}{2Mc}[L(2L+1)]^{1/2} \langle r^{L-1} \rangle \left[\frac{2g_l}{L+1} \langle l'\tfrac{1}{2}j' \| (Y^{L-1}, \boldsymbol{l})^L \| l\tfrac{1}{2}j \rangle \right. \\ &\quad \left. + \mu \langle l'\tfrac{1}{2}j' \| (Y^{L-1}, \boldsymbol{\sigma})^L \| l\tfrac{1}{2}j \rangle \right]. \end{aligned} \tag{III.42}$$

In the first reduced matrix element, we substitute $(\mathbf{j} - \frac{1}{2}\boldsymbol{\sigma})$ for $\mathbf{l}$ and obtain the expression within the square brackets equal to

$$\frac{2g_l}{L+1}\langle l'\tfrac{1}{2}j'\|(Y^{L-1}, \mathbf{j})^L\|l\tfrac{1}{2}j\rangle + (\mu - \frac{g_l}{L+1})\langle l'\tfrac{1}{2}j'\|(Y^{L-1}, \boldsymbol{\sigma})^L\|l\tfrac{1}{2}j\rangle.$$

Remembering that $\mathbf{j}$ cannot change the state $|l\frac{1}{2}j\rangle$, we obtain, from (BIII.11) in Appendix B,

$$\begin{aligned}\langle l'\tfrac{1}{2}j'\|(Y^{L-1}, \mathbf{j})^L\|l\tfrac{1}{2}j\rangle &= U(L-1, 1, j', j; Lj)\langle l'\tfrac{1}{2}j'\|Y^{L-1}\|l\tfrac{1}{2}j\rangle\langle l\tfrac{1}{2}j\|\mathbf{j}\|l\tfrac{1}{2}j\rangle \\ &= U(L-1, 1, j', j; Lj)\tfrac{1}{2}[1-(-1)^{l+l'+L}](-1)^{L-1} \\ &\quad\times \sqrt{\frac{2L-1}{4\pi}}\begin{bmatrix} j' & L-1 & j \\ \frac{1}{2} & 0 & \frac{1}{2}\end{bmatrix}\sqrt{j(j+1)}.\end{aligned} \tag{III.43a}$$

In the final step, the reduced matrix elements of Y^{L-1} and $\mathbf{j}$ have been written by the application of (BIII.3) and (BIII.7) from Appendix B. Similarly, by an application of (BIII.16) from Appendix B, we get

$$\begin{aligned}\langle l'\tfrac{1}{2}j'\|(Y^{L-1}, \boldsymbol{\sigma})^L\|l\tfrac{1}{2}j\rangle &= \begin{bmatrix} l & \frac{1}{2} & j \\ L-1 & 1 & L \\ l' & \frac{1}{2} & j'\end{bmatrix}\langle l'\|Y^{L-1}\|l\rangle\langle\tfrac{1}{2}\|\boldsymbol{\sigma}\|\tfrac{1}{2}\rangle \\ &= \begin{bmatrix} l & \frac{1}{2} & j \\ L-1 & 1 & L \\ l' & \frac{1}{2} & j'\end{bmatrix}\left\{\frac{[l](2L-1)}{4\pi[l']}\right\}^{1/2}\begin{bmatrix} l & L-1 & l' \\ 0 & 0 & 0\end{bmatrix}\sqrt{3}.\end{aligned} \tag{III.43b}$$

Once again, the reduced matrix elements of Y^{L-1} and $\boldsymbol{\sigma}$ have been obtained respectively from (BIII.3) and (BIII.6) in Appendix B.

For the special case of M1-transition, the result is better derived by directly using the simplified form of the M1-operator, given by (III.36b), rather than specializing the general results (III.42) and (III.43) with $L = 1$. By an application of (BIII.13), (BIII.6), and (BIII.7) from Appendix B, we directly obtain

$$\begin{aligned}\langle n'l'\tfrac{1}{2}j'\|\Omega^1(\mathrm{M})\|nl\tfrac{1}{2}j\rangle &= \frac{e\hbar}{2Mc}\left(\frac{3}{4\pi}\right)^{1/2}\delta_{nn'}\delta_{ll'}[\langle l\tfrac{1}{2}j'\|\mathbf{j}\|l\tfrac{1}{2}j\rangle g_l + (\mu - \tfrac{1}{2}g_l)\langle l\tfrac{1}{2}j'\|\boldsymbol{\sigma}\|l\tfrac{1}{2}j\rangle] \\ &= \frac{e\hbar}{2Mc}\left(\frac{3}{4\pi}\right)^{1/2}\delta_{nn'}\delta_{ll'}[g_l\sqrt{j(j+1)}\delta_{jj'} + U(l\tfrac{1}{2}j'1; j\tfrac{1}{2})\sqrt{3}(\mu - \tfrac{1}{2}g_l)].\end{aligned} \tag{III.44}$$

Selection Rules

The general selection rules on the total angular momentum and parity have been described in Section 27A. For single-particle wavefunctions, the initial and final parities are determined by $(-1)^l$ and $(-1)^{l'}$, respectively. Therefore, for the EL-transition, $(-1)^L = (-1)^{l+l'}$, and for the ML-transition, $(-1)^L = (-1)^{l+l'+1}$. These requirements are obvious from the factor $\frac{1}{2}[1 + (-1)^{l+l'+L}]$ in (III.38a), $\frac{1}{2}[1 - (-1)^{l+l'+L}]$ in (III.43a), and the Clebsch-Gordon coefficient in (III.43b). According to the general angular momentum selection rule, j, j', and L must satisfy angular momentum coupling rules. In our special case of single-particle wavefunctions, this requirement is incorporated in the Clebsch-Gordon coefficient in (III.38a), in the first U-function in (III.43a), and in the 9j-symbol in (III.43b). From our earlier discussion, it is also clear that if $L_{\min} = |j - j'|$ and $(-1)^{l+l'} = (-1)^{L_{\min}}$, then the most predominant transition is $B(\mathrm{E}, L_{\min})$. On the other hand, if $(-1)^{l+l'} = (-1)^{L_{\min}+1}$, then the predominant transition is a mixture of $B(\mathrm{E}, L_{\min} + 1)$ and $B(\mathrm{M}, L_{\min})$. These statements have to be modified if the single nucleon undergoing the transition is a neutron. This point is discussed later in this section.

Some more selection rules can also be derived in the single-particle case. For example, due to the presence of Y^L in (III.38a), it is clear that, for the EL-transition, l, l', and L must satisfy the angular momentum coupling rule in addition to the parity rule. A little reflection would, however, show that if the requirements on parity and total angular momenta, already mentioned, are satisfied, then the equalities $l = j \pm \frac{1}{2}$ and $l' = j' \pm \frac{1}{2}$ automatically take care of the requirement $|l - l'| \leqslant L \leqslant l + l'$. On the other hand, the operators in $\Omega^L(\mathrm{M})$ for the magnetic transition [see (III.42)] contain a Y^{L-1}, and hence we get an additional requirement $|l - l'| \leqslant (L - 1) \leqslant (l + l')$. In the special case of M1-transition, i.e., $L = 1$, this requirement becomes identical to $l = l'$ [the $\delta_{ll'}$-factor in (III.44) should be noted]. This additional requirement for magnetic transitions is quite nontrivial, as is apparent from the example that follows.

Let us consider a transition between $(0d_{3/2})$ and $(1s_{1/2})$. In this case, $(-1)^{l+l'} = +1$, $L_{\min} = \frac{3}{2} - \frac{1}{2} = 1$, and hence $(-1)^{l+l'} = (-1)^{L_{\min}+1}$. Thus, it is expected to be a case of parity-unfavoured transition comprising a mixture of M1 and E2. However, the additional selection rule on the orbital angular momentum for magnetic transitions, which we have just stated, forbids an M1-transition between the state $l = 2$ and the state $l' = 0$. Thus, *according to our single-particle picture*, the transition under consideration is pure E2. This point is discussed later in this section in the context of experimental data.

Approximate Estimate and Weisskopf Units

We know that if (EL) is allowed according to selection rules, then $\frac{1}{2}[1 + (-1)^{l+l'+L}] = 1$. Using (III.37), together with (III.38) in (III.29a) for the transition probability $\lambda(\mathrm{E}L)$, we derive the *exact* result for a single-particle type transition:

$$\lambda(\mathrm{E}L) = \frac{2(L+1)}{L[(2L+1)!!]^2}\left(\frac{E_\gamma}{\hbar c}\right)^{2L+1}\frac{e^2}{\hbar c}c\langle r^L\rangle^2 S(jj'L). \tag{III.45}$$

Here

$$S(jj'L) = \frac{[L][j']}{[j]}\begin{bmatrix} j' & L & j \\ \frac{1}{2} & 0 & \frac{1}{2} \end{bmatrix}^2, \tag{III.46}$$

$E_\gamma(=\hbar\omega)$ is the energy of the γ-ray, and hence $E_\gamma/(\hbar c) = \omega/c \equiv k$.

An approximate estimate, based only on the magnitudes of kR and v/c, can now be made for the single-particle transition probability, which is somewhat better than the very rough order of magnitude estimate made in Section 27A. The quantity S is called the statistical factor and is replaced by unity for our estimate. The radial integral of r^L is estimated by replacing the wavefunctions $\mathcal{R}_{nl}$ $(=R_{nl}/r)$ by a constant magnitude in a sphere of radius R, roughly equal to the nuclear radius. The constant magnitude N of the wavefunction is obtained by integrating its square over the entire sphere, and requiring the resultant probability to be unity. Thus,

$$1 = \int_0^R \mathcal{R}_{nl}^2 r^2\, dr = \tfrac{1}{3}N^2R^3$$

or

$$N = (\frac{3}{R^3})^{1/2}.$$

The radial wavefunctions $\mathcal{R}_{nl}$ and $\mathcal{R}_{n'l'}$ have the aforestated magnitude between 0 and R, after which they drop sharply to zero. Thus, (III.38b) yields

$$\langle r^L \rangle = \frac{3}{R^3}\int_0^R r^{L+2}\,dr = \frac{3}{L+3}R^L. \tag{III.47}$$

It should be observed that this procedure gives a gross overestimate of the radial integral because a perfect overlap has been assumed between the two radial wavefunctions R_{nl} and $R_{n'l'}$. In general, the overlap between them may be far from perfect.

Substituting (III.47) and $S(jj'L) = 1$—together with $\hbar c = 1.97 \times 10^{-11}$ MeV cm, $e^2/(\hbar c) = 1/137$, and $c = 3 \times 10^{10}$ cm/sec—in the exact equation (III.45), we obtain the rough estimate

$$\lambda(\mathrm{E}L) \approx \frac{4.4(L+1)}{L[(2L+1)!!]^2}(\frac{E_\gamma}{197})^{2L+1}(\frac{3}{L+3})^2 R^{2L}(10^{21}\ \mathrm{sec}^{-1}), \tag{III.48}$$

where E_γ is understood to be in MeV, and R in units of fm ($=10^{-13}$ cm). The quantity 4.4 in the numerator must *not* be confused with 4×4. This rough value of $\lambda(\mathrm{E}L)$ was first given by Weisskopf and is very often referred to as the *Weisskopf unit* for the probability of electric multipole transition. A similar rough estimate can be obtained for the ML-transition as well. In this case, we extract the factors $e\hbar/(Mc)$, $\langle r^{L-1}\rangle$, and $(4\pi)^{-1/2}$ from (III.42) and denote the rest of the expression by $S(ll'jj'L)$. Then (III.29b) yields

$$\lambda(\mathrm{M}L) = \frac{2(L+1)}{L[(2L+1)!!]^2}(\frac{E_\gamma}{\hbar c})^{2L+1}\frac{e^2}{\hbar c}c\langle r^{L-1}\rangle^2(\frac{\hbar}{Mc})^2 S(ll'jj'L). \tag{III.49}$$

The factor $(4\pi)^{-1/2}$ was actually contained in the two matrix elements occurring in (III.42), as is evident from the expressions (III.43). In the present case, Weisskopf argued that $S(ll'jj'L)$, which contains the square of the magnetic moment of the nucleons, may be much larger than unity, and he used a rough value 10 for making the estimate. If we take the ratio of (III.49) and (III.45) with the aforementioned rough values of the quantities S occurring in them, then

$$\lambda(\mathrm{M}L)/\lambda(\mathrm{E}L) = 10[\hbar/(McR)]^2. \tag{III.50}$$

We recall that, according to our estimate in Section 27A where v/c and kR were used, the same quantity was found to be roughly equal to $(v/c)kR$. The discrepancy between the two estimates arises because in Section 27A we have replaced the value of the momentum operator between the initial and final states by its average value in the nuclear ground state; this procedure is obviously not strictly accurate. Going back to (III.49) with $S(ll'jj'L) = 10$ and substituting numbers, we obtain

$$\lambda(\mathrm{M}L) = \frac{1.9(L+1)}{L[(2L+1)!!]^2}(\frac{E_\gamma}{197})^{2L+1}(\frac{3}{L+2})^2 R^{2L-2}(10^{21}\ \mathrm{sec}^{-1}), \tag{III.51}$$

which is the Weisskopf unit for measuring the probability of magnetic multipole transition. As before, E_γ is in MeV, R in fm, and the number 1.9 must not be confused with 1×9. The substitution of $L - 1$ for L in (III.47) yields $(L + 2)$ in the denominator instead of $L + 3$, and R^{L-1} in the numerator instead of R^L; this accounts for the difference between the last two factors in (III.48) and (III.51). In making the numerical evaluation, we have used the nucleon Compton wavelength $\hbar/(Mc) = 0.21$ fm.

According to our previous observation, the Weisskopf unit is usually an overestimate of the *single-particle* transition probability due to the assumption of a complete overlap between the two radial functions. Therefore, most of the single-particle type transition data are expected to be a fraction of the Weisskopf unit. On the other hand, if a particular transition is not of the single-particle type, and many nucleons participate in it in a 'collective' manner, then the observed probability may be much larger than the Weisskopf unit.

The reader must remember that the exact formulas derived in this section enable him to obtain the *exact* value of the single-particle transition probability, and it is this exact value that should be compared with the observed one to check the success or shortcoming of the single-particle shell model. The Weisskopf unit gives only a rough guide as to whether a particular transition is of the collective type or the single-particle type.

Special Consideration of Single-Particle Neutron Transition

If the single particle undergoing the transition is a neutron, then the matrix elements Q^L_M and $\mathcal{M}^L_M$ are identically zero because they are associated with the charge of the particle. The matrix elements Q'^L_M and $\mathcal{M}'^L_M$, associated with the magnetic moment of the particle, will, however, be nonvanishing. The derivation of the matrix elements of these operators, using the techniques of Appendix B (Section III), is left as an exercise. The derivation is of academic interest only. In practice, the odd-neutron nuclei behave similar to the odd-proton ones, and their cases have to be treated by using Q^L_M and $\mathcal{M}^L_M$ with appropriate effective charges. The situation is comparable to that noticed for the static moments, and points out, once again, that the core of an odd-mass nucleus plays a role in electromagnetic transition as well.

Application of Single-Particle Results

The major uncertainty in trying to apply the single-particle results in a serious quantitative manner is in the radial integrals $\langle r^L \rangle$. A reliable quantitative estimate of this integral, which we shall call the L-th moment, presupposes a very detailed knowledge of the single-particle radial wavefunctions. In most shell-model work, the radial functions for an infinite harmonic oscillator well are used. A Woods-Saxon type well has also sometimes been applied to get the radial functions. Any error in the radial function manifests itself rather crucially in the determination of a moment of large order L. Since the square of the moment enters the transition probability, the error is still further magnified.

Under the most ideal conditions, the single-particle results would apply only to the closed-shell-plus-one-nucleon or closed-shell-minus-one-nucleon case. The simplest wavefunctions for the closed-shell-plus-one-nucleon case are given by $|\Phi_0, j:j\rangle$, where Φ_0 is the closed-shell core state, and j is the single-particle state. The transition probability between any two states j and j' of this type is given entirely by the single-particle result of this section. The same observation is true for one-hole states. In connection with the static moments, we have observed that considerable core polarization exists in these nuclei, which makes the wavefunctions more complicated. For each single-particle state j, the wavefunction is a mixture of the type

$$\Psi(j) = C_0^{(j)}|\Phi_0, j:j\rangle + \sum_{J, j''} C^{(j)}_{j'', J}|\Phi_J, j'':j\rangle, \tag{III.52a}$$

where Φ_J is an excited core state of angular momentum J. Since the multipole operators are of the one-body type, the core states Φ_J important for our purpose are those that can be obtained by exciting one particle out of Φ_0. The transition matrix element between $\Psi(j)$ and

$\Psi(j')$ is given, to the first order in the small numbers $C^{(j)}_{j'',J}$ and $C^{(j')}_{j'',J}$, by the expression

$$\langle\Psi(j')|\,\Omega^L_M\,|\Psi(j)\rangle = C^{(j)}_0 C^{(j')}_0 \langle\Phi_0, j' : j'|\,\Omega^L_M\,|\Phi_0, j : j\rangle + \sum_J C^{(j)}_{j',J} C^{(j')}_0 \langle\Phi_0, j' : j'|\,\Omega^L_M\,|\Phi_J, j' : j\rangle$$
$$+ \sum_J C^{(j)}_0 C^{(j')}_{j,J} \langle\Phi_J, j : j'|\,\Omega^L_M\,|\Phi_0, j : j\rangle. \qquad \text{(III.52b)}$$

The first term is the usual matrix element between closed-shell-plus-one-nucleon states. The second and third terms represent the effects of core polarization. Each of these matrix elements can be trivially shown, by the results of Appendix B (Section III), to be equal to $\delta_{J,L}\langle\Phi_L|\,\Omega^L_M\,|\Phi_0\rangle$. The reason that the single-particle states in the core-polarization terms have been put equal is that Φ_J already contains one particle state different from that in Φ_0. If the single-particle state h in Φ_0 is changed to p in Φ_L, then the core-matrix element is given by $\langle p|\,\Omega^L_M\,|h\rangle$. Since any of the single-particle states in Φ_0 could change to something else, the state Φ_L can, in general, be a superposition of many 1h-1p type states; the core contribution to the transition matrix element is therefore a similar superposition of matrix elements of the type $\langle p|\,\Omega^L_M\,|h\rangle$. In most cases, the core contribution can thus be quite significant, and this is the second reason why the single-particle results given earlier compare very poorly with the observed data. In particular, the core contribution to the E2- and M1-transition (which are the most frequently observed ones) may have a very large vitiating effect from the single-particle values.

A discussion similar to the one just concluded can be given in the case of a nonclosed-shell odd-mass nucleus in terms of its seniority-zero core state Φ_0. The transition between the states $|\Phi_0, j : j\rangle$ and $|\Phi_0, j' : j'\rangle$ now gives essentially the single-particle result; the only modification is due to the antisymmetry requirement of the last nucleon with the seniority-zero nucleons of the partially filled levels. The argument on core-polarization effects is also similar.

The experimental data on odd-mass nuclei is shown in Figs. III.7 and III.8. The horizontal lines in these diagrams correspond to the Weisskopf estimates of (III.48) and (III.51) with $R = 1.2A^{1/3}$ fm.

Finally, a few remarks on the trends displayed in Figs. III.7 and III.8 are pertinent. The ordinate here is the log of the *reduced lifetime*, defined as $\tau(EL)(E_\gamma^{2L+1}A^{2L/3})$ and $\tau(ML)(E_\gamma^{2L+1}A^{(2L-2)/3})$ for the electric and magnetic transitions, respectively. The mean lifetime τ is obtained from the probability λ by $\tau(L) = \lambda^{-1}(L)$. The multiplying quantities containing gamma-energy and the nucleon number erase the k^{2L+1}-dependence and nucleon-number dependence through the factors R^{2L} and R^{2L-2}, respectively. It is clear from the two diagrams that the observed lifetimes are usually too long (i.e., the probability is too small) as compared with the Weisskopf estimate. This is in agreement with our earlier expectation. On the other hand, most E2-transitions are seen to have lifetimes that are too short (i.e., the probability is too large) as compared with the Weisskopf estimate. For spherical nuclei, this is due to the strong quadrupole-type core-polarization effect, whereas for deformed nuclei this is a direct consequence of the charge deformation. Both these effects may be referred to as 'collective' effects.

28. EXACT TREATMENT OF TWO NUCLEONS BY SHELL MODEL

We have so far confined our discussion of the shell model to single-particle type states. In this section, we shall consider the case of two nucleons and do a complete shell-model calculation. The first step will be to write down all the antisymmetric wavefunctions in a convenient

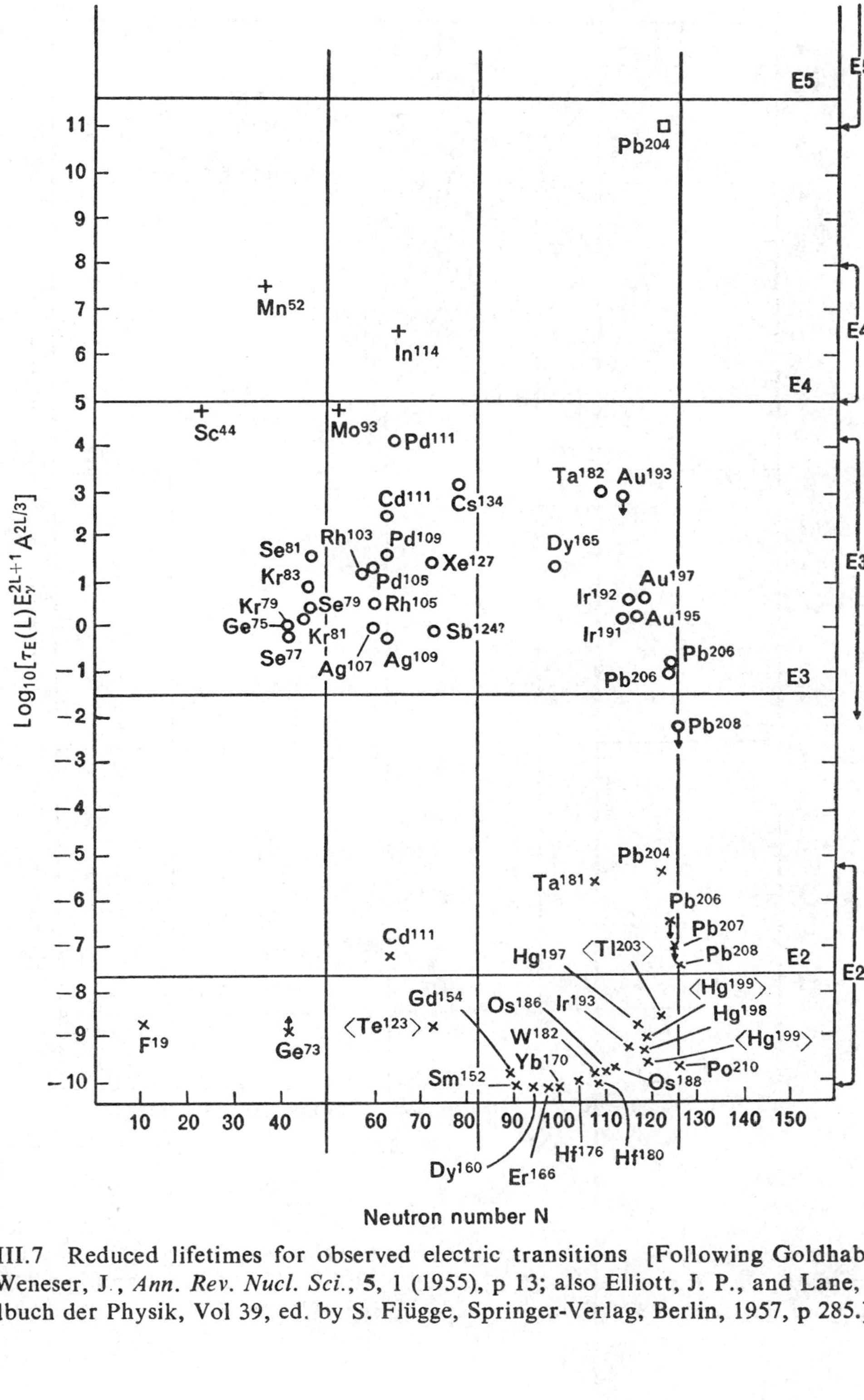

Fig. III.7 Reduced lifetimes for observed electric transitions [Following Goldhaber, M., and Weneser, J., *Ann. Rev. Nucl. Sci.*, 5, 1 (1955), p 13; also Elliott, J. P., and Lane, A. M., Handbuch der Physik, Vol 39, ed. by S. Flügge, Springer-Verlag, Berlin, 1957, p 285.]

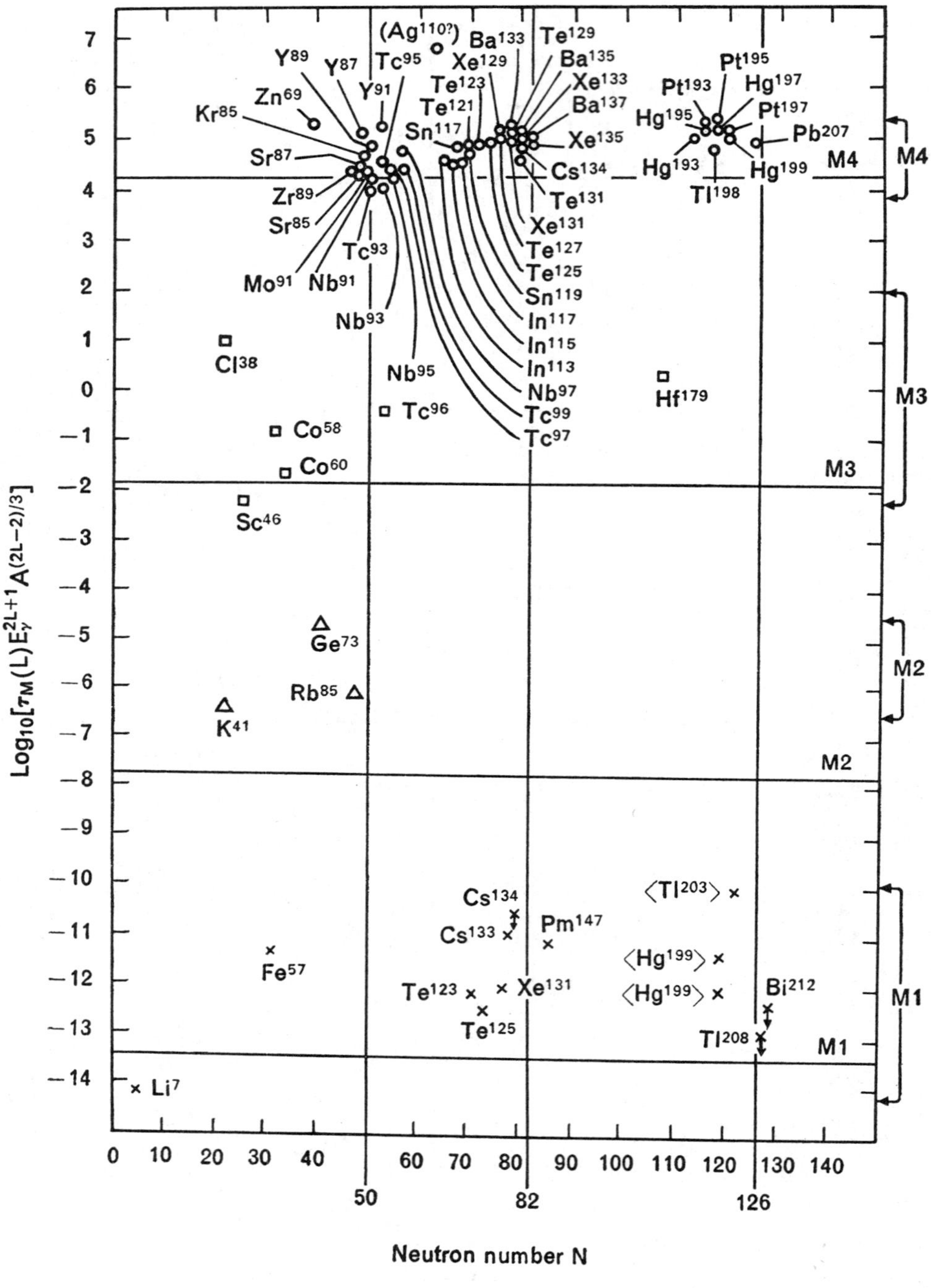

Fig. III.8 Reduced lifetimes for observed magnetic transitions. [Following Goldhaber, M., and Weneser, J., *Ann. Rev. Nucl. Sci.*, **5**, 1 (1955), p 12; also Elliott, J. P., and Lane, A. M., Handbuch der Physik, Vol 39, ed. by S. Flügge, Springer-Verlag, Berlin, 1957, p 286.]

manner, and the next step will be the evaluation of matrix elements of various physical observables using these wavefunctions. This prepares the groundwork for the treatment, in Section 29, of a general case of more than two nucleons.

A. TWO-NUCLEON WAVEFUNCTIONS

We start with the simplest case, namely, that of two identical nucleons (i.e., either protons or neutrons). If a set of single-particle states $(nljm)$, $(n'l', j'm')$, $(n''l'', j''m'')$, . . . is available to these two nucleons beyond the closed-shell core, then the two nucleons may distribute themselves in several ways amongst the available levels. If the total number of states is N, then obviously ${}^NC_2 = \frac{1}{2}N(N-1)$ is the total number of two-particle states permitted by the Pauli exclusion principle. According to this principle, the two nucleons must occupy two different single-particle states. By definition, the two single-particle states $nljm$ and $n'l'j'm'$ are considered to be different if at least one of the quantum numbers in $nljm$ differs from the corresponding quantum number in $n'l'j'm'$. Thus, the total number of antisymmetric states is equal to the total number of ways of picking out two *different* states from a set N, and hence this is equal to ${}^NC_2$, as just stated. With two such single-particle states, for example, $nljm$ and $n'l'j'm'$, we can construct a 2×2 determinant

$$\frac{1}{\sqrt{2}}\begin{vmatrix} \phi_{nljm}(1) & \phi_{n'l'j'm'}(1) \\ \phi_{nljm}(2) & \phi_{n'l'j'm'}(2) \end{vmatrix},$$

which is obviously a normalized antisymmetric state of the two nucleons (see Section 15). Since an actual nuclear state always corresponds to a good total angular momentum J, it is more convenient for our shell-model work to use a set of wavefunctions (for the two- as well as the several-nucleon case) in which the coupling of the individual particle angular momenta has already been done in order to produce the given total angular momenta. In so doing, each wavefunction has also to remain antisymmetric under the exchange of any two nucleons (guarantees the Pauli principle). One way to obtain antisymmetric wavefunctions with good total angular momentum is to use suitable linear combinations of determinantal wavefunctions multiplied by appropriate Clebsch-Gordon coefficients so as to produce the required total angular momentum. Although this straightforward procedure using determinants is quite feasible in the two-nucleon case, it becomes extremely cumbersome as the number of nucleons increases. We shall therefore follow an alternative method which directly uses angular momentum coupled states and guarantees antisymmetry by a suitable linear combination of such states. The angular momentum coupled states can be used in a straightforward manner in evaluating matrix elements of physical observables by the standard results of tensor operator algebra listed in Appendix B (Section III). At no stage do we then need to break up the angular momentum coupling with the explicit use of the Clebsch-Gordon coefficients. The determinantal method and the alternative method just described lead to identical states in the case of two nucleons; in the case of three or more nucleons, however, the two methods yield different sets of states. The set of states obtained by either method is complete but, as mentioned before, the states of the determinantal method are exceedingly cumbrous to construct and use in actual calculations.

In the two-nucleon case, we first distinguish between two situations: (i) The sets of quantum numbers nlj and $n'l'j'$ of the two nucleons are the same; in this situation, the two nucleons are said to be *equivalent*. (ii) *At least* one quantum number of the set nlj is different from the corresponding quantum number in $n'l'j'$; here the two nucleons are said to be *inequivalent*. It should be noticed that the projection quantum numbers were not introduced in these definitions because when the angular momenta of the nucleons are coupled the reference to their individual projection quantum numbers is lost.

In situation (i), the angular momentum coupling produces a two-nucleon state given by

$$|(nlj),(nlj):JM\rangle = \sum_{m,m'} \begin{bmatrix} j & j & J \\ m & m' & M \end{bmatrix} |nljm\rangle_1 |nljm'\rangle_2. \tag{III.53a}$$

The subscripts 1 and 2 on the right-hand side refer to the two nucleons. As already mentioned, we shall never use the *explicit* forms of the coupled wavefunction given by the right-hand side of this *definition*; the compact notation on the left-hand side, in which the coupling has been kept *implicit*, suffices for our work.

In situation (ii), the coupling produces the state

$$|(nlj),(n'l'j'):JM\rangle = \sum_{m,m'} \begin{bmatrix} j & j' & J \\ m & m' & M \end{bmatrix} |nljm\rangle_1 |n'l'j'm'\rangle_2. \tag{III.53b}$$

We remind the reader that the general notation nlj and $n'l'j'$ for the quantum numbers does not imply that all the three quantum numbers in one set need *necessarily* be different from the corresponding ones in the other set; to agree with our definition of the inequivalent nucleons, it would suffice if only one of the quantum numbers in one set is different from the corresponding one in the other set. We draw the reader's attention also to our convention that, in an angular momentum coupled state, the first state always goes with the first particle, and the second with the second particle. In the case of more than two nucleons also, the same convention will be followed; namely, the third state goes with the third nucleon, and so on.

The wavefunctions (III.53) are characterized by a total angular momentum J and its projection M. But we have also to ensure the antisymmetry of these wavefunctions under the exchange of the two nucleons. Let us first examine the equivalent case (III.53a). Exchanging the nucleon indices 1 and 2 on the right-hand side, we get

$$\begin{aligned} P_{12}|(nlj),(nlj):JM\rangle &= \sum_{m,m'} \begin{bmatrix} j & j & J \\ m & m' & M \end{bmatrix} |nljm'\rangle_1 |nljm\rangle_2 \\ &= \sum_{m,m'} (-1)^{j+j-J} \begin{bmatrix} j & j & J \\ m' & m & M \end{bmatrix} |nljm'\rangle_1 |nljm\rangle_2 \\ &= -(-1)^J \sum_{m,m'} \begin{bmatrix} j & j & J \\ m & m' & M \end{bmatrix} |nljm\rangle_1 |nljm'\rangle_2 \\ &= -(-1)^J |(nlj),(nlj):JM\rangle, \end{aligned} \tag{III.54}$$

where P_{12} is the exchange operator. In the second step, we have performed the first symmetry operation (AIV.4) from Appendix A on the Clebsch-Gordon coefficient; in the third step, we have used the fact that $2j$ is always an odd integer, and hence $(-1)^{2j} = -1$; finally, the definition (III.53a) has been used to replace the m, m' summations by the coupled state in the fourth step; since m, m' are *summation symbols*, we have simply interchanged them in the third step so that the identification with (III.53a) becomes clear. The result (III.54) tells us that the angular momentum coupled wavefunctions (III.53a) for the *equivalent case* are antisymmetric if we choose the total angular momentum to be even integers. According to the angular momentum coupling rule, J could be any integral quantity between $0\,(=j-j)$ and $2j\,(=j+j)$;

but the antisymmetry requirement rules out every alternate value (the odd integers), the permissible values being 0, 2, 4, . . . , $(2j-1)$. This completes our discussion on the wavefunctions of two *equivalent*, identical (i.e., either protons or neutrons) nucleons.

In the inequivalent case (III.53b), we perform the exchange operation P_{12} and obtain

$$\begin{aligned} P_{12}|(nlj), (n'l'j') : JM\rangle &= \sum_{m,m'} \begin{bmatrix} j & j' & J \\ m & m' & M \end{bmatrix} |n'l'j'm'\rangle_1 |nljm\rangle_2 \\ &= (-1)^{j+j'-J} \sum_{m,m'} \begin{bmatrix} j' & j & J \\ m' & m & M \end{bmatrix} |n'l'j'm'\rangle_1 |nljm\rangle_2 \\ &= (-1)^{j+j'-J} |(n'l'j'), (nlj) : JM\rangle. \end{aligned} \tag{III.55}$$

In this case also, we have used the first symmetry relation of (AIV.4) from Appendix A and, in the final step, the definition (III.53b); since, however, the state $|n'l'j'm'\rangle$ now goes with the first particle, and $|nljm\rangle$ with the second particle, we have written, in the final step, $n'l'j'$ first and then nlj. It should be noticed that this reversed ordering of the states of 1 and 2 demands the first two columns of the Clebsch-Gordon coefficient in the reversed order, and we therefore used the symmetry relation (AIV.4) from Appendix A to get the Clebsch-Gordon coefficient to conform to the new order of coupling. It is clear from (III.55) that, since the sets $n'l'j'$ and nlj are inequivalent, the right-hand side does not have a simple relationship with the state $|(nlj), (n'l'j') : JM\rangle$. Therefore, the antisymmetry of this state now has to be ensured by an explicit antisymmetrization procedure. Using the two-particle antisymmetrizer $(1/\sqrt{2})(\mathbb{1} - P_{12})$ from Section 15 and the result (III.55), we obtain

$$\begin{aligned} |(nlj), (n'l'j') : JM) &= \frac{1}{\sqrt{2}}(\mathbb{1} - P_{12})|(nlj), (n'l'j') : JM\rangle \\ &= \frac{1}{\sqrt{2}}[|(nlj), (n'l'j') : JM\rangle - (-1)^{j+j'-J}|(n'l'j'), (nlj) : JM\rangle]. \end{aligned} \tag{III.56}$$

We draw the reader's attention to another convention in our notation: the *parenthesis* at the end of the state on the left-hand side indicates an *antisymmetrized*, normalized state, whereas the *ket* notation at the ends of the states on the right-hand side denotes states with *only* angular momentum coupling and *no antisymmetrization*. We once again remind the reader that, in the first state on the right-hand side, nlj goes with the first particle, and $n'l'j'$ with the *second* particle, whereas in the second state the situation is just the opposite.

Using the same parenthetic notation in the *equivalent* case too, we simply have

$$\begin{aligned} |(nlj), (nlj) : JM) &= \tfrac{1}{2}[1 + (-1)^J]|(nlj), (nlj) : JM\rangle \\ &= \tfrac{1}{2}[|(nlj), (nlj) : JM\rangle - (-1)^{j+j-J}|(nlj), (nlj) : JM\rangle]. \end{aligned} \tag{III.57}$$

The factor $\frac{1}{2}[1 + (-1)^J]$ takes care of the requirement that J has to be an even integer. It should be observed that, except for the normalization constant, the right-hand side of (III.57) could be obtained from (III.56) by using $n = n'$, $l = l'$, and $j = j'$. The correct normalization can be taken care of by writing the normalization constant as $[2(1 + \delta_{nn'}\delta_{ll'}\delta_{jj'})]^{-1/2}$, and then the single general expression

$$\begin{aligned} |(nlj), (n'l'j') : JM) &= \frac{1}{2[(1 + \delta_{nn'}\delta_{ll'}\delta_{jj'})]^{1/2}} \\ &\times [|(nlj), (n'l'j') : JM\rangle - (-1)^{j+j'-J}|(n'l'j'), (nlj) : JM\rangle] \end{aligned} \tag{III.58}$$

is sufficient for the equivalent as well as the inequivalent case.

We have already answered the question: what are the permitted values of J in the equivalent case ? The same question applies to the inequivalent case. It is clear from the way the states (III.56) have been constructed that all possible values of J permitted by the angular momentum coupling rule, i.e., those lying between $|j-j'|$ and $(j+j')$, are allowed in this case.

Finally, we have to extend our discussion to the general case of *nonidentical* nucleons (i.e., neutron and proton). There are two ways of doing this. First, we can simply couple the angular momenta of a neutron and a proton and use all possible states of the type $|(nlj)_\nu, (n'l'j')_\pi : JM\rangle$, where ν and π refer to neutron and proton, respectively. The angular momentum J now ranges from $|j-j'|$ to $(j+j')$. Since the neutron-proton wavefunction does not have to be antisymmetric under exchange, the angular momentum coupled states serve our purpose. We do not therefore have to distinguish between equivalent and inequivalent cases.

The second method is to use the isospin as an additional quantum number and demand antisymmetry of the wavefunction under an exchange of *any two* nucleons. We then do not have to specify whether (nlj), $(n'l'j')$ correspond to ν or π. Neither do we have to discuss the cases of identical and nonidentical nucleons separately. All such information is contained in the two-nucleon isospin T and its projection M_T. We have discussed (see Section 2C) that both these are good quantum numbers for a two-nucleon system; T is conserved due to the charge-independence of nuclear forces, and M_T because of the charge conservation of the two-nucleon system.

The antisymmetric wavefunctions specified by JM, TM_T may be constructed by following a procedure similar to that already described for states of good JM. In this way, the general expression that covers the equivalent, as well as the inequivalent, cases is

$$|(nlj), (n'l'j') : JM; \tfrac{1}{2}, \tfrac{1}{2} : TM_T) = \frac{1}{[2(1+\delta_{nn'}\delta_{ll'}\delta_{jj'})]^{1/2}}[|(nlj), (n'l'j') : JM; \tfrac{1}{2}, \tfrac{1}{2} : TM_T\rangle \\ -(-1)^{j+j'-J}(-1)^{1/2+1/2-T}|(n'l'j'), (nlj) : JM; \tfrac{1}{2}, \tfrac{1}{2} : TM_T\rangle]. \tag{III.58}$$

This expression now replaces (III.57), given for identical nucleons. The part of each two-nucleon wavefunction following JM shows the isospin coupling of the two particles. The second term on the right-hand side is the result of $-P_{12}$ operating on the first term. In the same way as that for (III.55), we study this effect of P_{12}. Obviously, the interchange of the particles 1 and 2 in the angular momentum part would once again give the same result, i.e., (III.55); but the same exchange in the isospin part of the wavefunction would give, according to (I.5b), the additional phase factor $(-1)^{1/2+1/2-T}$. This accounts for the structure of the second term in (III.58). The δ-term in the normalization factor takes care, as before, of the special case of equivalent particles $n=n'$, $l=l'$, $j=j'$. For such particles, the two angular momentum coupled states are obviously identical, and (III.58) reduces to

$$|(nlj), (nlj) : JM; \tfrac{1}{2}, \tfrac{1}{2} : TM_T) = \tfrac{1}{2}[1-(-1)^{j+j-J+1-T}]|(nlj), (nlj) : JM; \tfrac{1}{2}, \tfrac{1}{2} : TM_T\rangle.$$

Since $(2j+1)$ is an even integer, it is necessary, in order that this expression be nonvanishing, that

$$J+T = \text{odd integer.}$$

Thus,

$$J = 0, 2, 4, \ldots, (2j-1) \quad \text{for } T=1, \tag{III.59a}$$

$$J = 1, 3, 5, \ldots, 2j \quad \text{for } T = 0. \tag{III.59b}$$

In the case of inequivalent nucleons, however, there is no such restriction on the value of J. For each T, J can have any value between $|j - j'|$ and $(j + j')$. Since two identical nucleons always correspond to $T = 1$ ($M_T = 1$ for two neutrons, and $M_T = -1$ for two protons), the result (III.59a) agrees with our earlier result, as it should. The states (III.59b), having $M_T = 0$ (since $T = 0$), are possible only in the neutron-proton system.

B. MATRIX ELEMENTS OF ONE-BODY OPERATORS

In shell-model calculations, we usually confront two types of operators: (i) the single-particle type operators, such as the static and transition electromagnetic moments, the kinetic energy operator, and the one-body average potential $\mathcal{V}$; and (ii) the two-body type operators; the only operator of this type that concerns us is the two-nucleon interaction potential V. In this section, we shall derive the matrix elements of a general one-body type operator,

$$F = \sum_{i=1}^{n} f(i),$$

using the two-nucleon wavefunctions from Section 28A. In Section 28C, the same work is accomplished for the two-body potential V.

Before proceeding with the actual evaluation, we shall derive a very important result which is applicable to any symmetric operator, i.e., any operator that does not change under an exchange of particle coordinates. The operator F is one of this kind. Further, in the proof, we shall use general antisymmetric wavefunctions of *any number* of particles, say, n. Let the two states be denoted by $\Psi(1, 2, \ldots, k, \ldots, n)$ and $\Psi'(1, 2, \ldots, k, \ldots, n)$ and the operator by

$$F = \sum_{i=1}^{n} f(i),$$

where the numbers $1, 2, \ldots, k, \ldots, n$ denote the particle coordinates.

We have

$$\begin{aligned} &\langle \Psi(1, 2, \ldots, k, \ldots, n)|\, f(1)\, |\Psi'(1, 2, \ldots, k, \ldots, n)\rangle \\ &= \langle \Psi(k, 2, \ldots, 1, \ldots, n)|\, f(k)\, |\Psi'(k, 2, \ldots, 1, \ldots, n)\rangle \\ &= \langle \Psi(1, 2, \ldots, k, \ldots, n)|\, f(k)\, |\Psi'(1, 2, \ldots, k, \ldots, n)\rangle. \end{aligned} \tag{III.60a}$$

In the first step, we have relabelled in the *entire expression* the particle coordinates 1 and k; the relabelling is legitimate because the particle coordinates are *dummy* integration variables in the *entire expression*. In the last step, we have interchanged the particles 1 and k *in the wavefunctions only*. Each interchange gives only a minus sign to the wavefunction, due to its antisymmetry under exchange. The two minus signs resulting from the interchanges in the two wavefunctions cancel each other, and hence the final result follows. Since k could be any particle index from $1, 2, \ldots, n$, the result, just proved, can be stated as follows: in the matrix element $\langle \Psi|\, F\, |\Psi'\rangle$, each term of the symmetric operator F contributes an equal result, and hence $\langle \Psi|\, F\, |\Psi'\rangle = n\langle \Psi|\, f(k)\, |\Psi'\rangle$, where k can be any particle. In shell-model work, usually the last particle $k = n$ is chosen in this expression.

For the sake of completeness, we derive also the result for the two-body type symmetric operator $G = \sum_{i<j} g(i, j)$. Proceeding in the same manner as in (III.60a), we obtain

$$\begin{aligned}&\langle\Psi(1, 2, \ldots, k, \ldots, l, \ldots, n)|\ g(1, 2)\ |\Psi'(1, 2, \ldots, k, \ldots, l, \ldots, n)\rangle \\ &\quad = \langle\Psi(k, l, \ldots, 1, \ldots, 2, \ldots, n)|\ g(k, l)\ |\Psi'(k, l, \ldots, 1, \ldots, 2, \ldots, n)\rangle \\ &\quad = \langle\Psi(1, 2, \ldots, k, \ldots, l, \ldots, n)|\ g(k, l)\ |\Psi'(1, 2, \ldots, k, \ldots, l, \ldots, n)\rangle.\end{aligned}$$

Once again, the first step follows by relabelling the dummy integration variables $1 \leftrightarrow k$, $2 \leftrightarrow l$ throughout the whole expression; the second step follows by making the same exchanges in the wavefunctions alone. Thus, we have proved that every term in G gives equal matrix elements, and hence

$$\langle\Psi|\ G\ |\Psi'\rangle = \tfrac{1}{2}n(n-1)\langle\Psi|\ g(k, l)\ |\Psi'\rangle, \tag{III.60b}$$

where k, l is any pair. In shell-model work, the pair $(n-1, n)$ or the pair $(1, 2)$ is usually chosen.

We now return to the problem of evaluating F between shell-model two-body states. We first define an abbreviated notation for the rest of our shell-model work. The sets of quantum numbers $nlj, n'l'j', \ldots$ of a single nucleon are, from now on, denoted by single letters $a, a', \ldots$. The product of δ-functions in the normalization constant $\delta_{nn'}\delta_{ll'}\delta_{jj'}$ is therefore denoted by the notation $\delta_{aa'}$. Since the single-nucleon isospin is always $\frac{1}{2}$, the isospin coupling $\frac{1}{2}, \frac{1}{2} : TM_T$ is denoted simply by TM_T for brevity. Whenever we need to show the quantum numbers n, l, j explicitly, we shall do so with the appropriate subscripts $a, a', \ldots$. Thus, n_a, l_a, j_a denotes these quantum numbers for a single-particle state whose abbreviated notation is a. In our new notation, we use two-nucleon wavefunctions of the type (III.58), utilize the general result (III.60a), and obtain

$$\begin{aligned}&(ab : JMTM_T|\ F\ |a'b' : J'M'T'M_T') \\ &\quad = \{(1+\delta_{ab})(1+\delta_{a'b'})\}^{-(1/2)}[\langle ab : JMTM_T| + (-1)^{j_a+j_b-J-T}\langle ba : JMTM_T|]f(2) \\ &\qquad \times[|a'b' : J'M'T'M_T'\rangle + (-1)^{j_{a'}+j_{b'}-J-T}|b'a' : J'M'T'M_T'\rangle] \\ &\quad = \{(1+\delta_{ab})(1+\delta_{a'b'})\}^{-(1/2)}\{1+P(abJT)\}\{1+P(a'b'J'T')\} \\ &\qquad \times[\langle ab : JMTM_T|\ f(2)\ |a'b' : J'M'T'M_T'\rangle] \\ &\quad = \begin{bmatrix} J' & K & J \\ M' & q & M \end{bmatrix} f_\tau(TM_T, T'M_T')\{(1+\delta_{ab})(1+\delta_{a'b'})\}^{-(1/2)}\{1+P(abJT)\} \\ &\qquad \times\{1+P(a'b'J'T')\}[\delta_{aa'}U(a'b'JK; J'b)\langle b|\ |f^k|\ |b'\rangle].\end{aligned} \tag{III.61}$$

In the first step, we have explicitly expressed the two terms in each antisymmetric wavefunction. The four matrix elements of $f(2)$ contained in that expression have been denoted in the next step with the help of the two operators $(1 + P)$ acting on the single matrix element. Each of the operators P is defined by

$$P(abJT) = (-1)^{j_a+j_b-J-T}P_{ab},$$

where P_{ab} exchanges a and b. In the final step, we have extracted the isospin part of the operator f, and denoted its matrix element by $f_\tau(TM_T, T'M_T')$. Since the isospin matrix element in all the four terms is the same, it has been written as a common factor to the left of the P-operators. Given the explicit form of the isospin operator, this matrix element can be written down with the help of the formulas given in Appendix B (Section III). After the extraction of the isospin part, we are left with the space-spin part of f. Assuming this to be of the form f_q^K, i.e., a tensor of rank K and component q (the general form will obviously be a sum of such tensors), we have written down, in the final step, its matrix element with the help of (BIII.13) from Appendix B.

If the single-particle states a, b are equivalent ($a = b$), then

$$(1 + \delta_{ab})^{-(1/2)}\{1 + P(abJT)\} = \frac{1}{\sqrt{2}}\{1 - (-1)^{J+T}\}$$

$$= \begin{cases} \sqrt{2} & \text{for } J + T = \text{odd integer (i.e., the states allowed for equivalent nucleons)} \\ 0 & \text{otherwise} \end{cases}$$

A similar observation holds for the case $a' = b'$. If both the states ab and $a'b'$ correspond to equivalent nucleons, then the product of the $(1 + P)$-operators gives a simple numerical factor with which the expression within the square brackets in the final step of (III.61) has to be multiplied. If any one of the states ab and $a'b'$ corresponds to equivalent particles, then the corresponding $(1 + P)$-operator reduces to a number, whereas the operator $(1 + P)$, for the inequivalent pair, operates on the expression within the square brackets. In the most general case of inequivalent nucleons in both ab and $a'b'$, the product of the two $(1 + P)$-factors,

$$1 + P(abJT) + P(a'b'J'T') + P(abJT)P(a'b'J'T'),$$

operates on the expression within the square brackets and gives rise to four different terms. However, some of these terms may turn out to be zero through the δ-factor in the expression within the square brackets. Since the δ-factor is also affected by the P-operations, we should let the operators first act on the expression, *as it appears in the final step of (III.61)*, and then put special values of a, b, a', b'.

The foregoing statements become clear if we consider an example: the special case of $a' = a$ and $b' = b$ with $a \neq b$. In this case, the term $P(abJT)$ gives rise to the δ-factor $\delta_{ba'}$ after the exchange $a \leftrightarrow b$ is done. But since $a' = a$ and $a \neq b$, this δ-factor gives zero, and hence the $P(abJT)$-term vanishes. Similarly, the $P(a'b'J'T')$-term produces the δ-factor $\delta_{ab'}$, and hence a zero result. The term that goes with 1 contains $\delta_{aa'}$ which is obviously non-zero, and so is the term produced by $P(abJT)P(a'b'J'T')$, the latter containing a δ-factor $\delta_{bb'}$.

Since F is a one-body type operator, we know that, acting on $a'b'$, it can produce a state ab in which *at the most* one single-particle state has changed. Thus, the nonvanishing matrix elements in (III.61) are obtained by connecting the states ab with the states $a'b'$ of five different types:

(i) $(aa) \leftarrow (aa)$,

(ii) $(aa) \leftarrow (ab)$ or $(aa) \leftarrow (ba)$,

(iii) $(ab) \leftarrow (ab)$,

(iv) $(ab) \leftarrow (ab')$ or $(ab) \leftarrow (a'b)$,

(v) $(ab) \leftarrow (aa)$ or $(ab) \leftarrow (bb)$.

It should be noticed that $(aa) \leftarrow (ab)$ or $(aa) \leftarrow (ba)$ represents the same matrix element with a possible difference in sign because

$$\begin{aligned}|ba:JMTM_T) &= \frac{1}{\sqrt{2}}[|ba:JMTM_T\rangle + (-1)^{j_b+j_a-J-T}|ab:JMTM_T\rangle] \\ &= \frac{1}{\sqrt{2}}(-1)^{j_b+j_a-J-T}[|ab:JMTM_T\rangle + (-1)^{j_a+j_b-J-T}|baJM, TM_T\rangle] \\ &= (-1)^{j_b+j_a-J-T}|ab:JMTM_T). \end{aligned} \tag{III.62}$$

For the same reason, in (iv), if the case $(ab) \leftarrow (ab')$ is worked out with suitable relabelling and phase factors, the result for the case $(ab) \leftarrow (a'b)$ will follow. Finally, the type (v) is already covered by (ii), the initial and final states being simply interchanged in the two cases. Thus, every conceivable result can be worked out if we specialize the expression (III.61) to the cases

(i) $(aa) \leftarrow (aa)$,

(ii) $(aa) \leftarrow (ab)$,

(iii) $(ab) \leftarrow (ab)$,

(iv) $(ab) \leftarrow (ab')$.

We have already explained how the P-operators in (III.61) should be handled in the equivalent and inequivalent cases. By following the instructions given there, specialization of the formula to the cases (i)–(iv) can easily be done.

C. MATRIX ELEMENTS OF TWO-BODY POTENTIAL

We shall consider the central and noncentral potentials discussed in Section 3, first with a well-behaved shape, say, any of the forms listed in (I.45b). The way to use the t-matrix elements in this method, in the case of a strong-core potential, is described in the course of our discussion in this section.

Since the state $|ab:JMTM_T)$ has a specified isospin, the result of the isospin exchange operator P_τ acting on it is very simply given by the multiplicative constant $(-1)^{T+1}$, which is 1 for $T=1$ and -1 for $T=0$. This means that the general Wigner and Heisenberg exchange-dependence of the noncentral and central forces immediately reduces to simple numbers. This is not true for the Majorana and Bartlett exchanges of the central force, and the noncentral operators $\mathcal{T}$, $\boldsymbol{l}\cdot\mathbf{S}$, and $(\boldsymbol{l}\cdot\mathbf{S})^2$. The Majorana operator P_M, being equal to $-P_\sigma P_\tau$, gives $(-1)^T P_\sigma$. The P_σ occurring here, and also in the Bartlett exchange term, could be replaced by $(-1)^{S+1}$ if it acted on a state of specified two-body spin quantum number S. Similarly, from our work in Section 4, we know that the evaluation of $\mathcal{T}$, $\boldsymbol{l}\cdot\mathbf{S}$, and $(\boldsymbol{l}\cdot\mathbf{S})^2$ also requires the knowledge of the spin S. They vanish for $S=0$, and have nonvanishing values for $S=1$. In the present case, our jj-coupled shell-model state $|ab:JMTM_T)$ does not as such contain this information on spin.

The term *jj-coupling* used here has been explained at the beginning of Section 25, where we also described the alternative way of coupling the angular momenta, called LS-coupling. In the case of two nucleons, this alternative method entails (i) first coupling $\boldsymbol{l}_a$, $\boldsymbol{l}_b$ to produce the total orbital angular momentum $\mathbf{L}$, then (ii) coupling the individual spins, each equal to $\frac{1}{2}$, to produce the total spin $\mathbf{S}$, and finally (iii) coupling $\mathbf{L}$ and $\mathbf{S}$ to produce the total $\mathbf{J}$. The mathematical transformation between the two-nucleon jj-coupled and LS-coupled states is presented in (AVI.2) of Appendix A. The transformation coefficients are the $9j$-symbols.

The information that we need for obtaining the values of P_σ and the noncentral operators is clearly contained in the LS-coupled state. Hence, for the purpose of calculating the matrix elements, a transformation of the state $|ab : JMTM_T)$ into the LS-coupled states is desirable. It should be noticed, however, that (AVI.2) from Appendix A applies to each of the two non-antisymmetrized parts of $|ab : JMTM_T)$ separately. Thus,

$$|ab : JMTM_T\rangle = \sum_{LS} \begin{bmatrix} l_a & \frac{1}{2} & j_a \\ l_b & \frac{1}{2} & j_b \\ L & S & J \end{bmatrix} |(n_a l_a, n_b l_b)L, (\tfrac{1}{2}\,\tfrac{1}{2})S : JM;\ TM_T\rangle, \tag{III.63a}$$

$$(-1)^{j_a+j_b-J-T}|ba : JMTM_T\rangle = (-1)^{j_a+j_b-J-T} \sum_{LS} \begin{bmatrix} l_b & \frac{1}{2} & j_b \\ l_a & \frac{1}{2} & j_a \\ L & S & J \end{bmatrix}$$
$$\times |(n_b l_b, n_a l_a)L, (\tfrac{1}{2}\,\tfrac{1}{2})S : JM;\ TM_T\rangle. \tag{III.63b}$$

According to the symmetry property of the $9j$-symbols, stated in Appendix A (Section VI), we can convert the $9j$-symbol of (III.63b) into that of (III.63a) by multiplying it with the phase factor $(-1)^{\Sigma}$, where $\Sigma = l_a + l_b + 1 + j_a + j_b + J + L + S$. Combining this phase factor with the phase $(-1)^{j_a+j_b-J-T}$ already present in (III.63b), we obtain

$$(-1)^{j_a+j_b-J-T}|ba : JMTM_T\rangle = (-1)^{l_a+l_b+L}(-1)^{1+S+T} \sum_{LS} \begin{bmatrix} l_a & \frac{1}{2} & j_a \\ l_b & \frac{1}{2} & j_b \\ L & S & J \end{bmatrix}$$
$$\times |(n_b l_b, n_a l_a)L, (\tfrac{1}{2}\,\tfrac{1}{2})S : JM;\ TM_T\rangle. \tag{III.63c}$$

Adding this to (III.63a) and multiplying by the normalization factor $[2(1 + \delta_{ab})]^{-1/2}$, we obtain

$$|ab : JMTM_T) = [2(1 + \delta_{ab})]^{-1/2} \sum_{LS} \begin{bmatrix} l_a & \frac{1}{2} & j_a \\ l_b & \frac{1}{2} & j_b \\ L & S & J \end{bmatrix} [|(n_a l_a, n_b l_b)L, (\tfrac{1}{2}\,\tfrac{1}{2})S : JM;\ TM_T\rangle$$
$$- (-1)^{l_a+l_b+L+S+T}|(n_b l_b, n_a l_a)L, (\tfrac{1}{2}\,\tfrac{1}{2})S : JM;\ TM_T\rangle]. \tag{III.64}$$

If we use this type of expression for the wavefunctions, then all the exchange operators immediately reduce to numbers dependent on the quantum numbers S and T, as explained. Therefore, the entire problem of evaluating the matrix element of the central potential reduces to dealing with the shape function $f(r/a)$ of the potential sandwiched between two states of the type (III.64). For the noncentral potential, there is the extra complication of inserting the matrix elements of the noncentral operators, as given in Section 4.

Since the two-nucleon potential depends on the relative coordinate $\mathbf{r}$ of the two nucleons, it is advantageous to introduce the Moshinsky transformation (II.85) while writing the *spatial part* of the wavefunction (III.64), i.e., $|(n_a l_a, n_b l_b)L\rangle$ or $|(n_b l_b, n_a l_a)L\rangle$. The first term in (III.64) gives rise to the harmonic oscillator bracket $\langle \mathcal{NL}, nl : L|\,|n_a l_a, n_b l_b : L\rangle$, whereas the second term gives $\langle \mathcal{NL}, nl : L|\,|n_b l_b, n_a l_a : L\rangle$. These two brackets are related to each other (see Appendix F) through a phase factor equal to $(-1)^{l_a+l_b+L+l}$. Together with the phase already present in the second term, we get a net phase factor equal to $(-1)^{l+S+T}$. Thus,

$$|ab : JMTM_T\rangle = [2(1 + \delta_{ab})]^{-1/2} \sum_{LS} \sum_{\mathcal{NL},nl} \begin{bmatrix} l_a & \frac{1}{2} & j_a \\ l_b & \frac{1}{2} & j_b \\ L & S & J \end{bmatrix} \langle \mathcal{NL}, nl : L| \, |n_a l_a, n_b l_b : L\rangle$$
$$\times [1 - (-1)^{l+S+T}] |(\mathcal{NL}, nl)L, (\tfrac{1}{2}\,\tfrac{1}{2})S : JM; TM_T\rangle. \qquad \text{(III.65)}$$

The quantum numbers $\mathcal{NL}$ now refer to the harmonic oscillator state for the centre-of-mass motion, and nl to the harmonic oscillator state for the *relative* coordinate of the two nucleons. The reader is cautioned not to confuse the latter with the single-particle oscillator function. It is understood that the Moshinsky coordinates $\bar{\mathbf{R}}$ and $\bar{\mathbf{r}}$, defined in (II.44), are being used in the spatial wavefunction $|(\mathcal{NL}, nl)L\rangle$ of (III.65). Therefore, while evaluating the matrix element, the shape function $f(r/a)$ should be suitably transformed into the appropriate function of $\bar{r}$. This is done later in this section.

The factor $1 - (-1)^{l+S+T}$ guarantees the antisymmetry of the nonvanishing terms on the right-hand side of (III.65). Since the centre-of-mass coordinate remains unchanged under an exchange of the two nucleons, the centre-of-mass wavefunction specified by $(\mathcal{NL})$ also remains unchanged under this operation. On the other hand, since the relative coordinate changes sign under an exchange of the nucleons, the part nl of the wavefunction gives a multiplicative factor $(-1)^l$ from the property of the spherical harmonic. The exchange on the spin and isospin parts of the wavefunction produces respectively $(-1)^{1+S}$ and $(-1)^{1+T}$. Putting all these phase factors together, we obtain $(-1)^{l+S+T}$. Thus, $1 - (-1)^{l+S+T}$ is equivalent to the operator $(\mathbb{1} - P_{12})$, which is the antisymmetrizer for the two-body wavefunction. The occurrence of antisymmetrized states on the right-hand side of (III.65) is to be naturally expected because the state on the left-hand side, which has here been expanded, is antisymmetric under the exchange of nucleons.

One final step remains to be incorporated in (III.65) before we can use the expression for the evaluation of matrix elements. The state on the right-hand side contains the couplings of three angular momenta in the order (i) $\mathcal{L} + \mathbf{l} = \mathbf{L}$ and (ii) $\mathbf{L} + \mathbf{S} = \mathbf{J}$. A reference to Appendix A (Section V) will instruct the reader on an alternative scheme of coupling: (i) $\mathbf{l} + \mathbf{S} = \mathcal{J}$, $\mathcal{J}$ being the total angular momentum of the relative two-nucleon state, and (ii) $\mathcal{L} + \mathcal{J} = \mathbf{J}$. The states corresponding to this second manner of coupling are more convenient to use because here the relative state is *completely* specified by its total angular momentum $\mathcal{J}$. We denote these states by the obvious notation $|\mathcal{NL}, (nl, S)\mathcal{J} : JM; TM_T\rangle$, and use them together with the six j-coefficients of Appendix A (Section V) to replace the states appearing in (III.65). We thus obtain

$$|ab : JMTM_T\rangle = [2(1 + \delta_{ab})]^{-1/2} \sum_{LS} \sum_{\mathcal{NL},nl} \begin{bmatrix} l_a & \frac{1}{2} & j_a \\ l_b & \frac{1}{2} & j_b \\ L & S & J \end{bmatrix} \langle \mathcal{NL}, nl : L| \, |n_a l_a, n_b l_b : L\rangle$$
$$\times [1 - (-1)^{l+S+T}] \sum_{\mathcal{J}} U(\mathcal{L}lJS; L\mathcal{J}) |\mathcal{NL}, (nl, S)\mathcal{J} : JM; TM_T\rangle. \qquad \text{(III.66)}$$

The wavefunction of the relative coordinate, specified by the quantum numbers l, S, $\mathcal{J}$, and the isospin wavefunction described by TM_T are exactly the same as those occurring in Chapter I. The only difference is in the radial function. In Chapter I, we have determined them by solving the exact Schrödinger equation for the relative coordinate r; here the radial dependence is that of the harmonic oscillator, given by R_{nl} (see Section 17C) and expressed as a function of $\bar{r} = r/\sqrt{2}$. The reader's attention is drawn also to the distinction between the orbital angular momenta L and l. L denotes the *total* two-nucleon orbital angular momentum, whereas l denotes the same for the *relative* motion.

Since V conserves JM, TM_T, we shall evaluate its matrix element between the two states $(a'b' : JMTM_T|$ and $|ab : JMTM_T)$. For both these wavefunctions, we shall use expansions of the type (III.66). V is a scalar operator for the relative wavefunction specified by $(nl, \mathcal{J})$ and the isospin wavefunction given by TM_T, and we know from our discussion in Section 4 that it keeps the quantum numbers S, $\mathcal{J}$, T, M_T unaltered. Further, being independent of the centre-of-mass coordinate, it does not affect the centre-of-mass wavefunction specified by $\mathcal{NL}$. As far as the relative orbital angular momentum l is concerned, we know from Section 4 that the noncentral part of V causes an admixture of $l = \mathcal{J} \pm 1$ in the spin-triplet state, and keeps the $(S = 1)$-state with $l = \mathcal{J}$ unmixed. The central part of V, on the other hand, conserves l in addition to S, $\mathcal{J}$, T, ... Therefore, in general, the potential V, acting on $|\mathcal{NL}, (nl, S)\mathcal{J} : JM; TM_T\rangle$, changes it to states of the type $|\mathcal{NL}, (n'l', S)\mathcal{J} : JM; TM_T\rangle$. Conservation laws similar to those applicable to l hold for the total angular momentum L as well. The central potential, which is a scalar in the two-body space, requires L to remain unchanged, whereas the non-central potential, containing tensors of higher rank, can change L to L' having the same J, S, and parity.

The foregoing conservation requirements enable us to write down the general result

$$
\begin{aligned}
&(a'b' : JMTM_T|\ V\,|ab : JMTM_T) \\
&= \tfrac{1}{2}[(1+\delta_{ab})(1+\delta_{a'b'})]^{-1/2} \sum_{L,L',S}\sum_{\mathcal{NL}}\sum_{nl,n'l'}\sum_{\mathcal{J}} \begin{bmatrix} l_a & \frac{1}{2} & j_a \\ l_b & \frac{1}{2} & j_b \\ L & S & J \end{bmatrix} \begin{bmatrix} l_{a'} & \frac{1}{2} & j_{a'} \\ l_{b'} & \frac{1}{2} & j_{b'} \\ L' & S & J \end{bmatrix} \\
&\quad\times\langle \mathcal{NL}, nl : L|\ |n_a l_a, n_b l_b : L\rangle\langle \mathcal{NL}, n'l' : L'|\ |n_{a'} l_{a'}, n_{b'} l_{b'} : L'\rangle \\
&\quad\times[1-(-1)^{l+S+T}][1-(-1)^{l'+S+T}]U(\mathcal{L}lJS; L\mathcal{J})U(\mathcal{L}l'JS; L'\mathcal{J}) \\
&\quad\times\langle \mathcal{NL}, (n'l', S)\mathcal{J} : JM; TM_T|\ V\,|\mathcal{NL}, (nl, S)\mathcal{J} : JM; TM_T\rangle. \qquad \text{(III.67)}
\end{aligned}
$$

Since l, l' have the same parity, we may write

$$[1-(-1)^{l+S+T}][1-(-1)^{l'+S+T}] = 2[1-(-1)^{l+S+T}].$$

Further, since V is a scalar operator, the application of the Wigner-Eckart theorem, and (BIII.13) from Appendix B, yields

$$
\begin{aligned}
&\langle \mathcal{NL}, (n'l', S)\mathcal{J} : JM; TM_T|\ V\,|\mathcal{NL}, (nl, S)\mathcal{J} : JM; TM_T\rangle \\
&= \langle (n'l', S)\mathcal{J}; T|\ |V|\ |(nl, S)\mathcal{J}; T\rangle \\
&= \langle (n'l', S)\mathcal{J}\mathcal{M}; TM_T|\ V\,|(nl, S)\mathcal{J}\mathcal{M}; TM_T\rangle. \qquad \text{(III.68a)}
\end{aligned}
$$

Since the Clebsch-Gordon coefficient in the Wigner-Eckart theorem is 1 for the present case, there is no difference between the full matrix element and the reduced matrix element. $\mathcal{M}$ here is the projection of $\mathcal{J}$. If the shape function of V is given by $f(r/a) \equiv f(\sqrt{2}\,\bar{r}/a) = \bar{f}(\bar{r})$, then the radial integral contained in (III.68a) is given by

$$\int_0^\infty R_{n'l'}(\bar{r})\bar{f}(\bar{r})R_{nl}(\bar{r})\,d\bar{r}. \qquad \text{(III.68b)}$$

The rest of the matrix element in (III.68a) connecting $\langle l'S\mathcal{J}\mathcal{M}; TM_T|$ and $|lS\mathcal{J}\mathcal{M}; TM_T\rangle$ is the same matrix element of the central and noncentral potentials evaluated in Section 4. Therefore, (III.68a) and (III.68b), together with (III.67), give a complete evaluation of the shell-model matrix elements of the two-body potential.

In view of the expression (II.81b) of the harmonic oscillator radial function, it is clear that the radial integral (III.68b) is a linear sum of integrals of the type

$$I_p = \int_0^\infty \exp(-\bar{r}^2/b^2)\bar{r}^{2(p+1)}\bar{f}(\bar{r})\, d\bar{r}, \tag{III.68c}$$

where $b[=\hbar/(M\omega)]^{1/2}$ is the harmonic oscillator parameter, and the integer p takes values between $\frac{1}{2}(l + l')$ and $\frac{1}{2}(l + l') + n + n' = \frac{1}{2}(\lambda + \lambda')$; λ is defined by (II.78). Since the relative states must have the same parity, $\frac{1}{2}(l + l')$ is guaranteed to be an integer. The radial integrals (III.68c) are called *Talmi integrals* for the potential function f. If f represents a Gaussian shape function $\exp[-(1/2)r^2/a^2]$, then the Talmi integrals are very easy to evaluate. Introducing $c^{-2} = a^{-2} + b^{-2}$, we then obtain

$$I_p = \int_0^\infty e^{-\bar{r}^2/c^2}\bar{r}^{2p+2}\, d\bar{r} = \tfrac{1}{2}c^{2p+3}\Gamma(p + \tfrac{3}{2}).$$

For an exponential $e^{-r/a}$ or a Yukawa shape function $e^{-r/a}/(r/a)$, the Talmi integral is more difficult to evaluate. This is left as an exercise. The final results are expressed in terms of the error function

$$\operatorname{erf}(x) = \int_0^x e^{-y^2}\, dy,$$

which can be numerically evaluated.

We now point out how the G-matrix elements of the actual two-nucleon potential with a strongly repulsive core (see Section 22C) could be used in the foregoing calculation. In the roughest approximation, discussed in Section 22, the relative t-matrix elements used in computing G do not at all depend on the centre-of-mass states. These matrix elements are actually of the type of those on the right-hand side of (III.68a), with V being replaced by t, and hence can be directly used in (III.67) in conjunction with (III.68a). In a more refined calculation, the t-matrix elements are considered to be diagonal in $\mathcal{N}\mathcal{L}$, but dependent on these quantum numbers through the (Q/e)-operator in the definition of t. These more refined matrix elements are of the type of those on the left-hand side of (III.68a), with V being replaced by t. These $(\mathcal{N}\mathcal{L})$-dependent matrix elements can be directly inserted in (III.67).

Alternative Method Using Slater Integrals

The foregoing method using the harmonic oscillator brackets is almost universally used at present. In earlier shell-model work, a method originally used in atomic physics was very popular. We shall demonstrate this method for a central potential.

When we use the $(LS\text{-}jj)$-transformation (III.63a), the values of the exchange operators can be immediately written down, as discussed earlier in this section, and then the problem reduces to treating only $V_0 f(r/a)$. We can expand the shape function $f(r/a)$ in a complete set of Legendre polynomials of the angle θ_{12} between the vectors $\mathbf{r}_1$ and $\mathbf{r}_2$, where $\mathbf{r} = \mathbf{r}_1 - \mathbf{r}_2$. The expansion coefficients will obviously be functions of the magnitudes r_1, r_2 of the two vectors, and we shall denote them by $f_K(r_1, r_2)$. Thus,

$$f(r/a) = \sum_{K=0}^{\infty} f_K(r_1, r_2) P_K(\cos \theta_{12})$$
$$= \sum_{K=0}^{\infty} f_K(r_1, r_2) y^K(1) \cdot y^K(2). \tag{III.69}$$

The Legendre polynomial has been substituted in terms of the spherical harmonics of the directions of $\mathbf{r}_1$ and $\mathbf{r}_2$, according to the spherical harmonic addition theorem (BI.21) from Appendix B. The unknown quantities $f_K(r_1, r_2)$ are given in terms of $f(r/a)$ by the integral

$$f_K(r_1, r_2) = \frac{2K+1}{2} \int_0^{\infty} f(r/a) P_K(\cos \theta_{12})\, d(\cos \theta_{12}). \tag{III.70}$$

We have multiplied the first line of (III.69) by $P_{K'}$ $(\cos \theta_{12})$ and integrated over the angle; the right-hand side yields a nonvanishing result only for the term $K = K'$, by virtue of (AI.8) from Appendix A. The final result, which is true for all K', has been rewritten in terms of K in (III.70). Using the last line of the operator (III.69) between the LS-coupled states of (III.63a), we get

$$\langle (n_{a'} l_{a'}, n_{b'} l_{b'}) L, (\tfrac{1}{2}\, \tfrac{1}{2}) S : JM, TM_T |\; f(r/a)\; | (n_a l_a, n_b l_b) L, (\tfrac{1}{2}\, \tfrac{1}{2}) S : JM, TM_T \rangle$$
$$= \sum_{K=0}^{\infty} F_K(n_{a'} l_{a'}, n_{b'} l_{b'}; n_a l_a, n_b l_b) \langle l_{a'} l_{b'} L |\, | y^K(1) \cdot y^K(2) |\, | l_a l_b L \rangle, \tag{III.71}$$

where the radial integral F_K is called the *Slater integral* and is defined by

$$F_K(n_{a'} l_{a'}, n_{b'} l_{b'}; n_a l_a, n_b l_b) = \int_0^{\infty} dr_1 \int_0^{\infty} dr_2 R_{n_{a'} l_{a'}}(r_1) R_{n_{b'} l_{b'}}(r_2) f_K(r_1, r_2) R_{n_a l_a}(r_1) R_{n_b l_b}(r_2). \tag{III.72}$$

In writing down (III.71), we have made use of the standard result (BIII.14) from Appendix B. To further simplify (III.71), we can express the reduced matrix element, by using (BIII.15) from Appendix B, as

$$\langle l_{a'} l_{b'} L |\, | y^K(1) \cdot y^K(2) |\, | l_a l_b L \rangle = (-1)^{l_{a'} + l_b - L} \{[l_{a'}][l_{b'}]\}^{1/2} W(l_a l_b l_{a'} l_{b'}; LK)$$
$$\times \langle l_{a'} |\, | y^K |\, | l_a \rangle \langle l_{b'} |\, | y^K |\, | l_b \rangle. \tag{III.73}$$

The reduced matrix elements in this final expression can finally be written down by using (BIII.3) from Appendix B.

It should be remembered that the antisymmetric state $|ab : JM, TM_T)$ contains two LS-coupled terms, as shown in (III.64). Thus, the matrix element between antisymmetric states is a linear combination of four matrix elements of the type (III.71). The general expression (III.77b) will enable the reader to reduce these four terms to two terms and evaluate them separately.

In a slight variant of the method, we can avoid expressing the jj-coupled states in terms of the LS-coupled states by suitably rewriting the operator. For a central potential, we have already noticed that the operator $\boldsymbol{\sigma}_1 \cdot \boldsymbol{\sigma}_2$, contained in the exchange operator P_σ, is troublesome to handle between the jj-coupled states. We shall show how this can be done by manipulating the operators. We have

$$\sum_{K=0}^{\infty} f_K(r_1, r_2) y^K(1) \cdot y^K(2) \boldsymbol{\sigma}_1 \cdot \boldsymbol{\sigma}_2 = \sum_{K=0}^{\infty} f_K(r_1, r_2) \sum_{q=-K}^{K} (-1)^q y_q^K(1) y_{-q}^K(2)$$
$$\times \sum_{\mu=-1}^{1} (-1)^\mu (\sigma_1)_\mu (\sigma_2)_{-\mu}. \tag{III.74}$$

Using the definition (BII.8) from Appendix B, we can rewrite the operator (III.74) as

$$-\sum_{K=0}^{\infty} f_K(r_1, r_2) \sum_{K'=|K-1|}^{K+1} (-1)^{K-K'} \sum_{Q} (-1)^Q (y^K(1), \boldsymbol{\sigma}_1)_Q^{K'} (y^K(2), \boldsymbol{\sigma}_2)_{-Q}^{K'}$$

$$= -\sum_{K=0}^{\infty} f_K(r_1, r_2) \sum_{K'=|K-1|}^{K+1} (-1)^{K-K'} (y^K(1), \boldsymbol{\sigma}_1)^{K'} \cdot (y^K(2), \boldsymbol{\sigma}_2)^{K'}. \tag{III.75}$$

Thus, the matrix element of (III.75) between the jj-coupled states is given directly, by using (BIII.15) from Appendix B, as

$$\langle a'b' : JM, TM_T| f(r/a)\boldsymbol{\sigma}_1 \cdot \boldsymbol{\sigma}_2 |ab : JM, TM_T\rangle$$
$$= -(-1)^{j_{a'}+j_b-J}\sqrt{[j_{a'}][j_{b'}]} \sum_{K=0}^{\infty} \sum_{K'} (-1)^{K-K'} W(j_{a'}j_{b'}j_aj_b; JK') F_K(n_{a'}l_{a'}, n_{b'}l_{b'}; n_al_a, n_bl_b)$$
$$\times \langle l_{a'}\tfrac{1}{2}j_{a'}| |(y^K, \boldsymbol{\sigma})^{K'}| |l_a\tfrac{1}{2}j_a\rangle \langle l_{b'}\tfrac{1}{2}j_{b'}| |(y^K, \boldsymbol{\sigma})^{K'}| |l_b\tfrac{1}{2}j_b\rangle. \tag{III.76a}$$

Each of the reduced matrix elements in the final expression can be obtained by an application of (BIII.16), followed by that of (BIII.3) and (BIII.6)—all from Appendix B.

For the Wigner and Heisenberg exchange terms, we have to handle the simpler expression (III.69) between the jj-coupled states and, using (BIII.15) from Appendix B directly, we obtain

$$\langle a'b' : JM, TM_T| f(r/a) |ab : JM, TM_T\rangle$$
$$= (-1)^{j_{a'}+j_b-J}\sqrt{[j_{a'}][j_{b'}]} \sum_{K=0}^{\infty} W(j_{a'}j_{b'}, j_aj_b; JK) F_K(n_{a'}l_{a'}, n_{b'}l_{b'}; n_al_a, n_bl_b)$$
$$\times \langle l_{a'}\tfrac{1}{2}j_{a'}| |y^K| |l_a\tfrac{1}{2}j_a\rangle \langle l_{b'}\tfrac{1}{2}j_{b'}| |y^K| |l_b\tfrac{1}{2}j_b\rangle. \tag{III.76b}$$

The reduced matrix elements in this expression are given by (BIII.17) in Appendix B.

Once again we draw the reader's attention to the fact that the expressions (III.76) represent matrix elements between simple angular momentum coupled states. There will therefore be four such matrix elements in the expression using the antisymmetric states $(a'b' : JM, TM_T|$ and $|ab : JM, TM_T)$. For the entire two-body potential V, these four terms are given, according to (III.58), by

$$(a'b' : JM, TM_T| V |ab : JM, TM_T)$$
$$= \tfrac{1}{2}[(1+\delta_{ab})(1+\delta_{a'b'})]^{-(1/2)}[\langle a'b' : JM, TM_T| V |ab : JM, TM_T\rangle$$
$$+ (-1)^{j_a+j_b+j_{a'}+j_{b'}}\langle b'a' : JM, TM_T| V |ba : JM, TM_T\rangle$$
$$+ (-1)^{j_a+j_b-J-T}\langle a'b' : JM, TM_T| V |ba : JM, TM_T\rangle$$
$$+ (-1)^{j_{a'}+j_{b'}-J-T}\langle b'a' : JM, TM_T| V |ab : JM, TM_T\rangle] \tag{III.77a}$$
$$= [(1 + \delta_{ab})(1 + \delta_{a'b'})]^{-(1/2)}[\langle a'b' : JM, TM_T| V |ab : JM, TM_T\rangle$$
$$+ (-1)^{j_a+j_b-J-T}\langle a'b' : JM, TM_T| V |ba : JM, TM_T\rangle]. \tag{III.77b}$$

It needs some detailed explanation to understand how the final step (III.77b) follows from (III.77a). First, it should be noticed that, in the step (III.77a), the first matrix element corresponds to a transition of the first nucleon (one in the first location in the two-body states, according to our convention) from $a \to a'$, and that of the second nucleon from $b \to b'$. Then it should be observed that, in the second matrix element, the second nucleon goes from $a \to a'$ and the first from $b \to b'$. Since the nucleon variables are dummy integration variables in the matrix element, we can relabel them by exchanging $1 \leftrightarrow 2$ throughout. V is not affected by the exchange because it is a symmetric operator, whereas the states change to $|b(2)a(1):JM,\ TM_T\rangle$ and $\langle b'(2)a'(1):JM,\ TM_T|$, respectively. If we reverse the order of coupling in these two states, then we obtain the phase factors $(-1)^{j_a+j_b-J+1-T}$ and $(-1)^{j_{a'}+j_{b'}-J+1-T}$, respectively, and the states become $|a(1)b(2):JM,\ TM_T\rangle$ and $\langle a'(1)b'(2):JM,\ TM_T|$. The phase factors cancel out with $(-1)^{j_a+j_b+j_{a'}+j_{b'}}$, already contained in the second term, and the matrix element between the new states is obviously the same as the matrix element in the first term. [It should be remembered that the nucleon coordinates (1) and (2) are understood in the first term.] Therefore, the first two terms in (III.77a), which are equal, add up to cancel the factor $\frac{1}{2}$ at the beginning and give rise to the first term in (III.77b). In an analogous manner, it can be shown that the third and fourth terms in (III.77a) are equal, and hence add up to produce the second term in (III.77b). The first term in (III.77b) is usually called the *direct term*, and the second term the *exchange term*. The matrix element of the two-body potential between antisymmetric states is therefore a combination of the direct and exchange terms. The results already obtained will enable the reader to evaluate each of these terms separately, and thus obtain the matrix element between antisymmetric states.

We now discuss the evaluation of the Slater integrals F_K. In atomic physics, the interaction between two electrons is given by the Coulomb potential e^2/r_{12}. In any textbook on analysis, the Legendre polynomial $P_K(\cos\theta_{12})$ is actually *defined* from $1/r_{12}$ by

$$\frac{1}{r_{12}} = (r_1^2 + r_2^2 - 2r_1r_2\cos\theta_{12})^{-(1/2)} = \sum_{K=0}^{\infty} \frac{r_<^K}{r_>^{K+1}} P_K(\cos\theta_{12}),$$

where $r_<$ and $r_>$ are respectively the smaller and the larger of the quantities r_1 and r_2. Comparing this expression with (III.69), we obtain, for the Coulomb interaction,

$$f_K(r_1, r_2) = \frac{r_<^K}{r_>^{K+1}}.$$

Thus, f_K is a very simple function of r_1, r_2 which can be directly used in the definition (III.72) to obtain the Slater integrals. In the two-nucleon case, the function $f_K(r_1, r_2)$ is not at all a simple function for any of the familiar shape functions of finite range. In fact, a straightforward integration of (III.70) is not possible in terms of known analytic functions. Because of this, the Slater integral method does not have in the two-nucleon case any *natural* advantage similar to that in the two-electron case.

To evaluate the Slater integrals in the two-nucleon case, the definition (III.70) has to be inserted in (III.72) and then a straightforward calculation of the resultant triple integral has to be accomplished. The expression we have to evaluate is given by

$$\frac{2K+1}{2}\int_0^\infty dr_1 \int_0^\infty dr_2 \int_{-1}^{+1} d(\cos\theta_{12})\, R_{n_{a'}l_{a'}}(r_1)R_{n_{b'}l_{b'}}(r_2)f(r/a)P_K(\cos\theta_{12})R_{n_al_a}(r_1)R_{n_bl_b}(r_2). \tag{III.78}$$

The product of the radial functions of r_1 occurring here can be written as

$$R_{n_al_a}(r_1)R_{n_{a'}l_{a'}}(r_1) = b^{-1}\sum_{i=0}^{N} C_i(n_al_a,\ n_{a'}l_{a'})\rho_1^{2i+L+2}\exp(-\rho_1^2), \tag{III.79a}$$

where $N = n_a + n_{a'}$, $L = l_a + l_{a'}$, $\rho_1 = r_1/b$. The coefficients C_i are easily obtained with the help of (II.81b). A similar expression can be written for the radial functions of r_2 appearing in (III.78) as

$$R_{n_b l_b}(r_2) R_{n_{b'} l_{b'}}(r_2) = b^{-1} \sum_{i'=0}^{N'} C_{i'}(n_b l_b, n_{b'} l_{b'}) \rho_2^{2i'+L'+2} \exp(-\rho_2^2) \qquad \text{(III.79b)}$$

with $N' = n_b + n_{b'}$, $L' = l_b + l_{b'}$, and $\rho_2 = r_2/b$. The factors b^{-1} appearing in (III.79) convert $dr_1\, dr_2$ into $d\rho_1\, d\rho_2$. The factors ρ_1^2 and ρ_2^2 can also be extracted from (III.79) to yield the elementary integration volume $\rho_1^2\, d\rho_1\, \rho_2^2\, d\rho_2\, d(\cos\theta_{12})$.

We shall next transform the dimensionless variables ρ_1, ρ_2, and $\cos\theta_{12}$ as follows:

$$\bar{\boldsymbol{\rho}} = \frac{1}{\sqrt{2}}(\boldsymbol{\rho}_1 - \boldsymbol{\rho}_2), \qquad \boldsymbol{\rho} = \frac{1}{\sqrt{2}}(\boldsymbol{\rho}_1 + \boldsymbol{\rho}_2)$$

so that

$$\boldsymbol{\rho}_1 = \frac{1}{\sqrt{2}}(\boldsymbol{\rho} + \bar{\boldsymbol{\rho}}), \qquad \boldsymbol{\rho}_2 = \frac{1}{\sqrt{2}}(\boldsymbol{\rho} - \bar{\boldsymbol{\rho}}).$$

Therefore,

$$\boldsymbol{\rho}_1 \cdot \boldsymbol{\rho}_2 = \rho_1 \rho_2 \cos\theta_{12} = \tfrac{1}{2}(\rho^2 - \bar{\rho}^2), \qquad \text{(III.80a)}$$

$$\rho_1^2 + \rho_2^2 = \tfrac{1}{2}[(\boldsymbol{\rho} + \bar{\boldsymbol{\rho}})^2 + (\boldsymbol{\rho} - \bar{\boldsymbol{\rho}})^2] = \rho^2 + \bar{\rho}^2 \qquad \text{(III.80b)}$$

and

$$\rho_1^2 = \tfrac{1}{2}(\rho^2 + \bar{\rho}^2 + 2\rho\bar{\rho}\cos\alpha), \qquad \text{(III.80c)}$$

$$\rho_2^2 = \tfrac{1}{2}(\rho^2 + \bar{\rho}^2 - 2\rho\bar{\rho}\cos\alpha). \qquad \text{(III.80d)}$$

Here α denotes the angle between the vectors $\boldsymbol{\rho}$ and $\bar{\boldsymbol{\rho}}$, which are respectively the relative and centre-of-mass coordinates in units of b (and hence dimensionless). As a result of these transformations,

$$\rho_1^2\, d\rho_1\, \rho_2^2\, d\rho_2\, d(\cos\theta_{12}) \to \rho^2\, d\rho\, \bar{\rho}^2\, d\bar{\rho}\, d(\cos\alpha).$$

The potential function $f(r/a)$ transforms into $f(\sqrt{2}\,\bar{r}/a) = f(\sqrt{2}\,\bar{\rho}b/a) = \bar{f}(\bar{\rho})$. The factor $\exp[-(\rho_1^2 + \rho_2^2)]$ in the product of (III.79a) and (III.79b) transforms, by virtue of (III.80b), into $\exp[-(\rho^2 + \bar{\rho}^2)]$. The factors involving the powers of ρ_1, ρ_2, and $\cos\theta_{12}$ (contained in the Legendre polynomial) can be transformed by using (III.80a), (III.80c), and (III.80d). We first write the Legendre polynomial $P_K(z)$, where $z = \cos\theta_{12}$, in powers of z, as

$$P_K(z) = \sum_{\substack{t=0 \\ (\text{parity of } K)}}^{K} c_t(K) z^t. \qquad \text{(III.81a)}$$

The coefficients $c_t(K)$ can be obtained from the definition of the Legendre polynomial given in Appendix A (Section I). The easiest way, however, is to make use of the recursion relation of the Legendre polynomials, i.e.,

$$P_{K+1}(z) = \frac{K}{K+1} P_{K-1}(z) - \frac{2K+1}{K+1} z P_K(z), \qquad \text{(III.81b)}$$

together with the initial expressions for P_0 and P_1, namely,

$$P_0(z) = 1, \qquad P_1(z) = z. \qquad \text{(III.81c)}$$

It is clear from (III.81b) that $P_2(z)$, given in terms of $zP_1(z)$ and $P_0(z)$, contains z^0 and z^2. Similarly, $P_3(z)$, a linear combination of $P_1(z)$ and $zP_2(z)$, contains z and z^3. In general, the recurrence relation (III.81b), together with (III.81c), guarantees that $P_K(z)$ has a maximum power of z equal to K, and that only even or odd powers of z appear in the sum, depending on whether K is even or odd. The latter fact is denoted in (III.81a) by specifying that the summation index t has the same parity as K. Further, since $P_K(\cos\theta_{12}) = y^K(1)\cdot y^K(2)$, and its matrix element was evaluated between the states l_a, $l_{a'}$ of the first nucleon and l_b, $l_{b'}$ of the second nucleon, it is guaranteed that (i) both $l_a + l_{a'} + K$ and $l_b + l_{b'} + K$ are equal to even integers; (ii) $|l_a - l_{a'}| \leqslant K \leqslant l_a + l_{a'}$ and $|l_b - l_{b'}| \leqslant K \leqslant l_b + l_{b'}$. These facts ensure that, for any t in the expansion (III.81a), the quantities $l_a + l_{a'} - t$ and $l_b + l_{b'} - t$ are either a *positive even* integer or zero. Using all these results in (III.78), and denoting $\cos\alpha$ by x, we get

$$F_K(n_{a'}l_{a'}, n_{b'}l_{b'}; n_a l_a, n_b l_b) = \frac{2K+1}{2}\int_0^\infty \bar{\rho}^2\, d\bar{\rho}\int_0^\infty \rho^2\, d\rho \int_{-1}^{+1} dx \sum_{i=0}^{N}\sum_{i'=0}^{N'}\sum_{\substack{t=0\\ (\text{parity of } K)}}^{K} C_i(n_a l_a, n_{a'} l_{a'})$$
$$\times C_{i'}(n_b l_b, n_{b'} l_{b'})c_t(K)\exp[-(\rho^2+\bar{\rho}^2)]\bar{f}(\bar{\rho})\{\rho_1^{2i+L}\rho_2^{2i'+L'}z^t\}. \tag{III.82}$$

We next consider the transformation of the part of (III.82) enclosed within the braces. The quantity $(\rho_1\rho_2 z)^t \equiv (\rho_1\rho_2\cos\theta_{12})^t$ can be rewritten with the help of (III.80a), and then ρ_1^{2i+L-t} and $\rho_2^{2i'+L'-t}$ can be transformed by (III.80c) and (III.80d), respectively. From what we have discussed, we know $L - t$ and $L' - t$ are positive even integers or zero, and hence both $\frac{1}{2}(L-t)$ and $\frac{1}{2}(L'-t)$ are integers greater than or equal to 0. Thus,

$$\rho_1^{2i+L-t} = (\rho_1^2)^{i+(1/2)(L-t)} = \frac{1}{2^{i+(1/2)(L-t)}}(\rho^2+\bar{\rho}^2+2\rho\bar{\rho}x)^{i+(1/2)(L-t)},$$

$$\rho_2^{2i'+L'-t} = (\rho_2^2)^{i'+(1/2)(L'-t)} = \frac{1}{2^{i'+(1/2)(L'-t)}}(\rho^2+\bar{\rho}^2-2\rho\bar{\rho}x)^{i'+(1/2)(L'-t)}.$$

We shall use these results in a typical term of (III.82) and evaluate the resultant triple integral as

$$\frac{1}{2^{i+i'+(1/2)(L+L')}}\int_0^\infty \bar{\rho}^2\, d\bar{\rho}\int_0^\infty \rho^2\, d\rho\int_{-1}^{+1} dx\, \exp[-(\rho^2+\bar{\rho}^2)]\{\rho^2-\bar{\rho}^2\}^t \bar{f}(\bar{\rho})$$
$$\times\{(\rho^2+\bar{\rho}^2)+2\rho\bar{\rho}x\}^{i+(1/2)(L-t)}\{(\rho^2+\bar{\rho}^2)-2\rho\bar{\rho}x\}^{i'+(1/2)(L'-t)}.$$

To abbreviate the notation, let us introduce

$$\mu = 2i + L, \qquad \nu = 2i' + L',$$

and denote the foregoing expression by $\mathcal{F}(\mu, \nu, t)$. To simplify $\mathcal{F}(\mu, \nu, t)$, we have to make binomial expansions of the three quantities enclosed within the braces. The first integration to be performed is that of x, which will give the nonvanishing result for the *even powers* of x *only* as

$$\int_{-1}^{+1} dx\, x^{2s} = \frac{2}{2s+1}.$$

The next integration to be done is that of ρ, which will have the general form

$$\int_0^\infty d\rho\, \rho^{2s}\exp(-\rho^2) = \tfrac{1}{2}\Gamma(s+\tfrac{1}{2}).$$

The last integration we have to do will be that of $\bar{\rho}$, and clearly the integrals will have the general form

$$\int_0^\infty d\bar{\rho}\, \bar{\rho}^{2(s+1)}\exp(-\bar{\rho}^2)\bar{f}(\bar{\rho}),$$

41 (45-122/1977)

which is, by the definition (III.68c), the Talmi integral I_s. Therefore, we anticipate the final expression of $\mathcal{F}(\mu, \nu, t)$ to be a linear sum of Talmi integrals. Carrying out the binomial expansions, and the integrations in the indicated order, we easily obtain the result

$$\mathcal{F}(\mu, \nu, t) = 2^{-(1/2)(\mu+\nu)} \sum_{\eta=0}^{\eta_m} \frac{2^{2\eta}}{2\eta+1} \left[\sum_x \sum_y \binom{\frac{1}{2}(\mu+\nu)-t-2\eta}{x} \binom{t}{y} \Gamma(x+y+\eta+\tfrac{3}{2}) I_s \right.$$
$$\left. \times \sum_\xi \sum_{\xi'} \left\{ \binom{\frac{1}{2}\mu - t}{\xi} \binom{\frac{1}{2}\nu - t}{\xi'} \right\} \right], \tag{III.83}$$

where $s = \frac{1}{2}(\mu + \nu) - x - y$, $\xi + \xi' = 2\eta$, and the maximum value of η, namely, η_m, is given by the condition of nonvanishing of the first binomial coefficient. The requirement for nonvanishing is that $\frac{1}{2}(\mu + \nu) - t - 2\eta$ must be greater than or equal to 0, and hence, if $\frac{1}{2}(\mu + \nu) - t$ is an even integer, $2\eta_m = \frac{1}{2}(\mu + \nu) - t$; on the other hand, if it is an odd integer, then $2\eta_m = \frac{1}{2}(\mu + \nu) - t - 1$. The limits of the summation of x and y are restricted, once again, by the appropriate binomial coefficient; the limits on ξ and ξ' are determined by the binomial coefficients, as well as the condition $\xi + \xi' = 2\eta$.

In terms of (III.83), the final expression for the Slater integral, (III.82), is given by

$$F_K(n_{a'}l_{a'}, n_{b'}l_{b'}; n_a l_a, n_b l_b) = \sum_{i=0}^{N} \sum_{i'=0}^{N'} \sum_{\substack{t=0 \\ (\text{parity of } K)}}^{K} \frac{2K+1}{2} C_i(n_a l_a, n_{a'} l_{a'})$$
$$\times C_{i'}(n_b l_b, n_{b'} l_{b'}) c_t(K) \mathcal{F}(\mu, \nu, t), \tag{III.84}$$

where $N = n_a + n_{a'}$, $L = l_a + l_{a'}$, $N' = n_b + n_{b'}$, $L' = l_b + l_{b'}$, $\mu = 2i + L$, and $\nu = 2i' + L'$. The expression (III.83) explicitly shows that $\mathcal{F}(\mu, \nu, t)$ is *symmetric in* μ *and* ν; further, since L, L', t have the same parity, we have to consider only the odd values of μ, ν for $t =$ odd, and the even values for $t =$ even. The quantities $\mathcal{F}(\mu, \nu, t)$ can be independently tabulated with the help of (III.83), and then used in (III.84) with appropriate values of K, $n_a l_a$, $n_b l_b$, ... together with the coefficients C_i and $c_t(K)$.

D. δ-FUNCTION POTENTIAL

The Slater integral method is very easy to apply in the case of a zero range potential because the radial functions $f_K(r_1, r_2)$ are extremely simple and the Slater integrals can be evaluated in a trivial manner for such a potential. Further, the summation over K (the order of the Legendre polynomial) which appears in the matrix element can be carried out in a closed form, yielding very simple final expressions for the matrix elements.

The spatial part of the potential is written as

$$V_0\delta(\mathbf{r}) = V_0\delta(\mathbf{r}_1 - \mathbf{r}_2), \tag{III.85}$$

where the subscripts 1 and 2 refer to the two interacting nucleons. The three-dimensional δ-function is actually a product of three one-dimensional δ-functions corresponding to the three coordinates. In terms of the cartesian coordinates,

$$\delta(\mathbf{r}_1 - \mathbf{r}_2) = \delta(x_1 - x_2)\delta(y_1 - y_2)\delta(z_1 - z_2), \tag{III.86a}$$

whereas in terms of the polar coordinates,

$$\delta(\mathbf{r}_1 - \mathbf{r}_2) = \frac{\delta(r_1 - r_2)}{r_1^2}\delta(\cos\theta_1 - \cos\theta_2)\delta(\phi_1 - \phi_2). \tag{III.86b}$$

The correctness of these assertions can be verified by integrating (III.86a) over one set of coordinates, say, $dx_1\, dy_1\, dz_1$, and integrating (III.86b) over $r_1^2\, dr_1\, d(\cos\theta_1)\, d\phi_1$; the result in both the cases is clearly equal to unity. Incidentally, the occurrence of r_1^2 in the elementary integration volume in terms of polar coordinates demands that an r_1^2 be put in the denominator of (III.86b).

From now on, we shall work in terms of polar coordinates only. The two δ-functions in the angle space can be replaced by the closure relation for the complete set of eigenfunctions of orbital angular momentum, i.e., the spherical harmonics. We thus have

$$\delta(\mathbf{r}) = \frac{\delta(r_1 - r_2)}{r_1^2} \sum_{K=0}^{\infty} \sum_{q=-K}^{K} Y_q^K(\theta_1, \phi_1) Y_q^{K*}(\theta_2, \phi_2). \tag{III.87}$$

Since $Y_q^{K*} = (-1)^q Y_{-q}^K$, we can replace the q-summation by our scalar product notation. Further, we switch over to the renormalized spherical harmonics of (AI.7a) from Appendix A, and obtain

$$\delta(\mathbf{r}) = \frac{\delta(r_1 - r_2)}{r_1^2} \sum_{K=0}^{\infty} \frac{[K]}{4\pi}\, y^K(1) \cdot y^K(2). \tag{III.88}$$

Comparing this expression with (III.69), we conclude

$$f_K(r_1, r_2) = \frac{[K]}{4\pi} \frac{\delta(r_1 - r_2)}{r_1^2}. \tag{III.89}$$

Thus, the K-dependence of the Slater integral of the δ-function potential will be through the factor $[K]$ only. Using the definition (III.72b), we have

$$\begin{aligned} F_K(n_{a'}l_{a'}, n_{b'}l_{b'}; n_a l_a, n_b l_b) &= \frac{[K]}{4\pi} \int_0^\infty dr_1 \int_0^\infty dr_2\, R_{n_{a'}l_{a'}}(r_1) R_{n_{b'}l_{b'}}(r_2) \frac{\delta(r_1 - r_2)}{r_1^2} R_{n_a l_a}(r_1) R_{n_b l_b}(r_2) \\ &= \frac{[K]}{4\pi} \int_0^\infty dr_2\, \frac{1}{r_2^2} R_{n_{a'}l_{a'}}(r_2) R_{n_{b'}l_{b'}}(r_2) R_{n_a l_a}(r_2) R_{n_b l_b}(r_2) \\ &= \frac{[K]}{4\pi} \int_0^\infty r_2^2\, dr_2\, \mathcal{R}_{n_{a'}l_{a'}}(r_2) \mathcal{R}_{n_{b'}l_{b'}}(r_2) \mathcal{R}_{n_a l_a}(r_2) \mathcal{R}_{n_b l_b}(r_2) \\ &= [K] \mathcal{J}(n_{a'}l_{a'}, n_{b'}l_{b'}; n_a l_a, n_b l_b), \end{aligned} \tag{III.90}$$

where $\mathcal{R}_{nl}$ is the full radial function, equal to R_{nl}/r, and the quantity $\mathcal{J}$ has the obvious definition of $(4\pi)^{-1}$ times the radial integral.

A peculiar fact about the depth V_0 of the δ-function potential follows if we integrate (III.85) over the entire three-dimensional volume corresponding to $\mathbf{r}$. Since $\int \delta(\mathbf{r})\, d^3r = 1$, the result of the integration is V_0. And since the entire potential $V_0\delta(\mathbf{r})$ has the dimension of energy, the volume integration must yield a quantity having the dimension of energy multiplied by the volume. Thus, the depth of the δ-function potential cannot be specified in MeV only; it has to be specified in MeV cm^3. Its difference from the depth V_0 of a finite-ranged potential $V_0 f(r/a)$ should be noted. As $f(r/a)$ is a dimensionless quantity (r/a being dimensionless), V_0 has the dimension of energy, and hence it is quoted in MeV. To avoid the distinction between the depth of a δ-function potential and that of a finite-ranged potential, the former is sometimes written as $V_0 R^3 \delta(\mathbf{r})$, where R is a length of nuclear dimension (i.e., 10^{-13} cm). Since the dimension of volume has been explicitly put into R^3, V_0 is now in MeV.

A simplified feature of the δ-function potential is the treatment of its exchange-dependence. Since an antisymmetric *spatial* wavefunction of two nucleons vanishes when $\mathbf{r}_1 = \mathbf{r}_2$, it is *automatically* rejected by the δ-function potential which is nonvanishing only for $\mathbf{r}_1 = \mathbf{r}_2$. Hence, after the δ-function potential has operated on a two-nucleon state, the value of the Majorana operator P_M is very easy to write down. Since the operation of the δ-function has rejected the antisymmetric spatial wavefunctions, and picked up only the symmetric parts, the value of P_M is equal to $+1$. For the same reason, the Bartlett exchange operator P_σ, which is identically equal to $-P_M P_\tau$, will have the value $-P_\tau = (-1)^T$ for the δ-function potential. If we denote the general exchange-dependence by

$$g_W + g_M P_M - g_H P_\tau + g_B P_\sigma, \tag{III.91a}$$

then, for the δ-function potential, it is equivalent to

$$(g_W + g_M) - (g_H + g_B)P_\tau,$$

and has the values

$$g_W + g_M - (g_H + g_B) \qquad (\text{for } T = 1), \tag{III.91b}$$

$$g_W + g_M + g_H + g_B \qquad (\text{for } T = 0). \tag{III.91c}$$

Since the nonvanishing contributions come from space-symmetric states, the combined spin isospin part of these states has to be antisymmetric, and hence (i) $T = 1$ will go with $S = 0$, and (ii) $T = 0$ will go with $S = 1$. Therefore, for the states $T = 1$, V_0 times (III.91b) is the even, spin-singlet state depth. We shall use the symbol $V_0^{\mathrm{e,s}}$ to denote this quantity. Similarly, for the states $T = 0$, V_0 times (III.91c) will be denoted by $V_0^{\mathrm{e,t}}$, where e, t stand for an even, spin-triplet state.

We now proceed to evaluate the direct and exchange terms of (III.77b). The effect of the depth V_0 times the exchange operators will be introduced at the very end of our derivation in terms of the depth parameters $V_0^{\mathrm{e,t}}$ and $V_0^{\mathrm{e,s}}$. For the present, we handle only $\delta(\mathbf{r})$, given by (III.88). Putting the Slater integral of (III.90) into (III.76b), we get, for the direct term,

$$\begin{aligned}
&\langle a'b' : JM, TM_T|\, \delta(\mathbf{r})\, |ab : JM, TM_T\rangle \\
&= \mathcal{I}(n_{a'}l_{a'}, n_{b'}l_{b'}; n_a l_a, n_b l_b) \sum_{K=0}^{\infty} [K](-1)^{j_{a'}+j_b-J}\{[j_{a'}][j_{b'}]\}^{1/2} W(j_{a'}j_{b'}j_a j_b; JK) \\
&\quad \times \langle j_{a'}\| y^K \| j_a\rangle \langle j_{b'}\| y^K \| j_b\rangle \\
&= \mathcal{I}(n_{a'}l_{a'}, n_{b'}l_{b'}; n_a l_a, n_b l_b) \sum_{K=0}^{\infty} [K](-1)^{j_{a'}+j_b-J}\left\{\frac{[j_{a'}][j_{b'}]}{[J][K]}\right\}^{1/2} \\
&\quad \times U(j_{a'}j_{b'}j_a j_b; JK)\tfrac{1}{2}[1 + (-1)^{l_{a'}+l_a+K}]\begin{bmatrix} j_{a'} & K & j_a \\ \frac{1}{2} & 0 & \frac{1}{2}\end{bmatrix}\begin{bmatrix} j_{b'} & K & j_b \\ \frac{1}{2} & 0 & \frac{1}{2}\end{bmatrix}.
\end{aligned} \tag{III.92}$$

We have converted the W-function into the U-function by using (AV.3) from Appendix A, and have written the values of the two reduced matrix elements from (BIII.17) in Appendix B. The parity factors $\frac{1}{2}[1 + (-1)^{l_a+l_{a'}+K}]$ and $\frac{1}{2}[1 + (-1)^{l_b+l_{b'}+K}]$ have to be simultaneously non-zero in order that the result is nonvanishing. This therefore requires the parity of $l_a + l_{a'}$ to be the same as the parity of $l_b + l_{b'}$, i.e., $l_a + l_{a'} + l_b + l_{b'} =$ even integer. It should be noticed that this requirement guarantees that the parity of the two two-body states on the left- and right-hand side of (III.92) is the same. Therefore, it is clearly enough to retain one of the parity factors, as done in (III.92). This expression contains two kinds of K-summation, one with and the other without $(-1)^K$ from the parity factor. We shall show how these two kinds of K-summation can be carried out. First, consider

$$\sum_K [K]^{1/2} U(j_{a'} j_{b'} j_a j_b; JK) \begin{bmatrix} j_{a'} & K & j_a \\ \frac{1}{2} & 0 & \frac{1}{2} \end{bmatrix} \begin{bmatrix} j_{b'} & K & j_b \\ \frac{1}{2} & 0 & \frac{1}{2} \end{bmatrix}. \qquad \text{(III.93a)}$$

Then, by using symmetry relations, rewrite the product of the two Clebsch-Gordon coefficients, as

$$(-1)^{j_{a'}-(1/2)} \sqrt{\frac{[j_a]}{[K]}} \begin{bmatrix} j_{a'} & j_a & K \\ \frac{1}{2} & -\frac{1}{2} & 0 \end{bmatrix} (-1)^{j_{b'}-j_b} \sqrt{\frac{[j_b]}{[j_{b'}]}} \begin{bmatrix} K & j_b & j_{b'} \\ 0 & -\frac{1}{2} & -\frac{1}{2} \end{bmatrix}. \qquad \text{(III.93b)}$$

The product of the two Clebsch-Gordon coefficients in (III.93b) is now replaced by (AV.10) from Appendix A, yielding

$$(-1)^{j_{a'}-(1/2)+j_{b'}-j_b} \left\{\frac{[j_a][j_b]}{[j_{b'}][K]}\right\}^{1/2} \sum_{J'} U(j_{a'} j_a j_{b'} j_b; KJ') \begin{bmatrix} j_a & j_b & J' \\ -\frac{1}{2} & -\frac{1}{2} & -1 \end{bmatrix} \begin{bmatrix} j_{a'} & J' & j_{b'} \\ \frac{1}{2} & -1 & -\frac{1}{2} \end{bmatrix}. \qquad \text{(III.93c)}$$

We have to replace the two Clebsch-Gordon coefficients in (III.93a) by this expression. The factor $[K]^{1/2}$ is seen to cancel out, and the K-dependence is now contained entirely in the two U-functions. We use from Appendix A the symmetry relations (AV.6) of the U-function to convert $U(j_{a'} j_a j_{b'} j_b; KJ')$ into $U(j_{a'} j_{b'} j_a j_b; J'K)$; then, according to (AV.5), also from Appendix A, the K-summation of the two U-functions yields $\delta_{JJ'}$. Thus, the summation over J' gets eliminated, and we finally simplify (III.93a) to

$$(-1)^{j_{a'}-(1/2)+j_{b'}-j_b} \left\{\frac{[j_a][j_b]}{[j_{b'}]}\right\}^{1/2} \begin{bmatrix} j_a & j_b & J \\ -\frac{1}{2} & -\frac{1}{2} & -1 \end{bmatrix} \begin{bmatrix} j_{a'} & J & j_{b'} \\ \frac{1}{2} & -1 & -\frac{1}{2} \end{bmatrix}. \qquad \text{(III.93d)}$$

We recall that this is the result of summing over K the K-dependent quantities in (III.92) without $(-1)^K$ from the parity factor. Therefore, substituting it in (III.92), we obtain, after converting the Clebsch-Gordon coefficients into more symmetric forms, the result

$$\text{the term of } \langle a'b' : JM, TM_T | \, \delta(\mathbf{r}) \, | ab : JM, TM_T \rangle \text{ without } (-1)^{l_a + l_{a'} + K}$$
$$= \tfrac{1}{2} \mathcal{I}(n_{a'} l_{a'}, n_{b'} l_{b'}; n_a l_a, n_b l_b) f(j_a j_b J) f(j_{a'} j_{b'} J), \qquad \text{(III.94a)}$$

where

$$f(j_a j_b J) = (-1)^{j_a + j_b} \left\{\frac{[j_a][j_b]}{[J]}\right\}^{1/2} \begin{bmatrix} j_a & j_b & J \\ \frac{1}{2} & \frac{1}{2} & 1 \end{bmatrix}. \qquad \text{(III.94b)}$$

The K-summation of the K-dependent quantities in (III.92) with the phase factor $(-1)^{l_{a'}+l_a+K}$ can be carried out in a similar manner. (III.93a) is now replaced by another expression which has the extra factor $(-1)^{l_{a'}+l_a+K}$. At step (III.93b), we can dispense with this extra K in the phase factor by changing the sign of the projection quantum numbers of the second Clebsch-Gordon coefficient, which gives an extra phase $(-1)^{K+j_b-j_{b'}}$. The product of these two phases gives $(-1)^{l_{a'}+l_a+j_b-j_{b'}}$, which goes into (III.93c). Further, since the second Clebsch-Gordon coefficient in (III.93b) got modified to

$$\begin{bmatrix} K & j_b & j_{b'} \\ 0 & \frac{1}{2} & \frac{1}{2} \end{bmatrix},$$

the two Clebsch-Gordon coefficients appearing in (III.93c) also change to

$$\begin{bmatrix} j_a & j_b & J' \\ -\frac{1}{2} & \frac{1}{2} & 0 \end{bmatrix} \begin{bmatrix} j_{a'} & J' & j_{b'} \\ \frac{1}{2} & 0 & \frac{1}{2} \end{bmatrix}.$$

The K-summation of the product of the two U-functions once again gives $\delta_{JJ'}$, and the step (III.93d) gets replaced by

$$(-1)^{l_{a'}+l_a+j_{a'}-(1/2)}\left\{\frac{[j_a][j_b]}{[j_{b'}]}\right\}^{1/2}\begin{bmatrix} j_a & j_b & J \\ -\frac{1}{2} & \frac{1}{2} & 0 \end{bmatrix}\begin{bmatrix} j_{a'} & J & j_{b'} \\ \frac{1}{2} & 0 & \frac{1}{2} \end{bmatrix}.$$

Once again, we bring the Clebsch-Gordon coefficients to a more symmetric form and substitute in (III.92), obtaining the final result

$$\text{the term of } \langle a'b' : JM,\ TM_T|\ \delta(\mathbf{r})\ |ab : JM,\ TM_T\rangle \text{ with } (-1)^{l_a+l_{a'}+K}$$
$$= \tfrac{1}{2}\mathcal{I}(n_{a'}l_{a'},\ n_{b'}l_{b'};\ n_a l_a,\ n_b l_b) g(j_a j_b J) g(j_{a'} j_{b'} J), \tag{III.95a}$$

where

$$g(j_a j_b J) = (-1)^{j_a-(1/2)+l_a}\left\{\frac{[j_a][j_b]}{[J]}\right\}^{1/2}\begin{bmatrix} j_a & j_b & J \\ \frac{1}{2} & -\frac{1}{2} & 0 \end{bmatrix}. \tag{III.95b}$$

Combining (III.94a) and (III.95a), we have

$$\langle a'b' : JM,\ TM_T|\ \delta(\mathbf{r})\ |ab : JM,\ TM_T\rangle$$
$$= \tfrac{1}{2}\mathcal{I}(n_{a'}l_{a'},\ n_{b'}l_{b'};\ n_a l_a,\ n_b l_b)[f(j_a j_b J) f(j_{a'} j_{b'} J) + g(j_a j_b J) g(j_{a'} j_{b'} J)]. \tag{III.96a}$$

The expression for the exchange term can be easily obtained from (III.96) with suitable relabelling. According to (III.90), the radial integral $\mathcal{I}$ remains unchanged when $a \leftrightarrow b$. Therefore, changes take place only in the factors $f(j_a j_b J)$ and $g(j_a j_b J)$. From the definitions (III.94b) and (III.95b), and the symmetry properties of the Clebsch-Gordon coefficients, we have

$$f(j_b j_a J) = (-1)^{j_a+j_b-J} f(j_a j_b J),$$
$$g(j_b j_a J) = -\pi(-1)^{j_a+j_b} g(j_a j_b J).$$

Here π denotes the parity of the two-body states, i.e., $\pi = (-1)^{l_a+l_b} = (-1)^{l_{a'}+l_{b'}}$. Therefore, the exchange term is given by

$$(-1)^{j_a+j_b-J-T}\langle a'b' : JM,\ TM_T|\ \delta(\mathbf{r})\ |ba : JM,\ TM_T\rangle$$
$$= \tfrac{1}{2}\mathcal{I}(n_{a'}l_{a'},\ n_{b'}l_{b'};\ n_a l_a,\ n_b l_b)[(-1)^T f(j_a j_b J) f(j_{a'} j_{b'} J) - \pi(-1)^{J+T} g(j_a j_b J) g(j_{a'} j_{b'} J)]. \tag{III.96b}$$

We put (III.96a) and (III.96b) in (III.77b), use explicit values of T, and now bring in the appropriate strengths $V_0^{e,s}$ and $V_0^{e,t}$. We thus obtain

$$(a'b' : JM,\ TM_T|\ V\ |ab : JM,\ TM_T)$$
$$= [(1 + \delta_{ab})(1 + \delta_{a'b'})]^{-(1/2)}\mathcal{I}(n_{a'}l_{a'} n_{b'}l_{b'};\ n_a l_a,\ n_b l_b)$$
$$\times \begin{cases} V_0^{e,s}\tfrac{1}{2}[1 + (-1)^J\pi] g(j_a j_b J) g(j_{a'} j_{b'} J) & \text{for } T = 1 \\ V_0^{e,t}[f(j_a j_b J) f(j_{a'} j_{b'} J) + \tfrac{1}{2}\{1 - (-1)^J\pi\} g(j_a j_b J) g(j_{a'} j_{b'} J)] & \text{for } T = 0 \end{cases} \tag{III.97}$$

where $V_0^{e,s}$ and $V_0^{e,t}$ are the depths of the potential in even, spin-singlet and even, spin-triplet states, respectively.

E. SHELL-MODEL DIAGONALIZATION

The formulas derived in this section will enable us to do a detailed shell-model diagonalization of the Hamiltonian matrix for a special class of nuclei, namely, those that consist of two nucleons outside a set of completely filled levels. Examples are He^6, Li^6, O^{18}, F^{18}; Ca^{42}, Sc^{42}

(with the reservation that Ca^{40} has been found to be not a very good closed-shell nucleus); Ni^{58}, Co^{58} (once again, for Ni^{56}, there is the same reservation as for Ca^{40}); Zr^{92}, Nb^{92}; and Po^{210}, Bi^{210}. We can pick out several other clear examples where the core of the two outer particles, instead of consisting of only closed shells, has also some closed sublevels in a major shell, the latter being fairly well-separated in energy from the other partially filled sublevels (which receive the two outermost nucleons).

If the picture of a core comprising completely filled levels is good, then the core has a very simple role to play in the process of diagonalization of the Hamiltonian matrix. This is true in the general case of any number of nucleons (i.e., not necessarily two, as in our present case) distributed in the partially filled levels. We shall first prove this statement.

Let the single-particle states in the core be denoted by the symbols $c, c', \ldots,$ and those above the closed-shell core by $p, p', \ldots$. We shall prove our statement for simple determinantal states of all the nucleons. Once the proof is completed, a little reflection will convince the reader that the same results would have been true had we considered states that are a superposition of several determinants coupling the states of the outer nucleons to a good total angular momentum J. As discussed in Section 25, the core state is unique, has $J = 0$, and has no role to play in this angular momentum coupling.

Let us denote the determinantal states by $\Psi, \Psi', \ldots$. Then, according to (II.23) and (II.25), any diagonal matrix element of the Hamiltonian (II.3) is given by

$$\langle\Psi| H |\Psi\rangle = \sum_{\xi}^{\text{occ}} \langle\xi| T |\xi\rangle + \sum_{\xi<\eta}^{\text{occ}} (\xi\eta| V |\xi\eta), \tag{III.98}$$

where the summations go over the occupied single-particle states in the determinant Ψ, and the antisymmetrized matrix element of V contains 'direct' minus 'exchange' parts. Keeping in mind the meaning of the notation $c, c', \ldots, p, p', \ldots,$ as just described, we split up the summations as

$$\langle\Psi| H |\Psi\rangle = \sum_{c} \langle c| T |c\rangle + \sum_{p} \langle p| T |p\rangle + \sum_{c<c'} (cc'| V |cc') + \sum_{p}\sum_{c} (pc| V |pc) + \sum_{p<p'} (pp'| V |pp'). \tag{III.99}$$

Let us introduce

$$E_c = \sum_{c} \langle c| T |c\rangle + \sum_{c<c'} (cc'| V |cc'), \tag{III.100}$$

which is clearly the energy of the closed-shell core. We also use the definition (II.95) of the overall potential $\mathcal{V}$, set up by the closed-shell core on a *single* nucleon outside the core, and rewrite the fourth term of (III.99) as

$$\sum_{c} (pc| V |p'c) = \langle p| \mathcal{V} |p'\rangle. \tag{III.101}$$

The shell-model single-particle energy $\epsilon(p)$ is clearly the eigenvalue of $T + \mathcal{V}$ for the state p. So we write

$$\epsilon(p) = \langle p| T |p\rangle + \langle p| \mathcal{V} |p\rangle. \tag{III.102a}$$

In general, if we consider two different single-particle states, p and p', we have

$$\epsilon(p)\delta_{pp'} = \langle p| (T + \mathcal{V}) |p'\rangle. \tag{III.102b}$$

Using (III.100), (III.101), and (III.102a), we get, from (III.99),

$$\langle\Psi| H |\Psi\rangle = E_c + \sum_{p} \epsilon(p) + \sum_{p<p'} (pp'| V |pp'). \tag{III.103}$$

There are two classes of nondiagonal matrix elements: (i) when Ψ and Ψ' differ through one single-particle state; let these single-particle states in Ψ and Ψ' be denoted by p_1 and p_1'; (ii) when Ψ and Ψ'' differ through two single-particle states; let these be (p_1, p_2) in Ψ and (p_1', p_2') in Ψ''. In case (i), according to (II.24) and (II.26), we have

$$\begin{aligned} \langle\Psi| H |\Psi'\rangle &= \langle p_1| T |p_1'\rangle + \sum_\xi (\xi p_1| V |\xi p_1') \\ &= \langle p_1| T |p_1'\rangle + \sum_c (cp_1| V |cp_1') + \sum_p (pp_1| V |pp_1') \\ &= \langle p_1| (T + \mathcal{V}) |p_1'\rangle + \sum_p (pp_1| V |pp_1') \\ &= \sum_p (pp_1| V |pp_1'). \end{aligned} \tag{III.104}$$

In the second step, we have split up the summation over ξ, as before, and then replaced the summation over c, according to the definition (III.101); finally, (III.102b) has been used to drop the $(T + \mathcal{V})$-term.

In case (ii), according to the discussion in Section 15, we have a zero matrix element for T, and the result for the potential energy can be had from (II.27). We obtain, in this way,

$$\langle\Psi| H |\Psi''\rangle = (p_1p_2| V |p_1'p_2'). \tag{III.105}$$

We now interpret the results contained in (III.103)-(III.105). It is clear from these expressions that the Hamiltonian matrix has a potential energy part, which *could be obtained by calculating with states for the outer nucleons only*. The effect of the closed-shell core is entirely confined to the diagonal matrix elements (III.103). Firstly, the core contributes its energy E_c to all the *diagonal* matrix elements. Secondly, it produces the shell-model single-particle energies in the *diagonal* matrix elements, and thus removes any explicit occurrence of the kinetic energy operator T. Since a constant quantity E_c, occurring in all the diagonal matrix elements, can be removed from the matrix without altering the eigenvectors, we shall do so, and then remember that the eigenvalues of the modified matrix give the energies of the nucleus, with the zero of the scale fixed at the core energy E_c. After the elimination of E_c, we are left with a diagonalization problem which corresponds to the shell-model (SM) Hamiltonian *for the outer nucleons only* and is described as

$$H_{\text{SM}} = \sum_i h_i + \sum_{i<j} V_{ij}, \tag{III.106}$$

where the single-particle Hamiltonian h is equal to $(T + \mathcal{V})$ and has eigenvalues $\epsilon(p)$, defined according to (III.102a). We finally emphasize that the single-particle energy, as defined by (III.102a) and (III.101), is the energy of a *single* nucleon in state p outside the closed-shell core. Thus, even though we may be doing the diagonalization for O^{18} or F^{18}, the single-particle energies that enter the diagonal matrix elements must be taken from O^{17} or F^{17}. Usually, the numbers are taken from *experimental data* on the appropriate closed shell plus a single-nucleon nucleus.

F. EXAMPLE OF SHELL-MODEL DIAGONALIZATION FOR TWO NUCLEONS

We shall describe the example of O^{18} and F^{18} in some detail. The single-particle levels above the closed shells of O^{16} are $(0d_{5/2})$, $(1s_{1/2})$, and $(0d_{3/2})$. The single-particle energies of these levels, as determined from O^{17} and F^{17} spectra, are given by

$$O^{17} : 0d_{5/2}(0), \quad 1s_{1/2}(0.87 \text{ MeV}), \quad 0d_{3/2}(5.08 \text{ MeV}),$$

F^{17} : $0d_{5/2}(0)$, $1s_{1/2}(0.5\ \text{MeV})$, $0d_{3/2}(4.7\ \text{MeV})$.

In both cases, we have taken the lowest level to be the zero of the scale, which eventually fixes the reference point of the eigenvalues of the Hamiltonian matrix at $[E_c + N_p\epsilon_p(0d_{5/2}) + N_n\epsilon_n(0d_{5/2})]$, where the subscripts n and p denote neutron and proton, and N is the number of nucleons outside the O^{16}-core. For any given nucleus, N_p and N_n are fixed numbers, and hence this fixed reference point of energy measurement need not be known while calculating the *spectra*, which consists of energies of the different eigenstates, as measured from the lowest (i.e., the ground state) of them. In the difference of the calculated energy of any state from that of the lowest one, the constant reference energy simply cancels out. A knowledge of this number is necessary only if we want to calculate the absolute binding energy of the ground state of a nucleus or determine the difference in binding energies of neighbouring nuclei.

The possible two-nucleon configurations and the total J- and T- values of the corresponding two-nucleon states are listed in Table III.4 in accordance with the considerations in Section 28A.

Table III.4 Two-nucleon states in $(1s, 0d)$-shell

$(0d_{5/2})^2$	$T = 1$	$J = 0, 2, 4$
	$T = 0$	$J = 1, 3$
$(0s_{1/2})^2$	$T = 1$	$J = 0$
	$T = 0$	$J = 1$
$(0d_{3/2})^2$	$T = 1$	$J = 0, 2$
	$T = 0$	$J = 1, 3$
$(0d_{5/2}, 0s_{1/2})$	$T = 1, 0$	$J = 2, 3$
$(0d_{5/2}, 0d_{3/2})$	$T = 1, 0$	$J = 1, 2, 3, 4$
$(0d_{3/2}, 0s_{1/2})$	$T = 1, 0$	$J = 1, 2$

Since O^{18} comprises two neutrons outside the closed shells, its states correspond to $T = 1$ only, whereas F^{18}, having a neutron and a proton, has both the states $T = 0$ and $T = 1$. According to the discussions on isotopic spin in Section 14, in the context of nuclear symmetry energy, we expect the ground state of F^{18} to arise from $T = 0$ (because $N = Z$).

The list in Table III.4 shows that, for $T = 1$, $J = 0$, there are three possible states: (i) $|(0d_{5/2})^2\, J = 0, T = 1\rangle$, (ii) $|(0s_{1/2})^2\, J = 0, T = 1\rangle$, and (iii) $|(0d_{3/2})^2\, J = 0, T = 1\rangle$. Therefore, for the calculation of the states $J = 0$ of O^{18}, we have to set up the 3×3 matrix for the Hamiltonian (III.106) with these three states as the basis, and then diagonalize. In a similar manner, the list gives five basis states corresponding to $J = 2$ and $T = 1$. The programme includes setting up the matrices for the various (J, T)-values with the appropriate basis states picked up from the list. The diagonalization of each matrix yields a set of eigenvalues and eigenstates with the appropriate J and T quantum numbers. The lowest energies for $T = 1$ and $T = 0$ obtained in this way will give respectively the ground states of O^{18} and F^{18}. The sequence

42 (45-122/1977)

of the excited levels in each nucleus can also be obtained in this manner. The eigenvector in each case is obtained as a string of numbers, $c_1, c_2, \ldots$, and the corresponding two-nucleon eigenstate is given by $\sum_i c_i|i)$, where $|i)$ is the i-th basis state of the particular energy matrix.

The two-body effective potential V is usually assumed to be of the Yukawa form with a general type of exchange-dependence. The depths of the potential in the four possible spin(S)-isospin(T) states are treated as parameters in the calculation. The range parameter is usually given a fixed value. The only other unknown constant, namely, the harmonic oscillator parameter b contained in the wavefunctions, is determined by fitting the experimental root-mean-square radius of the nucleus. The total number of energy levels of O^{18} and F^{18} fitted by this procedure is larger than the total number (the four depth constants) of adjustable parameters. The final values of these parameters, together with the calculated and experimental energy levels, are shown in Table III.5. The same effective potential has been successfully used

Table III.5 Energy states in (1*s*, 0*d*)-shell with Yukawa potential

State		−Energy (MeV)	
Isospin T	Total angular momentum J	Experimental	Calculated
	0	3.96	4.02
1	2	1.98	1.81
	4	0.41	0.52
	1	5.00	5.11
0	3	4.06	3.97
	5	3.87	3.77

Harmonic oscillator parameter = 1.72 fm.
Parameters of Yukawa potential—range parameter = 1.42 fm; depth parameter $V_{T,S}$ for the various two-nucleon isospin (T) and spin (S) states:

$V_{0,0} = 40.86$ MeV,
$V_{0,1} = -46.91$ MeV,
$V_{1,1} = 19.36$ MeV,
$V_{1,0} = -34.43$ MeV.

in structural calculations on many other nuclei in the (0*d*, 1*s*)-shell. From our work in Section 28C it is clear that more general potentials, containing tensor and two-body spin-orbit terms, could also have been used in such calculations. The only restriction is on the shape of the potential, which has to be of a smooth well-behaved form. In particular, the realistic shape of the two-body potential having a strongly repulsive core cannot be *directly* used in such calculations.

In Section 28C, we have mentioned what has to be done for the realistic hard-core potential. The effective G-matrix elements, including the core-polarization effect, have to be substituted for the matrix elements of V in (III.67), and then the corresponding two-body matrix elements can be used in the manner described for matrix diagonalization. A summary of the G-matrix calculations for the nuclei F^{18} and O^{18}, indicating success and failure, has been given in Section 22C under "Finite nuclei".

29. MORE THAN TWO NUCLEONS IN UNFILLED SHELL

In this section, we shall consider the case where more than two nucleons are distributed amongst a few partially filled levels, outside a set of completely filled shells. The case of three nucleons in the partially filled levels is first treated in detail to elucidate all the essential points; finally, a generalization is made to the case of n nucleons, where n is less than the number of available single-particle states.

A. THREE NUCLEONS

In analogy with the case of two nucleons, we have two types of problem to solve: (i) finding all the allowed J- and T-quantum numbers (and also additional quantum numbers, if necessary) of the *antisymmetric* three-nucleon states, and the detailed construction of such states in terms of angular momentum coupled states of the three nucleons; (ii) calculating the matrix elements of one- and two-body type operators, using the antisymmetric states obtained in (i).

Construction of Antisymmetric States

If several single-particle levels $a, b, c, \ldots$ are available to the nucleons, then we can distribute the three nucleons amongst the levels to form configurations of the distinctive types

(i) a^3;

(ii) a^2, c;

(iii) a, b, c.

These different types are characterized by the fact that, in (i), all the three nucleons occupy a single level a (equivalent nucleons); in (ii), two of them are in the same level a, but the third one is in a different level c; and in (iii), all the three nucleons are in different single-particle levels a, b, c.

Let us consider *any* antisymmetric wavefunction $\Phi(1, 2)$ of the first two nucleons, and multiply it by any single-particle wavefunction $\phi(3)$ of the third nucleon. The product wavefunction $\Phi(1, 2)\phi(3)$ is, by definition, antisymmetric under the exchange P_{12} of the first two nucleons, but it does not have any definite behaviour under the exchange operation P_{23} and P_{13}. Let us now consider the operator

$$\mathcal{A} = (1 - P_{23} - P_{13}) \tag{III.107}$$

and let it operate on $\Phi(1, 2)\phi(3)$. The wavefunction obtained in this way will be called $\Psi(1, 2, 3)$ and is given by

$$\begin{aligned}\Psi(1, 2, 3) &= (1 - P_{23} - P_{13})[\Phi(1, 2)\phi(3)] \\ &= \Phi(1, 2)\phi(3) - \Phi(1, 3)\phi(2) - \Phi(3, 2)\phi(1).\end{aligned} \tag{III.108}$$

Operating in turn with P_{12}, P_{23}, and P_{13} on the last step, we can very easily show that $\Psi(1, 2, 3)$ is indeed antisymmetric under all these exchanges. So $\Psi(1, 2, 3)$ is an acceptable antisymmetric wavefunction of the three-nucleon system which has been obtained by means of a general prescription: take *any* antisymmetric state of nucleons 1 and 2, multiply it by *any* single-particle wavefunction for nucleon 3, and then antisymmetrize the product by the antisymmetrizer (III.107). We shall see later in this section that an antisymmetric wavefunction of *any* number (n) of nucleons can be obtained from the product of an antisymmetric wavefunction of the first $(n-1)$ nucleons and the single-particle wavefunction of the last nucleon by the application of a suitable antisymmetrizer.

The technique just described can be directly applied while constructing antisymmetric states out of configurations of the type (a^3), (a^2, c), and (a, b, c). We shall consider first the case of (a^2, c), followed by (a, b, c), and then the case of *equivalent* nucleons (a^3). We know, from Section 28A, all the antisymmetric states of a^2; they are of the type $|a^2J_0T_0\rangle$ with the allowed J_0 and T_0 being given by $J_0 + T_0 =$ odd integer, $T_0 = 0$ or $T_0 = 1$, and $0 \leqslant J_0 \leqslant 2a$. Let us associate this type of antisymmetric state with the nucleons 1 and 2, i.e., it is equivalent to $\Phi(1, 2)$ of (III.108). We put the third nucleon in the single-particle state c and couple J_0, T_0 with j_0 and $\frac{1}{2}$, respectively, to produce a three-nucleon state of total angular momentum J and total isospin T. Thus, $|a^2(1, 2)J_0T_0; c(3): JT\rangle$ is the analogue of $\Phi(1, 2)\phi(3)$ of (III.108). To produce a three-nucleon antisymmetric state, we therefore operate on this state by $\mathcal{A}$ and obtain

$$\begin{aligned}\mathcal{A}|a^2(1, 2)J_0T_0; c(3): JT\rangle &= (1 - P_{23} - P_{13})|a^2(1, 2)J_0T_0; c(3): JT\rangle \\ &= |a(1)a(2)J_0T_0, c(3): JT\rangle - |a(1)a(3)J_0T_0; c(2): JT\rangle \\ &\quad -|a(3)a(2)J_0T_0; c(1): JT\rangle. \end{aligned} \tag{III.109}$$

Following the convention described in Section 28, we shall use angular momentum coupled states in which the first, second, and third single-particle states from the left correspond to the first, second, and third nucleons, respectively. Whereas the sequence is already correct for the first state on the right-hand side of (III.109), the other two states require appropriate rearrangements. In the second state, for example, the coupling order, according to the aforementioned convention, must be changed to $|a(1), c(2)\mathcal{J}\mathcal{T}, a(3): JT\rangle$, where $\mathcal{J}$, $\mathcal{T}$ are the two-body angular momentum and isospin obtained by the coupling of $a(1)$ and $c(2)$. This new arrangement can be achieved in three steps: (i) interchange the coupling order of $|a(1), a(3)J_0T_0\rangle$ and $|a(3), a(1)J_0T_0\rangle$ which gives rise to a phase factor according to the first relation of (AIV.4) in Appendix A; (ii) rearrange the angular momentum coupling with the $6j$-symbol of Appendix A (Section V), and get the state $|a(3), a(1)c(2)\mathcal{J}\mathcal{T}: JT\rangle$; and (iii) interchange the order of coupling between $a(3)$ and $(\mathcal{J}\mathcal{T})$ so as to obtain $|a(1)c(2)\mathcal{J}\mathcal{T}, a(3): JT\rangle$. The third step once again brings in a phase factor. At every stage we have to remember that there are two types of coupling involved: one between the angular momenta and the other between the isospins. Following the steps indicated, we obtain

$$\begin{aligned}|a(1)a(3)J_0T_0, c(2): JT\rangle &= (-1)^{2j_a+1-J_0-T_0} \sum_{\mathcal{J},\mathcal{T}} (-1)^{j_a+\mathcal{J}-J+1/2+\mathcal{T}-T} \\ &\quad \times U(j_aj_aJj_c; J_0\mathcal{J})U(\tfrac{1}{2}\,\tfrac{1}{2}T\tfrac{1}{2}; T_0\mathcal{T})|a(1)c(2)\mathcal{J}\mathcal{T}, a(3): JT\rangle. \end{aligned} \tag{III.110}$$

In an exactly similar manner, the last term of (III.109) can be rearranged as

$$\begin{aligned}|a(3)a(2)J_0T_0, c(1): JT\rangle &= \sum_{\mathcal{J},\mathcal{T}} (-1)^{j_a+j_c-\mathcal{J}+1-\mathcal{T}}(-1)^{j_a+\mathcal{J}-J+1/2+\mathcal{T}-T} \\ &\quad \times U(j_aj_aJj_c; J_0\mathcal{J})U(\tfrac{1}{2}\,\tfrac{1}{2}T\tfrac{1}{2}; T_0\mathcal{T})|c(1)a(2)\mathcal{J}\mathcal{T}; a(3): JT\rangle. \end{aligned} \tag{III.111}$$

We recall that the allowed values of J_0, T_0 satisfy $(-1)^{J_0+T_0} = -1$, and since $(2j_a + 1)$ is an even integer, the first phase factor in (III.110) yields -1. Thus, combining the last two terms of (III.109), i.e., the expressions (III.110) and (III.111), we get

$$\sum_{\mathcal{J},\mathcal{T}} (-1)^{j_a+\mathcal{J}-J+1/2+\mathcal{T}-T}\sqrt{2}U(j_aj_aJj_c;\, J_0\mathcal{J})U(\tfrac{1}{2}\,\tfrac{1}{2}T\tfrac{1}{2},\, T_0\mathcal{T})|\{ac(1,2)\mathcal{J}\mathcal{T}\}_A,\, a(3):JT\rangle, \tag{III.112}$$

where $\{ac(1, 2)\mathcal{J}\mathcal{T}\}_A$ stands for the *normalized antisymmetric* two-nucleon state defined in Section 28A. Substituting (III.112) for the last two terms of (III.109), we obtain

$$\mathcal{A}|a^2(1,2)J_0T_0,\, c(3):JT\rangle = |a^2(1,2)J_0T_0,\, c(3):JT\rangle + \sum_{\mathcal{J},\mathcal{T}} (-1)^{j_a+\mathcal{J}-J+1/2+\mathcal{T}-T}\sqrt{2}$$
$$\times\, U(j_aj_aJj_c;\, J_0\mathcal{J})U(\tfrac{1}{2}\,\tfrac{1}{2}T\tfrac{1}{2},\, T_0\mathcal{T})|\{ac(1,2)\mathcal{J}\mathcal{T}\}_A,\, a(3):JT\rangle. \tag{III.113a}$$

The first state on the right-hand side is orthogonal to any of the states appearing in the summation for the simple reason that nucleon 3 belongs to the different states c and a. Any two different states inside the summation are also orthogonal to each other because they have different values of $(\mathcal{J}, \mathcal{T})$ for the first two nucleons. Therefore, the normalization factor can be worked out from the sum of the squares of the coefficients of all these states:

$$1 + \sum_{\mathcal{J},\mathcal{T}} 2U^2(j_aj_aJj_c;\, J_0\mathcal{J})U^2(\tfrac{1}{2}\,\tfrac{1}{2}T\tfrac{1}{2};\, T_0\mathcal{T}) = 3.$$

The square of each U-function, when summed, produces unity, according to (AV.5) from Appendix A, and hence this result. The normalization factor is therefore $3^{-1/2}$. We shall denote the normalized antisymmetrized state by $|a^2J_0T_0, c:JT)$, where the parenthesis, as usual, denotes antisymmetry, and the particle labels 1, 2, 3 have been omitted because, in the antisymmetrized state, these coordinates cannot be attached to specific single-particle states. Thus, (III.113a) yields

$$|a^2J_0T_0,\, c:JT) = \frac{1}{\sqrt{3}}|a^2(1,2)J_0T_0,\, c(3):JT\rangle + \sqrt{\tfrac{2}{3}} \sum_{\mathcal{J},\mathcal{T}} (-1)^{j_a+\mathcal{J}-J+1/2+\mathcal{T}-T}$$
$$\times\, U(j_aj_aJj_c;\, J_0\mathcal{J})U(\tfrac{1}{2}\,\tfrac{1}{2}T\tfrac{1}{2};\, T_0\mathcal{T})|\{ac(1,2)\mathcal{J}\mathcal{T}\}_A,\, a(3):JT\rangle. \tag{III.113b}$$

Each state of this expansion has a notable characteristic: it has been obtained by the angular momentum coupling of a single-particle state (either c or a) of the last nucleon (number 3) with an antisymmetric state of the remaining nucleons. Each such state is superposed in the formation of the antisymmetrized state with a definite numerical coefficient, which is called the *coefficient of fractional parentage* (*cfp*). Even though we started by coupling $c(3)$ with a given two-body state $|a^2(1, 2)J_0T_0\rangle$, on the right-hand side we not only obtained this state, but also a whole set of other states, the latter being the consequence of antisymmetrization. An expansion of this type for an antisymmetrized state is called a fractional parentage expansion.

The same type of procedure as for (a^2, c) can be easily repeated for states of the configuration (a, b, c). Here we have to start with an antisymmetric state of the first two nucleons given by

$$|ab(1,2)J_0T_0) = \frac{1}{\sqrt{2}}[|a(1)b(2)J_0T_0\rangle + (-1)^{j_a+j_b+J_0+T_0}|b(1)a(2)J_0T_0\rangle],$$

then couple the single-particle state $c(3)$ to produce the final J and T. The same antisymmetrizer $\mathcal{A}$ has then to be applied to this state. Following the same type of rearrangements, and finally carrying out the normalization of the state, we obtain the fractional parentage expansion in this case as

$$|(ab)J_0T_0, c:JT) = \frac{1}{\sqrt{3}}|\{ab(1,2)J_0T_0\}_A, c(3):JT\rangle + \frac{1}{\sqrt{3}} \sum_{\mathcal{J},\mathcal{T}} (-1)^{\mathcal{J}+\mathcal{T}-J-T}$$
$$\times U(\tfrac{1}{2}\tfrac{1}{2}T\tfrac{1}{2}; T_0\mathcal{T})[(-1)^{j_a+1/2}U(j_aj_bJj_c; J_0\mathcal{J})|\{bc(1,2)\mathcal{J}\mathcal{T}\}_A, a(3):JT\rangle$$
$$+ (-1)^{j_b+1/2}U(j_bj_aJj_c; J_0\mathcal{J})|\{ac(1,2)\mathcal{J}\mathcal{T}\}_A, b(3):JT\rangle]. \quad \text{(III.114)}$$

In both (III.113b) and (III.114), the first state, whose antisymmetrization gave rise to the entire expansion, is called the *principal parent*. It is clear from this procedure that all possible values of J, T consistent with the angular momentum coupling rules will be allowed in the expression (III.113b) of $|a^2J_0T_0, c:JT)$ and the expression (III.114) of $|(ab)J_0T_0, c:JT)$.

States of three equivalent nucleons The case of the configuration (a^3) of equivalent nucleons which we shall deal with now is very distinctive in several respects. It is possible to construct an antisymmetric state in the manner of (III.113b) by substituting the single-particle state c with a, but we have to be careful with the normalization. Therefore, we first go back to the steps (III.110) and (III.111) and notice that

$$(-1)^{2j_a+1-J_0-T_0} = -1, \qquad (-1)^{2j_a+1-\mathcal{J}-\mathcal{T}} = (-1)^{\mathcal{J}+\mathcal{T}}.$$

Hence, (III.112), which was the sum of these two expressions, is now replaced by

$$\sum_{\mathcal{J},\mathcal{T}} (-1)^{j_a+1/2-J-T}(-1)^{\mathcal{J}+\mathcal{T}}[1-(-1)^{\mathcal{J}+\mathcal{T}}]U(j_aj_aJj_a; J_0\mathcal{J})U(\tfrac{1}{2}\tfrac{1}{2}T\tfrac{1}{2}; T_0\mathcal{T})$$
$$\times |a^2(1,2)\mathcal{J}\mathcal{T}; a(3):JT\rangle.$$

Thus, $(\mathcal{J}+\mathcal{T})$ is required to be an odd integer, and consequently the allowed $(\mathcal{J}, \mathcal{T})$-states of the configuration $a^2(1, 2)$ appearing in the foregoing expression are antisymmetric. The expression (III.109) therefore yields

$$|a^2J_0T_0, a:JT) = N \sum_{\mathcal{J},\mathcal{T}} [\delta_{\mathcal{J},J_0}\delta_{\mathcal{T},T_0} - (-1)^{j_a+1/2-J-T}\{1-(-1)^{\mathcal{J}+\mathcal{T}}\}$$
$$\times U(j_aj_aJj_a; J_0\mathcal{J})U(\tfrac{1}{2}\tfrac{1}{2}T\tfrac{1}{2}; T_0\mathcal{T})]|\{a^2(1,2)\mathcal{J}\mathcal{T}\}_A, a(3):JT\rangle, \quad \text{(III.115a)}$$

where N is the normalization constant. N can be determined by equating to unity N^2 times the sum over $(\mathcal{J}, \mathcal{T})$ of the square of the expression within the square brackets:

$$1 = N^2 \sum_{\mathcal{J},\mathcal{T}} [\delta_{\mathcal{J},J_0}\delta_{\mathcal{T},T_0} - 2\delta_{\mathcal{J},J_0}\delta_{\mathcal{T},T_0}(-1)^{j_a+1/2-J-T}\{1-(-1)^{\mathcal{J}+\mathcal{T}}\}$$
$$\times U(j_aj_aJj_a; J_0\mathcal{J})U(\tfrac{1}{2}\tfrac{1}{2}T\tfrac{1}{2}; T_0\mathcal{T}) + 2\{1-(-1)^{\mathcal{J}+\mathcal{T}}\}U^2(j_aj_aJj_a; J_0\mathcal{J})U^2(\tfrac{1}{2}\tfrac{1}{2}T\tfrac{1}{2}; T_0\mathcal{T})]$$
$$= N^2[3 - 6(-1)^{j_a+1/2-J-T}U(j_aj_aJj_a; J_0J_0)U(\tfrac{1}{2}\tfrac{1}{2}T\tfrac{1}{2}; T_0T_0)].$$

The tricky part of the foregoing summation is the handling of the phase factor $(-1)^{\mathcal{J}+\mathcal{T}}$ times the U^2-functions. This has been done by the application of (AV.13) from Appendix A to the summation over $\mathcal{J}$ of $(-1)^{\mathcal{J}}U^2(j_aj_aJj_a; J_0\mathcal{J})$ and also to the summation over $\mathcal{T}$ of $U^2(\tfrac{1}{2}\tfrac{1}{2}T\tfrac{1}{2}; T_0\mathcal{T})$. The symmetry relations (AV.6) from Appendix A of the W-function, and the conversion equation from the U-function to the W-function, namely, (AV.3) from Appendix A, are also needed in this algebraic manipulation which is left as an exercise. The normalization constant N is then given by

$$N = \frac{1}{\sqrt{3}}[1 - 2(-1)^{j_a+1/2-J-T}U(j_aj_aJj_a; J_0J_0)U(\tfrac{1}{2}\tfrac{1}{2}T\tfrac{1}{2}; T_0T_0)]^{-1/2},$$

and the fractional parentage expansion (III.115a) finally reduces to

$$|a^2(J_0T_0), a : JT) = \frac{1}{\sqrt{3}}[1 - 2(-1)^{j_a+1/2-J-T}U(j_aj_aJj_a;\ J_0J_0)U(\tfrac{1}{2}\ \tfrac{1}{2}T\tfrac{1}{2};\ T_0T_0)]^{-1/2}$$

$$\times \sum_{\mathcal{J},\mathcal{T}} [\delta_{\mathcal{J},J_0}\delta_{\mathcal{T},T_0} - (-1)^{j_a+1/2-J-T}2U(j_aj_aJj_a;\ J_0\mathcal{J})U(\tfrac{1}{2}\ \tfrac{1}{2}T\tfrac{1}{2};\ T_0\mathcal{T})]$$

$$\times |\{a^2(1, 2)\mathcal{J}, \mathcal{T}\}_A, a(3) : JT\rangle. \tag{III.115b}$$

The most important distinction of the equivalent configuration (a^3) is now stated. In the other two cases, (a^2, c) and (a, b, c), we found that any antisymmetric two-nucleon state (J_0, T_0) of (a^2) or of (a, b), when coupled to c and antisymmetrized, yields a set of three-nucleon states specified by the quantum numbers J, T; all the values of J, T consistent with angular momentum coupling rules are admissible. In the equivalent case, (a^3), this is not true. Beginning with a given (J_0T_0)-state of (a^2) and coupling the last single-particle state a to it, we may find that the moment the product is antisymmetrized, all the cfp's turn out to be zero for many combinations of J, T which are otherwise permitted by the angular momentum coupling rules. That is to say, three-nucleon states in (a^3) having such (J, T)-quantum numbers are ruled out by the Pauli principle. The situation is analogous to that observed in the two-nucleon case for the equivalent configuration (a^2).

There is a second important distinctive feature of the equivalent configuration (a^3). Suppose we use the foregoing procedure to construct an antisymmetric state for a *given* (J, T) from all the different sets of (J_0, T_0) which are consistent with angular momentum coupling; for example,

$$|(\tfrac{5}{2})^2 01;\ \tfrac{5}{2} : \tfrac{5}{2}\ \tfrac{3}{2}), \qquad |(\tfrac{5}{2})^2\ 21;\ \tfrac{5}{2} : \tfrac{5}{2}\ \tfrac{3}{2}).$$

We ask ourselves: are all these states independent of one another? The answer is, in general, "No". In the foregoing example, the two states are actually identical to each other, and this can be verified by working out their cfp's and checking that the numerical values of the two sets of cfp's are the same.

Instead of realizing all these unpleasant facts at the *end* of a detailed computing of the cfp's, it is clearly desirable to have *independent* methods of reckoning the *permissible independent* states of the equivalent configuration (a^3). We now proceed to outline such methods.

For simplicity, let us first consider the case of three neutrons or three protons. The isospin projection being $\pm\frac{3}{2}$ in these cases, the three-nucleon states we are looking for must have $T = \frac{3}{2}$. As a matter of fact, using the isospin quantum number is now an unnecessary luxury. It is enough if we merely omit the isospin part of the states and investigate the permissible J-quantum numbers occurring in completely antisymmetric states of (a^3). The single-particle state j_a has $(2j_a + 1)$ projection quantum numbers. We have to find out in how many different ways these various projection quantum numbers can be assigned to the three nucleons; in any given assignment, all the three projection quantum numbers have to differ according to the Pauli principle.

As a specific example, let us consider the case $j = \frac{5}{2}$ (we deliberately delete the subscript a to j) which has the projection quantum number $m = \pm\frac{5}{2}, \pm\frac{3}{2}, \pm\frac{1}{2}$. The assignment that makes the total projection M (sum of the three individual m-values) maximum is clearly given by $(+\frac{5}{2}, +\frac{3}{2}, +\frac{1}{2})$ with $M = \frac{9}{2}$. This state could arise from $J \geqslant \frac{9}{2}$; but we can permit only $J = \frac{9}{2}$ because were J greater than $\frac{9}{2}$, projection values M larger than $\frac{9}{2}$ would have arisen; such values are however not permitted by the Pauli principle. The next lower value of M, which is

$\frac{7}{2}$, can be had from only one assignment, viz., $(+\frac{5}{2}, +\frac{3}{2}, -\frac{1}{2})$. This projection is already contained in the state $J = \frac{9}{2}$, and hence this new assignment does not indicate the presence of any new J-state. Let us consider then the states $M = \frac{5}{2}$, which can arise from two and only two assignments, viz., $(+\frac{5}{2}, +\frac{3}{2}, -\frac{3}{2})$ and $(+\frac{5}{2}, +\frac{1}{2}, -\frac{1}{2})$. Since there are two states of projection $\frac{5}{2}$, one of them has to arise from $J = \frac{5}{2}$, whereas the other is accountable by the state $J = \frac{9}{2}$ Continuing in this manner with $M = \frac{3}{2}$ and $M = \frac{1}{2}$, we obtain the assignments

$$M = \tfrac{3}{2}: (+\tfrac{5}{2}, +\tfrac{3}{2}, -\tfrac{5}{2}), \quad (+\tfrac{3}{2}, +\tfrac{1}{2}, -\tfrac{1}{2}), \quad (+\tfrac{5}{2}, +\tfrac{1}{2}, -\tfrac{3}{2}),$$

$$M = \tfrac{1}{2}: (+\tfrac{5}{2}, +\tfrac{1}{2}, -\tfrac{5}{2}), \quad (+\tfrac{3}{2}, +\tfrac{1}{2}, -\tfrac{3}{2}), \quad (+\tfrac{5}{2}, -\tfrac{1}{2}, -\tfrac{3}{2}).$$

In the first case, the states $J = \frac{9}{2}$ and $J = \frac{5}{2}$, found earlier, account for two of the states, and the third one then demands the presence of a $(J = \frac{3}{2})$-state. In the case of $M = \frac{1}{2}$, the three possible assignments are accountable in terms of $J = \frac{9}{2}, \frac{5}{2}, \frac{3}{2}$, already found, and hence no new J-value is warranted. The same method can be applied with other j-values, and also for nucleon numbers exceeding 3.

The allowed J-values of completely antisymmetric states of identical nucleons (i.e., either neutrons or protons), prepared in the foregoing manner, are given in Table III.6; the allowed values of J for one nucleon and two nucleons are also included for the sake of completeness. For each j, the total number of nucleons n goes up to $\frac{1}{2}(2j + 1)$. The states of any n in this range are the same as those of the nucleon number $(2j + 1 - n)$. This can be understood in terms of the equivalence between hole and particle states. The nucleon number $(2j + 1 - n)$ can be interpreted as n holes in the level j, and hence the states are identical to those of n particles in the same level. The superscripts to many of the integer values of J, and to the parentheses enclosing many of the half-integer values of J, denote the number of states with that value of J occurring in the corresponding j^n-configuration. Clearly, in such cases, additional quantum numbers (to be discussed later in this section) are necessary for distinguishing the various states of the same J.

For the present, however, we are interested in only the part $n = 3$ of Table III.6. We notice that in this case there is no repetition of the same J for $j \leqslant \frac{7}{2}$; repeated states of the same J start for $j = \frac{9}{2}$ (only once for the state $J = \frac{9}{2}$), and continue to occur for $j = \frac{11}{2}$ (and also for higher values of j not included in this table).

In a shell where both neutrons and protons are present, the enumeration of states in Table III.6 is not adequate. In such instances, we have to bring in the T-quantum number also. The method we have given can now be repeated with the inclusion of the projection of isospin as well. To demonstrate the added complexity that we shall have to face, let us consider, in the case of $j = \frac{5}{2}$, the assignment $(+\frac{5}{2}, +\frac{5}{2}, +\frac{3}{2})$. As already noted, this is not allowed by the Pauli principle when isospin is not considered. Now the projection $+\frac{5}{2}$ of j can go with $+\frac{1}{2}$ and $-\frac{1}{2}$ projections of isospin, and hence $(+\frac{5}{2}^+, +\frac{5}{2}^-, +\frac{3}{2}^+)$ is an allowed assignment of projection quantum numbers having $M = \frac{13}{2}$ and $M_T = +\frac{1}{2}$. (In our new notation, the superscripts + and − denote respectively the two isospin projections $+\frac{1}{2}$ and $-\frac{1}{2}$.) It is obvious that the bookkeeping necessary for this kind of 'direct attack' would be quite cumbersome. We therefore describe an alternative method which relies essentially on standard techniques of group theory; the basic rules can, however, be stated and understood in an intuitive manner.

While explaining the alternative method, we shall consider isospin states and space-spin states separately and then suitably combine them such that the complete state is antisymmetric. Let us denote the isospin or the state j, m of a single nucleon by a single box (a) in Fig. III.9.

Table III.6 List of total angular momentum J of antisymmetric states in configuration j^n of identical nucleons (i.e., either neutrons or protons)

j	n	J
	1	$\frac{1}{2}$
$\frac{1}{2}$	2	0
	1	$\frac{3}{2}$
$\frac{3}{2}$	2	0, 2
	1	$\frac{3}{2}$
$\frac{3}{2}$	2	0, 2, 4
	3	$\frac{3}{2}, \frac{5}{2}, \frac{9}{2}$
	1	$\frac{7}{2}$
	2	0, 2, 4, 6
$\frac{7}{2}$	3	$\frac{3}{2}, \frac{5}{2}, \frac{7}{2}, \frac{9}{2}, \frac{11}{2}, \frac{15}{2}$
	4	$0, 2^2, 4^2, 5, 6, 8$
	1	$\frac{9}{2}$
	2	0, 2, 4, 6, 8
$\frac{9}{2}$	3	$\frac{3}{2}, \frac{5}{2}, \frac{7}{2}, (\frac{9}{2})^2, \frac{11}{2}, \frac{13}{2}, \frac{15}{2}, \frac{17}{2}, \frac{21}{2}$
	4	$0^2, 2^2, 3, 4^3, 5, 6^3, 7, 8^2, 9, 10, 12$
	5	$\frac{1}{2}, \frac{3}{2}, (\frac{5}{2})^2, (\frac{7}{2})^2, (\frac{9}{2})^3, (\frac{11}{2})^2, (\frac{13}{2})^2, (\frac{15}{2})^2, (\frac{17}{2})^2, \frac{19}{2}, \frac{21}{2}, \frac{25}{2}$
	1	$\frac{11}{2}$
	2	0, 2, 4, 6, 8, 10
	3	$\frac{3}{2}, \frac{5}{2}, \frac{7}{2}, (\frac{9}{2})^2, (\frac{11}{2})^2, \frac{13}{2}, (\frac{15}{2})^2, \frac{17}{2}, \frac{19}{2}, \frac{21}{2}, \frac{23}{2}, \frac{27}{2}$
$\frac{11}{2}$	4	$0^2, 2^2, 3, 4^4, 5^2, 6^4, 8^4, 9^2, 10^3, 11, 12^3, 13, 14, 16$
	5	$\frac{1}{2}, (\frac{3}{2})^2, (\frac{5}{2})^3, (\frac{7}{2})^4, (\frac{9}{2})^4, (\frac{11}{2})^5, (\frac{13}{2})^4, (\frac{15}{2})^5, (\frac{17}{2})^4, (\frac{19}{2})^4, (\frac{21}{2})^3, (\frac{23}{2})^3, (\frac{25}{2})^2, (\frac{27}{2})^2, \frac{29}{2}, \frac{31}{2}, \frac{35}{2}$
	6	$0^3, 2^4, 3^3, 4^6, 5^3, 6^7, 7^4, 8^6, 9^4, 10^5, 11^2, 12^4, 13^2, 14^2, 15, 16, 18$

43 (45-122/1977)

Clearly, this box corresponds to isospin $\frac{1}{2}$ and angular momentum j. Then consider the product of two single nucleon wavefunctions which, as we know, can be split up into parts that are symmetric and antisymmetric under the exchange of the two nucleons. This split is represented by (b) in Fig. III.9. By our convention, the two boxes in the same row represent the symmetric state, and the two boxes in the same column represent the antisymmetric state. The symmetric diagram for the case of isospin represents $T = 1$, whereas for the space-spin function it stands for $J = 1, 3, \ldots, 2j$. Similarly, the antisymmetric diagram corresponds to $T = 0$ for isospin and $J = 0, 2, \ldots, (2j - 1)$ for space-spin.

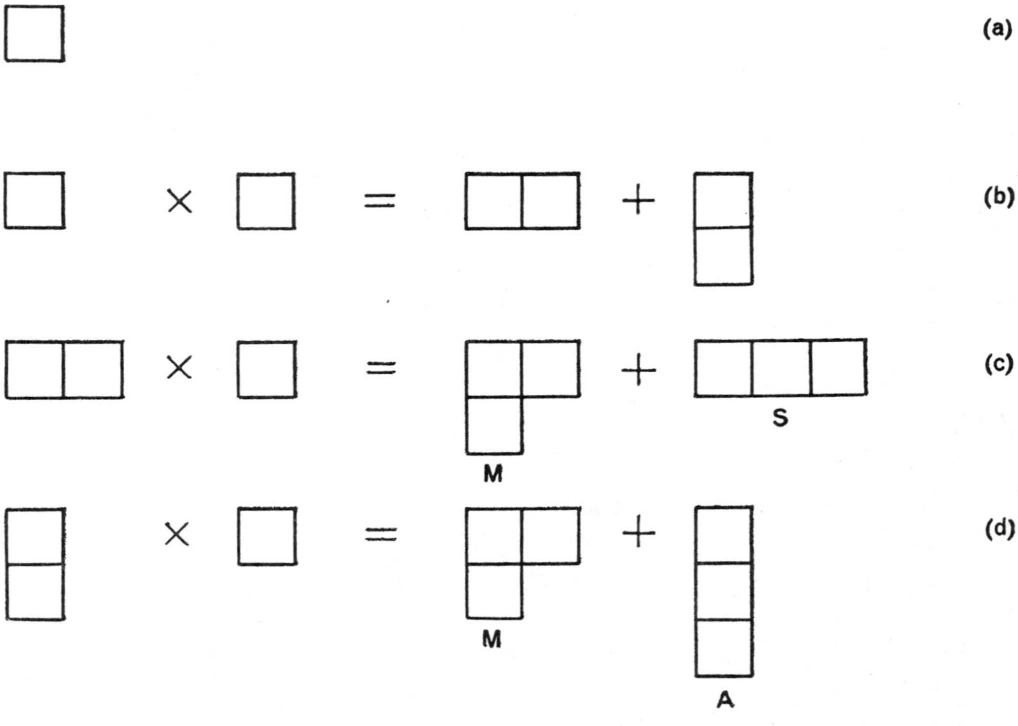

Fig. III.9 Young's tableau diagrams for two- and three-particle system.

Now let us consider the process of multiplying the two-nucleon symmetric and antisymmetric functions by the wavefunction of the third nucleon; these two cases are represented by (c) and (d) respectively in Fig. III.9. When we multiply the symmetric two-nucleon function by the wavefunction of the third nucleon, it should be possible to symmetrize it with respect to all the three nucleons by multiplying with $(1 + P_{13} + P_{12})$, and this completely symmetric function is represented by the three boxes in one row, labelled S in Fig. III.9. It is impossible to produce a completely antisymmetric function of three nucleons from the aforementioned product by multiplying with $(1 - P_{12} - P_{13})$ because the first two nucleons are already present in a symmetric state and the operation of this antisymmetrizer produces a zero result. In a similar way, if we consider multiplying the antisymmetric two-nucleon function by the wavefunction of the third nucleon, it is possible, in general, to produce a completely antisymmetric function of three nucleons by an antisymmetrizer (already explicitly demonstrated earlier in this section). However, it is impossible to produce a completely symmetric function from such a product because nucleons 1 and 2 in the product are already antisymmetric. In Fig. III.9, the completely antisymmetric function is represented by the three boxes in the same column, labelled A. The extra pattern, labelled M, comprising two boxes in the first row and one box in the second row, which appears both in (c) and (d) of Fig. III.9 cannot, however, be understood so intuitively. All we can deduce is that the product of a two-nucleon symmetric function with a wavefunction of the third nucleon cannot be identically equal to the completely symmetric function of the three nucleons; there must be some other type of function occurring along with it. Similar observations hold for the right-hand side of (d) in Fig. III.9. The fact that the symmetry character of this extra function in (c) and (d) is the same cannot be established intuitively; neither can we assume that there will be no function of any other symmetry character in these results. These are precisely the points clarified by some knowledge of the theory of groups, namely, the group of all permutation operators acting on the nucleons. Without going into the group theoretical details, we now simply state the working rules. The rule for multiplying any pattern by a single box is to attach this extra box to that pattern in as many different ways as possible, keeping in mind that the number of boxes in any row should not exceed that in any of the preceding rows. Further, the total number of rows in the resultant patterns must not exceed $(2j+1)$ in the case of space-spin functions, and 2 in the case of isospin.

If we keep the working rules in mind, then, in the case of isospin states, the completely antisymmetric pattern A is ruled out because it consists of three rows. The other pattern, labelled M in (d) of Fig. III.9, will be loosely called 'mixed symmetry'. In the case of isospin, therefore, only the pattern M remains on the right-hand side of (d) in Fig. III.9, and the three-nucleon isotopic spin T, corresponding to this diagram, must be obtainable from the isotopic spin content of the left-hand side. The latter corresponds to an isospin zero of the antisymmetric two-nucleon diagram and isospin $\frac{1}{2}$ of the single box; the product implies that all total isospins available from the coupling of the two must be present in the result. The coupling in this case produces a unique value $T=\frac{1}{2}$ for the left-hand side, and hence the same value of T is associated with the mixed symmetry pattern M. In a similar manner, we couple the isospin 1 of the symmetric two-nucleon diagram, and $\frac{1}{2}$ for a single box, and obtain $T=\frac{1}{2}$ and $T=\frac{3}{2}$ as the isospins on the left-hand side of (c) in Fig. III.9. By this argument, the same isospins must occur on the right-hand side also. But the pattern M has already been shown to contain only $T=\frac{1}{2}$; therefore, the pattern S contains only $T=\frac{3}{2}$.

In a manner analogous to reckoning the isospin T, we can try to find out the permitted J-values in the patterns M, S, and A of Fig. III.9 for various values of j. If $j=\frac{1}{2}$, the pattern A can be dropped as in the case of isospin, and we get the pattern M going with $J=\frac{1}{2}$, and the pattern S with $J=\frac{3}{2}$. If $j=\frac{3}{2}$, then the three nucleons in this level are equivalent to a single hole, and hence the permitted J-value for the completely antisymmetric case is, by the equivalence of a hole and a particle, given by $J=j=\frac{3}{2}$. So the pattern A goes with $J=\frac{3}{2}$. The total angular momenta on the left-hand side of (d) in Fig. III.9 are obtained from the coupling of 0 and 2 (which are contained in the antisymmetric two-nucleon pattern) with $\frac{3}{2}$ (corresponds to the single box), i.e., $J=\frac{3}{2}, \frac{1}{2}, \frac{3}{2}, \frac{5}{2}, \frac{7}{2}$. Since the pattern A contains $J=\frac{3}{2}$, we take away this value from the J's just listed, and hence associate the pattern M with $J=\frac{1}{2}, \frac{3}{2}, \frac{5}{2}, \frac{7}{2}$. It should be noticed that in preparing the list of J we must be careful to keep any repeated value of J as many times as it is produced (in the present case, $\frac{3}{2}$ occurred twice—once in the coupling of 0 with $\frac{3}{2}$ and once in the coupling of 2 with $\frac{3}{2}$—and we retained it twice). The same method for the determination of the allowed J-values of mixed symmetry can be applied to higher values of j, in which case we shall have to use Table III.6 (the $n = 3$ rows) to eliminate the J-values of the completely antisymmetric pattern. Having obtained the J's contained in the mixed symmetry pattern, we can now proceed to (c) in Fig. III.9, find out the list of angular momentum J by angular momentum coupling on the left-hand side, and finally remove the J's of the pattern M from this list to obtain the allowed J-values of the symmetric three-nucleon diagram S. The J-structure of the symmetric pattern is, however, unnecessary for our purpose, as shown later in this section.

Let us next consider the construction of antisymmetric states from the product of isospin and space-spin functions. We have specified the symmetry characteristic of these two types of function *separately*, and also found out the allowed T- and J-values corresponding to each symmetry pattern. When we take their product, we can, however, admit only products that are completely antisymmetric under the combined exchange of space, spin, and isospin coordinates of any two nucleons. In the case of the completely antisymmetric space-spin function A, it is clear that it must be multiplied by the completely symmetric isospin function S, which has $T = \frac{3}{2}$, in order that the product is antisymmetric under the combined exchange of space-spin and isospin. It is, however, not intuitively clear what is to be done for the mixed symmetry functions M. Once again, group theory gives the answer, and the rule can be very simply stated: the space-spin function of any symmetry pattern must be multiplied by the isospin function of the *conjugate* symmetry pattern (which has to be obtained from the former by the interchange of the rows and columns) The mixed symmetry pattern is clearly the conjugate of itself. Thus, the mixed symmetry isospin state $T = \frac{1}{2}$ must be multiplied by the mixed symmetry space-spin functions whose J-values have already been determined. It should be observed that the conjugate of the completely symmetric pattern is the completely antisymmetric pattern, and hence the general group-theoretical rule actually includes the statement made about the ($T = \frac{3}{2}$)-states. By the same rule, the completely symmetric space-spin state would have to be multiplied by the completely antisymmetric isospin state, but we have already found that this pattern does not exist for isospin. Hence, even though the J-structure of the completely symmetric space-spin functions can be determined in principle, it is entirely unnecessary.

Let us now ask why the completely antisymmetric isospin state of three nucleons is non-existent. The answer is very simple. The single-particle isospin state has two possible projections, $\pm\frac{1}{2}$. If we have more than two particles, then more than one particle must necessarily be given the same isospin projection quantum number. An attempt to antisymmetrize such an isospin wavefunction with respect to the nucleons that have the same projection would clearly produce a zero result. Thus, it is impossible to produce a completely antisymmetric isospin state of more than two nucleons. This explanation throws light also on the requirement that there cannot be more than two rows in an isospin symmetry pattern because the existence of more than two rows implies that at least one column has more than two boxes, and a completely antisymmetric isospin state represented by that column is impossible.

Arguments similar to the foregoing can be advanced for the space-spin functions as well. A single-particle state j has $(2j + 1)$ projection quantum numbers. As long as we are dealing with a nucleon number less than or equal to $(2j + 1)$, we can put all the nucleons in *different* projection states of j, and hence produce a completely antisymmetric function by a suitable antisymmetrizer. The moment we have more than $(2j + 1)$ nucleons, more than one nucleon has to be put in the same projection state, and then an attempt at antisymmetrization necessarily produces zero. This explains why we cannot admit columns of more than $(2j + 1)$ boxes [i.e., more than $(2j + 1)$ rows] in the symmetry pattern of space-spin functions.

We have already listed (see Table III.6) the J-values of the completely antisymmetric states. For three nucleons, these states correspond to $T = \frac{3}{2}$ (completely symmetric in isospin), and can be read from the $(n = 3)$ rows of the table. The preceding discussion enables us to find out, for three equivalent nucleons, the allowed J-values for mixed symmetry that have $T = \frac{1}{2}$. These states for several values of j are listed in Table III.7.

Table III.7 J-values of mixed symmetry states ($T = \frac{1}{2}$) of three nucleons in level j

j	J
$\frac{3}{2}$	$\frac{1}{2}, \frac{3}{2}, \frac{5}{2}, \frac{7}{2}$
$\frac{5}{2}$	$\frac{1}{2}, \frac{3}{2}, (\frac{5}{2})^2, (\frac{7}{2})^2, \frac{9}{2}, \frac{11}{2}, \frac{13}{2}$
$\frac{7}{2}$	$\frac{1}{2}, \frac{3}{2}, (\frac{5}{2})^2, (\frac{7}{2})^3, (\frac{9}{2})^2, (\frac{11}{2})^2, (\frac{13}{2})^2, \frac{15}{2}, \frac{17}{2}, \frac{19}{2}$
$\frac{9}{2}$	$\frac{1}{2}, \frac{3}{2}, (\frac{5}{2})^2, (\frac{7}{2})^3, (\frac{9}{2})^3, (\frac{11}{2})^3, (\frac{13}{2})^3, (\frac{15}{2})^2, (\frac{17}{2})^2, (\frac{19}{2})^2, \frac{21}{2}, \frac{23}{2}, \frac{25}{2}$
$\frac{11}{2}$	$\frac{1}{2}, \frac{3}{2}, (\frac{5}{2})^2, (\frac{7}{2})^3, (\frac{9}{2})^3, (\frac{11}{2})^4, (\frac{13}{2})^4, (\frac{15}{2})^3, (\frac{17}{2})^3, (\frac{19}{2})^3, (\frac{21}{2})^2, (\frac{23}{2})^2, (\frac{25}{2})^2, \frac{27}{2}, \frac{29}{2}, \frac{31}{2}$

Seniority and reduced isotopic spin of three-nucleon states The concept of seniority of shell-model states of equivalent nucleons is based on the addition of a specially defined 'pair' of nucleons. Such a pair is, by definition, coupled to the angular momentum state $J_P = 0$; for the antisymmetry of this 'paired' state, its isotopic spin has necessarily to be $T_P = 1$. In the case of identical nucleons, the isotopic spin of a pair is automatically unity and may be altogether omitted from the discussion.

In the case of two nucleons in the level j, the quantum numbers of the states are known. They are

$$J = 0, 2, 4, \ldots, (2j - 1) \quad \text{with} \quad T = 1,$$
$$J = 1, 3, 5, \ldots, 2j \quad \text{with} \quad T = 0.$$

In the first group, the state $J = 0$ $(T = 1)$ is identical to the special pair defined. This two-nucleon state can therefore be regarded as having been obtained by the addition of one specially defined pair to the state of *zero* nucleons. We shall call it a state of seniority *zero* $(v = 0)$, where seniority stands for the number of 'unpaired' nucleons. All the other two-nucleon states are of seniority *two* $(v = 2)$ because in these states there are *two* unpaired nucleons, a pair being reckoned in the special sense, as just mentioned. The two-nucleon states of seniority two have two different values of isotopic spin, namely, $T = 1$ and $T = 0$.

In the case of three equivalent nucleons, we can (as with the two-nucleon states) consider obtaining a special set of states by adding one pair to the state $(j, t = \frac{1}{2})$ of a single nucleon. The resultant state clearly has a total angular momentum $J = j$ because $J_P = 0$; on the other hand, the total isotopic spin T of such a state is the result of coupling $t = \frac{1}{2}$ with $T_P = 1$, i.e., $T = \frac{1}{2}$ or $T = \frac{3}{2}$. Both these states are said to have a *reduced* isotopic spin $t = \frac{1}{2}$, which is, by definition, equal to the isotopic spin of the *parent* state (in this case, the one-nucleon state) to which a pair has been added to obtain the resultant state. The states obtained in this manner have one unpaired nucleon and a pair, and their seniority, which is equal to the number of unpaired nucleons, is then equal to one ($v = 1$). Instead of defining seniority as the number of unpaired nucleons in a state, we can alternatively define it as the number of nucleons in the *parent* state to which ($J_P = 0$) pairs have been added to obtain the final state. In the case of three identical nucleons (i.e., only neutrons or only protons), T has the unique value $\frac{3}{2}$, and hence there is only one state $J = j$ of seniority one.

If we now consider the ($n = 3$) rows of the list in Table III.6, which refers to identical nucleons, we notice that up to $j = \frac{7}{2}$ there is only one state with $J = j$. This single state can then be constructed in the manner already described. All the other three-nucleon states cannot be obtained by adding a pair to the single-nucleon state because they have $J \neq j$. In these states, therefore, all the three nucleons are unpaired and have, by definition, seniority $v = 3$. Now, examining the case of $j = \frac{9}{2}$ in Table III 6, we have two states with the same $J = j$. One of these states can clearly be obtained by adding a pair to a single nucleon $j = \frac{9}{2}$, which then has seniority $v = 1$. The other state with $J = \frac{9}{2}$, *which can be made orthogonal* to the state ($v = 1$, $J = \frac{9}{2}$) (the process of orthogonalization is indicated presently), is then called a state of seniority $v = 3$. The other states with $J \neq \frac{9}{2}$ in the list are also of seniority three, according to our earlier definition. They are *automatically* orthogonal to the state of seniority one because they have $J \neq j$.

The case of $j = \frac{11}{2}$ (in Table III.6) is more complex. Here the two states of $J = \frac{11}{2}$ are clearly of seniority one and three, and both of them can be constructed in the manner indicated for the ($j = \frac{9}{2}$)-case. All the other states are of seniority three, but there are two values of J, namely, $\frac{9}{2}$ and $\frac{15}{2}$, for which there are two states with the same value of J. In each of these cases then, seniority fails to distinguish between the two states of the same J. This is the complexity just mentioned.

Now, while describing the general procedure for constructing the three-nucleon state of good seniority, we shall show how two states of the same J and seniority can be explicitly constructed and used in calculations. The expression (III.115b) has to be generally used for writing the states. When there is only one state with the given value of J ($T = \frac{3}{2}$ in the case under consideration), this state can be constructed from any value of J_0 ($T_0 = 1$) consistent with the angular momentum coupling rule. In such a case, it is automatically guaranteed that, if we use two different values of J_0 in constructing the state, the resultant fractional parentage expansion (III.115b) differs, at the most, through an overall phase factor. In the case of more than one state with the same J, we have to use different values of J_0 in setting up the fractional parentage expansion (III.115b). If we deal with two states, then two different values of J_0 suffice. If we deal with two states of $J = j$, then we know that they are of seniority one and three, respectively. The states with seniority one can be obtained from (III.115b) by using $J_0 = 0$ (recall the definition of a seniority-one state), and the other state obtained with a different J_0 is not, in general, automatically orthogonal to this (seniority-one, $J = j$)-state. However, it can be made orthogonal by following a general procedure: take any two normalized states $|P\rangle$ and $|Q\rangle$, where $|Q\rangle$ is not orthogonal to $|P\rangle$; $|\bar{Q}\rangle$, as now defined, can be easily checked to be orthogonal to $|P\rangle$:

$$|\bar{Q}\rangle = |Q\rangle - |P\rangle\langle P \mid Q\rangle.$$

$|\bar{Q}\rangle$, however, is not normalized to unity, but can be made so very easily. The same procedure can be used in dealing with the two states $J = \frac{9}{2}$ and $J = \frac{15}{2}$ arising from $j = \frac{11}{2}$. Two states have to be constructed according to (III.115b) by using two different values J_0, each of which is not necessarily equal to zero. These two states can then be orthogonalized in the manner just described.

The procedure for constructing the states of mixed symmetry, listed in Table III.7, is now described. All these states have $T = \frac{1}{2}$. When we wish to construct a $(J = j, T = \frac{1}{2})$-state of seniority one and reduced isotopic spin $\frac{1}{2}$, we must choose $J_0 = 0$, $T_0 = 1$ in (III.115b). In the case of $J = \frac{5}{2}$, the other $(J = \frac{5}{2})$-state is clearly of seniority three and can be constructed by choosing any $J_0 \neq 0$ and the commensurate value of T_0. In all the other cases in this table, where the same J occurs more than once, it is evident that the seniority quantum number by itself is not adequate to distinguish between the states. The reduced isotopic spin also does not help because, for the seniority-one state, it has the unique value $t = \frac{1}{2}$. Then the procedure for obtaining the different states is to use different values of J_0 and T_0 in (III.115b) and in the subsequent orthogonalization of the states. The orthogonalization procedure, described in the case of two states, can be very easily generalized to the case of more than two states.

Calculation of Matrix Elements

The advantage of the fractional parentage expansion of a state is now made explicit by working out the expressions for the matrix elements of one- or two-particle type operators using these expansions. The earlier notation is made simpler before writing the desired expressions. We now denote the n-nucleon states by $|\Psi_1\rangle, |\Psi_2\rangle, \ldots$, and the $(n-1)$-nucleon states by $|\overline{\Psi}_1\rangle, |\overline{\Psi}_2\rangle, \ldots$. Thus, the fractional parentage expansion of an n-particle state, which is an expansion in terms of the states of the $(n-1)$ nucleons, coupled to the state of the last nucleon, can be schematically written as

$$|\Psi_1) = \sum_{\overline{\Psi}_1, j_1} C_1(\overline{\Psi}_1, j_1)|\overline{\Psi}_1(1, 2, \ldots, n-1), j_1(n) : \Psi_1\rangle, \tag{III.116a}$$

$$|\Psi_2) = \sum_{\overline{\Psi}_2, j_2} C_2(\overline{\Psi}_2, j_2)|\overline{\Psi}_2(1, 2, \ldots, n-1), j_2(n) : \Psi_2\rangle. \tag{III.116b}$$

As usual, the parentheses on the left-hand side signify that these states are antisymmetric in all the n nucleons; C_1, C_2 are the cfp's which depend on the quantum numbers of the states shown within the parentheses; $\overline{\Psi}_1$, $\overline{\Psi}_2$ are states of the nucleons $1, 2, \ldots, n-1$, as shown; and the last nucleon n is in a single-particle state specified by $j_1, j_2, \ldots$. The comma separating $\overline{\Psi}_1$ from j_1 denotes coupling of angular momentum and isotopic spin. The angular brackets on the states on the right-hand side indicate that there is only angular momentum and isotopic spin coupling of the state of the last nucleon with that of the first $(n-1)$ nucleons; that is to say, there is no antisymmetry in *each* basis state on the right-hand side under the exchange P_{in}, where $i = 1, 2, \ldots, n-1$. In fact, the *full* expansion with the cfp's, by definition, builds up this antisymmetry requirement and makes $|\Psi_1)$ or $|\Psi_2)$ completely antisymmetric.

Let F and G denote a one-particle type and a two-particle type operator, respectively. According to the general results (III.60), we have

$$(\Psi_1| F |\Psi_2) = n(\Psi_1| F(n) |\Psi_2), \tag{III.117}$$

$$(\Psi_1| G |\Psi_2) = \tfrac{1}{2}n(n-1)(\Psi_1| G(1, 2) |\Psi_2) \tag{III.118a}$$

$$= \tfrac{1}{2}n(n-1)(\Psi_1| G(n-1, n) |\Psi_2). \tag{III.118b}$$

In the three-nucleon case, we shall make use of (III.118a) for the matrix element of G. In the case of $n > 3$, it is more advantageous to use the alternative expression (III.118b), although (III.118a) can also be applied.

We substitute the fractional parentage expansions (III.116) in (III.117), and note that the operator belongs to the particle n, and hence, in the resultant summations over $\overline{\Psi}_1$ and $\overline{\Psi}_2$, which are states of the first $(n-1)$ nucleons, only the terms with $\overline{\Psi}_1 = \overline{\Psi}_2$ contribute a non-zero result. In this way,

$$(\Psi_1|\, F\, |\Psi_2) = n \sum_{\overline{\Psi}, j_1, j_2} C_1(\overline{\Psi}, j_1) C_2(\overline{\Psi}, j_2) \times \langle \overline{\Psi}(1, 2, \ldots, n-1), j_1(n) : \Psi_1|\, F(n)\, |\overline{\Psi}(1, 2, \ldots, n-1), j_2(n) : \Psi_2\rangle. \tag{III.119a}$$

The matrix element is now in the form of the standard Racah result (BIII.13) from Appendix B, and can therefore be expressed in terms of the single-particle matrix element $\langle j_1|\,|F|\,|j_2\rangle$. To write out the final expression in detail, we now need the explicit symbols $(\overline{J}, \overline{T})$, which denote the angular momentum and isotopic spin quantum numbers of the $(n-1)$-nucleon state $\overline{\Psi}$; (J_1, T_1) and (J_2, T_2), which denote the angular momentum and isotopic spin quantum numbers for the n-nucleon states Ψ_1 and Ψ_2, respectively; and the detailed structure of the operator F. Let F be a product of a tensor F_q^K acting on the angular momentum states, and $\mathcal{T}_{q'}^{K'}$ acting on the isotopic spin states. Introducing all the notation explicitly, and applying the standard results (BIII.1) and (BIII.13) from Appendix B, we finally obtain

$$(\Psi_1|\, F\, |\Psi_2) = n \sum_{\overline{\Psi}, j_1, j_2} C_1(\overline{\Psi}, j_1) C_2(\overline{\Psi}, j_2) \begin{bmatrix} J_2 & K & J_1 \\ M_2 & q & M_1 \end{bmatrix} \begin{bmatrix} T_2 & K' & T_1 \\ \tau_2 & q' & \tau_1 \end{bmatrix} \times U(\overline{J} j_2 J_1 K; J_2 j_1) U(\overline{T}\tfrac{1}{2} T_1 K'; T_2 \tfrac{1}{2}) \langle j_2|\,|F^K|\,|j_1\rangle \langle \tfrac{1}{2}|\,|\mathcal{T}^{K'}|\,|\tfrac{1}{2}\rangle. \tag{III.119b}$$

Here M_1, M_2, τ_1, and τ_2 are given projection quantum numbers corresponding to J_2, J_1, T_2, and T_1, respectively.

In the case of the two-body operator, we write the result for $n = 3$. We exactly follow the foregoing procedure, starting with the expression (III.118a), and obtain

$$(\Psi_1|\, G\, |\Psi_2) = 3 \sum_{\overline{\Psi}_1, \overline{\Psi}_2, j} C_1(\overline{\Psi}_1, j) C_2(\overline{\Psi}_2, j) \times \langle \overline{\Psi}_1(1, 2), j(3) : \Psi_1|\, G(1, 2)\, |\overline{\Psi}_2(1, 2), j(3) : \Psi_2\rangle. \tag{III.120a}$$

In this case, the operator is independent of the last particle, and hence the matrix element is diagonal with respect to j_1 and j_2; we have denoted the common value by j. Once again, the matrix element of $G(1, 2)$ in the final expression is clearly in the standard Racah form (BIII.14) from Appendix B, and can be easily evaluated in terms of $\langle \overline{\Psi}_1|\,|G|\,|\overline{\Psi}_2\rangle$. The last quantity is a two-body matrix element, whose evaluation has been shown in detail in Section 28. The detailed expression, analogous to (III.119b), will be written only where G is the two-body potential V. In this case, the operator has a rank zero in both space-spin and isotopic spin space, and hence the Clebsch-Gordon coefficients, the U-functions, and the phase factors in the standard Racah result all reduce to unity; there is no difference between the reduced matrix element $\langle \overline{\Psi}_1|\,|V|\,|\overline{\Psi}_2\rangle$ and the full matrix element $\langle \overline{\Psi}_1|\, V\, |\overline{\Psi}_2\rangle$. In this way,

$$(\Psi_1|\, V\, |\Psi_2) = 3 \sum_{\overline{\Psi}_1, \overline{\Psi}_2, j} C_1(\overline{\Psi}_1, j) C_2(\overline{\Psi}_2, j) \langle \overline{\Psi}_1(1, 2)|\, V(1, 2)\, |\overline{\Psi}_2(1, 2)\rangle. \tag{III.120b}$$

The cfp's of all the three-nucleon states can be picked out of the expressions (III.113b), (III.114), and (III.115b). Therefore, (III.119b) with $n = 3$, and (III.120b) enable us to evaluate all the necessary matrix elements in the case of three nucleons for a shell-model calculation.

An evaluation of the two-body type matrix element for $n > 3$ is now detailed. When we use (III.118b), it becomes clear that (III.120a) gets replaced by

$$(\Psi \qquad \Psi_2)$$
$$= \tfrac{1}{2}n(n-1) \sum_{\overline{\Psi}_1, \overline{\Psi}_2, j_1, j_2} C_1(\overline{\Psi}_1, j_1)C_2(\overline{\Psi}_2, j_2)$$
$$\times \langle \overline{\Psi}_1(1, 2, \ldots, n-1), j_1(n) : \Psi_1 |\; G(n-1, n)\; | \overline{\Psi}_2(1, 2, \ldots, n-1), j_2(n) : \Psi_2 \rangle. \tag{III.121}$$

The operator now belongs to the last two nucleons, and hence we need another fractional parentage expansion of the states $\overline{\Psi}_1$, $\overline{\Psi}_2$ of $(n-1)$ nucleons, which will separate the wave-function of the first $(n-2)$ nucleons from the $(n-1)$-th nucleon. Let these expansions be written as

$$|\overline{\Psi}_1) = \sum_{\overline{\Psi}'_1, j'_1} d_1(\overline{\Psi}'_1, j'_1)|\overline{\Psi}'_1(1, 2, \ldots, n-2), j'_1(n-1) : \overline{\Psi}_1\rangle, \tag{III.122a}$$

$$|\overline{\Psi}_2) = \sum_{\overline{\Psi}'_2, j'_2} d_2(\overline{\Psi}'_2, j'_2)|\overline{\Psi}'_2(1, 2, \ldots, n-2), j'_2(n-1) : \overline{\Psi}_2\rangle. \tag{III.122b}$$

Substituting these expressions in (III.121), we obtain

$$(\Psi_1 |\; G\; | \Psi_2)$$
$$= \tfrac{1}{2}n(n-1) \sum_{\overline{\Psi}_1, \overline{\Psi}_2, j_1, j_2} \; \sum_{\overline{\Psi}'_1, \overline{\Psi}'_2, j'_1, j'_2} C_1(\overline{\Psi}_1, j_1)C_2(\overline{\Psi}_2, j_2)d_1(\overline{\Psi}'_1, j'_1)d_2(\overline{\Psi}'_2, j'_2)$$
$$\times \langle \overline{\Psi}'_1(1, 2, \ldots, n-2), j'_1(n-1) : \overline{\Psi}_1, j_1(n) : \Psi_1 |\; G(n-1, n)\; | \overline{\Psi}'_2(1, 2, \ldots, n-2),$$
$$j'_2(n-1) : \overline{\Psi}_2, j_2(n) : \Psi_2\rangle. \tag{III.123}$$

To proceed, we need the information on the angular momentum and isotopic spin of the states $\overline{\Psi}'_1$ and $\overline{\Psi}'_2$, which we denote by $\bar{J}'_1$, $\bar{T}'_1$, and $\bar{J}'_2$, $\bar{T}'_2$. Using the U-function to change the order of coupling, we then write

$$|\overline{\Psi}'_1(1, 2, \ldots, n-2), j'_1(n-1) : \overline{\Psi}_1, j_1(n) : \Psi_1\rangle$$
$$= \sum_{\mathcal{J}_1, \mathcal{T}_1} U(\bar{J}'_1 j'_1 J_1 j_1;\; \bar{J}_1 \mathcal{J}_1) U(\bar{T}'_1 \tfrac{1}{2} T_1 \tfrac{1}{2};\; \bar{T}_1 \mathcal{T}_1)$$
$$\times |\overline{\Psi}_1(1, 2, \ldots, n-2), \{j'_1(n-1), j_1(n) : \mathcal{J}_1 \mathcal{T}_1\} : \Psi_1\rangle.$$

Here the braces indicate the coupling of the states of the last two nucleons to $\mathcal{J}_1\mathcal{T}_1$ and the subsequent coupling of $\overline{\Psi}_1$ of the first $(n-2)$ nucleons to form the final state Ψ_1. An analogous expression can be written for the state on the right-hand side of $G(n-1, n)$ in (III.123). Substituting these expressions in (III.123) and then noting that the matrix element in the new expression is in the standard form (BIII.13) of Appendix B, we obtain

$$(\Psi_1 |\; G\; | \Psi_2) = \tfrac{1}{2}n(n-1) \sum_{\overline{\Psi}_1, \overline{\Psi}_2, j_1, j_2} \; \sum_{\overline{\Psi}', j'_1, j'_2} C_1(\overline{\Psi}_1, j_1)C_2(\overline{\Psi}_2, j_2)d_1(\overline{\Psi}', j'_1)d_2(\overline{\Psi}', j'_2)$$
$$\times \sum_{\mathcal{J}\mathcal{T}} U(\bar{J}' j'_1 J j_1;\; \bar{J}_1 \mathcal{J}) U(\bar{T}' \tfrac{1}{2} T \tfrac{1}{2};\; \bar{T}_1 \mathcal{T}) U(\bar{J}' j'_2 J j_2;\; \bar{J}_2 \mathcal{J}) U(\bar{T}' \tfrac{1}{2} T \tfrac{1}{2};\; \bar{T}_2 \mathcal{T})$$
$$\times \langle j'_1(n-1), j_1(n) : \mathcal{J}\mathcal{T} |\; G(n-1, n)\; | j'_2(n-1), j_2(n) : \mathcal{J}\mathcal{T}\rangle. \tag{III.124}$$

This final expression has been written on the assumption that the operator G is a scalar, i.e., the result is applicable to the two-body potential. All the equalities $J_1 = J_2 = J$, $T_1 = T_2 = T$, $\mathcal{J}_1 = \mathcal{J}_2 = \mathcal{J}$, and $\mathcal{T}_1 = \mathcal{T}_2 = \mathcal{T}$, used in writing (III.124), are actually attributable to this special assumption on the rank of the operator G. The equality $\overline{\Psi}_1' = \overline{\Psi}_2' = \overline{\Psi}'$, and the equalities of the corresponding quantum numbers, namely, $\bar{J}_1' = \bar{J}_2' = \bar{J}'$ and $\bar{T}_1' = \bar{T}_2' = \bar{T}'$, are however quite general and follow from the fact that the operator $G(n-1, n)$ belongs to the last two nucleons, whereas the wavefunctions $\overline{\Psi}_1'$, $\overline{\Psi}_2'$ belong to the first $(n-2)$ nucleons.

Before closing this section, it may be worthwhile to draw attention to a particular fact: in the fractional parentage expansion of a state of n nucleons, we get the states $\overline{\Psi}$ of the first $(n-1)$ nucleons coupled to the state j of the last nucleon, and the summation is, *in general*, over both $\overline{\Psi}$ and j; in the special case of equivalent nucleons (i.e., all in one shell-model orbital), the last nucleon in the fractional parentage break-up is necessarily in the same given orbital, and the summation is therefore only over all possible states $\overline{\Psi}$ of the first $(n-1)$ nucleons. The points just mentioned will be appreciated if the reader refers to the expression (III.115b), valid for the configuration (a^3) of three equivalent nucleons, and the expressions (III.113b) and (III.114), which apply to the case of three non-equivalent nucleons distributed over mixed configurations of the type (a^2, b) and (a, b, c).

B. MORE THAN THREE NUCLEONS

The detailed shell-model calculation for a given number of nucleons distributed in several orbitals entails setting up the matrix for the Hamiltonian by using the angular momentum coupled states as the basis; the resultant eigenvalues should be compared with the observed spectrum, and the eigenstates can be used in calculating other observed quantities, such as static electromagnetic moments and dynamic transition matrix elements. The general results already given in the calculation of the one- and two-body type matrix element are clearly sufficient for carrying out this part of the programme.

It is clear then that the first step in a shell-model calculation for any number of nucleons distributed over several single-particle levels is to find out all the configurations and the possible states (their seniority, J, and T, and other quantum numbers, if necessary) in the various configurations; the second task entails obtaining the fractional parentage expansion of each n-particle state in terms of the states of $(n-1)$ particles coupled to the state of the last particle. Once these two steps are executed, the actual shell-model diagonalization programme and the calculation of the other properties can be carried out in a straightforward manner. The complexity of the programme is contained entirely in the two steps just mentioned. We have shown (in Section 29A) the procedure for the entire programme in the case of three nucleons. That the complexity increases with an increase in the number of nucleons and in the number of available single-particle orbitals (which is so in the big major shells) can be very easily appreciated.

As a precursor to the problems, let us consider first the case of four nucleons distributed amongst four or more orbitals. We immediately notice that the configurations can be of five types, namely, a^4, (a^3, b), (a^2, b, c), (a^2, b^2), and (a, b, c, d), in contrast to only the three types possible for configurations of three nucleons. The quantum numbers J, T of the possible states in the last four configurations can be found out fairly easily from straightforward coupling rules; the information needed for this purpose consists of the possible angular momentum and isotopic quantum numbers in the configurations a^2 and a^3. This is already known to us. The nontrivial problem of finding the states is faced in the case of the configuration a^4 of equivalent nucleons. Once again, for the completely antisymmetric states, the quantum numbers J, T

can be found by counting the various possible distributions of nucleons amongst the distinct combinations of the single-particle projection quantum numbers m, m_t. In view of the large number of particles, it is clear that the counting will be quite tedious. Moreover, there will be states of other symmetry types, which may be obtained by adding a fourth box to the three-nucleon symmetry diagrams of Fig. III.9, following the rules described in Section 29A. For the reasons stated there, we have to consider, for the isotopic spin states, the symmetry patterns consisting of up to two rows and the conjugate symmetry patterns (which necessarily cannot have more than two columns) of up to $(2a + 1)$ rows. Any such symmetry pattern for isotopic spin corresponds to a unique value of T equal to $\frac{1}{2}(n_1 - n_2)$, where n_1 and n_2 are the number of boxes in the two rows of the pattern. The possible seniority quantum numbers and J-quantum numbers in the conjugate symmetry patterns are not, however, trivial to find out. In the case of three nucleons, we indicated a fairly straightforward procedure, which depends on the coupling of the angular momenta of the two-nucleon patterns with the angular momentum of the last nucleon. *In general*, for four or more nucleons, this simple approach does not always give information on each individual symmetry pattern. To solve this problem, we need more sophisticated knowledge of the group of permutations. The representations of this group have a one-to-one correspondence with those of the group of unitary transformations. The latter contains the symplectic group and the rotation group as subgroups. The seniority quantum number corresponds to a reduction of the representations of the unitary group in terms of those of the symplectic group; similarly, the quantum number J corresponds to a reduction in terms of the representations of the three-dimensional rotation group. These details are outside the scope of our discussion. A pragmatic reader, who should be satisfied if he can do a calculation using standard tables, may take consolation in the fact that this group-theoretical problem of finding the quantum numbers of allowed states in the configuration of n equivalent nucleons has been solved by Jahn et al[5], and that published lists with the required information exist in the literature on the subject; tables on the corresponding cfp's are also available.

The cfp's for states belonging to mixed configurations (inequivalent nucleons are to be treated in these cases) can be calculated in the manner already described for three nucleons. Let us consider a state $\overline{\Psi}_C(1, 2, \ldots, n-1)$ of the $(n-1)$ nucleons belonging to a configuration labelled C. An antisymmetric state belonging to the configuration (C, j) of n nucleons can be written with the help of the antisymmetrizer $\mathcal{A}$ as

$$|\Psi) = \mathcal{A}|\overline{\Psi}_C(1, 2, \ldots, n-1), j(n) : \Psi\rangle, \tag{III.125}$$

where the comma before j denotes angular momentum and isotopic spin coupling, and the antisymmetrizer $\mathcal{A}$ is explicitly given in terms of the exchange operators P_{in} as

$$\mathcal{A} = 1 - \sum_{i=1}^{n-1} P_{in}. \tag{III.126}$$

Let us first examine what $P_{n-1,n}$ does. We have

$$\begin{aligned} P_{n-1,n}&|\overline{\Psi}_C(1, 2, \ldots, n-1), i(n) : \Psi\rangle \\ &= |\overline{\Psi}_C(1, 2, \ldots, n-2, n), j(n-1) : \Psi\rangle \\ &= \sum_{\overline{\Psi}', j'} C(\overline{\Psi}', j')|\overline{\Psi}'(1, 2, \ldots, n-2), j'(n) : \overline{\Psi}_C, j(n-1) : \Psi\rangle \\ &= \sum_{\overline{\Psi}', j'} C(\overline{\Psi}', j') \sum_{\mathcal{J}, \mathcal{T}} U(j'\bar{J}'Jj; \bar{J}\mathcal{J})U(\tfrac{1}{2}\bar{T}'T\tfrac{1}{2}; \bar{T}\mathcal{T}) \\ &\quad \times(-1)^{\bar{J}'-\bar{J}+\bar{T}'-\bar{T}-\mathcal{J}-J-\mathcal{T}-T}|\overline{\Psi}'(1, 2, \ldots, n-2), j(n-1) : \mathcal{J}\mathcal{T}, j'(n) : \Psi\rangle. \end{aligned} \tag{III.127}$$

We have here first exchanged $(n-1)$ with n, then expanded $\overline{\Psi}(\bar{J}, \bar{T})$ in terms of the cfp $C(\overline{\Psi}', j')$; the $(n-2)$-particle state, $\overline{\Psi}'(\bar{J}', \bar{T}')$, couples to j' of the particle n in this break-up. Finally, we have used the U-functions to so rearrange the coupling that the particle n is coupled last. In doing so, the order of coupling $\overline{\Psi}'$ with j' was first reversed, giving $|j'(n), \overline{\Psi}' : \overline{\Psi}_C, j(n-1) : \Psi\rangle$; the recoupled state was then in the order $|j'(n), \{\overline{\Psi}', j(n-1) : \mathcal{J}\mathcal{T}\} : \Psi\rangle$ in which the coupling of $j'(n)$ with $(\mathcal{J}\mathcal{T})$ was finally reversed to obtain the state in the last step. The two reversals of coupling account for the phase factor present in (III.127).

In the same manner as for $P_{n-1,n}$, the effect of P_{in}, with $i \neq n-1$, can be easily shown to produce the same coefficients and summations, but gives the state $|\overline{\Psi}'(1, 2, \ldots, n-1, \ldots, n-2), j(i) : \mathcal{J}\mathcal{T} : j'(n) : \Psi\rangle$ *with a minus sign,* where $n-1$ has taken the place of the i-th nucleon. To understand this result, we note that P_{in} puts n in the position of i in the wavefunction $\overline{\Psi}$ and then n and $n-1$ in $\overline{\Psi}$ are interchanged so that the subsequent fractional parentage expansion separates the last particle n from this wavefunction. This additional interchange in the antisymmetric wavefunction $\overline{\Psi}$ accounts for the extra minus sign, and also for the fact that $(n-1)$ occupies the position of i in $\overline{\Psi}'$ in the final result. Putting together the result of operating with $P_{n-1,n}$ and P_{in} for all $i \neq n-1$, we obtain, along with the coefficients in (III.127), the state $|\bar{\mathcal{A}}\{\overline{\Psi}'(1, 2, \ldots, n-2), j(n-1) : \mathcal{J}\mathcal{T}\}, j'(n) : \Psi\rangle$, where $\bar{\mathcal{A}} = (1 - \sum_{i=1}^{n-2} P_{i,n-1})$, and $\bar{\mathcal{A}}$ operates on the state enclosed within the braces. A comparison with (III.126) tells us that $\bar{\mathcal{A}}$ is the antisymmetrizer for any wavefunction of $(n-1)$ nucleons written as a product of an antisymmetric wavefunction of the first $(n-2)$ nucleons and the wavefunction of the last nucleon. Let the normalization constant of this $(n-1)$-nucleon state be $N^{-1}(\overline{\Psi}', j : \mathcal{J}\mathcal{T})$. Using the aforestated results along with the coefficients of (III.127) and the first term unity of $\bar{\mathcal{A}}$ of (III.126), we obtain, for the wavefunction (III.125), the expression

$$|\overline{\Psi}) = |\overline{\Psi}_C(1, 2, \ldots, n-1), j(n) : \Psi\rangle - \sum_{\overline{\Psi}', j'} \sum_{\mathcal{J}, \mathcal{T}} (-1)^{\bar{J}'-\bar{J}-\mathcal{J}-J+\bar{T}'-\bar{T}-\mathcal{T}-T}$$
$$\times U(j'\bar{J}'Jj; \bar{J}\mathcal{J}) U(\tfrac{1}{2}\bar{T}'T\tfrac{1}{2}; \bar{T}\mathcal{T}) N(\overline{\Psi}', j : \mathcal{J}\mathcal{T})$$
$$\times |\{\overline{\Psi}(1, 2, \ldots, n-2), j(n-1) : \mathcal{J}\mathcal{T}\}_A, j'(n) : \Psi\rangle. \qquad \text{(III.128)}$$

This is the required fractional parentage expansion of the particular n-particle state; of course, we have first to find out the normalization constant, and then the cfp's have to be written with this constant suitably incorporated.

The right procedure, it must be obvious to the reader, is to move step by step, starting with the wavefunctions $n = 2$. That is to say, build up the wavefunctions $n = 3$ from $n = 2$ in the manner of Section 29A; then proceed to the wavefunctions $n = 4$, utilizing the knowledge of the wavefunctions $n = 3$, and so on. In this way, at the time of evaluating the cfp's of the n-nucleon states from the expansion (III.128), complete information on the antisymmetric states of the $(n-1)$ nucleons appearing therein is already at our disposal, and hence there will be no problem with the normalization constants $N(\overline{\Psi}', j : \mathcal{J}\mathcal{T})$, and the permitted values of the summation labels in (III.128). The cfp's of the n nucleons can thus be explicitly evaluated.

30. USE OF SECOND-QUANTIZATION TECHNIQUE IN SHELL-MODEL PROBLEMS

The second-quantization technique of dealing with the antisymmetrization problem in the many-body theory can be very profitably used in some shell-model calculations. We shall demonstrate a few such applications in this section. The technique itself has been presented in Section 16.

A. RELATIONSHIP BETWEEN HOLE STATE AND PARTICLE STATE

A *particle* state outside a closed shell $|\Phi_0\rangle$ is described in the second-quantization technique by the creation operator $C_\alpha^\dagger$ according to

$$C_\alpha^\dagger|\Phi_0\rangle,$$

where α stands for the set of quantum numbers defining the state of the particle. Since $|\Phi_0\rangle$ is the closed-shell state, it already contains a set of single-particle states occupied in it and, according to the work in Section 16, any occupied state from $|\Phi_0\rangle$ can be destroyed by an annihilation operator C_β, where β is one such state. The state

$$C_\beta|\Phi_0\rangle$$

describes a *hole* in the single-particle state β of $|\Phi_0\rangle$.

In the nuclear shell model, each of the labels $\alpha, \beta, \ldots$ stands for the set of quantum numbers n, l, j, m. While referring to any such state, we shall omit the quantum numbers n, l and use only j, m since only the latter are relevant to most of our discussions. Further, we shall establish a relationship between the particle and hole states corresponding to the *same* (n, l, j)-quantum numbers, and hence very frequently the creation and destruction operators will bear only the projection quantum number m as a label; whenever a knowledge of the angular momentum is explicitly required, the symbol j will be used to denote it.

The antisymmetric state $|\Phi_0; jm)$ formed with one particle in the single-particle state j, m outside the closed-shell state $|\Phi_0\rangle$ is described by

$$|\Phi_0; jm) = C_m^\dagger|\Phi_0). \tag{III.129}$$

This is a state of angular momentum j and projection m, and hence, under a rotation of the coordinate system, it transforms according to

$$|\Phi_0; jm) \to \overline{|\Phi_0; jm)} = \sum_{m'} \mathcal{D}^j_{mm'}|\Phi_0; jm') = \sum_{m'} \mathcal{D}^j_{mm'}C_{m'}^\dagger|\Phi_0).$$

Thus, $C_m^\dagger$ transforms according to

$$\sum_{m'} \mathcal{D}^j_{mm'}C_{m'}^\dagger.$$

The annihilation operator C_m then transforms according to

$$\sum_{m'} \mathcal{D}^{j*}_{mm'}C_{m'}$$

or, using (BI.16) from Appendix B, according to

$$\sum_{m'} (-1)^{m-m'}\mathcal{D}^j_{-m,-m'}C_{m'}.$$

Therefore, the operator $(-1)^{j-m}C_{-m}$ transforms as

$$\sum_{m'} (-1)^{j-m}(-1)^{-m+m'}\mathcal{D}^j_{mm'}C_{-m'} = \sum_{m'} \mathcal{D}^j_{mm'}(-1)^{j-m'}C_{-m'}.$$

Thus, the quantity $(-1)^{j-m}C_{-m}$ has the same transformation property as a state of angular momentum j and projection m. The state generated by this operator from a closed-shell state $|\Psi_0\rangle$, which already has the single-particle state $j, -m$ occupied, will therefore be called a *hole* state of angular momentum j and projection m. We use the notation $|\Psi_0; (jm)^{-1})$ to denote such a hole state, which in the second-quantized notation becomes

$$|\Psi_0; (jm)^{-1}) = (-1)^{j-m}C_{-m}|\Psi_0). \tag{III.130}$$

We have deliberately used two different closed-shell states Φ_0 and Ψ_0 in (III.129) and (III.130) while describing the *particle* state *jm*, and the hole state of the same quantum numbers. This is because the single particle denoted by *jm* has to be created by starting with the closed shell just outside which the single-particle level *jm* occurs. When all the $(2j+1)$-substates of this particular *j*-level are filled with nucleons, we reach another closed-shell nucleus, which has been denoted here by Ψ_0. Obviously, the hole state with the quantum numbers *jm* has to be defined by removing particles from the state Ψ_0 in which the states *jm* with all *m* are occupied. The relationship between Φ_0 and Ψ_0 is clearly

$$|\Psi_0\rangle = \prod_{m=-j}^{+j} C_m^\dagger|\Phi_0) \tag{III.131}$$

wherein the creation operator corresponding to each *m*-quantum number belonging to the given *j* occurs.

The antisymmetric states of more than one particle or hole in the level *j* can also be similarly described. As an example, consider the case of two *particles*; such a state, normalized to unity, is given by

$$C_{m'}^\dagger C_m^\dagger|\Phi_0).$$

The normalization of this state can be checked by working out $(\Phi_0|\, C_m C_{m'} C_{m'}^\dagger C_m^\dagger\, |\Phi_0)$ with the help of the anticommutators (II.30) from Section 16. The clue is to take the destruction operators to the extreme right, past the creation operators, by using this anticommutator relation, and finally noting $C_m|\Phi_0) = C_{m'}|\Phi_0) = 0$ because all the substates of this *j* are, by definition, unoccupied in Φ_0. We can equivalently use angular momentum coupled two-particle states $|j^2JM)$ which are given by

$$|j^2JM) = A^\dagger(j^2JM)|\Phi_0), \tag{III.132a}$$

where

$$A^\dagger(j^2JM) = \frac{1}{\sqrt{2}} \sum_{m,m'} \begin{bmatrix} j & j & J \\ m & m' & M \end{bmatrix} C_{m'}^\dagger C_m^\dagger. \tag{III.132b}$$

Here the numerical factor $1/\sqrt{2}$ has to be introduced so that the state (III.132a) is normalized to unity; the normalization can be checked in the manner just described, together with the property of the Clebsch-Gordon coefficients. The angular momentum coupling of the creation operators has been achieved in (III.131) with the Clebsch-Gordon coefficient because these operators do indeed transform as states of angular momenta *jm* and *jm'*, respectively. According to our convention, the operators act on $|\Phi_0)$ *starting* from the right, and hence the first angular momentum in the coupling (the first column of the Clebsch-Gordon coefficient) is *jm* and the second is *jm'*, although, while writing the operators in (III.132b), $C_{m'}^\dagger$ occurs first and $C_m^\dagger$ second.

We now wish to construct the operator that produces from $|\Psi_0)$ the two-hole states of angular momentum *J* and projection *M*. Clearly, the Clebsch-Gordon coefficient of (III.132b) has now to be used with the operators $(-1)^{j-m}C_{-m}$ and $(-1)^{j-m'}C_{-m'}$ in view of the behaviour of these annihilation operators under rotation, as already described. Thus, the two-hole states are given by

$$\begin{aligned} |j^{-2}JM) &= \frac{1}{\sqrt{2}} \sum_{m,m'} \begin{bmatrix} j & j & J \\ m & m' & M \end{bmatrix} (-1)^{j-m}C_{-m}(-1)^{j-m'}C_{-m'}|\Psi_0) \\ &= (-1)^{J-M}A(j^2, J, -M)|\Psi_0), \end{aligned} \tag{III.132c}$$

where the pair annihilation operator $A(j^2JM)$ is defined as the Hermitean conjugate of the pair creation operator $A^\dagger(j^2JM)$ of (III.132b).

The observant reader may have noted a flaw in the first step of (III.132c). According to this step, from $|\Psi_0)$ is produced first the hole state $(jm')^{-1}$ and then the hole state $(jm)^{-1}$, and hence, in the Clebsch-Gordon coefficient, we should have taken jm' as the first column and jm as the second column. However, this discrepancy, as is clear, could have been corrected simply by interchanging the two destruction operators, which would have given rise to, by virtue of the anticommutator (II.30a), an extra minus sign in the definition of the two-hole states. Since an overall phase of a state is irrelevant in computation, we adhere to the definition just given. *This merely fixes our phase convention for the hole states.* The convenience in choosing the phase convention in (III.132c) is that it relates the operator for the two-*hole* state, namely, $(-1)^{J-M}A(j^2J, -M)$, to the corresponding two-*particle* state creation operator, $A^\dagger(j^2JM)$, in a manner that is exactly analogous to the relationship between a single-hole and a single-particle state. For the many-hole states also, we shall use the same phase convention, namely, the n-hole state of angular momentum J and M with other quantum numbers α (if necessary) will be obtained from $|\Psi_0)$ through the operator $(-1)^{J-M}\mathscr{B}(j^n\alpha J, -M)$, where $\mathscr{B}^\dagger(j^n\alpha JM)$ creates the corresponding n-particle state from $|\Phi_0)$. It is clear that $\mathscr{B}^\dagger(j^n\alpha JM)$ comprises a string of single-particle creation operators which are coupled with suitable Clebsch-Gordon coefficients starting from the right. To illustrate this statement, let us consider a three-particle state where the quantum number J_0, obtained after coupling the first two creation operators from the extreme right, plays the role of α. Thus,

$$|j^3, J_0JM) = \mathscr{B}^\dagger(j^3, J_0JM)|\Phi_0)$$
$$= \sum_{m,m',m''} \begin{bmatrix} j & j & J_0 \\ m & m' & M_0 \end{bmatrix} \begin{bmatrix} J_0 & j & J \\ M_0 & m'' & M \end{bmatrix} C^\dagger_{m''} C^\dagger_{m'} C^\dagger_m |\Phi_0). \tag{III.133}$$

This antisymmetric three-nucleon state is, by definition, identical to the expression (III.115b). We have noted in Section 29 that all possible values of J are not permitted. In the second-quantized version (III.133), this will automatically follow from the anticommuting property (II.30a) of the creation operators. For the forbidden values of J, this expression will produce terms that are all zero due to either the occurrence of two creation operators in the anti-commutator combination or the presence of more than one creation operator having the same particle label.

We now emphasize the fact that the foregoing relationship between the operators that generate the antisymmetric states of n holes and of n particles sets up a one-to-one correspondence between the two sets of states. That is to say, all the states in the n-particle system will have their counterparts, and only these, in the n-hole system. We shall prove that the matrix element of an operator connecting any two n-hole states is also very closely related to the matrix elements of the same operator taken between the two corresponding n-particle states. Once this is done, we can conclude that the shell-model calculations in the first half of a j-shell, i.e., j^n with n up to $(j + \frac{1}{2})$, will enable us to reproduce, with very trivial modifications, the results in the second half, i.e., n ranging from $(j + \frac{3}{2})$ to $(2j + 1)$. It should be noted that j^n is the configuration of the n particles, and j^{2j+1-n} is the configuration of the n holes. If n lies in the first half of the shell, i.e., between 1 and $(j + \frac{1}{2})$, the corresponding $(2j + 1 - n)$ lies in the second half of the shell.

The relationship between the matrix elements will now be established. Let us first consider a one-body type tensor operator $\sum_{I=1}^{n} T_q^K(I)$. Then, taking its second-quantized form (II.31a), we get

$$(j^{-n}\alpha JM| \sum_{I=1}^{2j+1-n} T_q^K(I)\,|j^{-n}\alpha' J'M')$$
$$= \sum_{\mu,\nu} \langle\mu|\, T_q^K\,|\nu\rangle(-1)^{J-M}(\Psi_0|\, \mathcal{B}^\dagger(j^n, \alpha J, -M)C_\mu^\dagger C_\nu \mathcal{B}(j^n, \alpha' J', -M')\,|\Psi_0)(-1)^{J'-M'}. \tag{III.134}$$

In the same manner, the corresponding particle state matrix element is given by

$$(j^n\alpha JM| \sum_{I=1}^{n} T_q^K(I)\,|j^n\alpha' J'M') = \sum_{\mu,\nu} \langle\mu|\, T_q^K\,|\nu\rangle(\Phi_0|\, \mathcal{B}(j^n\alpha JM)C_\mu^\dagger C_\nu \mathcal{B}^\dagger(j^n\alpha' J'M')\,|\Phi_0). \tag{III.135}$$

Here μ, ν each corresponds to a set of single-particle quantum numbers. To simplify (III.134) and (III.135), we must break up each of the operators $\mathcal{B}^\dagger$, $\mathcal{B}$, ... into the products of n creation and annihilation operators, and then proceed as follows. A typical term in (III.135) will be of the type

$$\sum_{\mu,\nu} \langle\mu|\, T_q^K\,|\nu\rangle(\Phi_0|\, C_1C_2 \ldots C_nC_\mu^\dagger C_\nu C_{n'}^\dagger \ldots C_{2'}^\dagger C_{1'}^\dagger\,|\Phi_0), \tag{III.136}$$

where we have abbreviated the labels (recall that these are the various m-values corresponding to the given j) on the creation and annihilation operators by the numbers $(1, 2, \ldots, n)$ and $(1', 2', \ldots, n')$. Writing

$$C_\mu^\dagger C_\nu = \delta_{\mu\nu} - C_\nu C_\mu^\dagger, \tag{III.137}$$

we obtain, from (III.136),

$$\sum_{\mu}^{\text{all}} \langle\mu|\, T_q^K\,|\mu\rangle(\Phi_0|\, C_1C_2 \ldots C_nC_{n'}^\dagger \ldots C_{2'}^\dagger C_{1'}^\dagger\,|\Phi_0) - \sum_{\mu\neq(1',2',\ldots,n')}\ \sum_{\nu\neq(1,2,\ldots,n)} \langle\mu|\, T_q^K\,|\nu\rangle$$
$$\times (\Phi_0|\, C_1C_2 \ldots C_nC_\nu C_\mu^\dagger C_{n'}^\dagger \ldots C_{2'}^\dagger C_{1'}^\dagger\,|\Phi_0). \tag{III.138}$$

The restrictions on μ and ν in the second term arise from the fact that the product of two creation operators or two destruction operators with the same label is zero by virtue of the anticommutation properties. We next notice that the matrix elements of the creation and destruction operators occurring in the first term of (III.138) are nonvanishing only if the quantum number on each destruction operator is identical to the corresponding quantum number on one of the creation operators to its right, that is to say, the destruction and creation operators must occur in pairs. The reason for this is evident. If any destruction operator does not find its counterpart amongst the creation operators, it can be easily permuted to the extreme right; the process of permuting gives, by virtue of the anticommutation property, $(-1)^N$, where N is the number of operators across which it has been shifted. When the particular destruction operator has been taken to the extreme right, it directly operates on $|\Phi_0)$ and produces zero because each of the state labels $(1, 2, \ldots, n)$ on the destruction operators corresponds to a single-particle state unoccupied in $|\Phi_0)$. Therefore, we conclude that the first term of (III.138) contributes only in the diagonal matrix element, i.e., when the set of states $(1, 2, \ldots, n)$ is the same as the set $(1', 2', \ldots, n')$, except of course for a permutation of these labels. If a permutation P is needed to change the latter set into the former, then

$$C^{\dagger}_{n'} \ldots C^{\dagger}_{2'} C^{\dagger}_{1'} = (-1)^P C^{\dagger}_{n} \ldots C^{\dagger}_{2} C^{\dagger}_{1},$$

the sign $(-1)^P$ having arisen from the anticommuting property. Therefore,

$$\text{first term of (III.138)} = (-1)^P \sum_{\mu}^{\text{all}} \langle\mu|\, T^K_q \,|\mu\rangle (\Phi_0|\, C_1 C_2 \ldots C_n C^{\dagger}_n \ldots C^{\dagger}_2 C^{\dagger}_1 \,|\Phi_0)$$

$$= (-1)^P \sum_{\mu}^{\text{all}} \langle\mu|\, T^K_q \,|\mu\rangle. \tag{III.139a}$$

The last step follows by noting that $C_n C^{\dagger}_n = 1 - C^{\dagger}_n C_n$ and the term $C^{\dagger}_n C_n$ produces zero in the manner described when C_n is taken to the extreme right, past the other creation operators, all of which have quantum number labels different from that of n; so we replace $C_n C^{\dagger}_n$ by 1 and then tackle $C_{n-1} C^{\dagger}_{n-1}$, occurring next to each other, which can again be replaced by unity, and so on.

We have now to treat the second term of (III.138). Here we first note that $C^{\dagger}_{\mu}$ can be taken to the extreme right and C_ν to the extreme left, the two signs due to the permutation of operator cancelling each other. The second fact to observe is that $C^{\dagger}_{\mu}|\Phi_0)$ and $(\Phi_0|C_\nu$ are nonvanishing if μ, ν *are both states unoccupied in* Φ_0. Therefore, here too we conclude that the set of states $(\mu, 1', 2', \ldots, n')$ and $(\nu, 1, 2, \ldots, n)$ must be identical, except for a permutation of labels in one set. If $\mu = \nu$, then of course this requirement implies that $(1, 2, \ldots, n)$ and $(1', 2', \ldots, n')$ must be identical, except for a permutation P. So $\mu = \nu$ in the second term once again corresponds to the case of a diagonal matrix element, and we obtain

diagonal contribution of second term of (III.138)

$$= (-1)^P \sum_{\mu \neq (1, \ldots, n)}^{\text{unocc in } \Phi_0} \langle\mu|\, T^K_q \,|\mu\rangle (\Phi_0|\, C_\mu C_1 C_2 \ldots C_n C^{\dagger}_n \ldots C^{\dagger}_2 C^{\dagger}_1 C^{\dagger}_\mu \,|\Phi_0)$$

$$= (-1)^P \sum_{\mu \neq (1, \ldots, n)}^{\text{unocc in } \Phi_0} \langle\mu|\, T^K_q \,|\mu\rangle. \tag{III.139b}$$

The matrix element of the creation and destruction operators has reduced to unity in the manner described after (III.139a). Subtracting (III.139b) from (III.139a), according to (III.138), we get

$$\text{diagonal matrix element} = (-1)^P \Big(\sum_{\mu=1}^{n} \langle\mu|\, T^K_q \,|\mu\rangle + \sum_{\xi}^{\text{occ in } \Phi_0} \langle\xi|\, T^K_q \,|\xi\rangle \Big). \tag{III.139c}$$

The second term of (III.138) gives rise to a nondiagonal matrix element as well, i.e., it produces a nonvanishing result when the sets $(1, 2, \ldots, n)$ and $(1', 2', \ldots, n')$ differ. In our way of reckoning, this contribution will be picked up by admitting the possibility $\mu \neq \nu$ and then requiring the set $(\mu, 1', 2', \ldots, n')$ to be identical to the set $(\nu, 1, 2, \ldots, n)$, except for a permutation. We have already considered the consequence of this requirement for the case $\mu = \nu$. Now, if $\mu \neq \nu$, the two sets can clearly be identical only if μ is equal to one of the labels $(1, 2, \ldots, n)$, and ν is equal to one of the labels $(1', 2', \ldots, n')$; the remaining labels in $(1, 2, \ldots, n)$ and $(1', 2', \ldots, n')$ must of course be identical, except for a permutation. Thus, the only type of nondiagonal matrix element we obtain corresponds to the case where the sets $(1, 2, \ldots, n)$ and $(1', 2', \ldots, n')$ differ through only one single-particle state. Denoting the single-particle states different in the two sets by m and m', we conclude $\mu = m$ and $\nu = m'$. In this case, while permuting $(1', 2', \ldots, n')$, we place m' in the location where m occurs in the set $(1, 2, \ldots, n)$; then the remaining labels which are the same in the two sets are permuted to identical locations. Denoting this permutation by P', we then have

nondiagonal contribution of second term of (III.138)

$$= -(-1)^{P'}\langle m|\ T_q^K\ |m'\rangle(\Phi_0|\ C_{m'}C_1C_2\ldots C_m\ldots C_nC_n^\dagger\ldots C_{m'}^\dagger\ldots C_2^\dagger C_1^\dagger C_m^\dagger\ |\Phi_0)$$

$$= (-1)^{P'}\langle m|\ T_q^K\ |m'\rangle. \tag{III.139d}$$

Here the first minus sign has been erased while switching $C_{m'}$ and C_m; after the switch over has been effected, the destruction and creation operators come in pairs one after another and are replaced by unity at each stage, as already described.

Let us now examine the counterpart of the term (III.136) contained in the hole matrix element (III.134). We recall that the break-up of the $\mathscr{B}$-operators for the hole states gives rise to the same Clebsch-Gordon coefficients as contained in the corresponding particle states; but a creation operator $C_m^\dagger$ is to be replaced by $(-1)^{j-m}C_{-m}$, and the string of annihilation operators has to be written in the reverse order, when compared with the string of creation operators in $\mathscr{B}^\dagger$, because $\mathscr{B}$ is the Hermitean conjugate of $\mathscr{B}^\dagger$. Thus, the counterpart of (III.136) is, in the case of holes, given by

$$\sum_{\mu,\nu} s_1s_2\ldots s_n\langle\mu|\ T_q^K\ |\nu\rangle(\Psi_0|\ C_{-n}^\dagger\ldots C_{-2}^\dagger C_{-1}^\dagger C_\mu^\dagger C_\nu C_{-1'}C_{-2'}\ldots C_{-n'}\ |\Psi_0)s_{1'}s_{2'}\ldots s_{n'}. \tag{III.140}$$

Here we have used abbreviated notation: the minus sign in front of a label denotes that the corresponding m-quantum number be written with a minus; and the symbol s with a subscript stands for the phase factor $(-1)^{j-m}$ corresponding to the m-quantum number of that subscript. It should be noted that the closed-shell state in (III.140) is Ψ_0, which is obtained from Φ_0 by filling up all the $(2j+1)$ sublevels of the single-particle state j immediately outside Φ_0. Therefore, in this case, any of the creation operators $C_{-1}^\dagger, C_{-2}^\dagger, \ldots, C_{-n}^\dagger$ acting on $|\Psi_0)$ gives zero because we then create a particle in a single-particle state which is already occupied. This fact is useful to us while simplifying (III.140).

As a beginning to the process of simplification, we first permute $C_\mu^\dagger$ to the extreme left and C_ν to the extreme right, and then note that $C_\nu|\Psi_0)$ and $(\Psi_0|C_\mu^\dagger$ are nonvanishing only if μ, ν are states occupied in Ψ_0. Further, μ cannot be equal to any of the states in the set $(-1, -2, \ldots, -n)$, and ν cannot be equal to any of the states in the set $(-1', -2', \ldots, -n')$ by virtue of the basic anticommutation property. The next fact we notice is that the set of states $(-1', -2', \ldots, -n', \nu)$ and $(-1, -2, \ldots, -n, \mu)$ must be identical, except for a permutation of the labels in one set. As before, we consider two cases. First, the diagonal matrix element corresponding to the case when $(-1, -2, \ldots, -n)$ is identical to $(-1', -2', \ldots, -n')$, except for a permutation P; the labels μ and ν have to be equal. The phase factors $s_1s_2\ldots s_n$ and $s_{1'}s_{2'}\ldots s_{n'}$ cancel because each factor occurs in a pair. In this case, considering all the aforementioned facts, we obtain

diagonal matrix element in (III.140)

$$= (-1)^P \sum_{\mu\neq(-1,-2,\ldots,-n)}^{\text{occ in }\Psi_0} \langle\mu|\ T_q^K\ |\mu\rangle(\Psi_0|\ C_\mu^\dagger C_{-n}^\dagger\ldots C_{-2}^\dagger C_{-1}^\dagger C_{-1}C_{-2}\ldots C_{-n}C_\mu\ |\Psi_0)$$

$$= (-1)^P \sum_{\mu\neq(-1,-2,\ldots,-n)}^{\text{occ in }\Psi_0} \langle\mu|\ T_q^K\ |\mu\rangle$$

$$= (-1)^P\Big(\sum_{\xi}^{\text{occ in }\Psi_0} \langle\xi|\ T_q^K\ |\xi\rangle - \sum_{\mu=1}^{n} \langle-\mu|\ T_q^K\ |-\mu\rangle\Big). \tag{III.141a}$$

Each expression such as $C^{\dagger}_{-1}C_{-1}$ and $C^{\dagger}_{-2}C_{-2}$ is replaceable by unity in this matrix element, and hence the result.

The nondiagonal matrix element can also be treated in the manner described for the case of particle matrix elements. Defining the permutation P' as before, and denoting the single-particle states that are different in the two sets $(-1, -2, \ldots, -n)$ and $(-1', -2', \ldots, -n')$ by $-m$ and $-m'$, respectively, we now obtain $\mu = -m'$ and $\nu = -m$; all but the phase factors $s_m s_{m'}$ cancel out and we finally obtain

$$\begin{aligned}&\text{nondiagonal matrix element in (III.140)}\\&= (-1)^{P'}\langle -m'|\, T^K_q\,|-m\rangle(\Psi_0|\, C^{\dagger}_{-m'}C^{\dagger}_{-n}\ldots C^{\dagger}_{-m}\ldots C^{\dagger}_{-2}C^{\dagger}_{-1}\\&\quad\times C_{-1}C_{-2}\ldots C_{-m'}\ldots C_{-n}C_{-m}\,|\Psi_0)s_m s_{m'}\\&= -(-1)^{P'}\langle -m'|\, T^K_q\,|-m\rangle s_m s_{m'}.\end{aligned} \tag{III.141b}$$

The minus sign in the front arises from the interchange of $C^{\dagger}_{-m'}$ and $C^{\dagger}_{-m}$. Noting that Ψ_0 consists of Φ_0 and all the occupied single-particle states $m = -j$ to $m = +j$, we write

$$\sum_{\xi}^{\text{occ in }\Psi_0}\langle\xi|\, T^K_q\,|\xi\rangle = \sum_{\xi}^{\text{occ in }\Phi_0}\langle\xi|\, T^K_q\,|\xi\rangle + \sum_{m=-j}^{+j}\langle jm|\, T^K_q\,|jm\rangle$$

or, by virtue of the proof given in Section 26A concerning the sum over all the sublevels of a closed shell,

$$\delta_{K0}\delta_{q0}\sum_{\xi}^{\text{occ in }\Psi_0}\langle\xi|\, T^K_q\,|\xi\rangle = \delta_{K0}\delta_{q0}\Big[\sum_{\xi}^{\text{occ in }\Phi_0}\langle\xi|\, T^K_q\,|\xi\rangle + [j]\langle j|\,|T^K_q|\,|j\rangle\Big]. \tag{III.142a}$$

The Clebsch-Gordon coefficient in $\langle jm|\, T^0_0\,|jm\rangle$ is identically equal to unity, and hence the sum over m has yielded the factor $(2j+1)$.

To demonstrate the relationship between the other terms in (III.139c) and (III.141a), we recall that the labels $\mu, -\mu, \ldots$ describe the projection quantum numbers corresponding to the angular momentum j, and hence the application of the Wigner-Eckart theorem yields

$$\begin{aligned}-\langle -\mu|\, T^K_q\,|-\mu\rangle &= -\begin{bmatrix} j & K & j\\ -\mu & q & -\mu\end{bmatrix}\langle j|\,|T^K|\,|j\rangle\delta_{q0}\\&= -\begin{bmatrix} j & K & j\\ \mu & q & \mu\end{bmatrix}(-1)^{j+K-j}\langle j|\,|T^K|\,|j\rangle\delta_{q0}\\&= -\delta_{q0}\langle\mu|\, T^K_q\,|\mu\rangle(-1)^K.\end{aligned} \tag{III.142b}$$

We have used here the symmetry relation of the Clebsch-Gordon coefficient to change the $-\mu$ to $+\mu$, and then the Wigner-Eckart theorem to recover the matrix element in the final step. (III.142) can now be summarized as follows: the diagonal matrix element of a single-particle operator T^K_q between any two hole states can be obtained from the same matrix element between the corresponding particle states by changing its sign and multiplying it with $(-1)^K$ [see (III.142b)]; in the special case of the zero-rank tensor operator, the matrix elements have an extra contribution from the closed shells of Φ_0, and the result for the hole states in the level j is obtained by subtracting the contribution of the corresponding particle states in the level j from the contribution $(2j+1)\langle j|\,|T^0_0|\,|j\rangle$ of all the nucleons in the level j [see (III.142a)].

In the case of the nondiagonal matrix elements, once again the result is nonvanishing only if one single-particle state is different. As for (III.142b), so too here we can establish

$$-s_m s_{m'}\langle -m'|\, T^K_q\, |-m\rangle = -\begin{bmatrix} j & K & j \\ -m & -m'+m & -m \end{bmatrix}(-1)^{j-m}(-1)^{j-m'}\langle j||T^K||j\rangle$$

$$= -\delta_{q,m-m'}\begin{bmatrix} j & K & j \\ m' & m-m' & m \end{bmatrix}(-1)^K\langle j||T^K||j\rangle$$

$$= -\delta_{q,m-m'}(-1)^K\langle m|\, T^K_q\, |m'\rangle. \tag{III.143}$$

Therefore, we conclude that the relationship between the nondiagonal matrix elements is also given by a reversal of sign and the multiplicative factor $(-1)^K$.

We have now succeeded in relating the particle matrix element (III.136) with the hole matrix element (III.140). We, however, recall that these expressions are typical terms contained in (III.135) and (III.134), respectively, after angular momentum decoupling is effected in the $\mathcal{B}^\dagger$- and $\mathcal{B}$-operators. We further recall that the Clebsch-Gordon coefficients in this decoupling in the case of particles were the same as those in the case of holes. In view of this, the relationship between the particle and hole matrix elements (III.136) and (III.140) [i.e., through the phase factor $-(-1)^K$, where K is the rank of the tensor operator of the single-particle type] remains unchanged even after the angular momentum coupling has been redone.

We shall next derive the relationship between the particle and hole matrix elements for a two-body type operator; the two-body potential being the only such operator of concern to us, we shall restrict ourselves to this operator.

The two-body potential energy operator is given by [see (II.31b)]

$$V = \tfrac{1}{2}\sum_{\alpha\beta\gamma\delta}\langle\alpha\beta|\, V\,|\gamma\delta\rangle C^\dagger_\alpha C^\dagger_\beta C_\delta C_\gamma = \tfrac{1}{4}\sum_{\alpha\beta\gamma\delta}(\alpha\beta|\, V\,|\gamma\delta)C^\dagger_\alpha C^\dagger_\beta C_\delta C_\gamma, \tag{III.144}$$

where the antisymmetrized matrix element of V is, as usual, the direct term minus the exchange term, i.e.,

$$(\alpha\beta|\, V\,|\gamma\delta) = \langle\alpha\beta|\, V\,|\gamma\delta\rangle - \langle\alpha\beta|\, V\,|\delta\gamma\rangle.$$

For treating the hole matrix element, the form (III.144) is desirable; on the other hand, the particle matrix element can be better tackled by using the alternative form

$$C^\dagger_\alpha C^\dagger_\beta C_\delta C_\gamma = \delta_{\alpha\gamma}\delta_{\beta\delta} - \delta_{\alpha\delta}\delta_{\beta\gamma} + \delta_{\beta\gamma}C_\delta C^\dagger_\alpha - \delta_{\beta\delta}C_\gamma C^\dagger_\alpha + \delta_{\alpha\delta}C_\gamma C^\dagger_\beta - \delta_{\alpha\gamma}C_\delta C^\dagger_\beta + C_\delta C_\gamma C^\dagger_\alpha C^\dagger_\beta$$

which follows from successive use of the anticommutation rule. Substituting this expression in (III.144), and suitably relabelling the summation symbols, we can easily show that the first two terms give identical contribution in the sum, and so do the next four terms containing one destruction and one creation operator. In this way,

$$V = \tfrac{1}{2}\sum_{\alpha\beta}(\alpha\beta|\, V\,|\alpha\beta) - \sum_{\alpha\beta\gamma}(\alpha\beta|\, V\,|\gamma\beta)C_\gamma C^\dagger_\alpha + \tfrac{1}{4}\sum_{\alpha\beta\gamma\delta}(\alpha\beta|\, V\,|\gamma\delta)C_\delta C_\gamma C^\dagger_\alpha C^\dagger_\beta. \tag{III.145}$$

In the case of particles, this operator has to be inserted between the states $(\Phi_0|C_1C_2\ldots C_n$ and $C^\dagger_{n'}\ldots C^\dagger_{2'}C^\dagger_{1'}|\Phi_0)$, whereas in the case of holes, the expression (III.144) has to be inserted between the states $(\Psi_0|C^\dagger_{-n}\ldots C^\dagger_{-2}C^\dagger_{-1}s_1s_2\ldots s_n$ and $s_{1'}s_{2'}\ldots s_{n'}C_{-1'}C_{-2'}\ldots C_{-n'}|\Psi_0)$. The arguments and procedure for simplification of the resultant expressions are identical to that given in connection with the single-particle operator T^K_q. The reader may go through the necessary exercise, if he so desires. We quote the final results for the complete Hamiltonian $T+V$. The definitions (III.101) and (III.102) have to be used in the derivation in order to obtain the single-nucleon energies ϵ. In both the diagonal and nondiagonal matrix elements, the hole matrix elements of the two-body potential are identical to those of the particles. For the diagonal matrix element, there is a constant additive part which is equal to the energy E_c of the closed-shell core Φ_0 *plus* the sum of the single-particle energies ϵ in the field of Φ_0 in the case of the particle matrix element; the same constant additive part, in the case of the hole matrix element, consists of the energy $\tilde{E}_c$ of the closed-shell state Ψ_0 *minus* the sum of the energy ϵ of the hole states, the latter being reckoned in the field of Ψ_0.

To properly understand Φ_0 and Ψ_0, let us give an illustration. The nuclei He^6 and Li^6 consist of two particles in the $0p$-shell just outside the core He^4, which here is the state Φ_0; on the other hand, O^{14} and N^{14} comprise two holes each in the $0p$-shell, the holes being reckoned with reference to the closed-shell state Ψ_0 of O^{16}.

B. STATES OF HOLE-PARTICLE EXCITATION AND CORE POLARIZATION

The ground state Ψ_0 of a closed-shell nucleus comprises a set of completely occupied single-particle levels. The simplest type of excitation of such a nucleus is caused by promoting a nucleon from one of these occupied single-particle levels to an unoccupied level. The resultant state is thus of the 1h-1p type. These states can be conveniently expressed in the second-quantized terminology. Let the vacancy be in the state $(j_h, -m_h)$; the corresponding hole state of angular momentum j_h and projection m_h is then given by the operator $(-1)^{j_h-m_h}C_{j_h,\,-m_h}$. The particle state (j_p, m_p) is created by $C^\dagger_{j_p,\,m_p}$. Hence, the 1h-1p state of total angular momentum J_c and projection M_c is given by

$$|j_h^{-1}j_pJ_cM_c) = A^\dagger(j_h^{-1}j_pJ_cM_c)|\Psi_0), \tag{III.146a}$$

where

$$A^\dagger(j_h^{-1}j_pJ_cM_c) = \sum_{m_h,\,m_p} (-1)^{j_h-m_h}\begin{bmatrix} j_h & j_p & J_c \\ m_h & m_p & M_c \end{bmatrix} C^\dagger_{j_pm_p}C_{j_h,\,-m_h}. \tag{III.146b}$$

Here the operators occur in an order in which the hole state is created first and the particle state next; the two columns in the Clebsch-Gordon coefficients corresponding to these angular momenta have therefore been arranged in the same order.

In view of (II.31) and (III.146), typical terms in the matrix elements of T and V connecting hole-particle states are of the form

$$\sum_{\alpha,\,\beta} (\Psi_0|\, C^\dagger_{j_h',\,-m_h'}C_{j_p'm_p'}C^\dagger_\alpha C_\beta C^\dagger_{j_pm_p}C_{j_h,\,-m_h}\,|\Psi_0)\langle\alpha|\,T\,|\beta\rangle, \tag{III.147a}$$

$$\tfrac{1}{4}\sum_{\alpha\beta\gamma\delta} (\Psi_0|\, C^\dagger_{j_h',\,-m_h'}C_{j_p'm_p'}C^\dagger_\alpha C^\dagger_\beta C_\delta C_\gamma C^\dagger_{j_pm_p}C_{j_h,\,-m_h}\,|\Psi_0)(\alpha\beta|\,V\,|\gamma\delta). \tag{III.147b}$$

These simplified results are to be multiplied by the appropriate Clebsch-Gordon coefficients and phase factors occurring in (III.146b). The necessary summations over the projection quantum numbers are then to be carried out. The details of this derivation are given in Appendix G (Section III). Here we shall quote only the final result, namely,

$$(j_h'^{-1}j_p'MJ_c|\,(T+V)\,|j_h^{-1}j_pJ_cM_c) = \delta_{j_hj_h'}\delta_{j_pj_p'}[E_0 + \epsilon(j_p) - \epsilon(j_h)] + F(j_h'j_p'j_hj_p;\,J_c), \tag{III.148}$$

where the hole-particle matrix element F of the two-body potential V is given by (GIII.9b) and (GIII.7) in Appendix G, and the quantity E_0 is the energy of the state Ψ_0 defined after (GIII.3) in Appendix G.

As a result of the diagonalization of the matrix defined by (III.148), we obtain the eigenvalues E_e and the corresponding eigenvectors with the elements x^e_{hp}. The eigenstate belonging to E_e can clearly be written as

$$|\Psi_e) = \sum_{j_h, j_p} x^e_{hp}|j_h^{-1}j_pJ_cM_c)$$
$$= \sum_{j_h, j_p} x^e_{hp}A^\dagger(j_h^{-1}j_pJ_cM_c)|\Psi_0). \tag{III.149}$$

The matrix element for the transition of this state to the ground state Ψ_0 through the multipole operator Ω^K_q is given by (the summation I runs over all the nucleons in the nucleus)

$$(\Psi_0| \sum_{I=1}^{A} \Omega^K_q(I)\, |\Psi_e) = \sum_{\alpha, \beta} \langle\alpha|\, \Omega^K_q\, |\beta\rangle(\Psi_0|\, C^\dagger_\alpha C_\beta\, |\Psi_e). \tag{III.150a}$$

Here we have simply used the second-quantized expression for the single-particle type operator. Explicitly writing Ψ_e from (III.149) and (III.146b) and then manipulating the creation and annihilation operators in the familiar way, we get the result

$$\text{transition matrix element} = \sum_{j_h, j_p} x^e_{hp} \sum_{m_h, m_p} (-1)^{j_h - m_h}\begin{bmatrix} j_h & j_p & J_c \\ m_h & m_p & M_c \end{bmatrix}$$
$$\times \langle j_h, -m_h|\, \Omega^K_q\, |j_p, m_p\rangle$$
$$= \delta_{K, J_c}\delta_{q, -M_c}(-1)^{J_c + M_c} \sum_{j_h, j_p} x^e_{hp}\sqrt{\frac{[j_h]}{[J_c]}}\langle j_h|\, |\Omega^{J_c}|\, |j_p\rangle. \tag{III.150b}$$

We have here used the Wigner-Eckart theorem to write down the matrix element, and the Clebsch-Gordon coefficients have been summed in the usual way by using their symmetry and orthogonality properties. A more detailed derivation and discussion on the transition matrix element is given in Appendix G, which should be read while studying Chapter VI. For the present, a few brief remarks suffice.

We notice that in (III.150b) the contribution of the various hole-particle states occurs in a linear sum multiplied by their respective amplitudes x^e_{hp} in the eigenstate. In most cases (especially when $J_c = 2$, and the hole and particle states are of the same parity; the multipole operator Ω^{J_c} is then of the E2-type), one of the eigenstates Ψ_e has the amplitudes x^e_{hp} occurring with suitable relative phases, such that the contributions of the various hole-particle states in (III.150b) add up coherently. A state Ψ_e of this special type then has a large probability of transition to the ground state. The coherence in the phases of the amplitudes x^e_{hp} for this special state also has a marked consequence on its eigenvalue E_e; the particular eigenvalue E_e is found to be significantly removed from the unperturbed energy $(E_0 + \epsilon_p - \epsilon_h)$ of the group of hole-particle states. As already mentioned, more insight into all these features is given in Chapter VI and Appendix G. For the present, the reader may just accept the validity of the statements made. In this section, we shall refer to this special Ψ_e and its energy E_e as Ψ_c and E_c, respectively, where the label c denotes "collective" (the large enhancement of the transition probability is a collective effect of the various hole-particle excitations).

We next turn our attention to a single nucleon outside the closed shell Ψ_0 If this nucleon belongs to a single-particle level jm, then the resultant nuclear state, which has angular momentum j and projection m, is given by

$$|\Phi_{jm}) = C^\dagger_{jm}|\Psi_0). \tag{III.151a}$$

In such a nucleus, more complicated states may arise if the closed-shell core gets excited to the various possible eigenstates Ψ_e. States with total angular momentum j and projection m can then be produced by coupling the excited state Ψ_e of the core with the last nucleon. Denoting the angular momentum of the last nucleon in such a case by $j'm'$, we then have complicated states of the type

$$|\Phi(eJ_cj' : jm)) = \sum_{M_c,\, m'} \begin{bmatrix} J_c & j' & j \\ M_c & m' & m \end{bmatrix} C^\dagger_{j'm'}|\Psi_e(J_cM_c))$$

$$= \sum_{M_c,\, m'} \begin{bmatrix} J_c & j' & j \\ M_c & m' & m \end{bmatrix} \sum_{hp} x^e_{hp} C^\dagger_{j'm'} A^\dagger(j_h^{-1} j_p J_c M_c)|\Psi_0). \qquad \text{(III.151b)}$$

$j'm'$ is, by definition, a state above the occupied states in Ψ_0, i.e., it is a particle state, and hence the state (III.151b) is of the two-particle one-hole type. These states *per se* are not very interesting. However, the effect of their admixture in the state (III.151a), though small, has very important consequences, as now discussed.

We expect that the ground state of the nucleus with one nucleon outside the closed shell is *predominantly* of the type (III.151a). The residual interaction causes a slight admixture of the states (III.151b) in the state (III.151a). The detailed procedure for calculating the admixture entails diagonalizing the matrix of the Hamiltonian using as basis the state (III.151a) and the states (corresponding to different values of e, J_c, and j') given by (III.151b). An evaluation of the typical matrix elements for this purpose is quite straightforward. However, in our work here we shall compute the admixture in the first-order perturbation theory. That is to say, we write the ground-state wavefunction as

$$\Psi_{jm} = \Phi_{jm} + \sum_{e,\, J_c,\, j'} y_{eJ_cj'} \Phi(eJ_cj' : jm), \qquad \text{(III.152a)}$$

where the admixture coefficient $y_{eJ_cj'}$ is given by the first-order perturbation expression

$$y_{eJ_cj'} = \frac{(\Phi(eJ_cj' : jm)|\, H_I \,|\Phi_{jm})}{(E_0 + \epsilon_j) - (E_e + \epsilon_{j'})}. \qquad \text{(III.152b)}$$

The perturbation H_I in this expression is given, as usual, by

$$H_I = -\sum_{i=1}^{A+1} \mathcal{V}_i + \sum_{i<j}^{A+1} V_{ij}$$

$$= -\sum_{\alpha,\beta} \langle\alpha|\, \mathcal{V} \,|\beta\rangle C^\dagger_\alpha C_\beta + \tfrac{1}{4} \sum_{\alpha\beta\gamma\delta} (\alpha\beta|\, V \,|\gamma\delta) C^\dagger_\alpha C^\dagger_\beta C_\delta C_\gamma, \qquad \text{(III.152c)}$$

where $\mathcal{V}_i$ is the average shell model potential on the nucleon i, and V_{ij} is the two-body potential acting between the nucleons i and j.

Since H_I is a Hermitean operator, the matrix element occurring in (III.152b) is equal to $\langle\Phi_{jm}|\, H_I \,|\Phi(eJ_cj' : jm)$, which we can evaluate by using (III.151) and (III.152c) along with the usual artifices on the creation and destruction operators. Some of the relevant details are now stated. A typical expression contained in the matrix element arising from the $\mathcal{V}$-term of (III.152c) is given by

$$-\sum_{\alpha,\beta} \langle\alpha|\, \mathcal{V} \,|\beta\rangle (\Psi_0|\, C_{jm} C^\dagger_\alpha C_\beta C^\dagger_{j'm'} C^\dagger_{j_pm_p} C_{j_h, -m_h} \,|\Psi_0)$$

$$= \sum_{\alpha,\beta} \delta_{\alpha;\, (j_h,\, -m_h)} [\delta_{\beta;\, (j'm')} \delta_{jm;\, j_pm_p} - \delta_{\beta;\, (j_pm_p)} \delta_{jm;\, j'm'}] \langle\alpha|\, \mathcal{V} \,|\beta\rangle$$

$$= \langle j_h, -m_h|\, \mathcal{V} \,|j'm'\rangle \delta_{jm;\, j_pm_p} - \langle j_h, -m_h|\, \mathcal{V} \,|j_pm_p\rangle \delta_{jm;\, j'm'}. \qquad \text{(III.153a)}$$

In this expression, $\delta_{jm;\,j_pm_p}$ stands for $\delta_{jj_p}\delta_{mm_p}$, and $\delta_{\beta;\,(j'm')}$ is a notation that requires the angular momentum and projection quantum numbers contained in the symbol β to be equal to j' and m', respectively. In the same manner, the typical expression contained in the matrix element arising from the V-term of (III.152c) is given by

$$\tfrac{1}{4}\sum_{\alpha\beta\gamma\delta}(\alpha\beta|\,V\,|\gamma\delta)(\Psi_0|\,C_{jm}C_\alpha^\dagger C_\beta^\dagger C_\delta C_\gamma C_{j'm'}^\dagger C_{j_pm_p}^\dagger C_{j_h,-m_h}\,|\Psi_0)$$

$$= -\sum_\beta^{\text{occ}}(j_h,\,-m_h;\,\beta|\,V\,|j'm';\,\beta)\delta_{jm;\,j_pm_p}$$

$$+\sum_\beta^{\text{occ}}(j_h,\,-m_h;\,\beta|\,V\,|j_pm_p;\,\beta)\delta_{jm;\,j'm'} - (j_h,\,-m_h;\,jm|\,V\,|j'm';\,j_pm_p). \qquad \text{(III.153b)}$$

When we add (III.153a) and (III.153b), the first and second terms of (III.153a) combine with the same terms of (III.153b) and produce zero by virtue of the definition (III.102b) of $\mathcal{V}$. Thus, we are left with only the last term of (III.153b). This term has now obviously to be multiplied by the Clebsch-Gordon coefficients and the phase factor occurring in (III.151b) and (III.146b), and then summed over the magnetic quantum numbers; that will produce the final result for the matrix element of H_{I} occurring in (III.152b). In this way,

$$y_{eJ_cj'} = -[(E_0+\epsilon_j)-(E_c+\epsilon_{j'})]^{-1}\sum_{hp}x_{hp}^{e}$$

$$\times\sum_{M_c,\,m'}\ \sum_{m_h,\,m_p}(-1)^{j_h-m_h}\begin{bmatrix} j_h & j_p & J_c\\ m_h & m_p & M_c\end{bmatrix}\begin{bmatrix} J_c & j' & j\\ M_c & m' & m\end{bmatrix}$$

$$\times(j_h,\,-m_h;\,jm|\,V\,|j'm';\,j_pm_p). \qquad \text{(III.154)}$$

The matrix element of V can be replaced by

$$\sum_{J'}\begin{bmatrix} j_h & j & J'\\ -m_h & m & m-m_h\end{bmatrix}\begin{bmatrix} j' & j_p & J'\\ m' & m_p & m'+m_p\end{bmatrix}(j_hjJ'|\,V\,|j_h'j_pJ'),$$

where the magnetic quantum number of J' has been omitted in the matrix element because the angular momentum coupled matrix element of V is actually independent of this quantum number. Substituting this expression in (III.154), we find four Clebsch-Gordon coefficients to be summed over the projection quantum numbers M_c, m', m_h, m_p, only two of which are independent, as per the property of the Clebsch-Gordon coefficients. A summation over the two independent quantum numbers then produces a U-function in the manner of Appendix A (Section V). The final result is given by

$$y_{eJ_cj'} = [(E_c+\epsilon_{j'})-(E_0+\epsilon_j)]^{-1}\sum_{hp}x_{hp}^{e}\sqrt{\frac{[J_c]}{[j]}}F(j_hj_pj'j;\,J_c), \qquad \text{(III.155)}$$

where F is defined by (GIII.7) in Appendix G.

We next evaluate the static moment of the state (III.152a) due to the multipole operator

$$\sum_{l=1}^{A+1}\Omega_q^K(l).$$

Up to the first order in the admixture coefficients, the expression is given by

$$\langle \Psi_{jm}| \sum_{i=1}^{A+1} \Omega_0^K(I)\, |\Psi_{jm}\rangle = \langle \Phi_{jm}| \sum_{I=1}^{A+1} \Omega_0^K(I)\, |\Phi_{jm}\rangle$$

$$+ \Big[\sum_{eJ_c j'} y_{eJ_c j'} \langle \Phi_{jm}| \sum_{I=1}^{A+1} \Omega_0^K(I)\, |\Phi(eJ_c j' : jm)\rangle$$

$$+ \text{complex conjugate}\Big]. \qquad \text{(III.156)}$$

Using the expressions (III.151a) and (III.150a), we simplify the first term of (III.156) to

$$\langle \Phi_{jm}| \sum_{I=1}^{A+1} \Omega_0^K(I)\, |\Phi_{jm}\rangle = \sum_{\mu}^{\text{occ in } \psi_0} \langle \mu|\, \Omega_0^K\, |\mu\rangle + \langle jm|\, \Omega_0^K\, |jm\rangle$$

$$= \langle jm|\, \Omega_0^K\, |jm\rangle = \begin{bmatrix} j & K & j \\ m & 0 & m \end{bmatrix} \langle j\| \Omega^K \| j\rangle. \qquad \text{(III.157a)}$$

Here the summation μ over all the occupied states in the closed shell Ψ_0 yields zero, according to the proof in Section 26A, except in the special case $K = 0$. The final result (III.157a) for non-zero multipoles is equal to the moment of the single nucleon outside the closed shell. We shall therefore refer to this part of (III.156) as the single-particle contribution to the moment.

The implication of the second term of (III.156) is now explored. The expression derived for the $\mathcal{V}$-part of $\langle \Phi_{jm}|\, H_I\, |\Phi(eJ_c j' : jm)\rangle$ clearly suffices in dealing with the matrix element of Ω_0^K. We replace $-\mathcal{V}$ by Ω_0^K in the expression (III.153a), multiply by the necessary Clebsch-Gordon coefficients and the phase factor, and obtain

$$\langle \Phi(eJ_c j' : jm)| \sum_{I=1}^{A+1} \Omega_0^K(I)\, |\Phi_{jm}\rangle$$

$$= \sum_{hp} x_{hp}^{e} \sum_{M_c,\, m'} \sum_{m_h,\, m_p} (-1)^{j_h - m_h} \begin{bmatrix} J_c & j' & j \\ M_c & m' & m \end{bmatrix} \begin{bmatrix} j_h & j_p & J_c \\ m_h & m_p & M_c \end{bmatrix}$$

$$\times \left(-\langle j_h, -m_h|\, \Omega_0^K\, |j'm'\rangle \delta_{jm;\, j_p m_p} + \langle j_h, -m_h|\, \Omega_0^K\, |j_p m_p\rangle \delta_{jm;\, j'm'}\right)$$

$$= \sum_{hp} x_{hp}^{e} \sum_{m_h} (-1)^{j_h - m_h} \begin{bmatrix} J_c & j & j \\ 0 & m & m \end{bmatrix} \begin{bmatrix} j_h & j_p & J_c \\ m_h & -m_h & 0 \end{bmatrix} \begin{bmatrix} j_p & K & j_h \\ -m_h & 0 & -m_h \end{bmatrix} \langle j_h\| \Omega^K \| j_p\rangle$$

$$- \sum_{h} x_{h;\, jm}^{e} \sum_{m_h} (-1)^{j_h - m_h} \begin{bmatrix} J_c & j' & j \\ m + m_h & -m_h & m \end{bmatrix} \begin{bmatrix} j_h & j & J_c \\ m_h & m & m + m_h \end{bmatrix}$$

$$\times \begin{bmatrix} j' & K & j_h \\ -m_h & 0 & -m_h \end{bmatrix} \langle j_h\| \Omega^K \| j'\rangle. \qquad \text{(III.157b)}$$

In the second step, we have explicitly put in the requirements due to the Kronecker deltas and the relationship between the various projection quantum numbers required by the Clebsch-Gordon coefficients. The summation over m_h in the first term is trivial to carry out. Using the symmetry and orthogonality property of the Clebsch-Gordon coefficients, we get

$$\sum_{m_h} \begin{bmatrix} j_h & j_p & J_c \\ m_h & -m_h & 0 \end{bmatrix} \begin{bmatrix} j_p & K & j_h \\ -m_h & 0 & -m_h \end{bmatrix} (-1)^{j_h - m_h} = (-1)^K \sqrt{\frac{[j_h]}{[K]}}\, \delta_{K,\, J_c}.$$

46 (45-122/1977)

Therefore,

$$\text{first term of (III.157b)} = \delta_{K,J_c}\begin{bmatrix} j & J_c & j \\ m & 0 & m \end{bmatrix}\delta_{jj'}\sum_{hp} x^{\mathrm{e}}_{hp}\sqrt{\frac{[j_h]}{[J_c]}}\langle j_h|\,|\Omega^{J_c}|\,|j_p\rangle. \qquad \text{(III.157c)}$$

The close relationship of this expression with (III.150b) is apparent. In fact, we can write

$$\text{first term of (III.157b)} = \delta_{K,J_c}\begin{bmatrix} j & K & j \\ m & 0 & m \end{bmatrix}(-1)^{J_c}\delta_{jj'}\langle\Psi_0|\sum_{I=1}^{A}\Omega^K_0(I)\,|\Psi_{\mathrm{e}}\rangle. \qquad \text{(III.157d)}$$

The Clebsch-Gordon coefficient in this expression merely confirms the validity of the Wigner-Eckart theorem, and is identical to that occurring in (III.157a) which gives the single-particle contribution to the moment. The remaining factor in (III.157d) is the transition matrix element in the closed-shell nucleus from its state Ψ_{e} to the ground state Ψ_0. We had discussed after (III.150b) that this transition matrix element is large for one special state Ψ_{c}, in which the coefficients x^{c}_{hp} bear a suitable phase relationship and make the summation over (h, p) coherent. Even though the admixture coefficient $y_{cJ_cj'}$ is small, the large transition matrix element from Ψ_{c} to Ψ_0 makes the contribution of the corresponding term in (III.156) comparable to, and sometimes even larger than, the single-particle contribution (III.157a).

The entire expression (III.157d), especially the occurrence of the phase factor $(-1)^{J_c}$, can be explained as follows. We first claim that if we had ignored the antisymmetrization of the last nucleon with the core particles and evaluated the contribution of the electromagnetic moment due to the core particles alone, then the result would have been identical to (III.157d). To see this, let us write the angular momentum coupled states (without the antisymmetrization) as $\langle\Psi_0(0), j:jm|$ and $|\Psi_{\mathrm{e}}(J_c), j':jm\rangle$, where the quantities within the parentheses following Ψ_0 and Ψ_{e} denote the angular momenta of these states. Since the operator

$$\sum_{I=1}^{A}\Omega^K_0(I)$$

acts on the core part of these states, we have to apply the standard result (BIII.14) from Appendix B, which yields

$$\begin{aligned}
&\langle\Psi_0(0), j:jm|\sum_{I=1}^{A}\Omega^K_0(I)\,|\Psi_{\mathrm{e}}(J_c), j':jm\rangle \\
&= \begin{bmatrix} j & K & j \\ m & 0 & m \end{bmatrix}U(jJ_cjK; j0)(-1)^{J_c}\delta_{jj'}\langle\Psi_0(0)|\,|\sum_{I=1}^{A}\Omega^K_0(I)|\,|\Psi_{\mathrm{e}}(J_c)\rangle \\
&= \begin{bmatrix} j & K & j \\ m & 0 & m \end{bmatrix}\left\{(-1)^{J_c}\delta_{KJ_c}\sqrt{\frac{[j]}{[K][j]}}\right\}(-1)^{J_c}\begin{bmatrix} J_c & J_c & 0 \\ 0 & 0 & 0 \end{bmatrix}^{-1}\begin{bmatrix} J_c & J_c & 0 \\ 0 & 0 & 0 \end{bmatrix} \\
&\quad\times\langle\Psi_0(0)|\,|\sum_{I=1}^{A}\Omega^K_0(I)|\,|\Psi_{\mathrm{e}}(J_c)\rangle \\
&= \begin{bmatrix} j & K & j \\ m & 0 & m \end{bmatrix}(-1)^{J_c}\langle\Psi_0|\sum_{I=1}^{A}\Omega^K_0(I)\,|\Psi_{\mathrm{e}}\rangle.
\end{aligned}$$

In the second step, the expression within the braces is the value of the U-function. The Clebsch-Gordon coefficient and its inverse have been introduced such that the former, along with the *reduced* matrix element, produces the matrix element in the last step. The inverse of the Clebsch-Gordon coefficient has the value $(-1)^{J_c}[J_c]^{1/2}$; this value has been used in obtaining the final expression, which incidentally checks our assertion about (III.157d).

We shall next investigate the effect of the second term of (III.157b). The summation over m_h can be carried out by using the standard result (AV.12) from Appendix A, leading to the expression

$$-\begin{bmatrix} j & K & j \\ m & 0 & m \end{bmatrix} \sum_h x^{\mathrm{e}}_{h,j}(-1)^{j_h+j+J_c+K}\sqrt{\frac{[j_h]}{[K]}}U(jj_hjj';\, J_cK)\langle j_h|\,|\Omega^K|\,|j'\rangle. \tag{III.158}$$

We have already mentioned that the first term of (III.157b), as given by (III.157c) or (III.157d), is large in the special case of the collective core state (e = c) due to the coherence in the sum over the hole-particle states. The expression (III.158) is usually much smaller, by comparison, for several reasons: (i) the summation does not proceed over the full set of hole-particle states, as in (III.157c); (ii) the various terms in the summation may not coherently add up; and (iii) the transition matrix element of Ω^K is reduced by the presence of the U-function. Although the effect of (III.158) can be retained in any detailed numerical calculation, it is, for the reasons just stated, usually dropped in a discussion of the core-polarization effect (i.e., hole-particle excitation from the core) on the electromagnetic moments of a single nucleon outside the closed shell.

We reject the second term of (III.157b) and get the final expression for the electromagnetic moment from the parent expression (III.156) after substituting from (III.157a), (III.157d), and (III.155). We keep only the contribution of the collective state c, dropping the other terms in the summation over the various eigenstates e. The transition matrix element of the closed-shell nucleus for the collective state, occurring in (III.157d), is denoted by m_K for brevity; the moment of the state Ψ_{jm} appearing on the left-hand side of (III.156) is also replaced by the notation $\langle \Omega_0^K \rangle_{jm}$. In this way,

$$\langle \Omega_0^K \rangle_{jm} = \begin{bmatrix} j & K & j \\ m & 0 & m \end{bmatrix}\{\langle j|\,|\Omega^K|\,|j\rangle + 2(E_c - E_0)^{-1}(-1)^K m_K \sum_{hp} x^{\mathrm{c}}_{hp}\sqrt{\frac{[K]}{[j]}}F(j_hj_pjj;\, K)\}. \tag{III.159}$$

The factor of 2 here takes care of the 'complex conjugate' term in (III.156) in view of the reality of all the quantities appearing in the expression. The second term within the braces is the contribution of the collective state of core polarization to the single-particle electromagnetic moment.

The effect of the second term of (III.159) may be viewed in an alternative manner. If we can express this term as a number multiplying the matrix element of an operator connecting $\langle j|$ with $|j\rangle$, then the whole expression (III.159) becomes the single-particle matrix element of an effective electromagnetic operator of which the first term is the actual electromagnetic operator Ω^K, and the second term is a consequence of core polarization. With this programme in view, let us examine, from Appendix G, the expression (GIII.7) of the F-matrix element. The two-body potential can, in general, be expanded in the form

$$V(\mathbf{r}, \boldsymbol{\sigma}_1, \boldsymbol{\sigma}_2) = \sum_L F_L(r_1, r_2)T^L(\hat{\mathbf{r}}_1, \boldsymbol{\sigma}_1)\cdot T^L(\hat{\mathbf{r}}_2, \boldsymbol{\sigma}_2), \tag{III.160}$$

where the unit vectors $\hat{\mathbf{r}}_1, \hat{\mathbf{r}}_2$ stand for the (θ, ϕ)-angles of $\mathbf{r}_1$ and $\mathbf{r}_2$, respectively, and T^L is a tensor of rank L constructed out of the spherical harmonics of these angles and the spin operator shown within the parentheses. If the isotopic spin has to be included, then the whole expression (III.160) has to be multiplied by $(a_0 + a_\tau \boldsymbol{\tau}_1\cdot\boldsymbol{\tau}_2)$; the reader may do this generalization, if he so wishes. Using the foregoing general expression for the two-body potential, and the standard result (BIII.15) from Appendix B, we obtain, for the direct term and the exchange term, respectively, of F [see (GIII.7) in Appendix G], the expressions

$$F_D(abcd; J_c) = -(-1)^{a+b+c+d} \sum_{J'} \sqrt{\frac{[J']}{[J_c]}} U(abdc; J_cJ')\langle cbJ'| V |adJ'\rangle$$
$$= (-1)^{a+b+c+d} \sum_{J'} \sqrt{\frac{[J']}{[J_c]}} U(abdc; J_cJ') \sum_L (-1)^L \sqrt{\frac{[b]}{[d]}}$$
$$\times U(LabJ'; cd)\langle c| |T^L| |a\rangle\langle b| |T^L| |d\rangle \mathcal{J}_L(abcd) \qquad \text{(III.161a)}$$

and

$$F_E(abcd; J_c) = (-1)^{a+b+c+d} \sum_{J'} (-1)^{a+d+J'} \sqrt{\frac{[J']}{[J_c]}} U(abdc; J_cJ')\langle cbJ'| V |daJ'\rangle$$
$$= (-1)^{a+b+c+d} \sum_{J'} \sqrt{\frac{[J']}{[J_c]}} U(abdc; J_cJ') \sum_L (-1)^{L+a+d+J'} \sqrt{\frac{[b]}{[a]}}$$
$$\times U(LdbJ'; ca)\langle c| |T^L| |d\rangle\langle b| |T^L| |a\rangle \mathcal{J}_L(abcd). \qquad \text{(III.161b)}$$

The quantity $\mathcal{J}_L(abcd)$ is the radial integral of $F_L(r_1, r_2)$. The summation over J' in the exchange matrix element (III.161b) can be very easily carried out by using the standard symmetry and orthogonality relations of the U-function given in Appendix A (Section V). This summation produces δ_{LJ_c}, and hence the summation over L in (III.161b) disappears, yielding

$$F_E(abcd; J_c) = (-1)^{a+b} \frac{\sqrt{[bc]}}{[J_c]} \langle c| |T^{J_c}| |d\rangle\langle b| |T^{J_c}| |a\rangle \mathcal{J}_{J_c}(abcd). \qquad \text{(III.161c)}$$

A similar attempt at carrying out the summation over J' in (III.161a) with the help of the result (AV.13) from Appendix A produces the U-function $U(abcd; J_cL)$, and the summation over L remains. For simplicity, the entire direct term (III.161a) is dropped in many discussions, and the F-matrix element is approximated by (III.161b) or, equivalently, its simplified version (III.161c). It is also usual to approximate the radial integral $\mathcal{J}_{J_c}(abcd)$ by a constant $\mathcal{J}$. Substituting the simplified version of F in the second term of (III.159), we obtain

$$-2(E_c - E_0)^{-1}(-1)^K m_K \mathcal{J} \sum_{hp} x^c_{hp} \sqrt{\frac{[K]}{[j]}} (-1)^{j_h+j_p} \sqrt{\frac{[jj_p]}{[K]}} \langle j| |T^K| |j\rangle\langle j_p| |T^K| |j_h\rangle.$$

Therefore, according to the scheme stated just before (III.160), we introduce the effective multipole operator Ω^K_{eff} and write

$$\langle \Omega^K_0\rangle_{jm} = \begin{bmatrix} j & K & j \\ m & 0 & m \end{bmatrix} \langle j| |\Omega^K_{\text{eff}}| |j\rangle,$$

where

$$\Omega^K_{\text{eff}} = \Omega^K - [2(E_c - E_0)^{-1}(-1)^K m_K \mathcal{J} \sum_{hp} x^c_{hp}(-1)^{j_h+j_p}\sqrt{[j_p]}\langle j_p| |T^K| |j_h\rangle]T^K. \qquad \text{(III.161d)}$$

The case of the magnetic moment is now specially discussed. Here

$$\Omega^1 = g_l \boldsymbol{l} + g_s \mathbf{s}.$$

The hole and particle states that contribute to m_1 through the matrix elements $\langle j_h| |g_l\boldsymbol{l} + g_s\mathbf{s}| |j_p\rangle$ must therefore be of the same parity and have the same orbital angular momentum, and

$(j_h, j_p, 1)$ must satisfy angular momentum addition rules. The operator T^1, by virtue of the matrix element $\langle j_p| |T^1| |j_h\rangle$, should then be able to connect hole and particle states with these characteristics. We recall that T^K occurring in the two-body potential has the general structure $(Y^{K'}, \sigma)^K$. Thus, the simplest T^1 that conserves parity and orbital angular momentum is $(Y^0, \sigma)^1$, i.e., σ itself (because Y^0 is a constant). According to angular momentum coupling rules, K' can have two other possible values, namely, 1 and 2. The former is ruled out because it can connect only states of opposite parity. Thus, the only other term is of the type $(Y^2, \sigma)^1$. With this discussion on the case of the magnetic moment, the reader should refer back to Section 26A, where the particular structure derived here was assumed.

C. SENIORITY AND FRACTIONAL PARENTAGE BY SECOND-QUANTIZATION TECHNIQUE

Seniority

The concept of seniority has been introduced in Section 29A in terms of the addition of 'pairs' coupled to zero angular momentum. The same concept is now discussed in terms of creation and annihilation operators. The addition of a zero angular momentum pair to any state is achieved by operating on it with the operator $A^\dagger(jjJ = 0, M = 0)$, defined according to (III.132b). Since we shall consider, in this section, nucleons belonging to a single value of j only, this quantum number will be frequently omitted from all labels; the projection quantum number m will be explicitly used for the purpose of labelling the operators $C^\dagger$ and C. The pair operator also will be denoted by $A_0^\dagger$ for brevity. Thus,

$$A_0^\dagger = \frac{1}{\sqrt{2}} \sum_m \begin{bmatrix} j & j & 0 \\ m & -m & 0 \end{bmatrix} C_{-m}^\dagger C_m^\dagger$$

$$= \frac{1}{\sqrt{2[j]}} \sum_m (-1)^{j-m} C_{-m}^\dagger C_m^\dagger. \qquad \text{(III.162a)}$$

The Hermitean conjugate of the operator A_0 is then given by

$$A_0 = \frac{1}{\sqrt{2[j]}} \sum_m (-1)^{j-m} C_m C_{-m}. \qquad \text{(III.162b)}$$

We first evaluate the commutator of these two operators with the help of the known anti-commuting property of $C^\dagger$, C, ... Commencing with the expression $C_{-m}^\dagger C_m^\dagger C_{m'} C_{-m'}$, we take the creation operators to the extreme right, repeatedly using the anticommutators. Thus,

$$\begin{aligned} C_{-m}^\dagger C_m^\dagger C_{m'} C_{-m'} &= C_{-m}^\dagger C_{-m'} \delta_{mm'} - C_{-m}^\dagger C_{m'} C_m^\dagger C_{-m'} \\ &= C_{-m}^\dagger C_{-m'} \delta_{mm'} - C_{-m}^\dagger C_{m'} \delta_{m,-m'} + C_{-m}^\dagger C_{m'} C_{-m'} C_m^\dagger \\ &= C_{-m}^\dagger C_{-m'} \delta_{mm'} - C_{-m}^\dagger C_{m'} \delta_{m,-m'} + \delta_{-m,m'} C_{-m'} C_m^\dagger - C_{m'} C_{-m}^\dagger C_{-m'} C_m^\dagger \\ &= C_{-m}^\dagger C_{-m'} \delta_{mm'} - C_{-m}^\dagger C_{m'} \delta_{m,-m'} + \delta_{-m,m'} C_m^\dagger C_{-m'} - \delta_{-m,m'} \\ &\quad + C_m^\dagger C_{m'} \delta_{-m,-m'} - \delta_{-m',-m} + C_{m'} C_{-m'} C_{-m}^\dagger C_m^\dagger. \end{aligned}$$

Transferring the last term to the left-hand side, we obtain the desired commutator

$$[C_{-m}^\dagger C_m^\dagger, C_{m'} C_{-m'}] = (\delta_{mm'} - \delta_{m,-m'})(C_m^\dagger C_m + C_{-m}^\dagger C_{-m} - 1). \qquad \text{(III.163)}$$

Making use of this result, we easily obtain

$$[A_0^\dagger, A_0] = \frac{1}{2[j]} \sum_{m,m'} (-1)^{j-m}(-1)^{j-m'}[C_{-m}C_m^\dagger, C_{m'}C_{-m'}]$$

$$= \frac{1}{[j]} \sum_m (C_m^\dagger C_m + C_{-m}^\dagger C_{-m} - 1)$$

$$= \frac{2\hat{N}}{[j]} - 1, \tag{III.164a}$$

where

$$\hat{N} = \sum_m C_m^\dagger C_m = \sum_m C_{-m}^\dagger C_{-m}. \tag{III.164b}$$

The term $\delta_{mm'}$ produces 1 out of $(-1)^{j-m}(-1)^{j-m'}$; on the other hand, these phase factors produce -1 with the term $\delta_{m,-m'}$. Therefore, $(\delta_{mm'} - \delta_{m,-m'})$ of (III.163) gives a factor of 2 in (III.164a).

We next evaluate the commutator of $\hat{N}$ with $A_0^\dagger$ and A_0. Using the same procedure as just described, we get

$$[C_m^\dagger C_m, C_{-m'}^\dagger C_{m'}] = \delta_{m,-m'} C_m^\dagger C_{m'}^\dagger - \delta_{mm'} C_m^\dagger C_{-m'}^\dagger,$$

and hence

$$[\hat{N}, A_0^\dagger] = \frac{1}{\sqrt{2[j]}} \sum_{m,m'} (\delta_{m,-m'} - \delta_{mm'}) C_m^\dagger C_{-m}^\dagger (-1)^{j-m}$$

$$= \sqrt{\frac{2}{[j]}} \sum_m C_{-m}^\dagger C_m^\dagger (-1)^{j-m} = 2A_0^\dagger. \tag{III.165a}$$

By taking the Hermitean conjugate ($\hat{N}$ is Hermitean) of (III.165a), we have

$$[\hat{N}, A_0] = -2A_0. \tag{III.165b}$$

The expressions (III.165) imply that if we multiply $A_0^\dagger$ and A_0 by any arbitrary constant C, then

$$[\tfrac{1}{2}\hat{N}, CA_0^\dagger] = CA_0^\dagger, \tag{III.166a}$$

$$[\tfrac{1}{2}\hat{N}, CA_0] = -CA_0. \tag{III.166b}$$

By the same multiplication, (III.164a), however, changes to

$$[CA_0^\dagger, CA_0] = C^2\left(\frac{2\hat{N}}{[j]} - 1\right). \tag{III.166c}$$

Let us now compare (III.166a), (III.166b), and (III.166c) with the standard commutators of the components of *any* angular momentum operator **J**:

$$[J_z, J_+] = J_+, \tag{III.167a}$$

$$[J_z, J_-] = -J_-, \tag{III.167b}$$

$$[J_+, J_-] = 2J_z, \tag{III.167c}$$

where

$$J_\pm = J_x \pm iJ_y.$$

To make their similarity obvious, we first choose C^2 in such a way that the $\hat{N}$-dependent term of (III.166c) becomes equal to $\hat{N}$, i.e.,

$$C^2 = \frac{[j]}{2} \quad \text{or} \quad C = \sqrt{\frac{[j]}{2}}. \tag{III.168}$$

This choice is dictated by the fact that the right-hand side of (III.167c) is $2J_z$ when the left-hand side of (III.167a) and (III.167b) contains J_z; since the left-hand side of (III.166a) and (III.166b) contains $\frac{1}{2}\hat{N}$, we make the corresponding term on the right-hand side of (III.166c) equal to twice $\frac{1}{2}\hat{N}$. The presence of the constant term $-C^2$ on the right-hand side of (III.166c) presents no trouble. We simply include $-\frac{1}{2}C^2$ along with $\frac{1}{2}\hat{N}$ in (III.166a) and (III.166b). It should be noticed that this inclusion does not invalidate (III.166a) and (III.166b) because the commutator of the constant with $A_0^\dagger$ or A_0 is zero.

We now introduce the new nomenclature $S_\pm$, S_z according to the relations

$$S_+ = CA_0^\dagger = \frac{1}{2}\sum_m (-1)^{j-m} C_{-m}^\dagger C_m^\dagger, \tag{III.169a}$$

$$S_- = CA_0 = \frac{1}{2}\sum_m (-1)^{j-m} C_m C_{-m}, \tag{III.169b}$$

$$S_z = \tfrac{1}{2}\hat{N} - \tfrac{1}{2}C^2 = \tfrac{1}{2}\left(\hat{N} - \frac{[j]}{2}\right). \tag{III.169c}$$

According to (III.166) and the value of C given in (III.168), these operators satisfy

$$[S_z, S_\pm] = \pm S_\pm, \qquad [S_+, S_-] = 2S_z. \tag{III.170}$$

Now the identity of the commutators with (III.167) is fully transparent. The operator **S** whose components are $S_\pm$ and S_z can therefore be identified as an angular momentum operator in a fictitious space. It has been called *quasispin* in the literature (see Kerman[6]). The simultaneous eigenstates of S^2 and S_z can then be investigated in accordance with the standard procedure of angular momentum algebra.

The eigenvalues M_S of S_z are very easy to find. From (III.169c), we have

$$M_S = \tfrac{1}{2}\left(n - \frac{[j]}{2}\right), \tag{III.171a}$$

where n is the eigenvalue of the number operator $\hat{N}$, i.e., n is the number of nucleons belonging to the level j. From the structure of (III.169a) and (III.169b), we notice that S_+ and S_- respectively add and remove two particles, i.e., they change n to $n + 2$ and $n - 2$, respectively. From (III.171), it is then clear that S_+ and S_- respectively step up and step down M_S by unity, as they should. According to the well-known property of the step-up and step-down operators, the quantum number S of a state does not change by their operation. S here gives the eigenvalue of S^2, in the usual way, as $S(S + 1)$. We therefore notice that the $(2S + 1)$-fold multiplet corresponding to the various M_S-quantum numbers of a given quasispin S represents levels of different nuclei, each nucleus differing from the next in the number of nucleons n by 2. The minimum value of M_S is $-S$ and the maximum is $+S$. The state $|S, -S\rangle$ produces zero when S_- acts on it and the state $|S, +S\rangle$ produces zero under the operation of S_+. (III.171a) says that the minimum value M_S in a multiplet corresponds to a minimum value of n; let us denote this $n_{\min}$ by v. All the other states belonging to the same S, but with higher values of the M_S-quantum number, correspond to $n = v + 2,\ v + 4, \ldots$. Clearly, the S-quantum number of a multiplet of levels, which are found starting with $n = v$, is given by

$$-S = \tfrac{1}{2}\left(v - \frac{[j]}{2}\right) \quad \text{or} \quad S = \tfrac{1}{2}\left(\frac{[j]}{2} - v\right). \tag{III.171b}$$

The quantity v is the seniority quantum number of this set of levels. The state $|S, -S\rangle$ is then identical to the state $|j^v vJ\rangle$, and hence we have

$$S_-|j^v vJ\rangle = 0. \tag{III.172}$$

By the repeated application of S_+, i.e., the repeated addition of zero-coupled pairs to the state $|j^v vJ\rangle$, we generate all the states of the multiplet. Thus, with $n > v$, we have

$$|j^n vJ\rangle = N_P(S_+)^P|j^v vJ\rangle, \tag{III.173a}$$

where N_P is the normalization factor, and P denotes the number of zero-coupled pairs, i.e., $P = \tfrac{1}{2}(n - v)$. The normalization constant N_P can be worked out as follows. We have

$$S_-S_+ = S_x^2 + S_y^2 - S_z = S^2 - S_z^2 - S_z$$

and, therefore, its matrix element connecting two normalized states of seniority v and number of particles ν is given, according to (III.171a) and (III.171b), by

$$\begin{aligned}\langle j^\nu vJ|\, S_-S_+ \,|j^\nu vJ\rangle &= S(S+1) - M_S(M_S+1) - M_S \\ &= (S - M_S)(S + M_S + 1) \\ &= \tfrac{1}{4}([j] - \nu - v)(\nu - v + 2).\end{aligned} \tag{III.173b}$$

At the stage of adding one pair to the state $\nu = v$, we have

$$|j^{v+2} vJ\rangle = N_1 S_+|j^v vj\rangle \tag{III.174a}$$

such that the normalization to unity demands

$$1 = (N_1)^2\langle j^v vJ|\, S_-S_+ \,|j^v vJ\rangle$$

or

$$(N_1)^{-2} = \tfrac{1}{4}([j] - 2v)\cdot 2 = \left(\frac{[j]}{2} - v\right)\cdot 1.$$

We have here substituted from the expression (III.173b) with $\nu = v$. In the same manner, if we write

$$|j^{v+4} vJ\rangle = \mathcal{N}_2 S_+|j^{v+2} vJ\rangle, \tag{III.174b}$$

then

$$(\mathcal{N}_2)^{-2} = \tfrac{1}{4}([j] - 2v - 2)\cdot 4 = \left(\frac{[j]}{2} - v - 1\right)\cdot 2.$$

The normalization constant N_2 occurring in (III.173a) for $P = 2$ is defined by

$$|j^{v+4} vJ\rangle = N_2(S_+)^2|j^v vJ\rangle$$

and is given, according to (III.174a) and (III.174b), by

$$N_2 = N_1\mathcal{N}_2.$$

Therefore,

$$(N_2)^{-2} = (\frac{[j]}{2} - v - 1)(\frac{[j]}{2} - v) \cdot 2!$$

The pattern for proceeding towards N_P is now obvious. A step-by-step approach clearly produces

$$(N_P)^{-2} = P!\{\frac{[j]}{2} - v - (P - 1)\} \ldots (\frac{[j]}{2} - v)$$

$$= \frac{P!(\frac{[j]}{2} - v)!}{(\frac{[j]}{2} - v - P)!}. \quad \text{(III.174c)}$$

The correspondence (III.171b) of the seniority quantum number v with the quasispin S is one-to-one. Thus, states of different seniority but of the same n and J belong to different values of the S-quantum number; according to the general property of angular momentum eigenfunctions, the states of different S must be orthogonal to each other. Thus, we have proved

$$\langle j^n v' J \mid j^n v J \rangle = \delta_{vv'}. \quad \text{(III.175)}$$

Incidentally, the maximum M_S $(= +S)$ for a given S occurs for a maximum n given by

$$S = \tfrac{1}{2}(n_{\max} - \frac{[j]}{2}).$$

Using the value of S given by (III.171b), we then have

$$n_{\max} = \frac{[j]}{2} - v + \frac{[j]}{2} = [j] - v = (2j + 1 - v).$$

Now a configuration of $(2j + 1 - v)$ nucleons can be regarded as that of v holes in the j-level and, according to our discussion in Section 30A, the states of v holes are the same as those of v particles. Since the multiplet under discussion commences with $n = v$, and no level of this particular multiplet is to be found for a smaller number of *particles*, the aforementioned correspondence between the hole states and particle states, in fact, leads us to expect that no levels of the same seniority will be found for a number of *holes* smaller than v. The derivation of $n_{\max}$ merely verifies this fact.

We shall next derive an interesting result on the pairing interaction (V_P) which is defined in a general manner by (VI.93). Since we are concerned here with levels of a single j, we express the pairing interaction in this special case as

$$V_P = \tfrac{1}{4}G \sum_{m, m'} (-1)^{j-m} C_m^\dagger C_{-m}^\dagger C_{-m'} C_{m'} (-1)^{j-m'}, \quad \text{(III.176a)}$$

where the constant G denotes the strength of the interaction. Referring to the expressions (III.169a) and (III.169b), we immediately obtain

$$V_P = G S_+ S_- = \tfrac{1}{2}G[j] A_0^\dagger A_0. \quad \text{(III.176b)}$$

The operator identity

$$S_+ S_- = S_x^2 + S_y^2 + S_z = S^2 - S_z^2 + S_z$$

then clarifies that V_P is diagonal with respect to states labelled by the quasispin S and its projection M_S or, equivalently, those labelled by seniority and the particle number n. In fact, using the values of S and M_S from (III.171), we obtain

$$\langle j^n vJ| V_P |j^n vJ\rangle = G(S + M_S)(S - M_S + 1)$$
$$= \tfrac{1}{4}G(n - v)(2j - n - v + 3). \tag{III.177a}$$

In particular, for an even value of n, and the state of lowest seniority ($v = 0$), (III.177a) yields the value

$$\tfrac{1}{4}Gn(2j - n - 3). \tag{III.177b}$$

A somewhat more general two-body potential can also be exactly treated on the basis of seniority wavefunctions. Any two-body potential V_{12} can be expanded in the manner of (III.160), where the sum L proceeds over all possible integer values. The special potential to be treated here contains only the odd values of L in this sum, and will be called the odd-tensor interaction. It is given by

$$V_{12} = \sum_L^{\text{odd}} F_L(r_1, r_2)T^L(\hat{\mathbf{r}}_1, \boldsymbol{\sigma}_1) \cdot T^L(\hat{\mathbf{r}}_2, \boldsymbol{\sigma}_2). \tag{III.178}$$

We begin with the second-quantized version of a completely general two-body potential, and then point out, at a suitable stage, that we need only the odd-tensor terms in the expansion in order to achieve a closed expression for the matrix element. At that stage, we fall back on the special potential (III.178). The second-quantized form of the potential will be written by using the angular momentum coupled pair operators (III.132b) and their Hermitean conjugate. Further, since we are concerned with a single j, the summation will be on the projection quantum numbers alone. Thus, keeping the j-quantum number understood, we have

$$V = \tfrac{1}{2} \sum_{m_1m_2m_3m_4} \langle m_1m_2| V |m_3m_4\rangle C^\dagger_{m_1}C^\dagger_{m_2}C_{m_4}C_{m_3}$$
$$= \tfrac{1}{2} \sum_{J,M} \sum_{m_1m_2m_3m_4} \begin{bmatrix} j & j & J \\ m_1 & m_2 & M \end{bmatrix}\begin{bmatrix} j & j & J \\ m_3 & m_4 & M \end{bmatrix}\langle jjJ| V |jjJ\rangle C^\dagger_{m_1}C^\dagger_{m_2}C_{m_4}C_{m_3}$$
$$= \sum_{J,M} V_J A^\dagger_{JM} A_{JM}. \tag{III.179}$$

The operator $A^\dagger_{JM}$ is the same as $A^\dagger(j^2JM)$ of (III.132b). The quantity V_J is defined by

$$V_J \equiv \langle jjJ| V | jjJ\rangle$$
$$= \sum_L (-1)^L U(LjjJ; jj)\langle j| |T^L| |j\rangle^2. \tag{III.180}$$

Here we have used the multipole expansion (III.178) of the two-body potential and the standard result (BIII.15) from Appendix B.

We next want to evaluate the commutator of S_- with V. With this aim in view, we first work out, in the usual way,

$$[A_0, A^\dagger_{JM}] = \delta_{J0}\delta_{M0} - \frac{2}{\sqrt{[j]}}\hat{N}_{JM}, \tag{III.181a}$$

where the operator $\hat{N}_{JM}$ is defined by

$$\hat{N}_{JM} = \sum_{m,m'} \begin{bmatrix} j & j & J \\ m & m' & M \end{bmatrix} C^\dagger_m C_{-m'}(-1)^{j-m'}. \tag{III.181b}$$

Using the expressions (III.181a) and (III.179) and the relation

$$S_- = \sqrt{\frac{[j]}{2}} A_0,$$

we get

$$[S_-, V] = \sqrt{\frac{[j]}{2}} \sum_{J,M} V_J(\delta_{J0}\delta_{M0} - \frac{2}{\sqrt{[j]}} \hat{N}_{JM}) A_{JM}$$

$$= V_0 S_- - \sqrt{2} \sum_{JM} V_J \hat{N}_{JM} A_{JM}. \qquad \text{(III.182)}$$

We subsequently see that the second term of (III.182) creates complications, and hence we now look for a special two-body potential for which this term is zero. This investigation leads us to the odd-tensor interaction (III.178). We write $\hat{N}_{JM}A_{JM}$ explicitly and, after using the expression (III.180), obtain for the second term the result

$$-\sqrt{2} \sum_{JM} \sum_{L} (-1)^L U(LjjJ; jj) \langle j||T^L||j\rangle^2 \sum_{m_1m'_1} \sum_{m_2m'_2} \begin{bmatrix} j & j & J \\ m_1 & m'_1 & M \end{bmatrix}$$

$$\times C^\dagger_{m'_1} C_{-m_1} (-1)^{j-m_1} \begin{bmatrix} j & j & J \\ m_2 & m'_2 & M \end{bmatrix} C_{m_2} C_{m'_2}. \qquad \text{(III.183a)}$$

Here J occurs in the U-function and the two Clebsch-Gordon coefficients. The summation over J of these factors, according to the standard result (AV.11) in Appendix A, will give rise to two new Clebsch-Gordon coefficients. In this way, we finally obtain, for the second term of (III.182), the expression

$$-\sqrt{2} \sum_{L,M} \langle j||T^L||j\rangle^2 \frac{[j]}{[L]} \hat{N}_{LM} A_{L,-M}. \qquad \text{(III.183b)}$$

Because of the anticommuting property of the single-particle destruction operators, the quantity L in (III.183b) labelling $A_{L,-M}$ can have only even values. Therefore, we conclude that the second term of (III.182) is zero if the even-L tensor components of the potential are absent from the beginning. We have thus established our assertion following (III.178). For the odd-tensor interaction, we then have the simple result

$$[S_-, V] = V_0 S_-. \qquad \text{(III.184)}$$

We now use (III.184) in evaluating the matrix element $\langle j^n v' J| V |j^n v J\rangle$. According to (III.173a), we have

$$\langle j^n v' J| V |j^n v J\rangle = N_P N_{P'} \langle j^{v'} v' J| S_-^{P'} V S_+^P |j^v v J\rangle,$$

where

$$P = \tfrac{1}{2}(n - v), \qquad P' = \tfrac{1}{2}(n - v').$$

Now we commute the factors S_- from the left of V to its right by using the anticommutator (III.184). After commuting the first S_-, we obtain, for example,

$$\langle j^n v' J| V | j^n v J\rangle = N_P N_{P'} \langle j^v v' J| (S_-)^{P'-1}([S_-, V] + VS_-)(S_+)^P |j^v v J\rangle$$

$$= V_0 N_P N_{P'} \langle j^{v'} v' J| (S_-)^{P'} (S_+)^P |j^v v J\rangle$$

$$+ N_P N_{P'} \langle j^{v'} v' J| (S_-)^{P'-1} V S_- (S_+)^P |j^v v J\rangle.$$

In the first term, we get back $V_0\langle j^n v' J \mid j^n vJ\rangle$ [according to the definition (III.173a)] which is $V_0\delta_{vv'}$. The second term can be handled by repeating the procedure of commuting another S_- from the left to the right of V, and so on. In this way, every time one S_- is taken to the right of V, we pick up a term $V_0\delta_{vv'}$, getting altogether $PV_0\delta_{vv'}$ (notice that $v = v'$ makes $P = P'$). When all the S_--factors have shifted to the right of V, we are also left with the term $N_P N_{P'}\langle j^{v'} v' J| V(S_-)^{P'}(S_+)^P |j^v vJ\rangle$. Therefore, the final expression is given by

$$\langle j^n v' J| V |j^n vJ\rangle = PV_0\delta_{vv'} + N_P N_{P'}\langle j^{v'} v' J| V(S_-)^{P'}(S_+)^P |j^v vJ\rangle. \tag{III.185}$$

We, however, know that the operators S_+ and S_- cannot change the quasispin quantum number S or, equivalently, the quantum number v. The potential V, on the other hand, cannot change the number of nucleons in a state. Thus, the second term in (III.185) can be written as

$$N_P N_{P'}\langle j^{v'} v' J| V |j^{v'} vJ\rangle\langle j^{v'} vJ| (S_-)^{P'}(S_+)^P |j^v vJ\rangle. \tag{III.186}$$

If the state $|j^{v'} vJ\rangle$ exists, then of course $v' \geqslant v$. However, it can be shown that the matrix element $\langle j^{v'} v' J| V |j^{v'} vJ\rangle$ is nonvanishing only if $v' = v$. In that case, $P' = P$ and the remaining factors of (III.186) reduce to

$$N_P^2\langle j^v vJ| (S_-)^P(S_+)^P |j^v vJ\rangle \equiv \langle j^n vJ |j^n vJ\rangle = 1.$$

Therefore, from (III.186) and (III.185) for the odd-tensor interaction, we obtain the final result

$$\langle j^n v' J| V |j^n vJ\rangle = \delta_{vv'}(PV_0 + \langle j^v vJ| V |j^v vJ\rangle). \tag{III.187}$$

Coefficient of Fractional Parentage (cfp)

Our discussion on the cfp is based mainly on the work of Macfarlane[7]. Let us start with any antisymmetric state $|j^n vJM)$ of n equivalent particles. By definition, $|j^{n-1}v'J', j_n' : JM\rangle$ is a state obtained by coupling the angular momentum of the n-th nucleon with the antisymmetric state $|j^{n-1}, v'J'M')$ of the first $(n - 1)$ nucleons; in this state, there is no antisymmetry between the n-th nucleon and any of the first $(n - 1)$ nucleons. The state $|j^n vJM)$ can be expanded in terms of the states $|j^{n-1}v'J', j_n' : JM\rangle$; the summation will be over the quantum numbers $v'J'$. The coefficients in this expansion are the cfp's. We write the expansion as

$$|j^n vJ) = \sum_{v'J'} c(n - 1, v'J'; n, vJ)|j^{n-1}v'J', j_n' : JM\rangle \tag{III.188a}$$

such that the cfp $c(n - 1, v'J'; n, vJ)$ is identically equal to the overlap of the two states

$$c(n - 1, v'J'; n, vJ) = \langle j^{n-1}v'J', j_n : JM| j^n vJM). \tag{III.188b}$$

From the transformation property of the two states under rotation, it can be shown that this overlap is indeed independent of the quantum number M.

In view of the antisymmetry of $|j^n vJM)$, we have

$$(1 - \sum_{i=1}^{n-1} P_{in})|j^n vJM) = n|j^n vJM).$$

Therefore,

$$c(n - 1, v'J'; n, vJ) = \frac{1}{n}\langle j^{n-1}v'J', j_n : JM| (1 - \sum_{i=1}^{n-1} P_{in}) |j^n vJM). \tag{III.188c}$$

We now let the antisymmetrizer operator

$$(1 - \sum_{i=1}^{n-1} P_{in})$$

operate to its left in (III.188c). The result is a completely antisymmetric state in which the $(n - 1)$ nucleons have a parent state $(v'J')$; the state, however, is unnormalized, and requires a normalization factor to be normalized to unity. We introduce the parenthetical notation to denote the completely antisymmetric *unnormalized* state:

$$\langle j^{n-1}v'J', j_n : JM|(1 - \sum_{i=1}^{n-1} P_{in}) = (j^{n-1}v'J', j : JM|. \tag{III.189}$$

From this point onwards, it is convenient for us to apply second-quantization algebra. In this terminology, the antisymmetric state (III.189) is written as

$$|j^{n-1}v'J', j : JM) = \sqrt{n} \sum_{M'm} \begin{bmatrix} J' & j & J \\ M' & m & M \end{bmatrix} C_{jm}^{\dagger}|j^{n-1}v'J'M') \tag{III.190a}$$

or

$$(j^{n-1}v'J', j : JM| = \sqrt{n} \sum_{M',m} \begin{bmatrix} J' & j & J \\ M' & m & M \end{bmatrix} (j^{n-1}v'J'M'|C_{jm}. \tag{III.190b}$$

The numerical factor $\sqrt{n}$ in this expression originates from the basic definition of the effect of the $C_{jm}^{\dagger}$-operator on determinantal states. We note that

$$\frac{1}{\sqrt{n}}(1 - \sum_{i=1}^{n-1} P_{in})$$

operating on $\phi_{jm}(n)D$, where D stands for a *normalized* determinantal state of $(n - 1)$ nucleons, produces a *normalized* n-particle determinant of which the last row and last column are labelled by the particle coordinate n and the single-particle state jm, respectively. By definition, $C_{jm}^{\dagger}|D)$ is identical to this normalized n-particle determinant. If we identify the determinant $|D)$ with any of the normalized $(n - 1)$ particle determinantal states whose linear superposition builds up $|j^{n-1}v'J'M')$, we find that the factor $\sqrt{n}$ follows.

We now substitute (III.190b) in the definition (III.188c) and obtain

$$c(n - 1, v'J'; nvJ) = \frac{1}{\sqrt{n}} \sum_{M',m} \begin{bmatrix} J' & j & J \\ M' & m & M \end{bmatrix} (j^{n-1}v'J'M'|\, C_{jm}\, |j^n vJM). \tag{III.191}$$

The operator $(-1)^{j-m}C_{j,-m}$ is a tensor of rank j and projection m with respect to the rotations in the ordinary three-dimensional space. We may also investigate its tensorial property with respect to the rotations in the quasispin space. For this purpose, we record the criterion in the case of any angular momentum. A tensor T_q^K of rank K and component q in the space of $\mathbf{J}$ is defined by

$$\begin{aligned} &[J_z, T_q^K] = qT_q^K, \\ &[J_{\pm}, T_q^K] = [(K \mp q)(K \pm q + 1)]^{1/2} T_{q\pm 1}^K. \end{aligned} \tag{III.192}$$

To analyze the tensorial character of $(-1)^{j+m}C_{jm}$ in the quasispin space, we therefore need its commutators with $S_{\pm}$ and S_z. In the usual way, we derive

$$\begin{aligned} &[S_z, C_m](-1)^{j-m} = -\tfrac{1}{2}(-1)^{j-m}C_m, \\ &[S_+, C_m](-1)^{j-m} = C_{-m}^{\dagger}, \\ &[S_-, C_m](-1)^{j-m} = 0. \end{aligned}$$

Similarly,

$$[S_z, C^\dagger_{-m}] = +\tfrac{1}{2}C^\dagger_{-m},$$
$$[S_+, C^\dagger_{-m}] = 0,$$
$$[S_-, C^\dagger_{-m}] = C_m(-1)^{j-m}.$$

These commutation relations are comparable with the general results (III.192), provided we put in them $K = \frac{1}{2}$ and $q = \pm\frac{1}{2}$. Thus, the operators $(-1)^{j-m}C_m$ and $C^\dagger_{-m}$ are respectively the $(-\frac{1}{2})$- and $(+\frac{1}{2})$-component of a tensor of rank $\frac{1}{2}$ in the quasispin space.

We go back to (III.191) and write the matrix element occurring there in terms of S and M_S; that is, we write

$$(-1)^{j-m}(j^{n-1}v'J'M'|\, C_{jm}\,|j^n vJM) \equiv \langle S'M'_S|\, T^{1/2}_{-1/2}\,|SM_S\rangle, \tag{III.193a}$$

where

$$T^{1/2}_{-1/2} = (-1)^{j-m}C_{jm}, \qquad S = \tfrac{1}{2}(\frac{[j]}{2} - v), \qquad M_S = \tfrac{1}{2}(n - \frac{[j]}{2}),$$
$$S' = \tfrac{1}{2}(\frac{[j]}{2} - v'), \qquad M'_S = \tfrac{1}{2}(n - 1 - \frac{[j]}{2}). \tag{III.193b}$$

By the Wigner-Eckart theorem, we then obtain

$$(-1)^{j-m}(j^{n-1}v'J'M'|\, C_{jm}\,|j^n vJM) = \begin{bmatrix} S & \frac{1}{2} & S' \\ M_S & -\frac{1}{2} & M'_S \end{bmatrix}\langle S'|\, |T^{1/2}|\, |S\rangle. \tag{III.194a}$$

The Clebsch-Gordon coefficients require that this result be nonvanishing only if

$$S' = S \pm \tfrac{1}{2}, \quad \text{i.e.,} \quad v' = v \mp 1, \tag{III.194b}$$

$$M'_S = M_S - \tfrac{1}{2}. \tag{III.194c}$$

(III.194c) is already satisfied according to (III.193b). The reduced matrix element occurring in (III.194a) depends on S and S' only, i.e., on v and v' only. Let us consider (III.194a) in the two cases $v' = v \pm 1$.

Case 1 ($v' = v - 1$ or $S' = S + \frac{1}{2}$) The smallest n that produces a state of seniority v is equal to v. So we consider the state $|j^v vJM)$. If C_{jm} operates on it, the number of particles is reduced by 1, and hence we reach a state j^{v-1}. This configuration can have $v' = v - 1$. Hence, we evaluate

$$(-1)^{j-m}(j^{v-1}, v - 1, J'M'|\, C_{jm}\,|j^v vJM) = (-1)^{j-m}\langle S + \tfrac{1}{2}, \bar{M}'_S|\, C_{jm}\,|S, \bar{M}_S\rangle, \tag{III.195a}$$

where

$$\bar{M}_S = \tfrac{1}{2}(v - \frac{[j]}{2}), \qquad \bar{M}'_S = \tfrac{1}{2}(v - 1 - \frac{[j]}{2}). \tag{III.195b}$$

Thus, the Wigner-Eckart theorem produces

$$(-1)^{j-m}(j^{v-1}, v - 1, J'M'|\, C_{jm}\,|j^v vJM)$$
$$= \begin{bmatrix} S & \frac{1}{2} & S + \frac{1}{2} \\ \bar{M}_S & -\frac{1}{2} & \bar{M}'_S \end{bmatrix}\langle S + \tfrac{1}{2}|\, |T^{1/2}|\, |S\rangle(-1)^{j-m}. \tag{III.195c}$$

Eliminating the reduced matrix element from (III.194a) with the help of (III.195c), we obtain

$$
\begin{aligned}
&(-1)^{j-m}(j^{n-1}, v-1, J'M'|\, C_{jm}\, |j^n vJM) \\
&= \begin{bmatrix} S & \frac{1}{2} & S+\frac{1}{2} \\ \bar{M}_S & -\frac{1}{2} & \bar{M}'_S \end{bmatrix}^{-1} \begin{bmatrix} S & \frac{1}{2} & S+\frac{1}{2} \\ M_S & -\frac{1}{2} & M'_S \end{bmatrix} (-1)^{j-m}(j^{v-1}, v-1, J'M'|\, C_{jm}\, |j^v vJM).
\end{aligned}
\tag{III.196}
$$

Making use of the basic definition (III.191) of the cfp, we then have, from (III.196),

$$
\begin{aligned}
c(n-1, v-1, J'; nvJ) &= \sqrt{\frac{v}{n}} \begin{bmatrix} S & \frac{1}{2} & S+\frac{1}{2} \\ \bar{M}_S & -\frac{1}{2} & \bar{M}'_S \end{bmatrix}^{-1} \begin{bmatrix} S & \frac{1}{2} & S+\frac{1}{2} \\ M_S & -\frac{1}{2} & M'_S \end{bmatrix} \\
&\quad \times c(v-1, v-1, J'; vvJ) \\
&= \sqrt{\frac{v}{n}} \sqrt{\frac{S - M'_S + \frac{1}{2}}{S - \bar{M}'_S + \frac{1}{2}}}\, c(v-1, v-1, J'; vvJ) \\
&= \sqrt{\frac{v(2j+3-n-v)}{n(2j+3-2v)}}\, c(v-1, v-1, J'; vvJ).
\end{aligned}
\tag{III.197}
$$

We have here made use of the explicit expression of the Clebsch-Gordon coefficient from Table AVII.1 in Appendix A, and then substituted the values of S, M'_S, and $\bar{M}'_S$ from (III.193b) and (III.195b). The expression (III.197) is a recurrence relation for obtaining the cfp connecting j^n with j^{n-1} from that connecting j^v with j^{v-1}.

Case 2 ($v' = v + 1$ or $S' = S - \frac{1}{2}$) Here if we take the state $|j^v vJM)$ in (III.194a), C_{jm} takes us to j^{v-1}, which cannot have a state of seniority ($v' = v + 1$). To have ($v' = v + 1$), we need j^{n-1} on the left-hand side of (III.194a) to be at least j^{v+1}. We reach this j^{v+1} only if C_{jm} operates on $|j^{v+2}vJM)$. Therefore, in this case we evaluate

$$
(-1)^{j-m}(j^{v+1}, v+1, J'M'|\, C_{jm}\, |j^{v+2}vJM) = \langle S - \tfrac{1}{2}, \bar{M}'_S|\, T^{1/2}_{-1/2}\, |S, \bar{M}_S\rangle,
$$

where

$$
\bar{M}_S = \tfrac{1}{2}\left(v + 2 - \frac{[j]}{2}\right), \qquad \bar{M}'_S = \tfrac{1}{2}\left(v + 1 - \frac{[j]}{2}\right).
$$

Proceeding exactly as in the case of $v' = v - 1$, we have

$$
\begin{aligned}
c(n-1, v+1, J'; n, v, J) &= \sqrt{\frac{v+2}{n}} \sqrt{\frac{S + M'_S + \frac{1}{2}}{S + \bar{M}'_S + \frac{1}{2}}}\, c(v+1, v+1, J'; v+2, v, J) \\
&= \sqrt{\frac{(v+2)(n-v)}{2n}}\, c(v+1, v+1, J'; v+2, v, J).
\end{aligned}
\tag{III.198}
$$

In this recurrence relation, the state $|v + 2, v, JM)$ can be further reduced as follows. Using (III.173a) and (III.174c), we have

$$
|j^{v+2}, v, JM) = \frac{1}{\left(\frac{[j]}{2} - v\right)^{1/2}} S_+ |j^v vJM),
$$

and hence

$$c(v+1, v+1, J'; v+2, v, J)$$

$$= \frac{1}{\sqrt{v+2}} \sum_{M'm} (j^{v+1}, v+1, J'M'|\, C_{jm}\, |j^{v+2}, vJM) \begin{bmatrix} J' & j & J \\ M' & m & M \end{bmatrix}$$

$$= \frac{1}{\sqrt{v+2}} \left(\frac{[j]}{2} - v\right)^{-1/2} \sum_{M', m} \begin{bmatrix} J' & j & J \\ M' & m & M \end{bmatrix} (j^{v+1}, v+1, J'M'|\, C_{jm}S_+\, |j^v vJM). \quad \text{(III.199)}$$

Since $(j^{v+1}, v+1, J'M'|S_+ = 0$, according to the Hermitean conjugate of (III.172), the matrix element on the right-hand side of (III.199) is in fact equal to that of $[C_{jm}, S_+]$, which is equal to $(-1)^{j+m}C^\dagger_{j,-m}$, according to the results given after (III.192). Considering all these facts, we get

$$(j^{v+1}, v+1, J'M'|\, C_{jm}S_+\, |j^v vJM) = (-1)^{j+m}(j^{v+1}, v+1, J'M'|\, C^\dagger_{j,-m}\, |j^v, vJM).$$

Since the matrix element on the right-hand side of this equation is real, we replace it by its complex conjugate and obtain

$$(j^{v+1}, v+1, J'M'|\, C_{jm}S_+\, |j^v vJM) = (-1)^{j+m}(j^v vJM|\, C_{j,-m}\, |j^{v+1}, v+1, J'M')$$

$$= -\begin{bmatrix} J' & j & J \\ M' & m & M \end{bmatrix} (j^v vJ|\, |C_j|\, |j^{v+1}, v+1, J'). \quad \text{(III.200)}$$

In the last step, we have used the tensorial property of $(-1)^{j-m}C_{j,-m}$ (a tensor of rank j and projection m) under ordinary rotation and applied the Wigner-Eckart theorem with respect to the angular momenta JM and $J'M'$ of the two states. We substitute (III.200) in (III.199) and carry out the sum of the product of the two Clebsch-Gordon coefficients over the magnetic quantum numbers in the usual way, thus obtaining

$$c(v+1, v+1, J'; v+2, v, J) = -\sqrt{2}[(v+2)(2j+1-2v)]^{-1/2} \times (j^v vJ|\, |C_j|\, |j^{v+1}, v+1, J'). \quad \text{(III.201a)}$$

In the same manner, we handle the expression of the cfp $c(v, v, J; v+1, v+1, J')$ and obtain

$$c(v, v, J; v+1, v+1, J')$$

$$= \frac{1}{\sqrt{v+1}} \sum_{M,m} \begin{bmatrix} J & j & J' \\ M & m & M' \end{bmatrix} (j^v vJM|\, C_{jm}\, |j^{v+1}, v+1, J'M')$$

$$= \frac{1}{\sqrt{v+1}} \sum_{M,m} \begin{bmatrix} J & j & J' \\ M & m & m' \end{bmatrix} \begin{bmatrix} J' & j & J \\ M' & -m & M \end{bmatrix} (-1)^{j+m}(j^v vJ|\, |C_j|\, |j^{v+1}, v+1, J')$$

$$= -(-1)^{J+j-J'} \sqrt{\frac{[J]}{(v+1)[J']}} (j^v vJ|\, |C_j|\, |j^{v+1}, v+1, J'). \quad \text{(III.201b)}$$

Comparing (III.201a) and (III.201b), we finally obtain

$$c(v+1, v+1, J'; v+2, v, J) = (-1)^{J+j-J'}\left[\frac{2(v+1)(2J'+1)}{(v+1)(2j+1-2v)(2J+1)}\right]^{1/2} \times c(vvJ; v+1, v+1, J'). \quad \text{(III.202)}$$

This is the final recurrence relation of the cfp's that we wish to derive. Together, the three recurrence relations (III.197), (III.198), and (III.202) enable us to calculate all the cfp's in the seniority scheme.

The cfp's are now explicitly calculated from the recurrence relations in two special cases: first, the $(v = 0, J = 0)$-state for an even value of n and, second, the $(v = 1, J = j)$-state for an odd value of n. These two cases correspond to the lowest seniority state of even and odd n, respectively. In the first case, the value of v' in the j^{n-1}-configuration, which is normally equal to $v + 1$, can only be equal to $v + 1$, i.e., 1. Here $v - 1$ happens to be -1, which is clearly not a permissible value for the seniority v'. We therefore take the recurrence relation (III.198) and derive

$$c(n-1, 1, J' = j; n, 0, 0) = \sqrt{\frac{2n}{2n}}\, c(1, 1, j; 2, 0, 0).$$

The value of J' corresponding to $v' = 1$ is unique, and equal to j, as shown. The cfp on the right-hand side can be easily seen to be equal to unity, by following its definition and then constructing the state $J = 0$ of j^2 by coupling the j of the second nucleon to the angular momentum j of the first nucleon. Thus, the only nonvanishing cfp in the first case is unity, and we get

$$|j^n v = 0, J = 0) = |j^{n-1}, v' = 1, J' = j, j_n : J = 0\rangle,$$

which is a rather surprising result because the single angular momentum coupled state on the right-hand side is already completely antisymmetric in all the n nucleons.

In the second case, both the values of v', equal to 0 and 1, are permitted. When $v' = 0$, the corresponding J' has the unique value 0, whereas for the case $v' = 2$ the corresponding J' are clearly the other permitted values of J' in the two-nucleon system, i.e., J' = even integers (> 0). We use all the three recurrence relations and obtain, by virtue of $c(0, 0, J' = 0; 1, 1, J = j) = 1$ and $c(1, 1, j; 2, 2, J') = 1$, the results

$$c(n-1, 0, 0; n, 1, j) = \left[\frac{2j + 2 - n}{n(2j + 1)}\right]^{1/2}, \tag{III.203a}$$

$$c(n-1, 2, J'; n, 1, j) = -\left[\frac{2(n-1)(2J'+1)}{(n2j-1)(2j+1)}\right]^{1/2} \text{ for } J' = \text{even integers } (>0). \tag{III.203b}$$

The cfp's can be used to write out the expansion for the lowest seniority state of odd n as

$$|j^n, v = 1, J = j) = \left[\frac{2j + 2 - n}{n(2j + 1)}\right]^{1/2} |j^{n-1}, v' = 0, J' = 0, j_n : J = j\rangle$$

$$- \sum_{J' = \text{even integers } (>0)} \left[\frac{2(n-1)(2J'+1)}{n(2j-1)(2j+1)}\right]^{1/2} |j^{n-1}, v' = 2, J', j_n : J = j\rangle. \tag{III.204}$$

The value of a single-particle operator $\sum_l T_q^K(l)$ using the foregoing expansion may be easily worked out. For some interesting results concerning these expectation values, see Problems 5 and 6 at the end of this chapter.

31. SUMMARY OF SHELL-MODEL CALCULATIONS AND RESULTS

Without attempting to be comprehensive, we shall review a few typical shell-model calculations and their results.

A. 0*p*-SHELL WORK

The earliest calculations were for 0*p*-shell nuclei. An even or odd number of nucleons in the 0*p*-level give rise to nuclear states of even or odd parity, respectively. In some nuclei of this region of the periodic table ($A = 5$ to $A = 16$), states of the alternative parity (that is, odd parity for even nuclei and even parity for odd-mass nuclei) are located fairly close to the ground state. Obviously, to enable us to explain these states of 'non-normal' parity, 0*p*-nucleons need to be excited to higher shells.

The calculations on 'normal' parity states, which belong to configurations of the type p^n, were done as follows. There is evidence in this shell that the single-particle spin-orbit coupling term of the shell-model Hamiltonian may be comparable with the residual two-body type interaction amongst the nucleons. The particular single-particle term $\sum_{i=1}^{n} \xi(r_i)\boldsymbol{l}_i \cdot \mathbf{s}_i$ and the residual two-body interaction $\sum_{i<j} V_{ij}$ are taken together in setting up the matrix of the Hamiltonian connecting the different basis states with the same *J*- and *T*-quantum number.

After the initial work by Inglis[8], exhaustive calculations for all the nuclei in the 0*p*-shell were done by Kurath[9] with a fair amount of success in respect of agreement with the observed spectra. The reader is referred to the original work for detailed numerical results.

More recently, Cohen and Kurath[10] extensively reinvestigated the 0*p*-nuclei. They considered only states of normal parity belonging to configurations of the type p^n. Their work lays more emphasis on determining the *p*-shell effective interaction. To do this, all the energy matrices for various values of *n* are expressed, in the manner of (III.124), in terms of cfp's and two-body matrix elements of the potential; the latter are then determined by a least-squares fit to a large number of energy levels belonging to nuclei spread over the entire shell. (They deliberately excluded the nuclei with the mass numbers 6 and 7 while picking up the data for a least-squares fit.) The two-body matrix elements used in the fitting procedure can be in either the *LS*-representation or the *jj*-representation; these matrix elements are related to each other through the basic relation (III.65) between the states. The number of independent matrix elements in the 0*p*-shell, in the two representations, is necessarily equal (=15, in fact). However, if the *LS*-representation is used, four of the nondiagonal matrix elements connect the state $L = 1$ with a state of even *L*; the orbital wavefunction of $L = 1$ is antisymmetric under the exchange of space coordinates of the two nucleons, whereas the orbital wavefunctions with even *L* are symmetric. Such nondiagonal matrix elements vanish for phenomenological potentials. Therefore, the number of independent matrix elements in the *LS*-representation reduces to 11, which are determined from the least-squares fit. They are then reproduced from a phenomenological two-body potential consisting of central, tensor, *two-body type* spin-orbit coupling, and quadratic spin-orbit coupling terms (all described in Section 3). In fact, by determining only the radial-reduced matrix elements with the help of the full two-body matrix elements, after inserting the assumed form of the phenomenological potential, Cohen and Kurath bypassed the specification of the radial shapes of the various terms in the potential. Once this information on the potential is derived, levels of any nucleus (including $A = 6$ and $A = 7$ which they had deliberately left out of the analysis) can be calculated. The energy levels and wavefunctions so determined have been extensively checked against experimental data. For checking the wavefunctions, the authors calculated the magnetic dipole moment, M1-transition strength, and beta-decay probability in all relevant cases. Detailed results are available in the original paper.

B. (1*s*, 0*d*)-SHELL NUCLEI

One of the first sets of calculations was done up to the mass number 19 by Elliott and Flowers[11]. (The experimentally observed energies of the single-particle levels of this shell have been quoted in Section 28.) The splitting between the $(0d_{5/2})$-level and $(0d_{3/2})$-level enables an estimation of where the 0*d*-level should have been when there is no spin-orbit coupling. For the $(1s_{1/2})$-level, there is no energy due to the spin-orbit interaction, and hence the level 1*s*, in the absence of spin-orbit coupling, occurs exactly at its experimentally observed position. In the (1*s*, 0*d*)-shell calculation, the energies of the 1*s* level and the 0*d*-level are required as input data.

In the calculations by Elliott and Flowers, the Rosenfeld exchange mixture with a Yukawa-type radial shape was used for the central two-nucleon interaction. The range parameter of the potential was chosen to be 1.37 fm, whereas the harmonic oscillator parameter was taken as 1.64 fm. The strength of the spin-orbit coupling was fixed by the observed $0d_{5/2}$-$0d_{3/2}$ splitting in $A = 17$. Calculations were done by varying the strength parameter of the central potential which effectively plays the role of the intermediate coupling parameter. The exchange mixture was found to crucially determine the relative separation of the $(T = 1)$-group of levels from the $(T = 0)$-group of levels. However, the separation inside each of these groups was rather insensitive to the exchange mixture. The central force strength for which the spectra are well-reproduced was observed to be in the vicinity of 40–42 MeV.

Elaborate shell-model calculations for $A = 17$ to $A = 22$ were done in very recent times by Halbert et al[12]. Several types of two-body interactions were used in their calculations, which take into account the complete configuration mixing effect amongst the $(d_{5/2})$-, $(d_{3/2})$-, and $(s_{1/2})$-orbital. The two-body matrix elements, renormalized with core-polarization effects, and estimated in the manner described in Sections 22B and 22C for the realistic Hamada-Johnston potential by both Kuo and Brown and by Kuo alone were used in the work of Halbert et al. The three single-particle energies were either taken from experimental data on $A = 17$ or treated as free parameters in making a least-squares fit to a large number of energy level data in the range $A = 17$ to $A = 22$. These authors also attempted a parametrization of the two-body interaction in terms of independent combinations of Talmi integrals. For detailed results, the reader is referred to the original article. The main feature of the results is a prediction of levels that appear surprisingly close to a rotational spectrum (see Chapter IV).

In fact, many nuclei in the mass range $A = 19$ to $A = 25$ have been treated by various authors from the point of view of the rotational model described in Chapter IV. In Chapter V, we shall describe the Hartree-Fock type theory for the same nuclei, which also predicts a rotational-type spectrum of many of these nuclei. The nuclei from Si^{28} to the neighbourhood of A^{36} belonging to the (*s*, *d*)-shell are still not very well-understood in terms of any of the theoretical approaches. Direct shell-model diagonalization around Si^{28} is also numerically prohibitive, and has not yet been done.

C. (0*f*, 1*p*)-SHELL NUCLEI

The earliest calculation was on Ca^{43} by Levinson and Ford[13]. This work involved several simplifying assumptions on the spectra of Ca^{41} and Ca^{43}. The authors finally calculated the complete spectra of Ca^{43} by taking advantage of the assumed data on Ca^{41} and Ca^{43}. Talmi did an extensive shell-model analysis of all the Ca isotopes from $A = 41$ to $A = 48$, using pure $(f_{7/2})$-configurations for the neutrons outside the closed-shell core Ca^{40}. A review of this work may be found in the book by de-Shalit and Talmi[14]. Configuration-mixing calculations were recently done by McGrory et al[15] for the nuclei Ca^{42}–Ca^{50}.

D. Ni ISOTOPES

Complete configuration-mixing calculations on the Ni isotopes $A = 56$ to $A = 60$ were done by Gambhir and Ram Raj[16] by using several different types of two-nucleon interaction. Similar calculations were worked out for all the Ni isotopes up to $A = 64$ by Auerbach[17] who treated the two-nucleon matrix elements as parameters in fitting the spectral data of all the isotopes. Lawson et al[18] covered the same ground, treating the two-body matrix elements as parameters, and also with the matrix elements computed in Kuo's work as described in Sections 22B and 22C.

E. Pb-REGION

The earliest calculation on Pb^{206} was by True and Ford[19]. The single-particle and single-hole levels in this region have been very extensively investigated experimentally. Of utmost interest in recent times is a set of seven close-lying levels (the septuplet) in Bi^{209}, which is interpreted as resulting from coupling a $(h_{9/2})$-proton (the single-particle state that gives the ground state of Bi^{209}) to the excited 3^- state of the Pb^{208} core.

REFERENCES

1. Mayer, M. G., and Jensen, J. H. D., Elementary Theory of Nuclear Shell Structure, Wiley, New York, 1955.
2. Gomes, L. C., Walecka, J. D., and Weisskopf, V. F., *Ann. Phys.* (N.Y.), **3**, 241 (1958).
3. de-Shalit, A., *Phys. Rev.*, **91**, 1479 (1953).
4. Arima, A., and Horie, H., *Prog. Theor. Phys.*, **11**, 509 (1954); Blinstoyle, R. J., and Perks, M. A., *Proc. Phys. Soc.* (London), **A67**, 885 (1954); *Revs. Mod. Phys.*, **28**, 75 (1956).
5. Jahn, H. A., *Proc. Roy. Soc.* (London), **A205**, 192 (1951); Jahn, H. A., and van Wieringen, H., *Proc. Roy. Soc.* (London), **A209**, 502 (1951); Elliott, J. P., Hope, J., and Jahn, H. A., *Phil. Trans. Roy. Soc.* (London), **A246**, 241 (1953); Flowers, B. H., *Proc. Roy. Soc.* (London), **A210**, 497 (1952); *ibid*, **A212**, 248 (1952); Edmonds, A. R., and Flowers, B. H., *Proc. Roy. Soc.* (London), **A214**, 515 (1952). See also Auerbach, N., *Nucl. Phys.*, **76** (1966).
6. Kerman, A. K., *Ann. Phys.* (N.Y.), **12**, 300 (1961).
7. Macfarlane, M. H., Lectures in Theoretical Physics, University of Colorado Press, Boulder, 1966, p 583.
8. Inglis, D. R., *Revs. Mod. Phys.*, **25**, 390 (1953).
9. Kurath, D., *Phys. Rev.*, **101**, 216 (1956).
10. Cohen, S., and Kurath, D., *Nucl. Phys.*, **A101**, 1 (1967).
11. Elliott, J. P., and Flowers, B. H., *Proc. Roy. Soc.* (London), **A242**, 57 (1957).
12. Halbert, E. C., McGrory, J. B., Wildenthal, B. H., and Pandya, S. P., *Adv. Nucl. Phys.*, **4**, 315 (1971).
13. Levinson, C. A., and Ford, K. W., *Phys. Rev.*, **99**, 792 (1952); *ibid*, **100**, 13 (1955).
14. de-Shalit, A., and Talmi, I., Nuclear Shell Theory, Academic Press, New York, 1963, pp 342–57.

15. McGrory, J. B., Wildenthal, B. H., and Halbert, E. C., *Phys. Rev.*, **C2**, 186 (1970).
16. Gambhir, Y. K., and Ram Raj, *Phys. Rev.*, **161**, 1125 (1967).
17. Auerbach, N., *Nucl. Phys.*, **76**, 321 (1966).
18. Cohen, S., Lawson, R. D., Macfarlane, M. H., Pandya, S. P., and Soga, M., *Phys. Rev.*, **160**, 903 (1967).
19. True, W., and Ford, K. W., *Phys. Rev.*, **109**, 1675 (1958).

PROBLEMS

1. Derive the expression (III.75) from (III.74).

2. Evaluate completely the Slater integrals (III.72) when all the single-particle radial functions belong to the $0p$-harmonic oscillator level, and the potential has a Gaussian shape.

3. Derive (III.183b) from (III.183a).

4. Prove that $\langle j^{v'}v'J|\, V\, |j^{v'}vJ\rangle = 0$ unless $v' = v$ for the odd-tensor interaction which satisfies the commutator relation (III.184). [Hint: Write

$$|j^{v'}vJ\rangle = N_{P''}(S_+)^{P''}|j^{v}vJ\rangle$$

with a suitable definition of P'' and use the Hermitean conjugates of (III.184) and (III.172).]

5. Use the fractional parentage expansion (III.204) and the method described in Section 29 to write down the expectation value of the single-particle operator

$$\sum_{I=1}^{n} T_q^K(I)$$

for the state $|j^n, v = 1, J = j, m\rangle$. Express the result in terms of the reduced matrix element $\langle j|\, |T^K|\, |j\rangle$ by using the standard results from Appendix B (Section III). Explicitly carry out the summation over J' in the final expression and obtain the required expectation value in the form

$$\langle T_q^K\rangle_{jm}\left[\frac{2j+2-n}{2j+1} - \frac{2(n-1)}{(2j-1)(2j+1)} + \frac{n-1}{2j-1}\{(-1)^{K+1} + \delta_{K,0}(2j+1)\}\right].$$

Note The expression within the square brackets is equal to (i) 1 for $K =$ odd integer, (ii) n for $K = 0$, and (iii) $(2j + 1 - 2n)/(2j - 1)$ which is less than 1 for $n > 1$ in the case $K =$ even integers greater than zero. Here $\langle T_q^K\rangle_{jm}$ is the expectation value for the single-particle state $|j, m\rangle$.

6. (a) Show that the second-quantized expression for the single-particle operator in Problem 5 for a single j-shell is given by $(-1)^K\langle j|\, |T^K|\, |j\rangle\hat{N}_{Kq}$, where $\hat{N}$ is as defined by (III.181b).

(b) Work out the commutator and prove that $[S_-, \hat{N}_{Kq}] = (-1)^q A_{-q}^K$. Use the property of the operator occurring on the right-hand side to show that the commutator is zero for $K =$ odd integer.

(c) Use the commutator in Problem 6b and the hint given in Problem 4 to work out

$$\langle j^n v'J'M'|\sum_{I=1}^{n} T_q^K(I)\, |j^n vJM\rangle$$

for (i) $K =$ odd integer and (ii) $K = 0$.

Note The result in both cases contains $\delta_{vv'}$ and the matrix element connecting the two states of the configuration j^v. The extra numerical factor is 1 in case (i) and n/v in case (ii). In the case of $K =$ even integers (>0), the matrix element is nonvanishing for both $v' = v$ and $v' \neq v$. Work out the case $v' = v$ and show that the numerical factor is $(2j + 1 - 2n)/(2j + 1 - 2v)$ for this matrix element.

IV Phenomenological Collective and Unified Model

32. COLLECTIVE MODEL AND ROTATION OF DEFORMED NUCLEI

A. PARAMETRIZATION OF NUCLEAR SURFACE AND BASIC FACTS ABOUT THE ELLIPSOID

The phenomenological collective model (see Bohr and Mottelson[1]) is based on the concept of a nuclear surface defined by the surface coordinate R in the direction (θ, ϕ) as

$$R(\theta, \phi) = R_0[1 + \sum_{\lambda=0}^{\infty} \sum_{\mu=-\lambda}^{\lambda} \alpha^*_{\lambda\mu} Y^\lambda_\mu(\theta, \phi)]. \tag{IV.1}$$

Here R_0 is the radius of a spherical nucleus, and the extra terms containing $\alpha^*_{\lambda\mu}$ represent the expansion of any general function of the angles (θ, ϕ) in terms of the complete set of spherical harmonics $Y^\lambda_\mu(\theta, \phi)$. With *small* nonvanishing values of the coefficients $\alpha^*_{\lambda\mu}$, the angle-dependent terms represent a *small* departure of the nuclear surface from a spherical shape. If the quantities $\alpha^*_{\lambda\mu}$ are time-dependent, then (IV.1) obviously represents the instantaneous orientation of the surface. The motion of the surface with respect to time is then delineated through the exact nature of the time-dependence of $\alpha^*_{\lambda\mu}$. If the equilibrium shape of the nucleus is spherical, then the equilibrium values of the deformation parameters $\alpha^*_{\lambda\mu}$ are zero, and they develop nonvanishing values when the surface moves about the spherical equilibrium shape. If, on the other hand, the equilibrium shape is nonspherical, then the deformation parameters initially have a set of nonvanishing equilibrium values; as the nonspherical surface moves, the quantities $\alpha^*_{\lambda\mu}$ undergo changes about their non-zero equilibrium values.

Certain very general properties of the coefficients $\alpha^*_{\lambda\mu}$ can be immediately derived. Since $R(\theta, \phi)$ has to be real, we must impose the condition

$$R^*(\theta, \phi) = R(\theta, \phi). \tag{IV.2}$$

Taking the complex conjugate of (IV.1), we have

$$\begin{aligned} R^*(\theta, \phi) &= R_0[1 + \sum_{\lambda=0}^{\infty} \sum_{\mu=-\lambda}^{\lambda} \alpha_{\lambda\mu} Y^{\lambda *}_\mu(\theta, \phi)] \\ &= R_0[1 + \sum_{\lambda,\mu} \alpha_{\lambda\mu}(-1)^\mu Y^\lambda_{-\mu}(\theta, \phi)] \\ &= R_0[1 + \sum_{\lambda,\mu} \alpha_{\lambda,-\mu}(-1)^\mu Y^\lambda_\mu(\theta, \phi)]. \end{aligned} \tag{IV.3}$$

We have first used $Y^{\lambda *}_\mu = (-1)^\mu Y^\lambda_{-\mu}$, a well-known property of the spherical harmonic, and then relabelled the summation index μ as $-\mu$. Comparing (IV.3) and (IV.1), we obtain, by virtue of the requirement (IV.2),

$$\alpha^*_{\lambda\mu} = (-1)^\mu \alpha_{\lambda,-\mu}. \tag{IV.4}$$

For brevity, we shall very often write (IV.1) as

$$R(\theta, \phi) = R_0[1 + \xi(\theta, \phi)], \tag{IV.5a}$$

where

$$\xi(\theta, \phi) = \sum_{\lambda, \mu} \alpha^*_{\lambda\mu} Y^\lambda_\mu(\theta, \phi) \tag{IV.5b}$$

and is a small quantity if the coefficients $\alpha^*_{\lambda\mu}$ are small. If the nucleus is deformed from the spherical equilibrium shape of the radius R_0, then its volume prior to deformation must have been $(4\pi/3)R_0^3$. After the deformation has occurred, the volume changes to

$$\tau = \int d\omega \int_0^R r^2\, dr, \tag{IV.6a}$$

where $d\omega = \sin\theta\, d\theta\, d\phi$, and the limits of the (θ, ϕ)-integration are, as usual, from $0 \to \pi$ and $0 \to 2\pi$. Since the maximum value R of the radial coordinate r is dependent on (θ, ϕ), it is absolutely essential, as is evident from (IV.6a), that we first carry out the indicated integration of r and then obtain the final volume

$$\begin{aligned} \tau &= \tfrac{1}{3} \int_0^\pi \sin\theta\, d\theta \int_0^{2\pi} d\phi\, [R(\theta, \phi)]^3 \\ &= \tfrac{1}{3} R_0^3 \int_0^\pi \sin\theta\, d\theta \int_0^{2\pi} d\phi\, (1 + \xi)^3 \\ &= \frac{4\pi}{3} R_0^3 + \tfrac{1}{3} R_0^3 \int_0^\pi \sin\theta\, d\theta \int_0^{2\pi} d\phi\, (3\xi + 3\xi^2 + \xi^3). \end{aligned} \tag{IV.6b}$$

In view of the smallness of ξ, we reject the ξ^3-term. The ξ-term contributes only through the $(\lambda = 0, \mu = 0)$-term of (IV.5b); the higher order spherical harmonics produce zero when integrated over the angles. The ξ^2-term contributes from the spherical harmonics of all orders (λ, μ). The nuclear fluid is thought to be highly incompressible, which means that the total nuclear volume must not change as a result of deformation. The change in volume is represented by the integral term of (IV.6b). If we ignore the contribution of ξ^3 in this term, then the condition of volume conservation is clearly given by

$$0 = \int d\omega\, (\xi + \xi^2). \tag{IV.7}$$

An interesting consequence of this condition is that α^*_{00} becomes a second-order quantity if the coefficients $\alpha^*_{\lambda\mu}$ with higher values of λ, μ are of first-order smallness. The proof follows.

We have, using (IV.5b),

$$\int d\omega\, \xi = \alpha^*_{00} \frac{1}{\sqrt{4\pi}} \int d\omega = \sqrt{4\pi}\, \alpha^*_{00}. \tag{IV.8a}$$

On the other hand,

$$\begin{aligned} \int d\omega\, \xi^2 &= \sum_{\lambda, \mu} \sum_{\lambda', \mu'} \alpha^*_{\lambda\mu} \alpha^*_{\lambda'\mu'} \int d\omega\, Y^\lambda_\mu Y^{\lambda'}_{\mu'} \\ &= \sum_{\lambda, \mu} \alpha^*_{\lambda\mu} \alpha^*_{\lambda, -\mu} (-1)^\mu = \sum_{\lambda, \mu} |\alpha_{\lambda\mu}|^2. \end{aligned} \tag{IV.8b}$$

In deriving this result, we have used the orthogonality integral of two spherical harmonics and the reality condition (IV.4). Using (IV.8a) and (IV.8b), in conjunction with the requirement

(IV.7), we derive the result

$$\sqrt{4\pi}\alpha_{00}^{*} = -\sum_{\lambda,\mu} |\alpha_{\lambda\mu}|^2. \tag{IV.8c}$$

Another important and interesting property of the coefficients $\alpha_{\lambda\mu}^{*}$ follows from the physical requirement that the distance of a point R on the surface should not depend on the coordinate system chosen to specify the direction of **R** in terms of the polar and azimuthal angles (θ, ϕ). Let us choose another coordinate system, oriented with respect to the initial system, through the Eulerian angles $(\theta_1, \theta_2, \theta_3)$. In this new coordinate system, the polar and azimuthal angles of **R** are (θ', ϕ'), and hence

$R(\theta, \phi)$ of (IV.1) $\equiv R(\theta', \phi')$ of the new coordinate system

$$= R_0[1 + \sum_{\lambda=0}^{\infty} \sum_{\mu=-\lambda}^{\lambda} a_{\lambda\mu}^{*} Y_{\mu}^{\lambda}(\theta', \phi')], \tag{IV.9}$$

where $a_{\lambda\mu}^{*}$ are the expansion parameters in the new coordinate frame. The spherical harmonics $Y_{\mu}^{\lambda}(\theta, \phi)$ of the unrotated frame and $Y_{\mu}^{\lambda}(\theta', \phi')$ of the rotated frame are related to each other by (BI.13) of Appendix B, i.e.,

$$Y_{\mu}^{\lambda}(\theta, \phi) = \sum_{\nu} \mathcal{D}_{\mu\nu}^{\lambda}(\theta_1, \theta_2, \theta_3) Y_{\nu}^{\lambda}(\theta', \phi'). \tag{IV.10a}$$

Using this expression in (IV.1) and comparing the resultant expression with the right-hand side of (IV.9), we get

$$\sum_{\mu} \alpha_{\lambda\mu}^{*} \mathcal{D}_{\mu\nu}^{\lambda}(\theta_1, \theta_2, \theta_3) = a_{\lambda\nu}^{*}. \tag{IV.10b}$$

Taking the complex conjugate of this expression, and then using, from Appendix B, the unitarity relation (BI.17) of the $\mathcal{D}^{\lambda}$-matrix, we easily obtain

$$\alpha_{\lambda\mu} = \sum_{\nu} \mathcal{D}_{\mu\nu}^{\lambda}(\theta_1, \theta_2, \theta_3) a_{\lambda\nu}. \tag{IV.10c}$$

Comparing this relation with (IV.10a), we conclude that, under a rotation of the coordinate frame, the expansion parameters $\alpha_{\lambda\mu}$ transform in the same way as the spherical harmonics Y_{μ}^{λ}, for example, $\alpha_{\lambda\mu}$ is the μ-component of a spherical tensor of rank λ. Therefore, according to (BII.10) of Appendix B,

$$\sum_{\mu=-\lambda}^{\lambda} \alpha_{\lambda\mu}^{*} Y_{\mu}^{\lambda}(\theta, \phi) \equiv \boldsymbol{\alpha}_{\lambda} \cdot Y^{\lambda}.$$

In this scalar product form, the rotational invariance of the expression (IV.1) becomes completely transparent.

We next want to examine the physical meaning of some of the lowest multipole terms $(\lambda = 1, 2)$ in the expression of $R(\theta, \phi)$. To understand the $(\lambda = 1)$-term, let us first consider the translation of a sphere of radius R_0, *as a whole*, through a distance $\boldsymbol{\alpha}'$, as depicted in Fig. IV.1. The displacement $\boldsymbol{\alpha}'$ is taken to be infinitesimal. As a vector, it has three cartesian components, which are related to the spherical components α_{μ}', according to (AI.9) of Appendix A. Under this displacement, the centre of the sphere moves from C to C', and a point P on the sphere moves to P'. Both CC' and PP' are equal to $\boldsymbol{\alpha}'$, by definition. With respect to the new origin C', the point P' still has a radial coordinate R_0, whereas with respect to the old origin C its distance R is clearly a function of the polar and azimuthal angles (θ, ϕ) of the direction CP'. To find R as a function of (θ, ϕ), we use the fact that, vectorially, **R** in the direction CP' is the

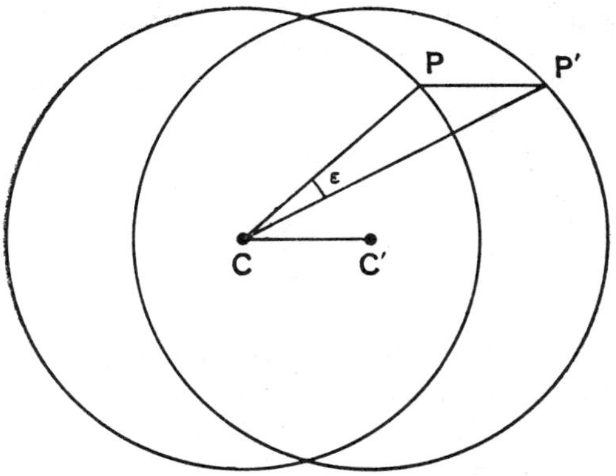

Fig. IV.1 Diagram showing correspondence of dipole term in radius R (a function of angles) of nuclear surface to a motion of nucleus as a whole without any accompanying deformation.

sum of a vector $\mathbf{R}_0$ along CP and the vector $\boldsymbol{\alpha}'$ along PP', i.e.,

$$\mathbf{R} = \mathbf{R}_0 + \boldsymbol{\alpha}'.$$

Multiplying this equation by the unit vector $\mathbf{e_R}$ along $\mathbf{R}$, we get the magnitude R of CP' on the left-hand side, whereas on the right-hand side we get R_0 times the cosine of the angle ϵ between CP and CP', plus $\boldsymbol{\alpha}' \cdot \mathbf{e_R}$. Thus,

$$R(\theta, \phi) = R_0 \cos \epsilon + \boldsymbol{\alpha}' \cdot \mathbf{e_R}. \tag{IV.11a}$$

According to (AI.9) in Appendix A, the unit vector $\mathbf{e_R}$ has cartesian components e_X, e_Y, e_Z which are related to its spherical components as

$$(\mathbf{e_R})_{\pm 1} = \mp \frac{1}{\sqrt{2}}(e_X \pm ie_Y) = \mp \frac{1}{\sqrt{2}} \sin \theta \, e^{\pm i\phi} = \sqrt{\frac{4\pi}{3}} Y^1_{\pm 1}(\theta, \phi),$$

$$(\mathbf{e_R})_0 = e_Z = \cos \theta = \sqrt{\frac{4\pi}{3}} Y^1_0(\theta).$$

Therefore, according to the definition (BII.10)—in Appendix B—of the scalar product, we have

$$\boldsymbol{\alpha}' \cdot \mathbf{e_R} = \sqrt{\frac{4\pi}{3}} \sum_\mu (-1)^\mu \alpha'_{-\mu} Y^1_\mu(\theta, \phi) = \sqrt{\frac{4\pi}{3}} \sum_\mu \alpha'^*_\mu Y^1_\mu(\theta, \phi). \tag{IV.11b}$$

Here we have used the fact that, for any vector $\mathbf{A}$ having real cartesian components, we have $(-1)^\mu A_{-\mu} = A^*_\mu$ according to (AI.9) in Appendix A. Thus, if we identify $\sqrt{(4\pi/3)}\alpha'^*_\mu$ with $\alpha^*_{1\mu}$, the expression (IV.11b), which is the second term of $R(\theta, \phi)$ as given in (IV.11a), exactly agrees with the ($\lambda = 1$)-term of the general multipole expansion (IV.1) for $R(\theta, \phi)$. The first term of (IV.11a) is $R_0 + \mathcal{O}(\epsilon^2)$ because, for $\boldsymbol{\alpha}'$ infinitesimal, the angle ϵ is also infinitesimally small. This agrees with the fact that, in the general expression (IV.1), if we decide to terminate the series at $\lambda = 1$, we still have, in addition to the R_0-term, a term $\alpha^*_{00}R_0$ in which α^*_{00} is of second-order smallness.

To sum up, we have shown that the ($\lambda = 1$)-term of the multipole expansion (IV.1) corresponds to a displacement of the nucleus as a whole, and hence it is of no interest whatsoever in nuclear structure study. We shall omit this term in our subsequent work.

We now investigate the ($\lambda = 2$)-terms of (IV.1). Let us consider a point P on the surface of an ellipsoid. The coordinates (x, y, z) of P, by the definition of the ellipsoid, satisfy the equation

$$\frac{x^2}{a^2} + \frac{y^2}{b^2} + \frac{z^2}{c^2} = 1, \tag{IV.12}$$

where a, b, and c are the half lengths of the three principal axes of the ellipsoid. The section of the ellipsoid on each of the three principal planes xy, yz, and zx is, by definition, an ellipse with the semi-axis (a, b), (b, c), and (c, a), respectively. If the ellipsoid has a rotational symmetry about an axis—for instance, the z-axis—that is to say, if the ellipsoid has been generated by rotating the ellipse in the yz-plane (or the zx-plane) around the z-axis, then such an ellipsoid of revolution has a special name; it is called a spheroid. The cross-section of this spheroid perpendicular to the symmetry axis, i.e., in the present case, the cross-section in the xy-plane, is clearly a circle. Any two mutually perpendicular directions in this plane can therefore serve as the principal x- and y-axis of the spheroid. Further, in such a case, we have $a = b$ in (IV.12). In the case of a spheroid, we can distinguish between two physically different shapes: one corresponds to $c > a(=b)$ and the other to $c < a(=b)$. In the former case, the semi-axis along the symmetry direction is longer than the semi-axes contained in the perpendicular plane; such a spheroid is called a prolate spheroid, whereas in the latter case it is called an oblate spheroid, which is shorter in the symmetry direction than in the other two directions.

In the special case of the spheroid, let us put $a = b$ in (IV.12) and make the substitutions

$$a = b = R_0 e^{-\delta}, \tag{IV.13a}$$

$$c = R_0 e^{2\delta}. \tag{IV.13b}$$

It should be noticed that, for a positive δ, these expressions define a prolate spheroid, whereas for a negative δ they define an oblate one. The δ is called the deformation parameter. If we restrict ourselves to the case of a small deformation, then

$$a^{-1} = b^{-1} = R_0^{-1} e^{\delta} \approx R_0^{-1}(1 + \delta),$$

$$c^{-1} = R_0^{-1} e^{-2\delta} \approx R_0^{-1}(1 - 2\delta),$$

and hence (IV.12) reduces to

$$1 = \frac{x^2 + y^2}{R_0^2}(1 + 2\delta) + \frac{z^2}{R_0^2}(1 - 4\delta)$$

$$= \frac{R^2}{R_0^2} - \frac{2\delta}{R_0^2}(3z^2 - R^2) = \frac{R^2}{R_0^2}\left[1 - 4\delta\sqrt{\frac{4\pi}{5}}Y_0^2(\theta)\right].$$

Here we have used $z = R\cos\theta$, and then replaced $(3\cos^2\theta - 1)$ in terms of $Y^2(\theta)$ with the help of Table AI.1 in Appendix A. In view of the smallness of δ, the foregoing equation yields

$$R(\theta, \phi) = R_0\left[1 - 4\delta\sqrt{\frac{4\pi}{5}}Y_0^2(\theta)\right]^{-1/2} = R_0\left[1 + 2\delta\sqrt{\frac{4\pi}{5}}Y_0^2(\theta)\right]. \tag{IV.13c}$$

A comparison of this expression with the ($\lambda = 2$, $\mu = 0$)-term of (IV.1) results in the identification

$$4\sqrt{\frac{\pi}{5}}\delta = \alpha_{20}^* \tag{IV.13d}$$

and the conclusion that this particular term leads to the surface equation of a spheroid. The spheroid is prolate or oblate, depending on the sign of α_{20}^*.

In the more general case of the ellipsoid, let us substitute

$$a = R_0 e^{-\delta_1}, \qquad b = R_0 e^{-\delta_2}, \qquad c = R_0 e^{\delta_3}. \tag{IV.14a}$$

The volume of this ellipsoid is

$$\frac{4\pi}{3}abc = \frac{4\pi}{3}R_0^3 e^{\delta_3 - (\delta_1 + \delta_2)}.$$

If this volume has to be the same as that of the original sphere of radius R_0, which got deformed into the ellipsoidal shape, then we must have

$$\delta_3 = \delta_1 + \delta_2. \tag{IV.14b}$$

It should be noticed that, in the case of the spheroid, it is the volume-conservation requirement that demanded 2δ in (IV.13b) when $-\delta$ was used in (IV.13a). Let us now make the substitution (IV.14a) and (IV.14b) in (IV.12) and follow steps similar to those already applied for a spheroid. In this way, we obtain

$$R_0^2 = R^2 + 2x^2\delta_1 + 2y^2\delta_2 - 2z^2(\delta_1 + \delta_2). \tag{IV.14c}$$

We should now observe the algebraic identities

$$(x + iy)^2 = x^2 - y^2 + 2ixy, \tag{IV.15a}$$

$$(x - iy)^2 = x^2 - y^2 - 2ixy, \tag{IV.15b}$$

$$(x + iy)(x - iy) = x^2 + y^2. \tag{IV.15c}$$

Since the cross term xy is absent in (IV.14c), we conclude that (IV.15a) and (IV.15b) must always appear in a sum with equal weight in any alternative expression of (IV.14c). We have

$$\tfrac{1}{2}[(x + iy)^2 + (x - iy)^2] = x^2 - y^2, \tag{IV.15d}$$

and hence, from (IV.15c) and (IV.15d),

$$x^2 = \tfrac{1}{2}(x + iy)(x - iy) + \tfrac{1}{4}[(x + iy)^2 + (x - iy)^2],$$

$$y^2 = \tfrac{1}{2}(x + iy)(x - iy) - \tfrac{1}{4}[(x + iy)^2 + (x - iy)^2].$$

Using these results in (IV.14c), we obtain

$$R_0^2 = R^2 - (\delta_1 + \delta_2)(3z^2 - R^2) + \tfrac{1}{2}(\delta_1 - \delta_2)[(x + iy)^2 + (x - iy)^2]$$

$$= R^2\left[1 - 2\sqrt{\frac{4\pi}{5}}(\delta_1 + \delta_2)Y_0^2 + \sqrt{\tfrac{2}{3}}\sqrt{\frac{4\pi}{5}}(\delta_1 - \delta_2)(Y_2^2 + Y_{-2}^2)\right].$$

To introduce spherical harmonics in this expression, we have used the results of Table AI.1 from Appendix A. Inverting this relation for small δ_1 and δ_2, we obtain

$$R = R_0\left[1 + \sqrt{\frac{4\pi}{5}}\{(\delta_1 + \delta_2)Y_0^2 + \tfrac{1}{2}\sqrt{\tfrac{2}{3}}(\delta_2 - \delta_1)(Y_2^2 + Y_{-2}^2)\}\right]. \tag{IV.16}$$

A comparison of this expression and the ($\lambda = 2$)-terms of (IV.1) yields

$$\alpha_{20}^* = \sqrt{\frac{4\pi}{5}}(\delta_1 + \delta_2), \qquad \alpha_{2,2}^* = \alpha_{2,-2}^* = \sqrt{\frac{2\pi}{15}}(\delta_2 - \delta_1). \tag{IV.17a}$$

The fact that α_{21}^* and $\alpha_{2,-1}^*$ are zero and $\alpha_{22}^* = \alpha_{2,-2}^*$ in (IV.16) is a consequence of not having any cross terms xy, yz, zx in the expression (IV.12) for the ellipsoid. This, in turn, resulted from our choice of the coordinate axes, while writing (IV.12), along the principal axes of the ellipsoid. It has been explicitly pointed out after (IV.15) that the absence of the cross term xy demands the sum of $(x + iy)^2$ and $(x - iy)^2$ with equal weight. These are the terms that yielded Y_2^2 and Y_{-2}^2, respectively, in (IV.16). Therefore, $\alpha_{22}^* = \alpha_{2,-2}^*$ is a consequence of the absence of the term xy in (IV.12). Table AI.1 in Appendix A will convince the reader that $Y_{\pm 1}^2$ is proportional to $z(x \pm iy)$. Thus, the absence of the cross terms zx and yz in (IV.12) is responsible for $\alpha_{21}^* = \alpha_{2,-1}^* = 0$.

We shall now slightly change our notation. When the ellipsoid is referred to its principal axes as the coordinate axes, the corresponding surface parameters $\alpha_{2\mu}^*$ are simply denoted by a_μ. Thus, we have

$$a_1 = a_{-1} = 0, \qquad a_2 = a_{-2} \neq 0, \qquad a_0 \neq 0. \tag{IV.17b}$$

In view of (IV.4) and the equality $a_2 = a_{-2}$, we conclude that the two independent nonvanishing parameters are real, and hence they can be rewritten in terms of two other real parameters, β and γ, defined as

$$a_0 = \beta \cos \gamma, \qquad a_2 = a_{-2} = \frac{1}{\sqrt{2}}\beta \sin \gamma. \tag{IV.18}$$

We have already seen that, in the case of a spheroid having z as its symmetry axis, only $a_0 \neq 0$, and all the other a_μ-parameters ($\mu \neq 0$) vanish. From (IV.18), therefore, such a situation corresponds to $\gamma = 0$. Positive and negative values of β for this γ will then correspond respectively to a prolate and an oblate spheroid having z as its symmetry axis.

In general, if we refer the ellipsoid to any arbitrary coordinate system, the equation of the surface contains cross terms in x, y, z, and hence all the five coefficients are nonvanishing. Thus, we usually require five parameters to specify the ellipsoidal surface; when we refer the body to its principal axes as the coordinate system, only two parameters, β and γ, are enough to specify its shape. But then we also have to define the orientation of the principal axes system with reference to an arbitrary coordinate system. This, as we know, requires the specification of the three Eulerian angles θ_1, θ_2, θ_3, that is, the three independent rotations needed to move from the arbitrary coordinate system to the principal axes system. Thus, the five parameters $\alpha_{2\mu}^*$ are now replaced by the alternative set (β, γ) and (θ_1, θ_2, θ_3). The arbitrary coordinate system, which is fixed with respect to an observer, is called the *stationary* frame, and the principal axes system, which is necessarily fixed in the ellipsoidal body but goes on changing its orientation as the body moves, is referred to as the *body-fixed* frame. For both these coordinate frames, we shall conventionally adhere to a right-handed system (i.e., a right-handed screw driven from x to y moves along the z-axis).

As a part of the general discussion on the ellipsoidal surface, we shall now point out the arbitrariness we have in going from the stationary frame to the body-fixed frame. If we have a given ellipsoidal surface, and a given stationary frame, the values of the expansion parameters $\alpha_{2\mu}$ are uniquely determined. This statement, however, is not true for the parameters a_μ, i.e., (β, γ) and (θ_1, θ_2, θ_3). This is because of the arbitrariness involved in choosing the body-fixed frame along the principal axes. In fact, each principal axis goes in both directions through the origin, which is chosen to be the 'centre' of the ellipsoidal body; hence, each coordinate axis can be chosen in either of these two directions. In the case of a spheroid, the arbitrariness in this choice is even infinitely many because, as already pointed out, the coordinate axes perpendicular to the symmetry axis could be chosen along any two mutually perpendicular directions; the only restriction in *labelling* these axes is that with the symmetry axis they should satisfy the right-handed convention of our coordinate system.

In order to systematically study this arbitrariness in the case of an ellipsoid, let us first introduce three rotation operators:

(i) $\mathcal{R}_1$ [this operator produces a right-handed rotation through π about the body-fixed y-axis];

(ii) $\mathcal{R}_2$ [this operator produces a right-handed rotation through $\pi/2$ about the body-fixed z-axis];

(iii) $\mathcal{R}_3$ [this operator produces the Eulerian rotations $(\pi/2, \pi/2, \pi)$ starting with a given orientation of the body-fixed frame].

The effect of these three rotations is shown in Fig. IV.2, where the solid lines x, y, z denote the initial situations. The position of the axes after each subsequent rotation is shown by the

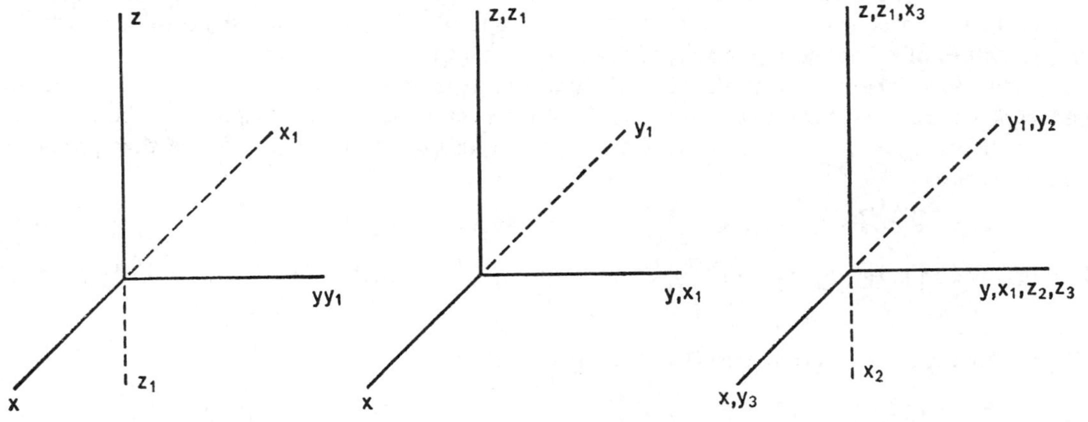

(a) One rotation about y-axis through angle π

(b) One rotation about z-axis through $\pi/2$

(c) Three Eulerian rotations $(\pi/2, \pi/2, \pi)$

Fig. IV.2 Three symmetry operations (rotations) of ellipsoidal body.

corresponding subscripted letter. The subscripts 1, 2, and 3 refer to the situation after the first, second, and third rotations, respectively. Figures IV.2a and IV.2b relate to only one rotation, and hence only the subscript 1 occurs. The reader's attention is drawn to the final situation (x_3, y_3, z_3) in Fig. IV.2c, where the axes are coincident with the original axes x, y, z taken in a different cyclic order, namely, y, z, x. Thus, the rotation $\mathcal{R}_3$ allows a convenient relabelling of the body-fixed axes while maintaining the cyclic order. A second operation with $\mathcal{R}_3$, commencing with the situation (x_3, y_3, z_3), would clearly allow the relabelling $(x, y, z) \rightarrow (z, x, y)$.

Now let us consider the degree of arbitrariness in the choice of the body-fixed frame. Because of the special symmetries of the ellipsoidal body, the eight octants into which the body can be divided by the three principal planes are identical, and hence the body-fixed frame could be chosen along the three edges of each of these octants. Once this is done, the coordinate axes in each case may have three possible labels: (x, y, z), (y, z, x), or (z, x, y). Thus, the actual choice of the body-fixed frame can be done in $(8 \times 3 =)$ 24 different ways.

In terms of the operators $\mathcal{R}_1$, $\mathcal{R}_2$, $\mathcal{R}_3$, we can generate the 24 different choices as follows. First, let us start with the initial situation depicted in Fig. IV.2b, where the z-axis is pointing 'up'. There are four octants in the 'upper' half of the ellipsoid, which corresponds to the 'up' position of the z-axis. In passing from one such octant to the next, it is clear, from Fig. IV.2b, that all we need is the rotation $\mathcal{R}_2$. Three such rotations, in succession, take us from one starting octant to the other three. In order to follow the same procedure in the 'lower' half of the ellipsoid, we first need to reverse the direction of the z-axis from its 'up' position to the 'down' position, which can clearly be done by the rotation $\mathcal{R}_1$, as depicted in Fig. IV.2a. Once we have entered the lower half of the ellipsoid, the subsequent rotations with $\mathcal{R}_2$ again enable us to move into all the four 'lower-half' octants. Finally, as already discussed, the three cyclic orders in which the coordinate axes can be labelled, for each of the eight orientations of the body-fixed frame in the eight octants, are generated by the successive applications of $\mathcal{R}_3$.

The foregoing discussion clarifies that it is not really necessary to study what happens to (β, γ) and $(\theta_1, \theta_2, \theta_3)$ in all the 24 possible cases. It is sufficient if this effect is examined with respect to the three fundamental rotation operations $\mathcal{R}_1$, $\mathcal{R}_2$, $\mathcal{R}_3$. The effects of these rotations on the values of a_μ or, equivalently, (β, γ) can be examined as follows.

We have already seen that, when the coordinate system is rotated, the surface parameters undergo the transformation (IV.10b). In the present case, since both frames are body-fixed we denote the unrotated and rotated coefficients by a_μ and $\bar{a}_\mu$, respectively. Since they are all real, (IV.10b) reads

$$\bar{a}_\mu = \sum_\nu \mathcal{D}^{2*}_{\mu\nu}(\bar{\theta}_1, \bar{\theta}_2, \bar{\theta}_3) a_\nu. \tag{IV.19}$$

For $\mathcal{R}_1$, we have $(\bar{\theta}_1, \bar{\theta}_2, \bar{\theta}_3) = (0, \pi, 0)$ and, according to (BI.12a) and (BI.15) of Appendix B,

$$\mathcal{D}^{2*}_{\mu\nu}(0, \pi, 0) = \delta_{\nu, -\mu}(-1)^\mu. \tag{IV.20a}$$

Thus, under $\mathcal{R}_1$, we have from (IV.19) and (IV.20a)

$$\bar{a}_\mu = (-1)^\mu a_{-\mu},$$

which, written explicitly, establishes

$$\bar{a}_2 = a_{-2} = \frac{\beta}{\sqrt{2}} \sin \gamma, \qquad \bar{a}_{-2} = a_2 = \frac{\beta}{\sqrt{2}} \sin \gamma, \qquad \bar{a}_0 = a_0 = \beta \cos \gamma.$$

Thus, the new values $(\bar{\beta}, \bar{\gamma})$ corresponding to $\bar{a}_\mu$ are the same as the old ones. In other words, $\mathcal{R}_1$ keeps (β, γ) unchanged.

In a similar manner, $\mathcal{R}_2$ corresponds to the Eulerian angles $(\pi/2, 0, 0)$ in which case

$$\mathcal{D}^{2*}_{\mu\nu}(\pi/2, 0, 0) = e^{-i\mu(\pi/2)}\delta_{\mu\nu} = (-i)^\mu \delta_{\mu\nu}, \tag{IV.20b}$$

and hence, according to (IV.19),

$$\bar{a}_\mu = (-i)^\mu a_\mu.$$

Therefore, under $\mathcal{R}_2$, we get

$$\bar{a}_2 = \bar{a}_{-2} = -a_2 = -\frac{\beta}{\sqrt{2}} \sin \gamma, \qquad \bar{a}_0 = a_0 = \beta \cos \gamma.$$

These equations imply $\bar{\beta} = \beta$, $\bar{\gamma} = -\gamma$. Thus, $\mathcal{R}_2$ transforms (β, γ) into $(\beta, -\gamma)$.

Finally, in the case of $\mathcal{R}_3$, we have to evaluate $\mathcal{D}^{2*}_{\mu\nu}(\pi/2, \pi/2, \pi)$. According to (BI.12a) of Appendix B,

$$\mathcal{D}^{2*}_{\mu\nu}(\pi/2, \pi/2, \pi) = e^{-i\mu(\pi/2)} d^2_{\mu\nu}(\pi/2)\, e^{-i\nu\pi}$$
$$= (-i)^{\mu}(-1)^{\nu} d^2_{\mu\nu}(\pi/2). \tag{IV.21}$$

For the evaluation of the last factor, we need to consider (BI.12c) in Appendix B for the d-function. The general result for any θ is given in Table IV.1. Using this table with $\theta = \pi/2$,

Table IV.1 Matrix elements of $d^2(\theta)$

$\mu \backslash \nu$	$+2$	$+1$	0	-1	-2
$+2$	$\cos^4 \frac{\theta}{2}$	$-\sin\theta \cos^2 \frac{\theta}{2}$	$\frac{\sqrt{6}}{4}\sin^2\theta$	$-\sin\theta \sin^2\frac{\theta}{2}$	$\sin^4\frac{\theta}{2}$
$+1$	$\sin\theta\cos^2\frac{\theta}{2}$	$\cos\theta\cos^2\frac{\theta}{2} - \frac{1}{2}\sin^2\theta$	$-\sqrt{\frac{3}{2}}\sin\theta\cos\theta$	$\cos\theta\sin^2\frac{\theta}{2} + \frac{1}{2}\sin^2\theta$	$-\sin\theta\sin^2\frac{\theta}{2}$
0	$\frac{\sqrt{6}}{4}\sin^2\theta$	$\sqrt{\frac{3}{2}}\sin\theta\cos\theta$	$1 - \frac{3}{2}\sin^2\theta$	$-\sqrt{\frac{3}{2}}\sin\theta\cos\theta$	$\frac{\sqrt{6}}{4}\sin^2\theta$
-1	$\sin\theta\sin^2\frac{\theta}{2}$	$\cos\theta\sin^2\frac{\theta}{2} + \frac{1}{2}\sin^2\theta$	$\sqrt{\frac{3}{2}}\sin\theta\cos\theta$	$\cos\theta\cos^2\frac{\theta}{2} - \frac{1}{2}\sin^2\theta$	$-\sin\theta\cos^2\frac{\theta}{2}$
-2	$\sin^4\frac{\theta}{2}$	$\sin\theta\sin^2\frac{\theta}{2}$	$\frac{\sqrt{6}}{4}\sin^2\theta$	$\sin\theta\cos^2\frac{\theta}{2}$	$\cos^4\frac{\theta}{2}$

and the expressions (IV.21) and (IV.19), we obtain, in a straightforward manner,

$$\bar{a}_2 = \bar{a}_{-2} = -\tfrac{1}{4}(a_2 + a_{-2}) - \frac{\sqrt{6}}{4} a_0,$$
$$\bar{a}_1 = \bar{a}_{-1} = 0,$$
$$\bar{a}_0 = \frac{\sqrt{6}}{4}(a_2 + a_{-2}) - \tfrac{1}{2}a_0.$$

Since $\cos 60° = \frac{1}{2}$ and $\sin 60° = \sqrt{3}/2$, these relations can be rewritten as

$$\bar{a}_2 = \bar{a}_{-2} \equiv \frac{\bar{\beta}}{\sqrt{2}} \sin\bar{\gamma} = -\frac{\beta}{\sqrt{2}}(\sin\gamma\cos 60° + \cos\gamma\sin 60°) = -\frac{\beta}{\sqrt{2}}\sin(\gamma + 60°), \tag{IV.22a}$$

$$\bar{a}_0 \equiv \frac{\bar{\beta}}{\sqrt{2}}\cos\bar{\gamma} = \beta(\sin\gamma\sin 60° - \cos\gamma\cos 60°) = -\beta\cos(\gamma + 60°). \tag{IV.22b}$$

The overall minus sign in (IV.22a) has arisen from the factor $(-i)^{\mu}$ in the expression (IV.21). Since $-\sin(\gamma + 60°) = \sin(\gamma - 120°)$ and $-\cos(\gamma + 60°) = \cos(\gamma - 120°)$, we conclude from (IV.22) that, under $\mathcal{R}_3$, the parameters (β, γ) go to $(\beta, \gamma - 120°)$.

We summarize all the foregoing results as

$$\begin{aligned} &(\beta, \gamma) \to (\beta, \gamma) && \text{under } \mathcal{R}_1, \\ &(\beta, \gamma) \to (\beta, -\gamma) && \text{under } \mathcal{R}_2, \\ &(\beta, \gamma) \to (\beta, \gamma - 120°) && \text{under } \mathcal{R}_3. \end{aligned} \tag{IV.23}$$

These results are very important and are used subsequently.

To study what happens to the Eulerian angles θ_1, θ_2, θ_3 because of the rotations given by (IV.23), we first note that we are now considering two successive rotations. The first rotation $\mathcal{R}$ through the Eulerian angles θ_1, θ_2, θ_3 brings us from the stationary frame to a particular orientation of the body-fixed frame, let us say the (x, y, z)-frame of Fig. IV.2. The second rotation $\overline{\mathcal{R}}$ then takes us from this particular body-fixed frame to any of the 23 other alternatives of the body-fixed frame. We shall refer to this frame as the *second* body-fixed frame. The rotation $\overline{\mathcal{R}}$, as already defined, is through the Eulerian angles $\bar{\theta}_1$, $\bar{\theta}_2$, $\bar{\theta}_3$. The values of these angles, when $\overline{\mathcal{R}} = \mathcal{R}_1$, $\mathcal{R}_2$, $\mathcal{R}_3$, have also been stated. We now want to find the Eulerian rotation R which takes us directly from the stationary frame to the second body-fixed frame. By definition,

$$R = \overline{\mathcal{R}}\mathcal{R}, \tag{IV.24}$$

and let the Eulerian angles for R be denoted by Θ_1, Θ_2, Θ_3. In (BI.6) and (BI.7b) of Appendix B, we see that the rotation operator $\mathcal{R}(\theta_1, \theta_2, \theta_3)$ has two equivalent forms:

$$\mathcal{R}(\theta_1, \theta_2, \theta_3) = \begin{cases} \exp(-i\theta_3 L_{Z_2}) \exp(-i\theta_2 L_{Y_1}) \exp(-i\theta_1 L_Z) \\ \exp(-i\theta_1 L_Z) \exp(-i\theta_2 L_Y) \exp(-i\theta_3 L_Z) \end{cases}$$

In the first form, the subsequent Eulerian rotations about the Z-axis through θ_1, about the Y_1-axis (i.e., the Y-axis after the first rotation) through θ_2, and the final rotation θ_3 about the Z_2-axis (i.e., the Z-axis after the second rotation) appear from right to left as they should. In its second form, all the rotations are about the *initial* axes—first through θ_3 about Z, then through θ_2 about Y, and finally through θ_1 about Z. What is to be remembered as a lesson from this equivalence is that the Euler rotations $\theta_1, \theta_2, \theta_3$ get reversed to the order $\theta_3, \theta_2, \theta_1$ when the *stationary* axes Z, Y, and Z are chosen as the subsequent axes of rotation. We must remember this while writing the operator for $\overline{\mathcal{R}}\mathcal{R}$. If we wish to use the stationary axes as the axes of rotation, then, instead of giving the rotations in the order θ_1, θ_2, θ_3 followed by $\bar{\theta}_1$, $\bar{\theta}_2$, $\bar{\theta}_3$, we must reverse the entire order of the six rotations, that is, write

$$\begin{aligned} \overline{\mathcal{R}}(\bar{\theta}_1, \bar{\theta}_2, \bar{\theta}_3)\mathcal{R}(\theta_1, \theta_2, \theta_3) = {} & \exp(-i\theta_1 L_Z) \exp(-i\theta_2 L_Y) \exp(-i\theta_3 L_Z) \exp(-i\bar{\theta}_1 L_Z) \\ & \times \exp(-i\bar{\theta}_2 L_Y) \exp(-i\bar{\theta}_3 L_Z). \end{aligned} \tag{IV.25a}$$

In terms of the single rotation R, we have

$$R(\Theta_1, \Theta_2, \Theta_3) = \exp(-i\Theta_1 L_Z) \exp(-i\Theta_2 L_Y) \exp(-i\Theta_3 L_Z). \tag{IV.25b}$$

Thus, taking the matrix elements of (IV.25), and using the equality (IV.24) and, from Appendix B, the basic definition (BI.12a) of the $\mathcal{D}$-function, we get

$$\mathcal{D}^{\lambda *}_{\mu\nu}(\Theta_1, \Theta_2, \Theta_3) = \sum_{\mu'} \mathcal{D}^{l*}_{\mu\mu'}(\theta_1, \theta_2, \theta_3)\mathcal{D}^{l*}_{\mu'\nu}(\bar{\theta}_1, \bar{\theta}_2, \bar{\theta}_3). \tag{IV.26}$$

In obtaining the right-hand side from (IV.25a), we have inserted the complete set of spherical harmonic states

$$\sum_{\lambda'\mu'} |\lambda'\mu'\rangle\langle\lambda'\mu'|$$

between $\exp(-i\theta_3 L_Z)$ and $\exp(-i\bar{\theta}_1 L_Z)$. The equation (IV.26) gives us the required relationship between the set of Euler angles Θ_1, Θ_2, Θ_3 and the sets $(\theta_1, \theta_2, \theta_3)$, $(\bar{\theta}_1, \bar{\theta}_2, \bar{\theta}_3)$. For example, in the case of $\mathcal{R}_1$, using (IV.20a) for $\mathcal{D}^{2*}_{\mu\nu}(\bar{\theta}_1, \bar{\theta}_2, \bar{\theta}_3)$, we obtain from (IV.26)

$$\mathcal{D}^{2*}_{\mu\nu}(\Theta_1, \Theta_2, \Theta_3) = (-1)^{\nu}\mathcal{D}^{2*}_{\mu,-\nu}(\theta_1, \theta_2, \theta_3).$$

According to the definition (BI.12a) in Appendix B, this relation leads to the identification

$$\Theta_1 = \theta_1, \qquad \Theta_2 = \theta_2 + \pi, \qquad \Theta_3 = -\theta_3. \tag{IV.27a}$$

In a similar manner, the expression (IV.20b) for $\mathcal{R}_2$ leads, with the help of the basic equation (IV.26), to the result

$$\mathcal{D}^{2*}_{\mu\nu}(\Theta_1, \Theta_2, \Theta_3) = (-i)^{\nu}\mathcal{D}^{2*}_{\mu\nu}(\theta_1, \theta_2, \theta_3).$$

Once again, this relation implies

$$\Theta_1 = \theta_1, \qquad \Theta_2 = \theta_2, \qquad \Theta_3 = \theta_3 + \frac{\pi}{2}. \tag{IV.27b}$$

For the rotation $\mathcal{R}_3$, (IV.21) and Table IV.1 enable us to write down the basic equation (IV.26) explicitly. This equation does not lead to very simple solutions of $(\Theta_1, \Theta_2, \Theta_3)$ in terms of $(\theta_1, \theta_2, \theta_3)$. However, in our subsequent work, the equation connecting the $\mathcal{D}$-functions of these two sets of angles is adequate.

B. DERIVATION OF COLLECTIVE MODEL HAMILTONIAN

In this chapter, we shall consider the collective rotational and vibrational motions of the nuclear surface, restricting ourselves to the $\lambda = 2$ (quadrupole)-terms in the expansion (IV.1). However, for the present, let us keep the formulation general by considering the complete multipole expansion (IV.1) and then deriving the kinetic and potential energies of a nuclear fluid of uniform density ρ_0 confined within this surface. Since we have already imposed the volume-conservation condition (IV.7), the density ρ_0 remains unchanged during the deformation.

Calculation of Hamiltonian

The first part of the calculation is classical. The second part entails taking the classical expression, quantizing it, and writing the Schrödinger equation. The Hamiltonian consists of kinetic energy and potential energy. The latter has two classical types of contribution, one due to the Coulomb repulsion of the charge contained in the nuclear drop, and the other accounting for the change in potential energy as a result of alteration in the nuclear surface area. In analogy with classical hydrodynamics, the latter effect is calculated in terms of an empirically defined surface tension S of the nuclear drop. By definition, S represents the surface energy per unit area. The value of this quantity may be estimated in the manner described in Section 14.

Kinetic energy In deriving the kinetic energy, we shall assume an irrotational motion of the fluid, and take the velocity of the fluid at the nuclear surface to be in the radial direction.

Denoting the velocity by $\mathbf{v}$ and the velocity potential by χ, we note they are related to each other through

$$\mathbf{v} = \text{grad}\, \chi. \tag{IV.28a}$$

Due to the irrotational nature of the fluid motion, we have

$$0 = \text{div}\, \mathbf{v} = \nabla^2 \chi,$$

which has solutions for χ of the type $r^{\lambda} Y^{\lambda}_{\mu}(\theta, \phi)$. The general solution is a sum of all multipolarities, and is given by

$$\chi(r, \theta, \phi) = \sum_{\lambda=0}^{\infty} \sum_{\mu=-\lambda}^{\lambda} r^{\lambda} \beta^{*}_{\lambda\mu} Y^{\lambda}_{\mu}(\theta, \phi), \tag{IV.28b}$$

where the parameters $\beta^*_{\lambda\mu}$ are constants to be determined. The velocity at the nuclear surface $r = R$ can therefore be written as

$$(\text{grad } X)_{r=R} = \sum_{\lambda=0}^{\infty} \sum_{\mu=-\lambda}^{\lambda} \beta^*_{\lambda\mu}[\nabla\{r^\lambda Y^\lambda_\mu(\theta, \phi)\}]_{r=R}.$$

As per our assumption, we take this velocity as pointing in the radial direction; hence, its magnitude is given by

$$v_{r=R} = \sum_{\lambda=0}^{\infty} \sum_{\mu=-\lambda}^{\lambda} \beta^*_{\lambda\mu}[\frac{\partial}{\partial r}\{r^\lambda Y^\lambda_\mu(\theta, \phi)\}]_{r=R}$$

$$= \sum_{\lambda=0}^{\infty} \sum_{\mu=-\lambda}^{\lambda} \beta^*_{\lambda\mu}\lambda R^{\lambda-1} Y^\lambda_\mu(\theta, \phi). \tag{IV.29a}$$

An alternative expression for the same velocity is obtained by taking the time derivative of (IV.1), which yields

$$v_{r=R} = \dot{R}(\theta, \phi) = R_0 \sum_{\lambda=0}^{\infty} \sum_{\mu=-\lambda}^{\lambda} \dot{\alpha}^*_{\lambda\mu}(t) Y^\lambda_\mu(\theta, \phi). \tag{IV.29b}$$

Replacing $R^{\lambda-1}$ in (IV.29a) by the approximate expression $R_0^{\lambda-1}$, and then equating (IV.29a) and (IV.29b), we easily obtain

$$\beta^*_{\lambda\mu} = \lambda^{-1} R_0^{2-\lambda} \dot{\alpha}^*_{\lambda\mu}(t). \tag{IV.30}$$

We now have complete information on the velocity $\mathbf{v}$ at any point (r, θ, ϕ) through (IV.28) and (IV.30) in terms of the time derivatives of our old surface parameters $\alpha^*_{\lambda\mu}$. The kinetic energy of the entire drop can then be written as

$$T = \tfrac{1}{2}\rho_0 \int_\tau d\tau\, v^2,$$

where the quantity τ at the bottom of the integral sign denotes that the integration has to be carried over the nuclear volume τ enclosed within the surface defined by (IV.1). Substituting for v^2, we have

$$T = \tfrac{1}{2}\rho_0 \int d\omega \int_0^{R(\theta,\phi)} r^2\, dr \sum_{\lambda,\mu} \sum_{\lambda',\mu'} \beta^*_{\lambda\mu}\beta^*_{\lambda'\mu'}[\nabla r^\lambda Y^\lambda_\mu(\theta, \phi)] \cdot [\nabla r^{\lambda'} Y^{\lambda'}_{\mu'}(\theta, \phi)]. \tag{IV.31}$$

In this expression, the dr-integration *must* be carried out first, as indicated, because the upper limit of this integral depends on (θ, ϕ). The radial part of the scalar product is given by

$$(\frac{d}{dr} r^\lambda)(\frac{d}{dr} r^{\lambda'}) Y^\lambda_\mu Y^{\lambda'}_{\mu'} = \lambda\lambda' r^{\lambda+\lambda'-2} Y^\lambda_\mu Y^{\lambda'}_{\mu'}, \tag{IV.32a}$$

and the θ- and ϕ-part of the scalar product add to

$$(r^\lambda \frac{1}{r}\frac{\partial}{\partial\theta} Y^\lambda_\mu)(r^{\lambda'} \frac{1}{r}\frac{\partial}{\partial\theta} Y^{\lambda'}_{\mu'}) + (r^\lambda \frac{1}{r\sin\theta}\frac{\partial}{\partial\phi} Y^\lambda_\mu)(r^{\lambda'} \frac{1}{r\sin\theta}\frac{\partial}{\partial\phi} Y^{\lambda'}_{\mu'}). \tag{IV.32b}$$

It is clear from (IV.32a) and (IV.32b) that the radial function in all the three terms is $r^{\lambda+\lambda'-2}$, which changes to $r^{\lambda+\lambda'}$ when multiplied by the r^2 of the $(r^2\, dr)$. Thus, the radial integration in all the three terms gives

$$\int_0^{R(\theta,\phi)} dr\, r^{\lambda+\lambda'} = \frac{R^{\lambda+\lambda'+1}}{\lambda+\lambda'+1}$$

$$= \frac{R_0^{\lambda+\lambda'+1}}{\lambda+\lambda'+1}(1 + \sum_{\lambda''\mu''} \alpha^*_{\lambda''\mu''} Y^{\lambda''}_{\mu''})^{\lambda+\lambda'-1}$$

$$\approx \frac{R_0^{\lambda+\lambda'+1}}{\lambda+\lambda'+1} + \frac{\lambda+\lambda'-1}{\lambda+\lambda'+1} R_0^{\lambda+\lambda'+1} \sum_{\lambda''\mu''} \alpha^*_{\lambda''\mu''} Y^{\lambda''}_{\mu''}. \tag{IV.32c}$$

We have neglected all higher terms in the expansion, and in fact use only the first term of (IV.32c). Using this term, along with (IV.32a) and (IV.32b), in (IV.31), we get

$$T = \tfrac{1}{2}\rho_0 \sum_{\lambda,\mu} \sum_{\lambda',\mu'} \frac{R_0^{\lambda+\lambda'+1}}{\lambda+\lambda'+1} \beta^*_{\lambda\mu}\beta^*_{\lambda'\mu'}$$

$$\times \int d\omega\, [\lambda\lambda' Y^\lambda_\mu Y^{\lambda'}_{\mu'} + (\frac{\partial}{\partial\theta} Y^\lambda_\mu)(\frac{\partial}{\partial\theta} Y^{\lambda'}_{\mu'}) + \operatorname{cosec}^2\theta (\frac{\partial}{\partial\phi} Y^\lambda_\mu)(\frac{\partial}{\partial\phi} Y^{\lambda'}_{\mu'})]. \tag{IV.33}$$

The integration in the first term here is very easy to carry out with the orthogonality properties of the spherical harmonics:

$$\int d\omega\, Y^\lambda_\mu Y^{\lambda'}_{\mu'} = \delta_{\lambda\lambda'}\delta_{\mu,-\mu'}(-1)^\mu. \tag{IV.34a}$$

The integration of the other two terms requires a little more work. Using the identities (AI.10b) from Appendix A, we write

$$\frac{\partial}{\partial\theta} = \tfrac{1}{2}(L_+ e^{-i\phi} - L_- e^{i\phi}),$$

$$i \cot\theta \frac{\partial}{\partial\phi} = \tfrac{1}{2}(L_+ e^{-i\phi} + L_- e^{i\phi}).$$

Applying these relations, we get from the second and third terms of (IV.33)

$$(\frac{\partial}{\partial\theta} Y^\lambda_\mu)(\frac{\partial}{\partial\theta} Y^{\lambda'}_{\mu'}) + (1 + \cot^2\theta)(\frac{\partial}{\partial\phi} Y^\lambda_\mu)(\frac{\partial}{\partial\phi} Y^{\lambda'}_{\mu'})$$

$$= \tfrac{1}{4}[\{(L_+ e^{-i\phi} - L_- e^{i\phi}) Y^\lambda_\mu\}\{(L_+ e^{-i\phi} - L_- e^{i\phi}) Y^{\lambda'}_{\mu'}\}$$

$$- \{(L_+ e^{-i\phi} + L_- e^{i\phi}) Y^\lambda_\mu\}\{(L_+ e^{-i\phi} + L_- e^{i\phi}) Y^{\lambda'}_{\mu'}\}] - \mu\mu' Y^\lambda_\mu Y^{\lambda'}_{\mu'}$$

$$= -\tfrac{1}{2}[(L_+ Y^\lambda_\mu)(L_- Y^{\lambda'}_{\mu'}) + (L_- Y^\lambda_\mu)(L_+ Y^{\lambda'}_{\mu'})] - \mu\mu' Y^\lambda_\mu Y^{\lambda'}_{\mu'}. \tag{IV.34b}$$

The last term in this expression has arisen from $[(\partial/\partial\phi) Y^\lambda_\mu][(\partial/\partial\phi) Y^{\lambda'}_{\mu'}]$ in view of the $e^{im\phi}$-type dependence of Y^l_m. Since there is a summation over (λ, μ) and (λ', μ') in (IV.33), the two terms enclosed within the square brackets yield equal results. So we omit the factor $\frac{1}{2}$ and, using one of the terms, we write the final expression, with the help of (IV.34b) and (IV.33), as

$$T = \tfrac{1}{2}\rho_0 \sum_{\lambda,\mu} \sum_{\lambda',\mu'} \frac{R_0^{\lambda+\lambda'+1}}{\lambda+\lambda'+1} \beta^*_{\lambda\mu}\beta^*_{\lambda'\mu'} \int d\omega\, [(\lambda\lambda' - \mu\mu') Y^\lambda_\mu Y^{\lambda'}_{\mu'} - (L_+ Y^\lambda_\mu)(L_- Y^{\lambda'}_{\mu'})]. \tag{IV.34c}$$

Using the result (AI.3) from Appendix A in the L_+- and L_--term, and then the result (IV.34a), we finally obtain

$$T = \tfrac{1}{2}\rho_0 \sum_{\lambda\mu} \frac{R_0^{2\lambda+1}}{2\lambda+1}|\beta_{\lambda\mu}|^2[\lambda(2\lambda+1) - \mu]$$

$$= \tfrac{1}{2}\rho_0 R_0^5 \sum_{\lambda\mu} \lambda^{-1}|\dot{\alpha}_{\lambda\mu}|^2.$$

In getting the final line from its predecessor, we have used (IV.30) and further taken cognizance of the fact that the μ-term within the square brackets vanishes because, in the sum, the contribution of each positive μ-term is exactly cancelled by the corresponding negative μ-term. The change from $+\mu$ to $-\mu$ leaves $|\beta_{\lambda\mu}|^2$ unaffected. We rewrite this result in a more compact form as

$$T = \tfrac{1}{2} \sum_{\lambda,\mu} B_\lambda |\dot{\alpha}_{\lambda\mu}|^2, \tag{IV.35a}$$

where

$$B_\lambda = \lambda^{-1}\rho_0 R_0^5 = \frac{3}{4\pi}\lambda^{-1} MAR_0^2, \tag{IV.35b}$$

MA being equal to $(4\pi/3)\rho_0 R_0^3$, i.e., the total mass of the nucleus. B_λ is called the *inertial parameter*. Since the expression for T, as given by (IV.35a), is already of the second order in $\dot{\alpha}_{\lambda\mu}$, we do not work out the consequences of the second term of (IV.32c) which already contains one $\alpha^*_{\lambda''\mu''}$. However, if this higher-order term is desired, it can very easily be worked out by using the results (BIII.1) and (BIII.3) from Appendix B for the integral of the product of three spherical harmonics. This correction term in T is obviously of the type $\dot{\alpha}\dot{\alpha}\alpha$.

Surface energy In terms of the empirical surface tension parameter S, the surface energy is given by

$$V_S = S\int_A dA,$$

where the integral has to be extended over the whole area of the nuclear surface defined by (IV.1). To express the area element dA on the deformed surface in terms of the polar coordinates R, θ, ϕ of a point on the surface is a little intricate. It is *not* simply $R^2\, d\omega$, which is the area element on a *spherical* surface of radius R contained within the solid angle $d\omega$ ($=\sin\theta\, d\theta\, d\phi$). To express dA, we proceed as follows. We first note that the normal to the surface, defined by (IV.1), which relates the radial coordinate R to the other two coordinates θ, ϕ, is given by the gradient of the expression

$$f(R, \theta, \phi) \equiv R - R_0[1 + \sum_{\lambda\mu} \alpha^*_{\lambda\mu} Y^\lambda_\mu(\theta, \phi)] = 0$$

in the usual way. The gradient, when resolved along the three coordinate axes, yields

$$R\text{-component} = \frac{\partial}{\partial R} f(R, \theta, \phi) = 1,$$

$$\theta\text{-component} = \frac{1}{R}\frac{\partial}{\partial\theta} f(R, \theta, \phi) = -\frac{1}{R}\frac{\partial R(\theta, \phi)}{\partial\theta},$$

$$\phi\text{-component} = \frac{1}{R\sin\theta}\frac{\partial}{\partial\phi} f(R, \theta, \phi) = -\frac{1}{R\sin\theta}\frac{\partial R(\theta, \phi)}{\partial\phi}.$$

The direction cosines l, m, n of the gradient, i.e., the normal to the surface, at R, θ, ϕ are clearly proportional to these quantities. If the constant of proportionality is k, then k can be determined from

$$1 = l^2 + m^2 + n^2 = k^2[1 + \frac{1}{R^2}(\frac{\partial R}{\partial \theta})^2 + \frac{1}{R^2 \sin^2 \theta}(\frac{\partial R}{\partial \phi})^2].$$

That is,

$$k = [1 + \frac{1}{R^2}(\frac{\partial R}{\partial \theta})^2 + \frac{1}{R^2 \sin^2 \theta}(\frac{\partial R}{\partial \phi})^2]^{-1/2},$$

$$l = k, \qquad m = -\frac{k}{R}\frac{\partial R}{\partial \theta}, \qquad n = -\frac{k}{R \sin \theta}\frac{\partial R}{\partial \phi}.$$

The projection of the area element dA (whose normal is defined by the direction cosines l, m, n) on the (θ, ϕ)-plane is clearly given by $l\,dA$ and, by definition, this has to be equal to the elementary area $R^2 \sin \theta\, d\theta\, d\phi$ in the (θ, ϕ)-plane. Thus,

$$\begin{aligned} dA &= l^{-1}R^2 \sin \theta\, d\theta\, d\phi \\ &= [1 + \frac{1}{R^2}(\frac{\partial R}{\partial \theta})^2 + \frac{1}{R^2 \sin^2 \theta}(\frac{\partial R}{\partial \phi})^2]^{1/2} R^2 \sin \theta\, d\theta\, d\phi. \end{aligned}$$

We have here used $l = k$, and the expression of k as just given. It is advantageous now to write R in the form (IV.5a) and retain the terms up to the second order in ξ in the expression dA. In this way,

$$\begin{aligned} dA &= R_0^2(1 + 2\xi + \xi^2)[1 + (1 + \xi)^{-2}\{(\frac{\partial \xi}{\partial \theta})^2 + \frac{1}{\sin^2 \theta}(\frac{\partial \xi}{\partial \phi})^2\}]^{1/2} \\ &\approx R_0^2(1 + 2\xi + \xi^2)[1 + \tfrac{1}{2}(\frac{\partial \xi}{\partial \theta})^2 + \frac{1}{2 \sin^2 \theta}(\frac{\partial \xi}{\partial \phi})^2] \\ &\approx R_0^2[1 + 2\xi + \xi^2 + \tfrac{1}{2}(\frac{\partial \xi}{\partial \theta})^2 + \frac{1}{2 \sin^2 \theta}(\frac{\partial \xi}{\partial \phi})^2]. \end{aligned}$$

It should be noticed that, although $(1 + \xi)^{-2} = (1 - 2\xi + 3\xi^2 + \ldots)$, only the term 1 of this expansion contributes to our expression up to the second order in ξ because the quantities $(\partial\xi/\partial\theta)^2$ and $(\partial\xi/\partial\phi)^2$ are already of the second order in ξ.

From the foregoing expression of dA, we now obtain the surface energy

$$\begin{aligned} V_S &= SR_0^2 \int d\omega\, [1 + 2\xi + \xi^2 + \tfrac{1}{2}(\frac{\partial \xi}{\partial \theta})^2 + \frac{1}{2 \sin^2 \theta}(\frac{\partial \xi}{\partial \phi})^2] \\ &= SR_0^2 \int d\omega\, [1 + \xi + \tfrac{1}{2}\{(\frac{\partial \xi}{\partial \theta})^2 + \operatorname{cosec}^2 \theta (\frac{\partial \xi}{\partial \phi})^2\}]. \end{aligned} \qquad \text{(IV.36)}$$

We have used here the volume-conservation condition (IV.7) to delete $(\xi + \xi^2)$ from the integrand. The first term of this integral gives $4\pi R_0^2 S$, which is the surface energy V_S (sphere) of a sphere of radius R_0; the other terms represent the change in surface energy due to deformation. In our Hamiltonian, we shall omit V_S (sphere) and keep only the extra surface energy due to deformation. Using (IV.5b) for ξ, we get a contribution to the second term of the integral from the $(\lambda = 0, \mu = 0)$-term for which $Y_0^0 = (4\pi)^{-1/2}$. The integration over the solid angle gives 4π, and hence

$$R_0^2 S \int \xi \, d\omega = (4\pi)^{1/2} \alpha_{00}^* R_0^2 S = -R_0^2 S \sum_{\lambda, \mu} |\alpha_{\lambda\mu}|^2, \tag{IV.37a}$$

where we have used the result (IV.8c). The terms within the braces in (IV.36) yield, according to (IV.5b),

$$\tfrac{1}{2} R_0^2 S \sum_{\lambda\mu} \sum_{\lambda'\mu'} \alpha_{\lambda\mu}^* \alpha_{\lambda'\mu'}^* \int d\omega \, [(\frac{\partial Y_\mu^\lambda}{\partial\theta})(\frac{\partial Y_{\mu'}}{\partial\theta}) + \operatorname{cosec}^2 \theta (\frac{\partial Y_\mu^\lambda}{\partial\phi})(\frac{\partial Y_{\mu'}^{\lambda'}}{\partial\phi})]. \tag{IV.37b}$$

It should be observed that this integral has been encountered in (IV.33) while evaluating the kinetic energy. This is the integral in (IV.34c) with the omission of the $\lambda\lambda'$-term. Thus, the present integral is given by

$$\delta_{\lambda\lambda'} \delta_{\mu, -\mu'} (-1)^\mu [\mu^2 + (\lambda - \mu)(\lambda + \mu + 1)],$$

where the first term comes from the $\mu\mu'$-term of (IV.34c) and the second term from the (L_+, L_-)-terms of (IV.34c). We have

$$\mu^2 + (\lambda - \mu)(\lambda + \mu + 1) = \lambda(\lambda + 1) - \mu$$

and, as before, the $-\mu$-term here contributes nothing when summed over μ in (IV.37b). Thus, (IV.37b) reduces to

$$\tfrac{1}{2} R_0^2 S \sum_{\lambda, \mu} \lambda(\lambda + 1) |\alpha_{\lambda\mu}|^2.$$

Using this expression and (IV.37a) in (IV.36), we finally obtain

$$V_S = \tfrac{1}{2} \sum_{\lambda, \mu} C_\lambda^S |\alpha_{\lambda\mu}|^2, \tag{IV.38a}$$

where

$$C_\lambda^S = R_0^2 S (\lambda - 1)(\lambda + 2). \tag{IV.38b}$$

We recall that there is V_S(sphere) in the total surface energy, and the expressions (IV.38) give the surface energy due to deformation. For $\lambda = 1$, the latter is seen to be zero, whereas for all $\lambda > 1$, it is positive. Thus, the surface energy of a sphere increases in a volume-conserving deformation which is understandable because, for a given volume, the sphere has the least surface area.

Coulomb energy The Coulomb interaction energy between the charge located inside the volume elements of $d\tau_1 = r_1^2 \, dr_1 \, d\omega_1$ and $d\tau_2 = r_2^2 \, dr_2 \, d\omega_2$ is given by

$$\frac{\rho_0^2}{|\mathbf{r}_1 - \mathbf{r}_2|} r_1^2 \, dr_1 \, d\omega_1 \, r_2^2 \, dr_2 \, d\omega_2.$$

The total Coulomb energy of the entire charge distribution can be calculated by carrying out the integration over $d\tau_1$ and $d\tau_2$, where each integration proceeds over the entire volume of the charge distribution. However, we have to insert an extra factor of $\frac{1}{2}$ to cancel the double counting of each pair of elements $d\tau_1$ and $d\tau_2$ when *independent* integrations are carried out over these elements. Thus,

$$V_C = \tfrac{1}{2} \rho_0^2 \int d\omega_1 \int d\omega_2 \int_0^{R_0(1+\xi_1)} r_1^2 \, dr_1 \int_0^{R_0(1+\xi_2)} r_2^2 \, dr_2 \, \frac{1}{|\mathbf{r}_1 - \mathbf{r}_2|}. \tag{IV.39}$$

Here ξ_1 and ξ_2 are the quantity ξ defined by (IV.5), wherein the angles of the spherical harmonics should be put equal to (θ_1, ϕ_1) and (θ_2, ϕ_2).

In our earlier derivation of the kinetic and surface energies, we have evaluated every quantity to the second order in ξ. We adhere to the same order here. We first note the standard result

$$\frac{1}{|\mathbf{r}_1 - \mathbf{r}_2|} = \sum_{K=0}^{\infty} \frac{r_<^K}{r_>^{K+1}} P_K(\cos \omega_{12})$$

$$= \sum_{K=0}^{\infty} \frac{r_<^K}{r_>^{K+1}} \frac{4\pi}{2K+1} \sum_{q=-K}^{K} Y_q^{K*}(\theta_1, \phi_1) Y_q^K(\theta_2, \phi_2), \tag{IV.40}$$

where ω_{12} is the angle between the vectors $\mathbf{r}_1$ and $\mathbf{r}_2$, and $r_<$ and $r_>$ are respectively the smaller and the larger of the two *radial* coordinates r_1 and r_2.

In order to understand the complications involved through the occurrence of $r_<$ and $r_>$ in (IV.40), let us first evaluate (IV.39) for a sphere, i.e., in the case where the upper limit of each radial integral is R_0. Here we can first carry out the angle-integrations which lead to the requirement $\delta_{K,0}$, since the single spherical harmonic $Y_q^{K*}(\theta_1, \phi_1)$ has to be integrated over $d\omega_1$, and the single $Y_q^K(\theta_2, \phi_2)$ over $d\omega_2$. After completing the angle-integration, we get

$$V_C\,(\text{sphere}) = \tfrac{1}{2}(4\pi\rho_0)^2 \int_0^{R_0} r_1^2\, dr_1 \int_0^{R_0} r_2^2\, dr_2 \frac{1}{r_>}$$

$$= \tfrac{1}{2}(4\pi\rho_0)^2 \int_0^{R_0} r_1^2\, dr_1 \left(\frac{1}{r_1} \int_0^{r_1} r_2^2\, dr_2 + \int_{r_1}^{R_0} r_2\, dr_2\right)$$

$$= \tfrac{1}{2}(4\pi\rho_0)^2 \int_0^{R_0} r_1^2\, dr_1\, \tfrac{1}{2}(R_0^2 - \tfrac{1}{3}r_1^2)$$

$$= \tfrac{16}{15}\pi^2 R_0^5 \rho_0^2 = \tfrac{3}{5} \frac{(Ze)^2}{R_0}, \tag{IV.41}$$

where $Ze = (4\pi/3)R_0^3\rho_0$ is the total charge in the sphere. Because of the occurrence of $1/r_>$ in the integrand, we have had to split the integral over r_2 into two parts: one between $r_2 = 0 \to r_2 = r_1$, in which the range r_1 is $r_>$, and the other defined by $r_2 = r_1 \to r_2 = R_0$, where r_2 becomes $r_>$.

It may seem that we make a straightforward approach while evaluating (IV.39) for nonvanishing ξ_1 and ξ_2. We may like to split the r_2-integration in $0 \to r_1$ and $r_1 \to R_0(1 + \xi_2)$. But since the range of r_1 is $0 \to R_0(1 + \xi_1)$, there is no guarantee that, in the range $r_1 \to R_0(1 + \xi_2)$ of r_2, the quantity r_1 is always smaller than r_2. In fact, for any two given directions (θ_1, ϕ_1) and (θ_2, ϕ_2), the precise values of ξ_1 and ξ_2 determine when r_1 is smaller or larger than r_2. Because of this complicated interrelationship in the integration variables, we attempt an alternative method of working out (IV.39) to our desired degree of accuracy, i.e., to the second order in the quantities ξ_1 and ξ_2. These quantities occur as integration limits in (IV.39), and hence we make use of certain identities. If

$$F(x) = \int_0^x dx'\, f(x'),$$

then $F(x + \Delta)$, by definition, is given by

$$F(x + \Delta) = \int_0^{x+\Delta} dx'\, f(x').$$

If Δ is a small quantity, we can make a Taylor expansion of $F(x + \Delta)$ and keep terms up to, let us say, the second order. Then

$$F(x + \Delta) = F(x) + F'(x) + \tfrac{1}{2}\Delta^2 F''(x).$$

But

$$F'(x) = \frac{d}{dx}\int_0^x dx'\, f(x') = f(x),$$

and hence

$$F''(x) = \frac{df}{dx}.$$

Therefore, the final result for $F(x + \Delta)$ becomes

$$\int_0^{x+\Delta} f(x')\, dx' \equiv F(x + \Delta)$$

$$= \int_0^x f(x')\, dx' + \Delta f(x) + \tfrac{1}{2}\Delta^2\frac{df}{dx}.$$

Using this key relation to express the integral over r_2, we have

$$\int_0^{R_0(1+\xi_2)} dr_2\, r_2^2\frac{r_<^K}{r_>^{K+1}} = \int_0^{R_0} dr_2\, r_2^2\frac{r_<^K}{r_>^{K+1}} + R_0\xi_2R_0^2\left(\frac{r_<^K}{r_>^{K+1}}\right)_{r_2=R_0} + \tfrac{1}{2}(R_0\xi_2)^2\left(\frac{d}{dr_2}\frac{r_2^2 r_<^K}{r_>^{K+1}}\right)_{r_2=R_0}. \tag{IV.42}$$

We take the first term of (IV.42) and then move on to the r_1-integration. Making use of the key relation again, we obtain

$$\int_0^{R_0(1+\xi_1)} dr_1\, r_1^2\left(\int_0^{R_0} dr_2\, r_2^2\frac{r_<^K}{r_>^{K+1}}\right) = \int_0^{R_0} r_1^2\, dr_1\int_0^{R_0} r_2^2\, dr_2\frac{r_<^K}{r_>^{K+1}} + R_0\xi_1R_0^2\left(\int_0^{R_0} dr_2\, r_2^2\frac{r_<^K}{r_>^{K+1}}\right)_{r_1=R_0}$$

$$+ \tfrac{1}{2}(R_0\xi_1)^2\left(\frac{d}{dr_1}\int_0^{R_0} dr_2\, r_2^2\frac{r_1^2 r_<^K}{r_>^{K+1}}\right)_{r_1=R_0}. \tag{IV.43}$$

The first term of (IV.43), as is clear, taken with the other factors in (IV.40) and integrated over $d\omega_1$, $d\omega_2$, leads to the Coulomb energy of a sphere of radius R_0, evaluated in (IV.41). For the Coulomb energy of deformation, we shall use only the part that represents the departure from the spherical value. Hence, the first term of (IV.43) is omitted from our consideration in the subsequent derivation. The second and third terms in (IV.43) are now quite straightforward to evaluate. Because $r_1 = R_0$ in these terms, and r_2-integration goes from $0 \to R_0$, the role of $r_>$ is now taken by r_1, and that of $r_<$ is taken by r_2. Using these facts, we get the results

$$\text{second term of (IV.43)} = \frac{R_0^5}{K+3}\xi_1, \tag{IV.44a}$$

$$\text{third term of (IV.43)} = \tfrac{1}{2}R_0^5\frac{1-K}{K+3}\xi_1^2. \tag{IV.44b}$$

We have followed the consequence of the lowest-order term in (IV.42) right through our expressions (IV.43) and (IV.44). We next wish to apply the same procedure for the first- and second-order term in ξ_2 appearing in (IV.42). Since only those terms up to the second order in ξ_1 and ξ_2 are of interest to us in the subsequent integration over r_1 of the ξ_2-term in (IV.42), we need to use terms up to the order ξ_1; similarly, in the next integration over r_1 of the ξ_2^2-term in (IV.42), we need keep only the first term of the r_1-integration, which is independent of ξ_1. Keeping these facts in mind, we have, with the second term of (IV.42), the expression

$$R_0^3\xi_2 \int_0^{R_0(1+\xi_1)} dr_1\, r_1^2\Big(\frac{r_<^K}{r_>^{K+1}}\Big)_{r_2=R_0} = R_0^3\xi_2 \int_0^{R_0} dr_1\, r_1^2\Big(\frac{r_<^K}{r_>^{K+1}}\Big)_{r_2=R_0} + R_0^3\xi_2 R_0\xi_1 R_0^2\Big(\frac{r_<^K}{r_>^{K+1}}\Big)_{r_1=r_2=R_0}$$

$$= \frac{R_0^5}{K+3}\xi_2 + R_0^5\xi_1\xi_2. \tag{IV.44c}$$

In a similar manner, the third term of (IV.42) yields up to the second order

$$\tfrac{1}{2}(R_0\xi_2)^2 \int_0^{R_0} dr_1\, r_1^2\Big(\frac{d}{dr_2}\frac{r_2^2 r_<^K}{r_>^{K+1}}\Big)_{r_2=R_0} = \tfrac{1}{2}R_0^5\frac{1-K}{K+3}\xi_2^2. \tag{IV.44d}$$

We note the fact that, in the subsequent integration over $d\omega_1$ and $d\omega_2$, (IV.44a) and the first term of (IV.44c) lead to the same result; the same observation is true for (IV.44b) and (IV.44d). So, in the final expression of V_C, we keep (IV.44a) and (IV.44b) multiplied by 2 and the second term of (IV.44c). With the help of (IV.39) and (IV.40) and the fact just mentioned, we obtain

$$V_C = 4\pi\rho_0^2R_0^5 \int d\omega_1 \int d\omega_2 \sum_{K,q} \frac{1}{2K+1}\Big[\frac{\xi_1}{K+3} + \frac{1-K}{2(K+3)}\xi_1^2 + \tfrac{1}{2}\xi_1\xi_2\Big] Y_q^{K*}(\theta_1,\phi_1)Y_q^K(\theta_2,\phi_2). \tag{IV.45}$$

In the integration of the first two terms, the $d\omega_2$-integration of $Y_q^K(\theta_2, \phi_2)$ demands $K = q = 0$. We thus get, from these two terms,

$$\tfrac{1}{6}\int d\omega_1\,(2\xi_1 + \xi_1^2) = \tfrac{1}{6}\int d\omega_1\,\xi_1 = -\tfrac{1}{6}\sum_{\lambda,\mu} |\alpha_{\lambda\mu}|^2, \tag{IV.46a}$$

where we have used the volume-conservation conditions (IV.7) and (IV.8c). The last term in (IV.45) can be worked out as

$$\tfrac{1}{2}\sum_{K,q}\int d\omega_1 \int d\omega_2\, \xi_1\xi_2 Y_q^{K*}(\theta_1,\phi_1)Y_q^K(\theta_2,\phi_2)(2K+1)^{-1}$$

$$= \tfrac{1}{2}\sum_{K,q}\int d\omega_1 \sum_{\lambda\mu}\alpha_{\lambda\mu}^* Y_\mu^\lambda(\theta_1,\phi_1)Y_q^{K*}(\theta_1,\phi_1)\int d\omega_2 \sum_{\lambda'\mu'}\alpha_{\lambda'\mu'}^* Y_{\mu'}^{\lambda'}(\theta_2,\phi_2)Y_q^K(\theta_2,\phi_2)(2K+1)^{-1}$$

$$= \tfrac{1}{2}\sum_{\lambda,\mu}|\alpha_{\lambda\mu}|^2(2\lambda+1)^{-1}. \tag{IV.46b}$$

Using (IV.46) in (IV.45), we finally obtain

$$V_C = 4\pi\rho_0^2R_0^5 \sum_{\lambda\mu}|\alpha_{\lambda\mu}|^2\tfrac{1}{2}\Big(\frac{1}{2\lambda+1} - \tfrac{1}{3}\Big)$$

$$= -\frac{3}{4\pi}\frac{(Ze)^2}{R_0}\sum_{\lambda\mu}\frac{\lambda-1}{2\lambda+1}|\alpha_{\lambda\mu}|^2. \tag{IV.47}$$

We remind the reader, once again, that the actual V_C also contains V_C (sphere) of (IV.41); the expression (IV.47) gives the extra Coulomb energy due to deformation. For $\lambda = 1$, the latter is zero, whereas for all the higher multipoles it has a negative sign, signifying that the Coulomb energy decreases when the sphere gets deformed. This is in contrast to the behaviour of surface energy as stated after (IV.38). We rewrite (IV.47) in the form

$$V_C = \tfrac{1}{2}\sum_{\lambda,\mu} C_\lambda^C|\alpha_{\lambda\mu}|^2, \tag{IV.48a}$$

where

$$C_\lambda^{\mathrm{C}} = -\frac{3}{2\pi}\frac{(Ze)^2}{R_0}\frac{\lambda - 1}{2\lambda + 1}. \tag{IV.48b}$$

The total potential energy which is a sum of (IV.38a) and (IV.48a) is then given by

$$V = V_{\mathrm{S}} + V_{\mathrm{C}} = \tfrac{1}{2}\sum_{\lambda,\,\mu} C_\lambda|\alpha_{\lambda\mu}|^2 \tag{IV.49a}$$

with

$$C_\lambda = C_\lambda^{\mathrm{S}} + C_\lambda^{\mathrm{C}} = R_0^2 S(\lambda - 1)(\lambda + 2) - \frac{3}{2\pi}\frac{(Ze)^2}{R_0}\frac{\lambda - 1}{2\lambda + 1}. \tag{IV.49b}$$

Collecting the kinetic energy T from (IV.35a), we finally obtain the Hamiltonian as

$$H = T + V = \tfrac{1}{2}\sum_{\lambda\mu}(B_\lambda|\dot{\alpha}_{\lambda\mu}|^2 + C_\lambda|\alpha_{\lambda\mu}|^2), \tag{IV.50}$$

where

$$B_\lambda = \frac{3}{4\pi}\lambda^{-1}MAR_0^2.$$

Classically, this Hamiltonian corresponds to a vibrator for each (λ, μ), the quantity $\alpha_{\lambda\mu}$ being the vibrating coordinate; the classical frequency of vibration is given by $\omega_\lambda^2 = (C_\lambda/B_\lambda)$ which is independent of μ.

Transformation of Hamiltonian to Body-Fixed Frame

We shall continue to work with all the values of (λ, μ). Finally, we shall specialize to the case of quadrupole deformation only. The expansion coefficients in the body-fixed frame will be denoted by $a_{\lambda\nu}$, which is related to $\alpha_{\lambda\mu}$ by (IV.10b) and (IV.10c).

In general, we impose two kinds of symmetry on the body: (i) any point on the surface $R(\theta, \phi)$, when rotated through π about the body-fixed z-axis, maintains its old *value* of R at the changed angles; (ii) the observation in (i) is true also when the rotation π is given around the body-fixed y-axis. It should be noticed that an ellipsoidal body has both these symmetries. We are here *imposing* them for the higher multipoles. Further, we note the fact that now we are rotating a point on the surface, whereas in Section 32A we have considered rotations of the coordinate system. However, we know from Appendix B that the rotation of a point is given by the Hermitean conjugate of the operator that generates the same rotation for the coordinate frame. Thus, the rotations (i) and (ii) of a point on the surface are generated by $\exp(i\pi L_z)$ and $\exp(i\pi L_y)$, respectively. We therefore require the equality

$$\begin{aligned} R_0[1 + \sum_{\lambda\mu} a_{\lambda\mu}^* Y_\mu^\lambda(\theta, \phi)] &= R_0[1 + \sum_{\lambda\mu} a_{\lambda\mu}^* \exp(i\pi L_z) Y_\mu^\lambda(\theta, \phi)] \\ &= R_0[1 + \sum_{\lambda\mu} a_{\lambda\mu}^*(-1)^\mu Y_\mu^\lambda(\theta, \phi)], \end{aligned}$$

which implies $a_{\lambda\mu}^* = (-1)^\mu a_{\lambda\mu}^*$ or μ = even integers only; and the equality

$$\begin{aligned} R_0[1 + \sum_{\lambda\mu} a_{\lambda\mu}^* Y_\mu^\lambda(\theta, \phi)] &= R_0[1 + \sum_{\lambda\mu} a_{\lambda\mu}^* \exp(i\pi L_y) Y_\mu^\lambda(\theta, \phi)] \\ &= R_0[1 + \sum_{\lambda\mu} a_{\lambda\mu}^*(-1)^{\lambda+\mu} Y_{-\mu}^\lambda(\theta, \phi)], \end{aligned}$$

which implies $a_{\lambda\mu}^* = (-1)^{\lambda+\mu} a_{\lambda,-\mu}^*$. This relation, taken along with the previous requirement, namely, $\mu =$ even integers, demands

$$a^*_{\lambda\mu} = (-1)^\lambda a^*_{\lambda, -\mu}.$$

If we use this in conjunction with (IV.4), we obtain

$$a^*_{\lambda\mu} = (-1)^\lambda a_{\lambda\mu}.$$

In the special case of even λ, the last relation implies *real* coefficients $a_{\lambda\mu}$. It should be noticed that, in the case of an ellipsoidal body, these symmetry requirements become $a_\mu =$ real, and

$$a_1 = a_{-1} = 0, \qquad a_2 = a_{-2} \neq 0, \qquad a_0 \neq 0,$$

a fact that has been derived in Section 32A in a more straightforward manner from the equation of the ellipsoid.

Using the potential energy term of (IV.50) along with (IV.10c), we get

$$\begin{aligned} V &= \tfrac{1}{2} \sum_{\lambda\mu} C_\lambda |\alpha_{\lambda\mu}|^2 = \tfrac{1}{2} \sum_{\lambda\mu} C_\lambda \sum_{\nu} \sum_{\nu'} \mathcal{D}^\lambda_{\mu\nu}(\theta_1, \theta_2, \theta_3) \mathcal{D}^{\lambda *}_{\mu\nu'}(\theta_1, \theta_2, \theta_3) a_{\lambda\nu} a^*_{\lambda\nu'} \\ &= \tfrac{1}{2} \sum_{\lambda\nu} C_\lambda |a_{\lambda\nu}|^2, \end{aligned} \tag{IV.51}$$

where we have used the unitary property of the $\mathcal{D}^\lambda$-matrix in carrying out the summation over μ.

The transformation of the kinetic energy term of (IV.50) is a little more tedious. We first write from (IV.10c)

$$\dot{\alpha}_{\lambda\mu} = \sum_{\nu} [\mathcal{D}^\lambda_{\mu\nu}(\theta_1, \theta_2, \theta_3) \dot{a}_{\lambda\nu} + \sum_{j=1}^{3} a_{\lambda\nu} \dot{\theta}_j \frac{\partial}{\partial \theta_j} \mathcal{D}^\lambda_{\mu\nu}(\theta_1, \theta_2, \theta_3)], \tag{IV.52}$$

where, in the second term, we have noted the fact that the Eulerian angles are dependent on time; they change as the body-fixed frame moves with the movement of the body. In the kinetic energy, we need $\sum_{\mu} |\dot{\alpha}_{\lambda\mu}|^2$. The contribution to this from the first term of (IV.52) *alone* is quite easy to calculate. The work is similar to that involved in (IV.51). We call this part of the kinetic energy T_{vib}, where "vib" stands for vibration. Thus,

$$\begin{aligned} T_{\text{vib}} &= \tfrac{1}{2} \sum_{\lambda} B_\lambda \sum_{\mu} \sum_{\nu, \nu'} \mathcal{D}^\lambda_{\mu\nu} \mathcal{D}^{\lambda *}_{\mu\nu'} \dot{a}_{\lambda\nu} \dot{a}^*_{\lambda\nu'} \\ &= \tfrac{1}{2} \sum_{\lambda\nu} B_\lambda |\dot{a}_{\lambda\nu}|^2. \end{aligned} \tag{IV.53}$$

Due to the presence of the second term in (IV.52), there are additional contributions to $|\dot{\alpha}_{\lambda\mu}|^2$, one from the square of the second term and the other from the cross product of the first and second terms. We shall call the part of T that comes from the square of the second term T_{rot}, where "rot" stands for rotation. The part of T from the cross product of the first and second terms can be proved to be zero. To do this and to derive the expression for T_{rot}, we need a somewhat simplified form of the second term of (IV.52), which follows from the definition of the $\mathcal{D}$-function given in Appendix B. We shall first derive that result.

The definition given in Appendix B says

$$\mathcal{D}^{\lambda *}_{\mu\nu}(\theta_1, \theta_2, \theta_3) = \langle \lambda\mu | \exp(-i\theta_1 L_Z) \exp(-i\theta_2 L_Y) \exp(-i\theta_3 L_Z) | \lambda\nu \rangle \tag{IV.54a}$$

or

$$\mathcal{D}^\lambda_{\mu\nu}(\theta_1, \theta_2, \theta_3) = \langle \lambda\nu | \exp(+i\theta_3 L_Z) \exp(+i\theta_2 L_Y) \exp(+i\theta_1 L_Z) | \lambda\mu \rangle, \tag{IV.54b}$$

where X, Y, Z stand for the coordinate axes in the stationary frame. The results can be immediately established by a direct differentiation of (IV.54a) and the insertion of the full set of intermediate states at the appropriate point. Thus,

$$\frac{\partial}{\partial\theta_1}\mathcal{D}^{\lambda *}_{\mu\nu}(\theta_1, \theta_2, \theta_3) = -i\sum_{\mu'}\mathcal{D}^{\lambda *}_{\mu\mu'}(\theta_1, \theta_2, \theta_3)\langle\lambda\mu'|\,\hat{L}_Z\,|\lambda\nu\rangle,$$

$$\frac{\partial}{\partial\theta_2}\mathcal{D}^{\lambda *}_{\mu\nu}(\theta_1, \theta_2, \theta_3) = -i\sum_{\mu'}\mathcal{D}^{\lambda *}_{\mu\mu'}(\theta_1, \theta_2, \theta_3)\langle\lambda\mu'|\,\hat{L}_Y\,|\lambda\nu\rangle,$$

$$\frac{\partial}{\partial\theta_3}\mathcal{D}^{\lambda *}_{\mu\nu}(\theta_1, \theta_2, \theta_3) = -i\sum_{\mu'}\mathcal{D}^{\lambda *}_{\mu\mu'}(\theta_1, \theta_2, \theta_3)\langle\lambda\mu'|\,L_Z\,|\lambda\nu\rangle,$$

where

$$\begin{aligned}\hat{L}_Y &= \exp(i\theta_3 L_Z)L_Y\exp(-i\theta_3 L_Z)\\ &\equiv L_Y\cos\theta_3 + L_X\sin\theta_3,\\ \hat{L}_Z &= \exp(i\theta_3 L_Z)\exp(i\theta_2 L_Y)L_Z\exp(-i\theta_2 L_Y)\exp(-i\theta_3 L_Z)\\ &\equiv \exp(i\theta_3 L_Z)(L_Z\cos\theta_2 - L_X\sin\theta_2)\exp(-i\theta_3 L_Z)\\ &\equiv L_Z\cos\theta_2 - (L_X\cos\theta_3 - L_Y\sin\theta_3)\sin\theta_2.\end{aligned}$$

These final expressions for $\hat{L}_Y$ and $\hat{L}_Z$ follow from a straightforward application of the general result for *any* unitary transformation, namely,

$$\begin{aligned}\hat{A} &= e^{i\theta\Omega}Ae^{-i\theta\Omega}\\ &\equiv A + \frac{i\theta}{1!}[\Omega, A] + \frac{(i\theta)^2}{2!}[\Omega, [\Omega, A]] + \dots,\end{aligned}$$

and from the commutators of L_X, L_Y, L_Z amongst themselves. With the help of the foregoing expressions, we obtain

$$\sum_j \dot{\theta}_j\frac{\partial}{\partial\theta_j}\mathcal{D}^{\lambda *}_{\mu\nu}(\theta_1, \theta_2, \theta_3) = -i\sum_{\mu'}\mathcal{D}^{\lambda *}_{\mu\mu'}(\theta_1, \theta_2, \theta_3)\langle\lambda\mu'|\,\mathcal{L}\,|\lambda\nu\rangle, \tag{IV.55a}$$

where

$$\begin{aligned}\mathcal{L} &= \dot{\theta}_3 L_Z + \dot{\theta}_2(L_Y\cos\theta_3 + L_X\sin\theta_3) + \dot{\theta}_1[L_Z\cos\theta_2 - (L_X\cos\theta_3 - L_Y\sin\theta_3)\sin\theta_2]\\ &= \omega_x L_X + \omega_y L_Y + \omega_z L_Z.\end{aligned} \tag{IV.55b}$$

The components of the angular velocity $\boldsymbol{\omega}$ appearing here are along the body-fixed axes and given by

$$\begin{aligned}\omega_x &= \dot{\theta}_2\sin\theta_3 - \dot{\theta}_1\cos\theta_3\sin\theta_2,\\ \omega_y &= \dot{\theta}_2\cos\theta_3 + \dot{\theta}_1\sin\theta_3\sin\theta_2,\\ \omega_z &= \dot{\theta}_3 + \dot{\theta}_1\cos\theta_2,\end{aligned} \tag{IV.55c}$$

and hence the validity of the last line of (IV.55b) can be checked. As for the expressions (IV.55c) themselves, they follow from the fact that $\dot{\theta}_1$, $\dot{\theta}_2$, $\dot{\theta}_3$ are angular velocities about the stationary Z-axis, the Y_1-axis (i.e., the changed Y-axis after the first Euler rotation), and the Z_2-axis (the changed Z-axis after the second Euler rotation, which is the same as the final body-fixed z-axis), respectively. Therefore, $\dot{\theta}_3$ contributes entirely to ω_z. Since $\dot{\theta}_2$ is about the Y_1-axis, the connection between $\dot{\theta}_2$ and ω_x, ω_y, ω_z can be found by multiplying a vector

$$\begin{pmatrix}0\\ \dot{\theta}_2\\ 0\end{pmatrix}$$

by the third Euler rotation matrix, i.e.,

$$\begin{pmatrix} \cos\theta_3 & \sin\theta_3 & 0 \\ -\sin\theta_3 & \cos\theta_3 & 0 \\ 0 & 0 & 1 \end{pmatrix}.$$

This gives the part of ω_x and of ω_y in (IV.55c) dependent on $\dot{\theta}_2$. The contribution of $\dot{\theta}_1$ (which is about the Z-axis) can be similarly found from

$$\begin{pmatrix} 0 \\ 0 \\ \dot{\theta}_1 \end{pmatrix}$$

by multiplying with the product of the last two Eulerian matrices. The reader can verify that the $\dot{\theta}_1$-dependent parts of ω_x, ω_y, ω_z, obtained in this way, indeed agree with those shown in (IV.55c).

The expression (IV.55b) may at first sight appear a little misleading because the components of $\boldsymbol{\omega}$ along the body-fixed axes occur in combination with the components of the rotation operator $\mathbf{L}$ along the stationary axes. However, we shall show how L_X, L_Y, L_Z can be easily replaced by the corresponding body-fixed components. For this purpose, we make use of the geometrical fact mentioned in Appendix B, namely, that any number of rotations given in succession about coordinate axes at the successive stages of rotation is equivalent to the same rotations given in the reverse order about the corresponding stationary axes. In this way, we have the two alternative forms of the Eulerian rotations, namely,

$$R(\theta_1, \theta_2, \theta_3) = \begin{cases} \exp(-i\theta_3 L_{z_2}) \exp(-i\theta_2 L_{y_1}) \exp(-i\theta_1 L_Z) \\ \exp(-i\theta_1 L_Z) \exp(-i\theta_2 L_Y) \exp(-i\theta_3 L_Z) \end{cases}$$

We now wish to consider a further rotation $R_k(\theta)$, given about the body-fixed k-axis after the three Eulerian rotations. According to the geometry just stated,

$$R_k(\theta) \exp(-i\theta_3 L_{z_2}) \exp(-i\theta_2 L_{y_1}) \exp(-i\theta_1 L_Z) = \exp(-i\theta_1 L_Z) \exp(-i\theta_2 L_Y) \times \exp(-i\theta_3 L_Z) R_K(\theta),$$

where K is the stationary axis corresponding to the body-fixed axis k, i.e., it is the Z-axis if k stands for the z-axis, and so on. In the special case of an infinitesimal rotation (i.e., $\theta = \epsilon$), this relation leads to

$$L_k R(\theta_1, \theta_2, \theta_3) = R(\theta_1, \theta_2, \theta_3) L_K. \tag{IV.55d}$$

We note that (IV.55a) is a matrix element of $R(\theta_1, \theta_2, \theta_3)\mathcal{L}$ with respect to the spherical harmonic states $\langle\lambda\mu|$ and $|\lambda\nu\rangle$ in the stationary frame. So each term in (IV.55b) leads in (IV.55a) to

$$-i\omega_k \langle\lambda\mu| R(\theta_1, \theta_2, \theta_3) L_K |\lambda\nu\rangle = -i\omega_k \langle\lambda\mu| L_k R(\theta_1, \theta_2, \theta_3) |\lambda\nu\rangle, \tag{IV.55e}$$

the equality following from (IV.55d). The state $R(\theta_1, \theta_2, \theta_3)|\lambda\nu\rangle$ is, by definition, the spherical harmonic state in the body-fixed frame. Therefore,

$$\begin{aligned} \langle\lambda\mu| L_k R(\theta_1, \theta_2, \theta_3) |\lambda\nu\rangle &= \langle\lambda\mu| L_k |\lambda\nu\rangle_B \\ &= \sum_{\mu'} \langle\lambda\mu \mid \lambda\mu'\rangle_B \, {}_B\langle\lambda\mu'| L_k |\lambda\nu\rangle_B. \end{aligned}$$

The overlap $\langle\lambda\mu \mid \lambda\mu'\rangle_B$ is, by definition, equal to $\langle\lambda\mu| R(\theta_1, \theta_2, \theta_3) |\lambda\mu'\rangle$, i.e., $\mathscr{D}^{\lambda *}_{\mu\mu'}(\theta_1, \theta_2, \theta_3)$. Hence,

$$\langle\lambda\mu| L_k R(\theta_1, \theta_2, \theta_3) |\lambda\nu\rangle = \sum_{\mu'} \mathscr{D}^{\lambda *}_{\mu\mu'}(\theta_1, \theta_2, \theta_3)(L_k)^{\lambda}_{\mu'\nu}, \tag{IV.55f}$$

where $(L_k)^{\lambda}_{\mu'\nu}$ stands for the matrix elements ${}_B\langle\lambda\mu'| L_k |\lambda\nu\rangle_B$ of the body-fixed operator L_k with respect to the *spherical harmonics* ${}_B\langle\lambda\mu'|$ and $|\lambda\nu\rangle_B$ *of the body-fixed frame*. Using the results (IV.55d), (IV.55e), and (IV.55f) along with (IV.55a) and (IV.55b), we finally derive the important expression

$$\sum_j \dot{\theta}_j \frac{\partial}{\partial\theta_j} \mathscr{D}^{\lambda *}_{\mu\nu}(\theta_1, \theta_2, \theta_3) = -i \sum_{\mu'} \mathscr{D}^{\lambda *}_{\mu\mu'}(\theta_1, \theta_2, \theta_3) \sum_k \omega_k (L_k)^{\lambda}_{\mu'\nu}$$

or, taking the complex conjugate, we get

$$\sum_j \dot{\theta}_j \frac{\partial}{\partial\theta_j} \mathscr{D}^{\lambda}_{\mu\nu}(\theta_1, \theta_2, \theta_3) = i \sum_{\mu'} \mathscr{D}^{\lambda}_{\mu\mu'}(\theta_1, \theta_2, \theta_3) \sum_k \omega_k (L_k)^{\lambda}_{\nu\mu'} \tag{IV.56}$$

(in view of the hermiticity of the operators L_k).

We now use (IV.56) in the second term of (IV.52) and evaluate T_{rot} from the square of the modulus of this term. We have

$$\text{second term of } \dot{\alpha}_{\lambda\mu} = i \sum_{\nu} a_{\lambda\nu} \sum_{k=1}^{3} \omega_k \sum_{\mu'} \mathscr{D}^{\lambda}_{\mu\mu'}(L_k)^{\lambda}_{\nu\mu'}, \tag{IV.57}$$

and hence

$$T_{\text{rot}} = \tfrac{1}{2} \sum_{\lambda\mu} B_\lambda \sum_{\nu\nu_1} a_{\lambda\nu} a^{*}_{\lambda\nu_1} \sum_{kk_1} \omega_k \omega_{k_1} \sum_{\mu'\mu'_1} (L_k)^{\lambda}_{\nu\mu'} (L_{k_1})^{\lambda}_{\mu'_1\nu_1} \mathscr{D}^{\lambda}_{\mu\mu'} \mathscr{D}^{\lambda *}_{\mu\mu'_1}.$$

First, we carry out the summation over μ in the product of the $\mathscr{D}$-functions, and obtain $\delta_{\mu'\mu'_1}$; then we use this Krönecker delta to reduce the (μ', μ'_1)-summations to a single sum, which can be immediately worked out to yield $(L_k L_{k_1})^{\lambda}_{\nu\nu_1}$. Thus, the expression for T_{rot} reduces to

$$T_{\text{rot}} = \tfrac{1}{2} \sum_{\lambda} B_\lambda \sum_{\nu\nu_1} \sum_{kk_1} \omega_k \omega_{k_1} a_{\lambda\nu} a_{\lambda\nu_1} (-1)^{\lambda} (L_k L_{k_1})^{\lambda}_{\nu\nu_1}.$$

The terms that correspond to $k \neq k_1$ contain $(L_y L_z + L_z L_y)$, $(L_z L_x + L_x L_z)$, and $(L_x L_y + L_y L_x)$, where x, y, z now label the body-fixed axes. The first two cross terms obviously step up or step down the projection quantum number ν_1 by 1; hence, in a body-fixed system for a deformed nucleus having the two symmetries discussed in Section 32A, these terms reduce to zero by virtue of the fact that ν, ν_1 are even integers. The third cross term can be reduced to $(L_+L_+ - L_-L_-)$; since the matrix element $\nu\nu_1$ of L_+L_+ is exactly equal to the matrix element $\nu_1\nu$ of L_-L_-, this cross term also yields zero when the summations over ν and ν_1 are carried out. We are thus left with only the $(k = k_1)$-terms, and T_{rot} finally assumes the simple form

$$T_{\text{rot}} = \tfrac{1}{2} \sum_k \mathscr{J}_k \omega_k^2, \tag{IV.58a}$$

where

$$\mathscr{J}_k = \sum_{\lambda} B_\lambda \sum_{\nu\nu'} a_{\lambda\nu} a_{\lambda\nu'} (-1)^{\lambda} \langle\lambda\nu| L_k^2 |\lambda\nu'\rangle. \tag{IV.58b}$$

By similar arguments, it can be proved with the help of (IV.57) that the contribution to the kinetic energy from the cross product of the two terms in (IV.52) vanishes. This is left as an exercise.

Putting (IV.51), (IV.53), and (IV.58a) together, we obtain, for the transformed Hamiltonian, the expression

$$H = T_{\text{vib}} + T_{\text{rot}} + V = \tfrac{1}{2}\sum_{\lambda\mu}(B_\lambda|\dot{a}_{\lambda\mu}|^2 + C_\lambda|a_{\lambda\mu}|^2) + \tfrac{1}{2}\sum_k \mathcal{J}_k\omega_k^2. \tag{IV.59}$$

The terms enclosed within the parentheses once again represent a vibrator Hamiltonian, and the last term denotes a rotator with $\mathcal{J}_k$ ($k = x, y, z$) along the axes of the body-fixed frame. Thus, what was a pure vibrator Hamiltonian in the stationary frame has yielded place to a vibrator plus a rotator Hamiltonian in the body-fixed frame.

Classical expression of angular momentum The angular momentum of the body can be classically expressed as

$$\mathbf{M} = \rho_0 \int_\tau (\mathbf{r} \times \mathbf{v})\, d\tau,$$

where the velocity $\mathbf{v}$ is completely determined through (IV.28) and (IV.30). Thus,

$$\mathbf{M} = \rho_0 \sum_{\lambda\mu} \lambda^{-1} R_0^{2-\lambda} \dot{\alpha}_{\lambda\mu}^* \int d\omega \int_0^{R_0(1+\xi)} r^2\, dr\, (\mathbf{r} \times \nabla)[r^\lambda Y_\mu^\lambda(\theta, \phi)].$$

The rotation operator $\mathbf{L}$ is given by $-i(\mathbf{r} \times \nabla)$ and we know that it does not contain any $\partial/\partial r$. In other words, $\mathbf{L}$ operates only on $Y_\mu^\lambda(\theta, \phi)$ in the foregoing expression, and r^λ goes to its left where it can be immediately integrated on dr. In this way,

$$\mathbf{M} = i\rho_0 \sum_{\lambda\mu} \lambda^{-1}(\lambda + 3)^{-1} R_0^5 \dot{\alpha}_{\lambda\mu}^* \int d\omega\, (1 + \sum_{\lambda'\mu'} \alpha_{\lambda'\mu'}^* Y_{\mu'}^{\lambda'})^{\lambda+3} \mathbf{L} Y_\mu^\lambda.$$

We expand to the first order in $\alpha_{\lambda'\mu'}^*$, and note that the first term of this expansion gives the integral $\int d\omega\, \mathbf{L} Y_\mu^\lambda$, which is zero. This is because $\mathbf{L}$ operating on Y_μ^λ produces a spherical harmonic of the same order λ, whose integral is zero if $\lambda \neq 0$. For $\lambda = 0$, $\mathbf{L}Y_0^0$ itself is zero. The second term of the expansion, which is linear in $\alpha_{\lambda'\mu'}^*$, then produces the result

$$\mathbf{M} = i \sum_{\lambda\mu} B_\lambda \dot{\alpha}_{\lambda\mu'}^* \sum_{\mu'} \alpha_{\lambda\mu'} \int d\omega\, Y_{\mu'}^\lambda \mathbf{L} Y_\mu^\lambda. \tag{IV.60}$$

In getting this expression, we have used (IV.35b) for B_λ; we have also consistently replaced the summation label μ' by $-\mu'$, and thereby secured $\alpha_{\lambda'\mu'} Y_{\mu'}^{\lambda'*}$ out of $\alpha_{\lambda'\mu'}^* Y_{\mu'}^{\lambda'}$; finally, the summation label λ' has been put equal to λ in view of the fact that the last integral, which stands for the matrix element $\langle\lambda'\mu'|\, L_-\, |\lambda\mu\rangle$, demands $\delta_{\lambda\lambda'}$ from the property of the rotation operator $\mathbf{L}$. We have to remember that this is an equation connecting the vector $\mathbf{M}$ with the vector $\mathbf{L}$, and components can be taken with respect to either the body-fixed or the stationary axes. We shall take components along the body-fixed axes. Hence, we shall first transform the quantities $\dot{\alpha}_{\lambda\mu}^*$ and $\alpha_{\lambda\mu'}$ into the body-fixed frame, according to (IV.52), (IV.57), and (IV.10c). In this way,

$$M_k = i \sum_{\lambda\mu\mu'} B_\lambda \sum_{\nu\nu'} \mathcal{D}_{\mu'\nu'}^\lambda a_{\lambda\nu'} [\mathcal{D}_{\mu\nu}^{\lambda*} \dot{a}_{\lambda\nu}^* - i a_{\lambda\nu}^* \sum_{\mu_1} \mathcal{D}_{\mu\mu_1}^{\lambda*} \sum_{k_1} \omega_{k_1} (L_{k_1})_{\mu_1\nu}^\lambda] \langle\lambda\mu'|\, L_k\, |\lambda\mu\rangle. \tag{IV.61}$$

In further simplifying this expression, we write $\langle\lambda\mu'|\, L_k\, |\lambda\mu\rangle$ in terms of matrix elements by using states of the body-fixed frame. We also use the fact stated before (IV.55f), namely,

$$\langle\lambda\mu \mid \lambda\mu'\rangle_B = \mathcal{D}_{\mu\mu'}^{\lambda*}.$$

In this way,

$$\langle\lambda\mu'|\, L_k \,|\lambda\mu\rangle = \sum_{\mu_1\mu_2} \mathcal{D}^{\lambda *}_{\mu'\mu_1}\mathcal{D}^{\lambda}_{\mu\mu_2}\, {}_B\langle\lambda\mu_1|\, L_k \,|\lambda\mu_2\rangle_B$$

$$= \sum_{\mu_1\mu_2} \mathcal{D}^{\lambda *}_{\mu'\mu_1}\mathcal{D}^{\lambda}_{\mu\mu_2}(L_k)^{\lambda}_{\mu_1\mu_2}. \qquad \text{(IV.62)}$$

For the $\dot{a}_{\lambda\nu}$-term of (IV.61), the summation over μ and μ' can now be easily carried out, obtaining $\delta_{\mu_1\nu'}$ and $\delta_{\mu_2\nu}$ according to the unitarity of the $\mathcal{D}$-functions. Therefore, the $\dot{a}_{\lambda\nu}$-term yields

$$i \sum_{\lambda} B_\lambda \sum_{\nu\nu'} a_{\lambda\nu'}\dot{a}_{\lambda\nu}(-1)^{\lambda}(L_k)^{\lambda}_{\nu'\nu}.$$

Because both ν, ν' are even, this quantity vanishes. We shall not repeat the details of such proofs. We proceed with the simplification of the second term of (IV.61). In this term, we have $\mathcal{D}^{\lambda *}_{\mu\mu_1}$ instead of the $\mathcal{D}^{\lambda *}_{\mu\nu}$ of the first term. So the steps proceed in a similar manner, ultimately producing

$$M_k = \sum_{\lambda} B_\lambda \sum_{\nu\nu'} a_{\lambda\nu'}a^{*}_{\lambda\nu} \sum_{\mu_1} (L_k)^{\lambda}_{\nu'\mu_1} \sum_{k_1} \omega_{k_1}(L_{k_1})^{\lambda}_{\mu_1\nu}$$

$$= \sum_{\lambda} B_\lambda \sum_{\nu\nu'} a_{\lambda\nu}a_{\lambda\nu'}(-1)^{\lambda} \sum_{k_1} \omega_{k_1}(L_k L_{k_1})^{\lambda}_{\nu'\nu}.$$

Once again, through already familiar arguments, we can guarantee that only the $(k = k_1)$-term contributes since both ν, ν' are even. Hence,

$$M_k = \mathcal{J}_k\omega_k, \qquad \text{(IV.63a)}$$

where $\mathcal{J}_k$ has been defined in (IV.58b). This expression for angular momentum and the expression (IV.58a) for T_{rot} ensure that the quantities $\mathcal{J}_k$ are to be interpreted as the moment of inertia of the body along the three principal body-fixed axes. In terms of the angular momentum M_k and the moment of inertia $\mathcal{J}_k$, the expression for T_{rot} assumes the familiar form

$$T_{\text{rot}} = \sum_k \frac{M_k^2}{2\mathcal{J}_k}. \qquad \text{(IV.63b)}$$

Specialization to the case of quadrupole ($\lambda = 2$) motion In this case, we omit the label λ from every quantity that has it. The mass parameter B_λ becomes

$$B = \tfrac{1}{2}\rho_0 R_0^5. \qquad \text{(IV.64)}$$

The three nonvanishing deformation parameters a_0, a_2, and a_{-2} are given in terms of the two independent parameters β, γ by the relations (IV.18). The components of the moment of inertia are obtained from (IV.58b) by putting $\lambda = 2$, and for both ν and ν' the values 2, −2, 0; the matrix elements of L_x^2, L_y^2, and L_z^2 connecting the states $\langle 2\nu|$ with $|2\nu'\rangle$ can be obtained from

$$L_{\pm}|\lambda\nu\rangle = [(\lambda + \nu)(\lambda \mp \nu + 1)]^{1/2}|\lambda, \nu \mp 1\rangle, \qquad \text{(IV.65a)}$$

$$L_z|\lambda\nu\rangle = \nu|\lambda\nu\rangle. \qquad \text{(IV.65b)}$$

Here

$$L_{\pm} = L_x \pm iL_y. \qquad \text{(IV.65c)}$$

It should be noticed that (IV.65a) is different from the standard result (AI.3) of Appendix A for the step-up and step-down operators. The origin of this difference lies in the fact that, in the present case, we have taken the components of L along the axes of a coordinate system that is *moving* with the body; on the other hand, in deriving the general result (AI.3) in Appendix A, we have chosen components of **L** along the axes of a *fixed* coordinate system. The relations (IV.65) are discussed in Section 33A. Using (IV.65) explicitly and substituting for the values of a_0, a_2, and a_{-2} in terms of β and γ from (IV.18), we obtain

$$\mathcal{J}_k = 4B\beta^2 \sin^2 \left(\gamma - k\,\frac{2\pi}{3}\right), \tag{IV.66}$$

where k on the right-hand side has to be put equal to 1, 2, 3 for the components x, y, z, respectively. This expression and (IV.55c) enable us to write T_{rot} as a function of β, γ, and $\dot{\theta}_i$ ($i = 1, 2, 3$). We first write (IV.55c) as

$$\omega_k = \sum_{j=1}^{3} q_{kj}\dot{\theta}_j, \tag{IV.67a}$$

where the matrix q is given by

$$q = \begin{pmatrix} -\sin\theta_2\cos\theta_3 & \sin\theta_3 & 0 \\ \sin\theta_2\sin\theta_3 & \cos\theta_3 & 0 \\ \cos\theta_2 & 0 & 1 \end{pmatrix}. \tag{IV.67b}$$

We can then write

$$T_{\text{rot}} = \tfrac{1}{2}\sum_k \mathcal{J}_k\omega_k^2 = \tfrac{1}{2}\sum_k \mathcal{J}_k \sum_{jj'} q_{kj}q_{kj'}\dot{\theta}_j\dot{\theta}_{j'}. \tag{IV.68}$$

The expressions for T_{vib} and V, given by (IV.53) and (IV.51), respectively, can be very easily specialized to the case of $\lambda = 2$. In terms of the parameters β, γ defined in (IV.18), we obtain in a straightforward manner the expressions

$$T_{\text{vib}} = \tfrac{1}{2}B(\dot{\beta}^2 + \beta^2\dot{\gamma}^2), \tag{IV.69a}$$

$$V = \tfrac{1}{2}C\beta^2. \tag{IV.69b}$$

33. QUANTUM MECHANICAL TREATMENT OF COLLECTIVE MODEL

A. SCHRÖDINGER EQUATION AND ITS GENERAL SOLUTION

So far we have been treating the problem of the collective motion of a nuclear droplet classically. We now wish to quantize the quadrupole-type motion and write down the Schrödinger equation for the problem. The independent coordinates in terms of which the equation has to be written are five in number: β, γ and θ_1, θ_2, θ_3.

We first examine the expression for the kinetic energy T as a quadratic function of the time derivatives of these variables. The required expressions are given by (IV.68) and (IV.69a) which, when added, give the desired expressions of the kinetic energy T as a quadratic in the time derivatives of β, γ, θ_1, θ_2, θ_3. For convenience, we denote these five coordinates in the order shown here by β_μ with μ running from 1 to 5. In that case, the expression for T has the form

$$2T = \sum_{\mu,\nu} G_{\mu\nu}\dot{\beta}_\mu\dot{\beta}_\nu, \tag{IV.70}$$

where the coefficients $G_{\mu\nu}$ are functions of the variables β_μ. It is clear from (IV.69a) and (IV.68) that the coefficient matrix $G_{\mu\nu}$ actually consists of two submatrices, one in the space of $\dot{\beta}_1$, $\dot{\beta}_2$ (i.e., $\dot{\beta}$ and $\dot{\gamma}$) and the other in the space of $\dot{\beta}_3$, $\dot{\beta}_4$, $\dot{\beta}_5$ (i.e., $\dot{\theta}_1$, $\dot{\theta}_2$, $\dot{\theta}_3$); moreover, there

52 (45-122/1977)

being no cross term $\dot{\beta}\dot{\gamma}$ in (IV.69a), we conclude that the former submatrix is diagonal. Using the explicit expressions for q_{kj} from (IV.67b), we find it fairly straightforward to work out the elements $G_{\mu\nu}$ occurring in (IV.70), i.e., the coefficients $\sum_k \mathcal{J}_k q_{kj} q_{kj'}$. In this way, we obtain

$$G = \left(\begin{array}{c|c} g & 0 \\ \hline 0 & \mathcal{G} \end{array}\right), \tag{IV.71a}$$

where the 2×2 diagonal matrix g is given by

$$g = \begin{pmatrix} B & 0 \\ 0 & B\beta^2 \end{pmatrix}, \tag{IV.71b}$$

and the 3×3 matrix $\mathcal{G}$ is symmetric and is given by

$$\mathcal{G} = \begin{pmatrix} \sin^2\theta_2(\mathcal{J}_x\cos^2\theta_3 + \mathcal{J}_y\sin^2\theta_3) + \mathcal{J}_z\cos^2\theta_2 & \sin\theta_2\sin\theta_3\cos\theta_3(\mathcal{J}_y - \mathcal{J}_x) & \mathcal{J}_z\cos\theta_2 \\ & \mathcal{J}_x\sin^2\theta_3 + \mathcal{J}_y\cos^2\theta_3 & 0 \\ & & \mathcal{J}_z \end{pmatrix}. \tag{IV.71c}$$

The blank spaces in $\mathcal{G}$ can be filled up from its symmetry property. The implication of (IV.70) is that, in terms of the differential changes $d\beta_\mu$ in the variables, we can write the quantity $2T\,dt^2$ as

$$2T\,dt^2 = \sum_{\mu,\nu} G_{\mu\nu}(\beta)\,d\beta_\mu\,d\beta_\nu.$$

The left-hand side, by definition, describes the square of the length ds^2 of a line of infinitesimal length in the five-dimensional space of the coordinates β_μ. The fact that we have cross terms in $d\beta_\mu\,d\beta_\nu$ in this expression shows that the generalized coordinates β_μ are nonorthogonal, and the fact that the coefficient matrix $G_{\mu\nu}$ is a function of the coordinates β_μ implies that the coordinate system defined by the five β_μ is curvilinear.

The problem of setting up the Schrödinger equation in terms of generalized coordinates was tackled by Pauli. We merely quote the result. If the determinant of the matrix G is denoted by $|G|$, then the Schrödinger equation becomes

$$H\Psi = E\Psi, \tag{IV.72a}$$

where

$$H = T + V = -\tfrac{1}{2}\hbar^2 \sum_{\mu\nu} |G|^{-1/2} \frac{\partial}{\partial\beta_\mu} |G|^{1/2} (G^{-1})_{\mu\nu} \frac{\partial}{\partial\beta_\nu} + V. \tag{IV.72b}$$

Here $(G^{-1})_{\mu\nu}$ is the $\mu\nu$-th element of the inverse of the matrix G. The inverse matrix can be easily seen to have the structure

$$G^{-1} = \left(\begin{array}{c|c} g^{-1} & 0 \\ \hline 0 & \mathcal{G}^{-1} \end{array}\right), \tag{IV.73a}$$

where

$$g^{-1} = \begin{pmatrix} B^{-1} & 0 \\ 0 & B^{-1}\beta^{-2} \end{pmatrix} \tag{IV.73b}$$

and the $\mu\nu$-th element of $\mathcal{G}^{-1}$ is given by the usual formula

$$(\mathcal{G}^{-1})_{\mu\nu} = \frac{M_{\nu\mu}}{|\mathcal{G}|}. \tag{IV.73c}$$

$M_{\nu\mu}$ is the minor of the $\nu\mu$-th element of the matrix $\mathcal{G}$. The determinant $|G|$ of the whole matrix and the determinant $|\mathcal{G}|$ of the submatrix are clearly related by

$$|G| = B^2\beta^2|\mathcal{G}|. \tag{IV.74a}$$

The quantity $|\mathcal{G}|$ has to be evaluated in a straightforward manner from (IV.71c) and then we obtain

$$|G| = B^2\beta^2\mathcal{J}_x\mathcal{J}_y\mathcal{J}_z \sin^2\theta_2 = 4B^5\beta^8 \sin^2\theta_2 \sin^2 3\gamma. \tag{IV.74b}$$

We have here substituted $\mathcal{J}_x$, $\mathcal{J}_y$, $\mathcal{J}_z$ from (IV.66).

It is clear from the structure (IV.73a) of G^{-1} and the diagonal nature of g^{-1} that the derivatives of β_1 and β_2 (i.e., of β and γ) occur in the expression (IV.72b) of H only as $(\partial/\partial\beta \ldots \partial/\partial\beta)$ and $(\partial/\partial\gamma \ldots \partial/\partial\gamma)$. In particular, there is no cross term $(\partial/\partial\beta \ldots \partial/\partial\gamma)$, and no cross term multiplying $\partial/\partial\beta$ or $\partial/\partial\gamma$ by $\partial/\partial\theta_j$. The G_{11}^{-1}- and G_{22}^{-1}-term of (IV.72b) give rise in a straightforward manner to the expression

$$-\frac{\hbar^2}{2B}\left(\frac{1}{\beta^4}\frac{\partial}{\partial\beta}\beta^4\frac{\partial}{\partial\beta} + \frac{1}{\beta^2 \sin 3\gamma}\frac{\partial}{\partial\gamma}\sin 3\gamma\frac{\partial}{\partial\gamma}\right), \tag{IV.75}$$

and clearly have arisen respectively from the $\dot{\beta}^2$- and $\dot{\gamma}^2$-term of T_{vib}. The terms of H that arise from $(G^{-1})_{\mu\nu}$ with μ, ν lying between 3 and 5 will contain derivatives of the type $(\partial/\partial\theta_i \ldots \partial/\partial\theta_j)$ and can obviously be traced to T_{rot}. However, instead of writing the terms in H corresponding to T_{rot} in this form, we recall the expression (IV.63b), which is equivalent to (IV.68), and quantize the angular momentum operators M_k occurring in this expression in the usual quantum mechanical way. This means we have to write M_k in units of $\hbar$ as

$$M_k = \hbar R_k \tag{IV.76}$$

and then require appropriate commutation relations amongst the three components of R_k. In fact, the components R_X, R_Y, R_Z along the axes of a fixed coordinate system must satisfy the usual commutation rules

$$[R_X, R_Y] = iR_Z \text{ (and cyclic permutations of } X, Y, Z). \tag{IV.77a}$$

The components k, however, are along the x-, y-, z-axis of a coordinate system moving with the body. In such a coordinate system, the commutation relations acquire a minus sign on the right-hand side, that is,

$$[R_x, R_y] = -iR_z \text{ (and cyclic permutations of } x, y, z). \tag{IV.77b}$$

The proof of these unusual commutation properties is given after writing down the Schrödinger equation explicitly.

We have T_{vib} given by (IV.75), and T_{rot}, according to the foregoing statements, by

$$T_{\text{rot}} = \sum_k \frac{\hbar^2}{2\mathcal{J}_k}R_k^2. \tag{IV.78}$$

Putting together these two expressions and (IV.69b) for V, we obtain from (IV.72)

$$[-\frac{\hbar^2}{2B}(\frac{1}{\beta^4}\frac{\partial}{\partial\beta}\beta^4\frac{\partial}{\partial\beta}+\frac{1}{\beta^2\sin 3\gamma}\frac{\partial}{\partial\gamma}\sin 3\gamma\frac{\partial}{\partial\gamma})+\sum_k\frac{\hbar^2}{2\mathscr{I}_k}R_k^2+\tfrac{1}{2}C\beta^2]\Psi(\beta,\gamma,\theta_1,\theta_2,\theta_3)$$
$$=E\Psi(\beta,\gamma,\theta_1,\theta_2,\theta_3). \qquad \text{(IV.79)}$$

Using the expression (IV.66) for $\mathscr{I}_k$ and multiplying the equation throughout by β^2, we note that the term containing $\partial/\partial\gamma$ and the rotational terms are then independent of β. This means that the wavefunction is separable in the β-coordinate, and we can write

$$\Psi(\beta,\gamma,\theta_1,\theta_2,\theta_3)=f(\beta)\Phi(\gamma,\theta_1,\theta_2,\theta_3). \qquad \text{(IV.80)}$$

Substituting this expression and (IV.66) for $\mathscr{I}_k$ in (IV.79) and then multiplying from the left by $\beta^2[\Psi(\beta,\gamma,\theta_1,\theta_2,\theta_3)]^{-1}$, we easily obtain

$$-\frac{\hbar^2}{2B}[\frac{1}{f(\beta)}\frac{1}{\beta^2}\frac{d}{d\beta}\beta^4\frac{d}{d\beta}f(\beta)+\frac{1}{\sin 3\gamma}\frac{1}{\Phi}\frac{\partial}{\partial\gamma}\sin 3\gamma\frac{\partial}{\partial\gamma}\Phi]+\tfrac{1}{2}C\beta^4-E\beta^2$$
$$+\sum_k\frac{\hbar^2}{8B}\frac{1}{\Phi}\frac{R_k^2}{\sin^2(\gamma-k\frac{2\pi}{3})}\Phi=0$$

or

$$-\frac{\hbar^2}{2B}\frac{1}{f(\beta)}[\frac{1}{\beta^2}\frac{d}{d\beta}\beta^4\frac{d}{d\beta}f(\beta)]+\tfrac{1}{2}C\beta^4-E\beta^2$$
$$=-[\sum_k\frac{\hbar^2}{8B}\frac{1}{\Phi}\frac{R_k^2}{\sin^2(\gamma-k\frac{2\pi}{3})}\Phi-\frac{\hbar^2}{2B}\frac{1}{\sin 3\gamma}\frac{1}{\Phi}\frac{\partial}{\partial\gamma}\sin 3\gamma\frac{\partial}{\partial\gamma}\Phi]. \qquad \text{(IV.81)}$$

Here we have made use of the fact that the operators R_k^2 act only on a function of $\theta_1, \theta_2, \theta_3$ and not on a function of β. From the final step, by the usual arguments of the separation of variable method, we obtain each side of the equation as a constant. We write this constant as $[-\hbar^2/(2B)]\Lambda$. Then the left-hand side yields, for β-vibration, the equation

$$(-\frac{\hbar^2}{2B}\frac{1}{\beta^4}\frac{d}{d\beta}\beta^4\frac{d}{d\beta}+\tfrac{1}{2}C\beta^2+\frac{\Lambda\hbar^2}{2B}\frac{1}{\beta^2})f(\beta)=Ef(\beta). \qquad \text{(IV.82a)}$$

Similarly, the right-hand side of (IV.81) leads to the coupled equation for rotation and γ-motion, namely,

$$[-\frac{1}{\sin 3\gamma}\frac{\partial}{\partial\gamma}\sin 3\gamma\frac{\partial}{\partial\gamma}+\tfrac{1}{4}\sum_k\frac{R_k^2}{\sin^2(\gamma-k\frac{2\pi}{3})}]\Phi(\gamma,\theta_1,\theta_2,\theta_3)=\Lambda\Phi(\gamma,\theta_1,\theta_2,\theta_3). \qquad \text{(IV.82b)}$$

(IV.82a) is a second-order differential equation in one variable β, and hence could be solved by standard numerical methods even if the potential energy term were more complicated than $\tfrac{1}{2}C\beta^2$. This equation, as it stands, yields an analytic solution in terms of known functions and is treated in detail in Section 33B. The equation for γ-motion and rotation is more difficult to tackle because of the coupling of the γ-variable with the θ-variables. This equation can be simply treated by assuming the nucleus to be rigid against γ-vibration. This assumption is equivalent to dropping the first term of (IV.82b), and considering only the rotational part, namely,

$$\tfrac{1}{4}\sum_k\frac{R_k^2}{\sin^2(\gamma-k\frac{2\pi}{3})}\Phi(\theta_1,\theta_2,\theta_3)=\Lambda\Phi(\theta_1,\theta_2,\theta_3). \qquad \text{(IV.82c)}$$

The solution of this rotator equation is discussed later in this section.

There is a special case when (IV.82b) again becomes simple; this time it contains only the γ-motion and the *treatment is exact*. The special case we have in mind corresponds to a state Φ of zero angular momentum. Here the angular momentum operators R_k^2 produce zero, operating on Φ, and we have

$$\left(\frac{1}{\sin 3\gamma}\frac{\partial}{\partial\gamma}\sin 3\gamma\frac{\partial}{\partial\gamma}+\Lambda\right)\Phi_{I=0}(\gamma)=0. \tag{IV.82d}$$

If $\Phi(\gamma)$ has to be a single-valued function of 3γ, then in this equation we need

$$\Lambda = 9\lambda(\lambda+1), \qquad \lambda = 0, 1, 2, \ldots .$$

For these values of Λ, (IV.82d) becomes identical to the equation for the Legendre polynomial, and the solution $\Phi_{I=0}(\gamma)$ is given by

$$\Phi_{I=0}(\gamma) = P_\lambda(\cos 3\gamma).$$

Substituting this expression of Λ in (IV.82a), we obtain an oscillator equation for $f(\beta)$ which can be solved by the method described in Section 17 for the solution of the three-dimensional isotropic oscillator problem.

With this brief discussion of the simple cases, we now proceed to find the general solutions of (IV.82c) and, with their help, the general solutions of (IV.82b). An analysis of the properties of the R_k-operators, though a digression, is necessary before we can write down the solutions explicitly.

Let us first examine the operators R_k, which are components of the angular momentum operator $\mathbf{R}$ along the three body-fixed axes x, y, z. Since the body-fixed axes move with respect to a set of axes fixed in the laboratory, the properties of the operator R_k are somewhat different from those of the components of a general angular momentum operator $\mathbf{J}$, discussed in Appendix A. The components of $\mathbf{J}$, dealt with there, are along the axes of a *fixed* coordinate system. Let the unit vectors along the x-, y-, z-direction be denoted by $\hat{\mathbf{e}}_k$ $(k = 1, 2, 3)$. Denoting the components of $\mathbf{R}$ and $\hat{\mathbf{e}}_k$ along the fixed axes X, Y, Z by the subscripts I, J, . . . , we then write

$$R_k = \hat{\mathbf{e}}_k \cdot \mathbf{R} = \sum_I \hat{e}_{kI} R_I. \tag{IV.83}$$

For the components R_I along the fixed axes, we have the usual commutation relations

$$[R_I, R_J] = i \sum_K \epsilon_{IJK} R_K, \tag{IV.84}$$

where $\epsilon_{IJK} = +1$ or -1 depending on whether (I, J, K) is an even or odd permutation of (X, Y, Z), and $\epsilon_{IJK} = 0$ if any two or more of the labels are identical.

We use (IV.83) for R_k and $R_{k'}$, and obtain

$$\begin{aligned}[R_k, R_{k'}] &= \sum_{I,J} [\hat{e}_{kI} R_I, \hat{e}_{k'J} R_J] \\ &= \sum_{I,J} \{\hat{e}_{kI}\hat{e}_{k'J}[R_I, R_J] + \hat{e}_{kI}[R_I, \hat{e}_{k'J}]R_J + \hat{e}_{k'J}[\hat{e}_{kI}, R_J]R_I\}.\end{aligned} \tag{IV.85}$$

If k and k' were components along another *fixed* set of axes, then, of course, the quantities $\hat{e}_{kI}$, $\hat{e}_{k'J}$ would have been fixed quantities, and their commutators with R_I, R_J appearing in the last two terms of (IV.85) would have vanished. In the case under consideration, by virtue of the moving nature of the coordinate axes, the quantities $\hat{e}_{kI}$, $\hat{e}_{k'J}$ are the components of the vector *variables* $\hat{\mathbf{e}}_k$ and $\hat{\mathbf{e}}_{k'}$. A vector variable, *by definition*, satisfies the same commutation relations with the angular momentum as does the coordinate vector. Therefore, we have

$$[\hat{e}_{kI}, R_J] = i \sum_K \epsilon_{IJK} \hat{e}_{kK},$$

$$[\hat{e}_{k'J}, R_I] = i \sum_K \epsilon_{JIK} \hat{e}_{k'K} = -i \sum_K \epsilon_{IJK} \hat{e}_{k'K}.$$

Substituting these expressions and (IV.84) in (IV.85), we obtain

$$\begin{aligned}[R_k, R_{k'}] &= i \sum_{I,J,K} \epsilon_{IJK}(\hat{e}_{kI}\hat{e}_{k'J}R_K + \hat{e}_{k'K}\hat{e}_{kI}R_J + \hat{e}_{kK}\hat{e}_{k'J}R_I) \\ &= -i \sum_{I,J,K} \epsilon_{IJK}\hat{e}_{kI}\hat{e}_{k'J}R_K \\ &= -i \sum_K (\hat{\mathbf{e}}_k \times \hat{\mathbf{e}}_{k'})_K R_K \\ &= -i(\hat{\mathbf{e}}_k \times \hat{\mathbf{e}}_{k'}) \cdot \mathbf{R} = -i \sum_{k''} \epsilon_{kk'k''} R_{k''}. \end{aligned} \tag{IV.86}$$

In reaching the second step from the first, we have used the properties of ϵ_{IJK}. In the next step, we have carried out the summation over I, J and expressed the result in terms of the cross product of the two vectors, which again follows from the property of ϵ_{IJK}. Next, we have expressed the K-summation as a scalar product, and finally noted that $(\hat{\mathbf{e}}_k \times \hat{\mathbf{e}}_{k'})$ gives the unit vector along the third body-fixed axis k'' with a plus or minus sign, depending on whether (k, k', k'') represents an even or odd permutation of (x, y, z). Comparing (IV.84) with (IV.86), we notice that the components of an angular momentum in the moving coordinate frame satisfy commutation relations that have an extra minus sign on the right-hand side.

The fact that R_k is a *scalar* [see (IV.83)] formed from two vector variables immediately tells us that it will commute with all the components R_I $(I = X, Y, Z)$ of the angular momentum, and hence with R^2. The three components R_x, R_y, R_z, however, do not commute with one another. It is thus possible to set up a representation in which, besides the usual commuting operators R^2 and R_Z, a third commuting operator, taken to be one of the components of $\mathbf{R}$ in the moving frame, is diagonal. To be explicit, we choose R_z as the third commuting variable, and look for the simultaneous eigenfunctions of R^2, R_Z, and R_z corresponding to given eigenvalues of these operators. Let us consider an eigenfunction characterized by the eigenvalue $I(I+1)$ of R^2, the eigenvalue M of R_Z, and K of R_z. In the present case, the eigenfunction is obviously a function of the angles θ_1, θ_2, θ_3. The dependence of this function on θ_1 and θ_3 is very easy to calculate. The angle θ_1 denotes the rotation about the Z-axis; hence, the angular momentum operator R_Z (in units of $\hbar$) is given, according to (AI.10a) in Appendix A, by

$$R_Z = -i\frac{\partial}{\partial\theta_1},$$

and the corresponding eigenfunction by

$$\exp(iM\theta_1). \tag{IV.87a}$$

In the same manner, θ_3 being the rotation about the z-axis, we have

$$R_z = -i\frac{\partial}{\partial\theta_3},$$

and the corresponding eigenfunction is given by

$$\exp(iK\theta_3). \tag{IV.87b}$$

We draw the reader's attention to the fact that (IV.87a) and (IV.87b) are respectively the θ_1- and θ_3-part of the function $\mathcal{D}^I_{MK}(\theta_1, \theta_2, \theta_3)$, discussed in Appendix B (Section I). In fact, the eigenfunctions we are looking for are precisely the functions $\mathcal{D}^I_{MK}$. In order to prove this assertion, we have obviously to demonstrate that the θ_2-part of the eigenfunction is the same as $d^I_{MK}(\theta_2)$, defined by (BI.12b) and (BI.12c) in Appendix B; this proof is somewhat complicated and is not given here. This part of the eigenfunction arises because it is required to be a simultaneous eigenfunction of R^2 belonging to the eigenvalue $I(I+1)$. The reader interested in the details is referred to Landau and Lifshitz[2].

We now take it for granted that $\mathcal{D}^I_{MK}(\theta_1, \theta_2, \theta_3)$ are the desired eigenfunctions. From the definition of the eigenfunction, we then have

$$R^2\mathcal{D}^I_{MK}(\theta_1, \theta_2, \theta_3) = I(I+1)\mathcal{D}^I_{MK}(\theta_1, \theta_2, \theta_3), \tag{IV.88a}$$

$$R_Z\mathcal{D}^I_{MK}(\theta_1, \theta_2, \theta_3) = M\mathcal{D}^I_{MK}(\theta_1, \theta_2, \theta_3), \tag{IV.88b}$$

$$R_z\mathcal{D}^I_{MK}(\theta_1, \theta_2, \theta_3) = K\mathcal{D}^I_{MK}(\theta_1, \theta_2, \theta_3). \tag{IV.88c}$$

The operators $(R_X \pm iR_Y)$ will produce the usual stepping up and stepping down of the quantum numbers M. From the general results (AII.2c) of Appendix A, we write

$$(R_X \pm iR_Y)\mathcal{D}^I_{MK}(\theta_1, \theta_2, \theta_3) = [(I \mp M)(I \pm M + 1)]^{1/2}\mathcal{D}^I_{M\pm 1,\,K}(\theta_1, \theta_2, \theta_3). \tag{IV.89}$$

The corresponding results for $(R_x \pm iR_y)$, however, are not completely analogous. The difference is actually attributable to the minus sign on the right-hand side of the commutator (IV.86), as now shown. We have, according to this relation,

$$\begin{aligned}[(R_x \pm iR_y), R_z] &= iR_y \pm i(-i)R_x \\ &= \pm(R_x \pm iR_y).\end{aligned}$$

The first $\pm$ on the right-hand side would have been $\mp$ were the minus sign in (IV.86) absent. With the help of (IV.88c), we can easily establish that the foregoing equation implies

$$R_z[(R_x \pm iR_y)\mathcal{D}^I_{MK}] = (K \mp 1)[(R_x \pm iR_y)\mathcal{D}^I_{MK}].$$

Thus, $(R_x \pm iR_y)$ are eigenfunctions of R_z belonging to the eigenvalues $K \mp 1$. In other words, the role of stepping up and stepping down has now been interchanged between the two operators. Accordingly, we write

$$(R_x \pm iR_y)\mathcal{D}^I_{MK} = N\mathcal{D}^I_{M,\,K\mp 1}, \tag{IV.90}$$

where the multiplying constant N has to be determined as follows. We multiply this equation by its Hermitean conjugate and integrate over $d\theta_1 \sin\theta_2\, d\theta_2\, d\theta_3$. In this way,

$$\begin{aligned}&\int_0^{2\pi} d\theta_1 \int_0^{\pi} \sin\theta_2\, d\theta_2 \int_0^{2\pi} d\theta_3\, \mathcal{D}^{I*}_{MK}(\theta_1, \theta_2, \theta_3)(R_x \mp iR_y)(R_x \pm iR_y)\mathcal{D}^I_{MK}(\theta_1, \theta_2, \theta_3) \\ &\quad = |N|^2 \int_0^{2\pi} d\theta_1 \int_0^{\pi} \sin\theta_2\, d\theta_2 \int_0^{2\pi} d\theta_3\, \mathcal{D}^{I*}_{M,\,K\mp 1}(\theta_1, \theta_2, \theta_3)\mathcal{D}^I_{M,\,K\mp 1}(\theta_1, \theta_2, \theta_3).\end{aligned} \tag{IV.91}$$

We have

$$\begin{aligned}(R_x \mp iR_y)(R_x \pm iR_y) &= R_x^2 + R_y^2 \pm i[R_x, R_y] \\ &= R^2 - R_z^2 \pm R_z.\end{aligned}$$

Using this expression in (IV.91) and the results (IV.88a) and (IV.88c), and (BI.20b) from

Appendix B, we obtain

$$I(I+1) - K^2 \pm K = |N|^2$$

or

$$N = [(I \pm K)(I \mp K + 1)]^{1/2}.$$

Substituting this value of N back in (IV.90), we get the final results

$$(R_x \pm iR_y)\mathcal{D}^I_{MK}(\theta_1, \theta_2, \theta_3) = [(I \pm K)(I \mp K + 1)]^{1/2}\mathcal{D}^I_{M,\,K\mp1}(\theta_1, \theta_2, \theta_3). \tag{IV.92}$$

The reader should note that the $\pm$ signs on the right-hand side of this equation are the opposite of those occurring on the right-hand side of (IV.89).

We are now in a position to get back to the discussion of the solutions of (IV.82b) and (IV.82c). First, let us consider the case where the rotating body has a symmetry axis; we choose this axis to be the z-axis. The ellipsoidal body having this symmetry is a spheroid. In this case, any two mutually perpendicular directions in the plane perpendicular to the symmetry axis can be chosen as the x- and y-axis. The cross-section of the body perpendicular to the symmetry axis is a circle. By virtue of this symmetry, we have $\gamma = 0$, and hence

$$\mathcal{J}_x = \mathcal{J}_y, \qquad \mathcal{J}_z = 0, \tag{IV.93}$$

which follow from (IV.66) by putting this special value of γ. The second equation in (IV.93) is a general property which holds for the moment of inertia of a body about a symmetry axis; quantum mechanically, it is a consequence of the fact that rotation about a symmetry axis of the body cannot be detected. Classically, this property holds for the *irrotational* motion of a *fluid* about an axis of symmetry; recall that the classical expression (IV.66) was derived for such a motion. By virtue of $\mathcal{J}_z = 0$, the kinetic energy of rotation about the symmetry axis is also zero. Therefore, when the z-axis is an axis of symmetry, (IV.82c) reduces to

$$\tfrac{1}{3}(R_x^2 + R_y^2)\Phi(\theta_1, \theta_2, \theta_3) = \Lambda\Phi(\theta_1, \theta_2, \theta_3). \tag{IV.94a}$$

The last term on the left-hand side of (IV.82b) is also similarly reduced. Writing

$$R_x^2 + R_y^2 = R^2 - R_z^2,$$

we therefore note that $\mathcal{D}^I_{MK}(\theta_1, \theta_2, \theta_3)$ is a solution for the wavefunction $\Phi(\theta_1, \theta_2, \theta_3)$ appearing in (IV.94). With the help of (IV.88a) and (IV.88c), we have

$$\Lambda = \tfrac{1}{3}[I(I+1) - K^2]. \tag{IV.94b}$$

More details on the rotation of a spheroidal nucleus are discussed later in this section.

For the present, we consider the solutions of (IV.82b) and (IV.82c) in the more general case of an ellipsoidal body for which $\mathcal{J}_x \neq \mathcal{J}_y \neq \mathcal{J}_z$ and all the three principal moments of inertia are nonvanishing. Here the coefficients of R_x^2 and R_y^2 are *not* equal, and these operators have to be treated by the operator identities

$$R_x^2 = \tfrac{1}{4}(R_+ + R_-)^2, \qquad R_y^2 = -\tfrac{1}{4}(R_+ - R_-)^2, \tag{IV.95a}$$

where

$$R_\pm = R_x \pm iR_y. \tag{IV.95b}$$

According to (IV.95a), we now have the operators R_+R_+, R_-R_-, and $(R_+R_- + R_-R_+)$ to deal with. The last operator does not alter $\mathcal{D}^I_{MK}(\theta_1, \theta_2, \theta_3)$. However, R_+R_+ and R_-R_- change $\mathcal{D}^I_{MK}$ to respectively $\mathcal{D}^I_{M,\,K-2}$ and $\mathcal{D}^I_{M,\,K+2}$. (It should be noted that R_x, R_y commute with R_z, and hence R_+R_+ and R_-R_- leave the quantum number M unaltered.) Thus, in the present case, $\mathscr{D}^I_{MK}(\theta_1, \theta_2, \theta_3)$ by itself cannot be an eigenfunction. The actual eigenfunction for a given I and M is a mixture of all the $\mathscr{D}^I_{MK}$ with K differing by ± 2; the maximum and minimum values of K in the admixture are, however, required to be $+I$ and $-I$, respectively, because K is the projection quantum number corresponding to I.

If we are considering the simplified expression (IV.82c), then we have to set up the matrix of

$$\tfrac{1}{4} \sum_k R_k^2/\sin^2 (\gamma - k\frac{2\pi}{3})$$

in the basis of the $\mathscr{D}^I_{MK}$-functions. This can be easily done with the help of the standard results (IV.88c) and (IV.92). The diagonalization of the matrix gives the eigenstates and the eigenvalues.

The complete expression (IV.82b) in which the γ-motion is coupled to the rotational motion, described by the $\mathscr{D}^I_{MK}$-function, can also be solved in a manner similar to that for (IV.82c). Once again, the eigenfunctions are an admixture of many $\mathscr{D}^I_{MK}$ with K differing by ± 2. However, the γ-dependence of the equation now manifests itself by demanding that the coefficients of the various $\mathscr{D}^I_{MK}$-functions can be functions of γ. We therefore write

$$\Phi^I_M(\gamma, \theta_1, \theta_2, \theta_3) = \sum_K g^I_K(\gamma)\mathscr{D}^I_{MK}(\theta_1, \theta_2, \theta_3). \tag{IV.96}$$

We need to substitute this expression in (IV.82b), multiply by $\mathscr{D}^{I*}_{MK'}(\theta_1, \theta_2, \theta_3)$ from the left, and integrate over $d\theta_1 \sin\theta_2\, d\theta_2\, d\theta_3$. As a result, we obtain the matrix elements of R_k^2 connecting $\langle \mathscr{D}^I_{MK'}|$ with $|\mathscr{D}^I_{MK}\rangle$, which can be evaluated in the manner described. On the right-hand side, we are left with only the term $g^I_{K'}(\gamma)$ by virtue of the result (BI.20b) of Appendix B. On the left-hand side, the matrix elements of R_k^2, according to the selection rules described, lead to $g^I_{K'+2}(\gamma)$ and $g^I_{K'-2}(\gamma)$, in addition to the term $g^I_{K'}(\gamma)$. Thus, (IV.82b), for a given K', reduces to a coupled differential equation of these three functions, which describe the γ-motion. We have, however, to consider the coupled equation for all possible values of K' and, as a result, we are faced with a chain of coupled differential equations of $g^I_K(\gamma)$, with K differing by ± 2 and ranging between $-I$ and I. It is clear from this discussion that a solution of the γ-motion in the general ellipsoidal case is quite complicated.

We shall now discuss the consequences of the arbitrariness in labelling the x-, y-, z-axis. This arbitrariness has been discussed in detail in terms of the operators $\mathscr{R}_1$, $\mathscr{R}_2$, $\mathscr{R}_3$, as shown in Fig. IV.2 and explicitly described before this diagram. The five variables β, γ, θ_1, θ_2, θ_3 define a definite shape and orientation of the body. As long as the shape and orientation of the body remain the same, the *value* of the wavefunction must also remain unchanged. The relabelling of the principal axes caused by the operators $\mathscr{R}_1$, $\mathscr{R}_2$, $\mathscr{R}_3$ changes the values of the variables β, γ in accordance with (IV.23), and those of θ_i in accordance with (IV.27a) for $\mathscr{R}_1$, (IV.27b) for $\mathscr{R}_2$, and (IV.26) with $(\bar{\theta}_1, \bar{\theta}_2, \bar{\theta}_3) = (\pi/2, \pi/2, \pi)$ for $\mathscr{R}_3$. However, assuming that the orientation and deformation of the body remain the same, we require that the *value* of the wavefunction, evaluated with the new values of the variables, be the same as that evaluated with the old values of these variables. We make this requirement a basic criterion and then look for the consequences in each of the cases $\mathscr{R}_1$, $\mathscr{R}_2$, $\mathscr{R}_3$.

(i) Under $\mathscr{R}_1 : \gamma \to \gamma$, $\theta_1 \to \theta_1$, $\theta_2 \to \theta_2 + \pi$, $\theta_3 \to -\theta_3$. Hence, the aforementioned basic requirement demands

$$\sum_K g_K^I(\gamma)\mathcal{D}_{MK}^I(\theta_1, \theta_2, \theta_3) = \sum_{K'} g_{K'}^I(\gamma)\mathcal{D}_{MK'}^I(\theta_1, \theta_2 + \pi, -\theta_3)$$
$$= \sum_{K'} g_{K'}^I(\gamma)\mathcal{D}_{M,-K'}^I(\theta_1, \theta_2, \theta_3)(-1)^{I+K'}.$$

Here we have used the definition of the $\mathcal{D}$-function to reach the last step from the second. Now, equating the coefficients of a given $\mathcal{D}_{MK}^I$ on both sides of the foregoing equation, we obtain

$$g_K^I(\gamma) = g_{-K}^I(\gamma)(-1)^{I-K}. \tag{IV.97a}$$

In the case of $K = 0$, this equation has the interesting consequence that I must be equal to even integers only in order that $g_0^I(\gamma)$ remain nonvanishing.

(ii) Under $\mathcal{R}_2 : \gamma \to -\gamma$, $\theta_1 \to \theta_1$, $\theta_2 \to \theta_2$, $\theta_3 \to \theta_3 + \pi/2$. Hence, in the manner indicated in (i),

$$\sum_K g_K^I(\gamma)\mathcal{D}_{MK}^I(\theta_1, \theta_2, \theta_3) = \sum_{K'} g_{K'}^I(-\gamma)\mathcal{D}_{MK'}^I(\theta_1, \theta_2, \theta_3 + \frac{\pi}{2})$$
$$= \sum_{K'} g_{K'}^I(-\gamma)\mathcal{D}_{MK'}^I(\theta_1, \theta_2, \theta_3)i^{K'}$$

or

$$g_K^I(\gamma) = i^K g_K^I(-\gamma). \tag{IV.97b}$$

Applying (IV.97b) to express $g_K^I(-\gamma)$, we get the interesting equation

$$g_K^I(\gamma) = (-1)^K g_K^I(\gamma), \tag{IV.97c}$$

i.e., the quantum number K must be even integers only. Combining this requirement with (IV.97a), we then have

$$g_K^I(\gamma) = (-1)^I g_{-K}^I(\gamma), \qquad K = \text{even integers}. \tag{IV.97d}$$

(iii) Under $\mathcal{R}_3 : \gamma \to \gamma - 120°$, $(\theta_1, \theta_2, \theta_3) \to (\Theta_1, \Theta_2, \Theta_3)$ such that

$$\mathcal{D}_{MK}^I(\Theta_1, \Theta_2, \Theta_3) = \sum_{K'} \mathcal{D}_{MK'}^I(\theta_1, \theta_2, \theta_3)\mathcal{D}_{K'K}^I(\frac{\pi}{2}, \frac{\pi}{2}, \pi).$$

Therefore,

$$\sum_K g_K^I(\gamma)\mathcal{D}_{MK}^I(\theta_1, \theta_2, \theta_3) = \sum_{K'} g_{K'}^I(\gamma - 120°)\mathcal{D}_{MK'}^I(\Theta_1, \Theta_2, \Theta_3)$$
$$= \sum_{K, K'} g_{K'}^I(\gamma - 120°)\mathcal{D}_{KK'}^I(\frac{\pi}{2}, \frac{\pi}{2}, \pi)\mathcal{D}_{MK}^I(\theta_1, \theta_2, \theta_3)$$

or

$$g_K^I(\gamma) = \sum_{K'} \mathcal{D}_{KK'}^I(\frac{\pi}{2}, \frac{\pi}{2}, \pi)g_{K'}^I(\gamma - 120°). \tag{IV.97e}$$

The requirement (IV.97e) thus establishes a relationship between $g_K^I(\gamma)$ and $g_K^I(\gamma - 120°)$. Similarly, (IV.97b) establishes a very simple relationship between $g_K^I(\gamma)$ and $g_K^I(-\gamma)$. These two relationships, taken together, determine $g_K^I(\gamma)$ for all γ between 0° and 360° if the function is known only from 0° → 60°. For example, $\gamma = 0° \to 60°$ enables us to go to $\gamma = 0° \to -60°$ by virtue of (IV.97b). Then, starting with γ in the interval $-60° \to 60°$, the relation (IV.97e) takes us to values of γ lying between 60° and 180°. Since we know the latter, (IV.97e), if applied again, takes us to γ between 180° and 300°. The last angle is equivalent to $-60°$, and hence we have already covered the full range of γ.

Incorporating the property (IV.97d) in (IV.96), we then have

$$\Phi^I_M(\gamma, \theta_1, \theta_2, \theta_3) = \sum_K g^I_K(\gamma)[\mathcal{D}^I_{MK}(\theta_1, \theta_2, \theta_3) + (-1)^I \mathcal{D}^I_{M,-K}(\theta_1, \theta_2, \theta_3)], \qquad K = \text{even integers only.} \tag{IV.98}$$

This general expression can be very easily specialized in two respects: (i) when the γ-motion can be ignored, the quantities $g^I_K(\gamma)$ become independent of γ; (ii) if, in addition, the nucleus is spheroidal, then the summation over K is absent and the wavefunction requires the label K. Thus, the solution (with the incorporation of the proper symmetry) of the rotator equation (IV.82c) in the special case of a spheroidal body is given by

$$\Phi^I_{MK}(\theta_1, \theta_2, \theta_3) = g^I_K[\mathcal{D}^I_{MK}(\theta_1, \theta_2, \theta_3) + (-1)^I \mathcal{D}^I_{M,-K}(\theta_1, \theta_2, \theta_3)], \qquad K = \text{even integers only.} \tag{IV.99}$$

The quantity g^I_K in this expression plays the role of a normalization constant. If we demand the normalization

$$1 = \int_0^{2\pi} d\theta_1 \int_0^{\pi} \sin\theta_2\, d\theta_2 \int_0^{2\pi} d\theta_3\, \Phi^{I*}_{MK}(\theta_1, \theta_2, \theta_3)\Phi^I_{MK}(\theta_1, \theta_2, \theta_3),$$

then, with the help of (IV.99), and the standard result (BI.20b) in Appendix B, we easily establish that g^I_K is independent of K and has the value

$$g^I_K = \sqrt{\frac{2I+1}{16\pi^2}}.$$

Thus,

$$\Phi^I_{MK}(\theta_1, \theta_2, \theta_3) = \left(\frac{2I+1}{16\pi^2}\right)^{1/2}[\mathcal{D}^I_{MK}(\theta_1, \theta_2, \theta_3) + (-1)^I \mathcal{D}^I_{M,-K}(\theta_1, \theta_2, \theta_3)]. \tag{IV.100a}$$

The expression for g^I_K needs modification in the case of $K = 0$ because then both the terms in (IV.99) become $\mathcal{D}^I_{M0}$, and we have already proved that I is needed to be an even integer. Thus, in this special case,

$$\Phi^I_{M0}(\theta_1, \theta_2, \theta_3) = g^I_0 \mathcal{D}^I_{M0}(\theta_1, \theta_2, \theta_3),$$

where g^I_0 can be evaluated with the standard result cited, making

$$\Phi^I_{M0}(\theta_1, \theta_2, \theta_3) = \sqrt{\frac{2I+1}{8\pi^2}}\,\mathcal{D}^I_{M0}(\theta_1, \theta_2, \theta_3). \tag{IV.100b}$$

In fact, we now wish to show that the spheroidal symmetry demands *only* $K = 0$.

Let us refer to the derivation of (IV.97b) from the transformation of the wavefunction under $\mathcal{R}_2$, which denotes a rotation of $\pi/2$ about the z-axis. In the special case of a spheroid, the operator $\mathcal{R}_2$ should be reinterpreted as a rotation through *any* arbitrary angle ϕ about the z-axis because the principal axes system of a spheroid having z as the symmetry axis remains a principal axes system under this operation. Proceeding in the manner of deriving (IV.97b), we then prove

$$g_K^I(\gamma) = e^{iK\phi} g_K^I(-\gamma)$$

or, applying this relation again,

$$g_K^I(\gamma) = e^{2iK\phi} g_K^I(\gamma).$$

Since this equation has to be valid for any *arbitrary* ϕ, it proves that K must be zero.

B. ROTATION AND VIBRATION OF EVEN NUCLEI

Phenomenology

The collective model described so far can be applied to even nuclei of several different categories. For the description of odd-mass nuclei, we have to couple the last odd nucleon to the wavefunction of the even core, the latter being reckoned according to the work in Section 33C. The extension of the model to the case of odd-mass nuclei is dealt with in Section 33C.

The large quadrupole moment of the odd-mass nuclei having ground-state spin larger than $\frac{1}{2}$ in certain regions of the periodic table is taken as direct evidence of an equilibrium deformation of nuclei belonging to these mass ranges. These nuclei are found in the middle of the $(0d, 1s)$-shell in the mass number range $145 < A < 185$ (the two limits of this range are not very sharp though), the mass number $A > 226$. As a result of the equilibrium deformation, all the nuclei in these ranges exhibit spectra with the rotational $I(I+1)$-characteristics. The even nuclei have a ground-state spin $I = 0$, and hence they cannot exhibit a static quadrupole moment, but the effect of the deformation manifests itself in large E2-transition probability connecting the subsequent states of these nuclei.

As we proceed from the deformed regions to the neighbouring regions of the periodic table, we reach a point on either side where the rotational spectrum characterized by the $I(I+1)$-rule gives way to a vibrational spectrum, i.e., levels separated by approximately the same energy interval. The first excited state in almost all even nuclei is 2^+, and the ground state is always 0^+. In the rotational nuclei, the energy difference between these two levels is of the order of 100 keV, sometimes a little more and often a little less. When the vibrational regions of the periodic table are reached, this energy difference increases appreciably.

The transition from the rotational to the vibrational nuclei is not very sharp. Corresponding to every deformed region of the periodic table, we have at both ends of the region a few nuclei whose low-lying spectra and other properties are not characteristic of either pure rotation or pure vibration. These nuclei are called transitional.

In the case of transitional nuclei, we conclude that in (IV.82a) and (IV.82b) a clean separation between the rotational and vibrational modes is not possible. These equations have to be tackled in their existing form, and complicated solutions result at the end of fairly detailed work. We, however, observe that the potential energy term $\frac{1}{2}C\beta^2$ occurs only in the β-vibrational equation, i.e., (IV.82a), and there is no corresponding potential energy term in the γ-equation, i.e., (IV.82b). This, we recall, is a consequence of using the hydrodynamical model in calculating the collective model Hamiltonian. In principle, it is possible to have a potential energy function $\mathcal{V}(\beta, \gamma)$ of a more general type in these equations. In dealing with transitional nuclei, this freedom may also have to be utilized. We shall not discuss these nuclei any further. Suffice it to say that Kumar and Baranger[3] have extensively covered them, deriving a general potential energy function $\mathcal{V}(\beta, \gamma)$ from detailed microscopic structural calculations, and using the quantities B and $\mathcal{J}$, which are also dependent on (β, γ).

Before we discuss the application of the theory of the collective model to rotational and vibrational nuclei, we shall describe a few systematics of their spectra, already mentioned in this section. Figure IV.3 shows the systematics of the spectra of nuclei as we move away

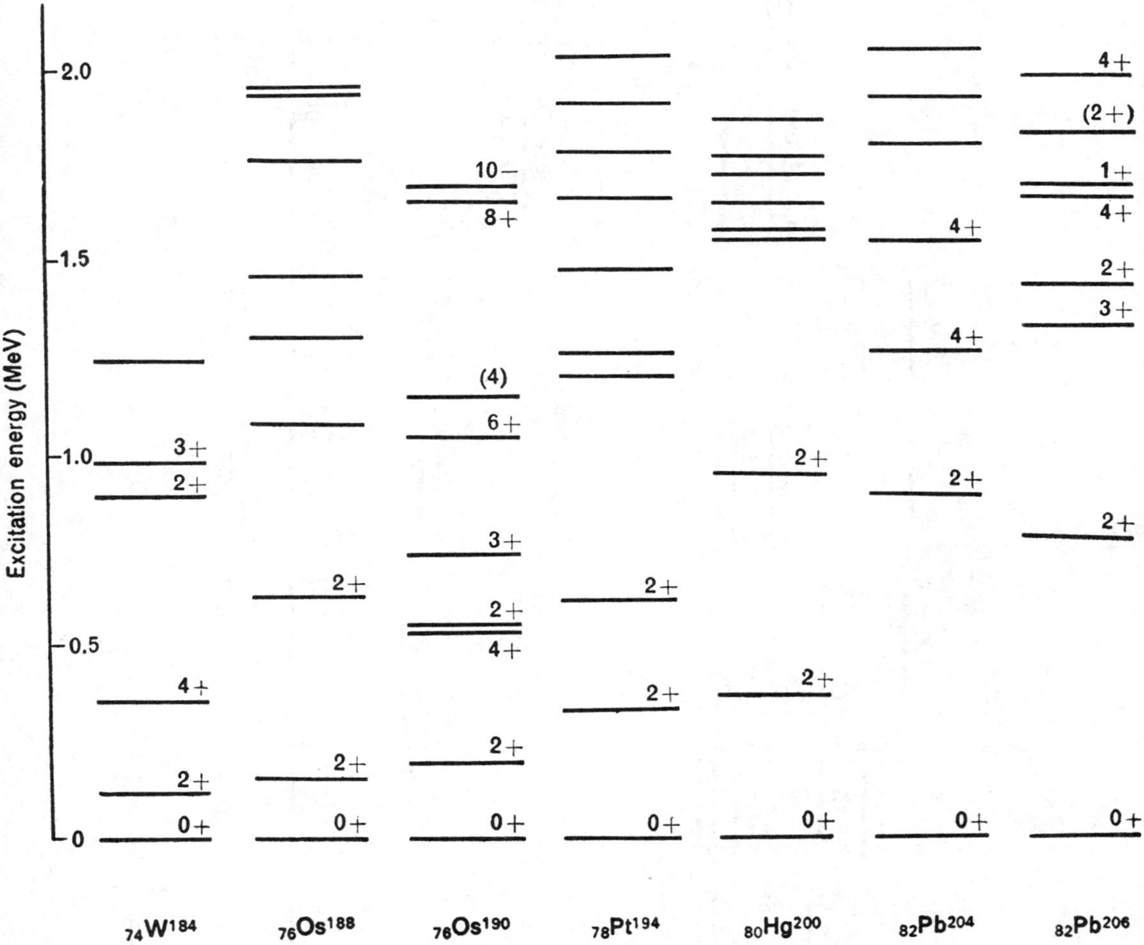

Fig. IV.3 Spectra of nuclei in range W^{184}–Pb^{206}. (From Bohr, A., and Mottelson, B. R., in Nuclear Spectroscopy, Part B, ed. by F. Ajzenberg-Selove, Academic Press, New York, 1960, p 1025.)

from the closed-shell nucleus $_{82}Pb^{208}$. As we go still further, the energy of the first excited 2^+ level diminishes and, finally, at the extreme left we have nuclei with the typical 0^+, 2^+, 4^+ level spectra satisfying approximately the $I(I+1)$-rule of a rotator. The spectra of nuclei to the right of those exhibiting the $I(I+1)$-type levels are interpreted as vibrational. Figure IV.4 shows the spectra of nuclei in the mass number range $A = 152$–192. At the two extremes are nuclei whose spectra are vibrational, the number of vibrational quanta (phonons) being shown by the value of n next to these levels. Between these two nuclei are a host of nuclei, all of which exhibit rotational spectra. Figure IV.5 shows the systematics of the first 2^+ excited level in vibrational even nuclei. The plots are against the proton number (Z) and the neutron

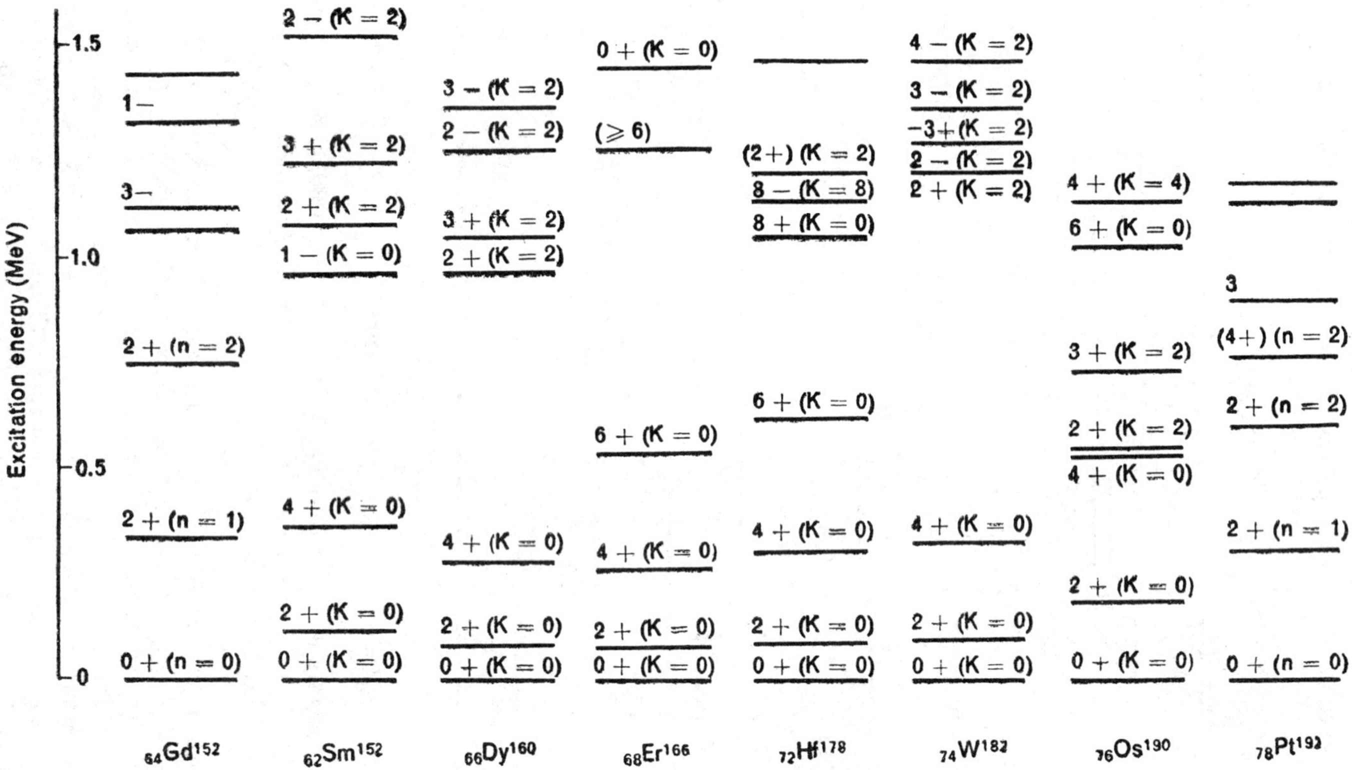

Fig. IV.4 Spectra of nuclei in range Gd^{152}–Pt^{192}. (From Bohr, A., and Mottelson, B. R., in Nuclear Spectroscopy, Part B, ed. by F. Ajzenberg-Selove, Academic Press, New York, 1960, p 1028.)

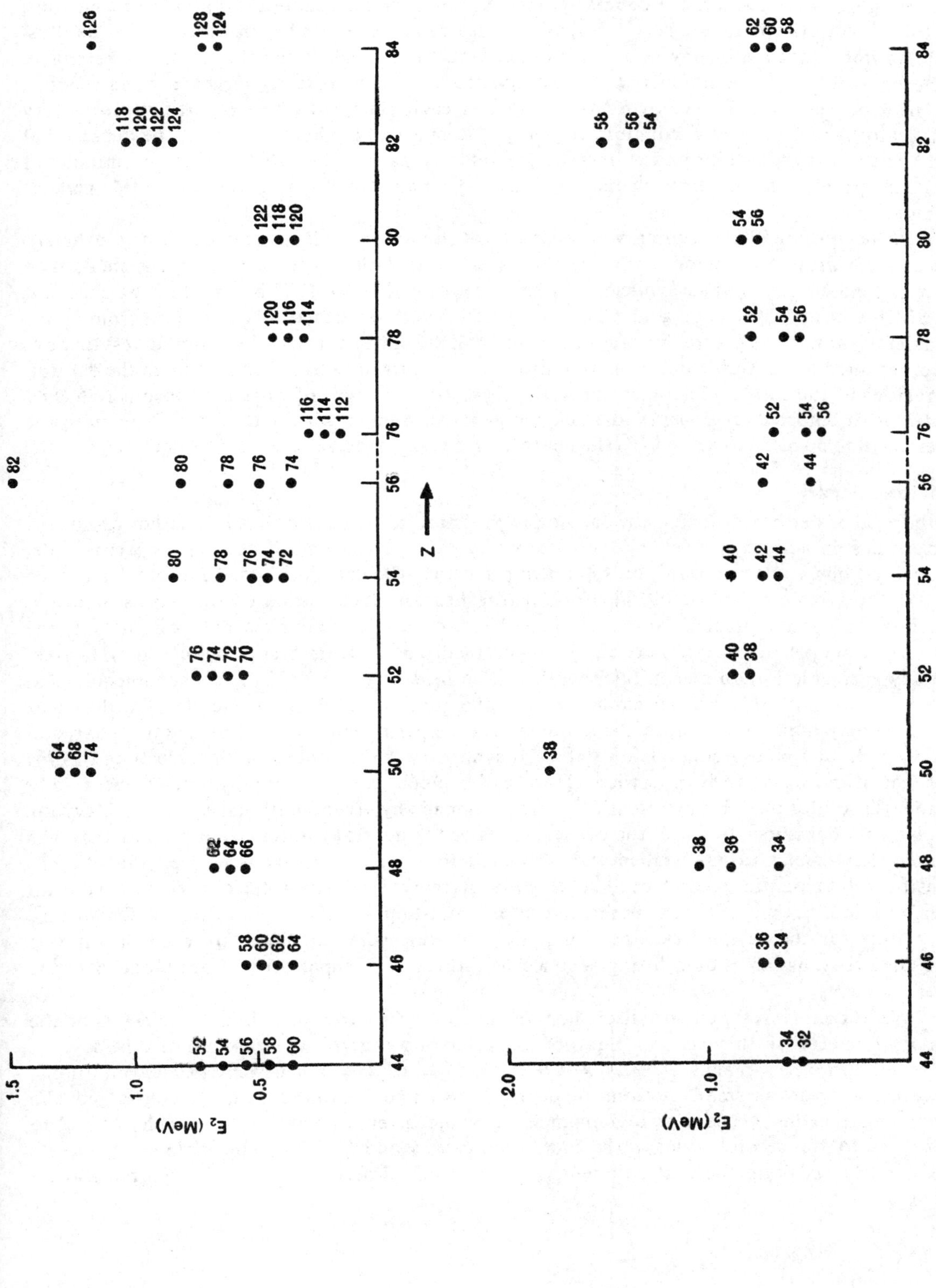

Fig. IV.5 Energy of first excited 2^+ level in even-even nuclei as function of proton number (upper figure) and neutron number (lower figure).

number (N). In the Z-plot, the figures against the solid circles denote the number of neutrons; in the N-plot, the figures against the solid circles denote the number of protons. Let us illustrate an important feature by referring to the points at $Z = 56$. When the neutrons belong to a closed shell (82), the 2^+ energy of that particular nucleus is fairly large; then, as we open the neutron shell too by removing two neutrons each time, the 2^+ energy at first drops very drastically, and then goes on diminishing very systematically. This means that the 'softness' of the nucleus towards deformation increases gradually as we proceed towards the middle of both the proton and neutron shells. In Fig. IV.5, there are many examples of this general phenomenon.

If we continue the foregoing process and create an appreciable neutron deficiency for nuclei whose Z lie around the middle of a shell so as to bring N also to the middle of the shell, then there is a possibility that such nuclei may show rotational spectra. This was first pointed out by Sheline et al[4]. Morinaga and Clark[5] very clearly demonstrated this transition from vibrational to rotational spectra by creating neutron-deficient isotopes of Xe. Similar results have been obtained for unstable nuclei in which there is a neutron excess. In addition to the deformed regions of the stable nuclei mentioned in this section, the type of work described here has revealed other deformed regions: (i) the neutron-deficient regions with both Z and N lying in the interval 28 to 50 or 50 to 82 and (ii) the neutron-rich nuclei with $28 < Z < 50$ and $50 < N < 82$.

Deformed Nuclei

In the case of deformed nuclei, the theoretical question to be settled first is whether or not the nucleus has an axis of symmetry. Historically, several applications of the theory were made, following Bohr and Mottelson[1], on the assumption that the deformed nucleus does have such an axis of symmetry. Most of the work in the heavier nuclei is based on this assumption. Amongst the lighter nuclei, Ne^{20} was treated as an axially symmetric nucleus. In fact, the answer to the question raised here can be theoretically given from first principles in a Hartree-Fock type calculation on nuclei. Such work is described in Chapter V. The phenomenological collective model of this chapter cannot answer the question from first principles. On the other hand, we may hope to obtain, in the systematics of experimental data, some definite characteristic which may be reproducible only if axial symmetry is introduced in our foregoing theory. The fact of the situation is otherwise. Historically, much after the application of the axially symmetric rotator model, a systematic attempt was made by several authors, especially Davydov and his collaborators[6], to check the consequences of the general rotator Hamiltonian that has no axis of symmetry (usually called an asymmetric rotator). For a time it was felt that the E2-transition data on the second excited 2^+ state of several deformed nuclei perhaps favour an interpretation in terms of the asymmetric rotator Hamiltonian. In a review article by Davidson[7], the results for the asymmetric and symmetric rotators were examined in detail and it was concluded that, as far as data fitting is concerned, there is no superiority of one model over the other.

We therefore restrict our discussion to axially symmetric nuclei. Let the z-axis be the axis of symmetry. In this case, the equilibrium value of γ is zero, whereas that of β is, say, β_0. The axial symmetry makes $\mathcal{J}_x = \mathcal{J}_y = 3B\beta_0^2$ and $\mathcal{J}_z = 0$, $K = 0$ *under equilibrium* conditions. However, as soon as small motions in β around β_0, and γ around zero are considered, the moment of inertia $\mathcal{J}_z$ acquires a nonvanishing value given by $4B\beta^2 \sin^2 \gamma$ which is equal to $4B\beta_0^2\gamma^2$ up to the second order in the small quantities γ and $(\beta - \beta_0)$. The components $\mathcal{J}_x$, $\mathcal{J}_y$ also, strictly speaking, become different; however, this difference is usually ignored and we continue to use their equilibrium values $3B\beta_0^2$. As a result of this assumption, the R_x^2- and R_y^2-term in (IV.82b) appear with the same coefficient, and the resultant combination $R_x^2 + R_y^2$ $(= R^2 - R_z^2)$ keeps K a good quantum number for the wavefunction. Therefore, the solution to (IV.82b), with this approximation, turns out to be of the type (IV.99), where g_K^I is now required to be a function of γ. The solution has thus separated into a pure rotational part $\mathscr{D}_{MK}^I(\theta_1, \theta_2, \theta_3)$ and a pure γ-vibrational part $g_K^I(\gamma)$. The equation for the latter is obtained by substituting in (IV.82b) the expressions

$$\left[\frac{1}{4}\frac{R_x^2}{\sin^2(\gamma - \frac{2\pi}{3})} + \frac{1}{4}\frac{R_y^2}{\sin^2(\gamma - \frac{4\pi}{3})}\right]\mathscr{D}_{MK}^I \approx \frac{1}{3}[I(I+1) - K^2]\,\mathscr{D}_{MK}^I$$

and

$$\frac{1}{4}\frac{R_z^2}{\sin^2\gamma}\,\mathscr{D}_{MK}^I \approx \frac{K^2}{4\gamma^2}\mathscr{D}_{MK}^I.$$

Multiplying the resultant equation from the left by $\mathscr{D}_{MK}^{I*}$ and integrating over the full range of $\theta_1, \theta_2, \theta_3$, we easily obtain

$$\left(-\frac{1}{\sin 3\gamma}\frac{d}{d\gamma}\sin 3\gamma\frac{d}{d\gamma} + \frac{K^2}{4\gamma^2}\right)g_K^I(\gamma) = [\Lambda - \tfrac{1}{3}\{I(I+1) - K^2\}]g_K^I(\gamma). \tag{IV.101}$$

As it stands, this equation does not describe a *vibrational* motion because it lacks the potential energy term of an oscillator. We have already mentioned that this is a result of our use of the hydrodynamical expression for potential energy. In general, it is possible to have a potential energy term in γ; in view of the small values of γ about the equilibrium value zero, this potential energy term is assumed to have the form $\frac{1}{2}C_\gamma\gamma^2$. In (IV.101), we insert this extra term and also replace $\sin 3\gamma$ by its approximate value 3γ for small γ. In this way, the γ-vibrational equation of a nucleus having a spheroidal equilibrium shape becomes

$$\left[-\frac{1}{\gamma}\frac{d}{d\gamma}\gamma\frac{d}{d\gamma} + \frac{(K/2)^2}{\gamma^2} + \tfrac{1}{2}C_\gamma\gamma^2\right]g_K^I(\gamma) = [\Lambda - \tfrac{1}{3}\{I(I+1) - K^2\}]g_K^I(\gamma). \tag{IV.102}$$

Let us refer to (IV.79) and the subsequent step for the separation of the β-equation. It becomes clear that the $\frac{1}{2}C_\gamma\gamma^2$-term in (IV.102) arises if we take a potential energy term in (IV.79) equal to $\frac{1}{2}\bar{C}_\gamma\gamma^2/\beta^2$ and substitute $\bar{C}_\gamma = [\hbar^2/(2B)]C_\gamma$. Further, the steps from (IV.79) to (IV.82b) will convince the reader that $[\hbar^2/(2B\beta^2)]\Lambda$ is equal to the sum of the rotational energy and the energy due to the γ-motion. Strictly speaking, β is a variable, and hence Λ determines the rotational plus γ-vibrational energy for each given value of β, as β oscillates around β_0. We shall, however, evaluate the rotational and γ-vibrational energy by using only the equilibrium value β_0 of β. Thus, $[\hbar^2/(2B\beta_0^2)]\Lambda$ is the energy of rotation plus γ-vibration. Of this total quantity, the contribution of the rotational energy in the case of axial symmetry is clearly the value of $[\hbar^2/(2\mathscr{I})](R_x^2 + R_y^2)$, where $\mathscr{I} = \mathscr{I}_x = \mathscr{I}_y$. The value of $\mathscr{I}$ is equal to $3B\beta_0^2$ for the hydrodynamical model in the equilibrium condition of the nucleus. The rotational energy for an axially symmetric nucleus is thus

$$E_{\text{rot}} = \frac{\hbar^2}{2\mathscr{I}}\langle R^2 - R_z^2\rangle = \frac{\hbar^2}{2\mathscr{I}}[I(I+1) - K^2] \tag{IV.103a}$$

which, with the hydrodynamical value of $\mathscr{I}$, reduces to

$$\frac{\hbar^2}{6B\beta_0^2}[I(I+1) - K^2]. \tag{IV.103b}$$

This value of E_{rot} and the meaning of $[\hbar^2/(2B\beta_0^2)]\Lambda$, just mentioned, then tell us that $[\hbar^2/(2B\beta_0^2)]$ times the eigenvalue on the right-hand side of (IV.102) is indeed the energy of the γ-motion.

(IV.102) can be solved in a straightforward manner by following the procedure described in Section 17 for the solution of an isotropic harmonic oscillator equation. The details are left as an exercise. In terms of the dimensionless coordinate (γ/b_γ), where $b_\gamma^2 = \hbar/(B\omega_\gamma)$, we write the wavefunction $g_K^I(\gamma)$ as

$$g_K^I(\gamma) = \gamma^m \exp(-\tfrac{1}{2}\gamma^2/b_\gamma^2)f(\gamma^2/b_\gamma^2), \qquad m = \tfrac{1}{2}K, \tag{IV.104a}$$

where the function f satisfies the confluent hypergeometric equation in Appendix C (Section II) with the parameters a and c given as

$$a = \tfrac{1}{2}(m + 1 - \mathcal{E}_\gamma), \qquad c = \tfrac{1}{2}(2m + 1). \tag{IV.104b}$$

The variable in the confluent hypergeometric equation is $x = \gamma^2/b_\gamma^2$. The quantity $\mathcal{E}_\gamma$ is the eigenvalue of γ-vibrational energy, expressed in units of $\hbar\omega_\gamma$, where $\omega_\gamma^2 = \bar{C}_\gamma/B$. The condition of well behaviour at $\gamma \to \infty$ of the solution (IV.104a) determines the eigenvalue $\mathcal{E}_\gamma$ in the usual way; we must reduce f to a polynomial by choosing a of (IV.104b) as a negative integer $(-n_\gamma)$. Thus,

$$\mathcal{E}_\gamma = 2n_\gamma + m + 1 = 2n_\gamma + \tfrac{1}{2}K + 1 = N_\gamma + 1. \tag{IV.104c}$$

The phonon quantum number N_γ is equal to $2n_\gamma + \frac{1}{2}K$. For the two lowest values of N_γ, we have

$$N_\gamma = 0, \qquad \text{hence } n_\gamma = 0,\ K = 0;$$

$$N_\gamma = 1, \qquad \text{hence } n_\gamma = 0,\ K = 2.$$

The first state corresponds to no γ-phonon, whereas the second state is the first excited γ-vibrational state having one phonon. The important point to note here is that this single γ-phonon state has a projection quantum number $K = 2$.

We next consider (IV.82a) for β-motion. Substituting

$$\bar{f}(\beta) = \beta^2 f(\beta),$$

$$\omega_\beta^2 = C/B,$$

$$b_\beta^2 = \hbar/(B\omega_\beta),$$

$$\hat{\beta} = \beta/b_\beta,$$

$$\Lambda = \lambda(\lambda + 3)$$

in (IV.82a), we obtain, in a straightforward manner,

$$\left[\frac{d^2}{d\hat{\beta}^2} - \frac{l(l+1)}{\hat{\beta}^2} - \hat{\beta}^2 + 2\mathcal{E}_\beta\right]\bar{f}(\hat{\beta}) = 0, \tag{IV.105a}$$

where

$$\begin{aligned} l &= \lambda + 1, \\ \mathcal{E}_\beta &= E_\beta/(\hbar\omega_\beta). \end{aligned} \tag{IV.105b}$$

(IV.105a) is identical to the isotropic oscillator equation of Section 17, and hence we obtain

$$\mathcal{E}_\beta = 2n_\beta + l + \tfrac{3}{2} = 2n_\beta + \lambda + \tfrac{5}{2} = N_\beta + \tfrac{5}{2}. \tag{IV.105c}$$

The function $\bar{f}$ is the same as the Laguerre polynomial defined in Section 17.

The total energy E can thus be represented as

$$E = \hbar\omega_\beta(N_\beta + \tfrac{5}{2}) + \hbar\omega_\gamma(2n_\gamma + \tfrac{1}{2}K + 1) + \frac{\hbar^2}{2\mathcal{J}}[I(I+1) - K^2]. \tag{IV.106}$$

The lowest set of states is obtained for $N_\beta = 0$ (no β-phonon), $n_\gamma = 0$, $K = 0$ (no γ-phonon). For $K = 0$, we can have only $I =$ even states, and these states ($I = 0, 2, 4, \ldots$), all of even parity, satisfy the energy equation

$$\tfrac{5}{2}\hbar\omega_\beta + \hbar\omega_\gamma + \frac{\hbar^2}{2\mathcal{J}}I(I+1). \tag{IV.107}$$

They are said to form the ground-state *rotational band*, with *the band quantum number* $K = 0$. The constant energy represented by the first two terms of (IV.107) is the zero-point energy of β- and γ-vibration. Initially, we had a five-dimensional oscillator corresponding to the five parameters, and the correct result should have been $\frac{5}{2}\hbar\omega$ for the zero-point energy. We recall that we introduced the γ-vibrational potential energy externally, and hence the extra zero-point energy $\hbar\omega_\gamma$ appears.

Another set of rotational states will arise for $N_\beta = 0$, but $N_\gamma = 1$ (one γ-phonon), i.e., $n_\gamma = 0$ and $K = 2$. This set of states can have any integral I consistent with $K = 2$, i.e., $I = 2, 3, 4, \ldots$, and it forms the first *γ-vibrational band*, with the band quantum number 2. The lowest state of this band is 2^+, and its energy is called the energy of the *band head*, which is above the ground-state band head by an amount $(\hbar\omega_\gamma + \hbar^2/\mathcal{J})$.

There is a rotational band of states built upon $N_\beta = 1$ (one β-phonon) and $N_\gamma = 0$ (no γ-phonon). Since there is no γ-vibration for this set of states, the coordinate γ always has the equilibrium value zero. Hence, of the three surface expansion parameters in the body-fixed frame, we have $a_2 = a_{-2} = (\beta/\sqrt{2}) \sin\gamma = 0$; on the other hand, $a_0 = \beta\cos\gamma = \beta$, and it is nonvanishing. Thus, the rotational band of β-vibration, just considered, corresponds to oscillations in a_0 and has the projection quantum number zero. Thus, the excitation of one β-phonon cannot change the projection quantum number $K = 0$ of the ground state. Therefore, we identify the band quantum number of the first band of β-vibrational states as $K = 0$ and, once again, the rotational states have the spin sequence $0^+, 2^+, 4^+, \ldots$. The band head is clearly at an energy $\hbar\omega_\beta$ above the ground-state band head.

A typical rotational spectrum in the nucleus Er^{166} is shown in Fig. IV.6. This figure gives the ground-state rotational band, a few levels of the γ-vibrational band, and only the band head of the β-vibrational band; the situation depicted for the β- and γ-band is somewhat typical. There are very few nuclei where many levels of these bands can be identified. The ground-state band is, however, very extensively and clearly observed; in many cases, it has been found to extend up to $I \approx 16$ and above. According to (IV.106), the same moment of inertia parameter $\mathcal{J}$ characterizes the ground-state rotational band, and the β- and γ-band. In practice, however, they have to be chosen different in order that the observed levels in the various bands can be reproduced. The moment of inertia $\mathcal{J}$ can be calculated from the hydrodynamical model formula $\mathcal{J} = 3B\beta_0^2$, with B determined from (IV.35b), and the deformation β_0 estimated from quadrupole transition data (see Section 34B). However, the rotational spectrum predicted with this hydrodynamical estimate usually turns out to be wrong. The usual practice is therefore to treat it as a parameter and determine it from the spacing between the first two levels of the band. Values of the parameter $(3\hbar^2/\mathcal{J})$ arrived at in this way are approximately equal to 95 keV for $150 < A < 200$, and approximately equal to 43 keV for $A > 220$. The hydrodynamical model gives smaller values of $\mathscr{I}$, whereas the rigid body

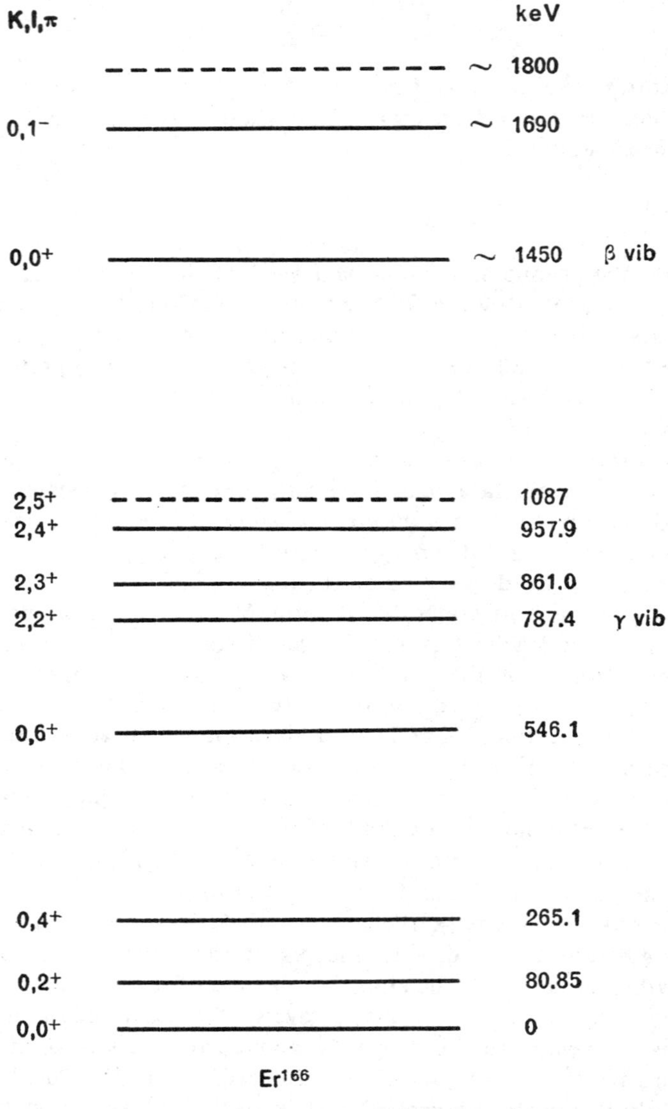

Fig. IV.6 Spectra of Er^{166} showing rotational levels in ground-state band ($K = 0^+$); several levels in γ-vibrational band ($K = 2^+$); and band head of β-vibration ($K = 0^+$). [From Sheline, R. K., *Revs. Mod. Phys.*, **32**, 1 (1950).]

moment of inertia $\mathscr{I}_{\text{rig}}$ of a spheroid of the same density and deformation turns out to be larger, as shown in Fig. IV.7. In this diagram, $\mathscr{I}$ has been plotted in units of $\mathscr{I}_{\text{rig}}$. The experimental values are given by the hollow circles, and the theoretical curve for irrotational fluid motion is shown by the dashed line. The rigid-body value is clearly given by the line (not shown)

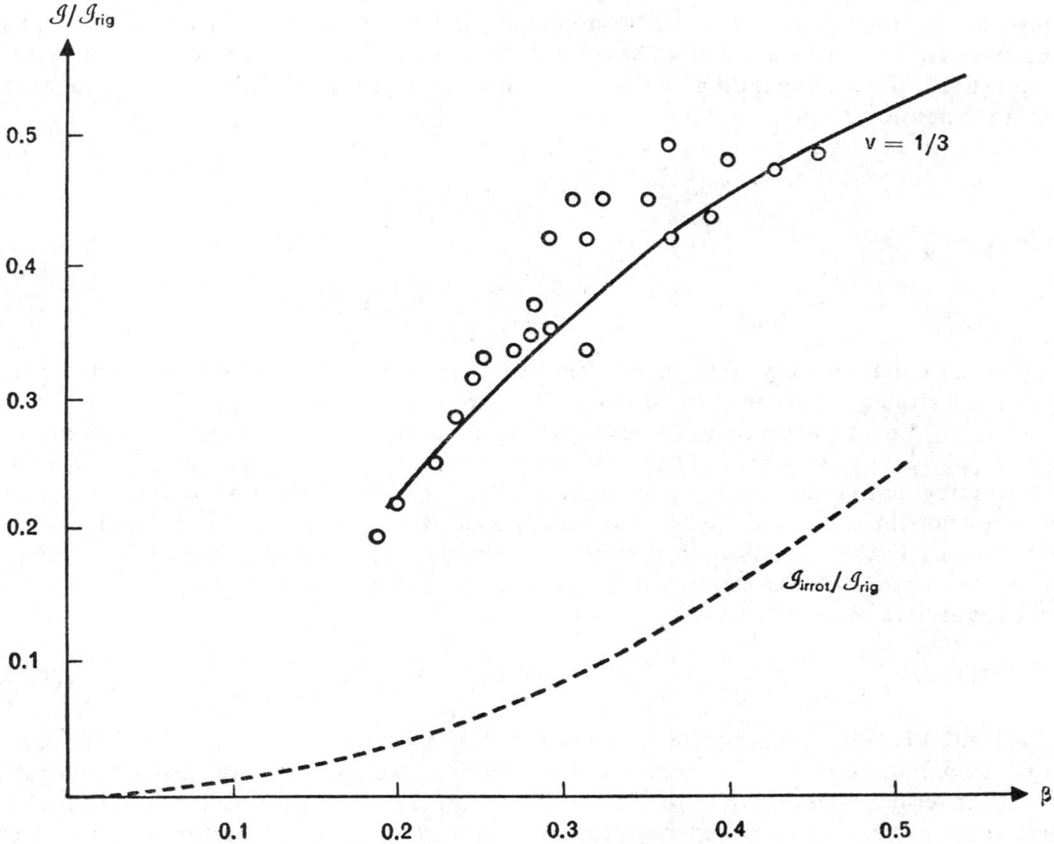

Fig. IV.7 Moments of inertia of nuclei (in rare-earth region) as function of deformation parameter. (Following Kerman, A. K., in Nuclear Reactions, Vol 1, ed. by P. M. Endt and M. Demeur, North-Holland, Amsterdam, 1959, p 480.)

parallel to the abscissa corresponding to $(\mathcal{J}/\mathcal{J}_{rig}) = 1$. All the experimental points lie between this line and the curve for irrotational motion. For the ground-state band, where levels with very large I have been observed, we can then check the validity of the $[\hbar^2/(2\mathcal{J})]I(I+1)$-rule for higher levels of the band. Usually, the first few levels are in accordance with this rule, but the discrepancy increases as we reach higher members of the band. One approach that explains this discrepancy is based on an empirical expansion of rotational energy in powers of the *operator* I^2, i.e., in powers of $I(I+1)$. Thus,

$$E_I = \frac{\hbar^2}{2\mathcal{J}} I(I+1) + b[I(I+1)]^2 + c[I(I+1)]^4 + \dots \qquad \text{(IV.108)}$$

The parameters $b, c, \dots$ can be determined by fitting the observed spectrum, and then we may try to understand them in terms of some theoretical model. One such theoretical approach to the understanding of the parameter b invokes the coupling between the rotation and the β-vibration, which arises in the way now outlined.

We recall that while writing the expression of rotational energy, we used the value of $\mathcal{I}$ corresponding to the equilibrium deformation β_0 and remarked that this is an approximate procedure. To be exact, we should have used, for a given β (which spreads around β_0 during the vibration), the corresponding $\mathcal{I}(\beta)$ in the rotational expression. However, we can at once make an expansion about $\beta = \beta_0$ and obtain

$$\mathcal{I}(\beta) = \mathcal{I}(\beta_0) + (\beta - \beta_0)(\frac{d\mathcal{I}}{d\beta})_{\beta_0} + \ldots,$$

and hence

$$\frac{\hbar^2}{2\mathcal{I}(\beta)}I(I+1) = \frac{\hbar^2}{2\mathcal{I}(\beta_0)}I(I+1) - \frac{\hbar^2}{2\mathcal{I}(\beta_0)}(\beta - \beta_0)(\frac{d\mathcal{I}}{d\beta})_{\beta_0}I(I+1) + \ldots. \qquad \text{(IV.109)}$$

The first term has already been considered. Since $(\beta - \beta_0)$ is small, we may work out the correction to the lowest order in $(\beta - \beta_0)$. The ground state, as discussed, corresponds to $n_\beta = 0$. According to a well-known result of harmonic oscillator theory, the oscillating coordinate $(\beta - \beta_0)$ can connect only the one-phonon state $n_\beta = 1$ with the state $n_\beta = 0$, and the corresponding matrix element is equal to $[\hbar\omega_\beta/(2C)]^{1/2}$ [see the discussion following (IV.122)]. The correction in energy of the ground state, due to the term $(\beta - \beta_0)$ in (IV.109), can therefore arise only in the second order. Inserting the energy denominator, which is the difference between the energy of the state $n_\beta = 0$ and that of the state $n_\beta = 1$, i.e., $-\hbar\omega_\beta$, we easily obtain the second-order correction

$$\Delta E = -\frac{1}{\hbar\omega_\beta}[\frac{\hbar^2}{2\mathcal{I}(\beta_0)}(\frac{d\mathcal{I}}{d\beta})_{\beta_0}]^2\frac{\hbar\omega_\beta}{2C}[I(I+1)]^2. \qquad \text{(IV.110)}$$

The last but one factor comes from the square of the matrix element, and the first from the energy denominator just mentioned. Strictly speaking, we should have also considered the next higher term containing $(\beta - \beta_0)^2$ in the expansion (IV.109). That term would have given a first-order correction to the energy proportional to $I(I+1)$; this therefore has the effect of renormalizing $\mathcal{I}(\beta_0)$ which is, in any case, treated as a parameter in the data fitting. The correction (IV.110) is in the form of the second term of (IV.108) and thus gives a theoretical estimate of the parameter b.

Several other approaches that explain the deviation of the higher members of a rotational band from the $I(I+1)$-rule are available. We shall not go into any more details here. A good reference on this subject is Mariscotti et al[8] and a summary of their work is given in Section 43D. For more details on the fitting of the β- and γ-vibrational band, the reader may refer to two review articles, one by Sheline[9] and the other by Davidson[7].

Spherical Vibrational Nuclei

In view of the absence of an equilibrium deformation of the spherical nuclei, all their moments of inertia under the equilibrium condition are zero, and hence the nuclei do not have any rotational energy. The vibrational energy due to β- and γ-vibration can be worked out from the appropriate equations. However, it is more convenient to treat the vibration in terms of the original Hamiltonian (IV.50).

We introduce the momentum π_μ canonically conjugate to the coordinate α_μ, defined according to

$$\pi_\mu = \frac{\partial T}{\partial \dot{\alpha}_\mu} = B\dot{\alpha}_\mu^* = (-1)^\mu B\dot{\alpha}_{-\mu}. \tag{IV.111}$$

We then impose the usual quantum mechanical commutation rule between the coordinates α_μ and the momenta π_μ, namely,

$$[\alpha_\mu, \pi_{\mu'}] = i\hbar\delta_{\mu\mu'}, \tag{IV.112a}$$

and make the coordinate α_μ dimensionless by dividing it with b, where $b^2 = \hbar/(B\omega)$ and $\omega^2 = C/B$. Thus,

$$\tilde{\alpha}_\mu = \left(\frac{B\omega}{\hbar}\right)^{1/2}\alpha_\mu \tag{IV.113a}$$

is dimensionless. We choose the dimensionless momentum $\tilde{\pi}_\mu$ in such a way that

$$[\tilde{\alpha}_\mu, \tilde{\pi}_{\mu'}] = i\delta_{\mu\mu'} \tag{IV.112b}$$

is satisfied. Comparing with (IV.112a), we then obtain

$$\tilde{\pi}_\mu = \hbar^{-1}\left(\frac{\hbar}{B\omega}\right)^{1/2}\pi_\mu. \tag{IV.113b}$$

Let us construct the operators

$$q_\mu^\dagger = \frac{1}{\sqrt{2}}[\tilde{\alpha}_\mu - i(-1)^\mu\tilde{\pi}_{-\mu}], \tag{IV.114a}$$

$$q_\mu = \frac{1}{\sqrt{2}}[(-1)^\mu\tilde{\alpha}_{-\mu} + i\tilde{\pi}_\mu]. \tag{IV.114b}$$

These operators satisfy the commutation rule [follows from the application of (IV.112b)]

$$[q_\mu, q_{\mu'}^\dagger] = \delta_{\mu\mu'}. \tag{IV.115}$$

We have already proved [see (IV.10c)] that α_μ transforms as the μ-component of a tensor of second rank. Hence, according to (IV.111), $(-1)^\mu\pi_{-\mu}$ must also transform similarly. (IV.114a) then guarantees a similar transformation property for the operator $q_\mu^\dagger$. In writing q_μ, we have made use of the property (IV.4) of α_μ and the corresponding property of π_μ derivable from (IV.111).

It can now be easily verified that the identity

$$\begin{aligned}\tfrac{1}{2}\hbar\omega \sum_\mu (q_\mu^\dagger q_\mu + q_\mu q_\mu^\dagger) &= \tfrac{1}{2}\hbar\omega \sum_\mu (|\tilde{\alpha}_\mu|^2 + |\tilde{\pi}_\mu|^2)\\ &= \frac{1}{2B}\sum_\mu |\pi_\mu|^2 + \tfrac{1}{2}C\sum_\mu |\alpha_\mu|^2\\ &\equiv H\end{aligned}$$

holds. But $q_\mu q_\mu^\dagger$, according to (IV.115), is equal to $(q_\mu^\dagger q_\mu + 1)$. Hence, we obtain

$$H = \hbar\omega \sum_\mu (q_\mu^\dagger q_\mu + 1). \tag{IV.116}$$

The commutator of H with $q_\mu^\dagger$ and q_μ can be very easily evaluated from (IV.116) and (IV.115). Thus,

$$[H, q_\mu^\dagger] = \hbar\omega q_\mu^\dagger, \tag{IV.117a}$$

$$[H, q_\mu] = -\hbar\omega q_\mu. \tag{IV.117b}$$

These equations tell us that if $|\psi\rangle$ is any eigenstate of H belonging to the eigenvalue E, then $q_\mu^\dagger|\psi\rangle$ is a state of eigenvalue $(E + \hbar\omega)$, and $q_\mu|\psi\rangle$ is one of eigenvalue $(E - \hbar\omega)$. Thus, $q_\mu^\dagger$ creates a vibrational quantum of energy $\hbar\omega$, and q_μ destroys the same. In particular, if $|\psi_0\rangle$ is the ground state of H, then

$$q_\mu|\psi_0\rangle = 0 \tag{IV.117c}$$

must hold because this state, if it exists, necessarily has an energy less than that of the ground state by an amount $\hbar\omega$. If $|\psi_0\rangle$ is the true ground state, this is, by definition, impossible. Denoting the vibrational quantum number of projection μ by n_μ, we can alternatively represent any eigenstate of H by $|n_\mu\rangle$ so that, in the same representation, $|\psi_0\rangle$ becomes $|0\rangle$. Starting with this state, we can then generate the states $|n_\mu\rangle$ by repeated application of the operator $q_\mu^\dagger$.

We have already proved that $q_\mu^\dagger|n_\mu\rangle$ is a state of energy larger than that of $|n_\mu\rangle$ by the amount $\hbar\omega$; that is to say, it is a state of $(n_\mu + 1)$ quanta. In general, however, the state produced through $q_\mu^\dagger|n_\mu\rangle$ may not be *automatically* normalized to unity. If $|n_\mu + 1\rangle$ denotes the normalized state, then it is possible to show

$$q_\mu^\dagger|n_\mu\rangle = \sqrt{n_\mu + 1}\,|n_\mu + 1\rangle. \tag{IV.118}$$

The proof of this statement can be given by constructing the states $|1\rangle, |2\rangle, \ldots$ in a chain, starting with $|0\rangle$. Thus, with the arbitrary constant N_1, to be presently determined, we have

$$q_\mu^\dagger|0\rangle = N_1|1_\mu\rangle.$$

Multiplying from the left by the Hermitean conjugate, we obtain

$$|N_1|^2 = \langle 0|\, q_\mu q_\mu^\dagger\, |0\rangle = \langle 0|\,(1 + q_\mu^\dagger q_\mu)\,|0\rangle = 1.$$

Here we have used (IV.115) and (IV.117c). Thus, $N_1 = 1$ and we have

$$q_\mu^\dagger|0\rangle = \sqrt{1}|1_\mu\rangle.$$

At the next step, we have

$$q_\mu^\dagger|1_\mu\rangle = N_2|2_\mu\rangle$$

or

$$|N_2|^2 = \langle 1_\mu|\, q_\mu q_\mu^\dagger\, |1_\mu\rangle = \langle 0|\, q_\mu q_\mu^\dagger q_\mu q_\mu^\dagger\, |0\rangle = 2$$

or

$$N_2 = \sqrt{2},$$

$$q_\mu^\dagger|1_\mu\rangle = \sqrt{2}\,|2_\mu\rangle.$$

Here we have simplified the matrix element by the repeated application of (IV.115) and (IV.117c). This chain procedure can be continued and (IV.118) established.

According to (IV.118), the *only* nonvanishing matrix element of $q_\mu^\dagger$ is of the type

$$\langle n'_\mu|\, q_\mu^\dagger\, |n_\mu\rangle = \delta_{n'_\mu,\, n_\mu+1}\sqrt{n_\mu + 1}. \tag{IV.119a}$$

Taking the complex conjugate of (IV.119a), we prove that the *only* nonvanishing matrix element of q_μ is of the type

$$\langle n_\mu|\, q_\mu\, |n'_\mu\rangle = \delta_{n'_\mu,\, n_\mu+1}\sqrt{n_\mu + 1} = \delta_{n_\mu,\, n'_\mu - 1}\sqrt{n'_\mu}. \tag{IV.119b}$$

This matrix element also establishes

$$q_\mu|n_\mu\rangle = \sqrt{n_\mu}|n_\mu - 1\rangle \tag{IV.120}$$

because (IV.119b) is the *only* nonvanishing matrix element of q_μ.

In the theory of vibrational nuclei, we confront the matrix elements of the coordinates α_μ, which can be very easily obtained by first noting, with the help of (IV.114), the result

$$\tilde{\alpha}_\mu = \frac{1}{\sqrt{2}}[q_\mu^\dagger + (-1)^\mu q_{-\mu}]$$

or

$$\alpha_\mu = \left(\frac{\hbar}{2B\omega}\right)^{1/2}[q_\mu^\dagger + (-1)^\mu q_{-\mu}]. \tag{IV.121}$$

Using this expression of α_μ with (IV.119), we then obtain the only two types of nonvanishing matrix elements of α_μ, namely,

$$\langle n_\mu + 1|\ \alpha_\mu\ |n_\mu\rangle = \left(\frac{\hbar}{2B\omega}\right)^{1/2}\langle n_\mu + 1|\ q_\mu^\dagger\ |n_\mu\rangle = \left(\frac{\hbar}{2B\omega}\right)^{1/2}\sqrt{n_\mu + 1}, \tag{IV.122a}$$

$$\langle n_{-\mu} - 1|\ \alpha_\mu\ |n_{-\mu}\rangle = \left(\frac{\hbar}{2B\omega}\right)^{1/2}(-1)^\mu\sqrt{n_{-\mu}}. \tag{IV.122b}$$

In the special case of $n_\mu = 0$, the matrix element (IV.122a) reduces to $[\hbar/(2B\omega)]^{1/2}$ or $[\hbar\omega/(2C)]^{1/2}$.

We have already stated the tensorial property of the operator $q_\mu^\dagger$. In the case of even nuclei, the ground state, which contains no vibrational quantum, has angular momentum 0 and parity +. Since $q_\mu^\dagger$ transforms as a tensor of second rank, the one-phonon state $q_\mu^\dagger|0\rangle$ must have angular momentum 2 and parity +. The parity of the quadrupole operator being even, our final statement follows. If we next consider states with two quadrupole phonons, the parity is still +, but the angular momentum of the two-phonon states is clearly the resultant of the coupling of two angular momenta, each of magnitude 2. Thus, the resultant states should normally have any integral angular momentum lying between 0 and 4. However, the odd values 1 and 3 are ruled out from the consideration that follows: each quadrupole phonon is a boson of integral spin, and hence, under an exchange of the two phonons, the two-phonon wavefunction must remain symmetric. From the property of the Clebsch-Gordon coefficient, we conclude that any state $|j_1 j_2 JM\rangle$ of total angular momentum J, obtained by coupling two angular momenta each of magnitude j, gets multiplied by the phase factor $(-1)^{j+j-J}$ under the exchange P_{12}. In the present case, $j = 2$, and hence this phase factor is +1 only if J = even integer. Since we require a wavefunction symmetric under P_{12}, we are tied down to this choice of J. Thus, the two-phonon states comprise a triplet 0^+, 2^+, 4^+.

A typical vibrational spectrum for an even nucleus is shown in Fig. IV.8. The energy of the first 2^+ state merely determines the parameter $\hbar\omega$. The two-phonon triplet, whenever observed, is split up in energy. Very frequently, only one or two members of the triplet are seen and levels of other spin occur in the vicinity of one or more members of the triplet. There is no systematics in the order and magnitude of the energy splitting of the triplet. In Fig. IV.8, the number of phonons is denoted by N, printed in the extreme left-hand column. The extreme right-hand column gives the ratio R of the energy of a level to that of the first 2^+ level. Although the nucleus Cd^{114} very clearly shows the two-phonon triplet, there is considerable departure from an ideal quadrupole vibrator. In particular, the first excited 2^+ state has been found to have a large static quadrupole moment.

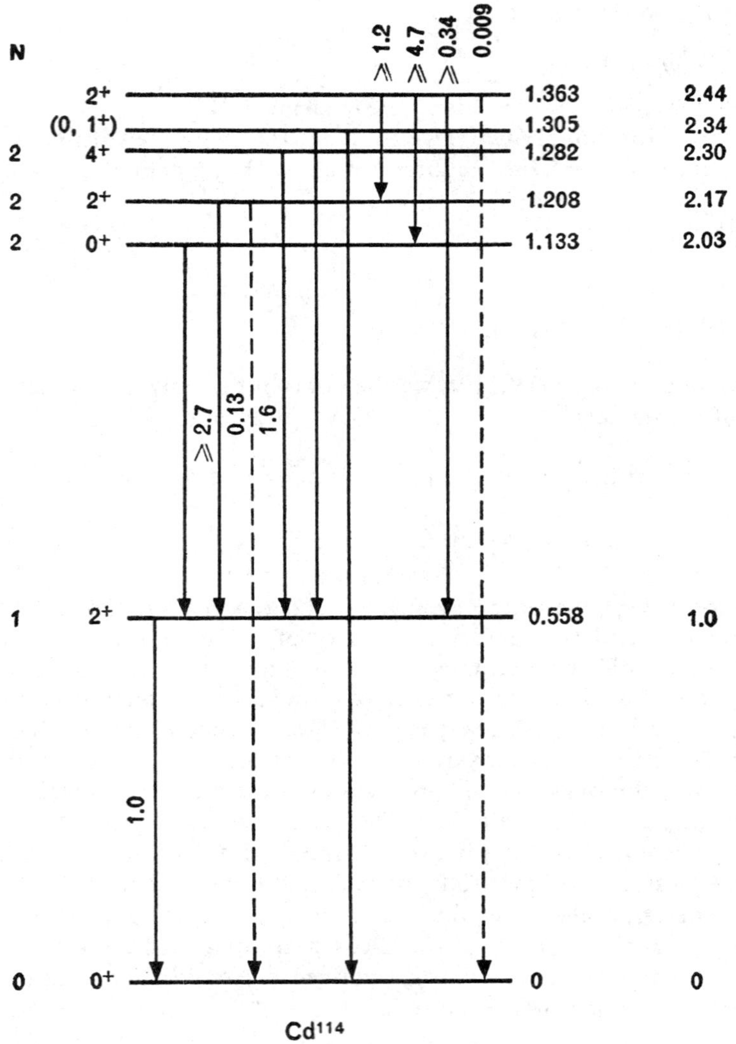

Fig. IV.8 Spectra of Cd^{114} showing one-phonon and two-phonon vibrational levels.

The electromagnetic properties of the first 2^+ level, and the levels of the triplet, are described in Section 34.

C. ODD-MASS NUCLEI: COUPLING OF PARTICLE TO EVEN-EVEN CORE

In the model we are considering here, we shall treat an odd-mass nucleus as the last odd nucleon coupled to the even-even core. For simplicity, we omit the β- and γ-vibrational part of the Hamiltonian already given for the even-even nucleus. Keeping only the rotational part for the even-even core and then adding to it the Hamiltonian of the last odd nucleon (H_p) and its coupling (H_{coupl}) with the core, we obtain

$$H = \sum_k \frac{\hbar^2 R_k^2}{2\mathcal{J}_k} + H_p + H_{coupl}. \tag{IV.123}$$

If we wish to specialize to the spheroidal deformation of the core, then we have to drop the $[R_z^2/(2\mathcal{J}_z)]$-term and put $\mathcal{J}_x = \mathcal{J}_y = \mathcal{J}$.

To find H_p and H_{coupl}, we formulate a procedure: the equipotential surfaces of a deformed nucleus are obtained by deforming those of the spherical nucleus to a shape identical to that of the deformed nucleus. Thus, the spherical equipotential surface of radius r_0 on which the value of the potential is $\mathcal{V}(r_0)$ gets deformed to an ellipsoid of radius $r(\theta, \phi)$ given, according to (IV.5a), by

$$r(\theta, \phi) = r_0[1 + \xi(\theta, \phi)], \tag{IV.124}$$

where $\xi(\theta, \phi)$ is a quantity of first-order smallness and is given by (IV.5b). The potential $\mathcal{V}_d(r)$ at each point on the surface of this ellipsoid is, according to our hypothesis, equal to $\mathcal{V}(r_0)$ of the undeformed nucleus. According to (IV.124), we have

$$r_0 = r[1 + \xi(\theta, \phi)]^{-1} \approx r[1 - \xi(\theta, \phi)] = r - r\xi(\theta, \phi),$$

and hence, by the foregoing hypothesis,

$$\mathcal{V}_d(\mathbf{r}) \equiv \mathcal{V}(r_0) = \mathcal{V}(r - r\xi) = \mathcal{V}(r) - r\frac{d\mathcal{V}}{dr}\xi$$

$$= \mathcal{V}(r) - r\frac{d\mathcal{V}}{dr} \sum_\mu \alpha_\mu^* Y_\mu^2(\theta, \phi). \tag{IV.125}$$

In the final step, we have used the expression (IV.5b) for $\xi(\theta, \phi)$ in the special case of quadrupole deformation. For $\mathcal{V}(r)$, we can take the central average potential of the spherical shell model (usually the harmonic oscillator potential). To obtain the complete expression for the particle Hamiltonian, we need the kinetic energy T_p, the spin-orbit coupling $\xi(r)\mathbf{l}\cdot\mathbf{s}$ of the shell model, and the potential energy as given by (IV.125). The second term in (IV.125) consists of a scalar product of the surface parameters α with the second-order spherical harmonic of the angle coordinates θ, ϕ of the particle. So this term can be interpreted as the surface-particle coupling term H_{coupl} occurring in (IV.123) and given by

$$H_{coupl} = -r\frac{d\mathcal{V}}{dr}\alpha \cdot Y^2(\theta, \phi). \tag{IV.126a}$$

According to the foregoing statements, we then have

$$H_p = T_p + \xi(r)\mathbf{l}\cdot\mathbf{s} + \mathcal{V}(r). \tag{IV.126b}$$

If $\mathcal{V}(r)$ is roughly represented by a square-well of radius R_0 and depth $-V_0$, then the strength $-r(d\mathcal{V}/dr)$ of the coupling term is clearly given by

$$-r\frac{d\mathcal{V}}{dr} = V_0 R_0 \delta(r - R_0). \tag{IV.126c}$$

On the other hand, with the more usual form of $\mathcal{V}(r)$, namely, the harmonic oscillator potential, $r(d\mathcal{V}/dr)$ is also a harmonic oscillator potential. The strength of the coupling appearing in (IV.126a) may be conveniently taken in either of these forms.

Spherical Vibrational Nuclei

In the case of spherical vibrational nuclei, the rotational part of the Hamiltonian can be put equal to zero. The ground and vibrational states of the even core have been described in Section 33B. The surface variables α_μ are treated there as vibrating coordinates. When we add H_p and H_{coupl} to the Hamiltonian of the core, the method of solving the problem clearly becomes dependent on how the magnitude of H_{coupl} compares with that of the other terms of the Hamiltonian, i.e., H_p and H_{core} given by (IV.126b) and (IV.116), respectively.

Since, by our hypothesis, the even core is spherical, the single-particle average potential created by it, and acting on the last odd nucleon, must be that of the spherical shell model. This is precisely what we have in (IV.126b). The single-particle levels in this potential are, we know, characterized by the quantum numbers *nljm*; the spin-orbit coupling term is known to be quite strong and is responsible in splitting the single-particle energy levels of the same l-value but different j-values ($j = l \pm \frac{1}{2}$).

The vibrating coordinates α_μ oscillate through small values about the equilibrium spherical shape for which all the $\alpha_\mu = 0$. The surface-particle coupling term H_{coupl} of (IV.126a) may therefore be taken to be small compared with, for example, the spin-orbit coupling term in (IV.126b). We therefore decide to work with the unperturbed Hamiltonian

$$\begin{aligned} H_0 &= H_{\text{core}} + H_p \\ &= [\tfrac{1}{2}B \sum_\mu |\dot{\alpha}_\mu|^2 + \tfrac{1}{2}C \sum_\mu |\alpha_\mu|^2] + [T_p + \xi(r)\mathbf{l}\cdot\mathbf{s} + \mathcal{V}(r)] \end{aligned} \tag{IV.127a}$$

and the corresponding unperturbed states

$$|n_\mu J_c, j : JM\rangle, \tag{IV.127b}$$

where n_μ, with the five possible values of μ, specifies the number of vibrational quanta for each μ present in the core; and J_c denotes the resultant angular momentum of the core (for example, $J_c = 2$ when $n_\mu = 1$); we have coupled J_c with the particle angular momentum j to produce the resultant J for the odd-mass nucleus under consideration. With these unperturbed states, we now take into account the effect of H_{coupl} through a perturbation treatment. The unperturbed ground state has $n_\mu = 0$ and $J_c = 0$, and hence $J = j$. When the perturbation H_{coupl} is switched on, this state mixes with other states of the type $|n'_\mu J'_c, j : J = j, m\rangle$. Since the resultant angular momentum is a good quantum number, we have used for this state the same value $J = j$ as that contained in the unperturbed state. We note that α is a tensor operator of second rank operating on the core state characterized by $(n_\mu J_c)$, and $Y^2(\theta, \phi)$ is a similar tensor acting on the single-particle state $j, j' \ldots$. Thus, the matrix element

$$\langle n'_\mu J'_c, j' : J = j, m|\; H_{\text{coupl}}\; |n_\mu = 0, J_c = 0, j : J = j, m\rangle,$$

which occurs in the first-order correction to the wavefunction of the ground state, can be easily evaluated by the standard Racah result (BIII.15) from Appendix B. We obtain

$$\begin{aligned} &\langle n'_\mu J'_c, j' : J = j, m|\; (-r\frac{d\mathcal{V}}{dr})\alpha \cdot Y^2\; |n_\mu = 0, J_c = 0, j : J = j, m\rangle \\ &= \left[\int_0^\infty r^2\, dr\, R_{n'l'}(-r\frac{d\mathcal{V}}{dr})R_{nl}\right](-1)^{J'_c}\{\frac{[j'][J'_c]}{5[j]}\}^{1/2}\langle n'_\mu J'_c\|\,\alpha\,\|n_\mu = 0, J_c = 0\rangle\langle j'\|\,Y^2\,\|j\rangle. \end{aligned} \tag{IV.128}$$

The particle-reduced matrix element can be substituted from the standard result (BIII.17) of

Appendix B. The quantity enclosed within the square brackets is the radial integral of the pure radial part of the operator, namely, $[-r(d\mathcal{V}/dr)]$, between the single-particle radial functions $R_{n'l'}$ and R_{nl} corresponding to the particle wavefunctions on the left- and right-hand side, respectively. To calculate the reduced matrix element of α between the core states, we first note that the only nonvanishing matrix elements of α_μ are of the type (IV.122). Thus, in the present case, n'_μ is uniquely determined to be 1. Using the projection quantum number of the core states explicitly, we then have, from (IV.122a),

$$\langle n'_\mu = 1, J'_c = 2, M'_c| \alpha_\mu |n_\mu = 0, J_c = 0, M_c = 0\rangle = (\frac{\hbar\omega}{2C})^{1/2}.$$

However, by the application of the Wigner-Eckart theorem, the same matrix element can be written as

$$\begin{bmatrix} 0 & 2 & 2 \\ 0 & \mu & M'_c \end{bmatrix} \langle n'_\mu = 1, J_c = 2| |\alpha| |n_\mu = 0, J_c = 0\rangle.$$

This Clebsch-Gordon coefficient happens to be unity, and hence we obtain

$$\langle n'_\mu = 1, J'_c = 2| |\alpha| |n_\mu = 0, J_c = 0\rangle = (\frac{\hbar\omega}{2C})^{1/2}. \tag{IV.129}$$

We denote the value of the radial integral by $-C$. Substituting (IV.129) and the particle-reduced matrix element explicitly, we then have

$$\langle n'_\mu J'_c, j' : J = j, m| H_{\text{coupl}} |n_\mu = 0, J_c = 0, j : J = j, m\rangle$$
$$= (\frac{\hbar\omega}{2C})^{1/2} \sqrt{\frac{5}{4\pi}}\, \tfrac{1}{2}[1 + (-1)^{l+l'}] \sqrt{\frac{[j']}{[j]}}\, C \begin{bmatrix} j' & 2 & j \\ \frac{1}{2} & 0 & \frac{1}{2} \end{bmatrix} \delta_{n'_\mu, 1}\delta_{J'_c, 2}. \tag{IV.130}$$

In order to obtain the admixture coefficient in the wavefunction, we have to divide this matrix element, according to the first-order perturbation theory, by the energy difference of the right- and left-hand state, i.e., $-\hbar\omega + \epsilon_j - \epsilon_{j'}$. In this way, we obtain, for the perturbed ground state, the expression

$$|\psi_0\rangle = |n_\mu = 0, J_c = 0, j : j, m\rangle + C(\frac{\hbar\omega}{2C})^{1/2} \sqrt{\frac{5}{4\pi}} \sum_{j'} \tfrac{1}{2}\{1 + (-1)^{l+l'}\} \begin{bmatrix} j' & 2 & j \\ \frac{1}{2} & 0 & \frac{1}{2} \end{bmatrix} \sqrt{\frac{[j']}{[j]}}$$
$$\times [\hbar\omega + \epsilon_{j'} - \epsilon_j]^{-1} |n'_\mu = 1, J'_c = 2, j' : J = j, m\rangle. \tag{IV.131}$$

The one-phonon admixture into the ground state has very important consequences on the electromagnetic properties of the states, which we shall discuss in Section 34.

Complete diagonalization calculations with the particle vibration coupling model have been done by many authors[10]. The necessary formulas were first derived by Choudhury[10]. Later, Lande and Brown[10], and then Sen et al[10] applied the model, including both the quadrupole and octupole phonons. Sen and his collaborators worked extensively on the odd-mass isotopes of Sb, In, Sr, Kr, Zr, Te, Rb, and Y. Besides obtaining the energy levels, they computed the electromagnetic transition strengths and spectroscopic factors of various levels.

Deformed Rotational Nuclei

In deformed rotational nuclei, we have the core states described by the rotator part of (IV.123). Strictly speaking, we have also the β- and γ-vibrational part of the Hamiltonian, which can be brought into the discussion if desired. For simplicity, we suppress these parts.

The question now is: is the surface-particle coupling weak or strong in comparison with H_p? In the applications of the model, the coupling has always been taken to be strong.

Since the nucleus has an equilibrium deformation, we express H_{coupl} in the body-fixed frame and obtain

$$\begin{aligned}(-r\frac{d\mathcal{V}}{dr})\alpha\cdot Y^2 &= (-r\frac{d\mathcal{V}}{dr})(a_0Y_0^2 + a_2Y_2^2 + a_{-2}Y_{-2}^2)\\ &= (-r\frac{d\mathcal{V}}{dr})[\beta\cos\gamma\ Y_0^2(\theta,\phi) + \frac{\beta}{\sqrt{2}}\sin\gamma\{Y_2^2(\theta,\phi) + Y_{-2}^2(\theta,\phi)\}]. \quad \text{(IV.132)}\end{aligned}$$

The projection quantum numbers on Y^2 now refer to the body-fixed z-axis.

If our assumption about strong coupling is correct, then we must consider the entire Hamiltonian at one time. In such a case, since H_{coupl} contains both the surface *variables* β, γ of the core and the particle variables r, θ, ϕ, it is almost impossible to solve the resultant coupled equations of the five surface variables and the three particle variables. In order to make the problem tractable, we make use of the adiabatic hypothesis which is as follows: assume that the motion of the surface variables β, γ, θ_1, θ_2, θ_3 is very slow compared with that of the particle. Further, since the even core has a fairly well-deformed equilibrium shape, and the motion of β, γ about this shape is slow, assume that solving the particle motion for a given value of β, γ is a good approximation; in other words, while solving the particle equation, treat β, γ in H_{coupl} as *fixed parameters* (and not variables) specifying a given shape of the nucleus.

Because of the strong coupling and the adiabatic hypothesis, we are entitled to solve the particle motion corresponding to the Hamiltonian

$$\begin{aligned}\mathcal{H}_p(\beta,\gamma) = H_p + H_{\text{coupl}} = T_p + \xi(r)\boldsymbol{l}\cdot\mathbf{s} + \mathcal{V}(r) - r\frac{d\mathcal{V}}{dr}[\beta\cos\gamma\ Y_0^2(\theta,\phi)\\ + \frac{\beta}{\sqrt{2}}\sin\gamma\{Y_2^2(\theta,\phi) + Y_{-2}^2(\theta,\phi)\}], \qquad \text{(IV.133)}\end{aligned}$$

which is a sum of (IV.126b) and (IV.132). In (IV.133), β, γ, as just mentioned, are to be treated as given parameters and not variables. The notation $\mathcal{H}_p(\beta,\gamma)$ on the left-hand side has been used to emphasize the fact that, even though the equation contains H_{coupl}, the adiabatic hypothesis has transformed the whole expression as the Hamiltonian for particle motion for a given β, γ; these quantities therefore legitimately occur within the parentheses following $\mathcal{H}_p$. When looked at in this way, $\mathcal{H}_p$ defines a single-particle Hamiltonian with a nonspherical potential. If the core always maintains axial symmetry about the z-axis, then $\gamma = 0$, and the nonspherical part of the potential in (IV.133) reduces to

$$-\beta r\frac{d\mathcal{V}}{dr}Y_0^2(\theta). \qquad \text{(IV.134)}$$

We shall consider the details of solving the single-particle wavefunctions for such a single-particle Hamiltonian in Section 33D where a minor modification is introduced in $\mathcal{H}_p$ and brought to the reader's attention at the appropriate point.

In this section, we shall use very general facts about the single-particle wavefunction just referred to. Let us denote it by χ. If the Y_2^2- and Y_{-2}^2-term are absent in (IV.133), i.e., if $\gamma = 0$, then the deformed potential has axial symmetry, and hence the projection quantum number Ω of the single-particle angular momentum will be a good quantum number for the specification of χ. The term (IV.134) of the Hamiltonian will mix wavefunctions of different n-, l-, and j-value and, as a result, χ_Ω will be of the form

$$\chi_\Omega = \sum_{n,l,j} C_{nlj\Omega}\phi(nlj\Omega); \tag{IV.135a}$$

the quantum numbers n, l, j, Ω are the usual single-particle quantum numbers specifying the single-particle states of H_p.

The effect of considering the more general form (IV.133), in which Y_2^2- and Y_{-2}^2-term are nonvanishing, is also fairly obvious. These terms of the potential clearly mix the wavefunctions $\phi(nlj\Omega)$ with values of Ω differing by ± 2. Thus, Ω can no longer be used as a labelling quantum number on χ, and we obtain

$$\chi = \sum_{nlj\Omega} C_{nlj\Omega}\phi(nlj\Omega). \tag{IV.135b}$$

It is obvious that, with suitable coefficients d_Ω, the wavefunction (IV.135b) can be expressed also in terms of χ_Ω of (IV.135a) with different values of Ω, that is,

$$\chi = \sum_\Omega d_\Omega \chi_\Omega. \tag{IV.135c}$$

The rotational part of the Hamiltonian (IV.123) has now to be written somewhat differently. This is because we would now like to characterize the states by the total angular momentum I and its projection M for the *whole* nucleus (i.e., even-even core plus the single nucleon) and not that of the core alone. Therefore, we introduce the corresponding total angular momentum operator $\mathbf{I}$, defined by

$$\mathbf{I} = \mathbf{R} + \mathbf{j}. \tag{IV.136}$$

For an axially symmetric nucleus, the rotational part of the Hamiltonian is given by $[\hbar^2/(2\mathcal{J})](R^2 - R_z^2)$, which can be written from (IV.136) in terms of $\mathbf{I}$ and $\mathbf{j}$ as

$$\begin{aligned} H_{\text{rot}} &= \frac{\hbar^2}{2\mathcal{J}}[(\mathbf{I} - \mathbf{j})^2 - (I_z - j_z)^2] \\ &= \frac{\hbar^2}{2\mathcal{J}}(I^2 - 2\mathbf{I}\cdot\mathbf{j} + j^2 - I_z^2 - j_z^2 + 2I_z j_z) \\ &= \frac{\hbar^2}{2\mathcal{J}}(I^2 - I_z^2 - (I_+ j_- + I_- j_+) + j^2 - j_z^2), \end{aligned} \tag{IV.137}$$

where $I_\pm = I_x \pm iI_y$ and $j_\pm = j_x \pm ij_y$. The terms j^2 and j_z^2 can be very easily transferred to the particle Hamiltonian $\mathcal{H}_p$. The rest of the expression (IV.137) is now treated.

We first note that $\mathbf{j}$ operates only on the single-particle wavefunction and leaves a $\mathcal{D}$-function unaffected. Hence, the result of operating on the $\mathcal{D}$-functions with the components of $\mathbf{I}$ is given, according to (IV.136), by the results for the same components of $\mathbf{R}$. Thus, all the results (IV.88), (IV.89), and (IV.92) hold if $\mathbf{R}$ is replaced in each equation by $\mathbf{I}$. Thus, the rotational wavefunction $\mathcal{D}_{MK}^I(\theta_1, \theta_2, \theta_3)$ remains an eigenfunction of the first two terms, namely, $(I^2 - I_z^2)$, occurring in (IV.137). The same statement is not applicable to the next term $-(I_+ j_- + I_- j_+)$, which is called the *Coriolis coupling* [the reader should distinguish it from the *surface*-particle coupling of (IV.126a)] or the *rotation*-particle coupling (RPC) because of the fact that it contains the product of the rotation operator $\mathbf{I}$ with the particle angular momentum $\mathbf{j}$. In view of (IV.92), the $(I_+ j_-)$-term of the RPC, operating on $\mathcal{D}_{MK}^I$, changes it to $\mathcal{D}_{M,K-1}^I$, and the (I_-j_+)-term changes the same to $\mathcal{D}^I_{M,K+1}$. The detailed effect of the RPC is treated later in this section. For the present, we take a rotator wavefunction $\mathcal{D}^I_{MK}(\theta_1, \theta_2, \theta_3)$, multiply it by the particle wavefunction $\chi_\Omega(\mathbf{r}_p)$, and then write our model wavefunction for the odd-mass nucleus with an axially symmetric core as

$$\Psi^I_{MK}(\theta_1, \theta_2, \theta_3; \mathbf{r}_p) = \mathcal{D}^I_{MK}(\theta_1, \theta_2, \theta_3)\chi_\Omega(\mathbf{r}_p).$$

This wavefunction has to be multiplied by $f(\beta)g^I_K(\gamma)$ if the effect of β- and γ-vibration has to be incorporated.

We have now to include the symmetry requirements due to the transformations $\mathcal{R}_1$, $\mathcal{R}_2$, $\mathcal{R}_3$ of the principal axes of an ellipsoid, just as we have done for the even nucleus in Section 33A [see (IV.97)]. All the discussions with respect to the functions of γ and $(\theta_1, \theta_2, \theta_3)$ hold here. The extra consideration, which we have now to incorporate, arises from the transformation of the particle wavefunction $\chi_\Omega(\mathbf{r}_p)$ under the rotations $\mathcal{R}_1$, $\mathcal{R}_2$, $\mathcal{R}_3$. We recall that $\chi_\Omega(\mathbf{r}_p)$ has been constructed with respect to a certain z-axis, along which the projection quantum number is Ω. Since the three transformations introduce new z-axes, $\chi_\Omega(\mathbf{r}_p)$ is certainly affected by each of these transformations.

We now write the desired transformation equations for χ_Ω as

$$\chi_\Omega \to \exp(-i\pi j_y)\chi_\Omega = (-1)^{j+\Omega}\chi_{-\Omega} \qquad \text{(under } \mathcal{R}_1\text{)}, \tag{IV.138a}$$

$$\chi_\Omega \to \exp\left(-i\frac{\pi}{2}j_z\right)\chi_\Omega = \exp\left(-i\frac{\pi}{2}\Omega\right)\chi_\Omega \qquad \text{(under } \mathcal{R}_2\text{)}, \tag{IV.138b}$$

$$\chi_\Omega \to \exp(-i\pi j_z)\exp\left(-i\frac{\pi}{2}j_y\right)\exp\left(-i\frac{\pi}{2}j_z\right)\chi_\Omega \qquad \text{(under } \mathcal{R}_3\text{)}. \tag{IV.138c}$$

In the case of axial symmetry of the core, we have to reinterpret $\mathcal{R}_2$ as mentioned after (IV.100b). In that case, the symmetry requirement applies to any arbitrary rotation ϕ about the z-axis, which produces

$$\chi_\Omega \to \exp(-i\phi j_z)\chi_\Omega = e^{-i\phi\Omega}\chi_\Omega. \tag{IV.138d}$$

The meaning of the quantum number j occurring in (IV.138a) has to be understood in the following sense: according to the work in Appendix B, $\exp(-i\pi j_y)$, operating on an eigenstate $|j\Omega\rangle$ of angular momentum, produces

$$\exp(-i\pi j_y)|j\Omega\rangle = \sum_{\Omega'} |j\Omega'\rangle\langle j\Omega'|\exp(-i\pi j_y)|j\Omega\rangle$$

$$= \sum_{\Omega'} d^{j*}_{\Omega\Omega'}(\pi)|j\Omega'\rangle = (-1)^{j+\Omega}|j, -\Omega\rangle \qquad (d^j_{\Omega\Omega'} \text{ is real}).$$

The wavefunction χ_Ω occurring in (IV.138a) is not an eigenfunction of angular momentum; however, it can be written as a superposition of these eigenfunctions, as shown in (IV.135a). The right-hand side of (IV.138a) denotes that *each term* of (IV.135a) gets the quantum number Ω occurring in the state (and *not* in the coefficients $C_{nlj\Omega}$) reversed in sign and multiplied by the *corresponding* phase factor $(-1)^{j+\Omega}$. Thus,

$$(-1)^{j+\Omega}\chi_{-\Omega} = \sum_{nlj} C_{nlj\Omega}(-1)^{j+\Omega}\phi(nlj, -\Omega) \tag{IV.138e}$$

when χ_Ω is given by (IV.135a).

Combining (IV.138a) and (IV.138b) with the transformed $\mathcal{D}^I_{MK}$-functions [the transformation of $\mathcal{D}$-functions under $\mathcal{R}_1$, $\mathcal{R}_2$ has been given when deriving (IV.97a) and (IV.97b)], we obtain

$$\mathcal{D}^I_{MK}(\theta_1, \theta_2, \theta_3)\chi_\Omega \to (-1)^{I+K}(-1)^{-j-\Omega}\chi_{-\Omega}\mathcal{D}^I_{M,-K} \quad \text{(under } \mathcal{R}_1\text{)}, \tag{IV.139a}$$

$$\mathcal{D}^I_{MK}(\theta_1, \theta_2, \theta_3)\chi_\Omega \to e^{i(\pi/2)(K-\Omega)}\chi_\Omega\mathcal{D}^I_{MK} \quad \text{(under } \mathcal{R}_2\text{)}. \tag{IV.139b}$$

In the special case of axial symmetry, $e^{i(\pi/2)(K-\Omega)}$ in (IV.139b) changes to $e^{i\phi(K-\Omega)}$ with ϕ completely arbitrary.

Applying $\mathcal{R}_2$ twice and then requiring that the *value* of the wavefunction does not change, we conclude from (IV.139b) that

$$K - \Omega = \text{even integer}. \tag{IV.140a}$$

In the special case of axial symmetry, in view of the arbitrariness of ϕ, we must require

$$K - \Omega = 0$$

or

$$\Omega = K. \tag{IV.140b}$$

(IV.139a) similarly tells us that the *value* of the wavefunction remains unaltered under $\mathcal{R}_1$, provided we consider the symmetrized wavefunction

$$\Psi^I_{MK}(\theta_1, \theta_2, \theta_3; \mathbf{r}_p) = \left(\frac{2I+1}{16\pi^2}\right)^{1/2}[\mathcal{D}^I_{MK}(\theta_1, \theta_2, \theta_3)\chi_K(\mathbf{r}_p) + (-1)^{I-j}\mathcal{D}^I_{M,-K}(\theta_1, \theta_2, \theta_3)\chi_{-K}(\mathbf{r}_p)]. \tag{IV.141}$$

We have already used the value of Ω, determined by the symmetry requirement $\mathcal{R}_2$, and have incorporated the normalization factor preceding (IV.100a). The special case of $K = 0$ does not arise here because, for odd-mass nuclei, $I =$ half-integer, and hence K is also necessarily a half-integral quantum number.

As in the case of even nuclei, we now obtain bands of rotational levels of various angular momenta I belonging to a given band quantum number K. Permitted values of I in a band must necessarily be $I \geqslant K$. For all the levels of a rotational band, the particle wavefunction χ_K remains the same; only the rotational part $\mathcal{D}^I_{MK}$ acquires different values of i. The contribution from $\mathcal{H}_p$ to the energy of a level therefore remains constant for all the levels of a band, and we denote this contribution by $E_0(K)$. The rotational energy gets added to it. In this way, the energy $E(I)$ of a level I is given by

$$E(I; K) = \frac{\hbar^2}{2\mathcal{J}}[I(I+1) - K^2] + E_0(K), \tag{IV.142}$$

where, in quoting the rotational energy, we have considered only the first two terms of (IV.137). The effect of j^2- and j_z^2-term of that equation is already contained in our definition of $E_0(K)$. Hence, the only additional contribution to the energy comes from the RPC term

$$-\frac{\hbar^2}{2\mathcal{J}}(I_+j_- + I_-j_+),$$

and we shall first consider its effect in the lowest-order perturbation theory. The first-order contribution to the energy of a perturbation H_I is its diagonal matrix element for the unperturbed state. Thus, the lowest-order change in energy due to the RPC is given by

$$\Delta E(I; K) = -\frac{\hbar^2}{2\mathcal{J}}\langle\Psi^I_{MK}|\,(I_+j_- + I_-j_+)\,|\Psi^I_{MK}\rangle.$$

Since $I_{\pm}$ change K to $K \mp 1$, we conclude that a nonvanishing contribution to this matrix element can arise only from the cross terms between $\mathcal{D}^I_{M,K}$ and $\mathcal{D}^I_{M,-K}$ and that too only if $K = \frac{1}{2}$. For this special value of K, $-K$ happens to be $K - 1$; for any other value of K, this condition between K and $-K$ does not hold. Therefore, the RPC contribution in the lowest order is given by

$$\begin{aligned}\Delta E(I; K) &= -\delta_{K,1/2}\frac{\hbar^2}{\mathcal{J}}\langle \mathcal{D}^I_{M,1/2}\chi_{1/2}|\, I_- j_+ \,|\mathcal{D}^I_{M,-1/2}\chi_{-1/2}\rangle(-1)^{I-j}\\ &= \delta_{K,1/2}\frac{\hbar^2}{\mathcal{J}}(-1)^{I+1/2}\langle \mathcal{D}^I_{M,1/2}|\, I_- \,|\mathcal{D}^I_{M,-1/2}\rangle\langle\chi_{1/2}|\, j_+ \,|\chi_{-1/2}\rangle(-1)^{j-1/2}\\ &= \delta_{K,1/2}(-1)^{I+1/2}(I+\tfrac{1}{2})a. \end{aligned} \tag{IV.143}$$

Here we have first noted that the other cross matrix element with $-\frac{1}{2}$ on the left and $+\frac{1}{2}$ on the right has the same value, and hence its effect has been taken into account by cancelling the 2 of $2\mathcal{J}$ in the denominator. Secondly, in the matrix element used here, we have noted that only the term $I_- j_+$ contributes a nonvanishing result; the other term $I_+ j_-$ would change $\mathcal{D}^I_{M,-1/2}$ to $\mathcal{D}^I_{M,-3/2}$. Finally, the value of the matrix element of I_- has been written with the help of (IV.92), a relation that holds good [as per our observation following (IV.137)] under the replacement of **R** by **I**. The quantity a can be evaluated with the help of the standard matrix element of j_+ from **Appendix A**, and the expressions (IV.135a) and (IV.138e). Thus,

$$\begin{aligned} a &= \frac{\hbar^2}{\mathcal{J}}\langle\chi_{1/2}|\, j_+ \,|\chi_{-1/2}\rangle(-1)^{j-1/2}\\ &= \frac{\hbar^2}{\mathcal{J}}\sum_{nlj}|C_{nlj(1/2)}|^2\langle nlj, +\tfrac{1}{2}|\, j_+ \,|nlj, -\tfrac{1}{2}\rangle(-1)^{j-1/2}\\ &= \frac{\hbar^2}{\mathcal{J}}\sum_{nlj}(-1)^{j-1/2}(j+\tfrac{1}{2})|C_{nlj(1/2)}|^2. \end{aligned} \tag{IV.144}$$

It is called the decoupling parameter in the literature on the collective model. The point to note about the expression (IV.144) is that the decoupling parameter does not depend on I, and is determined entirely by the deformed single-particle potential through the single-particle wavefunction $\chi_{1/2}$.

Collecting (IV.142) and (IV.143), we then write the general expression of energy as

$$E(I; K) = \mathcal{E}_0(K) + \frac{\hbar^2}{2\mathcal{J}}I(I+1) + \delta_{K,1/2}a(-1)^{I+1/2}(I+\tfrac{1}{2}), \tag{IV.145}$$

where

$$\mathcal{E}_0(K) = E_0(K) - \frac{\hbar^2}{2\mathcal{J}}K^2.$$

In the next higher order, the RPC can cause very interesting effects on the moment of inertia. We have already remarked that the RPC can give rise to nondiagonal matrix elements connecting different K's, i.e., different bands. The K-quantum numbers mixed in this way can differ by ± 1. The ground state of an odd-mass nucleus has $I_0 = K_0$, where K_0 is the band quantum number of the ground-state rotational band. There may be other excited bands in the same nucleus, and if the energy of the band head of an excited band is fairly close to the ground-state energy, there can be an appreciable mixing of the two bands through the RPC, provided, of course, the excited band has a band quantum number $K = K_0 \pm 1$. *Usually*, the close-lying excited bands have a band quantum number larger than K_0. As a result, we consider only the mixing of the ground-state band with a band of $K = K_0 + 1$. The angular momentum $I_0 (= K_0)$ is not contained in such an excited band because K is larger than I_0. Thus, the ground state of the nucleus remains unaffected by such band mixing. Each of the higher levels of the ground-state band having $I = I_0 + 1, I_0 + 2, \ldots$ finds a state of the same angular momentum in the excited band $K = K_0 + 1$, with which it therefore mixes through the RPC. The resultant states for a given I can clearly be obtained by diagonalizing a 2×2 matrix of the Hamiltonian, using the states $|I, K_0\rangle$ and $|I, K = K_0 + 1\rangle$ as the basis.

The diagonal matrix elements of our Hamiltonian, inclusive of the RPC term, are given by the general expression (IV.145). We have then to calculate only the nondiagonal matrix elements, which arise from the RPC term alone. In a straightforward manner,

$$-\frac{\hbar^2}{2\mathcal{J}}\langle I, K = K_0 + 1|\,(I_+j_- + I_-j_+)\,|I, K_0\rangle$$

$$= -\frac{\hbar^2}{2\mathcal{J}}\frac{2I+1}{16\pi^2}\langle \mathcal{D}^I_{MK}\chi_K + (-1)^{I-j}\mathcal{D}^I_{M,-K}\chi_{-K}|\,(I_+j_- + I_-j_+)\,|\mathcal{D}^I_{MK_0}\chi_{K_0}$$

$$+ (-1)^{I-j}\mathcal{D}^I_{M,-K_0}\chi_{-K_0}\rangle.$$

We have already taken $K = K_0 + 1$, and hence $-K = -K_0 - 1$. If the RPC operator has to connect the $\mathcal{D}^I_{MK_0}$-term with $\mathcal{D}^I_{M,-K}$, then clearly we need $K_0 = 0$, which is impossible for the odd-mass nucleus. Thus, the only nonvanishing terms are those connecting $\mathcal{D}^I_{MK_0}$ with $\mathcal{D}^I_{MK}$, and $\mathcal{D}^I_{M,-K_0}$ with $\mathcal{D}^I_{M,-K}$, and these two terms happen to be equal. We therefore evaluate only one term, incorporating a factor of 2, which converts $(2I+1)/(16\pi^2)$ into $(2I+1)/(8\pi^2)$, and the latter produces unity when taken with the integral of $\mathcal{D}^{I*}_{MK}\mathcal{D}^I_{MK}$ over all the angles.

Further, let us recall that only I_- can give a nonvanishing matrix element between $|\mathcal{D}^I_{MK_0}\rangle$ and $\langle\mathcal{D}^I_{M,K_0+1}|$, and hence

$$\langle I, K = K_0 + 1|\,\text{RPC}\,|I, K_0\rangle = -\frac{\hbar^2}{2\mathcal{J}}\frac{2I+1}{8\pi^2}\langle\mathcal{D}^I_{M,K_0+1}|\,I_-\,|\mathcal{D}^I_{M,K_0}\rangle\langle\chi_{K_0+1}|\,j_+\,|\chi_{K_0}\rangle$$

$$= A[(I - K_0)(I + K_0 + 1)]^{1/2}, \qquad \text{(IV.146a)}$$

where

$$A = -\frac{\hbar^2}{2\mathcal{J}}\langle\chi_{K_0+1}|\,j_+\,|\chi_{K_0}\rangle. \qquad \text{(IV.146b)}$$

Once again, A_K is determined entirely by the deformed single-particle wavefunction.

After the foregoing procedure, the matrix to be diagonalized has the structure

$$\begin{pmatrix} E(I, K_0) & [(I - K_0)(I + K_0 + 1)]^{1/2}A \\ [(I - K_0)(I + K_0 + 1)]^{1/2}A & E(I, K_0 + 1) \end{pmatrix}. \qquad \text{(IV.147)}$$

The eigenvalues of this 2×2 matrix can be easily found from the secular determinant equation, which yields

$$\mathcal{E}(I) = \tfrac{1}{2}[E(I, K_0) + E(I, K_0 + 1)]$$

$$\pm \tfrac{1}{2}[\{E(I, K_0 + 1) - E(I, K_0)\}^2 + 4A^2(I - K_0)(I + K_0 + 1)]^{1/2}. \qquad \text{(IV.148)}$$

In the limit of $A \to 0$, the minus sign separating the two terms leads to the energy $E(I, K_0)$ of the ground-state band. For small A, we make an expansion of the second term in powers of A^2, and collect only the lowest-order term; we also use the minus sign separating the two terms such that the resultant $E(I)$ may give the shifted position of the state I originally belonging to the ground-state band. Thus,

$$\mathcal{E}(I) \approx E(I, K_0) - \frac{[I(I+1) - K(K+1)]A^2}{E(I, K_0+1) - E(I, K_0)}. \tag{IV.149}$$

Up to the second order in A, the RPC has, according to (IV.149), the effect of renormalizing the moment of inertia parameter $[\hbar^2/(2\mathcal{J})]$ to a value $[\hbar^2/(2\mathcal{J}')]$ given by

$$\frac{\hbar^2}{2\mathcal{J}'} = \frac{\hbar^2}{2\mathcal{J}} - \frac{A^2}{E(I, K_0+1) - E(I, K_0)}. \tag{IV.150}$$

It should be noted that the numerical fitting of the levels in the ground-state band determines the parameter $[\hbar^2/(2\mathcal{J}')]$ and not the 'true' moment of inertia $[\hbar^2/(2\mathcal{J})]$.

Rotational spectra, and more than one band, have been observed in many odd-mass nuclei in the deformed regions. The case of Al^{25} is shown in Fig. IV.9. Here the different bands are shown horizontally displaced one from the other so that a clear visual identification becomes possible. The ratio of the energies of the first two states $I_0 + 1$ and $I_0 + 2$ measured

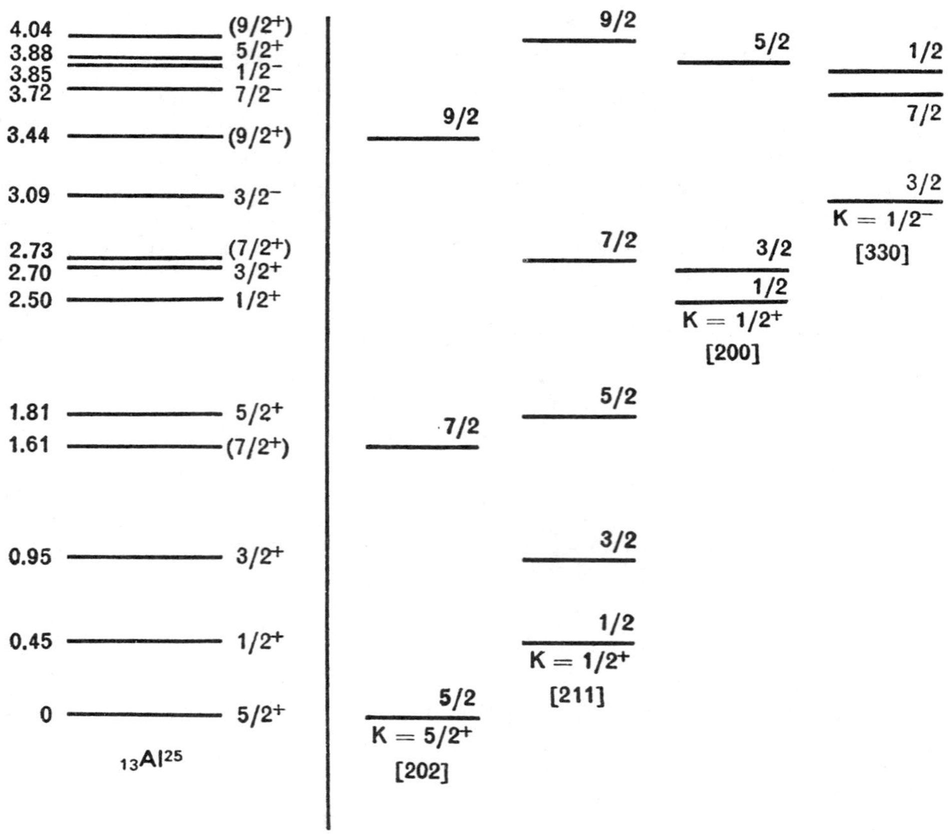

Fig. IV.9 Rotational spectra of Al^{25}. Data to right of vertical line denotes various rotational bands in different columns. (Following Bohr, A., and Mottelson, B. R., in Nuclear Spectroscopy, Part B, ed. by F. Ajzenberg-Selove, Academic Press, New York, 1960, p 1029.)

from the ground state I_0 is given, according to the rotational formula (when $K \neq \frac{1}{2}$), by

$$\frac{E(I_0 + 2)}{E(I_0 + 1)} = 1 + \frac{2}{I_0 + 1}.$$

Therefore, this experimental ratio determines I_0, and consequently the ground-state band quantum number K_0 $(=I_0)$. The excitation energy of the first state $I_0 + 1$ then determines the moment of inertia parameter. As stated, this parameter has to be identified with $\mathcal{J}'$ and not $\mathcal{J}$. Values of $(3\hbar^2/\mathcal{J}')$ obtained in this way are consistently larger than the value of $(3\hbar^2/\mathcal{J})$ determined for the neighbouring even nuclei. For example, ${}_{64}Gd^{155}$ has $(3\hbar^2/\mathcal{J}') \approx 120$ keV, whereas the neighbouring even nucleus yields $(3\hbar^2/\mathcal{J}) \approx 70$ keV. This difference has to be understood in terms of the RPC, as just shown. The decoupling parameter a in the special case of $K = \frac{1}{2}$ has also been determined by fitting the spectra of levels of the ground-state rotational band to the formula (IV.145). A few such nuclei are ${}_{69}Tm^{169}$, ${}_{73}Ta^{181}$, and ${}_{94}Pu^{239}$, all of which have a ground state $I_0 = \frac{1}{2}$ which belongs to a $(K = \frac{1}{2})$-band. The values of a found for the three nuclei are respectively -0.77, 0.19, and -0.58.

D. NILSSON MODEL FOR DEFORMED SINGLE-PARTICLE WAVEFUNCTIONS

We now discuss a generalization of the wavefunction (IV.141) which is applicable to both even- and odd-mass nuclei. Instead of considering χ_K as the wavefunction in a deformed potential of a *single* nucleon outside the even core, we take it as a normalized determinantal *many-nucleon* wavefunction for all the nucleons in the nucleus; the determinant is constructed out of single-particle wavefunctions generated by a deformed potential. Clearly, according to this extended interpretation of χ_K, we can write a wavefunction with the structure of (IV.141) for both the even- and the odd-mass nuclei. The determinantal wavefunction χ_K has reference to a set of principal axes moving with the rotating nucleus, and it describes the pure *intrinsic* motion of the nucleons. Therefore, it is usually termed the intrinsic wavefunction of the nucleus. We are still assuming the existence of axial symmetry, and this is why the intrinsic state χ_K carries the angular momentum projection quantum number K as a label. However, χ_K being a many-body determinant, the quantum number K here is to be reinterpreted as the projection quantum number of the entire determinantal wavefunction; that is to say, it is a sum of the individual particle projection quantum numbers Ω which label the wavefunctions of the individual particles in the determinant. In an expansion of χ_K in terms of the good angular momentum states χ_K^J, the quantum number J specifies the total angular momentum of the state χ_K^J of the many-nucleon system. Whenever an expansion is required, we shall write

$$\chi_K = \sum_J C_{JK} \chi_K^J. \tag{IV.151}$$

We now replace the single-particle angular momentum j in (IV.141) by the J just defined. Similarly, in the relation (IV.136), we have to replace **j** by **J**. The extended model described with the reinterpreted intrinsic wavefunctions is usually referred to as the *unified model.* This nomenclature relies on the fact that the model is a unification of the determinantal-type many-nucleon wavefunctions of the shell model (the only difference here is the deformation of the average single-particle potential in contradistinction to the spherical shell-model potential) and the collective rotational-type wavefunction $\mathcal{D}_{MK}^I$ of Bohr's collective model.

The calculation of the deformed single-particle wavefunctions with which an intrinsic wavefunction χ_K has to be constructed occupies a position of importance in the theory of the unified model. The work of Nilsson and Mottelson[11] is the most elaborate in this respect; their deformed single-particle potential and their calculation of the corresponding single-particle wavefunctions define what is popularly known as the *Nilsson model.*

The single-particle potential used in the Nilsson model is very similar to that used in (IV.133) with the omission of the Y_2^2- and Y_{-2}^2-term. This omission is because of the authors' use of a deformed potential having the z-axis as its symmetry axis. They were guided by the spherical harmonic oscillator potential used in the shell model. This potential is given by

$$\mathcal{V}(r) = \tfrac{1}{2}M\omega^2 r^2 = \tfrac{1}{2}M\omega^2(x^2 + y^2 + z^2)$$

and can be deformed by associating ω_x, ω_y, ω_z of different magnitudes with the x^2-, y^2-, z^2-term, respectively. Thus, the general deformed potential is given by

$$\mathcal{V}(\mathbf{r}) = \tfrac{1}{2}M(\omega_x^2 x^2 + \omega_y^2 y^2 + \omega_z^2 z^2). \tag{IV.152a}$$

In the special case of axial symmetry, considered by Mottelson and Nilsson, we have to make the coefficients of y^2 and x^2 the same. This we do by requiring

$$\omega_x = \omega_y = \omega_\perp,$$

and hence

$$\mathcal{V}(\mathbf{r}) = \tfrac{1}{2}M\omega_z^2 z^2 + \tfrac{1}{2}M\omega_\perp^2 \hat{\rho}^2, \tag{IV.152b}$$

where

$$\hat{\rho}^2 = x^2 + y^2.$$

The anharmonic oscillator has to be supplemented by other terms in order to become realistic enough for the average potential of a nucleus. Once again, we can take the clue from the shell model, where a spin-orbit coupling term $\xi(r)\mathbf{l}\cdot\mathbf{s}$ is needed in addition to the spherically symmetric oscillator potential. A term of this type is indeed taken in the single-particle potential of the Nilsson model. One more term is then added to the potential to simulate an important property of single-particle level energies. This term is now explained. The isotropic harmonic oscillator potential produces single-particle level energies having a high degree of degeneracy; in particular, all levels having n- and l-quantum number such that $(2n + l)$ is a given constant have the same energy. In the $(0d, 1s)$-shell, for example, the energies of $0d$- and $1s$-level before the switching on of the spin-orbit coupling are the same. The experimental data on the levels of O^{17}, however, tell us otherwise. The observed positions of the single-particle levels with respect to $(0d_{5/2})$ are

$$\epsilon(0d_{3/2}) - \epsilon(0d_{5/2}) = 5.08 \text{ MeV}, \qquad \epsilon(1s_{1/2}) - \epsilon(0d_{5/2}) = 0.87 \text{ MeV}.$$

The spin-orbit splitting of a level l is equal to

$$\epsilon(l - \tfrac{1}{2}) - \epsilon(l + \tfrac{1}{2}) = -\tfrac{1}{2}(2l + 1)\langle\xi(r)\rangle,$$

where $\langle\xi(r)\rangle$ is the radial integral of $\xi(r)$ connecting the single-particle radial wavefunction. $\xi(r)$ is attractive, and hence we shall denote $-\langle\xi(r)\rangle$ by a positive number C. Using 5.08 MeV as the observed splitting for $l = 2$, we obtain $C = 2.03$ MeV. The contribution of spin-orbit coupling to the energy of the $(0d_{5/2})$-level is given by

$$\langle\xi(r)\rangle\tfrac{1}{2}[j(j + 1) - l(l + 1) - \tfrac{3}{4}],$$

where $j = 5/2$, and $l = 2$. This amount, with our estimate of C, is equal to -2.03 MeV. Thus, the $0d$-level before the switching on of the spin-orbit coupling must have been above the observed $(0d_{5/2})$-level by 2.03 MeV. The $(1s_{1/2})$-level has no spin-orbit energy, and hence the same position of the $1s$-level is only 0.87 MeV above the level $0d_{5/2}$. Thus, we are led to the conclusion that a term in the single-particle Hamiltonian is needed to supplement the isotropic oscillator potential, which splits up the $0d$- and $1s$-level by the observed amount $(2.03 - 0.87 =)$ 1.16 MeV. In the Nilsson model, this role is played by a term in the Hamiltonian taken in the form Dl^2, where D is a parameter to be determined from experimental data. In our example, Dl^2 for the level $0d$ is $6D$, whereas the same for the level $1s$ is zero. Thus, $6D$ is required to be equal to 1.16 MeV, which determines $D = 0.193$ MeV. We shall return to a discussion of the terms $\xi(r)\boldsymbol{l}\cdot\mathbf{s}$ and Dl^2 in the Nilsson Hamiltonian later in this section. For the present, we go back to our treatment of $\mathcal{V}(\mathbf{r})$.

The Schrödinger equation for the anharmonic oscillator potential easily splits into three separate equations for the x-, y-, z-direction. We shall use a subscript $i\ (= x, y, z)$ to denote the wavefunction and energy ϵ for any of these coordinates. In the usual way, a dimensionless coordinate ξ_i is introduced, which is obtained from the corresponding coordinate x, y, or z by multiplying with the constant $(M\omega_i/\hbar)^{1/2}$. The energy ϵ_i is also taken in units of $\hbar\omega_i$. The separated equation for any of the three coordinates ξ_i is then given by

$$\left(-\frac{d^2}{d\xi_i^2} + \xi_i^2\right)\psi_i = 2\epsilon_i\psi_i. \tag{IV.153a}$$

The total energy $\mathcal{E}$ of the level is given by

$$\mathcal{E} = \sum_i \hbar\omega_i\epsilon_i, \tag{IV.153b}$$

and the single-particle wavefunction ψ by

$$\psi = \prod_i \psi_i. \tag{IV.153c}$$

The one-dimensional oscillator equation (IV.153a) can be solved in the usual way, leading to the eigenvalues $\epsilon_i = n_i + \frac{1}{2}$. The single-particle wavefunction is a product $\psi_x\psi_y\psi_z$ labelled by the quantum numbers n_x, n_y, n_z.

A more convenient representation for the present problem is obtained by treating the x- and y-equation together [in view of the symmetry about the z-axis of the Nilsson potential (IV.152b)] and the z-equation alone. The z-equation then produces the eigenvalues $\epsilon_z = n_z + \frac{1}{2}$ and the eigenfunction $\psi_{n_z}(\xi_z)$. The combined equation for the x-, y-direction is solved in terms of the variable $\hat{\rho}$ $(= \sqrt{x^2 + y^2})$ and the angle ϕ between $\hat{\rho}$ and the x-axis. Thus,

$$x = \hat{\rho}\cos\phi, \qquad y = \hat{\rho}\sin\phi.$$

The combined (x, y)-equation, given by

$$\left[-\frac{\hbar^2}{2M}\left(\frac{d^2}{dx^2} + \frac{d^2}{dy^2}\right) + \tfrac{1}{2}M\omega_\perp^2(x^2 + y^2)\right]\psi_\perp(x, y) = \mathcal{E}_\perp\psi_\perp(x, y),$$

easily transforms into

$$\left[-\left(\frac{\partial^2}{\partial\xi_x^2} + \frac{\partial^2}{\partial\xi_y^2}\right) + (\xi_x^2 + \xi_y^2)\right]\psi_\perp(\xi_x, \xi_y) = 2\epsilon_\perp\psi(\xi_x, \xi_y) \tag{IV.154a}$$

or

$$\left[-\left(\frac{\partial^2}{\partial\rho^2} + \frac{1}{\rho}\frac{\partial}{\partial\rho} + \frac{1}{\rho^2}\frac{\partial^2}{\partial\phi^2}\right) + \rho^2\right]\psi_\perp(\rho, \phi) = 2\epsilon_\perp\psi_\perp(\rho, \phi). \tag{IV.154b}$$

Here $\epsilon_\perp$ is the energy $\mathcal{E}_\perp$ in units of $\hbar\omega_\perp$, and ρ is the dimensionless variable $(M\omega_\perp/\hbar)^{1/2}\hat{\rho}$. This equation easily separates into two equations: one for the variable ρ and the other for the variable ϕ. The latter produces a solution $(1/\sqrt{2\pi})e^{i\Lambda\phi}$ in which Λ can be 0, ±1, ±2, ... in order that the function is single-valued in ϕ. The equation for ρ can then be obtained from (IV.154b) by substituting

$$\psi_\perp(\rho, \phi) = \frac{1}{\sqrt{2\pi}} e^{i\Lambda\phi} R(\rho) \tag{IV.155}$$

such that

$$[-(\frac{d^2}{d\rho^2} + \frac{1}{\rho}\frac{d}{d\rho} - \frac{\Lambda^2}{\rho^2}) + \rho^2]R(\rho) = \epsilon_\perp R(\rho). \tag{IV.156}$$

We have dealt with this equation earlier [see (IV.102)] and found that $\epsilon_\perp$ has the eigenvalues given by

$$\epsilon_\perp = (2n_\perp + \Lambda + 1) \quad \text{with } n_\perp = 0, 1, 2, \ldots; \Lambda = 0, \pm1, \pm2, \ldots. \tag{IV.157a}$$

The corresponding radial wavefunction $R(\rho)$ is labelled by the quantum numbers $n_\perp$ and Λ, and is given by [see (IV.104a) and (IV.104b)]

$$R_{n_\perp\Lambda}(\rho) = \rho^\Lambda \exp(-\tfrac{1}{2}\rho^2) F(-n_\perp, \tfrac{1}{2}(2\Lambda + 1); \rho^2), \tag{IV.157b}$$

where the confluent hypergeometric series F reduces to a polynomial by the choice (IV.157a).

The solutions of the nonisotropic oscillator problem can now be summarized. The eigenstates are of the type $|n_z, n_\perp, \Lambda\rangle$ and the eigenvalue corresponding to this eigenstate is given by

$$\mathcal{E}(n_z, n_\perp, \Lambda) = \hbar\omega_z(n_z + \tfrac{1}{2}) + \hbar\omega_\perp(2n_\perp + \Lambda + 1). \tag{IV.158}$$

We now write down the full Nilsson model Hamiltonian of the particle and try to solve the eigenvalues and eigenfunctions with the help of our knowledge of the solutions of the deformed oscillator problem. The particle Hamiltonian is taken in the form

$$\mathcal{H}_p = T_p + \tfrac{1}{2}M\omega_z^2 z^2 + \tfrac{1}{2}M\omega_\perp^2(x^2 + y^2) + C\mathbf{l}\cdot\mathbf{s} + D\mathbf{l}^2, \tag{IV.159}$$

where the quantities C and D are treated as parameters. In terms of ξ_i, we reexpress (IV.159) as

$$\mathcal{H}_p = [\tfrac{1}{2}\hbar\omega_z(-\frac{\partial^2}{\partial\xi_z^2} + \xi_z^2) + \tfrac{1}{2}\hbar\omega_\perp(-\frac{\partial^2}{\partial\xi_x^2} + \xi_x^2 - \frac{\partial^2}{\partial\xi_y^2} + \xi_y^2)] + C\mathbf{l}\cdot\mathbf{s} + D\mathbf{l}^2. \tag{IV.160}$$

Introducing the spin wavefunction $|\tfrac{1}{2}\Sigma\rangle$, where Σ denotes the projection $+\tfrac{1}{2}$ or $-\tfrac{1}{2}$ of the single-particle spin, we can characterize the eigenstates of the part of $\mathcal{H}_p$ enclosed within the square brackets by $|n_z, n_\perp, \Lambda, \Sigma\rangle$. It is possible to use these eigenstates as the basis, calculate the matrix elements of $C\mathbf{l}\cdot\mathbf{s}$- and $D\mathbf{l}^2$-term in that basis, and then diagonalize the matrix for $\mathcal{H}_p$. In his original paper, Nilsson did not do the diagonalization in this basis. He replaced $\mathbf{l}$ by the operator $\mathbf{l}_t$, where $\mathbf{l}_t$ is the orbital angular momentum operator constructed with the coordinates ξ_i and the corresponding momenta $-i(\partial/\partial\xi_i)$ in perfect analogy with the expression of $\mathbf{l}$ in terms of x, y, z and the momenta $-i(\partial/\partial x)$. For example, he used

$$(l_t)_z = -i\xi_x\frac{\partial}{\partial\xi_y} + i\xi_y\frac{\partial}{\partial\xi_x}.$$

The spin-orbit coupling and $\mathbf{l}^2$-terms were thus replaced by $C\mathbf{l}_t\cdot\mathbf{s}$ and $D\mathbf{l}_t^2$. The diagonal matrix elements of these two operators are easy to calculate in the deformed oscillator basis $|n_z, n_\perp, \Lambda, \Sigma\rangle$, and have been done in an appendix to Nilsson's paper. In the case of a very strong deformation, the terms $C\mathbf{l}_t\cdot\mathbf{s}$ and $D\mathbf{l}_t^2$ may be treated as small compared with the anisotropic oscillator Hamiltonian, enclosed within the square brackets in (IV.160). Thus, the first-order perturbation treatment of these two terms, which require only the diagonal matrix elements, is sufficient. Nilsson refers to the states $|n_z, n_\perp, \Lambda, \Sigma\rangle$, and the corresponding energies calculated in a first-order perturbation treatment of $C\mathbf{l}_t \cdot \mathbf{s} + D\mathbf{l}_t^2$, as asymptotic solutions for large deformation. We shall not enter into any more details on this subject.

For small and moderate deformations, the nondiagonal matrix elements of $C\mathbf{l}_t \cdot \mathbf{s}$ and $D\mathbf{l}_t^2$ should be kept, and we have also to take into account the effect of $C(\mathbf{l} - \mathbf{l}_t) \cdot \mathbf{s}$ and $D(\mathbf{l}^2 - \mathbf{l}_t^2)$ because the original Hamiltonian contained $(C\mathbf{l} \cdot \mathbf{s} + D\mathbf{l}^2)$ and *not* the terms with $\mathbf{l}_t$ in place of $\mathbf{l}$. Nilsson used a different set of basis states for a complete calculation. To introduce these states, we first use a deformation parameter ϵ, and define it in terms of $\omega_x, \omega_y, \omega_z$ as

$$\omega_z = \omega_0(\epsilon)(1 - \tfrac{2}{3}\epsilon), \tag{IV.161a}$$

$$\omega_x = \omega_y = \omega_\perp = \omega_0(\epsilon)(1 + \tfrac{1}{3}\epsilon). \tag{IV.161b}$$

The equipotential surfaces of the deformed potential

$$\tfrac{1}{2}M\omega_z^2 z^2 + \tfrac{1}{2}M\omega_\perp^2(x^2 + y^2)$$

are clearly spheroidal surfaces whose volume is equal to abc, where a, b, c are the lengths of the three semi-axes. If we are to consider an equipotential surface on which the potential is equal to N^2 everywhere, then we have

$$N^2 = \tfrac{1}{2}M\omega_z^2 z^2 + \tfrac{1}{2}M\omega_\perp^2(x^2 + y^2)$$

or

$$\frac{x^2}{a^2} + \frac{y^2}{b^2} + \frac{z^2}{c^2} = 1$$

with

$$a = N\omega_\perp^{-1}\sqrt{\frac{2}{M}} = b, \qquad c = N\omega_z^{-1}\sqrt{\frac{2}{M}}.$$

Thus, the volume enclosed by the equipotential surface is

$$N^3\omega_\perp^{-2}\omega_z^{-1}\left(\frac{2}{M}\right)^{3/2} = N^3\left(\frac{2}{M}\right)^{3/2}\omega_0^{-3}(\epsilon)(1 + \tfrac{1}{3}\epsilon)^{-2}(1 - \tfrac{2}{3}\epsilon)^{-1}.$$

If we require a constant volume for this surface as the deformation changes, we need

$$\omega_0^3(\epsilon)(1 + \tfrac{1}{3}\epsilon)^2(1 - \tfrac{2}{3}\epsilon) = \text{constant} = (\tilde{\omega}_0)^3,$$

where $(\tilde{\omega}_0)^3$ is the notation used for the deformation-independent constant. Therefore,

$$\omega_0(\epsilon) = \tilde{\omega}_0[(1 + \tfrac{1}{3}\epsilon)^2(1 - \tfrac{2}{3}\epsilon)]^{-1/3}$$

$$= \tilde{\omega}_0(1 - \tfrac{1}{3}\epsilon^2 - \tfrac{2}{27}\epsilon^3)^{-1/3}. \tag{IV.161c}$$

We now go back to (IV.160) and make the substitutions (IV.161a) and (IV.161b). The first terms of $\omega_x, \omega_y, \omega_z$ then combine to produce an isotropic oscillator Hamiltonian for the coordinates ξ_i, which we denote by H_0. The second terms of $\omega_x, \omega_y, \omega_z$ combine to produce an expression that will be denoted by $H'(\epsilon)$. Thus,

$$\mathcal{H}_p = H_0(\epsilon) + H'(\epsilon) + C\mathbf{l}_t \cdot \mathbf{s} + D\mathbf{l}_t^2 + C(\mathbf{l} - \mathbf{l}_t) \cdot \mathbf{s} + D(\mathbf{l}^2 - \mathbf{l}_t^2), \tag{IV.162a}$$

where

$$H_0(\epsilon) = \tfrac{1}{2}\hbar\omega_0(\epsilon)(-\nabla^2 + \rho^2), \tag{IV.162b}$$

$$H'(\epsilon) = \tfrac{1}{6}\epsilon\hbar\omega_0(\epsilon)\left[-2\left(-\frac{\partial^2}{\partial\xi_z^2} + \xi_z^2\right) + \left(-\frac{\partial^2}{\partial\xi_x^2} + \xi_x^2\right) + \left(-\frac{\partial^2}{\partial\xi_y^2} + \xi_y^2\right)\right]. \tag{IV.162c}$$

57 (45-122/1977)

The solution of the isotropic harmonic oscillator Hamiltonian (IV.162b) is already known from the work in Section 17. The quantum numbers lm occurring in that solution have now to be interpreted in terms of the operator l_t, i.e., $l(l+1)$ is the eigenvalue of l_t^2 and m is the eigenvalue of $(l_t)_z$. Thus, the eigenstates of $H_0(\epsilon)$, including the spin projection Σ, are $|nlm\Sigma\rangle$, and the corresponding eigenvalues are $\hbar\omega_0(\epsilon)(2n + l + \frac{3}{2})$. A treatment of the terms $H'(\epsilon)$, $Cl_t \cdot \mathbf{s}$, and Dl_t^2 occurring in (IV.162a) in this basis is now described. The last two terms of this expression, arising from the difference between l and l_t, have been proved by Nilsson to have very small effects and are therefore not considered any further.

The operator l_t^2 is clearly diagonal in the representation and has the matrix element $l(l+1)$. The spin-orbit coupling $l_t \cdot \mathbf{s}$ has both diagonal and nondiagonal (with respect to m, Σ) matrix elements, which can be written down with the help of the standard results given in Appendix A. We have

$$\begin{aligned}
&\langle nlm'\Sigma'|\, l_t \cdot \mathbf{s}\, |nlm\Sigma\rangle \\
&\quad = \delta_{mm'}\delta_{\Sigma\Sigma'} m\Sigma + \delta_{m',m+1}\delta_{\Sigma',\Sigma-1}\tfrac{1}{2}\langle nl, m+1, \Sigma - 1|\, (l_t)_+ s_-\, |nlm\Sigma\rangle \\
&\qquad + \delta_{m',m-1}\delta_{\Sigma',\Sigma+1}\tfrac{1}{2}\langle nl, m-1, \Sigma+1|\, (l_t)_- s_+\, |nlm\Sigma\rangle \\
&\quad = \delta_{mm'}\delta_{\Sigma\Sigma'} m\Sigma + \delta_{m',m+1}\delta_{\Sigma',\Sigma-1}\tfrac{1}{2}[(\tfrac{1}{2}+\Sigma)(\tfrac{1}{2}-\Sigma+1)(l-m)(l+m+1)]^{1/2} \\
&\qquad + \delta_{m',m-1}\delta_{\Sigma',\Sigma+1}\tfrac{1}{2}[(\tfrac{1}{2}-\Sigma)(\tfrac{1}{2}+\Sigma+1)(l+m)(l-m+1)]^{1/2}.
\end{aligned} \tag{IV.163}$$

Clearly, the second term requires Σ to be $+\frac{1}{2}$, and the last one requires Σ to be $-\frac{1}{2}$.

To complete the derivation, we now have to evaluate the necessary matrix element of $H'(\epsilon)$, as given by (IV.162c). We first notice that the total number of oscillator quanta N in an eigenstate of (IV.162b) is given in two ways: by $(2n + l)$ and by $(n_x + n_y + n_z)$. Thus,

$$2n + l = n_x + n_y + n_z = N. \tag{IV.164a}$$

Now, in any one term of (IV.162c), we have

$$\tfrac{1}{2}\hbar\omega_0(\epsilon)\left(-\frac{\partial^2}{\partial\xi_i^2} + \xi_i^2\right) \equiv \hbar\omega_0(\epsilon)(n_i + \tfrac{1}{2}), \tag{IV.164b}$$

which is diagonal with respect to the particular n_i. The operator is diagonal with respect to the other two n_i-quantum numbers too because the coordinates and derivatives occurring in it do not operate on the wavefunction of the other two coordinates. Thus, the entire operator (IV.162c) keeps each of n_x, n_y, n_z unaltered, and hence it is diagonal with respect to $(n_x + n_y + n_z)$ or $(2n + l)$, i.e., the number of oscillator quanta N. Further, we know that the diagonal matrix element of the kinetic energy operator and that of the potential energy operator of a harmonic oscillator are equal. Hence, the nonvanishing matrix elements of (IV.162c) can be written as

$$\begin{aligned}
\langle n'l'm'\Sigma'|\, H'(\epsilon)\, |nlm\Sigma\rangle &= \delta_{NN'}\delta_{\Sigma\Sigma'}\tfrac{1}{3}\epsilon\hbar\omega_0(\epsilon)\langle n'l'm'\Sigma'|\, (\xi_x^2 + \xi_y^2 - 2\xi_z^2)\, |nlm\Sigma\rangle \\
&= \delta_{NN'}\delta_{\Sigma\Sigma'}\frac{\epsilon}{3}\hbar\omega_0(\epsilon) 2\sqrt{\frac{4\pi}{5}}\, \langle n'l'm'\Sigma'|\, \rho^2 Y_0^2(\theta_\rho, \phi_\rho)\, |nlm\Sigma\rangle \\
&= \delta_{NN'}\delta_{\Sigma\Sigma'}\delta_{mm'}\tfrac{2}{3}\epsilon\hbar\omega_0(\epsilon)\sqrt{\frac{[l]}{[l']}}\begin{bmatrix} l & 2 & l' \\ m & 0 & m' \end{bmatrix}\begin{bmatrix} l & 2 & l' \\ 0 & 0 & 0 \end{bmatrix} \\
&\quad \times \int_0^\infty R_{n'l'}(\rho)\rho^4 R_{nl}(\rho)\, d\rho.
\end{aligned} \tag{IV.165}$$

In the first step, we have taken into account the kinetic energy terms [i.e., the derivatives occurring in (IV.162c)] by the inclusion of a factor of 2. In the second step, we have replaced $(\xi_x^2 + \xi_y^2 - 2\xi^2)$ in terms of the spherical harmonic of the angles θ_ρ, ϕ_ρ of the vector $\boldsymbol{\rho}$, whose components are ξ_x, ξ_y, ξ_z. In the last step, we have written down the matrix element by using the standard Racah results from Appendix B.

We notice from (IV.163) and (IV.165) that $\Omega = m + \Sigma$ is a conserved quantity in the matrix elements, i.e., $m + \Sigma = m' + \Sigma'$. (IV.165) conserves m and Σ independently, but (IV.163) does not.

The diagonalization programme then entails choosing a value of the oscillator quantum number N, and considering all the states in it with the various quantum numbers n, l, m. Let us take a given *total* projection quantum number Ω, and check in how many ways Ω can be formed from m and Σ. Taking all the basic states with these values of m and Σ and the permitted values of n, l consistent with m, we set up the matrix elements in the basis of all these states and diagonalize. The procedure has to be repeated for each N for all possible values of Ω. Nilsson obtained the single-particle eigenvalues and eigenfunctions in this way and extensively tabulated them. The level diagrams are reproduced in Fig. IV.10.

It is also possible to consider the angular momentum coupled states $|nlj\Omega\rangle$ as the basis for the Nilsson diagonalization. Here the resultant angular momentum $\mathbf{j}$ is the vector sum of $\boldsymbol{l}_t$ and $\mathbf{s}$. The operator $\boldsymbol{l}_t \cdot \mathbf{s}$ is diagonal in this representation and has the matrix element $\frac{1}{2}[j(j+1) - l(l+1) - \frac{3}{4}]$. The matrix element of $H'(\epsilon)$ in this new basis can also be easily written down with the help of the standard Racah result (BIII.17) from Appendix B as

$$\langle n'l'j'\Omega'|\, H'(\epsilon)\, |nlj\Omega\rangle = \delta_{NN'}\delta_{\Omega\Omega'}\tfrac{2}{3}\epsilon\hbar\omega_0(\epsilon)\begin{bmatrix} j & 2 & j' \\ \Omega & 0 & \Omega \end{bmatrix}\begin{bmatrix} j' & 2 & j \\ \frac{1}{2} & 0 & \frac{1}{2} \end{bmatrix}\tfrac{1}{2}[1 + (-1)^{l+l'}] \times \int_0^\infty R_{n'l'}\rho^4 R_{nl}\, d\rho. \quad \text{(IV.166)}$$

The level diagrams in Fig. IV.10 show the single-particle level energies as a function of a deformation parameter δ which is different from the deformation parameter ϵ and will shortly be defined. Each energy level is labelled on its right and left by the quantum number Ω and its parity. It is also marked by a serial number 2, 3, 4, ... on the extreme left; these serial numbers were used by Nilsson for the identification of the various levels. The abscissa point $\delta = 0$ corresponds to zero deformation of the potential, and at that point the positions of the energy levels are the same as those of the spherical shell-model. In fact, the parameters C and D were chosen by satisfying this condition as best as possible.

The deformation parameter δ occurs in Nilsson's original programme of diagonalization, and the parameter ϵ is in the programme described in an appendix to his paper. We have here followed the latter programme of diagonalization. In the original programme, δ is defined by the relations

$$\begin{aligned} \omega_z &= \omega_0(\delta)(1 - \tfrac{4}{3}\delta)^{1/2}, \\ \omega_x &= \omega_y = \omega_\perp = \omega_0(\delta)(1 + \tfrac{2}{3}\delta)^{1/2}. \end{aligned} \quad \text{(IV.167)}$$

The condition of the constant volume equipotential surfaces then gives

$$\omega_0^3(\delta)(1 + \tfrac{2}{3}\delta)(1 - \tfrac{4}{3}\delta)^{1/2} = \text{constant} = (\tilde{\omega}_0)^3$$

or

$$\begin{aligned} \omega_0(\delta) &= \tilde{\omega}_0[(1 + \tfrac{2}{3}\delta)^2(1 - \tfrac{4}{3}\delta)]^{-1/6} \\ &= \tilde{\omega}_0[1 - \tfrac{4}{3}\delta^2 + \tfrac{16}{27}\delta^3]^{-1/6}. \end{aligned} \quad \text{(IV.168)}$$

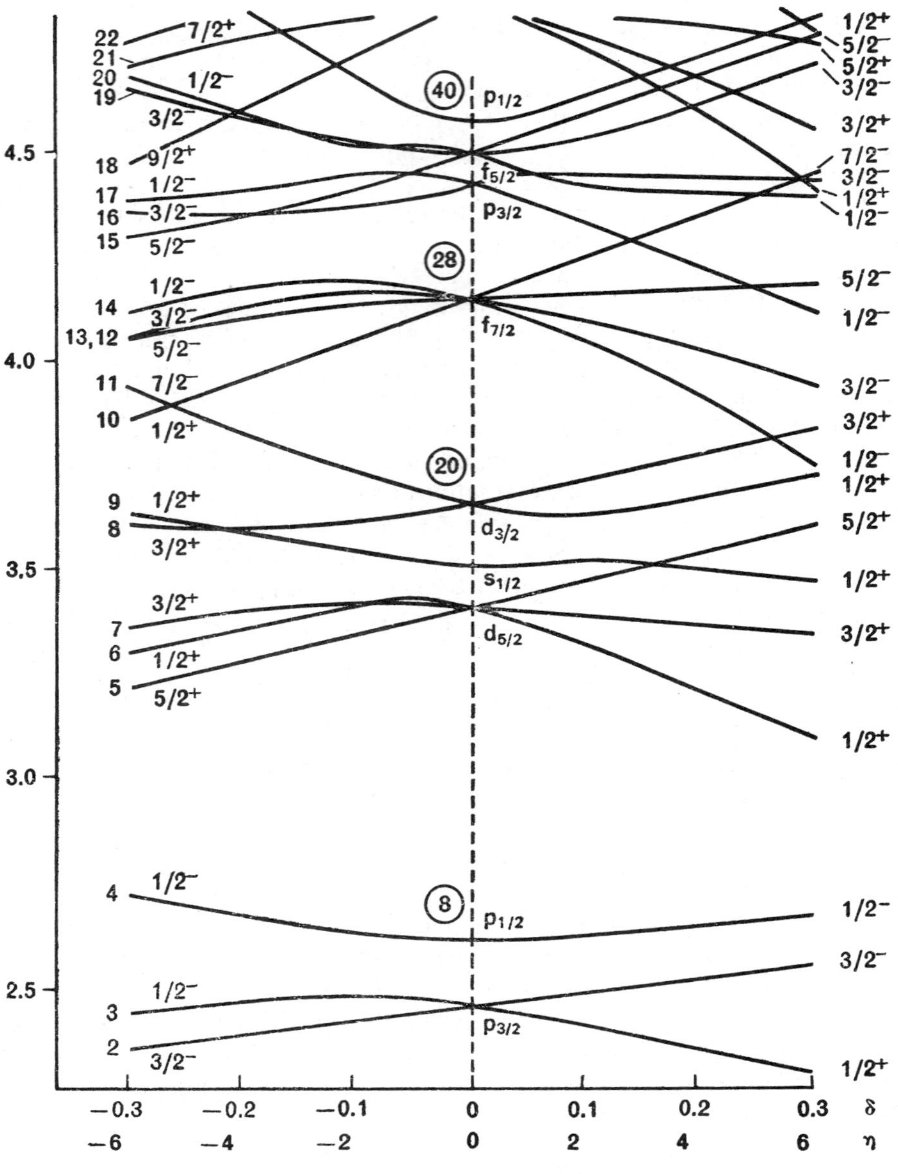

(a) First few major shells

Fig. IV.10 Nilsson levels calculated as function of deformation parameter. [Following Nilsson, S. G., *Kgl. Danske Videnskab. Selskab, Mat. Fys. Medd.*, **29**, No. 16, 68 (1955).]

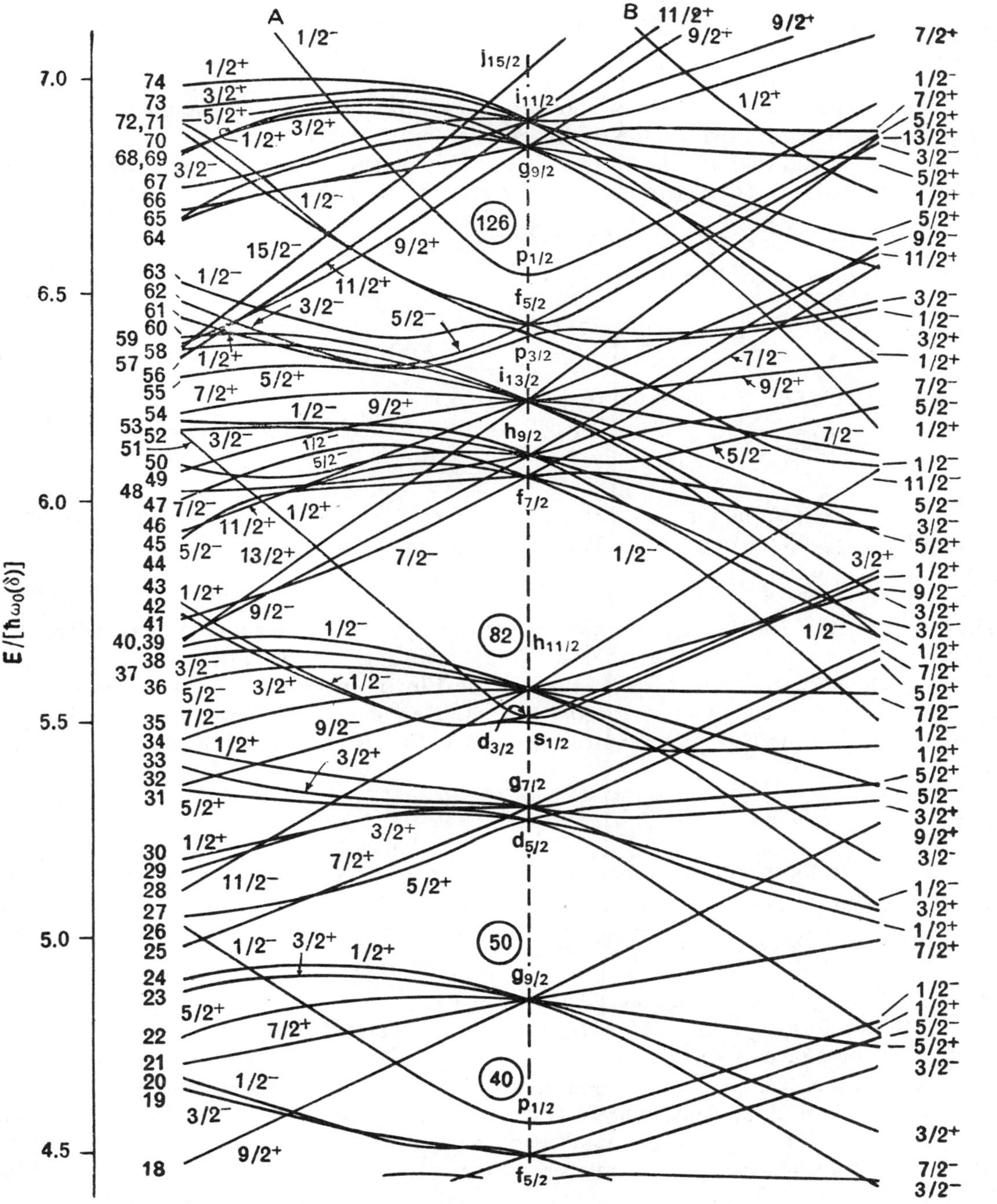

(b) Higher major shells

Fig. IV.10 Nilsson levels calculated as function of deformation parameter. [Following Nilsson, S. G., *Kgl. Danske Videnskab. Selskab, Mat. Fys. Medd.*, 29, No. 16, 68 (1955).]

It is left to the reader to establish that ϵ and δ are related by

$$\epsilon = \delta + \tfrac{1}{6}\delta^2 + O(\delta^3). \tag{IV.169}$$

The Nilsson results have been applied in computing many interesting physical properties of nuclei. We first note that each Nilsson level is doubly degenerate. This is because a level of projection Ω and that of $-\Omega$ result from the diagonalization of the same matrix. This can be checked from (IV.166) by comparing it with the matrix element

$$(-1)^{j'-\Omega'}\langle n'l'j', -\Omega'|\, H'(\epsilon)\, |nlj, -\Omega\rangle(-1)^{j-\Omega}.$$

The Clebsch-Gordon coefficient

$$\begin{bmatrix} j & 2 & j' \\ -\Omega & 0 & -\Omega' \end{bmatrix}$$

in this matrix element can be easily converted into

$$\begin{bmatrix} j & 2 & j' \\ \Omega & 0 & \Omega \end{bmatrix}$$

by including a phase factor $(-1)^{j+2-j'}$ which cancels the phase factors $(-1)^{j'-\Omega'}(-1)^{j-\Omega}$ by virtue of the $\delta_{\Omega\Omega'}$-factor in (IV.166). Thus, we have demonstrated

$$\langle n'l'j'\Omega'|\, H'(\epsilon)\, |nlj\Omega\rangle = (-1)^{j'-\Omega'}\langle n'l'j', -\Omega'|\, H'(\epsilon)\, |nlj, -\Omega\rangle(-1)^{j-\Omega}. \tag{IV.170}$$

The other terms of the Hamiltonian give matrix elements which are actually independent of Ω. For a negative projection $-\Omega$, we shall use $(-1)^{j-\Omega}|nlj, -\Omega\rangle$ as the basis states. In this basis, the Hamiltonian matrix is identical to that obtained for positive Ω in the basis $|nlj\Omega\rangle$. Thus, the eigenvalues and eigenvectors for the two cases are identical. Each eigenvector gives the expansion coefficients C_{nlj} in the expression of an eigenstate in terms of the basis states. Thus, if

$$\psi_\Omega = \sum_{nlj} C_{nlj}|nlj\Omega\rangle \tag{IV.171a}$$

is an eigenstate for positive Ω, then the state degenerates with it and, having the projection $-\Omega$, is given by

$$\psi_{-\Omega} = \sum_{nlj} C_{nlj}(-1)^{j-\Omega}|nlj, -\Omega\rangle \tag{IV.171b}$$

with the same expansion coefficients.

With the foregoing knowledge, let us now consider the intrinsic wavefunction in the ground state of a nucleus. We have to fill up the Nilsson levels at the appropriate deformation from the bottom of the level diagram, putting two neutrons (corresponding to $+\Omega$ and $-\Omega$) and two protons in each level. Thus, when a Nilsson level has its full quota of neutrons or protons, the sum of the projection quantum numbers of the nucleons in that level is zero. If both the neutron number and the proton number of the nucleus are even, then clearly the band quantum number K of the determinantal wavefunction χ_K, using the Nilsson orbitals in the manner described, must be zero. It should be noted that this is a definite prediction drawn from the Nilsson model. We repeat: according to this model, the band quantum number of the ground-state band of an even-even nucleus is zero. In an excited intrinsic state, we may promote one or more nucleons from its level (the hole level) in χ_K to another excited Nilsson level (particle level). Therefore, for such an excited intrinsic state, the band quantum number is $\Omega' - \Omega$, where Ω and Ω' are the projection quantum numbers of the hole and particle levels, respectively. If more than one particle has been excited, then all the corresponding $(\Omega' - \Omega)$ have to be summed to obtain the band quantum number of the excited intrinsic state.

In an odd-mass nucleus, if we are interested in the *ground-state band*, all the nucleons in the even-even core contribute zero to the band quantum number, and then the quantum number Ω of the last Nilsson level to which the last odd nucleon goes gives the band quantum number K of the ground-state band. The ground-state spin I of the odd-mass nucleus is also necessarily equal to this K. [For $K = \frac{1}{2}$, this may not always be true due to the decoupling term in (IV.145).] Thus, the Nilsson model can be, and has been, used in predicting the ground-state spins of odd-mass nuclei.

In order to get the sequence in which the Nilsson levels are to be occupied by nucleons, it is essential to know in advance the correct deformation of the nucleus. This is because the Nilsson levels do cross over one another when there are changes in the deformation parameter. The equilibrium deformation of a nucleus can be experimentally determined from its electromagnetic properties (see Section 34B) and used for the purpose just mentioned. However, an independent theoretical method can be tried with the Nilsson model itself. We know from (II.92) that the ground-state energy of a many-nucleon system described by a determinantal wavefunction is given by the sum of the kinetic energies of all the occupied single-nucleon states and the sum of the potential energies of all the occupied *pairs* of single-particle states (which interact through the two-nucleon interaction). In terms of the two-nucleon interactions, we can introduce an average one-body potential; then the total potential energy due to the two-body interaction becomes equal to half of the potential energy due to the average one-body potential. Thus, in the case of the Nilsson model, which is a model based on an average one-body potential, the total energy of the ground-state determinant is given by

$$E = \sum_{\alpha} (T_\alpha + \tfrac{1}{2}\mathcal{V}_\alpha),$$

where the sum is over all the occupied Nilsson levels, and $\mathcal{V}$ is the one-body potential in the Nilsson Hamiltonian. This quantity can be computed as a function of the deformation parameter using the Nilsson values of the energy levels for each deformation. When E is plotted as a function of the deformation parameter, it exhibits a minimum point, and the deformation corresponding to that point is then to be interpreted as the theoretically predicted equilibrium deformation. Figure IV.11 depicts the calculated equilibrium deformation for a large number of deformed nuclei and the comparison with experimental values derived from electromagnetic properties (see Section 34B). The hollow circles and the crosses represent the observed deformations for even-even and odd-mass nuclei, respectively. The solid line gives the calculated values.

The Nilsson model has also been applied to compute several intrinsic quantities which have already occurred in our calculations. For example, the decoupling parameter a, occurring in the expression of energy of a $(K = \frac{1}{2})$-band, and defined by (IV.144), can be very easily computed by taking the coefficients C_{nlj} from the Nilsson results. In Section 34, we shall encounter several such intrinsic quantities, which can be computed with the Nilsson wavefunctions. We shall draw the reader's attention to them at the appropriate stage.

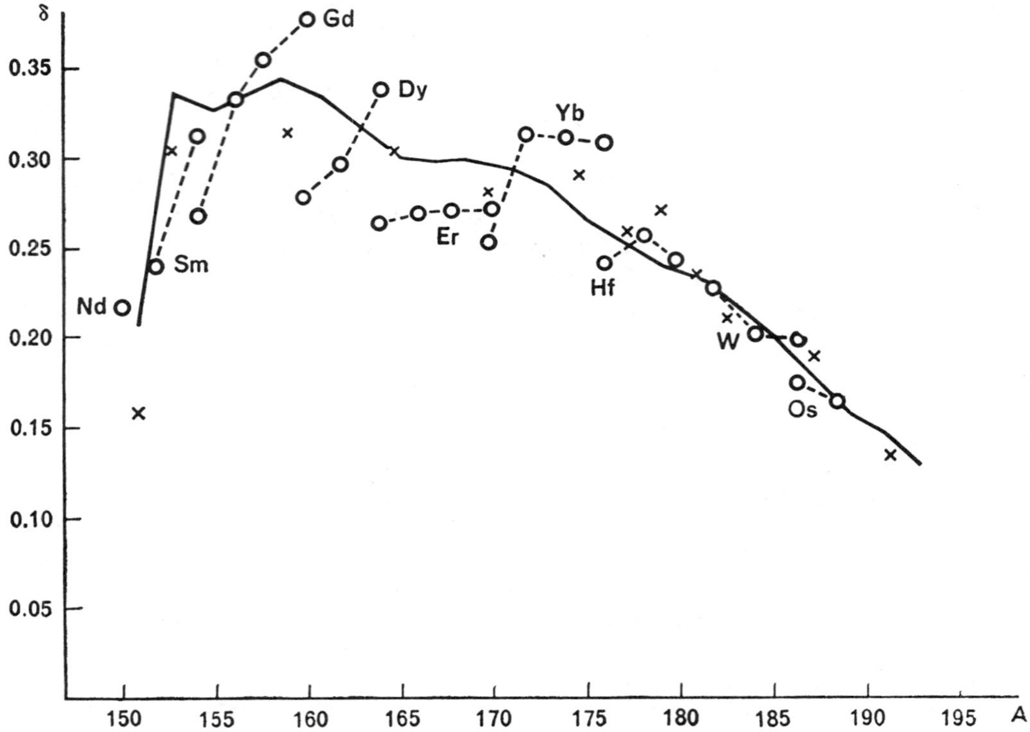

Fig. IV.11 Comparison of equilibrium deformations of rare-earth nuclei (calculated from Nilsson model) with experimental values. [Following Mottelson, B. R., and Nilsson, S. G., *Kgl. Danske Videnskab. Selskab, Mat. Fys. Skrifter*, **1**, No. 8, 76 (1959).]

34. ELECTROMAGNETIC PROPERTIES IN UNIFIED MODEL

A. SPHERICAL VIBRATIONAL NUCLEI

We shall consider for spherical vibrational nuclei the magnetic dipole operator and the electric quadrupole operator. If the even-even nucleus has the rotation operator **R**, then, associated with this angular momentum operator, we expect to get a magnetic dipole moment due to the rotational motion of the nuclear droplet as a whole. Let g_R be the gyromagnetic ratio for this magnetic moment. In the case of an odd-mass nucleus, there can be an additional magnetic moment arising from the angular momentum **j** of the last odd nucleon. Thus,

$$\boldsymbol{\mu} = g_R\mathbf{R} + g_j\mathbf{j}. \tag{IV.172}$$

The ground state of the even nucleus has $I = 0$, and hence the expectation value of $g_R\mathbf{R}$ in this state is zero. Therefore, this state does not have a static magnetic dipole moment. An excited state of the same nucleus having $I \neq 0$ can, however, have a nonvanishing static magnetic moment, provided $g_R \neq 0$. The value of the magnetic moment is given by

$$\langle I, M = I|\, g_R R_Z \,|I, M = I\rangle = g_R I. \tag{IV.173}$$

In the case of the odd-mass nucleus,

$$\mathbf{I} = \mathbf{R} + \mathbf{j}.$$

Therefore, for any state of total angular momentum I, obtained by coupling the core state J_c with the particle state j, we have the magnetic moment given by

$$\langle J_c, j : I, M = I | \, g_R R_Z + g_j j_Z \, | J_c, j : I, M = I \rangle$$

$$= I\left(g_R \frac{\langle \mathbf{R} \cdot \mathbf{I} \rangle}{\langle I^2 \rangle} + g_j \frac{\langle \mathbf{j} \cdot \mathbf{I} \rangle}{\langle I^2 \rangle}\right)$$

$$= \frac{1}{2(I+1)}[g_R\{I(I+1) + J_c(J_c+1) - j(j+1)\} + g_j\{I(I+1) + j(j+1) - J_c(J_c+1)\}]. \tag{IV.174}$$

We have used here the operator identity for the *diagonal* matrix element of any vector operator $\mathbf{A}$, which asserts

$$\langle \mathbf{A} \rangle = \langle \mathbf{J} \rangle \frac{\langle \mathbf{A} \cdot \mathbf{J} \rangle}{\langle J^2 \rangle},$$

where $\mathbf{J}$ is the angular momentum operator. (IV.174) easily simplifies to

$$\mu = \tfrac{1}{2}(g_R + g_j)I + \tfrac{1}{2}(g_R - g_j)\frac{1}{I+1}[J_c(J_c+1) - j(j+1)]. \tag{IV.175}$$

The gyromagnetic ratio g_R is usually treated as a parameter in data fitting; it can be roughly approximated by (Z/A) nuclear magnetons, where Z and A are respectively the charge and the mass number of the even core. This value is obtained from a particular consideration: when a proton of charge e and mass M rotates, it gives a magnetic moment $[e\hbar/(2Mc)]$, i.e., 1 nM; therefore, the nuclear core of charge Ze and mass MA gives rise to $[Ze\hbar/(2MAc)]$ or Z/A nM.

We next discuss the possibility of an M1-transition in the nuclei just mentioned. The M1-transition operator is equal to the operator $\boldsymbol{\mu}$ multiplied by a numerical factor $\sqrt{3/(4\pi)}$. For an even nucleus, the different excited states are, as per our assumption, *different* states of vibration, which are necessarily orthogonal to each other. The operator $\boldsymbol{\mu} = g_R \mathbf{R}$ cannot connect two such orthogonal states because the angular momentum operator $\mathbf{R}$ connects its eigenstate of angular momentum I and projection M with itself or other M-members of the same multiplet; it cannot connect two orthogonal states of the same I. Thus, even though the ordinary selection rules for an M1-transition (I_i, I_f, 1 satisfy the angular momentum coupling rule, and the parity of the initial and final states is the same) are satisfied between the two-phonon 2^+ level and the one-phonon 2^+ level, this transition is forbidden in our vibrational model.

The forbiddenness of the M1-transition holds good for the odd-mass nuclei as well. Here the initial and final states are of the type $|J_c, j : I_i M_i\rangle$ and $|J_c', j' : I_f M_f\rangle$. The operator $g_R \mathbf{R}$ of (IV.172), as argued, demands the same core state $J_c = J_c'$ for a nonvanishing matrix element; it also requires the same particle state $j = j'$ because this operator does not act on the particle coordinates. In the same manner, $g_j \mathbf{j}$ of (IV.172) also demands the same core and the same particle states. In other words, we have proved that it is possible to have only a nonvanishing diagonal matrix element of $\boldsymbol{\mu}$ (which gives the static magnetic dipole moment), but it is impossible to have a nondiagonal matrix element, which alone gives rise to the M1-transition.

58 (45-122/1977)

We next consider the electric quadrupole moment which is given by

$$\Omega_m(\mathrm{E2}) = \int_\tau \rho(\mathbf{r}) r^2 Y_m^2(\theta, \phi)\, d^3r, \tag{IV.176}$$

where the integration is over the nuclear volume τ. Our model of the core (with the quadrupole-type deformation only) contains a nuclear fluid of density ρ_0 confined within a surface defined by

$$R(\theta, \phi) = R_0[1 + \xi(\theta, \phi)] \tag{IV.177a}$$

with

$$\xi(\theta, \phi) = \sum_\mu \alpha_\mu^* Y_\mu^2(\theta, \phi). \tag{IV.177b}$$

In the case of spherical vibrational nuclei, the surface has an equilibrium spherical shape of radius R_0, and the deformation represented by $\xi(\theta, \phi)$ is acquired by the nucleus while it oscillates. Using this expression as the upper limit of the r-integration, we obtain, from (IV.176),

$$\begin{aligned}\Omega_m(\mathrm{E2}) &= \rho_0 \int_0^{2\pi} d\phi \int_0^\pi \sin\theta\, d\theta\, Y_m^2(\theta, \phi) \int_0^{R_0(1+\xi)} r^4\, dr \\ &= \rho_0 \frac{R_0^5}{5} \int_0^{2\pi} d\phi \int_0^\pi \sin\theta\, d\theta\, Y_m^2(\theta, \phi)[1 + \xi(\theta, \phi)]^5.\end{aligned} \tag{IV.178}$$

This expression can be evaluated to various orders in ξ, i.e., in the parameters α_μ. We expand

$$[1 + \xi(\theta, \phi)]^5 = 1 + 5\xi(\theta, \phi) + \ldots$$

and notice that the first term unity of this expansion produces zero in the integration over the angles contained in (IV.178). The second term of the expansion, together with (IV.177b), yields the lowest-order expression of $\Omega_m(\mathrm{E2})$, which is

$$\begin{aligned}\Omega_m(\mathrm{E2}) &= \rho_0 R_0^5 \sum_\mu \alpha_\mu \int_0^{2\pi} d\phi \int_0^\pi \sin\theta\, d\theta\, Y_m^2(\theta, \phi) Y_\mu^{2*}(\theta, \phi) \\ &= \rho_0 R_0^5 \alpha_m;\end{aligned} \tag{IV.179}$$

here the orthogonality integral of spherical harmonics has been used. The quantity ρ_0 in this expression can be eliminated by requiring the total charge in the nucleus to be Ze. Thus,

$$\begin{aligned}Ze &= \rho_0 \int_\tau d^3r \\ &= \rho_0 \int_0^{2\pi} d\phi \int_0^\pi \sin\theta\, d\theta \int_0^{R_0(1+\xi)} r^2\, dr \\ &= \rho_0 \frac{R_0^3}{3} \int_0^{2\pi} d\phi \int_0^\pi \sin\theta\, d\theta\, [1 + \xi(\theta, \phi)]^3.\end{aligned}$$

Proceeding as before, we get the lowest-order contribution to this integral from the term unity in the expansion

$$(1 + \xi)^3 = 1 + 3\xi + \ldots$$

and we obtain

$$Ze = \frac{4\pi}{3} \rho_0 R_0^3 \quad \text{or} \quad \rho_0 = \frac{3}{4\pi} Ze R_0^{-3}.$$

Going back to (IV.179), we then obtain

$$\begin{aligned}\Omega_m(\text{E2}) &= \frac{3Ze}{4\pi}R_0^2\alpha_m \\ &= \frac{3Ze}{4\pi}R_0^2\left(\frac{\hbar}{2B\omega}\right)^{1/2}[q_m^\dagger + (-1)^m q_{-m}] \\ &= \frac{3ZebR_0^2}{4\pi\sqrt{2}}[q_m^\dagger + (-1)^m q_{-m}], \end{aligned} \tag{IV.180}$$

where we have substituted for α_m from (IV.121) and have replaced $[\hbar/(B\omega)]^{1/2}$ by the oscillator parameter b.

The operators $q_m^\dagger$ and q_m respectively create and destroy a phonon. Therefore, the electric quadrupole operator (IV.180) cannot give a nonvanishing diagonal matrix element for a vibrational state, i.e., the *static quadrupole moment of any vibrational state is predicted to be zero*. The transition matrix element connecting a vibrational state of n-phonons with another of n'-phonons demands the selection rule $n - n' = \pm 1$. Thus, the two-phonon 2^+ level decays to the first 2^+, and is strictly forbidden to make an E2-transition to the ground state 0^+, which has no phonon. Without the phonon selection rule, which arises from the vibrational nature of the levels, the transition $2_2^+ \to 0^+$ would, from energetic considerations, have been faster than the transition $2_2^+ \to 2_1^+$. Here the subscripts to 2^+ denote the number of phonons.

For an odd-mass nucleus in the vibrational region, we have already computed an expression, namely (IV.131), of the ground-state wavefunction with an admixture of one-phonon states of the core brought in by the core-particle coupling. The structure (IV.180) of the quadrupole operator tells us that the static quadrupole moment $\langle\psi_0|\,\Omega_0(\text{E2})\,|\psi_0\rangle$ is nonvanishing and has the value

$$2C\left(\frac{\hbar\omega}{2C}\right)^{1/2}\sqrt{\frac{5}{4\pi}}\sum_{j'}\tfrac{1}{2}[1+(-1)^{l+l'}]\begin{bmatrix} j' & 2 & j \\ \frac{1}{2} & 0 & \frac{1}{2}\end{bmatrix}\sqrt{\frac{[j']}{[j]}}(\hbar\omega + \epsilon_{j'} - \epsilon_j)^{-1}$$
$$\times\langle n'_\mu = 1, J'_c = 2, j' : J = j, m|\,\Omega_0(\text{E2})\,|n_\mu = 0, J_c = 0, j : J = j, m\rangle. \tag{IV.181a}$$

The matrix element $\mathcal{M}$ occurring in this expression can be evaluated by using the standard Racah result (BIII.14) from Appendix B and (IV.129). Then

$$\mathcal{M} = \frac{3Ze}{4\pi}\left(\frac{\hbar\omega}{2C}\right)^{1/2}R_0^2(-1)^{j'-j}. \tag{IV.181b}$$

We have thus obtained the contribution of core excitation to the static quadrupole moment of the odd-mass nucleus. There can be an additional *particle* contribution to the quadrupole moment from the diagonal matrix element of the particle quadrupole moment operator for the first term of (IV.131). This is left as an exercise. The reader should note the similarity of these calculations and results with those derived in Section 30B from the shell model where the hole-particle excitations of the core were used.

The predictions made so far about the electromagnetic properties of the vibrational levels in even nuclei are seldom found to be exactly true in real nuclei. In recent years, some of the so-called vibrational nuclei, which exhibit a reasonably good vibrational-type spectra (showing even the two-phonon triplet), have been found to possess a large quadrupole moment in the first excited 2^+ state; yet, at the same time, the second 2^+ state decays predominantly to the first 2^+ (Cd^{114} of Fig. IV.8 is itself a good example). That is to say, the cross-over E2-transition from the second 2^+ to the ground state 0^+ is still found to be extremely weak. These two features make the theoretical explanation of such vibrational nuclei very difficult. Suppose a mixture of the one-phonon and two-phonon states is admitted through a residual interaction between the phonons. Then the one-phonon component of the first 2^+ can give a nonvanishing matrix element of the quadrupole moment operator with its two-phonon component. The experimentally observed large quadrupole moment of the 2^+ state requires the admixture coefficients of a one- and two-phonon state to be comparable with each other. The second 2^+ state being orthogonal to the first 2^+ then consists of an orthogonal combination of the one-phonon and two-phonon state. The one-phonon component of the second 2^+ state then causes a significant E2-transition to the ground state. Thus, an observed large quadrupole moment in the first 2^+ state and a weak cross-over E2-transition from the second 2^+ to the ground state are very difficult to understand from the phonon mixing model. For an idealized harmonic vibration, we obtain all the results derived in this section; when we switch on the interaction between phonons, the implication is that we are deviating from the situation of independent harmonic vibrational modes. Thus, the observed experimental data do point to the presence of anharmonicity in the vibration, and a detailed quantitative understanding of these effects has been obtained by various authors in some specific cases.

The electromagnetic data on the vibrational nuclei in most cases deviate from those of a pure harmonic vibrator. The energy-level data, especially on the two-phonon triplet, also show a somewhat whimsical trend. In view of all these facts, the general conclusion emerging in recent years is that there are very few 'good' vibrational nuclei in nature. Most of the so-called vibrational nuclei require an extremely elaborate theory for a proper understanding. This field is an active branch of present-day structure physics. Some trends in the new developments are given in Sections 42 and 43.

B. DEFORMED ROTATIONAL NUCLEI

For deformed rotational nuclei, we shall first discuss the static magnetic dipole moment and electric quadrupole moment. Then we shall consider the case of electromagnetic transition of a general multipole order. The wavefunctions used for the entire calculation are those given by the unified model, i.e., (IV.141) with the determinantal χ_K, and with the j replaced everywhere by J.

Any operator used for a dynamical variable is originally in the coordinate frame S fixed in the laboratory. However, the evaluation of the matrix elements becomes easier if we transform it into the body-fixed coordinate frame S'. The Euler angles specify the orientation of S' with respect to S; we then write

$$\Omega_m^L(S) = \sum_\mu \mathscr{D}_{m\mu}^L(\theta_1, \theta_2, \theta_3)\Omega_\mu^L(S'). \tag{IV.182}$$

The matrix element of this expression connecting $|\mathscr{D}_{MK}^I\chi_K\rangle$ with $\langle\mathscr{D}_{M'K'}^{I'}\chi_{K'}|$ is given by

$$\begin{aligned}&\langle\mathscr{D}_{M'K'}^{I'}\chi_{K'}|\ \Omega_m^L(S)\ |\mathscr{D}_{MK}^I\chi_K\rangle\\ &= \sum_\mu \int_0^{2\pi} d\theta_1 \int_0^{\pi} \sin\theta_2\, d\theta_2 \int_0^{2\pi} d\theta_3\ \mathscr{D}_{M'K'}^{I'*}(\theta_1, \theta_2, \theta_3)\mathscr{D}_{m\mu}^L(\theta_1, \theta_2, \theta_3)\mathscr{D}_{MK}^I(\theta_1, \theta_2, \theta_3)\\ &\quad \times \langle\chi_{K'}|\ \Omega_\mu^L(S')\ |\chi_K\rangle.\end{aligned} \tag{IV.183}$$

We recall that $\chi_K, \chi_{K'}, \ldots$ are determinantal wavefunctions in which the single-particle states have been calculated in the nuclear body-fixed frame S'. Thus, the matrix element of $\Omega_\mu^L(S')$

occurring in (IV.183) can be evaluated in a straightforward manner, using the standard result (II.23). For the present, we simply denote it by

$$\langle \chi_{K'}| \, \Omega^L_\mu(S') \, |\chi_K\rangle = \langle \Omega^L_\mu \rangle^{\text{intrinsic}}_{K',K}. \tag{IV.184}$$

The integral of the $\mathcal{D}$-functions occurring in (IV.183) can be written down from the standard result (BI.20c) of Appendix B. We thus obtain

$$\langle \mathcal{D}^{I'}_{M'K'}\chi_{K'}| \, \Omega^L_m(S) \, |\mathcal{D}^I_{MK}\chi_K\rangle = \frac{8\pi^2}{[I]} \begin{bmatrix} I & L & I' \\ M & m & M' \end{bmatrix} \begin{bmatrix} I & L & I' \\ K & \mu & K' \end{bmatrix} \langle \Omega^L_\mu \rangle^{\text{intrinsic}}_{K',K}. \tag{IV.185}$$

The summation over μ has been omitted because the Clebsch-Gordon coefficient determines it to be $K' - K$. (IV.185) has frequent use in this section.

We now consider the magnetic dipole operator

$$\boldsymbol{\mu} = g_R\mathbf{R} + \boldsymbol{\mu}^{\text{intrinsic}} = g_R\mathbf{I} + (\boldsymbol{\mu}^{\text{intrinsic}} - g_R\mathbf{J}), \tag{IV.186}$$

where the contribution of the rotational motion $\mathbf{R}$ of the nucleus is written in the manner of Section 34A; and the contribution of the intrinsic motion of the nucleons to the magnetic moment is represented by the second term, which, for the present, is just a notation. In the final expression, $\mathbf{R}$ is substituted from the relation $\mathbf{I} = \mathbf{R} + \mathbf{J}$. The static magnetic moment is the expectation value of the Z-component of this vector operator in the state of the highest projection quantum number. Thus, while using (IV.185), we must put $L = 1$, $m = 0$, $I' = I$, $M = M' = I$. In this way,

$$\begin{aligned} &\langle \mathcal{D}^I_{I,K'}\chi_{K'}| \, \mu^1_0(S) \, |\mathcal{D}^I_{IK}\chi_K\rangle \\ &= \frac{8\pi^2}{[I]}\left\{\delta_{KK'}g_RI + \begin{bmatrix} I & 1 & I \\ I & 0 & I \end{bmatrix} \begin{bmatrix} I & 1 & I \\ K & K'-K & K' \end{bmatrix} \langle(\boldsymbol{\mu}^{\text{intrinsic}} - g_R\mathbf{J})_{K'-K}\rangle_{K',K}\right\}. \end{aligned} \tag{IV.187}$$

Here we have directly written the value of the $\mathbf{I}$-term by noting the fact that $\mathcal{D}^I_{MK}$ is an eigenfunction of I_z belonging to the eigenvalue M. The result (IV.185) has been applied to work out the remainder of (IV.186); the value of μ has been explicitly put equal to $K' - K$.

Let us now use the expression (IV.141) for the wavefunction and evaluate its magnetic moment. Four terms of the type (IV.187) arise; of these, the terms K, K and $-K$, $-K$ are equal, and so are the two cross terms K, $-K$ and $-K$, K. For the first two terms, $\mu = K' - K = 0$, whereas for the cross terms, $\mu = K' - K = \pm 2K$. However, the rank of the operator being unity, $|\mu|$ can never exceed 1, and thus the only case where the cross terms contribute is $K = \frac{1}{2}$. Taking all these facts into consideration, we write the final result

$$\begin{aligned} \langle \Psi^I_{IK}| \, \mu_0 \, |\Psi^I_{IK}\rangle = g_RI &+ \begin{bmatrix} I & 1 & I \\ I & 0 & I \end{bmatrix} \begin{bmatrix} I & 1 & I \\ K & 0 & K \end{bmatrix} \langle(\boldsymbol{\mu}^{\text{intrinsic}} - g_R\mathbf{J})_0\rangle_{K,K} + \delta_{K,1/2} \begin{bmatrix} I & 1 & I \\ I & 0 & I \end{bmatrix} \\ &\times \begin{bmatrix} I & 1 & I \\ \frac{1}{2} & -1 & -\frac{1}{2} \end{bmatrix} (-1)^{I-j} \langle(\boldsymbol{\mu}^{\text{intrinsic}} - g_R\mathbf{J})_{-1}\rangle_{-1/2,+1/2}. \end{aligned} \tag{IV.188}$$

In view of our ignorance about $\boldsymbol{\mu}^{\text{intrinsic}}$, we *define* a gyromagnetic ratio g_K such that

$$\langle \boldsymbol{\mu}^{\text{intrinsic}} \rangle_{K,K} = g_K \langle \mathbf{J} \rangle_{K,K}, \tag{IV.189a}$$

where g_K may indeed depend on K. From this relation, we obtain

$$\langle \mu_0^{\text{intrinsic}} \rangle_{K,K} = g_K \langle J_0 \rangle_{K,K} = g_K \langle \chi_K| \, J_z \, |\chi_K\rangle = g_K K. \tag{IV.189b}$$

In the same way, we write

$$\langle\mu_{-1}^{\text{intrinsic}}\rangle_{-1/2,\,+1/2} = g_K\langle J_{-1}\rangle_{-1/2,\,+1/2} = g_K\langle\chi_{-1/2}|\,J_{-1}\,|\chi_{1/2}\rangle. \tag{IV.189c}$$

Using the definitions of (IV.189) in (IV.188) and substituting the Clebsch-Gordon coefficients by their explicit expressions from Table AVII.2 of Appendix A, we finally obtain the static magnetic moment of a deformed rotational nucleus:

$$\langle\Psi_{IK}^{I}|\,\mu_0\,|\Psi_{IK}^{I}\rangle = g_R I + (g_K - g_R)\Big[\frac{K^2}{I+1} + \delta_{K,\,1/2}(-1)^{I+1/2}\frac{[I]}{I+1}\tfrac{1}{4}b_0\Big], \tag{IV.190a}$$

where

$$b_0 = (-1)^{J-1/2}\langle\chi_{-1/2}|\,J_{-1}\,|\chi_{+1/2}\rangle. \tag{IV.190b}$$

The parameter b_0 can be calculated once $\chi_{1/2}$ and $\chi_{-1/2}$ have been explicitly built from the Nilsson model.

The final expressions (IV.190) tell us that the magnetic moment of deformed rotational nuclei can be reproduced by exploiting the three parameters g_R, g_K, and b_0 (in the case of $K = \frac{1}{2}$ only). For a discussion on the detailed fit to data, the reader is referred to the article by Bodenstedt and Rogers[12].

The derivation of the static quadrupole moment can now be done by following steps similar to those applied to the static magnetic moment. The static quadrupole moment operator is given by

$$Q_0 = \sum_i \sqrt{\frac{16\pi}{5}}\,r_i^2 Y_0^2(\theta_i, \phi_i),$$

where the summation is over all the protons in the nucleus. In this case, while using (IV.185), we must put $L = 2$, $m = 0$, $I = I'$, $M = M' = I$. The final result is given by

$$\langle\Psi_{IK}^{I}|\,Q_0\,|\Psi_{IK}^{I}\rangle = \begin{bmatrix} I & 2 & I \\ I & 0 & I\end{bmatrix}\begin{bmatrix} I & 2 & I \\ K & 0 & K\end{bmatrix}\langle Q_0^{\text{intrinsic}}\rangle_{K,\,K} + \begin{bmatrix} I & 2 & I \\ I & 0 & I\end{bmatrix}$$
$$\times\begin{bmatrix} I & 2 & I \\ K & -2K & -K\end{bmatrix}\langle Q_{-2K}^{\text{intrinsic}}\rangle_{-K,\,K}(-1)^{I-J}. \tag{IV.191}$$

The second term arises from the cross matrix elements connecting the term $\mathscr{D}_{M,\,-K}^{I}\chi_{-K}$ with the term $\mathscr{D}_{M,\,K}^{I}\chi_K$ of the wavefunction. From the fact that the ($\mu = -2K$)-component of the quadrupole moment occurs in this expression, we conclude that the second term of (IV.191) is non-vanishing only for $K = \frac{1}{2}$ and $K = 1$. If we are careless, it may appear that it occurs for $K = 0$ as well; but it should be recalled that for $K = 0$ there is only one term in Ψ_{IK}^{I}, and hence the cross terms of the type described do not arise at all. In view of the frequent occurrence of the intrinsic quadrupole moment contained in the first term of (IV.191), we shall denote it by the simpler notation Q_0 (even at the risk of confusion). Using an explicit expression of the Clebsch-Gordon coefficient, this term reduces to

$$\frac{[3K^2 - I(I+1)][3I^2 - I(I+1)]}{(2I-1)I(I+1)(2I+3)}Q_0, \tag{IV.192a}$$

which, in the special case of the ground state $K = I$, gives

$$\frac{I(2I-1)}{(I+1)(2I+3)}Q_0. \tag{IV.192b}$$

This expression is zero for $I = 0$ and $I = \frac{1}{2}$, a well-known result for the quadrupole moment of ground states.

The intrinsic quadrupole moment Q_0 occurring in (IV.192) is usually estimated as follows. Denoting an occupied Nilsson orbital in χ_K by ϕ_α, we obtain

$$Q_0 \equiv \langle \chi_K | Q_0^{\text{intrinsic}} | \chi_K \rangle = \sum_\alpha \langle \phi_\alpha | q_0 | \phi_\alpha \rangle,$$

where q_0 is the single-particle quadrupole moment operator, i.e.,

$$q_0 = \sqrt{\frac{16\pi}{5}} r^2 Y_0^2(\theta, \phi), \tag{IV.193}$$

(θ, ϕ) being the angles of $\mathbf{r}$ in the intrinsic frame. The summation α is over all the single-particle states of the proton occupied in χ_K. We replace $\sum_\alpha \phi_\alpha^*(\mathbf{r})\phi_\alpha(\mathbf{r})$ by the proton density $\rho(\mathbf{r})$ such that

$$\int_\tau d^3r \, \rho(\mathbf{r}) = Ze. \tag{IV.194a}$$

With this notation, Q_0 becomes

$$Q_0 = \int_\tau d^3r \, \rho(\mathbf{r}) q_0(r, \theta, \phi). \tag{IV.194b}$$

The integration is over the nuclear volume τ. We here adopt the same model as described after (IV.176): the charge density $\rho(\mathbf{r})$ is equal to a constant ρ_0 everywhere within the deformed nuclear surface of radius given by (IV.177). We thus obtain from (IV.194a)

$$\rho_0 = \frac{3}{4\pi} Ze R_0^{-3}. \tag{IV.195}$$

We substitute from (IV.193) in (IV.194b) and carry out the integrations. Then

$$Q_0 = \sqrt{\frac{16\pi}{5}} \rho_0 \tfrac{1}{5} R_0^5 \int_0^{2\pi} d\phi \int_0^\pi \sin\theta \, d\theta \, Y_0^2(\theta)(1 + \xi)^5.$$

We shall use the expression of ξ from (IV.177b) and specialize it to the axially symmetric case by keeping only the $(\mu = 0)$-term in the body-fixed frame, that is,

$$\xi(\theta, \phi) = a_0 Y_0^2(\theta) = \beta Y_0^2(\theta).$$

From the expansion of $(1 + \xi)^5$, we shall keep terms up to the second order in ξ. In this way,

$$Q_0 = \sqrt{\frac{16\pi}{5}} \frac{3}{4\pi} Ze \tfrac{1}{5} R_0^2 \int_0^{2\pi} d\phi \int_0^\pi \sin\theta \, d\theta \, Y_0^2(\theta)[1 + 5\beta Y_0^2(\theta) + 10\beta^2 Y_0^2(\theta) Y_0^2(\theta)].$$

The first term in the expansion produces zero after integration. The β- and β^2-term contribute nonvanishing results. The integral in the β^2-term is actually the matrix element of $Y_0^2(\theta)$ between two spherical harmonic states, and hence the result can be written down from the standard Racah expression (BIII.3) of Appendix B. Thus, we finally obtain

$$\begin{aligned} Q_0 &= \frac{3}{\sqrt{5\pi}} Ze R_0^2 \beta \left(1 + \beta \begin{bmatrix} 2 & 2 & 2 \\ 0 & 0 & 0 \end{bmatrix}^2 \sqrt{\frac{5}{4\pi}} \right) \\ &= \frac{3}{\sqrt{5\pi}} Z R_0^2 \beta \left(1 + \tfrac{4}{7} \sqrt{\frac{5}{4\pi}} \beta \right). \end{aligned} \tag{IV.196}$$

Let us now get back to the expression (IV.192b). This gives us the value of the ground-state quadrupole moment in *all* odd-mass nuclei. We recall that in odd-mass nuclei the second term of (IV.191) has to be used only if $K = \frac{1}{2}$, but in this band the ground state has $I = \frac{1}{2}$, which cannot have any quadrupole moment. The observed quadrupole moment of the odd-mass nucleus then determines its intrinsic quadrupole moment Q_0, which in its turn determines through (IV.196) the deformation parameter β of the spheroidal nucleus. The intrinsic quadrupole moment of the neighbouring even nucleus differs only very slightly from this Q_0, and hence the same value of β may be used in that even nucleus. Another method of determining β is mentioned in this section after our discussion on the electric quadrupole transition.

The derivation of the matrix elements of an electromagnetic transition operator Ω^L_m (precise expressions occur in Section 27) can also be given by using (IV.185) and the expression (IV.141) for the wavefunctions. We here denote all quantum numbers of the initial state by the subscript i and those for the final state by the subscript f. Once again, the matrix element connecting (K_i, K_f) is equal to the matrix element connecting $(-K_i, -K_f)$, and the cross element connecting $(K_i, -K_f)$ is equal to the cross element connecting $(-K_i, K_f)$. We keep one of each of these terms, and incorporate a factor of 2. In this way,

$$\langle \Psi^{I_f}_{M_f K_f} | \, \Omega^L_m(S) \, | \Psi^{I_i}_{M_i K_i} \rangle$$

$$= \frac{\{[I_i][I_f]\}^{1/2}}{16\pi^2} \frac{8\pi^2}{[I_f]} 2 \begin{bmatrix} I_i & L & I_f \\ M_i & m & M_f \end{bmatrix} \left\{ \begin{bmatrix} I_i & L & I_f \\ K_i & K_f - K_i & K_f \end{bmatrix} \langle \Omega^L_{K_f - K_i} \rangle^{\text{intrinsic}}_{K_f, K_i} \right.$$

$$\left. + (-1)^{I_f - J_f} \begin{bmatrix} I_i & L & I_f \\ K_i & -K_i - K_f & -K_f \end{bmatrix} \langle \Omega^L_{-K_i - K_f} \rangle^{\text{intrinsic}}_{-K_f, K_i} \right\}. \quad \text{(IV.197)}$$

The equality of the pairs of terms just mentioned, and also used in our derivation of the static moments earlier in this section, can be proved by using the definition

$$\exp(-i\pi J_y)\chi_K = (-1)^{J-K}\chi_{-K}$$

and the transformation of the tensor operator Ω^L_μ under a rotation π about the y-axis, namely,

$$\exp(i\pi J_y)\Omega^L_\mu \exp(-i\pi J_y) = (-1)^{L-\mu}\Omega^L_{\mu}.$$

The Clebsch-Gordon coefficient $\begin{bmatrix} I_i & L & I_f \\ -K_i & -\mu & -K_f \end{bmatrix}$ has also to be changed to $\begin{bmatrix} I_i & L & I_f \\ K_i & \mu & K_f \end{bmatrix}$ by multiplying with $(-1)^{I_i+L-I_f}$; similar changes have to be made in the Clebsch-Gordon coefficient $\begin{bmatrix} I_i & L & I_f \\ -K_i & K_i + K_f & K_f \end{bmatrix}$ of one of the cross terms to change it to $\begin{bmatrix} I_i & L & I_f \\ K_i & -K_i - K_f & -K_f \end{bmatrix}$. The details of this proof are left as an exercise.

The computation of the reduced transition strength from (IV.197) is straightforward: we have to take the square of the modulus of the matrix element, sum over M_f, and average over M_i. Since

$$\frac{1}{[I_i]} \sum_{M_i} \sum_{M_f} \begin{bmatrix} I_i & L & I_f \\ M_i & m & M_f \end{bmatrix}^2 = \frac{[I_f]}{[I_i]},$$

we obtain the reduced transition strength $B(L)$ as

$$B(L) = \left(\begin{bmatrix} I_i & L & I_f \\ K_i & K_f - K_i & K_f \end{bmatrix} \langle \Omega^L_{K_f - K_i} \rangle^{\text{intrinsic}}_{K_f, K_i} \right.$$

$$\left. + (-1)^{I_f - J_f} \begin{bmatrix} I_i & L & I_f \\ K_i & -K_i - K_f & -K_f \end{bmatrix} \langle \Omega^L_{-K_i - K_f} \rangle^{\text{intrinsic}}_{-K_f, K_i} \right)^2. \quad \text{(IV.198)}$$

In the special case of levels belonging to the same band, $K_i = K_f = K$, and we have

$$B(L) = \left(\begin{bmatrix} I_i & L & I_f \\ K & 0 & K \end{bmatrix} \langle \Omega_0^L \rangle_{K,K}^{\text{intrinsic}} + (-1)^{I_f - J_f} \begin{bmatrix} I_i & L & I_f \\ K & -2K & -K \end{bmatrix} \langle \Omega_{-2K}^L \rangle_{-K,K}^{\text{intrinsic}} \right)^2. \quad \text{(IV.199)}$$

For an M1-transition, the $g_R\mathbf{I}$ part of the operator cannot connect different states, and hence the effective part of the M1-operator, in units of nuclear magneton, is given by

$$\Omega_\mu^1 = \sqrt{\frac{3}{4\pi}}(g_K - g_R)J_\mu.$$

Thus, within the same band,

$$B(\text{M1}) = \frac{3}{4\pi}(g_K - g_R)^2 \left(\begin{bmatrix} I_i & 1 & I_f \\ K & 0 & K \end{bmatrix} K + \delta_{K,1/2}(-1)^{I_f + 1/2} \begin{bmatrix} I_i & 1 & I_f \\ \frac{1}{2} & -1 & -\frac{1}{2} \end{bmatrix} \frac{b_0}{\sqrt{2}} \right)^2, \quad \text{(IV.200)}$$

where b_0 is defined by (IV.190b).

For an E2-transition within the same band, the cross term contributes for only $K = \frac{1}{2}$ and $K = 1$. Since $K = 1$ is seldom encountered, we exclude this case and write B(E2) as

$$B(\text{E2}) = \frac{5}{16\pi}(Q_0)^2 \left(\begin{bmatrix} I_i & 2 & I_f \\ K & 0 & K \end{bmatrix} + \delta_{K,1/2}(-1)^{I_i - 1/2} \begin{bmatrix} I_i & 2 & I_f \\ \frac{1}{2} & -1 & -\frac{1}{2} \end{bmatrix} \frac{\tilde{Q}_1}{Q_0} \right)^2, \quad \text{(IV.201a)}$$

where

$$\tilde{Q}_1 = (-1)^{J_f - 1/2} \langle \chi_{-1/2} |\, Q_{-1} \,| \chi_{1/2} \rangle. \quad \text{(IV.201b)}$$

The quantity Q_0 is the intrinsic quadrupole moment defined by (IV.196). The operator Q_{-1} occurring on the right-hand side of (IV.201b) is the component -1 of the quadrupole moment operator, i.e.,

$$\sqrt{\frac{16\pi}{5}} \sum_i r_i^2 Y_{-1}^2(\theta_i, \phi_i),$$

the summation being over the protons. The factor $5/(16\pi)$ occurs in (IV.201a) through the definition of the operator Ω_μ^2(E2), which is $\sqrt{5/(16\pi)}$ times Q_μ. The parameter $\tilde{Q}_1$, as defined by (IV.201b), can also be computed once the Nilsson wavefunctions are known.

In even-even nuclei, the ground-state band has $K = 0$, and hence the measured B(E2)-value connecting any two consecutive states, i.e., $(2^+ \to 0^+)$, $(4^+ \to 2^+)$, . . . , determines $(Q_0)^2$, according to (IV.201a). The corresponding deformation parameter β can then be determined, except for its sign. It has already been mentioned in this section that Q_0 for the odd-mass nuclei and the corresponding β can be determined from static quadrupole moment measurements.

In the odd-mass nuclei, subsequent rotational levels differ in their angular momenta by ± 1. Thus, the E2- as well as the M1-transition are permitted between two such levels. The Coulomb excitation from the ground state to the first excited state $(I_i \to I_f)$ is predominantly of the E2-type because the Coulomb excitation takes place primarily through the electrostatic interaction of the projectile with the charge of the target nucleus. The B(E2)-value of $I_i \to I_f$ in the Coulomb excitation process indirectly determines the B(E2)-value of the γ-decay that takes place from the excited state I_f to the lower state I_i. According to the first term of (IV.201a) (assuming that the band does not correspond to $K = \frac{1}{2}$), the ratio of the two B(E2)-values is the ratio

59 (45-122/1977)

of the square of the two Clebsch-Gordon coefficients $\begin{bmatrix} I_i & 2 & I_f \\ K & 0 & K_f \end{bmatrix}^2$ and $\begin{bmatrix} I_f & 2 & I_i \\ K & 0 & K_i \end{bmatrix}^2$, which is $\dfrac{[I_f]}{[I_i]}$. This proves our assertion. Knowing the B(E2)-value of the γ-decay indirectly from the Coulomb excitation B(E2), and subtracting it from the observed reduced transition strength of the γ-decay, we determine the value of B(M1) for the γ-decay. According to the first term of (IV.200) (once again $K = \frac{1}{2}$ is excluded), this B(M1)-value determines the parameter $(g_K - g_R)^2$. If, in addition, we use the value of the static magnetic moment, which is linear in g_R and $(g_K - g_R)$, we obtain alternative choices for the values of g_K and g_R, corresponding to the two alternative signs of $(g_K - g_R)$. To remove this ambiguity, one more measurement is necessary to directly give the sign of $(g_K - g_R)$. The angular distribution of the γ-ray following the Coulomb excitation is such an experiment; it actually determines $(g_K - g_R)Q_0^{-1}$. Since the magnitude and sign of Q_0 are known from the static quadrupole moment measurement, the measurement just mentioned indeed determines the sign of $(g_K - g_R)$.

An important selection rule (called the K-selection rule) follows from (IV.198) for γ-transition between the two different bands K_i and K_f. The quantity $(K_f - K_i)$ occurring as the

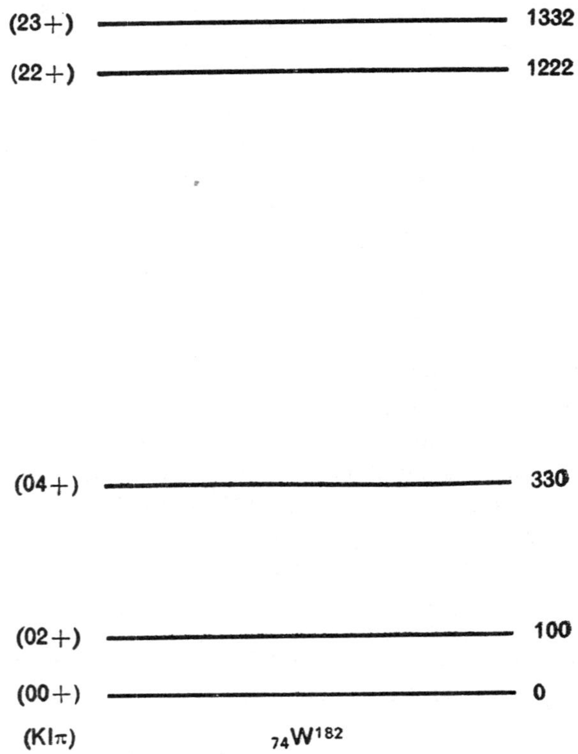

Fig. IV.12 Spectra of W^{182}. (From Kerman, A. K., in Nuclear Reactions, Vol 1, ed. by P. M. Endt and M. Demeur, North-Holland, Amsterdam, 1959, p 450.)

component of Ω^L in the first term must necessarily be smaller than or equal to L. It should be noted that $-K_i - K_f$ is larger in *magnitude* than $|K_f - K_i|$. Thus, if $|K_f - K_i| \leqslant L$ is not satisfied, the second term of (IV.198) also vanishes. Therefore, this inequality gives the desired K-selection rule.

Another important rule on the branching ratio of the transitions to *different* final levels of the *same band* from the *same* initial level follows from (IV.198). Once again, confining ourselves to the first term, we expect

$$\frac{B(L;\, I_i \to I_f)}{B(L;\, I_i \to I_f')} = \begin{bmatrix} I_i & L & I_f \\ K_i & K_f - K_i & K_f \end{bmatrix}^2 \begin{bmatrix} I_i & L & I_f' \\ K_i & K_f - K_i & K_f \end{bmatrix}^{-2}. \qquad \text{(IV.202)}$$

This ratio is simply that of the square of two Clebsch-Gordon coefficients. Amongst the spectra of deformed nuclei there are many interesting examples of the K-selection rule and the rule on the branching ratio. We show a typical case in Fig. IV.12 (see Kerman[13]). The spectrum is that of $_{74}W^{182}$. The numbers to the right of the levels denote their energies in keV, whereas those to the left denote the quantum numbers K, I, and $\pm$ denote parity. The first three levels form the ground-state rotational band, whereas the two upper ones form an excited band with $K = 2$. The (22+) level decays by the E2 γ-emission to two final states in the ground-state band—the ground state (00+) and the first excited state (02+). The branching ratio of the two B(E2)-values is found to be 0.62, whereas the simple intensity ratio (IV.202) predicts a value 0.70. In the same nucleus, we find the K-selection rule in operation in the case of the decay of the (23+)-level, which branches to the (02+)-level and the (04+)-level by E2-emission. From normal γ-decay selection rules on spin and parity, both these decays could have been M1. However, in this case, $K_f - K_i = -2$, and hence the K-selection rule is not satisfied for the M1-decay with $L = 1$, whereas it is indeed satisfied for the E2-transition, which is observed. The ratio of the two B(E2)-values also agrees well with (IV.202): the observed value is 2.1, and the predicted value is 2.5.

REFERENCES

1. Bohr, A., *Kgl. Danske Videnskab. Selskab, Mat. Fys. Medd.*, **26**, No. 14 (1952); Bohr, A., and Mottelson, B. R., *Kgl. Danske Videnskab. Selskab, Mat. Fys. Medd.*, **27**, No. 16 (1953).
2. Landau, L. D., and Lifshitz, E. M., Quantum Mechanics, Pergamon Press, Oxford, 1958, p 279.
3. Kumar, K., and Baranger, M., *Nucl. Phys.*, **A110**, 529 (1968); *ibid*, **A122**, 273 (1968); *ibid*, **A92**, 608 (1967); Kumar, K., *Nucl. Phys.*, **A92**, 653 (1967).
4. Sheline, R. K., Sikkeland, T. S., and Chadda, R. N., *Phys. Rev. Letters*, **7**, 446 (1961).
5. Morinaga, H., and Clark, N. L., *Nucl. Phys.*, **67**, 315 (1965).
6. Davydov, A. S., and Filipov, G. F., *Nucl. Phys.*, **8**, 237 (1958); Davydov, A. S., *Nucl. Phys.*, **16**, 597 (1960); *ibid*, **24**, 682 (1961); Davydov, A. S., and Chaban, A. A., *Nucl. Phys.*, **20**, 499 (1960); Davydov, A. S., Rostovsky, V. S., and Chaban, A. A., *Nucl. Phys.*, **27**, 134 (1961).
7. Davidson, J. P., *Rev. Mod. Phys.*, **37**, 105 (1965).
8. Mariscotti, M. A. J., Scharff-Goldhaber, G., and Buck, B., *Phys. Rev.*, **178**, 1864 (1959).
9. Sheline, R. K., *Revs. Mod. Phys.*, **32**, 1 (1950).

10. Choudhury, D. C., *Kgl. Danske Videnskab. Selskab, Mat. Fys. Medd.*, **28**, No. 4 (1954); *Nucl. Phys.*, **A93**, 300 (1967); *Phys. Rev.*, **C3**, 1619 (1971); Ford, K. W., and Levinson, C., *Phys. Rev.*, **100**, 1 (1955); Lande, A., and Brown, G. E., *Nucl. Phys.*, **75**, 344 (1966); Sen, S., et al, *Nucl. Phys.*, **A157**, 497 (1970); *ibid*, **A191**, 29 (1972); *ibid*, **A220**, 580 (1974); Sen, S., et al, *Phys. Rev.*, **C13**, 2055 (1976); *ibid*, **C14**, 758 (1976); Sen, S., et al, *J. Phys. G: Nucl. Phys.*, **1**, 286 (1975).

11. Nilsson, S. G., *Kgl. Danske Videnskab. Selskab, Mat. Fys. Medd.*, **29**, No. 16 (1955); Mottelson, B. R., and Nilsson, S. G., *Kgl. Danske Videnskab. Selskab, Mat. Fys. Skrifter*, **1**, 8 (1959); *Nucl. Phys.*, **13**, 281 (1959).

12. Bodenstedt, E., and Rogers, J. D., in Perturbed Angular Correlations, ed. by E. Karlsson, E. Matthias, and K. Siegbahn, North-Holland, Amsterdam, 1964, p 91.

13. Kerman, A. K., in Nuclear Reactions, Vol 1, ed. by P. M. Endt and M. Demeur, North-Holland, Amsterdam, 1959, p 427.

PROBLEMS

1. Prove that the part of T_{vib} from the product of the two terms in (IV.52) vanishes.

2. The Pauli recipe for T given in (IV.72b) corresponds to ∇^2 in a set of generalized coordinates. Derive it.

3. Solve (IV.82a) for β-vibration.

4. Prove that the solutions of (IV.82d) are single-valued functions of 3γ if $\Lambda = 9\lambda(\lambda + 1)$, $\lambda = 0, 1, 2, \ldots$, and that the equation then reduces to the equation of the Legendre polynomial (to be obtained from Appendix A).

5. Show that (IV.104a) and (IV.104b) represent the solution of (IV.102).

6. Work out the solution (IV.155)–(IV.157b) of (IV.154b).

7. Prove (IV.169).

8. Derive (IV.191).

V Microscopic Theory of Nuclear Structure

35. HARTREE-FOCK THEORY OF NUCLEAR SHAPE

A. DERIVATION OF HARTREE-FOCK EQUATIONS

The aim of microscopic theory of nuclear structure is to understand the various observed properties of nuclei, starting from a realistic many-body Hamiltonian, i.e., a Hamiltonian consisting of the kinetic energies of all the nucleons, and the sum of the *actual* two-nucleon interaction (as observed in the two-body problem) over all the pairs of nucleons. We have discussed in Chapter II that the strong repulsion in the core region of the realistic two-nucleon interaction has to be suitably treated and an effective two-body potential derived as a first step in this procedure. Here we assume that this has already been done and we have the Hamiltonian

$$H = \sum_{i=1}^{A} T_i + \sum_{i<j}^{A} V_{ij}, \tag{V.1}$$

where V is the *effective* two-nucleon potential. In practice, we calculate a set of matrix elements of the effective potential in a suitable basis, e.g., the harmonic oscillator basis. Although we have used a configuration space representation of V in (V.1), in our subsequent work the matrix elements of V in a convenient basis (e.g., harmonic oscillator) suffice.

In the phenomenological shell model, we *assumed* a form for the average potential field in which the nucleons move. This was taken to be the harmonic oscillator potential with the oscillator parameter chosen to fit the experimentally observed root-mean-square radius of the nucleus, and a spin-orbit coupling whose strength was so determined as to produce a fairly good fit to the observed spin-orbit splitting of the single-particle levels. In microscopic theory, this average potential field is *derived* by commencing with the Hamiltonian (V.1). The method for doing this is the well-known Hartree-Fock (HF) method of atomic physics.

There are many different ways of presenting the HF formalism. We shall give a derivation that is best suited for the purpose of nuclear calculations. Let us use the symbols $|i\rangle, |j\rangle, \ldots$ to denote the correct single-particle states corresponding to the best average potential $\mathcal{V}$ which is to be determined by the HF method. We shall also use a set of known basic single-particle states, $|\alpha\rangle, |\beta\rangle, \ldots$. These basic states may, for example, be taken to be the harmonic oscillator states (in the presence of spin-orbit coupling) used in ordinary shell-model calculations. The states $|i\rangle$ can, in general, be expanded in terms of the basic states $|\alpha\rangle$ as

$$|i\rangle = \sum_{\alpha} x_{\alpha}^{i} |\alpha\rangle, \tag{V.2}$$

where the coefficients x^i_α are to be determined, subject to the normalization requirement

$$\sum_\alpha |x^i_\alpha|^2 = 1. \tag{V.3}$$

The principle, followed for the determination of the coefficients x^i_α, is the well-known quantum mechanical variational principle. We allow the set of A single-particle states $|i\rangle, |j\rangle, \ldots$ to be occupied in a normalized determinantal many-body state Φ for A nucleons. The parameters x^i_α, occurring in Φ through the single-particle states, are then treated as variational parameters in the minimization of the expectation value of the Hamiltonian. That is, we require

$$\langle\Phi| H |\Phi\rangle = \text{minimum} \tag{V.4}$$

under arbitrary variations of x^i_α and x^{i*}_α, subject to the constraint (V.3). To carry out this programme, we first work out the left-hand side of (V.4), with H given by (V.1), by the application of the standard results of Section 15. Thus,

$$\langle\Phi| H |\Phi\rangle = \sum_i^{\text{occ}} \langle i| T |i\rangle + \sum_{i<j}^{\text{occ}} (ij| V |ij). \tag{V.5a}$$

Next, we substitute the expansion (V.2) for each single-particle state on the right-hand side and obtain

$$\langle\Phi| H |\Phi\rangle = \sum_{i=1}^{\text{occ}} \sum_{\alpha,\beta} x^{i*}_\alpha x^i_\beta \langle\alpha| T |\beta\rangle + \tfrac{1}{2} \sum_{i,j}^{\text{occ}} \sum_{\alpha\beta\gamma\delta} x^{i*}_\alpha x^{j*}_\beta x^i_\gamma x^j_\delta (\alpha\beta| V |\gamma\delta). \tag{V.5b}$$

The equation of constraint (V.3) for all i can be taken into account by the usual method of Lagrange multipliers. Calling the Lagrange multipliers ϵ_i, we are then required to minimize (V.5b) minus $\sum_i \epsilon_i \sum_\alpha |x^i_\alpha|^2$; that is, we require the variations of

$$F(x^i, x^{i*}) \equiv -\sum_i^{\text{occ}} \epsilon_i \sum_\alpha x^{i*}_\alpha x^i_\alpha + \sum_i^{\text{occ}} \sum_{\alpha,\beta} x^{i*}_\alpha x^i_\beta \langle\alpha| T |\beta\rangle + \tfrac{1}{2} \sum_{i,j}^{\text{occ}} \sum_{\alpha\beta\gamma\delta} x^{i*}_\alpha x^{j*}_\beta x^i_\gamma x^j_\delta (\alpha\beta| V |\gamma\delta) \tag{V.5c}$$

to be zero when x^i_α or x^{i*}_α (all i and all α) are arbitrarily varied. Let us consider the variation of F through a particular x-coefficient, for instance, x^{k*}_μ, and equate the coefficient of δx^{k*}_μ to zero to obtain the general condition of minimum. In this way,

$$0 = -\epsilon_k x^k_\mu + \sum_\beta x^k_\beta \langle\mu| T |\beta\rangle + \sum_j^{\text{occ}} \sum_{\beta\gamma\delta} x^{j*}_\beta x^k_\gamma x^j_\delta (\mu\beta| V |\gamma\delta)$$

or

$$\epsilon_k x^k_\mu = \sum_\gamma x^k_\gamma (\langle\mu| T |\gamma\rangle + \langle\mu| \mathcal{V} |\gamma\rangle), \tag{V.6a}$$

where

$$\langle\mu| \mathcal{V} |\gamma\rangle = \sum_{\beta\delta} (\mu\beta| V |\gamma\delta) \rho_{\delta\beta} \tag{V.6b}$$

with

$$\rho_{\delta\beta} = \sum_j^{\text{occ}} x^j_\delta x^{j*}_\beta. \tag{V.6c}$$

The factor $\frac{1}{2}$ appearing in (V.5c) has disappeared in (V.6a) because we have picked up two equal terms corresponding to the two possibilities for the summation labels, namely,

$$i = k, \alpha = \mu \quad \text{and} \quad j = k, \beta = \mu.$$

By letting μ proceed over all the basis states, we obtain, from (V.6a), a system of linear equations for the unknown coefficients x_μ^k. The solution of this system of equations clearly corresponds to diagonalizing the matrix of $(T + \mathcal{V})$ in the representation spanned by our basis states $|\alpha\rangle$, $|\beta\rangle, \ldots$. The eigenvalues of the matrix are the quantities ϵ_k. For each ϵ_k, the corresponding eigenvector determines the coefficients x_μ^k, μ running over all the basis states. The operator $(T + \mathcal{V})$ can be clearly interpreted as a single-particle Hamiltonian, of which T and $\mathcal{V}$ are kinetic and potential energies, respectively. The one-body potential $\mathcal{V}$, as is apparent from (V.6b), has been derived by averaging the given two-body potential V in a certain manner.

The potential $\mathcal{V}$ is called the Hartree-Fock potential, and the quantities ϵ_k are the HF single-particle energies. The states $|i\rangle$, $|j\rangle$, which are known as soon as the coefficients x_μ^k are determined, are the HF single-particle states.

It is clear from (V.6b) and (V.6c) that there is a very serious complication involved in the solution of the system of equations (V.6a). The potential $\mathcal{V}$ is not completely known to start with. The matrix elements of V appearing in (V.6b) are known, but the matrix elements $\rho_{\delta\beta}$ given by (V.6c) are not known because they involve the coefficients x_δ^j, x_β^{j*}, which are available only after the diagonalization of the matrix for $(T + \mathcal{V})$. For this reason, the solution of the HF equations (V.6a) can be achieved only by an iterative procedure. To start with, we have to *guess* a set of values for the x-coefficients, and calculate the matrix elements of $\mathcal{V}$ with these values. Then we need to diagonalize $(T + \mathcal{V})$ in the chosen representation, and obtain a new set of x-coefficients, and a set of HF energies. The new set of x-coefficients will, in general, not agree with the starting set. Therefore, we repeat the entire procedure with the new set, and once again compare the last available set of x-coefficients with those belonging to the previous set. This iteration procedure is ended when the x-coefficients in two successive iterations agree within the desired degree of accuracy. The check can be equivalently applied to the set of HF energies in successive iterations. When the final goal has been achieved, we say that the HF equations have been *self-consistently* solved; this is because in our procedure we have used a set of single-particle states (determined with the x-coefficients) that determined an average potential $\mathcal{V}$, which in its turn regenerated the same set of single-particle states. In this sense, the procedure achieved self-consistency between the single-particle states and the average potential.

The expression (V.6b), which defines the HF potential in conjunction with (V.6c), can be rewritten as follows. Since

$$\sum_\delta x_\delta^j |\delta\rangle = |j\rangle, \tag{V.7a}$$

$$\sum_\beta \langle\beta| x_\beta^{j*} = \langle j|, \tag{V.7b}$$

we have

$$\langle\mu|\, \mathcal{V}\, |\gamma\rangle = \sum_j^{\text{occ}} (\mu j|\, V\, |\gamma j). \tag{V.8}$$

The quantity $\rho_{\delta\beta}$, defined by (V.6c), is a matrix element of the single-particle density operator ρ connecting the single-particle states $\langle\delta|$ and $|\beta\rangle$. The reason for calling ρ the single-particle density operator is now explained.

From (V.7), we have

$$x_\delta^j = \langle \delta \mid j \rangle, \qquad x_\beta^{j*} = \langle j \mid \beta \rangle,$$

and hence (V.6c) reduces to

$$\rho_{\delta\beta} \equiv \langle \delta | \, \rho \, | \beta \rangle = \sum_j^{\text{occ}} \langle \delta \mid j \rangle \langle j \mid \beta \rangle.$$

Therefore,

$$\rho = \sum_j^{\text{occ}} |j\rangle\langle j|. \tag{V.9a}$$

In this form, ρ easily lends itself to an evaluation of its matrix element in the coordinate representation. We have

$$\begin{aligned} \langle \mathbf{r} | \, \rho \, | \mathbf{r}' \rangle &= \sum_j^{\text{occ}} \langle \mathbf{r} \mid j \rangle \langle j \mid \mathbf{r}' \rangle \\ &= \sum_j^{\text{occ}} \phi_j(\mathbf{r}) \phi_j^*(\mathbf{r}'), \end{aligned} \tag{V.9b}$$

where $\phi_j(\mathbf{r})$ is the single-particle wavefunction corresponding to the single-particle state $|j\rangle$. In particular, the diagonal matrix element of ρ in the coordinate space, to be denoted by $\rho(\mathbf{r})$, is given by

$$\langle \mathbf{r} | \, \rho \, | \mathbf{r} \rangle \equiv \rho(\mathbf{r}) = \sum_j^{\text{occ}} |\phi_j(\mathbf{r})|^2. \tag{V.9c}$$

The quantity $|\phi_j(\mathbf{r})|^2$ is, according to elementary quantum mechanics, the probability density at the point $\mathbf{r}$, corresponding to the single-particle state ϕ_j. The single-particle probability density corresponding to the many-body determinantal state Φ can be clearly obtained by summing the contribution of each single-particle state occupied in Φ. This expected result is embodied in (V.9c). Thus, the nomenclature chosen for the operator ρ is correct.

It is relevant here to give some algebraic details for the computation of the two-body matrix elements occurring in (V.6b) for the HF potential. Writing out in detail the quantum numbers *nljmt* contained in the abbreviated notation of each Greek letter, we have

$$(\mu\beta | \, V \, | \gamma\delta) \equiv (n_\mu l_\mu j_\mu m_\mu t_\mu ;\; n_\beta l_\beta j_\beta m_\beta t_\beta | \, V \, | n_\gamma l_\gamma j_\gamma m_\gamma t_\gamma ;\; n_\delta l_\delta j_\delta m_\delta t_\delta), \tag{V.9d}$$

where n is the radial quantum number, l and j are the orbital and total angular momenta of a single nucleon, m is the projection of j, and t is the isospin projection ($+\frac{1}{2}$ for neutron and $-\frac{1}{2}$ for proton). In Section 28C, we have given detailed calculations of the matrix elements of V between two-nucleon states coupled to the appropriate total angular momentum J, its projection M, and the total isospin T with its projection M_T. We shall therefore assume that such results are known, and rewrite (V.9d) in terms of the results. Using the Clebsch-Gordon coefficients for the coupling of angular momenta, we obtain

$$\begin{aligned} (\mu\beta | \, V \, | \gamma\delta) = \sum_{J,T} &\begin{bmatrix} j_\mu & j_\beta & J \\ m_\mu & m_\beta & m_\mu + m_\beta \end{bmatrix} \begin{bmatrix} j_\gamma & j_\delta & J \\ m_\gamma & m_\delta & m_\gamma + m_\delta \end{bmatrix} \begin{bmatrix} \frac{1}{2} & \frac{1}{2} & T \\ t_\mu & t_\beta & t_\mu + t_\beta \end{bmatrix} \\ &\times \begin{bmatrix} \frac{1}{2} & \frac{1}{2} & T \\ t_\gamma & t_\delta & t_\gamma + t_\delta \end{bmatrix} (n_\mu l_\mu j_\mu, \, n_\beta l_\beta j_\beta : JT | \, V \, | n_\gamma l_\gamma j_\gamma, \, n_\delta l_\delta j_\delta : JT). \end{aligned} \tag{V.9e}$$

The Clebsch-Gordon coefficients require $m_\mu + m_\beta = m_\gamma + m_\delta$ and $t_\mu + t_\beta = t_\gamma + t_\delta$. The two-body matrix elements on the right-hand side of (V.9e) are diagonal in the quantum numbers of J and T, and independent of the projection quantum numbers. These facts have been utilized in writing (V.9e).

B. HARTREE-FOCK CALCULATION AND VARIOUS OBSERVED QUANTITIES

The physical quantities that can be calculated immediately after a self-consistent calculation are now described.

Energy for the State Φ

The energy for the state Φ is given by (V.5a) which can be written in alternative ways, using the definitions (V.6) together with the fact that ϵ_i is the energy of the self-consistent state $|i\rangle$, i.e.,

$$\begin{aligned}\epsilon_i &= \langle i|\,(T+\mathcal{V})\,|i\rangle \\ &= \langle i|\,T\,|i\rangle + \sum_j^{\text{occ}} (ij|\,V\,|ij). \qquad\qquad (\text{V.10})\end{aligned}$$

We have, according to (V.5a),

$$\begin{aligned}\langle\Phi|\,H\,|\Phi\rangle &= \sum_i^{\text{occ}} \langle i|\,T\,|i\rangle + \tfrac{1}{2}\sum_{i,j}^{\text{occ}} (ij|\,V\,|ij) \\ &= \tfrac{1}{2}\sum_i^{\text{occ}} \langle i|\,(T+\mathcal{V})\,|i\rangle + \tfrac{1}{2}\sum_i^{\text{occ}} \langle i|\,T\,|i\rangle \\ &= \tfrac{1}{2}\sum_i^{\text{occ}} \epsilon_i + \tfrac{1}{2}\sum_{\alpha,\beta} \langle\alpha|\,T\,|\beta\rangle \sum_i^{\text{occ}} x_\alpha^{i*} x_\beta^i \qquad\qquad (\text{V.11a}) \\ &= \tfrac{1}{2}\sum_i^{\text{occ}} \epsilon_i + \tfrac{1}{2}\sum_{\alpha,\beta} \langle\alpha|\,T\,|\beta\rangle \rho_{\beta\alpha} \qquad\qquad (\text{V.11b}) \\ &= \tfrac{1}{2}\sum_i^{\text{occ}} \epsilon_i + \tfrac{1}{2}\,\mathrm{Tr}\,(T\rho), \qquad\qquad (\text{V.11c})\end{aligned}$$

where Tr denotes the trace of the matrix following it. Alternatively, we could write

$$\tfrac{1}{2}\sum_{i,j}^{\text{occ}} (ij|\,V\,|ij) = \tfrac{1}{2}\sum_i^{\text{occ}} \langle i|\,\mathcal{V}\,|i\rangle,$$

and hence

$$\begin{aligned}\langle\Phi|\,H\,|\Phi\rangle &= \sum_i^{\text{occ}} \langle i|\,(T+\tfrac{1}{2}\mathcal{V})\,|i\rangle \\ &= \sum_{\alpha,\beta} \langle\alpha|\,(T+\tfrac{1}{2}\mathcal{V})\,|\beta\rangle \sum_i^{\text{occ}} x_\alpha^{i*} x_\beta^i \\ &= \sum_{\alpha,\beta} \langle\alpha|\,(T+\tfrac{1}{2}\mathcal{V})\,|\beta\rangle \rho_{\beta\alpha} = \mathrm{Tr}\,[(T+\tfrac{1}{2}\mathcal{V})\rho]. \qquad\qquad (\text{V.11d})\end{aligned}$$

The total energy calculated in this way does not, *in general*, correspond to the energy of an observed state of the nucleus because, *in general*, Φ does not have a fixed angular momentum, whereas any physically observed state always has a conserved value of total angular momentum. The method of obtaining states of good angular momentum from the HF state Φ is described in Section 36A. The energy calculation for the observed states is also described in that context.

60 (45-122/1977)

The reader should note that the HF state Φ, in this sense, plays the role of the intrinsic state described in Section 33D.

Deformation of the State Φ

The expectation value of any single-particle operator

$$\sum_{I=1}^{A} \Omega(I)$$

in the determinantal state Φ is given by

$$\langle\Phi| \sum_{I=1}^{A} \Omega(I) |\Phi\rangle = \sum_{i}^{\text{oc}} \langle i| \Omega |i\rangle. \tag{V.12a}$$

In the special case of the μ-component of the quadrupole operator, we have

$$\langle Q_\mu\rangle = \sum_{i}^{\text{occ}} \langle i| Q_\mu |i\rangle = \text{Tr}\,(Q_\mu \rho), \tag{V.12b}$$

where the symbol $\langle\ \rangle$ on the left-hand side is an abbreviated notation for the expectation value. If (V.12b) turns out to be zero for all μ, then we infer that the state Φ does not have a quadrupole deformation. If, on the other hand, $\langle Q_0\rangle$ turns out to be non-zero and $\langle Q_\mu\rangle = 0$ for $\mu \neq 0$, then the state Φ has a quadrupole deformation with symmetry about the z-axis. In such a case, the value of $\langle Q_0\rangle$, calculated from (V.12b), can be used in conjunction with the expression (IV.196) to define a phenomenological deformation parameter β corresponding to the HF solution Φ. In the more general case when (V.12b) yields non-zero values for all $\langle Q_\mu\rangle$, we can, by a suitable rotation of the coordinate frame (the method for guaranteeing such a frame, to start with, is described in Section 35C), make $\langle Q_1\rangle = \langle Q_{-1}\rangle = 0$ and $\langle Q_2\rangle = \langle Q_{-2}\rangle$. The nonvanishing values of $\langle Q_0\rangle$ and $\langle Q_2\rangle$ or, equivalently, $\langle Q_{-2}\rangle$ can then be used along with (IV.18) and (IV.196) to define two quadrupole deformation parameters β and γ for the HF state Φ.

As already mentioned, when the HF state Φ has a deformation it does not correspond to any physically observed state of the nucleus because Φ does not have a given angular momentum. It then plays the role of an intrinsic state, with its deformation observable only to an observer in the intrinsic coordinate frame. The values of the quadrupole moments $\langle Q_0\rangle$, $\langle Q_2\rangle, \ldots$ are intrinsic quantities, and not observable moments of the ground state or any other physical state.

Single-Particle Energies

The self-consistent single-particle energies ϵ_i corresponding to the determinantal state Φ are the observable energies of single nucleons in the nucleus, provided the state Φ has been very carefully determined. The question of careful determination arises because the solutions of the HF equations, as discussed in detail in Section 35C, are not unique. We may obtain different sets of solutions for the single-particle energies and wavefunctions, which will lead to different HF determinants Φ. Corresponding to each determinant, we have a value of $\langle\Phi| H |\Phi\rangle$. If enough care is exercised to explore all the different solutions, then the particular Φ that leads to the lowest value of $\langle\Phi| H |\Phi\rangle$ is the best HF solution for the ground state of the nucleus. The corresponding single-particle energies of the states occupied in the determinant should, in principle, be the observable energies of single-particle states measured in a nucleon pick-up or a (p, 2p)-type experiment conducted on the nucleus.

Several attempts have been made to give a more extended meaning and interpretation to the HF single-particle energies. Such attempts are based on the arguments that follow. Suppose the HF calculation has been carried out for an even nucleus having A nucleons and has yielded a determinant Φ with a set of occupied single-particle states (to be denoted by h) and also a set of unoccupied single-particle states (to be denoted by p) lying above the occupied ones. Consider now a determinant for $(A - 1)$ nucleons which has been obtained from Φ by omitting the row (column) corresponding to the A-th nucleon, and the column (row) corresponding to one of the occupied states, say, h. The expectation value of the Hamiltonian for the new determinant Φ_h is obviously related to the old determinant Φ as

$$\begin{aligned}\langle\Phi| H |\Phi\rangle &= \sum_i{}' \langle i| T |i\rangle + \langle h| T |h\rangle + \sum_{i<j}{}' (ij| V |ij) + \sum_i{}' (ih| V |ih)\\ &= \langle\Phi_h| H |\Phi_h\rangle + \langle h| T |h\rangle + \sum_i{}' (ih| V |ih)\\ &= \langle\Phi_h| H |\Phi_h\rangle + \epsilon_h.\end{aligned}$$

Here the summations with prime on them go over all the occupied states, with the exception of h. The first and third terms in the first line therefore add up to $\langle\Phi_h| H |\Phi_h\rangle$ in the second line. The other two terms in the first and second lines add up to ϵ_h, according to (V.10). It may appear at first sight that it is not so in view of the prime on the last summation. But the term missing from this summation, and apparently present in (V.10), is $(hh| V |hh)$, which is identically zero. The foregoing equation therefore reduces to

$$\langle\Phi| H |\Phi\rangle - \langle\Phi_h| H |\Phi_h\rangle = \epsilon_h. \tag{V.13a}$$

This is a perfectly valid *mathematical* relation. The determinant Φ_h is clearly *not* the self-consistent solution for the nucleus having $(A - 1)$ nucleons. However, if h is the uppermost occupied state in Φ and it is *assumed* that Φ_h is very close to the ground state of the nucleus $(A - 1)$, then the left-hand side of (V.13a) represents the difference in the binding energies of the two nuclei having $(A - 1)$ and A nucleons, respectively. This difference in the two binding energies, according to the derivation (V.13a), is equal to the calculated HF energy ϵ_h of the single-particle state h. For the simple reason just mentioned, this interpretation of ϵ_h may, however, be grossly wrong.

In the same manner as for Φ_h, by comparing a determinant Φ_p for $(A + 1)$ nucleons, having an extra row and column, with Φ, which is labelled by the coordinate of the $(A + 1)$-th nucleon and the single-particle state p (which is unoccupied in Φ), we can establish the identity

$$\langle\Phi_p| H |\Phi_p\rangle - \langle\Phi| H |\Phi\rangle = \epsilon_p. \tag{V.13b}$$

Once again, this is a valid mathematical relation and, in order to give any physical meaning to its left-hand side, we have to introduce assumptions about the interpretation of Φ_p. When p is the lowest unoccupied state, we may *assume* that Φ_p is a good approximation to the ground state of the nucleus having $(A + 1)$ nucleons, and the difference in the binding energies of the nuclei A and $(A + 1)$ is, according to (V.13b), equal to the HF energy ϵ_p. In the same manner, we may also identify p with any single-particle state above the lowest unoccupied one, and use, on the excited state of the nucleus $(A + 1)$, the same sort of approximation as that just mentioned. The left-hand side will then be the observed difference in the binding energies plus the excitation energy of the particular excited state. The equality of this number with the calculated value of the respective HF single-particle energy can be checked. Once again, this kind of interpretation of the HF particle energies may be grossly violated in practice.

C. PRACTICAL ASPECTS OF HARTREE-FOCK CALCULATION

We have mentioned in Section 35B that the self-consistent solution of the HF equations is not unique. In fact, the nonuniqueness has its origin in the quantum mechanical variation principle on which the derivation of the HF equations is based. It is well-known that the minimization of the expectation value of the Hamiltonian with respect to a variational wavefunction gives the best wavefunction of the particular *class* of functions used in the variation process. It may be possible to obtain another variational wavefunction Φ of a different *class* which, at the end of the minimization programme, yields a lower value of $\langle\Phi| H |\Phi\rangle$. If it does, then the second wavefunction is better than the first. In practice, we therefore have to use some intuition, try several classes of variational wavefunctions, and choose the solution for which $\langle\Phi| H |\Phi\rangle$ has the lowest value.

In a practical HF calculation, the possibility of different classes of solution arises in three different ways. It is obvious that, in order to make the computations practicable, we cannot take an unlimited number of basis states $|\alpha\rangle, |\beta\rangle, \ldots$ for setting up the HF matrix of $(T + \mathcal{V})$. If we decide to truncate the number of basis states to a reasonable size, then different classes of solution result from the various sets of basis states that may be chosen. Let us write the quantum numbers contained in any basis state $|\alpha\rangle$ as $nljm$, where l is the orbital angular momentum, j the total angular momentum, m the projection of the latter, and n the number of nodes in the radial wavefunction. Even if we restrict ourselves to the same set of basis states $|nljm\rangle$, we have an alternative way of confronting various classes of HF solution. This arises from the arbitrary choice of the radial wavefunctions. For example, we may choose the basis states corresponding to the solutions of a harmonic oscillator potential or of a Saxon-Woods type potential. The basis states in the two cases, although labelled by the same set of quantum numbers $nljm$, produce different classes of HF wavefunction because the radial wavefunctions are different in the two cases.

Even after we have resolved the arbitrariness due to the foregoing two reasons and have limited ourselves to a given number of basis states corresponding to a given potential, there is a third possibility of producing different classes of HF solution. One possible class is obtained as follows. From the given set of states we make several subsets, each labelled by a *given* projection quantum number m. The states within each subset differ one from the other through one or more of the quantum numbers n, l, j. Now let us consider an HF state defined by a superposition of the type (V.2), where the sum proceeds over the basis states in a subset. Each state $|i\rangle$ then carries a given quantum number m, but is a superposition of the states having different n-, l-, j-value. The set of all variational wavefunctions $|i\rangle$, defined in this manner, forms one class of HF functions. We shall call this class *axially symmetric*. If we lift the restriction of ensuring m as a good quantum number for the states $|i\rangle$, the summation (V.2) can go over *all* the states of our chosen set. This new set of functions $|i\rangle$ comprises a more general class of HF functions. We shall describe this class as *triaxial*. On the other hand, a third class of functions, more restricted than the axially symmetric ones, can be obtained by requiring l, j, m to be good quantum numbers. In that case, our set of basis states divides itself into subsets, each having a given l, j, m quantum number, and the functions within each subset differ one from the other through the n-quantum number. By restricting the superposition (V.2) within these subsets, we generate a new class of $|i\rangle$, each of which is a superposition of several states with different n-quantum numbers but with the same quantum numbers l, j, m. We shall call this restricted class of functions *spherically symmetric*.

In the case of the axially symmetric class of functions, the reason for the choice of nomenclature is obvious. Single-particle wavefunctions having a good projection quantum number do correspond to a potential that has rotational symmetry about the z-axis. From the general expression of the density matrix, it is obvious that in this case $\rho_{\alpha\beta}$ is nonvanishing only if the projection quantum numbers m_α and m_β are equal. Thus, the matrix ρ splits up into several smaller submatrices, each contained within the subset of basis states of a given projection quantum number, as

$\rho =$ 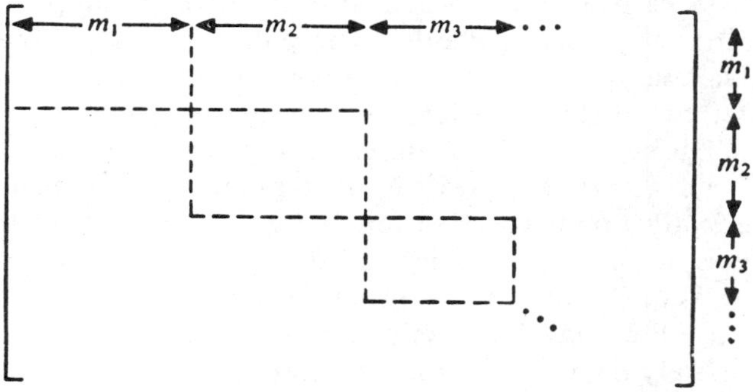 (V.14)

Nonvanishing matrix elements occur only inside the dashed boxes along the leading diagonal. It is also clear from the definition (V.8) that the matrix for $\mathcal{V}$ has the same kind of structure; this follows from the fact that the total projection quantum number in the matrix element of the two-body potential V must be conserved, i.e., $m_\mu + m_j = m_\gamma + m_j$ or $m_\mu = m_\gamma$. Thus, the symmetry about the z-axis is a self-consistent symmetry; put into the single-particle wavefunction (or, equivalently, into the density matrix), it generates itself in the HF potential $\mathcal{V}$. The solution of the HF equations, that is, the diagonalization of $(T + \mathcal{V})$ with $\mathcal{V}$, of such a form, then regenerates wavefunctions of the initial type (i.e., conserved m-quantum number). The axially symmetric class of trial functions in the HF calculation cannot therefore lead, during the process of iteration, to the more general class of triaxial functions. The reverse, however, is not true. If we start with the triaxial type of functions, then ρ has nonvanishing matrix elements everywhere, inside as well as outside the dashed boxes in (V.14). The corresponding $\mathcal{V}$ also has a general structure of this kind. However, as the iteration proceeds, the self-consistency may be achieved finally for single-particle states of the same general structure or for special values of the x-coefficients pertinent to the axially symmetric case.

Let us examine the spherically symmetric class of functions in some detail. Writing out the quantum numbers explicitly, we are here concerned with single-particle functions of the type

$$\psi(ljm) = \sum_n x_n^{ljm} \phi(nljm),$$

where ϕ is a basis wavefunction. Since the states on both sides have the same angular momentum j, it can be proved, by a well-known procedure (see Problem 4 of Appendix A), that the expansion coefficients x are necessarily independent of m. The density matrix generated by such coefficients splits into submatrices, each corresponding to a given l, j, m; but, what is more important, the *submatrices for a given l, j, m are independent of the value of m.* Now let us refer to (V.6b). The property of $\rho_{\delta\beta}$, already mentioned, ensures $m_\beta = m_\delta$, $l_\beta = l_\delta$, and $j_\beta = j_\delta$. From the property of the matrix elements of V, we immediately conclude that $m_\mu = m_\gamma$. In order to prove $j_\mu = j_\gamma$, we need to couple the angular momenta in the matrix elements of V and then carry out the summation over $m_\beta (= m_\delta)$ occurring through the Clebsch-Gordon coefficients. Because $\rho_{\delta\beta}$ is independent of this quantum number, the summation can be carried out. The details of the proof are left to the reader. Once $j_\mu = j_\gamma$ has been proved, the equality of l_μ and l_γ follows from the parity conservation in the matrix elements of V. Thus, the matrix for $\mathcal{V}$ also subdivides into the same structure as that of ρ. But there is a very important point involved in this derivation, namely, the possibility of carrying out a *complete* summation over m_β. This summation in (V.6b) appears through $\rho_{\delta\beta}$. The foregoing complete summation is justified only if all the magnetic substates m_β for a given j_β are occupied in the determinant Φ. Therefore, a starting spherical density matrix generates a similar HF potential only if the number of nucleons is such that all the magnetic substates of each energy level are occupied with particles. If the number of nucleons falls short of this criterion, then, even if we start with a spherical class of functions, the resultant HF potential $\mathcal{V}$ acquires matrix elements dependent on the projection quantum numbers and connects states of different angular momentum. Such a potential is obviously not spherically symmetric. During the course of further iteration, the nonspherical $\mathcal{V}$ generates a nonspherical set of single-particle functions, and eventually the iteration, in general, achieves self-consistency for a nonspherical solution.

One advantage of doing an HF calculation with a class of functions restricted by axial or spherical symmetry is immediately obvious. Because the matrix for $(T + \mathcal{V})$ splits into a set of smaller submatrices, the work involved in calculating the matrix elements appreciably reduces in comparison with what is involved in setting up the matrix for the general triaxial case. Moreover, iterative diagonalization of the smaller submatrices takes much less machine time than the time needed for one large matrix. The main disadvantage of such restricted calculations is that the best solution may be overlooked when it happens to be of a genuine triaxial type.

In view of the saving in time that can be achieved by a symmetry requirement, it is worthwhile to look for some other fairly general types of symmetry. A symmetry that is very generally imposed is that due to good parity of the single-particle states. The most general single-particle wavefunction does not necessarily need good parity, but there is some evidence from calculations using mixed parity functions that iteration with such functions and *reasonable* two-body potentials ultimately leads to self-consistent solutions having good parity. If the good parity requirement is imposed from the very beginning, then each matrix, or submatrix mentioned in the foregoing discussions, further splits into two matrices corresponding to the basis states of even and odd parity. The basic definitions make it clear that this symmetry is also a self-consistent symmetry, i.e., when put into the single-particle wavefunction it automatically reproduces itself in the potential $\mathcal{V}$ and vice versa.

Many authors who are reluctant to restrict themselves to axial symmetry have used a symmetry under a rotation of π about the z-axis. If we write the potential $\mathcal{V}$, which is a function of the coordinates r, θ, ϕ, σ of a single nucleon, as a general multipole expansion

$$\mathcal{V}(r, \theta, \phi, \sigma) = f(r)[1 + \sum_{\lambda, \mu} \alpha_{\lambda\mu} T^\lambda_\mu(\theta, \phi, \sigma)], \tag{V.15a}$$

then the imposition of the aforementioned symmetry on $\mathcal{V}$ demands

$$\begin{aligned}\mathcal{V} &\equiv \exp(-i\pi j_z)\mathcal{V}\exp(i\pi j_z) = f(r)[1 + \sum_{\lambda, \mu} \alpha_{\lambda\mu} \exp(-i\pi j_z) T^\lambda_\mu \exp(i\pi j_z)] \\ &= f(r)[1 + \sum_{\lambda, \mu} \alpha_{\lambda\mu}(-1)^\mu T^\lambda_\mu(\theta, \phi, \sigma)].\end{aligned} \tag{V.15b}$$

The last step follows from the fact that any tensor operator T^λ_μ under the unitary transformation $\exp(-i\pi j_z)$ gets multiplied by $\exp(-i\pi\mu) = (-1)^\mu$. Comparing (V.15a) and (V.15b), we conclude

$$\alpha_{\lambda\mu} = (-1)^\mu \alpha_{\lambda\mu}. \tag{V.15c}$$

This general relation demands that all $\alpha_{\lambda\mu}$, where μ is an odd integer, be zero. Since the potential $\mathcal{V}$ contains the tensors T^λ_μ, where μ is an even integer only, the nonvanishing matrix elements of $\mathcal{V}$ with respect to the basis states $|nljm\rangle$ exist between states differing in m by an even integer. If $m_{\max}$ is the maximum projection quantum number available for the basis states, then the nonvanishing matrix elements clearly exist within the subspace of states labelled by

$$m = m_{\max}, (m_{\max} - 2), \ldots, (-m_{\max} + 1).$$

States within the other subspace

$$m = (m_{\max} - 1), (m_{\max} - 3), \ldots, -m_{\max}$$

are then interconnected. We shall refer to these subspaces as number 1 and number 2, respectively. We thus find that the imposition of symmetry under a rotation of π about the z-axis, even in the case of the most general class of functions, helps to produce two submatrices of half the dimension of the entire space. This symmetry is also of the self-consistent type. Moreover, as in the case of parity, some authors have done specimen calculations without imposing this symmetry and have eventually arrived at solutions of this special type at the end of iterations.

Finally, let us consider a symmetry under a rotation of π about the y-axis. To do this, we require $\exp(-i\pi j_y)\mathcal{V}\exp(i\pi j_y) = \mathcal{V}$. Letting $\exp(-i\pi j_y)$ operate on (V.15a), and using the general result (BI.15) from Appendix B, we get

$$\begin{aligned}\exp(-i\pi j_y)\mathcal{V}\exp(i\pi j_y) &= f(r)[1 + \sum_{\lambda,\mu} \alpha_{\lambda\mu}(-1)^{\lambda-\mu} T^\lambda_{-\mu}(\theta, \phi, \sigma)] \\ &= f(r)[1 + \sum_{\lambda,\mu} \alpha_{\lambda,-\mu}(-1)^{\lambda+\mu} T^\lambda_\mu(\theta, \phi, \sigma)].\end{aligned} \tag{V.16a}$$

Equating this expression with (V.15a), we obtain

$$(-1)^{\lambda+\mu}\alpha_{\lambda,-\mu} = \alpha_{\lambda,\mu}. \tag{V.16b}$$

Combining (V.16b) with (V.15c), we conclude that, in the case of even λ, only combinations such as $(T^\lambda_\mu + T^\lambda_{-\mu})$ and, in the case of odd λ, combinations such as $(T^\lambda_\mu - T^\lambda_{-\mu})$ are allowed in $\mathcal{V}$. In both the cases, μ is an even integer. Under the symmetry operation $\exp(-i\pi j_y)$, a basis state $|nljm\rangle$ goes to $(-1)^{j-m}|nlj, -m\rangle$, and hence a single-particle state

$$|i\rangle = \sum_{nljm} x^i_{nljm}|nljm\rangle \tag{V.16c}$$

goes to $|\bar{i}\rangle$, where

$$|\bar{i}\rangle = \sum_{nljm} x^i_{nljm}(-1)^{j-m}|nlj, -m\rangle. \tag{V.16d}$$

If $\mathcal{V}$ is invariant under $\exp(-i\pi j_y)$, both $|i\rangle$ and $|\bar{i}\rangle$ must be degenerate. Therefore, if we combine this symmetry requirement with the preceding one, we conclude that it is necessary to diagonalize the matrix for $(T + \mathcal{V})$ only once in the subspace number 1 of

$$m = m_{\max}, (m_{\max} - 2), \ldots, (-m_{\max} + 1).$$

The basis states of subspace number 2 are obtained from those of subspace number 1 through the symmetry operation $\exp(-i\pi j_y)$, and hence the result of diagonalization in subspace number 2 produces the same eigenvalues and the corresponding eigenvectors given by (V.16d).

In the case of axial symmetry, the imposition of the additional symmetry under $\exp(-i\pi j_y)$ guarantees that each eigenvalue for m is degenerate with an eigenvalue corresponding to $-m$; the eigenvectors are given by expressions such as (V.16c) and (V.16d) by omitting the summation over m.

If we begin with a set of doubly-degenerate single-particle functions, such as (V.16c) and (V.16d), the density operator ρ remains, according to the definition (V.9a), invariant under $\exp(-i\pi j_y)$, provided we let both the states $|i\rangle$ and $|\bar{i}\rangle$ be occupied in the determinant Φ. This presupposes that both the neutron number and proton number at our disposal are even. Since the matrix elements of the two-body potential are invariant under $|i\rangle \rightarrow |\bar{i}\rangle$, as soon as the invariance of ρ is ascertained in the aforementioned way, the invariance of $\mathcal{V}$ under $\exp(-i\pi j_y)$ is automatically guaranteed. If, on the other hand, the number of either neutrons or protons is odd, then even after starting with a doubly-degenerate set of states, such as (V.16c) and (V.16d), we will have to put the last odd nucleon in a certain state while keeping its partner state unoccupied. The resultant ρ, and consequently $\mathcal{V}$, then fails to have the symmetry under $\exp(-i\pi j_y)$, and we are faced with the necessity of diagonalizing the matrices in both the subspaces number 1 and number 2, which no longer lead to identical eigenvalues. The particular symmetry under discussion is therefore not a self-consistent symmetry; it has to be ensured at every stage of the iteration by defining the occupancy of the single-particle states and their partners in pairs.

We next enquire into the physical meaning of the HF solution of different classes. We have already stated that the particular solution that produces the lowest expectation value of H should be accepted, according to the definition of the ground state, as the best intrinsic wavefunction from which the ground state with the appropriate angular momentum should be projected by the techniques described in Section 36A. The solutions of the other classes represent excited intrinsic states of the nucleus having different equilibrium shapes. For example, any solution of the axially symmetric class can, according to (V.12b), have only the moment $\langle Q_0\rangle$; the state $|i\rangle$ has a definite projection quantum number which demands only $\mu = 0$. The magnitude of $\langle Q_0\rangle$ then defines, for such a solution, the equilibrium deformation parameter β. On the other hand, a triaxial solution, with the restriction imposed by $\exp(-i\pi j_z)$ symmetry, has in $|i\rangle$ a superposition of m-values differing by even integers. Hence, according to (V.12b), $\langle Q_0\rangle$ as well as $\langle Q_2\rangle$ and $\langle Q_{-2}\rangle$ are nonvanishing. Under the further restriction due to $\exp(-i\pi j_y)$ symmetry, $\langle Q_2\rangle$ and $\langle Q_{-2}\rangle$ are equal because the summation in (V.12b) then includes $|i\rangle$ and $|\bar{i}\rangle$ in pairs, and $\langle i|\, Q_2\, |i\rangle$ is identically equal to $\langle \bar{i}|\, Q_{-2}\, |\bar{i}\rangle$. Therefore, for an HF solution of this class, we can immediately define the equilibrium deformation parameters β and γ. As long as the parity is taken to be a good quantum number, the HF states cannot have a nonvanishing value for the octupole moment, or any other moment of an odd order. In general, the states can have even moments; the maximum order clearly depends on the maximum l- and j-value of the basis states contained in the states $|i\rangle$.

In general, it is possible to have several HF solutions of the same shape, i.e., axially symmetric or triaxial. As before, the one with the lowest expectation value of H contains the ground state. The possibility of having different solutions of the same shape arises from the choice in defining the occupancy of the single-particle states while forming the variational wavefunction Φ. The reason becomes obvious (in the case of axial symmetry) when the Nilsson diagram (see Fig. IV.10) is considered. Here we notice that the single-particle levels cross each other as we go from a small to a large deformation parameter. To the left of such a cross-over, one set of single-particle states is occupied in the lowest energy Φ, whereas to the right of the cross-over a different set of single-particle states is occupied. Therefore, if we do the HF calculation with the occupancy of states as on the left of the cross-over, we obtain a self-consistent solution having an equilibrium deformation at an appropriate point to the left of the cross-over. On the other hand, an equilibrium deformation lying to the right of the cross-over point arises for the self-consistent solution obtained with the occupancy of states in Φ corresponding to that part of the single-particle level diagram. Although we have illustrated our point with reference to the Nilsson diagram, the validity of these arguments is quite general. The HF single-particle levels also can be plotted on a similar diagram; however, the lines in this case are not continuous because the solutions correspond only to a few equilibrium points of deformation.

We have now to determine the nature of the energy $E = \langle\Phi| H |\Phi\rangle$ as a function of the deformation parameters. To simplify matters, let us consider solutions of the axially symmetric class alone. As just mentioned, we generally have a few solutions of this class, corresponding to different equilibrium values of β. So we know $E(\beta)$ only at these equilibrium values of β. Since the variational method in determining the solution at each equilibrium deformation (say, β_0) ensures a local extremum of $E(\beta)$ at β_0, we know that the next term of $E(\beta)$, expanded near β_0, is quadratic in $(\beta - \beta_0)$. The linear term in $(\beta - \beta_0)$ in this expansion has the coefficient $[\partial E(\beta)/\partial\beta]_{\beta=\beta_0}$, which is equal to zero from the definition of the local extremum at β_0. The coefficient of the next term $\frac{1}{2}(\beta - \beta_0)^2$ is $[\partial^2 E(\beta)/\partial\beta^2]_{\beta=\beta_0}$, and its sign obviously determines whether $E(\beta)$ has a local minimum or local maximum at β_0.

Let us see how we can trace out the $E(\beta)$-versus-β-curve in the vicinity of the equilibrium point β_0. The HF calculation, described so far, always leads to $\beta = \beta_0$. The mathematically satisfactory way of achieving this aim is to minimize $\langle\Phi| H |\Phi\rangle$ with respect to the variational parameters x^i_α in the single-particle states, subject to the constraint (V.3), already used, and the additional constraint that $\langle\Phi| Q_0 |\Phi\rangle$ have a fixed value. The final equation of constraint is written as

$$\begin{aligned} \text{constant} &= \langle Q_0\rangle = \sum_i^{\text{occ}} \langle i| Q_0 |i\rangle \\ &= \sum_{\alpha,\beta} \langle\alpha| Q_0 |\beta\rangle \sum_i^{\text{occ}} x^{i*}_\alpha x^i_\beta. \end{aligned} \tag{V.17}$$

According to the well-known method of Lagrange multipliers, the expression to be minimized is (V.5c) minus η_0 times (V.17), where η_0 is the Lagrange multiplier. If we minimize the resultant expression in the usual way, (V.6a) gets replaced by

$$\epsilon_k x^k_\mu = \sum_\gamma x^k_\gamma \langle\mu| (T + \mathcal{V} - \eta_0 Q_0) |\gamma\rangle. \tag{V.18}$$

The self-consistent solutions of this system of equations for any value of η_0 give a set of eigenvalues $\epsilon_k(\eta_0)$, coefficients $x^k_\gamma(\eta_0)$, and a self-consistent potential $\mathcal{V}(\eta_0)$. The total energy $\langle\Phi(\eta_0)| H |\Phi(\eta_0)\rangle$, calculated from (V.11a), then determines $E(\eta_0)$. If we were to take the starting set of x^k_γ as corresponding to the $(\beta = \beta_0)$-solution of (V.6a), then the new solution, for small values of η_0, would lie in the vicinity of this deformation; the value of β corresponding to the solution of (V.18) can be determined in the usual way from the calculated value of $\langle\Phi(\eta_0)| Q_0 |\Phi(\eta_0)\rangle$.

Although we have described the procedure for axially symmetric solutions only, the

61 (45-122/1977)

generalization to the triaxial case is obvious. We have to recast the equation of constraint (V.17) in terms of $\langle Q_\mu \rangle$, and then replace the $\eta_0 Q_0$-term of (V.18) by $\sum_\mu \eta_\mu Q_\mu$. With the $\exp(-i\pi j_z)$ and $\exp(-i\pi j_y)$ symmetries, this sum reduces to $(\eta_0 Q_0 + \eta_2 Q_2 + \eta_2 Q_{-2})$.

The practical procedure for exploring a large class of HF solutions can now be appreciated by the reader: make axially symmetric calculations by using different possible types of occupancy of single-particle states. Then, corresponding to each solution, do the calculation based on (V.18). The new energy $E(\beta)$, when compared with $E(\beta_0)$ based on the solution of (V.6a), tells whether β_0 is a minimum point of the energy versus the β-deformation curve. If it is found to be a minimum, then, commencing with the $(\beta = \beta_0)$-solution, do a calculation based on the aforementioned extension of (V.18) for the triaxial case. With a nonvanishing value of η_2, the new HF solution has finite values for both $\langle Q_0 \rangle$ and $\langle Q_2 \rangle$. The new energy $E(\beta, \gamma)$ may be larger or smaller than $E(\beta_0)$. In the first case, $E(\beta_0)$ is a genuine equilibrium solution, whereas in the latter case β_0 is merely a saddle point for the surface representing $E(\beta, \gamma)$ plotted as a function of the two variables β and γ. Faced with such a situation, we can continue generating solutions with increasing γ, and finally reach a genuine minimum point of $E(\beta, \gamma)$ for a triaxial shape.

36. HARTREE-FOCK INTRINSIC STATE AND ACTUAL NUCLEAR STATES OF GOOD ANGULAR MOMENTUM

A. ANGULAR MOMENTUM PROJECTION

In this section, we describe the method of obtaining states of good angular momentum from a deformed HF determinant. The method has been successfully applied to the axially symmetric case. The derivation given here is general enough and applies to the triaxial case as well. However, the triaxial computations become a lot more involved for reasons that are indicated at the appropriate point.

In the axially symmetric case, each single-particle state has a given projection quantum number, and hence the HF determinant Φ has a given total projection quantum number K, which is the sum of the individual particle projection quantum numbers. In analogy with the collective model terminology, we shall refer to K, if necessary, as the band quantum number, and use it as a labelling quantum number for Φ. The determinant Φ_K can obviously be written as a linear superposition of states of the type $\Phi_K^{\alpha J}$, where J is the angular momentum of the state, and α stands for one or more quantum numbers needed to distinguish between different states with the same J and K. We do not know $\Phi_K^{\alpha J}$ to start with. However, if we write

$$\Phi_K = \sum_{\alpha J} C_{\alpha J} \Phi_K^{\alpha J}, \tag{V.19a}$$

it is possible to determine

$$\sum_\alpha C_{\alpha J} \Phi_K^{\alpha J}$$

in a straightforward manner from the given determinant Φ_K. When this is achieved, we say that the part of Φ_K that has the total angular momentum J has been projected from it. We shall denote this projected part by Ψ_K^J; that is, in general,

$$\Psi_K^J = \sum_\alpha C_{\alpha J} \Phi_K^{\alpha J} \tag{V.19b}$$

and it may not be normalized to unity if $\Phi_K^{\alpha J}$ is. In special circumstances, there may be only one state of a given J contained in Φ_K; in that case, α on $\Phi_K^{\alpha J}$ is a redundant label and Ψ_K^J, $\Phi_K^{\alpha J}$ are simply proportional to each other.

The general method of obtaining Ψ_K^J from Φ_K can be understood by considering a rotation of the coordinate system $R(\phi, \theta, \psi)$, given through the Eulerian angles ϕ, θ, ψ. Applying this rotation to both sides of (V.19a), we have

$$R(\phi, \theta, \psi)\Phi_K = \sum_{\alpha J'} C_{\alpha J'} R(\phi, \theta, \psi)\Phi_K^{\alpha J'}$$

$$= \sum_{\alpha J'} C_{\alpha J'} \sum_{M'} \mathscr{D}_{M'K}^{J'*}(\phi, \theta, \psi)\Phi_{M'}^{\alpha J'}.$$

Multiplying both sides by $\{[J]/(8\pi^2)\}\mathscr{D}_{MK}^J(\phi, \theta, \psi)$ and integrating over the Eulerian angles, we obtain, with the use of the standard result (BI.20b) of Appendix B and the notation $d(\phi, \theta, \psi)$ explained prior to (BI.20a) in Appendix B,

$$\Psi_M^J \equiv \sum_{\alpha} C_{\alpha J}\Phi_M^{\alpha J} = \frac{[J]}{8\pi^2}\int d(\phi, \theta, \psi)\mathscr{D}_{MK}^J(\phi, \theta, \psi)[R(\phi, \theta, \psi)\Phi_K]. \qquad \text{(V.20)}$$

We have to put $M = K$ in this expression to obtain Ψ_K^J, the projected part of Φ_K with the angular momentum J. The state Ψ_M^J with a general value M for the magnetic quantum number, as given by (V.20), is clearly related to Ψ_K^J through the operation of the required set of step-up (J_+) or step-down (J_-) operators.

According to Appendix B (Section I), the rotation operator $R(\phi, \theta, \psi)$ is given by

$$R(\phi, \theta, \psi) = \exp(-iJ_z\phi) \exp(-iJ_y\theta) \exp(-iJ_z\psi).$$

The operator $\exp(-iJ_z\psi)$, when substituted in (V.20), can be immediately replaced by $\exp(-iK\psi)$ because Φ_K is to its right. This simplification would not have followed in the case of a triaxial Φ for which K is not a good quantum number. The operator $\exp(-iJ_z\phi)$ to the left of $R(\phi, \theta, \psi)$ also gets similarly replaced by $\exp(-iK\phi)$ when we use (V.20) in the evaluation of $\langle\Psi_K^J|\, H\, |\Psi_K^J\rangle$ and $\langle\Psi_K^J \mid \Psi_K^J\rangle$. These are the expressions that we need in evaluating the energy E_J of the projected state Ψ_K^J, which is given by

$$E_J = \frac{\langle\Psi_K^J|\, H\, |\Psi_K^J\rangle}{\langle\Psi_K^J \mid \Psi_K^J\rangle}$$

$$= \frac{\langle\Phi_K|\, P_J H P_J\, |\Phi_K\rangle}{\langle\Phi_K|\, P_J^2\, |\Phi_K\rangle}. \qquad \text{(V.21a)}$$

Here P_J is the operator that projects Ψ_K^J out of the determinant Φ_K. According to (V.20), this operator is given by

$$P_J = \frac{[J]}{8\pi^2}\int d(\phi, \theta, \psi)\mathscr{D}_{KK}^J(\phi, \theta, \psi)R(\phi, \theta, \psi). \qquad \text{(V.21b)}$$

It can be verified in a straightforward manner that P_J acting on Ψ_K^J gives back Ψ_K^J. Thus, $P_J^2 = P_J$, in conformity with the fact that P_J is a projection operator. Since H is a scalar operator, and cannot change the angular momentum J of a state on which it operates, P_J to the left of H in (V.21a) can be shifted to its right, and then the property $P_J^2 = P_J$ can be used. In this way, (V.21a) yields

$$E_J = \frac{\langle\Phi_K|\, H P_J\, |\Phi_K\rangle}{\langle\Phi_K|\, P_J\, |\Phi_K\rangle}. \qquad \text{(V.21c)}$$

In this form, it is clear that $\exp(-iJ_z\phi)$ and $\exp(-iJ_z\psi)$ contained in $R(\phi, \theta, \psi)$ of (V.21b) get replaced by $\exp(-iK\phi)$ and $\exp(-iK\psi)$, respectively, when substituted in (V.21c). For the denominator, this statement about $\exp(-iJ_z\phi)$ is obvious, whereas for the numerator it follows, provided it is remembered that the scalar operator H cannot change the magnetic quantum number of Φ_K when it operates on it. Once again, we point out that these simplifications could not have been achieved in the case of a triaxial determinant Φ.

We next make use of the form (BI.12a) of $\mathscr{D}^I_{KK}(\phi, \theta, \psi)$ given in Appendix B. This lets us promptly carry out the ϕ- and ψ-integration in (V.21c), appearing through (V.21b). The result of each integration is clearly a factor of 2π. Thus, only the θ-integration of (V.21b) remains to be carried out in (V.21c). We finally obtain

$$E_J = \frac{\int_0^\pi \sin\theta\, d\theta\, d^J_{KK}(\theta)\langle\Phi_K|\, H \exp(-i\theta J_y)\,|\Phi_K\rangle}{\int_0^\pi \sin\theta\, d\theta\, d^J_{KK}(\theta)\langle\Phi_K|\exp(-i\theta J_y)\,|\Phi_K\rangle}. \tag{V.21d}$$

We denote the expectation values with respect to the determinantal state Φ_K appearing in (V.21d) by $H(\theta)$ and $I(\theta)$, i.e.,

$$H(\theta) = \langle\Phi_K|\, H \exp(-i\theta J_y)\,|\Phi_K\rangle, \tag{V.22a}$$

$$I(\theta) = \langle\Phi_K|\exp(-i\theta J_y)\,|\Phi_K\rangle. \tag{V.22b}$$

Since $d^J_{KK}(\theta)$ is a known function [see (BI.12c) in Appendix B], the expression (V.21d) can be numerically evaluated, provided we can calculate $H(\theta)$ and $I(\theta)$. This is done as follows.

J_y is a many-body operator obtained by summing j_y of an individual nucleon over all the particles. Thus,

$$\exp(-i\theta J_y) = \exp[-i\theta \sum_{I=1}^{A} j_y(I)];$$

that is, the exponential is a product of factors such as $\exp[-i\theta j_y(I)]$, there being one factor for each nucleon. When such a product of factors operates on the determinant Φ_K, each factor $\exp(-i\theta j_y)$ operates on the column in the determinant that corresponds to the coordinates of this particular nucleon. (We adhere to the convention of labelling the columns of the determinant by the particle coordinates, and the rows by the single-particle states, as done in Chapter II.) Thus, $\exp(-i\theta J_y)\Phi_K$ is a new determinantal state in which the single-nucleon states have been transformed through the operation of $\exp(-i\theta j_y)$. Let the result of this transformation on a single-particle state $|i\rangle$ be denoted by $|i_\theta\rangle$. If the axially symmetric state $|i\rangle$ having the magnetic quantum number m_i is given by

$$|i\rangle = \sum_{nlj} x^i_{nlj}|nljm_i\rangle, \tag{V.23a}$$

then $|i_\theta\rangle$ can be calculated from

$$\begin{aligned}|i_\theta\rangle \equiv \exp(-i\theta j_y)|i\rangle &= \sum_{nlj} x^i_{nlj} \exp(-i\theta j_y)|nljm_i\rangle \\ &= \sum_{nlj} x^i_{nlj} \sum_{m'} d^j_{m'm_i}(\theta)|nljm'\rangle.\end{aligned} \tag{V.23b}$$

The last step follows from the definition (BI.12b) of Appendix B. Subsequently, we shall need the overlap between the state $\langle i|$ and the rotated state $|k_\theta\rangle$ corresponding to *any other* state $|k\rangle$.

Using (V.23a) and (V.23b), we easily obtain

$$\langle i \mid k_\theta \rangle = \sum_{nlj} x^{i*}_{nlj} x^{k}_{nlj} d^{j}_{m_i m_k}(\theta). \tag{V.23c}$$

The task that remains now is to give explicit expressions for (V.22) in terms of functions such as (V.23a) and (V.23b). First, let us examine (V.22b). We write the determinants Φ_K and $\exp(-i\theta J_y)\Phi_K$ as

$$\Phi_K = \frac{1}{\sqrt{A!}} \det \{i, j, k, \ldots, p\}, \tag{V.24a}$$

$$\exp(-i\theta J_y)\Phi_K = \frac{1}{\sqrt{A!}} \det \{i_\theta, j_\theta, k_\theta, \ldots, p_\theta\}, \tag{V.24b}$$

where $1/\sqrt{A!}$ is the normalization factor and $i, j, k, \ldots, p$ is the list of occupied single-particle states. If P and P' denote arbitrary permutation of these state labels, then

$$\begin{aligned} I(\theta) &\equiv \langle \Phi_K | \exp(-i\theta J_y) | \Phi_K \rangle \\ &= \frac{1}{A!} \sum_{P, P'} (-1)^{P+P'} \langle i(1), j(2), k(3), \ldots, p(A) | \, PP' \, | i_\theta(1), j_\theta(2), k_\theta(3), \ldots, p_\theta(A) \rangle, \end{aligned}$$

where P operates to the left, P' operates to the right, and $1, 2, \ldots, A$ stand for the particle coordinates. This expression follows from the well-known expression of a determinant in terms of permutation operators. The right-hand side can be symbolically rewritten as

$$\frac{1}{A!} \sum_{P, P'} (-1)^{P+P'} PP' \langle i \mid i_\theta \rangle \langle j \mid j_\theta \rangle, \ldots, \langle p \mid p_\theta \rangle, \tag{V.24c}$$

where the angular brackets stand for single-particle overlap integrals, which no longer depend on the coordinate labels $1, 2, \ldots, A$. We have to remember that P operates on $i, j, k, \ldots, p$ and P' on $i_\theta, j_\theta, k_\theta, \ldots, p_\theta$. For any given P, the summation over P' clearly produces the determinant D, defined as

$$D = \begin{vmatrix} \langle i \mid i_\theta \rangle & \langle i \mid j_\theta \rangle & \cdots & \langle i \mid p_\theta \rangle \\ \langle j \mid i_\theta \rangle & \langle j \mid j_\theta \rangle & \cdots & \langle j \mid p_\theta \rangle \\ \langle k \mid i_\theta \rangle & \langle k \mid j_\theta \rangle & \cdots & \langle k \mid p_\theta \rangle \\ \vdots & \vdots & & \vdots \\ \langle p \mid i_\theta \rangle & \langle p \mid j_\theta \rangle & \cdots & \langle p \mid p_\theta \rangle \end{vmatrix}. \tag{V.25a}$$

The summation over P then produces the same determinant $A!$ times which removes $1/A!$ from (V.24c). Thus, we finally have

$$I(\theta) \equiv \langle \Phi_K | \exp(-i\theta J_y) | \Phi_K \rangle = D \text{ of (V.25a)}. \tag{V.25b}$$

It is a little more involved to handle the quantity $H(\theta)$ of (V.22a) than $I(\theta)$. Since H is a sum of kinetic and potential energies, we write

$$H(\theta) = T(\theta) + V(\theta), \tag{V.26a}$$

where

$$T(\theta) = \langle \Phi_K | \, (\sum_{J=1}^{A} T_I) \exp(-i\theta J_y) \, | \Phi_K \rangle, \tag{V.26b}$$

$$V(\theta) = \langle \Phi_K | \, (\sum_{I<J}^{A} V_{IJ}) \exp(-i\theta J_y) \, | \Phi_K \rangle. \tag{V.26c}$$

To evaluate (V.26b), we make use of the well-known expansion of the determinant Φ_K in terms of the elements of the first column labelled by the coordinates of the first nucleon:

$$\Phi_K = \frac{1}{\sqrt{A}} \sum_s (-1)^{N_s+1} |s(1)\rangle \Phi_K^s(2, 3, \ldots, A), \tag{V.27a}$$

where Φ_K^s is the *normalized* [normalization factor $= 1/\sqrt{(A-1)!}$] determinant of the $(A-1)$ nucleons indicated within the parentheses following it, obtained by suppressing the first column and the row in position N_s corresponding to the single-particle state $|s\rangle$. The summation over s in (V.27a) proceeds over all the occupied states in Φ_K. The state $\exp(-i\theta J_y)\Phi_K$ can be similarly expanded in terms of the rotated states $|t_\theta(1)\rangle$ and the corresponding determinants $\Phi_K^{t_\theta}(2, 3, \ldots, A)$ as

$$\exp(-i\theta J_y)\Phi_K = \frac{1}{\sqrt{A}} \sum_t (-1)^{N_t+1} |t_\theta(1)\rangle \Phi_K^{t_\theta}(2, 3, \ldots, A). \tag{V.27b}$$

The matrix element of the one-body type operator

$$\sum_{I=1}^{A} T_I$$

between antisymmetric states is equal, according to (III.60a), to A times the matrix element of T_1. Thus,

$$T(\theta) = \sum_{s,t} (-1)^{N_s+N_t} \langle s| T |t_\theta\rangle \langle \Phi_K^s \mid \Phi_K^{t_\theta}\rangle. \tag{V.28a}$$

The overlap of *any* two determinantal states using different sets of single-particle wavefunctions has already been worked out and given by a determinant such as (V.25a). The same general result can obviously be used to express the overlap $\langle \Phi_K^s \mid \Phi_K^{t_\theta}\rangle$ of (V.28a). A little reflection shows that the $(A-1)$-dimensional determinant for this overlap, with the phase factor $(-1)^{N_s+N_t}$, is the minor of the element $\langle s \mid t_\theta\rangle$ of (V.25a). This minor can be very simply evaluated from D and the element $\langle s \mid t_\theta\rangle$ (see Problem 2 at the end of this chapter).

The algebra involved in the evaluation of $V(\theta)$ of (V.26c) is very similar to that of $T(\theta)$. Starting with the expression (V.27a), we extract the states belonging to the coordinates of the second nucleon from each $\Phi_K^s(2, 3, \ldots, A)$ by expanding it in the usual way in terms of the elements of its first column (which is labelled by 2). A similar expansion of $\Phi_K^{t_\theta}(2, 3, \ldots, A)$ is introduced in (V.27b). We also use the fact [see (III.60b)] that the matrix element of $\sum_{I<J} V_{IJ}$ is equal to $\frac{1}{2}A(A-1)$ times the matrix element of V_{12}. In this way, we finally obtain

$$V(\theta) = \sum_{r<s} \sum_{t<u} (-1)^{N_r+N_s+N_t+N_u} (rs| V |tu) \langle \Phi_K^{rs} \mid \Phi_K^{t_\theta u_\theta}\rangle, \tag{V.28b}$$

where a determinant labelled by two superscripts is obtained from Φ_K (or its rotated form) by omitting the two rows labelled by the superscripts and omitting also the columns 1 and 2. The $(A-2)$-dimensional determinant which represents the overlap $\langle \Phi_K^{rs} \mid \Phi_K^{t_\theta u_\theta}\rangle$ is also a minor of the determinant D, and an expression that should be proved (see Problem 3 at the end of this chapter) can be used for its evaluation.

Approximation Valid for Small θ

The quantity $I(\theta)$ represents the overlap of the same determinantal state referred to two different coordinate frames oriented with respect to each other through the Eulerian angle θ. If θ is zero, then the overlap is complete, and $I(\theta)$ has the value unity. As θ increases through very small values, the overlap falls increasingly below unity, and finally it may be expected to be nearly zero for large θ. If only very small values of θ are important for the functions $H(\theta)$ and $I(\theta)$, then a useful approximation can be developed as follows. For small θ, $\exp(-i\theta J_y)$ can be replaced by $(1 - i\theta J_y - \frac{1}{2}\theta^2 J_y^2)$ to get

$$\begin{aligned} d^J_{KK}(\theta) &\equiv \langle JK|\exp(-i\theta J_y)|JK\rangle \\ &\approx 1 - \tfrac{1}{2}\theta^2\langle JK|J_y^2|JK\rangle \\ &= 1 - \tfrac{1}{4}\theta^2\langle JK|(J^2 - J_z^2)|JK\rangle \\ &= 1 - \tfrac{1}{4}\theta^2\{J(J+1) - K^2\}. \end{aligned} \tag{V.29}$$

In view of the axial symmetry, we have here used the fact that J_x^2 and J_y^2 have the same expectation value. The term $-i\theta J_y$ in the expansion of $\exp(-i\theta J_y)$ cannot connect $|JK\rangle$ with $\langle JK|$, and hence it does not appear in $d^J_{KK}(\theta)$.

Using (V.29) in (V.21d), we obtain

$$E_J \approx \frac{\int_0^\pi \sin\theta\, d\theta\, [1 - \frac{1}{4}\theta^2\{J(J+1) - K^2\}]H(\theta)}{\int_0^\pi \sin\theta\, d\theta\, [1 - \frac{1}{4}\theta^2\{J(J+1) - K^2\}]I(\theta)}.$$

The expressions for $H(\theta)$ and $I(\theta)$, which contain $\exp(-i\theta J_y)$, can also be similarly simplified. We shall deal with that presently. For now, we simplify the expression for E_J by assuming that the integrals containing θ^2 are of one order of smallness compared with the integrals without this factor. In such a case, keeping terms up to the first order of smallness, we get

$$E_J \approx \frac{\int_0^\pi \sin\theta\, d\theta\, H(\theta)}{\int_0^\pi \sin\theta\, d\theta\, I(\theta)} + \tfrac{1}{4}[J(J+1) - K^2]\left[\frac{\int_0^\pi \sin\theta\, d\theta\, \theta^2 I(\theta)}{\int_0^\pi \sin\theta\, d\theta\, I(\theta)} - \frac{\int_0^\pi \sin\theta\, d\theta\, \theta^2 H(\theta)}{\int_0^\pi \sin\theta\, d\theta\, H(\theta)}\right]. \tag{V.30}$$

To this approximation, therefore, the energies of the angular momentum projected states fit into a rotational spectrum given by the $J(J+1)$-dependence. In general, the good angular momentum states projected *accurately* from the intrinsic state Φ_K do not have the exact rotational-type spectrum. Here we have merely shown the approximate conditions under which they are expected to do so.

We next indicate the approximate forms that $I(\theta)$ and $H(\theta)$ assume if only small θ is important. Once again, the term $-i\theta J_y$ acting on Φ_K changes the band quantum number, and hence

$$\begin{aligned} I(\theta) &\approx \langle\Phi_K|(1 - \tfrac{1}{2}\theta^2 J_y^2)|\Phi_K\rangle \\ &= 1 - \tfrac{1}{2}\theta^2\langle\Phi_K|J_y^2|\Phi_K\rangle \approx \exp(-\tfrac{1}{2}\theta^2\langle\Phi_K|J_y^2|\Phi_K\rangle). \end{aligned} \tag{V.31a}$$

Similarly,

$$\begin{aligned} H(\theta) &\approx \langle\Phi_K|H(1 - \tfrac{1}{2}\theta^2 J_y^2)|\Phi_K\rangle \\ &= \langle\Phi_K|H|\Phi_K\rangle\left(1 - \tfrac{1}{2}\theta^2\frac{\langle\Phi_K|HJ_y^2|\Phi_K\rangle}{\langle\Phi_K|H|\Phi_K\rangle}\right) \end{aligned}$$

$$\approx \langle \Phi_K | H | \Phi_K \rangle \exp [-\tfrac{1}{2}\theta^2 \frac{\langle \Phi_K | HJ_y^2 | \Phi_K \rangle}{\langle \Phi_K | H | \Phi_K \rangle}]. \tag{V.31b}$$

According to these approximate expressions, $I(\theta)$ and $H(\theta)$ decrease very rapidly in a Gaussian manner with increasing θ. The quantity $\langle \Phi_K | H | \Phi_K \rangle$ is already known from the initial stage of the HF calculation [see (V.11a)]. To compute (V.31a) and (V.31b), we therefore need to calculate the expectation values of the operators J_y^2 and HJ_y^2 for the determinantal state Φ_K, which can be done in a straightforward manner.

B. CALCULATION OF MOMENT OF INERTIA

It is clear from the derivation in Section 36A that the calculation of nuclear spectra by projecting out good angular momentum states from the HF determinant is a very tedious procedure, and only under very special circumstances does this procedure lead to a rotational-type spectrum described by

$$E_J = E_0 + \mathcal{A}J(J+1), \tag{V.32}$$

where E_0 is the energy of the state $J = 0$, and $\mathcal{A}$ is related to the moment of inertia $\mathcal{J}$ by $\mathcal{A} = \hbar^2/(2\mathcal{J})$.

If it is known from experimental data that a nucleus has a rotational-type spectrum, described by (V.32), then a short cut to calculating $\mathcal{A}$ from the HF determinant can be designed as follows. For simplicity, let us assume that the HF determinant Φ_0 has spheroidal symmetry. If z is the axis of symmetry, then the value of J_x or J_y, where $\mathbf{J}$ is the total angular momentum operator, vanishes for the state Φ_0 because the latter has a good band quantum number. In order to generate a nonvanishing value for J_x, for instance, let us consider, instead of H, the Hamiltonian

$$H(\omega) = H - \omega J_x, \tag{V.33a}$$

where ω is a parameter. If we want to solve this Hamiltonian self-consistently, then it clearly amounts to doing an HF type diagonalization of the single-particle Hamiltonian

$$T + \mathcal{V} - \omega j_x, \tag{V.33b}$$

where $\mathbf{j}$ is the single-particle angular momentum operator. The resultant HF determinant may be denoted by $\Phi(\omega)$. The quantity $\langle \Phi(\omega) | J_x | \Phi(\omega) \rangle$, which is the induced value of the x-component of angular momentum, is then equated with $\omega\mathcal{J}_x$. For a spheroidal nucleus, $\mathcal{J}_x = \mathcal{J}_y = \mathcal{J}$. This procedure then determines $\mathcal{J}_x$ or, equivalently, $\mathcal{J}$. The main drawback of this type of derivation is the lack of rigour in equating $\langle \Phi(\omega) | J_x | \Phi(\omega) \rangle$ with $\omega\mathcal{J}_x$. It is not clear from (V.33a) whether the parameter ω has anything to do with an imposed angular velocity. Self-consistent diagonalization of $H(\omega)$ indeed makes the expectation value of H minimum subject to the constraint that $\langle J_x \rangle$ have a given value. The quantity ω in this derivation is therefore merely a Lagrange multiplier for the equation of constraint.

In general, the value of the angular momentum $\langle \Phi(\omega) | J_x | \Phi(\omega) \rangle$ is not linear in ω, and hence $\mathcal{J}_x$ determined through the foregoing procedure may indeed depend on the value chosen for this parameter. Only in the special case, where we treat ωJ_x of (V.33a) in the first-order perturbation theory, can the linearity with respect to ω hold. When such an approximate treatment of (V.33a) is done, we obtain an expression for the moment of inertia which, although very approximate, has been very widely used in nuclear calculations. This is the celebrated cranking formula for the moment of inertia, first derived by Inglis, by a somewhat different method. The first-order solution and the cranking formula follow.

Let Φ_0 be the lowest solution of H and the excited states be denoted by Φ_n. Then the perturbed state $\Phi(\omega)$ to the first order in the perturbation is given by

$$\Phi(\omega) = \Phi_0 + \omega \sum_n \frac{\langle\Phi_n| J_x |\Phi_0\rangle}{E_n - E}\Phi_n, \tag{V.34a}$$

where E_n and E are the unperturbed energies of the states Φ_n and Φ_0, respectively. Since $\Phi_n \neq \Phi_0$ and J_x is a one-body type operator, according to Section 15, Φ_n can differ from Φ_0 only through one single-particle state, i.e., it is a one-hole one-particle type state with reference to the ground state Φ_0. Let the occupied state h in Φ_0 be replaced by an unoccupied state p to form a particular Φ_n. Then

$$\langle\Phi_n| J_x |\Phi_0\rangle \equiv \langle p| j_x |h\rangle, \tag{V.34b}$$

$$E_n - E = \epsilon_p - \epsilon_h, \tag{V.34c}$$

where ϵ is the HF single-particle energy. Using the expressions (V.34) together, and working to the first order in ω, we easily obtain

$$\begin{aligned}\langle\Phi(\omega)| J_x |\Phi(\omega)\rangle &= 2\omega \sum_n \frac{|\langle\Phi_n| J_x |\Phi_0\rangle|^2}{E_n - E}\\ &= 2\omega \sum_{h,\,p} \frac{|\langle p| j_x |h\rangle|^2}{\epsilon_p - \epsilon_h}.\end{aligned} \tag{V.35a}$$

The term $\langle\Phi_0| J_x |\Phi_0\rangle$ is zero because J_x acting on Φ_0 changes the band quantum number. Equating (V.35a) with $\omega\mathcal{J}_x$, we then obtain the cranking formula

$$\mathcal{J} = 2 \sum_{h,\,p} \frac{|\langle p| j_x |h\rangle|^2}{\epsilon_p - \epsilon_h}. \tag{V.35b}$$

A better way of obtaining the moment of inertia is now described. If the energy of an eigenstate Ψ^J of angular momentum J is really given by (V.32), then

$$H\Psi^J = E_J\Psi^J = [E_0 + \mathcal{A}J(J+1)]\Psi^J. \tag{V.36a}$$

The HF determinantal wavefunction Φ_0 can be expanded in terms of the eigenstates Ψ^J such that

$$\Phi_0 = \sum_J C_J\Psi^J, \tag{V.36b}$$

where

$$\sum_J |C_J|^2 = 1, \qquad \langle\Psi^J \mid \Psi^J\rangle = 1.$$

Then the expectation value of H for the determinant Φ_0 is given by

$$\begin{aligned}\langle\Phi_0| H |\Phi_0\rangle &= \sum_J |C_J|^2\langle\Psi^J| H |\Psi^J\rangle\\ &= \sum_J |C_J|^2[E_0 + \mathcal{A}J(J+1)]\\ &= E_0 + \mathcal{A} \sum_J |C_J|^2 J(J+1).\end{aligned}$$

Similarly, the expectation value of $\mathcal{A}J^2$ is easily found to be

$$\langle\Phi_0| \mathcal{A}J^2 |\Phi_0\rangle = \mathcal{A} \sum_J |C_J|^2 J(J+1).$$

Therefore,

$$\langle \Phi_0 | (H - \mathcal{A}J^2) | \Phi_0 \rangle = E_0. \tag{V.36c}$$

The method referred to is based on this result.

If the spectrum is really given by (V.32), then our aim in an HF calculation should obviously be to determine E_0 (which is the energy of the ground state having $J = 0$) and *not* $\langle \Phi_0 | H | \Phi_0 \rangle$. Thus, instead of minimizing the expectation value of H with respect to the HF determinant, we should minimize the expectation value of $H - \mathcal{A}J^2$, where $\mathcal{A}$ is an unknown and variable parameter to start with. According to our discussions on the method of Lagrange multipliers, it is clear that the present procedure amounts to minimizing the expectation value of H subject to the constraint of requiring a specified value of $\langle \Phi_0 | J^2 | \Phi_0 \rangle$. The parameter $\mathcal{A}$ then plays the role of the Lagrange multiplier for the equation of constraint.

In view of the result

$$J^2 = \sum_{I=1}^{A} j_I^2 + 2 \sum_{I<K}^{A} \mathbf{j}_I \cdot \mathbf{j}_K, \tag{V.37}$$

where the summations go over the A particles, we easily conclude that the minimization of $\langle (H - \mathcal{A}J^2) \rangle$ is equivalent to the self-consistent diagonalization of the single-particle Hamiltonian

$$T + \mathcal{V} - 2\mathcal{A}\mathcal{J} - \mathcal{A}j^2,$$

where the last term is a consequence of the first term (*one-body* term) of (V.37), and the third term is a consequence of the second term (*two-body* term) of (V.37). The quantity $\mathcal{J}$ is clearly defined, analogous to $\mathcal{V}$, as

$$\langle \alpha | \mathcal{J} | \gamma \rangle = \sum_{\beta, \delta} (\alpha\beta | \mathbf{j}_I \cdot \mathbf{j}_K | \gamma\delta) \rho_{\delta\beta}.$$

The numerical programme for obtaining the best value of the parameter $\mathcal{A}$ is as follows: carry out the self-consistent diagonalization for a range of values of $\mathcal{A}$ and, in each case, obtain the corresponding value of $E_0(\mathcal{A})$ according to

$$E_0(\mathcal{A}) \equiv \langle \Phi_0(\mathcal{A}) | (H - \mathcal{A}J^2) | \Phi_0(\mathcal{A}) \rangle$$
$$= \sum_{i}^{\text{occ}} \langle i(\mathcal{A}) | (T - \mathcal{A}j^2) | i(\mathcal{A}) \rangle + \sum_{i<k}^{\text{occ}} (i(\mathcal{A})k(\mathcal{A}) | (V - 2\mathbf{j}_1 \cdot \mathbf{j}_2) | i(\mathcal{A})k(\mathcal{A})),$$

where $i(\mathcal{A})$, $k(\mathcal{A})$ are the HF single-particle states for a chosen value of the parameter $\mathcal{A}$. Then we use (V.32) to obtain $E_J(\mathcal{A})$ for the levels of different J. The parameter $\mathcal{A}$ can be finally determined by a least-squares fit of the exact experimental spectra E_J^{exact} with the theoretically calculated values. That is, minimize

$$\sum_{J} [-E_J(\mathcal{A}) + E_J^{\text{exact}}]^2 W_J \tag{V.38}$$

with respect to the parameter $\mathcal{A}$, where W_J is a suitably chosen weight factor for the level J, and the summation goes over all the observed levels of the ground-state rotational band. The best choice for W_J is, as is evident in the proof that follows, $|C_J|^2$, which is the probability of the state Ψ^J in the determinant Φ_0.

Since Φ_0 is not an exact eigenstate of $(H - \mathcal{A}J^2)$, we can write

$$(H - \mathcal{A}J^2)\Phi_0 = E_0\Phi_0 + \chi, \tag{V.39a}$$

where X is a quantity with a positive norm. We have

$$X = [(H - \mathcal{A}J^2) - E_0]\Phi_0,$$

and hence

$$\langle X \mid X\rangle = \langle\Phi_0|\,[(H - \mathcal{A}J^2) - E_0]^2\,|\Phi_0\rangle.$$

Substituting the expansion (V.36b) for Φ_0, and the exact eigenvalue E_J^{exact} of H for Ψ^J, we easily obtain

$$\begin{aligned}\langle X \mid X\rangle &= \sum_J |C_J|^2[E_J^{\text{exact}} - \mathcal{A}J(J+1) - E_0]^2\\ &= \sum_J |C_J|^2[E_J^{\text{exact}} - E_J(\mathcal{A})]^2.\end{aligned} \tag{V.39b}$$

According to the definition (V.39a), it is clear that the lower the value of (V.39b), the better the solution and the corresponding Φ_0. So the theoretical criterion for choosing $\mathcal{A}$ is to minimize (V.39b) or, equivalently, (V.38) with the choice of W_J equal to $|C_J|^2$, as already asserted.

It is possible to calculate $\langle X \mid X\rangle$ without using the exact observed eigenvalue E_J^{exact}. We write

$$\begin{aligned}\langle X \mid X\rangle &= \langle\Phi_0|\,[(H - \mathcal{A}J^2)^2 + E_0^2 - 2E_0(H - \mathcal{A}J^2)]\,|\Phi_0\rangle\\ &= \sum_n \langle\Phi_0|\,(H - \mathcal{A}J^2)\,|\Phi_n\rangle\langle\Phi_n|\,(H - \mathcal{A}J^2)\,|\Phi_0\rangle - E_0^2\\ &= \sum_n{}' \,|\langle\Phi_0|\,(H - \mathcal{A}J^2)\,|\Phi_n\rangle|^2,\end{aligned} \tag{V.39c}$$

where the summation with the prime excludes the state Φ_0 from the summation over the intermediate states Φ_n. In reaching the final expression, we have used the definition $E_0 = \langle\Phi_0|\,(H - \mathcal{A}J^2)\,|\Phi_0\rangle$. Since $(H - \mathcal{A}J^2)$ contains only one- and two-body type operators, the states Φ_n could be 1h-1p and 2h-2p type states with respect to Φ_0. The former type of states does not contribute because of the well-known result (see Problem 7 at the end of this chapter) of the HF theory, which states that the matrix element of the Hamiltonian, minimized in the HF calculation (in the present case $H - \mathcal{A}J^2$), connecting the HF state with any 1h-1p type state, is identically zero.

37. HARTREE-FOCK CALCULATIONS AND RESULTS

A. TWO-BODY POTENTIAL

Many different types of two-body potential have been used in HF calculations. Levinson and his collaborators[1] used a central potential with the Rosenfeld exchange-dependence and Yukawa shape:

$$V(r) = \tfrac{1}{3}V_0\boldsymbol{\tau}_1\cdot\boldsymbol{\tau}_2(0.3 + 0.7\boldsymbol{\sigma}_1\cdot\boldsymbol{\sigma}_2)\frac{e^{-r/a}}{r/a}, \tag{V.40}$$

where V_0 and a are the depth and range parameters and are taken to be

$$V_0 = -42.5 \text{ MeV}, \qquad a = 1.37 \text{ fm};$$

the subscripts 1 and 2 refer to the two interacting nucleons with the relative coordinate **r**, and **σ**, **τ** are the spin and isospin operators.

Baranger and his coworkers[2,3] used several different kinds of two-body potential. In one of their earliest studies (see Muthukrishnan and Baranger[2]), they applied a separable nonlocal potential to the relative S-state of the two nucleons:

$$V(r) = V_0 \exp[-\{r/(2a)\}^2] \exp[-\{r'/(2a)\}^2] \tag{V.41a}$$

with $V_0 = -0.362$ MeV fm^{-3} and $a = 1.175$ fm. This potential gives a binding energy of 15.7 MeV per nucleon, and a saturation density corresponding to the fermi momentum $k_F = 1.42$ fm^{-1} in the infinite nuclear matter calculation. This group all along adopted the following attitude: instead of trying to use a two-body potential that fits all two-nucleon data perfectly and yet fails in the binding energy calculation in nuclear matter, it is more advisable in HF calculations (which is essentially a binding energy calculation) to use a two-body potential that succeeds well in the latter calculation and reproduces *some* low-energy (such as scattering length and effective range) two-nucleon data fairly well.

In another set of calculations (see Davies et al[3]), the same group used a velocity-dependent central potential acting only in the even spatial states:

$$V(r, p) = \frac{\hbar^2}{M}[U(r) + \nabla^2 W(r) + W(r)\nabla^2]. \tag{V.41b}$$

Here M is the nucleon mass, and $-\hbar^2\nabla^2$ is the operator standing for the square of the relative momentum. The shape functions $U(r)$ and $W(r)$ are taken in the form

$$U(r) = -A \exp(-\tfrac{1}{2}r^2/\alpha^2), \qquad W(r) = B \exp(-\tfrac{1}{2}r^2/\beta^2)$$

with $\alpha = \beta$. The values of the depth parameters A, B, of the range parameter $\alpha(=\beta)$, and of the scattering length (a) and the effective range (r_0) that this potential reproduces are given in Table V.1a.

Table V.1 Parameter of velocity-dependent potential used in Hartree-Fock calculations
(a) Velocity-dependent central potential

Spin State	A (fm)$^{-2}$	B	α (fm)	a (fm)	r_0 (fm)
Singlet	0.835	0.60	1.41	−22.95	4.52
Triplet	2.56	0.50	1.01	5.76	2.13

Nestor et al[3] supplemented the velocity-dependent potential with the tensor and spin-orbit potentials, and used nonvanishing potentials in *all* the four spin-isospin states. The extra terms in the potential are

$$V_T(r)S_{12} + V_{LS}(r)\mathbf{L}\cdot\mathbf{S}$$

with

$$V_T(r) = -A_T(\tfrac{1}{2}r^2/\alpha_T^2)\exp(-\tfrac{1}{2}r^2/\alpha_T^2).$$

A typical set of values of the parameters is given in Table V.1b.

Table V.1 Parameter of velocity-dependent potential used in Hartree-Fock calculations (b) Velocity-dependent central potential plus tensor and spin-orbit coupling

State	A (fm)$^{-2}$	α (fm)	B	A_T (fm)$^{-2}$	α_T (fm)	A_{LS} (fm)$^{-2}$	α_{LS} (fm)
Triplet-even	7.227	0.575	1.0	0.490	1.20		
Singlet-even	2.127	0.903	0.6				
Triplet-odd	0.219	1.40	0.128	−0.35	1.0	1.40	0.8
Singlet-odd	−0.128	1.40	0.307				

Another type of potential was pioneered by Volkov[4] and has been used by several authors. In his $0p$-shell work, Volkov chose

$$V(r) = f(r)(1 - m + mP_M) \tag{V.42a}$$

with

$$f(r) = V_a \exp[-(r/\alpha)^2] + V_r \exp[-(r/\rho)^2].$$

The criteria he applied for the determination of the parameters m, V_a, V_r, α, and ρ are:

(i) the two-nucleon S-wave scattering length should agree reasonably well with the experimental singlet and triplet values;

(ii) the two-nucleon effective range should also be reproduced fairly well; and

(iii) the binding energies of He^4 and O^{16} should be well-reproduced with a shell-model Slater determinant.

Eight different sets of parameters were obtained in this way. A typical set of values is given by

$$m = 0.6, \qquad V_a = -83.34 \text{ MeV}, \qquad V_r = 144.86 \text{ MeV},$$

$$\alpha = 1.60 \text{ fm}, \qquad \rho = 0.82 \text{ fm}.$$

In later calculations, Volkov used a general exchange-dependence given by

$$V(r) = f(r)(0.29 + 0.2P_\sigma - 0.05P_\tau + 0.71P_M), \tag{V.42b}$$

where

$$f(r) = -78 \exp[-(r/1.5)^2] + 82.5 \exp[-(r/0.8)^2].$$

Here the depths 78 and 82.5 are in MeV, r and the range parameters 1.5 and 0.8 are in fm, and P_σ, P_τ, P_M are the usual two-nucleon exchange operators. The Gaussian radial-dependence of this potential makes it specially suitable for use with basic single-particle wavefunctions corresponding to a deformed harmonic oscillator potential.

Several different potentials have been used by Brink and Boeker[5], one of which is a slight variant of the Volkov potential just quoted. The criteria they imposed on their two-body potential are:

(i) ^4He ground-state energy (−28.2 MeV) be correctly reproduced;

(ii) ^4He root-mean-square radius (1.62 fm) be correctly given;

(iii) the calculated binding energy per nucleon in nuclear matter be 15.75 MeV; and

(iv) the saturation density, calculated for nuclear matter, should lead to a fermi momentum $k_F = 1.45$ fm^{-1}.

Brink and Boeker used three different types of potential satisfying the criteria just mentioned. These types are now described.

(1) *Volkov-type Gaussian potential*

$$V(r) = V_1(1 - m_1 + m_1 P_M) \exp[-(r/\mu_1)^2] + V_2(1 - m_2 + m_2 P_M) \exp[-(r/\mu_2)^2]. \tag{V.43a}$$

Reasonable sets of values were chosen for the range parameters μ_1, μ_2 and then four other parameters, namely, V_1, V_2, m_1, m_2, were determined to fulfil the criteria (i)–(iv). Of the several sets of parameters calculated by Brink and Boeker, a typical set of values is given by

$$\mu_1 = 1.4 \text{ fm}, \qquad \mu_2 = 0.7 \text{ fm},$$
$$V_1 = -140.6 \text{ MeV}, \qquad V_2 = 389.5 \text{ MeV},$$
$$m_1 = 0.4864, \qquad m_2 = 0.529.$$

(2) *Gaussian plus delta-function potential*

$$V(r) = V_0(1 - m + mP_M) \exp[-(r/\mu)^2] + A\delta(r). \tag{V.43b}$$

The range parameter μ was fixed to a reasonable value, and then V_0, m, and A determined to fit the criteria (i), (ii), and (iv); the value of the binding energy in criterion (iii) then automatically follows. A typical set of values of the parameters is given by

$$\mu = 1.12 \text{ fm}, \qquad V_0 = -301.0 \text{ MeV},$$
$$m = 0.6169, \qquad A = 1011 \text{ MeV fm}^3.$$

(3) *Velocity-dependent delta-function interaction*

$$V(r) = A\delta(r) - B[\nabla^2\delta(r) + \delta(r)\nabla^2] - 2C\nabla \cdot \delta(r)\nabla, \tag{V.43c}$$

which in the momentum space reads as

$$\langle \mathbf{k}| V |\mathbf{k}'\rangle = A + B(k^2 + k'^2) + 2C\mathbf{k} \cdot \mathbf{k}'. \tag{V.43c$'$}$$

In actual application, a more specialized version of this equation was used; this consists of only an S-state interaction in the form

$$V(r) = A\delta(r) - B[\nabla^2\delta(r) + \delta(r)\nabla^2] + D\nabla^2\delta(r)\nabla^2 \tag{V.43d}$$

or

$$\langle \mathbf{k}| V |\mathbf{k}'\rangle = A + B(k^2 + k'^2) + Dk^2k'^2. \tag{V.43d$'$}$$

The values of A, B, D were determined to fit the criteria (ii)–(iv). The value of ^{4}He binding energy [criterion (i)] then automatically follows. A typical set of values in MeV fm^3 is given by

$$A = -923.5, \qquad B = 226.5, \qquad D = 166.9.$$

Besides the foregoing two-body potentials, all of which [except (V.40)] were specially designed for HF work, realistic two-body potentials have also been used in self-consistent calculations. The Tabakin potential has been used by Kerman et al[6] in both first- and second-order energy calculations. In another set of calculations by Shakin et al[7], effective two-body matrix elements of the Yale potential have been applied. Similar matrix elements of the Hamada-Johnston potential, calculated by Kuo, have been used in HF calculations by Davies et al[8].

B. RESULTS OF HARTREE-FOCK CALCULATIONS

Work by Levinson and Collaborators

Several distinctly different types of HF calculation have been done by various groups of workers. For nuclei in the ($1s$, $0d$)-shell, Levinson and his collaborators[1] treated O^{16} as an inert core, and worked out self-consistent calculations for only the (1*s*, 0*d*)-nucleons, using spherical harmonic oscillator wavefunctions as the basis. The two-body potential used by this group has already been given (see Section 37A). The formalism in Section 35A has to be slightly modified in such calculations; because the core nucleons do not explicitly enter into the calculation, their effect has to be incorporated through the replacement of the kinetic energy operator of the (1*s*, 0*d*)-nucleons by a suitably defined set of single-particle energies, following the principle explained in Section 28E. The exact values of these single-particle energies are obtained from experimental data. The basis wavefunctions in these calculations are the various projection states $k = \frac{5}{2}, \frac{3}{2}, \frac{1}{2}$, arising from the $(1s_{1/2})$-, $(0d_{5/2})$-, and $(0d_{3/2})$-harmonic oscillator states. Their single-particle wavefunctions corresponding to spheroidal self-consistent solutions were therefore very similar to the Nilsson wavefunctions (described in Section 33D). However, the large energy gap between the occupied and unoccupied single-particle states rendered their single-particle energy spectrum distinctly different from the Nilsson spectrum.

The single-particle spectra for various nuclei are shown in Fig. V.1. The quantity θ denotes the degree of filling of each level denoted by the subscripts.

Some qualitative understanding of the existence of the energy gap in the single-particle spectrum is now given, following the work of Bar-Touv and Levinson[9]. For simplicity, these authors assumed the unperturbed single-particle energies to be degenerate, and considered a central two-body potential with a general exchange-dependence but infinite range. In other words, the radial shape of the potential can be taken to be a constant. Such a potential can be written as

$$V = a_W + a_M P_M + a_\sigma P_\sigma - a_\tau P_\tau, \tag{V.44}$$

where P_M, P_σ, $-P_\tau$ are respectively the Majorana, Bartlett, and Heisenberg exchange operators; a_W, a_M, a_σ, a_τ are the strength parameters. Levinson et al assumed an infinite-range potential because they found that, by varying the range of the two-body potential, the *relative* level spacing remains nearly the same. Bar-Touv and Levinson therefore argued that the spectrum obtained by using the infinite-range potential can, by multiplication with a constant, be changed to the actual spectrum for a potential having a more realistic value of the range.

The contribution to the single-particle energy due to the two-body potential (V.44) is given, according to the definition of the HF potential, by

$$\langle i|\mathcal{V}|i\rangle = \sum_j^{\text{occ}} (i, j|\, V\, |i, j). \tag{V.45}$$

We shall work out the right-hand side for each term of (V.44) separately. While doing so, we remember that each state $i, j, \ldots$ consists of a space ϕ, spin χ, and isospin ρ part. Thus, $(\phi_i\chi_i\rho_i)$ stands for the state i, and so on. The summation over j can similarly be split into summations over ϕ_j, χ_j, and ρ_j. The Wigner term of (V.44) therefore contributes to (V.45) the result

$$\begin{aligned} a_W & \sum_{\phi_j}^{\text{occupied}}\sum_{\chi_j}\sum_{\rho_j} (\langle\phi_i\chi_i\rho_i;\, \phi_j\chi_j\rho_j \mid \phi_i\chi_i\rho_i;\, \phi_j\chi_j\rho_j\rangle - \langle\phi_i\chi_i\rho_i;\, \phi_j\chi_j\rho_j \mid \phi_j\chi_j\rho_j;\, \phi_i\chi_i\rho_i\rangle) \\ &= a_W\Big(\frac{n}{4} \times 2 \times 2 - \sum_{\phi_j}^{\text{occupied}}\sum_{\chi_j}\sum_{\rho_j} \delta_{\phi_i\phi_j}\delta_{\chi_i\chi_j}\delta_{\rho_i\rho_j}\Big) \\ &= a_W(n - \theta_i), \end{aligned} \tag{V.46a}$$

where n is the total number of nucleons and θ_i is equal to 1 or 0 when the state i is occupied or unoccupied. The two terms on the left-hand side of (V.46a) are the direct and exchange terms.

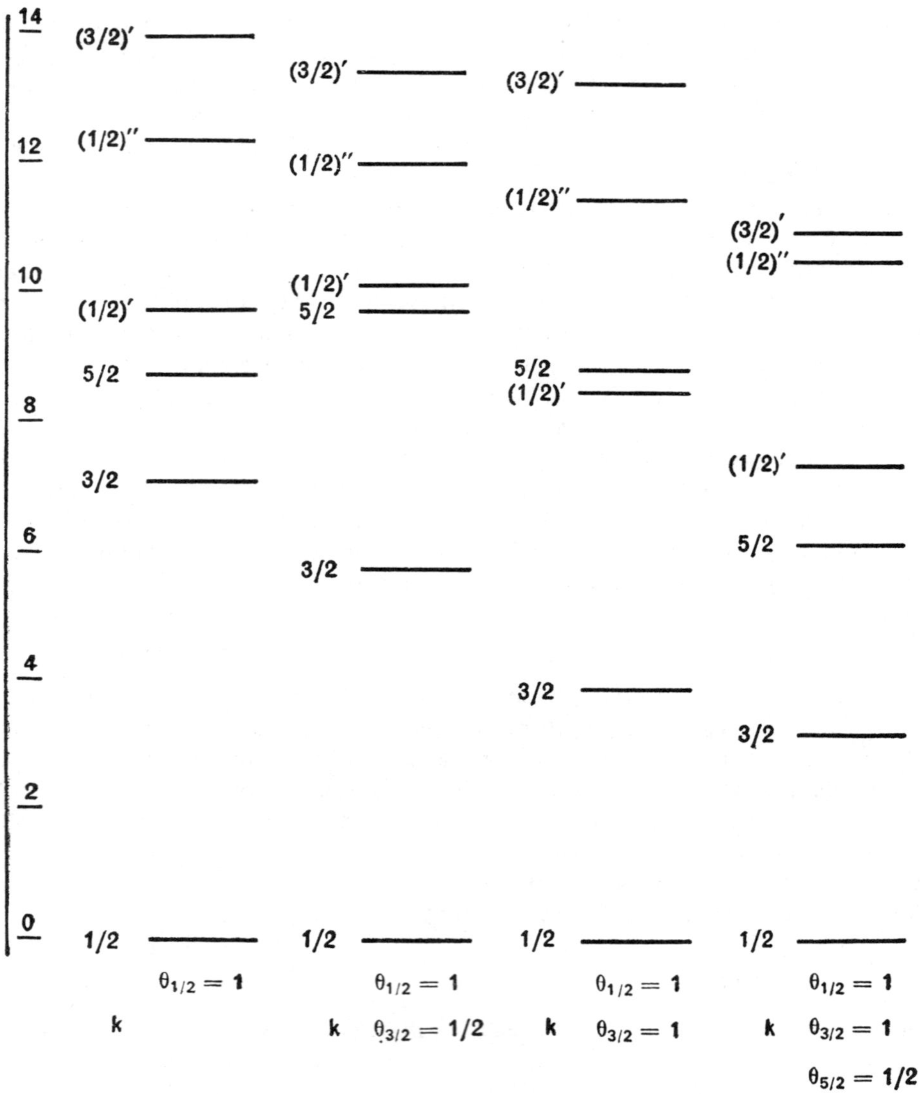

Fig. V.1 Single-particle Hartree-Fock spectra for various (0*d*, 1*s*)-shell nuclei. [Following Kelson, I., and Levinson, C. A., *Phys. Rev.*, **134**, B274 (1964).]

Each spatial state has four nucleons for the two spin states and two isospin states. This accounts for the factors 2×2 on the right-hand side of (V.46a), whereas the factor $n/4$ represents the total number of occupied spatial states. The occupation probability θ_i appears in the final step by virtue of the Krönecker deltas and the fact that j stands for the occupied states.

In a manner similar to that for the Wigner term, the Majorana term yields, from its space-exchange property, the result

$$a_{\mathrm{M}} \sum_{\phi_j}^{\text{occupied}} \sum_{\chi_j} \sum_{\rho_j} (\langle \phi_i\chi_i\rho_i;\ \phi_j\chi_j\rho_j \mid \phi_j\chi_i\rho_i;\ \phi_i\chi_j\rho_j\rangle - \langle \phi_i\chi_i\rho_i;\ \phi_j\chi_j\rho_j \mid \phi_i\chi_j\rho_j;\ \phi_j\chi_i\rho_i\rangle)$$

$$= a_{\mathrm{M}}\Big(4 \sum_{\phi_j}^{\text{occ}} \delta_{\phi_i\phi_j} - \frac{n}{4} \sum_{\chi_j}^{\text{occupied}} \sum_{\rho_j} \delta_{\chi_i\chi_j}\delta_{\rho_i\rho_j}\Big)$$

$$= a_{\mathrm{M}}(4\theta_i - \tfrac{1}{4}n). \tag{V.46b}$$

The evaluation of the Bartlett and Heisenberg terms is simple and is therefore left as an exercise. The results are given by

$$a_{\mathrm{B}}(\tfrac{1}{2}n - 2\theta_i) \qquad \text{(Bartlett)}, \tag{V.46c}$$

$$-a_{\mathrm{H}}(\tfrac{1}{2}n - 2\theta_i) \qquad \text{(Heisenberg)}. \tag{V.46d}$$

Putting together the expressions (V.46), we get the expression for the energy of the single-particle state i in the model under consideration as

$$\epsilon_i = (a_{\mathrm{W}} - \tfrac{1}{4}na_{\mathrm{M}} + \tfrac{1}{2}a_{\mathrm{B}} - \tfrac{1}{2}a_{\mathrm{H}})n + (-a_{\mathrm{W}} + 4a_{\mathrm{M}} - 2a_{\mathrm{B}} + 2a_{\mathrm{H}})\theta_i. \tag{V.47}$$

In view of the occupation probability θ_i in the second term, its coefficient $(-a_{\mathrm{W}} + 4a_{\mathrm{M}} - 2a_{\mathrm{B}} + 2a_{\mathrm{H}})$ is the energy gap between the occupied and unoccupied states in this model. The reader can easily convince himself from (V.46a), (V.46b), and the steps in the derivation of (V.46c) and (V.46d) that the θ_i-terms of (V.47) appear whenever the spatial wavefunctions ϕ_i and ϕ_j exchange places. If we use a coordinate space representation of the HF potential, then we see that the terms in which ϕ_i and ϕ_j interchange lead to nonlocal contributions to the potential. For details of the coordinate space representation, the reader should refer to Problem 1 at the end of this chapter.

We next summarize some of the other important features of the pioneering calculations by Levinson et al[1]. For even nuclei, these authors calculated the moments of inertia by applying the Inglis cranking formula, i.e., (V.35b), and also the alternative method described following it. They used the calculated moments of inertia to predict the spectra of these nuclei by means of the usual rotational formula. In Ne^{20} and Mg^{24}, they tried to calculate the position of the band heads of the various known excited bands. The procedure entails exciting a nucleon from the uppermost occupied single-particle level to one of the unoccupied levels so that the resultant determinantal state can have the appropriate band quantum number K. For example, in the case of Ne^{20}, while calculating the excited band $K = 0$, we have to excite a nucleon (see Fig. V.1) from the topmost occupied levels $k = \frac{1}{2}, -\frac{1}{2}$ (which are degenerate) to the other two unoccupied levels $k = (1/2)', (1/2)''$ [and $k = (-1/2)', (-1/2)''$]. This way the band quantum number of the excited determinant remains the same (i.e., $K = 0$) as that of the ground-state determinant. Since there are several possible determinants to consider, the actual calculation involves diagonalizing the two-body interaction plus the unperturbed energy, using all these excited determinantal states (which are of the one-hole one-particle type) as the basis. Such calculations are described in detail in Section 42B. The qualitative fact that the 0^+ excited band of Ne^{20} occurs at 6.75 MeV gives general confirmation of the existence of a large energy gap between the occupied and unoccupied levels (the gap is the minimum value of the unperturbed energy of the one-hole one-particle type determinants). Without such an energy gap (i.e., with the Nilsson spectrum), the calculated value of the energy of the excited band becomes too low. In the calculations of Levinson et al, the intrinsic energy of the lowest eigenstate (which is a mixture of several one-hole one-particle type determinants) was first calculated through the diagonalization procedure just mentioned. Then the position of the band head E_0 was determined by first *assuming* that the band gives rise to pure rotational states, and next applying the method described (see Section 36B) in connection with the calculation of the moment of inertia of such intrinsic states. The actual numerical agreements are of a qualitative nature, and hence are not reproduced here.

For odd nuclei, Levinson et al[1] used a hybrid method based on the HF intrinsic determinants and the collective Hamiltonian (IV.137) described in Section 33C. The procedure entails using the collective model wavefunction (IV.141) with χ_K determined by the HF method. Then, as far as the energy calculation is concerned, we may use the expression (IV.145), but we should also introduce the values of the parameters $\mathcal{J}, a, \ldots$ calculated with the HF χ_K. Levinson et al[1] treated also the nondiagonal part, $(I_+j_- + I_-j_+)$, of the rotation-particle coupling (RPC) term $\mathbf{I}\cdot\mathbf{j}$ of the collective model Hamiltonian (IV.137) by working out a complete diagonalization of the RPC term. Once again, the collective model expressions (IV.146) were used for this purpose along with the HF wavefunctions. The results obtained for the various bands of Ne^{21}, Na^{23}, Mg^{23}, Mg^{25}, Al^{25}, Al^{27}, Si^{27} are really outstanding. In the case of $A = 21, 23$, the odd particle belongs to the level $k = 3/2$ (see Fig. V.1) in the ground-state band. The neighbouring excited bands correspond to the levels $k = 5/2$, $(1/2)'$ of this nucleon. The higher levels $k = (1/2)''$, 3/2 were ignored in the calculation. The RPC diagonalization thus incurred mixing the bands $K = 3/2, 5/2, 1/2$, where both 1/2 and 5/2 bands mix with the 3/2 band via the RPC term for which $\Delta K = 1$ is obviously a necessity. In the case of $A = 25, 27$, on the other hand, the ground-state band corresponds to the last odd nucleon going to the level $k = 5/2$ (see Fig. V.1); the next excited band corresponds to $k = (1/2)'$, and hence the RPC causes no mixing, unless the high-lying levels $k = (1/2)''$, $(3/2)'$ are also brought into the calculation. This was not done in the work described here; hence, for $A = 25, 27$, the results are really based on the diagonal matrix elements of RPC. As a typical illustration of the numerical agreement, we have presented the results for $A = 21$ in Fig. V.2.

Unrestricted HF Calculations for Nuclei up to Ca^{40} (Deformed self-consistent solutions)

The second distinctly different type of calculation is unrestricted in the sense that it achieves self-consistency for *all* the nucleons in the nucleus using the Hamiltonian (V.1), which consists of the *kinetic energy* and the two-body interaction. When applied to nuclei between two magic numbers, these calculations use all the harmonic oscillator states up to a certain major shell. In the earliest calculations of this type for the $(1s, 0d)$-shell nuclei, all the projection sublevels resulting from $0s_{1/2}$, $0p_{3/2}$, $0p_{1/2}$, $1s_{1/2}$, $0d_{5/2}$, and $0d_{3/2}$ were used as the basis. In later calculations on the same nuclei, the basis was extended to include all the sublevels from the $(0f, 1p)$-shell as well.

A typical set of single-particle states for $N = Z$ even nuclei has been depicted by Ripka[10], whose work can be consulted for further reading on the subject of HF calculation.

The most important results of the unrestricted type of HF calculation are on the (i) absolute binding energy of nuclei and (ii) deformation parameters and quadrupole moment of nuclei. In calculations that omit the core nucleons, the computation of (i) does not rely entirely on the two-body potential; the single-particle energies of the extra core nucleons and the binding energy of the closed-shell core are to be treated as empirical input data. For a deformed nucleus, the core is also deformed, and hence these input data which concern a spherical core may lead to serious errors in the calculated value of the binding energy in a 'limited' calculation of the type discussed in Section 36B. Similar observations hold for the

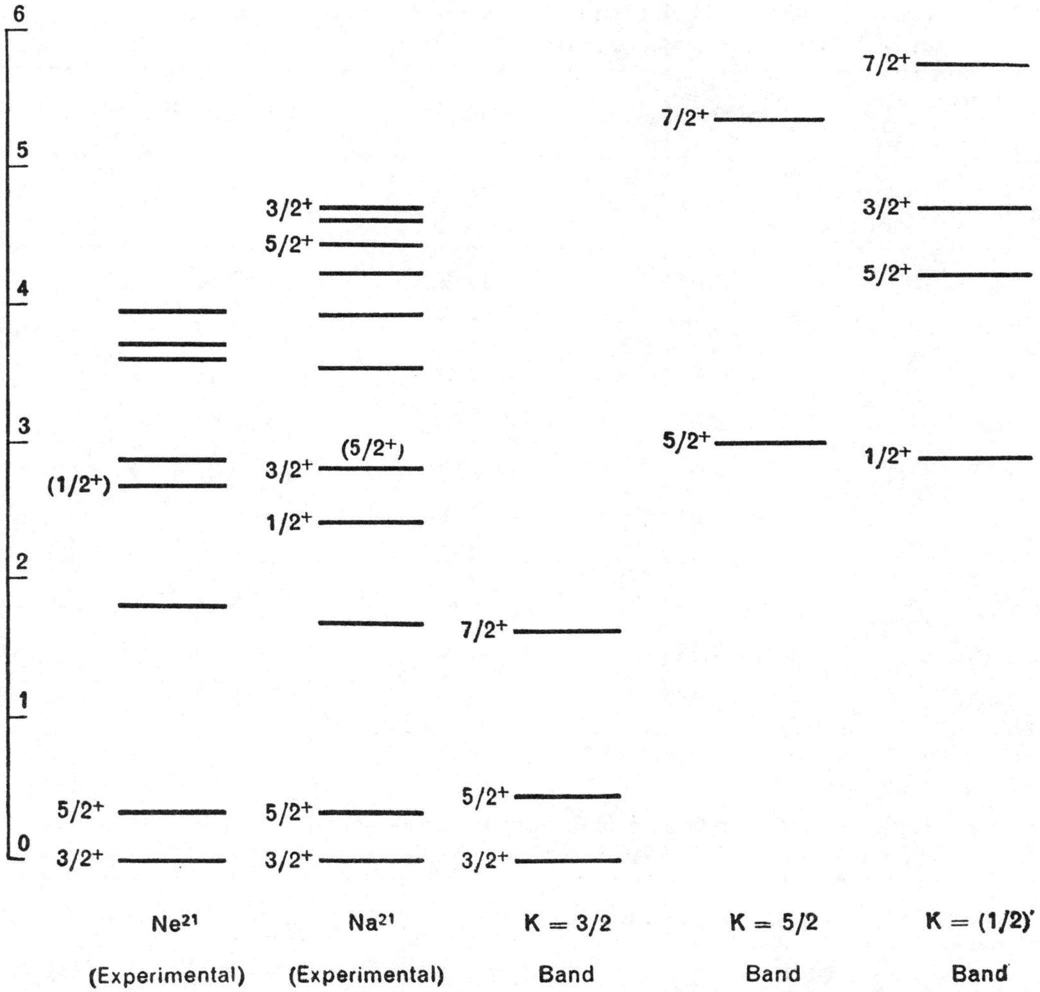

Fig. V.2 Spectra of ($A = 21$)-nuclei, using Hartree-Fock results for even core and coupling single particle to core, according to unified model described in Section 33. [Following Kelson, I., and Levinson, C. A., *Phys. Rev.*, **134,** B276 (1964).]

values of the deformation parameter and quadrupole moment in the 'limited' calculations.

In general, several deformed shapes lead to the extremum in HF calculations. Table V.2a shows several HF solutions for Mg^{24}; the single-particle energy levels are labelled by the basic state having the largest weight in the corresponding self-consistent single-particle wavefunction. The subscript and superscript in each basic state refer to the values of j and projection k, respectively. The different solutions for Ne^{20} are given in Table V.2b. In Mg^{24}, the 'triaxial' solution is seen to have the lowest energy, whereas in Ne^{20} the prolate spheroidal solution is the lowest. For the latter case, the last column of Table V.2b shows the energy gap between the last occupied and first unoccupied levels for HF solutions of the various shapes. It is

Table V.2 Results of Hartree-Fock calculations in ($1s$, $0d$)-shell
(a) Different HF solutions of Mg^{24}

Single-Particle Level (MeV) \ Energy per Nucleon (MeV)	Spherical	Oblate	Prolate	Triaxial
	-4.79	-5.20	-5.40	-5.54
$0s_{1/2}^{\pm 1/2}$	-58.04	-56.79	-59.24	-58.94
$0p_{3/2}^{\pm 3/2}$	-30.39	-33.72	-27.68	-29.42
$0p_{3/2}^{\pm 1/2}$	-30.39	-24.51	-35.85	-35.35
$0p_{1/2}^{\pm 1/2}$	-31.63	-32.54	-26.75	-24.85
$0d_{5/2}^{\pm 5/2}$	-8.01	-12.98	-4.97	-5.62
$0d_{5/2}^{\pm 3/2}$	-8.01	-10.54	-9.50	-12.13
$0d_{5/2}^{\pm 1/2}$	-8.01	-4.97	-14.96	-14.95
$1s_{1/2}^{\pm 1/2}$	-7.17	-9.63	-4.25	-6.75
$0d_{3/2}^{\pm 1/2}$	-7.58	-5.48	-1.85	-0.99
$0d_{3/2}^{\pm 1/2}$	-7.58	-1.69	-9.04	-3.78

Source: Stamp, A. P., *Nucl. Phys.*, **A105**, 627 (1967).

Table V.2 Results of Hartree-Fock calculations in ($1s$, $0d$)-shell
(b) Different HF solutions of Ne^{20}

Shape	$(\sqrt{\pi/5})\langle r^2 Y_0^2\rangle$ (fm)^2	$(\sqrt{\pi/20})\langle r^2(Y_2^2 + Y_{-2}^2)\rangle$ (fm)^2	Energy per Nucleon (MeV)	Energy Gap (MeV)
Spherical	0.0	0.0	-4.53	
Oblate	-6.54	0.0	-4.66	0.10
Prolate	12.89	0.0	-5.07	5.81
Triaxial	3.19	3.91	-4.80	2.95

Source: Stamp, A. P., *Nucl. Phys.*, **A105**, 627 (1967).

interesting to note that the lowest solution has the largest value of the energy gap. This means that the probability of exciting hole-particle pairs across the energy gap is the least for this state. In other words, the lowest HF solution is also the most stable against small perturbations. In most calculations of this type (see Ripka[10], Stamp[11], and Bassichis et al[6]), all the even-even ($N = Z$)-nuclei, starting from He^4 and ending with Ca^{40}, have been investigated. With the exception of the closed-shell nuclei He^4, O^{16}, and Ca^{40}, all the other nuclei were found to be deformed. With the exception of C^{12}, Mg^{24}, and S^{32} which were found to be triaxial, all the other deformed nuclei showed spheroidal symmetry. Of these spheroidal nuclei, Be^{8} and Ne^{20} were observed to be prolate, whereas C^{12} and Si^{28} produced the lowest minimum for an oblate shape.

A summary of the deformation parameters is given in Table V.2c, where the first set of effective two-body matrix elements (unpublished) of the Yale potential calculated by Shakin et al[7] are used. Stamp[11] chose the final set of matrix elements of Shakin and his coworkers.

Table V.2 Results of Hartree-Fock calculations in (1*s*, 0*d*)-shell
(c) Summary of deformation parameters

	Be^{8}	C^{12}	O^{16}	Ne^{20}	Mg^{24}	Si^{28}	S^{32}	Ca^{40}
β	0.786	−0.343	0	0.335	0.324	−0.282	−0.190	0
γ	0	0	0	0	17°52′	0	23°33′	0

Source: Pal, M. K., and Stamp, A. P., *Phys. Rev.*, **158**, 924 (1967).

Several aspects of the results presented in Table V.2 deserve mention here.

(i) The calculated binding energies with the realistic two-body potential are somewhat lower than the observed values. This is true also of the binding energy of nuclear matter, determined with this kind of two-body potential.

(ii) The values of the quadrupole moments are also somewhat lower than the observed values. This is due partly to the defect in the radial part of the wavefunction, and partly to the noninclusion of the (0*f*, 1*p*)-shell states in the calculations reported in Table V.2. The quadrupole deformation of the 0*p*-core nucleons has the immediate effect of mixing in (0*f*, 1*p*)-states into the 0*p*-wavefunction. The exclusion of the latter states therefore amounts to an incomplete reckoning of the core-polarization effect.

(iii) In the spherical solution of Mg^{24} (see Table V.2a), the spin-orbit splitting between the ($0p_{3/2}$)-level and the ($0p_{1/2}$)-level is the opposite to that required by the shell model, whereas the correct order of this splitting appears in the asymmetrical solution. Thus, an incorrect equilibrium shape of the nucleus is found to lead to a faulty order of spin-orbit splitting.

In the calculations of Bassichis et al[6] and those by Ripka[10], the complete core-polarization effect was included by using the (0*f*, 1*p*)-shell states also in the basis wavefunctions. These calculations still suffer from the drawback that the radial part of the wavefunction is not too well-determined. To improve the radial wavefunction, we have obviously to include, in the basis, harmonic oscillator states with a fairly large number of radial nodes. Such calculations are tractable (described in Section 37B) only if spherical symmetry is imposed when l, j, k become good quantum numbers and the HF matrices (independent of k) connect oscillator states of different radial quantum number n. In the deformed nuclei now being discussed, the HF matrices in the expanded space with several (some 3 or 4) n-quantum numbers present tremendous computational problems.

Ripka[10] used a potential of the type (V.42b), whereas Bassichis et al[6] chose the Tabakin separable nonlocal potential (see Tabakin[27], Chapter I). The HF calculation with the latter potential, reported by Kerman et al[6], leads to too small a binding energy in closed-shell nuclei such as O^{16} and Ca^{40}; Kerman and Pal[6] therefore did a calculation for these two nuclei, incorporating the second-order corrections to the binding energy. The same approach was maintained by Bassichis et al[6] while calculating all the even-even ($N = Z$)-nuclei from He^4 to Ca^{40}. A summary of their results is presented in Fig. V.3. It is seen that the second-order correction

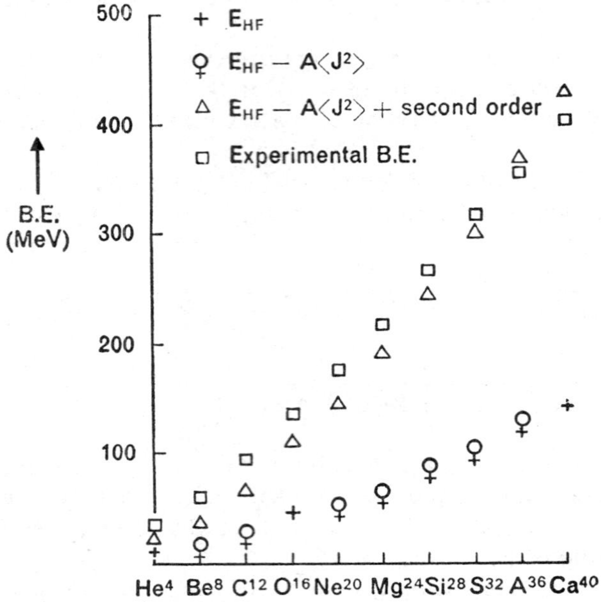

Fig. V.3 Binding energies of even-even ($N = Z$)-nuclei in range He^4–Ca^{40}, calculated with Tabakin two-nucleon potential including second-order contribution (E_2). [Following Bassichis, W. H., Kerman, A. K., and Svenne, J. P., *Phys. Rev.*, **160**, 746 (1967).]

leads to a very close agreement with the experimental results in spite of a *systematic* disagreement (the calculated values are below the observed ones for the lighter nuclei, increase systematically, and then cross over on the higher side of the observed values for A^{36} and Ca^{40}).

Deformed HF Calculations (Nuclei heavier than Ca^{40})

Parikh and Svenne[12] did calculations on several even-even nuclei in the range Ca^{42} to Ge^{72}. They used Ca^{40} as the core, and experimental single-particle energies, rather than the kinetic energies, for the extra core nucleons in the manner of Levinson et al[1]. However, since Ca^{40} is not a very good closed-shell core, its use alone can be regarded a serious drawback of this work. The two-body matrix elements were taken from Shakin et al[7], who derived the reaction matrix elements for the Yale hard-core potential. However, since the core nucleons are suppressed for the extra core nucleons in the calculations by Parikh and Svenne, the renormalized values (which are not yet available) of the reaction matrix elements of Shakin et al, which take into account the core-polarization effects, would have been more suitable.

The nuclei calculated by Parikh and Svenne have $Z = 20$ (Ca), 22 (Ti), 24 (Cr), 26 (Fe), 28 (Ni), 30 (Zn), 32 (Ge). The single-particle energies used by them are listed in Table V.3a. Except for Ca^{48} and Ge^{72}, deformed solutions were obtained in all the even-even nuclei calcu-

Table V.3 Single-particle energies in (0*f*, 1*p*)-shell
(a) Single-particle energies of Parikh and Svenne

Single-Particle State	Unperturbed Energy (MeV)	
	Proton	Neutron
$f_{7/2}$	−1.09	−8.36
$p_{3/2}$	0.70	−6.29
$f_{5/2}$	4.41	−2.86
$p_{1/2}$	3.04	−4.23

Source: Parikh, J. C., and Svenne, J. P., *Phys. Rev.*, **174**, 1343 (1968).

lated. For the Ca isotopes, the deformation was found to be rather small; this is because of the presence of the ($f_{7/2}$)-shell, whose filling gives rise to the fairly good closed-shell nucleus Ca^{48}. For Cr^{48}, Fe^{56}, and Ni^{58}, the lowest solutions were of the triaxial type. All the other nuclei showed deformed solutions of the axially symmetric type, with a preponderance of the prolate shape; the oblate shape was found only amongst heavier nuclei, such as Zn^{68}, Ge^{64}, and Ge^{70}. The energy gap between the last occupied and the first unoccupied levels is about one-third of that observed in the (s, d)-shell. As a result, the excitation of particles from the HF ground state across this gap becomes comparatively easier, and may lead to large corrections in the HF determinantal wavefunction.

Another calculation by Harishchandra[13] on even Ti, Cr, and Fe isotopes reaches almost identical conclusions. Like Parikh and Svenne, he too treated Ca^{40} as the closed-shell core, but used slightly different unperturbed single-particle energies of the extra core nucleons. Since he ignored the Coulomb interaction between protons, he opted for a common set of single-particle energies for both kinds of nucleons. The values are given in Table V.3b. The zero of the single-particle energies has been arbitrarily chosen at the location of the level $f_{5/2}$. Since calculations

Table V.3 Single-particle energies in (0*f*, 1*p*)-shell
(b) Single-particle energies of Harishchandra

Single-Particle State	Unperturbed Energy (MeV)
$f_{7/2}$	−6.5
$p_{3/2}$	−4.5
$f_{5/2}$	0
$p_{1/2}$	−2.5

Source: Harishchandra, *Phys. Rev.*, **185**, 1320 (1969).

of the type reported by Parikh and Svenne and by Harishchandra do not attempt to get the absolute value of the binding energy, it is immaterial where the zero of the energy scale is fixed. The two-body potential used by Harishchandra consists of a central potential of the Yukawa type with the depths in various spin-isospin states as listed in Table III.5. The major difference between this work and that of Parikh and Svenne concerns the relative positions of the

$(k = \frac{7}{2})$-level and the second $(k = \frac{1}{2})$-level. This is important because the shape of Ni^{56}, for example, in Parikh and Svenne is very greatly dependent on the fact that the $(k = \frac{7}{2})$-level rises very high. This discrepancy between the two works may be connected with the different unperturbed single-particle energies in Tables V.3a and V.3b.

Another important result obtained by Harishchandra relates to the special status of the $(N = 28)$-nuclei. Although an HF calculation in the even nuclei with this neutron number produces a deformed solution, the spherical shape is restored as soon as the correction to the HF ground state due to the pairing type excitation across the energy gap is taken into account. This has been achieved by Harishchandra by means of a Hartree-Fock-Bogoliubov (HFB) type calculation, which is discussed in Chapter VI. In contrast with the $(N = 28)$-nuclei, the other even-even nuclei treated in this calculation retain their deformed shape even when the effect of the pairing type excitations is included by applying the HFB method. This result, once again, marks another major departure from the conclusion of Parikh and Svenne on the importance of the pairing effects in modifying the HF ground state of Ti^{44}. In Harishchandra's work, the HF and HFB solutions of this nucleus show very little difference.

Faessler et al[14] worked on Fe^{56}, Ge^{72}, Se^{78}, and $_{46}Pd^{110}$. In contrast with Parikh and Svenne, and Harishchandra, these authors did an unrestricted calculation (i.e., they did not use the concept of a core) for all the nucleons, using the original Hamiltonian consisting of the *kinetic* and two-body interaction energy. They used the Brink and Boeker two-body potential given by (V.43a), and all the harmonic oscillator states up to the levels $0g$, $1d$, $2s$ as the basis for HF diagonalization. This basis is large enough to accommodate 70 nucleons of each kind. These authors, like Harishchandra, did both HF and HFB calculations and several minor variants ranging between these two methods. For Fe^{56}, Se^{78}, and Pd^{110}, they conclude, in the manner of Harishchandra, that HF and HFB solutions are almost identical, in spite of the small energy gap in the HF spectrum. On the other hand, in the nucleus Ge^{72}, the HF calculation yields two deformed solutions, which in the HFB calculation make place for a *spherical* solution. (Incidentally, Ge^{72} has a neutron number $N = 40$, a semi-magic number in this range of the periodic table.) This result for Ge^{72} is therefore very similar to that obtained in Harishchandra's work for the $N = 28$ (a semi-magic number) Ti, Cr, and Fe isotopes.

Calculations with Deformed Harmonic Oscillator States

The first set of calculations with deformed harmonic oscillator wavefunctions is due to Volkov[4]. Strictly speaking, these calculations for the $0p$-shell nuclei are not of the HF type. In the simplest form, Volkov used the deformed $0s$-wavefunction as

$$\phi_s = C_s \,\mathrm{expf} \tag{V.48a}$$

and the deformed $0p$-wavefunctions of projections 0 and ± 1 as

$$\phi_0 = C_0(bz)\,\mathrm{expf}, \tag{V.48b}$$

$$\phi_{\pm 1} = C_{\pm 1}a(x \pm iy)\,\mathrm{expf}, \tag{V.48c}$$

where expf is the deformed Gaussian factor occurring in all the harmonic oscillator functions, i.e.,

$$\mathrm{expf} = \exp\,[-\tfrac{1}{2}\{a(x^2 + y^2) + bz^2\}]. \tag{V.48d}$$

The quantities C_s, C_0, and $C_{\pm 1}$ are the normalization constants. If the parameters a and b are made equal, all these deformed wavefunctions reduce to spherical harmonic oscillator functions.

The same parameter a is used for both the x- and y-directions, implying that these deformed single-particle wavefunctions retain spheroidal symmetry with z as the symmetry axis. The ratio of the parameters a and b specifies the extent of the deformation. For this purpose, a deformation parameter ϵ is used, defined by

$$\frac{a}{b}=\frac{1+\frac{1}{3}\epsilon}{1-\frac{1}{3}\epsilon}. \tag{V.49}$$

The calculation entails constructing a determinantal wavefunction for a given p-shell nucleus with the aforementioned single-particle states, assuming that each such spatial state accepts two protons and two neutrons with opposite spin projection. The expectation value of the Hamiltonian is then calculated in a straightforward manner. For the potential energy, we obtain space integrals of the type

$$\int d^3r_1 \int d^3r_2\, \phi_\alpha(1)\phi_\beta(2)V(r_{12})[\phi_\alpha(1)\phi_\beta(2)-\phi_\beta(1)\phi_\alpha(2)].$$

The Gaussian form of the potential (V.42) makes the evaluation of these integrals quite simple *with the use of cartesian coordinates*. It should be observed that a Yukawa shape of the potential, which is $e^{-r/a}/(r/a)$, i.e., $a[x^2+y^2+z^2]^{-1/2}\exp[-a^{-1}(x^2+y^2+z^2)^{1/2}]$, would have given rise to a very complicated space integral in terms of cartesian coordinates. The choice of the cartesian coordinates for the purpose of integration becomes necessary in view of the expressions (V.48) of the deformed wavefunctions. After the expectation value of the Hamiltonian is calculated, the result is minimized with respect to the parameters a and b or, equivalently, the nuclear size and deformation. In this sense, Volkov's calculation is therefore a more restricted variational calculation than the self-consistent HF calculation. In one practical way, however, this type of calculation very quickly achieves what sometimes becomes a tremendous computational problem in the self-consistent work. In a deformed self-consistent work, the inclusion of basis wavefunctions with a large number of radial nodes sometimes poses a big computing problem; on the other hand, the simple deformed functions (V.48a) to (V.48c) already contain higher radial nodes, when analyzed in terms of spherical harmonic oscillator solutions.

The minimization with respect to a and b has been tackled numerically as follows. First, a choice is made of the deformation parameter ϵ, as given by (V.49), and then the energy minimized with respect to a (the size parameter). This process is repeated for a large number of values of ϵ, both positive (prolate) and negative (oblate).

In a more extended calculation, the parameters a and b were attributed three different sets of values in the three wavefunctions (V.48a) to (V.48c); however, the equality of the ratio a/b for the three wavefunctions was maintained. This means, according to (V.49), that the deformations of the orbits ϕ_s, ϕ_0, and $\phi_{\pm 1}$ are given the same value but their size parameters are allowed to be different. Numerically, we now need a minimization at each ϵ with respect to three size parameters. To avoid the complexity of this three-parameter numerical search, certain assumptions were introduced. At zero deformation, the size parameters of ϕ_0 and $\phi_{\pm 1}$ must be the same, whereas that of ϕ_s could be different. Therefore, at zero deformation, the minimization is with respect to two independent size parameters. In the subsequent minimization at non-zero deformation, the value of the size parameter for ϕ_s is retained. One of the size parameters, that of ϕ_0 or $\phi_{\pm 1}$, is also retained at the same value; when $\epsilon > 0$, the ϕ_0 size parameter is chosen for this purpose, whereas that of $\phi_{\pm 1}$ is chosen when $\epsilon < 0$. The remaining size

64 (45-122/1977)

parameter is then determined by the minimization of the energy. The results for He^6 and Be^8 are shown in Fig. V.4 as representative of Volkov's findings. Curve I represents the case of the

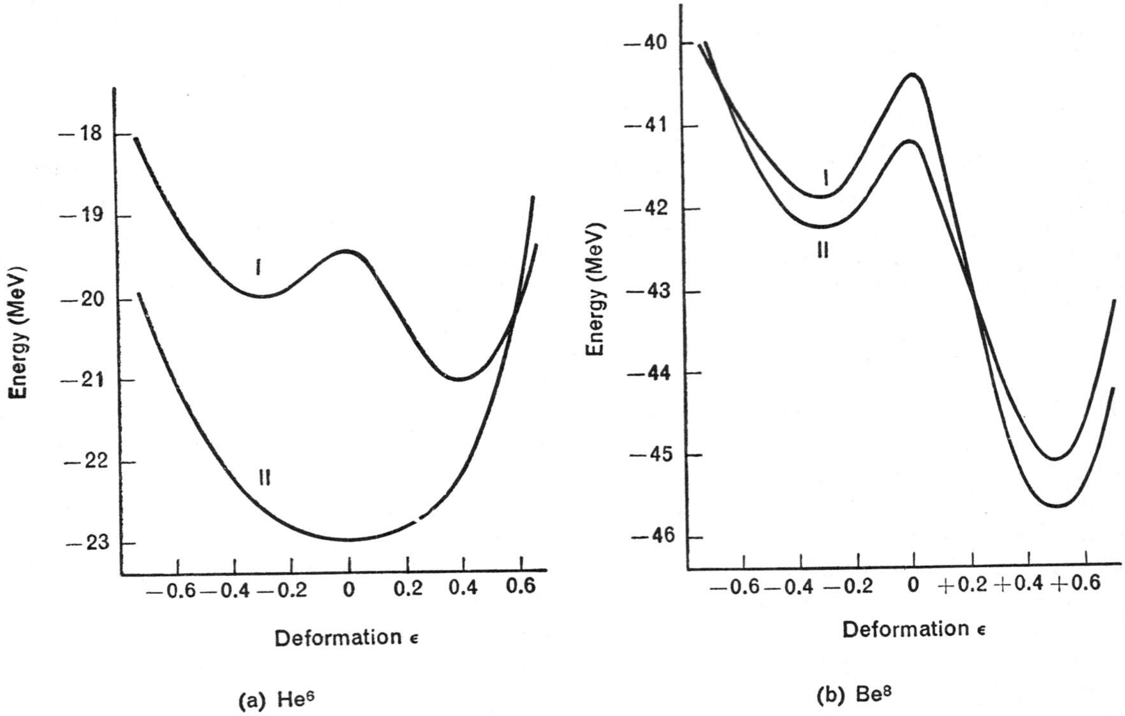

Fig. V.4 Ground-state binding energy, using different oscillator constants for various single-particle orbitals. [Following Volkov, A. B., *Nucl. Phys.*, **74**, 33 (1965).]

same size parameter and displays two pronounced minima; the one at a fairly large prolate deformation, $\epsilon \approx 0.4$–0.6, corresponds to the intrinsic ground state. Curve II shows the effect of choosing different size parameters for the single-particle states. In the case of He^6, it has a very pronounced effect, namely, it leads to an almost spherical solution. In the case of Be^8, however, no qualitative change of this type takes place.

All the $0p$-shell nuclei between $A = 5$ and $A = 13$ were found to be deformed. Up to $A = 8$, there is a definite preference for the prolate shape; beyond that the nuclei tend to be oblate. $A = 14$ and $A = 15$ have almost spherical shapes.

Brink and Boeker[5], and Boeker[15], did calculations similar to those by Volkov, using harmonic oscillator wavefunctions in the cartesian basis. Let these states, corresponding to a *spherical* harmonic oscillator of energy $\hbar\omega$, be denoted by $|n_x, n_y, n_z\rangle_\omega$. Brink and Boeker used the states $|0, 0, 0\rangle_\omega$, $|1, 0, 0\rangle_\omega$, $|0, 1, 0\rangle_\omega$, $|0, 0, 1\rangle_\omega$, . . . as occupied states in setting up determinantal wavefunctions for He^4 (only the first state is occupied), Be^8 (the first and second states are occupied), C^{12} (the first three states are occupied), O^{16} (all the four states are occupied) . . . . The quantity $\hbar\omega$ was treated as a variational parameter for the minimization of energy. The main purpose was to set up good two-nucleon potentials [described by (V.43a)–(V.43d)] for HF work; therefore, the actual energy calculation was rendered as simple as possible. An important expression obtained in this work, which facilitates an evaluation of the matrix elements of the Gaussian potential in the cartesian basis, is now detailed. A product of two harmonic oscillator wavefunctions, given for *any* of the three coordinates (say, x) of two nucleons, can be expanded in terms of relative and centre-of-mass coordinates with a Moshinsky-like transformation of a much simpler type. Let

$$x = \frac{1}{\sqrt{2}}(x_1 - x_2), \qquad X = \frac{1}{\sqrt{2}}(x_1 + x_2), \tag{V.50}$$

where the subscripts refer to the two nucleons, and x and X refer to the relative and centre-of-mass x-coordinates. Then

$$|n_1, n_2\rangle = \sum_{n, N} \langle n, N \mid n_1, n_2\rangle |n, N\rangle, \tag{V.51a}$$

where $\langle n, N \mid n_1, n_2\rangle$ is the required transformation bracket and is given by

$$\langle n, N \mid n_1, n_2\rangle = \left(\frac{n_1!\, n_2!\, n!\, N!}{2^{n+N}}\right)^{1/2} \delta(n_1 + n_2 - n - N) \times \sum_{\nu} (-1)^{\nu} [(n_1 - n + \nu)!\, (n_2 - \nu)!\, (n - \nu)!\, \nu!]^{-1}. \tag{V.51b}$$

The derivation of this simple transformation is left as an exercise.

Boeker[15] attempted the same calculation as that in Brink and Boeker[5] with a deformed basis and in a self-consistent manner. However, the deformed states were used only indirectly. If we denote the harmonic oscillator frequency ω_x of a harmonic oscillator state $|n_x\rangle_{\omega_x}$ as $\omega_x = \omega e^{-2\alpha}$, then the state $|n_x\rangle_{\omega_x}$ can, in general, be expanded in terms of a complete set of harmonic oscillator states $|n_x\rangle_{\omega}$ corresponding to the harmonic oscillator parameter ω. The quantity α in the definition of ω_x may be considered a deformation parameter for the direction x. *In the lowest order* expansion, we have

$$|n_x\rangle_{\omega_x} = |n_x\rangle_{\omega} + \tfrac{1}{2}\alpha[\sqrt{(n_x + 1)(n_x + 2)}\,|n_x + 2\rangle_{\omega} - \sqrt{n_x(n_x - 1)}\,|n_x - 2\rangle_{\omega}]. \tag{V.52}$$

The derivation of this result is also left as an exercise. In the same manner, we can consider the harmonic oscillator wavefunctions for the directions y and z, namely, $|n_y\rangle_{\omega_y}$ and $|n_z\rangle_{\omega_z}$, with different harmonic oscillator parameters, namely, ω_y and ω_z. Let these be related to ω through the two deformation parameters β and γ as

$$\omega_y = \omega e^{-2\beta}, \qquad \omega_z = \omega e^{-2\gamma}.$$

Once again, $|n_y\rangle_{\omega_y}$ and $|n_z\rangle_{\omega_z}$ can be expanded in a manner similar to (V.52). In the full three-dimensional deformed wavefunction $|n_x, n_y, n_z\rangle_{\omega_x, \omega_y, \omega_z}$, we therefore have states of the spherical harmonic oscillator of the type

$$|n_x, n_y, n_z\rangle_{\omega}, |n_x + 2, n_y, n_z\rangle_{\omega}, |n_x - 2, n_y, n_z\rangle_{\omega}, |n_x, n_y + 2, n_z\rangle_{\omega} \ldots .$$

Boeker therefore decided to do an approximate deformed calculation, using all these spherical states as the basis. In his evaluation of two-body matrix elements, the results (V.51) are obviously very handy. To fix the ideas, let us consider the lowest harmonic oscillator state $|0, 0, 0\rangle_{\omega}$. In Boeker's calculation, according to our discussion, this state has to be mixed with $|2, 0, 0\rangle_{\omega}$, $|0, 2, 0\rangle_{\omega}$, $|0, 0, 2\rangle_{\omega}$. The admixture of these four states in He^4 is then determined by a regular HF calculation with the four states as the basis. Similar HF diagonalization has been done for Be^8, C^{12}, ... using the higher states: (i) $|1, 0, 0\rangle_{\omega}$, (ii) $|0, 1, 0\rangle_{\omega}$, ... and the states that mix with them, i.e.,

$$\text{(i)}\ |1, 0, 0\rangle_\omega,\quad |3, 0, 0\rangle_\omega,\quad |1, 2, 0\rangle_\omega,\quad |1, 0, 2\rangle_\omega;$$

$$\text{(ii)}\ |0, 1, 0\rangle_\omega,\quad |0, 3, 0\rangle_\omega,\quad |2, 1, 0\rangle_\omega,\quad |0, 1, 2\rangle_\omega.$$

Calculations for Closed-Shell Nuclei

All calculations for closed-shell nuclei have been done by mixing wavefunctions with different radial nodes n, and taking the basis wavefunctions ϕ_{nljm} to be spherical harmonic oscillator functions.

In their calculation on O^{16} and Ca^{40}, Kerman et al[6] used the two-body matrix elements of the Tabakin nonlocal potential. They also varied the harmonic oscillator parameter b and the number of radial nodes mixed in the single-particle ($s_{1/2}$)-, ($p_{3/2}$)-, and ($p_{1/2}$)-wavefunction. Their argument is that if a sufficient number of radial nodes is mixed, then the self-consistent wavefunctions *should* eventually be independent of the harmonic oscillator parameter, which merely characterizes the basis functions. In practice, only a finite number of radial nodes can be mixed, and for four radial nodes a fairly flattish dependence of the HF total energy on the parameter b results. We may, of course, choose to use the value of b that corresponds to the minimum of this curve. The single-particle levels obtained by these authors for O^{16} are given in Table V.4 together with the experimental values. The numbers do not compare too well (this is a general characteristic of all HF calculations). One more aspect of the energy levels that deserves mention is the very large binding of the lowest ($s_{1/2}$)-level. If the energy levels near the top of the fermi sea obtained with a *local* potential are fit to the observed levels, then the lowermost levels of the same local potential do not go as deep down as with the HF potential. This trait of the HF spectrum arises from the nonlocal nature of the HF potential. In Table V.4, the single-particle levels are denoted by the usual shell-model n-quantum number. For the HF states, this quantum number merely identifies the lowest energy level with the particular l, j and the next higher level of the same l, j, and so on. It is clear from this table that there are two major differences between the calculated numbers and the experimental ones: (i) the extent of spin-orbit splitting and (ii) the unbound nature of the first unoccupied ($0d_{5/2}$)-HF level in contrast with the bound nature of the experimentally observed ($0d_{5/2}$)-level.

Table V.4 Calculated single-particle levels of O^{16}

Level	Hartree-Fock Energy (MeV)	Experimental Energy (MeV)
$0s_{1/2}$	−48.72	
$0p_{3/2}$	−19.65	−21.81
$0p_{1/2}$	−9.45	−15.65
$0d_{5/2}$	2.31	−4.15
$1s_{1/2}$	6.12	−3.28
$0d_{3/2}$	10.74	0.93

The total binding energy of O^{16}, found by Kerman and his colleagues[6], is 38.5 MeV, the corresponding experimental number being 127 MeV. Kerman and Pal[6] later obtained a second-order correction to the energy which has improved the binding energy to 107.5 MeV.

Baranger et al[3] did an exhaustive set of calculations for the closed-shell nuclei, using the set of potentials described earlier in this section. The nuclei calculated by these authors are O^{16}, Ca^{40}, Ca^{48}, Ni^{56}, Sr^{88}, Zr^{90}, and Pb^{208}. Some of these nuclei have a rather doubtful closed-shell character. The single-particle energy levels, once again, have a rather nonspectacular agreement with experimental values. The most important result obtained by this group concerns the bumps in the mass and charge-density distributions due to the closure of the inner shells. The typical cases of Ca^{40}, Ca^{48}, and Pb^{208} are shown in Fig. V.5. The dashed curves labelled P correspond to the densities calculated with pure harmonic oscillator wavefunctions. The solid curves labelled 3×3 correspond to the wavefunctions obtained by mixing, for each l, the first three harmonic oscillator radial functions, i.e., those with a radial quantum number $n = 0, 1, 2$. In the case of Pb^{208}, the curves labelled $P + N$ $(N = 1, 2)$ correspond to the wavefunctions obtained by mixing, for a given l, the first $P + N$ harmonic oscillator radial functions, where P is the highest radial wavefunction for that l in Pb^{208}, according to the pure shell model. The bumps mentioned earlier distinguish the HF density distributions from the empirical fermi-type flat distribution in the core region, which is followed by the decaying tail region. The HF density distribution can be regarded as a superposition of the fermi-type distribution with an oscillating density due to shell effects. There is experimental confirmation of this general qualitative fact in the 750-MeV electron-scattering experiment of Bellicard et al[16]

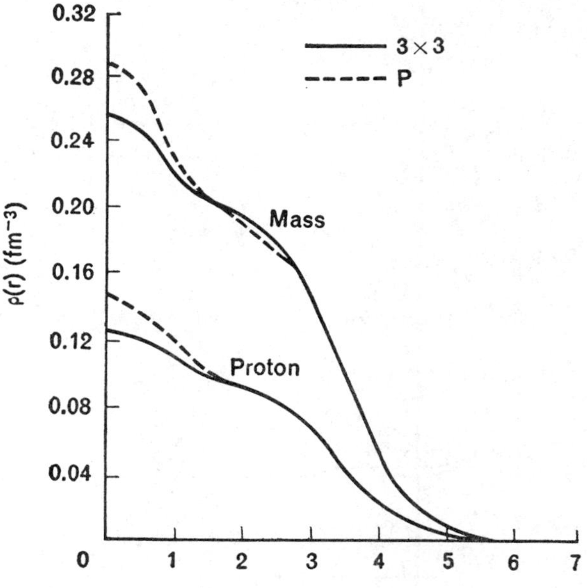

(a) Ca^{40}

Fig. V.5 Proton and mass densities. [Following Turbutton, R. M., and Davies, K. T. R., *Nucl. Phys.*, **A120**, 1 (1968).] (cont.)

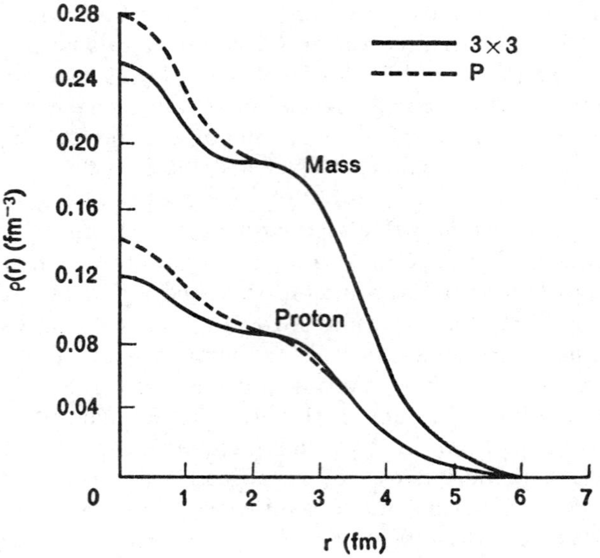

(b) Ca^{48}

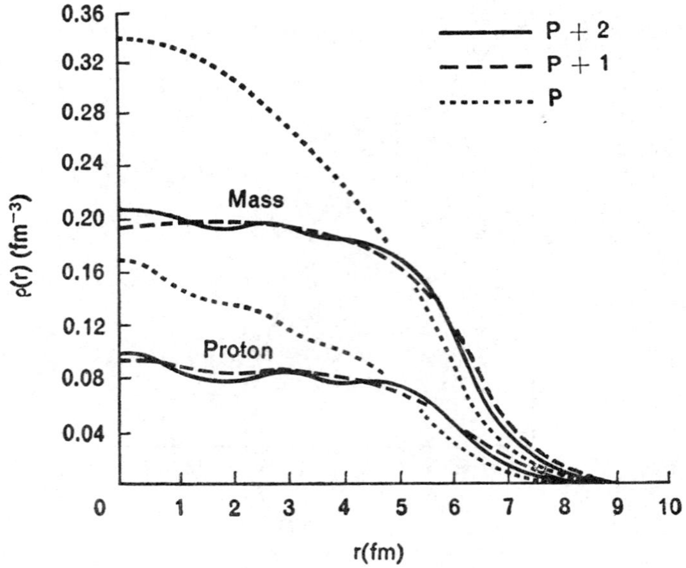

(c) Pb^{208}

Fig. V.5 Proton and mass densities. [Following Turbutton, R. M., and Davies, K. T. R., *Nucl. Phys.*, **A120**, 1 (1968).]

on the nuclei Ca^{40} and Ca^{48}. Their results are shown in Fig. V.6 for the purpose of comparison. In Fig. V.6a, the experimental data are denoted by the solid circles and the theoretical curves by the solid and dashed lines. The dashed lines have been calculated with a parabolic fermi-type charge distribution, whereas the solid lines contain the additional contribution of an oscillating correction $\Delta\rho(r)$ to the density. In Fig. V.6b, the dashed lines correspond to a parabolic fermi-type charge distribution [i.e., without the $\Delta\rho(r)$-term]. The dotted curve labelled "Shell model" gives the pure shell-model charge distribution. All these curves have been plotted with reference to the left-hand ordinate. The right-hand ordinate has been used to plot the difference (the dashed curve) in the parabolic fermi-type distributions in Ca^{40} and Ca^{48}. With reference to the same ordinate, the distributions $\Delta\rho(r)$ are represented by the solid curves running close to the abscissa line. The oscillating part of the density, found in the

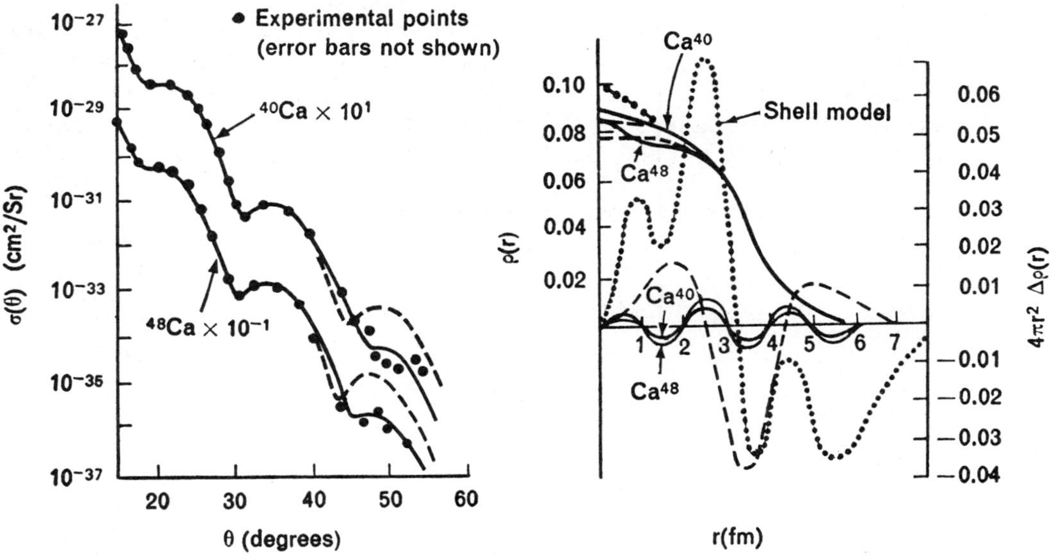

(a) Observed by Stanford group of Bellicard et al

(b) Charge distributions (solid lines) determined from fitting electron scattering data shown in Fig. V.6a

Fig. V.6 Differential cross-section of electron scattering and charge distributions of Ca^{40} and Ca^{48}. [Following Proceedings of International Conference on Nuclear Structure, ed. by J. Sanada, Tokyo, 1967, p 539; also Bellicard, J. B., Bounin, P., Frosch, R. F., Hofstadter, R., McCarthy, J. S., Uhrhane, E. J., Yearian, M. R., Clark, B. C., Herman, R., and Ravenhall, D. G., *Phys. Rev. Letters*, **19**, 527 (1967).]

experiment, is not very uniquely determined. The authors first took an empirical form of the extra addition $\Delta F(q)$ needed for the form factor so that the shape of the cross-section curve at large q can be reproduced. The form chosen for this purpose is arbitrary and is given by

$$\Delta F(q) = A \exp [(q - q_0)^2/p^2]$$

with A, q_0, and p treated as parameters. The Fourier transform of $\Delta F(q)$ was then taken for the change $\Delta\rho(r)$ needed in the density distribution. Thus,

$$\Delta\rho(r) = \frac{ZeApq_0^2}{2\pi^{3/2}}\left(\frac{\sin q_0 r}{q_0 r} + \frac{p^2}{2q_0^2}\cos q_0 r\right)e^{-(1/4)p^2r^2}.$$

This oscillating part of the density is shown in Fig. V.6b. Since the addition $\Delta F(q)$ to the form factor is a Gaussian centred at q_0, a suitable choice of the latter ensures that there is very little effect of this oscillating part of the density on the small q-part of the cross-section curve, which was fitted earlier for the 250-MeV electron scattering data with a parabolic type fermi distribution. This distribution contains three parameters—c, z, and w—and is given by

$$\rho(r) = \rho_0(1 + wr^2/c^2)[1 + \exp\{(r - c)/z\}]^{-1}.$$

It can be seen, by comparison with the fermi distribution (II.58), that this distribution has the additional quadratic term wr^2/c^2 in the numerator. The quantity ρ_0 was determined from the value of the total charge Ze of the nucleus. The values of the parameters c, z, and w and the parameters p, q, and A (needed for the oscillating part of the density) for the two nuclei are given in Table V.5. It is clear that the 'half-density' radius c is larger for Ca^{48}, whereas the skin thickness z for it is smaller than that of Ca^{40}. (It should be noted that, because of the presence of the w-term, strictly speaking ρ no longer becomes $\frac{1}{2}\rho_0$ at $r = c$.) The root-mean-square

Table V.5 Parameters of charge density

	Ca^{40}	Ca^{48}
c (fm)	3.6685	3.7369
z (fm)	0.5839	0.5245
w	−0.1017	−0.0300
p (fm)$^{-1}$	0.5	0.5
q_0 (fm)$^{-1}$	3.0	3.0
$A \times 10^3$	0.5	0.8

radii with the complete experimental charge distributions turn out to be 3.50 fm and 3.49 fm, respectively, for Ca^{40} and Ca^{48}. The values determined from the HF charge distributions are 3.303 fm and 3.337 fm. There is therefore a qualitative discrepancy here in the relative magnitude of the root-mean-square radii of Ca^{40} and Ca^{48}. The other discrepancy relates to the magnitude of the oscillations in the density, which are somewhat large compared with the oscillating part found by fitting the experimental data. Of course, more reliable experimental data at larger electron energies are needed to settle these questions conclusively.

C. RESULTS OF ANGULAR MOMENTUM PROJECTION

In this section, we discuss some typical numerical results of spectroscopic calculations based on the angular momentum projection from HF intrinsic states, described in Section 36A. Many different workers have obtained results on the (s, d)-shell nuclei; some of them used HF determinants for the (s, d)-shell nucleons only (i.e., treating O^{16} as a core), and others the determinantal state for all the nucleons in the nucleus.

A typical set of results obtained by Ripka[10] is shown in Fig. V.7; the p-shell nucleus C^{12} is also included. The number to the right of each level is its spin, whereas the number to the left of each projected level gives its energy with reference to the energy of the determinantal state. It is clear that the projected spectra are in obvious disagreement with the experimental

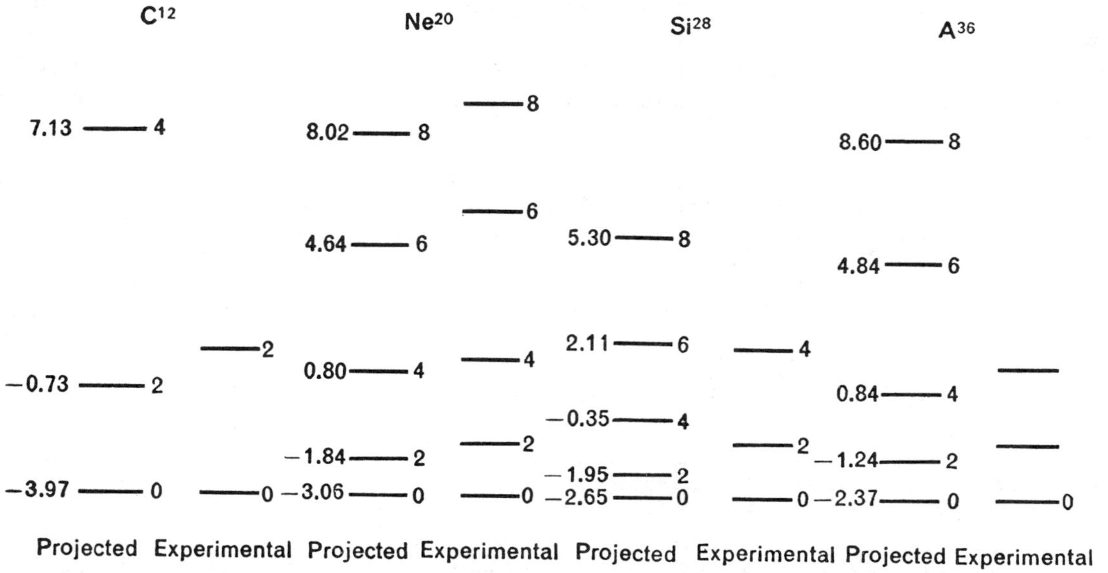

Fig. V.7 Spectra of several even-even ($N = Z$)-nuclei, projected from deformed Hartree-Fock determinants. [Following Ripka, G., in Advances in Nuclear Physics, Vol 1, ed. by M. Baranger, Plenum Press, New York, 1968, p 208.]

data for Si^{28} and A^{36}. It is quite well-known that the nuclei from Si^{28} onwards in the (s, d)-shell do not possess collective rotational spectra, whereas the projection from the HF determinant roughly produces the rotational features.

In the case of Si^{28}, the results shown refer to the oblate lowest-energy HF determinant. There is a prolate solution whose energy is somewhat higher than the lowest solution. Many authors have tried to utilize the existence of this solution in securing agreement with experimental data. Tewari and Grillot[17] mixed the prolate and oblate solutions, but the mixing was found to be very small. They also tried to correct the intrinsic HF state by mixing it with two-hole two-particle (2h-2p) states [the two particles being in a pair of time-reversed states, such as (p, $\bar{p}$), and the two holes also in a pair of states (h, $\bar{h}$)]. Mixing of this type is also found to be small, and hence the projected spectrum changes very little through this type of correction.

Angular momentum projection has been worked out by Bouten et al[18] for the $0p$-shell nuclei.

Considerable work on angular momentum projection and on properties of projected states has been done by Warke et al[19–22]. They were amongst the early workers who derived the angular momentum projection algebra given in Section 36A. They obtained special symmetry relations for the functions $I(\theta)$ and $H(\theta)$, defined by (V.25b and (V.26a). Further, they derived the behaviour of $E(J)$, the energy of the projected state of angular momentum J, as a function of J. These authors worked out the values of the magnetic dipole and electric quadrupole moments of projected states and the E2-transition strengths connecting neighbouring states of the band. In addition, electron scattering, β-decay probability, etc. were also calculated. While applying their theoretical expressions to ($1s$, $0d$)-nuclei, Warke and Gunye[19] used the Rosenfeld interaction, given by (V.40) with $V_0 = -45$ MeV and $a = 1.37$ fm. O^{16} was treated as the core producing single-particle energies for the ($1s$, $0d$)-nucleons, as described in Section 37B. The aforementioned energies of the projected states and their properties were quite satisfactory. A few typical cases of their results on the electromagnetic properties are shown in Table V.6.

Table V.6 Magnetic dipole moment μ (nM) and electric quadrupole moment Q ($e \times 10^{-24}$ cm^2)

Nucleus	J	μ^*_{sp}	μ_{calc}	μ_{expt}	Q^{**}_{calc}	Q_{expt}
Ne^{21}	3/2	0.85	−0.57	−0.66	0.097	0.093
Na^{23}	3/2	0.12	2.11	2.22	0.10	0.097
Mg^{25}	5/2	−1.91	−0.58	−0.85	0.18	0.22
Si^{29}	1/2	−1.91	−0.38	−0.55	0	0

*μ_{sp} denotes the Schmidt value of the magnetic moment.
**In calculating Q, effective charges of $1.5e$ for the proton and $0.5e$ for the neutron have been used.

Dworzecka and Warke[20] calculated the binding energy for ($1s$, $0d$)-shell nuclei using the same interaction as in Pal and Stamp[7] and in Stamp[11]. Instead of doing an HF calculation with all the nucleons, they developed an approximate method, treating the core nucleons and extra core nucleons separately. In addition, they calculated the binding energy for the projected ground state, unlike Pal and Stamp who computed only the energy of the HF state.

The HF potential is so defined that the matrix element of the total Hamiltonian connecting the HF ground state with any one-hole one-particle (1h-1p) excited state is zero (see Problem 7 at the end of this chapter). Thus, the lowest set of excited states that can mix with the HF state is the 2h-2p states. It can therefore be argued that an intrinsic state that is better than the HF state can be obtained by diagonalizing the total Hamiltonian, using the HF state and the 2h-2p states as basis states. However, because of the energy gap in the single-particle spectrum, the 2h-2p states have unperturbed energies very much higher than the HF energy, and hence their admixture into the HF state as a result of the diagonalization may turn out to be small. This idea was first explored by Pal and Stamp[23], using 2p-2h excitations caused by a special pairing interaction active between pairs of identical nucleons in states that are time-reversed of each other [see (VI.92)]. Padjen and Ripka[23] derived a self-consistent method for taking into account this admixture of 2h-2p states, and also emphasized that a realistic two-nucleon interaction containing both the components $T = 0$ and $T = 1$ be used. Earlier, Goodman et al[24] had also made the point that, for a ($1s$, $0d$)-shell $N = Z$ even nucleus, such as Mg^{24}, the two-nucleon ($T = 0$)-interaction makes the 2h-2p admixture important. In Mg^{24}, the HF calculations lead to a prolate, an oblate, and a triaxial minima, of which the last one is lowermost in energy. By means of a self-consistent Hartree-Fock-Bogoliubov (HFB) calculation (this method is described in Appendix G, Section II), Goodman et al[24] demonstrated that, after the pair excitations due to the ($T = 0$)-interaction are included, both the prolate and triaxial states are pushed down, but the prolate is lowered 4.3 MeV more than the oblate, thereby becoming the lower of the two.

Gunye et al[21] have also dealt with the foregoing problem. They showed that the detailed self-consistent treatment of Padjen and Ripka[23] is not essential, whereas a perturbation treatment of the admixture is quite adequate. In their detailed calculations on Ne^{20}, Mg^{24}, and Si^{28}, they used the two-nucleon interaction matrix elements as given by Elliott et al[25], which is more popularly known as the Sussex interaction. These authors verified that the prolate solution for Mg^{24} indeed comes below the triaxial, the relative lowering being 3.35 MeV; however, the contributions of the components $T = 1$ and $T = 0$ of the interaction in the relative lowering are 2.25 MeV and 1.10 MeV, respectively. They ascribed the large ($T = 0$)-contribution found by Goodman et al[24] to the nonconservation of the nucleon number in the HFB method used by these authors.

Warke[22] explored in detail the dependence of the energy $E(J)$ of the projected state on the angular momentum J. He used (see the first paper of Warke[22]) Lanczo's algorithm for the eigenvalue problem of a matrix and obtained an expression of the form

$$E(J) = E(\beta) + \sum_{n=1}^{N-1} A_n(\beta)[J(J+1)]^n,$$

where the HF determinant contains N states of different angular momenta J, and β is the deformation parameter for the HF determinant. Expressions have been obtained for the coefficients $A_n(\beta)$, of which $A_1(\beta)$ is obviously related to the moment of inertia. The expression for $A_1(\beta)$, obtained by Warke, agrees with the formula for the moment of inertia derived earlier by Skyrme[26], and his expression for $E(J)$, keeping only the ($n = 1$)-term, agrees with that obtained by Dasgupta and Van Ginneken[27] by using an approximate form of the projection operator. Warke again derived $E(J)$ (see the second paper of Warke[22]), by using the projection operator given in Section 36A and following the spin-matrix polynomial expansion method of an arbitrary function of J_y [in this case, $\exp(-i\theta J_y)$]. He concluded that the expressions of $E(J)$ arrived at by the various methods have different forms which suggest nonuniqueness in the construction of the projection operator.

Castel et al[28] calculated the single-particle occupation probability in the ground states of even nuclei of the ($1s$-$0d$)-shell and compared the results with stripping data. Parikh et al[28] made calculations which critically examine the nature of the energy gap, in particular its relationship with the nonlocality of the HF potential as proposed by Bar-Touv and Levinson[9] (described in Section 37B). An alternative explanation of the energy gap, which has been verified by Parikh and his coworkers, is as follows. The quadrupole part of the two-body interaction produces the deformed HF solution and breaks the spherical symmetry of the original Hamiltonian in the resultant wavefunction. The energy gap is a result of this symmetry breaking, and hence related to the proportion of quadrupole and hexadecapole components of the interaction. However, these authors argued that, in a limited model space calculation, it is difficult to conclude in favour of either of the two alternative explanations (i.e., nonlocality or quadrupole interaction).

In a series of papers, Parikh et al[29] investigated the question of minimizing the variance in energy σ, defined by

$$\sigma^2 = \langle H^2 \rangle - \langle H \rangle^2,$$

where $\langle\ \rangle$ indicates the expectation value. Obviously, the minimization of $\langle H \rangle$ in the HF calculation produces a better value of the energy; on the other hand, the minimization of σ (which is related to the spread of the state over the actual eigenstates of H) produces a better wavefunction. In order that such a wavefunction does not lead to a very poor value of the energy, we have to look for a Ψ to minimize σ^2 in the neighbourhood of the HF wavefunction Φ. These authors showed that such solutions exist and are not very different from Φ. The spread σ at the minimum and also that of the HF solution are quite large by any standards. For example, the HF state in O^{16} may have a spread as large as 24.3 MeV.

Elaborate HF calculations have been done as a replacement of the shell-model configuration-mixing calculations. Angular momentum states projected from HF determinants, corresponding to several intrinsic shapes, and hole-particle excited determinants have been used in such work for diagonalizing the Hamiltonian. Nair and Satpathy[30] did such calculations for Ne^{22}, whereas Khadkikar, Nair, and Pandya[30] and, later, Khadkikar, Kulkarni, and Pandya[30] applied the method to Ne^{24}.

Very elaborate work of the foregoing type on the $(0f\text{-}1p)$-shell nuclei has been done by Dhar et al[31]. Earlier calculations by Parikh and Svenne[12] on the $(0f\text{-}1p)$-shell nuclei have been described in Section 37B. Work involving angular momentum projection has also been done by Khadkikar and Gunye[32], and Khadkikar and Banerjee[33]. The nuclei, treated by Dhar and his coworkers, are V^{48}, Ti^{45-51} (odd A isotopes), Cr^{49}, V^{51}, and Mn^{53}. The method depends on first doing an axially symmetric HF calculation for the extra core nucleons (Ca^{40} is treated as the core), using the modified Kuo-Brown two-body matrix elements as proposed by McGrory et al[34]. In all the cases, prolate and oblate solutions are found. In addition to the angular momentum states projected from these two intrinsic wavefunctions, the authors used also states projected from 1p-1h determinants based on the two HF solutions. For each J, a set of nonorthogonal states was thus obtained. Taking the overlap of pairs of these states and diagonalizing the resultant overlap matrix, they constructed a set of orthogonalized basis states with the help of the eigenvector of the overlap matrix. The Hamiltonian matrix was then set up with the orthogonalized basis and diagonalized. The energy eigenvalues and the corresponding eigenstates were obtained in this way. The authors also computed various electromagnetic properties of the resultant states using the effective charges $e_p = 1.32e$ and $e_n = 0.89e$ in most cases; although, in a few cases, the values $e_p = 1.5e$ and $e_n = 0.5e$ produce slightly better agreement. For details, the reader is referred to the original papers.

A simple interaction, originally proposed by Skyrme[35] and subsequently modified by Vautherin and Brink[36], has been extensively used in HF calculations. The Skyrme interaction consists of a two-nucleon part V_{12} and a three-nucleon part V_{123}. The part V_{12} is conveniently defined in the momentum space by

$$\langle \mathbf{k} | V_{12} | \mathbf{k}' \rangle = t_0(1 + x_0 P_\sigma) + t_1(k^2 + k'^2) + t_2 \mathbf{k} \cdot \mathbf{k}' + iW(\boldsymbol{\sigma}_1 + \boldsymbol{\sigma}_2) \cdot (\mathbf{k} \times \mathbf{k}'),$$

whereas V_{123} is taken to be a three-nucleon contact interaction

$$V_{123} = t_3 \delta(\mathbf{r}_1 - \mathbf{r}_2)\delta(\mathbf{r}_2 - \mathbf{r}_3).$$

The modification introduced by Vautherin and Brink[36] entails replacing V_{123} by a density-dependent two-nucleon interaction v_{12} (the two forms being equivalent in the HF energy calculations of even nuclei) given by

$$v_{12} = \tfrac{1}{6} t_3 (1 + P_\sigma) \delta(\mathbf{r}_1 - \mathbf{r}_2) \rho(\mathbf{R}).$$

Here P_σ is the Bartlett exchange operator, and the density ρ is to be taken at $\mathbf{R}$, the centre-of-mass of the two nucleons, i.e., $\frac{1}{2}(\mathbf{r}_1 + \mathbf{r}_2)$. The complete density-dependent Skyrme interaction, $V_{12} + v_{12}$, thus contains the six parameters t_i ($i = 0, 1, 2, 3$), x_0, and W. Earlier values of the parameters, suggested by Skyrme, have been revised by Vautherin and Brink[36] by fitting the binding energies of O^{16} and Pb^{208}, and by Beiner et al[37] by fitting the binding energies and charge radii of a large number of magic nuclei. Moszkowski[38] used a somewhat simpler form of a density-dependent interaction in his many-body calculations. In his many-body G-matrix work, Bethe[39] proposed a density-dependent effective interaction, and Nemeth and Bethe[40] applied it to self-consistent energy calculations. Negele[41] extensively computed finite nuclei using the density-dependent interaction, and Negele and Vautherin[42] tried to link the G-matrix version of the density-dependent interaction with the phenomenological Skyrme version.

Khadkikar and Kamble[43] applied the Skyrme interaction to angular momentum projected HF calculations on Li^8, Be^8, B^8, C^{12}, and Ne^{20}. In some of this work, the density-dependent part of the interaction has been omitted and the results compared with those obtained in the same calculations using the Sussex interaction (Elliott et al[25]). While applying the density-dependent interaction to deformed nuclei, the authors had to use an averaging procedure to circumvent the difficulty relating to the nonscalar nature of v_{12}, caused by the angle-dependence of the density ρ.

The HF energy needs correction due to the centre-of-mass motion of the nucleus, and the Coulomb energy before it is compared with experimental values. The Coulomb interaction between pairs of protons can be included in the total Hamiltonian while doing the HF self-consistent calculation. An approximate calculation of the Coulomb energy can also be done by using the HF wavefunction obtained without the Coulomb force. The question of centre-of-mass motion plagues all many-body wavefunctions based on individual particle states. This question was first discussed by Elliott and Skyrme[44] in the context of the harmonic oscillator shell-model wavefunctions. The problem involved is as follows: the centre-of-mass of a nucleus can be displaced through a distance $\mathbf{a}$ with the help of the operator $\exp[-(i/\hbar)\mathbf{P}\cdot\mathbf{a}]$, where $\mathbf{P} = \sum_{i=1}^{A} \mathbf{p}_i$ is the total momentum of the A nucleons in the nucleus. Under this displacement, the nuclear wavefunction should, at the most, get multiplied by a numerical factor $\exp[i\delta(\mathbf{a})]$, where $\delta(\mathbf{a})$ is real. This implies that the nuclear wavefunction be an eigenfunction of the operator $\mathbf{P}$, and the centre-of-mass part of the wavefunction be given by $\exp(i\mathbf{K}\cdot\mathbf{R})$, where $\mathbf{R}$ is the centre-of-mass coordinate vector and $\hbar\mathbf{K}$ is the eigenvalue of $\mathbf{P}$. It is difficult to ensure in a determinantal HF state that the centre-of mass dependence of the wavefunction automatically becomes a factor of the type $\exp(i\mathbf{K}\cdot\mathbf{R})$. Various approximate ways of dealing with the centre-of-mass motion have therefore been suggested. Kerman et al[6] proposed that the intrinsic energy be computed by eliminating the kinetic energy of the centre-of-mass from the total Hamiltonian. In other words, an HF calculation is done with $H - P^2/(2MA)$, where M is the nucleon mass. Since

$$H - \frac{P^2}{2MA} = \sum_{i=1}^{A} \frac{p_i^2}{2M} + \sum_{i<j} V_{ij} - \frac{\sum_i \mathbf{p}_i \cdot \sum_j \mathbf{p}_j}{2MA}$$

$$= \sum_{i=1}^{A} (1 - \frac{1}{A})\frac{p_i^2}{2M} + \sum_{i<j} (V_{ij} - \frac{\mathbf{p}_i \cdot \mathbf{p}_j}{MA}),$$

the HF calculation with this modified Hamiltonian can be carried out in the usual way. Some questions relating to the centre-of-mass motion with the Hamiltonian just given and an equivalent form of it, namely,

$$\sum_{i<j} [V_{ij} + \frac{(\mathbf{p}_i - \mathbf{p}_j)^2}{2MA}],$$

have been investigated by Khadkikar and Kamble[43].

REFERENCES

1. Kelson, I., *Phys. Rev.*, **132**, 2189 (1963); Kelson, I., and Levinson, C. A., *Phys. Rev.*, **134**, B269 (1964); Kelson, I., *Phys. Rev.*, **134**, B267 (1964); Bassichis, W., Kelson, I., and Levinson, C. A., *Phys. Rev.*, **136**, B380 (1964); Bar-Touv, J., and Kelson, I., *Phys. Rev.*, **138**, B1035 (1965).
2. Muthukrishnan, R., and Baranger, M., *Phys. Letters*, **18**, 160 (1965).
3. Davies, K. T. R., Krieger, S. J., and Baranger, M., *Nucl. Phys.*, **84**, 545 (1966); Nestor (Jr.), C. W., Davies, K. T. R., Krieger, S. J., and Baranger, M., *Nucl. Phys.*, **A113**, 14 (1968); Turbutton, R. M., and Davies, K. T. R., *Nucl. Phys.*, **A120**, 1 (1968).
4. Volkov, A. B., *Nucl. Phys.*, **74**, 33 (1965).
5. Brink, D. M., and Boeker, E., *Nucl. Phys.*, **A91**, 1 (1967).
6. Kerman, A. K., Svenne, J. P., and Villars, F. M. H., *Phys. Rev.*, **147**, 710 (1966); Kerman, A. K., and Pal, M. K., *Phys. Rev.*, **162**, 970 (1967); Bassichis, W. H., Kerman, A. K., and Svenne, J. P., *Phys. Rev.*, **160**, 746 (1967).
7. Shakin, C. M., Waghmare, Y. R., and Hull (Jr.), M. H., *Phys. Rev.*, **161**, 1006 (1967); Shakin, C. M., Waghmare, Y. R., Tomaselli, M., and Hull (Jr.), M. H., *Phys. Rev.*, **161**, 1015 (1967); Shakin, C. M., and Waghmare, Y. R., *Phys. Rev. Letters*, **16**, 403 (1966); Pal, M. K., and Stamp, A. P., *Phys. Rev.*, **158**, 924 (1967).
8. Davies, K. T. R., Baranger, M., Turbutton, R. M., and Kuo, T. T. S., *Phys. Rev.*, **177**, 1519 (1969).
9. Bar-Touv, J., and Levinson, C. A., *Phys. Rev.*, **153**, 1099 (1967).
10. Ripka, G., in Advances in Nuclear Physics, Vol 1, ed. by M. Baranger, Plenum Press, New York, 1968, p 183.
11. Stamp, A. P., *Nucl. Phys.*, **A105**, 627 (1967).
12. Parikh, J. C., and Svenne, J. P., *Phys. Rev.*, **174**, 1343 (1968).
13. Harishchandra, *Phys. Rev.*, **185**, 1320 (1969).
14. Faessler, A., Sauer, P. U., and Stingl, N. M., *Phys. Letters*, **26B**, 71 (1967).
15. Boeker, E., *Nucl. Phys.*, **A91**, 27 (1967).
16. Bellicard, J. B., Bounin, P., Frosch, R. F., Hofstadter, R., McCarthy, J. S., Uhrhane, E. J., Yearian, M. R., Clark, B. C., Herman, R., and Ravenhall, D. G., *Phys. Rev. Letters*, **19**, 527 (1967).
17. Tewari, S. N., and Grillot, D., *Phys. Rev.*, **177**, 1717 (1969).
18. Bouten, M., in Theory of Nuclear Structure: Trieste Lectures, 1969, IAEA, Vienna, 1970, p 361.
19. Warke, C. S., and Gunye, M. R., *Phys. Rev.*, **155**, 1084 (1967); Gunye, M. R., and Warke, C. S., *Phys. Rev.*, **156**, 1087 (1967); *ibid*, **159**, 885 (1967); *ibid*, **164**, 1264 (1967).
20. Dworzecka, M., and Warke, C. S., *Phys. Rev.*, **175**, 1346 (1968).
21. Gunye, M. R., Warke, C. S., and Khadkikar, S. B., *Phys. Rev.*, **C3**, 1936 (1971).
22. Warke, C. S., *Phys. Rev*, **C10**, 418 (1974); *Pramāṇa*, **5**, 33 (1975).
23. Pal, M. K., and Stamp, A. P., *Nucl. Phys.*, **A99**, 228 (1967); Padjen, R., and Ripka, G., *Nucl. Phys.*, **A149**, 273 (1970).

24. Goodman, A. L., Strubble, G. L., and Goswami, A., *Phys. Letters*, **26B**, 260 (1968).
25. Elliott, J. P., Jackson, A. D., Mavromatis, H. A., Sanderson, E. A., and Singh, B., *Nucl. Phys.*, **A121**, 241 (1968).
26. Skyrme, T. H. R., *Proc. Roy. Soc.* (London), **A70**, 433 (1957).
27. Dasgupta, S., and Van Ginneken, A., *Phys. Rev.*, **164**, 1320 (1967).
28. Castel, B., Johnstone, I. P., Singh, B. P., and Parikh, J. C., *Nucl. Phys.*, **A157**, 137 (1970); Parikh, J. C., Castel, B., Stewart, K. W. C., and Lande, A., *Nucl. Phys.*, **A156**, 501 (1970).
29. Parikh, J. C., and Svenne, J. P., *Phys. Rev.*, **C6**, 34 (1972); Bhatt, K. H., and Parikh, J. C., *Phys. Letters*, **44B**, 338 (1973); Satyan, V., and Parikh, J. C., *Phys. Rev.*, **C14**, 1198 (1976).
30. Satpathy, L., and Nair, S. C. K., *Phys. Letters*, **26B**, 716 (1968); Khadkikar, S. B., Nair, S. C. K., and Pandya, S. P., *Phys. Letters*, **36B**, 290 (1971); Khadkikar, S. B., Kulkarni, D. R., and Pandya, S. P., *Pramāṇa*, **2**, 259 (1974).
31. Dhar, A. K., Kulkarni, D. R., and Bhatt, K. H., *Nucl. Phys.*, **A238**, 340 (1975); Dhar, A. K., and Bhatt, K. H., *Nucl. Phys.*, **A271**, 36 (1976); Dhar, A. K., Khadkikar, S. B., Kulkarni, D. R., and Bhatt, K. H., in Proc. Int. Conf. on Nuclear Self-Consistent Fields, ICTP, Trieste (1975), ed. by G. Ripka and M. Porneuf, North-Holland, Amsterdam, 1976; Bhatt, K. H., and Dhar, A. K., *Phys. Rev.*, **C16**, 792 (1977).
32. Khadkikar, S. B., and Gunye, M. R., *Nucl. Phys.*, **A110**, 472 (1968).
33. Khadkikar, S. B., and Banerjee, B., *Nucl. Phys.*, **A129**, 220 (1969).
34. McGrory, J. B., Wildenthal, B. H., and Halbert, E. C., *Phys. Rev.*, **C2**, 186 (1970).
35. Skyrme, T. H. R., *Phil. Mag.*, **1**, 1043 (1956); *Nucl. Phys.*, **9**, 615 (1959).
36. Vautherin, D., and Brink, D. M., *Phys. Rev.*, **C5**, 626 (1972); Vautherin, D., *Phys. Rev.*, **C7**, 296 (1973).
37. Beiner, M., Flocard, H., Giai, N. V., and Quentin, P., *Nucl. Phys.*, **A238**, 29 (1975).
38. Moszkowski, S. A., in Heavy-Ion, High Spin States and Nuclear Structure, IAEA, Vienna, 1975, p 131.
39. Bethe, H. A., *Phys. Rev.*, **167**, 879 (1968).
40. Nemeth, J., and Bethe, H. A., *Nucl. Phys.*, **A116**, 241 (1968); Nemeth, J., *Nucl. Phys.*, **A156**, 183 (1970); Nemeth, J., and Ripka, G., *Nucl. Phys.*, **A194**, 329 (1972).
41. Negele, J. W., *Phys. Rev.*, **C1**, 1260 (1970).
42. Negele, J. W., and Vautherin, D., *Phys. Rev.*, **C5**, 1472 (1972); *Nucl. Phys.*, **A207**, 298 (1973).
43. Khadkikar, S. B., and Kamble, V. B., *Phys. Letters*, **64B**, 131 (1976).
44. Elliott, J. P., and Skyrme, T. H. R., *Proc. Roy. Soc.* (London), **A232**, 561 (1955).

PROBLEMS

1. Use the definitions (V.6b) and (V.6c) to rewrite the HF potential in the coordinate space. Show that it consists of a local and a nonlocal part. Assume V to be a well-behaved central potential with a general exchange-dependence, and derive explicit expressions for the strengths of the local and nonlocal parts.

2. It is well-known that a determinant D having the elements d_{ij} can be expanded as $\sum_j d_{ij}M_{ij}$, where M_{ij} is the minor of the element d_{ij}. Prove that $\sum_i d_{ij}M_{ik} = 0$ when $j \neq k$. [Hint: Use the property of a determinant that states that whenever two rows or two columns are identical, the determinant vanishes.] Thus, we can write $D\delta_{jk} = \sum_i d_{ij}M_{ik}$. Use this relation to obtain an expression of the minor in terms of D and the elements d_{ij}.

3. Generalize the results of Problem 2 to obtain an expression of the minor $M_{ij;\,kl}$, obtained by suppressing the rows i, j and the columns k, l of the determinant and multiplying by the appropriate phase factor.

4. Derive (V.51b).

5. Prove (V.52).

6. Obtain an expression for the expectation value of $\sum_{i=1}^{A} T_q^K(i)$ by using an angular momentum projected state.

7. Consider the HF ground state Φ_0 and the 1h-1p state Φ_{hp}. Use the Hamiltonian (V.1) and the standard results (II.24) and (II.26) to write down the matrix element $\langle\Phi_0| H |\Phi_{hp}\rangle$. Prove that it vanishes by virtue of the definition (V.8) and the fact that h and p are orthogonal eigenstates of the HF Hamiltonian.

VI Microscopic Theory of Nuclear Structure—Vibrational States

38. INTRODUCTION

The phenomenon of nuclear vibration, which is the subject of this chapter, has been introduced in Chapter IV in a phenomenological manner. Here we are concerned with a microscopic description of the phenomenon.

The nucleus is a many-body system, which is made up of nucleons moving in an average potential field and interacting weakly through a residual interaction. It is obvious that such a many-body system has excited states corresponding to the excitation of one or a few particles. This type of excitation mode of the nucleus is well-known near closed shells. An altogether different mode of excitation is that in which many particles participate in a coherent manner. Such an excitation is well-known in the many-electron problem, where there is a coherent plasma mode of oscillation. Similar oscillation modes have been observed in nuclei also; in fact, we have a very rich variety here, one distinguished from the other by the angular momentum (J), parity (π), and isospin (T) of the phonon associated with the vibrational mode.

The most important nuclear vibrations, classified with J^π and T, are (i) the giant dipole oscillation ($J^\pi = 1^-$, $T = 1$), (ii) quadrupole vibration (2^+, $T = 0$), and (iii) octupole vibration (3^-, $T = 0$).

For theoretical purposes, it is convenient to classify vibrational states according to the type of nuclei they belong to. In so doing, we have the theory of vibration of (i) closed-shell nuclei, (ii) spherical nonclosed-shell nuclei, and (iii) deformed nuclei, which differ from each other in some details. The basic concepts (now outlined) are however the same for all these different cases.

The single-particle level schemes for the three cases and the highest occupied level λ (fermi level) are shown in Fig. VI.1. In the case of a closed-shell nucleus, the level immediately above λ is separated from it by a large interval (the spacing between two shells), and hence the excitation of a nucleon from the occupied levels to the lowest unoccupied level requires a fairly large energy. The ground state therefore is a fairly good closed-shell state having an occupation probability P, below and above λ, as depicted in Fig. VI.2a. The state obtained by exciting a nucleon from an occupied to an unoccupied level is usually called a state of the one-hole one-particle (1h-1p) type because, in this process of excitation, a vacancy (hole) is left in the occupied levels and a particle is produced instead in one of the unoccupied levels. This type of excitation is a possible elementary mode of excitation of a closed-shell nucleus, and such a mode has an energy very nearly equal to the separation of the two shells at the top of the fermi sea. However, if each of the nucleons in the fermi sea is allowed to get excited in turn above the sea, there will be a whole set of 1h-1p type states of the closed-shell nucleus. The question is whether these independent particle modes of the different fermi-sea nucleons

66 (45-122/1977)

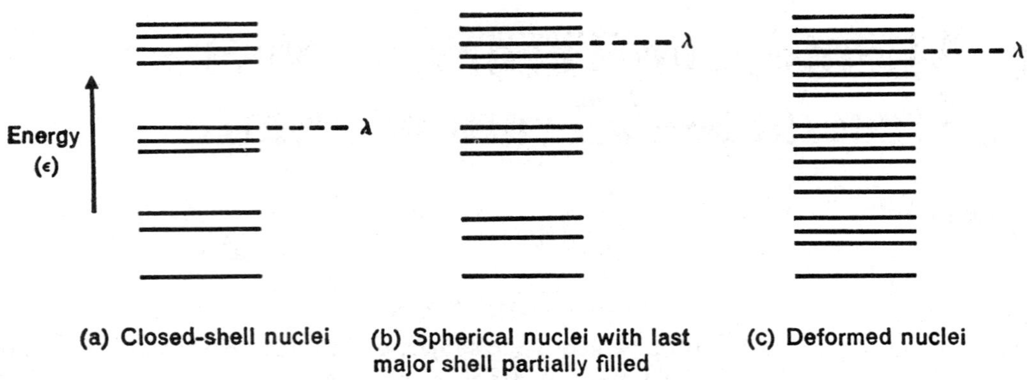

Fig. VI.1 Single-particle level schemes.

act coherently and build up a collective type of excitation mode. That this is so becomes evident in this chapter, and the resultant coherent state is, for all intents and purposes, a vibrational state of the closed-shell nucleus. The exact meaning of the word "vibration" in this context is made more quantitative as we go along.

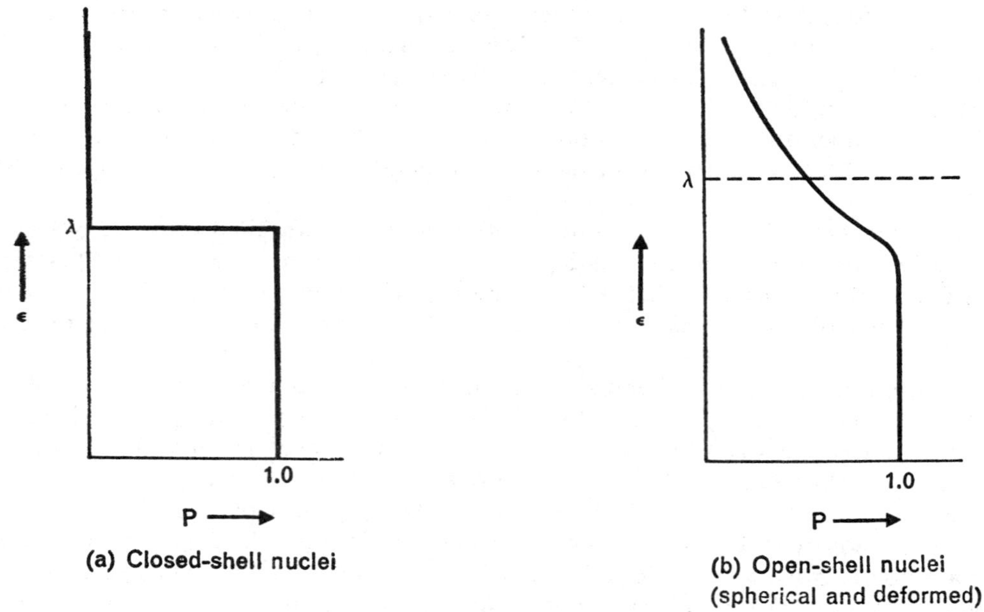

Fig. VI.2 Occupation probability of single-particle levels as function of single-particle energy.

Let us now go back to Figs. VI.1b, VI.1c, and VI.2b and examine what happens in the case of nonclosed-shell spherical and deformed nuclei. The only difference between these two cases is the nature of the single-particle states. In the case of spherical nuclei (Fig. VI.1b), the states are labelled by the quantum numbers *nljm*, and the separation between the major shells is still large. On the other hand, the single-particle states of deformed nuclei (Fig. VI.1c) are a superposition of several sets of *nljm*; for axially symmetric nuclei, only m is a good quantum number. In this case, the levels belonging to various major shells get closer to each other, and can even penetrate each other for a large deformation. The main point of interest, at present, is what occurs near the fermi level λ. In both the cases, there are several single-particle energy levels which crowd together in this energy region. As a consequence, our picture of a sharp fermi sea as the ground state of these nuclei is sure to be rather poor. Since it costs very little energy to promote a nucleon from the fermi sea to one of these crowded levels immediately above λ, we expect the ground state to contain such excitations as make the occupation probability of levels immediately below and above λ depart from their ideal values 1 and 0, respectively. This picture of the ground state having a diffuse single-particle probability distribution is shown in Fig. VI.2b. Since the ground state is already somewhat complicated for these nuclei, the type of elementary excitations that will take place in them is not immediately clear. We shall show in Section 41 that it is still possible to find elementary excitation modes, to be called *quasiparticles*. In this approximate theory, the ground state of Fig. VI.2b is characterized by the absence of any quasiparticle; the states of odd-mass and even nuclei correspond respectively to an odd and even number of quasiparticles. For an even nucleus, the elementary excitation modes of the lowest energies thus correspond to a *pair* of quasiparticles. This is closely analogous to the hole-particle *pair* excitation mode of a closed-shell nucleus. Therefore, we can formulate a microscopic theory of vibration corresponding to Fig. VI.2b by a coherent superposition of quasiparticle pair states. This is the core of the microscopic theory of vibration developed in our subsequent discussion.

The microscopic picture of vibration, as just outlined, apparently seems rather distinct from the classical description of vibration of a dynamical variable. In reality, however, there is an underlying link between the two pictures, and it is possible to show that the nuclear vibrational state, which is a coherent superposition of many elementary excitation modes of the hole-particle (or pair of quasiparticles) type, corresponds to a single-particle density variable that indeed oscillates with time. The equation of motion for the oscillating density is exactly identical to the system of linear equations that connects the elementary excitation modes.

In our discussion of the Hartree-Fock (HF) theory in Section 35, we have noted that the single-particle density is self-consistently connected to the average nuclear field. Therefore, the oscillations we are dealing with can equivalently be looked upon as oscillations of the average HF potential. We should keep in mind, however, that this kind of vibration does not exhaust all the possibilities for a nucleus. An example of states corresponding to vibrations of a different type of dynamical variables has been recently encountered; this is the case of pairing vibration. The dynamical quantity that vibrates here is the pairing density or, equivalently, the pairing potential.

We shall begin by describing a general method of treating vibration. Several methods have been applied by various authors in deriving the equations for nuclear vibrations. Needless to say, the final equations derived are independent of the method used. Basically, the derivation can be given in two different ways. The first (see Section 40) is a time-dependent HF treatment, and features the semiclassical picture of an oscillating density distribution; the second derivation (Section 41) is more formal, and is based on the transformation of the Hamiltonian of the many-body system to a form that clearly shows the part responsible for the vibrations, and what remains after such a treatment. The second method therefore allows a natural extension, which enables the incorporation of the effects of the residual parts of the

Hamiltonian. Such an extension gives rise to anharmonic effects in nuclear vibration. As mentioned in the phenomenology of Section 33B, there is ample experimental evidence of such anharmonicity in nuclear quadrupole vibration.

Towards the end of this chapter we shall describe the application of the theory to the vibrational states in (i) closed-shell nuclei, (ii) spherical nonclosed-shell nuclei, and (iii) deformed nuclei. A preview of these may be pertinent here.

Closed-Shell Nuclei

The nuclei that have been exhaustively treated by many authors are O^{16}, Ca^{40}, and Pb^{208}. In all these nuclei, the giant dipole resonance states ($J^\pi = 1^-, T = 1$) and the octupole states ($J^\pi = 3^-$, $T = 0$) are very well-reproduced by the theory based on the 1h-1p type excitations. The agreement of the transition strength in the case of octupole states is very good, whereas there are some minor discrepancies in the giant dipole states. The latter states are around 20–25 MeV in the lighter nuclei, whereas in Pb^{208} they are in the neighbourhood of 14–15 MeV. It has been found that, for these states, the excitation of a nucleon to the continuum states is also important; an extension of the kind of theory we shall describe here, so as to incorporate the effect of the continuum states, truly belongs to the realm of the microscopic theory of nuclear reactions.

The octupole states are very strongly collective, have a large E3-transition probability to the ground state, and are very low-lying in energy. The type of theory we derive is very appropriate for describing these two features.

For a time, there were many calculations on the states 2^+, $T = 0$ in closed-shell nuclei using the method based on the 1h-1p excitations. However, it is now conclusively established that getting a *low-lying* 2^+ state by this method is difficult. Experimentally also, the low-lying 2^+ states of closed-shell nuclei are found to have very little collective vibrational character—the E2-transition strength connecting such a state to the ground state is rather small.

The most enigmatic state in closed-shell nuclei (and also many spherical even nuclei) is the low-lying 0^+ state. In early calculations on vibrational states, this state also was unsuccessfully attributed a 1h-1p type character. However, the experimentally observed small transition strength to the ground state rules out such a treatment of the 0^+ state. In O^{16} and Ca^{40}, this state has been established to have a deformation, even though the ground states are spherical. However, Pb^{208} is a very stable closed-shell nucleus, and does not tend to acquire a deformation in any low-lying excited state. The 0^+ state in this nucleus is now understood to be a member of the chain of levels in the Pb region connected to each other by the pairing vibration. In O^{16} and Ca^{40}, there are states other than the 0^+ state that correspond to a deformed shape.

Spherical Nonclosed-Shell Nuclei

In contrast with closed-shell nuclei, spherical nonclosed-shell nuclei have a strongly collective quadrupole vibrational level (2^+) as their first excited state. The octupole level (3^-) has also been observed in many of these nuclei. The main point of interest centres around the group of excited states above the first excited 2^+. Phenomenological details of the two-phonon triplet, the presence of anharmonicity, and the case of transitional nuclei have already been described in Sections 33B and 34A.

Deformed Nuclei

As described in Section 33B, in deformed nuclei the levels near the ground state belong to a rotational band. However, they exhibit excited states of the vibrational type, the most important ones being the quadrupole β- and γ-vibrational state, and the octupole vibrational states. In deformed nuclei, the phonon of the vibration does not, strictly speaking, have a definite angular momentum; yet it is customary to consider these vibrational states as 'quadrupole' or 'octupole', in keeping with the semiclassical picture of the shape oscillation of the nucleus starting from an *equilibrium* prolate spheroidal shape in the manner of Section 33B, where the cases of β- and γ-vibration have been described in detail. When the deformed shape during the oscillation is of the octupole type, we may have states with the projection 0^-, 1^-, 2^-, 3^-, all of which comprise the octupole vibrational states. Only a brief summary of the existing results is given in this chapter.

Very recently, a fresh resurgence of interest has occurred in the area of giant multipole resonances. Experimental studies have been made by using high-energy electron scattering and the scattering of high-energy protons and alpha-particles, and also heavy ions such as Li^6. The energy region of the nucleus where these resonances have been found is roughly given by $60A^{-1/3}$ MeV. Multipole excitations studied in this way are E2, E0, M1, and M2, and, in small proportions, E4 and E6 as well. The experimental findings, the theoretical analysis, and further details on giant multipole resonances are available in current literature*.

39. GENERAL COMMUTATOR METHOD

A. SIMPLE CASE

In its simplest form, the commutator method entails looking for an operator $Q^\dagger$, whose commutator with the Hamiltonian H is a numerical mutiple of itself, i.e.,

$$[H, Q^\dagger] = \hbar\omega Q^\dagger, \tag{VI.1}$$

where $\hbar\omega$ is a number. This equation automatically guarantees (take the Hermitean conjugate and reverse the sign)

$$[H, Q] = -\hbar\omega Q. \tag{VI.2}$$

Without any loss of generality, we may assume $\hbar\omega$ to be positive; because, if $\hbar\omega$ is not positive, then $-\hbar\omega$ is, and hence all that becomes necessary is to reverse the roles of Q and $Q^\dagger$.

Now, if Ψ is an eigenfunction of H belonging to the eigenvalue E, then (VI.1) guarantees that $Q^\dagger\Psi$ is also an eigenfunction of H belonging to the eigenvalue $(E + \hbar\omega)$. The proof follows.

Given $H\Psi = E\Psi$, (VI.1) ensures

$$HQ^\dagger\Psi - Q^\dagger H\Psi = \hbar\omega Q^\dagger\Psi$$

or

$$H(Q^\dagger\Psi) - E(Q^\dagger\Psi) = \hbar\omega(Q^\dagger\Psi)$$

or

$$H(Q^\dagger\Psi) = (E + \hbar\omega)(Q^\dagger\Psi),$$

*See, for example, Youngblood, D. H., Moss, J. M., Rosza, C. M., Bronson, J. D., Bacher, A. D., and Brown, D. R., *Phys. Rev.*, C13, 994 (1976); Halbert, E. C., McGrory, J. B., Satchler, G. R., and Speth, J., *Nucl. Phys.*, **A245**, 189 (1975); Speth, J., Werner, E., and Wild, W., *Phys. Rep.*, **33C**, 128 (1977).

which proves the required result. In a similar manner, we can prove, with the help of (VI.2),

$$H(Q\Psi) = (E - \hbar\omega)(Q\Psi).$$

Thus, $Q^\dagger$ acts as the step-up operator for energy, and Q as the step-down operator. In particular, if Ψ_0 denotes the ground state of H, then

$$Q\Psi_0 = 0 \tag{VI.3}$$

because it would be essential for $Q\Psi_0$ to have an energy lower than that of the ground state by $\hbar\omega$, which is, by definition of the ground state, impossible.

It is clear that the method so far described is applicable to *any* Hamiltonian, and not necessarily to the Hamiltonian of an oscillator. We have found that, for any general Hamiltonian, if $Q^\dagger$, as defined by (VI.1), exists, then the spectrum of H is given by

$$E_n = E_0 + n\hbar\omega, \qquad n = 0, 1, 2, \ldots. \tag{VI.4}$$

Clearly, it is impossible to know what E_0 is without specifying the Hamiltonian H in detail.

B. GENERALIZATION

We now generalize the commutator method as follows. Suppose we have found a set of operators $A_i^\dagger$ ($i = 1, 2, \ldots, N$) which satisfy

$$[H, A_i^\dagger] = \sum_{j=1}^{N} M_{ji} A_j^\dagger = (\tilde{M} A^\dagger)_i \tag{VI.5}$$

for every $i = 1, 2, \ldots, N$. Here M is a numerical matrix and $\tilde{M}$ is its transpose.

The question that now arises is whether the spectrum of H can be calculated. To obtain the answer, we first diagonalize the matrix M, and find its eigenvalues $\mathcal{E}_\alpha$ ($\alpha = 1, 2, \ldots, N$) and the corresponding eigenvectors $X^{(\alpha)}$ which have the components $X_1^{(\alpha)}, X_2^{(\alpha)}, \ldots, X_N^{(\alpha)}$. By definition,

$$\sum_{j=1}^{N} M_{ij} X_j^{(\alpha)} = \mathcal{E}_\alpha X_i^{(\alpha)}, \qquad i = 1, 2, \ldots, N. \tag{VI.6}$$

Now we construct the operators with these eigenvectors:

$$Q_\alpha^\dagger = \sum_{i=1}^{N} X_i^{(\alpha)} A_i^\dagger, \qquad i = 1, 2, \ldots, N. \tag{VI.7}$$

Substituting (VI.5) and (VI.6) in the commutator, we obtain

$$\begin{aligned}[H, Q_\alpha^\dagger] &= \sum_{i=1}^{N} X_i^{(\alpha)} [H, A_i^\dagger] \\ &= \sum_{j=1}^{N} \Big[\sum_{i=1}^{N} M_{ji} X_i^{(\alpha)}\Big] A_j^\dagger \\ &= \mathcal{E}_\alpha \sum_{j=1}^{N} X_j^{(\alpha)} A_j^\dagger = \mathcal{E}_\alpha Q_\alpha^\dagger.\end{aligned} \tag{VI.8}$$

Comparing this expression with the results proved in Section 39A, we conclude that, in the present case, there is a set of step-up operators $Q_\alpha^\dagger$, $\alpha = 1, 2, \ldots, N$, which steps up the energy by the respective $\mathcal{E}_\alpha$. The corresponding Hermitean conjugate operators Q_α step down the energy by $\mathcal{E}_\alpha$ and, acting on the ground state Ψ_0 of H, produce a zero result, i.e.,

$$Q_\alpha \Psi_0 = 0, \qquad \alpha = 1, 2, \ldots, N. \tag{VI.9}$$

We go back to (VI.5) and multiply from the left and the right by $\langle\Psi|$ and $|\Psi_0\rangle$, respectively, where Ψ is *any* excited state and Ψ_0 the ground state. Since

$$[H, A_i^\dagger] = HA_i^\dagger - A_i^\dagger H,$$

we let H operate on $\langle\Psi|$ in the first term and on $|\Psi_0\rangle$ in the second term, thereby obtaining the energies E and E_0 of the states Ψ and Ψ_0, respectively. Thus, (VI.5) yields, with all these manipulations, the result

$$(E - E_0)\langle\Psi|\, A_i^\dagger\, |\Psi_0\rangle = \sum_{j=1}^{N} M_{ji}\langle\Psi|\, A_j^\dagger\, |\Psi_0\rangle. \tag{VI.10}$$

Quantities such as $\langle\Psi|\, A_i^\dagger\, |\Psi_0\rangle$ give the probability amplitude X_i of a basis state $A_i^\dagger|\Psi_0\rangle$ contained in the excited state Ψ. The set of equations (VI.10) then defines the eigenvalue equations for the amplitudes X_i, where $(E - E_0)$ is the eigenvalue, obtained by diagonalizing the coefficient matrix $\tilde{M}$. This set of equations is trivially satisfied if all the amplitudes X_i are identically zero for a given Ψ and Ψ_0. Nontrivial values of these amplitudes correspond to the eigenvectors of $\tilde{M}$ belonging to the various eigenvalues of $\tilde{M}$. Thus, we conclude that we get a set of nontrivial amplitudes connecting the two nuclear states Ψ and Ψ_0 only if these states differ in energy through an eigenvalue of $\tilde{M}$ (which is the same as that of M), which is incidentally the energy of a vibrational quantum. Pairs of nuclear states differing in energy by two or more vibrational quanta have all the amplitudes X_i equal to zero.

The general method described in this section is well suited to the treatment of the nuclear Hamiltonian. In view of the general nature of this method, it is capable of yielding both the single 'quasiparticle'-type and the 'vibrational'-type solutions of the many-body Hamiltonian. We have seen in Section 16 that H is very conveniently expressed in the second-quantized form. If we use the single-particle creation operators and the destruction operators for $A_i^\dagger$ in the general relation (VI.5), then we do *not* get a set of linear relations connecting these quantities. The application to the nuclear 'quasiparticle' mode therefore relies on *approximations* that reduce these commutators to a set of linear relations for the single-particle creation and destruction operators. In the same way, if we use the 1h-1p creation operators, or the quasiparticle pair creation operators, for the $A_i^\dagger$, and introduce suitable approximations to obtain a set of linear relations such as (VI.5), then the resultant $Q^\dagger$-operators give rise to the so-called vibrational states. It is clear that in both cases we have approximations, and hence in practical applications to nuclei we have to calculate the corrections as well. We therefore emphasize that the general method as described here does not give *exact* results when applied to the nuclear many-body Hamiltonian.

40. TIME-DEPENDENT HARTREE-FOCK THEORY

If $\Psi(t)$ is the time-dependent many-body determinantal state, then the variational equation

$$\langle\delta\Psi(t)|\, (H - i\hbar\frac{\partial}{\partial t})\, |\Psi(t)\rangle = 0 \tag{VI.11}$$

can be used to determine the single-particle states $\psi_h(t)$ comprising the determinant. The variation $\delta\Psi(t)$ actually consists of variations in the occupied single-particle states $\psi_h(t)$. The solution to this variational problem is given by

$$(T + \mathcal{C}\hat{V})\psi_h(t) = i\hbar\frac{\partial\psi_h(t)}{\partial t}, \tag{VI.12}$$

where the time-dependent HF potential $\hat{\mathcal{V}}$ is defined by the matrix elements

$$\langle A|\, \hat{\mathcal{V}}\, |B\rangle = \sum_{h'} (A\psi_{h'}|\, V\, |B\psi_{h'}). \tag{VI.13}$$

By convention, the single-particle states denoted by the letter h stand for the occupied states in the determinant $\Psi(t)$, whereas the unoccupied states are indicated by the letter p. The basis states $\langle A|$, $|B\rangle$ in (VI.13) can be *any* two single-particle states.

We shall denote the determinantal state obtained through the time-independent HF treatment by Φ, and the corresponding single-particle states by $\phi_h, \phi_p, \ldots$. The time-dependent single-particle states $\psi_h(t)$ can be clearly expressed in terms of the stationary single-particle states as

$$\psi_h(t) = \phi_h \exp\left(-\frac{i}{\hbar}\epsilon_h t\right) + \sum_p C_{hp}(t)\phi_p \exp\left(-\frac{i}{\hbar}\epsilon_p t\right). \tag{VI.14}$$

We shall consider the coefficients $C_{hp}(t)$ to be of the first order of smallness, and work out all the results to the first order in these quantities. Thus,

$$\langle A|\, \hat{\mathcal{V}}\, |B\rangle = \sum_{h'} (A\phi_{h'}|\, V\, |B\phi_{h'}) + \sum_{h',\, p'} \Big[(A\phi_{h'}|\, V\, |B\phi_{h'})C_{h'p'} \exp\left(-\frac{i}{\hbar}\epsilon_{p'h'}t\right)$$
$$+ (A\phi_{p'}|\, V\, |B\phi_{h'})C^*_{h'p'} \exp\left(\frac{i}{\hbar}\epsilon_{h'p'}t\right)\Big], \tag{VI.15}$$

where

$$\epsilon_{p'h'} = \epsilon_{p'} - \epsilon_{h'}. \tag{VI.16}$$

We now express the $\hat{\mathcal{V}}$-term of (VI.12) in the complete set of basis states $|\phi_A\rangle$, where A runs over the unoccupied as well as the occupied single-particle states. Using (VI.15) in the resultant expression, we obtain

$$\hat{\mathcal{V}}|\psi_h\rangle = \sum_A |\phi_A\rangle\langle\phi_A|\, \hat{\mathcal{V}}\, |\psi_h\rangle$$
$$= \sum_A |\phi_A\rangle\Big[\sum_{h'} (\phi_A\phi_{h'}|\, V\, |\psi_h\phi_{h'}) + \sum_{h',\, p'} \Big\{(\phi_A\phi_{h'}|\, V\, |\phi_h\phi_{p'})C_{h'p'} \exp\left(-\frac{i}{\hbar}\epsilon_{p'h'}t\right)$$
$$+ (\phi_A\phi_{p'}|\, V\, |\phi_h\phi_{h'})C^*_{h'p'} \exp\left(\frac{i}{\hbar}\epsilon_{p'h'}t\right)\Big\} \exp\left(-\frac{i}{\hbar}\epsilon_h t\right)\Big]. \tag{VI.17}$$

Since the expression within the braces is already of the first order in the coefficients $C_{h'p'}$, we have used only the ϕ_h-term of ψ_h. The other term contains ψ_h and is simplified as

$$\sum_A |\phi_A\rangle \sum_{h'} (\phi_A\phi_{h'}|\, V\, |\psi_h\phi_{h'}) = \sum_A |\phi_A\rangle \sum_{h'} \Big[(\phi_A\phi_{h'}|\, V\, |\phi_h\phi_{h'}) \exp\left(-\frac{i}{\hbar}\epsilon_h t\right)$$
$$+ \sum_p (\phi_A\phi_{h'}|\, V\, |\phi_p\phi_{h'})C_{hp} \exp\left(-\frac{i}{\hbar}\epsilon_p t\right)\Big]$$
$$= \mathcal{V}\Big[|\phi_h\rangle \exp\left(-\frac{i}{\hbar}\epsilon_h t\right) + \sum_p |\phi_p\rangle C_{hp} \exp\left(-\frac{i}{\hbar}\epsilon_p t\right)\Big]. \tag{VI.18}$$

Here we have used the definition of the time-independent HF potential $\mathcal{V}$, and the completeness of the set of states ϕ_A.

We substitute (VI.18) in (VI.17) and then write (VI.12) in detail. We further note that

$$(T + \mathcal{V})\phi_h = \epsilon_h\phi_h, \qquad (T + \mathcal{V})\phi_p = \epsilon_p\phi_p$$

and the terms arising from the left-hand side of (VI.12) cancel out with the corresponding terms on the right-hand side of the same equation, where $\partial/\partial t$ operates on the exponential functions of the time. On the right-hand side, we are thus left with only the terms where $\partial/\partial t$ operates on the coefficients $C_{hp}(t)$; and on the left-hand side there remain only the terms of (VI.17) enclosed within the braces. In this way,

$$\sum_p i\hbar \frac{\partial C_{hp}}{\partial t} \exp\left(-\frac{i}{\hbar}\epsilon_p t\right)|\phi_p\rangle = \sum_A |\phi_A\rangle \sum_{h'p'} \Big[(\phi_A\phi_{h'}|\, V\, |\phi_h\phi_{p'})C_{h'p'} \exp\left(-\frac{i}{\hbar}\epsilon_{p'h'}t\right) + (\phi_A\phi_{p'}|\, V\, |\phi_h\phi_{h'})C^*_{h'p'} \exp\left(\frac{i}{\hbar}\epsilon_{p'h'}t\right)\Big] \exp\left(-\frac{i}{\hbar}\epsilon_h t\right). \quad \text{(VI.19)}$$

Multiplying with a given $\langle\phi_p|$, we then obtain

$$i\hbar\frac{\partial C_{hp}}{\partial t} \exp\left(-\frac{i}{\hbar}\epsilon_{ph}t\right) = \sum_{h'p'} \Big[(\phi_p\phi_{h'}|\, V\, |\phi_h\phi_{p'})C_{h'p'} \exp\left(-\frac{i}{\hbar}\epsilon_{p'h'}t\right) + (\phi_p\phi_{p'}|\, V\, |\phi_h\phi_{h'})C^*_{h'p'} \exp\left(\frac{i}{\hbar}\epsilon_{p'h'}t\right)\Big]. \quad \text{(VI.20)}$$

We next break up the time-dependent quantities $C_{hp}(t)$ as

$$C_{hp}(t) \exp\left(-\frac{i}{\hbar}\epsilon_{ph}t\right) = X_{hp}e^{-i\omega t} + Y^*_{hp}e^{i\omega t}, \quad \text{(VI.21)}$$

where X_{hp} and Y_{hp} are independent of time. Substituting (VI.21) in (VI.20) and equating the coefficients of $e^{-i\omega t}$, we obtain

$$(-\epsilon_{ph} + \hbar\omega)X_{hp} = \sum_{h'p'} [(\phi_p\phi_{h'}|\, V\, |\phi_h\phi_{p'})X_{h'p'} + (\phi_p\phi_{p'}|\, V\, |\phi_h\phi_{h'})Y_{h'p'}]. \quad \text{(VI.22a)}$$

In the same manner, the coefficients of $e^{i\omega t}$ yield

$$(-\epsilon_{ph} - \hbar\omega)Y_{hp} = \sum_{h'p'} [(\phi_h\phi_{h'}|\, V\, |\phi_p\phi_{p'})X_{h'p'} + (\phi_h\phi_{p'}|\, V\, |\phi_p\phi_{h'})Y_{h'p'}]. \quad \text{(VI.22b)}$$

The set of simultaneous equations (VI.22) determines the coefficients X_{hp} and Y_{hp} and the vibrational frequency ω. These equations are called the random phase approximation (RPA equations of vibration. In the rest of this section, we shall omit the letter ϕ and use only h, p, . . . in the matrix elements while quoting (VI.22). The origin of the label RPA may be traced to a set of similar equations derived for the plasma oscillation mode of an electron gas where the assumption was that the excitation amplitudes of a pair of electrons corresponding to different momentum transfers have a completely random phase with respect to each other.

It is easy to rewrite the system of equations (VI.22) in terms of matrices by introducing the definitions of the two matrices A and B as

$$A_{hp,\, h'p'} = \epsilon_{ph}\delta_{hh'}\delta_{pp'} - (ph'|\, V\, |p'h), \quad \text{(VI.23a)}$$

$$B_{hp,\, h'p'} = (pp'|\, V\, |hh'). \quad \text{(VI.23b)}$$

Each hole-particle state defines a row or a column of these matrices. In view of the definition of the antisymmetrized matrix element, we have, in the two-body matrix element in (VI.23a), interchanged, with a reversal of sign, the original ordering of h and p' as it occurred in (VI.22a). We also introduce the notation of a column vector X whose elements are X_{hp}, and a similar column vector Y whose elements are Y_{hp}. Then it is easy to verify that (VI.22) can be replaced

67 (45–122/1977)

by the matrix equation

$$\begin{pmatrix} A & B \\ B^* & A^* \end{pmatrix}\begin{pmatrix} X \\ Y \end{pmatrix} = \hbar\omega\begin{pmatrix} X \\ -Y \end{pmatrix}. \tag{VI.24}$$

We denote the matrix on the left-hand side by M, i.e.,

$$M = \begin{pmatrix} A & B \\ B^* & A^* \end{pmatrix}. \tag{VI.25}$$

It is clear from the definitions (VI.23) that A is Hermitean, and B symmetric. Using these properties, we have

$$M^\dagger = \begin{pmatrix} A^\dagger & \tilde{B} \\ B^\dagger & \tilde{A} \end{pmatrix} = \begin{pmatrix} A & B \\ B^* & A^* \end{pmatrix} = M. \tag{VI.26}$$

That is, M is a Hermitean matrix. However, we must notice that (VI.24), because of the presence of the minus sign with Y on the right-hand side, is *not an eigenvalue equation for the matrix M.*

Since (VI.24) contains the two matrix equations

$$AX + BY = \hbar\omega X, \tag{VI.27a}$$

$$B^*X + A^*Y = -\hbar\omega Y, \tag{VI.27b}$$

we can very easily rewrite (VI.27b) as

$$-B^*X - A^*Y = \hbar\omega Y \tag{VI.27c}$$

and then replace the set (VI.27a), (VI.27b) by (VI.27a), (VI.27c). The latter set can then be re-written as

$$M'\begin{pmatrix} X \\ Y \end{pmatrix} = \hbar\omega\begin{pmatrix} X \\ Y \end{pmatrix} \tag{VI.28a}$$

with

$$M' = \begin{pmatrix} A & B \\ -B^* & -A^* \end{pmatrix}. \tag{VI.28b}$$

It should be observed that (VI.28a) is indeed the eigenvalue equation of the new matrix M'. But because of the appearance of the minus signs in (VI.28b), we can easily verify that M' is *not a Hermitean matrix.* Thus, the RPA equations represent the eigenvalue problem for a non-Hermitean matrix.

Several important mathematical properties of the matrix M' are now noted.

(i) The Hermitean matrix M has real eigenvalues. It is possible to prove (see Appendix G) that these real eigenvalues are positive definite if our equilibrium state corresponds to a minimum of energy as a function of the deformation. From this property of the eigenvalues of M, it is then possible to ascertain that *the eigenvalues of the RPA matrix M' are real.* This proof is also given in Appendix G. (The student must read Appendix G as a sequel to this section.)

(ii) Equations (VI.28a) and (VI.28b) are equivalent to (VI.27a) and (VI.27c). Taking the complex conjugate of the latter set and keeping in mind that $\hbar\omega$ is real, we obtain

$$A^*X^* + B^*Y^* = \hbar\omega X^*,$$

$$-BX^* - AY^* = \hbar\omega Y^*.$$

By reversing the sign, we rewrite these equations as

$$AY^* + BX^* = -\hbar\omega Y^*,$$
$$-B^*Y^* - A^*X^* = -\hbar\omega X^*.$$

These two equations can be clearly written in matrix form as

$$M'\begin{pmatrix} Y^* \\ X^* \end{pmatrix} = -\hbar\omega\begin{pmatrix} Y^* \\ X^* \end{pmatrix}. \tag{VI.29}$$

This proves that the eigenvalues of M' occur in pairs $\pm\hbar\omega$; if $\begin{pmatrix} X \\ Y \end{pmatrix}$ is a vector for the eigenvalue $\hbar\omega$, then $\begin{pmatrix} Y^* \\ X^* \end{pmatrix}$ is, according to (VI.29), an eigenvector belonging to the eigenvalue $-\hbar\omega$.

(iii) From the positive definiteness of the eigenvalues of M,

$$\hbar\omega(|X|^2 - |Y|^2) \geqslant 0,$$

i.e., in more detail,

$$\hbar\omega \sum_{h,\,p} (|X_{hp}|^2 - |Y_{hp}|^2) \geqslant 0. \tag{VI.30}$$

This property provides us with a convenient normalization of the eigenvector $\begin{pmatrix} X \\ Y \end{pmatrix}$. For a positive energy $\hbar\omega$, the summation in (VI.30) is positive definite, and hence we can normalize the corresponding eigenvectors to

$$\sum_{h,\,p} (|X_{hp}|^2 - |Y_{hp}|^2) = 1. \tag{VI.31}$$

For a negative energy $-\hbar\omega$, the corresponding normalization is to minus unity. According to property (ii), X for a negative energy solution is Y^* for the corresponding positive energy solution, and Y for the same negative energy solution is X^* for the corresponding positive energy solution. Hence, the normalization of the negative energy solution is automatically guaranteed the moment the normalization (VI.31) for the positive energy solution is assumed.

A detailed interpretation of the RPA equations and amplitudes is given in Section 42B, after we derive these equations by an alternative method.

41. GENERAL MICROSCOPIC THEORY—QUASIPARTICLE MODE

A. BASIC CONCEPT OF QUASIPARTICLES

In this section, we shall derive a general microscopic theory of nuclear states in a manner that will make it applicable to all the three cases mentioned in Section 38. For this purpose, we shall use the second-quantized version of the many-body Hamiltonian

$$H = \sum_{\alpha,\,\beta} \langle\alpha|\, T\, |\beta\rangle C_\alpha^\dagger C_\beta + \tfrac{1}{2} \sum_{\alpha,\,\beta,\,\gamma,\,\delta} \langle\alpha\beta|\, V\, |\gamma\delta\rangle C_\alpha^\dagger C_\beta^\dagger C_\delta C_\gamma, \tag{VI.32}$$

where $C_\alpha^\dagger$, C_β, . . . are the creation and destruction operators for a set of single-particle basis states. The factor $\frac{1}{2}$ in front of the second term has to be changed to $\frac{1}{4}$ if we wish to use in (VI.32) the antisymmetrized matrix element $(\alpha\beta|\, V\, |\gamma\delta)$. However, this is not necessary. The form (VI.32) automatically gives rise to the direct minus exchange terms of the two-body potential in actual calculations.

The destruction operator used in (VI.32) has an important property: the physical vacuum state $|0\rangle$ does not contain any nucleons by definition, and hence it should be impossible to destroy any nucleon from this state, that is,

$$C_\beta|0\rangle = 0 \quad \text{for all } \beta. \tag{VI.33}$$

In order to generalize the concept just stated, let us first examine what happens in the HF state, that is, the situation depicted in Figs. VI.1a and VI.2a. The HF state $|\Phi_0\rangle$ contains a set of occupied single-particle states $h, h', \ldots$, and hence it is impossible to *create* another particle in an occupied state h without violating the Pauli exclusion principle. Therefore,

$$C_h^\dagger|\Phi_0\rangle = 0 \quad \text{for all } h \text{ occupied in } \Phi_0. \tag{VI.34a}$$

The state $|\Phi_0\rangle$ does not contain any of the unoccupied single-particle states $p, p', \ldots$, and hence it should be impossible to *destroy* any such state from $|\Phi_0\rangle$. Thus,

$$C_p|\Phi_0\rangle = 0 \quad \text{for all } p \text{ unoccupied in } \Phi_0. \tag{VI.34b}$$

If we now define a *new* destruction operator for the self-consistent single-particle states by the relation

$$\begin{aligned} b_i &= C_i^\dagger \quad &&\text{(when } i = h\text{, i.e., any of the occupied states)} \\ &= C_i \quad &&\text{(when } i = p\text{, i.e., any of the unoccupied states),} \end{aligned} \tag{VI.35}$$

then (VI.34a) can be combined with (VI.34b) and written as

$$b_i|\Phi_0\rangle = 0 \quad \text{for all } i. \tag{VI.36}$$

Now the formal similarity between (VI.33) and (VI.36) should be observed. We can then interpret the HF state Φ_0 as the vacuum state for the new objects whose destruction operators are defined by (VI.35). These new objects are created by the operators Hermitean conjugate to (VI.35), that is,

$$\begin{aligned} b_i^\dagger &= C_i \quad &&\text{(when } i \text{ is any of the occupied states)} \\ &= C_i^\dagger \quad &&\text{(when } i \text{ is any of the unoccupied states).} \end{aligned} \tag{VI.37}$$

We call these new objects the 'quasiparticles' and the HF state Φ_0 plays the role of the vacuum for these quasiparticles. The quasiparticle states are certainly distinct from the particle states we started with because Φ_0 *does* contain many *particles*, even though it does *not* contain any *quasiparticle*. In fact, the quasiparticle in the present example is either a 'hole' in the HF state or a 'particle' above the occupied states in $|\Phi_0\rangle$. This is clear from (VI.37), the first line of which indicates that if we destroy a particle state in the fermi sea (i.e., the set of occupied states), and thereby create a 'hole' in that state, this becomes equivalent to creating a 'quasiparticle'; on the other hand, the second line of (VI.37) gives an alternative way of getting a quasiparticle, namely, by creating a 'particle' above the fermi sea.

In the general microscopic theory, which we shall now describe, it is our aim to obtain a ground state satisfying the property (VI.36) in the more complicated cases depicted in Figs. VI.1b, VI.1c, and VI.2b. Since the probability distribution near the top of the fermi sea in Fig. VI.2b is very much spread out, in this more general case we cannot talk of a pure 'hole' state inside the fermi sea or a pure 'particle' state above the fermi sea. The quasiparticle therefore obviously loses the simple definitions (VI.35) and (VI.37). The main object of the theory is to *find* the 'quasiparticles' *suitably* so that the ground state in the more general situation can also behave as a vacuum state for those quasiparticles.

B. QUASIPARTICLE TRANSFORMATION OF HAMILTONIAN

Even without writing down the quasiparticle operators explicitly, it is possible to derive a general form for the transformed Hamiltonian. For this purpose, we only assume that there exists a ground state which is a vacuum state for the quasiparticles, and that there is a transformation from the particle creation and destruction operators to the corresponding operator for the quasiparticles.

According to Wick's theorem, which holds for any product of a set of creation and destruction operators, we can write the identities

$$C_\alpha^\dagger C_\beta = :C_\alpha^\dagger C_\beta: + \langle C_\alpha^\dagger C_\beta \rangle, \tag{VI.38}$$

$$\begin{aligned} C_\alpha^\dagger C_\beta^\dagger C_\delta C_\gamma = {} & :C_\alpha^\dagger C_\beta^\dagger C_\delta C_\gamma: \\ & + :C_\alpha^\dagger C_\beta^\dagger: \langle C_\delta C_\gamma \rangle + \langle C_\alpha^\dagger C_\beta^\dagger \rangle :C_\delta C_\gamma: \\ & + :C_\alpha^\dagger C_\gamma: \langle C_\beta^\dagger C_\delta \rangle + \langle C_\alpha^\dagger C_\gamma \rangle :C_\beta^\dagger C_\delta: \\ & - :C_\alpha^\dagger C_\delta: \langle C_\beta^\dagger C_\gamma \rangle - \langle C_\alpha^\dagger C_\delta \rangle :C_\beta^\dagger C_\gamma: \\ & + \langle C_\alpha^\dagger C_\beta^\dagger \rangle \langle C_\delta C_\gamma \rangle + \langle C_\alpha^\dagger C_\gamma \rangle \langle C_\beta^\dagger C_\delta \rangle - \langle C_\alpha^\dagger C_\delta \rangle \langle C_\beta^\dagger C_\gamma \rangle. \end{aligned} \tag{VI.39}$$

In these expressions, $\langle\ \rangle$ denotes an expectation value with respect to the ground state Ψ_0 which is the vacuum state of the quasiparticles. This is usually called the *contraction* of the pair of operators enclosed by the angular brackets. The symbol : : denotes the *normal product* of the operators appearing between the dots. To evaluate the normal product, we first have to express all the particle creation and destruction operators in terms of the quasiparticle creation and destruction operators, and then permute all the *quasiparticle* destruction operators to the right of all the *quasiparticle* creation operators; the sign is plus or minus depending on the even or odd nature of the permutation needed. Since we have not yet defined the transformation from the particle operators to the quasiparticle operators, we have to keep the normal products in their symbolic forms for the present.

The explicit evaluation of the contractions also involves substituting the particle operators by the quasiparticle operators and then simplifying the expression with the help of the basic property of the ground state, namely, a quasiparticle destruction operator produces zero on it. This explicit evaluation also has to be postponed. For now, we proceed to transform the Hamiltonian (VI.32) with the standard results (VI.38) and (VI.39). Even though we have merely quoted these standard results, the enterprising reader may have already recognized the rules of the procedure: first, write down the normal product of the whole expression [the first terms of (VI.38) and (VI.39)]; next, contract all the possible pairs of operators and multiply by the normal product of any operator that may be left [this gives the second term of (VI.38) and the second, third, and fourth lines of (VI.39)]; then continue the contraction process with an *additional* pair and multiply by the normal product of any operator that may be left [this gives the last line of (VI.39)]; repeat these steps until all the operators have been contracted pairwise. In every term, take the sign as plus or minus depending on the even or odd nature of the permutation of operators that has taken place in the particular term with respect to the ordering of the operators on the left-hand side. This explains the minus sign in the fourth line and the last term of (VI.39).

We substitute the expressions (VI.38) and (VI.39) in the original Hamiltonian (VI.32). We get terms that contain only contractions, and hence those terms represent a pure number. We denote this part of H by H_0 (*zero* denotes that there are *no* operators in this part of H).

Then there is a second set of terms which contains the normal product of a pair of operators. Such terms are denoted by H_2 (2 indicates that there are two operators). Finally, we have a part, H_4, coming from the first line of (VI.39). We write these different parts of H one after another. First,

$$H_0 = \sum_{\alpha,\beta} \langle\alpha| T |\beta\rangle\langle C^\dagger_\alpha C_\beta\rangle + \tfrac{1}{2} \sum_{\alpha,\beta,\gamma,\delta} \langle\alpha\beta| V |\gamma\delta\rangle$$
$$\times (\langle C^\dagger_\alpha C^\dagger_\beta\rangle\langle C_\delta C_\gamma\rangle + \langle C^\dagger_\alpha C_\gamma\rangle\langle C^\dagger_\beta C_\delta\rangle - \langle C^\dagger_\alpha C_\delta\rangle\langle C^\dagger_\beta C_\gamma\rangle). \quad \text{(VI.40)}$$

Since γ and δ are summation symbols, we can interchange them in the last term, which then becomes

$$-\tfrac{1}{2} \sum_{\alpha\beta\gamma\delta} \langle\alpha\beta| V |\delta\gamma\rangle\langle C^\dagger_\alpha C_\gamma\rangle\langle C^\dagger_\beta C_\delta\rangle.$$

This term can then be put together with its predecessor to obtain

$$\tfrac{1}{2} \sum_{\alpha\beta\gamma\delta} (\alpha\beta| V |\gamma\delta)\langle C^\dagger_\alpha C_\gamma\rangle\langle C^\dagger_\beta C_\delta\rangle. \quad \text{(VI.41)}$$

The appearance of the antisymmetrized matrix element replacing the direct term minus the exchange term should be noted. If we use the definition (V.6c) for the density operator (which can be shown to be identical to $\langle C^\dagger_\beta C_\delta\rangle$ for any general ground-state wavefunction Ψ_0), and the definition (V.6b) for the self-consistent potential $\mathcal{V}$, then the kinetic energy term of (VI.40) and the last two terms [given by (VI.41)] give rise to

$$\sum_{\alpha,\beta} \langle\alpha| T |\beta\rangle\langle\beta| \rho |\alpha\rangle + \tfrac{1}{2} \sum_{\alpha\beta\gamma\delta} (\alpha\beta| V |\gamma\delta)\langle\gamma| \rho |\alpha\rangle\langle\delta| \rho |\beta\rangle$$
$$= \sum_{\alpha,\beta} \langle\alpha| (T + \tfrac{1}{2}\mathcal{V}) |\beta\rangle\langle\beta| \rho |\alpha\rangle. \quad \text{(VI.42)}$$

This expression agrees with (V.11d), which was the ground-state energy for an HF state. In the present case, H_0 is indeed the energy of the ground state because the terms H_2 and H_4 of the Hamiltonian contain the normal product of the operators in which the quasiparticle destruction operators, by definition, appear on the right, and hence produce zero when operating on the ground state Ψ_0. We notice that the expression (VI.40) for the energy of the ground state, which is more general than the HF state, contains an extra term, namely, $\langle C^\dagger_\alpha C^\dagger_\beta\rangle\langle C_\delta C_\gamma\rangle$. To study the consequence of this term, we introduce the definition of the *pairing density* κ as

$$\kappa_{\delta\gamma} = \langle C_\delta C_\gamma\rangle \quad \text{(VI.43a)}$$

such that

$$\kappa^*_{\beta\alpha} = \langle C_\beta C_\alpha\rangle^* = \langle C^\dagger_\alpha C^\dagger_\beta\rangle. \quad \text{(VI.43b)}$$

Since the destruction operators anticommute with each other, we have

$$\kappa_{\gamma\delta} = \langle C_\gamma C_\delta\rangle = -\langle C_\delta C_\gamma\rangle = -\kappa_{\delta\gamma}. \quad \text{(VI.43c)}$$

We therefore write the extra term of (VI.40) as

$$\tfrac{1}{2} \sum_{\alpha\beta\gamma\delta} \langle\alpha\beta| V |\gamma\delta\rangle\kappa^*_{\beta\alpha}\tfrac{1}{2}(\kappa_{\delta\gamma} - \kappa_{\gamma\delta}) = \tfrac{1}{4} \sum_{\alpha\beta\gamma\delta} (\alpha\beta| V |\gamma\delta)\kappa^*_{\beta\alpha}\kappa_{\delta\gamma}. \quad \text{(VI.44)}$$

In writing the last step, we have interchanged γ, δ in the term with $\kappa_{\gamma\delta}$, thereby picking up the antisymmetrized matrix element of V.

Now we introduce the definition of the pairing potential as

$$\Delta_{\alpha\beta} = \tfrac{1}{2} \sum_{\gamma,\,\delta} (\alpha\beta|\, V\, |\gamma\delta)\kappa_{\delta\gamma}, \tag{VI.45}$$

and then (VI.44) simplifies to

$$\tfrac{1}{2} \sum_{\alpha,\,\beta} \Delta_{\alpha\beta}\kappa^{*}_{\beta\alpha}. \tag{VI.46}$$

Adding (VI.46) to (VI.42), we get the final expression of the ground-state energy H_0, given by (VI.40), as

$$H_0 = \sum_{\alpha,\,\beta} [\langle\alpha|\,(T + \tfrac{1}{2}\mathcal{V})\,|\beta\rangle\langle\beta|\,\rho\,|\alpha\rangle + \tfrac{1}{2}\Delta_{\alpha\beta}\kappa^{*}_{\beta\alpha}]. \tag{VI.47}$$

We next collect, via (VI.38) and (VI.39), all the terms of (VI.32) that contain the normal product of a pair. Thus,

$$\begin{aligned} H_2 = {} & \sum_{\alpha,\,\beta} \langle\alpha|\, T\, |\beta\rangle{:}C^{\dagger}_{\alpha}C_{\beta}{:} + \tfrac{1}{2} \sum_{\alpha\beta\gamma\delta} \langle\alpha\beta|\, V\, |\gamma\delta\rangle \\ & \times ({:}C^{\dagger}_{\alpha}C^{\dagger}_{\beta}{:}\langle C_{\delta}C_{\gamma}\rangle + \langle C^{\dagger}_{\alpha}C^{\dagger}_{\beta}\rangle{:}C_{\delta}C_{\gamma}{:} + {:}C^{\dagger}_{\alpha}C_{\gamma}{:}\langle C^{\dagger}_{\beta}C_{\delta}\rangle \\ & + \langle C^{\dagger}_{\alpha}C_{\gamma}\rangle{:}C^{\dagger}_{\beta}C_{\delta}{:} - {:}C^{\dagger}_{\alpha}C_{\delta}{:}\langle C^{\dagger}_{\beta}C_{\gamma}\rangle - \langle C^{\dagger}_{\alpha}C_{\delta}\rangle{:}C^{\dagger}_{\beta}C_{\gamma}{:}). \end{aligned} \tag{VI.48}$$

Since $\langle\alpha\beta|\, V\, |\gamma\delta\rangle = \langle\gamma\delta|\, V\, |\alpha\beta\rangle$, a straightforward interchange of summation labels verifies that the first two terms in the potential energy are Hermitean conjugates of each other. Similarly, since $\langle\beta\alpha|\, V\, |\delta\gamma\rangle = \langle\alpha\beta|\, V\, |\gamma\delta\rangle$, the next two terms in the potential energy can be proved to be equal to each other; for the same reason, the last two terms of (VI.48) are also equal. Further, by the interchange of summation labels, the last term can be converted into $-\langle\alpha\beta|\, V\, |\delta\gamma\rangle\langle C^{\dagger}_{\alpha}C_{\gamma}\rangle{:}C^{\dagger}_{\beta}C_{\delta}{:}$. If this is taken with the corresponding positive term in (VI.48), we obtain the antisymmetrized matrix element of V. As far as the first term in the potential energy is concerned, we can get the antisymmetrized matrix element by replacing $\kappa_{\delta\gamma}$ by $\tfrac{1}{2}(\kappa_{\delta\gamma} - \kappa_{\gamma\delta})$ and then using the artifice employed in (VI.44). With all these simplifications, (VI.48) reads as

$$\begin{aligned} H_2 = {} & \sum_{\alpha,\,\beta} \langle\alpha|\,(T + \mathcal{V})\,|\beta\rangle{:}C^{\dagger}_{\alpha}C_{\beta}{:} + \sum_{\alpha\beta\gamma\delta} (\alpha\beta|\, V\, |\gamma\delta)\langle\delta|\,\rho\,|\beta\rangle{:}C^{\dagger}_{\alpha}C_{\gamma}{:} \\ & + \tfrac{1}{4} \sum_{\alpha\beta\gamma\delta} [(\alpha\beta|\, V\, |\gamma\delta)\kappa_{\delta\gamma}{:}C^{\dagger}_{\alpha}C^{\dagger}_{\beta}{:} + \text{Hermitean conjugate}]. \end{aligned}$$

Once again, using the definitions of $\mathcal{V}$ and Δ, we get

$$H_2 = \sum_{\alpha,\,\beta} \langle\alpha|\,(T + \mathcal{V})\,|\beta\rangle{:}C^{\dagger}_{\alpha}C_{\beta}{:} + \tfrac{1}{2} \sum_{\alpha,\,\beta} (\Delta_{\alpha\beta}{:}C^{\dagger}_{\alpha}C^{\dagger}_{\beta}{:} + \text{Hermitean conjugate}). \tag{VI.49}$$

Finally, the expression for H_4 comes from the $({:}C^{\dagger}_{\alpha}C^{\dagger}_{\beta}C_{\delta}C_{\gamma}{:})$-term of (VI.39). From (VI.32), it is given by

$$\begin{aligned} H_4 & = \tfrac{1}{2} \sum_{\alpha\beta\gamma\delta} \langle\alpha\beta|\, V\, |\gamma\delta\rangle{:}C^{\dagger}_{\alpha}C^{\dagger}_{\beta}C_{\delta}C_{\gamma}{:} \\ & = \tfrac{1}{4} \sum_{\alpha\beta\gamma\delta} (\alpha\beta|\, V\, |\gamma\delta){:}C^{\dagger}_{\alpha}C^{\dagger}_{\beta}C_{\delta}C_{\gamma}{:}\,. \end{aligned} \tag{VI.50}$$

The last step follows after writing $C_{\delta}C_{\gamma} = \tfrac{1}{2}(C_{\delta}C_{\gamma} - C_{\gamma}C_{\delta})$ and then interchanging the summation indices γ, δ in the negative term.

C. SELF-CONSISTENT HARTREE-FOCK AND HARTREE-FOCK-BOGOLIUBOV THEORY

We have now come to a stage where we can try to determine the ground state Ψ_0 by minimizing H_0, which is the expectation value of the Hamiltonian with respect to this state. For this purpose, let us first note the structure of the pairing density (VI.43a) and the pairing potential (VI.45). In (VI.43a), we commence with the ground state Ψ_0 on the right, destroy two particles in the states γ and δ, and then try to reach the state Ψ_0 on the left. Such a quantity can clearly be nonvanishing *only* if Ψ_0 contains wavefunctions with a different number of particles. If the conservation of the number particles is strictly true for Ψ_0, then after $C_\delta C_\gamma$ operates on it we reach a state where the nucleus has two nucleons less, and hence the overlap of a number conserving $\langle\Psi_0|$ with $C_\delta C_\gamma|\Psi_0\rangle$ is zero.

For simplicity, let us first consider a ground-state wavefunction Ψ_0 for which the conservation of the number of particles is strictly valid. The terms containing the pairing potential in (VI.47) and (VI.49) are then absent. As far as the determination of Ψ_0 is concerned, we may apply the variational principle, and require that

$$\bar{H}_0 = \sum_{\alpha,\beta} \langle\alpha|\, (T + \tfrac{1}{2}\mathcal{V})\, |\beta\rangle\langle\beta|\, \rho\, |\alpha\rangle \tag{VI.51}$$

be minimized. In Chapter V, we have seen that this is exactly what the HF programme does. We recall that the solution to this variational problem is as follows. We define a set of single-particle states

$$|i\rangle = \sum_\alpha x^i_\alpha|\alpha\rangle \quad \text{or} \quad C^\dagger_i = \sum_\alpha x^i_\alpha C^\dagger_\alpha; \tag{VI.52a}$$

then the variational parameters x^i_α which make $\bar{H}_0$ a minimum satisfy

$$\sum_\beta \langle\alpha|\, (T + \mathcal{V})\, |\beta\rangle x^i_\beta = \epsilon_i x^i_\alpha. \tag{VI.52b}$$

The system of equations for the determination of x^i_α is the HF equations; these have to be solved *self-consistently* because $\mathcal{V}$ already contains ρ which, in its turn, includes the unknown coefficients x^i_α.

If we want to minimize the complete H_0, with the pairing potential term included, then we know that a variational state Ψ_0, in which the number of particles is conserved, is not adequate. We therefore consider a generalized variational state Ψ_0, which contains a superposition of the wavefunctions of nuclei having different numbers of nucleons. If we want to apply such a theory to a *given* nucleus, then, of course, we have to work out the variational calculation on H_0, subject to the constraint that the average value of the number of nucleons in the state Ψ_0 be equal to the given number of nucleons of the nucleus under consideration. The constraint in a variational problem, as is well-known, is specified in terms of a Lagrange multiplier. Therefore, instead of minimizing $\langle\Psi_0|\, H\, |\Psi_0\rangle$, i.e., H_0 of (VI.47), we have to minimize, in this case, $\langle\Psi_0|\, (H - \lambda\hat{N})\, |\Psi_0\rangle$, where λ is the Lagrange multiplier and $\hat{N}$ is the operator for the total number of nucleons

$$\hat{N} = \sum_\alpha C^\dagger_\alpha C_\alpha.$$

If we had transformed $H - \lambda\hat{N}$, instead of H, into H_0, H_2, and H_4, then it would have been clear from the form of $\hat{N}$, namely,

$$\hat{N} = \sum_{\alpha} C_\alpha^\dagger C_\alpha = \sum_{\alpha} [\langle C_\alpha^\dagger C_\alpha \rangle + :C_\alpha^\dagger C_\alpha:], \tag{VI.53}$$

that the term $-\lambda\hat{N}$ contributes a term to H_0 and another term to H_2, leaving H_4 unaltered. In analogy with the kinetic energy term, it is obvious that these extra terms are obtained by replacing, in both H_0 and H_2, the operator T by $T - \lambda\mathbb{1}$, where $\mathbb{1}$ is the unity operator. We use (VI.47) and (VI.49), together with the fact just mentioned, to write the expressions

$$H_0(\lambda) = \sum_{\alpha,\beta} [\langle\alpha| (T - \lambda\mathbb{1} + \mathcal{V}) |\beta\rangle\langle\beta| \rho |\alpha\rangle + \tfrac{1}{2}\Delta_{\alpha\beta}\kappa^*_{\beta\alpha}], \tag{VI.54a}$$

$$H_2(\lambda) = \sum_{\alpha,\beta} \langle\alpha| (T - \lambda\mathbb{1} + \mathcal{V}) |\beta\rangle :C_\alpha^\dagger C_\beta: + \tfrac{1}{2} \sum_{\alpha,\beta} (\Delta_{\alpha\beta} :C_\alpha^\dagger C_\beta^\dagger: + \text{Hermitean conjugate}). \tag{VI.54b}$$

In analogy with the transformation (VI.52a), we now use a generalized transformation

$$b_i^\dagger = \sum_{\alpha} (x_\alpha^i C_\alpha^\dagger + y_\alpha^i C_\alpha), \tag{VI.55}$$

where the coefficients x_α^i and y_α^i are the variational parameters. The operators $b_i^\dagger$ are the creation operators for our quasiparticles and they will be completely determined when the minimization programme for $H_0(\lambda)$ is carried out. This procedure gives rise, in analogy with the HF case, to a set of coupled equations for the variational parameters which have once again to be solved self-consistently. This self-consistent method of finding the quasiparticles is called the Hartree-Fock-Bogoliubov (HFB) method, and the quasiparticle equation (VI.55) is called the HFB quasiparticle transformation. The self-consistent equations of this method are shown in Appendix G (Section II). An observant reader may have already noticed that the HFB transformation equation (VI.55) is indeed based on the concept of particle number non-conservation in the states. The first part of $b_i^\dagger$ adds a particle, whereas the second part removes a particle operating on any state; thus, $b_i^\dagger$, acting on any wavefunction, produces wavefunctions with a mixed number of nucleons.

We shall follow a two-step minimization programme of $H_0(\lambda)$ instead of the fully self-consistent HFB procedure.

D. SIMPLE PAIRING THEORY OF THE BCS TYPE

In the simple pairing theory of the Bardeen-Cooper-Schriffer (BCS) type, we omit the pairing potential term of (VI.54a) at the first step of minimizing $H_0(\lambda)$. The resultant minimization programme is clearly identical to what we have done in connection with (VI.51). The only difference now is the extra term $-\lambda\mathbb{1}$. But the diagonalization problem of $(T + \mathcal{V})$, as shown in (VI.52b), is not really changed by this extra term. The eigenvectors of $(T + \mathcal{V})$ and $(T + \mathcal{V} - \lambda\mathbb{1})$ are obviously the same, except that the energies are shifted from ϵ_i to $(\epsilon_i - \lambda)$. Therefore, at this stage of dealing with $H_0(\lambda)$, we have the set of single-particle self-consistent states of (VI.52b) and the corresponding energies $(\epsilon_i - \lambda)$. At the next step, we make use of the states $|i\rangle$ as the basis in the rewriting of $H_0(\lambda)$ as

$$H_0(\lambda) = \sum_{i,j} \langle i| (T - \lambda\mathbb{1} + \tfrac{1}{2}\mathcal{V}) |j\rangle\langle j| \rho |i\rangle + \tfrac{1}{2} \sum_{i,j} \Delta_{ij}\kappa^*_{ji}, \tag{VI.56}$$

where

$$\langle j| \rho |i\rangle = \langle C_i^\dagger C_j \rangle, \tag{VI.57a}$$

68 (45-122/1977)

$$\kappa_{ij} = \langle C_i C_j \rangle, \tag{VI.57b}$$

$$\Delta_{ij} = \tfrac{1}{2} \sum_{k,l} (ij|\, V\, |kl)\kappa_{lk}, \tag{VI.57c}$$

$$\langle i|\, \mathcal{V}\, |j\rangle = \sum_{k,l} (ik|\, V\, |jl)\langle l|\, \rho\, |k\rangle. \tag{VI.57d}$$

We then do a fresh minimization for $H_0(\lambda)$, using a quasiparticle transformation that is much simpler than (VI.55). This simpler transformation, which we introduce in (VI.62), is called the BCS transformation, after the name of the people who first applied a similar transformation to the theory of superconductivity of metals.

To prepare the ground for the BCS transformation, let us first examine the characteristics of the self-consistent single-particle states $|i\rangle$. In a spherical nucleus, such a state is specified by the quantum numbers l, j, m, and also the number of nodes n in the radial wavefunction. All the states having the same (nlj) but different m $(= -j, \ldots, +j)$ are degenerate. The state $|nljm\rangle$ has a time-reversed *partner* which is given by

$$(-1)^{j-m}|nlj, -m\rangle. \tag{VI.58}$$

We shall denote these two partner states as $|i\rangle$ and $|\bar{i}\rangle$; sometimes $|\bar{i}\rangle$ will also be written as $s_i|-i\rangle$, where s_i is the phase factor of (VI.58) and $|-i\rangle$ stands for the same quantum numbers (nlj) as in $|i\rangle$ but with the projection quantum number reversed in sign. In a more general situation where the single-particle states correspond to a nonspherical $\mathcal{V}$, $|i\rangle$ and $|\bar{i}\rangle$ are not as simple as just stated. But a state $|i\rangle$ in this more complicated case can still be expanded in terms of states of the type $|nljm\rangle$ as

$$|i\rangle = \sum_{nljm} x^i_{nljm}|nljm\rangle. \tag{VI.59a}$$

In this case, the time-reversed partner $|\bar{i}\rangle$ is obtained by using (VI.58) for every term in the summation as

$$|\bar{i}\rangle = \sum_{nljm} x^i_{nljm}(-1)^{j-m}|nlj, -m\rangle. \tag{VI.59b}$$

If $\mathcal{V}$ has a spheroidal symmetry, then m is a good quantum number, and there is no summation over m in (VI.59), and the states $|i\rangle$ and $|\bar{i}\rangle$ have m and $-m$, respectively, as labelling quantum numbers. Even in a more complicated case such as (VI.59b), we shall denote $|\bar{i}\rangle$ symbolically as $s_i|-i\rangle$.

We shall now deal with a result of prime importance—that relating to the time-reversed partner $|\bar{i}\rangle$. Let us do a second time-reversal on the state (VI.58). This clearly gives

$$|\bar{\bar{i}}\rangle = (-1)^{j-m}(-1)^{j+m}|nljm\rangle = -|nljm\rangle = -|i\rangle. \tag{VI.60a}$$

Similarly, for the more complicated state (VI.59b), we get

$$|\bar{\bar{i}}\rangle = \sum_{nljm} x^i_{nljm}(-1)^{j-m}(-1)^{j+m}|nljm\rangle = -|i\rangle. \tag{VI.60b}$$

In both the cases, the minus sign is a consequence of the phase factor $(-1)^{2j}$, where $2j$ is always an odd integer. The explicit general results (VI.60) explain that our symbolic quantity s_i has the property

$$s_{-i} = -s_i. \tag{VI.61}$$

The steps in proving this important result are as follows. By definition,

$$|\bar{i}\rangle = s_i|-i\rangle,$$

and hence

$$|\bar{\bar{i}}\rangle = s_i s_{-i}|i\rangle$$

or, by comparison with (VI.60),

$$s_i s_{-i} = -1$$

which is the same as (VI.61).

In an HF calculation for even nuclei, a state $|i\rangle$ is usually found to be degenerate with its time-reversed partner $|\bar{i}\rangle = s_i|-i\rangle$. Let us now examine the consequence of creating a particle in the single-particle state $|i\rangle$ and of destroying its time-reversed partner $|\bar{i}\rangle$. The first process is achieved by the operator $C_i^\dagger$, and the second by the operator $s_i C_{-i}$. It was shown after (III.129), from the angular momentum property of the second-quantized operators, that, under a rotation of the coordinate system, both the operator $C_{nljm}^\dagger$ and the operator $(-1)^{j-m}C_{nlj,-m}$ transform with the same rotation matrix $\mathscr{D}^j_{mm'}$.

Using these basic properties, and the expansions (VI.59), we can directly establish that $C_i^\dagger$ and $s_i C_{-i}$ behave similarly under rotation. Therefore, instead of using the most general HFB transformation of (VI.55), we now use the simplified BCS transformation given by

$$b_i^\dagger = u_i C_i^\dagger - v_{-i} s_i C_{-i}. \tag{VI.62}$$

The choice of the minus sign in the second term simply defines the phase convention for v_{-i}. We can create a particle in the state $|i\rangle$ only if it is empty, and we can destroy a particle from a certain state only if it is occupied. Therefore, u_i must be the probability *amplitude* of getting the state i empty (probability of nonoccupation), whereas v_{-i} must be the probability *amplitude* for the occupation of the corresponding time-reversed state. (VI.62) automatically gives the quasiparticle destruction operator

$$b_i = u_i C_i - v_{-i} s_i C_{-i}^\dagger \tag{VI.63}$$

or

$$s_i b_{-i} = v_i C_i^\dagger + s_i u_{-i} C_{-i}, \tag{VI.64}$$

where we have used the property (VI.61).

The formal expression (VI.62) can clearly be used even in the HF case. There, however, u_i and v_i have known values, i.e.,

$$\begin{aligned} u_i &= 1 \quad \text{for a particle state} \\ &= 0 \quad \text{for a hole state,} \\ v_i &= 1 \quad \text{for a hole state} \\ &= 0 \quad \text{for a particle state.} \end{aligned} \tag{VI.65}$$

That is to say, in the HF case, even if we formally write two terms in (VI.62), one or the other usually survives. An exception to this occurs whenever we encounter either of two situations in the HF ground state: (i) a state i is unoccupied but its time-reversed partner is occupied; in this case, both the terms in (VI.62) are present; (ii) a state i is occupied and its time-reversed partner unoccupied; then both the terms in (VI.62) become zero. These exceptional situations are usually not encountered in *even* nuclei, in which either both $|i\rangle$ and $|\bar{i}\rangle$ are occupied or both are unoccupied.

In the cases depicted in Fig. VI.2b, the occupation probability and nonoccupation probability of states near the fermi surface are nonvanishing fractional quantities, and hence both the terms in the BCS transformation (VI.62) survive. Therefore, in such cases, the transformation (VI.62) indeed gives rise to a nucleon number nonconservation, and hence it is able to take into account the consequences of the pairing potential in $H_0(\lambda)$ and $H_2(\lambda)$.

We now require that the quasiparticle operators behave as fermions. Thus, the fermion anticommutation properties

$$\{b_i, b_j\} = \{b_i^\dagger, b_j^\dagger\} = 0, \tag{VI.66a}$$

$$\{b_i, b_j^\dagger\} = \delta_{ij} \tag{VI.66b}$$

have to be ensured. From (VI.62), we have

$$\begin{aligned}\{b_i^\dagger, b_j^\dagger\} &= [(u_i C_i^\dagger - v_{-i}\delta_i C_i), (u_j C_j^\dagger - v_{-j} s_j C_{-j})] \\ &= -\delta_{i,-j} u_i v_{-j} s_j - \delta_{-i,j} v_{-i} u_j s_i \\ &= (u_i v_i - u_{-i} v_{-i}) s_i.\end{aligned}$$

Here we have used the standard anticommutation rules of $C_i^\dagger, C_j, \ldots$. In the HF case, this result is always zero because a state is either occupied or unoccupied, and hence either u or v for any state has to be zero. But, in the more general BCS case, the anticommutation demands

$$u_i v_i = u_{-i} v_{-i}. \tag{VI.67a}$$

Next, we apply (VI.66b) together with (VI.63) and (VI.62). This gives

$$\begin{aligned}\delta_{ij} = \{b_i, b_j^\dagger\} &= [(u_i C_i - v_{-i} s_i C_{-i}^\dagger), (u_j C_j^\dagger - v_{-j} s_j C_{-j})] \\ &= u_i u_j \delta_{ij} + v_{-i} v_{-j} s_i s_j \delta_{-i,-j} \\ &= (u_i^2 + v_{-i}^2)\delta_{ij}.\end{aligned}$$

Thus, the requirement is

$$u_i^2 + v_{-i}^2 = 1, \qquad u_{-i}^2 + v_i^2 = 1. \tag{VI.67b}$$

Taking (VI.67b) with (VI.67a), we prove

$$u_i/v_{-i} = u_{-i}/v_i \quad \text{or} \quad u_i^2/(u_i^2 + v_{-i}^2) = u_{-i}^2/(u_{-i}^2 + v_i^2)$$

or

$$u_i^2 = u_{-i}^2, \qquad v_i^2 = v_{-i}^2. \tag{VI.68}$$

Thus, a state i and its time-reversed partner $\bar{i}$ must have the same value for their occupation probability and nonoccupation probability. Then, (VI.67a) and (VI.68) together require either

$$u_i = u_{-i}, \qquad v_i = v_{-i} \tag{VI.69a}$$

or

$$u_i = -u_{-i}, \qquad v_i = -v_{-i}. \tag{VI.69b}$$

We choose the phase convention of (VI.69a). The statement in italics is *usually* taken as an assumption to start with, in which case we write, from the very beginning, v_i instead of v_{-i} in the BCS transformation (VI.62). The present approach is more physical and rigorous. Now that we have proved (VI.69a), we shall no longer use a negative subscript in the defining equations, namely, (VI.62) and (VI.64), which become

$$b_i^\dagger = u_i C_i^\dagger - v_i s_i C_{-i}, \tag{VI.70a}$$

$$s_i b_{-i} = v_i C_i^\dagger + u_i s_i C_{-i}. \tag{VI.70b}$$

It is easy to invert these relations with the help of (VI.67b). We then obtain

$$C_i^\dagger = u_i b_i^\dagger + v_i s_i b_{-i}, \tag{VI.71a}$$

and hence

$$C_i = u_i b_i + v_i s_i b_{-i}^\dagger. \tag{VI.71b}$$

We are now fully equipped to go back to (VI.56) and (VI.57a), (VI.57b), and (VI.57c) and do the minimization of $H_0(\lambda)$, treating u_i and v_i as the variational parameters subject to the constraint $u_i^2 + v_i^2 = 1$. We shall use the basic definition

$$b_i|\Psi_0\rangle = 0 \quad \text{for all } i, \tag{VI.72}$$

and evaluate (VI.57a) and (VI.57b) with the help of (VI.71). Thus,

$$\begin{aligned}\langle j|\,\rho\,|i\rangle &= \langle \Psi_0|\,(u_i b_i^\dagger + v_i s_i b_{-i})(u_j b_j + v_j s_j b_{-j}^\dagger)\,|\Psi_0\rangle \\ &= \langle \Psi_0|\,b_{-i} b_{-j}^\dagger\,|\Psi_0\rangle v_i v_j s_i s_j = \delta_{ij} v_i^2,\end{aligned} \tag{VI.73a}$$

$$\begin{aligned}\kappa_{ij} &= \langle \Psi_0|\,(u_i b_i + v_i s_i b_{-i}^\dagger)(u_j b_j + v_j s_j b_{-j}^\dagger)\,|\Psi_0\rangle \\ &= \langle \Psi_0|\,b_i b_{-j}^\dagger\,|\Psi_0\rangle u_i v_j s_j \\ &= \delta_{i,-j} u_j v_j s_j.\end{aligned} \tag{VI.73b}$$

In simplifying (VI.73), we have used (VI.72) and its Hermitean conjugate, and then the anticommutator

$$\{b_i, b_j^\dagger\} = \delta_{ij}, \quad \text{i.e.,} \quad b_i b_j^\dagger = -b_j^\dagger b_i + \delta_{ij}.$$

Using the expression of κ_{ij} given by (VI.73b), we get, for the pairing potential of (VI.57c), the result

$$\begin{aligned}\Delta_{ij} &= \tfrac{1}{2} \sum_{k,l} (ij|\,V\,|kl) u_k v_k s_k \delta_{k,-l} \\ &= \tfrac{1}{2} \sum_k (ij|\,V\,|k, \bar{k}) u_k v_k.\end{aligned}$$

It is easy to show, from the time-reversal invariance of the two-body potential V, that the state j in this expression must be the time-reversed partner of i. Thus,

$$\Delta_{ij} = \tfrac{1}{2}\delta_{j,\bar{i}} \sum_k (i, \bar{i}|\,V\,|k, \bar{k}) u_k v_k. \tag{VI.73c}$$

Similarly, from (VI.73a) and (VI.57d),

$$\langle i|\,\mathcal{V}\,|j\rangle = \sum_k (ik|\,V\,|jk) v_k^2. \tag{VI.73d}$$

Inserting (VI.73) in (VI.56), we obtain

$$H_0(\lambda) = \sum_i \langle i|\,(T - \lambda\mathbb{1})\,|i\rangle v_i^2 + \tfrac{1}{2} \sum_i \Delta_{i,\bar{i}} u_i v_i + \tfrac{1}{2} \sum_i \langle i|\,\mathcal{V}\,|i\rangle v_i^2 \tag{VI.74a}$$

$$= \sum_i \langle i|\,(T - \lambda\mathbb{1})\,|i\rangle v_i^2 + \tfrac{1}{4} \sum_{i,k} (i, \bar{i}|\,V\,|k, \bar{k}) u_k v_k u_i v_i + \tfrac{1}{2} \sum_{i,k} (ik|\,V\,|ik) v_i^2 v_k^2. \tag{VI.74b}$$

We now minimize (VI.74b) by requiring

$$\frac{\partial H_0(\lambda)}{\partial v_a} = 0$$

for every a, and using

$$u_a^2 = 1 - v_a^2$$

or

$$\frac{\partial u_a}{\partial v_a} = -(v_a/u_a)$$

while carrying out the differentiation. We thus obtain

$$\begin{aligned} 0 &= 2(\epsilon_a - \lambda)v_a + \tfrac{1}{2}\sum_k (a, \bar{a}|V|k, \bar{k})u_k v_k \frac{\partial}{\partial v_a}(u_a v_a) \\ &= 2(\epsilon_a - \lambda)v_a + \Delta_{a,\bar{a}}(u_a - \frac{v_a^2}{u_a}) \\ &= 2u_a^{-1}[(\epsilon_a - \lambda)u_a v_a + \tfrac{1}{2}\Delta_{a,\bar{a}}(u_a^2 - v_a^2)] \end{aligned}$$

or, since u_a^{-1} is nonvanishing,

$$0 = (\epsilon_a - \lambda)u_a v_a + \tfrac{1}{2}\Delta_{a,\bar{a}}(u_a^2 - v_a^2). \qquad \text{(VI.75)}$$

In view of $u_a^2 + v_a^2 = 1$, we can easily solve (VI.75) by putting

$$v_a = \cos\theta_a, \qquad u_a = \sin\theta_a.$$

(VI.75) then yields

$$(\epsilon_a - \lambda)\sin 2\theta_a = \Delta_a \cos 2\theta_a$$

or

$$\tan 2\theta_a = \frac{\Delta_a}{\epsilon_a - \lambda}$$

or

$$\begin{aligned} 2u_a v_a &\equiv \sin 2\theta_a = \Delta_a[(\epsilon_a - \lambda)^2 + \Delta_a^2]^{-1/2}, \\ u_a^2 - v_a^2 &\equiv \cos 2\theta_a = (\epsilon_a - \lambda)[(\epsilon_a - \lambda)^2 + \Delta_a^2]^{-1/2}. \end{aligned} \qquad \text{(VI.76)}$$

In these equations, we have used the definition

$$\begin{aligned} \Delta_a &= -\Delta_{a,\bar{a}} \\ &= -\tfrac{1}{2}\sum_k (a, \bar{a}|V|k, \bar{k})u_k v_k \end{aligned} \qquad \text{(VI.77)}$$

which is a positive quantity by virtue of the special matrix element of V being attractive. This matrix element of V which takes a *paired* state (a state and its time-reversed partner) to another paired state is called the *pairing* matrix element of the potential. Using the second relation of (VI.76) with $u_a^2 + v_a^2 = 1$, we obtain

$$u_a^2 = \tfrac{1}{2}(1 + \frac{\epsilon_a - \lambda}{E_a}), \qquad v_a^2 = \tfrac{1}{2}(1 - \frac{\epsilon_a - \lambda}{E_a}), \qquad \text{(VI.78)}$$

where

$$E_a = \sqrt{(\epsilon_a - \lambda)^2 + \Delta_a^2}. \qquad \text{(VI.79)}$$

Similarly, substituting the first relation of (VI.76) in (VI.77), we get

$$\Delta_a = -\tfrac{1}{4}\sum_k (a, \bar{a}|V|k, \bar{k})\frac{\Delta_k}{E_k}. \qquad \text{(VI.80)}$$

Further, from (VI.73a) we have

$$\sum_i \langle i|\, \rho \,|i\rangle \equiv \langle \Psi_0|\, C_i^\dagger C_i \,|\Psi_0\rangle \equiv \langle \Psi_0|\, \hat{N} \,|\Psi_0\rangle = \Sigma\, v_i^2,$$

where $\hat{N}$ is the total number operator. According to our earlier stipulation, this expression should be equated to the given number of nucleons (say, A) in the nucleus we want to calculate. Thus,

$$A = \sum_a v_a^2 = \tfrac{1}{2} \sum_a \left(1 - \frac{\epsilon_a - \lambda}{E_a}\right). \tag{VI.81}$$

(VI.80) and (VI.81) are a system of coupled nonlinear equations [due to definition (VI.79) of E_a] in the unknown quantity λ and the set of Δ_a. The total number of the latter depends on the total number of single-particle states admitted into the calculation. These equations can be solved for the unknown quantities, starting with a *given* two-body potential. Only a very special type of matrix element of the potential, namely, the pairing matrix element, is, however, relevant at this stage of the calculation. Since we know λ and the Δ_a's, it is trivial to compute the occupation probabilities of the various states (v_a^2) from (VI.78) and the energies E_a of (VI.79).

To help the reader in his assimilation of the facts, we mention that the key equations in the lengthy derivation just given are (i) (VI.70), defining the quasiparticle transformation; (ii) (VI.78), which determine the transformation coefficients u_a and v_a in the quasiparticle; (iii) (VI.79), which determines E_a; and (iv) (VI.80) and (VI.81), whose solutions determine λ and the set of Δ_a. In fact, we have to start the computation by solving these two equations and then we go to the equations mentioned in (i), (ii), and (iii) for the determination of the various quantities.

Treatment of $H_2(\lambda)$ and Interpretation of E_a

We next show that the minimization of $H_0(\lambda)$ has a very interesting consequence on $H_2(\lambda)$ which is given by (VI.54b). First, we rewrite this equation in terms of the HF representation, in which the first term $(T - \lambda\mathbb{1} + \mathcal{V})$ is diagonal and Δ has nonvanishing elements of the type $\Delta_{i,\bar{i}}$. Therefore,

$$H_2(\lambda) = \sum_i (\epsilon_i - \lambda){:}C_i^\dagger C_i{:} + \tfrac{1}{2} \sum_i (\Delta_{i,\bar{i}} s_i {:}C_i^\dagger C_{-i}^\dagger{:} + \text{Hermitean conjugate}). \tag{VI.82}$$

We proceed to evaluate the normal products from their definition and (VI.71). Thus,

$$\begin{aligned} {:}C_i^\dagger C_i{:} &= {:}(u_i b_i^\dagger + v_i s_i b_{-i})(u_i b_i + v_i s_i b_{-i}^\dagger){:} \\ &= u_i^2 b_i^\dagger b_i - v_i^2 b_{-i}^\dagger b_{-i} + u_i v_i s_i (b_i^\dagger b_{-i}^\dagger + b_{-i} b_i). \end{aligned} \tag{VI.83a}$$

Only in one term, namely, $v_i^2 b_{-i} b_{-i}^\dagger$, we needed to take the destruction operator to the right of the creation operator through *one* permutation, and hence this term has a minus sign. With similar algebra,

$$\begin{aligned} {:}C_i^\dagger C_{-i}^\dagger{:} &= {:}(u_i b_i^\dagger + v_i s_i b_{-i})(u_i b_{-i}^\dagger - v_i s_i b_i){:} \\ &= u_i^2 b_i^\dagger b_{-i}^\dagger - v_i^2 b_{-i} b_i - u_i v_i s_i (b_i^\dagger b_i + b_{-i}^\dagger b_{-i}). \end{aligned} \tag{VI.83b}$$

Using (VI.83) in (VI.82), we get

$$H_2(\lambda) = H_{11}(\lambda) + H_{20}(\lambda) + H_{02}(\lambda), \tag{VI.84}$$

where the two subscripts denote respectively the number of *quasiparticle* creation and destruction operators. Thus, picking up the appropriate terms from (VI.83), we get

$$H_{11}(\lambda) = \sum_i (\epsilon_i - \lambda)(u_i^2 b_i^\dagger b_i - v_i^2 b_{-i}^\dagger b_{-i}) - \sum_i \Delta_{i,\bar{i}} u_i v_i (b_i^\dagger b_i + b_{-i}^\dagger b_{-i}). \tag{VI.85}$$

We can further simplify this expression by noting that ϵ_i is equal to ϵ_{-i}, the energy for the time-reversed state. Hence, changing $i \to -i$ in the summation of the term containing $b_{-i}^\dagger b_{-i}$, we get the first term simplified to

$$\sum_i (\epsilon_i - \lambda)(u_i^2 - v_i^2) b_i^\dagger b_i. \tag{VI.86a}$$

Similar stratagems should be resorted to with the $b_{-i}^\dagger b_{-i}$ occurring with the pairing potential. Here $\Delta_{i,\bar{i}} \equiv s_i \Delta_{i,-i}$ changes to $s_{-i}\Delta_{-i,i}$ when $-i \to i$. The latter is equal to $-s_i\Delta_{-i,i} = -\Delta_{\bar{i},i}$. However, according to the definition (VI.73c), $-\Delta_{\bar{i},i} = \Delta_{i,\bar{i}}$. Therefore, under $-i \to i$, the term $b_{-i}^\dagger b_{-i}\Delta_{i,\bar{i}}$ simply changes to $\Delta_{i,\bar{i}} b_i^\dagger b_i$ and the second term of (VI.85) simplifies to

$$-2 \sum_i \Delta_{i,\bar{i}} u_i v_i b_i^\dagger b_i. \tag{VI.86b}$$

Adding (VI.86a) and (VI.86b), we obtain

$$\begin{aligned} H_{11}(\lambda) &= \sum_i [(\epsilon_i - \lambda)(u_i^2 - v_i^2) - 2\Delta_{i,\bar{i}} u_i v_i] b_i^\dagger b_i \\ &= \sum_i [(\epsilon_i - \lambda)\cos 2\theta_i + \Delta_i \sin 2\theta_i] b_i^\dagger b_i \\ &= \sum_i E_i b_i^\dagger b_i. \end{aligned} \tag{VI.87}$$

In simplifying this expression, we have used the first line of (VI.77), and (VI.76) and (VI.79).

We next select the appropriate terms from (VI.83), put them in (VI.82), and obtain

$$\begin{aligned} H_{20}(\lambda) + H_{02}(\lambda) &= \sum_i (\epsilon_i - \lambda) u_i v_i s_i (b_i^\dagger b_{-i}^\dagger + b_{-i} b_i) \\ &\quad + \tfrac{1}{2} \sum_i [\Delta_{i,\bar{i}} s_i (u_i^2 b_i^\dagger b_{-i}^\dagger - v_i^2 b_{-i} b_i) + \text{Hermitean conjugate}] \\ &= \sum_i [(\epsilon_i - \lambda) u_i v_i + \tfrac{1}{2}\Delta_{i,\bar{i}}(u_i^2 - v_i^2)] s_i (b_i^\dagger b_{-i}^\dagger + b_{-i} b_i). \end{aligned} \tag{VI.88}$$

Comparing the quantity enclosed within the square brackets with the expression occurring in the minimization condition (VI.75), we find them identical. Hence, the minimization condition *automatically* guarantees that (VI.88) is zero, i.e.,

$$H_{20}(\lambda) + H_{02}(\lambda) = 0. \tag{VI.89}$$

This is the interesting consequence we mentioned.

Now we can summarize the transformed Hamiltonian as

$$H = H_0(\lambda) + \sum_i E_i b_i^\dagger b_i + H_4(\lambda), \tag{VI.90a}$$

where $H_0(\lambda)$ and $H_4(\lambda)$ are given by (VI.74a) and (VI.50), respectively. The term $H_0(\lambda)$ is the ground-state energy and, when we use the expression for $u_i v_i$ from (VI.76), it simplifies to

$$H_0(\lambda) = \sum_i \langle i| (T - \lambda \mathbb{1} + \tfrac{1}{2}\mathcal{V}) |i\rangle - \tfrac{1}{4} \sum_i \frac{\Delta_i^2}{E_i}, \tag{VI.90b}$$

where

$$\langle i | \mathcal{V} | i \rangle = \sum_k (ik| V |ik) v_k^2. \tag{VI.90c}$$

The second term of (VI.90a) which is the sole survivor of $H_2(\lambda)$ obviously represents the energy of single quasiparticles. This can be seen as follows.

For the present, we ignore $H_4(\lambda)$, and consider a state having one quasiparticle, i.e., a state $b_k^\dagger |\Psi_0\rangle$. The energy of this state is given by

$$\langle \Psi_0 | b_k [H_0(\lambda) + \sum_i E_i b_i^\dagger b_i] b_k^\dagger |\Psi_0\rangle = H_0(\lambda) \langle \Psi_0 | b_k b_k^\dagger |\Psi_0\rangle + \sum_i E_i \langle \Psi_0 | b_k b_i^\dagger b_i b_k^\dagger |\Psi_0\rangle$$

$$= H_0(\lambda) + E_k. \tag{VI.91}$$

The clue in carrying out the algebra here is to permute the destruction operators to the extreme right by using the anticommutator relations, and then dropping terms in which a destruction operator appears next to $|\Psi_0\rangle$, or a creation operator next to $\langle \Psi_0 |$. The energy of the state $|\Psi_0\rangle$ is $H_0(\lambda)$, and the energy of the state having one quasiparticle in the state k in addition to Ψ_0 is found to be $H_0(\lambda) + E_k$. Therefore, the natural interpretation of E_k is that it is the energy carried by the single quasiparticle.

The term $H_4(\lambda)$ actually contains the interaction between quasiparticles, and a proper calculation of the consequences of $H_4(\lambda)$ is the major task of the general microscopic theory. The development so far has led us only to independent quasiparticle excitations of the nucleus. The treatment of $H_4(\lambda)$ gives rise to other modes of excitation (see Section 42).

Interpretation of λ

The quantity λ used in the theory is called the chemical potential and it is interpreted as follows. Let us consider the expression (VI.78) for v_a^2. The fact that this is the occupation probability can be verified by evaluating $\langle \Psi_0 | C_a^\dagger C_a |\Psi_0\rangle$; the result is indeed v_a^2, as demonstrated by (VI.73a). Now (VI.78) reveals that

$$v_a^2 > 0.5 \qquad (\text{when } \epsilon_a < \lambda),$$

$$v_a^2 < 0.5 \qquad (\text{when } \epsilon_a > \lambda),$$

$$v_a^2 = 0.5 \qquad (\text{when } \epsilon_a = \lambda).$$

Let us now look at the curve of Fig. VI.2b. The occupation probability of deep-down states is ≈ 1, which also follows from (VI.78) by making ϵ_a very small, i.e., $\epsilon_a - \lambda \approx -\lambda$ and $E_a \approx \lambda$. As ϵ_a becomes comparable with λ, v_a^2, according to (VI.78), starts decreasing from 1 and approaches 0.5 for $\epsilon_a \to \lambda$. On the other side of λ, ϵ_a decreases below 0.5 and, according to (VI.78), reaches zero value when $\epsilon_a \to \infty$. Thus, λ is the single-particle energy, on both sides of which the occupation probability departs significantly from the ideal HF values. It therefore belongs to the same energy region where the fermi level would have been located had we ignored the existence of close-lying levels near the fermi surface and allowed the nucleons to go to the lowest available states. The quantity λ is therefore very often loosely called the fermi level.

Interpretation of Δ_a

From the expression of the quasiparticle energy E_a, it is clear that, even if the HF energy ϵ_a of the state a is coincident with the fermi level λ, there is a minimum value of E_a (given by the positive quantity Δ_a) that owes its origin entirely to the pairing matrix elements of the two-

69 (45-122/1977)

body potential. For reasons explained later in this section, the even nuclei in the quasiparticle theory are associated with a ground state Ψ_0, which behaves as the quasiparticle vacuum, and the excited states of such nuclei are taken to have 2, 4, . . . quasiparticles. The lowest excited states of such nuclei have at least two quasiparticles, and hence an energy that is at least twice the parameter Δ. These parameters are therefore called the *energy gap* parameters, the gap being roughly a half of the energy difference between the ground state and the first set of excited states of even nuclei. However, we must keep in mind the interaction term $H_4(\lambda)$ of the Hamiltonian. As we have stated qualitatively in Section 38, the interaction between pairs of quasiparticles may build up collective effects, and produce a coherent state that may very well differ markedly in energy from the unperturbed energies of the quasiparticle pair. If this happens, then the *observed* first excited state of an even nucleus definitely does not show a very pronounced energy gap with respect to the ground state.

The odd-mass nuclei in this theory are associated with an odd number of quasiparticles, the lowest states being 1 and 3 quasiparticle-type states. All the 1 quasiparticle states have a minimum energy Δ, and hence the levels near the ground state of odd-mass nuclei are never expected to show an energy gap *with respect to the ground state.*

Special Pairing Force

In our derivation of the quasiparticles where we used a general potential, we saw that the special BCS transformation picked up only the pairing part of it in the fundamental energy gap equations of the theory. For this reason, sometimes the entire theory is worked out with a model *two-body* potential, called the *pairing potential* (not to be confused with the earlier Δ_{ij}, which is certainly not the two-body potential). This idealized potential is *defined* to have nonvanishing matrix elements of the following type with a constant magnitude:

$$(a, \bar{a}|\, V_P\, |b, \bar{b}) = -G. \tag{VI.92}$$

In the special case of *spherical* single-particle states, we have

$$(-1)^{j-m}(jm;\, j,\, -m|\, V_P\, |j'm';\, j',\, -m')(-1)^{j'-m'} = -G. \tag{VI.93}$$

In this second form, the pairing potential has the important property of being nonvanishing only for the two-body ($J = 0$)-state. To prove this statement, we proceed as follows:

$$\begin{aligned}
&(j_1 j_2 JM|\, V_P\, |j_1' j_2' J' M') \\
&= \sum_{m_1 m_2}\sum_{m_1' m_2'} \begin{bmatrix} j_1 & j_2 & J \\ m_1 & m_2 & M \end{bmatrix}\begin{bmatrix} j_1' & j_2' & J' \\ m_1' & m_2' & M' \end{bmatrix}(j_1 m_1,\, j_2 m_2|\, V_P\, |j_1' m_1',\, j_2' m_2') \\
&= -G \sum_{m_1, m_2}\sum_{m_1', m_2'} \begin{bmatrix} j_1 & j_2 & J \\ m_1 & m_2 & M \end{bmatrix}\begin{bmatrix} j_1' & j_2' & J' \\ m_1' & m_2' & M' \end{bmatrix}\delta_{j_1 j_2}\delta_{m_2, -m_1}\delta_{j_1' j_2'}\delta_{m_2', -m'} \\
&\quad \times (-1)^{j_1 - m_1}(-1)^{j_1' - m_1'} \\
&= -G \sum_{m_1} (-1)^{j_1 - m_1}\begin{bmatrix} j_1 & j_1 & J \\ m_1 & -m_1 & 0 \end{bmatrix} \sum_{m_1'} \begin{bmatrix} j_1' & j_1' & J' \\ m_1' & -m_1' & 0 \end{bmatrix}(-1)^{j_1' - m_1'}\delta_{M0}\delta_{M'0}\delta_{j_1 j_2}\delta_{j_1' j_2'}.
\end{aligned}$$

In this expression, we have merely used the definition (VI.93) for the nonvanishing matrix elements of V_P. The summation in the final step can be carried out with the standard orthogonality property of the Clebsch-Gordon coefficients. The result is given by

$$\sum_m (-1)^{j-m}\begin{bmatrix} j & j & J \\ m & -m & 0 \end{bmatrix} = \sum_m \begin{bmatrix} j & j & J \\ m & -m & 0 \end{bmatrix}(-1)^{j-m}\begin{bmatrix} j & 0 & j \\ m & 0 & m \end{bmatrix}$$

$$= \sum_m \begin{bmatrix} j & j & J \\ m & -m & 0 \end{bmatrix}\begin{bmatrix} j & j & 0 \\ m & -m & 0 \end{bmatrix}\sqrt{2j+1}$$

$$= \delta_{J0}\sqrt{2j+1}. \tag{VI.94a}$$

The extra Clebsch-Gordon coefficient introduced in the derivation is identically equal to unity. Using this standard result, we finally have

$$(j_1 j_2 JM|\, V_P \,|\, j_1' j_2' J'M') = -\delta_{j_1 j_2}\delta_{j_1' j_2'}\delta_{M0}\delta_{M'0}\delta_{J0}G\sqrt{(2j_1+1)(2j_1'+1)}. \tag{VI.94b}$$

For the special pairing force (VI.92) or (VI.93), the energy gap equations (VI.80) become very simple. We have, according to (VI.92),

$$\Delta_a = \tfrac{1}{4}G \sum_k (\Delta_k / E_k).$$

Since the summation on the right-hand side is independent of the state a, it is clear that Δ_a now is actually independent of a; there is only one energy gap Δ for all the single-particle states. *There is only one gap equation*, and that is

$$\Delta = \tfrac{1}{4}G \sum_k (\Delta / E_k)$$

or

$$1 = \tfrac{1}{4}G \sum_k [(\epsilon_k - \lambda)^2 + \Delta^2]^{-1/2}. \tag{VI.95}$$

We have to solve only this equation and (VI.81) simultaneously.

Ground-State Wavefunction

Finally, we wish to show that a ground-state wavefunction satisfying

$$b_i|\Psi_0\rangle = 0 \quad \text{for all } i \tag{VI.96}$$

actually exists, and it can be constructed in terms of the particle creation operators as

$$|\Psi_0\rangle = \prod_k (u_k + v_k s_k C_k^\dagger C_{-k}^\dagger)|0\rangle, \tag{VI.97}$$

where $|0\rangle$ is the vacuum state for the particle operators and satisfies

$$C_i|0\rangle = 0 \quad \text{for all } i. \tag{VI.98}$$

To show that $|\Psi_0\rangle$ satisfies (VI.96), we write

$$b_i = (u_i C_i - v_i s_i C_{-i}^\dagger)$$

and let it operate from the left on (VI.97). It is clear that C_i and $C_{-i}^\dagger$ can be commuted past all the factors in (VI.97) for which $k \neq i$. When we encounter the factor for $k = i$, we have to examine what happens to it when b_i operates on it from the left. We have

$$(u_i C_i - v_i s_i C_{-i}^\dagger)(u_i + v_i s_i C_i^\dagger C_{-i}^\dagger) = u_i^2 C_i - u_i v_i s_i C_{-i}^\dagger + u_i v_i s_i C_i C_i^\dagger C_{-i}^\dagger - v_i^2 C_{-i}^\dagger C_i^\dagger C_{-i}^\dagger. \tag{VI.99}$$

The last term is zero because it contains two $C_{-i}^{\dagger}$, and hence it violates the Pauli exclusion principle. The first term can be commuted past all the other factors which are $k \neq i$, and eventually it operates on $|0\rangle$ and produces zero according to (VI.98). The third term can be written as

$$u_i v_i s_i C_{-i}^{\dagger} + u_i v_i s_i C_i^{\dagger} C_{-i}^{\dagger} C_i,$$

where we have used $\{C_i, C_i^{\dagger}\} = 1$. The first term here cancels out with the second term of (VI.99), and the second term produces zero when acting on $|0\rangle$. Thus, we have established the property (VI.96) for our ground-state wavefunction (VI.97).

The ground-state wavefunction clearly contains states of 0, 2, 4, 6, . . . particles when all the factors are multiplied. The process of fixing λ in the calculation ensures that the strength of the A-particle wavefunction in this superposition has a maximum. Because of this form of Ψ_0, it is necessary that we restrict the quasiparticle ground state to the description of the ground state of an even nucleus. The moment we let $b_i^{\dagger} = u_i C_i^{\dagger} - v_i s_i C_{-i}$ operate on Ψ_0, the new wavefunction $b_i^{\dagger}|\Psi_0\rangle$ clearly becomes a superposition of 1, 3, 5, . . . particles. This is the reason odd-mass nuclei are described in terms of an odd number of quasiparticle excitation. In the same way, an even number of quasiparticle operators, operating on Ψ_0, always generates a wavefunction having a superposition of an even number of nucleons. Thus, in this theory, the excited states of even nuclei are associated with an even number of quasiparticle excitation.

Special Case of Spherical Nuclei

We have already shown that considerable simplification takes place if we use the special pairing force of constant strength G. *Even without this special approximation about the two-body force*, the general equations (VI.80) take a simpler form for spherical nuclei. For such nuclei, the single-particle states are of the type $(nljm)$. For simplicity, we omit the quantum numbers (nl) in the derivation that follows. For spherical nuclei, it is not physically possible to make any special choice of the Z-axis, all directions in space being equivalent for them. Therefore, we know that the quantities $u_a, v_a, E_a, \Delta_a, \ldots$ cannot depend on the projection quantum number m_a of the state a. We keep this fact in mind and write (VI.80) as

$$\Delta_j = -\tfrac{1}{4} \sum_{j', m'} (-1)^{j-m}(-1)^{j'-m'}(jm; j, -m|\, V\, |j'm'; j', -m')\frac{\Delta_{j'}}{E_{j'}}. \qquad \text{(VI.100)}$$

We replace the two-body states in this expression by the corresponding angular momentum coupled states, i.e.,

$$|j'm'; j', -m'\rangle = \sum_{J'} \begin{bmatrix} j' & j' & J' \\ m' & -m' & 0 \end{bmatrix} |j'j'J'0\rangle. \qquad \text{(VI.101a)}$$

In the exchange term of the matrix element in (VI.100), we have the state

$$|j', -m'; j'm'\rangle = \sum_{J'} \begin{bmatrix} j' & j' & J' \\ -m' & m' & 0 \end{bmatrix} |j'j'J'0\rangle$$

$$= -\sum_{J'} (-1)^{J'} \begin{bmatrix} j' & j' & J' \\ m' & -m' & 0 \end{bmatrix} |j'j'J'0\rangle. \qquad \text{(VI.101b)}$$

In the final step, we have used a standard symmetry relation of the Clebsch-Gordon coefficient.

Using (VI.101), we have

$$(jm; j, -m| V |j'm'; j', -m') \equiv \langle jm; j, -m| V |j'm'; j', -m'\rangle - \langle jm; j, -m| V |j', -m'; j'm'\rangle$$
$$= \sum_J \begin{bmatrix} j & j & J \\ m & -m & 0 \end{bmatrix} \begin{bmatrix} j' & j' & J \\ m' & -m' & 0 \end{bmatrix} \langle jjJ0| V |j'j'J0\rangle \times [1 + (-1)^J].$$

Here we have used the property of the potential V which requires J to be equal to J'. The last factor tells us that J must be an even integer, which in turn guarantees the antisymmetry of the state on the right-hand side [it is known that the antisymmetric states of two equivalent nucleons in the same (nlj)-orbit can have only even J-values]. Therefore, we replace the factor $\frac{1}{2}[1 + (-1)^J]$ times the two-body matrix element by

$$(jjJ0| V |j'j'J0),$$

where the parentheses as usual denote the matrix element with respect to *antisymmetric* two-body angular momentum coupled states. When the expression for the matrix element of V is used in (VI.100), we obtain

$$\Delta_j = -\tfrac{1}{2} \sum_{Jj'm'} (-1)^{j-m}(-1)^{j'-m'} \begin{bmatrix} j' & j' & J \\ m' & -m' & 0 \end{bmatrix} \begin{bmatrix} j & j & J \\ m & -m & 0 \end{bmatrix} (jjJ0| V |j'j'J0) \frac{\Delta_{j'}}{E_{j'}}.$$

In this expression, the summation over m' can be immediately carried out by using the standard result (VI.94a). Thus,

$$\Delta_j = -\tfrac{1}{2}(-1)^{j-m} \begin{bmatrix} j & j & 0 \\ m & -m & 0 \end{bmatrix} \delta_{J0} \sum_{j'} (jjJ0| V |j'j'J0)\sqrt{2j'+1}\frac{\Delta_{j'}}{E_{j'}}$$

$$= -\tfrac{1}{2} \sum_{j'} \sqrt{\frac{2j'+1}{2j+1}} (jjJ=0, M=0| V |j'j'J=0, M=0)\frac{\Delta_{j'}}{E_{j'}}. \tag{VI.102}$$

In obtaining the final expression, we have substituted for the special Clebsch-Gordon coefficient its value $(-1)^{j-m}(2j+1)^{-1/2}$. Thus, we find that, for spherical nuclei, even though we have a set of Δ_j's coupled through (VI.102), the part of the interaction potential V responsible for this coupling is merely the interaction in the two body state of angular momentum $J = 0$.

42. MICROSCOPIC THEORY—VIBRATIONAL MODE

A. GENERAL EXPRESSION OF QUASIPARTICLE INTERACTION

In Section 41, we began transforming the many-body Hamiltonian and showed that a part (H_0) of it represents the ground-state energy, and another part (H_{11}) represents the energy of single quasiparticles. What remains of the Hamiltonian after this is the term H_4 [of (VI.50)], which actually represents the interaction of the quasiparticles. In view of the importance of this term in building up collective vibrational states, we shall devote this section solely to its treatment.

We have seen in Section 41 that a BCS type quasiparticle transformation, given by (VI.70), *where both* u_i *and* v_i *are nonvanishing* and computed from (VI.78), describes the case of nuclei having a diffuse fermi surface. Such a quasiparticle is *partly* a 'hole' and *partly* a 'particle'. On the other hand, we found that the same transformation equations can be *formally* used to describe the quasiparticles also in a nucleus having a sharp fermi sea (i.e., a closed-shell nucleus or a nucleus where the HF calculation produces a large energy gap between the occupied and unoccupied single-particle states). In this case, however, we must remember to put special values of u_i and v_i, namely, $u_i = 1$ and $v_i = 0$ if i is an unoccupied (i.e., a 'particle') state, and $u_i = 0$ and $v_i = 1$ if i is an occupied (i.e., a 'hole') state. In other words, the quasiparticles here are either *pure* 'particle' or *pure* 'hole' states. Keeping this fact in mind, we shall write down the general expression for H_4 with the substitutions (VI.71) and then apply the same formal expression with the appropriate values of u_i and v_i to describe both the types of nuclei.

Since the BCS transformation was done with HF type single-particle states, we first rewrite H_4 of (VI.50) in the HF representation as

$$H_4 = \tfrac{1}{4} \sum_{ijkl} (ij|\, V\, |kl) {:} C_i^\dagger C_j^\dagger C_l C_k {:}\,. \tag{VI.103}$$

We have to substitute from (VI.71) and then permute in each term the destruction operators to the right of the creation operators; finally, the appropriate sign of each term has to be incorporated according to the rules for forming the normal product described after (VI.39). It is clear from (VI.71) that there will be one term in (VI.103) which contains four quasiparticle creation operators, and one term with four destruction operators; these terms obviously have the coefficients $u_i u_j v_l v_k$ and $v_i v_j u_l u_k$ respectively. We shall denote these terms by H_{40} and H_{04} where, as usual, the two subscripts specify the number of creation and destruction operators, respectively. Then we easily obtain

$$H_{40} = \tfrac{1}{4} \sum_{ijkl} (ij|\, V\, |\bar{k}, \bar{l}) u_i u_j v_l v_k b_i^\dagger b_j^\dagger b_l^\dagger b_k^\dagger, \tag{VI.104a}$$

$$H_{04} = \text{Hermitean conjugate of } H_{40}. \tag{VI.104b}$$

Here a state with a bar above denotes the corresponding time-reversed state, i.e., $|\bar{k}\rangle = s_k|-k\rangle$, where s_k is the phase factor appearing in (VI.71). Since k, l are summation indices, we have made the legitimate replacements $k \to -k$ and $l \to -l$ in obtaining (VI.104a).

In a similar manner, it is clear from (VI.71) that there are four terms in (VI.103) having one quasiparticle destruction operator and three creation operators (H_{31}); the number of terms with the roles of destruction and creation operators reversed (H_{13}) is also four. It is possible to relabel the summation indices in the different terms of H_{31} and H_{13} and then obtain the final expressions

$$H_{31} = \tfrac{1}{2} \sum_{ijkl} [u_i u_j v_k u_l (ij|\, V\, |\bar{k}, l) + u_i v_j v_k v_l (i, \bar{l}|\, V\, |\bar{j}, \bar{k})] b_i^\dagger b_j^\dagger b_k^\dagger b_l, \tag{VI.104c}$$

$$H_{13} = \text{Hermitean conjugate of } H_{31}. \tag{VI.104d}$$

The last set of terms (H_{22}) contained in (VI.103) has two quasiparticle creation and two quasiparticle destruction operators. Their total number is six, but once again the summation indices can be relabelled to write H_{22} in the compact form

$$H_{22} = \tfrac{1}{2} \sum_{ijkl} [(u_i u_j u_k u_l + v_i v_j v_k v_l)\tfrac{1}{2}(ij|\, V\, |kl) - (u_i v_j u_k v_l + v_i u_j v_k u_l)(j, \bar{k}|\, V\, |l, \bar{i})] b_i^\dagger b_j^\dagger b_l b_k. \tag{VI.104e}$$

In fact, if we compare (VI.104e) with the second term of (VI.32), we are immediately struck by their similarity. Such a comparison leads us to interpret the expression enclosed within the square brackets as $\langle ij|\, V_Q\, |kl\rangle$, where V_Q is the effective interaction potential of two quasiparticles. Such a clear identification of the other terms, namely, (VI.104a) to (VI.104d), is not very obvious, because these terms do not conserve the number of quasiparticles, whereas (VI.104e) does. (VI.32) conserves the number of particles, and hence the formal similarity is evident only for the term (VI.104e) which conserves the number of quasiparticles.

B. APPLICATION TO HARTREE-FOCK CASE

Diagrammatic Representation of Interaction

By the HF case here we mean those nuclei where an HF-type ground state having a sharp fermi sea is a good description. The quasiparticles are either pure holes or pure particles. The coefficients u_i, v_i, . . . in (VI.104) indicate whether a given state has to be a hole or a particle. For example, in the term H_{40} of (VI.104a), the factors $u_i u_j$ and $v_l v_k$ tell us that i, j are particle states and k, l are hole states. Therefore, $b_i^\dagger b_j^\dagger b_l^\dagger b_k^\dagger$ occurring in this expression describes the creation of two hole-particle pairs. In the matrix element of V, the state on the right is the initial state, and that on the left is the final state. Hence, as a result of the two-body interaction, the hole state k proceeds to the particle state i while the hole state l goes to the particle state j. The appearance in the two-body matrix element of a time-reversed state corresponding to a hole has its origin in the transformation property of a hole state under rotation.

All the foregoing facts can be illustrated by drawing a dashed horizontal line to represent V, and solid lines moving towards or away from points to represent the various states. To distinguish between holes and particles, we attach an arrow to a line, and make the conventions described in Fig. VI.3. Figures VI.3a and VI.3b have *upward* arrows and are taken to represent *particle* states, whereas Figs. VI.3c and VI.3d have *downward* arrows representing

(a) Particle creation (b) Particle destruction (c) Hole creation (d) Hole destruction

Fig. VI.3 Creation and destruction of particle and hole.

hole states. For particles, an arrow moving *away* from a point, as in Fig. VI.3a, denotes *creation* at that point, whereas an arrow moving *towards* a point, as in Fig. VI.3b, represents *destruction* of the particle at that point. In the case of holes, the convention is just the opposite: Fig. VI.3c having a hole line moving *towards* a point denotes the *creation* of the hole, whereas Fig. VI.3d with a hole line moving *away* from a point denotes the *destruction* of the hole. Keeping these conventions in mind, we describe the interaction term H_{40} as in Fig. VI.4.

It should be observed that the direction of the arrows indeed tells us that the state k goes to i, and the state l to j. According to Figs. VI.3a and VI.3c, we indeed have in Fig. VI.4 the creation of the holes k and l, and the creation of the particles i and j. Each 'hen track' in Fig. VI.4 corresponds to a hole-particle pair, and the entire diagram corresponds to the creation of two such pairs.

In a similar manner, the Hermitean conjugate of H_{40}, i.e., H_{04} of (VI.104b), describes the destruction of two hole-particle pairs, and is represented as in Fig. VI.5.

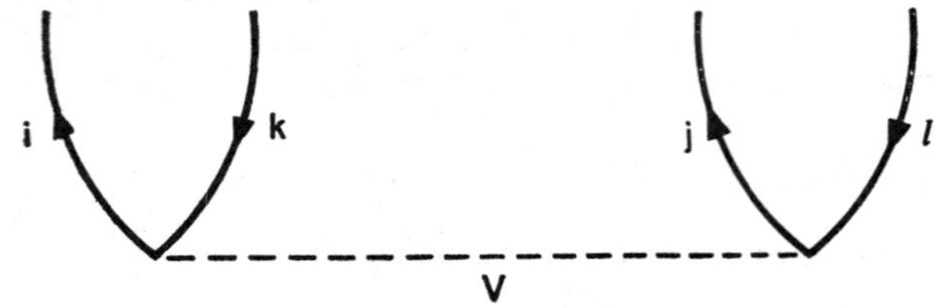

Fig. VI.4 Representation of term H_{40} corresponding to the part of two-nucleon interaction responsible for exciting a pair of holes and particles.

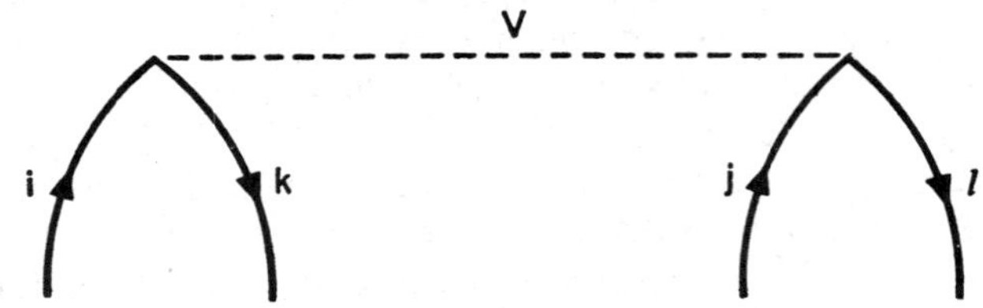

Fig. VI.5 Representation of term H_{04} corresponding to absorption of pair of holes and particles.

We next examine H_{31}. Of the two terms in (VI.104c), the coefficients u, v in the first term indicate that H_{31} corresponds to the creation of the two particles i, j and a hole k and the destruction of a particle l. The matrix element of V appearing in this term shows that k goes to i, and l to j. This is depicted in the diagram of Fig. VI.6, which explicitly shows that this

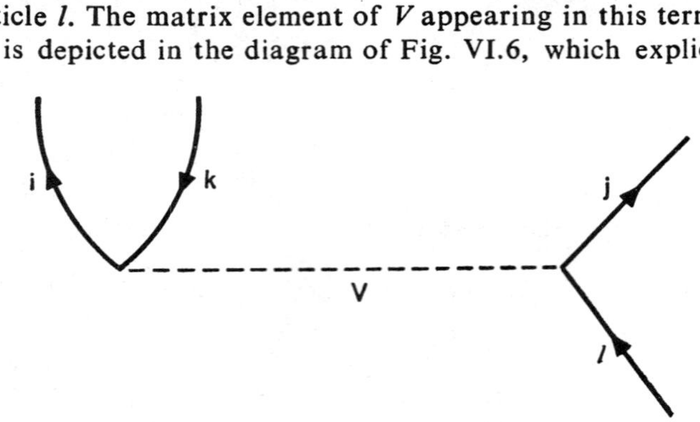

Fig. VI.6 First term of H_{31} representing interaction of particle with hole-particle pair.

term represents the interaction of a hole-particle pair with a *particle*. In the same manner, the reader can convince himself that the second term of H_{31} represents the interaction of a hole-particle pair with a *hole* state. As an exercise, he may draw the diagrams for H_{13} of (VI.104d) and interpret them appropriately.

Finally, we consider the expression H_{22} of (VI.104e) which has three different types of terms. The first type comes from the term $u_i u_j u_k u_l$ of the first line. Clearly, all the states i, j, k, l are required to be particle states of which i and j are created and k and l are destroyed (because of $b_i^\dagger b_j^\dagger b_l b_k$). With proper recognition of the two-body matrix elements, we can easily draw the diagram of Fig. VI.7 which depicts the interaction between two particles, and is

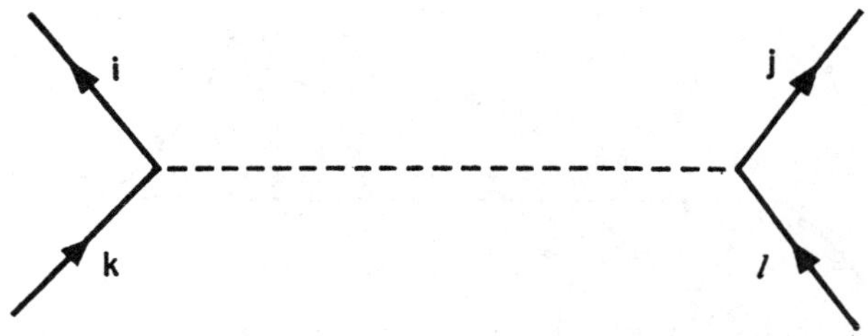

Fig. VI.7 Particle-particle interaction term of H_{22}.

usually called the *particle-particle* interaction term. In a similar manner, we examine the second term in the first line of (VI.104e), where the coefficients v tell us that i, j, k, l are all hole states. This term therefore represents the *hole-hole* interaction, and is given by the diagram of Fig. VI.8. The third type of term in H_{22} corresponds to the second line of (VI.104e). From

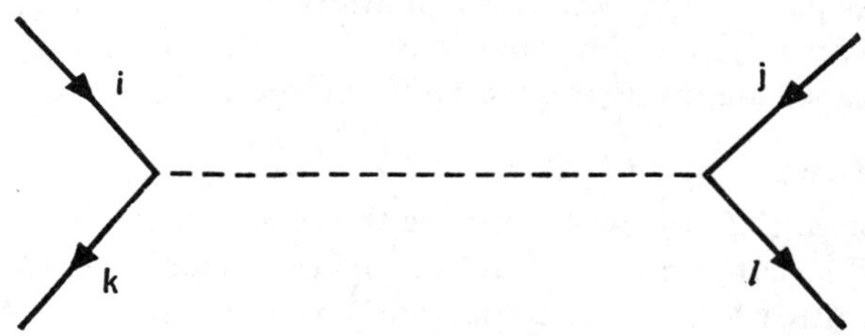

Fig. VI.8 Hole-hole interaction term of H_{22}.

arguments similar to those just repeated several times, we conclude that this term corresponds to the interaction of a hole and a particle and can be represented by the diagram of Fig. VI.9.

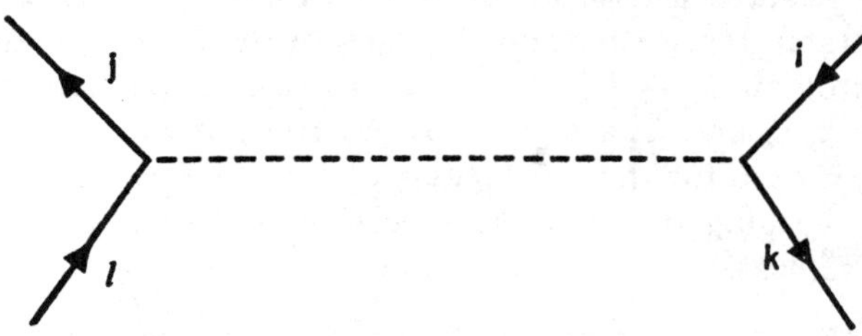

Fig. VI.9 Hole-particle interaction term of H_{22}.

Finally, we must point out that we have followed here the convention that allows only one diagram corresponding to an antisymmetric matrix element of V. Many authors split such a matrix element explicitly into the direct and the exchange parts and represent these processes by separate diagrams. To give an example, let us consider the hole-particle term, which contains the matrix element $(j, \bar{k}| V |l, \bar{i}) \equiv \langle j, \bar{k}| V |l, \bar{i}\rangle - \langle j, \bar{k}| V |\bar{i}, l\rangle$. The direct term here still corresponds to a diagram such as the one given by Fig. VI.9, whereas in the exchange term the state i goes to j, and the state l to k. Hence, if we decide to have a separate diagram for the exchange process, then it would be as in Fig. VI.10. Those who follow the convention

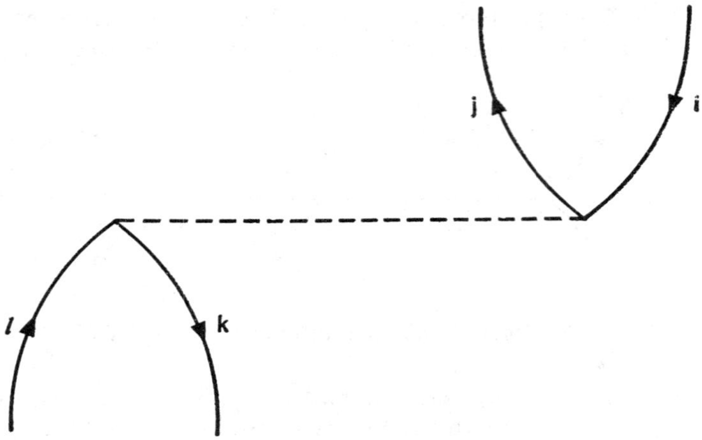

Fig. VI.10 Exchange part of hole-particle interaction.

of illustrating the direct part of the hole-particle interaction (as in Fig. VI.9) and drawing the exchange part of the same interaction separately (as in Fig. VI.10) refer to these diagrams as the 'ladder' and 'hen track' diagrams, respectively, because of their visual appearance.

Tamm-Dancoff Calculation for 1h-1p States

A glance at the expressions (VI.104) will convince the reader that all the parts of the interaction Hamiltonian, except H_{40}, contain a destruction operator for a quasiparticle (in the present case, a quasiparticle is either a 'hole' or a 'particle') to the extreme right. Thus, if we operate on the ground state $|\Phi_0\rangle$, only the H_{40}-part can produce a nonvanishing result; in fact, the discussion in Section 42B tells us that the result is a 2h-2p type state. Therefore, when the interaction is switched on, the ground state acquires an admixture of 2h-2p type states in the lowest approximation. What is important is that there is no term in the interaction Hamiltonian that can mix a 1h-1p type state into $|\Phi_0\rangle$. This is a well-known result of the HF theory.

The 1h-1p type states represent excited states of the Hamiltonian, and it follows, from our unperturbed Hamiltonian $H_0 + H_{11}$, that such a two-quasiparticle state has an unperturbed energy $E_h + E_p$, where E_h and E_p are given by (VI.86a), and *not* by (VI.79). This is because the last term of (VI.85) is zero for the HF theory, the product uv vanishing for both occupied and unoccupied states. Keeping in mind the special values of u, v for the hole and particle states, we have, from (VI.86a),

$$E_h = -(\epsilon_h - \lambda) = \lambda - \epsilon_h, \tag{VI.105a}$$

$$E_p = \epsilon_p - \lambda, \tag{VI.105b}$$

where λ is the energy of the fermi level. These results can be inferred if we remember that a particle state is *above* the fermi level, whereas a hole state is below it. Finally, from (VI.105), we have the unperturbed energy of a 1h-1p type state, given by

$$E_h + E_p = \epsilon_p - \epsilon_h. \tag{VI.106}$$

All the 1h-1p type states that are obtained by lifting a nucleon across one major shell have their unperturbed energies (VI.106) lying within a few MeV of each other. The unperturbed energies of the 2h-2p type states are, on the other hand, roughly double this magnitude.

We next examine what happens to the 1h-1p type states when the interaction H_4 is switched on. It is clear from the term H_{22} in (VI.104e) that the hole-particle interaction part of it can give a nonvanishing matrix element connecting the 1h-1p states with each other. Similarly, the term H_{31} in (VI.104c) can produce a nonvanishing matrix element between a 1h-1p type state and a 2h-2p type state. The easiest way to see this is that H_{31} has one destruction operator to its extreme right, and hence the state it operates on must have at least one quasiparticle; each of the 1h-1p type states has two quasiparticles, and hence meets this criterion. Further, H_{31} creates three quasiparticles after destroying one, that is, it produces two additional quasiparticles; 2h-2p states have four quasiparticles, whereas 1h-1p states have two, and hence, after the production of two additional quasiparticles, the 1h-1p state changes naturally to a 2h-2p state. This kind of argument, even though elementary, has been given here in detail so that a beginner in this subject understands the logic behind similar statements, subsequently made, without an elaborate explanation.

According to the perturbation theory, therefore, the effect of the interaction is to produce a ground state and a set of excited states with the structure

$$|\Phi_0\rangle \to |\Phi_0'\rangle = |\Phi_0\rangle + \sum_{hph'p'} f_{hp,\,h'p'}|\Phi(hp,\, h'p')\rangle, \tag{VI.107a}$$

$$|\Phi(hp)\rangle \to |\Phi'(hp)\rangle = |\Phi(hp)\rangle + \sum_{h'p'} g_{hp,\,h'p'}|\Phi(h'p')\rangle. \tag{VI.107b}$$

We have written here only the lowest-order perturbation terms. Hole-particle indices within the parentheses following a Φ specify the hole-particle excitations present in that particular basis state. The quantities f and g contain, according to the perturbation theory, the appropriate matrix element of H_4 in the numerator and an energy denominator which is equal to $(E_h + E_p + E_{h'} + E_{p'})$ for f and $(E_{h'} + E_{p'} - E_h - E_p)$ for g. Since the energy denominator of g is much smaller than that of f, we can consider an approximate treatment of H_4 as follows. We neglect any modification of the ground state through the interaction, but consider the admixture of the 1h-1p type states with each other. A calculation based on this point of view is called the *Tamm-Dancoff Approximation* (TDA).

In fact, the energy denominator $(E_{h'} + E_{p'} - E_h - E_p)$ can, in principle, even vanish when the energies $E_{h'} + E_{p'}$ and $E_h + E_p$ are accidentally degenerate. At any rate, all the 1h-1p type states across a given major shell form a near-degenerate group of levels, and hence their admixture is better taken into account by a matrix diagonalization (degenerate perturbation theory), rather than a perturbation treatment such as (VI.107b). The calculation in the TDA therefore entails setting up the matrix of the hole-particle part of H_{22} in the 1h-1p space, adding the unperturbed energies along the diagonal and then diagonalizing the resultant Hamiltonian matrix. We proceed to write down the expression for a typical matrix element.

The $(u_iv_ju_kv_l)$-term in the hole-particle part of (VI.104e) corresponds to the identification

$$i = \text{particle}, \qquad j = \text{hole}, \qquad k = \text{particle}, \qquad l = \text{hole}; \tag{VI.108a}$$

whereas the $(v_iu_jv_ku_l)$-term in the same expression needs

$$i = \text{hole}, \qquad j = \text{particle}, \qquad k = \text{hole}, \qquad l = \text{particle}. \tag{VI.108b}$$

We want the matrix element of these terms connecting the hole-particle state $b_p^\dagger b_h^\dagger|\Phi_0\rangle$ with another hole-particle state, namely, $\langle\Phi_0|b_{h'}b_{p'}$. In the usual shell-model notation, these states are denoted by $|\bar{h}^{-1}p\rangle$ and $\langle\bar{h}'^{-1}p'|$, respectively. It should be observed that, according to (VI.70a), the hole creation operator $b_h^\dagger$ corresponds to $s_hC_{-h} \equiv C_{\bar{h}}$, and hence it corresponds to a vacancy in the shell-model state $\bar{h}$ (and *not* h). This fact accounts for the equivalent shell-model notation just mentioned. We record these correspondences for subsequent use:

$$|\bar{h}^{-1}p\rangle = b_p^\dagger b_h^\dagger|\Phi_0\rangle, \tag{VI.109a}$$

$$\langle\bar{h}'^{-1}p'| = \langle\Phi_0|b_{h'}b_{p'}. \tag{VI.109b}$$

We now take the operator $b_i^\dagger b_j^\dagger b_l b_k$ from (VI.104e) and easily arrive at the results

$$\begin{aligned} b_l b_k|\bar{h}^{-1}p\rangle &= b_l b_k b_p^\dagger b_h^\dagger|\Phi_0\rangle \\ &= (\delta_{kp}\delta_{lh} - \delta_{kh}\delta_{lp})|\Phi_0\rangle, \end{aligned} \tag{VI.110a}$$

$$\begin{aligned} \langle\bar{h}'^{-1}p'|b_i^\dagger b_j^\dagger &= \langle\Phi_0|b_{h'}b_{p'}b_i^\dagger b_j^\dagger \\ &= \langle\Phi_0|(\delta_{ip'}\delta_{jh'} - \delta_{ih'}\delta_{jp'}). \end{aligned} \tag{VI.110b}$$

In deriving these results, we have made use of the anticommutators of the creation and destruction operators to permute the destruction operators on to $|\Phi_0\rangle$ and the creation operators on to $\langle\Phi_0|$, both of which are zero according to (VI.36). When we multiply (VI.110b) by (VI.110a), the requirements (VI.108a) and (VI.108b) imply that two of the four terms are non-vanishing. Thus, putting (VI.110b), (VI.110a), (VI.108), and (VI.104e) together, we finally obtain

$$\langle\bar{h}'^{-1}p'|\, H_{22}\, |\bar{h}^{-1}p\rangle = -\tfrac{1}{2} \sum_{ijkl} (\delta_{kp}\delta_{lh}\delta_{ip'}\delta_{jh'} + \delta_{kh}\delta_{lp}\delta_{ih'}\delta_{jp'})(j, \bar{k}|\, V\, |l, \bar{i}).$$

Using the time-reversal invariance of V, we can easily convince ourselves that the two terms in this expression are equal, and the final result is given by

$$\langle\bar{h}'^{-1}p'|\, H_{22}\, |\bar{h}^{-1}p\rangle = -(p', \bar{h}|\, V\, |p, \bar{h}'). \tag{VI.111}$$

It has already been mentioned that between the same two states the unperturbed part $H_0 + H_{11}$ of the Hamiltonian gives, according to (VI.106),

$$\langle\bar{h}'^{-1}p'|\, (H_0 + H_{11})\, |\bar{h}^{-1}p\rangle = \delta_{hh'}\delta_{pp'}(\epsilon_p - \epsilon_h). \tag{VI.112}$$

The details of the angular momentum coupling in (VI.111) and (VI.112) are given in Appendix G (Section III).

Alternative Derivations of TDA Equations

There are several ways of deriving TDA equations. We shall describe two such methods in this section.

Commutator method with linearization In this method, we evaluate the commutator of H with $b_p^\dagger b_h^\dagger$. Being a number, H_0 commutes with it. We next evaluate the commutator

$$[b_i^\dagger b_i,\, b_p^\dagger b_h^\dagger] \equiv b_i^\dagger[b_i,\, b_p^\dagger b_h^\dagger].$$

Since

$$\begin{aligned} b_i b_p^\dagger b_h^\dagger &= \delta_{ip} b_h^\dagger - b_p^\dagger b_i b_h^\dagger \\ &= \delta_{ip} b_h^\dagger - \delta_{ih} b_p^\dagger + b_p^\dagger b_h^\dagger b_i, \end{aligned}$$

we have

$$[b_i,\, b_p^\dagger b_h^\dagger] = \delta_{ip} b_h^\dagger - \delta_{ih} b_p^\dagger. \tag{VI.113}$$

Therefore,

$$[b_i^\dagger b_i, b_p^\dagger b_h^\dagger] = \delta_{ip} b_i^\dagger b_h^\dagger - \delta_{ih} b_i^\dagger b_p^\dagger,$$

and hence

$$\begin{aligned}[H_{11}, b_p^\dagger b_h^\dagger] &\equiv \sum_i E_i [b_i^\dagger b_i, b_p^\dagger b_h^\dagger] \\ &= (E_p + E_h) b_p^\dagger b_h^\dagger \\ &= (\epsilon_p - \epsilon_h) b_p^\dagger b_h^\dagger. \end{aligned} \tag{VI.114a}$$

Finally, as a step towards evaluating the commutator with the hole-particle part of H_{22}, we work out, with the help of (VI.113), the result

$$\begin{aligned}[b_i^\dagger b_j^\dagger b_l b_k, b_p^\dagger b_h^\dagger] &= b_i^\dagger b_j^\dagger [b_l b_k, b_p^\dagger b_h^\dagger] \\ &= b_i^\dagger b_j^\dagger (b_l [b_k, b_p^\dagger b_h^\dagger] + [b_l, b_p^\dagger b_h^\dagger] b_k) \\ &= b_i^\dagger b_j^\dagger (\delta_{kp} b_l b_h^\dagger - \delta_{kh} b_l b_p^\dagger + \delta_{lp} b_h^\dagger b_k - \delta_{lh} b_p^\dagger b_k) \\ &= b_i^\dagger b_j^\dagger (\delta_{kp}\delta_{lh} - \delta_{kh}\delta_{lp} + \delta_{kh} b_p^\dagger b_l - \delta_{kp} b_h^\dagger b_l + \delta_{lp} b_h^\dagger b_k - \delta_{lh} b_p^\dagger b_k). \end{aligned} \tag{VI.114b}$$

We are finally going to use the manipulation mentioned after (VI.9). When $|\Phi_0\rangle$ appears to the right of (VI.114b), the last four terms obviously produce zero. This is why, in the subsequent steps, we keep only the first two terms of (VI.114b) and obtain, from (VI.104e) and (VI.108),

$$\begin{aligned}[H_{22}, b_p^\dagger b_h^\dagger] &= -\tfrac{1}{2} \sum_{i,j} [u_i v_j b_i^\dagger b_j^\dagger (j, \bar{p}| V |h, \bar{i}) - v_i u_j b_i^\dagger b_j^\dagger (j, \bar{h}| V |p, \bar{i})] \\ &= -\tfrac{1}{2} \sum_{h', p'} b_{p'}^\dagger b_{h'}^\dagger [(h', \bar{p}| V |h, \bar{p}) + (p', \bar{h}| V |p, \bar{h}')] \\ &= - \sum_{h', p'} (p', \bar{h}| V |p, \bar{h}') b_{p'}^\dagger b_{h'}^\dagger. \end{aligned} \tag{VI.115}$$

Once again, we have used the time-reversal invariance of V to equate the two matrix elements; further, we have taken note of the fact that $u_i v_j$ in the first term requires i to be a particle state p' and j to be a hole state h', whereas $v_i u_j$ in the second term imposes just the opposite requirement. In the second term, $b_{h'}^\dagger b_{p'}^\dagger$ has been rearranged to $-b_{p'}^\dagger b_{h'}^\dagger$. The artifice of putting $|\Phi_0\rangle$ to the right and $\langle\Phi|$ to the left of the commutator has enabled us to discard terms that are not linear in the lh-lp creation operator $b_{p'}^\dagger b_{h'}^\dagger$. Therefore, this procedure is usually called the *method of linearization*. It should be noticed that the commutator method of diagonalizing a Hamiltonian, described in Section 39, succeeds *only* if it is possible to obtain a closed system of linear equations connecting a given set of operators. This has been achieved in (VI.115). Putting (VI.114a) and (VI.115) together, we are once again led, according to the general derivation of Section 39, to the diagonalization of the same hole-particle matrix of the Hamiltonian which we have described in (VI.111) and (VI.112).

Quasi-boson method In this method, we first introduce the operator $A_{hp}^\dagger$ which creates a hole-particle pair:

$$A_{hp}^\dagger = b_p^\dagger b_h^\dagger. \tag{VI.116a}$$

The Hermitean conjugate of this operator destroys a hole-particle pair and is given by

$$A_{hp} = b_h b_p. \tag{VI.116b}$$

It can be easily verified from the anticommutators of $b_h, b_p, b_h^\dagger, b_p^\dagger, \ldots$ that

$$A_{hp}^\dagger = -A_{ph}^\dagger, \qquad A_{hp} = -A_{ph}, \tag{VI.116c}$$

$$[A_{hp}, A_{h'p'}] = [A_{hp}^\dagger, A_{h'p'}^\dagger] = 0. \tag{VI.117a}$$

The commutator of a pair-destruction operator with a pair-creation operator can be similarly worked out with the help of (VI.113). We obtain

$$\begin{aligned}
[A_{hp}, A_{h'p'}^\dagger] &= [b_h b_p, b_{p'}^\dagger b_{h'}^\dagger] \\
&= b_h[b_p, b_{p'}^\dagger b_{h'}^\dagger] + [b_h, b_{p'}^\dagger b_{h'}^\dagger]b_p \\
&= \delta_{pp'} b_h b_{h'}^\dagger - \delta_{hh'} b_{p'}^\dagger b_p \\
&= \delta_{hh'}\delta_{pp'} - \delta_{pp'} b_{h'}^\dagger b_h - \delta_{hh'} b_{p'}^\dagger b_p.
\end{aligned} \tag{VI.117b}$$

Introducing the abbreviated notation

$$\hat{N}_{i',i} = b_{i'}^\dagger b_i, \tag{VI.118}$$

we rewrite (VI.117b) as

$$[A_{hp}, A_{h'p'}^\dagger] = \delta_{hh'}\delta_{pp'} - \delta_{pp'}\hat{N}_{h'h} - \delta_{hh'}\hat{N}_{p'p}. \tag{VI.119}$$

Once again, if we use this commutator with the state $|\Phi_0\rangle$ to its right, then the last two terms produce zero and we obtain

$$[A_{hp}, A_{h'p'}^\dagger]|\Phi_0\rangle = \delta_{hh'}\delta_{pp'}|\Phi_0\rangle. \tag{VI.120}$$

Therefore, as long as we have the state $|\Phi_0\rangle$ to the right, the creation and destruction operators of the hole-particle pairs satisfy the commutation relations (VI.117a) and (VI.120) which are merely the commutation relations satisfied by a set of boson operators. The operators A_{hp}, $A_{h'p'}^\dagger, \ldots$ are called quasi-boson operators; the adjective "quasi" here is a reminder of the fact that (VI.120) is not an exact operator equation for the commutator, its validity being restricted to the presence of $|\Phi_0\rangle$ to the right.

The commutator of these quasi-boson operators with the operator $\hat{N}_{i'i}$ of (VI.118) is also straightforward to derive. We have, using (VI.113), the result

$$\begin{aligned}
[\hat{N}_{i'i}, A_{hp}^\dagger] &= [b_{i'}^\dagger b_i, b_p^\dagger b_h^\dagger] \\
&= b_{i'}^\dagger[b_i, b_p^\dagger b_h^\dagger] \\
&= \delta_{ip} b_{i'}^\dagger b_h^\dagger - \delta_{ih} b_{i'}^\dagger b_p^\dagger \\
&= \delta_{ip} A_{hi'}^\dagger + \delta_{ih} A_{i'p}^\dagger.
\end{aligned} \tag{VI.121a}$$

From the Hermitean conjugate of this equation, we have

$$[\hat{N}_{ii'}, A_{hp}] = -(\delta_{ih}A_{i'p} + \delta_{ip}A_{hi'}). \tag{VI.121b}$$

The next step in the derivation is to evaluate the commutator of $A_{hp}^\dagger$ with the part of the Hamiltonian that is treated by the TDA. The part H_0 gives zero to the commutator; the parts H_{11} and the hole-particle part of H_{22} are rewritten in terms of the new operators as

$$H_{11} = \sum_i E_i \hat{N}_{ii}, \tag{VI.122a}$$

$$\text{hole-particle part of } H_{22} = -\tfrac{1}{2}\sum_{h'p'}\sum_{h''p''}[(h'\bar{p}''|\,V\,|h''\bar{p}')A^{\dagger}_{h'p'}A_{h''p''} + (p'\bar{h}''|\,V\,|p''\bar{h}')A^{\dagger}_{p'h'}A_{p''h''}]$$

$$= -\sum_{h'p'}\sum_{h''p''}(p'\bar{h}''|\,V\,|p''\bar{h}')A^{\dagger}_{h'p'}A_{h''p''}. \tag{VI.122b}$$

We have used here the time-reversal invariance of V and (VI.116c). The commutators are now easy to evaluate with the help of (VI.117a), (VI.119), and (VI.121a). Once again, we shall multiply the commutator by $|\Phi_0\rangle$ from the right and $\langle\Phi|$ from the left in accordance with the artifice described in Section 39B, and hence the simpler result (VI.120), instead of (VI.119), can be used. In this way, we have

$$[H_{11}, A^{\dagger}_{hp}] = (E_h + E_p)A^{\dagger}_{hp} = (\epsilon_p - \epsilon_h)A^{\dagger}_{hp}, \tag{VI.123a}$$

$$[H_{22}, A^{\dagger}_{hp}] = -\sum_{h'p'}(p'\bar{h}|\,V\,|p\bar{h}')A^{\dagger}_{h'p'}. \tag{VI.123b}$$

These equations are identical to (VI.114a) and (VI.115). Thus, the linearization described in the commutator method is exactly equivalent to using the quasi-boson commutation rule (VI.120), instead of the exact result (VI.119). Once again, when we multiply the commutators from the right by $|\Phi_0\rangle$ and from the left by $\langle\Phi|$, the equations (VI.123a) plus (VI.123b) become a system of linear equations for the amplitudes $\langle\Phi|\,A^{\dagger}_{hp}\,|\Phi_0\rangle$ with all possible h, p admitted into the calculation. The solution to this system of equations is obtained by diagonalizing the TDA matrix in the hole-particle space. The amplitudes $\langle\Phi|\,A^{\dagger}_{hp}\,|\Phi_0\rangle$ are given by the elements of the eigenvector, and they clearly correspond to the amplitude of the basic hole-particle states present in the expression of the excited state $|\Phi\rangle$.

RPA Calculation for 1h-1p States

It is clear from the discussion so far that the TDA calculation diagonalizes only a limited part of the interaction. The improvement over the TDA could be achieved by two alternative methods. (i) In the first method, we introduce states with a larger number of hole-particle pairs, or at least the 2h-2p pairs. Then the parts H_{31} and H_{13} become effective in connecting the 1h-1p states with the 2h-2p states and vice versa; there is a second important consequence, namely, H_{40} connects the state $|\Phi_0\rangle$ with the 2h-2p states, and H_{04} does the opposite. Therefore, according to this approach, we must set up the matrix in the space of $|\Phi_0\rangle$, all the 1h-1p states, and the 2h-2p states. $|\Phi_0\rangle$ does not mix directly with the 1h-1p states but does so through the intermediary of the 2h-2p states. Such a calculation is usually called the *Higher Tamm-Dancoff Approximation* (HTDA). Care has to be exercised in constructing the 2h-2p states in such a calculation, because $A^{\dagger}_{hp}A^{\dagger}_{h'p'}|\Phi_0\rangle$ for all h, p and h', p' are not nontrivial states; whenever $h = h'$ and $p = p'$, such basic states do not exist due to the Pauli principle (that is, due to the anticommutation of fermion operators). While considering angular momentum coupling in these operators, we have to ensure that we keep only states in which hh and pp couple to even values of J for $T = 1$ and odd values of J for $T = 0$, where J and T are respectively the resultant angular momentum and isospin of the two holes or the two particles. (ii) In the second method, we do not *explicitly* introduce the states with a higher number of hole-particle pairs. Their effect is indirectly taken into account by including *correlations* in the ground state.

We shall now describe the second method. To understand the basic principle, let us write down the structure of the correlated ground state. In the lowest order of approximation, this is given by (VI.107a). In a similar manner, let us consider what happens to the mixed 1h-1p states of the TDA in the next higher approximation when they mix with the 2h-2p states. For simplicity, we write a mixed 1h-1p TDA state as Φ(1h-1p). This state changes to $|\Phi'\rangle$, which is given by

$$|\Phi'\rangle = |\Phi(1\text{h-}1\text{p})\rangle + \sum_{h'p'} \sum_{h''p''} \bar{f}_{h'p',\,h''p''} |\Phi(h'p', h''p'')\rangle. \tag{VI.124}$$

It is clear from (VI.107a) and (VI.124) that the amplitude $\langle\Phi'|\, A^{\dagger}_{hp}\, |\Phi'_0\rangle$ is large and of the order of unity. Because of the second term (which represents correlation in the ground state) in (VI.107a), an amplitude of the type $\langle\Phi'|\, A_{hp}\, |\Phi'_0\rangle$ is also nonvanishing and arises from the connection of $\Phi(hp, h'p')$ of (VI.107a) with Φ(1h-1p) of (VI.124) via the pair destruction operator. Such an amplitude is clearly of the order of $f_{hp,\,h'p'}$, i.e., of the first order of smallness compared with $\langle\Phi'|\, A^{\dagger}_{hp}\, |\Phi'_0\rangle$. In the same manner, it is clear from (VI.107a) and (VI.124) that an amplitude of the type $\langle\Phi'_0|\, \hat{N}_{i'i}\, |\Phi'_0\rangle$ is quadratic in f, whereas $\langle\Phi'|\, \hat{N}_{i'i}\, |\Phi'_0\rangle$ is bilinear in f and $\bar{f}$. Both these amplitudes are therefore of the second order of smallness when compared with the large amplitude $\langle\Phi'|\, A^{\dagger}_{hp}\, |\Phi'_0\rangle$, and one order smaller than $\langle\Phi'|\, A_{hp}\, |\Phi'_0\rangle$. Therefore, the natural extension of the TDA method is to construct a set of equations in which the amplitudes $\langle\Phi'|\, A^{\dagger}_{hp}\, |\Phi'_0\rangle$ and $\langle\Phi'|\, A_{hp}\, |\Phi'_0\rangle$ are retained while amplitudes of the type $\langle\Phi'|\, \hat{N}_{i'i}\, |\Phi'_0\rangle$ are neglected. For simplicity, we shall omit in our notation the primes on the corrected ground and excited states, and denote them by Ψ_0 and Ψ.

The algebraic procedure, once again, is to evaluate the commutator of H with $A^{\dagger}_{hp}$, and sandwich the result between $\langle\Psi|$ and $|\Psi_0\rangle$. From the resultant expression, we have to drop all the terms that are of a smaller order of magnitude than the amplitude $\langle\Psi|\, A_{hp}\, |\Psi_0\rangle$. In order to quickly recognize the terms that will be finally ignored, we carry out the following book-keeping type of work by suppressing the hole-particle subscripts on all $A^{\dagger}$- and A-operators. The $\hat{N}$-operators carry either two hole or two particle labels, and these too will be suppressed. An exception is made in the case of the hole-hole and particle-particle part of H_{22} [the first line of (VI.104e)], where the subscripts are explicitly retained on $A^{\dagger}$ and A, because, in this case, these operators will carry *two* particle or *two* hole labels, as distinguished from $A^{\dagger}_{hp}$ and A_{hp}.

The terms H_{40} and H_{04}, as given by (VI.104a) and (VI.104b), have the structures

$$H_{40} \rightarrow A^{\dagger}A^{\dagger}, \tag{VI.125a}$$

$$H_{04} \rightarrow AA. \tag{VI.125b}$$

Similarly, H_{31} and H_{13} of (VI.104c) and (VI.104d) have the schematic composition

$$H_{31} \rightarrow A^{\dagger}\hat{N}, \tag{VI.125c}$$

$$H_{13} \rightarrow \hat{N}A. \tag{VI.125d}$$

The hole-particle part of H_{22} has already been explicitly written down in (VI.122b) and has the structure

$$H_{22}\ (\text{hole-particle}) \rightarrow A^{\dagger}A. \tag{VI.125e}$$

In contrast to this, the hole-hole and particle-particle part of H_{22} [the first line of (VI.104e)] appear as

$$H_{22}\ (\text{hole-hole}) \rightarrow A^{\dagger}_{h_1h_2}A_{h_3h_4}, \tag{VI.125f}$$

$$H_{22}\ (\text{particle-particle}) \to A^{\dagger}_{p_1p_2}A_{p_3p_4}. \qquad \text{(VI.125g)}$$

We next use the commutators (VI.119) and (VI.121a), and obtain from (VI.125a) to (VI.125e) the schematic expressions

$$[H_{40}, A^{\dagger}] \to 0, \qquad \text{(VI.126a)}$$

$$[H_{04}, A^{\dagger}] \to A + \hat{N}A + A\hat{N}, \qquad \text{(VI.126b)}$$

$$[H_{31}, A^{\dagger}] \to A^{\dagger}A^{\dagger}, \qquad \text{(VI.126c)}$$

$$[H_{13}, A^{\dagger}] \to A^{\dagger}A + \hat{N} + \hat{N}\hat{N}, \qquad \text{(VI.126d)}$$

$$[H_{22}\ (\text{hole-particle}), A^{\dagger}] \to A^{\dagger} + A^{\dagger}\hat{N}. \qquad \text{(VI.126e)}$$

Evaluating the commutators of (VI.125f) and (VI.125g) with $A^{\dagger}$ requires a little extra work. First, we have to evaluate the commutators of $A_{h_3h_4}$ and $A_{p_3p_4}$ with $A^{\dagger}_{hp}$, which can be done in a manner analogous to (VI.117b). This work is left as an exercise. Finally, we obtain

$$[H_{22}\ (\text{hole-hole}), A^{\dagger}] \to A^{\dagger}\hat{N}, \qquad \text{(VI.126f)}$$

$$[H_{22}\ (\text{particle-particle}), A^{\dagger}] \to A^{\dagger}\hat{N}. \qquad \text{(VI.126g)}$$

If we now put $\langle\Psi|$ to the left and $|\Psi_0\rangle$ to the right of (VI.126), we obtain, besides the amplitudes $\langle\Psi|\, A^{\dagger}\, |\Psi_0\rangle$ and $\langle\Psi|\, A\, |\Psi_0\rangle$, the quantities

$$\langle\Psi|\, \hat{N}\, |\Psi_0\rangle, \quad \langle\Psi|\, \hat{N}\hat{N}\, |\Psi_0\rangle, \quad \langle\Psi|\, \hat{N}A\, |\Psi_0\rangle,$$

$$\langle\Psi|\, A\hat{N}\, |\Psi_0\rangle, \quad \langle\Psi|\, A^{\dagger}\hat{N}\, |\Psi_0\rangle, \quad \langle\Psi|\, A^{\dagger}A\, |\Psi_0\rangle, \quad \langle\Psi|\, A^{\dagger}A^{\dagger}\, |\Psi_0\rangle.$$

It has already been remarked after (VI.124) that $\langle\Psi|\, \hat{N}\, |\Psi_0\rangle$ is of a smaller order than $\langle\Psi|\, A\, |\Psi_0\rangle$. To examine the other amplitudes which are bilinear in the operators $A^{\dagger}$, A, and $\hat{N}$, we follow the practice of introducing a complete set of states $\sum_n |\Psi_n\rangle\langle\Psi_n|$ between the two operators. As a demonstration of the type of arguments that follow this procedure, let us consider the case of $\langle\Psi|\, A^{\dagger}\hat{N}\, |\Psi_0\rangle$. After the introduction of the intermediate states, this term yields

$$\langle\Psi|\, A^{\dagger}\hat{N}\, |\Psi_0\rangle = \sum_n \langle\Psi|\, A^{\dagger}\, |\Psi_n\rangle\langle\Psi_n|\, \hat{N}\, |\Psi_0\rangle. \qquad \text{(VI.127)}$$

We recall that by our choice of notation the present Ψ_0 is actually the same as Φ_0' of (VI.107a), and the first term in that equation represents the states with no hole or no particle. Thus, the second factor $\langle\Psi_n|\, \hat{N}\, |\Psi_0\rangle$ of (VI.127) gets, with the choice of the two-hole two-particle states for Ψ_n, a nonvanishing contribution only from the second term of (VI.107a). In other words, this factor of (VI.127) was already linear in the f-coefficients of (VI.107a), and hence of the same order as $\langle\Psi|\, A\, |\Psi_0\rangle$. The other factor $\langle\Psi|\, A^{\dagger}\, |\Psi_n\rangle$ of (VI.127) under the same circumstances (i.e., for Ψ_n equal to a two-hole two-particle state) is extremely small (in fact, zero in the lowest approximation), because Ψ is predominantly a linear combination of one-hole one-particle type states. Thus, the term $A^{\dagger}\hat{N}$ in the commutator of H with $A^{\dagger}_{hp}$ can be ignored in comparison with the $A^{\dagger}$- and A-term. In the same manner, it is possible to argue that all the other bilinear terms in $A^{\dagger}$, A, $\hat{N}$, with the exception of $\langle\Psi|\, A\hat{N}\, |\Psi_0\rangle$ and $\langle\Psi|\, A^{\dagger}A^{\dagger}\, |\Psi_0\rangle$, can be ignored. We shall now see that the contribution of $\langle\Psi|\, A\hat{N}\, |\Psi_0\rangle$ is kept in the RPA theory, whereas that of $\langle\Psi|\, A^{\dagger}A^{\dagger}\, |\Psi_0\rangle$ is usually disregarded.

71 (45-122/1977)

Going back to (VI.126), we conclude that, in the RPA derivation, only the commutator with H_{04} and the $A^{\dagger}$-term in the commutator of H_{22} (hole-particle) need be worked out in detail. In fact, the latter term has already been carefully evaluated in the TDA theory; hence, all we need to do here is evaluate (VI.126b) in detail. For this purpose, we go back to (VI.104b) which we write as

$$H_{04} = -\tfrac{1}{4} \sum_{h'p'} \sum_{h''p''} (p''p'|\, V\, |\bar{h}'\bar{h}'')A_{h'p'}A_{h''p''}$$
$$= \tfrac{1}{4} \sum_{h'p'} \sum_{h''p''} (p''p'|\, V\, |\bar{h}''\bar{h}')A_{h'p'}A_{h''p''}. \qquad \text{(VI.128)}$$

Next, we evaluate its commutator with $A^{\dagger}_{hp}$, using (VI.119). If we decide to keep only the terms containing one A-operator, we omit the last two terms of (VI.119) and obtain

$$[H_{04}, A^{\dagger}_{hp}] = \tfrac{1}{2} \sum_{h'p'} (pp'|\, V\, |\bar{h}\bar{h}')A_{h'p'}. \qquad \text{(VI.129)}$$

This result is, however, not quite right. The factor $\frac{1}{2}$ is actually absent in a more careful derivation. The source of this error lies in the use of the quasi-boson commutation relation [the first term of (VI.119)]. In fact, if we had used the entire expression (VI.119), we would have obtained terms with the structure of the last two terms of (VI.126b).

We shall now show that terms of the type $\langle\Psi|\, A\hat{N}\, |\Psi_0\rangle$ have some $\langle\Psi|\, A\, |\Psi_0\rangle$ concealed in them. The terms that have to be added to the right-hand side of (VI.129) are obtained from (VI.128) and the last two terms of (VI.119). These are given by

$$-\tfrac{1}{4} \sum_{h'p'} \sum_{h''p''} (p''p'|\, V\, |\bar{h}''h')[A_{h'p'}(\delta_{pp''}\hat{N}_{hh''} + \delta_{hh''}\hat{N}_{pp''}) + A_{h''p''}(\delta_{pp'}\hat{N}_{hh'} + \delta_{hh'}\hat{N}_{pp'})]. \qquad \text{(VI.130)}$$

Let us examine the first term, $A_{h'p'}\hat{N}_{hh''} \equiv b_{h'}b_{p'}b^{\dagger}_{h}b_{h''}$. When we insert $\langle\Psi|$ and $|\Psi_0\rangle$ to the left and the right and $\sum_n |\Psi_n\rangle\langle\Psi_n|$ in the middle, it *appears* that this term is always of a smaller order of magnitude than $\langle\Psi|\, A_{h'p'}\, |\Psi_0\rangle$, because of the second factor $\langle\Psi_0|\, b^{\dagger}_{h}b_{h''}\, |\Psi_0\rangle$ (we are here considering the case $\Psi_n = \Psi_0$). It is, however, not apparent that there is a possibility of obtaining a nonvanishing result for the case $\Psi_n = \Psi$ as well. To see this, all that we need is to rewrite, by permuting operators,

$$A_{h'p'}\hat{N}_{hh''} \equiv b_{h'}b_{p'}b^{\dagger}_{h}b_{h''} = -b_{h'}b^{\dagger}_{h}b_{p'}b_{h''} = b_{h'}b^{\dagger}_{h}b_{h''}b_{p'}. \qquad \text{(VI.131)}$$

With the choice $\Psi_n = \Psi$, this expression yields $\langle\Psi|\, b_{h'}b^{\dagger}_{h}\, |\Psi\rangle\langle\Psi|\, b_{h''}b_{p'}\, |\Psi_0\rangle$. The first factor is large when $h = h'$ (actually equal to $\delta_{hh'}$ when Ψ is normalized to unity), whereas the second factor is the amplitude $\langle\Psi|\, A_{h''p'}\, |\Psi_0\rangle$. This establishes the assertion we made regarding the concealed presence of $\langle\Psi|\, A\, |\Psi_0\rangle$ in $\langle\Psi|\, A\hat{N}\, |\Psi_0\rangle$. The second term of (VI.130) can also be similarly handled to yield an amplitude of the type $\langle\Psi|\, A\, |\Psi_0\rangle$. An analogous attempt with the third and fourth terms of (VI.130), however, fails to yield anything. In all this derivation, we have consistently to make the assumption that $\langle\Psi|\, \hat{N}\, |\Psi\rangle$, which is the probability of getting a particle or a hole in the excited state, is of an order of magnitude smaller than $\langle\Psi|\, A^{\dagger}\, |\Psi_0\rangle$. The details of the algebra indicated, to be applied to the first two terms of (VI.130), are left as an exercise. With this extra contribution, we finally obtain

$$[H_{04}, A^{\dagger}_{hp}] = \sum_{h'p'} (pp'|\, V\, |\bar{h}\bar{h}')A_{h'p'}. \qquad \text{(VI.132)}$$

Putting together (VI.123) and (VI.132), we have

$$[H, A_{hp}^\dagger] = (\epsilon_p - \epsilon_h)A_{hp}^\dagger - \sum_{h'p'} (p'\bar{h}| V |p\bar{h}')A_{h'p'}^\dagger + \sum_{h'p'} (pp'| V |\bar{h}\bar{h}')A_{h'p'}. \quad \text{(VI.133a)}$$

Taking the Hermitean conjugate and reversing the sign, we immediately obtain

$$[H, A_{hp}] = -(\epsilon_p - \epsilon_h)A_{hp} + \sum_{h'p'} (p\bar{h}'| V |p'\bar{h})A_{h'p'} - \sum_{h'p'} (\bar{h}\bar{h}'| V |pp')A_{h'p'}^\dagger. \quad \text{(VI.133b)}$$

In the usual way, the system of equations (VI.133) gives rise to a set of linear equations for the amplitudes $\langle\Psi| A_{hp}^\dagger |\Psi_0\rangle \equiv x_{hp}$ and $\langle\Psi| A_{hp} |\Psi_0\rangle \equiv y_{hp}$. The matrix to be diagonalized is clearly of the form (VI.28b), with the definitions of A and B given by

$$A_{hp;\,h'p'} = (\epsilon_p - \epsilon_h)\delta_{hh'}\delta_{pp'} - (p\bar{h}'| V |p'\bar{h}), \quad \text{(VI.134a)}$$

$$B_{hp;\,h'p'} = (pp'| V |\bar{h}\bar{h}'). \quad \text{(VI.134b)}$$

Except for the phase factor, and the sign of the projection quantum numbers of the hole states, these definitions agree with (VI.23). This slight difference in the hole states originates from the definition $b_h^\dagger = s_h C_{-h}$.

C. THEORY OF VIBRATION IN BCS CASE

The treatment of the quasiparticle interaction given in Section 42A in the case of a BCS type nucleus is now a straightforward generalization of what we have done in Section 42B.

Definitions

In analogy with the hole-particle pair and the number operators of Section 42B, we now introduce the quasiparticle pair and number operators

$$A_{mn}^\dagger = b_n^\dagger b_m^\dagger, \qquad A_{mn} = b_m b_n, \qquad \hat{N}_{mn} = b_m^\dagger b_n. \quad \text{(VI.135)}$$

Using the basic anticommutation rules, we obtain

$$[b_i, b_n^\dagger b_m^\dagger] = \delta_{in} b_m^\dagger - \delta_{im} b_n^\dagger. \quad \text{(VI.136)}$$

By a straightforward application of (VI.136), we then derive the commutator relations

$$\begin{aligned}[A_{ij}, A_{mn}^\dagger] &= [b_i b_j, b_n^\dagger b_m^\dagger] \\ &= b_i[b_j, b_n^\dagger b_m^\dagger] + [b_i, b_n^\dagger b_m^\dagger]b_j \\ &= \delta_{jn} b_i b_m^\dagger - \delta_{jm} b_i b_n^\dagger + \delta_{in} b_m^\dagger b_j - \delta_{im} b_n^\dagger b_j \\ &= \delta_{jn} b_i b_m^\dagger - \delta_{jm} b_i b_n^\dagger + \delta_{in}\hat{N}_{mj} - \delta_{im}\hat{N}_{nj} \qquad \text{(VI.137a)} \\ &= (\delta_{im}\delta_{jn} - \delta_{in}\delta_{jm}) + \delta_{jm}\hat{N}_{ni} - \delta_{jn}\hat{N}_{mi} + \delta_{in}\hat{N}_{mj} - \delta_{in}\hat{N}_{nj}. \qquad \text{(VI.137b)}\end{aligned}$$

The quasi-boson approximation entails dropping all the terms containing the number operator in (VI.137b). *In this approximation*, we have, by virtue of the Krönecker deltas within the parentheses in (VI.137b),

$$[A_{ij}, A_{ij}^\dagger] = 1, \qquad [A_{ij}, A_{ji}^\dagger] = -1. \quad \text{(VI.138)}$$

From the definition (VI.135), we have

$$A_{ji}^\dagger = -A_{ij}^\dagger,$$

and hence the two equations in (VI.138) are one and the same. Therefore, in spite of the *two* terms with the Krönecker deltas in (VI.137b), the operators A_{ij}, $A^{\dagger}_{ij}$ indeed behave as boson operators under the quasi-boson approximation.

Once again, applying (VI.136), we can derive the commutator

$$[\hat{N}_{ij}, A^{\dagger}_{mn}] = \delta_{jn}A^{\dagger}_{mi} - \delta_{jm}A^{\dagger}_{ni}. \tag{VI.139}$$

Finally, we write down the obvious results

$$[A^{\dagger}_{ij}, A^{\dagger}_{mn}] = [A_{ij}, A_{mn}] = 0. \tag{VI.140}$$

TDA Equations

In the manner of Section 42B, we evaluate the commutator of the pair creation operator $A^{\dagger}_{mn}$ with H_{11} and H_{22}. Using (VI.139), we have

$$[H_{11}, A^{\dagger}_{mn}] = \sum_{i} E_i[\hat{N}_{ii}, A^{\dagger}_{mn}] = (E_m + E_n)A^{\dagger}_{mn}, \tag{VI.141}$$

where E_m is the energy of the quasiparticle m.

The operator in H_{22} is clearly $A^{\dagger}_{ji}A_{lk}$, and its commutator with $A^{\dagger}_{mn}$, according to the quasi-boson commutation relation, is given by

$$\begin{aligned}[A^{\dagger} A_{lk}, A^{\dagger}_{mn}] &= A^{\dagger}_{ji}[A_{lk}, A^{\dagger}_{mn}] \\ &= A^{\dagger}_{ji}(\delta_{lm}\delta_{kn} - \delta_{ln}\delta_{km}).\end{aligned} \tag{VI.142}$$

Using (VI.142), together with the coefficient in (VI.104e), we obtain

$$\begin{aligned}[H_{22}, A^{\dagger}_{mn}] = \tfrac{1}{2} \sum_{ij} [&(u_iu_ju_mu_n + v_iv_jv_mv_n)(ij| V |nm) \\ &+ u_iv_j\{u_nv_m(j\bar{n}| V |m\bar{i}) - u_mv_n(j\bar{m}| V |n\bar{i})\} \\ &+ v_iu_j\{u_mv_n(j\bar{n}| V |m\bar{i}) - u_nv_m(j\bar{m}| V |n\bar{i})\}]A^{\dagger}_{ij}.\end{aligned} \tag{VI.143}$$

By interchanging the labels i, j and using $A^{\dagger}_{ji} = -A^{\dagger}_{ij}$ together with the time-reversal invariance of the matrix elements of V, we can easily establish that in (VI.143) (i) the first term in the third line is equal to the second term in the second line, and (ii) the first term in the second line is equal to the second term in the third line. Therefore, we retain the second terms in the second and third lines of (VI.143) and delete the factor $\frac{1}{2}$ from these two lines. Thus,

$$[H_{22}, A^{\dagger}_{mn}] = \sum_{ij} [(u_iu_ju_mu_n + v_iv_jv_mv_n)\tfrac{1}{2}(ij| V |mn) - (u_iv_ju_mv_n + v_iu_jv_mu_n)(j\bar{m}| V |n\bar{i})]A^{\dagger}_{ij}. \tag{VI.144}$$

(VI.141) plus (VI.144) give rise to a linear set of equations for the amplitudes $\langle\Psi| A^{\dagger}_{ij} |\Psi_0\rangle$. The matrix to be diagonalized is therefore the coefficient matrix on the right-hand side of (VI.144) together with the quasiparticle pair energies $(E_m + E_n)$ of (VI.141) along the leading diagonal. As in the hole-particle case of Section 42B, this TDA calculation is identical to the diagonalization of the unperturbed Hamiltonian plus the quasiparticle interaction in the space of all the quasiparticle pair states of the type $A^{\dagger}_{mn}|\Psi_0\rangle$. Finally, in perfect analogy with the TDA theory of Section 42B, the use of the quasi-boson commutation rule in writing (VI.142) is justified by the fact that the remaining terms of (VI.137b) produce zero when acting on the BCS state $|\Psi_0\rangle$.

RPA Equations

As in Section 42B, the part H_{40} of the quasiparticle interaction can mix four quasiparticle states into the BCS ground state Ψ_0. The two quasiparticle excited states can also mix with the four quasiparticle states via H_{31}. A higher TDA calculation can therefore be done by mixing zero, two, and four quasiparticle states. An alternative way to treat the effect of correlation in the BCS ground state is the RPA method, which admits two types of amplitudes $\langle\Psi| A^\dagger_{mn} |\Psi_0\rangle$ and $\langle\Psi| A_{mn} |\Psi_0\rangle$, Ψ_0 being the correlated ground state.

By arguments similar to that presented in Section 42B, we can show that the RPA treatment, in the present case also, requires the detailed evaluation of only one more commutator, namely, that of H_{04} with $A^\dagger_{mn}$.

According to (VI.104b), the expression for H_{04} is given by

$$H_{04} = \tfrac{1}{4} \sum_{ijkl} (k, \bar{l}| V |ij) u_i u_j v_l v_k A_{kl} A_{ji}. \tag{VI.145}$$

As in Section 42B, if we keep only the quasi-boson terms of (VI.137b), we obtain

$$[H_{04}, A^\dagger_{mn}] = \tfrac{1}{2} \sum_{kl} (\bar{k}\bar{l}| V |nm)(u_m u_n v_k v_l + v_m v_n u_k u_l) A_{kl}. \tag{VI.146a}$$

This, however, is not the correct result because the other terms in (VI.137b) give rise to terms in the commutator of the type $A\hat{N}$ and $\hat{N}A$ of which the former yields some extra A_{kl} type terms in the final result. Terms of the type $\hat{N}A$, on the other hand, yield $\langle\Psi| \hat{N} |\Psi\rangle\langle\Psi| A |\Psi_0\rangle$ which is of a smaller order of magnitude than $\langle\Psi| A |\Psi_0\rangle$ because the factor $\langle\Psi| \hat{N} |\Psi\rangle$, representing the probability of getting a *given* quasiparticle in the excited state Ψ, must be very small compared with unity (Ψ being a coherent superposition of *many* quasiparticle pair states in the lowest approximation). Therefore, we write only the extra terms of the type $A\hat{N}$, and derive the extra contribution to the right-hand side of (VI.146a) as

$$\tfrac{1}{4} \sum_{ijkl} (\bar{k}\bar{l}| V |ij) u_i u_j v_l v_k A_{kl}(\delta_{jm}\hat{N}_{nl} - \delta_{jn}\hat{N}_{ml} + \delta_{ln}\hat{N}_{mj} - \delta_{im}\hat{N}_{nj}). \tag{VI.146b}$$

For example, the first term of (VI.146b) yields, when the summation indices k and l, in turn, are equal to n, the result

$$A_{kl}\hat{N}_{nl} = b_k b_l b^\dagger_n b_i \rightarrow \delta_{ln} b_k b_i + \delta_{kn} b_i b_l. \tag{VI.146c}$$

It should be noticed that, when $l = n$, k is necessarily not equal to l or n (because if $k = l$, $b_k b_l = 0$), and hence b_k can be taken to the right of $b^\dagger_n$; the same observation holds for the case $k = n$, when b_l can be moved to the right of $b^\dagger_n$ without trouble. In fact, in (VI.146c), we have substituted $b_l b^\dagger_n$ and $b_k b^\dagger_n$ in the two cases by δ_{ln} and δ_{kn} and omitted the other terms, namely, $-b^\dagger_n b_l$ and $-b^\dagger_n b_k$. These terms, together with the other two destruction operators, have the structure $\hat{N}A$, and we have already argued that they lead to terms smaller in order of magnitude than $\langle\Psi| A |\Psi_0\rangle$. The type of work carried out in (VI.146c) can be repeated for all the terms in (VI.146b) which then yields a new set of A_{kl}-terms. Adding these to (VI.146a), we finally obtain

$$[H_{04}, A^\dagger_{mn}] = \sum_{k,l} [\tfrac{1}{2}(\bar{k}, \bar{l}| V |nm)(u_m u_n v_k v_l + v_m v_n u_k u_l) + (\bar{k}\bar{n}| V |lm) u_m v_n v_k u_l$$
$$- (\bar{k}\bar{m}| V |ln) v_m u_n v_k u_l] A_{kl}. \tag{VI.147}$$

(VI.141), (VI.144), and (VI.147) together give the commutator $[H, A^{\dagger}_{mn}]$ in the RPA approximation. The commutator $[H, A_{mn}]$ can then be written by taking the Hermitean conjugate and reversing the sign. Together they provide us with the required system of linear equations for the amplitudes $\langle\Psi| A^{\dagger}_{kl} |\Psi_0\rangle$ and $\langle\Psi| A_{kl} |\Psi_0\rangle$. The task of angular momentum coupling is tackled in Appendix G (Section III); the transition probability of the TDA and RPA states is also calculated there. For a complete understanding of these methods, this particular appendix *must* be read.

43. REVIEW OF APPLICATIONS OF THEORY OF VIBRATION

In this section, we shall review some of the calculations based on the theory developed in the preceding discussion. The purpose of this treatment being pedagogic, no attempt is made to survey the existing literature in a *complete* and comprehensive manner. In conformity with the attitude adopted in the introduction to this chapter, the material is divided into three categories: (i) hole-particle states and their coherent superposition in closed-shell nuclei; (ii) quasi-particle states and their coherent superposition in spherical heavy nuclei, where pairing effects are important; and (iii) vibrational states in deformed nuclei, where too the pairing effects are quite significant. Some more details of the extension of the theory are also presented in the appropriate context while discussing specific applications.

A. CLOSED-SHELL NUCLEI

The nucleus that has been the testing ground of many theoretical calculations is O^{16}. Some calculations have been done for Ca^{40} and Pb^{208} as well.

The single-particle and single-hole energies needed in the calculation are invariably taken from the neighbouring odd nuclei (e.g., O^{17}, F^{17} and O^{15}, N^{15}). In the Ca^{40} region, it is very uncertain whether the low-lying levels of the neighbouring odd nuclei can ever be interpreted as single-particle or single-hole levels. To that extent, the treatment of Ca^{40} as a closed-shell nucleus is also doubtful. However, standard hole-particle diagonalization calculations for this nucleus have been carried out by several authors.

The other necessary ingredient for the calculations is the two-nucleon potential. In the earliest work on the subject, Elliott and Flowers[1] have used the Rosenfeld exchange mixture with a Yukawa-type radial-dependence:

$$V = \frac{V_0}{3}\frac{e^{-r/a}}{r/a}\boldsymbol{\tau}_1 \cdot \boldsymbol{\tau}_2(0.3 + 0.7\boldsymbol{\sigma}_1 \cdot \boldsymbol{\sigma}_2), \tag{VI.148}$$

where $\boldsymbol{\tau}$ and $\boldsymbol{\sigma}$ are respectively the isospin and spin operators of the nucleon labelled by the subscript. These authors have used the *LS*-coupling representation for their calculation, and hence the spin-orbit coupling of the shell-model potential, namely, $\xi \boldsymbol{l} \cdot \mathbf{s}$, was also explicitly used as a part of the interaction while doing the diagonalization. The radial integrals of ξ were, however, chosen to reproduce the experimentally observed single-particle splitting of $(d_{5/2})$- and $(d_{3/2})$-level in $O^{17}(F^{17})$ and the splitting of the hole levels $p_{3/2}$ and $p_{1/2}$ in $N^{15}(O^{15})$.

The schematic δ-function potential described in Appendix G (Section III) has also been used with various types of exchange-dependence by several authors. One of the favourite formulas is the Soper mixture

$$0.3 + 0.43P_M + 0.27P_B, \tag{VI.149}$$

where P_M and P_B are respectively the Majorana and Bartlett exchange operators.

The most extensive calculations have been performed by Gillet et al[2] and Green[3] who determined the potential by the least-squares fit to several levels. Their potential has a Gaussian shape and is given by

$$V_0 \exp[-(r/\mu)^2](W + BP_B + HP_H + MP_M), \tag{VI.150a}$$

where P_H is the Heisenberg exchange operator. The values of the various parameters, as determined by the least-squares fit, are given by

$$M - W = 0, \quad M + W - B - H = 0.4, \quad H = 0.4, \tag{VI.150b}$$

$$V_0 = -40 \text{ MeV}, \quad \mu/b = 1.0, \tag{VI.150c}$$

where b is the harmonic oscillator parameter, and was chosen to be 1.68 fm from the observed root-mean-square radius of O^{16}. The four parameters W, M, B, H obey the normalization

$$W + M + B + H = 1. \tag{VI.150d}$$

In recent years, effective shell-model matrix elements of the Tabakin potential, and the hard-core Hamada-Johnston (HJ) potential have also been evaluated. Vinh-Mau has used the Tabakin potential to calculate the giant resonance of O^{16} (J. P. Svenne, private communication). Brown has referred to the calculation on the same states of Pb^{208}, using the effective matrix elements of the HJ potential [see Proceedings of the Dubna Symposium on Nuclear Structure (1968), IAEA].

The structure of the various levels of O^{16} that has emerged from all these calculations, and also from experimental observations, is summarized in Table VI.1. It is clear from this table that there are many excited levels in this nucleus which are not of the 1h-1p type. The experimental evidence on the deformation of some of these states and the several rotational bands

Table VI.1 States observed in O^{16}

Energy (MeV)	J	T	Proposed Description
6.06	0^+	0	Deformed
6.14	3^-	0	1h-1p
6.92	2^+	0	Rotational on 6.06 MeV state
7.12	1^-	0	1h-1p
8.88	2^-	0	3h-3p
9.59	1^-	0	3h-3p (deformed)
9.85	2^+	0	Deformed
10.36	4^+	0	Rotational on 6.06 MeV state
10.95	0^-	0	1h-1p
11.26	0^+	0	Deformed
11.62	3^-	0	Rotational on 9.59 MeV state
12.43	1^-	0	1h-1p
12.52	2^-	0	1h-1p
12.78	0^-	1	1h-1p
12.96	2^-	1	1h-1p
13.10	1^-	1	1h-1p

Table VI.1 States observed in O^{16} (cont.)

Energy (MeV)	J	T	Proposed Description
13.26	3^-	1	1h-1p
13.98	2^-	0	
16.3	0^-	0	
17.3	1^-	1	1h-1p
19.5	$2^{+,-}$		
22.4	1^-	1	1h-1p
24.5	1^-	1	1h-1p

Source: Green, A. M., *Rep. Prog. Phys.*, **28**, 113 (1965).

they form are summarized in Fig. VI.11, which is taken from Carter et al[4]. In this figure, the solid and hollow circles correspond to positive parity levels, and the rectangles stand for odd-parity levels. Actually, the 1h-1p type calculations that we are now describing do not apply to the levels of this category. Setting them aside, we are left with the ($T = 1$)- and ($T = 0$)-state of $J^\pi = 1^-, 2^-, 3^-$. With the exception of the 2^+ state at 19.5 MeV, all the other positive parity states listed in the table belong to the rotational bands and need a different kind of treatment.

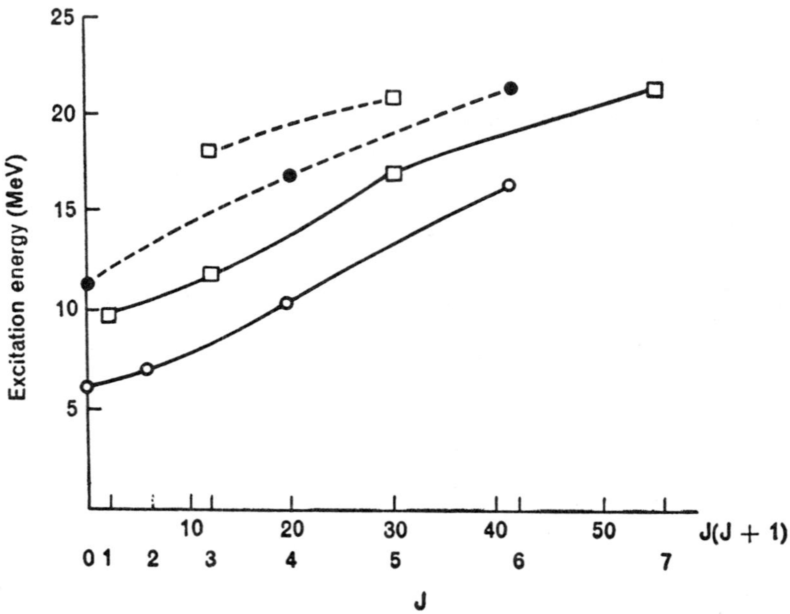

Fig. VI.11 Experimental results on rotational bands in O^{16}. [Following Green, A. M., *Rep. Prog. Phys.*, **28**, 135 (1965).]

Amongst the most notable deformed odd-parity states are 9.59 MeV (1^-) and 11.62 MeV (3^-), which are regarded as the first two levels of a rotational band. The other notable odd-parity level in this table that does not conform to the 1h-1p picture is the state 8.88 MeV (2^-); there is direct experimental evidence that this state cannot be excited in inelastic proton scattering which rules out its having a 1h-1p type structure.

We shall now discuss the states $J = 1^-, 2^-, 3^-$ of $T = 1$ and $T = 0$, which are fairly well-described by a 1h-1p type TDA and an RPA calculation. The general observations are: (i) both the TDA and RPA type theories explain the observed energies equally well, provided we adjust the overall strength V_0 of the two-nucleon interaction in the two cases separately; (ii) the agreement in the transition probability is better with the RPA theory in the cases of the collective states 3^- $(T = 0)$ and 1^- $(T = 1)$, which occur at 6.14 MeV (3^-) and 22–25 MeV (1^-); for the other odd-parity states, which do not show a collective enhancement of the transition probability, there is not much difference between the TDA and RPA predictions; and (iii) the agreement in energy in the hole-particle calculation is somewhat better for the $(T = 1)$-states than for the $(T = 0)$-states.

There are two important discrepancies in the results of the calculations on the dipole states. An experimental situation is shown in Fig. VI.12, where the ordinate gives the (γ, n)

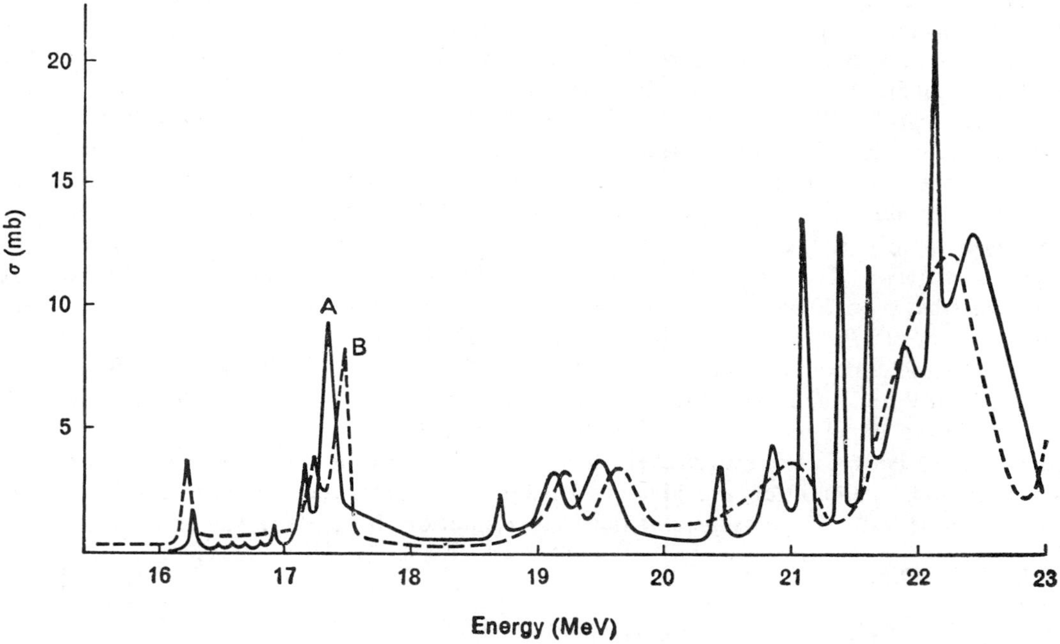

Fig. VI.12 Experimental results on giant dipole states in O^{16}. [Following Green, A. M., *Rep. Prog. Phys.*, **28**, 127 (1965).]

cross-section (curve A) and the (γ, p) cross-section (curve B) in millibarns. Most of the dipole cross-section is found to be located in the energy range 21–25 MeV. Theoretically, we get two strong dipole states in this energy region from the following mechanism. Let us consider an idealized dipole state $\mathbf{D}|\Phi_0\rangle$, where Φ_0 is the ground-state wavefunction, and $\mathbf{D}$ the dipole operator

$$\mathbf{D} = \sum_{p=1}^{Z} (\mathbf{r}_p - \mathbf{R}) = \frac{N}{A} \sum_{p=1}^{Z} \mathbf{r}_p - \frac{Z}{A} \sum_{n=1}^{N} \mathbf{r}_n. \tag{VI.151a}$$

72 (45-122/1977)

Here **R** is the coordinate of the centre-of-mass of the nucleus, i.e., $\mathbf{R} = A^{-1} \sum_{i=1}^{A} \mathbf{r}_i$; p denotes a proton and n a neutron; N, Z, and A are respectively the number of neutrons, protons, and nucleons. In the special case of $N = Z$ (for instance, O^{16} and Ca^{40}), (VI.151a) reduces to

$$\mathbf{D} = -\tfrac{1}{2} \sum_{i=1}^{A} \mathbf{r}_i \tau_3(i), \tag{VI.151b}$$

τ_3 being the third component of isospin of a nucleon. In analogy with the state $D|\Phi_0\rangle$, we can also consider an idealized state $\mathbf{D}_\sigma|\Phi_0\rangle$, where $\mathbf{D}_\sigma$ is given by

$$\mathbf{D}_\sigma = -\tfrac{1}{2} \sum_{i=1}^{A} (\mathbf{r}_i, \boldsymbol{\sigma}_i)^1 \tau_3(i). \tag{VI.151c}$$

Here $\boldsymbol{\sigma}$ is the nucleon spin operator and $(\mathbf{r}, \boldsymbol{\sigma})^1$ is the tensor of rank 1, obtained by compounding $\mathbf{r}$ with $\boldsymbol{\sigma}$ (it is actually proportional to $\mathbf{r} \times \boldsymbol{\sigma}$). Acting on the ground state Φ_0, $\mathbf{D}_\sigma$ generates a state of total angular momentum $J = 1$, parity $\pi = -1$, and the total intrinsic spin $S = 1$ [by virtue of the presence of the spin vector $\boldsymbol{\sigma}$ in (VI.151c) and the fact that Φ_0 has $S = 0$]. We therefore refer to the state $\mathbf{D}_\sigma|\Phi_0\rangle$ as the spin-flip dipole state. It is clear that if $\mathbf{D}|\Phi_0\rangle$ is an exact eigenstate of the Hamiltonian, then all the dipole transition probability is exhausted by this state; and all the other states, by virtue of their orthogonality with this state, have zero E1-transition probability to the ground state. However, in a more realistic situation (i.e., for the exact nuclear Hamiltonian), this *idealized* state may be shared between several *actual* states, all these states then producing a nonvanishing E1-transition probability. In particular, the spin-flip dipole state $\mathbf{D}_\sigma|\Phi_0\rangle$ can mix fairly well with $\mathbf{D}|\Phi_0\rangle$ through the spin-orbit coupling; we then expect the dipole transition probability to be large for the two states that are mixtures of $\mathbf{D}|\Phi_0\rangle$ and $\mathbf{D}_\sigma|\Phi_0\rangle$. This idea of a splitting of the giant dipole resonance into two strong components between 20 MeV and 25 MeV was first proposed by Ferrell, and later substantiated by Fallieros et al[5], Pal and Lee[5], and Lee[5]. In the detailed calculation by Elliott and Flowers[1] and the schematic model calculations of other authors, the splitting into two strong peaks, one around 22 MeV and the other around 25 MeV, has been well-established. However, one of the discrepancies in experimental data, already mentioned, has reference to the distribution of dipole strength between these two levels. The results of the Elliott-Flowers calculation and the schematic model calculation are given in the first line and second line, respectively, of each datum in Table VI.2. The closeness of the two calculations is obvious. The experimental values

Table VI.2 Dipole states of O^{16}

Calculated Energy	13.1	17.3	20.4	22.6	25.2
	13.7	17.6	20.0	22.2	25.0
Transition Strength (%)	0	1	0	67	32
	1	1	1	68	29

of the strength for the energy region around 22 MeV (the integrated area under the curve representing strength as a function of energy) and for the region around 25 MeV are more in the ratio 1 : 1 in contrast to the calculated ratio 2 : 1.

The other discrepancy concerns the dipole sum rule. We shall derive the sum rule before discussing the discrepancy. The transition strength $B(EL)$ for the multipole transition of order L (in the present case, $L = 1$) is defined by

$$B_{J_i \to J_f}(EL) = \frac{1}{2J_i + 1} \sum_{M_i, M_f} \sum_{M} |\langle \alpha J_f M_f | \Omega_M^L | J_i M_i \rangle|^2. \tag{VI.152}$$

Here (J_iM_i) are the ground-state angular momentum and its projection; Ω_M^L is the multipole operator, and $(\alpha J_f M_f)$ specify the state that is decaying by the multipole transition. Only two of the summation indices M_i, M_f, M are actually independent. $B(EL)$ is an experimentally measurable quantity, and sum rules can be derived by summing it over all the possible states (αJ_f). A more useful sum rule is obtained by weighting (VI.152) with the energy involved in the transition, i.e., $(E_f - E_i)$, and then carrying out the summation over (αJ_f). In this way,

$$\sum_{\alpha J_f} (E_f - E_i) B_{J_i \to J_f}(EL) = \frac{1}{2J_i + 1} \sum_{M_i, M} \sum_{\alpha J_f M_f} (E_f - E_i) |\langle \alpha J_f M_f | \Omega_M^L | J_i M_i \rangle|^2. \tag{VI.153}$$

The left-hand side can be evaluated from experimental data, and the right-hand side can be estimated from theoretical considerations as follows. The factor $(E_f - E_i)$ in (VI.153) can be automatically produced from the commutator of Ω_M^L or Ω_M^{L*} with H. Thus,

$$\begin{aligned}
&\sum_{\alpha J_f M_f} (E_f - E_i) |\langle \alpha J_f M_f | \Omega_M^L | J_i M_i \rangle|^2 \\
&= \sum_{\alpha J_f M_f} [\tfrac{1}{2}(E_f - E_i) \langle J_i M_i | \Omega_M^{L*} | \alpha J_f M_f \rangle \langle \alpha J_f M_f | \Omega_M^L | J_i M_i \rangle \\
&\quad - \tfrac{1}{2}(E_i - E_f) \langle J_i M_i | \Omega_M^{L*} | \alpha J_f M_f \rangle \langle \alpha J_f M_f | \Omega_M^L | J_i M_i \rangle] \\
&= \tfrac{1}{2} \sum_{\alpha J_f M_f} [\langle J_i M_i | \Omega_M^{L*} | \alpha J_f M_f \rangle \langle \alpha J_f M_f | (H\Omega_M^L - \Omega_M^L H) | J_i M_i \rangle \\
&\quad - \langle J_i M_i | (H\Omega_M^{L*} - \Omega_M^{L*} H) | \alpha J_f M_f \rangle \langle \alpha J_f M_f | \Omega_M^L | J_i M_i \rangle] \\
&= \tfrac{1}{2} \langle J_i M_i | \Omega_M^{L*}[H, \Omega_M^L] - [H, \Omega_M^{L*}]\Omega_M^L | J_i M_i \rangle.
\end{aligned} \tag{VI.154a}$$

The operator H produces E_f and E_i, by definition, when it operates on the states $(\alpha J_f M_f)$ and $(J_i M_i)$, respectively. The replacement of $(E_f - E_i)$ and $(E_i - E_f)$ by the commutators is thus explained. In the final step of (VI.154a), we have used the closure relation to carry out the summation over $(\alpha J_f M_f)$. Next, we recall that there is a sum over M in (VI.153) and, clearly,

$$\begin{aligned}
\sum_M [H, \Omega_M^{L*}]\Omega_M^L &\equiv \sum_M (-1)^M [H, \Omega_{-M}^L]\Omega_M^L \\
&= \sum_M (-1)^M [H, \Omega_M^L]\Omega_{-M}^L = \sum_M [H, \Omega_M^L]\Omega_M^{L*}.
\end{aligned} \tag{VI.154b}$$

All we have done here is replace the summation index label M by $-M$. Using (VI.154) in (VI.153), we finally obtain

$$\sum_{\alpha J_f} (E_f - E_i) B_{J_i \to (\alpha J_f)}(EL) = \frac{1}{2(2J_i + 1)} \sum_{M_i, M_f} \langle J_i M_i | [\Omega_M^{L*}, [H, \Omega_M^L]] | J_i M_i \rangle. \tag{VI.155}$$

If there is no velocity dependence and space-exchange in the nuclear potential, then $[H, \Omega_M^L]$ can be substituted with $[T, \Omega_M^L]$, where T is the kinetic energy operator. In the special case of dipole transition, Ω_M^L is given, according to (VI.151a), by

$$D_M = \frac{N}{A} \sum_{p=1}^{Z} r_M(p) - \frac{Z}{A} \sum_{n=1}^{N} r_M(n),$$

where the M-component of any vector $\mathbf{A}$ is defined by

$$A_0 = A_z, \qquad A_{\pm 1} = \mp \frac{1}{\sqrt{2}}(A_x \pm iA_y).$$

The kinetic energy operator T is equal to the sum over all the nucleons of the operator $[-\hbar^2/(2m)]\nabla^2$, m being the nucleon mass. Using the basic commutation relations between the components of momentum and coordinate, we can easily evaluate $\sum_M [D_M^*, [T, D_M]]$, which is equal to $(NZ/A)(\hbar^2/m)$. Thus, the theoretical estimate of the sum rules (VI.155), *with the approximation of no velocity dependence and space-exchange in the nuclear potential*, is given by

$$\sum_{\alpha J_f} (E_f - E_i) B_{J_i \to (\alpha J_f)}(EL) \approx \frac{NZ}{A} \frac{\hbar^2}{m}. \tag{VI.156}$$

The actual two-nucleon potential, contained in H, has some velocity dependence and space-exchange. This part of V gives a nonvanishing commutator with **D**, and hence extra contribution to the sum rule (VI.156). It is possible to make a reasonable estimate of this additional contribution. According to the hole-particle calculation of the giant resonance, we should therefore expect the sum on the left-hand side of (VI.156), carried over all the dipole states (which are all below 25–26 MeV), to be fairly close to the estimate just mentioned. In practice, the experimental value of the sum is about a half of the estimated sum rule, if the sum on the left-hand side is restricted to states up to 25–26 MeV. The remaining half of the sum rule is therefore contained in rather high-lying states. In the hole-particle calculation, however, we have no such states. A way out of this discrepancy is to extend the hole-particle calculation by admitting states with a larger number of hole-particle pairs. Such states mix with the dipole state and are very high-lying in energy. The little transition strength that they acquire in this way, when weighted with the factor $(E_f - E_i)$, can very well account for the transfer of about a half of the sum rule to states above 25–26 MeV.

The results for Ca^{40} and Pb^{208} will now be briefly described. Table VI.3 lists the states

Table VI.3 States observed in ^{40}Ca

Energy (MeV)	J^π	Proposed Description
3.35	0^+	Deformed
3.73	3^-	1h-1p
3.90	2^+	Rotational on 3.35 MeV state
4.48	5^-	1h-1p
5.27	3^- or 1^-	3h-3p basis of an odd-parity rotational band
5.62	4^+	Rotational on 3.35 MeV state
5.89	1^-	
6.29	3^-	Rotational on 5.27 MeV state and 1h-1p
6.56	3^-	
6.94	1^-	1h-1p
7.12	$(\geqslant 6^+)$	Rotational on 3.35 MeV state (?)
7.91	4^+	
8.09	2^+	
8.37	5^-	1h-1p or rotational on 5.27 MeV state

Source: Green, A. M., *Rep. Prog. Phys.*, **28**, 113 (1965).

of Ca^{40} and their structure. The situation here is very similar to O^{16}, namely, there are some deformed states and some states that can be explained in terms of the 1h-1p type configurations. In the case of Pb^{208}, there are two rather interesting results. (i) The energy of the octupole state (2.6 MeV) can be fairly well-reproduced only if we mix a large number (about 45) of hole-particle configurations having unperturbed energies up to 15 MeV. If, on the other hand, the number of basis states is cut down by restricting ourselves to a lower unperturbed energy (about 6 MeV), the energy of the collective octupole state is very unsatisfactory. This result is shown in Fig. VI.13, where the upper line of numbers along the abscissa gives the number of states, the lower line of numbers gives the maximum unperturbed energy in MeV spanned by the corresponding basis states, and the solid and dashed curves correspond to the TDA and RPA calculations, respectively. (ii) Experimentally, the giant dipole state occurs in the

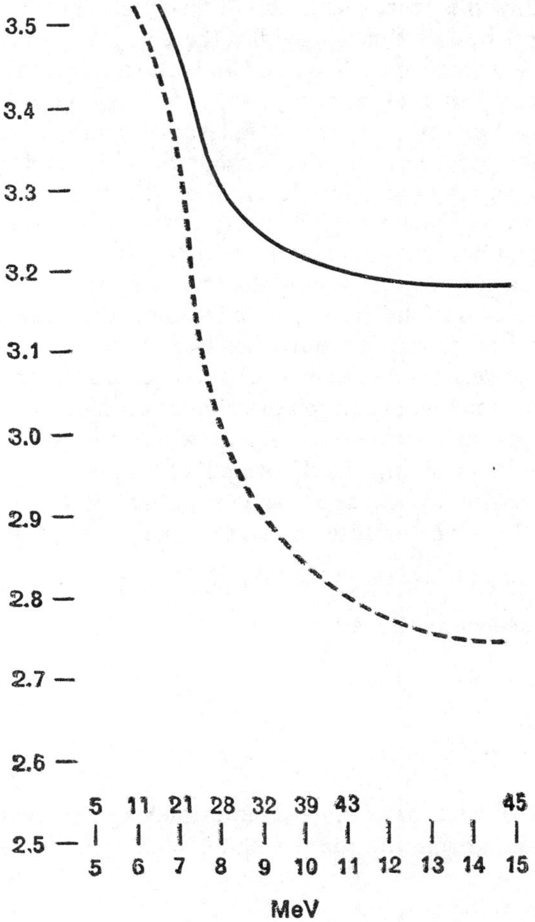

Fig. VI.13 Energy of octupole state of Pb^{208} as function of number of hole-particle basis states used in setting up energy matrix. [Following Gillet, V., and Sanderson, E. A., *Nucl. Phys.*, **54**, 472 (1964).]

energy region 13.5–15 MeV. The theoretical calculations using Gillet's potential or the effective potential derived from the HJ potential (see Brown cited earlier) push this state only up to about 11 MeV. To achieve a satisfactory result, we need an artificial bolstering of the overall strength of the two-nucleon potential by about a factor of two.

Deformed States of Closed-Shell Nuclei

The deformed states of closed-shell nuclei need special treatment, and the existing theoretical understanding of them is rather unsatisfactory. Almost all the experiments have so far been restricted to the deformed 0^+ (6.06 MeV) state in O^{16}. At one time, 1h-1p type calculations were done by several workers on this state. But it was soon realized that the monopole transition from this state to the ground state had no collective enhancement. Therefore, it was surmised for some time that this state has predominantly 2h-2p type configurations, although it was rather poorly understood at that stage why the 2h-2p type states should be so low in energy. Then there came a suggestion from Bohr and Mottelson that this state, and the similar low-lying 0^+ state in many other even nuclei, may correspond to a deformed shape of the nucleus. If we consider the Nilsson single-particle level diagram, we immeditaely realize that, as we move on the positive deformation side, a single-particle level from the upper shell comes down close to the uppermost single-particle level from the lower shell, and the gap between the major shells is therefore almost destroyed as a result of the deformation. Since the 6.06 MeV state has to have more than one hole-particle pair, it was suggested that it comprise primarily 2h-2p pairs, and the reason for the state being so low was attributed to the tendency of the 2h-2p state to acquire a deformation; as soon as the nucleus becomes deformed, according to the foregoing statement, very little energy is required to transfer a pair of nucleons from the topmost level of the p-shell to the lowest level of the (s, d)-shell. Based on this suggestion, Brown and Green[6] have done a phenomenological type of calculation by mixing the deformed 2h-2p and 4h-4p $(J = 0)$-states with the spherical $(J = 0)$-state (which has no hole-particle pair). They also considered the $(J = 2)$-state by mixing the deformed 2h-2p and 4h-4p states. Let us denote the number of hole-particle pairs by N, and the various states by $|N, J\rangle$. Then, according to Brown and Green (BG), there will be three matrix elements of the nucleon-nucleon potential:

$$M_1 = \langle 0, 0|\, V\,|2, 0\rangle, \qquad M_2 = \langle 2, 0|\, V\,|4, 0\rangle, \qquad M_3 = \langle 2, 2|\, V\,|4, 2\rangle.$$

The hole state of the deformed nuclei is taken to be

$$x'|p_{3/2}, \tfrac{1}{2}\rangle + y'|p_{1/2}, \tfrac{1}{2}\rangle, \tag{VI.157a}$$

and the particle state to be

$$x|d_{5/2}, \tfrac{1}{2}\rangle + y|s_{1/2}, \tfrac{1}{2}\rangle + z|d_{3/2}, \tfrac{1}{2}\rangle, \tag{VI.157b}$$

where the second number in each basis state denotes the angular momentum projection. Brown and Green have taken the deformation for the 2h-2p state as $\beta = 0.3$, and for the 4h-4p state as $\beta = 0.5$. The values of x, y, z for a given value of the deformation are known from Nilsson's work. For example, for $\beta = 0.3$, we have

$$x = 0.828, \qquad y = 0.537, \qquad z = -0.160.$$

Actually, for the hole state, instead of (VI.157a), Brown and Green have used the following approximation. The hole state was taken to be a pure $|p_{1/2}, \tfrac{1}{2}\rangle$ state, but then the matrix elements were multiplied in the end by an overlap factor ($=0.75$ for $\beta = 0.3$) to take into account the deformation of the state. In this way, they obtained the estimate

$$M_1 = -4.3 \text{ MeV}.$$

For the other two matrix elements, namely, M_2 and M_3, very rough estimates, based on the SU_3-type wavefunctions, were used:

$$M_2 = -1.8 \text{ MeV}, \qquad M_3 = -1.6 \text{ MeV}.$$

The unperturbed energies of the states appearing in the matrix elements were also treated as empirically adjustable quantities. Denoting these by $E_J(N)$, Brown and Green chose the values

$$E_0(0) = 2.31 \text{ MeV}, \qquad E_0(2) = 8.51 \text{ MeV}, \qquad E_0(4) = 6.51 \text{ MeV},$$
$$E_2(2) = 10.83 \text{ MeV}, \qquad E_2(4) = 7.61 \text{ MeV}.$$

The final results they obtained are summarized in Table VI.4.

Table VI.4 Deformed states of O^{16}
(a) Energy and wavefunction

State (J_n)*	Energy (MeV)	Amplitudes of Basis State $\|N, J\rangle$		
		$N=0$	$N=2$	$N=4$
0_1^+	0	0.874	0.469	0.130
0_2^+	6.07	−0.262	0.229	0.937
0_3^+	11.36	−0.410	0.853	−0.323
2_1^+	6.92		0.377	0.923
2_2^+	11.52		0.923	−0.377

*J_n denotes the n-th state of angular momentum J.

Table VI.4 Deformed states of O^{16}
(b) B(E2)-values [in units of e^2 (fm)^4]

Transition	B(E2)-Value	
	Calculated	Experimental
2_1^+ (6.93 MeV) $\rightarrow 0_1^+$ (0 MeV)	5.3	4.6 ± 1.0
2_1^+ (6.93 MeV) $\rightarrow 0_2^+$ (6.06 MeV)	10.3	40
4_1^+ (10.36 MeV) $\rightarrow 2_1^+$ (6.93 MeV)	152	117 ± 10

In spite of the highly successful nature of the results in Table VI.4, we need a more fundamental calculation based on a lesser number of empirically determined quantities. The most interesting approach of this kind started with the initial work by Ripka and Bassichis[7]. These authors considered states of the type (VI.157) and determined the coefficients x', y', x, y, and z by Hartree-Fock (HF) calculations. The closed-shell ground-state wavefunction, say, $|\Phi_0\rangle$, was taken as an inert core, over which 2h-2p and 4h-4p type excitations were allowed. The unperturbed single-particle and single-hole energies (ϵ) were chosen in such a way that an HF calculation yielded the correct single-particle spectrum of O^{17} and O^{15}. The two-nucleon interaction (V) used for this purpose was taken to be a well-behaved Gaussian potential with a Rosenfeld exchange mixture, the same interaction that had already been used in the HF calculations on (s, d)-shell nuclei by other authors. The HF Hamiltonian Γ, for the self-consistent calculation of Nh-Np states, is defined in the harmonic oscillator representation by

$$\langle nljm\tau_3|\,\Gamma\,|n'l'j'm\tau_3\rangle = \epsilon_{nlj}\delta_{nn'}\delta_{ll'}\delta_{jj'} + \sum_{i=1}^{N}\langle nljm\tau_3;\,p_i|\,V\,|n'l'j'm\tau_3;\,p_i\rangle - \sum_{i=1}^{N}\langle nljm\tau_3;\,h_i|\,V\,|n'l'j'm\tau_3;\,h_i\rangle, \tag{VI.158}$$

where τ_3 is the projection of isospin; ($nljm$) is a spin-orbit coupled harmonic oscillator state; and p_i and h_i stand for the quantum numbers of the particle and hole states. In the case of 4h-4p states, the determinantal wavefunction to be considered has four holes in the state (VI.157a) and its time-reversed state, and four particles in the state (VI.157b) and its time-reversed state. Such a determinant automatically has a total isospin $T = 0$. In the case of 2h-2p type states, this is not so, and Ripka and Bassichis have ensured the correct total isospin by suitable coupling in the determinantal wavefunctions. There are clearly three possibilities. (i) The two hole states are coupled to $T' = 1$ and the two particle states are coupled to the same isospin, and finally these isospins are coupled to the total $T = 0$; the state of two holes or two particles with $T' = 1$ has to be antisymmetric in the space-spin components, and hence we have only one possibility, namely, the two holes must be in *different* space-spin states [the state (VI.157a) and its time-reversed conjugate], and the two particles also must be in different space-spin states [the state (VI.157b) and its time-reversed conjugate]. (ii) In the case of $T = 0$, on the other hand, the space-spin part of the two holes as well as of the two particles must be *symmetric*, and hence two alternatives exist: first, the possibility just mentioned, and second, the possibility of the two holes going to the same state (VI.157a) and the two particles going to the same state (VI.157b). For 2p-2h states, Ripka and Bassichis denote the possibility in (i) as a, and the two possibilities in (ii) as b and c.

The self-consistent energy of the 2p-2h type states (cases a, b, c) and the energy of the 4p-4h state are shown as a function of the strength of the two-body interaction in Fig. VI.14. These energies correspond to the energies of the intrinsic states. Assuming that every intrinsic state gives rise to a rotational band, we can find the energy of the observed 0^+ state from the calculated energy of the intrinsic state by subtracting the value of $\mathcal{A}J^2$, where the parameter $\mathcal{A}$ is obviously related to the moment of inertia of the deformed intrinsic state. $\mathcal{A}$ was calculated by Ripka and Bassichis from the cranking model expression, and then the energy of the 0^+ state was obtained from the solid line marked 4p-4h in Fig. VI.14 by subtracting $\mathcal{A}\langle J^2\rangle$. The dashed line labelled E_0 was obtained in this way.

It is clear from Fig. VI.14 that, for suitable values of the interaction strength, the deformed 4h-4p state comes very much below the deformed 2h-2p type states. This explains why the 6.06 MeV state of O^{16} is composed predominantly of 4h-4p states, rather than 2h-2p states. The main interest that now remains in the calculation of the 6.06 MeV state is whether or not a *reasonable* two-body interaction really pushes down the 0^+ state to 6.06 MeV. From

Fig. VI.14, it is clear that the Ripka-Bassichis potential with a strength around 32 MeV can do the job. However, it is currently believed that more realistic two-body potentials do not satisfy this criterion too well.

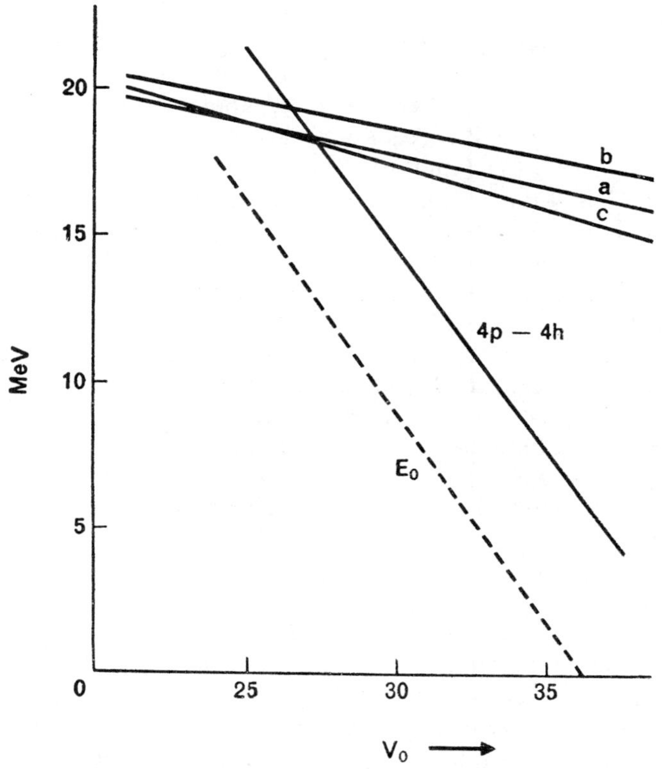

Fig. VI.14 Energy of ground state and 4p-4h states as function of strength of two-body interaction. [Following Ripka, G., and Bassichis, W., *Phys. Letters*, **15**, 320 (1965).]

There is an interesting calculation by Stephenson and Banèrjee[8] who have shown that it is likely that the 0^+ state of O^{16} has a deformed triaxial shape, rather than the axially symmetric shape assumed by earlier workers. The 4h-4p type state in a triaxial HF potential goes much lower than a similar state in an axially symmetric potential. Once again, such calculations need to be worked out with a more realistic two-body interaction.

B. SPHERICAL VIBRATIONAL NUCLEI

Input Data

Single-*particle* energies are usually taken from experimental data on nuclei with a closed-shell-plus-one-nucleon state in the neighbourhood of the nuclei being calculated. A smooth variation of these energies with the mass number (A), of the type $A^{-1/3}$ (i.e., the energy is inversely proportional to the radius of the nucleus), is also allowed in many calculations. Wherever experimental data are not available, theoretical values of the single-particle energies calculated with the Saxon-Woods type average potential should be used; one set of energies and wave-functions is given by Blomqvist and Wahlborn[9].

The other type of input data concerns the two-body potential. Very little information was available from shell-model type calculations. The potential used in the earliest work, namely, that of Kisslinger and Sorensen[10], consisted of the idealized pairing interaction V_P, defined by (VI.94b), and the schematic quadrupole potential, defined by (GIII.43a) in Appendix G, with $K = 2$. The BCS equations are solved only with V_P to obtain the independent quasiparticles, and the vibrational equations in Section 42 are solved only with the model expression (GIII.43a). In their first paper, dealing with single closed-shell nuclei (i.e., nuclei in which either the neutrons or the protons are in closed shells), Kisslinger and Sorensen used an adiabatic method, originally suggested by Bohr and Mottelson, to obtain the vibrational energy. In their second paper, dealing with nuclei having both neutrons and protons in unfilled shells (the unfilled levels of neutrons were, however, much above those of the protons so that, at the stage of calculating the quasiparticles, the neglect of the pairing interaction between neutrons and protons could be justified), they used a more sophisticated technique, very much akin but not exactly analogous to that described in Section 42. The strength of the pairing interaction used in their calculations was obtained by fitting the odd-even mass difference on the assumption that this is almost entirely determined by the pairing effect. To elucidate the procedure, let us consider the binding energies (BE) of three neighbouring nuclei and compute

$$\mathrm{BE}(Z, N-1) - 2\mathrm{BE}(Z, N) + \mathrm{BE}(Z, N+1),$$

where the letters within the parentheses denote the proton and neutron numbers of the nuclei. The nucleus (Z, N) is even-even, whereas the other two are even-odd (i.e., odd mass). If the effect of the residual quasiparticle interaction is ignored, then this quantity should be equated to $2E_{\min}$, where $E_{\min}$ is the energy of the single neutron quasiparticle state that represents the ground state of the two odd-mass nuclei. The strength of the pairing force determines the energy gap and through that the quantity $E_{\min}$, thereby making possible a determination of the former. Kisslinger and Sorensen found that the strength of the pairing force G is given by $(19/A)$ MeV and by $(23/A)$ MeV in the neighbourhood of Sn and Pb, respectively. The strength of the quadrupole force X was determined by fitting the energy of the first excited 2^+ state. Instead of treating X as an adjustable parameter for each individual nucleus, they tried to attribute a smooth A-dependence to this parameter so that the trend of the 2^+ level of a whole set of nuclei in a particular mass number region is fairly well-reproduced. The final value obtained by them is given by $[5/(4\pi)](n + \frac{3}{2})^2 b^4 X = 110A^{-1}$, where b is the harmonic oscillator constant and n is the number of harmonic oscillator quanta associated with most of the levels (i.e., with the exception of the level that comes down from the upper shell through the spin-orbit coupling) of the unfilled major shell.

In their second paper, Kisslinger and Sorensen made the pairing force strength of the neutrons G_n equal to that of the protons G_p. They ignored the pairing effects between a neutron and a proton as per the arguments already stated. The quadrupole force between two neutrons, between two protons, and between a neutron and a proton was assumed to have the same strength, i.e., $X_n = X_p = X_{np}$. They took estimates of these quantities and G_n, G_p from their first paper. In addition, they used a short-range (δ-function type) potential between a neutron and a proton, and explicitly considered the effect of this interaction in changing the proton single-particle energy as the neutron number increases, and vice versa. This, in fact, amounts to considering only the contribution of the δ-function potential to the average potential acting on a proton or a neutron. Apart from this specific change in the single-particle energies with mass number, the smooth $A^{-1/3}$-type dependence of χ_n, χ_p, χ_{np} mentioned earlier, and another smooth A-dependent term (representing the change in spin-orbit coupling effects with A) were also considered. For the details, the reader is referred to Appendix II of the second paper by Kisslinger and Sorensen[10].

Many of the later workers used more detailed forms of the two-nucleon potential. Arvieu and his colleagues[11] used a spin-dependent Gaussian potential, whose range, depth, and ratio of singlet-to-triplet strengths were treated as parameters in a least-squares fit to the energy levels of odd-mass nuclei. The same potential was then used to predict the 2^+ state of the even nuclei. The work was done in the Sn- and Pb-region.

In the Ni- and Sn-region, many different well-behaved central potentials have been used by Pal et al[12] and Savoia et al[13]. A set of phenomenologically determined two-body matrix elements due to the Argonne group have also been used by Pal et al[12].

Two-body matrix elements, starting from the realistic HJ two-nucleon interaction, have been calculated in the Ni-region by Kuo[14] and calculated and applied in the Sn-region by Gmitro and his coworkers[15]. The latter team used also the separable nonlocal Tabakin interaction. In both Lawson et al[14] and Gmitro et al[15], the effect of excitation of the core nucleons, in renormalizing the two-body matrix elements, has been taken into account.

Inverse Gap Equation Method

Gillet and Rho[16] suggested a new way of supplying input to the quasiparticle calculation. This is now known as the inverse gap equation (IGE) method. To understand the principle of this approach, let us consider (VI.102) and explicitly pull out the attractive strength $-V_0$ of the potential; then we denote the rest of the two-body matrix element by $m_0(jj')$, where the subscript refers to the total J. In this manner, (VI.102) can be rewritten as

$$\left(\frac{1}{V_0}\right)\Delta_j = \sum_{j'} M_{jj'}\Delta_{j'}, \tag{VI.159a}$$

where

$$M_{jj'} = \frac{1}{2E_{j'}}\sqrt{\frac{2j'+1}{2j+1}}\, m_0(jj'), \tag{VI.159b}$$

$$-V_0 m_0(jj') = (jjJ=0, M=0|\, V\, |j'j'J=0, M=0). \tag{VI.159c}$$

If the strength of the potential V_0 is treated as an unknown quantity at the beginning of the calculation, and $m_0(jj')$ and $E_{j'}$, for all j and j', are taken to be known, then (VI.159a) becomes the eigenvalue-eigenvector equation for the *known* matrix M, of which V_0^{-1} is the eigenvalue and the quantities Δ_j are *proportional* to the elements of the eigenvector. In general, M has many eigenvalues; however, the eigenvalue for which the elements of the eigenvector are all real positive quantities alone serves our purpose because the energy gap parameters must necessarily be real and positive. Since the matrix elements on the right-hand side of (VI.159c) are attractive, the matrix M is positive definite and, by a theorem due to Frobenius, such a matrix has one, and only one, eigenvalue for which the eigenvector has all the elements positive. This theorem guarantees the type of solution we want and, in view of the uniqueness of the solution, there is no ambiguity in the method.

It is clear, however, that the quasiparticle energies E_j for all j must be known at the beginning of an IGE calculation. Usually, we assume that the few lowest levels of the odd-mass nuclei are very nearly pure single-quasiparticle type states. Then their observed energies can be directly fed into the IGE method as input data for E_j. The results will be unreliable to the extent that this starting hypothesis about the nature of the levels of the odd-mass nuclei is erroneous.

The consistency of the method is usually checked as follows. We have already remarked that the elements of the special eigenvector of M, in which we are interested, are *proportional* to the energy gap parameters Δ_j; this is because the system of equations (VI.159a), for all j, forms a set of *homogeneous* equations for the unknown quantities Δ_j and, as such, remains unaltered through the multiplication of any overall constant, say, ξ. Let us denote the elements of the eigenvector by $\bar{\Delta}_j$ and then $\Delta_j = \xi\bar{\Delta}_j$, where ξ has to be determined. For this purpose, we first write down $(\epsilon_j - \lambda)$ from E_j and Δ_j as

$$(\epsilon_j - \lambda)^2 + \Delta_j^2 = E_j^2$$

or

$$(\epsilon_j - \lambda) = \pm(E_j^2 - \Delta_j^2)^{1/2} = \pm(E_j^2 - \xi^2\bar{\Delta}_j^2)^{1/2}. \tag{VI.160}$$

To settle the sign of this expression, we have once again to take recourse to experimental data. The information needed is whether or not the occupation probability v_j^2, of the level j, as measured in nucleon stripping or pick-up experiments, exceeds $\frac{1}{2}$. According to the interpretation of λ in Section 41D, we know that $(\epsilon_j - \lambda)$ is negative or positive, depending on whether v_j^2 is greater or less than $\frac{1}{2}$, respectively. Once the sign is settled in this manner, the expression (VI.160) for $(\epsilon_j - \lambda)$ becomes specified to the extent of an undetermined ξ. The latter is then determined by using (VI.160) in (VI.81). Since E_j has to be larger than Δ_j, i.e., $\xi\bar{\Delta}_j$, we conclude that the positive number ξ must be less than $(E_j/\bar{\Delta}_j)$. Very often, (VI.81) cannot be exactly satisfied with this restriction on ξ. In practice, we therefore use the value of ξ, subject to the aforementioned restriction, such that the departure of the right-hand side of (VI.81) from the given nucleon number A be a minimum. When $(\epsilon_j - \lambda)$ have been completely determined for all the levels in this manner, we can easily calculate the energies ϵ_j of all the single *particles* (not *quasiparticles*) with respect to one of them, say, $\epsilon_{j'}$, which will be treated as the reference state. If the whole procedure is repeated for a few odd-mass nuclei in a given region, then the values of ϵ_j relative to the reference state can be plotted as a function of the mass number. For the internal consistency of the whole approach, we must expect that the dependence on the mass number of this curve be as smooth and as slow as possible. The result of this consistency check (see Gambhir[17]) for the odd-mass Ni isotopes is rather disturbing; this in turn may imply that some of the actual levels of the odd-mass nuclei used in the calculation cannot be interpreted as fairly good one-quasiparticle states.

Salient Features of Numerical Calculations

Both the IGE method and the straightforward solution of the BCS equations yield a set of values for the quasiparticle energies E_j and the corresponding transformation coefficients u_j and v_j. With these quantities, we then solve the vibrational equation for the even nuclei. The single-quasiparticle energies can be checked against the energy levels of the odd-mass nuclei, and the lowest eigenvalue of the vibrational equation gives the energy of the first 2^+ state of the even nuclei. This general programme was carried out by Kisslinger and Sorensen[10] and

Arvieu et al[11]. The B(E2)-values were also calculated for the decay of the 2^+ state to the ground state. For the odd-mass nuclei, Kisslinger and Sorensen have calculated magnetic and quadrupole moments. The expression (GIII.48b), in Appendix G, for any single-particle type operator can be used to evaluate its expectation value in a one-quasiparticle state. Only the first line of this expression contributes because the second line changes the number of quasiparticles acting on any state. We can easily verify that the magnetic and quadrupole moments of a single-quasiparticle state are equal to that of a single-particle state. Thus, the quasiparticle theory cannot provide any agreement to the observed values of these quantities, unless we use effective gyromagnetic ratios and an effective charge for the nucleon. An *understanding* of the latter quantities is, however, beyond the capability of a theory that tries to interpret the observed ground state of odd-mass nuclei as a one-quasiparticle state. For the magnetic moment, Kisslinger and Sorensen have used a δ-function interaction to produce an admixture of other shell-model configurations which could then cause additional contribution to the magnetic moment. The switching on of the quasiparticle interaction is not of great help. This produces a mixture of the three-quasiparticle states with the one-quasiparticle state. There is a coherent mixture of two-quasiparticle states which is identified as the vibrational 2^+ state (phonon state) of the even nucleus. So there is a special three-quasiparticle state that can be considered a single-quasiparticle state coupled to the phonon. The admixture of this state into the one-quasiparticle state was taken into consideration by Kisslinger and Sorensen. This type of admixture changes the electric quadrupole properties, but does not appreciably affect the magnetic moment. For all the detailed results, refer to the original papers of Kisslinger and Sorensen[10] and Arvieu et al[11].

Extension of RPA Work: Modified Tamm-Dancoff Method

The theory just reviewed is nearly powerless when we attempt to explain the levels above the first 2^+ state in the even nuclei and the levels above the first few states in the odd-mass nuclei. According to the simple RPA theory, we expect to get states, at roughly twice the energy of the calculated 2^+ state, which have a two-phonon (i.e., a special admixture of four-quasiparticle states) character. The other two-quasiparticle states, which are orthogonal to the one-phonon state, also occur in this general neighbourhood of energy. Hence, it becomes imperative to develop a theory that allows the admixture of two- and four-quasiparticle states. Since the zero-quasiparticle state is coupled by the part H_{40} of the quasiparticle residual interaction to the four-quasiparticle states, we are led to the diagonalization of the quasiparticle Hamiltonian in the space of zero-, two-, and four-quasiparticle states. This procedure is usually called the modified Tamm-Dancoff approximation (MTDA) because it is a natural extension of the diagonalization in the two-quasiparticle space which is known as the Tamm-Dancoff approximation (TDA). Just as the TDA can be generalized to the RPA by allowing both the creation and the destruction of quasiparticle pairs, the MTDA method can also be generalized by allowing the destruction of two- and four-quasiparticle states in addition to their creation. Such a treatment is called the higher random phase approximation (HRPA). For details of these methods, refer to Pal et al[12] and Sawicki et al[13]. In the review presented here, we shall mention only a few important points of the MTDA approach.

For odd-mass nuclei, the MTDA calculation should admix one-, three-, and five-quasiparticle states. In the calculations done so far, only the admixture of one- and three-quasiparticle states has been taken into account.

One of the important points about the three- and four-quasiparticle basis states was first pointed out by Sawicki[13], who was the originator of the movement towards the HRPA and MTDA methods and calculations. It is well-known from the shell-model classification of states that antisymmetric states of three and four Fermions in the level span only a subspace of all possible states. The same result is true also for quasiparticles because they are also Fermions. Let us consider a three-quasiparticle state of the type $[(b_j^\dagger, b_j^\dagger)^{J'}, b_j^\dagger]_M^J$, where J and M are the total angular momentum and its projection, J' is the angular momentum of the first two quasiparticles, and the third quasiparticle has then been coupled to J' to produce the total J. According to our prior observation, many such states identically vanish, and sometimes two or more states with different J' but the same J are linearly dependent on each other. The same kind of observation holds for four-quasiparticle states. If we recall our work on the TDA, then the procedure for deriving the MTDA equations becomes pretty obvious: we have to evaluate the commutator of H with a one- and three-quasiparticle creation operator in the case of odd nuclei, and a two- and four-quasiparticle operator in the case of even nuclei. Since all three-quasiparticle creation operators of the foregoing type and similar four-quasiparticle creation operators are not independent, it is clear that we have to exercise caution in obtaining a set of equations involving a set of *nonredundant* operators.

In the work by Pal et al[12], a complete set of antisymmetric states of three- and four-quasiparticle states has been constructed by standard shell-model methods, and the matrix elements of the quasiparticle interaction connecting different states have been obtained by a combination of shell-model and second-quantized techniques. Sawicki et al[13] followed a somewhat different technique, based on the Schmidt orthogonalization procedure, to construct their nonredundant set of states.

Another important caution that has to be exercised in this kind of work was also first pointed out by Sawicki and his collaborators. This relates to the elimination of spurious states. The existence of a spurious 0^+ state in the two-quasiparticle space was fairly well-known to earlier workers. This spurious state can be derived as follows. Let us consider the effect of the total number operator $\hat{N}$, operating on the ground state Ψ_0. Since

$$\hat{N} = \sum_a C_a^\dagger C_a = \sum_a v_a^2 + \sum_a (u_a^2 - v_a^2) b_a^\dagger b_a + \sum_a u_a v_a s_a (b_a^\dagger b_{-a}^\dagger + b_{-a} b_a)$$

and $b_a^\dagger b_a$ as well as $b_{-a} b_a$ produce zero acting on Ψ_0, we have

$$\hat{N}|\Psi_0\rangle = \sum_a v_a^2 |\Psi_0\rangle + \sum_a u_a v_a s_a b_a^\dagger b_{-a}^\dagger |\Psi_0\rangle.$$

In the quasiparticle theory, we equate $\sum_a v_a^2$ with the nucleon number A. Had there been a strict conservation of the number of nucleons, we would have obtained only the first term on the right-hand side. The second term of this expression, namely,

$$|\Psi\rangle = \sum_a u_a v_a s_a b_a^\dagger b_{-a}^\dagger |\Psi_0\rangle,$$

is thus an entirely spurious state. By doing the angular momentum coupling for spherical nuclei, where u_a, v_a do not depend on the projection quantum number, we can easily show that this represents an angular momentum zero state.

Any four-quasiparticle state that is obtained by coupling a two-quasiparticle state to the $|\Psi\rangle$ just mentioned is therefore spurious. A similar observation holds for three-quasiparticle states. There can also be more complicated spurious states in the four-quasiparticle case; for example, we can examine the result of $\hat{N}^2$ on Ψ_0, and identify an additional spurious state and so on. Pal et al[12] explicitly projected out the spurious states arising from $|\Psi\rangle$, as just defined. Sawicki and his team took care to eliminate some more spurious states. For all the mathematical details and numerical results, refer to Pal et al[12], Sawicki et al[13], and Sawicki[18]. Some important aspects of the results have been summarized by Pal[19].

C. VIBRATIONAL LEVELS IN DEFORMED NUCLEI

In deformed nuclei, the single-particle, as well as the single-quasiparticle, states are not specified by *given* angular momentum quantum numbers. In the case of deformation with axial symmetry, only the projection quantum number is a good quantum number for these states, whereas for deformation having no axis of symmetry even this is not true. However, under very general circumstances, a single-particle state of a deformed nucleus and its time-reversed conjugate are usually found to be degenerate. This has been verified, for example, in all the HF type calculations done to date. So, we take for granted the degeneracy of a quasiparticle state m, and its time-reversed state $\bar{m}$. Further, we keep in mind the general result (VI.60b), namely, $|\bar{\bar{m}}\rangle = -|m\rangle$.

Since the single-quasiparticle states do not have given angular momenta, we do not do the angular momentum coupling in the quasiparticle pair creation and destruction operators, as we did for spherical nuclei. Instead, we use the RPA equations without this angular momentum coupling, derived in Section 42 and embodied in (VI.141), (VI.144), and (VI.147). Because of the presence of a summation over k, l in (VI.147), we can clearly rewrite the right-hand side, changing the summation indices to $\bar{k}$, $\bar{l}$ throughout. The equation for the commutator $[H, A_{mn}]$ which can be easily derived from $[H, A^\dagger_{mn}]$, as mentioned after (VI.147), can also be changed to an equation for the commutator $[H, A_{\bar{m}\bar{n}}]$, where

$$A_{\bar{m}\bar{n}} = b_{\bar{m}}b_{\bar{n}}.$$

In the usual way, we define the amplitudes

$$Z^{(\pm)}_{mn} = \langle\Psi|\ (A^\dagger_{mn} \pm A_{\bar{m}\bar{n}})\ |\Psi_0\rangle. \tag{VI.161}$$

Adding and subtracting the equations for $[H, A^\dagger_{mn}]$ and $[H, A_{\bar{m}\bar{n}}]$, placed between $\langle\Psi|$ and $|\Psi_0\rangle$, we easily obtain the coupled equations [satisfied by the amplitudes (VI.161)]

$$\begin{aligned} EZ^{(+)}_{mn} = (E_m + E_n)Z^{(-)}_{mn} + \sum_{(i,j)} [(ij|\ V\ |mn)(u_iu_j + v_iv_j)(u_mu_n + v_mv_n) \\ - \{(j\bar{m}|\ V\ |n\bar{i}) + (i\bar{m}|\ V\ |n\bar{j})\}(u_mv_n - v_mu_n)(u_iv_j - v_iu_j)]Z^{(-)}_{ij}, \end{aligned} \tag{VI.162a}$$

$$\begin{aligned} EZ^{(-)}_{mn} = (E_m + E_n)Z^{(+)}_{mn} + \sum_{(i,j)} [(ij|\ V\ |mn)(u_iu_j - v_iv_j)(u_mu_n - v_mv_n) \\ - \{(j\bar{m}|\ V\ |n\bar{i}) - (i\bar{m}|\ V\ |n\bar{j})\}(u_mv_n + v_mu_n)(u_iv_j + v_iu_j)]Z^{(+)}_{ij}. \end{aligned} \tag{VI.162b}$$

Here the parentheses enclosing the summation symbols i, j denote that a given pair (i, j) is to be counted only once in the summation, and not twice as (i, j) and (j, i).

In conformity with what was done for spherical nuclei, we can introduce models for V in order to simplify the general equations (VI.162). The usual assumption is to ignore the matrix elements $(ij|\ V\ |mn)$, and substitute the matrix element $(j\bar{m}|\ V\ |n\bar{i})$ by a factorable form

$$-(j\bar{m}|\ V\ |n\bar{i}) = \sum_{\mu} \kappa_\lambda\langle j|\ \Omega^\lambda_\mu\ |\bar{i}\rangle\langle\bar{m}|\ \Omega^{\lambda *}_\mu\ |n\rangle. \tag{VI.163}$$

This expression follows from a multipole potential of the type

$$V_{12} = \kappa_\lambda \sum_\mu \Omega_\mu^\lambda(1)\Omega_\mu^{\lambda *}(2), \tag{VI.164}$$

where 1 and 2 denote nucleon coordinates, κ_λ denotes the strength of the interaction, and the multipole operator Ω_μ^λ is usually taken to be

$$\Omega_\mu^\lambda(\mathbf{r}) = r^\lambda Y_\mu^\lambda(\theta, \phi). \tag{VI.165}$$

It is clear that, when we use the potential (VI.164), the right-hand side of (VI.163) actually corresponds to only the exchange part of the matrix element on the left-hand side of (VI.163).

The model potential, described by (VI.163), converts (VI.162) into equations of the type (GIII.44d) and (GIII.44e) in Appendix G. (This task is left as a simple exercise.) The solutions of such equations are given by (GIII.46) (see Appendix G). For this, a sum of quadrupole-plus-octupole-type potential for V is usually taken. The matrix elements are then obtained from (VI.163) by adding the results for $\lambda = 2$ and $\lambda = 3$. The parity of these multipoles tells us that, in the case of $\lambda = 2$, quasiparticle pair states such as (m, n) and (i, j) which are coupled through (VI.162a) and (VI.162b) have even parity, whereas in the case of $\lambda = 3$ the pair states have odd parity. Thus, the vibrational equations corresponding to these two multipoles in the potential are completely decoupled from each other, and can be solved separately. For details of the theoretical derivation and the properties of the nuclei calculated, refer to the work by Solóviev[20].

In this chapter, we have so far presented a systematic method of calculating vibrational states based on an expansion in terms of the number of quasiparticles; in applications, the expansion has been cut off at four quasiparticles for even nuclei, and three quasiparticles for odd-mass nuclei. No investigation has been made into the rapidity of convergence of this expansion. There is an alternative approach based on an expansion in terms of the number of phonons. Some of the pertinent references are given at the end of the chapter (see Belyaev and Zelevinsky[21]).

D. MOMENT OF INERTIA, HIGH ANGULAR MOMENTUM STATES, AND PHASE TRANSITION

The moment of inertia formula (V.35b) needs to be changed in two ways in the quasiparticle formalism. First, the energy denominator $(\epsilon_p - \epsilon_h)$ should be substituted by the energy $(E_\mu + E_\nu)$ of the state that has the two quasiparticles μ and ν. Second, the numerator should contain the matrix element of the many-body operator J_x connecting the ground state with the two-quasiparticle state just mentioned. Writing

$$J_x = \sum_{\alpha, \beta} \langle\alpha| j_x |\beta\rangle C_\alpha^\dagger C_\beta \tag{VI.166}$$

and then transforming $C_\alpha^\dagger$ and C_β in terms of quasiparticle operators, we immediately obtain

$$\begin{aligned}\langle\Psi_{\mu,\nu}| J_x |\Psi_0\rangle &= \sum_{\alpha, \beta} \langle\alpha| j_x |\beta\rangle s_\beta u_\alpha v_\beta(\delta_{\mu\alpha}\delta_{\nu,-\beta} - \delta_{\nu,\alpha}\delta_{\mu,-\beta})\\ &= \langle\mu| j_x |\bar{\nu}\rangle(-u_\mu v_\nu) + \langle\nu| j_x |\bar{\mu}\rangle(u_\nu v_\mu)\\ &= \langle\mu| j_x |\bar{\nu}\rangle(u_\nu v_\mu - u_\mu v_\nu).\end{aligned} \tag{VI.167}$$

Here we have used the fact that the operator j_x changes sign under time-reversal. Thus,

(V.35b) gets replaced by

$$\mathcal{J} = 2 \sum_{(\mu,\nu)} \frac{|\langle\mu|\, j_x \,|\bar{\nu}\rangle|^2}{E_\mu + E_\nu}(u_\nu v_\mu - u_\mu v_\nu)^2. \tag{VI.168}$$

The summation is over all the pairs (μ, ν) of quasiparticles. This expression gives values of $\mathcal{J}$ smaller than those given by (V.35b), which is based on holes and particles because of two reasons: (i) the quasiparticle energy denominator is larger by virtue of the energy gap Δ and (ii) the numerator is smaller due to the overlap factor containing the coefficients u, v. Thus, the present theory modifies the cranking model expression for the moment of inertia in the right direction, i.e., more towards the experimental results. For detailed numerical agreement, refer to the original work of Nilsson and Prior[22].

It has been mentioned in Section 33B that the $I(I+1)$-dependence of the energy with a constant moment of inertia fails to fit the data as the angular momentum I increases. At the initial stage, several workers proposed various models containing an I-dependence of the moment of inertia. The most popular approach of this type is by Mariscotti et al[23]. These authors proposed an empirical expression for the energy of a level I to be given by

$$E_I = \tfrac{1}{2}C(\mathcal{J}_I - \mathcal{J}_0)^2 + \frac{\hbar^2}{2\mathcal{J}_I}I(I+1). \tag{VI.169}$$

Here the first term represents a kind of vibrational energy due to shape changes in the nucleus with changing I, leading in its turn to changes in the moment of inertia $(\mathcal{J}_I - \mathcal{J}_0)$ as shown; C is a constant parameter representing the spring constant of the vibration; and the second term stands for the rotational energy with a variable moment of inertia (VMI). The authors then prescribed that the dependence of $\mathcal{J}_I$ on I be such that the expression (VI.169) is minimized. This requirement leads to the equation

$$0 = \frac{\partial E_I}{\partial \mathcal{J}_I} = C(\mathcal{J}_I - \mathcal{J}_0) - \frac{\hbar^2}{2\mathcal{J}_I^2}I(I+1).$$

This is a cubic equation for $\mathcal{J}_I$ which can be solved in terms of the two parameters C and $\mathcal{J}_0$. The value of $\mathcal{J}_I$ thus obtained can be substituted in the parent expression (VI.169) and the energy levels fitted in terms of the aforementioned parameters. For the detailed data fitting, refer to Mariscotti et al[23].

In recent years, the data on rotational nuclei have been extended to very high angular momentum states, $20\hbar$ and beyond and, in some cases, to much higher values near 60–$70\hbar$. Even a Mariscotti-type parametrization has now been found to be inadequate. The experiment entails bombarding suitable target nuclei with α-particles or heavy ions having energy much above the Coulomb barrier so that a compound nucleus with a large angular momentum and excitation energy may be formed. Subsequently, the compound system de-excites by evaporating a few neutrons, each carrying 1 or 2 units ($\hbar$) of angular momentum. A typical case using A^{40} as the heavy-ion projectile and four emitted neutrons is shown in Fig. VI.15. For each angular momentum I, the state of lowest energy is called the *yrast* state. The figure depicts the yrast line which passes through the energies of the yrast states for different values of I. With the emission of neutrons, the residual compound system gradually approaches this yrast line until it comes so close in energy to it that the emission of any more neutrons is not energetically possible. The nucleus then emits one or two γ-rays and de-excites to the yrast state at the angular momentum concerned. Under suitable circumstances, this yrast state may be an excited member of the rotational band and, from then on, a chain of γ-rays is emitted due to

74 (45-122/1977)

the transition to the successively lower members of the rotational band until the nucleus ends up in its ground state. Ideally, each of these γ-transitions is expected to be of the E2-type and strongly enhanced due to collective effects.

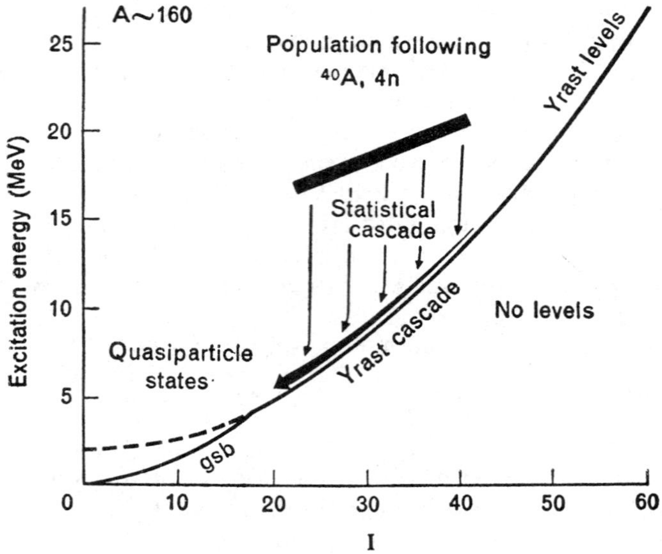

Fig. VI.15 Diagram showing final arrival of product of (^{40}A, 4n) reaction into yrast band of levels. [Following Stephens, F. S., *Revs. Mod. Phys.*, **47**, 57 (1975).]

The experiment entails making various measurements on the γ-rays, such as the measurement of (i) the γ-yield as a function of the bombarding energy (the so-called excitation function); (ii) the γ-yield at various angles with the incoming beam; (iii) the half-lives of the γ-transitions; and (iv) the γ-γ-coincidence.

On the basis of the measured energies and multipolarity of the γ-transitions, a level scheme of the nucleus is ultimately built up.

E. IMPORTANT EXPERIMENTAL FEATURES

Even-Even Nuclei

Using an angular momentum dependent moment of inertia $\mathcal{I}_I$, we can write the energy E_I of the state of angular momentum I as

$$E_I = \frac{\hbar^2}{2\mathcal{I}_I} I(I+1),$$

and hence

$$\frac{\hbar^2}{2\mathcal{I}_I} = \frac{\partial E_I}{\partial[I(I+1)]}. \tag{VI.170a}$$

The derivative in (VI.170a) can be evaluated from the experimental data only approximately in terms of finite differences as

$$\frac{\hbar^2}{2\mathcal{I}_I} = \frac{E_I - E_{I-2}}{I(I+1) - (I-2)(I-1)} = \frac{E_I - E_{I-2}}{2(2I-1)}. \tag{VI.170b}$$

In the same way, the angular frequency ω_I, for a given I, can be defined by

$$\mathcal{J}_I \omega_I = \hbar \sqrt{I(I+1)}. \qquad \text{(VI.171)}$$

Denoting $\sqrt{I(I+1)}$ by $\hat{I}$, we easily get

$$\frac{\partial E_I}{\partial \hat{I}} = \frac{\hbar^2}{2\mathcal{J}_I} 2\hat{I} = \frac{\hbar^2 \hat{I}}{\mathcal{J}_I} = \hbar\omega_I, \qquad \text{(VI.172a)}$$

where, in the last step, the definition (VI.171) has been used. The quantity ω_I can, once again, by taking the finite differences, be approximately evaluated from the data

$$\hbar\omega_I = \frac{E_I - E_{I-2}}{\sqrt{I(I+1)} - \sqrt{(I-2)(I-1)}} \approx \tfrac{1}{2}(E_I - E_{I-2}). \qquad \text{(VI.172b)}$$

For moderately large I, the difference in the denominator works out to be almost 2, and hence the last step. As long as $\mathcal{J}_I$ is roughly a constant, the energy difference in (VI.172b) increases with I, and hence ω_I also increases. The experimental data show this behaviour for not too large values of I. As I increases, the experimental level spacing starts falling below that given by the $I(I+1)$-rule for E_I, and hence $\mathcal{J}_I$, evaluated according to (VI.170b), starts showing an increase with I and, at the same time, ω_I, evaluated with (VI.172b), shows a retarded growth. Eventually, a stage may come when the energy difference in (VI.172b) starts diminishing from I to $(I+2)$. At this time, ω_I will, in fact, decrease with I, and $\mathcal{J}_I$, according to (VI.170b), will simultaneously show a fast increase.

This feature of the experimental data is shown in Fig. VI.16. The data may be displayed simply on an E_I-versus-$I(I+1)$ plot, as shown in the curves in Fig. VI.16a. This plot must

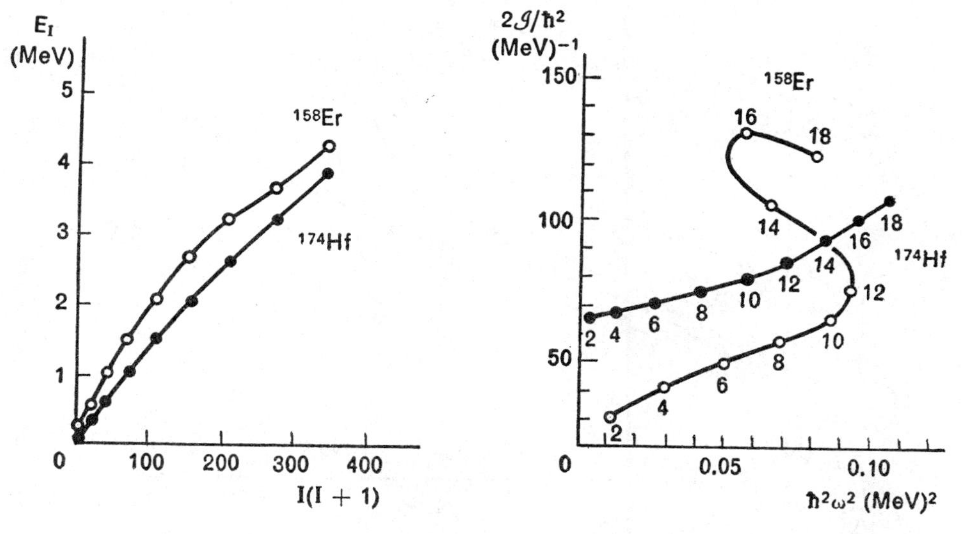

Fig. VI.16 Energy levels and backbending in ^{158}Er and ^{174}Hf. [Following Faessler, A., KFA (Jülich) preprint, XIth International School on Nuclear Physics, Predeal, Roumania, 1976.]

give a straight line with the slope equal to $\hbar^2/(2\mathcal{J})$ if the moment of inertia is a constant. A slight departure of the plot Hf^{174} from the straight line indicates a dependence of $\mathcal{J}_I$ on I which is not too drastic. On the other hand, the curve for Er^{158} indicates a marked change in the curvature at the values of I in the neighbourhood of 12. This fact has been demonstrated more emphatically in the $(2\mathcal{J}_I/\hbar^2)$-versus-$(\hbar\omega_I)^2$ plot in Fig. VI.16b. In the case of Hf^{174}, the increase in ω_I with increasing I has been retarded and the moment of inertia has started growing somewhat faster. In the case of Er^{158}, we notice that ω_I has even started decreasing at $I = 12$ while the moment of inertia has grown extraordinarily fast to a final value at $I = 16$–18, which is two-to-three times larger than the earlier value around $I = 10$–12. This particular S-shaped $\mathcal{J}_I$-versus-ω_I^2 curve is called the phenomenon of "backbending" and has been noticed in some cases.

Odd-Mass Nuclei

The backbending situation in odd-mass nuclei is somewhat similar to what is noticed for even nuclei, namely, it is observed only in some of the cases. The situation is also interesting when compared with neighbouring even nuclei. In Fig. VI.17, the case of Er^{159} is compared with

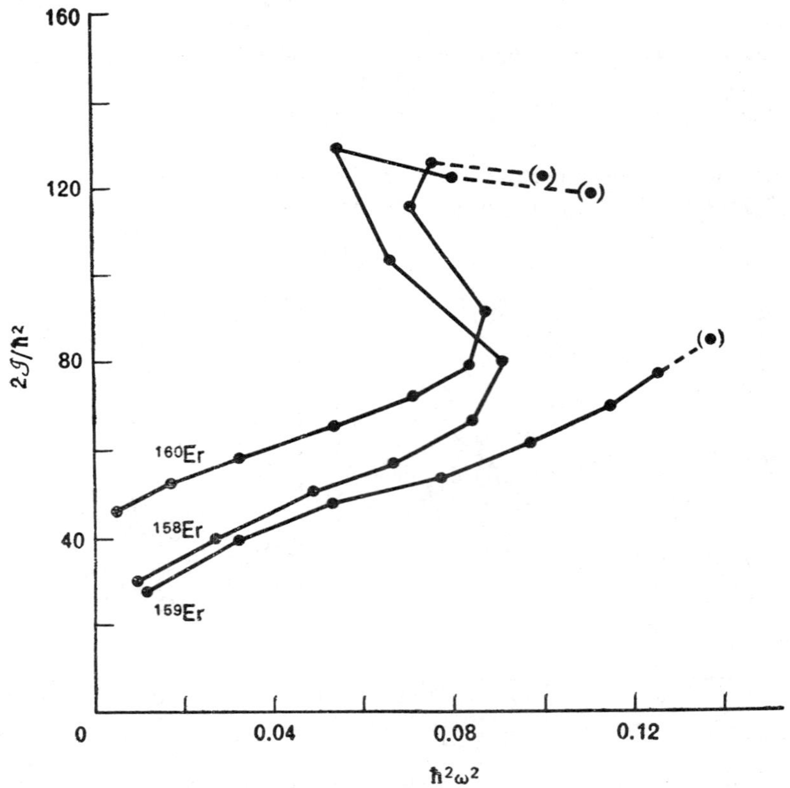

Fig. VI.17 Comparison of backbending plot for $^{158,160}Er$ with that of ^{159}Er. [Following Stephens, F. S., *Revs. Mod. Phys.*, **47**, 64 (1975).]

$Er^{158, 160}$. Although these even nuclei backbend, the odd neutron in Er^{159} has the effect of washing out the backbending trend. On the other hand, each of the odd-proton nuclei $Ho^{157, 159, 161}$ shows backbending, as is apparent from Fig. VI.18. The neighbouring even nuclei (Dy^{156}, Er^{158}), (Dy^{158}, Er^{160}), and (Dy^{160}, Er^{162}) corresponding to these three odd-proton nuclei also backbend.

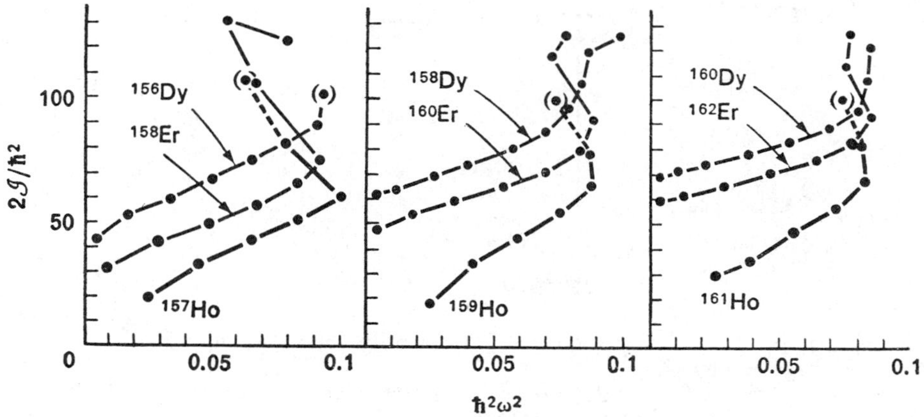

Fig. VI.18 Comparison of backbending plots of even- and odd-mass nuclei. [Following Stephens, F. S., *Revs. Mod. Phys.*, **47**, 64 (1975).]

In addition to the observations on backbending, there are several other interesting features of the data on odd-mass nuclei. As an example, the level spectra of $Er^{161, 163, 165}$ are shown in Fig. VI.19. It is clear that the ($\Omega = 5/2$)-band (Ω is the projection of **I** on the symmetry axis) is normal in Er^{165}, whereas in Er^{161} neighbouring pairs of levels are inverted. The energy spacings also are quite different from the pure rotational ones.

Another interesting feature of the data on odd-mass nuclei is shown in Fig. VI.20. Here the negative parity bands of $La^{125-133}$ are compared with the levels of the neighbouring even Ba-nuclei in the mass range $A = 124$–132. The levels of La are plotted relative to the $(11/2)^-$ level (which is not always the ground state). Two facts are to be noted here: first, in the rotational sequence of these odd-mass La isotopes, every alternate member is missing and, second, the spectra of each La isotope bears a close agreement with the $I = 0, 2, 4$ spectra of the preceding even Ba isotope.

F. THEORETICAL INTERPRETATION OF HIGH-SPIN STATES

In Chapter IV, we have explained that a rotational band represents the collective rotation of a deformed nucleus as a whole, maintaining the intrinsic state of the nucleons. The same nucleus may display several rotational bands built on different intrinsic states of the nucleons. To understand the moment of inertia of the ground-state band, we have therefore to know the nature of the corresponding intrinsic state of the nucleons. Suppose there is an excited intrinsic state corresponding to a second band having a larger moment of inertia; the level spacings in this band are then smaller than those in the ground-state band. As a result, the two bands cross each other at some value of angular momentum I_c, as shown in Fig. VI.21a. After the crossing point, the levels of the ground-state band are higher than those of the excited band. If we concentrate on the yrast states, then we may conclude that, for these states, the moment

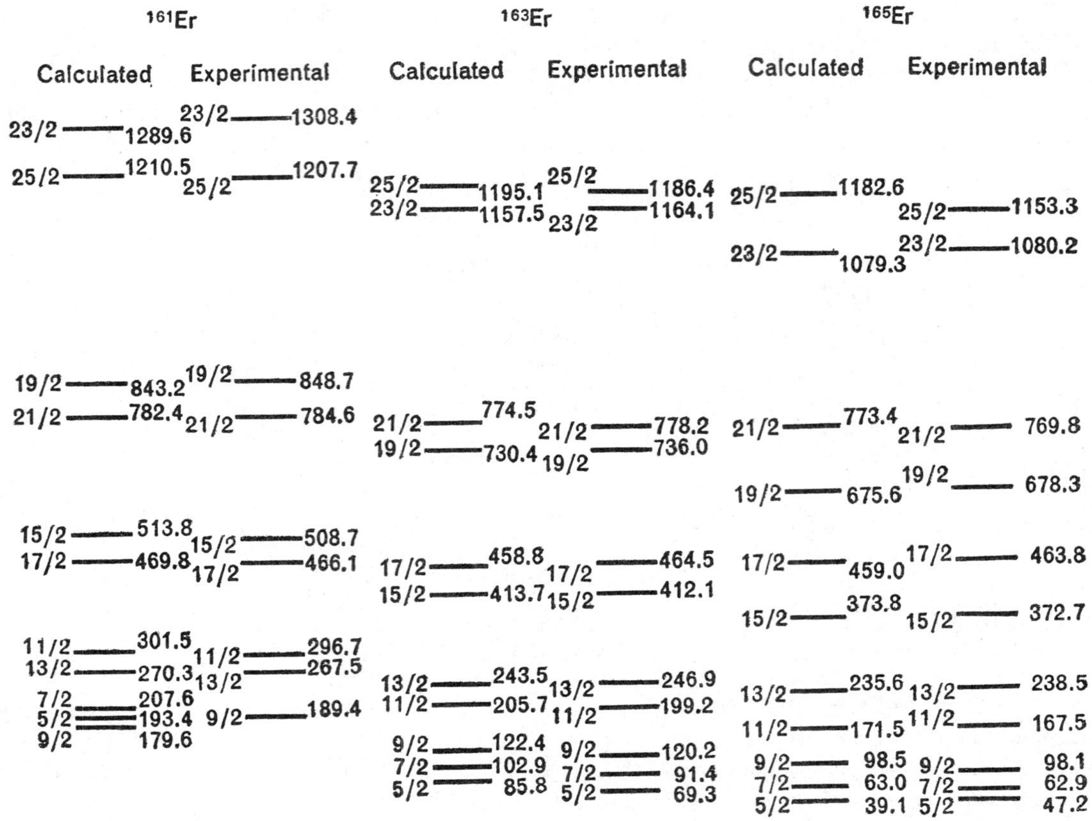

Fig. VI.19 Experimental and calculated positive parity levels in odd-mass Er-nuclei. [Following Stephens, F. S., *Revs. Mod. Phys.*, **47**, 48 (1975).]

of inertia before and after the crossing point is different; the levels after the crossing point have a larger moment of inertia. At the crossing point, the angular momentum I_c on the two bands is identical and, since it is equal to $(\mathcal{J}\omega)$ for both the bands, the value of ω for the ground-state band at I_c is larger than that for the excited band at the same angular momentum. This idealized picture of band crossing gives a very preliminary but qualitatively satisfying explanation of backbending, as depicted by the two lines with the sharp break (dashed line) between them in Fig. VI.21b.

In the nonidealized version, we have to put in ingredients that explain the mild changes in $\mathcal{J}_I$ as a function of I before the backbending point, and mix the two bands near I_c so as to reproduce a smoother change from one band to the other and, what is more important, to keep open the possibility of gamma-transition between the two levels on the two sides of I_c. In Fig. VI.21a, the dotted and dashed curves represent the lower mixed band for strong and weak mixing, respectively. Depending on the strength of the admixture in Fig. VI.21a, the transition in the backbending region is less distinct, as shown by the curves A and B in Fig. VI.21b, and may even be completely washed out (curve C) for very strong mixing.

The main point in the theoretical understanding is then the nature of the two intrinsic

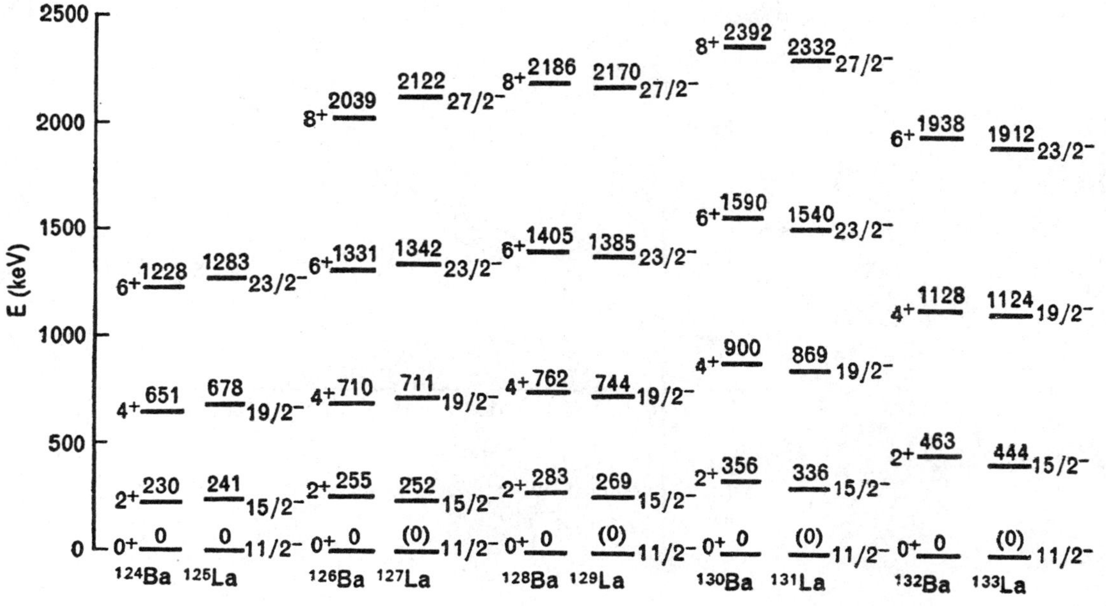

Fig. VI.20 Comparison of negative parity bands in odd-mass La isotopes with ground band of Ba-nuclei. [Following Stephens, F. S., *Revs. Mod. Phys.*, **47**, 54 (1975).]

states that give rise to the crossing bands. The ground-state band is pretty well understood. It corresponds to an intrinsic state of the Hartree-Fock-Bogoliubov (HFB) type [see Appendix G (Section II)] in which pairs of neutrons and protons are strongly correlated through a strong two-body pairing interaction. We have already learnt that the pairing correlation reduces the moment of inertia to a value that is half to one-third of the rigid-body value in keeping with observations. The mild variation of $\mathcal{J}_I$ in the ground-state band as a function of I is a matter of detail that has to be reproduced by a self-consistent HFB treatment varying the deformation and the pairing energy gap with changing I.

Several alternatives have been suggested with respect to the excited band which has to have a larger moment of inertia than that of the ground-state band. Earlier treatments used an explanation emanating from a suggestion by Mottelson and Valatin[24], proposed several years before the experiments on high-spin states. This explanation is now sketched. According to the pairing ideas, the nucleus is excited by creating pairs of quasiparticles for which a minimum energy 2Δ is needed, where Δ is the energy gap. We have also explained that it is this energy gap that accounts for a twofold-to-threefold reduction in the value of the moment of inertia, as compared with the value for a shell-model state. When the nucleus rotates very fast, the individual nucleon angular momentum **j** interacts with the angular momentum **I** of the spinning nucleus through an interaction proportional to $\mathbf{I}\cdot\mathbf{j}$. This interaction has been popularly called the Coriolis coupling. It tends to bring the **j** of every individual nucleon into alignment with **I**, and hence the strong pairing correlation between pairs of nucleons having the angular momentum projection m and $-m$ gets destroyed. As I increases, this destruction

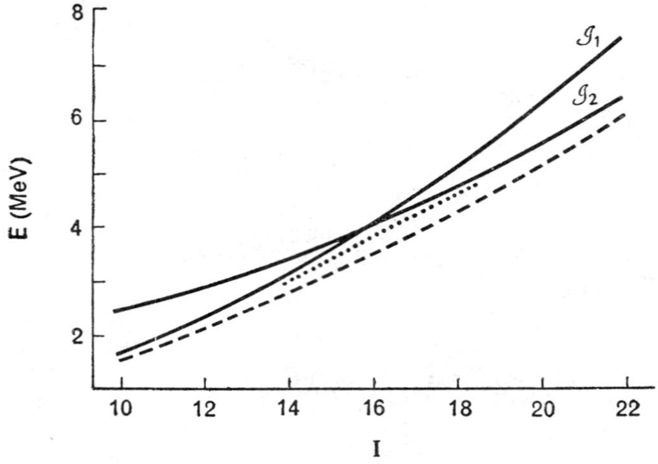

(a) Crossing of two bands with moments of inertia $\mathcal{J}_1$ and $\mathcal{J}_2$ at $I_c = 16$

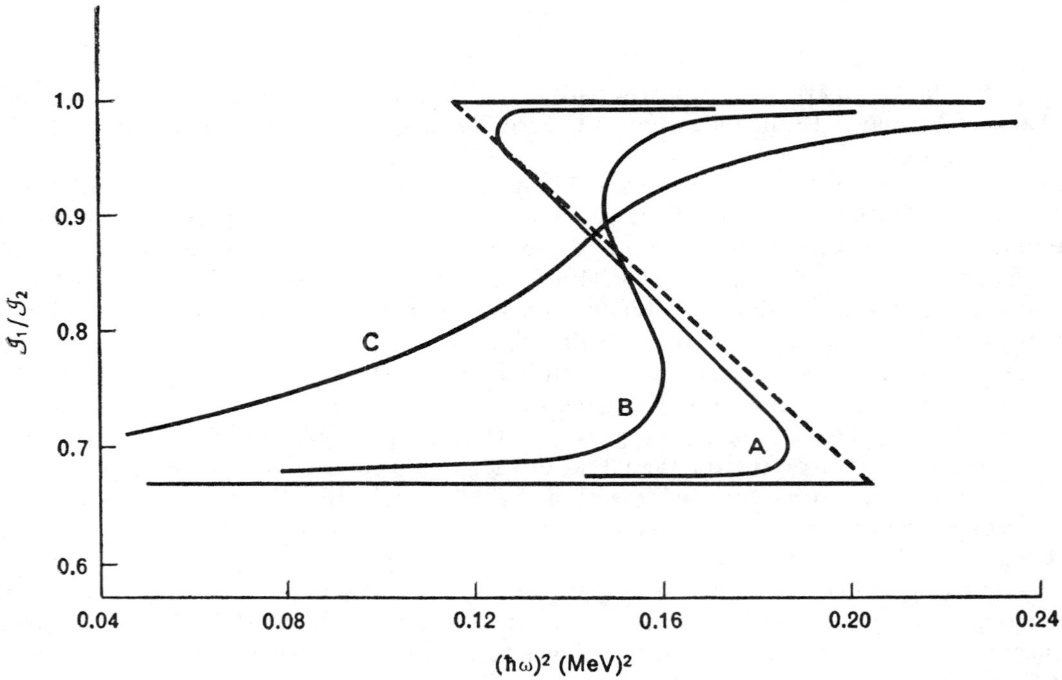

(b) Backbending plot for yrast band as per idealized situation represented in Fig. VI.21a

Fig. VI.21 Band crossing and backbending plot—idealized case. [Following Stephens, F. S., *Revs. Mod. Phys.*, **47**, 58, 59 (1975).]

of pairing correlation goes on, and ultimately we reach a completely unpaired intrinsic state for which the moment of inertia has the rigid-body value, i.e., two-to-three times the value of the initial paired intrinsic state. Thus, according to this model, the pair correlated BCS type ground state undergoes a phase transition to the normal state that has no pair correlation.

Another explanation, which now seems most often applicable, was first suggested by Stephens et al[25]. To understand it, let us consider the core-particle coupling model of Section 33C described by the Hamiltonian (IV.123) whose coupling term is given by (IV.132). For an axially symmetric core, the coupling term simplifies to (IV.134) and the rotational part to (IV.137). Under the same circumstances, I_z and j_z have the same well-defined value, say, Ω.

In the old treatment of the collective model described in Chapter IV, the coupling term was assumed to be very strong (the strong coupling assumption), and the motion of the collective variables β, γ, and of the rotation angles, was taken to be slower than the particle motion (adiabatic assumption). Hence, the particle motion was solved by the Nilsson model for given values of the collective deformation variables. The effect of the Coriolis coupling term $(I_+ j_- + I_- j_+)$, present in (IV.137), was then included in some cases, especially in the case of the $(k = \frac{1}{2})$-bands, by admixing the $(k = \frac{3}{2})$-band through this interaction.

Stephens et al[25] pointed out that the relative magnitude of the coupling and the Coriolis term should be carefully considered in every case. It is clear that in instances where $\mathcal{J}$ is small and I and j large, the Coriolis term can be quite large as compared with the coupling term. In such cases, a detailed diagonalization has to be done to obtain reliable results, using a large number of bands. The experimental data in such a situation are shown in Fig. VI.19, which also gives the theoretical results of the band-mixing calculation.

An interesting fact pointed out by Stephens and his colleagues relates to those nuclei where the last odd nucleon belongs to the high j-orbital which intrudes into the major oscillator shell from the shell above and has a parity opposite to that of the other levels of the major shell, for example, $h_{11/2}$ in the major shell comprising $(g_{7/2}, d_{5/2}, d_{3/2}, s_{1/2})$-levels, and $i_{13/2}$ in the shell comprising $(h_{9/2}, f_{7/2}, f_{5/2}, p_{3/2}, p_{1/2})$-levels. In the Nilsson treatment, such a level remains quite pure, except that only the various projection components split under the influence of the deformation term $k\beta Y_0^2$. For the projection state Ω of angular momentum j, the matrix element of this deformation term is given by

$$k\beta\left[\frac{3\Omega^2 - j(j+1)}{4j(j+1)}\right] \tag{VI.173a}$$

and the $(I_z^2 + j_z^2)$-term from (IV.137) has the value

$$-\frac{\hbar^2}{\mathcal{J}}\Omega^2. \tag{VI.173b}$$

It is clear from (VI.173a) that, for prolate (β = positive) deformations, smaller Ω-components of j lie lower, whereas the reverse is the case for oblate (β = negative) deformations. Moreover, by taking the Ω^2-terms of (VI.173) together [the other term of (VI.173a) is a constant for a given j], it is interesting to explore whether or not, in the prolate case, the coefficients can mutually cancel out each other. In that case, we conclude that the Coriolis term becomes very important and has to be treated carefully. Since the body has symmetry about the z-axis, there cannot be any rotation about this axis. The Coriolis term, which is equal to $(I_x j_x + I_y j_y)$, can therefore be replaced by $I\mathbf{j}\cdot\mathbf{n}$, where $\mathbf{n}$ is the rotation axis. Thus, the single-particle eigenstates of the Coriolis term will be states of the type $|j, \alpha\rangle$, where α denotes the projection of $\mathbf{j}$ along

75 (45-122/1977)

the rotation axis $\mathbf{n}$. This fact has been called the *rotation alignment effect*, which is the complete decoupling of the particle from the rotation of the core. It has been shown, from symmetry arguments, that $(I - \alpha)$ must be even and α, of course, has values starting from $\frac{1}{2}$ and ending with j. The energy of a nuclear state specified by (I, j, α) is given by

$$E(I, j, \alpha) = e_j + \frac{\hbar^2}{2\mathcal{J}}[(I - \alpha)(I - \alpha + 1) + j(j + 1) - \alpha^2]. \tag{VI.174}$$

Thus, the highest α-states have the lowest energy. Further, for each α, a set of rotational levels is generated with the moment of inertia $\mathcal{J}$ of the core and the quantum number $(I - \alpha)$ playing the role of the rotation angular momentum. Since $(I - \alpha)$ is even, we get a set of rotational levels in the band specified by α with the angular momentum $I = \alpha, \alpha + 2, \alpha + 4, \ldots$. For the lowest lying band, $\alpha = j$ and we have $I = j, j + 2, j + 4, \ldots$. All these observations explain the data in Fig. VI.20.

In the case of even nuclei, the ground-state band is still the completely paired state. However, the excited band in this model has to be secured by considering excitations of two quasiparticles in the high j-orbital. The Coriolis interaction mixes the ground state and all these two-quasiparticle states with each other. The fit to the data is then a matter of detailed computation. At large values of I, the excited band is predominantly of two decoupled quasiparticles with their α-quantum number equal to j and $j - 1$ (because of the Pauli principle). In the odd-mass nuclei, the ground band at high spin has a decoupled quasiparticle $\alpha = j$, and the excited band has three quasiparticles $\alpha = j, j - 1, j - 2$. Referring back to Fig. VI.17, we note that the odd-neutron nucleus ^{159}Er has an ($\alpha = 13/2$, $i_{13/2}$)-neutron quasiparticle in the ground band, whereas in the excited band it has the ($\alpha = 13/2, 11/2, 9/2$)-neutron quasiparticles decoupled from the core. Both the even nuclei ^{158,160}Er, which have been compared with ^{159}Er in this figure, have a paired ground-state band and the excited band of ($\alpha = 13/2, 11/2$)-decoupled quasiparticles. The energy difference between the aforementioned one- and three-quasiparticle states is higher than that between the zero- and two-quasiparticle states. Thus, the band crossing in the odd-mass nucleus occurs at higher energies and angular momenta as compared with the even ones. This qualitatively explains the absence of backbending of ^{159}Er. The odd-proton nuclei $^{157-161}$Ho of Fig. VI.18, on the other hand, have a proton quasiparticle in the ground state and their excited bands can be had by exciting two neutron quasiparticles with $\alpha = 13/2, 11/2$ without violating the Pauli principle. Therefore, in this case, the energy difference between the one- and three-quasiparticle states for the odd-mass nucleus is identical to that between the zero- and two-quasiparticle states of the even nuclei. Thus, the backbending behaviour of all the three nuclei is expected to be identical.

The Coriolis decoupling model is suited to all the cases where a high j single-particle level is involved. In other cases, the Mottelson-Valatin effect may still have to be invoked. The estimates of the critical angular momentum made by Mottelson and Valatin were around 14. Later sophisticated estimates place this value higher. If the later estimates are reliable, this explanation may still be invoked in cases where backbending (if found at all) occurs at $I = 20$ or higher.

There are many detailed calculations (see Kumar[26]) on the high-spin states based on the self-consistent HFB method. The projection of angular momentum, the correct number of particles, the various self-consistency requirements on the deformation and pairing parameters involved in this method are very cumbersome, and almost every group working on it has made some compromise. In an approximate version of this method, the Hamiltonian $H - \omega J_x$ is self-consistently solved, treating ω as a Lagrange multiplier. It is determined for each I by requiring the given value $\sqrt{I(I+1)}$ of the angular momentum $\langle J_x \rangle$, where x is the rotation axis. This simplified microscopic method is similar to the Coriolis decoupling model in many of its physical contents.

G. CURRENT TRENDS

The Coriolis decoupling model has been extended and applied to cases where the departure from axial symmetry is important (see Meyer-ter-Vehn[27]).

There have been speculations that, at very high spins $I \approx 60$–$70\hbar$, where the rotation is nearly classical, the nuclear rotation may give rise to an appearance of the oblate shape. This classical problem was studied long ago by Chandrasekhar[28], and the backbending to be expected as a result of this sudden change in nuclear shape at very high spin is called *giant backbending* in recent work.

At very high spin, the levels form a near continuum, and hence the experimental study of these levels in the (γ-γ)-coincidence work is difficult. The problem arises when we aim to proceed very much beyond the ($I = 24$–26)-levels. Some progress has been made by Stephens and Diamond[29] and by Herskind et al[30].

Several good review articles exist on the subject of high-spin states. See, for example, Johnson and Szymanski[31], Stephens[31], and Faessler[31].

REFERENCES

1. Elliott, J. P., and Flowers, B. H., *Proc. Roy. Soc.* (London), **A242**, 57 (1957).
2. Gillet, V., *Nucl. Phys.*, **51**, 410 (1964); Gillet, V., and Sanderson, E. A., *Nucl. Phys.*, **54**, 472 (1964); Gillet, V., and Vinh-Mau, N., *Nucl. Phys.*, **54**, 321 (1964).
3. Green, A. M., *Rep. Prog. Phys.*, **28**, 113 (1965).
4. Carter, E. B., Mitchell, G. E., and Davis, R. H., *Phys. Rev.*, **133**, B1421 (1964).
5. Fallieros, S., Ferrell, R. A., and Pal, M. K., *Nucl. Phys.*, **15**, 363 (1960); Pal, M. K., and Lee, Y. C., *Bull. Am. Phys. Soc.*, **4**, 407 (1959); Lee, Y. C., Ph.D. thesis, Department of Physics, University of Maryland, 1963 (unpublished).
6. Brown, G. E., and Green, A. M., *Nucl. Phys.*, **75**, 401 (1966).
7. Ripka, G., and Bassichis, W., *Phys. Letters*, **15**, 320 (1965).
8. Stephenson (Jr.), G. J., and Banerjee, M. K., *Phys. Letters*, **24B**, 209 (1967).
9. Blomqvist, I., and Wahlborn, S., *Arkiv fur Fysik*, **16**, No. 46 (1959).
10. Kisslinger, L. S., and Sorensen, R. A., *Kgl. Danske Videnskab. Selskab, Mat. Fys. Medd.*, **32**, No. 9 (1960); *Revs. Mod. Phys.*, **35**, 853 (1963).
11. Arvieu, R., Baranger, E., Baranger, M., Gillet, V., and Veneroni, M., *Phys. Letters*, **4**, 119 (1963); Arvieu, R., and Veneroni, M., *Phys. Letters*, **5**, 142 (1963); Arvieu, R., Salusti, E., and Veneroni, M., *Phys. Letters*, **8**, 334 (1964).
12. Pal, M. K., Gambhir, Y. K., and Ram Raj, *Phys. Rev.*, **155**, 1144 (1967); *ibid*, **162**, 1139 (1967); *ibid*, **163**, 1004 (1967).
13. Savoia, M., Sawicki, J., and Tomasini, A., *Nuovo Cimento*, **32**, 991 (1964); Ottaviani, P. L., Savoia, M., Sawicki, J., and Tomasini, A., *Phys. Rev.*, **153**, 1138 (1967); Ottaviani, P. L., Savoia, M., and Sawicki, J., *Phys. Letters*, **24B**, 353 (1967); Gmitro, M., Hendekovic, J., and Sawicki, J., *Phys. Letters*, **26B**, 252 (1968).

14. Lawson, R. D., Macfarlane, M. H., and Kuo, T. T. S., *Phys. Letters*, **22**, 168 (1966); Kuo, T. T. S., *Nucl. Phys.*, **A90**, 199 (1967); *ibid*, **A103**, 71 (1967); Kuo, T. T. S., *Phys. Letters*, **26B**, 63 (1967).

15. Gmitro, M., Hendekovic, J., and Sawicki, J., *Phys. Rev.*, **169**, 983 (1968); Gmitro, M., Rimini, A., Sawicki, J., and Weber, T., *Phys. Rev.*, **175**, 1243 (1963).

16. Gillet, V., and Rho, M., *Phys. Letters*, **21**, 82 (1966); Gillet, V., Giraud, B., and Rho, M., *Nucl. Phys.*, **A103**, 257 (1967).

17. Gambhir, Y. K., *Nucl. Phys.*, **A120**, 193 (1968).

18. Sawicki, J., in Proceedings of the 12th International Summer Meeting in Physics, Herceg Novi, Yugoslavia, 1967, p 281.

19. Pal, M. K., in Theory of Nuclear Structure: Trieste Lectures 1969, IAEA, Vienna (1970), pp 547–695.

20. Soloviev, V. G., *Atom. Energy Rev.*, **3**, 117 (1965); in Proceedings of the Dubna Symposium on Nuclear Structure, IAEA, Vienna, 1968, p 101.

21. Belyaev, S. T., and Zelevinsky, V. G., *Nucl. Phys.*, **39**, 582 (1962); Sorensen, B., *Suppl. to J. Phys. Soc.* (Japan), **24**, 588 (1968).

22. Nilsson, S. G., and Prior, O., *Dan. Mat. Fys. Medd.*, **32**, No. 16 (1961); Griffin, J. J., and Rich, M., *Phys. Rev.*, **118**, 850 (1960).

23. Mariscotti, M. A. J., Scharff-Goldhaber, G., and Buck, B., *Phys. Rev.*, **178**, 1864 (1969).

24. Mottelson, B. R., and Valatin, J. G., *Phys. Rev. Letters*, **5**, 511 (1960).

25. Stephens, F. S., Diamond, R. M., and Nilsson, S. G., *Phys. Letters*, **44B**, 429 (1973).

26. Kumar, K., *Physica Scripta* (Sweden), **6**, 270 (1972); Banerjee, B., Mang, H. J., and Ring, P., *Nucl. Phys.*, **A215**, 366 (1973); Ansari, A., and Nair, S. C. K., *Nucl. Phys.*, **A283**, 326 (1977); Faessler, A., Sandhya Devi, K. R., Grümmer, F., Schmid, K. W., and Hilton, R. R., *Nucl. Phys.*, **A256**, 106 (1976); Warke, C. S., and Gunye, M. R., *Phys. Rev.*, **C12**, 1647 (1975); *ibid*, **C13**, 859 (1976).

27. Meyer-ter-Vehn, J., *Nucl. Phys.*, **A249**, 111, 141 (1975); Toki, H., and Faessler, A., *Nucl. Phys.*, **A253**, 231 (1975); *Phys. Letters*, **59B**, 211 (1975).

28. Chandrasekhar, S., Ellipsoidal Figures of Equilibrium, Yale Press, Yale, Connecticut, 1969.

29. Stephens, F. S., in Proc. Int. Symp. on Highly Excited States in Nuclei (Jülich), 1975, Vol 2, ed. by A. Faessler, 1976, p 12.

30. Hagemann, G. B., Broda, R., Herskind, B., Ishihara, M., Ogaza, S., and Ryde, H., *Nucl. Phys.*, **A245**, 166 (1975).

31. Johnson, A., and Szymanski, Z., *Phys. Rep.*, **7**, 181 (1973); Stephens, F. S., *Revs. Mod. Phys.*, **47**, 43 (1975); Faessler, A., KFA (Jülich) preprint, XIth International School on Nuclear Physics, Predeal, Roumania, 1976.

PROBLEMS

1. Derive the completeness relation for the eigenvectors of the RPA-matrix defined by (VI.28).

2. Define the pair addition and pair removal operators as $A^\dagger_{pp'} = C^\dagger_p C^\dagger_{p'}$ and $A_{hh'} = C_{h'}C_h$, and use the commutator method to obtain the linearized system of equations satisfied by them.

3. Show, by semiclassical arguments, that $\partial E_J/\partial\sqrt{J(J+1)}$ is the square of the angular velocity of a rotational nucleus in the angular momentum state J.

Appendix A

Angular Momentum

I. ORBITAL ANGULAR MOMENTUM

Let $\mathbf{L}\hbar$ denote the orbital angular momentum operator. The properties of the components of $\mathbf{L}$ and of the spherical harmonics $Y^l_m(\theta, \phi)$, which are the eigenfunctions of L^2 and L_z, are discussed in standard texts on quantum mechanics (see, for example, Schiff[1]). Some important results are listed here:

$$L^2 Y^l_m(\theta, \phi) = l(l+1) Y^l_m(\theta, \phi), \qquad l = 0, 1, 2, \ldots, \tag{AI.1}$$

$$L_z Y^l_m(\theta, \phi) = m Y^l_m(\theta, \phi), \qquad m = -l \rightarrow +l \text{ (integrally spaced)}, \tag{AI.2}$$

$$(L_x \pm iL_y) Y^l_m(\theta, \phi) = [(l \mp m)(l \pm m + 1)]^{1/2} Y^l_{m\pm 1}(\theta, \phi). \tag{AI.3}$$

The step-up and step-down operators $L_x \pm iL_y$ will frequently be denoted by $L_\pm$. The orthonormality condition is given by

$$\int_0^{2\pi} d\phi \int_0^{\pi} \sin\theta \, d\theta \; Y^{l'*}_{m'}(\theta, \phi) Y^l_m(\theta, \phi) = \delta_{ll'}\delta_{mm'}. \tag{AI.4}$$

The spherical harmonic can be broken up into a θ-part and a ϕ-part as

$$Y^l_m(\theta, \phi) = P^l_m(\cos\theta)\Phi_m(\phi), \tag{AI.5}$$

where

$$\Phi_m(\phi) = \frac{1}{\sqrt{2\pi}} e^{im\phi}, \tag{AI.6a}$$

$$P^l_m(\zeta) = \left\{\frac{[l](l+m)!}{2(l-m)!}\right\}^{1/2} \frac{1}{2^l l!}(1-\zeta^2)^{-m/2} \frac{d^{l-m}}{d\zeta^{l-m}}(\zeta^2 - 1)^l. \tag{AI.6b}$$

Here $\zeta = \cos\theta$ and $[l] = 2l + 1$; the function P^l_m is called the associated Legendre polynomial.

Renormalized Spherical Harmonics and Legendre Polynomial

Renormalized spherical harmonics and Legendre polynomials are defined respectively by

$$y^l_m(\theta, \phi) = \sqrt{\frac{4\pi}{[l]}}\, Y^l_m(\theta, \phi) \tag{AI.7a}$$

and

$$P_l(\cos\theta) = y^l_0(\theta) = \frac{1}{2^l l!} \frac{d}{d\zeta^l}(\zeta^2 - 1)^l. \tag{AI.7b}$$

(AI.4) yields

$$\int_{-1}^{+1} d\zeta \, P_l(\zeta) P_{l'}(\zeta) = \frac{2}{[l]} \delta_{ll'}. \tag{AI.8}$$

The specific expressions for $l = 0, 1, 2$ are listed in Table AI.1.

Table AI.1 Renormalized spherical harmonics

l	m	$y_m^l(\theta, \phi)$
0	0	1
1	± 1	$\mp \frac{1}{\sqrt{2}} \sin\theta \, e^{\pm i\phi}$
	0	$\cos\theta$
2	± 2	$\frac{1}{2}\sqrt{\frac{3}{2}} \sin^2\theta \, e^{\pm 2i\phi}$
	± 1	$\mp \sqrt{\frac{3}{2}} \cos\theta \sin\theta \, e^{\pm i\phi}$
	0	$\frac{1}{2}(3\cos^2\theta - 1)$

It is clear from Table AI.1 that

$$r_{\pm 1} \equiv r y_{\pm 1}^1(\theta, \phi) = \mp \frac{1}{\sqrt{2}}(x + iy), \tag{AI.9a}$$

$$r_0 \equiv r y_0^1(\theta, \phi) = z. \tag{AI.9b}$$

Expression for Components

$$L_z = -i\frac{\partial}{\partial\phi}, \tag{AI.10a}$$

$$L_{\pm} = e^{\pm i\phi}\left(\pm \frac{\partial}{\partial\theta} + i \cot\theta \frac{\partial}{\partial\phi}\right), \tag{AI.10b}$$

$$-L^2 = \frac{1}{\sin\theta}\frac{\partial}{\partial\theta}\left(\sin\theta \frac{\partial}{\partial\theta}\right) + \frac{1}{\sin^2\theta}\frac{\partial^2}{\partial\phi^2}. \tag{AI.10c}$$

Relationship with ∇^2

The operator L^2 is related to ∇^2 as

$$\nabla^2 = \nabla_r^2 - \frac{1}{r^2}L^2, \tag{AI.11a}$$

where

$$\nabla_r^2\left(\frac{R}{r}\right) = \frac{1}{r^2}\frac{\partial}{\partial r}\left(r^2\frac{\partial}{\partial r}\right)\left(\frac{R}{r}\right) = \frac{1}{r}\frac{d^2R}{dr^2}. \tag{AI.11b}$$

II. ANGULAR MOMENTUM (General)

The details on angular momentum may be obtained from Condon and Shortley[1] or any other standard text on the subject (see, for example, Rose[1]). The angular momentum operator is denoted by $\mathbf{J}\hbar$, and its components satisfy the commutation relations

$$[J_x, J_y] = iJ_z \text{ (and cyclic permutations of } x, y, z). \tag{AII.1}$$

The simultaneous eigenstates of J^2 and J_z are denoted by $|jm\rangle$ such that

$$J^2|jm\rangle = j(j+1)|jm\rangle, \tag{AII.2a}$$

$$J_z|jm\rangle = m|jm\rangle, \tag{AII.2b}$$

where j has either integral or half-integral positive values and m varies between $-j$ and $+j$, integrally spaced. The step-up and step-down operators $J_\pm$ yield

$$J_\pm|jm\rangle = [(j \mp m)(j \pm m + 1)]^{1/2}|j, m \pm 1\rangle, \tag{AII.2c}$$

$$J_+|j, m = j\rangle = 0, \qquad J_-|j, m = -j\rangle = 0. \tag{AII.2d}$$

III. SPIN ANGULAR MOMENTUM

Spin angular momentum is a special case of angular momentum (with $j = \frac{1}{2}$) for which we use the symbol $\mathbf{s}$ instead of $\mathbf{J}$. The Pauli spin operator $\boldsymbol{\sigma}$ is defined by

$$\boldsymbol{\sigma} = 2\mathbf{s}. \tag{AIII.1}$$

The spin eigenstates are denoted by $\chi_m^{1/2}$ instead of $|\frac{1}{2}, m\rangle$, where the superscript refers to $j = \frac{1}{2}$. The two substates $m = \pm\frac{1}{2}$ are called the spin-up and spin-down states. The results (AII.1) and (AII.2) apply. With the help of (AII.2a), (AII.2b), and (AII.2c), we can easily construct the matrix elements $\langle\chi_m^{1/2}|\, s_i\, |\chi_{m'}^{1/2}\rangle$, where i denotes any of the components x, y, z. It is more convenient to write the matrices for the components of $\boldsymbol{\sigma}$ which are given by

$$\sigma_x = \begin{pmatrix} 0 & 1 \\ 1 & 0 \end{pmatrix}, \qquad \sigma_y = \begin{pmatrix} 0 & -i \\ i & 0 \end{pmatrix}, \qquad \sigma_z = \begin{pmatrix} 1 & 0 \\ 0 & -1 \end{pmatrix}. \tag{AIII.2}$$

Some Properties of Pauli Spin Matrices

$$\sigma_k\sigma_l + \sigma_l\sigma_k = 2\delta_{kl}\mathbb{1}, \tag{AIII.3a}$$

$$\sigma_x\sigma_y\sigma_z = i\mathbb{1}, \tag{AIII.3b}$$

$$\mathrm{Tr}\,\sigma_k = 0. \tag{AIII.3c}$$

Here k, l are any two of the components x, y, z, and $\mathbb{1}$ is the 2×2 unit matrix.

Arbitrary Spin State

In the two-dimensional representation, used for the Pauli matrices, the basis vectors are the two eigenstates $\chi_m^{1/2}$ with $m = \pm\frac{1}{2}$. These states themselves are represented by $\begin{pmatrix} 1 \\ 0 \end{pmatrix}$ and $\begin{pmatrix} 0 \\ 1 \end{pmatrix}$. σ_z has the value $+1$ and -1 for these two states; therefore, these two states correspond to spin alignment along the z-axis and along the direction opposite to the z-axis, respectively. It will now be seen that an arbitrary spin state $\begin{pmatrix} a \\ b \end{pmatrix}$ with

$$|a|^2 + |b|^2 = 1 \tag{AIII.4}$$

has spin alignment in a direction (θ, ϕ) in space, determined in a particular way by the values of a and b. Let $\mathbf{n}$ be the unit vector in the direction (θ, ϕ) such that

$$n_x = \sin\theta\cos\phi, \qquad n_y = \sin\theta\sin\phi, \qquad n_z = \cos\theta. \tag{AIII.5}$$

By our assertion, $(\boldsymbol{\sigma}\cdot\mathbf{n})$ has the value unity for the state $\begin{pmatrix}a\\b\end{pmatrix}$, that is,

$$(\boldsymbol{\sigma}\cdot\mathbf{n})\begin{pmatrix}a\\b\end{pmatrix} = \begin{pmatrix}a\\b\end{pmatrix}. \tag{AIII.6}$$

It is a straightforward exercise to obtain, from this equation, with the help of (AIII.2) and (AIII.5), the result

$$\frac{a}{b} = e^{-i\phi}\cot\frac{\theta}{2}. \tag{AIII.7}$$

Using (AIII.4), we get

$$b = e^{i\xi}\sin\frac{\theta}{2}, \qquad a = e^{-i(\phi-\xi)}\cos\frac{\theta}{2}, \tag{AIII.8}$$

where ξ is an arbitrary phase. For convenience, we choose $\xi = \phi/2$, and then

$$a = \cos\frac{\theta}{2}\,e^{-i(\phi/2)}, \qquad b = \sin\frac{\theta}{2}\,e^{i(\phi/2)}. \tag{AIII.9}$$

IV. COUPLING OF TWO ANGULAR MOMENTA

If $\mathbf{J}_1$ and $\mathbf{J}_2$ denote two independent (i.e., commuting) angular momenta, each satisfying (AII.1), then the resultant $\mathbf{J}$ defined by

$$\mathbf{J}_1 + \mathbf{J}_2 = \mathbf{J}$$

also satisfies (AII.1), and hence denotes an angular momentum operator in the quantum mechanical sense. In keeping with Condon and Shortley[1], the product states $|j_1m_1\rangle|j_2m_2\rangle$ are denoted by $|j_1j_2m_1m_2\rangle$ and the eigenstates of the coupled angular momentum by $|j_1j_2jm\rangle$. Using the completeness of each of these two sets of state vectors, we obtain

$$|j_1j_2m_1m_2\rangle = \sum_{jm} |j_1j_2jm\rangle\langle j_1j_2jm\,|\,j_1j_2m_1m_2\rangle, \tag{AIV.1a}$$

$$|j_1j_2jm\rangle = \sum_{m_1m_2} |j_1j_2m_1m_2\rangle\langle j_1j_2m_1m_2\,|\,j_1j_2jm\rangle. \tag{AIV.1b}$$

With a suitable choice of phase of the states, the transformation coefficients in (AIV.1) can be made real, and then they become identical. Therefore, a common symbol $\begin{bmatrix} j_1 & j_2 & j \\ m_1 & m_2 & m \end{bmatrix}$ is used to denote them. This coefficient is called the Clebsch-Gordon coefficient, and it is nonvanishing only when

$$m_1 + m_2 = m, \tag{AIV.2a}$$

$$|j_1 - j_2| \leqslant j \leqslant (j_1 + j_2). \tag{AIV.2b}$$

The requirement (AIV.2b) is often called the triangular ($\triangle$) inequality of j_1, j_2, and j. A few important properties of the Clebsch-Gordon coefficients are now listed.

Orthogonality Propertie

$$\sum_{m_1m_2}\begin{bmatrix} j_1 & j_2 & j \\ m_1 & m_2 & m \end{bmatrix}\begin{bmatrix} j_1 & j_2 & j' \\ m_1 & m_2 & m' \end{bmatrix} = \delta_{jj'}\delta_{mm'}, \tag{AIV.3a}$$

$$\sum_{jm}\begin{bmatrix} j_1 & j_2 & j \\ m_1 & m_2 & m \end{bmatrix}\begin{bmatrix} j_1 & j_2 & j \\ m_1' & m_2' & m \end{bmatrix} = \delta_{m_1m_1'}\delta_{m_2m_2'}. \tag{AIV.3b}$$

Symmetry Properties

$$\begin{bmatrix} j_1 & j_2 & j \\ m_1 & m_2 & m \end{bmatrix} = (-1)^{j_1+j_2-j}\begin{bmatrix} j_2 & j_1 & j \\ m_2 & m_1 & m \end{bmatrix} = (-1)^{j_1+j_2-j}\begin{bmatrix} j_1 & j_2 & j \\ -m_1 & -m_2 & -m \end{bmatrix}$$
$$= (-1)^{j_1-m_1}\sqrt{\frac{[j]}{[j_2]}}\begin{bmatrix} j_1 & j & j_2 \\ m_1 & -m & -m_2 \end{bmatrix} = (-1)^{j_2+m_2}\sqrt{\frac{[j]}{[j_1]}}\begin{bmatrix} j & j_2 & j_1 \\ -m & m_2 & -m_1 \end{bmatrix}. \tag{AIV.4}$$

Here, and subsequently, $[a] = 2a + 1$ and $[abc\ldots] = [a][b][c]\ldots$.

Special Clebsch-Gordon Coefficients

$$\begin{bmatrix} j & 0 & j \\ m & 0 & m \end{bmatrix} = \begin{bmatrix} 0 & j & j \\ 0 & m & m \end{bmatrix} = 1. \tag{AIV.5a}$$

$$\begin{bmatrix} j & j & 0 \\ m & -m & 0 \end{bmatrix} = (-1)^{j-m}[j]^{-1/2}. \tag{AIV.5b}$$

If $m_1 = j_1$, $m_2 = j_2$, $j = m = j_1 + j_2$, then

$$\text{Clebsch-Gordon coefficient} = 1. \tag{AIV.5c}$$

When $m_1 = m_2 = m = 0$, the Clebsch-Gordon coefficient is nonvanishing only if $j_1 + j_2 + j$ is an even integer ($=2g$ with g an integer), and in this case it is given by

$$(-1)^{j_1-j_2+g}[j]^{1/2}\frac{g!}{(g-j_1)!\,(g-j_2)!\,(g-j)!}\Delta(j_1j_2j), \tag{AIV.6a}$$

where

$$\Delta(j_1j_2j) = \left[\frac{(j_1+j_2-j)!\,(j+j_1-j_2)!\,(j_2+j-j_1)!}{(j_1+j_2+j+1)!}\right]^{1/2}. \tag{AIV.6b}$$

Formula for Numerical Computation of Clebsch-Gordon Coefficient

$$\begin{bmatrix} j_1 & j_2 & j \\ m_1 & m_2 & m \end{bmatrix} = [j]^{1/2}\Delta(j_1j_2j)[(j_1+m_1)!\,(j_1-m_1)!\,(j_2+m_2)!\,(j_2-m_2)!\,(j+m)!$$
$$\times (j-m)!]^{1/2}\sum_n (-1)^n[n!\,(j_1+j_2-j-n)!\,(j-j_2+m_1+n)!$$
$$\times (j-j_1-m_2+n)!\,(j_1-m_1-n)!\,(j_2+m_2-n)!]^{-1}. \tag{AIV.7}$$

In this summation, the range of n, which is allowed to be zero or positive integers, is determined by the fact that the arguments of none of the factorial functions be negative.

V. COUPLING OF THREE ANGULAR MOMENTA

The scheme of coupling, described in the preceding section, may now be further extended by

76 (45-122/1977)

considering the coupling of three angular momenta

$$\mathbf{J} = \mathbf{J}_1 + \mathbf{J}_2 + \mathbf{J}_3.$$

Since the coupling of any two of these angular momenta can be done by the method given in Section IV of this appendix, we shall constantly deal in terms of such coupled states.

Two types of states can be constructed: $|(j_1j_2)j_{12}, j_3 : jm\rangle$ and $|j_1, (j_2j_3)j_{23} : jm\rangle$, wherein the first pair to be coupled has been enclosed within the parentheses. Thus, the first state denotes a state of total angular momentum j and projection m obtained by *first* coupling (j_1j_2) to obtain the resultant j_{12} and *then* coupling j_{12} and j_3 to form the final j. On the other hand, the second state corresponds to the *first* coupling of (j_2j_3) to form j_{23} and the *second* coupling of j_1 and j_{23} to form j and projection m. In this section, we shall consider the transformation coefficients between these two types of states.

The coefficients, which are specified by six angular momentum quantum numbers, are real and denoted by U. They are defined by

$$|(j_1j_2)j_{12}, j_3 : jm\rangle = \sum_{j_{23}} U(j_1j_2jj_3; j_{12}j_{23})|j_1, (j_2j_3)j_{23} : jm\rangle, \tag{AV.1}$$

$$|j_1, (j_2j_3)j_{23} : jm\rangle = \sum_{j_{12}} U(j_1j_2jj_3; j_{12}j_{23})|(j_1j_2)j_{12}, j_3 : jm\rangle. \tag{AV.2}$$

The angular momentum couplings contained in (AV.1) and (AV.2) are

$$\mathbf{J}_1 + \mathbf{J}_2 = \mathbf{J}_{12}, \qquad \mathbf{J}_{12} + \mathbf{J}_3 = \mathbf{J}$$

$$\mathbf{J}_2 + \mathbf{J}_3 = \mathbf{J}_{23}, \qquad \mathbf{J}_1 + \mathbf{J}_{23} = \mathbf{J}.$$

The triangular inequalities for these four vector couplings must be satisfied so that the transformation coefficients U (originally due to Jahn[2]) may be nonvanishing. U is related to W (see Racah[3]) by

$$U(j_1j_2jj_3; j_{12}j_{23}) = [j_{12}j_{23}]^{1/2} W(j_1j_2jj_3; j_{12}j_{23}). \tag{AV.3}$$

Orthogonality Properties

$$\sum_{j_{12}} U(j_1j_2jj_3; j_{12}j_{23})U(j_1j_2jj_3; j_{12}j'_{23}) = \delta_{j_{23}, j'_{23}}, \tag{AV.4}$$

$$\sum_{j_{23}} U(j_1j_2jj_3; j_{12}j_{23})U(j_1j_2jj_3; j'_{12}j_{23}) = \delta_{j_{12}, j'_{12}}. \tag{AV.5}$$

Symmetry Properties

$$\begin{aligned} W(j_1j_2jj_3; j_{12}j_{23}) &= W(j_2j_1j_3j; j_{12}j_{23}) = W(jj_3j_1j_2; j_{12}j_{23}) = W(j_1jj_2j_3; j_{23}j_{12}) \\ &= (-1)^{j_2+j-j_{12}-j_{23}} W(j_1j_{12}j_{23}j_3; j_2j) \\ &= (-1)^{j_1+j_3-j_{12}-j_{23}} W(j_{12}j_2jj_{23}; j_1j_3). \end{aligned} \tag{AV.6}$$

Specific Values for a Few Special Cases

U(0 in any of the first four numbers; $j_{12}j_{23}$) = 1 (provided $\triangle$-inequalities are satisfied), (AV.7)

$$\begin{aligned} W(j_1j_2jj_3; 0j_{23}) &= \delta_{j_1j_2}\delta_{jj_3}(-1)^{j_1+j_3-j_{23}}[j_1j_3]^{-1/2}, \\ W(j_1j_2jj_3; j_{12}0) &= \delta_{j_1j}\delta_{j_2j_3}(-1)^{j_1+j_3-j_{12}}[j_1j_3]^{-1/2}. \end{aligned} \tag{AV.8}$$

It should be noted that the δ's arise from the $\triangle$-inequality requirements.

Clebsch-Gordon Coefficients and 6*j*-Symbol

$$\sum_{m_1 m_2} \begin{bmatrix} j_1 & j_2 & j_{12} \\ m_1 & m_2 & m_1+m_2 \end{bmatrix} \begin{bmatrix} j_{12} & j_3 & j \\ m_1+m_2 & m-m_1-m_2 & m \end{bmatrix} \begin{bmatrix} j_2 & j_3 & j_{23} \\ m_2 & m-m_1-m_2 & m-m_1 \end{bmatrix}$$

$$\times \begin{bmatrix} j_1 & j_{23} & j \\ m_1 & m-m_1 & m \end{bmatrix}$$

$$= U(j_1 j_2 j j_3; j_{12} j_{23}), \tag{AV.9}$$

$$\begin{bmatrix} j_1 & j_2 & j_{12} \\ m_1 & m_2 & m_1+m_2 \end{bmatrix} \begin{bmatrix} j_{12} & j_3 & j \\ m_1+m_2 & m-m_1-m_2 & m \end{bmatrix}$$

$$= \sum_{j_{23}} \begin{bmatrix} j_2 & j_3 & j_{23} \\ m_2 & m-m_1-m_2 & m-m_1 \end{bmatrix} \begin{bmatrix} j_1 & j_{23} & j \\ m_1 & m-m_1 & m \end{bmatrix} U(j_1 j_2 j j_3; j_{12} j_{23}), \tag{AV.10}$$

$$\begin{bmatrix} j_2 & j_3 & j_{23} \\ m_2 & m-m_1-m_2 & m-m_1 \end{bmatrix} \begin{bmatrix} j_1 & j_{23} & j \\ m_1 & m-m_1 & m \end{bmatrix}$$

$$= \sum_{j_{12}} \begin{bmatrix} j_1 & j_2 & j_{12} \\ m_1 & m_2 & m_1+m_2 \end{bmatrix} \begin{bmatrix} j_{12} & j_3 & j \\ m_1+m_2 & m-m_1-m_2 & m \end{bmatrix} U(j_1 j_2 j j_3; j_{12} j_{23}), \tag{AV.11}$$

$$\sum_{m_2} \begin{bmatrix} j_1 & j_2 & j_{12} \\ m_1 & m_2 & m_1+m_2 \end{bmatrix} \begin{bmatrix} j_{12} & j_3 & j \\ m_1+m_2 & m-m_1-m_2 & m \end{bmatrix} \begin{bmatrix} j_2 & j_3 & j_{23} \\ m_2 & m-m_1-m_2 & m-m_1 \end{bmatrix}$$

$$= \begin{bmatrix} j_1 & j_{23} & j \\ m_1 & m-m_1 & m \end{bmatrix} U(j_1 j_2 j j_3; j_{12} j_{23}). \tag{AV.12}$$

Special Sum Rules

$$\sum_e (2e+1)(-1)^{a+b+e} W(abcd; ef) W(bacd; eg) = W(agfb; cd), \tag{AV.13}$$

$$\sum_g (2g+1) W(a'gdc; ac') W(bgec'; b'c) W(a'gfb; ab') = W(abde; cf) W(a'db'e; c'f). \tag{AV.14}$$

Numerical Computation

Numerical values of the Racah coefficients are obtained on the computing machine by using the expression (due to Racah[3])

$$\begin{aligned} W(j_1 j_2 j j_3; j_{12} j_{23}) = {} & (-1)^{j_1+j_2+j+j_3} \Delta(j_1 j_2 j_{12}) \Delta(j_{12} j_3 j) \Delta(j_2 j_3 j_{23}) \Delta(j_1 j_{23} j) \\ & \times \sum_n (-1)^n (n+1)! \, [(n-j_1-j_2-j_{12})! \, (n-j_{12}-j_3-j)! \\ & \times (n-j_2-j_3-j_{23})! \, (n-j_1-j_{23}-j)! \, (j_1+j_2+j+j_3-n)! \\ & \times (j_2+j_{12}+j+j_{23}-n)! \, (j_1+j_{12}+j_3+j_{23}-n)!]^{-1}. \end{aligned} \tag{AV.15}$$

The Δ-functions appearing here have already been defined by (AIV.6b). The summation index n, as in the case of the Clebsch-Gordon coefficient, goes over zero and positive integral values, subject to the restriction that the arguments of none of the factorial functions are negative.

VI. COUPLING OF FOUR ANGULAR MOMENTA

Continuing our scheme of coupling more and more angular momenta, we now consider coupling four of them as

$$\mathbf{J} = \mathbf{J}_1 + \mathbf{J}_2 + \mathbf{J}_3 + \mathbf{J}_4.$$

The alternative schemes of coupling, to be considered now, are given by

$$\mathbf{J}_1 + \mathbf{J}_2 = \mathbf{J}_{12}, \qquad \mathbf{J}_3 + \mathbf{J}_4 = \mathbf{J}_{34}, \qquad \mathbf{J}_{12} + \mathbf{J}_{34} = \mathbf{J},$$

$$\mathbf{J}_1 + \mathbf{J}_3 = \mathbf{J}_{13}, \qquad \mathbf{J}_2 + \mathbf{J}_4 = \mathbf{J}_{24}, \qquad \mathbf{J}_{13} + \mathbf{J}_{24} = \mathbf{J}.$$

The quantum numbers corresponding to different angular momenta are denoted by the corresponding lowercase letters. The transformation coefficient connecting the two sets of states is called the $9j$-symbol and is defined by

$$|(j_1j_2)j_{12}, (j_3j_4)j_{34} : jm\rangle = \sum_{j_{13},\, j_{24}} \begin{bmatrix} j_1 & j_2 & j_{12} \\ j_3 & j_4 & j_{34} \\ j_{13} & j_{24} & j \end{bmatrix} |(j_1j_3)j_{13}, (j_2j_4)j_{24} : jm\rangle. \tag{AVI.1}$$

The expression defining a state on the right-hand side of (AVI.1) in terms of the set of states on the left-hand side contains the same $9j$-symbol as in (AVI.1) and a summation over j_{12}, j_{34}.

A useful case is the transformation of an LS-coupling wavefunction to a jj-coupling wavefunction or vice versa for a pair of particles. In this specific case, we have the orbital, spin, and total angular momenta of the two particles. In obvious notation, we have

$$|(l_1\tfrac{1}{2})j_1, (l_2\tfrac{1}{2})j_2 : JM\rangle = \sum_{LS} \begin{bmatrix} l_1 & \frac{1}{2} & j_1 \\ l_2 & \frac{1}{2} & j_2 \\ L & S & J \end{bmatrix} |(l_1l_2)L, (\tfrac{1}{2}\,\tfrac{1}{2})S : JM\rangle, \tag{AVI.2}$$

$$|(l_1l_2)L, (\tfrac{1}{2}\,\tfrac{1}{2})S : JM\rangle = \sum_{j_1j_2} \begin{bmatrix} l_1 & \frac{1}{2} & j_1 \\ l_2 & \frac{1}{2} & j_2 \\ L & S & J \end{bmatrix} |(l_1\tfrac{1}{2})j_1(l_2\tfrac{1}{2})j_2 : JM\rangle. \tag{AVI.3}$$

S can have two possible values, namely, $S = 0$ and $S = 1$. These two $9j$-symbols are very simply expressed in terms of the $6j$-symbol, i.e., the W-function, as

$$\begin{bmatrix} l & \frac{1}{2} & j \\ l' & \frac{1}{2} & j' \\ L & 0 & J \end{bmatrix} = \delta_{LJ}(-1)^{L+l'+j-1/2}\sqrt{\frac{[jj']}{[\frac{1}{2}]}}\, W(ll'jj'; L\tfrac{1}{2}), \tag{AVI.4}$$

$$\begin{bmatrix} l & \frac{1}{2} & j \\ l' & \frac{1}{2} & j' \\ L & 1 & J \end{bmatrix} = (-1)^{\Sigma}\sqrt{\frac{[jj'L]}{[1]}}\, W^{-1}(JL\tfrac{1}{2}\,\tfrac{1}{2}; 1K)$$
$$\times\, [W(jj'K\tfrac{1}{2}; Jl')W(ll'K\tfrac{1}{2}; Lj) + \delta_{LJ}\frac{(-1)^{K-1/2}}{2[J]}W(Jl'j\tfrac{1}{2}; jj')], \tag{AVI.5}$$

where Σ is the sum of all the nine angular momenta and $K = \frac{1}{2}(L + J)$ if $L \neq J$ and $K = L + \frac{1}{2}$ if $L = J$.

Orthogonality Relations

$$\sum_{j_{12},\, j_{34}} \begin{bmatrix} j_1 & j_2 & j_{12} \\ j_3 & j_4 & j_{34} \\ j_{13} & j_{24} & j \end{bmatrix} \begin{bmatrix} j_1 & j_2 & j_{12} \\ j_3 & j_4 & j_{34} \\ j'_{13} & j'_{24} & j \end{bmatrix} = \delta_{j_{13},\, j'_{13}} \delta_{j_{24},\, j'_{24}}, \tag{AVI.6}$$

$$\sum_{j_{13},\, j_{24}} \begin{bmatrix} j_1 & j_2 & j_{12} \\ j_3 & j_4 & j_{34} \\ j_{13} & j_{24} & j \end{bmatrix} \begin{bmatrix} j_1 & j_2 & j'_{12} \\ j_3 & j_4 & j'_{34} \\ j_{13} & j_{24} & j \end{bmatrix} = \delta_{j_{12},\, j'_{12}} \delta_{j_{34},\, j'_{34}}. \tag{AVI.7}$$

Symmetry Property

The symmetry property is better stated in terms of an associated $9j$-symbol defined as

$$\begin{bmatrix} j_1 & j_2 & j_{12} \\ j_3 & j_4 & j_{34} \\ j_{13} & j_{24} & j \end{bmatrix} = [j_{12} j_{34} j_{13} j_{24}]^{1/2} \begin{Bmatrix} j_1 & j_2 & j_{12} \\ j_3 & j_4 & j_{34} \\ j_{13} & j_{24} & j \end{Bmatrix}. \tag{AVI.8}$$

In this new $9j$-symbol, enclosed within the braces, the exchange of any two rows or columns gives rise to a phase factor $= (-1)^{\Sigma}$, where $\Sigma =$ the sum of the nine j's.

Expansion in Terms of $6j$-Symbol

$$\begin{bmatrix} j_1 & j_2 & j_{12} \\ j_3 & j_4 & j_{34} \\ j_{13} & j_{24} & j \end{bmatrix} = (-1)^{j_1 - j_{12} - j_{13}} \sum_r (-1)^r U(j_{12} j_3 j j_4;\, r j_{34}) U(j_{13} j_2 j j_4;\, r j_{24}) U(j_3 j_1 r j_2;\, j_{13} j_{12}). \tag{AVI.9}$$

Here the summation index r takes all possible values consistent with the various angular momentum couplings in the U-functions. This expression, supplemented by a code for the computation of the Racah coefficient, enables us to evaluate the $9j$-symbol.

In particular, when $j = 0$, we have, from (AVI.9),

$$\begin{bmatrix} j_1 & j_2 & j_{12} \\ j_3 & j_4 & j_{34} \\ j_{13} & j_{24} & 0 \end{bmatrix} = \delta_{j_{12} j_{34}} \delta_{j_{13} j_{24}} (-1)^{j_1 + j_4 - j_{12} - j_{13}} U(j_1 j_2 j_3 j_4;\, j_{12} j_{13}). \tag{AVI.10}$$

VII. TABLES OF ANGULAR MOMENTUM COUPLING COEFFICIENTS

Square of Clebsch-Gordon Coefficient $\begin{bmatrix} j & j' & J \\ m & m' & M \end{bmatrix}$

To evaluate a Clebsch-Gordon coefficient, we should *first* write the square-root of the corresponding expression and *then* insert numerals. Finally, we should insert a multiplicative factor of -1 if an $*$ appears in the table.

Coefficients will now be tabulated only for the positive values of m'. Those for $-m'$ will be obtained from the values of m' by the symmetry relation (AIV.4).

Table AVII.1 $j' = \frac{1}{2}$

$J \backslash m'$	$\frac{1}{2}$
$j + \frac{1}{2}$	$\frac{j + M + \frac{1}{2}}{2j + 1}$
$j - \frac{1}{2}$	$\frac{j - M + \frac{1}{2}}{2j + 1}$

Table AVII.2 $j' = 1$

$J \backslash m'$	1	0
$j + 1$	$\frac{(j + M)(j + M + 1)}{(2j + 1)(2j + 2)}$	$\frac{(j - M + 1)(j + M + 1)}{(2j + 1)(j + 1)}$
j	$* \frac{(j + M)(j - M + 1)}{2j(j + 1)}$	$\frac{M^2}{j(j + 1)}$
$j - 1$	$\frac{(j - M)(j - M + 1)}{2j(2j + 1)}$	$* \frac{(j - M)(j + M)}{j(2j + 1)}$

Table AVII.3 $j' = \frac{3}{2}$

$J \backslash m'$		$\frac{1}{2}$
$j + \frac{3}{2}$	$\frac{(j + M - \frac{1}{2})(j + M + \frac{1}{2})(j + M + \frac{3}{2})}{(2j + 1)(2j + 2)(2j + 3)}$	$\frac{3(j + M + \frac{1}{2})(j + M + \frac{3}{2})(j - M + \frac{3}{2})}{(2j + 1)(2j + 2)(2j + 3)}$
$j + \frac{1}{2}$	$* \frac{3(j + M - \frac{1}{2})(j + M + \frac{1}{2})(j - M + \frac{3}{2})}{2j(2j + 1)(2j + 3)}$	$* \frac{(j - 3M + \frac{3}{2})^2(j + M + \frac{1}{2})}{2j(2j + 1)(2j + 3)}$
$j - \frac{1}{2}$	$\frac{3(j + M - \frac{1}{2})(j - M + \frac{1}{2})(j - M + \frac{3}{2})}{(2j - 1)(2j + 1)(2j + 2)}$	$* \frac{(j + 3M - \frac{1}{2})^2(j - M + \frac{1}{2})}{(2j - 1)(2j + 1)(2j + 2)}$
$j - \frac{3}{2}$	$* \frac{(j - M - \frac{1}{2})(j - M + \frac{1}{2})(j - M + \frac{3}{2})}{2j(2j - 1)(2j + 1)}$	$\frac{3(j + M - \frac{1}{2})(j - M - \frac{1}{2})(j - M + \frac{1}{2})}{2j(2j - 1)(2j + 1)}$

Table AVII.4 $i' = 2$

J \ m'	2	1	0
$j+2$	$\frac{(j+M-1)(j+M)(j+M+1)(j+M+2)}{(2j+1)(2j+2)(2j+3)(2j+4)}$	$\frac{(j-M+2)(j+M+2)(j+M+1)(j+M)}{(2j+1)(j+1)(2j+3)(j+2)}$	$\frac{3(j-M+2)(j-M+1)(j+M+2)(j+M+1)}{(2j+1)(2j+2)(2j+3)(j+2)}$
$j+1$	$*\frac{(j+M-1)(j+M)(j+M+1)(j-M+2)}{2j(j+1)(j+2)((2j+1)}$	$*\frac{(j-2M+2)^2(j+M+1)(j+M)}{2j(2j+1)(j+1)(j+2)}$	$\frac{3M^2(j-M+1)(j+M+1)}{j(2j+1)(j+1)(j+2)}$
j	$\frac{3(j+M-1)(j+M)(j-M+1)(j-M+2)}{(2j-1)(2j)(j+1)(2j+3)}$	$\frac{(1-2M)^2 3(j-M+1)(j+M)}{(2j-1)j(2j+2)(2j+3)}$	$\frac{[3M^2-j(j+1)]^2}{(2j-1)j(j+1)(2j+3)}$
$j-1$	$*\frac{(j+M-1)(j-M)(j-M+1)(j-M+2)}{2(j-1)j(j+1)(2j+1)}$	$\frac{(j+2M-1)^2(j-M+1)(j-M)}{(j-1)j(2j+1)(2j+2)}$	$*\frac{3M^2(j-M)(j+M)}{(j-1)j(2j+1)(j+1)}$
$j-2$	$\frac{(j-M-1)(j-M)(j-M+1)(j-M+2}{(2j-2)(2j-1)2j(2j+1)}$	$*\frac{(j-M+1)(j-M)(j-M-1)(j+M-1)}{(j-1)(2j-1)j(2j+1)}$	$\frac{3(j-M)(j-M-1)(j+M)(j+M-1)}{(2j-2)(2j-1)j(2j+1)}$

Square of 6j-Coefficient $W(abJj; cJ')$

In evaluating the coefficients, we should first take the square-root of the expressions and then insert numerals. The $*$ appearing in front of some terms corresponds to a multiplication by -1 *after* the evaluation of the coefficient. For any given value j, the W-coefficients that do not appear in the table can be evaluated from the tabulated ones by using the symmetry relations (AV.6). In all the tables that follow, $s = a + b + c$ and $X = b(b+1) + c(c+1) - a(a+1)$.

Table AVII.5 $j = \frac{1}{2}$

J	J'	
$c - \frac{1}{2}$	$b - \frac{1}{2}$	$\dfrac{(s+1)(s-2a)}{2b(2b+1)2c(2c+1)}$
	$b + \frac{1}{2}$	$\dfrac{(s-2b)(s-2c+1)}{(2b+1)(2b+2)2c(2c+1)}$

Table AVII.6 $j = 1$

J	J'	
$c - 1$	$b - 1$	$\dfrac{s(s+1)(s-2a-1)(s-2a)}{(2b-1)2b(2b+1)(2c-1)2c(2c+1)}$
	b	$\dfrac{2(s+1)(s-2a)(s-2b)(s-2c+1)}{2b(2b+1)(2b+2)(2c-1)2c(2c+1)}$
	$b + 1$	$\dfrac{(s-2b-1)(s-2b)(s-2c+1)(s-2c+2)}{(2b+1)(2b+2)(2b+3)(2c-1)2c(2c+1)}$
c	b	$\dfrac{2X^2}{b(2b+1)(2b+2)2c(2c+1)(2c+2)}$

Table AVII.7 $j = \frac{3}{2}$

J	J'	
$c - \frac{3}{2}$	$b - \frac{3}{2}$	$\frac{(s-1)s(s+1)(s-2a-2)(s-2a-1)(s-2a)}{(2b-2)(2b-1)2b(2b+1)(2c-2)(2c-1)2c(2c+1)}$
	$b - \frac{1}{2}$	$\frac{3s(s+1)(s-2a-1)(s-2a)(s-2b)(s-2c+1)}{(2b-1)2b(2b+1)(2b+2)(2c-2)(2c-1)2c(2c+1)}$
	$b + \frac{1}{2}$	$\frac{3(s+1)(s-2a)(s-2b-1)(s-2b)(s-2c+1)(s-2c+2)}{2b(2b+1)(2b+2)(2b+3)(2c-2)(2c-1)2c(2c+1)}$
	$b + \frac{3}{2}$	$\frac{(s-2b-2)(s-2b-1)(s-2b)(s-2c+1)(s-2c+2)(s-2c+3)}{(2b+1)(2b+2)(2b+3)(2b+4)(2c-2)(2c-1)2c(2c+1)}$
$c - \frac{1}{2}$	$b - \frac{1}{2}$	$*\frac{[2(s-2b)(s-2c)-(s+2)(s-2a-1)]^2(s+1)(s-2a)}{(2b-1)2b(2b+1)(2b+2)(2c-1)2c(2c+1)(2c+2)}$
	$b + \frac{1}{2}$	$*\frac{[(s-2b-1)(s-2c)-2(s+2)(s-2a)]^2(s-2b)(s-2c+1)}{2b(2b+1)(2b+2)(2b+3)2c(2c+1)(2c+2)(2c-1)}$

Table AVII.8 $j=2$

J	J'	
$c-2$	$b-2$	$\frac{(s-2)(s-1)s(s+1)(s-2a-3)(s-2a-2)(s-2a-1)(s-2a)}{(2b-3)(2b-2)(2b-1)2b(2b+1)(2c-3)(2c-2)(2c-1)2c(2c+1)}$
	$b-1$	$\frac{4(s-1)s(s+1)(s-2a-2)(s-2a-1)(s-2a)(s-2b)(s-2c+1)}{(2b-2)(2b-1)2b(2b+1)(2b+2)(2c-3)(2c-2)(2c-1)2c(2c+1)}$
	b	$\frac{6s(s+1)(s-2a-1)(s-2b-1)(s-2a)(s-2b)(s-2c+1)(s-2c+2)}{(2b-1)2b(2b+1)(2b+2)(2b+3)(2c-3)(2c-2)(2c-1)2c(2c+1)}$
	$b+1$	$\frac{4(s+1)(s-2a)(s-2b-2)(s-2b-1)(s-2b)(s-2c+1)(s-2c+2)(s-2c+3)}{2b(2b+1)(2b+2)(2b+3)(2b+4)(2c-3)(2c-2)(2c-1)2c(2c+1)}$
	$b+2$	$\frac{(s-2b-3)(s-2b-2)(s-2b-1)(s-2b)(s-2c+1)(s-2c+2)(s-2c+3)(s-2c+4)}{(2b+1)(2b+2)(2b+3)(2b+4)(2b+5)(2c-3)(2c-2)(2c-1)2c(2c+1)}$
$c-1$	$b-1$	$*\frac{16[(a+b)(a-b+1)-(c-1)(c-b+1)]^2s(s+1)(s-2a-1)(s-2a)}{(2b-2)(2b-1)2b(2b+1)(2b+2)(2c-2)(2c-1)2c(2c+1)(2c+2)}$
	b	$*\frac{24[(a+b+1)(a-b)-c^2+1]^2(s+1)(s-2a)(s-2b)(s-2c+1)}{(2b-1)2b(2b+1)(2b+2)(2b+3)(2c-2)(2c-1)2c(2c+1)(2c+2)}$
	$b+1$	$*\frac{16[(a+b+2)(a-b-1)-(c-1)(b+c+2)]^2(s-2b-1)(s-2b)(s-2c+1)(s-2c+2)}{2b(2b+1)(2b+2)(2b+3)(2b+4)(2c-2)(2c-1)2c(2c+1)(2c+2)}$
c	b	$\frac{4[3X(X-1)-4b(b+1)c(c+1)]^2}{(2b-1)2b(2b+1)(2b+2)(2b+3)(2c-1)2c(2c+1)(2c+2)(2c+3)}$

The algebraic tables (i.e., the type given in this appendix) of the Clebsch-Gordon coefficients have been given for $j' = \frac{5}{2}$ by Melvin and Swamy[4], for $j' = 3$ by Falkoff et al[5], and for $j' = \frac{7}{2}$ by Edmonds and Flowers[6]. *Numerical* tables for the Clebsch-Gordon coefficients have been given by Alder[7], Simon[8], and Rotenberg et al[9].

Algebraic tables for the $6j$-coefficients $W(abJj;\, cJ')$ corresponding to $c = 3, 4$ are contained in the *Annals of the Tokyo Astronomical Observatory*[10]. The same annals contain *numerical* tables too. Numerical tables have been given also by Sharp[11], Biedenharn[12], and Rotenberg et al[9].

Numerical tables of the $9j$-coefficients have been given by Sharp[11], Kennedy[13], Matsunobu and Takebe[14], Stephenson and Smith[15], and Rotenberg et al[9].

REFERENCES

1. Schiff, L. I., Quantum Mechanics, 2nd edn., McGraw-Hill, New York, 1955, p 70; Condon, E. U., and Shortley, G. H., Theory of Atomic Spectra, Macmillan, New York, 1935, p 45; Rose, M. E., Elementary Theory of Angular Momentum, Wiley, New York, 1957.
2. Jahn, H. A., *Proc. Roy. Soc.* (London), **A205**, 192 (1951).
3. Racah, G., *Phys. Rev.*, **62**, 438 (1942).
4. Melvin, M. A., and Swamy, M. V. V. J., *Phys. Rev.*, **107**, 186 (1957).
5. Falkoff, D. L., Colladay, G. S., and Sells, R. E., *Can. J. Phys.*, **30**, 253 (1952).
6. Edmonds, A. R., and Flowers, B. H., *Proc. Roy. Soc.* (London), **A214**, 515 (1952).
7. Alder, K., *Helv. Phys. Acta*, **25**, 235 (1952).
8. Simon, A., ORNL–1718, Oak Ridge National Laboratory, Tennessee, 1954 (unpublished).
9. Rotenberg, M., Bivins, R., Metropolis, N., and Wooten (Jr.), J. K., The $3j$- and $6j$-Symbols, Technology Press, Massachusetts Institute of Technology, Cambridge, Massachusetts, 1959.
10. *Annals of the Tokyo Astronomical Observatory*, *University of Tokyo* (Second Series), **III**, No. 3 (1954); *ibid*, **IV**, No. 1 (1954).
11. Sharp, W. T., CRT–556, Atomic Energy of Canada Limited, Chalk River, Canada, 1953.
12. Biedenharn, L. C., ORNL–1098, Oak Ridge National Laboratory, Tennessee, 1952.
13. Kennedy, J. M., CRT–569, Atomic Energy of Canada Limited, Chalk River, Canada, 1954.
14. Matsunobu, H., and Takebe, H., *Prog. Theor. Phys.*, **14**, 589 (1955).
15. Stephenson, J. W., and Smith, K., *Proc. Phys. Soc.* (London), **A70**, 571 (1957).

PROBLEMS

1. Construct the Pauli matrices of (AIII.2) with the help of the general results (AII.2a), (AII.2b), and (AII.2c).

2. Evaluate the trace of the product of any three and any four Pauli matrices.

3. Derive the values of a and b by solving (AIII.6) and (AIII.4), where the components of $\mathbf{n}$ are given by (AIII.5), and hence verify the results of (AIII.8).

4. If $\Psi^J_M = \sum_\alpha C_\alpha(J, M)\Phi^{\alpha J}_M$, prove that $C_\alpha(J, M)$ is independent of M. [Hint: Operate with $J_\pm$.]

Note The result proved in Problem 4 explains why the U-coefficient in (AV.1) and (AV.2) and the $9j$-symbol in (AVI.1) have been taken as independent of the projection quantum number.

Appendix B

Rotation and Irreducible Tensors

I. ROTATION

When a coordinate system is rotated, the coordinates x, y, z of a point change to x', y', z'. Denoting the three components by a subscript which can take the values 1, 2, 3, we write the transformation equation as

$$x_i \rightarrow x_i' = \sum_j a_{ij} x_j, \tag{BI.1}$$

where a denotes the transformation matrix. In the case of rotation, it is a real orthogonal matrix with

$$\tilde{a}a = a\tilde{a} = \mathbb{1}, \tag{BI.2a}$$

$$\det a = 1. \tag{BI.2b}$$

Here $\tilde{a}$ is the transpose of the matrix a, and $\mathbb{1}$ is the unit matrix.

The most general rotation of the coordinate frame may be described by the three Eulerian rotations: first, a rotation α about the z-axis; then, a rotation β about the new y-axis (to be called y_1-axis); and, lastly, a rotation γ about the final z-axis (i.e., the z'-axis).

If we refer a body to a coordinate frame and rotate the latter about any axis in a certain direction through an angle α, the coordinates x, y, z of a point on the body undergo the transformation (BI.1). It is geometrically obvious that, if we keep the coordinate frame fixed and rotate the body about the same axis in the opposite direction through the same angle α (i.e., give the inverse rotation to the body), the point on the body reaches the same final orientation with respect to the coordinate frame. Thus, any rotation of the coordinate frame is equivalent to an equal and opposite rotation of the body. In what follows, we shall confine ourselves to transformations under a rotation of the coordinate frame. If the transformation corresponding to the rotation of a body in a fixed coordinate frame is required, all the results we give will be applicable with a change in the sign of the rotation angles.

Under a rotation of the coordinate frame, any function $F(x, y, z)$ undergoes the change

$$F(x, y, z) \rightarrow F(x', y', z') \equiv RF(x, y, z). \tag{BI.3}$$

This relation defines the rotation operator R.

An infinitesimal rotation ϵ about any axis $\mathbf{n}$ is produced by an infinitesimal rotation operator

$$R_{\mathbf{n}}(\epsilon) = 1 - i\epsilon\mathbf{n} \cdot \mathbf{L}, \tag{BI.4}$$

where $\mathbf{L}$ is the angular momentum operator. A finite rotation α can be generated by applying this infinitesimal operator N times in succession such that $\epsilon = \alpha/N$. By letting $N \rightarrow \infty$, ϵ

indeed becomes infinitesimal. In this way,

$$R_{\mathbf{n}}(\alpha) = \operatorname*{Lt}_{N\to\infty} \left(1 - i\frac{\alpha}{N}\mathbf{n}\cdot\mathbf{L}\right)^N = e^{-i\alpha\mathbf{n}\cdot\mathbf{L}}. \tag{BI.5}$$

In particular, the spherical harmonics $Y^l_m(\theta, \phi)$ when subjected to the three successive Eulerian rotations α, β, γ undergo the transformation

$$\begin{aligned} Y^l_m(\theta', \phi') &\equiv RY^l_m(\theta, \phi) \\ &= \exp(-i\gamma L_{z'}) \exp(-i\beta L_{y_1}) \exp(-i\alpha L_z) Y^l_m(\theta, \phi). \end{aligned} \tag{BI.6}$$

From the geometry of Eulerian rotations, it can be established that the three rotations α, β, γ as given in (BI.6) are exactly equivalent to rotating first through γ about the z-axis, then through β about the y-axis, and finally through α about the z-axis. (It should be noted that all these rotations are about the old axes z and y.) Therefore, we replace (BI.6) by

$$Y^l_m(\theta', \phi') \equiv RY^l_m(\theta, \phi), \tag{BI.7a}$$

where

$$R = \exp(-i\alpha L_z) \exp(-i\beta L_y) \exp(-i\gamma L_z). \tag{BI.7b}$$

The rotation operator R is clearly unitary, that is,

$$R^\dagger R = RR^\dagger = \mathbb{1}, \qquad R^{-1} = R^\dagger. \tag{BI.8}$$

Using the completeness relation of the spherical harmonics, we then obtain, from (BI.7a),

$$\begin{aligned} Y^l_m(\theta', \phi') &= \sum_{l'm'} Y^{l'}_{m'}(\theta, \phi) \int_0^{2\pi} d\phi \int_{-1}^{+1} d(\cos\theta)\, Y^{l'*}_{m'}(\theta, \phi) R Y^l_m(\theta, \phi) \\ &= \sum_{l'm'} Y^{l'}_{m'}(\theta, \phi) \langle l'm'|\, R\, |lm\rangle \\ &= \sum_{m'} Y^l_{m'}(\theta, \phi) \langle lm'|\, R\, |lm\rangle \\ &= \sum_{m'} \mathcal{D}^{l*}_{m'm}(\alpha, \beta, \gamma) Y^l_{m'}(\theta, \phi), \end{aligned} \tag{BI.9}$$

where

$$\mathcal{D}^{l*}_{m'm}(\alpha, \beta, \gamma) = \langle lm'|\, R\, |lm\rangle. \tag{BI.10}$$

In an intermediate step in (BI.9), we have omitted the sum on l' because the operators L_y and L_z contained in R are diagonal in the quantum number l. All these results are applicable to the case of a general angular momentum state $|jm\rangle$ when $\mathbf{L}$ is replaced everywhere by $\mathbf{J}$, and $Y^l_m(\theta, \phi)$, i.e., the state $|lm\rangle$, by $|jm\rangle$. The state after rotation is then denoted by $|jm\rangle_R$. Thus,

$$R = \exp(-i\alpha J_z) \exp(-i\beta J_y) \exp(-i\gamma J_z), \tag{BI.11a}$$

$$|jm\rangle_R = R|jm\rangle \tag{BI.11b}$$

$$= \sum_{m'} \mathcal{D}^{j*}_{m'm}(\alpha, \beta, \gamma)|jm'\rangle, \tag{BI.11c}$$

where

$$\mathcal{D}^{j*}_{m'm}(\alpha, \beta, \gamma) = \langle jm'|\, R\, |jm\rangle. \tag{BI.11d}$$

This expression for the $\mathcal{D}^{j*}$-matrix can, by using (BI.7b), be simplified to

$$\mathcal{D}^{j*}_{m'm}(\alpha, \beta, \gamma) = \langle jm'|\exp(-i\alpha J_z) \exp(-i\beta J_y) \exp(-i\gamma J_z)\,|jm\rangle = e^{-im'\alpha} d^j_{m'm}(\beta) e^{-im\gamma}, \tag{BI.12a}$$

where

$$d^j_{m'm}(\beta) = \langle jm' | \exp(-i\beta J_y) | jm \rangle. \tag{BI.12b}$$

The expression that follows, which is due to Wigner, shows that $d^j_{m'm}(\beta)$ is a real function:

$$d^j_{m'm}(\beta) = [(j+m)!\,(j-m)!\,(j+m')!\,(j-m')!]^{1/2} \times \sum_r \frac{(-1)^{r+m-m'}(\cos\frac{1}{2}\beta)^{2j-2r+m'-m}(\sin\frac{1}{2}\beta)^{2r+m-m'}}{r!\,(j+m'-r)!\,(j-m-r)!\,(r+m-m')!}. \tag{BI.12c}$$

In view of the reality of $d^j_{m'm}(\beta)$, we obtain

$$\mathcal{D}^j_{m'm}(\alpha, \beta, \gamma) = e^{im'\alpha} d^j_{m'm}(\beta) e^{im\gamma}. \tag{BI.12d}$$

Using the unitarity property (BI.8) of R and the definition (BI.12d) of the $\mathcal{D}^j$-matrix, we can easily invert the relation (BI.11b) and obtain

$$|jm\rangle = R^\dagger |jm\rangle_R = \sum_{m'} \mathcal{D}^j_{mm'}(\alpha, \beta, \gamma) |jm'\rangle_R. \tag{BI.13}$$

Properties of $\mathcal{D}$-Function

$$d^j_{mm'}(\beta) = (-1)^{m-m'} d^j_{m'm}(\beta) = (-1)^{m-m'} d^j_{mm'}(-\beta) = (-1)^{m-m'} d^j_{-m,-m'}(\beta), \tag{BI.14}$$

$$d^j_{mm'}(\pi) = (-1)^{j-m'} \delta_{m',-m}, \tag{BI.15}$$

$$\mathcal{D}^{j*}_{mm'}(\alpha, \beta, \gamma) = (-1)^{m-m'} \mathcal{D}^j_{-m,-m'}(\alpha, \beta, \gamma), \tag{BI.16}$$

$$\sum_{m'} \mathcal{D}^{j*}_{m'm} \mathcal{D}^j_{m'm''} = \sum_{m'} \mathcal{D}^{j*}_{mm'} \mathcal{D}^j_{m''m'} = \delta_{m,m''} \qquad \text{(unitarity)}, \tag{BI.17}$$

$$\mathcal{D}^l_{m,0}(\alpha, \beta, \gamma) = \sqrt{\frac{4\pi}{[l]}}\, Y^l_m(\beta, \alpha), \tag{BI.18a}$$

$$\mathcal{D}^l_{0,m}(\alpha, \beta, \gamma) = (-1)^m \sqrt{\frac{4\pi}{[l]}}\, Y^l_m(\beta, \gamma), \tag{BI.18b}$$

$$\mathcal{D}^l_{0,0}(\alpha, \beta, \gamma) = \sqrt{\frac{4\pi}{[l]}}\, Y^l_0(\beta) = P_l(\cos\beta). \tag{BI.18c}$$

Coupling of $\mathcal{D}$-Functions

$$\mathcal{D}^{j_1}_{m_1m_1'} \mathcal{D}^{j_2}_{m_2m_2'} = \sum_j \begin{bmatrix} j_1 & j_2 & j \\ m_1 & m_2 & m \end{bmatrix} \begin{bmatrix} j_1 & j_2 & j \\ m_1' & m_2' & m' \end{bmatrix} \mathcal{D}^j_{mm'}, \tag{BI.19a}$$

$$\mathcal{D}^j_{mm'} = \sum_{m_1m_1'} \begin{bmatrix} j_1 & j_2 & j \\ m_1 & m-m_1 & m \end{bmatrix} \begin{bmatrix} j_1 & j_2 & j \\ m_1' & m'-m_1' & m' \end{bmatrix} \mathcal{D}^{j_1}_{m_1m_1'} \mathcal{D}^{j_2}_{m-m_1,\,m'-m_1'}. \tag{BI.19b}$$

All these $\mathcal{D}$-functions refer to the same angles α, β, γ.

Integrals of $\mathcal{D}$-Functions

Let us denote the integration

$$\int_0^{2\pi} d\gamma \int_{-1}^{+1} d(\cos\beta) \int_0^{2\pi} d\alpha$$

by $\int d(\alpha, \beta, \gamma)$ for brevity. Then

$$\int d(\alpha, \beta, \gamma)\mathscr{D}^{j}_{mm'}(\alpha, \beta, \gamma) = \delta_{m,0}\delta_{m',0}\delta_{j,0}8\pi^2, \tag{BI.20a}$$

$$\int d(\alpha, \beta, \gamma)\mathscr{D}^{j_1*}_{m_1m_1'}(\alpha, \beta, \gamma)\mathscr{D}^{j_2}_{m_2m_2'}(\alpha, \beta, \gamma) = \delta_{j_1j_2}\delta_{m_1m_2}\delta_{m_1'm_2'}\frac{8\pi^2}{[j_1]}, \tag{BI.20b}$$

$$\int d(\alpha, \beta, \gamma)\mathscr{D}^{j_1*}_{m_1m_1'}(\alpha, \beta, \gamma)\mathscr{D}^{j}_{mm'}(\alpha, \beta, \gamma)\mathscr{D}^{j_2}_{m_2m_2'}(\alpha, \beta, \gamma)$$

$$= \frac{8\pi^2}{[j_1]}\begin{bmatrix} j_2 & j & j_1 \\ m_2 & m & m_1 \end{bmatrix}\begin{bmatrix} j_2 & j & j_1 \\ m_2' & m' & m_1' \end{bmatrix}. \tag{BI.20c}$$

Spherical Harmonics Addition Theorem

$$\sum_m Y^l_m(\theta_1, \phi_1)Y^{l*}_m(\theta_2, \phi_2) = \frac{[l]}{4\pi}P_l(\cos\theta_{12}), \tag{BI.21}$$

where (θ_1, ϕ_1) and (θ_2, ϕ_2) denote any two directions in space and θ_{12} denotes the angle between these two directions. A proof of this theorem is based on the transformation of the left-hand side of (BI.21) to another arbitrary coordinate frame with the help of the $\mathscr{D}$-functions, and then showing that the form remains invariant. We thus note that the left-hand side of (BI.21) can be evaluated in any coordinate frame. For a convenient frame leading to the expression on the right-hand side of (BI.21), we have to choose the direction (θ_1, ϕ_1) to be coincident with the z-axis, and fix the x-axis in such a way that ϕ_2 becomes zero.

II. IRREDUCIBLE TENSORS

Under a rotation of the coordinate frame, the coordinates undergo the transformation (BI.1) subject to the property (BI.2) of the transformation matrix. A reflection of either (i) all the three or (ii) one of the axes of the coordinate frame is also described by a relation such as (BI.1), where the transformation matrix a is much simpler. All the nondiagonal components of a are zero, whereas a diagonal component is -1 or $+1$, depending on whether or not the corresponding axis has been reflected. It is clear that the matrix a is still real orthogonal, but for both the cases of reflection just described

$$\det a = -1 \qquad \text{(reflection)}. \tag{BII.1}$$

It should be noted that if two of the axes are reflected, leaving the third unchanged, $\det a = +1$; this is because this particular reflection is actually a case of rotation through π about the unchanged axis.

A vector **A** is defined to have the three components A_i $(i = 1, 2, 3)$ transforming, under the rotation and reflection of the coordinate frame, in the same manner as the three components x_i $(i = 1, 2, 3)$. However, the vector product $\mathbf{P} = (\mathbf{A} \times \mathbf{B})$ of two vectors does not change sign under reflection simply because each of the vectors **A** and **B** does. Such a three-component quantity **P** is called an *axial vector* or a *pseudovector*. It can be shown, from the orthogonality property of the transformation matrix a, that a quantity, e.g., $S = \mathbf{A} \cdot \mathbf{B}$, remains invariant under rotation and reflection; it is called a *scalar*. On the other hand, the scalar product $\mathbf{A} \cdot \mathbf{P}$ of a vector **A** and an axial vector **P** obviously changes sign under reflection (because **A** does so, whereas **P** does not); under rotation, however, $\mathbf{A} \cdot \mathbf{P}$ remains unchanged. Such an entity is called a *pseudoscalar*.

A vector and a scalar are called tensors of rank 1 and 0, respectively. In general, a tensor of rank K has 3^K components, where 3 denotes the dimension of the space. The components are therefore labelled by K subscripts. If we are given the transformation property (BI.1) for the coordinate vector, the components of the tensor of rank K may be *defined* to obey the transformation law

$$T'_{ij\ldots k} = \sum_{i'j'\ldots k'} a_{ii'}a_{jj'}\ldots a_{kk'}T_{i'j'\ldots k'} \quad (K \text{ labels in all}), \tag{BII.2}$$

where there are K factors such as $a_{ii'}$, $a_{jj'}$, ... corresponding to the K labels of the tensor component. Basically, any component of a tensor of rank K is a product of the components of K vectors. Thus, products such as A_iB_j of the components of the *two* vectors $\mathbf{A}$ and $\mathbf{B}$ are components of a *second*-rank tensor.

Example of Reduction

Taking the foregoing example of a second-rank tensor constructed from the two vectors $\mathbf{A}$ and $\mathbf{B}$, we note that the products A_iB_j are nine in number. These are the 3^2 components of the tensor of rank 2. We now make linear combinations of the nine components:

(i) $\sum_i A_iB_i \equiv \mathbf{A}\cdot\mathbf{B}$ (a scalar).

(ii) $(A_iB_j - A_jB_i)$ with $i \neq j$ (an axial vector); there will be three such combinations and they are clearly the three components of the vector product $(\mathbf{A}\times\mathbf{B})$.

(iii) $\frac{1}{2}(A_iB_j + A_jB_i)$; there will be three such nondiagonal components with $i \neq j$, and three diagonal components with $i = j$. Of these six components, five are independent because the sum of the diagonal components (called the trace) is equal to $\mathbf{A}\cdot\mathbf{B}$ and is already included in (i). Thus, of the nine components of a second-rank tensor, the five components of the *traceless symmetric* second-rank tensor

$$\tfrac{1}{2}(A_iB_j + A_jB_i) - \tfrac{1}{3}\delta_{ij}\mathbf{A}\cdot\mathbf{B} \tag{BII.3}$$

cannot be further reduced to tensors of lower rank. These five components define an irreducible tensor of second rank.

General Case

In the general case of a tensor of any rank K, it can be shown that linear combinations can be obtained from the 3^K components to form tensors of lower rank. However, there always remain $(2K + 1)$ components which behave as a tensor of rank K and cannot be further reduced. These $(2K + 1)$ components are symmetric in any two indices, and also traceless with respect to any two indices (i.e., if we make any two indices identical and sum over that index, the result is zero), and they form the components of an irreducible tensor of rank K.

Spherical Components

So far, we have defined vector, tensor, ... in terms of their cartesian components. As far as the foregoing process of constructing irreducible tensors is concerned, it is much more convenient to work in terms of the spherical components, which are defined for the coordinate vector $\mathbf{r}$ by the three components r_m, $m = \pm 1$, and 0, in accordance with the expressions (AI.9). Since the magnitude r does not change under rotation, the components r_m clearly transform as the spherical harmonic $Y^1_m(\theta, \phi)$, i.e., according to

$$r'_m = \sum_{m'} \mathcal{D}^{1*}_{m'm} r_{m'}. \tag{BII.4}$$

The spherical components of any vector **A** are defined to transform in accordance with (BII.4). The quantity $A_m B_n$, where m and n denote spherical components of the vectors **A** and **B**, then transforms according to

$$A'_m B'_n = \sum_{m'} \mathcal{D}^{1*}_{m'm} A_{m'} \sum_{n'} \mathcal{D}^{1*}_{n'n} B_{n'}.$$

Using (BI.16) and (BI.19a), and the new labels $M = m + n$, $M' = m' + n'$, we directly obtain

$$A'_m B'_{M-m} = \sum_{J'=0}^{2} \sum_{M',m'} \begin{bmatrix} 1 & 1 & J' \\ m' & M'-m' & M' \end{bmatrix} \begin{bmatrix} 1 & 1 & J' \\ m & M-m & M \end{bmatrix} \mathcal{D}^{J'*}_{M'M} A_{m'} B_{M'-m'}.$$

Multiplying both sides by $\begin{bmatrix} 1 & 1 & J \\ m & M-m & M \end{bmatrix}$, summing over m, and using (AIV.3a), we get

$$(\mathbf{A}', \mathbf{B}')^J_M = \sum_{M'} \mathcal{D}^{J*}_{M'M} (\mathbf{A}, \mathbf{B})^J_{M'}, \tag{BII.5}$$

where

$$(\mathbf{A}, \mathbf{B})^J_M = \sum_m \begin{bmatrix} 1 & 1 & J \\ m & M-m & M \end{bmatrix} A_m B_{M-m}. \tag{BII.6}$$

A similar definition holds when **A**, **B** are substituted by **A′**, **B′**. Using explicit values of the appropriate Clebsch-Gordon coefficients from Table AVII.2, we easily verify

$$(\mathbf{A}, \mathbf{B})^0_0 = -\frac{1}{\sqrt{3}} \mathbf{A} \cdot \mathbf{B}, \tag{BII.7a}$$

$$(\mathbf{A}, \mathbf{B})^1_m = -\frac{i}{\sqrt{2}} (\mathbf{A} \times \mathbf{B})_m. \tag{BII.7b}$$

The linear combinations of the second-rank tensor components, as defined by (BII.6), therefore contain the scalar ($J = 0$) and the axial vector ($J = 1$). The remaining five components corresponding to $J = 2$ belong to the irreducible tensor of second rank. The five spherical components of this irreducible tensor can be related to the five cartesian components, earlier defined by (BII.3), through a detailed algebra which uses explicit values of the Clebsch-Gordon coefficients.

In general, we can construct spherical components of tensors of arbitrary rank K, and subdivide them into groups of irreducible tensors ranging in rank from $J = 0$ to $J = K$. In analogy with (BII.5), the irreducible tensor of any rank J has $(2J + 1)$ components transforming amongst themselves with the $\mathcal{D}^{J*}$-matrix, i.e., in the manner of the spherical harmonic of the same rank. From this definition, it follows that, given two *irreducible* tensors T^{K_1} and U^{K_2}, it is generally possible to form an irreducible compounded tensor according to the definition

$$(T^{K_1}, U^{K_2})^K_Q = \sum_{q_1 q_2} \begin{bmatrix} K_1 & K_2 & K \\ q_1 & q_2 & Q \end{bmatrix} T^{K_1}_{q_1} U^{K_2}_{q_2}. \tag{BII.8}$$

A special case of (BII.8) corresponds to $K_1 = K_2 = k$ and $K = 0$, when, using the explicit value of the Clebsch-Gordon coefficient,

$$(T^k, U^k)^0_0 = (-1)^k [k]^{-1/2} \sum_q (-1)^q T^k_q U^k_{-q}. \tag{BII 9}$$

The scalar product of the two tensors T^k and U^k is defined by

$$T^k \cdot U^k = \sum_q (-1)^q T_q^k U_{-q}^k. \tag{BII.10}$$

It should be observed that, when $k = 1$, this definition agrees with the usual definition of the scalar product of two vectors.

Vector Spherical Harmonics

The spherical components of the unit vector **e** can be defined in accordance with (AI.9). We can combine the spherical harmonic $Y_q^k(\theta, \phi)$ with the vector **e** to form a compounded tensor according to the general definition (BII.8), namely,

$$\mathbf{T}_Q^{(k1)K} \equiv (Y^k, \mathbf{e})_Q^K = \sum_\mu \begin{bmatrix} k & 1 & K \\ Q-\mu & \mu & Q \end{bmatrix} Y_{Q-\mu}^k(\theta, \phi)\mathbf{e}_\mu. \tag{BII.11}$$

The tensors $\mathbf{T}_Q^{(k1)K}$ are called vector spherical harmonics.

III. MATRIX ELEMENTS OF IRREDUCIBLE TENSOR OPERATORS

It follows from some algebra involving the $\mathscr{D}$-matrices that the matrix element of an irreducible tensor operator connecting eigenstates of angular momentum is given by

$$\langle jm|\, T_q^k \,|j'm'\rangle = \begin{bmatrix} j' & k & j \\ m' & q & m \end{bmatrix} \langle j|\, |T^k|\, |j'\rangle, \tag{BIII.1}$$

where the double-barred quantity $\langle j|\,|T^k|\,|j'\rangle$ is independent of the projection quantum numbers m, m' and the component label q of the tensor. This result is popularly known as the Wigner-Eckart theorem. The Clebsch-Gordon coefficient immediately yields, for the nonvanishing of the matrix element, the selection rules

$$m' + q = m, \qquad |j' - k| \leqslant j \leqslant (j' + k). \tag{BIII.2}$$

In practice, the reduced matrix element is worked out, once and for all, by directly evaluating one matrix element with convenient values of m, m', and q and then equating that result with the right-hand side of (BIII.1).

Examples

Spherical harmonic between orbital angular momentum states Use $m = m' = q = 0$, and then

$$\langle l0|\, Y_0^k \,|l'0\rangle = \int_0^\pi \sin\theta\, d\theta \int_0^{2\pi} d\phi\, Y_0^{l*}(\theta, \phi) Y_0^k(\theta, \phi) Y_0^{l'}(\theta, \phi)$$

$$= \left\{\frac{[l][k][l']}{(4\pi)^3}\right\}^{1/2} 2\pi \int_0^\pi \sin\theta\, d\theta\, P_l(\cos\theta) P_k(\cos\theta) P_{l'}(\cos\theta).$$

The integral of the Legendre polynomials can be directly evaluated. Using that value and the Wigner-Eckart theorem, we then obtain

$$\langle l|\, |Y^k|\, |l'\rangle = \left\{\frac{[l'][k]}{4\pi[l]}\right\}^{1/2} \begin{bmatrix} l' & k & l \\ 0 & 0 & 0 \end{bmatrix}. \tag{BIII.3}$$

The same expression follows from using (BI.18a) for each of the three spherical harmonics, and then using (BI.20c) for the value of the integral.

78 (45-122/1977)

The matrix element of the spherical harmonics can be used, together with their completeness relation, to establish

$$Y^l_m Y^{l'}_{m'} = \sum_{LM} Y^L_M \langle LM|\, Y^l_m\, |l'm'\rangle$$

$$= \sum_L \begin{bmatrix} l' & l & L \\ m' & m & m+m' \end{bmatrix} \left\{\frac{[l'][l]}{4\pi[L]}\right\}^{1/2} \begin{bmatrix} l' & l & L \\ 0 & 0 & 0 \end{bmatrix} Y^L_{m+m'}. \tag{BIII.4}$$

Therefore, using the expressions (BII.8) and (BIII.4), we get

$$(Y^l, Y^{l'})^L_M \equiv \sum_\mu \begin{bmatrix} l & l' & L \\ \mu & M-\mu & M \end{bmatrix} Y^l_\mu Y^{l'}_{M-\mu}$$

$$= \left\{\frac{[l'][l]}{4\pi[L]}\right\}^{1/2} \begin{bmatrix} l' & l & L \\ 0 & 0 & 0 \end{bmatrix} Y^L_M$$

or

$$Y^L_M = \left\{\frac{[L]4\pi}{[l][l']}\right\}^{1/2} \begin{bmatrix} l' & l & L \\ 0 & 0 & 0 \end{bmatrix}^{-1} \sum_\mu \begin{bmatrix} l & l' & L \\ \mu & M-\mu & M \end{bmatrix} Y^l_\mu Y^{l'}_{M-\mu}. \tag{BIII.5}$$

The results (BIII.4) and (BIII.5) are very useful when working with spherical harmonics.

Spin operator between spin states Use $m = m' = \frac{1}{2}$ and $q = 0$, and note that $\sigma^1_0 = \sigma_z = 2s_z$; then

$$\langle \tfrac{1}{2}\, \tfrac{1}{2}|\, \sigma_0\, |\tfrac{1}{2}\, \tfrac{1}{2}\rangle = 1.$$

The application of the Wigner-Eckart theorem then yields

$$\langle \tfrac{1}{2}|\, |\boldsymbol{\sigma}|\, |\tfrac{1}{2}\rangle = \sqrt{3}. \tag{BIII.6}$$

Angular momentum J between angular momentum states Once again, using $m = m' = j$ and $q = 0$, we obtain

$$\langle j|\, |\mathbf{J}|\, |j\rangle = \sqrt{j(j+1)}. \tag{BIII.7}$$

Operator ∇ between orbital angular momentum states For this differential operator, we have

$$\nabla_0 \equiv \frac{\partial}{\partial z} = \cos\theta \frac{\partial}{\partial r} - \frac{1}{r}\sin\theta\frac{\partial}{\partial\theta}. \tag{BIII.8}$$

Using $m = m' = 0$, we obtain

$$\langle l0|\, \nabla_0\, |l'0\rangle = \int_0^\pi \sin\theta\, d\theta \int_0^{2\pi} d\phi\, Y^{l*}_0(\theta)\left(\cos\theta\frac{\partial}{\partial r} - \frac{1}{r}\sin\theta\frac{\partial}{\partial\theta}\right) Y^{l'}_0(\theta).$$

This integral can be directly evaluated. Then, with the Wigner-Eckart theorem and an explicit value of the relevant Clebsch-Gordon coefficient,

$$\langle l|\, |\nabla|\, |l'\rangle = (-1)^{g+l}\left\{\frac{g}{[l]}\right\}^{1/2}\left[\frac{\partial}{\partial r} + \frac{1+(-1)^{g+l'}g}{r}\right], \tag{BIII.9}$$

where

$$2g = l + l' + 1.$$

Matrix element of compounded tensor By the Wigner-Eckart theorem,

$$\langle jm|\,(T^{k_1}, U^{k_2})^k_q\,|j'm'\rangle = \begin{bmatrix} j' & k & j \\ m' & q & m \end{bmatrix} \langle j|\,|(T^{k_1}, U^{k_2})^k|\,|j'\rangle, \qquad \text{(BIII.10)}$$

where

$$\langle j|\,|(T^{k_1}, U^{k_2})^k|\,|j'\rangle = \sum_{j''} (-1)^{k_1+k_2-k} U(k_1k_2jj'; kj'')\langle j|\,|T^{k_1}|\,|j''\rangle\langle j''|\,|U^{k_2}|\,|j'\rangle. \qquad \text{(BIII.11)}$$

The result (BIII.11) is derived by using, on the left-hand side of (BIII.10), the expression (BII.8) for the operators, then a complete set of states

$$\sum_{j''m''} |j''m''\rangle\langle j''m''|$$

between the two operators, and finally the Wigner-Eckart theorem for each of the two matrix elements. The sum over the appropriate projection quantum number of the product of the three Clebsch-Gordon coefficients is carried out with the help of (AV.12).

Matrix Elements between Angular Momentum Coupled States

Let the angular momentum coupled states be denoted by $|j_1j_2jm\rangle$. In what follows, the number 1 or 2 enclosed within the parentheses after the operator denotes whether the operator acts in the space of j_1 or j_2, respectively. The application of the Wigner-Eckart theorem yields a reduced matrix element connecting the coupled states. The main point lies in simplifying this reduced matrix element in terms of a simpler reduced matrix element in the space of the appropriate angular momentum to which the operator belongs. For example,

$$\langle j_1j_2jm|\; T^k_q(2)\;|j_1'j_2'j'm'\rangle = \begin{bmatrix} j' & k & j \\ m' & q & m \end{bmatrix} \langle j_1j_2j|\;|T^{k_2}(2)|\;|j_1'j_2'j'\rangle, \qquad \text{(BIII.12)}$$

where

$$\langle j_1j_2j|\;|T^k(2)|\;|j_1'j_2'j'\rangle = \delta_{j_1j_1'} U(j_1'j_2'jk; j'j_2)\langle j_2|\;|T^k|\;|j_2'\rangle. \qquad \text{(BIII.13)}$$

The simplification leading to (BIII.13) will not be explicitly demonstrated. The result similar to (BIII.13) for the operator $T^k(1)$ is

$$\langle j_1j_2j|\;|T^k(1)|\;|j_1'j_2'j'\rangle = \delta_{j_2j_2'}(-1)^{j_1'-j_1+j-j'} U(j_2'j_1'jk; j'j_1)\langle j_1|\;|T^k|\;|j_1'\rangle. \qquad \text{(BIII.14)}$$

The results for more complicated operators are

$$\langle j_1j_2jm|\; T^k(1)\cdot U^k(2)\;|j_1'j_2'j'm'\rangle = \delta_{jj'}\delta_{mm'}(-1)^{j_1+j_2'-j}\sqrt{[j_1][j_2]}\,W(j_1'j_2'j_1j_2; jk)$$
$$\times\langle j_1|\;|T^k|\;|j_1'\rangle\langle j_2|\;|U^k|\;|j_2'\rangle, \qquad \text{(BIII.15)}$$

$$\langle j_1j_2j|\;|(T^{k_1}(1), U^{k_2}(2))^k|\;|j_1'j_2'j'\rangle = \begin{bmatrix} j_1' & j_2' & j' \\ k_1 & k_2 & k \\ j_1 & j_2 & j \end{bmatrix} \langle j_1|\;|T^{k_1}|\;|j_1'\rangle\langle j_2|\;|U^{k_2}|\;|j_2'\rangle. \qquad \text{(BIII.16)}$$

Examples

Matrix elements of spin operator connecting states $|l\tfrac{1}{2}jm\rangle$ Here the operator belongs to the space of the second angular momentum, and hence we have a case for applying (BIII.12) and (BIII.13).

Matrix elements of spherical harmonic connecting states $|l\tfrac{1}{2}jm\rangle$ This is a case for applying

(BIII.14). It so happens that here the properties of the particular U-coefficient lead to the simplified result

$$\langle l\tfrac{1}{2}j\|\,y^k\,\|l'\tfrac{1}{2}j'\rangle = \tfrac{1}{2}\{1 + (-1)^{l+l'+k}\}(-1)^k\begin{bmatrix} j & k & j' \\ \tfrac{1}{2} & 0 & \tfrac{1}{2} \end{bmatrix}. \tag{BIII.17}$$

We have used the renormalized spherical harmonic y^k instead of the spherical harmonic Y^k.

FURTHER READING

Fano, U., and Racah, G., Irreducible Tensorial Sets, Academic Press, New York, 1959.

Rose, M. E., Elementary Theory of Angular Momentum, Wiley, New York, 1957.

Wigner, E., Group Theory and its Application to the Quantum Mechanics of Atomic Spectra, Academic Press, New York, 1959.

PROBLEMS

1. An orthogonal transformation keeps $\sum_i x_i^2$ invariant. Use this fact in (BI.1) to prove (BI.2a).

2. Write down the transformation matrix a corresponding to a rotation through α about the Z-axis.

3. Construct the matrix a for the full Eulerian rotation (α, β, γ), as defined before (BI.3), by multiplying the matrices for the three rotations in succession.

4. Obtain the transformation matrix for the spherical harmonic $Y_m^1(\theta, \phi)$ by applying to (x, y, z) the transformation matrix obtained in Problem 3. [Hint: Use (AI.9).] Work out $\mathcal{D}_{mm'}^{1*}(\alpha, \beta, \gamma)$ from (BI.12c) and (BI.12d) and verify its equality with what you have derived.

5. Derive $\mathcal{D}_{mm'}^{\frac{1}{2}*}(\alpha, \beta, \gamma)$ without using the standard expression (BI.12c).

6. Prove (BI.19a), (BI.20a), (BI.20b), and (BI.20c).

7. Derive (BI.21) by following the directions given after those equations.

8. If a tensor of rank K has to be symmetric in any two of its K indices, and also traceless with respect to any two indices (i.e., if we make any two indices identical and sum over the common index, the result is zero), show that it can then have only $(2K + 1)$ independent components.

9. Use the definition of an irreducible tensor given before (BII.8) and show that the left-hand side of this expression indeed transforms as an irreducible tensor of rank K. [Hint: Use the properties of the $\mathcal{D}$-matrices.]

10. Prove the Wigner-Eckart theorem (BIII.1). [Hint: Insert $R^\dagger R$, where R is an arbitrary rotation operator on both sides of T_q^K, and then use the transformation properties of the states and the operator under R.]

11. Derive (BIII.11), (BIII.13), (BIII.14), (BIII.15), (BIII.16), and (BIII.17).

Appendix C

Solutions of Special Radial Schrödinger Equations

I. RADIAL EQUATION FOR $r \to \infty$: SPHERICAL BESSEL, NEUMANN, AND HANKEL FUNCTIONS

In this appendix, we shall list the properties of the solutions of the radial Schrödinger equation (I.49) for a particle moving freely in space (i.e., no potential). A short-range potential vanishes at large r, and hence the Schrödinger equation in such a case, for large r, has the desired property. The solutions are (see Schiff[1])

$$krj_l(kr), \qquad krn_l(kr),$$

where $j_l(kr)$ and $n_l(kr)$ are respectively the spherical Bessel and spherical Neumann functions. These functions are related to Bessel functions of positive and negative half-integral order as

$$j_l(x) = \left(\frac{\pi}{2x}\right)^{1/2} J_{l+1/2}(x), \tag{CI.1}$$

$$n_l(x) = (-1)^{l+1}\left(\frac{\pi}{2x}\right)^{1/2} J_{-l-1/2}(x). \tag{CI.2}$$

The *exact* expressions for the spherical Bessel and spherical Neumann functions for $l = 0, 1, 2$ are

$$\begin{aligned}
j_0(x) &= \frac{\sin x}{x}, & n_0(x) &= -\frac{\cos x}{x}, \\
j_1(x) &= \frac{\sin x}{x^2} - \frac{\cos x}{x}, & n_1(x) &= -\frac{\cos x}{x^2} - \frac{\sin x}{x}, \\
j_2(x) &= \left(\frac{3}{x^3} - \frac{1}{x}\right) \sin x - \frac{3}{x^2} \cos x, & n_2(x) &= -\left(\frac{3}{x^3} - \frac{1}{x}\right) \cos x - \frac{3}{x^2} \sin x.
\end{aligned} \tag{CI.3}$$

The behaviour of these functions for $x \to 0$ and $x \to \infty$ is very important and is given by

$$\begin{aligned}
i_l(x) &= \frac{x^l}{(2l+1)!!}, & x &\to 0, \\
n_l(x) &= \frac{(2l-1)!!}{x^{l+1}}, & x &\to 0, \\
j_l(x) &= x^{-1} \sin (x - \tfrac{1}{2} l\pi), & x &\to \infty, \\
n_l(x) &= -x^{-1} \cos (x - \tfrac{1}{2} l\pi), & x &\to \infty.
\end{aligned} \tag{CI.4}$$

Therefore, the solution $j_l(kr) \to 0$ as $r \to 0$, whereas $n_l(kr)$ blows up at the origin.

Some standard results in terms of $f_l(x)$, where $f_l(x)$ can be both $j_l(x)$ and $n_l(x)$, are

$$f_{l-1}(x) + f_{l+1}(x) = \frac{2l+1}{x} f_l(x),$$
$$\frac{d}{dx} f_l(x) = \frac{1}{2l+1}[l f_{l-1}(x) - (l+1) f_{l+1}(x)],$$
$$\frac{d}{dx}[x^{l+1} f_l(x)] = x^{l+1} f_{l-1}(x),$$
$$\frac{d}{dx}[x^{-l} f_l(x)] = -x^{-l} f_{l+1}(x). \tag{CI.5}$$

The spherical Hankel functions of the first and second kind are defined by

$$h_l^{(1)}(x) = j_l(x) + i n_l(x),$$
$$h_l^{(2)}(x) = j_l(x) - i n_l(x). \tag{CI.6}$$

For large x, these functions behave as

$$h_l^{(1)}(x) = -i x^{-1} e^{i[x-(1/2)l\pi]}, \qquad x \to \infty,$$
$$h_l^{(2)}(x) = i x^{-1} e^{-i[x-(1/2)l\pi]}, \qquad x \to \infty. \tag{CI.7}$$

These functions are useful in writing the solutions of the radial Schrödinger equation in the absence of a potential for a particle of negative energy. In other words, the equation we have in mind is that of a bound state in the region of large r where the potential goes to zero. In this case, $k^2 = -\gamma^2$ or $k = i\gamma$. Therefore, for large r, the two solutions $h_l^{(1)}(i\gamma r)$ and $h_l^{(2)}(i\gamma r)$ behave, according to (CI.7), as $e^{-\gamma r}$ and $e^{\gamma r}$, respectively. The solution $h_l^{(1)}(i\gamma r)$ is well-behaved at $r \to \infty$. For $l = 0, 1, 2$, the *exact* expressions of this function are obtained from (CI.6) and (CI.3) as

$$i\gamma r h_0^{(1)}(i\gamma r) = -i e^{-\gamma r},$$
$$i\gamma r h_1^{(1)}(i\gamma r) = -(1 + \frac{1}{\gamma r}) e^{-\gamma r},$$
$$i\gamma r h_2^{(1)}(i\gamma r) = i[1 + \frac{3}{\gamma r} + \frac{3}{(\gamma r)^2}] e^{-\gamma r}. \tag{CI.8}$$

Expansion of Plane Wave in Terms of Partial Wave

The Schrödinger equation for a free particle of positive energy is given by

$$(\nabla^2 + k^2)\psi = 0$$

which immediately yields the plane-wave solution

$$\psi(\mathbf{r}) = e^{i\mathbf{k}\cdot\mathbf{r}} = e^{ikr\xi}, \tag{CI.9}$$

where $\xi = \cos\theta$, and θ is the angle between the vectors **k** and **r**. Clearly, this function can be expanded in a complete set of the Legendre polynomials $P_l(\xi)$, and the radial function for the l-th partial wave, namely, $j_l(kr)$, described earlier in this appendix. Thus,

$$e^{i\mathbf{k}\cdot\mathbf{r}} \equiv e^{ikr\xi} = \sum_{l=0}^{\infty} c_l j_l(kr) P_l(\xi), \tag{CI.10}$$

where the coefficients c_l are to be determined. Using the orthogonality property (AI.8) of the

Legendre polynomials, we immediately obtain

$$c_l j_l(kr) = \frac{2l+1}{2} \int_{-1}^{+1} e^{ikr\xi} P_l(\xi)\, d\xi. \tag{CI.11}$$

This result, being true for all values of r, yields, in the limit of $r \to 0$, with the help of the first expression of (CI.4),

$$c_l \frac{(kr)^l}{(2l+1)!!} = \frac{2l+1}{2} \underset{r\to 0}{\text{Lt}} \sum_{n=0}^{\infty} \frac{(ikr)^n}{n!} \int_{-1}^{+1} \xi^n P_l(\xi)\, d\xi. \tag{CI.12}$$

The integral vanishes when $n < l$. In view of the limit $r \to 0$, we need only the leading non-vanishing term, i.e., the term for $n = l$. The value of the corresponding integral is given by

$$\int_{-1}^{+1} \xi^l P_l(\xi)\, d\xi = \frac{2(l!)}{(2l+1)!!}. \tag{CI.13}$$

Substituting (CI.13) in (CI.12), we have the final result

$$c_l = i^l(2l+1), \tag{CI.14}$$

and hence

$$e^{i\mathbf{k}\cdot\mathbf{r}} = \sum_{l=0}^{\infty} i^l(2l+1) j_l(kr) P_l(\cos\theta). \tag{CI.15}$$

II. CONFLUENT HYPERGEOMETRIC EQUATION AND ITS SOLUTIONS

The radial Schrödinger equation for a Coulomb potential is solved in terms of the confluent hypergeometric function, some standard results of which are summarized here. For more details, see Morse and Feshbach[2].

The confluent hypergeometric equation is given by

$$\frac{d^2F}{dz^2} + \frac{1}{z}(c - z)\frac{dF}{dz} - \frac{a}{z}F = 0 \tag{CII.1}$$

of which $z = 0$ is a regular singular point and $z = \infty$, an irregular singular point. A series solution around the point $z = 0$ is valid up to (but not including) $z = \infty$. In the usual way, the indicial equation and the recurrence relation lead to the two solutions

$$F(a, c; z), \qquad z^{1-c}F(a+1-c, 2-c; z), \tag{CII.2}$$

where

$$F(m, n; z) = 1 + \frac{m}{n}z + \frac{m(m+1)}{n(n+1)}\frac{z^2}{2!} + \ldots \tag{CII.3a}$$

$$= \frac{\Gamma(n)}{\Gamma(m)} \sum_{r=0}^{\infty} \frac{\Gamma(m+r)}{\Gamma(n+r)} \frac{z^r}{r!}. \tag{CII.3b}$$

Clearly, the infinite series reduces to a polynomial when m is zero or a negative integer, whereas the series blows up (hence is undefined) when n is zero or a negative integer. Therefore, one of the two solutions (CII.2) becomes undefined when c is zero or an integer (positive or negative); in the special case of $c = 1$, although both solutions remain well-defined, they are obviously identical. Thus, for integral values of c, we have to find another linearly independent solution by some other means. Another problem faced in the series solutions (CII.2) is that, although their values for $z \to 0$ are well-defined, the behaviour for $z \to \infty$ is difficult to determine from the infinite series.

Two other linearly independent solutions of (CII.1) are best written in the integral forms as

$$U_1(a, c; z) = \frac{1}{\Gamma(c-a)} z^{a-c} e^z \int_0^\infty du\, e^{-u} u^{c-a-1} (1 - \frac{u}{z})^{a-1} = z^{a-c} e^z, \qquad z \to \infty, \tag{CII.4a}$$

$$U_2(a, c; z) = \frac{1}{\Gamma(a)} z^{-a} \int_0^\infty du\, e^{-u} u^{a-1} (1 + \frac{u}{z})^{c-a-1} = z^{-a}, \qquad z \to \infty. \tag{CII.4b}$$

The asymptotic forms for $z \to \infty$ of these two functions, as just shown, follow easily from the integral expressions by noting that the integrals are respectively $\Gamma(c-a)$ and $\Gamma(a)$ in that limit.

Similarly, an integral representation for $F(a, c; z)$ is given by

$$F(a, c; z) = \frac{\Gamma(c)}{\Gamma(a)\Gamma(c-a)} \int_0^1 e^{zt} t^{a-1} (1-t)^{c-a-1}\, dt \tag{CII.5a}$$

$$\equiv \frac{\Gamma(c)}{\Gamma(a)} U_1(a, c; z) + \frac{\Gamma(c)}{\Gamma(c-a)} e^{\pm i\pi a} U_2(a, c; z). \tag{CII.5b}$$

(CII.5b) follows from (CII.5a) by suitably deforming the contour of integration and then using (CII.4). The two signs $+$ and $-$ in the phase factor $e^{\pm i\pi a}$ in (CII.5b) hold for z lying in the upper and lower half-planes, respectively. The asymptotic form for $z \to \infty$ is then given by

$$F(a, c; z) = \frac{\Gamma(c)}{\Gamma(a)} z^{a-c} e^z + \frac{\Gamma(c)}{\Gamma(c-a)} e^{\pm i\pi a} z^{-a}, \qquad z \to \infty. \tag{CII.5c}$$

A second linearly independent solution to go with $F(a, c; z)$ can then be chosen as

$$G(a, c; z) = \frac{\Gamma(c)}{\Gamma(a)} U_1(a, c; z) - \frac{\Gamma(c)}{\Gamma(c-a)} e^{\pm i\pi a} U_2(a, c; z) \tag{CII.6a}$$

$$= \frac{\Gamma(c)}{\Gamma(a)} z^{a-c} e^z - \frac{\Gamma(c)}{\Gamma(c-a)} e^{\pm i\pi a} z^{-a}, \qquad z \to \infty. \tag{CII.6b}$$

The final problem that remains is to find what happens to $G(a, c; z)$ in the limit $z \to 0$ for integral values of c. We have to first express $z^{1-c}F(a+1-c, 2-c; z)$ in terms of U_1 and U_2, using (CII.5b), then solve for U_1 and U_2 in terms of the two solutions (CII.2), and finally substitute the results in (CII.6a). Some further tricky manipulations then lead to the final expression

$$\begin{aligned} G(a, n; z) = \pm \frac{e^{\pm 2i\pi a} - 1}{2\pi i} & [\{2 \ln z + \pi \cot \pi a \mp i\pi\} F(a, n; z) \\ & - 2 \sum_{r=0}^{\infty} \{\psi(r+1) + \psi(n+r) - \psi(a+r)\} \frac{\Gamma(a+r)\Gamma(n)}{\Gamma(a)\Gamma(n+r)} \frac{z^r}{r!}] \\ & - 2 e^{\pm i\pi a} \frac{\Gamma(n)}{\Gamma(a)} \sum_{s=1}^{n-1} \frac{\Gamma(s)}{\Gamma(n-s)\Gamma(s-a-1)} z^{-s}. \end{aligned} \tag{CII.7}$$

The $+$ and $-$ signs are applicable for z in the upper and lower half-planes, respectively. The function $\psi(x)$ is the log-derivative of the gamma-function, i.e.,

$$\psi(x) = \frac{1}{\Gamma(x)} \frac{d}{dx} \Gamma(x) \equiv -\gamma + \sum_{r=0}^{\infty} (\frac{1}{r+1} - \frac{1}{r+x}) \tag{CII.8}$$

such that $\psi(1) = -\gamma$, where γ is the Euler constant $= 0.5772\ldots$. The expression (CII.7) enables us to write $G(a, n; z)$ in the limit $z \to 0$.

III. PARTIAL-WAVE EXPANSION OF COULOMB WAVEFUNCTION

We shall here give all the algebraic steps in the derivation of C_l appearing in the partial-wave expansion (I.133) for the Coulomb wavefunction. We start with the expression for C_l, given in (I.135) as

$$C_l = \sqrt{\pi[l]}\,\exp(i\eta_l)\frac{\Gamma(1+i\alpha)}{\Gamma(l+1+i\alpha)}\,\frac{\Gamma(2l+2)}{(2k)^l}$$

$$\times \operatorname*{Lt}_{r\to 0} r^{-l}\int_{-1}^{+1} dx\; e^{-ikr(1-x)}F(-i\alpha,\,1;\,ikr(1-x))P_l(x). \tag{CIII.1}$$

For some time, we shall consider only the integral appearing in (CIII.1). Making power series expansions of the exponential and the confluent hypergeometric function [see (CII.3b)] and denoting the integral by I, we obtain

$$I = \int_{-1}^{+1} dx\; P_l(x)\sum_{m=0}^{\infty}\frac{(-ikr)^m(1-x)^m}{m!}\,\frac{\Gamma(1)}{\Gamma(-i\alpha)}\sum_{n=0}^{\infty}\frac{\Gamma(-i\alpha+n)}{\Gamma(1+n)n!}(ikr)^n(1-x)^n.$$

As argued after (CI.12), the lowest power of x that gives a nonvanishing integral with $P_l(x)$ is l. Therefore, the lowest value of $m+n$ is l. The power of r multiplying that term is also seen to be $m+n=l$. Higher values of $m+n$ also produce a nonvanishing integral, but the power of r with such terms is larger than l. Since we have to evaluate the limit $\operatorname*{Lt}_{r\to 0} r^{-l}I$, only the leading r^l-term of I can contribute a nonvanishing result in (CIII.1). Thus, we evaluate I for only $m+n=l$, and get

$$I = \frac{(-1)^l i^l (kr)^l}{\Gamma(-i\alpha)}\int_{-1}^{+1} dx\; P_l(x)(1-x)^l\sum_{n=0}^{l}(-1)^n\frac{\Gamma(-i\alpha+n)}{(l-n)!\,(n!)^2}.$$

Since only the $(-1)^l x^l$-term of $(1-x)^l$ contributes to the integral, we obtain, with the help of the standard result (CI.13),

$$I = \frac{i^l(kr)^l}{\Gamma(-i\alpha)}\,\frac{2}{(2l+1)!!}\,l!\sum_{n=0}^{l}(-1)^n\frac{\Gamma(-i\alpha+n)}{(l-n)!\,(n!)^2}.$$

Substituting this result in (CIII.1), we get

$$C_l = i^l\sqrt{4\pi[l]}\,\exp(i\eta_l)\Big[\frac{\Gamma(1+i\alpha)}{\Gamma(-i\alpha)\Gamma(l+1+i\alpha)}\sum_{n=0}^{l}(-1)^n({}^lC_n)^2(l-n)!\,\Gamma(-i\alpha+n)\Big], \tag{CIII.2}$$

where we have already used the expressions

$$\Gamma(2l+2) = (2l+1)! = 2^l(2l+1)!!\,l!,$$

$$\frac{l!}{(l-n)!\,n!} = {}^lC_n \qquad \text{(the binomial coefficient).}$$

The expression enclosed within the square brackets in (CIII.2) can easily be seen to be equal to unity for the two lowest values of l, i.e., $l = 0, 1$. Actually, it is so for any arbitrary value of l (see Problem 5 of Appendix C), and hence we write the final result as

$$C_l = i^l\sqrt{4\pi[l]}\,\exp(i\eta_l). \tag{CIII.3}$$

REFERENCES

1. Schiff, L. I., Quantum Mechanics, 2nd edn., McGraw-Hill, New York, 1955, p 77.

79 (45-122/1977)

2. Morse, F. M., and Feshbach, H., Methods of Theoretical Physics, McGraw-Hill, New York, 1953, pp 551, 604.

PROBLEMS

1. Derive the result (CI.13). [Hint: $\xi = \cos\theta = y_0^1(\theta)$, hence $\xi^l = y_0^1 y_0^1 \ldots y_0^1$ (l-factors); use (BIII.4) repeatedly in this expression to find the coefficient of y_0^l in the final result.]

2. Obtain the series solutions (CII.2) of the confluent hypergeometric equation.

3. With the help of the integral expressions (CII.4), prove that

$$z^{1-c}U_i(a + 1 - c, 2 - c; z) = U_i(a, c; z),$$

where $i = 1, 2$. [Hint: Make a binomial expansion, express the binomial coefficients in terms of Γ-functions, and use the standard result $\Gamma(x)\Gamma(1 - x) = \pi/(\sin \pi x)$.]

4. Express $z^{1-c}F(a + 1 - c, 2 - c; z)$ in terms of $U_1(a, c; z)$ and $U_2(a, c; z)$.

5. Prove, by the method of induction, that the expression within the square brackets in (CIII.2) is equal to unity for any value of l (i.e., assume it to be true for $l = L$, and then prove the validity for $l = L + 1$).

Appendix D

Time-Reversal in Nonrelativistic Quantum Mechanics

Some basic results on time-reversal are summarized in this appendix. The reader is referred to Sachs[1] for more details.

From the Schrödinger equation

$$i\hbar\frac{\partial\Psi(\mathbf{r},\, t)}{\partial t} = H\Psi(\mathbf{r},\, t), \tag{D.1}$$

it is easy to see that $\Psi(\mathbf{r}, -t)$ and $\Psi^*(\mathbf{r}, t)$ satisfy the same equation. Thus, the effect of time-reversal on the wavefunction can be described by complex conjugation. Under time-reversal, it is expected, from common sense, that the operator $\mathbf{r}$ should remain invariant, whereas the operator $\mathbf{p}$ which contains $d\mathbf{r}/dt$ must reverse its sign. Since $\mathbf{p} = -i\hbar\boldsymbol{\nabla}$ and $\boldsymbol{\nabla}$ is real, we notice that this anticipation is true as long as time-reversal produces complex conjugates of these operators. Similarly, we expect the angular momenta $\boldsymbol{l}$, $\boldsymbol{\sigma}$, and $\mathbf{j}$ to change sign under time-reversal. The case of $\boldsymbol{l}$ is easy to check because it is equal to $\mathbf{r} \times \mathbf{p}$. However, if we take merely the complex conjugates of the Pauli matrices, we notice that only σ_y changes sign but σ_x, σ_z remain the same. Thus, we conclude that in the case of spin the process of time-reversal must do more than complex conjugation. In general, we then write the time-reversal operator T as

$$T = UK, \tag{D.2}$$

where K produces complex conjugation, and U operates only on the spin wavefunctions and spin operators. Since K keeps $\Psi^*\Psi$ unchanged, it is unitary; thus, T will be unitary, provided we choose U to be so. According to our requirement,

$$T\boldsymbol{\sigma}T^{-1} = -\boldsymbol{\sigma}$$

or

$$U\boldsymbol{\sigma}^*U^{-1} = -\boldsymbol{\sigma}. \tag{D.3}$$

In terms of the components, (D.3) yields

$$U\sigma_xU^{-1} = -\sigma_x, \qquad U\sigma_yU^{-1} = \sigma_y, \qquad U\sigma_zU^{-1} = -\sigma_z. \tag{D.4}$$

Clearly, the operator $c\sigma_y$ for U, where c is an arbitrary constant, satisfies the three equations in (D.4). The choice $c = -i$ leads to the unitary operator

$$U = -i\sigma_y = \exp\left(-\frac{i}{2}\pi\sigma_y\right) = \exp\left(-i\pi s_y\right). \tag{D.5}$$

Operating on the eigenfunction $\chi_\mu^{1/2}$ of spin, the time-reversal operator then produces

$$T\chi_\mu^{1/2} = U\chi_\mu^{1/2} = \exp\left(-i\pi s_y\right)\chi_\mu^{1/2} = (-1)^{1/2-\mu}\chi_{-\mu}^{1/2}. \tag{D.6}$$

Here the reality of the spin state and the result (BI.15) have been used.

The orbital angular momentum eigenfunction $Y^l_m(\theta, \phi)$, on the other hand, transforms under T as

$$TY^l_m(\theta, \phi) = Y^{l*}_m(\theta, \phi) = (-1)^m Y^l_{-m}(\theta, \phi). \tag{D.7}$$

In order to make (D.6) and (D.7) look identical, we need an additional phase $(-1)^l$ in the latter equation. This can be obtained if we work with the redefined angular momentum eigenfunctions

$$\Phi^l_m(\theta, \phi) \equiv i^l Y^l_m(\theta, \phi). \tag{D.8}$$

In that case,

$$T\Phi^l_m(\theta, \phi) \equiv \Phi^{l*}_m(\theta, \phi) = (-1)^{l-m}\Phi^l_{-m}(\theta, \phi). \tag{D.9}$$

Since i^l in (D.8) is an overall phase, Φ^l_m can be coupled with the spin function $\chi^{1/2}_\mu$ in the usual way with a Clebsch-Gordon coefficient, and the resultant states denoted by $|l\frac{1}{2}jm\rangle_p$, where p is a reminder of the extra phase factor i^l in the orbital angular momentum part of this state. Under time-reversal, this state can be proved, with the help of (D.6) and (D.9), to undergo the transformation

$$T|l\tfrac{1}{2}jm\rangle_p = (-1)^{j-m}|l\tfrac{1}{2}j, -m\rangle_p. \tag{D.10}$$

It should be noticed that, for the state $|l\frac{1}{2}jm\rangle$, where the orbital angular momentum part is Y^l_m without the phase factor i^l, the same kind of transformation is produced by a rotation π about the y-axis:

$$\exp(-i\pi j_y)|l\tfrac{1}{2}jm\rangle = (-1)^{j-m}|l\tfrac{1}{2}j, -m\rangle. \tag{D.11}$$

In nuclear literature, the transformation $\exp(-i\pi j_y)$ of the states $|l\frac{1}{2}jm\rangle$ is sometimes loosely called the *time-reversal operation*.

The effect of time-reversal on isospin is once again deduced from common sense. Since τ_z represents the charge of the particle, it is required to remain invariant under T. Therefore, T can be only a rotation about the z-axis in the isospin space. In that case, operators such as $(\boldsymbol{\tau}_1 \cdot \boldsymbol{\tau}_2)$ and $(\boldsymbol{\tau}_1 \times \boldsymbol{\tau}_2)_z$ remain invariant under time-reversal.

REFERENCE

1. Sachs, R. G., Nuclear Theory, Addison-Wesley, Reading, Mass., 1953, p 353.

PROBLEMS

1. Change $t \to -t$ in (D.1), and also write the complex conjugate of the same equation, assuming H to be real. Verify that $\Psi(\mathbf{r}, -t)$ and $\Psi^*(\mathbf{r}, t)$ satisfy the same equation.

2. Simplify $\exp(-\frac{1}{2}\theta\sigma_y)$, where θ is any arbitrary angle. [Hint: Expand in a power series and group the odd and even powers of θ separately; simplify, using the properties of σ_y.] Hence, verify $-i\sigma_y = \exp(-\frac{1}{2}i\pi\sigma_y)$.

3. Write $|l\frac{1}{2}jm\rangle_p$ of (D.10) in terms of the Clebsch-Gordon coefficients and verify this equation by using (D.6) and (D.9) and the properties of the Clebsch-Gordon coefficients.

4. In the same way as in Problem 3, expand $|l\frac{1}{2}jm\rangle$ of (D.11) in terms of the Clebsch-Gordon coefficients and verify (D.11).

Appendix E

Nuclear Size and Charge Distribution

I. ELECTRON SCATTERING BY NUCLEI

The scattering amplitude $f(\theta)$ for the scattering of a particle of mass μ by a potential $V(\mathbf{r})$ is given by (see Schiff[1])

$$f(\theta) = \frac{1}{4\pi} \int \exp(-i\mathbf{k}_f \cdot \mathbf{r}) v(\mathbf{r}) \psi(\mathbf{r})\, d^3r, \tag{EI.1a}$$

where

$$v(\mathbf{r}) = -\frac{2\mu}{\hbar^2} V(\mathbf{r}) \tag{EI.1b}$$

and $\psi(\mathbf{r})$ is the exact wavefunction. The quantity $\mathbf{k}_f$ represents the final momentum of the particle. In the Born approximation (see Schiff[1]), the exact wavefunction is substituted by the incident plane wave $\exp(i\mathbf{k}_i \cdot \mathbf{r})$, where $\mathbf{k}_i$ is the initial momentum of the particle, and we obtain

$$f_B(\theta) = \frac{1}{4\pi} \int d^3r \exp(-i\mathbf{k}_f \cdot \mathbf{r}) v(\mathbf{r}) \exp(i\mathbf{k}_i \cdot \mathbf{r}). \tag{EI.1c}$$

We first apply the Born approximation formula (EI.1c) to the scattering of an electron by a point charge (subscript or superscript P) Ze. The attractive Coulomb potential is given by $-Ze^2/r$, and hence

$$v_P(r) = \frac{2\mu}{\hbar^2} \frac{Ze^2}{r}.$$

When this potential is substituted in (EI.1c), it gives rise to the Fourier transform of $1/r$:

$$f_B^P(\theta) = \frac{1}{4\pi} \frac{2\mu Ze^2}{\hbar^2} \int d^3r \frac{\exp(i\mathbf{q} \cdot \mathbf{r})}{r}, \tag{EI.2}$$

where $\mathbf{q}$ is the momentum transfer $(\mathbf{k}_i - \mathbf{k}_f)$. When $|\mathbf{k}_i| = |\mathbf{k}_f| = k$, we easily obtain

$$q = 2k \sin \tfrac{1}{2}\theta, \tag{EI.3}$$

where θ is the angle between $\mathbf{k}_i$ and $\mathbf{k}_f$, i.e., the scattering angle.

Let us denote the Fourier transform by $\mathcal{F}(\mathbf{q})$, that is,

$$\mathcal{F}(\mathbf{q}) = \int d^3r \frac{\exp(i\mathbf{q} \cdot \mathbf{r})}{r}.$$

By the inverse Fourier transformation,

$$\frac{1}{r} = \frac{1}{(2\pi)^3} \int d^3q\, \mathcal{F}(\mathbf{q}) \exp(-i\mathbf{q} \cdot \mathbf{r}).$$

Therefore,

$$\nabla^2\left(\frac{1}{r}\right) = \frac{1}{(2\pi)^3}\int d^3q\ \mathcal{F}(\mathbf{q})\nabla^2[\exp(-i\mathbf{q}\cdot\mathbf{r})]$$

$$= -\frac{1}{(2\pi)^3}\int d^3q\ q^2\mathcal{F}(\mathbf{q})\exp(-i\mathbf{q}\cdot\mathbf{r}). \tag{EI.4a}$$

The left-hand side of (EI.4a) is also directly given by Poisson's equation for the electrostatic potential at a distance r from a unit point charge as

$$\nabla^2\left(\frac{1}{r}\right) = -4\pi\delta(\mathbf{r}).$$

Using the familiar integral expression of $\delta(\mathbf{r})$, we get

$$\nabla^2\left(\frac{1}{r}\right) = -\frac{4\pi}{(2\pi)^3}\int d^3q\ \exp(-i\mathbf{q}\cdot\mathbf{r}). \tag{EI.4b}$$

Equating the right-hand sides of (EI.4a) and (EI.4b), we easily obtain

$$\mathcal{F}(\mathbf{q}) = \frac{4\pi}{q^2}.$$

Therefore, (EI.2), with the expression (EI.3) for q, yields the final result

$$f_B^P(\theta) = \frac{2\mu Ze^2}{\hbar^2}\frac{1}{q^2} = \frac{\mu Ze^2}{2\hbar^2k^2}\operatorname{cosec}^2\tfrac{1}{2}\theta$$

$$= \frac{Ze^2}{2\mu v^2}\operatorname{cosec}^2\tfrac{1}{2}\theta. \tag{EI.5}$$

It should be noticed that $\sigma(\theta)$, computed from this expression, is identical with the Rutherford expression (I.122) derived in Section 8 from an *exact* solution of the Schrödinger equation. While making this comparison, we have to remember to replace e^2 and M of (I.122) by Ze^2 and 2μ, respectively, because that equation was derived for the case of p-p scattering. The exact expression (I.121) of $f(\theta)$ obtained there, however, differs from the Born approximation expression (EI.5) through the energy-dependent phase factor $\exp(2i\eta_0)\exp[-i\alpha\ln\sin^2(\theta/2)]$. It is just an accident that the *exact* and the *approximate* Born approximation results for the scattering amplitude of a Coulomb potential differ *only* through a phase factor, and hence, perhaps a bit paradoxically, give rise to the same expression for the differential cross-section.

We next derive the Born approximation scattering amplitude of an electron scattered by the Coulomb potential exerted by an extended charge distribution of the nucleus. The charge density at any point $\mathbf{r}'$ of this distribution is denoted by $Ze\rho(\mathbf{r}')$. Since the total charge of the nucleus is Ze, we obtain the normalization of our density distribution function ρ as

$$Ze\int_\tau \rho(\mathbf{r}')\,d^3r' = Ze$$

or

$$\int_\tau \rho(\mathbf{r}')\,d^3r' = 1, \tag{EI.6}$$

where the volume τ denotes the extent of the nuclear charge distribution. The potential energy of the electron located at $\mathbf{r}$ is evidently given by

$$-Ze^2 \int_\tau d^3r' \frac{\rho(\mathbf{r}')}{|\mathbf{r}-\mathbf{r}'|} = V(\mathbf{r})$$

or

$$v(\mathbf{r}) = \frac{2\mu Ze^2}{\hbar^2} \int_\tau d^3r' \frac{\rho(\mathbf{r}')}{|\mathbf{r}'-\mathbf{r}|}. \tag{EI.7}$$

We substitute (EI.7) in (EI.1c) and obtain

$$f_B(\theta) = \frac{1}{4\pi}\frac{2\mu Ze^2}{\hbar^2} \int d^3r \exp(i\mathbf{q}\cdot\mathbf{r}) \int_\tau d^3r' \frac{\rho(\mathbf{r}')}{|\mathbf{r}-\mathbf{r}'|}. \tag{EI.8}$$

This expression takes a more convenient form if we make the coordinate transformation

$$\mathbf{R} = \mathbf{r} - \mathbf{r}', \qquad \boldsymbol{\xi} = \mathbf{r}'.$$

The Jacobian J of this transformation can be easily shown to be unity. Therefore,

$$d^3r\, d^3r' \equiv J\, d^3R\, d^3\xi = d^3R\, d^3\xi.$$

Introducing the foregoing transformation in (EI.8), we easily obtain

$$f_B(\theta) = \frac{1}{4\pi}\frac{2\mu Ze^2}{\hbar^2} \int d^3R \frac{\exp(i\mathbf{q}\cdot\mathbf{R})}{R} \int_\tau d^3\xi\, \rho(\boldsymbol{\xi}) \exp(i\mathbf{q}\cdot\boldsymbol{\xi}). \tag{EI.9}$$

Since $\mathbf{R}$ is a dummy integration variable, the first integral, multiplied by the constant factor in the front, exactly agrees with the expression (EI.2) for the point-nucleus scattering. Therefore, (EI.9) finally reduces to

$$f_B(\theta) = f_B^P(\theta) F(\mathbf{q}), \tag{EI.10a}$$

where the *nuclear form factor* $F(\mathbf{q})$ is given by

$$F(\mathbf{q}) = \int_\tau d^3r\, \rho(\mathbf{r}) \exp(i\mathbf{q}\cdot\mathbf{r}). \tag{EI.10b}$$

The differential cross-section is therefore given by

$$\sigma_B(\theta) = \sigma_R(\theta)|F(\mathbf{q})|^2, \tag{EI.11}$$

where the subscript R stands for Rutherford scattering.

The expression (EI.11) is, however, not quite right for the scattering of a very high-energy (several hundred MeV) electron. The electron has to have a high energy so that its De Broglie wavelength can be very small compared with the nuclear dimension; then, and then only, the electron penetrating the nucleus is able to probe the details of the nuclear charge distribution. The electrons of several hundred MeV are highly relativistic, and hence we should have applied the Dirac equations to describe their scattering. In the Born approximation, instead of using the Schrödinger plane waves $\exp(i\mathbf{k}\cdot\mathbf{r})$, we should have applied the four-component Dirac wavefunctions for a free particle. We shall not give the details of the relativistic Born approximation calculation, but we shall state only the results. For a point nucleus, the Rutherford expression $\sigma_R(\theta)$ gets multiplied by a factor of $\cos^2(\theta/2)$; the result quoted in (II.48) is called the Mott scattering cross-section. For an extended nuclear distribution and relativistic electron energy, (EI.11) gets modified by the replacement of $\sigma_R(\theta)$ by $\sigma_M(\theta)$, M standing for Mott.

For very heavy nuclei, even the relativistic Born approximation turns out to be inadequate. The most rigorous method entails considering the Dirac equation for the electron in the presence of the potential due to an extended charge distribution, and then computing the scattering cross-section by detailed calculation of the phase shifts.

Theoretical and Experimental Results

Experiments have been done on many nuclei, for instance, p, n, d, α, Li, Be, C, Mg, Si, S, A, Sr, Ca, V, Co, In, Sb, Hf, Ta, W, Au, Bi, Th, and U.

As a typical example, the experimental curve for C^{12}, together with the theoretical Born approximation curve, and the exact phase shift results, are shown in Fig. E.1. It can be easily derived from the harmonic oscillator radial functions in Section 17C that any $0p$-shell nucleus with charge Ze, having a ground-state spin equal to zero, has a charge distribution given by

$$\rho(r) = \frac{2}{Z(\pi^{1/2}b)^3} \exp(-r^2/b^2)[1 + \tfrac{1}{3}(Z-2)\frac{r^2}{b^2}], \tag{EI.12}$$

where b is the harmonic oscillator parameter that makes r dimensionless. With this charge density, the form factor $F(\mathbf{q})$ in the Born approximation works out to be

$$\begin{aligned} F(\mathbf{q}) &= \int d^3r \exp(i\mathbf{q}\cdot\mathbf{r})\rho(r) \\ &= 4\pi\frac{2}{Z(\pi^{1/2}b)^3}\int_0^\infty r^2\,dr\,\frac{\sin qr}{qr}\exp(-r^2/b^2)[1+\tfrac{1}{3}(Z-2)\frac{r^2}{b^2}] \\ &= \frac{8}{Z\pi^{1/2}\bar{q}}\int_0^\infty d\bar{r}\,\sin\bar{q}\bar{r}\,\exp(-\bar{r}^2)[\bar{r}+\tfrac{1}{3}(Z-2)\bar{r}^3], \end{aligned} \tag{EI.13}$$

where $\bar{r} = r/b$ and $\bar{q} = bq$. Working out the integral explicitly, we obtain a very simple expression, namely,

$$F(\mathbf{q}) = (1 - \frac{Z-2}{6Z}\bar{q}^2)\exp(-\tfrac{1}{4}\bar{q}^2). \tag{EI.14}$$

This expression vanishes at

$$\bar{q}^2 = \frac{6Z}{Z-2}.$$

Thus, the Born approximation cross-section for the $0p$-shell nuclei becomes exactly zero at some angle. We shall refer to this as a *diffraction minimum*. The Born approximation results for heavier nuclei are characterized by several such diffraction minima. The experimental data for C^{12} and the phase shift fit (solid line) to it are also shown in Fig. E.1. It is clear that the exact zero at the diffraction minimum, predicted by the Born approximation (dashed line), is in reality considerably filled up, and only a dip in the cross-section appears in the experimental or the exact phase shift curve. The failure of the Born approximation is very much marked at the diffraction minima and becomes more spectacular for heavier nuclei.

An important effect which we have to mention concerns the angular deformation in the charge distribution of certain nuclei. The experimental data for several nonspherical nuclei are shown in Fig. E.2. In these curves, the diffraction minima are almost non-existent. These deformed nuclei appear so when viewed in a coordinate system attached to the surface of the nucleus. With respect to a coordinate system fixed in the laboratory, the nuclear surface rotates

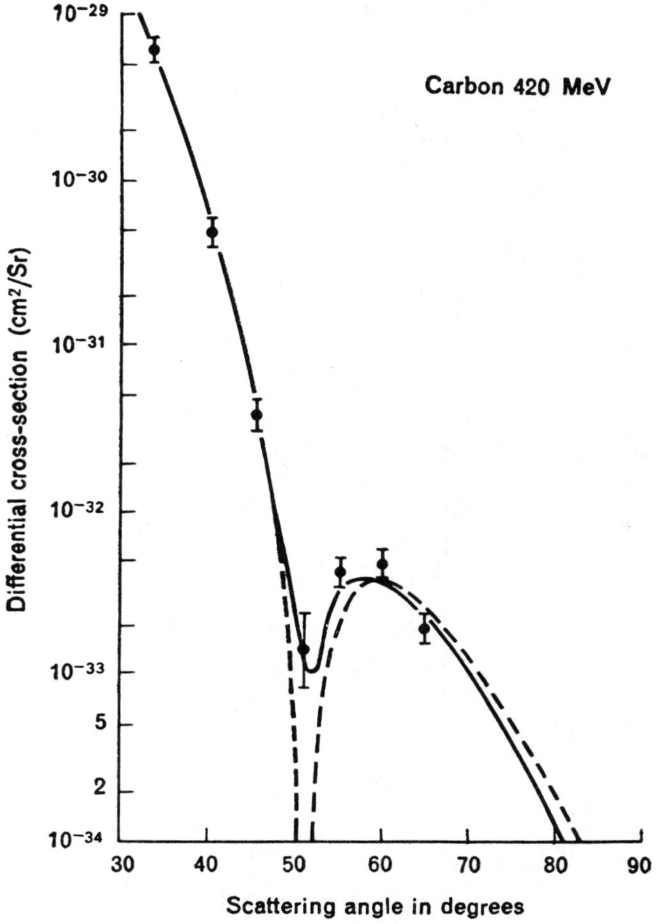

Fig. E.1 Differential cross-section for scattering of 420 MeV electrons on C^{12}. [Following Hofstadter, R., *Ann. Rev. Nucl. Sc.*, 7, 231 (1957).]

with time, and the effect of averaging over all orientations is a spherical density with a surface region much thicker than the usual extent of the surface of a spherical nucleus.

There are other nuclei whose charge distribution is deformed even with respect to a coordinate system fixed in the laboratory. A nucleus such as N^{14}, for example, has a ground-state spin 1, and hence the static ground-state charge density can have a quadrupole component. The scattering due to the quadrupole part of the charge distribution makes itself strongly felt in the region of the diffraction minimum, and has, once again, the effect of filling up the expected diffraction dip. Such analyses of N^{14} data, which give a good idea about the quadrupole moment of this nucleus, have been made (see Pal[2]). The charge density, calculated according to the shell model using the $0p$-level only, has a quadrupole moment that is too

80 (45-122/1977)

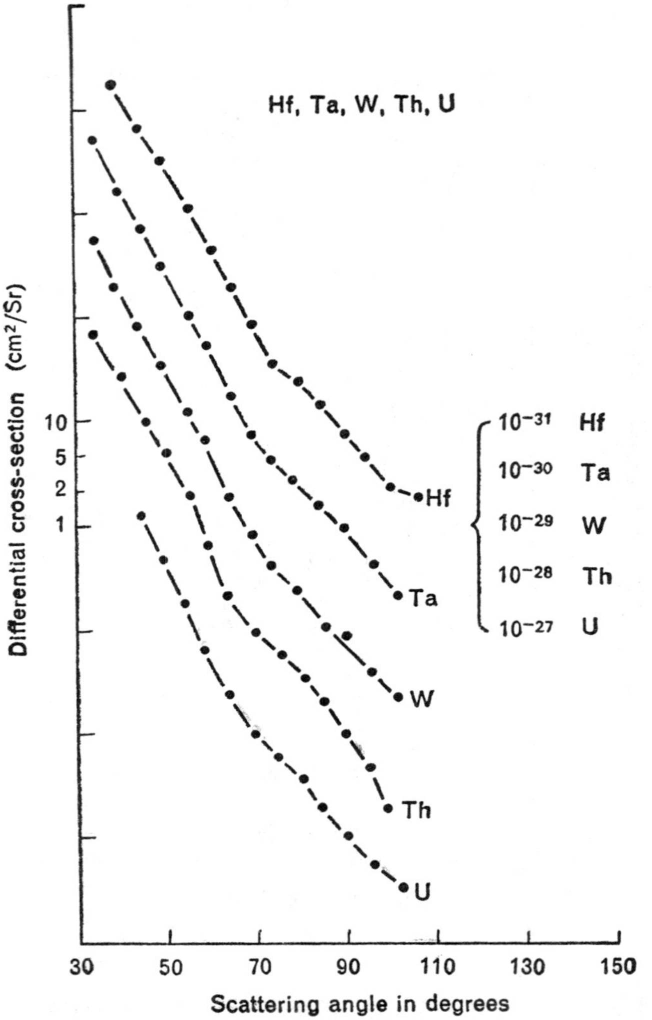

Fig. E.2 Experimental data for nonspherical nuclei. [Following Hofstadter, R., *Revs. Mod. Phys.*, **28**, 214 (1956).]

small to explain the observed extent of filling-up of the diffraction minimum. The extra quadrupole moment necessary for this purpose has been estimated, and attributed to the effect of excitation of the nucleus from the lowest shell-model configuration to higher orbitals (see Pal[2]).

As a general summary of the charge distribution of nuclei, obtained from the analyses of electron scattering data, we mention that (i) the light nuclei do not show a marked dependence on the model of the charge distribution, and usually good fits are produced with a charge density calculated from harmonic oscillator wavefunctions; and (ii) for heavier nuclei, there is a definite preference for the fermi-type charge distribution [see (II.58)]. For a summary of charge distribution, the reader is referred to Hofstadter[3,4].

Finally, we give a special description of the information revealed in the scattering of electrons by the nucleons themselves, i.e., proton and neutron. For the nucleons, the scattering due to their intrinsic magnetic moments becomes quite important, especially at large angles where scattering due to the charge becomes small. The expression of square of the form factor is given by

$$F_1^2 + \frac{\hbar^2 q^2}{4M^2c^2}[2(F_1 + \mu_a F_2)^2 \tan^2 \tfrac{1}{2}\theta + \mu_a^2 F_2^2], \tag{EI.15}$$

where F_1 and F_2 are the charge and magnetic form factors, respectively, and μ_a is the anomalous part of the nucleon magnetic moment, i.e., $\mu_a = 1.78$ nM for the proton and $\mu_a = -1.91$ nM for the neutron. Although the charge of the neutron is zero, its charge form factor does not vanish because of the meson cloud surrounding it. The total charge contained in the cloud and inside the core region of a neutron must be zero; but since there is a distribution of charge in space, and the high-energy electron sees the details of this distribution, there is a corresponding non-zero form factor even for the neutron. Experimental data at small values of q^2 determine only the root-mean-square radii of the charge and magnetic moment distributions of neutron and proton; the data at higher values of q^2 give more details about the distributions.

A theoretical model has helped the proper interpretation of experimental results. The picture that emerges is one of a core region of the nucleon, in which a part of the charge and magnetic moment is located, followed by a meson cloud which contains the remaining charge and magnetic moment. The theory considers, instead of F_1^n, F_2^n, F_1^p, and F_2^p (the superscripts n and p stand for neutron and proton, respectively), the more fundamental quantities F_1^S, F_2^S, F_1^V, and F_2^V, where S and V refer to the isotopic scalar and the third component of the isotopic vector, respectively. In terms of these quantities, the nucleon form factors are given by

$$\begin{aligned} &F_1^p = \tfrac{1}{2}(F_1^S + F_1^V), \qquad F_1^n = \tfrac{1}{2}(F_1^S - F_1^V), \\ &1.79F_2^p = \tfrac{1}{2}(-0.06F_2^S + 1.85F_2^V), \\ &1.91F_2^n = \tfrac{1}{2}(0.06F_2^S + 1.85F_2^V). \end{aligned} \tag{EI.16}$$

The opposite signs of F_1^V and F_2^V occurring in the neutron and proton form factors are due to the fact that these quantities originate from the third component of the isotopic spin vector. The isoscalar and isovector form factors owe their origin to meson clouds. The mesons for the two types are respectively $T = 0$ (isoscalar) and $T = 1$ (isovector), and both have $J = 1$. The theory suggests that the former is a resonance state of three pions, whereas the latter is a resonance state of two pions. The theory gives, for each of the scalar and vector form factors, the functional form

$$F(q^2) = (1 - \alpha) + \frac{\alpha}{1 + \dfrac{1}{6\alpha}a^2 q^2}, \tag{EI.17}$$

where a^2 is the mean-square radius of the corresponding meson cloud, and α and $(1 - \alpha)$ are the fraction of charge or magnetic moment contained in the cloud and in the core, respectively. It is further assumed in the form (EI.17) that the fraction $(1 - \alpha)$ belonging to the core is concentrated at a point. The accuracy of the experimental data also is not sufficient to make a finer study of the core region; it cannot be conclusively said whether the charge or magnetic moment of the core is located at a point or distributed over a distance of the order of the Compton wavelength of the nucleon.

The experimental values of the parameters α and a for the scalar and vector form factors are listed in Table EI.1. If we make use of an expression such as (EI.17), with the appropriate values of the parameters from this table, and substitute in (EI.16), then it is clear that each of the proton and neutron form factors consists of one constant term plus two terms of the type $\alpha[1 + (6\alpha)^{-1}a^2q^2]^{-1}$. The Fourier transform of the constant term in the coordinate space is a delta function at the origin, and the Fourier transform of each of the other terms is a Yukawa function. Thus, the foregoing analysis of the form factors implies that a nucleon consists of a localized charge and magnetic moment surrounded by an isovector and an isoscalar meson cloud, each of the Yukawa-type spatial distribution.

Table EI.1 Parameters of nucleon form factors

Form Factor	Measured Parameter*	Column 1**	Column 2†
F_1^S	α_1^S	0.56	0.58
	a_1^S	1.13 fm	1.16 fm
F_1^V	α_1^V	1.20	1.10
	a_1^V	0.77 fm	0.85 fm
F_2^S	α_2^S	−3.0	−1.5
	a_2^S	1.13 fm	1.16 fm
F_2^V	α_2^V	1.20	1.14
	a_2^V	0.77 fm	0.85 fm
F_1^n	a_1^n‡	0	0
F_1^p	a_1^p	0.85 fm	0.88 fm
F_2^n	a_2^n	0.76 fm	0.87 fm
F_2^p	a_2^p	0.94 fm	0.95 fm

*Parameters are described by the theoretical model of Bergia, S., Stanghellini, A., Fubini, S., and Villi, C., *Phys. Rev. Letters*, **6**, 367 (1961).

Data correspond to the experimental determinations in Hofstadter, R., and Herman, R., *Phys. Rev. Letters*, **6, 293 (1961).

†Data correspond to the experimental determinations in Littauer, R. M., Schopper, H. F., and Wilson, R. R., *Phys. Rev. Letters*, **7**, 144 (1961).

‡This quantity was required to be zero in the fitting process.

Because of the finite size of the nucleon charge distribution, the analysis of the experimental data, in terms of the form factor of various nuclei, has to be slightly extended. In practice, however, this modification is important only for light nuclei; for the heavier nuclei, the nucleon size as compared with the nuclear size is insignificant, and practically no difference is obtained by including the nucleon-size effect in the computation. In the case of nuclei, for instance, deuteron, alpha-, and the $0p$-shell nuclei, the nucleon-size effect produces appreciable changes in the interpretation of data.

First, let us consider how the experimental analysis goes. The differential cross-section is experimentally measured, and then the experimental quantities are divided by the Mott cross-section; in this way, experimental values of the form factor $F(q^2)$ are obtained. At small values of q^2, the form of $F(q^2)$ is given by

$$F(q^2) \approx 1 - \tfrac{1}{6}a^2q^2,$$

where a^2 is the mean-square radius of the nuclear charge distribution. We shall refer to a^2, obtained in this way, as the experimental mean-square radius. If we now take a theoretical expression, such as (EI.12), for the nuclear density, and compute $F(q^2)$ corresponding to it [see (EI.14)], the corresponding expression will also yield a value of a^2. But this value should not be directly compared with the experimental value of a^2 just described because, in constructing the theoretical expression for the charge density, we have assumed point nucleons; on the other hand, the actual charge density, whose Fourier transform is the experimental $F(q^2)$, is set up by nucleons having a finite size. We shall therefore put a subscript P on the charge density, form factors, mean-square radius, etc., calculated for point nucleons to distinguish them from the same quantities (without the subscript) pertaining to real nucleons.

By the definition of charge density, the many-body *operator* standing for it is given by

$$\hat{\rho}_{\mathrm{P}}(\mathbf{r}) = Z^{-1} \sum_{i=1}^{Z} \delta(\mathbf{r} - \mathbf{r}_i), \tag{EI.18}$$

where a point nucleon is located at the various points $\mathbf{r}_1, \mathbf{r}_2, \ldots, \mathbf{r}_i, \ldots$. If the nucleon has a finite extent defined by a density operator $\hat{\rho}_\nu$, where ν stands for a nucleon, then the nucleon centred at $\mathbf{r}_i$ contributes $\hat{\rho}_\nu(\mathbf{r} - \mathbf{r}_i)$ to the density at a point $\mathbf{r}$ inside a nucleus. Summing the contribution of all the nucleons, we get

$$\begin{aligned}\hat{\rho}(\mathbf{r}) &= Z^{-1} \sum_{i=1}^{Z} \hat{\rho}_\nu(\mathbf{r} - \mathbf{r}_i)\\ &= Z^{-1} \int d^3r'\, \hat{\rho}_\nu(\mathbf{r} - \mathbf{r}') \sum_{i=1}^{Z} \delta(\mathbf{r}' - \mathbf{r}_i)\\ &= \int d^3r'\, \hat{\rho}_\nu(\mathbf{r} - \mathbf{r}')\hat{\rho}_{\mathrm{P}}(\mathbf{r}'). \end{aligned}\tag{EI.19}$$

That the second step here is equal to the first can be checked by doing the d^3r'-integration with the well-known property of the Dirac delta function. In the last step, we have made use of (EI.18). The charge densities corresponding to the operators $\hat{\rho}_{\mathrm{P}}(\mathbf{r})$ and $\hat{\rho}(\mathbf{r})$ are the expectation values of the operators. We denote these expectation values by removing the ^. Thus,

$$\begin{aligned}\rho(\mathbf{r}) = \langle\hat{\rho}(\mathbf{r})\rangle &= \langle\int d^3r'\, \hat{\rho}_\nu(\mathbf{r} - \mathbf{r}')\hat{\rho}_{\mathrm{P}}(\mathbf{r}')\rangle\\ &\approx \int d^3r'\, \rho_\nu(\mathbf{r} - \mathbf{r}')\rho_{\mathrm{P}}(\mathbf{r}'). \end{aligned}\tag{EI.20}$$

Clearly, an approximation is involved at the last step; we have replaced the expectation value of a product of two density operators by the product of their expectation values. The Fourier transform of $\rho(\mathbf{r})$, namely, $F(q^2)$, can be identified with the experimentally observed $F(q^2)$, whereas the Fourier transform $F_P(q^2)$ of $\rho_P(\mathbf{r}')$ is what we evaluated in (EI.14). Taking the Fourier transform in (EI.20), we get

$$F(q^2) = \int \exp(i\mathbf{q}\cdot\mathbf{r})\rho(\mathbf{r})\, d^3r = \int d^3r \int d^3r' \exp[i\mathbf{q}\cdot(\mathbf{r}-\mathbf{r}')]\rho_\nu(\mathbf{r}-\mathbf{r}')\exp(i\mathbf{q}\cdot\mathbf{r}')\rho_P(\mathbf{r}').$$

Making the same coordinate transformation as described following (EI.8), we easily obtain

$$F(q^2) = F_P(q^2)F_\nu(q^2), \tag{EI.21}$$

where $F_\nu(q^2)$ is the charge form factor of the nucleon. Making an expansion for small q^2, we immediately prove

$$1 - \tfrac{1}{6}a^2q^2 = (1 - \tfrac{1}{6}a_P^2q^2)(1 - \tfrac{1}{6}a_\nu^2q^2) \approx 1 - \tfrac{1}{6}(a_P^2 + a_\nu^2)q^2$$

or

$$a^2 = a_P^2 + a_\nu^2. \tag{EI.22}$$

(EI.22) enables us to determine the mean-square radius a_P^2 of the nucleus, corresponding to point nucleons, from the experimental value a^2, and the nucleon mean-square radius a_ν^2. The value of a_P^2, obtained in this way, must then be understood from model charge distributions, such as (EI.12), which are valid for point nucleons. Actually, the mean-square radius of the nucleus determines the value of the harmonic oscillator parameter b, and the important point to remember is that the mean-square radius a_P^2, determined from (EI.22), must be used for this purpose, and not the a^2 that the experimental data directly give.

II. ALTERNATIVE METHODS OF NUCLEAR SIZE DETERMINATION

The electron scattering method, described in Section I of this appendix, provides the most elaborate information so far available on the size and detailed shape of the charge distribution of nuclei. There are several other methods, most of which are historically older than the electron scattering method. These methods usually determine the value of only one parameter—the radius—of the distribution. Some of them do not directly determine the charge or matter distribution inside the nuclei; the parameters that they determine pertain to the average nuclear *potential* inside the nucleus.

The oldest determination of nuclear size was, of course, done by Rutherford and his collaborators in their α-scattering experiments, which revealed that the nucleus occupies a dimension $\sim 10^{-12}$ cm at the centre of the atom. In later α-scattering experiments, done with higher-energy alpha-particles, much more information has been revealed on the shape and parameters of the optical potential. Similar scattering experiments done with neutrons and protons have also given some information on the potential. A summary of this type of information on the potential has been given in Section 17.

Another old method, which is applicable to naturally radioactive α-emitting nuclei, is based on the barrier penetration probability of the α-particles. The barrier is determined by the combined Coulomb and nuclear potentials, and the penetrability gives one radius parameter of this potential. The relationship of this radius parameter, or that for the optical potential, and the root-mean-square radius determined in electron scattering experiments, is, however, quite obscure.

In Section 14, we have also discussed the Coulomb energy of nuclei. In particular, the difference in Coulomb energy between mirror nuclei has been treated with the specific aim of determining one radius parameter for the nucleus.

There are several methods based on atomic spectroscopy (optical and X-rays). In the usual spectroscopic calculations, the electron energy levels in the field of a point nucleus are calculated. The extended charge distribution of the nucleus causes a shift in these energy levels with respect to their positions for a point nucleus. Since different electron energy levels are affected to a different extent, it is possible to get information on the nuclear charge distribution from the energies of X-rays or optical transitions that correspond to the differences in energy between two such electron levels. Usually, such measurements determine the expectation value $\langle r^{2\sigma}\rangle$ for the nuclear charge distribution, where $\sigma^2 \approx 1 - (Z/137)^2$. For nuclei with small Z, $\langle r^{2\sigma}\rangle$ actually becomes almost the same as the mean-square radius. The value of r_0 determined from this mean-square radius, however, has a rather small magnitude as compared with the value determined by electron scattering and other methods. The reason for this is understood to be a significant correction due to the finite nuclear compressibility.

A method that compares very favourably with the electron scattering method is the one based on the observation of the X-ray energies emitted by μ-mesonic atoms. The only interaction that the negative μ-meson undergoes with the atomic nucleus is of the electromagnetic type. When the μ-meson is absorbed by an atom, it is usually captured at first in one of the outer orbits having a large value of n and l. From this level, it starts cascading to the lower levels by the emission of Auger electrons from the atom. Finally, it reaches the innermost orbits $2P$, $1S$, . . . , at which stage the transition between the levels is predominantly radiative. The transition from the $2P$- to the $1S$-level, in particular, corresponds to an X-ray frequency. The most important point here is that these orbits have significantly smaller radii as compared with the radii of the corresponding orbits of an atomic electron. The radius of the orbit is inversely proportional to the mass of the particle, and hence more than 200 times smaller for the μ-meson. As a matter of fact, the radius of a μ-meson orbit with the quantum number n is given by $[m_e\alpha^2 n^2/(m_\mu Z)]$ cm, where the subscripted m's denote the corresponding masses, Z denotes the atomic number of the absorbing atom, and α the fine-structure constant. A rough calculation will show that for Pb the radius of the orbit for $n = 1$ is about 3 fm, and that for $n = 2$ is roughly 12 fm, the latter being just outside the nuclear surface. The μ-meson in the state $1S$ therefore spends most of its time right inside the Pb-nucleus. The predicted $2P_{1/2} \rightarrow 1S$ transition for the μ-meson in Pb corresponds to an X-ray of energy 16.4 MeV, whereas the value expected for a point nucleus is only about 6 MeV. Thus, the discrepancy is so large that a precise measurement of it, experimentally, can ultimately lead to a great precision in the determination of the nuclear charge distribution. In the early experiments for light nuclei, only one parameter of the distribution used to be determined, whereas in later experiments on comparatively heavy nuclei detailed information on the actual distribution $\rho(r)$ has been secured. The analysis rests heavily on the theoretically computed energies of the $1S$- and $2P$-level. The first step corresponds to numerically solving the potential $V(r)$ on the μ-meson due to the extended nuclear charge distribution. Then this $V(r)$ is used in the Dirac equation, which leads to two coupled radial equations in the usual way. Finally, the numerical solution determines the energy levels.

REFERENCES

1. Schiff, L. I., Quantum Mechanics, 2nd edn., McGraw-Hill, New York, 1955, p 161.

2. Pal, M. K., *Phys. Rev.*, **117**, 566 (1960).
3. Hofstadter, R., *Ann. Rev. Nucl. Sc.*, **7**, 231 (1957); *Revs. Mod. Phys.*, **28**, 214 (1956).
4. Hofstadter, R., ed., Nuclear and Nucleon Structure, Benjamin, New York, 1963.

PROBLEMS

1. Derive (EI.12) for a $0p$-shell nucleus of charge Ze with the help of the harmonic oscillator radial functions given in Section 17C.

2. (a) Work out the integral

$$\int_0^\infty \cos qr \, e^{-r^2} \, dr.$$

[Hint: Use the gamma-function, expand the cosine in a power series, and integrate term by term. Sum the resultant series.]

(b) Evaluate

$$\int_0^\infty r^{2n} \cos qr \, e^{-r^2} \, dr, \qquad \int_0^\infty r^{2n+1} \sin qr \, e^{-r^2} \, dr,$$

where n is a positive integer. [Hint: Differentiate the integral in Problem 2a successively with respect to q.]

3. Use the Dirac plane-wave function and derive the Mott scattering formula for an electron on a point nucleus.

Appendix F

Harmonic Oscillator Brackets

Harmonic oscillator transformation brackets are defined by (II.85). In this appendix, we shall list the necessary expressions for their numerical evaluation. The detailed formula will be given first for the case $n_1 = n_2 = 0$. The case of higher values of n_1 can be tackled by the use of a recurrence relation that connects several transformation brackets for any given (n_1, n_2) with a single bracket for $(n_1 + 1, n_2)$. The stepping up of the value of n_2 can be done by the combined use of the same recurrence relation and a symmetry relation that allows interchanging the quantum numbers n_1 and n_2 within the bracket. For the derivation of all these expressions, refer to Moshinsky et al[1] (see, however, the problems at the end of this appendix for hints on the derivation).

In the case of $n_1 = n_2 = 0$, the expression is given by

$$\begin{aligned}
\langle \mathcal{N}\mathcal{L}, nl : L|\, |0l_1, 0l_2 : L\rangle &= (-1)^{\mathcal{N}+n}\left[\frac{\mathcal{N}!\, n!\, \Gamma(\mathcal{N} + \mathcal{L} + \frac{3}{2})\Gamma(n + l + \frac{3}{2})}{2^{l_1+l_2}\Gamma(l_1 + \frac{3}{2})\Gamma(l_2 + \frac{3}{2})}\right]^{1/2} \\
&\quad \times \left\{\frac{[l_1][l_2]}{[\mathcal{L}][l]}\right\}^{1/2} \sum_{i=0}^{l_1} \sum_{j=0}^{l_2} \delta_{(i+j),\,(2n+l)}\delta_{(i+j),\,(l_1+l_2-2\mathcal{N}-\mathcal{L})} \\
&\quad \times (-1)^j B(l_1, i) B(l_2, j) \begin{bmatrix} l_1 - i & l_2 - j & \mathcal{L} \\ 0 & 0 & 0 \end{bmatrix} \begin{bmatrix} i & j & l \\ 0 & 0 & 0 \end{bmatrix} \\
&\quad \times \begin{bmatrix} l_1 - i & i & l_1 \\ l_2 - j & j & l_2 \\ \mathcal{L} & l & L \end{bmatrix},
\end{aligned} \tag{F.1}$$

where

$$B(l, i) = \frac{(2l)!}{(2i)!\,(2l - 2i)!}. \tag{F.2}$$

The recurrence relation is given by

$$\begin{aligned}
&\langle \mathcal{N}\mathcal{L}, nl : L)|\, |(n_1 + 1)l_1, n_2 l_2 : L\rangle[(n_1 + 1)(n_1 + l_1 + \tfrac{3}{2})]^{1/2} \\
&= - \sum_{\mathcal{N}'\mathcal{L}'} \sum_{n'l'} \langle \mathcal{N}\mathcal{L}, nl : LM|\, \tfrac{1}{2}(R^2 + r^2 + 2\mathbf{R} \cdot \mathbf{r})\, |\mathcal{N}'\mathcal{L}', n'l' : LM\rangle \\
&\quad \times \langle \mathcal{N}'\mathcal{L}', n'l' : L|\, |n_1 l_1, n_2 l_2 : L\rangle,
\end{aligned} \tag{F.3}$$

81 (45-122)/1977

where $\mathbf{R}$ and $\mathbf{r}$ are the centre-of-mass and relative coordinates. The nonvanishing matrix elements on the right-hand side of (F.3) are

$$-\langle \mathcal{N}\mathcal{L}, nl : LM|\ \tfrac{1}{2}(R^2 + r^2 + 2\mathbf{R}\cdot\mathbf{r})\ |\mathcal{N}'\mathcal{L}', n'l' : LM\rangle$$

$$= \begin{cases} \tfrac{1}{2}[\mathcal{N}(\mathcal{N} + \mathcal{L} + \tfrac{1}{2})]^{1/2} & \text{for } \mathcal{N}' = \mathcal{N} - 1,\ \mathcal{L}' = \mathcal{L},\ n' = n,\ l' = l \\ \tfrac{1}{2}[n(n + l + \tfrac{1}{2})]^{1/2} & \text{for } \mathcal{N}' = \mathcal{N},\ \mathcal{L}' = \mathcal{L},\ n' = n - 1,\ l' = l \\ (-1)^{\mathcal{L}+l-L}W(\mathcal{L} + 1, l - 1, \mathcal{L}, l; L, 1)[\mathcal{N}(\mathcal{L} + 1)(n + l + \tfrac{1}{2})l]^{1/2} & \\ \qquad \text{for } \mathcal{N}' = \mathcal{N} - 1,\ \mathcal{L}' = \mathcal{L} + 1,\ n' = n,\ l' = l - 1 & \\ (-1)^{\mathcal{L}+l-L}W(\mathcal{L} + 1, l + 1, \mathcal{L}, l; L, 1)[\mathcal{N}(\mathcal{L} + 1)n(l + 1)]^{1/2} & \\ \qquad \text{for } \mathcal{N}' = \mathcal{N} - 1,\ \mathcal{L}' = \mathcal{L} + 1,\ n' = n - 1,\ l' = l + 1 & \\ (-1)^{\mathcal{L}+l-L}W(\mathcal{L} - 1, l - 1, \mathcal{L}, l; L, 1)[(\mathcal{N} + \mathcal{L} + \tfrac{1}{2})\mathcal{L}(n + l + \tfrac{1}{2})l]^{1/2} & \\ \qquad \text{for } \mathcal{N}' = \mathcal{N},\ \mathcal{L}' = \mathcal{L} - 1,\ n' = n,\ l' = l - 1 & \\ (-1)^{\mathcal{L}+l-L}W(\mathcal{L} - 1, l + 1, \mathcal{L}, l; L, 1)[(\mathcal{N} + \mathcal{L} + \tfrac{1}{2})\mathcal{L}n(l + 1)]^{1/2} & \\ \qquad \text{for } \mathcal{N}' = \mathcal{N},\ \mathcal{L}' = \mathcal{L} - 1,\ n' = n - 1,\ l' = l + 1 & \end{cases} \tag{F.4}$$

Symmetry Relations

Before stating the symmetry relations, we emphasize that there are two concepts behind the Moshinsky brackets: (i) the *quantum numbers* for the two particles and their centre-of-mass and relative motion, and (ii) the *coordinates* of the particles $\mathbf{r}_1$, $\mathbf{r}_2$ and their centre-of-mass and relative coordinates $\mathbf{R}$ and $\mathbf{r}$. The brackets being overlap integrals, the coordinates present in them have already been integrated upon. Considered in this way, the bracket stands for the expression

$$\langle \mathcal{N}\mathcal{L}, nl : L|\ |n_1l_1, n_2l_2 : L\rangle$$

$$= \sum_m \sum_{m_1} \begin{bmatrix} \mathcal{L} & l & L \\ M - m & m & M \end{bmatrix} \begin{bmatrix} l_1 & l_2 & L \\ m_1 & M - m_1 & M \end{bmatrix}$$

$$\times \int d^3r_1 \int d^3r_2\ \phi^*_{\mathcal{N}\mathcal{L}(M-m)}(\mathbf{R})\phi^*_{nlm}(\mathbf{r})\phi_{n_1l_1m_1}(\mathbf{r}_1)\phi_{n_2l_2(M-m_1)}(\mathbf{r}_2), \tag{F.5}$$

where each ϕ stands for the harmonic oscillator radial function times the appropriate spherical harmonic. The integration in (F.5) could have been equally well taken over $\int d^3R \int d^3r$. In writing (F.5), when we associate the coordinates with the four states appearing in the bracket notation we have kept a definite convention in mind, namely, that the first and the second state within the *ket* correspond to the two particle coordinates $\mathbf{r}_1$ and $\mathbf{r}_2$, respectively, whereas the first and the second state within the *bra* correspond respectively to the coordinates $\mathbf{R}$ and $\mathbf{r}$. This association has been explicitly used in deriving the algebraic expressions for the brackets, and hence it must be kept in mind while interpreting the symmetry relations which we shall derive. To illustrate what we mean by this statement, let us consider the bracket $\langle nl, \mathcal{N}\mathcal{L} : L|\ |n_1l_1, n_2l_2 : L\rangle$. This bracket means that the *quantum numbers nl*, associated in (F.5) with $\mathbf{r}$, are now associated with $\mathbf{R}$, whereas the reverse is true for $\mathcal{N}\mathcal{L}$. In a similar manner, the bracket $\langle n_1l_1, n_2l_2 : L|\ |\mathcal{N}\mathcal{L}, nl : L\rangle$ corresponds to the case where $\mathbf{r}_1$, $\mathbf{r}_2$ are associated with the quantum numbers $\mathcal{NL}$ and nl because they now appear in the *ket*, whereas n_1l_1 and n_2l_2 having occurred in the *bra* go with the coordinates $\mathbf{R}$ and $\mathbf{r}$, respectively.

We now try to get the bracket $\langle \mathcal{NL}, nl : L| |n_2l_2, n_1l_1 : L\rangle$ from the bracket occurring in (F.5). The new bracket is also given by an integral of the type (F.5), but now

$$\phi_{n_1l_1m_1}(\mathbf{r}_2)\phi_{n_2l_2(M-m_1)}(\mathbf{r}_1)\begin{bmatrix} l_2 & l_1 & L \\ M-m_1 & m_1 & M \end{bmatrix}$$

replaces the corresponding factors on the right-hand side of (F.5). The Clebsch-Gordon coefficient can be immediately brought to the form of (F.5) by inserting a phase factor $(-1)^{l_1+l_2-L}$ or, equivalently, $(-1)^{\mathcal{L}+l-L}$. Since $\mathbf{r}_1$ and $\mathbf{r}_2$ are dummy integration variables, we can interchange them, which gives back $\phi_{n_1l_1m_1}(\mathbf{r}_1)\phi_{n_2l_2(M-m_1)}(\mathbf{r}_2)$; but, at the same time, this interchanging changes $\mathbf{r}$ to $-\mathbf{r}$, leaving $\mathbf{R}$ unaltered. Next, $-\mathbf{r}$ can be brought back to $\mathbf{r}$, whereupon the spherical harmonic contained in $\phi^*_{nlm}(-\mathbf{r})$ gives a factor $(-1)^l$. Putting all these results together, we obtain

$$\langle \mathcal{NL}, nl : L| |n_2l_2, n_1l_1 : L\rangle = (-1)^{\mathcal{L}-L}\langle \mathcal{NL}, nl : L| |n_1l_1, n_2l_2 : L\rangle. \tag{F.6a}$$

This is the symmetry relation we mentioned while describing the use of the recurrence relation.

The bracket $\langle nl, \mathcal{NL} : L| |n_1l_1, n_2l_2 : L\rangle$ can be similarly related to that in (F.5). Here the change needed is $\mathbf{r}_2 \to -\mathbf{r}_2$, which gives a factor $(-1)^{l_2}$ from the state of $\mathbf{r}_2$; this transformation simultaneously changes $\mathbf{R} \to \mathbf{r}$ and $\mathbf{r} \to \mathbf{R}$, according to their definitions, in terms of $\mathbf{r}_1$ and $\mathbf{r}_2$. An additional phase factor $(-1)^{\mathcal{L}+l-L} = (-1)^{l_1+l_2-L}$ comes from the Clebsch-Gordon coefficient. Thus,

$$\langle nl, \mathcal{NL} : L| |n_1l_1, n_2l_2 : L\rangle = (-1)^{l_1-L}\langle \mathcal{NL}, nl : L| |n_1l_1, n_2l_2 : L\rangle. \tag{F.6b}$$

Finally, since the transformation

$$\mathbf{r}_1 \to \mathbf{R}, \qquad \mathbf{r}_2 \to \mathbf{r}$$

automatically implies

$$\mathbf{R} \to \mathbf{r}_1, \qquad \mathbf{r} \to \mathbf{r}_2,$$

we can easily show, by such relabelling of the variables, that

$$\langle n_1l_1, n_2l_2 : L| |\mathcal{NL}, nl : L\rangle = \langle \mathcal{NL}, nl : L| |n_1l_1, n_2l_2 : L\rangle. \tag{F.6c}$$

Once again, we remind the reader that the bracket on the left is *not* to be interpreted as the complex conjugate of that on the right; it is interpreted according to (F.5) by letting $(\mathcal{NL})$, (nl) go with $\mathbf{r}_1$, $\mathbf{r}_2$ and (n_1l_1), (n_2l_2) go with $\mathbf{R}$, $\mathbf{r}$ respectively.

REFERENCES

1. Moshinsky, M., *Nucl. Phys.*, **13**, 104 (1959); Brody, T. A., Jacob, G., and Moshinsky, M., *Nucl. Phys.*, **17**, 16 (1960).

PROBLEMS

1. Let A be any vector with the polar coordinates A, θ, ϕ. Prove that

$$[Ay_1^1(\theta, \phi)]^\lambda = [\frac{\lambda!}{(2\lambda-1)!!}]^{1/2}A^\lambda y_\lambda^\lambda(\theta, \phi).$$

Here y is the renormalized spherical harmonic defined by (AI.7a). [Hint: Multiply $Ay_1^1(\theta, \phi)$ in succession λ times; at each stage, use (BIII.4) and note that, since the projection quantum number is the maximum, there is only one spherical harmonic on the right-hand side at each stage. Further, the Clebsch-Gordon coefficient

$$\begin{bmatrix} 1 & K & K+1 \\ 1 & K & K+1 \end{bmatrix}$$

has the value unity.]

2. Use the basic definitions of $\mathbf{r}_1$, $\mathbf{r}_2$, $\mathbf{R}$, and $\mathbf{r}$ to obtain

$$r_1 y_1^1(\theta_1, \phi_1) = \frac{1}{\sqrt{2}}[R y_1^1(\Theta, \Phi) + r y_1^1(\theta, \phi)],$$

where the angles in each spherical harmonic correspond to the vector whose magnitude appears in the corresponding term.

3. Use the result of Problem 1 to write $[r_1 y_1^1(\theta_1, \phi_1)]^{l_1}$ directly. Also, use the binomial expansion on the right-hand side of the expression in Problem 2, followed again by the result of Problem 1 for the powers of $R y_1^1(\Theta, \Phi)$ and $r y_1^1(\theta, \phi)$. In this way, prove

$$r_1^{l_1} y_{l_1}^{l_1}(\theta_1, \phi_1) = \frac{1}{\sqrt{2^{l_1}}}\left\{\frac{(2l_1-1)!!}{l_1!}\right\}^{1/2} \sum_{i=0}^{l_1} {}^{l_1}C_i \left\{\frac{(l_1-i)!\, i!}{(2l_1-2i-1)!!\,(2i-1)!!}\right\}^{1/2}$$
$$\times R^{l_1-i} r^i [y^{l_1-i}(\Theta, \Phi), y^i(\theta, \phi)]_{l_1}^{l_1},$$

where the last expression within the square brackets is a compounded tensor of rank l_1 and component l_1.

Note The last expression being a relation connecting the l_1-component of tensors, it should hold in general for any component m.

4. Derive a relationship similar to that in Problem 3 for $r_2^{l_2} y_{l_2}^{l_2}(\theta_2, \phi_2)$ and again note that the result should be true for any component of the tensor.

5. Multiply the expressions from Problems 3 and 4 and do the necessary angular momentum coupling on the left-hand side to obtain

$$r_1^{l_1} r_2^{l_2} [y^{l_1}(\theta_1, \phi_1), y^{l_2}(\theta_2, \phi_2)]_M^L.$$

On the right-hand side, recouple the angular momenta using a $9j$-symbol such that the two spherical harmonics of (Θ, Φ) are coupled together and those of (θ, ϕ) are coupled together. Convert the left-hand side to $|0l_1, 0l_2 : LM\rangle$ by multiplying with suitable factors.

Note The right-hand side now consists of a sum over terms such as

$$e^{(-1/2)(R^2+r^2)} R^{l_1+l_2-i-j} r^{i+j} |\mathcal{L} l : LM\rangle,$$

where i and j are summation indices. Further, this expression has to be written in terms of $|\mathcal{N}\mathcal{L}, nl : LM\rangle$ and then their coefficients can be identified as the harmonic oscillator bracket (F.1). For this purpose, we proceed to Problem 6.

6. Let $|n\lambda\mu\rangle$ denote a harmonic oscillator state of radial function $\mathcal{R}_{n\lambda}(\rho)$ and angle function Y_μ^λ, to be written simply as $|\lambda\mu\rangle$. Then, expanding in terms of the complete set, we obtain

$$e^{(-1/2)\rho^2} \rho^K |\lambda\mu\rangle = \sum_{n\lambda'\mu'} |n\lambda'\mu'\rangle\langle n\lambda'\mu'|\, e^{(-1/2)\rho^2}\rho^K\, |\lambda\mu\rangle$$

$$= \sum_n |n\lambda\mu\rangle \int_0^\infty e^{(-1/2)\rho^2} \rho^{K+2} \mathcal{R}_{n\lambda}(\rho)\, d\rho.$$

Use the radial function ($\mathcal{R}_{n\lambda}$) from Section 17C to evaluate the integral appearing in this expression.

7. (a) If $|nl\rangle$ denotes the radial function $\mathcal{R}_{nl}(r)$ for the harmonic oscillator, prove that

$$r^2|nl\rangle = (2n + l + \tfrac{3}{2})|nl\rangle + [(n+1)(n+l+\tfrac{3}{2})]^{1/2}|n+1, l\rangle$$
$$+ [n(n+l+\tfrac{1}{2})]^{1/2}|n-1, l\rangle.$$

[Hint: Use the Laguerre polynomials and the normalization constant for the radial functions appropriately.]

(b) Evaluate the matrix elements $\langle n, l|\, r\, |n', l+1\rangle$ and $\langle n, l|\, r\, |n', l-1\rangle$.

8. Write the expression for $-r_1^2|n_1l_1, n_2l_2 : LM\rangle$ with the help of the result in Problem 7a. Multiply from the left by $\langle \mathcal{NL}, nl : LM|$, where $\mathcal{NL}$ and nl are such that

$$(2\mathcal{N} + \mathcal{L} + 2n + l) = 2(n_1 + 1) + l_1 + 2n_2 + l_2.$$

In this way, obtain the relation

$$\langle \mathcal{NL}, nl : L|\,|(n_1+1)l_1, n_2l_2 : L\rangle[(n_1 + l_1)(n_1 + l_1 + \tfrac{3}{2})]^{1/2}$$
$$= - \sum_{\mathcal{N}'\mathcal{L}'} \sum_{n'l'} \langle \mathcal{NL}, nl : LM|\, \tfrac{1}{2}(R^2 + r^2 + 2\mathbf{R}\cdot\mathbf{r})\, |\mathcal{N}'\mathcal{L}', n'l' : LM\rangle$$
$$\times \langle \mathcal{N}'\mathcal{L}', n'l' : L|\,|n_1l_1, n_2l_2 : L\rangle.$$

Use the result from Problems 7a, 7b and the standard result (BIII.15) to evaluate the matrix element on the right-hand side.

Note The final result thus obtained is the recurrence relation (F.4).

Appendix G

Some Algebraic Details of Microscopic Theory of Structure

I. CONDITION FOR A HARTREE-FOCK MINIMUM AND POSITIVE DEFINITENESS OF M

If the equilibrium Hartree-Fock (HF) state corresponds to a genuine minimum point on the energy-versus-deformation curve, then, after the perturbation has been introduced, the ground-state energy must, *by definition*, be larger than the equilibrium HF energy. We shall calculate the change in energy, requiring that it be positive.

The expression for the ground-state energy is given by [see (V.5a)]

$$\langle H\rangle = \sum_{h} \langle h|\, T\, |h\rangle + \tfrac{1}{2} \sum_{h,\, h'} (hh'|\, V\, |hh'), \tag{GI.1}$$

where the notation $h, h' \ldots$ stands for occupied states. If we use the arbitrary representation (V.2), then this expression can be easily rewritten in a more convenient and general form [see (V.11d) and (V.6b)] as

$$\langle H\rangle = \sum_{\alpha,\, \beta} \langle \alpha|\, T\, |\beta\rangle\langle\beta|\, \rho\, |\alpha\rangle + \tfrac{1}{2} \sum_{\alpha,\, \beta,\, \gamma,\, \delta} (\alpha\beta|\, V\, |\gamma\delta)\langle\delta|\, \rho\, |\beta\rangle\langle\gamma|\, \rho\, |\alpha\rangle. \tag{GI.2}$$

If we use the equilibrium value of ρ in this expression, we get the equilibrium value of the ground-state energy. On the other hand, using the perturbed density

$$\hat{\rho} = \rho + \hat{\rho}^{(1)} + \hat{\rho}^{(2)} + \ldots$$

in the same expression, we get the energy of the perturbed state. We denote this energy by $\langle\hat{H}\rangle$ and obtain

$$\begin{aligned}\langle\hat{H}\rangle = &\sum_{\alpha,\, \beta} \langle \alpha|\, T\, |\beta\rangle\langle\beta|\, \{\rho + \hat{\rho}^{(1)} + \hat{\rho}^{(2)}\}\, |\alpha\rangle \\ &+ \tfrac{1}{2} \sum_{\alpha,\, \beta,\, \gamma,\, \delta} (\alpha\beta|\, V\, |\gamma\delta)\langle\delta|\, \{\rho + \hat{\rho}^{(1)} + \hat{\rho}^{(2)}\}\, |\beta\rangle\langle\gamma|\, \{\rho + \hat{\rho}^{(1)} + \hat{\rho}^{(2)}\}\, |\alpha\rangle.\end{aligned} \tag{GI.3}$$

We have used up to the second-order terms in $\hat{\rho}$, and hence are satisfied by evaluating $\langle\hat{H}\rangle$ to the second order. The zero-order term in (GI.3) comes everywhere from ρ, and hence exactly agrees with the equilibrium value $\langle H\rangle$, as given by (GI.2). We shall explicitly write the first- and second-order terms of (GI.3). Denoting the order by the corresponding superscripts, we have

$$\langle\hat{H}^{(1)}\rangle = \sum_{\alpha,\, \beta} \langle \alpha|\, T\, |\beta\rangle\langle\beta|\, \hat{\rho}^{(1)}\, |\alpha\rangle + \sum_{\alpha,\, \beta,\, \gamma,\, \delta} (\alpha\beta|\, V\, |\gamma\delta)\langle\delta|\, \rho\, |\beta\rangle\langle\gamma|\, \hat{\rho}^{(1)}\, |\alpha\rangle. \tag{GI.4}$$

In writing the potential energy term in this equation, we notice that there are two first-order terms, namely, $\langle\delta|\, \hat{\rho}^{(1)}\, |\beta\rangle\langle\gamma|\, \rho\, |\alpha\rangle$ and $\langle\delta|\, \rho\, |\beta\rangle\langle\gamma|\, \hat{\rho}^{(1)}\, |\alpha\rangle$. Because of the summation over

$\alpha, \beta, \gamma, \delta$, these two terms lead to identical results. We have therefore kept one of these factors in (GI.4) and have cancelled the factor $\frac{1}{2}$ in (GI.3) with the factor 2 acquired from the identity of the two terms. In a similar manner, we collect the second-order terms of the type $\rho\hat{\rho}^{(2)}$, $\hat{\rho}^{(2)}\rho$, and $\hat{\rho}^{(1)}\hat{\rho}^{(1)}$ from (GI.3) and obtain

$$\langle \hat{H}^{(2)} \rangle = \sum_{\alpha, \beta} \langle \alpha | T | \beta \rangle \langle \beta | \hat{\rho}^{(2)} | \alpha \rangle + \sum_{\alpha, \beta, \gamma, \delta} (\alpha\beta | V | \gamma\delta) \langle \delta | \rho | \beta \rangle \langle \gamma | \hat{\rho}^{(2)} | \alpha \rangle + \frac{1}{2} \sum_{\alpha, \beta, \gamma, \delta} (\alpha\beta | V | \gamma\delta) \langle \delta | \hat{\rho}^{(1)} | \beta \rangle \langle \gamma | \hat{\rho}^{(1)} | \alpha \rangle. \tag{GI.5}$$

We shall first simplify (GI.4) and show that the first-order change in energy is zero. To do this, we first insert the definition of the HF potential $\mathcal{V}$ from (V.6b) in the second term of (GI.4) and obtain

$$\langle \hat{H}^{(1)} \rangle = \sum_{\alpha, \beta} \langle \alpha | T | \beta \rangle \langle \beta | \hat{\rho}^{(1)} | \alpha \rangle + \sum_{\alpha, \gamma} \langle \alpha | \mathcal{V} | \gamma \rangle \langle \gamma | \hat{\rho}^{(1)} | \alpha \rangle$$
$$= \sum_{\alpha, \beta} \langle \alpha | (T + \mathcal{V}) | \beta \rangle \langle \beta | \hat{\rho}^{(1)} | \alpha \rangle.$$

It should be noticed that this form is actually a trace of $(T + \mathcal{V})\hat{\rho}^{(1)}$ in our arbitrary representation $\alpha, \beta, \ldots$. Since the trace is invariant with respect to the representation, we examine this expression in the HF representation, where the HF single-particle Hamiltonian $(T + \mathcal{V})$ is diagonal, and $\hat{\rho}^{(1)}$ has only nondiagonal matrix elements connecting a hole state with a particle state (see Problem 1 at the end of this appendix). In the HF representation,

$$\langle \hat{H}^{(1)} \rangle = \sum_{i, j} \langle i | (T + \mathcal{V}) | j \rangle \langle j | \hat{\rho}^{(1)} | i \rangle$$
$$= \sum_{i, j} \epsilon_i \delta_{ij} \langle j | \hat{\rho}^{(1)} | i \rangle = 0 \tag{GI.6}$$

because the requirement δ_{ij} is contradictory to the property of $\hat{\rho}^{(1)}$ just mentioned. Here ϵ_i is the energy of the HF state i.

We next examine the second-order expression (GI.5). Once again, we use the definition (V.6b) of $\mathcal{V}$ and obtain

$$\langle H^{(2)} \rangle = \sum_{\alpha, \beta} \langle \alpha | (T + \mathcal{V}) | \beta \rangle \langle \beta | \hat{\rho}^{(2)} | \alpha \rangle + \frac{1}{2} \sum_{\alpha, \beta, \gamma, \delta} (\alpha\beta | V | \gamma\delta) \langle \delta | \hat{\rho}^{(1)} | \beta \rangle \langle \gamma | \hat{\rho}^{(1)} | \alpha \rangle.$$

Now we use the HF representation, together with the properties of the matrix elements of $\hat{\rho}^{(1)}$ and $\hat{\rho}^{(2)}$ (to be proved in Problems 1 and 2 at the end of this appendix). This gives, for the first term of $\langle \hat{H}^{(2)} \rangle$,

$$\sum_{h} \langle h | (T + \mathcal{V}) | h \rangle \langle h | \hat{\rho}^{(2)} | h \rangle + \sum_{p} \langle p | (T + \mathcal{V}) | p \rangle \langle p | \hat{\rho}^{(2)} | p \rangle$$
$$= \sum_{h, p} (\epsilon_p - \epsilon_h) \langle h | \hat{\rho}^{(1)} | p \rangle \langle p | \hat{\rho}^{(1)} | h \rangle,$$

where $(\epsilon_p - \epsilon_h)$ is the energy of the unperturbed hole-particle state with respect to the ground state. Combining this with the second term of $\langle \hat{H}^{(2)} \rangle$, we obtain the final expression

$$\begin{aligned}\langle \hat{H}^{(2)} \rangle = &\sum_{h,p} \langle h | \hat{\rho}^{(1)} | p \rangle \langle p | \hat{\rho}^{(1)} | h \rangle (\epsilon_p - \epsilon_h) \\ &+ \tfrac{1}{2} \sum_{h,p} \sum_{h',p'} [(pp'| V | h'h) \langle h | \hat{\rho}^{(1)} | p \rangle \langle h' | \hat{\rho}^{(1)} | p' \rangle \\ &+ (h'h | V | p'p) \langle p | \hat{\rho}^{(1)} | h \rangle \langle p' | \hat{\rho}^{(1)} | h' \rangle + (h'p | V | p'h) \langle h | \hat{\rho}^{(1)} | p \rangle \langle p' | \hat{\rho}^{(1)} | h' \rangle \\ &+ (p'h | V | h'p) \langle p | \hat{\rho}^{(1)} | h \rangle \langle h' | \hat{\rho}^{(1)} | p' \rangle]. \end{aligned} \tag{GI.7}$$

The results (GI.6) and (GI.7) have an interesting physical meaning. We notice that, when we disturb the nucleus from its equilibrium, the first-order change in density cannot generate any first-order change in energy. The lowest-order change in energy is given by (GI.7), and *it is quadratic in the density fluctuation* $\hat{\rho}^{(1)}$. This has a classical analogy in terms of what happens to a vibrator.

We introduce the simplified notation

$$\langle p | \hat{\rho}^{(1)} | h \rangle = x_{hp},$$

and hence

$$\langle h | \hat{\rho}^{(1)} | p \rangle = x^*_{hp}$$

since $\hat{\rho}^{(1)}$ is Hermitean. Using the definitions (VI.23), we can write (GI.7) as

$$\begin{aligned}\langle \hat{H}^{(2)} \rangle &= \tfrac{1}{2} \sum_{h,p} \sum_{h',p'} (x^*_{hp} A_{hp,h'p'} x_{h'p'} + x_{hp} A^*_{hp,h'p'} x^*_{h'p'} + x^*_{hp} B_{hp,h'p'} x^*_{h'p'} + x_{hp} B^*_{hp,h'p'} x_{h'p'}) \\ &= \tfrac{1}{2} (x^* \quad x) \begin{pmatrix} A & B \\ B^* & A^* \end{pmatrix} \begin{pmatrix} x \\ x^* \end{pmatrix},\end{aligned} \tag{GI.8}$$

where x and x^* are vectors having the components x_{hp} and x^*_{hp}, respectively. *The condition for a genuine HF minimum then guarantees that (GI.8) be greater than zero, i.e.,*

$$(x^* \quad x) M \begin{pmatrix} x \\ x^* \end{pmatrix} > 0. \tag{GI.9}$$

If $\begin{pmatrix} x \\ y \end{pmatrix}$ be any eigenvector of M having the eigenvalue m, then

$$\begin{aligned} Ax + By &= mx, \\ B^*x + A^*y &= my. \end{aligned}$$

Since the eigenvalue m of the Hermitean matrix M is real, we get, by taking the complex conjugate of these two equations,

$$\begin{aligned} Ay^* + Bx^* &= my^*, \\ B^*y^* + A^*x^* &= mx^*. \end{aligned}$$

As a matrix equation, we therefore have

$$M \begin{pmatrix} y^* \\ x^* \end{pmatrix} = m \begin{pmatrix} y^* \\ x^* \end{pmatrix}.$$

So every eigenvalue of M is at least doubly-degenerate. The existence of an eigenvector $\begin{pmatrix} x \\ y \end{pmatrix}$ automatically ensures the existence of another eigenvector $\begin{pmatrix} y^* \\ x^* \end{pmatrix}$, as just proved. Any linear combination of these two vectors is also an eigenvector of M belonging to the same eigenvalue. Thus, in particular,

$$\begin{pmatrix} x + y^* \\ y + x^* \end{pmatrix}, \qquad \begin{pmatrix} i(x - y^*) \\ -i(y - x^*) \end{pmatrix} \tag{GI.10}$$

are eigenvectors belonging to the same eigenvalue. Both these vectors have the form of the special vector appearing in (GI.9). Because of the different signs in the two vectors of (GI.10), at least one of them is nonvanishing. Therefore, we have proved, with the aid of (GI.9), where x is arbitrary, that for every eigenvalue m of the matrix M there is at least one eigenvector in the special form of (GI.9), and hence (GI.9) shows that *every eigenvalue of M is positive definite.*

Eigenvalues of M' Are Real

The positive definiteness of the eigenvalues of M enables us to show that the eigenvalues of M' are real. From the definition (VI.24), if we use an eigenvector $\begin{pmatrix} X \\ Y \end{pmatrix}$ of M' belonging to the eigenvalue $\hbar\omega$, we have

$$\begin{aligned} (X^* \quad Y^*)M\begin{pmatrix} X \\ Y \end{pmatrix} &= \hbar\omega(X^* \quad Y^*)\begin{pmatrix} X \\ -Y \end{pmatrix} \\ &= \hbar\omega(|X|^2 - |Y|^2). \end{aligned} \tag{GI.11}$$

But since all the eigenvalues of the matrix M are real and positive definite, the left-hand side of (GI.11), which is the expectation value of M for an arbitrary state (i.e., *not* an eigenstate of M), must also be real and positive definite. Therefore,

$$\hbar\omega(|X|^2 - |Y|^2) \text{ is real and } \geqslant 0. \tag{GI.12}$$

Since the quantity enclosed within the parentheses is explicitly real, we prove that $\hbar\omega$, the eigenvalue of M', is also real. The inequality in (GI.12) then requires that

$$\begin{aligned} |X|^2 - |Y|^2 &> 0 \qquad (\text{when } \hbar\omega > 0) \\ &< 0 \qquad (\text{when } \hbar\omega < 0). \end{aligned} \tag{GI.13}$$

II. HARTREE-FOCK-BOGOLIUBOV EQUATIONS

We shall now give the derivation of the Hartree-Fock-Bogoliubov (HFB) equations. These are the equations satisfied by the transformation coefficients x^i_α and y^i_α of (VI.55). One possible way of arriving at these equations is to minimize the expression (VI.54a) for the ground-state energy. A much simpler derivation can be obtained by posing the problem as follows: we want for our Hamiltonian (VI.32) an excitation mode of the type (VI.55), which is a linear superposition of the creation and destruction operators $C^\dagger_\alpha$ and C_α. According to the results in Section 39B, the existence of such a mode requires that the commutators of all the $C^\dagger_\alpha$ and C_α with the Hamiltonian (VI.32) give rise to a linear sum of these operators. If such results can be proved, then the quasiparticle operators $b^\dagger_i$ can be constructed from all the $C^\dagger_\alpha$ and C_α [according to (VI.7)]. This is the programme we shall follow in this section.

82 (45-122/1977)

By straightforward algebra, and using standard anticommutation rules, we get

$$\begin{aligned}[C_\alpha^\dagger C_\beta, C_\mu^\dagger] &= C_\alpha^\dagger C_\beta C_\mu^\dagger - C_\mu^\dagger C_\alpha^\dagger C_\beta \\ &= \delta_{\beta\mu} C_\alpha^\dagger - (C_\alpha^\dagger C_\mu^\dagger + C_\mu^\dagger C_\alpha^\dagger) C_\beta \\ &= \delta_{\beta\mu} C_\alpha^\dagger, \end{aligned} \tag{GII.1a}$$

$$\begin{aligned}[C_\alpha^\dagger C_\beta^\dagger C_\delta C_\gamma, C_\mu^\dagger] &= C_\alpha^\dagger C_\beta^\dagger [C_\delta C_\gamma, C_\mu^\dagger] \\ &= C_\alpha^\dagger C_\beta^\dagger (C_\delta C_\gamma C_\mu^\dagger - C_\mu^\dagger C_\delta C_\gamma) \\ &= C_\alpha^\dagger C_\beta^\dagger (\delta_{\gamma\mu} C_\delta - C_\delta C_\mu^\dagger C_\gamma - C_\mu^\dagger C_\delta C_\gamma) \\ &= C_\alpha^\dagger C_\beta^\dagger (\delta_{\gamma\mu} C_\delta - \delta_{\delta\mu} C_\gamma). \end{aligned} \tag{GII.1b}$$

Using these results with (VI.32), we obtain

$$\begin{aligned}[H, C_\mu^\dagger] &= \sum_\alpha \langle\alpha| T |\mu\rangle C_\alpha^\dagger + \tfrac{1}{4} \sum_{\alpha\beta\gamma\delta} (\alpha\beta| V |\gamma\delta) C_\alpha^\dagger C_\beta^\dagger (\delta_{\gamma\mu} C_\delta - \delta_{\delta\mu} C_\gamma) \\ &= \sum_\alpha \langle\alpha| T |\mu\rangle C_\alpha^\dagger + \tfrac{1}{2} \sum_{\alpha\beta\gamma} (\alpha\beta| V |\mu\gamma) C_\alpha^\dagger C_\beta^\dagger C_\gamma. \end{aligned} \tag{GII.2}$$

An interchange of the summation indices γ and δ in the $(\delta_{\delta\mu} C_\gamma)$-term would easily convince the reader that this term is equal to the $(\delta_{\gamma\mu} C_\delta)$-term. This accounts for the factor $\frac{1}{2}$ in the final step of (GII.2).

The exact result (GII.2) tells us that the commutator of $C_\mu^\dagger$ with the Hamiltonian does *not* give rise to a linear sum of *single* creation and *single* destruction operators only. Such a linear sum is a prerequisite for a quasiparticle excitation mode of the type (VI.55). Therefore, we conclude that the quasiparticle mode is an *approximate* mode of the Hamiltonian H. The approximation necessary for this purpose will now be discussed.

In analogy with the results contained in (VI.38) and (VI.39), we can write the *exact* expression

$$C_\alpha^\dagger C_\beta^\dagger C_\gamma = :C_\alpha^\dagger C_\beta^\dagger C_\gamma: + \langle C_\beta^\dagger C_\gamma\rangle C_\alpha^\dagger - \langle C_\alpha^\dagger C_\gamma\rangle C_\beta^\dagger + \langle C_\alpha^\dagger C_\beta^\dagger\rangle C_\gamma. \tag{GII.3}$$

The first term, by definition, contains *three* operators and *cannot* be reduced to *single* creation or destruction operators. On the other hand, the remaining terms in (GII.3) are linear in the operators $C_\alpha^\dagger, C_\gamma, \ldots$. Thus, the required approximation for the quasiparticle mode entails neglecting the first term in (GII.3). This approximation is often referred to as the linearization of (GII.3).

With the linearization approximation, (GII.2) reduces to

$$\begin{aligned}[H, C_\mu^\dagger] &= \sum_\alpha \langle\alpha| T |\mu\rangle C_\alpha^\dagger + \tfrac{1}{2} \sum_{\alpha\beta\gamma} (\alpha\beta| V |\mu\gamma)(\langle C_\beta^\dagger C_\gamma\rangle C_\alpha^\dagger - \langle C_\alpha^\dagger C_\gamma\rangle C_\beta^\dagger) \\ &\quad + \tfrac{1}{2} \sum_{\alpha\beta\gamma} (\alpha\beta| V |\mu\gamma)\langle C_\alpha^\dagger C_\beta^\dagger\rangle C_\gamma \\ &= \sum_\alpha \langle\alpha| T |\mu\rangle C_\alpha^\dagger + \sum_{\alpha\beta\gamma} (\alpha\beta| V |\mu\gamma)\langle C_\beta^\dagger C_\gamma\rangle C_\alpha^\dagger + \tfrac{1}{2} \sum_{\alpha\beta\gamma} (\alpha\beta| V |\mu\gamma)\langle C_\alpha^\dagger C_\beta^\dagger\rangle C_\gamma. \end{aligned} \tag{GII.4}$$

Once again, we have used the stratagem of interchanging the summation indices α and β to show that the term involving $-\langle C_\alpha^\dagger C_\gamma\rangle C_\beta^\dagger$ is equal to the term containing $\langle C_\beta^\dagger C_\gamma\rangle C_\alpha^\dagger$; hence, the final step in (GII.4) follows.

We next use the definition (V.6b) for $\mathcal{V}$, and the definition (VI.45) for Δ, and rewrite

(GII.4) as

$$[H, C_\mu^\dagger] = \sum_\alpha \langle\alpha|\, (T + \mathcal{V})\, |\mu\rangle C_\alpha^\dagger + \sum_\alpha \Delta_{\mu\alpha}^* C_\alpha. \tag{GII.5}$$

Since we are doing a particle number nonconserving theory (for the reasons explained in Section 41C), we should have actually worked with the Hamiltonian

$$H(\lambda) = H - \lambda \sum_\alpha C_\alpha^\dagger C_\alpha,$$

instead of H. According to (GII.1a), the extra term in $H(\lambda)$ contributes $-\lambda\delta_{\alpha\mu}C_\mu^\dagger$ to the commutator, and hence we finally obtain

$$\begin{aligned}[H(\lambda), C_\mu^\dagger] &= \sum_\alpha \langle\alpha|\, (T - \lambda\mathbb{1} + \mathcal{V})\, |\mu\rangle C_\alpha^\dagger + \sum_\alpha \Delta_{\mu\alpha}^* C_\alpha \\ &= \sum_\alpha \langle\mu|\, (T - \lambda\mathbb{1} + \mathcal{V})\, |\alpha\rangle^* C_\alpha^\dagger + \sum_\alpha \Delta_{\mu\alpha}^* C_\alpha. \end{aligned} \tag{GII.6a}$$

Here we have used the hermiticity of the operator $(T - \lambda\mathbb{1} + \mathcal{V})$.

To get a closed system of linear equations such as (VI.5), we need also the commutator of $H(\lambda)$ with C_μ, which is easily obtained by taking the Hermitean conjugate of (GII.6a) and reversing the sign. Therefore,

$$[H(\lambda), C_\mu] = -\sum_\alpha \Delta_{\mu\alpha} C_\alpha^\dagger - \sum_\alpha \langle\mu|\, (T - \lambda\mathbb{1} + \mathcal{V})\, |\alpha\rangle C_\alpha. \tag{GII.6b}$$

We now compare the set of equations (GII.6) with (VI.5). Let us assume that we have at our disposal N creation operators $C_\alpha^\dagger$ and N destruction operators C_α. These $2N$ operators here take the role of the N operators $A_i^\dagger$ ($i = 1, 2, \ldots, N$) of (VI.5). The matrix $\tilde{M}$ in the present case can be written down, with the help of (GII.6), as

$$\tilde{M} = \begin{pmatrix} (T - \lambda\mathbb{1} + \mathcal{V})^* & \Delta^* \\ -\Delta & -(T - \lambda\mathbb{1} + \mathcal{V}) \end{pmatrix}. \tag{GII.7a}$$

This is clearly a $2N \times 2N$ matrix, and each of the submatrices, Δ, Δ^*, $(T - \lambda\mathbb{1} + \mathcal{V})$, and $(T - \lambda\mathbb{1} + \mathcal{V})^*$ is $N \times N$. According to Section 39B, our immediate task is to diagonalize M which can be obtained from (GII.7a) by interchanging the rows and columns and then taking the transpose of each submatrix. Remembering that $\tilde{\Delta} = -\Delta$ and $\Gamma^\dagger = \Gamma$, i.e., $\tilde{\Gamma} = \Gamma^*$, where $\Gamma = (T - \lambda\mathbb{1} + \mathcal{V})$, we obtain

$$M = \begin{pmatrix} T - \lambda\mathbb{1} + \mathcal{V} & \Delta \\ -\Delta^* & -(T - \lambda\mathbb{1} + \mathcal{V})^* \end{pmatrix}. \tag{GII.7b}$$

From the properties of the matrices Γ and Δ just stated, we can show that M is a Hermitean matrix. While taking the Hermitean conjugate of M, we have first to interchange the rows and columns in (GII.7b) and then take the Hermitean conjugate of each submatrix. This Hermitean matrix has *real* eigenvalues.

Let us consider the eigenvector $\begin{pmatrix} x^i \\ y^i \end{pmatrix}$ corresponding to the eigenvalue E_i. Each of the quantities x^i and y^i consists of N elements of the type x_α^i, y_α^i . . . . Writing out the eigenvalue equation in detail, we obtain

$$M\begin{pmatrix} x^i \\ y^i \end{pmatrix} = E_i \begin{pmatrix} x^i \\ y^i \end{pmatrix} \tag{GII.8a}$$

or

$$\Gamma x^i + \Delta y^i = E_i x^i, \tag{GII.8b}$$

$$-\Delta^* x^i - \Gamma^* y^i = -E_i y^i. \tag{GII.8c}$$

If we take the complex conjugate of these two equations, and reverse their signs throughout, remembering that $E_i^* = E_i$, we then obtain two equations that are equivalent to

$$M\begin{pmatrix} y^{i*} \\ x^{i*} \end{pmatrix} = -E_i\begin{pmatrix} y^{i*} \\ x^{i*} \end{pmatrix}. \tag{GII.9}$$

Therefore, the eigenvalues of the matrix (GII.7b) occur in $\pm$ pairs. Given the vector for a positive eigenvalue, we can easily obtain the vector for the corresponding negative eigenvalue through the relationship expressed by (GII.8a) and (GII.9).

Let us consider a positive eigenvalue E_i and the corresponding eigenvector as appearing in (GII.8a). According to (VI.7), we can construct the step-up operator

$$b_i^\dagger = \sum_\alpha (x_\alpha^i C_\alpha^\dagger + y_\alpha^i C_\alpha) \tag{GII.10a}$$

and its Hermitean conjugate

$$b_i = \sum_\alpha (y_\alpha^{i*} C_\alpha^\dagger + x_\alpha^{i*} C_\alpha) \tag{GII.10b}$$

having the properties

$$H b_i^\dagger |\Psi_0\rangle = (E_0 + E_i) b_i^\dagger |\Psi_0\rangle, \tag{GII.11a}$$

$$b_i |\Psi_0\rangle = 0, \tag{GII.11b}$$

where $|\Psi_0\rangle$ is the ground state of H having the energy E_0. The state $b_i^\dagger|\Psi_0\rangle$ which has the energy E_i above the ground state is a one-quasiparticle state, the quasiparticle being in a state i. The operator b_i is the destruction operator for the quasiparticle, and (GII.11b) guarantees that the ground state $|\Psi_0\rangle$ does not contain any quasiparticle to start with. It should be noticed that the eigenvector corresponding to $-E_i$, as given by (GII.9), would have led, according to (VI.7), to the linear combination (GII.10b); therefore, in general, the linear combination (GII.10b), operating on any eigenstate of H, gives rise to another eigenstate with an energy stepped down by $-E_i$. This further confirms the interpretation of the operator b_i as the destruction operator of a quasiparticle of energy E_i. In general, the $2N \times 2N$ matrix M yields $2N$ eigenvalues, of which N are positive, and the remaining N, the negative partners of the positive set. Linear combinations such as (GII.10a) for the N positive eigenvalues are interpreted as N different quasiparticle creation operators, whereas similar linear combinations, namely, (GII.10b), corresponding to the negative eigenvalues give the destruction operators for the same set of N quasiparticles.

Written out in detail, (GII.8b), (GII.8c) for the quasiparticle transformation coefficients read respectively as

$$\sum_\beta (\langle\alpha|\,\Gamma\,|\beta\rangle x_\beta^i + \Delta_{\alpha\beta} y_\beta^i) = E_i x_\alpha^i, \tag{GII.12a}$$

$$\sum_\beta (\Delta_{\alpha\beta}^* x_\beta^i + \langle\alpha|\,\Gamma^*\,|\beta\rangle y_\beta^i) = -E_i y_\alpha^i. \tag{GII.12b}$$

These equations are called the Hartree-Fock-Bogoliubov (HFB) equations. Their appearance as linear equations is rather deceptive because, by definition, the matrix elements of Γ and Δ

contain the density and the pairing matrix, respectively, and the latter are quadratic in the transformation coefficients themselves. In close parallel with the HF case, the HFB equations also require a self-consistent treatment due to the properties of Γ and Δ, as just stated. We now proceed to derive the expressions for the density and pairing matrix elements such that the self-consistent method can be described in detail.

We first write down the orthogonality properties of the eigenvectors of the Hermitean matrix M. The orthogonality of an eigenvector of positive eigenvalue E_i and another of negative eigenvalue $-E_j$ is expressed, according to (GII.8a) and (GII.9), by

$$(y^j \quad x^j)\begin{pmatrix} x^i \\ y^i \end{pmatrix} = 0$$

or

$$\sum_\alpha (y^j_\alpha x^i_\alpha + x^j_\alpha y^i_\alpha) = 0. \tag{GII.13a}$$

Similarly, the scalar product of two eigenvectors, belonging to two *different* positive eigenvalues E_i and E_j, is also zero. However, when $E_i = E_j$, i.e., when the scalar product of an eigenvector of positive eigenvalue is considered with itself, the result can be normalized to unity. Thus,

$$(x^{j*} \quad y^{j*})\begin{pmatrix} x^i \\ y^i \end{pmatrix} = \delta_{ij}$$

or

$$\sum_\alpha (x^{j*}_\alpha x^i_\alpha + y^{j*}_\alpha y^i_\alpha) = \delta_{ij}. \tag{GII.13b}$$

It is trivial to verify from the definitions (GII.10) and the anticommutator relations of $C^\dagger_\alpha$, $C_\beta, \ldots$ that the summation on the left-hand side of (GII.13a) is equal to $\{b_j, b_i\}$ and that on the left-hand side of (GII.13b) is equal to $\{b^\dagger_j, b_i\}$. Thus, the orthogonality properties (GII.13) of the eigenvectors of M automatically guarantee that the quasiparticles also satisfy the Fermion anticommutation properties

$$\{b_j, b_i\} = 0, \qquad \{b^\dagger_j, b_i\} = \delta_{ij}. \tag{GII.13c}$$

(GII.13a) and (GII.13b) further guarantee the result

$$\begin{pmatrix} x & y \\ y^* & x^* \end{pmatrix}\begin{pmatrix} x & y \\ y^* & x^* \end{pmatrix}^\dagger = \begin{pmatrix} x & y \\ y^* & x^* \end{pmatrix}\begin{pmatrix} x^\dagger & \tilde{y} \\ y^\dagger & \tilde{x} \end{pmatrix} = \begin{pmatrix} xx^\dagger + yy^\dagger & x\tilde{y} + y\tilde{x} \\ y^*x^\dagger + x^*y^\dagger & y^*\tilde{y} + x^*\tilde{x} \end{pmatrix} = \begin{pmatrix} \mathbb{1} & 0 \\ 0 & \mathbb{1} \end{pmatrix}. \tag{GII.14}$$

In this equation, we have treated x and y as matrices having the elements x^i_α and y^i_α, where i stands for the index specifying the row and α for the one that specifies the column.

It should be noticed that the transformation equations (GII.10) can be written as

$$\begin{pmatrix} b^\dagger \\ b \end{pmatrix} = \begin{pmatrix} x & y \\ y^* & x^* \end{pmatrix}\begin{pmatrix} C^\dagger \\ C \end{pmatrix}. \tag{GII.15}$$

Here $b^\dagger$ and b stand for columns having the elements $b^\dagger_i$ ($i = 1, 2, \ldots, N$) and b_i ($i = 1, 2, \ldots, N$). Similarly, $C^\dagger$ and C also stand for columns having the elements $C^\dagger_\alpha$ ($\alpha = 1, 2, \ldots, N$)

and C_α ($\alpha = 1, 2, \ldots, N$). It is clear from (GII.14) that the inverse of this equation is given by

$$\begin{pmatrix} C^\dagger \\ C \end{pmatrix} = \begin{pmatrix} x & y \\ y^* & x^* \end{pmatrix}^\dagger \begin{pmatrix} b^\dagger \\ b \end{pmatrix} = \begin{pmatrix} x^\dagger & \tilde{y} \\ y^\dagger & \tilde{x} \end{pmatrix} \begin{pmatrix} b^\dagger \\ b \end{pmatrix},$$

i.e.,

$$C^\dagger = x^\dagger b^\dagger + \tilde{y} b$$

or

$$C^\dagger_\alpha = \sum_i \{(x^\dagger)_{\alpha i} b^\dagger_i + (\tilde{y})_{\alpha i} b_i\}$$

$$= \sum_i (x^{i*}_\alpha b^\dagger_i + y^i_\alpha b_i). \qquad \text{(GII.16a)}$$

Hence,

$$C_\alpha = \sum_i (y^{i*}_\alpha b^\dagger_i + x^i_\alpha b_i). \qquad \text{(GII.16b)}$$

(GII.16) and (GII.11b) now enable us to calculate the density and pairing matrices. In a straightforward manner, using (GII.13c) whenever necessary, we obtain

$$\langle\alpha|\, \rho\, |\beta\rangle = \langle\Psi_0|\, C^\dagger_\beta C_\alpha\, |\Psi_0\rangle$$

$$= \sum_i y^i_\beta y^{i*}_\alpha, \qquad \text{(GII.17a)}$$

$$\kappa_{\delta\gamma} = \langle\Psi_0|\, C_\delta C_\gamma\, |\Psi_0\rangle$$

$$= \sum_i x^i_\delta y^{i*}_\gamma. \qquad \text{(GII.17b)}$$

These expressions, together with the definitions for $\mathcal{V}$ and Δ, completely determine the HFB equations (GII.12).

A practical procedure for solving the HFB equations self-consistently is as follows:

(i) Solve the HF problem self-consistently.

(ii) Start the HFB solutions with a value of the chemical potential λ in the neighbourhood of the highest filled HF level; the starting density matrix elements are also taken from the HF results. Take some reasonable values for the pairing density $\kappa_{\delta\gamma}$.

(iii) Find $\mathcal{V}$ and Δ from the starting ρ, κ and the given two-body matrix elements. Diagonalize the HFB matrix M.

(iv) Recalculate ρ and κ from the eigenvectors. Repeat the entire procedure until the matrix elements of ρ and κ in two successive iterations remain unaltered to the desired degree of accuracy.

(v) Check, with the final self-consistent eigenvector, whether or not the equation for the chemical potential

$$A = \sum_\alpha \langle\alpha|\, \rho\, |\alpha\rangle = \sum_i |y^i_\alpha|^2 \qquad \text{(GII.18)}$$

is satisfied. Here A is the number of nucleons in the nucleus we want to calculate. In general, (GII.18) is not satisfied at this stage of the calculation.

(vi) Suitably change the value of λ, and repeat the entire procedure until (GII.18) is satisfied to the desired accuracy.

A complete HFB type calculation for a heavy nucleus is still computationally difficult on most of the present-day electronic computers, if we wish to keep all the nucleons in achieving the self-consistency. In most practical calculations, we therefore omit the nucleons in the filled major shells of the interior, and do the self-consistent calculation only for the outer nucleons in a few major shells near the fermi surface. In such calculations, the matrix element $\langle\alpha| T |\beta\rangle$ of the kinetic energy is to be replaced, for obvious reasons, by $\epsilon_\alpha\delta_{\alpha\beta}$, where ϵ_α is the single-particle energy of a *single* nucleon in a state outside the omitted major shells, and the quantity A of (GII.18) then stands for the number of such nucleons.

HFB calculations for determining the self-consistent energy of the nucleus as a function of deformation parameters have been done by Kumar and Baranger[1]. These authors used the self-consistent energy as the collective potential energy in computing the spectra of nuclei by the collective Hamiltonian method of Section 33A. The moment of inertia and the mass parameter needed for these calculations were also computed microscopically. Many other authors (see Warke and Khadkikar[2] and others[2]) did HFB calculations with angular momentum projection and sometimes also with the projection of the correct number of particles. In calculations for the high-spin states, the HFB method has been used as described in Section 43D.

III. ANGULAR MOMENTUM COUPLING IN TDA AND RPA THEORY

Angular Momentum Coupling in Hole-Particle Calculation

The hole-particle calculation has been dealt with in Section 30B. The expressions (III.147) have now to be simplified by commuting $C_p^\dagger C_{-h}$ (the notation $j_p m_p$ and $j_h, -m_h$ has been simplified with p and $-h$, respectively) from the extreme right to the extreme left and $C_{-h'}^\dagger C_{p'}$ in the opposite direction. We have to make use of the obvious results

$$C_{p'}|\Psi_0) = C_{-h'}^\dagger|\Psi_0) = 0, \qquad (\Psi_0|C_p^\dagger = (\Psi_0|C_{-h} = 0$$

in achieving the simplification. The final results are given by

$$(\text{III.147a}) = \delta_{hh'}\langle p'| T |p\rangle - \delta_{pp'}\langle -h| T |-h'\rangle + \delta_{hh'}\delta_{pp'}\langle\Psi_0| T |\Psi_0\rangle, \tag{GIII.1}$$

$$(\text{III.147b}) = -(p', -h| V |p, -h') + \delta_{hh'} \sum_\alpha^{\text{occ}} (p'\alpha| V |p\alpha) - \delta_{pp'} \sum_\alpha (\alpha, -h| V |\alpha, -h') + \delta_{hh'}\delta_{pp'}\langle\Psi_0| V |\Psi_0\rangle. \tag{GIII.2}$$

Making use of the definition of the single-nucleon energies given by (III.101) and (III.102), we easily obtain

$$(\text{III.147a}) + (\text{III.147b}) = \delta_{hh'}\delta_{pp'}[E_0 + \epsilon_p - \epsilon_h] - (p', -h| V |p, -h'), \tag{GIII.3}$$

where

$$E_0 = \langle\Psi_0| \sum_i T_i + \sum_{i<j} V_{ij} |\Psi_0\rangle.$$

It should be observed that when we include the phase factor of (III.146b), the states $-h$ and $-h'$ of (GIII.3) change to $\bar{h}$ and $\bar{h}'$ and then (GIII.3) agrees exactly with the sum of (VI.111) and (VI.112), except for the occurrence of the total energy E_0 of the closed-shell state Ψ_0 in (GIII.3). However, the constant diagonal term E_0 can be trivially dropped in the computation, meaning thereby that the computed energies are simply measured with reference to the energy of the closed-shell state.

Thus, the angular momentum coupling in the TDA expression (VI.111) plus (VI.112) is the same as that confronted in (III.147). Using (GIII.3) and the appropriate phase factors and the Clebsch-Gordon coefficients from (III.146b), we easily obtain

$$\langle h'^{-1}p'JM|\,(H_{11}+H_{22})\,|h^{-1}pJM\rangle = \sum_{m_h m_p}\sum_{m_{h'}m_{p'}} \begin{bmatrix} j_h & j_p & J \\ m_h & m_p & M \end{bmatrix}\begin{bmatrix} j_{h'} & j_{p'} & J \\ m_{h'} & m_{p'} & M \end{bmatrix}$$

$$\times \{\delta_{hh'}\delta_{pp'}(\epsilon_p - \epsilon_h) - (p', \bar{h}|\,V\,|p, \bar{h}')\}. \tag{GIII.4}$$

In view of the Krönecker deltas, and the standard result (AIV.3a), the first term reduces to

$$\langle h'^{-1}p'JM|\,H_{11}\,|h^{-1}pJM\rangle = \delta_{j_h j_{h'}}\delta_{j_p j_{p'}}(\epsilon_p - \epsilon_h). \tag{GIII.5}$$

To work out the summation over the magnetic quantum numbers in the second term of (GIII.4), we first make the replacement

$$(p', \bar{h}|\,V\,|p, \bar{h}') = (-1)^{j_h - m_h}(-1)^{j_{h'} - m_{h'}} \sum_{J'M'} \begin{bmatrix} j_p & j_{h'} & J' \\ m_p & -m_{h'} & M' \end{bmatrix}\begin{bmatrix} j_{p'} & j_h & J' \\ m_{p'} & -m_h & M' \end{bmatrix}$$

$$\times (j_{p'}j_h J'|\,V\,|j_p j_{h'} J'). \tag{GIII.6}$$

In (GIII.4), the quantum number M being given, only two of the summation symbols are independent; they satisfy $m_h + m_p = m_{h'} + m_{p'} = M$. Similarly, in (GIII.6) the summation over M' is actually redundant because M' is required to be equal to $(m_{p'} - m_h) = (m_p - m_{h'})$. The two-body matrix element in (GIII.6) is actually independent of M', and hence we have omitted M' in this matrix element. Substituting (GIII.6) in (GIII.4), and using (AV.9) to carry out the summation of the product of the four Clebsch-Gordon coefficients over the two independent magnetic quantum numbers, we obtain

$$\langle (h'^{-1}p')JM|\,H_{22}\,|(h^{-1}p)JM\rangle$$

$$= -(-1)^{j_h + j_{h'} + j_p + j_{p'}} \sum_{J'} (2J'+1) W(j_h j_p j_{p'} j_{h'};\, JJ')(j_h j_{p'} J'|\,V\,|j_{h'} j_p J'). \tag{GIII.7}$$

The result, with the isospin coupling, can be easily derived, and is almost obvious from (GIII.7). We simply quote it here:

$$\langle (h'^{-1}p')JM;\, TM_T|\,H_{22}\,|(h^{-1}p)JM;\, TM_T\rangle$$

$$= -(-1)^{j_h + j_{h'} + j_p + j_{p'}} \sum_{J'T'} (2J'+1)(2T'+1) W(j_h j_p j_{p'} j_{h'};\, JJ') W(\tfrac{1}{2}\,\tfrac{1}{2}\,\tfrac{1}{2}\,\tfrac{1}{2};\, TT')$$

$$\times (j_h j_{p'} JT|\,V\,|j_{h'} j_p J'T'). \tag{GIII.8}$$

So, given the two-nucleon matrix elements of V, we can easily carry out the summation shown in (GIII.8) and obtain the hole-particle matrix elements. Because of their very frequent occurrence in nuclear structure theory, they are denoted by the special symbols G and F which are defined as

$$G(abcd;\, JT) \equiv (abJT|\,V\,|cdJT), \tag{GIII.9a}$$

$$F(abcd;\, JT) \equiv \langle (a^{-1}b)JT|\,H_{22}\,|(c^{-1}d)JT\rangle. \tag{GIII.9b}$$

G can be calculated either by taking a well-behaved V or, in the case of a realistic two-nucleon potential, by assuming that it is equal to the reaction matrix elements dealt with in Sections 20 and 22.

We would also be led to the result (GIII.8) were we to use the more sophisticated equations (VI.123). We first introduce the angular momentum coupled pair creation operators

$$A^{\dagger}(hpJM;\, TM_T) = \sum_{m_h,\, m_p} \sum_{m_{h'},\, m_{p'}} \begin{bmatrix} j_h & j_p & J \\ m_h & m_p & M \end{bmatrix} \begin{bmatrix} \frac{1}{2} & \frac{1}{2} & T \\ \mu_h & \mu_p & M_T \end{bmatrix} b^{\dagger}_{j_p m_p \mu_p} b^{\dagger}_{j_h m_h \mu_h}, \tag{GIII.10}$$

where μ_h and μ_p are the projection of isospin for the hole and the particle. In defining $b^{\dagger}_{j_h m_h \mu_h}$, we shall now introduce an extra phase due to isospin, i.e.,

$$b^{\dagger}_{j_h m_h \mu_h} = (-1)^{j_h - m_h}(-1)^{1/2-\mu_h} C_{j_h,\, -m_h;\, 1/2,\, -\mu_h},$$

where C is the destruction operator for a state specified by the subscripts.

We shall now introduce the Clebsch-Gordon coefficients of (GIII.10) in (VI.123) and carry out the indicated summation. This produces $[H, A^{\dagger}(hpJM;\, TM_T)]$ when the left-hand side of (VI.123a) and that of (VI.123b) are added. On the right-hand side of (VI.123a), the summation produces $A^{\dagger}(hpJM;\, TM_T)$, provided of course the energies ϵ_p, ϵ_h are independent of the projection quantum numbers (true for spherical nuclei). On the right-hand side of (VI.123b), the summation yields the second term of (GIII.4), multiplied by $A^{\dagger}(h'p'JM;\, TM_T)$ with two analogous Clebsch-Gordon coefficients containing isospin. The rest of the procedure entails using (GIII.6), with of course the inclusion of isospin, and obviously the final result is $F(hph'p';\, JT)$ multiplied by $A^{\dagger}(h'p'JM;\, TM_T)$. We write all these results finally as

$$[H, A^{\dagger}(hpJM;\, TM_T)] = (\epsilon_p - \epsilon_h) A^{\dagger}(hpJM;\, TM_T) + \sum_{h'p'} F(hph'p';\, JT) A^{\dagger}(h'p'JM;\, TM_T). \tag{GIII.11}$$

Introducing $\langle\Psi|$ and $|\Psi_0\rangle$ to the left and right of the operators on both sides of (GIII.11), we obtain the system of linear equations for the amplitudes $\langle\Psi|\, A^{\dagger}(h'p'JM;\, TM_T)\, |\Psi_0\rangle$. The coefficient matrix to be diagonalized is clearly the hole-particle matrix F with $(\epsilon_p - \epsilon_h)$ along the diagonal.

We next move on to the angular momentum coupling in the RPA equations (VI.133). The first line of (VI.133a) obviously leads to (GIII.11) because this line is nothing but the TDA equation. To tackle the second line, we introduce the operator $A(hpJM;\, TM_T)$, which is just the Hermitean conjugate of (GIII.10):

$$A(hpJM;\, TM_T) = \sum_{m_h,\, m_p} \sum_{\mu_h,\, \mu_p} \begin{bmatrix} j_h & j_p & J \\ m_h & m_p & M \end{bmatrix} \begin{bmatrix} \frac{1}{2} & \frac{1}{2} & T \\ \mu_h & \mu_p & M_T \end{bmatrix} b_{j_h m_h \mu_h} b_{j_p m_p \mu_p}. \tag{GIII.12}$$

While indicating some of the steps in this angular momentum coupling, we shall first omit the isospin coupling, which can be introduced into the final result by analogy. As a result of the coupling $A^{\dagger}(hpJM)$, to be done on the left-hand side of (VI.133a), we acquire, for the second line on the right-hand side,

$$\sum_{j_{h'} j_{p'}} \sum_{m_{h'} m_{p'}} \sum_{m_h m_p} \begin{bmatrix} j_h & j_p & J \\ m_h & m_p & M \end{bmatrix} (pp'|\, V\, |\bar{h}\bar{h}') A_{h'p'}. \tag{GIII.13a}$$

Using the definition (GIII.12) in the reverse direction, we obtain (with the omission of isospin) the identity

$$A_{h'p'} = \sum_{J'} \begin{bmatrix} j_{h'} & j_{p'} & J' \\ m_{h'} & m_{p'} & M' \end{bmatrix} A(h'p'J'M'). \tag{GIII.13b}$$

83 (45-122/1977)

Since we have already coupled $A^{\dagger}_{hp}$ to JM on the left-hand side of (VI.133a), only the term ($J' = J$, $M' = -M$) in (GIII.13b) contributes to the final result (because H is a scalar). H times $A^{\dagger}(hpJM)$ transforms as a tensor of rank J and component M. The part of (GIII.13b) that has a similar transformation property actually corresponds to $J' = J$, $M' = -M$. That M' and M have opposite sign is due to the fact that $A(h'p'J'M')$ is a destruction operator whereas $A^{\dagger}(hpJM)$ is a creation operator. Using only the ($J' = J$, $M' = -M$)-term of (GIII.13b) in (GIII.13a), we obtain

$$\sum_{J'}\sum_{j_{h'}j_{p'}}\sum_{m_{h'}m_{p'}}\sum_{m_h m_p}\begin{bmatrix} j_h & j_p & J \\ m_h & m_p & M\end{bmatrix}\begin{bmatrix} j_{h'} & j_{p'} & J \\ m_{h'} & m_{p'} & -M\end{bmatrix}\begin{bmatrix} j_p & j_{p'} & J' \\ m_p & m_{p'} & M'\end{bmatrix}\begin{bmatrix} j_h & j_{h'} & J' \\ -m_h & -m_{h'} & M'\end{bmatrix}$$
$$\times(-1)^{j_h-m_h}(-1)^{j_{h'}-m_{h'}}(j_p j_{p'}J'|\, V\,|j_h j_{h'}J')A(h'p'J,\ -M). \tag{GIII.13c}$$

The third and fourth Clebsch-Gordon coefficients and the summation over J' in this expression have resulted from the replacement of the matrix element of V in (GIII.13a) by the angular momentum coupled matrix elements of V in (GIII.13c). The phase factors have resulted from the $\bar{h}$ and $\bar{h}'$ in (GIII.13a). Once again, M' is not independent, and only two of the magnetic quantum number summations in (GIII.13c) are independent. We carry out these summations by standard algebra and use the definition (GIII.9b) of the F-matrix to obtain

$$\text{(GIII.13c)} = -\sum_{j_{h'}j_{p'}} F(hpp'h';\ J)(-1)^{j_{h'}+j_{p'}-J}(-1)^{J+M}A(h'p'J,\ -M). \tag{GIII.13d}$$

The extension, which has to include isospin coupling, is obvious. The result obtained in this way has to be added to (GIII.11) to produce the final equation for RPA, namely,

$$[H,\ A^{\dagger}(hpJM;\ TM_T)] = (\epsilon_p - \epsilon_h)A^{\dagger}(hpJM;\ TM_T) + \sum_{j_{h'}j_{p'}}\{F(hph'p';\ JT)A^{\dagger}(h'p'JM;\ TM_T)$$
$$+ F(hpp'h';\ JT)(-1)^{j_{h'}+j_{p'}-J-T}(-1)^{J+M+T+M_T}$$
$$\times A(h'p'J,\ -M;\ T,\ -M_T)\}. \tag{GIII.14}$$

The equivalent of the phase factor $(-1)^{j_{h'}+j_{p'}-J}$, appearing in (GIII.13d) and (GIII.14) in the case of isospin coupling, is $(-1)^{1/2+1/2-T}$; this phase explains the change in the overall minus sign of (GIII.13d) to a plus sign of the corresponding term in (GIII.14), and the appearance of the phase factor $(-1)^{j_{h'}+j_{p'}-J-T}$. It should be noted that $(-1)^{J+M+T+M_T}A(h'p'J,\ -M;\ T,\ -M_T)$ appearing on the right-hand side of (GIII.14) and $A^{\dagger}(hpJM;\ TM_T)$ on the left-hand side of the same equation have identical transformation property under rotation.

(GIII.14) enables us to write down the expression for $[H,\ A(hpJ,\ -M;\ T,\ -M_T)] \times(-1)^{J+M+T+M_T}$ with the help of its Hermitean conjugate. Thus,

$$(-1)^{J+M+T+M_T}[H,\ A(hpJ,\ -M;\ T,\ -M_T)]$$
$$= (\epsilon_p - \epsilon_h)A(hpJ,\ -M;\ T,\ -M_T)(-1)^{J+M+T+M_T}$$
$$-\sum_{j_{h'}j_{p'}}\{F(hph'p';\ JT)A(h'p'J,\ -M;\ T,\ -M_T)(-1)^{J+M+T+M_T}$$
$$+ (-1)^{j_{h'}+j_{p'}-J-T}F(hpp'h';\ JT)A^{\dagger}(h'p'J,\ -M;\ T,\ -M_T)\}. \tag{GIII.15}$$

In the usual way, (GIII.14) and (GIII.15) give rise to a set of linear equations for the amplitudes $\langle\Psi|\ A^{\dagger}(hpJM;\ TM_T)\ |\Psi_0\rangle$ and $(-1)^{J+M+T+M_T}\langle\Psi|\ A(hpJ,\ -M;\ T,\ -M_T)\ |\Psi_0\rangle$. The matrix to be diagonalized can be identified from (GIII.14) and (GIII.15) and is given by

$$M' = \begin{pmatrix} \mathcal{E} + F & \bar{F} \\ -\bar{F} & -\mathcal{E} - F \end{pmatrix}, \tag{GIII.16a}$$

where $\mathcal{E}$ and $\bar{F}$ are matrices defined by

$$\mathcal{E}_{hp;\,h'p'} = (\epsilon_p - \epsilon_h)\delta_{hh'}\delta_{pp'}, \tag{GIII.16b}$$

$$\bar{F}(hph'p';\,JT) = (-1)^{j_{h'}+j_{p'}-J-T}F(hpp'h';\,JT). \tag{GIII.16c}$$

It is easy to establish that the matrix M' of (GIII.16a) has all the properties of M', defined by (VI.28b). If we denote the amplitudes with $A^\dagger$ and A by X_{hpJT} and Y_{hpJT}, it follows that these amplitudes satisfy the properties (VI.29)–(VI.31).

Angular Momentum Coupling in Quasiparticle Calculation

The angular momentum coupling in the equations of Section 42C is fairly straightforward. We shall specifically deal with the case of spherical nuclei where the coefficients u_i, v_j are independent of the projection quantum number. The quasiparticle energies also have the same property. In most applications of the theory derived here, the quasiparticle pair is taken to be either a pair of proton or a pair of neutron quasiparticles. This is because the special BCS type transformation we are considering treats only the $(J = 0)$-part of the two-nucleon interaction [which is the strong force for only a pair of protons or a pair of neutrons; for an n-p pair, the interaction in the $(J = 1,\, T = 0)$-state is stronger than that in the $(J = 0,\, T = 1)$-state] at the stage of producing the independent quasiparticles. Since a pair of proton or neutron quasiparticles always has $T = 1$, the isospin T becomes a redundant quantum number, and it is necessary and sufficient to couple the angular momenta j_m, j_n of the pair of quasiparticles.

We introduce the obvious definitions

$$A^\dagger(mnJM) = \sum_{\mu\nu} \begin{bmatrix} m & n & J \\ \mu & \nu & M \end{bmatrix} b^\dagger_{n\nu} b^\dagger_{m\mu}, \tag{GIII.17a}$$

$$A(mnJM) = \sum_{\mu\nu} \begin{bmatrix} m & n & J \\ \mu & \nu & M \end{bmatrix} b_{m\mu} b_{n\nu}. \tag{GIII.17b}$$

Following a procedure similar to that adopted for the hole-particle case, and using the definitions (GIII.17), we easily obtain (the details are left to the reader), from (VI.141), (VI.144), and (VI.147),

$$[H_{11}, A^\dagger(mnJM)] = (E_m + E_n)A^\dagger(mnJM), \tag{GIII.18a}$$

$$\begin{aligned}[H_{22}, A^\dagger(mnJM)] = \sum_{(a,\,b)} &[(u_a u_b u_m u_n + v_a v_b v_m v_n)G(mnab;\,J) \\ &+ (v_a u_b u_m v_n + u_a v_b v_m u_n)F(mnba;\,J)(-1)^{a+b-J} \\ &- (u_a v_b u_m v_n + v_a u_b v_m u_n)F(mnab;\,J)]A^\dagger(abJM),\end{aligned} \tag{GIII.18b}$$

$$\begin{aligned}[H_{04}, A^\dagger(mnJM)] = -\sum_{(a,\,b)} &[(u_a u_b v_m v_n + v_a v_b u_m u_n)G(abmn;\,J) \\ &- (u_a v_b u_n v_m + v_a u_b v_n u_m)F(abmn;\,J) \\ &+ (v_a u_b u_n v_m + u_a v_b v_n u_m)F(bamn;\,J)(-1)^{a+b-J}](-1)^{J+M}A(abJ,\,-M).\end{aligned} \tag{GIII.18c}$$

In (GIII.18b) and (GIII.18c), the summation (a, b) implies that it has to be carried out over a particular pair (a, b) only once, and not twice as a, b and b, a. In order to have a pair only once, we have used the obvious symmetry relations between $A^\dagger(abJM)$, $A^\dagger(baJM)$ and $A(abJ, -M)$, $A(baJ, -M)$, which follow from the property of the Clebsch-Gordon coefficient in (GIII.17a) and (GIII.17b). A similar symmetry relation between $G(abmn; J)$ and $G(bamn; J)$ has also been used. (GIII.18a), (GIII.18b), (GIII.18c), when added, give $[H, A^\dagger(mnJM)]$. As in the hole-particle case, we are led to the diagonalization of a matrix such as M' of (GIII.16a) in which $F(hph'p'; JT)$ and $\bar{F}(hph'p'; JT)$ are replaced respectively by the expressions enclosed within the square brackets in (GIII.18b) and (GIII.18c). Similarly, the diagonal matrix elements of $\mathcal{E}$, instead of being given by (GIII.16b), are given by $(E_m + E_n)$ as appearing in (GIII.18a). The eigenvectors of the matrix M' now determine the amplitudes X_{mnJ} and Y_{mnJ}, where

$$X_{mnJ} = \langle\Psi|\, A^\dagger(mnJM)\, |\Psi_0\rangle,$$
$$Y_{mnJ} = \langle\Psi|\, A(mnJ, -M)\, |\Psi_0\rangle(-1)^{J+M}.$$

Calculations Using Schematic Models for Two-Body Potential

Some general features of the theory of vibration, as worked out here, can be very elegantly demonstrated by using schematic models for the two-body potential. The oldest such model is the δ-function potential. The G-matrix elements of the δ-function potential are given by (III.97), where the f- and g-coefficient are defined by (III.94b) and (III.95b), respectively, and the radial integral $\mathcal{I}$ is given by (III.90).

The hole-particle matrix elements $F(hph'p'; JT)$ corresponding to the G-matrix elements of the δ-function potential can be computed by using the definitions (GIII.9b), (GIII.9a), and (GIII.8). For a given value of $T\,(=0$ or $1)$, the T'-summation in (GIII.8) can be explicitly written out, using known numerical values of $W(\tfrac12\,\tfrac12\,\tfrac12\,\tfrac12;\, TT')$ from (AV.8) and Table (AVII.5). Substituting the G-matrix elements for the two values of T' from (III.97) in (GIII.8), we confront three types of expressions to be summed over J'. These are

$$(-1)^{h+h'+p+p'} \sum_{J'} \sqrt{\frac{[J']}{[J]}}\, U(h'p'ph; JJ') \times \begin{cases} f(hp'J')f(h'pJ') & \text{(type 1)} \quad \text{(GIII.19a)} \\ g(hp'J')g(h'pJ') & \text{(type 2)} \quad \text{(GIII.19b)} \\ (-1)^{J'}g(hp'J')g(h'pJ') & \text{(type 3)} \quad \text{(GIII.19c)} \end{cases}$$

The steps involved in the J'-summation in the three cases will now be outlined.

Type 1 We use the symmetry relations (AIV.4) to rewrite the Clebsch-Gordon coefficient in $f(hp'J')$ as

$$\begin{bmatrix} h & p' & J' \\ \tfrac12 & \tfrac12 & 1 \end{bmatrix} = (-1)^{h-1/2}\sqrt{\frac{[J']}{[p]}}\begin{bmatrix} J' & h & p' \\ 1 & -\tfrac12 & -\tfrac12 \end{bmatrix}.$$

We keep the Clebsch-Gordon coefficient in $f(h'pJ')$ as it occurs, and then note that the dimensionality factor $[J']$ cancels out in (GIII.19a). We substitute the product

$$\begin{bmatrix} h' & p & J' \\ \tfrac12 & \tfrac12 & 1 \end{bmatrix}\begin{bmatrix} J' & h & p' \\ 1 & -\tfrac12 & \tfrac12 \end{bmatrix}$$

with the help of (AV.10) and then carry out the sum over J' in the product of the two U-functions by using (AV.5). In this way, we are left with the product of the two Clebsch-Gordon coefficients

$$\begin{bmatrix} p & h & J \\ \frac{1}{2} & -\frac{1}{2} & 0 \end{bmatrix}\begin{bmatrix} h' & J & p' \\ \frac{1}{2} & 0 & \frac{1}{2} \end{bmatrix}$$

which, by using (AIV.4), can be changed to

$$(-1)^{h'-1/2}\sqrt{\frac{[p']}{[J]}}\begin{bmatrix} h & p & J \\ \frac{1}{2} & -\frac{1}{2} & 0 \end{bmatrix}\begin{bmatrix} h' & p' & J \\ \frac{1}{2} & -\frac{1}{2} & 0 \end{bmatrix}.$$

Collecting the various factors as just outlined, we finally obtain

$$\text{(GIII.19a)} = \bar{g}(hpJ)\bar{g}(h'p'J), \tag{GIII.20a}$$

where

$$\bar{g}(abJ) = (-1)^{l_a}g(abJ). \tag{GIII.20b}$$

Type 2 In the same way as in Type 1, we change

$$\begin{bmatrix} h & p' & J' \\ \frac{1}{2} & -\frac{1}{2} & 0 \end{bmatrix}$$

of $g(hp'J')$ to

$$(-1)^{h-1/2}\sqrt{\frac{[J']}{[p']}}\begin{bmatrix} J' & h & p' \\ 0 & -\frac{1}{2} & -\frac{1}{2} \end{bmatrix}$$

and leave the Clebsch-Gordon coefficient of $g(h'pJ')$ unchanged. We proceed exactly as in Type 1, and, after summing the U-functions, obtain the product of the two Clebsch-Gordon coefficients

$$\begin{bmatrix} p & h & J \\ -\frac{1}{2} & -\frac{1}{2} & -1 \end{bmatrix}\begin{bmatrix} h' & J & p' \\ \frac{1}{2} & -1 & -\frac{1}{2} \end{bmatrix}$$

which can be changed by (AIV.4) to the form

$$(-1)^{h'-1/2}\sqrt{\frac{[p']}{[J]}}\begin{bmatrix} h & p & J \\ \frac{1}{2} & \frac{1}{2} & 1 \end{bmatrix}\begin{bmatrix} h' & p' & J \\ \frac{1}{2} & \frac{1}{2} & 1 \end{bmatrix}.$$

The final result, obtained in this way, is given by

$$\text{(GIII.19b)} = \bar{f}(hpJ)\bar{f}(h'p'J), \tag{GIII.21a}$$

where

$$\bar{f}(abJ) = (-1)^{l_a}f(abJ). \tag{GIII.21b}$$

Type 3 This is distinguished from Type 2 through the occurrence of the additional phase factor $(-1)^{J'}$. This phase factor is to be cancelled out by changing $\begin{bmatrix} J' & h & p' \\ 0 & -\frac{1}{2} & -\frac{1}{2} \end{bmatrix}$ of Type 2 to $\begin{bmatrix} J' & h & p' \\ 0 & \frac{1}{2} & \frac{1}{2} \end{bmatrix}(-1)^{J'+h-p'}$. Once the J'-dependent phase factor is cancelled out, the rest of the derivation will go as in Type 2. The final result thus obtained is then given by

$$\text{(GIII.19c)} = -(-1)^{J}\bar{g}(hpJ)\bar{g}(h'p'J). \tag{GIII.22}$$

Using the results (GIII.20), (GIII.21), and (GIII.22) in (GIII.8), we finally obtain

$$F(hph'p'; JT) = c_T\bar{f}(hpJ)\bar{f}(h'p'J) + c'_T\bar{g}(hpJ)\bar{g}(h'p'J), \quad \text{(GIII.23)}$$

where

$$c_{T=0} = -\tfrac{1}{4}\mathcal{I}(hph'p')[2V_t^e + (-1)^J\pi(V_t^e + 3V_s^e)], \quad \text{(GIII.24a)}$$

$$c'_{T=0} = \tfrac{1}{4}\mathcal{I}(hph'p')(3V_s^e - V_t^e), \quad \text{(GIII.24b)}$$

$$c_{T=1} = \tfrac{1}{4}\mathcal{I}(hph'p')[2V_t^e + (-1)^J\pi(V_t^e - V_s^e)], \quad \text{(GIII.24c)}$$

$$c'_{T=1} = \tfrac{1}{4}\mathcal{I}(hph'p')(V_t^e + V_s^e); \quad \text{(GIII.24d)}$$

for simplicity, $V_0^{e,s}$ and $V_0^{e,t}$ of (III.97) have been written as V_s^e and V_t^e.

From the symmetry property of the Clebsch-Gordon coefficients, we obtain

$$\bar{f}(p'h'J) = (-1)^{l_{h'}+l_{p'}+h'+p'-J}\bar{f}(h'p'J), \quad \text{(GIII.25a)}$$

$$\bar{g}(p'h'J) = -(-1)^{l_{h'}+l_{p'}}(-1)^{h'+p'+l_{h'}+l_{p'}}\bar{g}(h'p'J)$$
$$= -(-1)^{h'+p'}\bar{g}(h'p'J). \quad \text{(GIII.25b)}$$

Using the expressions (GIII.25) together with (GIII.16c), we can easily prove that

$$\bar{F}(hph'p'; JT) = c_T\pi(-1)^T\bar{f}(hpJ)\bar{f}(h'p'J) - c'_T(-1)^{J+T}\bar{g}(hpJ)\bar{g}(h'p'J). \quad \text{(GIII.26)}$$

The schematic model can be further simplified by following either of two assumptions: (i) Brown-Bolsterli assumption—the radial integral $\mathcal{I}$ is independent of the single-particle quantum numbers; (ii) Moszkowski's surface delta-interaction—assume that the radial part of the δ-function potential is such that it is nonvanishing only at the nuclear surface $r_1 = r_2 = R_0$, where R_0 is the nuclear radius. This implies that in the radial integral we have the additional factor $\delta(r - R_0)$ and the integral then reduces to $R_0^2R_h(R_0)R_p(R_0)R_{h'}(R_0)R_{p'}(R_0)$.

Under either of the foregoing two assumptions, the matrix elements (III.97) have the general form

$$G_{KL} = \chi_f f_K f_L + \chi_g g_K g_L, \quad \text{(GIII.27a)}$$

whereas the matrix elements (GIII.23) or (GIII.26) have the general form

$$F_{KL} = \chi'\bar{f}_K\bar{f}_L + \chi'_g\bar{g}_K\bar{g}_L, \quad \text{(GIII.27b)}$$

$$\bar{F}_{KL} = \bar{\chi}_f\bar{f}_K\bar{f}_L + \bar{\chi}_g\bar{g}_K\bar{g}_L. \quad \text{(GIII.27c)}$$

Here K and L each is the notation for an angular momentum coupled state such as (hpJ), and the strengths χ, χ', $\bar{\chi}$ can be either attractive (negative) or repulsive (positive), depending on the exact values of V_s^e, V_t^e and the values of J, π, and T.

Due to the factorable nature of (GIII.27), the diagonalization of the TDA or RPA matrix can be carried out in a somewhat closed form. To illustrate this, let us first diagonalize the TDA matrix $F(hph'p'; JT)$ *assuming* that the diagonal matrix $\mathcal{E}$, whose elements are $(\epsilon_p - \epsilon_h)$, is a constant times the unit matrix. $\mathcal{E}$ can then be left out of the diagonalization, the constant diagonal element simply giving the reference point for the eigenvalues of F.

The eigenvalue equation in the notation of (GIII.27b) is given by

$$\sum_L F_{KL}X_L = EX_K$$

or

$$\sum_L (\chi_f' \bar{f}_K \bar{f}_L + \chi_g' \bar{g}_K \bar{g}_L) X_L = E X_K. \tag{GIII.28}$$

To further simplify matters, let us *assume* that the $\bar{f}$-term is negligible compared with the $\bar{g}$-term. Then

$$\chi_g' \sum_L \bar{g}_K \bar{g}_L X_L \equiv \chi_g' \bar{g}_K (\sum_L \bar{g}_L X_L) = E X_K \tag{GIII.29a}$$

or

$$X_K = E^{-1} \chi_g' C \bar{g}_K, \tag{GIII.29b}$$

where

$$C = \sum_L \bar{g}_L X_L = E^{-1} \chi_g' C \sum_L (\bar{g}_L)^2 \qquad \text{[using (GIII.29b)]}$$

or

$$E = \chi_g' \sum_L (\bar{g}_L)^2. \tag{GIII.29c}$$

This gives us one eigenvalue of the problem; the K-th element of the corresponding eigenvector, X_K, is proportional, according to (GIII.29b), to $\bar{g}_K$. The constant of proportionality is fixed by the normalization of the vector. The eigenvector belonging to any other eigenvalue has to be orthogonal to this one. If the K-th element of such a vector is denoted by X_K', we then have

$$0 = \sum_K X_K X_K' = \sum_K \bar{g}_K X_K'. \tag{GIII.30}$$

The eigenvalue equation for this eigenvector is given by

$$\chi_g' \sum_L \bar{g}_K \bar{g}_L X_L' = E' X_K'.$$

But the summation $\sum_L \bar{g}_L X_L'$ on the left-hand side of this equation is zero according to (GIII.30). Hence, we conclude that $E' = 0$. Therefore, according to this model, only one eigenvalue, E of (GIII.29c), is displaced with respect to the unperturbed position of the hole-particle states (which are degenerate by our assumption about $\mathcal{E}$); all the other eigenvalues E' remain coincident with the unperturbed energy.

An extension of the foregoing derivation can now be made in two ways: (i) use the unperturbed energy matrix $\mathcal{E}$ with unequal elements; and (ii) use the complete expression (GIII.28) containing both the $\bar{f}$- and $\bar{g}$-term. In doing the first extension, we have, denoting $(\epsilon_p - \epsilon_h)$ by $\mathcal{E}_K$,

$$\sum_L (\mathcal{E}_K \delta_{KL} + \chi_g' \bar{g}_K \bar{g}_L) X_L = E X_K \tag{GIII.31}$$

or

$$(E - \mathcal{E}_K) X_K = \chi_g' C \bar{g}_K$$

or

$$X_K = \chi_g' C \frac{\bar{g}_K}{E - \mathcal{E}_K}, \tag{GIII.32a}$$

where

$$C = \sum_L \bar{g}_L X_L = \chi'_g C \sum_L \frac{(\bar{g}_L)^2}{E - \mathcal{E}_L}$$

or

$$1 = \chi'_g \sum_L \frac{(\bar{g}_L)^2}{E - \mathcal{E}_L}. \tag{GIII.32b}$$

This so-called dispersion equation then determines the energies E. Knowing E, we can determine the corresponding eigenvector up to a normalization factor by (GIII.32a). A graphical method of solution to (GIII.32b) is quite instructive (see Problem 6 at the end of this appendix).

Let us now do the second extension by using the entire expression (GIII.28). We have the equation

$$\sum_L (\mathcal{E}_K \delta_{KL} + \chi'_f \bar{f}_K \bar{f}_L + \chi'_g \bar{g}_K \bar{g}_L) X_L = E X_K \tag{GIII.33}$$

or

$$(E - \mathcal{E}_K) X_K = \chi'_f C_f \bar{f}_K + \chi'_g C_g \bar{g}_K, \tag{GIII.34a}$$

where

$$C_f = \sum_L \bar{f}_L X_L, \qquad C_g = \sum_L \bar{g}_L X_L. \tag{GIII.34b}$$

Using (GIII.34a) to yield X_K and then using the result in the two equations of (GIII.34b), we get two linear homogeneous equations for C_f and C_g. The determinant of the coefficient matrix, when equated to zero, gives the secular equation for E; this is left as an exercise.

The final stage in this schematic model is to solve the eigenvalues and eigenfunctions of the RPA matrix (GIII.16a). In the condensed notation, the equations are given by

$$\sum_L [(\mathcal{E}_K \delta_{KL} + F_{KL}) X_L + \bar{F}_{KL} Y_L] = E X_K, \tag{GIII.35a}$$

$$\sum_L \{-\bar{F}_{KL} X_L - (\mathcal{E}_K \delta_{KL} + F_{KL}) Y_L\} = E Y_K. \tag{GIII.35b}$$

We use (GIII.27c) for $\bar{F}_{KL}$. By a procedure similar to that already demonstrated, we have

$$(E - \mathcal{E}_K) X_K = (\chi'_f C_f + \bar{\chi}_f D_f) \bar{f}_K + (\chi'_g C_g + \bar{\chi}_g D_g) \bar{g}_K, \tag{GIII.36a}$$

$$-(E + \mathcal{E}_K) Y_K = (\bar{\chi}_f C_f + \chi'_f D_f) \bar{f}_K + (\bar{\chi}_g C_g + \chi'_g D_g) \bar{g}_K, \tag{GIII.36b}$$

where

$$C_f = \sum_L \bar{f}_L X_L, \qquad C_g = \sum_L \bar{g}_L X_L, \tag{GIII.36c}$$

$$D_f = \sum_L \bar{f}_L Y_L, \qquad D_g = \sum_L \bar{g}_L Y_L. \tag{GIII.36d}$$

Once again, we have to substitute from (GIII.36a) and (GIII.36b) for X_L, Y_L in (GIII.36c) and (GIII.36d) to obtain a set of four linear homogeneous equations in C_f, C_g, D_f, and D_g. The secular determinant equation of the coefficient matrix determines E.

In the two practical cases, namely, the giant dipole resonance ($J = 1$, $\pi = -1$, $T = 1$), and octupole vibration ($J = 3$, $\pi = -1$, $T = 0$), the equations (GIII.35) are considerably simplified. As an illustration, let us consider the latter case, for which, according to (GIII.26) and (GIII.23), we have

$$F_{KL} + \bar{F}_{KL} = 2c'_T\bar{g}_K\bar{g}_L, \tag{GIII.37a}$$

$$F_{KL} - \bar{F}_{KL} = 2c_T\bar{f}_K\bar{f}_L. \tag{GIII.37b}$$

Similar simplifications occur in the case of giant dipole resonance as well. It is now advantageous to add and subtract (GIII.35a) and (GIII.35b) and use $Z_K^{(+)} = X_K + Y_K$ and $Z_K^{(-)} = X_K - Y_K$. We thus have

$$EZ_K^{(+)} = \sum_L (\mathcal{E}_K\delta_{KL} + 2c_T\bar{f}_K\bar{f}_L)Z_L^{(-)}, \tag{GIII.38a}$$

$$EZ_K^{(-)} = \sum_L (\mathcal{E}_K\delta_{KL} + 2c'_T\bar{g}_K\bar{g}_L)Z_L^{(+)}. \tag{GIII.38b}$$

We substitute for $Z_K^{(+)}$ from (GIII.38a) in (GIII.38b), and vice versa. Then, introducing the notation

$$C^{(-)} = \sum_L \bar{f}_L Z_L^{(-)}, \qquad C^{(+)} = \sum_L \bar{g}_L Z_L^{(+)}, \tag{GIII.39}$$

we easily obtain

$$Z_K^{(+)} = 2(E^2 - \mathcal{E}_K^2)^{-1}(c'_T C^{(+)}\bar{g}_K + c_T C^{(-)}\bar{f}_K), \tag{GIII.40a}$$

$$Z_K^{(-)} = 2(E^2 - \mathcal{E}_K^2)^{-1}(c_T C^{(-)}\bar{f}_K + c'_T C^{(+)}\bar{g}_K). \tag{GIII.40b}$$

Substituting (GIII.40) back in (GIII.39), we obtain two linear homogeneous equations in $C^{(-)}$ and $C^{(+)}$ and then the solution proceeds as usual. Obtaining the final dispersion equation for E is left as an exercise.

Schematic Model for Quasiparticle Calculation

In order to be able to apply the factorable schematic model to (GIII.18), we must first of all verify that the products of u_a, v_b, ... also factorize automatically. In the usual way, we introduce

$$X_{ab} = \langle\Psi|\, A^\dagger(abJM)\, |\Psi_0\rangle,$$

$$Y_{ab} = (-1)^{J+M}\langle\Psi|\, A(abJ, -M)\, |\Psi_0\rangle,$$

$$Z_{ab}^{(+)} = X_{ab} + Y_{ab}, \qquad Z_{ab}^{(-)} = X_{ab} - Y_{ab}.$$

We then insert $\langle\Psi|$ and $|\Psi_0\rangle$ to the left and the right respectively of $[H, A^\dagger(mnJM)]$, which is the sum of (GIII.18a), (GIII.18b), and (GIII.18c), and write the result in terms of the amplitudes X and Y, as just defined. We next similarly work with the equation for $[H, A(mnJ, -M)]$. Adding and subtracting the two equations we thus get, we finally obtain the equations satisfied by $Z^{(\pm)}$:

$$EZ_{mn}^{(+)} = (E_m + E_n)Z_{mn}^{(-)} + \sum_{(a,b)} H(mnabJ)Z_{ab}^{(-)}, \tag{GIII.41a}$$

$$EZ_{mn}^{(-)} = (E_m + E_n)Z_{mn}^{(+)} + \sum_{(a,b)} K(mnabJ)Z_{ab}^{(+)}, \tag{GIII.41b}$$

where

$$\begin{aligned} H(mnabJ) &= G(mnabJ)(u_m u_n + v_m v_n)(u_a u_b + v_a v_b) \\ &\quad - [F(mnabJ) + (-1)^{a+b-J}F(mnbaJ)](u_m v_n - v_m u_n)(u_a v_b - v_a u_b), \end{aligned} \tag{GIII.41c}$$

84 (45-122/1977)

$$K(mnabJ) = G(mnabJ)(u_m u_n - v_m v_n)(u_a u_b - v_a v_b) - [F(mnabJ) - (-1)^{a+b-J}F(mnbaJ)](u_m v_n + v_m u_n)(u_a v_b + v_a u_b). \quad \text{(GIII.41d)}$$

(GIII.41c) and (GIII.41d) demonstrate that the factors involving the u-, v-coefficients do factorize automatically. Therefore, if we adopt the factorable forms of the F- and G-matrix elements already given for the schematic model, we can once again obtain the solutions of (GIII.41a) and (GIII.41b) by the same procedure.

We consider in the quasiparticle theory either a pair of proton quasiparticles or a pair of neutron quasiparticles, for which the isospin T is identically 1. Using $T = 1$ in (GIII.16c) and (GIII.26), we obtain

$$\bar{F}(mnabJ) = (-1)^{a+b-J}F(mnbaJ) = -c\pi\bar{f}(mnJ)\bar{f}(abJ) + c'(-1)^J\bar{g}(mnJ)\bar{g}(abJ).$$

Here we have suppressed the subscript $T = 1$ on c and c'. Using this expression in the case of the quadrupole vibration ($J = 2$, $\pi = +1$), for example, we have

$$F(mnabJ) + (-1)^{a+b-J}F(mnbaJ) = 2c\bar{f}(mnJ)\bar{f}(abJ), \quad \text{(GIII.42a)}$$

$$F(mnabJ) - (-1)^{a+b-J}F(mnbaJ) = 2c'\bar{g}(mnJ)\bar{g}(abJ). \quad \text{(GIII.42b)}$$

While applying the schematic model in this case, an excuse is usually given to ignore the $G(mnabJ)$-term in (GIII.41c) and (GIII.41d). If this is permitted, then it is immediately clear that (GIII.41a) and (GIII.41b) become exactly similar to (GIII.38a) and (GIII.38b), the only difference being that in the present case we have factors such as $\bar{f}(abJ)(u_a v_b - v_a u_b)$ and $\bar{g}(abJ)(u_a v_b + v_a u_b)$ instead of $\bar{f}(abJ)$ and $\bar{g}(abJ)$.

Another type of schematic model that has been used in this calculation is based on a multipole type two-body potential, namely,

$$V_{12} = \chi_K \Omega^K(1) \cdot \Omega^K(2),$$

where χ_K is the strength of the potential of multipolarity K. The operator Ω^K for the i-th nucleon ($i = 1, 2$ in the foregoing expression for V_{12}) is given by

$$\Omega_q^K(i) = r_i^K Y_q^K(\theta_i, \phi_i),$$

where r_i, θ_i, ϕ_i are the coordinates of the i-th nucleon, and Y_q^K is the normalized spherical harmonic. In particular, the multipole potentials corresponding to $K = 2$ (quadrupole) and $K = 3$ (octupole) have been extensively used for treating the quadrupole and octupole vibrational states, respectively. The schematic model of this type makes further approximations: (i) the G-type matrix elements can be ignored as compared with the F-type matrix elements, and (ii) in working out the F-matrix element, the direct part can be ignored as compared with the exchange part. We first use the definition (GIII.7) for the F-matrix element and replace $\langle anJ'|\, V\, |mbJ'\rangle$ by the exchange term, i.e., $-(-1)^{m+b-J'}\langle anJ'|\, V\, |bmJ'\rangle$; then we apply the standard result (BIII.15) to write down $\langle anJ'|\, \Omega^K(1)\cdot\Omega^K(2)\, |bmJ'\rangle$ explicitly, and use (AV.5) to carry out the summation of the product of two W-functions over J'. In this way, we obtain, in the present schematic model, the F-matrix element of the multipole potential of order K given by

$$F(mnabJ) = \delta_{KJ}\chi_J \bar{g}'(mnJ)\bar{g}'(abJ), \quad \text{(GIII.43a)}$$

where

$$\bar{g}'(abJ) = (4\pi)^{-1/2}\bar{g}(abJ) \int_0^\infty r^2\, dr\, R_a(r) r^K R_b(r). \quad \text{(GIII.43b)}$$

Thus, according to this model, the quadrupole and the octupole potential give nonvanishing matrix elements in ($J = 2$)- and ($J = 3$)-state, respectively. Using the definitions (GIII.43a) and (GIII.43b) and the fact that G is ignored in comparison with F, we obtain, from (GIII.41c) and (GIII.41d),

$$H(mnabJ) = -[1 - (-1)^J]\chi_J\bar{g}_1'(mnJ)\bar{g}_1'(abJ), \qquad \text{(GIII.44a)}$$

$$K(mnabJ) = -[1 + (-1)^J]\chi_J\bar{g}_2'(mnJ)\bar{g}_2'(abJ), \qquad \text{(GIII.44b)}$$

where

$$\begin{aligned} g_1'(abJ) &= g'(abJ)(u_a v_b - v_a u_b), \\ g_2'(abJ) &= g'(abJ)(u_a v_b + v_a u_b). \end{aligned} \qquad \text{(GIII.44c)}$$

It is clear from (GIII.44a) and (GIII.44b) that $H = 0$ and $K \neq 0$ for $J = 2$ (quadrupole vibration), whereas the reverse is true for the octupole state. In the case of quadrupole vibration, for example, (GIII.41a) and (GIII.41b) reduce to

$$EZ_{mn}^{(+)} = (E_m + E_n)Z_{mn}^{(-)}, \qquad \text{(GIII.44d)}$$

$$EZ_{mn}^{(-)} = (E_m + E_n)Z_{mn}^{(+)} - 2\chi_J\bar{g}_2'(mnJ) \sum_{(a,b)} \bar{g}_2'(abJ)Z_{ab}^{(+)}. \qquad \text{(GIII.44e)}$$

Substituting from (GIII.44d) into (GIII.44e), we obtain

$$[E^2 - (E_m + E_n)^2]Z_{mn}^{(-)} = -2\chi_J C^{(+)}\bar{g}_2'(mnJ), \qquad \text{(GIII.45a)}$$

where

$$\begin{aligned} C^{(+)} &= \sum_{(a,b)} E\bar{g}_2'(abJ)Z_{ab}^{(+)} \\ &= \sum_{(a,b)} \bar{g}_2'(abJ)(E_a + E_b)Z_{ab}^{(-)}. \end{aligned} \qquad \text{(GIII.45b)}$$

We solve (GIII.45a) for $Z_{mn}^{(-)}$ and substitute the solution for $Z_{ab}^{(-)}$ in (GIII.45b). This yields the dispersion equation for the energy E:

$$1 = -2\chi_J \sum_{(a,b)} [E^2 - (E_a + E_b)^2]^{-1}(E_a + E_b)[\bar{g}_2'(abJ)]^2. \qquad \text{(GIII.46)}$$

Calculation of Transition Probability

Any one-body operator

$$\sum_I M_q^K(I),$$

where the summation goes over all the particles, is given in the second-quantized notation by

$$\sum_I M_q^K(I) = \sum_{a,b} \langle a|\, M_q^K\, |b\rangle C_a^\dagger C_b. \qquad \text{(GIII.47)}$$

In the special case of electromagnetic transition, M_q^K is the electromagnetic transition operator of multipolarity K and component q. We shall consider only the electric multipole transition, in which case

$$M_q^K = \Omega_q^K = r^K Y_q^K(\theta, \phi),$$

the quantities r, θ, ϕ being the coordinates of a proton. In general,

$$\langle a|\, \Omega_q^K\, |b\rangle = \begin{bmatrix} b & K & a \\ m_b & q & m_a \end{bmatrix} \langle a|\, |\Omega^K|\, |b\rangle. \qquad \text{(GIII.48a)}$$

We substitute from (VI.71a) and (VI.71b) for $C_a^\dagger$ and C_b and write the product $C_a^\dagger C_b$ as

$$\begin{aligned} C_a^\dagger C_b &= (u_a b_a^\dagger + v_a s_a b_{-a})(u_b b_b + v_b s_b b_{-b}^\dagger) \\ &= (v_a^2 \delta_{ab} + u_a u_b b_a^\dagger b_b + v_a v_b s_a s_b b_{-b}^\dagger b_{-a}) \\ &\quad + (u_a v_b s_b b_a^\dagger b_{-b}^\dagger + v_a u_b s_a b_{-a} b_b). \end{aligned} \tag{GIII.48b}$$

To evaluate the transition matrix element between the states $\langle\Psi|$ and $|\Psi_0\rangle$, we have to substitute (GIII.48b) in (GIII.47) with $M_q^K = \Omega_q^K$ and then multiply the resultant expression from the left and the right by $\langle\Psi|$ and $|\Psi_0\rangle$, respectively. The last line of (GIII.48b) then gives rise to amplitudes of the type $\langle\Psi| A^\dagger |\Psi_0\rangle$ and $\langle\Psi| A |\Psi_0\rangle$; the first line of this equation, on the other hand, gives rise to amplitudes of the type $\langle\Psi| \hat{N} |\Psi_0\rangle$ from the second and third terms, which are smaller in order of magnitude as compared with $\langle\Psi| A |\Psi_0\rangle$, and hence will be consistently neglected. The first term in the first line of (GIII.48b) being a scalar number produces exactly zero when sandwiched between the two *different* states $\langle\Psi|$ and $|\Psi_0\rangle$. Therefore, from now on we shall work with only the last line of (GIII.48b) and do the angular momentum coupling. We have

$$b_a^\dagger b_{-b}^\dagger = -b_{-b}^\dagger b_a^\dagger = -\sum_J \begin{bmatrix} a & b & J \\ m_a & -m_b & q \end{bmatrix} A^\dagger(abJq), \tag{GIII.48c}$$

$$b_{-a} b_b = \sum_J \begin{bmatrix} a & b & J \\ -m_a & m_b & -q \end{bmatrix} A(abJ, -q). \tag{GIII.48d}$$

The Clebsch-Gordon coefficient in (GIII.48a) guarantees that $q = m_a - m_b$, and this fact has been utilized in the Clebsch-Gordon coefficients of (GIII.48c) and (GIII.48d). The summation in (GIII.47) consists of that over m_a, m_b and all other single-particle quantum numbers collectively designated by a and b. The sum over m_a, m_b can be easily carried out by picking up the Clebsch-Gordon coefficients of (GIII.48a), (GIII.48c), and (GIII.48d). A restriction $K = J$ then automatically appears in the result. Putting everything together, we finally obtain

$$\begin{aligned} \sum_I \Omega_q^J(I) &\equiv \sum_{a,b} \langle a| \Omega_q^J |b\rangle C_a^\dagger C_b \\ &= -\sum_{a,b} \bar{g}'(abJ)[u_a v_b A^\dagger(abJq) + v_a u_b (-1)^{J+q} A(abJ, -q)]. \end{aligned} \tag{GIII.49}$$

In the hole-particle case, because of the factors uv, the first term within the square brackets in (GIII.49) is nonvanishing when $a = p$, $b = h$ and the second term is nonvanishing when $a = h$, $b = p$. We also rewrite $A^\dagger(phJq)$ as $-(-1)^{h+p-J}A^\dagger(hpJq)$. Finally, we obtain

$$\langle\Psi| \sum_I \Omega_q^J(I) |\Psi_0\rangle = -\sum_{h,p} [\bar{g}'(hpJ) Y_{hp} - (-1)^{h+p-J} \bar{g}'(phJ) X_{hp}].$$

Using (GIII.25b) and (GIII.43b), we have

$$-(-1)^{h+p-J} \bar{g}'(phJ) = (-1)^J \bar{g}'(hpJ),$$

and hence

$$\langle\Psi| \sum_I \Omega_q^J(I) |\Psi_0\rangle = -\sum_{h,p} \bar{g}'(hpJ)[Y_{hp} + (-1)^J X_{hp}]. \tag{GIII.50}$$

In the quasiparticle case, we put together the terms a, b and b, a contained in (GIII.49) and in the summation count a pair (a, b) only once. In this way, we obtain

$$\langle\Psi|\sum_I \Omega_q^J(I)\,|\Psi_0\rangle = -\sum_{(a,b)} \bar{g}'(abJ)[u_a v_b + (-1)^J v_a u_b][X_{ab} + (-1)^J Y_{ab}]. \tag{GIII.51}$$

As an example, let us consider the quadrupole state with the schematic quadrupole potential model. Then (GIII.51) reduces to

$$-\sum_{(a,b)} \bar{g}_2'(abJ)Z_{ab}^{(+)} = -\sum_{a,b} E^{-1}(E_a + E_b)\bar{g}_2'(abJ)Z_{ab}^{(-)}. \tag{GIII.52}$$

The right-hand side of (GIII.52) follows from (GIII.44d). Although this expression can be evaluated exactly after solving (GIII.44d) and (GIII.44e), we would like to show here, by some rough work, that this transition probability is indeed very large for the state Ψ, which gets pushed down by a large amount from the vicinity of the unperturbed states. For this state, it will be a fairly good approximation to replace the pair energies $(E_a + E_b)$ by an average value $\mathcal{E}_0$ such that (GIII.52) becomes

$$\langle\Psi|\sum_I \Omega_q^J(I)\,|\Psi_0\rangle = -(\mathcal{E}_0/E)\sum_{(a,b)} \bar{g}_2'(abJ)Z_{ab}^{(-)}, \tag{GIII.53a}$$

while (GIII.45a) tells us that $Z_{mn}^{(-)}$ is now proportional to $\bar{g}_2'(mnJ)$. In terms of an overall normalization constant $\mathcal{N}$, we have

$$Z_{mn}^{(-)} = \mathcal{N}\bar{g}_2'(mnJ), \tag{GIII.53b}$$

where $\mathcal{N}$ is determined according to (VI.31) by

$$\begin{aligned} 1 &= \sum_{(m,n)} Z_{mn}^{(+)}Z_{mn}^{(-)} = (\mathcal{E}_0/E)\sum_{(m,n)} [Z_{mn}^{(-)}]^2 \\ &= \mathcal{N}^2(\mathcal{E}_0/E)\sum_{(m,n)} [\bar{g}_2'(mnJ)]^2. \end{aligned} \tag{GIII.53c}$$

Thus, the transition probability (GIII.53a) is given by

$$\begin{aligned} -(\mathcal{E}_0/E)\mathcal{N}\sum_{(a,b)} [\bar{g}_2'(abJ)]^2 &\equiv \mathcal{N}^{-1} \\ &= \{(\mathcal{E}_0/E)\sum_{(m,n)} [\bar{g}_2'(mnJ)]^2\}^{1/2}. \end{aligned} \tag{GIII.53d}$$

According to (GIII.46),

$$E^2 = \mathcal{E}_0^2 - 2\chi_J\mathcal{E}_0\sum_{(a,b)} [\bar{g}_2'(abJ)]^2$$

or

$$(\mathcal{E}_0/E) = \{1 - (2\chi_J/\mathcal{E}_0)\sum_{a,b} [\bar{g}_2'(abJ)]^2\}^{-1/2}.$$

Substituting this in (GIII.53d), we conclude that

$$|\langle\Psi|\sum_I \Omega_q^J(I)\,|\Psi_0\rangle|^2 > \{\sum_{m,n} [\bar{g}_2'(mnJ)]^2\}^{1/2}. \tag{GIII.53e}$$

It is interesting to compute the transition probability in the TDA approximation using the same model, and compare it with (GIII.53e). To do this, we use (GIII.18a) and (GIII.18b) with $G = 0$; assuming $(E_m + E_n) \approx \mathcal{E}_0$, we easily obtain

$$(E - \mathcal{E}_0)X_{mn} = -\sum_{a,b} (u_a v_b u_m v_n + v_a u_b v_m u_n)F(mnabJ)X_{ab}.$$

where the summation a, b goes over both (a, b) and (b, a). Using F from (GIII.43a) and (GIII.43b) and the symmetry relations (GIII.25b) and $X_{ba} = (-1)^{a+b-J} X_{ab}$, we rewrite the foregoing equation as

$$(E - \mathcal{E}_0) X_{mn} = -\chi \mathcal{C} \bar{g}_2'(mnJ) \tag{GIII.54a}$$

with

$$\mathcal{C} = \sum_{(a, b)} \bar{g}_2'(abJ) X_{ab}. \tag{GIII.54b}$$

Therefore,

$$X_{mn} = \mathcal{N}' \bar{g}_2'(mnJ), \tag{GIII.54c}$$

where $\mathcal{N}'$ is a normalization constant and determined by

$$1 = \sum_{mn} X_{mn}^2 = \mathcal{N}'^2 \sum_{mn} [\bar{g}_2'(mnJ)]^2. \tag{GIII.54d}$$

Putting $Y_{ab} = 0$ in (GIII.51), we obtain the transition probability in the TDA which is given by (for $J = 2$)

$$\begin{aligned} |\langle \Psi | \sum_I \Omega_q^J(I) | \Psi_0 \rangle|^2 &= \sum_{(a, b)} \bar{g}_2'(abJ) X_{ab} \\ &= \mathcal{N}' \sum_{(a, b)} [\bar{g}_2'(abJ)]^2 \equiv (\mathcal{N}')^{-1} \\ &= \{ \sum_{(a, b)} [\bar{g}_2'(abJ)]^2 \}^{1/2}. \end{aligned} \tag{GIII.54e}$$

In obtaining this result, we have successively used (GIII.54c) and (GIII.54d). Thus, the RPA transition probability (GIII.53e) is larger than the TDA estimate (GIII.54e). The TDA value is already much larger than a typical single-particle transition probability by virtue of the summation present in (GIII.54e). The same kind of demonstration can be given for the hole-particle case as well; the general result is that the TDA transition matrix element for the state that shows the collective type of lowering in energy (pushing up in the case of giant dipole resonance) is already very large as compared with the single-particle transition probability, and the RPA produces a further enhancement in this quantity.

REFERENCES

1. Kumar, K., and Baranger, M., *Nucl. Phys.*, **A110**, 529 (1968); *ibid*, **A122**, 273 (1968); *ibid*, **A92**, 608 (1967); Kumar, K., *Nucl. Phys.*, **A92**, 653 (1967).
2. Warke, C. S., and Khadkikar, S. B., *Phys. Rev.*, **170**, 1041 (1968); Satpathy, L., Goss, D., and Banerjee, M. K., *Phys. Rev.*, **183**, 887 (1969); Ansari, A., and Nair, S. C. K., *Nucl. Phys.*, **A163**, 56 (1971); *Phys. Letters*, **47B**, 200 (1973); Kumar, K., *Physica Scripta* (Sweden), **6**, 270 (1972); Banerjee, B., Mang, H. J., and Ring, P., *Nucl. Phys.*, **A215**, 366 (1973); Ansari, A., and Nair, S. C. K., *Nucl. Phys.*, **A283**, 326 (1977); Faessler, A., Sandhya Devi, K. R., Grümmer, F., Schmid, K. W., and Hilton, R. R., *Nucl. Phys.*, **A256**, 106 (1976); Warke, C. S., and Gunye, M. R., *Phys. Rev.*, **C12**, 1647 (1975); *ibid*, **C13**, 859 (1976).

FURTHER READING

Baranger, M., in 1962 Cargese Lectures in Theoretical Physics, ed. by M. Levy, Benjamin, New York, 1969.

PROBLEMS

1. The density operator corresponding to a set of occupied single-particle states in a many-body state has the property of its square being equal to itself. Use this property for the operator $\hat{\rho}$ defined after (GI.2) and equate separately the quantities in the zero-th, first, and second orders from the two sides of the equation. Use the HF representation in which ρ has the elements $\rho_{ij} = n_i\delta_{ij}$, where n_i is the occupation number (1 for an occupied state and 0 for an unoccupied state). Manipulate the first-order equation and prove that the only nonvanishing matrix elements of $\hat{\rho}^{(1)}$ are those connecting an occupied with an unoccupied state.

2. In the same way as in Problem 1, use the second-order equation and derive a relationship between the elements of $\hat{\rho}^{(2)}$ connecting two occupied states or two unoccupied states with the nonvanishing matrix elements of $\hat{\rho}^{(1)}$, mentioned in Problem 1.

3. Derive (GIII.7) by carrying out the summation over the magnetic quantum numbers described before that equation.

4. Derive (GIII.18) by completing the steps indicated before (GIII.18a).

5. Derive (GIII.23) by following the procedure indicated prior to this equation.

6. Take a set of close-lying fixed energies $\mathcal{E}_L$ and plot the function

$$F(E) = \sum_L \frac{(\bar{g}_L)^2}{E - \mathcal{E}_L}$$

schematically, noting that $F(E)$ goes to $+\infty$ and $-\infty$ in the immediate vicinities of each $\mathcal{E}_L$. Then draw a line corresponding to $(1/X_g')$. The intersection of the sets of curves $F(E)$ with this line determines the eigenvalues E of (GIII.32b). Show that the lowest of these eigenvalues is very much below the lowest $\mathcal{E}_L$ when X_g' is attractive, whereas the highest one is very much above the highest $\mathcal{E}_L$ when X_g' is repulsive. All the other eigenvalues lie within the range of the energies $\mathcal{E}_L$.

Index

85 (45-122/1977)